Lecture Notes in Computer Science 3410

Commenced Publication in 1973
Founding and Former Series Editors:
Gerhard Goos, Juris Hartmanis, and Jan van Leeuwen

Editorial Board

David Hutchison
　Lancaster University, UK
Takeo Kanade
　Carnegie Mellon University, Pittsburgh, PA, USA
Josef Kittler
　University of Surrey, Guildford, UK
Jon M. Kleinberg
　Cornell University, Ithaca, NY, USA
Friedemann Mattern
　ETH Zurich, Switzerland
John C. Mitchell
　Stanford University, CA, USA
Moni Naor
　Weizmann Institute of Science, Rehovot, Israel
Oscar Nierstrasz
　University of Bern, Switzerland
C. Pandu Rangan
　Indian Institute of Technology, Madras, India
Bernhard Steffen
　University of Dortmund, Germany
Madhu Sudan
　Massachusetts Institute of Technology, MA, USA
Demetri Terzopoulos
　New York University, NY, USA
Doug Tygar
　University of California, Berkeley, CA, USA
Moshe Y. Vardi
　Rice University, Houston, TX, USA
Gerhard Weikum
　Max-Planck Institute of Computer Science, Saarbruecken, Germany

Carlos A. Coello Coello
Arturo Hernández Aguirre
Eckart Zitzler (Eds.)

Evolutionary Multi-Criterion Optimization

Third International Conference, EMO 2005
Guanajuato, Mexico, March 9-11, 2005
Proceedings

Springer

Volume Editors

Carlos A. Coello Coello
Centro de Investigación y de Estudios Avanzados
del Instituto Politécnico Nacional (CINVESTAV-IPN)
Sección Computación
Av. IPN No. 2508, D.F. 07360, Col. San Pedro Zacatenco, Mexico
E-mail: ccoello@cs.cinvestav.mx

Arturo Hernández Aguirre
Centro de Investigación en Matemáticas (CIMAT)
A.P. 402, Guanajuato 36000, Mexico
E-mail: artha@cimat.mx

Eckart Zitzler
Swiss Federal Institute of Technology (ETH) Zurich
Gloriastraße 35, 8092 Zurich, Switzerland
E-mail: zitzler@tik.ee.ethz.ch

Library of Congress Control Number: 2005920590

CR Subject Classification (1998): F.2, G.1.6, G.1.2, I.2.8

ISSN 0302-9743
ISBN 3-540-24983-4 Springer Berlin Heidelberg New York

This work is subject to copyright. All rights are reserved, whether the whole or part of the material is concerned, specifically the rights of translation, reprinting, re-use of illustrations, recitation, broadcasting, reproduction on microfilms or in any other way, and storage in data banks. Duplication of this publication or parts thereof is permitted only under the provisions of the German Copyright Law of September 9, 1965, in its current version, and permission for use must always be obtained from Springer. Violations are liable to prosecution under the German Copyright Law.

Springer is a part of Springer Science+Business Media

springeronline.com

© Springer-Verlag Berlin Heidelberg 2005
Printed in Germany

Typesetting: Camera-ready by author, data conversion by Scientific Publishing Services, Chennai, India
Printed on acid-free paper   SPIN: 11393993   06/3142   5 4 3 2 1 0

Preface

Multicriterion optimization refers to problems with two or more objectives (normally in conflict with each other) which must be simultaneously satisfied. Multicriterion optimization problems have not one but a set of solutions (which represent trade-offs among the objectives), which are called Pareto optimal solutions. Thus, the main goal in multicriterion optimization is to find or to approximate the set of Pareto optimal solutions. Evolutionary algorithms have been used for solving multicriterion optimization problems for over two decades, gaining an increasing popularity over the last 10 years.

The 3rd International Conference on Evolutionary Multi-criterion Optimization (EMO 2005) was held during March 9–11, 2005, in Guanajuato, México. This was the third international conference dedicated entirely to this important topic, following the successful EMO 2001 and EMO 2003 conferences, which were held in Zürich, Switzerland in March 2001, and in Faro, Portugal in April 2003, respectively.

The EMO 2005 scientific program included two keynote addresses, one given by Peter Fleming on an engineering design perspective of many-objective optimization, and the other given by Milan Zeleny on the evolution of optimality. In addition, three tutorials were presented, one on metaheuristics for multiobjective combinatorial optimization by Xavier Gandibleux, another on multiobjective evolutionary algorithms by Gary B. Lamont, and a third one on performance assessment of multiobjective evolutionary algorithms by Joshua D. Knowles.

In response to the call for papers, 115 papers from 30 countries were submitted, each of which was independently reviewed by at least three members of the Program Committee. This volume contains the 59 papers that were accepted for presentation at the conference, together with contributions based on the invited talks and tutorials. It is worth noting that the number of submissions to the EMO conference has steadily increased over the years. For EMO 2001, 87 papers were submitted (from which 45 were accepted). For EMO 2003, 100 papers were submitted (from which 56 were accepted). This is a clear indication of the growing interest in this research field.

We would like to express our appreciation to the keynote and tutorial speakers for accepting our invitation. We also thank all the authors who submitted their work to EMO 2005, and the members of the Program Committee for their thorough reviews. The organizers are particularly thankful to the Honda Research Institute Europe for funding two student travel grants through Dr. Yaochu Jin to support students to attend the conference. Finally, we are also very thankful to Edgar Chávez and the Universidad Michoacana for providing us with technical support and for hosting the website that was used for submitting papers to the conference.

March 2005

Carlos A. Coello Coello,
Arturo Hernández Aguirre, and Eckart Zitzler

Organization

EMO 2005 was organized by the Centro de Investigación en Matemáticas(CIMAT) and the Centro de Investigación y de Estudios Avanzados del Instituto Politécnico Nacional (CINVESTAV-IPN) with the support of the Consejo de Ciencia y Tecnología del Estado de Guanajuato (CONCyTEG). The organizers also acknowledge support from the Mexican Consejo Nacional de Ciencia y Tecnología (CONACyT).

General Chairs

Carlos A. Coello Coello   CINVESTAV-IPN, México
Arturo Hernández Aguirre   CIMAT, México
Eckart Zitzler   ETH Zürich, Switzerland

Program Committee

Hussein Abbass   University of New South Wales, Australian Defence Force Academy Campus, Australia
Hernán E. Aguirre   Shinshu University, Japan
Johan Andersson   Linköping University, Sweden
Shapour Azarm   University of Maryland at College Park, USA
Jürgen Branke   University of Karlsruhe, Germany
Carlos A. Brizuela   CICESE, México
Dirk Büche   University of Applied Sciences, Aargau, Switzerland
Kay Chen Tan   National University of Singapore, Singapore
David W. Corne   University of Exeter, UK
Lino Costa   University of Minho, Portugal
Dragan Cvetković   Soliton Inc., Canada
Kalyanmoy Deb   Indian Institute of Technology Kanpur, India
Edwin D. de Jong   Utrecht University, The Netherlands
Rolf Drechsler   University of Bremen, Germany
Els Ducheyne   Institute of Tropical Medicine, Belgium
Joydeep Dutta   Indian Institute of Technology Kanpur, India
Marco Farina   STMicroelectronics, Italy
Jonathan E. Fieldsend   University of Exeter, UK
Mark A. Fleischer   Johns Hopkins University, USA
Peter Fleming   University of Sheffield, UK
Carlos M. Fonseca   Universidade do Algarve, Portugal
Tomonari Furukawa   University of New South Wales, Australia
Xavier Gandibleux   Université de Nantes, France
António Gaspar-Cunha   University of Minho, Portugal
Julia Handl   University of Manchester, UK

Thomas Hanne	Fraunhofer ITWM, Germany
Christian Haubelt	University of Erlangen-Nuremberg, Germany
Alberto Herreros López	Universidad de Valladolid, Spain
Tomo Hiroyasu	Doshisha University, Japan
Evan J. Hughes	Cranfield University, UK
Hisao Ishibuchi	Osaka Prefecture University, Japan
Andrzej Jaszkiewicz	Poznan University of Technology, Poland
Yaochu Jin	Honda Research Institute Europe, Germany
Joshua D. Knowles	University of Manchester, UK
Mario Köppen	Fraunhofer IPK Berlin, Germany
Rajeev Kumar	Indian Institute of Technology Kharagpur, India
Gary B. Lamont	AFIT, USA
Dario Landa Silva	University of Nottingham, UK
Marco Laumanns	ETH Zürich, Switzerland
Xiaodong Li	RMIT University, Australia
José Antonio Lozano	Universidad del País Vasco, Spain
S. Afshin Mansouri	Amirkabir University of Technology, Iran
Carlos E. Mariano Romero	Mexican Institute of Water Technology, México
Efrén Mezura Montes	CINVESTAV-IPN, México
Martin Middendorf	University of Leipzig, Germany
Sanaz Mostaghim	ETH Zürich, Switzerland
Tadahiko Murata	Kansai University, Japan
Shigeru Obayashi	Tohoku University, Japan
Pedro Oliveira	University of Minho, Portugal
Andrzej Osyczka	AGH University of Science and Technology, Poland
Geoffrey T. Parks	University of Cambridge, UK
Robin Purshouse	PA Consulting Group, UK
Ranji S. Ranjithan	North Carolina State University, USA
Tapabrata Ray	National University of Singapore, Singapore
Margarita Reyes Sierra	CINVESTAV-IPN, México
Katya Rodríguez-Vázquez	IIMAS-UNAM, México
Thomas Philip Runarsson	University of Iceland, Iceland
J. David Schaffer	Philips Research, USA
Pradyunm Kumar Shukla	Indian Institute of Technology Kanpur, India
Patrick Siarry	Université Paris 12 (LERISS), France
Thomas Stützle	Darmstadt University of Technology, Germany
El-Ghazali Talbi	University of Lille, France
Hisashi Tamaki	Kobe University, Japan
Jürgen Teich	University of Erlangen-Nuremberg, Germany
Lothar Thiele	ETH Zürich, Switzerland
Dirk Thierens	Utrecht University, The Netherlands

Andrea Toffolo	University of Padua, Italy
Gregorio Toscano Pulido	CINVESTAV-IPN, México
David Van Veldhuizen	Studies & Analyses, Air Mobility Command, USA
Gary G. Yen	Oklahoma State University, USA

Local Organizing Committee

Salvador Botello Rionda	CIMAT, México
Arturo Hernández Aguirre	CIMAT, México

EMO Steering Committee

David W. Corne	University of Exeter, UK
Kalyanmoy Deb	Indian Institute of Technology Kanpur, India
Peter J. Fleming	University of Sheffield, UK
Carlos M. Fonseca	Universidade do Algarve, Portugal
J. David Schaffer	Philips Research, USA
Lothar Thiele	ETH Zürich, Switzerland
Eckart Zitzler	ETH Zürich, Switzerland

Acknowledgments

Invited Speakers

We thank the keynote and tutorial speakers for their talks given at the conference.

Keynote Speakers

Milan Zeleny   Fordham University, USA
Peter Fleming   University of Sheffield, UK

Tutorial Speakers

Xavier Gandibleux   Université de Nantes, France
Gary B. Lamont   Air Force Institute of Technology, USA
Joshua D. Knowles   University of Manchester, UK

Local Sponsors

Support by the following organizations is gratefully acknowledged:

Centro de Investigación en Matemáticas (CIMAT)
Consejo de Ciencia y Tecnología del Estado de Guanajuato (CONCyTEG)
Universidad de Guanajuato
Gobierno del Estado de Guanajuato
Coordinadora de Turismo del Estado de Guanajuato
Centro de Investigación y de Estudios Avanzados del Instituto Politécnico Nacional (CINVESTAV-IPN)

Table of Contents

Invited Talks

The Evolution of Optimality: De Novo Programming
  Milan Zeleny 1

Many-Objective Optimization: An Engineering Design Perspective
  Peter J. Fleming, Robin C. Purshouse, Robert J. Lygoe 14

Tutorial

1984-2004 – 20 Years of Multiobjective Metaheuristics. But What About the Solution of Combinatorial Problems with Multiple Objectives?
  Xavier Gandibleux, Matthias Ehrgott 33

Algorithm Improvements

Omni-optimizer: A Procedure for Single and Multi-objective Optimization
  Kalyanmoy Deb, Santosh Tiwari 47

An EMO Algorithm Using the Hypervolume Measure as Selection Criterion
  Michael Emmerich, Nicola Beume, Boris Naujoks 62

The Combative Accretion Model – Multiobjective Optimisation Without Explicit Pareto Ranking
  Adam Berry, Peter Vamplew 77

Parallelization of Multi-objective Evolutionary Algorithms Using Clustering Algorithms
  Felix Streichert, Holger Ulmer, Andreas Zell 92

An Efficient Multi-objective Evolutionary Algorithm: OMOEA-II
  Sanyou Zeng, Shuzhen Yao, Lishan Kang, Yong Liu 108

Path Relinking in Pareto Multi-objective Genetic Algorithms
  Matthieu Basseur, Franck Seynhaeve, El-Ghazali Talbi 120

Dynamic Archive Evolution Strategy for Multiobjective
Optimization
   Yang Shu Min, Shao Dong Guo, Luo Yang Jie 135

Searching for Robust Pareto-Optimal Solutions in Multi-objective
Optimization
   Kalyanmoy Deb, Himanshu Gupta 150

Multi-objective MaxiMin Sorting Scheme
   E.J. Solteiro Pires, P.B. de Moura Oliveira,
   J.A. Tenreiro Machado 165

Multiobjective Optimization on a Budget of 250 Evaluations
   Joshua Knowles, Evan J. Hughes 176

Initial Population Construction for Convergence Improvement of
MOEAs
   Christian Haubelt, Jürgen Gamenik, Jürgen Teich 191

Multi-objective Go with the Winners Algorithm: A Preliminary Study
   Carlos A. Brizuela, Everardo Gutiérrez 206

Incorporation of Preferences

Exploiting Comparative Studies Using Criteria: Generating Knowledge
from an Analyst's Perspective
   Daniel Salazar, Néstor Carrasquero, Blas Galván 221

A Multiobjective Evolutionary Algorithm for Deriving Final Ranking
from a Fuzzy Outranking Relation
   Juan Carlos Leyva-Lopez, Miguel Angel Aguilera-Contreras 235

Performance Analysis and Comparison

Exploring the Performance of Stochastic Multiobjective Optimisers
with the Second-Order Attainment Function
   Carlos M. Fonseca, Viviane Grunert da Fonseca, Luís Paquete 250

Recombination of Similar Parents in EMO Algorithms
   Hisao Ishibuchi, Kaname Narukawa 265

A Scalable Multi-objective Test Problem Toolkit
   Simon Huband, Luigi Barone, Lyndon While, Phil Hingston 280

Extended Multi-objective fast messy Genetic Algorithm Solving
Deception Problems
  Richard O. Day, Gary B. Lamont 296

Comparing Classical Generating Methods with an Evolutionary
Multi-objective Optimization Method
  Pradyumn Kumar Shukla, Kalyanmoy Deb, Santosh Tiwari 311

A New Analysis of the LebMeasure Algorithm for Calculating
Hypervolume
  Lyndon While 326

Effects of Removing Overlapping Solutions on the Performance of the
NSGA-II Algorithm
  *Yusuke Nojima, Kaname Narukawa, Shiori Kaige,
  Hisao Ishibuchi* 341

Selection, Drift, Recombination, and Mutation in Multiobjective
Evolutionary Algorithms on Scalable MNK-Landscapes
  Hernán E. Aguirre, Kiyoshi Tanaka 355

Comparison Between Lamarckian and Baldwinian Repair on
Multiobjective 0/1 Knapsack Problems
  Hisao Ishibuchi, Shiori Kaige, Kaname Narukawa 370

The Value of Online Adaptive Search: A Performance Comparison of
NSGAII, ε-NSGAII and εMOEA
  Joshua B. Kollat, Patrick M. Reed 386

Uncertainty and Noise

Fuzzy-Pareto-Dominance and Its Application in Evolutionary
Multi-objective Optimization
  Mario Köppen, Raul Vicente-Garcia, Bertram Nickolay 399

Multi-objective Optimization of Problems with Epistemic Uncertainty
  Philipp Limbourg 413

Alternative Methods

The Naive MIDEA: A Baseline Multi–objective EA
  Peter A.N. Bosman, Dirk Thierens 428

New Ideas in Applying Scatter Search to Multiobjective Optimization
  Antonio J. Nebro, Francisco Luna, Enrique Alba 443

A MOPSO Algorithm Based Exclusively on Pareto Dominance Concepts
  Julio E. Alvarez-Benitez, Richard M. Everson, Jonathan E. Fieldsend 459

Clonal Selection with Immune Dominance and Anergy Based
Multiobjective Optimization
  Licheng Jiao, Maoguo Gong, Ronghua Shang, Haifeng Du, Bin Lu ... 474

A Multi-objective Tabu Search Algorithm for Constrained Optimisation
Problems
  Daniel Jaeggi, Geoff Parks, Timoleon Kipouros, John Clarkson 490

Improving PSO-Based Multi-objective Optimization Using Crowding,
Mutation and ϵ-Dominance
  Margarita Reyes Sierra, Carlos A. Coello Coello 505

DEMO: Differential Evolution for Multiobjective Optimization
  Tea Robič, Bogdan Filipič 520

Applications

Multi-objective Model Selection for Support Vector Machines
  Christian Igel 534

Exploiting the Trade-Off — The Benefits of Multiple Objectives in
Data Clustering
  Julia Handl, Joshua Knowles 547

Extraction of Design Characteristics of Multiobjective Optimization –
Its Application to Design of Artificial Satellite Heat Pipe
  Min Joong Jeong, Takashi Kobayashi, Shinobu Yoshimura 561

Gray Coding in Evolutionary Multicriteria Optimization: Application
in Frame Structural Optimum Design
  David Greiner, Gabriel Winter, José M. Emperador, Blas Galván 576

Multi-objective Genetic Algorithms to Create Ensemble of Classifiers
  Luiz S. Oliveira, Marisa Morita, Robert Sabourin, Flávio Bortolozzi .. 592

Multi-objective Model Optimization for Inferring Gene Regulatory
Networks
  Christian Spieth, Felix Streichert, Nora Speer, Andreas Zell 607

High-Fidelity Multidisciplinary Design Optimization of Wing Shape for
Regional Jet Aircraft
   *Kazuhisa Chiba, Shigeru Obayashi, Kazuhiro Nakahashi,
   Hiroyuki Morino* 621

Photonic Device Design Using Multiobjective Evolutionary Algorithms
   Steven Manos, Leon Poladian, Peter Bentley, Maryanne Large 636

Multiple Criteria Lot-Sizing in a Foundry Using Evolutionary
Algorithms
   Jerzy Duda, Andrzej Osyczka 651

Multiobjective Shape Optimization Using Estimation Distribution
Algorithms and Correlated Information
   *Sergio Ivvan Valdez Peña, Salvador Botello Rionda,
   Arturo Hernández Aguirre* 664

Evolutionary Multi-objective Environmental/Economic Dispatch:
Stochastic Versus Deterministic Approaches
   Robert T.F. Ah King, Harry C.S. Rughooputh, Kalyanmoy Deb 677

A Multi-objective Approach to Integrated Risk Management
   Frank Schlottmann, Andreas Mitschele, Detlef Seese 692

An Approach Based on the Strength Pareto Evolutionary Algorithm 2
for Power Distribution System Planning
   Francisco Rivas-Dávalos, Malcolm R. Irving 707

Proposition of Selection Operation in a Genetic Algorithm for a Job
Shop Rescheduling Problem
   Hitoshi Iima 721

A Two-Level Evolutionary Approach to Multi-criterion Optimization
of Water Supply Systems
   Matteo Nicolini 736

Evolutionary Multi-objective Optimization for Simultaneous Generation
of Signal-Type and Symbol-Type Representations
   Yaochu Jin, Bernhard Sendhoff, Edgar Körner 752

A Multi-objective Memetic Algorithm for Intelligent Feature Extraction
   Paulo V.W. Radtke, Tony Wong, Robert Sabourin 767

Solving the Aircraft Engine Maintenance Scheduling Problem Using a
Multi-objective Evolutionary Algorithm
  Mark P. Kleeman, Gary B. Lamont 782

Finding Pareto-Optimal Set by Merging Attractors for a Bi-objective
Traveling Salesmen Problem
  Weiqi Li 797

Multiobjective EA Approach for Improved Quality of Solutions for
Spanning Tree Problem
  Rajeev Kumar, P.K. Singh, P.P. Chakrabarti 811

Developments on a Multi-objective Metaheuristic (MOMH) Algorithm
for Finding Interesting Sets of Classification Rules
  Beatriz de la Iglesia, Alan Reynolds, Vic J Rayward-Smith 826

Preliminary Investigation of the 'Learnable Evolution Model' for
Faster/Better Multiobjective Water Systems Design
  Laetitia Jourdan, David Corne, Dragan Savic, Godfrey Walters 841

Particle Evolutionary Swarm for Design Reliability Optimization
  *Angel E. Muñoz Zavala, Enrique R. Villa Diharce,
  Arturo Hernández Aguirre* 856

Multiobjective Water Pinch Analysis of the Cuernavaca City Water
Distribution Network
  *Carlos E. Mariano-Romero, Víctor Alcocer-Yamanaka,
  Eduardo F. Morales* 870

Multi-objective Vehicle Routing Problems Using Two-Fold EMO
Algorithms to Enhance Solution Similarity on Non-dominated Solutions
  Tadahiko Murata, Ryota Itai 885

Multi-objective Optimisation of Turbomachinery Blades Using Tabu
Search
  *Timoleon Kipouros, Daniel Jaeggi, Bill Dawes, Geoff Parks,
  Mark Savill* 897

Author Index 911

The Evolution of Optimality:
De Novo Programming

Milan Zeleny

Fordham University, New York, USA
Tomas Bata University, Zlín, CR
Xidian University, Xi'an, China
mzeleny@fordham.edu
mzeleny@quick.cz

Abstract. Evolutionary algorithms have been quite effective in dealing with single-objective "optimization" while the area of Evolutionary Multiobjective Optimization (EMOO) has extended its efficiency to Multiple Criteria Decision Making (MCDM) as well. The number of technical publications in EMOO is impressive and indicative of a rather explosive growth in recent years. It is fair to say however that most of the progress has been in applying and evolving algorithms and their convergence properties, not in evolving the optimality concept itself, nor in expanding the notions of true optimization. Yet, the conceptual constructs based on evolution and Darwinian selection have probably most to contribute – at least in theory – to the evolution of optimality. They should be least dependent on a priori fixation of anything in problem formulation: constraints, objectives or alternatives. Modern systems and problems are typical for their *flexibility*, not for their fixation. In this paper we draw attention to the impossibility of optimization when crucial variables are given and present *Eight basic concepts of optimality*. In the second part of this contribution we choose a more realistic problem of linear programming where constraints are not "given" but flexible and to be optimized and objective functions are multiple: *De novo programming*.

1 Introduction

Evolutionary algorithms have been quite effective in dealing with single-objective "optimization" while the area of Evolutionary Multiobjective Optimization (EMOO) has extended its efficiency to Multiple Criteria Decision Making (MCDM) [10] as well. The number of technical publications in EMOO is impressive and indicative of a rather explosive growth in recent years [1]. It is fair to say however that most of the progress has been in applying and evolving algorithms and their convergence properties, not in evolving the optimality concept itself, nor in expanding the notions of true optimization. Yet, the conceptual constructs based on evolution and Darwinian selection have probably most to contribute – at least in theory – to the evolution of optimality. They should be least dependent on a priori fixation of anything in problem formulation: constraints, objectives or alternatives.

The notion of optimality and the process of optimization are pivotal to the areas of economics, engineering, as well as management and business. What does it mean to state that something is 'optimal'? If optimal means 'the best', then asking 'What is the best?' remains a legitimate and still mostly unanswered question.

Any maxima or minima could be declared optimal under specific circumstances, but optima are not necessarily maxima or minima. The two concepts are different: maximizing (or minimizing) is not optimizing.

Although dictionaries commonly use optimization as a synonym for maximization, we shall develop the concept of optimality in the *sense of balance* among multiple criteria or objectives.

When there is only a single dimension or attribute chosen to describe reality, then maximization or minimization with respect to constraints is sufficient. When there are multiple criteria (measures or yardsticks), as is true in most situations, then optimality and optimization (in the sense of balancing) need to be developed.

Optimization applies to an *economic problem* only: when scarce means (constraints) are used to satisfy alternative ends (multiple objectives). If the means are scarce, but there is only a single end, then the problem of how to use the means is a *technical problem*: no value judgments enter into its solution, no balancing is needed, and no optimization can take place. Only knowledge of physical and technical relationships is needed.

In other words, if all my constraints are "given" (fixed) and if my objective function is single, then the solution is fully defined and determined by mathematical problem formulation. The solution just needs to be revealed, explicated or computed by an algorithm. No optimization is possible: all is given and fully determined.

The technical problem is not what we wish to address when dealing with optimality and optimization.

2 The Evolution of Optimality

Balancing of multiple criteria is about optimization, not about "satisficing". Simon acknowledged this quite simply by saying: 'No one in his right mind will satisfice if he can just as well optimize.'

Surprisingly, multiple criteria or multiple objective functions - the necessary prerequisites for optimization - were not recognized and acknowledged by the optimization sciences until the early 1970s. Optimization in the sense of balancing multi-dimensionality is not compatible with the traditional concepts of "optimality" characterized by scalar or scalarized schemes, based on unique solutions under complete information. These are rather limited in capturing the richness and complexity of human problem solving, decision making and optimization.

We must strive to understand decision making not merely as computation of the given, already-constructed world, but as a way of constructing our local world, ordering of individual and collective experience. It is necessary to acknowledge multiple concepts of optimality.

3 Multiple Concepts of Optimality

There are some axiomatic prerequisites that must be at the base of any optimization scheme.

For example, what is determined or given a priori cannot be subject to subsequent optimization and thus, clearly, does not need to be optimized: it is given.

What is not given must be selected, chosen or identified and is therefore, by definition, subject to optimization.

Consequently, different optimality concepts can be derived from different distinctions between what is given and what is yet to be determined in problem-solving or decision-making formulations.

For example, if I determine the value of the objective function a priori, set it at a predetermined value, then I cannot optimize it (nor maximize or minimize). If I set a value of the constraint a priori, then I cannot optimize that constraint. Constraints have to become objectives in order to be optimized, see also [2]. Even if I do not determine the value of the objective a priori, but the constraints are fixed, I still cannot optimize it – it is strictly implied (given) by the constraints.

Traditionally, by optimal solution or optimization we implicitly understand maximizing (or minimizing) a single, pre-specified objective function (or criterion) with respect to a given, fixed set of decision alternatives (or situation constraints). Both criterion and decision alternatives are given, only the (optimal) solution itself remains to be calculated.

There are at least *eight distinct optimality concepts*, all mutually irreducible, all characterized by different applications, interpretations and mathematical formalisms. For details see [13, 14, and 15].

Single-Objective Optimality

This is not really optimization but refers to the conventional maximization (or 'optimization') problem. It should be included for the sake of completeness, out of respect for tradition and as a potential special case of bona fide optimization.

To maximize a single criterion it is fully sufficient to perform technical measurement and algorithmic search processes. Once X and f are formulated or specified, the 'optimum' (that is, maximum) is found by computation, not by decision processes or balancing. Search for optimality is reduced to 'scalarization': assigning each alternative a number (scalar) and then identifying the largest-numbered alternative.

Numerical example

Consider the following linear-programming problem with two variables and five constraints:

Max $f = 400x + 300y$
subj. to $4x \leq 20$
$2x + 6y \leq 24$

$12x + 4y \leq 60$
$3y \leq 10.5$
$4x + 4y \leq 26$

The maximal solution to the above problem is $x^* = 4.25$, $y^* = 2.25$, and $f^* = 2375$. Observe that all is given here and considering market prices of resources is unnecessary. However, if $p_1 = 30$, $p_2 = 40$, $p_3 = 9.5$, $p_4 = 20$ and $p_5 = 10$ were respective market prices (\$/unit) of the five respective resources, the total cost of current resource portfolio (20, 24, 60, 10.5, 26) would be B= \$2600.

Multi-objective Optimality

More generally, optimality, to be distinct from maximizing, should involve balancing and harmonizing multiple criteria. In the real world, people continually resolve conflicts among multiple criteria which are competing for their attention and assignments of importance. This corresponds to the vector optimization problem. This maximization of individual functions should be non-scalarized, separate and independent, that is, not subject to superfunctional aggregation which would effectively reduce multi-objective optimality to single-objective maximization: there would be no reason to consider multiple criteria other than for constructing the superfunction. Multiple criteria, if they are to be meaningful and functional, should be optimized (or balanced) in the non-scalarized vector sense, in mutual competition with each other.

Numerical example

Max $f_1 = 400x + 300y$
and $f_2 = 300x + 400y$
subj. to $4x \leq 20$
$2x + 6y \leq 24$
$12x + 4y \leq 60$
$3y \leq 10.5$
$4x + 4y \leq 26$

The maximal solution with respect to f_1 is $x^* = 4.25$ and $y^* = 2.25$, f_1 (4.25, 2.25) = 2375. The maximal solution with respect to f_2 is $x^* = 3.75$ and $y^* = 2.75$, f_2 (3.75, 2.75) = 2225. The set of optimal (non-dominated) solutions X^* includes the two maximal solutions (extreme points) and their connecting (feasible) line defined by $4x + 4y = 26$. For example, $0.5(4.25, 2.25) + 0.5(3.75, 2.75) = (4.0, 2.5)$ is another non-dominated point in the middle of the line. Total cost of resource portfolio remains B =\$2600.

Optimal System Design: Single Criterion

Instead of optimizing a given system with respect to selected criteria, humans often seek to form or construct an optimal system of decision alternatives (optimal feasible set), designed with respect to such criteria. Single-criterion design is the simplest of such concepts: it is analogous to single-criterion 'optimization', producing the best (optimal) set of alternatives X at which a given, single objective function f is maximized subject to the cost of design (affordability).

Numerical example

Max f = 400x + 300y
subj. to 4x ≤ 29.4
2x+ 6y ≤ 14.7
12x + 4y ≤ 88.0
3y ≤ 0
4x + 4y ≤ 29.4

where the right-hand sides (resource portfolio) have been optimally designed. Solving the above optimally designed system will yield x* = 7.3446, y*= 0 and f(x*) = 2937.84. If market prices of the five resources (p_1= 30, p_2= 40, p_3= 9.5, p_4= 20 and p_5 = 10) remain unchanged, then the total cost of the resource portfolio (29.4, 14.7, 88, 0, 29.4) is again B = $2600.

Optimal System Design: Multiple Criteria

As before, multiple criteria cannot be scalarized into a superfunction. Rather, all criteria compete independently or there would be no need for their separate treatment.

Numerical example

Max f_1 = 400x + 300y
and f_2 = 300x + 400y
subj. to 4x ≤ 16.12
2x + 6y ≤ 23.3
12x +4y ≤ 58.52
3y ≤ 7.62
4x + 4y ≤ 26.28

The above represents an optimally designed portfolio of resources: maximal solution with respect to both f_1 and f_2 is x* = 4.03 and y* =2.54, f_1(4.03, 2.54) = 2375 and f_2(4.03, 2.54) = 2225. This can be compared (for reference only) with the f_1 and f_2 performances in the earlier case of given right-hand sides. Assuming the same prices of resources, the total cost of this resource portfolio is B = $2386.74 ≤ $2600. One could therefore design even better performing portfolios by spending the entire budget of 2600 (or the additional $213.26).

Optimal Valuation: Single Criterion

All previously considered optimization forms assume that decision criteria are given *a priori*. However, in human decision making, different criteria are continually being tried and applied, some are discarded, new ones added, until an optimal (properly balanced) mix of both quantitative and qualitative criteria is identified. There is nothing more suboptimal than engaging perfectly good set of alternatives X towards unworthy, ineffective or arbitrarily determined criteria (goals or objectives).

If the set of alternatives X is given and fixed *a priori*, we face a problem of optimal *valuation:* According to what measures should the alternatives be evaluated or ordered? According to criterion f_1, f_2 or f_3? Which of the criteria captures best our values and purposes? What specific criterion engages the available means (X) in the most effective way?

Numerical example

In order to evaluate X, should we maximize f_1 or f_2? How do we select a criterion if only one is allowed (possible) or feasible?

Max $f_1 = 400x + 300y$
or $f_2 = 300x + 400y$
subj. to $4x \leq 20$
$2x + 6y \leq 24$
$12x + 4y \leq 60$
$3y \leq 10.5$
$4x + 4y \leq 26$

The maximal solution with respect to f_1 is $x^* = 4.25$, $y^* = 2.25$, $f_1 (4.25, 225) = 2375$.

Maximal solution with respect to f_2 is $x^* = 3.75$, $y^* = 2.75$, $f_2 (3.75, 2.75) = 2225$. Is 2375 of f_1 better than 2225 of f_2? Only one of these valuation schemes can be selected.

Optimal Valuation: Multiple Criteria

If the set of alternatives X is given and fixed a priori, but a set of multiple criteria is still to be selected for the evaluation and ordering of X, we have a problem of multiple-criteria valuation:

Which set of criteria best captures our value complex? Is it (f_1 and f_2)? Or (f_2 and f_3)? Or perhaps (f_1 and f_2 and f_3)? Or some other combination?

Numerical example

How do we select a set of criteria f_1 or f_2 or (f_1 and f_2) that would best express a given value complex?

Max $f_1 = 400x + 300y$
or/and Max $f_2 = 300x + 400y$
subj. to $4x \leq 20$

$$2x + 6y \le 24$$
$$12x + 4y \le 60$$
$$3y \le 10.5$$
$$4x + 4y \le 26$$

The maximal solution with respect to f_1 is $x^* = 4.25$, $y^* = 2.25$, f_1 (4.25, 2.25) = 2375. Maximal solution with respect to f_2 is $x^* = 3.75$, $y^* = 2.75$, f_2 (3.75, 2.75) = 2225. Should we use f_1 or f_2 or should we use both f_1 and f_2 to achieve the best valuation of X? Only one of possible (single and multiple criteria) valuation schemes is to be selected.

Optimal Pattern Matching: Single Criterion

All previously considered optimization concepts assume that relevant decision criteria are given and determined a priori. Yet, that is not how human decision-making processes are carried out: different criteria are being tried and applied, some are discarded, new ones added, until a proper balanced mix (or portfolio) of both quantitative and qualitative criteria is derived.

Like any other decision-problem factors, criteria should be determined and designed in an optimal fashion. There is nothing more wasteful than engaging perfectly good means and processes towards unworthy, ineffective or only arbitrarily determined criteria.

There is a problem formulation representing an 'optimal pattern' of interaction between alternatives and criteria. It is this optimal, ideal or balanced problem formulation or pattern that is to be approximated or matched by decision makers. Single-objective matching of such cognitive equilibrium [18] is once more the simplest special case.

Numerical example

Should we maximize f_1 or f_2? How do we select a single criterion if only one is allowed, possible or feasible?

Max $f_1 = 400x + 300y$
or Max $f_2 = 300x + 400y$
subj. to $4x \le 29.4$ or 0
$2x + 6y \le 14.7$ 41.27
$12x + 4y \le 88$ 27.52
$3y \le 0$ 20.63
$4x + 4y \le 29.4$ 27.52

The above presents two optimally designed portfolios of resources with respect to f_1 and f_2 respectively. Among the possible patterns are ($x^* = 7.3446$, $y^* = 0$, $f_1 (x^*) = 2937.84$, B= \$2600) and ($x^* = 0$, $y^* = 6.8783$, $f_2 (y^*) = 2751.32$, B = \$2600).

Suppose that the value complex requires that the chosen criterion should minimize the opportunity cost of the unchosen criteria, other things being equal. Choosing f_1 would make f_2 drop only to 80.08 per cent of the opportunity performance, whereas choosing f_2 would make f_1 drop to 70.24 per cent. So, f_1 has preferable opportunity impact, and the first pattern and its resource portfolio would be selected.

A value complex indicating that deployed resource quantities should be as small as possible would require choosing f_2 and thus the second pattern.

Optimal Pattern Matching: Multiple Criteria

Pattern matching with multiple criteria is more involved and the most complex optimality concept examined so far. In all/matching' optimality concepts there is a need to evaluate the closeness (resemblance or match) of a proposed problem formulation (single or multi-criterion) to the optimal problem formulation.

Numerical example

How do we select a set of criteria f_1, f_2 or (f_1, f_2) that would best express our current value complex?

Max f_1 = 400x+ 300y
or/and Max f_2 = 300x + 400y
subj. to 4x $\leq$ 29.4 or 0 or 19.98
2x+ 6y $\leq$ 14.7 41.27 28.78
12x + 4y $\leq$ 88 27.52 72.48
3y $\leq$ 0 20.63 9.39
4x + 4y $\leq$ 29.4 27.52 32.50

The above describes three optimally designed portfolios of resources with respect f_1, f_2 and (f_1, f_2) respectively. So, among the possible patterns are (x* =7.3446, y*=0, $f_1(x^*)$ = 2937.84, B=$2600), (x* = 0, y*= 6.8783, $f_2(x^*)$ = 2751.32, B = $2600) and (x* = 4.996, y* = 3.131, $f_1(x^*)$ = 2937.84, $f_2(x^*)$ = 2751.32, B = $2951.96).

If the value complex requires that B =2600 is not to be exceeded, we may 'match' the third optimal pattern to that level by scaling it down by the optimum-path ratio r =2600/2951.% = 0.88. The new pattern is (x* = 4.396, y* = 2.755, $f_1(x^*)$ = 2585.30, $f_2(x^*)$ = 2421.16, B = $2600). If producing both products is of value, then the choice could be maximization of both f_1 and f_2.

4 Summary of Eight Concepts

In Figure 1 we summarize the eight major optimality concepts according to a dual classification: single versus multiple criteria versus the extent of the 'given', ranging from 'all-but' to 'none except'. The traditional concept of optimality, charac-

Number of Criteria \ Given	Single	Multiple
Criteria & Alternatives	Traditional "Optimality"	MCDM
Criteria Only	Optimal Design (De Novo Programming)	Optimal Design (De Novo Programming)
Alternatives Only	Optimal Valuation (Limited Equilibrium)	Optimal Valuation (Limited Equilibrium)
"Value Complex" Only	Cognitive Equilibrium (Matching)	Cognitive Equilibrium (Matching)

Fig. 1. Eight concepts of optimality

terized by too many 'givens' and a single criterion, naturally appears to be the most remote from any sort of optimal conditions or circumstances for problem solving as represented by cognitive equilibrium (optimum) with multiple criteria.

The third row of Figure 1 can be solved by De novo programming in linear cases. In the next section we summarize the basic formalism of De Novo programming, as it applies to linear systems [11, 12, 16, and 17]. It is only with multiple objectives that optimal system design becomes fully useful, even though a single-objective formulation can also lead to performance improvements.

5 Formal Summary of De Novo Programming

Formulate linear programming problem:

$$\text{Max } Z = Cx \text{ s. t. } Ax - b \leq 0, \, pb \leq B, \, x \geq 0, \quad (1)$$

where $C \in \Re^{q \times n}$ and $A \in \Re^{m \times n}$ are matrices of dimensions $q \times n$ and $m \times n$, respectively, and $b \in \Re^m$ is m-dimensional *unknown* resource vector, $x \in \Re^n$ is n-dimensional vector of decision variables, $p \in \Re^m$ is the vector of the unit prices of m resources, and B is the given total available budget.

Solving problem (1) means finding the optimal allocation of B so that the corresponding resource portfolio b maximizes simultaneously the values $Z = Cx$ of the product mix x.

Obviously, we can transform problem (1) into:

$$\text{Max } Z = Cx \text{ s. t. } Vx \leq B, \, x \geq 0, \quad (2)$$

where $Z = (z_1, ..., z_q) \in \Re^q$ and $V = (V_1, ..., V_n) = pA \in \Re^n$.

Let $z_{k*} = \max z_k$, $k = 1, ..., q$, be the optimal value for kth objective of Problem (2) subject to $Vx \leq B$, $x \geq 0$. Let $Z^* = (z_{1*}, ..., z_{q*})$ be the q-objective value for the ideal system with respect to B. Then, a *metaoptimum problem* can be constructed as follows:

$$\text{Min } Vx \text{ s. t. } Cx \geq Z^*, \, x \geq 0. \quad (3)$$

Solving Problem (3) yields x^*, B^* ($= Vx^*$) and b^* ($= Ax^*$). The value B^* identifies the minimum budget to achieve Z^* through x^* and b^*.

Since $B^* \geq B$, the optimum-path ratio for achieving the ideal performance Z^* for a given budget level B is defined as:

$$r^* = B/B^* \quad (4)$$

and establish the optimal system design as (x, b, Z), where $x = r^*x^*$, $b = r^*b^*$ and $Z = r^*Z^*$. The optimum-path ratio r^* provides an effective and fast tool for efficient optimal redesign of large-scale linear systems.

Shi [8] observed that two additional types of budget (other than B and B^*) can be usefully introduced. One is B_j^k, the budget level for producing optimal x_j^k with respect to the kth objective, referring to a single-objective De Novo programming problem.

The other, B^{**}, refers to the case $q \leq n$ (the number of objectives smaller than the number of variables). If x^{**} is the degenerate optimal solution, then $B^{**} = Vx^{**}$ (See Shi [8]). It can be shown that $B^{**} \geq B^* \geq B \geq B_j^k$, for $k = 1, ..., q$.

Shi defines six types of optimum-path ratios:

$$r_1 = B^*/B^{**};\ r_2 = B/B^{**};\ r_3 = \Sigma\ \lambda_k B_j^k/B^{**};$$
$$r_4 = r^* = B/B^*;\ r_5 = \Sigma\ \lambda_k B_j^k/B^*;\ r_6 = \Sigma\ \lambda_k\ B_j^k/B,$$

leading to six different optimal system designs. Comparative economic interpretations of all optimum-path ratios are still to be fully researched.

The following numerical example is adapted from Zeleny [11]:

Max $z_1 = 50\ x_1 + 100\ x_2 + 17.5\ x_3$
$z_2 = 92\ x_1 + 75\ x_2 + 50\ x_3$
$z_3 = 25\ x_1 + 100\ x_2 + 75\ x_3$

subject to

$12\ x_1 + 17\ x_2 \leq b_1$
$3\ x_1 + 9\ x_2 + 8\ x_3 \leq b_2$
$10\ x_1 + 13\ x_2 + 15\ x_3 \leq b_3$ (5)
$6\ x_1 + 16\ x_3 \leq b_4$
$12\ x_2 + 7\ x_3 \leq b_5$
$9.5\ x_1 + 9.5\ x_2 + 4\ x_3 \leq b_6$

We assume that the objective functions z_1, z_2, and z_3 are equally important. We are to identify the optimal resource levels of b_1 through b_6 when the current unit prices of resources are $p_1 = 0.75$, $p_2 = 0.60$, $p_3 = 0.35$, $p_4 = 0.50$, $p_5 = 1.15$ and $p_6 = 0.65$. The initial budget $B = 4658.75$.

We calculate $Z^* = (10916.813;\ 18257.933;\ 12174.433)$ with respect to given $B\ (= 4658.75)$. The feasibility of Z^* can be only assured by the metaoptimum solution $x^* = (131.341,\ 29.683,\ 78.976)$ at the cost of $B^* = 6616.5631$.

Because the optimal-path ratio $r^* = 4658.75/6616.5631 = 70.41\%$, the resulting $x = (92.48,\ 20.90,\ 55.61)$ and $Z = (7686.87;\ 12855.89;\ 8572.40)$. It follows that the optimal portfolio b, with respect to $B = 4658.75$, can be calculated by substituting x into the constraints (5). We obtain:

$b_1 = 1465.06$
$b_2 = 910.42$
$b_3 = 2030.65$
$b_4 = 1444.64$ (6)
$b_5 = 640.07$
$b_6 = 1299.55$

If we spend precisely B = 4658.8825 (approx. 4658.75) the optimum portfolio of resources to be purchased at current market prices is displayed in (6), allowing us to produce **x** and realize **Z** in criteria performance. The analysis with respect to the additional Shi ratios would proceed along similar lines.

6 Extended De Novo Formulation

There are many interesting extensions of De Novo programming [3-8]. But the main extension always concerns the objective function [12]. The multiobjective form of Max (cx - pb) appears to be the right function to be maximized in a globally competitive economy. This is compatible with achieving long-term maximum sustainable yields from deployed resources. Another realistic feature would be multiple pricing and quantity discounts in both resources and products markets.

Searching for a better portfolio of resources (redefining the b_is of right-hand sides) is tantamount to the continuous reconfiguration and "reshaping" of systems boundaries. Such practical considerations lead to a more general programming formulation, starting to approximate the real concerns of free-market producers.

For example, the following optimal-design formulation of the production problem, although still quite incomplete, takes a full advantage of the De novo programming computational efficiency while delivering the necessary decision inputs:

Max $z = \Sigma_j\ c_j(x_j)x_j - [\Sigma_{i \in I_1}\ p_i b_i]\pi_1 - ... - [\Sigma_{i \in I_r}\ p_i b_i]\pi_r$
s.t $\Sigma_j\ a_{ij}x_j - b_i\ \leq 0\ i \in I$
$[\Sigma_{i \in I}\ p_i b_i]\beta \leq B,$
where $I = I_1 \cup ... \cup I_r,\ I_s \cap I_{s+1} = 0,\ 0 < \pi_s < 1,\ s = 1, ..., r,\ \beta \geq 1$

and

$c_{j1}\ x_j \leq x_{j1}$
$c_{j2}\ x_{j1} < x_j \leq x_{j2}$
.
$c_j(x_j) = .$
.
$c_{jkj}\ x_{jkj-1} < x_j,$
where $c_{jh} \geq c_{h+1},\ h = 1, ...,\ k_j.$

The formulation above is more practical than the traditional LP- systems, but perhaps still quite far away from the useful formulation of the real world-class management systems.

7 Conclusion

The challenge for EMOO is obvious: how to evolve its conceptual and technical capabilities to transform itself into a truly evolutionary vehicle that would encompass all practical concepts of optimality, including at least the eight basic ones, as outlined in this paper.

References

1. Coello Coello, C. A., A comprehensive survey of evolutionary-based multiobjective optimization techniques, *Knowledge and Information Systems* **1**(1999)3, pp. 269-308.
2. Coello Coello, C. A., Treating constraints as objectives for single-objective evolutionary optimization, *Engineering Optimization* **32**(2000)3, pp. 275-308.
3. Lee, E. S., Optimal-design models. *International Encyclopedia of Business & Management* (London: Routledge, London, 1996) pp. 3758-3766.
4. Lee, E. S., On Fuzzy De Novo Programming, in: *Fuzzy Optimization: Recent Advances*, edited by M. Delgado, J. Kacprzyk, J.-L. Verdegay and M. A. Vila, (Physica-Verlag, Heidelberg, 1994) pp. 33-44.
5. Li, R-J. and Lee, E. S, Fuzzy Approaches to Multicriteria De Novo Programs. *Journal of Mathematical Analysis and Applications* **153** (1990) pp. 97-111.
6. Li, R-J. and Lee, E. S, Multicriteria De Novo Programming with Fuzzy Parameters. *Computers and Mathematics with Applications* **19**(1990) pp. 13-20.
7. Li, R-J. and Lee, E. S., De Novo Programming with Fuzzy Coefficients and Multiple Fuzzy Goals. *Journal of Mathematical Analysis and Applications* **172**(1993) pp. 212-220.
8. Shi, Y., Studies on Optimum-Path Ratios in De Novo Programming Problems, *Computers and Mathematics with Applications* **29** (1995) pp. 43-50.
9. Zeleny, M, *IEBM Handbook of Information Technology in Business*, (Thomson, London, 2000)
10. Zeleny, M., *Multiple Criteria Decision Making* (McGraw-Hill, New York, 1982)
11. Zeleny, M., Optimal System Design with Multiple Criteria: De Novo Programming Approach. *Engineering Costs and Production Economics* **10** (1986) pp. 89-94.
12. Zeleny, M., Optimizing Given Systems vs. Designing Optimal Systems: The De Novo Programming Approach, *General Systems* **17**(1990) pp. 295-307.
13. Zeleny, M., Optimality and optimization. *International Encyclopedia of Business & Management* (Routledge, London, 1996) pp. 3767-3781.
14. Zeleny, M., Optimality and Optimization. *IEBM Handbook of Information Technology in Business*, ed. M. Zeleny, Thomson, London, 2000, pp. 392-409.
15. Zeleny, M., Multiple Criteria Decision Making: Eight Concepts of Optimality, *Human Systems Management* **17**(1998)2, pp. 97-107.
16. Zeleny, M., Multiobjective Design of High-Productivity Systems. *Proceedings of Joint Automatic Control Conference,* July 27-30, 1976, Purdue University, Paper APPL9-4, New York, 1976, pp. 297-300.
17. Zeleny, M., On the Squandering of Resources and Profits via Linear Programming. *Interfaces* **11** (1981) pp. 101-107.
18. Zeleny, M, Cognitive equilibrium: a knowledge-based theory of fuzziness and fuzzy sets, *General Systems* **19**(1991), pp. 359-81.

Many-Objective Optimization: An Engineering Design Perspective

Peter J. Fleming[1], Robin C. Purshouse[2], and Robert J. Lygoe[3]

[1] Department of Automatic Control and Systems Engineering,
University of Sheffield, Mappin Street,
Sheffield S1 3JD, UK
P.Fleming@sheffield.ac.uk
[2] PA Consulting Group,
123 Buckingham Palace Road,
London SW1W 9SR, UK
Robin.Purshouse@PACONSULTING.COM
[3] Powertrain Applications Product Development,
Ford Motor Company Ltd

Abstract. Evolutionary multicriteria optimization has traditionally concentrated on problems comprising 2 or 3 objectives. While engineering design problems can often be conveniently formulated as multiobjective optimization problems, these often comprise a relatively large number of objectives. Such problems pose new challenges for algorithm design, visualisation and implementation. Each of these three topics is addressed. Progressive articulation of design preferences is demonstrated to assist in reducing the region of interest for the search and, thereby, simplified the problem. Parallel coordinates have proved a useful tool for visualising many objectives in a two-dimensional graph and the computational grid and wireless Personal Digital Assistants offer technological solutions to implementation difficulties arising in complex system design.

1 Introduction

Real-world engineering design problems often involve the satisfaction of multiple performance measures, or objectives, which should be solved simultaneously. Automotive and aerospace examples provide illustrations of some typical design challenges and demonstrate that these problems often involve a large number of objectives. It is demonstrated how a typical set of engineering design specifications might be mapped onto a familiar formulation of an EMO problem. EMO research has, for the most part, focused on problems having 2 or 3 objectives; however, in recent years there has been growing interest in the area of *many-objective* optimization where the problem might consist of 4 – 20 objectives, for example.

Of the three key requirements for EMO solution set quality - proximity, diversity and pertinency - a case is made that pertinency, focussing on solutions in the designer's region of interest, has a special prominence in *many-objective* optimization studies. A method whereby the MOEA is operated in an interactive manner through progressive articulation of preferences is described and an example worked through to explore the potential of this approach.

The means of using the method of parallel coordinates to reduce the study of a *many-dimensional* Pareto front to a 2-D representation reveals a number of strengths and limitations. *Many-objective* optimization in an engineering design context is inevitably very compute-intensive and there is an expectation that a design procedure will often be time-consuming. Two schemes are introduced to deal with these demands. In one scheme, the MOEA is parallelised to execute effectively in a computational grid environment in reduced time. For the second scheme, it is shown that wireless PDAs can be effective tools for designers in the interactive computational steering of the design process.

2 Design Approaches

In this section we provide examples of conflicting objectives in two areas of engineering design and then provide some background to solution approaches used in the past.

2.1 Typical Engineering Design Optimization Problems

Automotive Engineering Examples

Historically, in automotive engineering, the process of establishing trade-offs has been to conduct parametric studies. That is, evaluating the conflicting objective functions at different values of the decision variables (parameters), comparing the results in objective space and then finally selecting a single trade-off solution. An example of such a parametric study is shown in an enumeration plot (see Fig. 1), where two conflicting objective functions (empirical models of NOx and Brake Torque) are plotted against each other, evaluated as a function of their input (decision) variables.

Brake Torque is a surrogate variable for fuel economy and is easily measurable on an engine dynamometer test rig; maximising brake torque is equivalent to optimizing fuel economy. NOx or Oxides of Nitrogen are one of the three legislated exhaust emission pollutants and in this case is measured pre-catalytic converter. Minimising NOx minimises the precious metal (e.g. Platinum or Rhodium) coating in the catalytic converter and thus cost.

In Fig. 1 the decision variable is EGR (Exhaust Gas Recirculation) rate, which gives a benefit in brake torque and NOx. Brake Torque maximises at moderate EGR rate, but NOx minimises at maximum EGR rate. Thus, the objectives conflict.

Since the optimization problem is to maximise Brake Torque and minimise NOx, trade-off solutions in the lower RH corner of Fig. 1 are preferred. Using a parametric approach, many objective function evaluations are required, which may be expensive, particularly if there are a large number of objective functions. (In automotive engineering it is not uncommon to have problems with 4-10 objectives.) Also, it is possible that suitable Pareto-optimal solutions will not be discovered and that a sub-optimal trade-off solution will be selected.

Fig. 1 is obtained from empirical models of Brake Torque and NOx. The models are based on automated test data arising from a designed experiment on an engine on an engine dynamometer test rig. The resulting models are then validated against independent test data and against known physical trends.

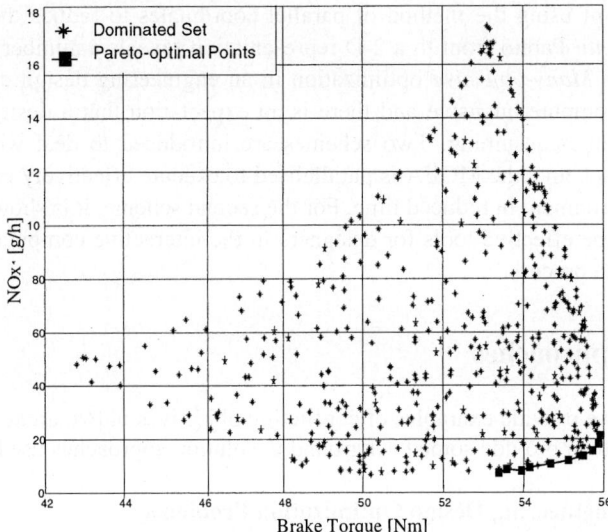

Fig. 1. Relationship between conflicting objective functions, NOx and Brake Torque

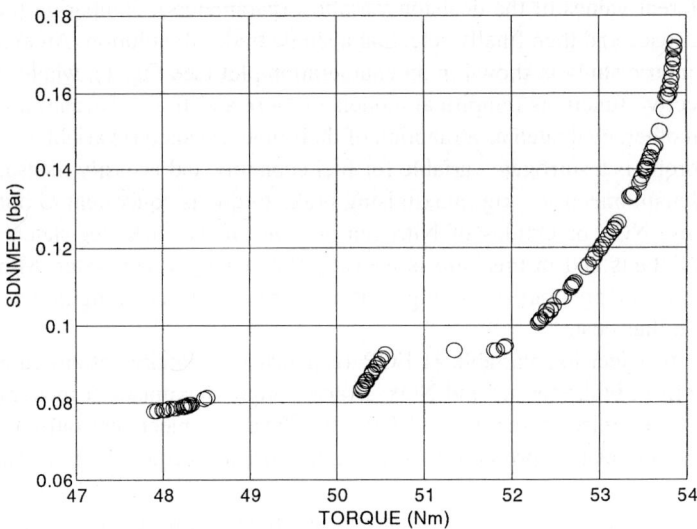

Fig. 2. Piece-wise continuous Pareto front obtained by comparing SDNMEP with torque

Such models offer practical advantages over an online approach (i.e. connecting an optimizer directly to the test rig) in terms of efficiency, re-use and a noise-free or repeatable objective function evaluation. Connecting the optimizer to these models provides an efficient and systematic search capability for Pareto-optimal solutions, which, by definition, represent optimal system capability that is of high engineering and business value.

Fig. 2 is an example of a piece-wise continuous Pareto front obtained by comparing SDNMEP (Standard Deviation of Net Mean Effective Pressure - a measure of combustion stability) with torque. Good combustion stability is necessary for engine smoothness, which is now a customer expectation of modern mass-produced engines. The decision variables, Intake Valve Opening (IVO) & Exhaust Valve Closing (EVC), are used to determine cam positions in a continuously variable Twin Independent Variable Cam Timing system. Torque maximises at moderate IVO + late EVC (medium overlap) or late IVO & EVC, whereas SDNMEP minimises at late IVO + early EVC (low overlap). Again, the objectives conflict.

Aerospace Engineering Examples

For different classes of gas turbine engines (GTEs) for aircraft propulsion, while the controller structure will often remain the same, there is a requirement for redesign of the control system for new engines and new performance requirements. (Besides aerospace, GTEs are also used for applications such as marine vehicle propulsion and power generation.) Each application requires the engine to operate in a specified manner and provide particular shaft velocities, efficiencies and thrust output. These depend on the purpose for which the engine is being used. One way of varying the performance of these engines is through the choice of a suitable control system that is able to regulate the fuel supplied to the combustion chamber of the engine and thus influences the performance of the engine.

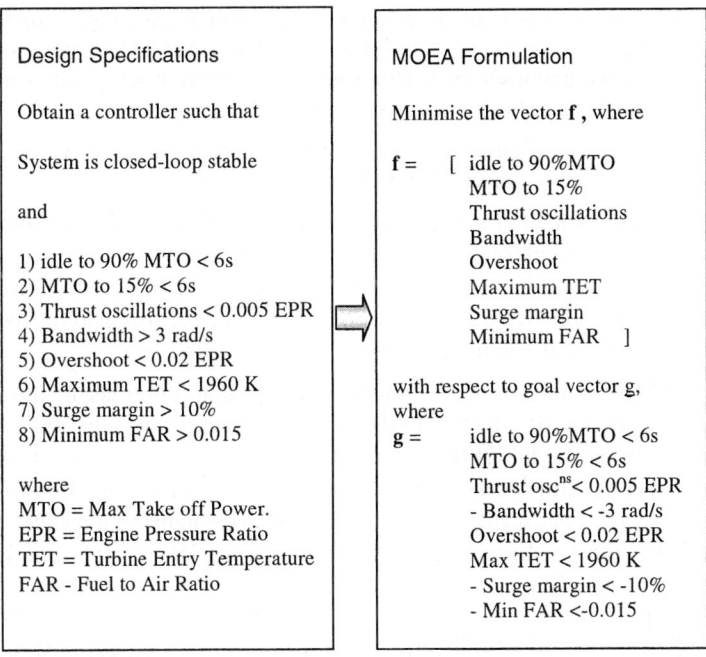

Fig. 3. Mapping of design specifications into a multiobjective evolutionary algorithm (MOEA) formulation

Fig. 3 provides a typical set of design specifications (and, for later discussion, a mapping to a specific EMO treatment is provided). For example, objective (1) represents a measure of rapid acceleration for go-around/aborted landing, objective (2) represents a measure of rapid deceleration for stopping on runway for aborted takeoff, and objective (8) prevents flameout, a factor which is relevant on rapid decelerations, e.g. when decelerating at high altitude at the end of cruise or an aborted takeoff.

High-fidelity nonlinear dynamic models of engine performance exist and controller design is realized by implementing alternative strategies on these models for performance evaluation and comparison. Until recently based designs were limited to scalar objectives and, often, addressed the design specifications indirectly by employing measures that required weight manipulation to shape the response into the desired form. The availability of tools to directly address *multiple* objectives is leading to shorter design times and improved system performance.

Fig. 3 demonstrates how a conventional design specification can be mapped into a multiobjective optimization (MO) formulation. Subsequent sections will describe and illustrate how such a MO formulation, complete with goals, can be treated in a multiobjective optimization evolutionary algorithm (MOEA) framework and the designs executed using progressive preference articulation.

2.2 Requirements of a Multi-objective Optimizer for Engineering Design

The globally optimal trade-off surface of a multi-objective optimization problem can contain a potentially infinite number of Pareto-optimal solutions. The task of a multi-objective optimizer is to provide an accurate and useful representation of the trade-off surface to the decision-maker. The set of solutions generated by the optimizer is known as an approximation set [32]. Three aspects of solution set quality can be considered. These are listed below, and shown graphically in Fig. 4.

Proximity. The approximation set should contain solutions whose corresponding objective vectors are close to the true Pareto front.

Diversity. The approximation set should contain a good distribution of solutions, in terms of both extent and uniformity. Good diversity is commonly of interest in objective-space, but may also be required in decision-space. In objective-space, the approximation set should extend across the entire range of the true Pareto front with a parametrically uniform distribution across the surface.

Pertinency. The approximation set should only contain solutions in the decision maker's (DM's) region of interest (ROI). In practice, and especially as the number of objectives increases, the DM is interested only in a sub-region of objective-space. Thus, there is little benefit in representing trade-off regions that lie outside the ROI. Focusing on pertinent areas of the search space helps to improve optimizer efficiency and reduces unnecessary information that the DM would otherwise have to consider.

2.3 A Brief Account of Multiobjective Optimization in Engineering Design

A typical control system design problem might be posed as follows. Given a system $\mathbf{x} = \Phi(\mathbf{x}, \mathbf{u}, t)$ where $\mathbf{x}$ and $\mathbf{u}$ are the system state and control vectors and Φ is a vector non-linear function, find a controller $\mathbf{u}$ such that the design specifications,

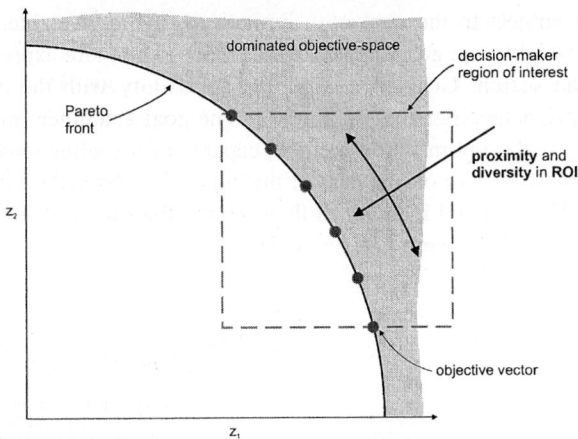

Fig. 4. The ideal solution to a multi-objective optimization problem

$$f_i(\mathbf{x},\mathbf{u},t) \le g_i, i = 1,\ldots,m, \tag{1}$$

are satisfied, where g_i are the design goals. Such problems were most often addressed via optimization by aggregating objectives directly (or by indirect means) into a weighted sum, such as the following objective function used in linear quadratic regulator design [1],

$$J = \int_0^\infty \{\mathbf{x}^T Q \mathbf{x} + \mathbf{u}^T R \mathbf{u}\} dt, \tag{2}$$

where Q and R are user-selected weighting matrices to guide the design. The well-known drawbacks of the weighted sum approach are the difficulty in setting values for the weights, and the fact that the method has been proved to be incapable of generating solutions in non-convex regions of the trade-off surface [4]. It is also a single solution method, requiring multiple starts in attempts to build up trade-off information.

Zakian and Al-Naib proposed a method for obtaining a control vector, **u**, of pre-specified structure [31], which satisfied the design specifications/constraints eqn. (1). However, this was a constraint satisfaction approach that (i) did not attempt to *optimize* the solution once one was found that satisfied the constraints, and (ii) had no recovery strategy, should the solution space prove to be null.

In the goal attainment method [18], the designer is required to specify a set of goals, let us also call them **g**, for the objective function vector, **f**. The nonlinear programming problem to be solved is:

$$\text{Min } \lambda, \text{ with } \lambda, \mathbf{p} \text{ subject to: } f_i - w_i \lambda \le g_i, i = 1,\ldots,m \tag{3}$$

where **p** is the decision variable vector (or controller parameters), $w_i \ge 0$ are weighting coefficients and $\ge$ is an unrestricted scalar variable. The quantity $w_i \lambda$ may thus be interpreted as the degree of under-attainment or over-attainment of the goal g_i. This method is not subject to the convexity limitations of the "weighted sum" approach and its use of weights and goal enables the designer to be more expressive and precise in directing the search. Goal expression has an affinity with the common form of engineering design specification (cf. Fig. 3). The goal attainment method is, though, irredeemably a scalar optimization method, capable of revealing only one solution on the Pareto front as a result of one pass of this algorithm. Nonetheless, it was to prove influential for Fonseca and Fleming as they refined their multiobjective genetic algorithm [14] for use in engineering design [15].

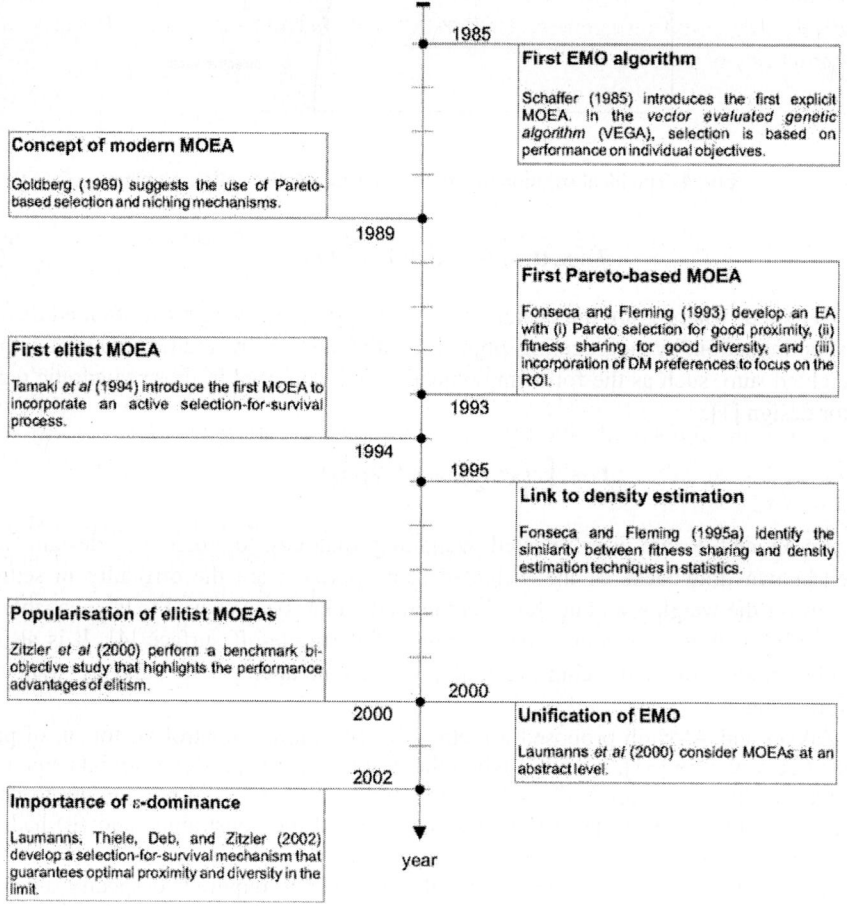

Fig. 5. Key developments in EMO history [24]

The population-based nature of evolutionary algorithms and their flexible selection mechanism have proved to be extremely successful for solving multi-objective optimization problems and for revealing a satisfactory approximation set to the desired globally optimal trade-off surface in a single execution of the algorithm. The fundamental benefit of this latter factor over multiple-start strategies is the potential for a cooperative search for ultimately different solutions, thus saving on the total number of solution evaluations required. Excellent descriptions of the history of EMO may be found, for example, in [8] and [6]. A timeline of key events is shown in Fig. 5.

3 *Many-Objective* Optimization

As we have seen in section 2, engineering design has a propensity to produce significant numbers of objectives, considerably in excess of 2 or 3 objectives commonly addressed by EMO researchers. This section looks at the new issues that arise and suggests way of tackling these.

3.1 Issues

Interactions often arise between objectives and these have been classified as *conflict* or *harmony* [23]. A relationship in which performance in one objective is seen to deteriorate as performance in another is improved is described as conflicting. A relationship in which enhancement of performance in an objective is witnessed as another objective is improved can be described as harmonious. The conflict that exists in a many-objective optimization task is a serious challenge for EMO researchers.

Given the typical numbers of requirements arising in engineering design, and elsewhere (see [6]), there is a very clear need to develop an understanding of the effects of increasing numbers of objectives on EMO. The phrase *many-objective* has been suggested in the OR community to refer to optimization problems with more than the standard two or three objectives [13].

For M conflicting objectives, an (M-1)-dimensional trade-off hypersurface exists in objective space. The number of samples required to achieve an adequate representation of the surface is exponential in M. In [9] it is shown that the proportion of locally non-dominated objective vectors in a finite randomly-generated sample becomes very large as the number of objectives increases. Since dominance is used to drive the search toward the true Pareto front, there may be insufficient selective pressure to make such progress. Of course, the use of a large population can help address this, but this is impractical for many engineering designs in which evaluation of objectives for a single candidate solution can be very compute-intensive.

Due to the 'curse of dimensionality' (the sparseness of data in high dimensions), the ability to fully explore surfaces in greater than five dimensions is regarded as highly limited [27]. Statisticians generally use dimensionality reduction techniques prior to application of the estimator. This assumes that the 'true' structure of the surface is of lower dimension, but the potential for reduction may be limited for a trade-off surface in which all objectives are in conflict with each other.

Possible measures that have been considered previously by the EMO community to address the issues arising from *many-objective* optimization include

- o the use of preferences,
- o aggregation,

The order in which the axes are set out in parallel coordinates does not have any bearing on the translation of the data in to parallel coordinates, although it is not entirely neutral to the technique, a point that will be expanded upon later.

It is not sufficient just to be able to display multivariate data in a 2-dimensional representation. The key requirement is to be able to easily interpret the relationships between the variables. It can be shown that the geometrical features of a surface in n-dimensional space are preserved in the parallel coordinates system. This is important because it allows these features to be easily identifiable when represented in parallel coordinates and therefore the relationship between the variables that give rise to these features can be visualised. For example, in Fig. 7, "crossing lines" indicates conflict between the two adjacent objectives. The degree of conflict is demonstrated by the intensity, or degree to which, the lines cross. Conversely, lines that do not cross demonstrate objectives which are in relative harmony with one another.

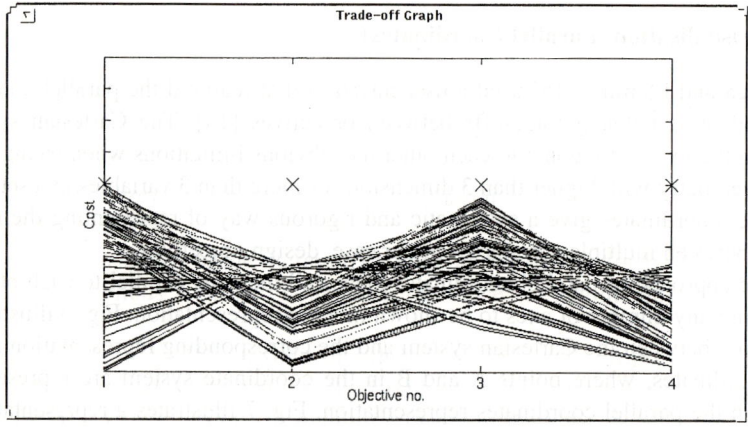

Fig. 7. Parallel coordinates for four objectives

Other requirements fulfilled by parallel coordinates are that there is no loss of data in the representation, which in turn ensures that there is a unique representation for each unique set of data. It also has a low representational complexity, $O(n)$, where n is the number of variables modelled, allowing the technique to scale well to large numbers of variables. Weaknesses of this visualisation method, however, are (i) that it requires multiple views (different orderings of objectives) to see different trade-offs and (ii) that it can be hard to see what is going on when many vectors are represented. Wegman [30] describes some countermeasures to these problems.

4 Facilitating the Engineering Design Exercise

4.1 Use of Preference Articulation

We will now illustrate the way in which preference articulation, as described in [15], can be used in a *many-objective* problem to focus on a specific region of interest (ROI) on the Pareto front and, ultimately to isolate a desired design solution. An

8 - objective flight control system design problem is used to illustrate the process; for reasons of clarity and commercial sensitivity, the objectives are unspecified, having titles such as Objective 1, etc. The preference articulation sequence is illustrated in Fig. 8 (a-f). Selection throughout the progress of the optimization uses the preferability operator defined in [15].

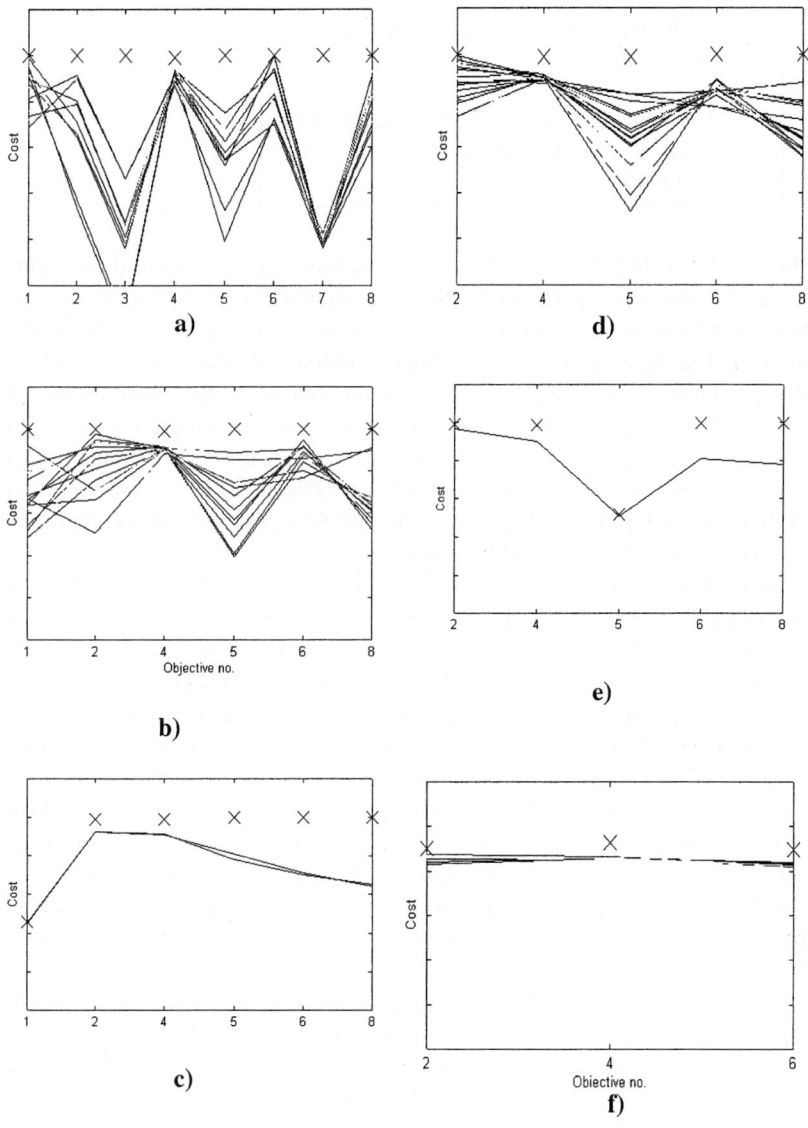

Fig. 8. Preference articulation sequence

Initially, the design is expressed as follows:

> 1. The designed controller seeks to simultaneously optimize *Objectives 1-8*
>
> 2. Each *Objective i* must satisfy *Goal i*, $i = 1,\ldots,8$
>
> 3. The controller has a prespecified structure with 7 variable parameters (*decision variables*)

Fig. 8(a) is a snapshot of the parallel coordinates representation of the eight objectives after a number of generations. Note that the initial design specification that each *Objective i* must satisfy *Goal i*, $i = 1\ldots 8$ immediately imposes a strict ROI for the optimizer. Fig. 8(a) shows Pareto-optimal solutions obtained after running the optimizer for a number of generations. Goal points for each of the Objectives are marked with an "x" in the plots. Since this example will exercise progressive articulation of preferences, these goal values will be subject to change during the design process.

From the plot we can immediately see that Objectives 2 and 3 are in "harmony" and that the goal of Objective 7 is easily satisfied. There is also a suggestion that Objectives 4 and 5 might be in "harmony".

The sequence that follows makes certain assumptions about the flight control system designer's preferences but serves to provide an example of how the interactive preference articulation process can reduce the ROI, focus on key Objectives and, ultimately, identify an acceptable solution - in this case, a flight controller that satisfies initial design goals and is "optimal" with respect to the design objectives.

The representation in Fig. 8(a) leads to the first interactive design decision: to remove Objective 3 from further consideration since it will benefit from improvements in Objective 2 and the latter is the more important of these two objectives. Objective 7 is converted to be a constraint; its value is constrained to be at least as good as the "worst" solution for that objective shown in Fig. 8(a). The optimizer is now released to cycle through more generations. Fig. 8(b) is a snapshot of the parallel coordinates representation of the remaining six objectives under consideration, after further runs of the optimizer.

A decision is made now to isolate the best solution so far with respect to Objective 1, see Fig. 8(c). The designer knows that this objective has a strong cost impact on the final solution although modest improvements beyond the best case here are likely to have little further impact. A decision is now made to convert Objective 1 to be a constraint, where its value must be at least as good as the isolated solution.

Fig. 8(d) is a snapshot of the parallel coordinates representation of the remaining five objectives under consideration after further runs of the optimizer. Now is the time to reduce the ROI still further. Isolating the best solution so far with respect to

Objective 5, see Fig. 8(e), and converting this objective into a constraint in a similar way to that of Objectives 1 and 7 achieves this. Observing this isolated solution, a decision is also made that Objective 8 can similarly be converted to a constraint, provided that its value is at least as good as this solution.

To arrive at Fig. 8(f), the goals of Objectives 2, 4 and 6 are progressively tightened until a very small set of solutions are obtained with little to discriminate them. At this point, the designer selects the best of these solutions with respect to Objective 2 (there is a slight preference with regard to the importance of this objective) as the desired solution.

4.2 Grid-Enabled EMO

The Grid computing paradigm is a recent development that enables complex system designers seeking to accelerate EMO solutions. The computational grid is a "hardware and software infrastructure that provides dependable, consistent, pervasive, and inexpensive access to high-end computational capabilities." [17]. Originally motivated by "big science" with high-performance computing requirements, the Grid lends itself to EMO applications whose objective function evaluations are sufficiently compute-intensive.

MOGA-G [25] is a grid-enabled framework for EMO. For non-trivial objective functions, EMO is compute-intensive since designs normally require a relatively large number of evaluations of the objective function to produce a satisfactory result. Furthermore, objectives arising in engineering designs often require considerable computational effort, for example involving nonlinear dynamic simulations. The population-based nature of evolutionary algorithms means that they are well suited for parallelism using the master-worker paradigm (see Fig. 9). Here, EMO operations (ranking, crossover, mutation, fitness sharing, etc.) are performed by the Master Node, and the evaluations of the objective function are executed in parallel on the Worker Nodes.

MOGA-G implements the master-worker paradigm in a Service Orientated Architecture (SOA). This is the view of grid computing taken by the Globus Project (globus.org) and focuses on providing access to the resources of the grid via services. The main advantages of using this approach are:

- suitability to the proposed form of parallelism,
- flexibility of use,
- interoperability with current (and, hopefully, future) standards, and
- the modular nature of the Globus Toolkit.

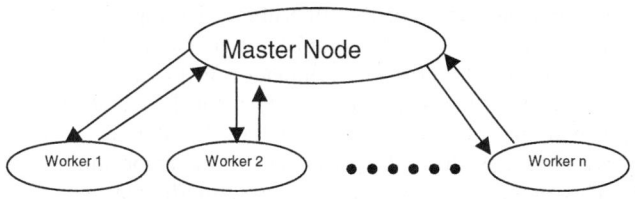

Fig. 9. Master-worker paradigm

Using the SOA approach, the client acts as the master node and the service acts as the worker. In the implementation of the MOGA-G framework (see Fig. 10) there are two different services. One service exposes the operations of the multi-objective evolutionary algorithm to the client, and the other provides operations for evaluating the objective function. Compute-intensive function evaluations can therefore be "farmed" out to a user-specified number of computing nodes on the Grid, often geographically distributed.

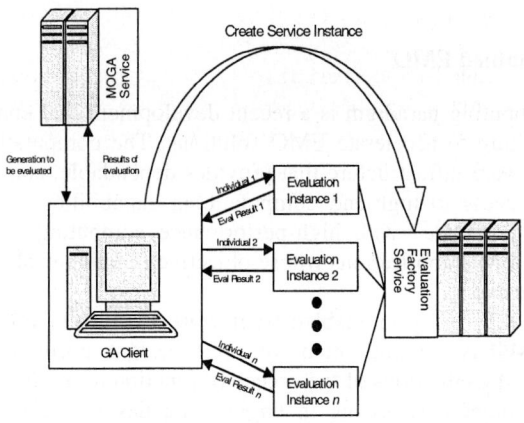

Fig. 10. MOGA-G implementation

The provision of these tools as services means that they can be accessed via the *http* protocol, and therefore via any device with a capable web browser such as a PDA Personal Digital Assistant (PDA). This flexibility stems from the use of easily accessible protocols like *http* and the loosely coupled nature of the SOA approach. For computationally trivial objective functions the communication overheads involved in executing the evaluations result in a decrease in performance. However, the framework shows significant performance advantages for more computationally complex objective functions such as nonlinear dynamic simulations.

4.3 Computational Steering of EMO-Based Engineering Designs Using a PDA

Large-scale, long-running, complex optimization routines, such as EMO, are usually run non-interactively. Typically, in an engineering design, the user will set the initial EMO parameters and then execute the algorithm. During this execution process, which can often take hours or days to complete, user interaction, if any, is limited to periodic interventions and the possible termination of the algorithm if it appears to have failed (for example, if the search process does not show convergence). When the execution is finished, the solutions produced by the algorithm are assessed and, if the design results are not satisfactory, the parameters of the algorithm are adjusted and it is run again. This process clearly leads to a very inefficient use of resources, and possibly, ultimately, to unsatisfactory solutions.

One solution to this problem is to allow the designer to interact with the optimization routine during execution (referred to as computational steering). This would allow the designer to influence the efficiency of the algorithm and the quality of the solutions that it produces. To enable interaction with the search process the user must be provided with an appropriate visualisation of the data so they can efficiently extract the relevant information [22].

A PDA-based client [26] has been developed to control this steering process (see Fig. 11). The client has to be stateless so that when there is no interaction from the user the optimization routine will run in batch mode. The PDA client provides an interface for observing the progress of the optimization routine (using a parallel coordinate plot, for example) and adjusting the parameters of the algorithm, if necessary. Due to issues related to scarcity of memory and computational power, plus small display size, a minimal interface has been developed, while still providing the desired functionality. It connects wirelessly to a web-service that allows the steering of the optimization algorithm. This steering web-service exposes methods for obtaining the current values of the candidate solutions and adjusting the parameters of the algorithm. Further development will allow the steering service to 'push' information to a client if one is connected.

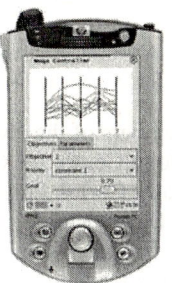 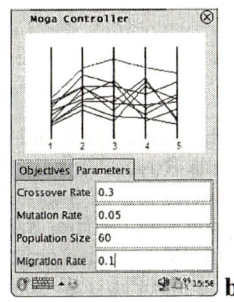 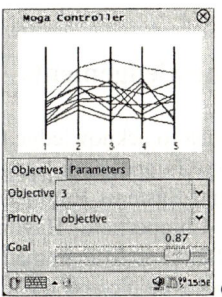

Fig. 11. a) PDA client for steering an EMO-based design; b) a parameter adjustment screen; c) a preference articulation screen

Steering of EMO can be performed in two main ways. Firstly, the internal parameters of the algorithm (such as crossover rate, mutation rate, etc.) can be adjusted from the steering client, see Fig. 11(b). Adjusting these parameters can alter the behaviour of the algorithm, such as speeding up or slowing down convergence. The second method of steering EMO is to alter the goal and priority information for the objectives, see Fig. 11(c). This information is used by the preferability operator [15]. Refining this preference information can help focus the algorithm on to a specific region of the non-dominated set. In this manner it is possible to guide the search and reduce the number of candidate solutions in the manner described in section 4.1.

5 Concluding Remarks

A particular application of EMO has been studied: design of engineering systems. These systems are often complex and, invariably, consist of many objectives. Large

27. Scott, D. W.: Multivariate Density Estimation: Theory, Practice, and Visualization, Wiley, New York (1992)
28. Tan, K. C., Khor, E. F., Lee, T. H., Sathikannan, R.: An evolutionary algorithm with advanced goal and priority specification for multi-objective optimization, Journal of Artificial Intelligence Research 18, 183–215 (2003)
29. Todd, D. S., Sen, P.: Directed multiple objective search of design spaces using genetic algorithms and neural networks, in W. Banzhaf, J. Daida, A. E. Eiben, M. H. Garzon, V. Honavar, M. Jakiela and R. E. Smith (eds), Proceedings of the 1999 Genetic and Evolutionary Computation Conference (GECCO 99), Vol. 2, Morgan Kaufmann Publishers, San Francisco, California, pp. 1738–1743 (1999)
30. Wegman, E. J.: Hyperdimensional data analysis using parallel coordinates, Journal of the American Statistical Association 85, 664–675 (1990)
31. Zakian V and Al-Naib U: Design of dynamical control systems by the method of inequalities, *Proc. IEE*, Vol 120, pp1421-1427 (1973)
32. Zitzler, E., Thiele, L., Laumanns, M., Fonseca, C. M., Grunert da Fonseca, V.: Performance assessment of multiobjective optimizers: An analysis and review, IEEE Transactions on Evolutionary Computation 7(2), 117–132 (2003)

1984-2004 – 20 Years of Multiobjective Metaheuristics. But What About the Solution of Combinatorial Problems with Multiple Objectives?

Xavier Gandibleux[1] and Matthias Ehrgott[2]

[1] LINA – FRE CNRS 2729
Université de Nantes, 2 rue de la Houssiniere BP 92208
F-44322 Nantes Cedex 03 - France
Xavier.Gandibleux@univ-nantes.fr
[2] Department of Engineering Science,
The University of Auckland, Private Bag 92019
Auckland - New Zealand
m.ehrgott@auckland.ac.nz

Abstract. After 20 years of development of multiobjective metaheuristics the procedures for solving multiple objective combinatorial optimization problems are generally the result of a blend of evolutionary, neighborhood search, and problem dependent components. Indeed, even though the first procedures were direct adaptations of single objective metaheuristics inspired by evolutionary algorithms or neighborhood search algorithms, hybrid procedures have been introduced very quickly. This paper discusses hybridations found in the literature and mentions recently introduced metaheuristic principles.

1 Multiobjective Optimization

A multiobjective optimization problem is defined as

$$\min_{x \in X}(z_1(x), \ldots, z_p(x)), \qquad \text{(MOP)}$$

where $X \subset \mathbb{R}^n$ is a feasible set in the *decision space*, and $z : \mathbb{R}^n \to \mathbb{R}^p$ is a vector valued objective function. By $Z = z(X) \subset \mathbb{R}^p$ we denote the image of the feasible set in the *objective space*. We consider optimal solutions of (MOP) in the sense of *efficiency* (or Pareto optimality), that is, a feasible solution $x \in X$ is called efficient if there does not exist $x' \in X$ such that $z(x') \leq z(x)$, i.e., $z_k(x') \leq z_k(x)$ for all $k = 1, \ldots, p$ and $z_j(x') < z_j(x)$ for some j. In other words, no solution is at least as good as x for all objectives, and strictly better for at least one.

Efficiency refers to solutions x in decision space. In terms of the objective space, with objective vectors $z(x) \in \mathbb{R}^p$ we use the notion of *non-dominance*: If x is an efficient solution then $z(x) = (z_1(x), \ldots, z_p(x))$ is a non-dominated vector (or point). The set of efficient solutions is X_E, the set of non-dominated vectors is Z_N. We may also refer to Z_N as the *non-dominated frontier* or the trade-off surface or the Pareto front. For $x^1, x^2 \in X$ we shall use the notation $x^1 \succ x^2$ if x^1 dominates x^2, i.e., if $z(x^1) \leq z(x^2)$.

In case of multiple feasible solutions $x, x' \in X$ mapping to the same non-dominated point $z(x) = z(x')$, the solutions are said to be *equivalent* [24]. A *complete set* X_E [24] is a set of efficient solutions such that all $x \in X \setminus X_E$ are either dominated or equivalent to at least one $x \in X_E$. I.e., for each nondominated point $z \in Z_N$ there exists at least one $x \in X_E$ such that $z(x) = z$. To solve a multiobjective optimization problem often means to find a complete set of efficient solutions. The computation of a set of efficient solutions is a major challenge of multiobjective optimization. But to precisely characterize the ability of an algorithm to solve an MOP the definition of complete set is refined as follows:

- [24] A *minimal complete set* X_{E_m} is a complete set without equivalent solutions. Any complete set contains a minimal complete set.
- [36] The *maximal complete set* X_{E_M} is a complete set including all equivalent solutions, i.e., all $x \in X \setminus X_{E_M}$ are dominated.

Multiobjective combinatorial optimization problems form a particular class of MOPs, which can be formulated as follows:

$$\min\{Cx : Ax \geq b, x \in \mathbb{Z}^n\}. \qquad \text{(MOCO)}$$

Here C is a $p \times n$ objective function matrix, where c_k denotes the k-th row of C. A is an $m \times n$ matrix of constraint coefficients and $b \in \mathbb{R}^m$. Usually the entries of C, A and b are integers. The feasible set $X = \{Ax \geq b, x \in \mathbb{Z}^n\}$ may describe a combinatorial structure such as, e.g., spanning trees of a graph, paths, matchings etc. We shall assume that X is a finite set. By $Z = CX$ we denote the image of X under C in $\mathbb{R}^p$.

2 Approximation Methods for MOCO

As in the single objective case, reasonable alternatives to exact methods for solving difficult MOCOs are approximation methods. An *approximation method* in a multiobjective optimization context is a method which finds either sets of locally potentially efficient solutions that are later merged to form a set of potentially efficient solutions – the approximation denoted by X_{PE} – or globally potentially efficient solutions according to the current approximation X_{PE}.

2.1 The Question of Quality of an Approximation

The quality of a solution of a combinatorial optimization problem can be estimated by comparing lower and upper bounds on the optimal objective function value. In multiobjective optimization the concept of bounds is not well developed. The best possible lower and upper bounds on values of all non-dominated points are given by the ideal and nadir point z^I and z^N defined by

$$z_k^I = \min_{x \in X} z_k(x), \quad k = 1, \ldots, p$$

and

$$z_k^N = \max_{x \in X_E} z_k(x), \quad k = 1, \ldots, p,$$

respectively. We sometimes refer to a utopian point $z^U = z^I - \varepsilon \mathbf{1}$, where **1** is a vector of all ones and ε is a small positive number. However, the ideal and nadir points are usually far away from non-dominated points and do not provide a good estimate of the non-dominated set. In addition, the nadir point is hard to compute for problems with more than two objectives, see [13].

To better capture the multiobjective nature of the problems and the fact that we are looking for a set of efficient solutions it is natural to generalize the notion of bounds to bound sets. Ehrgott and Gandibleux report first results on lower and upper bound sets in for the biobjective assignment, knapsack, traveling salesman, set covering, and set packing problems [11, 14]. Fernández and Puerto [15] use bound sets in their exact and heuristic methods to solve the multiobjective uncapacitated facility location problem.

There are a few other ideas in the literature. Kim et al. [30] propose a new measure, the integrated convex preference (ICP), to compare the quality of algorithms for MOCO problems with two objectives. Sayin [37] proposes the criteria of *coverage*, *uniformity*, and *cardinality* as quality measures. Although developed for continuous problems the ideas may be interesting for MOCO problems. However, the methods proposed in [37] can be efficiently implemented for linear problems only. Other authors propose *distance based measures* [44] and *visual comparisons* of the generated approximations. The latter are restricted to bi-objective problems. Jaszkiewicz [28] also distinguishes between *cardinal* and *geometric quality measures*. He gives further references and suggests preference-based evaluation of approximations of the non-dominated set using outperformance relations. Tenfelde-Podehl [42] proposes volume based measures.

None of these measures have been universally adopted in the multiobjective optimization literature, and further research is clearly needed.

2.2 Multiobjective Heuristics and Metaheuristics for MOP

Multiple objective heuristics (MOH) and *multiple objective metaheuristics (MOMH)* are methods that aim to provide a good tradeoff between an approximation of the set of efficient solutions and the time and memory requirements to obtain it. These methods may manipulate a complete or incomplete single solution or a collection of solutions at each iteration.

Heuristics are generally problem-specific, so that a method which works for one problem cannot be used to solve a different one. In contrast, metaheuristics are universal methods applicable to a large number of problems. A metaheuristic is a solution *concept*. The adaptation to a specific problem uses heuristics as solution *methods*. The family of metaheuristics includes, but is not limited to, constraint logic programming, genetic algorithms, evolutionary methods, neural networks, simulated annealing, tabu search, non-monotonic search strategies, greedy randomized adaptive search, ant colony systems, particle swarm optimization, noising methods, variable neighborhood search, scatter search, etc.

From a historical perspective, the pioneer approximation methods for multi-objective problems have appeared since 1984, in the following order: Genetic Algorithms (GA, Schaffer 1984 [38]), Artificial Neural Networks (ANN, Malakooti 1990 [32]), Simulated Annealing (SA, Serafini 1992 [39]), and Tabu Search (TS, Gandibleux 1996 [16]). The pioneer methods have three characteristics. First, they are inspired either by *Evolutionary Algorithms (EA)* or by *Neighborhood Search Algorithms (NSA)*. Second, the early methods are direct derivations of single objective optimization metaheuristics, incorporating small adaptations to integrate the concept of efficient solution for optimizing multiple objectives. Third, almost all methods were designed as a solution concept according to the principle of metaheuristics.

2.3 Evolutionary Algorithms Versus Neighborhood Search Algorithms

Evolutionary Algorithms manage a solution population $\mathcal{P}$ rather than a single feasible solution. In general, they start with an initial population and combine principles of self adaptation, i.e., independent evolution (such as the mutation strategy in genetic algorithms), and cooperation, i.e., the exchange of information between individuals (such as the "pheromone" used in ant colony systems), to improve approximation quality. The usual components of an evolutionnary algorithm are:

- a population of solutions
- evolutionary operators (crossover, mutation)
- an archive of elite solutions
- a ranking method
- a guiding method
- a clustering method
- a fitness measure
- a penalty strategy for infeasible solutions, etc.

Because the whole population contributes to the evolutionary process, the generation mechanism is parallel along the frontier, and thus these methods are also called *global convergence-based methods*. This characteristic makes population-based methods very attractive for solving multiobjective problems.

In *Neighborhood Search Algorithms*, the generation of solutions relies upon one individual, a current solution x_n, and its neighbors $\{x\} \subseteq \mathcal{N}(x_n)$. Using a local aggregation mechanism for the objectives (often based on a weighted sum), a weight vector $\lambda \in \Lambda$, and an initial solution x_0, the procedure iteratively projects the neighbors into the objective space in a search direction λ by optimizing the corresponding parametric single objective problem. A local approximation of the non-dominated frontier is obtained using archives of the successive potentially efficient solutions detected. This generation mechanism is sequential along the frontier, producing a local convergence to the non-dominated frontier, and so such methods are called *local convergence-based methods*. The principle is repeated for diverse search directions to completely approximate the non-dominated frontier. The elementary components of a neighborhood search algorithm are:

- a neighborhood structure (moves)
- an exploration strategy (partial, exhaustive)
- an acceptation rule (SA principle, TS principle)
- a list of candidates
- a scalarizing function
- an oscillation strategy
- a greedy (randomized) strategy
- a path-relinking strategy, etc.

NSAs are well-known for their ability to locate the non-dominated frontier, but they require more effort in diversification than EA in order to cover the efficient frontier completely.

While the first adaptation of metaheuristic techniques for the solution of multiobjective optimization problems has been introduced 20 years ago, the MOMH field has clearly mushroomed over the last ten years. The first approximation methods proposed for MOCO problems were "pure" NSA strategies and were straightforward extensions of well-known metaheuristics for dealing with the notion of non-dominated points. Simulated annealing (the MOSA method [43]), tabu search (the MOTS method [16], the method of Sun [41]), or GRASP (the VO-GRASP method [18]) are examples.

2.4 Hybrid Algorithms and Problem-Dependent Algorithms

The methods that followed the pioneer ones, designed to be more efficient algorithms in the MOCO context, have been influenced by two important observations.

The first observation is that on the one hand, NSAs focus on convergence to efficient solutions, but must be guided along the non-dominated frontier. On the other hand, EAs are very well able to maintain a population of solutions along the non-dominated frontier (in terms of diversity, coverage, etc.), but often converge too slowly to the non-dominated frontier. Naturally, methods have been proposed that try to take advantage of both EA and NSA features by combining components of both approaches, introducing *hybrid algorithms* for MOPs.

The second observation is that MOCO problems contain information deriving from their specific combinatorial structure, which can be advantageously exploited by the approximation process. Single objective combinatorial optimization is a very active field of research. Many combinatorial structures are very well understood. Thus combinatorial optimization represents a useful source of knowledge to be used in multiobjective optimization. This knowledge (e.g., cuts for reducing the search space) are more and more taken into account when designing a very efficient approximation method for a particular MOCO. It is not surprising to see an evolutionary algorithm – for global convergence – coupled advantage of an efficient heuristic known for the single objective problem in order to compute an initial set of very good solutions $\mathcal{P}_0$ in a first phase. The heuristic (a GRASP algorithm) is encapsulated in a basic generation procedure, for example using a convex combination of the objectives: λ-GRASP. The second phase works on a population of individuals $\mathcal{P} \supseteq \mathcal{P}_0$ derived from the initial set and performs an EA (a modified version of SPEA dealing with all potential efficient solutions and integrating a local search: A-SPEA) in order to consolidate the approximation of the non-dominated frontier (see Figure 3).

In Target Aiming Pareto Search (TAPaS) [29], the search directions of the procedure are given by the current set X_{PE}, similar to the principle of almost all the tabu search adaptations for MOP [16, 25, 41]. A series of goals is deduced from X_{PE} and a scalarizing function is used for guiding an NSA, defining a two phase strategy. This scheme has been applied to two vehicle routing problems. For one problem (the covering tour problem), TAPas is coupled with a EA plus a branch and cut algorithm specifically designed for the single objective version of the problem.

4. **EA + NSA + Problem Dependent Components**.
The most recent hybrid procedures integrate EA and NSA components as well as problem-dependent components in order to design a powerful approximation method for a MOCO problem. Gandibleux et al. [20, 21] propose a population based method where a crossover uses a "genetic" map of the population, and that includes a path-relinking operator. Path-relinking generates new solutions by exploring the trajectories that connect elite solutions. Starting from one solution – the initiating solution – a path is generated through the neighborhood space that leads to the other solution – the guiding solution [22] (see Figure 4). Only potentially efficient solutions compose the population at any time and bound sets limit the triggering of a local search. This procedure has been applied for approximating the non-dominated frontier of assignment and knapsack problems with two objectives.

5. **Approximation Method and Exact Procedure in a Hybrid Method**.
The combination of EA and NSA can be more accurately integrated than by a "simple" switch between the two mechanisms. Gandibleux and Fréville [19] propose a procedure for the biobjective knapsack problem combining an exact procedure for reducing the search space with a tabu search process for identifying the potentially efficient solutions. The reduction principle is based on cuts which eliminate parts of the decision space where (provably) no exact efficient solution exists (see Figure 5). The tabu search is triggered on the reduced space and dynamically updates the bounds in order to guarantee the tightest value at any time.

This category also applies if in an exact method for generating the non-dominated points the exact method needs bounds of good quality. For example in the *seek and cut method* for solving the assignment problem [36] the "seek" computes a local approximation of the non-dominated frontier (i.e., bounds are computed by a population-based algorithm coupled with path-

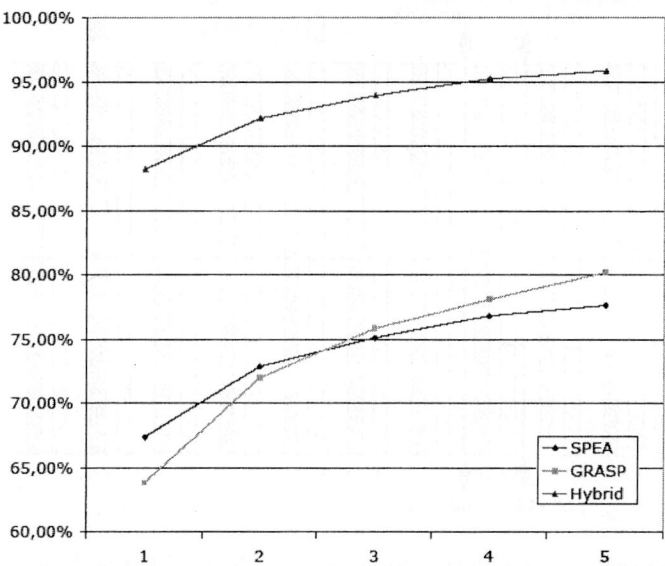

Fig. 3. The figure illustrates the average percentage of exact solutions found using λ-GRASP, A-SPEA, and the hybrid for the set packing problem with two objectives, when all three methods are allowed the same computational effort. (From Delorme et al. [8])

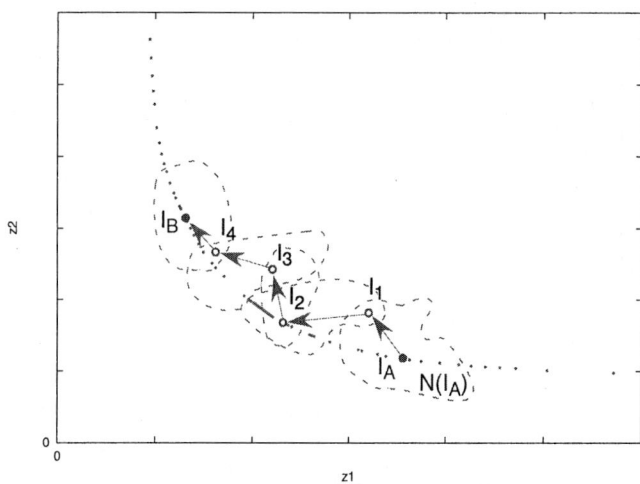

Fig. 4. Illustration of a possible path construction (see [21]). I_A and I_B are two individuals randomly selected from the current elite population (small bullets). I_A is the initiating solution, and I_B is the guiding solution. $\mathcal{N}(I_A)$ is the feasible neighborhood according to the move defined. $I_A - I_1 - I_2 - I_3 - I_4 - I_B$ is the path that is built

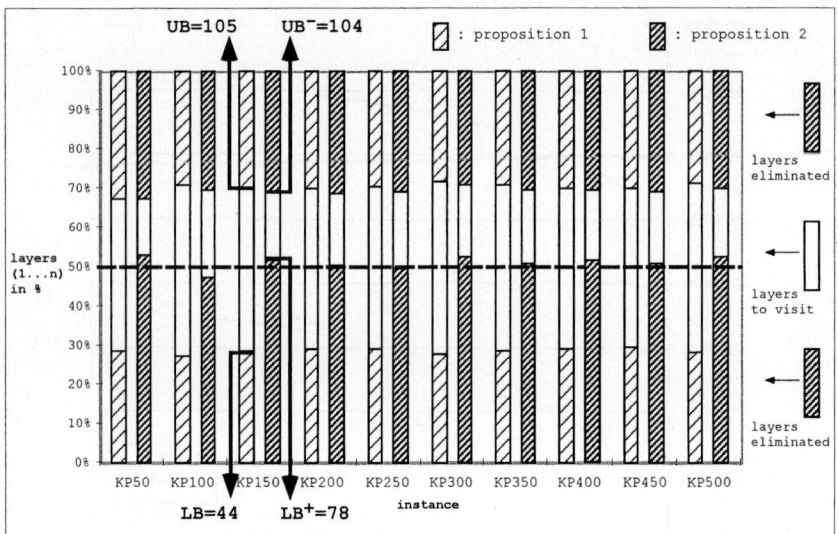

Fig. 5. A decision space reduction technique for the bi-objective knapsack problem uses an additional constraint on the cardinality of an optimal solution for computing a utopian reference point and an approximation set for verifying if the reference point is dominated. As output a *strategic map* is established eliminating all parts of the seach space where no efficient solution exists. A heuristic (a tabu search for example) can then be triggered inside the reduced decision space [19]

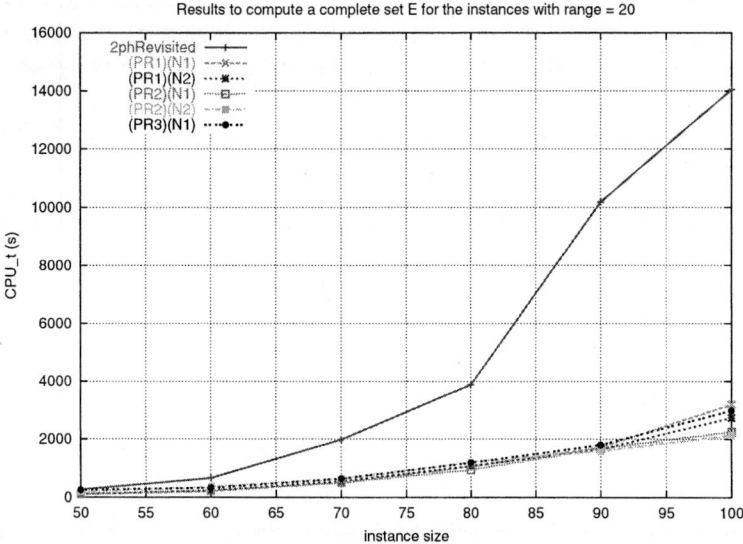

Fig. 6. CPU time used by an exact method for solving the assigment problem with two objectives without (the upper curve) and with (the lower curves) the use of approximate solutions for the pruning test inside the method [36]

relinking) which is then used for "cutting" the search space of an implicit enumeration scheme (see Figure 6).

Other well-known principles of metaheuristics have also been applied to MOP problems, but few concern MOCO problems:

1. The ant colony optimization principle, based on the behavior or real ants, is another EA principle (see [9, 23, 26, 40] for examples).
2. The scatter search principle is based on a linear combination of solutions selected from a candidate list, i.e., it uses a population of solutions [4, 33].
3. The particle swarm principle is based on the elementary moves of particles, i.e., a population of solutions [6, 34].
4. Constraint programming (CP) techniques have been introduced in the solution procedure of MOP problems with the PICPA method [2]. The main goal of CP in PICPA is to build a sharp bound in the the decision space around the efficient frontier using value propagation mechanisms over variables.

4 Conclusion

To be efficient, an approximation method for solving an MOP seems to be necessarily a hybrid algorithm, i.e., a combination of EA, NSA, and problem specific components. This is in particular true for MOCO, where the adapation of a universal method to a problem cannot compete with a method specifically designed for this problem. Because one main challenge is the scalability problem – the efficient solution of large scale problem instances – the future methods will be specifically designed methods, "recycling" the 50 years of knowledge of (single objective) optimization.

We are convinced that questions like how to reduce of the search space will become even more important than they are today. Constraint programming, cuts, and bounds appear to be possible answers as do the adaptation of intensification components identified as efficient for combinatorial problems, like path-relinking.

These challenges promise many future papers.

References

1. V. Barichard and J.K. Hao. Un algorithme hybride pour le problème de sac à dos multi-objectifs. In *JNPC'2002 Proceedings: Huitièmes Journées Nationales sur la Résolution Pratique de Problèmes NP-Complets*, Nice, France, 27–29 May 2002, pages 19–30.
2. V. Barichard and J.K. Hao. A population and interval constraint propagation algorithm. Volume 2632 of *Lecture Notes in Computer Science*, pages 88-101, Springer, 2003.
3. V. Barichard. *Approches hybrides pour les problèmes multiobjectifs*. PhD thesis, Université d'Angers, France, 2003.

4. R. Beausoleil. Multiple criteria scatter search. In J.P. de Sousa, editor, *MIC'2001 Proceedings of the 4th Metaheuristics International Conference, Porto, July 16-20, 2001*, volume 2, pages 539–543, 2001.
5. F. Ben Abdelaziz, J. Chaouachi, and S. Krichen. A hybrid heuristic for multiobjective knapsack problems. In S. Voss, S. Martello, I. Osman, and C. Roucairol, editors, *Meta-Heuristics: Advances and Trends in Local Search Paradigms for Optimization*, pages 205–212. Kluwer Academic Publishers, Dordrecht, 1999.
6. C. A. Coello Coello and M. S. Lechuga. MOPSO: A proposal for multiple objective particle swarm optimization. In *Congress on Evolutionary Computation* (CEC'2002), Vol. 2, pp. 1051–1056, IEEE Service Center, Piscataway, New Jersey, May 2002.
7. P. Czyzak and A. Jaszkiewicz. Pareto simulated annealing. In G. Fandel and T. Gal, editors, *Multiple Criteria Decision Making. Proceedings of the XIIth International Conference, Hagen (Germany)*, volume 448 of *Lecture Notes in Economics and Mathematical Systems*, pages 297–307, 1997.
8. X. Delorme, X. Gandibleux, F. Degoutin. Résolution approchée du problème de set packing bi-objectifs. *Ecole d'Automne en Recherche Opérationnelle*, Ecole Polytechnique, Université de Tours, October 2003, Tours, France.
9. K. Doerner, W.J. Gutjahr, R.F. Hartl, C. Strauss, and C. Stummer. Pareto ant colony optimization: A metaheuristic approach to multiobjective portfolio selection. *Annals of Operations Research*, 131:79–99, 2004.
10. M.P. Hansen. *Metaheuristics for multiple objective combinatorial optimization*. PhD thesis, Institute of Mathematical Modelling, Technical University of Denmark, Lyngby (Denmark), 1998. Report IMM-PHD-1998-45.
11. M. Ehrgott and X. Gandibleux. Bounds and bound sets for biobjective combinatorial optimization problems. In Murat Koksalan and Stan Zionts, editors, *Multiple Criteria Decision Making in the New Millennium*, volume 507 of *Lecture Notes in Economics and Mathematical Systems*, pages 241–253. Springer, 2001.
12. M. Ehrgott and X. Gandibleux, Multiobjective combinatorial optimization. In M. Ehrgott and X. Gandibleux, editors, *Multiple Criteria Optimization: State of the Art Annotated Bibliographic Survey*, volume 52 of *Kluwer's International Series in Operations Research and Management Science*, pages 369–444. Kluwer Academic Publishers, Boston, 2002.
13. M. Ehrgott and D. Tenfelde-Podehl. Computation of ideal and nadir values and implications for their use in MCDM methods. *European Journal of Operational Research*, 151(1):119–131, 2003.
14. M. Ehrgott and X. Gandibleux. Bound sets for biobjective combinatorial optimization problems. Technical report, Department of Engineering Science, The University of Auckland, 2004.
15. E. Fernández and J. Puerto. Multiobjective solution of the uncapacitated plant location problem. *European Journal of Operational Research*, 145(3):509–529, 2003.
16. X. Gandibleux, N. Mezdaoui, and A. Fréville. A tabu search procedure to solve multiobjective combinatorial optimization problems. In R. Caballero, F. Ruiz, and R. Steuer, editors, *Advances in Multiple Objective and Goal Programming*, volume 455 of *Lecture Notes in Economics and Mathematical Systems*, pages 291–300. Springer Verlag, Berlin, 1997.
17. X. Gandibleux, H. Morita, and N. Katoh. A genetic algorithm for 0-1 multiobjective knapsack problem. In *International Conference on Nonlinear Analysis and Convex Analysis (NACA98) Proceedings, July 28-31 1998, Niigata, Japan*, 1998.

18. X. Gandibleux, D. Vancoppenolle, and D. Tuyttens. A first making use of GRASP for solving MOCO problems. Technical report, University of Valenciennes, France, 1998. Paper presented at MCDM 14, June 8-12 1998, Charlottesville, VA.
19. X. Gandibleux and A. Fréville. Tabu search based procedure for solving the 0/1 multiobjective knapsack problem: The two objective case. *Journal of Heuristics*, 6(3):361–383, 2000.
20. X. Gandibleux, H. Morita, and N. Katoh. The supported solutions used as a genetic information in a population Heuristic. In E. Zitzler, K. Deb, L. Thiele, C. Coello, D. Corne, editors, *Evolutionary Multi-Criterion Optimization*, volume 1993 of *Lecture Notes in Computer Sciences*, pages 429–442, Springer Verlag, Berlin, 2001.
21. X. Gandibleux, H. Morita, and N. Katoh. A population-based metaheuristic for solving assignment problems with two objectives. Accepted for publication in Journal of Mathematical Modelling and Algorithms.
22. F. Glover and M. Laguna. *Tabu Search*. Kluwer Academic Publishers, Dordrecht, 1997.
23. M. Gravel, W.L. Price, and C. Gagné. Scheduling continuous casting of aluminium using a multiple objective ant colony optimization metaheuristic. *European Journal of Operational Research*, 143(1):218–229, 2002.
24. P. Hansen. Bicriterion path problems. In G. Fandel and T. Gal, editors, *Multiple Criteria Decision Making Theory and Application*, volume 177 of *Lecture Notes in Economics and Mathematical Systems*, pages 109–127. Springer Verlag, Berlin, 1979.
25. M.P. Hansen. Tabu search for multiobjective combinatorial optimization: TAMOCO. *Control and Cybernetics*, 29(3):799–818, 2000.
26. S. Iredi, D. Merkle, and M. Middendorf. Bi-criterion optimization with multi colony ant algorithms. In E. Zitzler, K. Deb, L. Thiele, C.A. Coello Coello, and D. Corne, editors, *First International Conference on Evolutionary Multi-Criterion Optimization*, volume 1993 of *Lecture Notes in Computer Science*, pages 359–372. Springer Verlag, Berlin, 2001.
27. A. Jaszkiewicz. Multiple objective genetic local search algorithm. In M. Köksalan and S. Zionts, editors, *Multiple Criteria Decision Making in the New Millennium*, volume 507 of *Lecture Notes in Economics and Mathematical Systems*, pages 231–240. Springer Verlag, Berlin, 2001.
28. A. Jaszkiewicz. *Multiple objective metaheuristic algorithms for combinatorial optimization*. Habilitation thesis, Poznan University of Technology, Poznan (Poland), 2001.
29. N. Jozefowiez. *Modélisation et résolution approchée de problèmes de tournées multi-objectif*. PhD thesis, Université de Lille 1, France, 2004.
30. B. Kim, E.S. Gel, W.M. Carlyle, and J.W. Fowler. A new technique to compare algorithms for bi-criteria combinatorial optimization problems. In M. Köksalan and S. Zionts, editors, *Multiple Criteria Decision Making in the New Millenium*, volume 507 of *Lecture Notes in Economics and Mathematical Systems*, pages 113–123. Springer Verlag, Berlin, 2001.
31. P. Lacomme, C. Prins, and M. Sevaux. Multi-objective capacitated arc routing problem. In C.M. Fonseca et al., editors, *Evolutionary multi-criterion optimization*, volume 2632 of *Lecture Notes in Computer Science*, pages 550–564. Springer, 2003.
32. B. Malakooti, J. Wang, and E.C. Tandler. A sensor-based accelerated approach for multi-attribute machinability and tool life evaluation. *International Journal of Production Research*, 28:2373–2392, 1990.

33. J. Molina, M. Laguna, R. Marti and R. Caballero. SSPMO: A scatter search procedure for non-linear multiobjective optimization. Working paper, Leeds School of Business, University of Colorado, USA.
34. S. Mostaghim and J. Teich. Strategies for finding good local guides in multi-objective particle swarm optimization (MOPSO). In *2003 IEEE Swarm Intelligence Symposium Proceedings*, IEEE Service Center, pages 26–33, Indianapolis, Indiana, USA, April 2003.
35. T. Murata and H. Ishibuchi. MOGA: Multi-objective genetic algorithms. In *Proceedings of the 2nd IEEE International Conference on Evolutionary Computing*, pages 289–294. IEEE Service Center, Piscataway, 1995.
36. A. Przybylski, X. Gandibleux and M. Ehrgott. Seek and cut algorithm computing minimal and maximal complete efficient solution sets for the biobjective assignment problem. *MOPGP'04 – 6th International Conference on Multi Objective Programming and Goal Programming*, April 14-16, 2004, Hammamet, Tunisia.
37. S. Sayin. Measuring the quality of discrete representations of efficient sets in multiple objective mathematical programming. *Mathematical Programming*, 87:543–560, 2000.
38. J.D. Schaffer. *Multiple Objective Optimization with Vector Evaluated Genetic Algorithms*. PhD thesis, Vanderbilt University, Nashville, TN (USA), 1984.
39. P. Serafini. Simulated annealing for multiobjective optimization problems. In *Proceedings of the 10th International Conference on Multiple Criteria Decision Making, Taipei-Taiwan*, volume I, pages 87–96, 1992.
40. P.S. Shelokar, S. Adhikari, R. Vakil, V. K. Jayaraman, and B.D. Kulkarni. Multiobjective ant algorithm: Combination of strength Pareto fitness assignment and thermodynamic clustering. *Foundations of Computing and Decision Sciences*, 25(4):213–230, 2000.
41. M. Sun. Applying tabu search to multiple objective combinatorial optimization problems. In *Proceedings of the 1997 DSI Annual Meeting, San Diego, California*, volume 2, pages 945–947. Decision Sciences Institute, Atlanta, GA, 1997.
42. D. Tenfelde-Podehl. *Facilities Layout Problems: Polyhedral Structure, Multiple Objectives and Robustness*, PhD thesis, University of Kaiserslautern, Germany, 2002.
43. E. L. Ulungu, *Optimisation combinatoire multicritère: Détermination de l'ensemble des solutions efficaces et méthodes interactives*, Université de Mons-Hainaut, Faculté des Sciences, 313 pages, 1993.
44. A. Viana and J. Pinho de Sousa. Using metaheuristics in multiobjective ressource constrained project scheduling. *European Journal of Operational Research*, 120(2):359–374, 2000.

Omni-optimizer: A Procedure for Single and Multi-objective Optimization

Kalyanmoy Deb and Santosh Tiwari

Kanpur Genetic Algorithms Laboratory (KanGAL),
Indian Institute of Technology Kanpur,
Kanpur, PIN 208 016, India
{deb, tiwaris}@iitk.ac.in
http://www.iitk.ac.in/kangal

Abstract. Due to the vagaries of optimization problems encountered in practice, users resort to different algorithms for solving different optimization problems. In this paper, we suggest an optimization procedure which specializes in solving multi-objective, multi-global problems. The algorithm is carefully designed so as to degenerate to efficient algorithms for solving other simpler optimization problems, such as single-objective uni-global problems, single-objective multi-global problems and multi-objective uni-global problems. The efficacy of the proposed algorithm in solving various problems is demonstrated on a number of test problems. Because of it's efficiency in handling different types of problems with equal ease, this algorithm should find increasing use in real-world optimization problems.

1 Introduction

With the advent of new and computationally efficient optimization algorithms, researchers and practitioners have been attempting to solve different kinds of search and optimization problems encountered in practice. One of the difficulties in solving real-world optimization problems is that they appear in different forms and types. Some optimization problems may have to be solved for only one objective, some other problems may have more than one conflicting objectives, some problems may be highly constrained, and some may have more than one optimal solutions. When faced with such problems, a user first analyzes the underlying problem and chooses a suitable algorithm for solving it. This is because an algorithm efficient for finding the sole optimum in a single-objective optimization problem cannot be adequately applied to find multiple optimal solutions present in another optimization problem. To solve different kinds of problems, a user needs to know different algorithms, each specialized in solving a particular class of optimization problem.

In this paper, we propose and evaluate a single optimization algorithm for solving different kinds of function optimization problems often encountered in practice. The proposed *omni-optimization* algorithm adapts itself to solve different kinds of problems – single or multi-objective problems and uni or multi-global problems. The motivation for developing such a generic procedure came from the generic programming practices. For example, if a programming task is to develop a code for adding a few integers, a generic approach would be to use the following strategy.

```
Add a few integers:
begin
    print 'enter number of integers to be added', read n
    sum = 0
    for i = 1 to n
        print 'enter integer i', read a[i]
        sum = sum + a[i]
    print 'Sum =' sum
end
```

Interestingly, the same code can be used for adding any number of integers initially defined by the variable n. If $n = 1$ is used (thereby trying to add only one number to zero), the code degenerates to printing the same integer as the outcome of the addition. On a similar vein, a generic optimization procedure should find optimal solutions for a multi-objective optimization problem and the same procedure should degenerate to solving a single-objective optimization problem if only one objective function is used. Similarly, our proposed approach can find multiple optimal solutions, if present in a problem, and will automatically degenerate to find the sole optimum of a uni-global optimization problem.

The proposed omni-optimizer is carefully designed to have various properties needed for solving different kinds of optimization problems and is also found to be computationally efficient. The simulation results on 12 test problems show the usefulness of the proposed algorithm and suggest more such studies in the near future.

2   Function Optimization Problems

A function optimization problem may be of different types, depending on the desired goal of the optimization task. The optimization problem may have only one objective function (known as a single-objective optimization problem), or it may have multiple conflicting objective functions (known as a multi-objective optimization problem). Some problems may have only one global optimum, thereby requiring the task of finding the global optimum[1]. Other problems may contain more than one global optima in the search space, thereby requiring the task of finding multiple such global optimal solutions. Although in some optimization

---

[1] However, in robust optimization tasks, instead of finding the global optimum, the emphasis is on finding a solution which is less sensitive to local perturbation of variables.

tasks, there may be a need of finding the local optimal solutions in addition to finding global optimum solutions, in this study we only concentrate in finding the global optima (one or more) of one or more objective functions.

We consider the following constrained M-objective ($M \geq 1$) minimization problem:

$$\begin{aligned}
\text{Minimize } & (f_1(\mathbf{x}), f_2(\mathbf{x}), \ldots, f_M(\mathbf{x})), \\
\text{Subject to } & g_j(\mathbf{x}) \geq 0, \quad j = 1, 2, \ldots, J, \\
& h_k(\mathbf{x}) = 0, \quad k = 1, 2, \ldots, K, \\
& x_i^{(L)} \leq x_i \leq x_i^{(U)}, \quad i = 1, 2, \ldots, n.
\end{aligned} \quad (1)$$

A n-variable solution vector $\mathbf{x}$ which satisfies all constraints and variable bounds shown above is called a *feasible* solution. The optimality of a solution depends on a number of KKT optimality conditions which involve finding the gradients of objective and constraint functions [2, 6, 11].

Here, we suggest an omni-optimizer which is capable of finding one or more near-optimal solutions for the following four types of optimization problems in a single simulation run of the algorithm:

1. Single-objective, uni-global optimization problems (the outcome is a single optimum solution),
2. Single-objective, multi-global optimization problems (the outcome is multiple optimal solutions),
3. Multi-objective, uni-global optimization problems (the outcome is multiple efficient points each corresponding to a single Pareto-optimal solution), and
4. Multi-objective, multi-global optimization problems (the outcome is multiple efficient points some of which may correspond to multiple Pareto-optimal solutions).

It is intuitive that the fourth type of optimization problem mentioned above is the most generic one. If designed carefully, an algorithm capable of solving the fourth type of problems can be made to solve other three types of problems in a degenerate sense. The developed algorithm should be capable of solving any of the above problems to its desired optimality without explicitly foretelling the type of problem it is handling. In the following section, we present one such omni-optimizer, which adapts itself to an efficient algorithm automatically for solving any of the above four types of problems.

3 Omni-optimizer

The optimization algorithm starts with an initial population P_0 of size N and an iteration counter t is initialized to zero. In each iteration, a bigger population R_t is constructed by two random orderings of the same population P_t. Thereafter, two binary-tournament selection operations are performed to select two parent solutions. The tournament selection operator prefers feasible solutions over infeasible solutions (constraint handling), non-dominated solutions over dominated solutions (multiple objective handling) and less-crowded solutions over more-crowded solutions (maintenance of diversity). The two parent solutions are then recombined and mutated to obtain two offspring solutions. Any standard genetic operator for each of these operations can be used here. The offspring solutions are included in the offspring population Q_t.

An elite-preservation is performed by combining both parent P_t and offspring Q_t populations together and then by ranking the combined population from best class of solutions to the worst. This way, a good parent solution is allowed to remain in the subsequent population, in case enough good offspring solutions are not created. The ranking procedure uses a modified domination principle (ϵ-domination) to classify the entire combined population into different classes. The best solutions of the population are stored in F_1, the next-best solutions are stored in F_2 and so on.

Now, to create the next population P_{t+1} of size N, we start accepting classes from the top of the list (F_1 onwards) and stop to the class (F_L) which cannot be completely accommodated to P_{t+1} due to size restriction. Then, based on crowding of the solutions of F_L in both objective and decision variable space, we select only those many solutions which will fill the population P_{t+1}. This completes one iteration of the proposed omni-optimizer.

Readers familiar with the elitist non-dominated sorting GA (NSGA-II) [5] may find the proposed omni-optimization procedure similar to that of NSGA-II with some differences. Thus, the proposed procedure is expected to solve multi-objective optimization problems in a manner similar to NSGA-II. The proposed procedure is also capable of solving other kinds of optimization problems mentioned in the previous section. Later, we shall discuss how this procedure degenerates to solve single-objective uni-global and multi-global problems. Here, we first present the omni-optimization procedure as a pseudo-code.

```
The omni-optimization procedure:
begin
   Initialize(P₀)
   t = 0   // iteration counter
   repeat
      Rₜ = Shuffle(Pₜ) ∪ Shuffle(Pₜ)
      for i = 1 to N - 1 step 2
         // two selection operations
         parent1 = tournament(Rₜ(2i - 1), Rₜ(2i))
         parent2 = tournament(Rₜ(2i + 1), Rₜ(2i + 2))
         // crossover and mutation operators
         (offspring1, offspring2) = variation(parent1, parent2)
         Qₜ(i)     = offspring1
         Qₜ(i + 1) = offspring2
      Rₜ = Pₜ ∪ Qₜ      // elite preservation
      (F₁, F₂, ...) = ranking(Rₜ) // best class F₁ and so on
      Pₜ₊₁ = ∅
      j = 1   // class number
      while |Pₜ₊₁ ∪ Fⱼ| ≤ N
         Pₜ₊₁ = Pₜ₊₁ ∪ Fⱼ  // include classes from best
```

```
            crowd_dist(F_j)    // crowding distance of each soln.
            j = j+1
            L = j      // last class to be included partially
            rem = N - |P_{t+1}|    // remaining solutions to be filled
            sorting(crowd_dist(F_L))  // sort F_L in decreasing order
                                                        of crowding distance
            P_{t+1} = P_{t+1} ∪ F_L(1:rem)  // include top solutions
            t = t+1    // increment iteration counter
      until (termination)
end
```

The operator `shuffle(P_t)` makes a random ordering of the population members of P_t. The `tournament(a,b)` operator compares two solutions a and b and declares the winner. The following algorithm is used for this purpose:

```
winner = tournament(a,b)
begin
   if a is feasible and b is infeasible, winner = a
   else if a is infeasible and b is feasible, winner = b
   else if both a and b are infeasible,
        winner = (if CV(a) < CV(b)) ? a : b
   else if both a and b are feasible,
        if a dominates b, winner = a
        else if b dominates a, winner = b
        else if crowd_dist(a) > crowd_dist(b), winner = a
        else if crowd_dist(a) < crowd_dist(b), winner = b
        else winner = random_choose(a,b)
end
```

The function `CV(a)` calculates the overall normalized constraint violation of solution a, as follows: $\text{CV}(\mathbf{a}) = \sum_{j=1}^{J} \langle \bar{g}_j(\mathbf{a}) \rangle + \sum_{k=1}^{K} |\bar{h}_k(\mathbf{a})|$, where $\langle \cdot \rangle$ is a bracket operator [2]. Here, the solution a dominates the solution b for an M-objective minimization problem, if following conditions are met:

1. $f_i^{\mathbf{a}} \leq f_i^{\mathbf{b}}$ for all $i = 1, 2, \ldots, M$,
2. $f_i^{\mathbf{a}} < f_i^{\mathbf{b}}$ for at least one $i \in \{1, M\}$.

This is the usual *domination* principle used in multi-objective optimization [11, 4]. The `random_choose(a,b)` selects a or b randomly. The procedure for computing the crowding distance metric `crowd_dist()` is discussed later.

The variation operator takes two solutions and performs genetic operations (such as crossover followed by mutation) and creates two offspring solutions.

The `ranking(R_t)` ranks the population R_t into different classes (from best to worst) depending on how good the solutions are. The following pseudo-code can be used for this purpose:

```
(F₁, F₂, ...) = ranking(Rₜ)
begin
    k = 1   // class counter
    repeat
        Fₖ = ∅
        for i = 1 to |Rₜ|
            for j = 1 to |Rₜ| and j ≠ i
                if Rₜ(j) ε-dominates Rₜ(i), break
            if (j = |Rₜ|)   // Rₜ(i) is ε-nondominated
                Fₖ = Fₖ ∪ Rₜ(i)
        Rₜ = Rₜ \ Fₖ
        k = k + 1
    until (all 2N solutions are classified)
end
```

It is not necessary that the repeat-until loop as described above continues till all $2N$ solutions are classified; the loop can be terminated as soon as the last class (L) which can be partially accommodated is encountered.

Solution a ϵ-dominates another solution b if the following conditions are true in an M-objective minimization problem:

1. $f_i^a \le f_i^b$ for all $i = 1, 2, \ldots, M$,
2. $f_i^a < f_i^b - \epsilon_i$ for at least one $i \in \{1, M\}$.

Here, ϵ_i is a quantity calculated from a user-defined parameter δ, described in equation 2. Figure 1 shows the region which is ϵ-dominated by solution a in a two-objective minimization problem. For simplicity, we use only one user-defined parameter δ as follows:

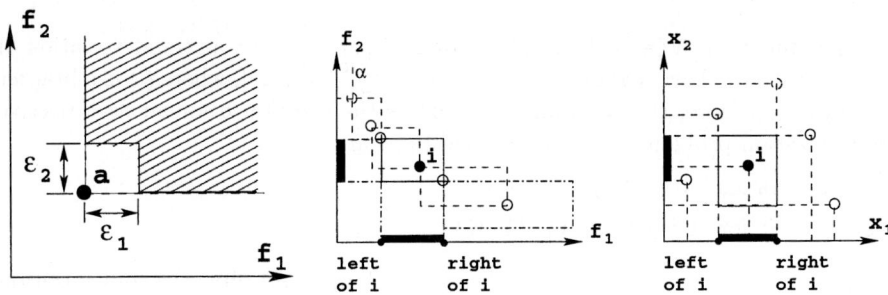

Fig. 1. The ϵ-dominance criterion. Point a ϵ-dominates the shaded region

Fig. 2. Objective space and variable space crowding distance computations

$$\epsilon_i = \delta(f_i^{\max} - f_i^{\min}), \qquad (2)$$

where $f_i^{\max}$ and $f_i^{\min}$ are the maximum and minimum value of the i-th objective in the population. After the ranking operation, the population is expected to be ranked from best solutions (class F_1) in a decreasing order of importance.

Omni-Optimizer: A Procedure for Single and Multi-objective Optimization 53

The `crowd_dist(F_j)` calculates a metric value of a solution in F_j providing an estimate of the neighboring members of F_j in both the objective and the decision variable space. The following pseudo-code describes the procedure:

```
crowd_dist(F_j)
begin
    // initialize all distances to zero
    for i = 1 to |F_j|
        crowd_dist_obj(i) = 0
        crowd_dist_var(i) = 0
    // objective space crowding
    for m = 1 to M
        for i = 1 to |F_j|
            if i is a minimum solution in m-th objective
                crowd_dist_obj(i) = ∞
            else crowd_dist_obj(i) += normalized_obj(i)
    // decision variable space crowding
    for j = 1 to n
        for i = 1 to |F_j|
            if i is a boundary solution in j-th variable
                crowd_dist_var(i) += 2×normalized_var(i)
            else crowd_dist_var(i) += normalized_var(i)
    // normalize distances and compute population average
    for i = 1 to |F_j|
        crowd_dist_obj(i) = crowd_dist_obj(i)/M
        crowd_dist_var(i) = crowd_dist_var(i)/n
    avg_crowd_dist_obj = ∑_{i=1:|F_j|} (crowd_dist_obj(i))/|F_j|
    avg_crowd_dist_var = ∑_{i=1:|F_j|} (crowd_dist_var(i))/|F_j|
    // if above average, assign larger of the two distances,
    // else assign smaller of the two distances
    for i = 1 to |F_j|
        if crowd_dist_obj(i) > avg_crowd_dist_obj or
           crowd_dist_var(i) > avg_crowd_dist_var
            crowd_dist(i) =
                max(crowd_dist_obj(i),crowd_dist_var(i))
        else crowd_dist(i) =
                min(crowd_dist_obj(i),crowd_dist_var(i))
end
```

To calculate the `normalized_obj(i)` of a solution i for the m-th objective, we first sort the population members in increasing order of the objective value and then calculate

$$\text{normalized\_obj}(i) = \frac{f_m(\text{right of } i) - f_m(\text{left of } i)}{f_m^{\max} - f_m^{\min}} \tag{3}$$

Similarly, the normalized_var(i) for the j-th variable is computed as follows:

$$\text{normalized\_var}(i) = \frac{x_j(\text{right of } i) - x_j(\text{left of } i)}{x_j^{\max} - x_j^{\min}} \quad (4)$$

Figure 2 illustrates the computation of objective space and decision variable space crowding distance. Note that if a solution is a boundary solution in the objective space, an infinite distance is assigned so as not to lose the solution. On the other hand, if a solution is a boundary solution in the decision variable space, the numerator in Equation 4 can be replaced by $(x_j(\text{right of } i) - x_j(i))$ or $(x_j(i) - x_j(\text{left of } i))$, as the case may be. Since this computes only a one-sided difference, the quantity is multiplied by a factor of two.

The sorting(A) sorts the population members of A into a decreasing order of crowding distance measure. Thus, the top-most member of the ordering has the maximum distance measure and the bottom-most member of the ordering has the least distance measure.

3.1 Computational Complexity

The ranking procedure involves $O(MN^2)$ computations, although a faster implementation can be achieved in $O(N \log^{M-2} N)$ computations [10, 9] for $M > 2$. The crowd_dist procedure involves $O(MN \log N)$ computations for calculating crowd_dist_obj values and $O(nN \log N)$ computations for calculating crowd_dist_var values. Thus, an efficient implementation of the algorithm requires an iteration-wise complexity of $O(N \log^{M-2} N)$ or $O(nN \log N)$, whichever is larger.

3.2 Single-Objective, Uni-global Optimizer

Now, let us analyze how the proposed omni-optimizer degenerates to different optimization algorithms of importance. First, we consider solving a single-objective minimization problem having one global optimum.

Here, the number of objectives is one, or $M = 1$. The tournament selection procedure described above degenerates to a constrained-tournament selection operator proposed in [3]. A check for dominance between two solutions degenerates to finding if a solution is better than the other or not. In the absence of constraints, the above procedure reduces to choosing the better of the two, and in case of a tie, the more diverse solution is chosen.

After the offspring population is created, it is combined with the parent population P_t, as it is done in standard single-objective EAs, such as in CHC [7] or in $(\mu+\lambda)$-ES [13]. The ranking procedure with a small ϵ parameter will assign each population member into a different class and the selection of N members from the combined population of size $2N$ degenerates to choosing the best N solutions of the combined parent-offspring population, a procedure followed in both CHC and $(\mu+\lambda)$-evolution strategies. If each class has only one or two members, the crowding distance computation assigns an infinite distance to them, thereby making no effect of crowding distance to the optimization procedure at all. However, in the presence of more than two solutions in a class which becomes the last class (class L) to be partially accepted, the extreme and sparsed solutions in the variable space are preferred. But we argue that such a decision, even though a low probability event, does not make much of a difference in obtaining the sole optimum of the problem.

Thus, if the underlying problem is a single-objective problem with one global optimum, the procedure is very similar to a standard EA procedure which uses tournament selection, genetic variation operators and an elite-preservation strategy along with a diversity preserving operator.

3.3 Single-Objective, Multi-global Optimizer

Often, there exist problems which have more than one global minimum solutions. The proposed omni-optimizer degenerates to finding multiple such global minima in a single simulation run.

The procedure becomes similar to the previous case, except that now there may exist multiple solutions in the top class (F_1), even towards the end of a simulation when the proposed method captures multiple optimum solutions. In such cases, the ranking procedure will put all such solutions in one class and the crowd_dist operation degenerates to a variable space crowding alone (since all these solutions will have identical function value or function values with a maximum difference of ϵ). In EA literature, such a variable space crowding operator emphasizing distant solutions in the variable space to remain in the population (often called a *niching* procedure) [8, 12] is used for finding multiple global optimum solutions. It is interesting to note how the proposed omni-optimization procedure degenerates to such a niching strategy automatically when there exist a number of global optimal solutions in a problem. Contrary to other niching methods, the proposed strategy does not require any additional parameter describing the niche size and it automatically adapts itself to find multiple optima offered by a problem.

3.4 Multi-objective Optimizer

In the presence of multiple objectives where each efficient solution in the objective space corresponds to a single Pareto-optimal solution in the decision space, the omni-optimization procedure degenerates to NSGA-II procedure [5] with the following modifications:

1. The ϵ-dominance criterion is used for classifying solutions into different fronts in the ranking procedure.
2. Both objective space and variable space crowding is performed for maintaining diversity among solutions of a single front.

The first modification makes the size of the non-dominated fronts larger than the non-dominated fronts obtained using the usual domination criterion. Since the selection begins from the top class and continues to worse classes, this modification does not make much of a difference in its working from that in NSGA-II. Due to the inclusion of variable-space niching in the omni-optimizer, the distribution of solutions is expected to be better in both objective and variable spaces.

However, a similar emphasis to that in NSGA-II has been followed in retaining extreme objective solutions, thereby ensuring a wide range of solutions in the objective space.

3.5 Multi-objective, Multi-global Optimizer

There exist some problems in which each efficient point in the objective space corresponds to a number of Pareto-optimal solutions in the decision variable space. In such problems, the task of an optimizer would be to find multiple such Pareto-optimal solutions corresponding to each efficient point. The proposed omni-optimizer can find such multiple solutions in a single simulation run.

The working of the proposed algorithm in such problems is similar to that in the previous case, except that in the crowd_dist operation, all such multiple solutions will be emphasized. These solutions will have identical objective function values, thereby making the crowd_dist_obj values to be zero (unless they are the extreme solutions), but their crowd_dist_var values will be non-zero. Since a solution's crowding distance value is chosen as the maximum of the two crowding distance values, these solutions will inherit the crowd_dist_var values. Thus, non-dominated solutions in a particular front can survive due to their sparsity either in the objective space or in the decision variable space. This allows not only sparsed efficient solutions to remain in the population, but any multiplicity of such efficient solutions in the decision variable space are also equally likely to survive in the population.

4 Simulation Results

In this section, we present simulation results of the omni-optimizer on various test problems chosen from the literature. In all problems, we use simulated binary crossover operator (with $\eta_c = 20$) and the polynomial mutation (with $\eta_m = 20$) [4] to handle real-valued variables. A crossover probability of 0.8 and a mutation probability of $1/n$ is used. For all problems, we use $\delta = 0.001$.

4.1 Single-Objective, Uni-global Test Problems

We choose 20-variable Rastrigin's function and 20-variable Schwefel's function. In both functions, there are many local minima, but there is only one global minimum and the corresponding function value is zero. Table 1 shows the best, median and worst number of evaluations needed in 10 runs of the proposed omni-optimizer to arrive at function value smaller than f^*. The seven-variable, four-constraint problem 5 described in [3] is solved next. With a population of size 70, the omni-optimizer is run for 5,000 generations and the best, median, and worst function values of 10 runs are found to be 680.216, 680.254, and 680.433, respectively. The best-known solution, as reported in [3], was $f(\mathbf{x}) = 680.630$ obtained for an identical number of function evaluations. For the welded beam design problem [3] having four variables and five non-linear inequality constraints, the omni-optimizer with a population size of 80 finds objective values of 2.381, 2.385,

Table 1. Results of omni-optimizer on single-objective, uni-global problems

Function	$f(\mathbf{x})$	n	Range	N	Target f^*	Func. Eval.		
						Best	Median	Worst
Rastrigin	$\sum_{i=1}^{n} x_i^2 + 10(1 - \cos(2\pi x_i))$	20	$[-10, 10]$	20	0.01	19,260	24,660	29,120
Schwefel	$418.9829n - \sum_{i=1}^{n} x_i \sin\sqrt{|x_i|}$	20	$[-500, 500]$	50	0.01	54,950	69,650	103,350

and 2.387 as best, median and worst of 10 runs of 4,000 generations each. The best result reported earlier with identical number of function evaluations also had a function value of 2.381.

4.2 Single-Objective, Multi-global Test Problems

We consider two test problems in this category. The first problem is a single-variable problem having 21 different global optimal solutions.

$$\text{Minimize} \quad f(x) = \sin^2(\pi x), \quad x \in [0, 20]. \tag{5}$$

Here, we use a population of size 100. Figure 3 shows the obtained solutions after 200 generations. It is clear that all 21 global minimum solutions are found by the omni-optimizer.

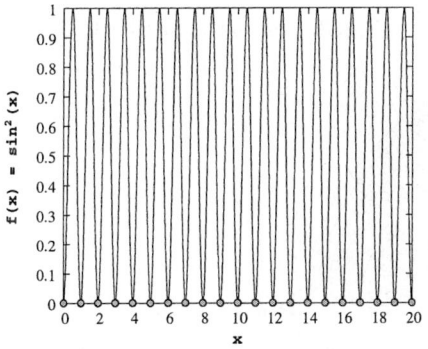

Fig. 3. All 21 minima are found for the $\sin^2(x)$ problem

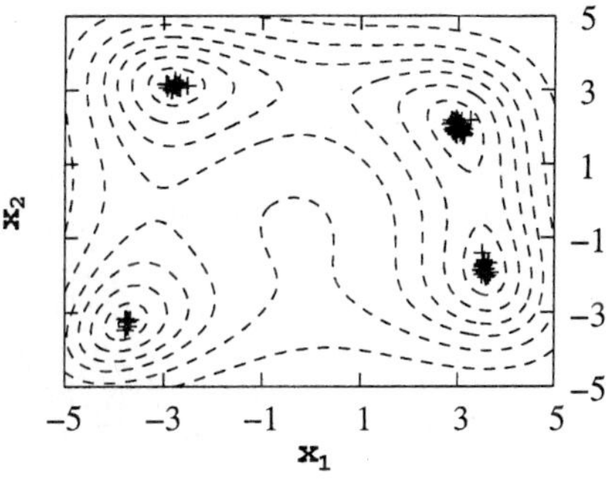

Fig. 4. All four minima are found for the Himmelblau's function

The second problem is the well-known Himmelblau's function [2]:

$$\text{Minimize } f(x_1, x_2) = (x_1^2 + x_2 - 11)^2 + (x_1 + x_2^2 - 7)^2, \quad -20 \leq x_1, x_2 \leq 20. \tag{6}$$

There are four minima, each having a function value equal to zero. Here, we use a population of size 100. Figure 4 shows the obtained solutions at the end of 100

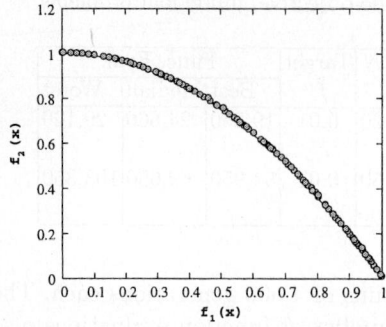

Fig. 5. Efficient points for ZDT2

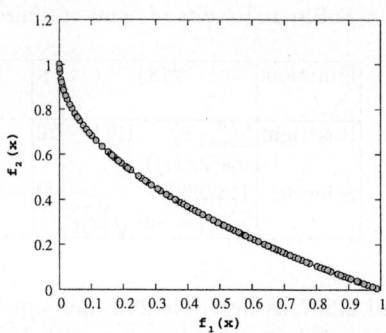

Fig. 6. Efficient points for ZDT4

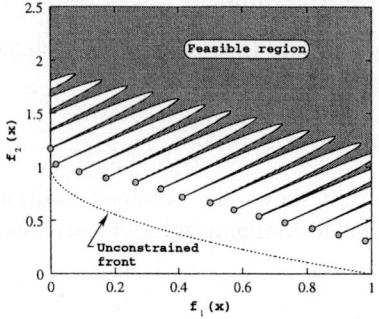

Fig. 7. Efficient points for CTP4

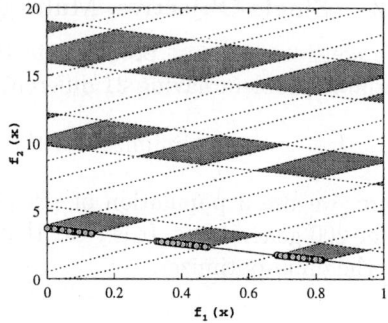

Fig. 8. Efficient points for CTP8

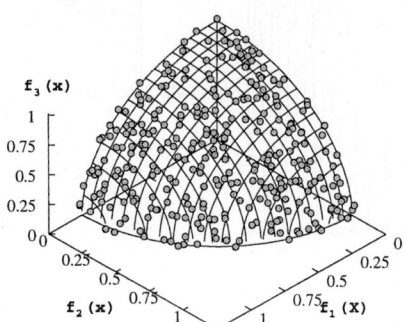

Fig. 9. Efficient points for DTLZ4S

generations. It is clear that the omni-optimizer is able to find all four minima in a single simulation run, similar to that reported in the literature [1] using a specialized niched EA.

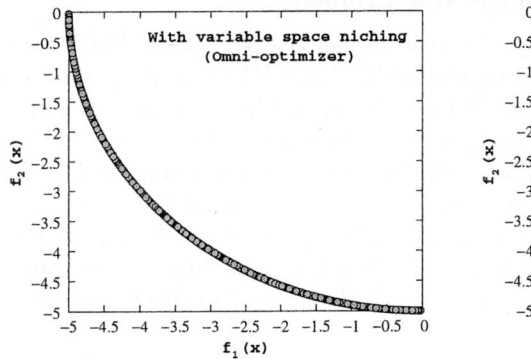

Fig. 10. Efficient points using omni-optimizer

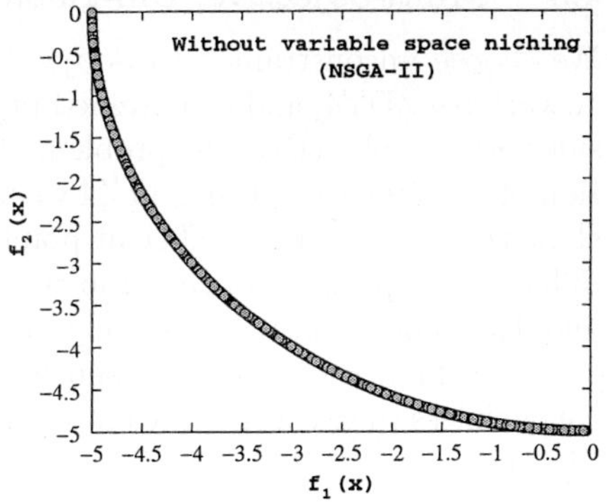

Fig. 11. Efficient points using NSGA-II

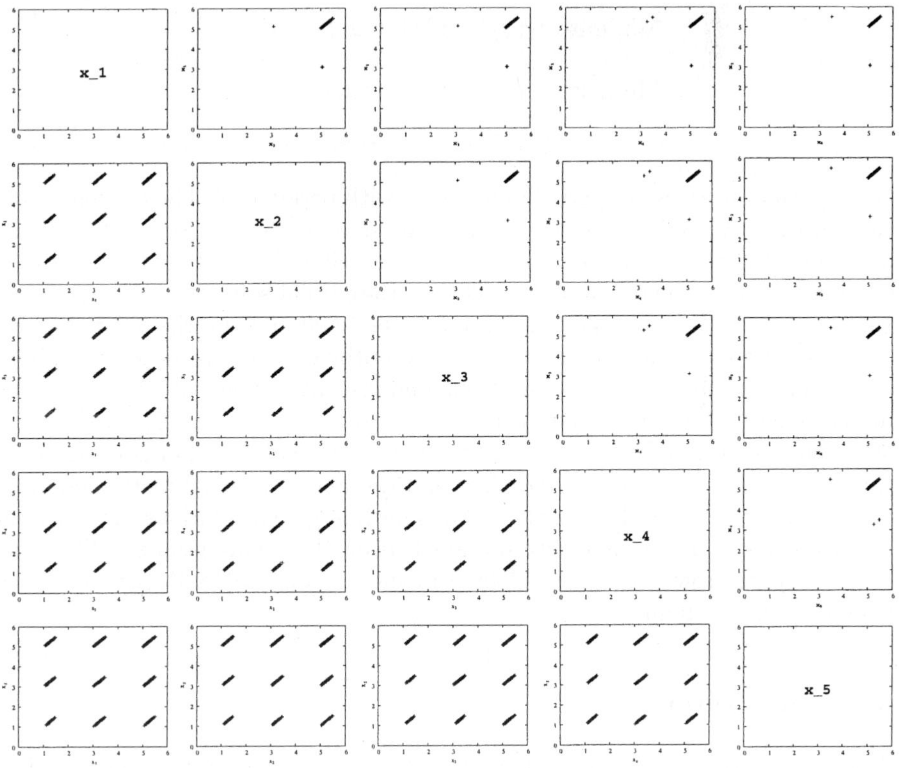

Fig. 12. Pareto-optimal solutions with omni-optimizer (left) and NSGA-II (right). The axes in a (i,j)-plot correspond to variables x_i and x_j

4.3 Multi-objective, Uni-global Test Problems

We choose unconstrained two-objective test problems – 30-variable ZDT2 and 10-variable ZDT4, and constrained problems CTP4 and CTP8 [4]. Also, we consider a three-objective test problem DTLZ4. For all problems, we use a population of size 100, except in DTLZ4 we use a population of size 300. Figures 5 till 9 show the corresponding efficient points obtained by the omni-optimizer. Due to different complexities of the problems, we have chosen to run till 100, 250, 7,000 and 100 generations, respectively, for ZDT2, ZDT4, CTP4 and CTP8. In the case of CTP4, the results are shown after 7,000 generations due to algorithm's sluggishness to reach the efficient points through the narrow feasible tunnels (see Figure 7).

4.4 Multi-objective, Multi-global Test Problem

We design the following problem:

$$\begin{aligned}
\text{Minimize } f_1(x) &= \sum_{i=1}^{n} \sin(\pi x_i), \\
\text{Minimize } f_2(x) &= \sum_{i=1}^{n} \cos(\pi x_i), \\
0 &\leq x_i \leq 6, \quad i = 1, 2, \ldots, n.
\end{aligned} \quad (7)$$

Here, both objectives are periodic functions with period of 2. The efficient frontier corresponds to the Pareto-optimal solutions $x_i \in [2m+1, 2m+3/2]$, where m is an integer. We choose $n = 5$. For every efficient point in the objective space there are in general $3 \times 5! = 360$ Pareto-optimal solutions in the decision variable space. We choose a population of size 1,000 and run the algorithm for 500 generations to capture as many Pareto-optimal solutions as possible. It is interesting to note that both algorithms find the entire range of efficient points in the objective space, as shown in Figures 10 and 11. However, the variable space plots show a different scenario. The lower diagonal plots in Figure 12 show the performance of the omni-optimizer and the upper diagonal plots show that of the original NSGA-II. It is clear that multiple global solutions for different combinations of x_i variables are obtained using the omni-optimizer. Since no variable space crowding was considered in the original NSGA-II, not all global combinations are found.

5 Conclusions

The optimization literature deals with single and multi-objective optimization problems and uni-global and multi-global problems differently.

In this paper, for the first time, we present an omni-optimizer which is designed to solve different types of optimization problems usually encountered in practice: multi-objective, multi-global problems requiring to find multiple efficient solutions each corresponding to multiple Pareto-optimal solutions. The algorithm is designed in a way so that it degenerates to an efficient procedure for solving an optimization problem with a simpler task of finding a single global minimum in a single-objective optimization problem or multiple global minima in a single-objective optimization problem or multiple efficient solutions in a multi-objective optimization problem. Proof-of-principle simulation results on 11 different test problems and one welded-beam design problem indicate the efficacy of the proposed omni-optimizer, suggest immediate further evaluation of the procedure, and stress the importance of more such studies in the near future. More research is needed to compare the proposed approach with dedicated single and multi-objective optimization procedures and on more synthetic and real-world problems.

References

1. K. Deb. Genetic algorithms in multi-modal function optimization. Master's thesis, Tuscaloosa, AL: University of Alabama, 1989.
2. K. Deb. *Optimization for Engineering Design: Algorithms and Examples*. New Delhi: Prentice-Hall, 1995.
3. K. Deb. An efficient constraint handling method for genetic algorithms. *Computer Methods in Applied Mechanics and Engineering*, 186(2–4):311–338, 2000.
4. K. Deb. *Multi-objective optimization using evolutionary algorithms*. Chichester, UK: Wiley, 2001.
5. K. Deb, S. Agrawal, A. Pratap, and T. Meyarivan. A fast and elitist multi-objective genetic algorithm: NSGA-II. *IEEE Transactions on Evolutionary Computation*, 6(2):182–197, 2002.
6. M. Ehrgott. *Multicriteria Optimization*. Berlin: Springer, 2000.
7. L. J. Eshelman. The CHC adaptive search algorithm: How to have safe search when engaging in nontraditional genetic recombination. In *Foundations of Genetic Algorithms 1 (FOGA-1)*, pages 265–283, 1991.
8. D. E. Goldberg and J. Richardson. Genetic algorithms with sharing for multimodal function optimization. In *Proceedings of the First International Conference on Genetic Algorithms and Their Applications*, pages 41–49, 1987.
9. M. T. Jensen. Reducing the run-time complexity of multiobjective EAs. *IEEE Transcations to Evolutionary Computation*, 7(5):503–515, 2003.
10. H. T. Kung, F. Luccio, and F. P. Preparata. On finding the maxima of a set of vectors. *Journal of the Association for Computing Machinery*, 22(4):469–476, 1975.
11. K. Miettinen. *Nonlinear Multiobjective Optimization*. Kluwer, Boston, 1999.
12. A. Pétrowski. A clearing procedure as a niching method for genetic algorithms. In *IEEE 3rd International Conference on Evolutionary Computation (ICEC'96)*, pages 798–803, 1996.
13. H.-P. Schwefel. *Evolution and Optimum Seeking*. New York: Wiley, 1995.

An EMO Algorithm Using the Hypervolume Measure as Selection Criterion

Michael Emmerich*, Nicola Beume+, and Boris Naujoks+

+University of Dortmund, Chair of Systems Analysis,
44221 Dortmund, Germany
*University of Leiden, Leiden Institute for Advanced Computer Science,
2333 CA Leiden, NL
emmerich@liacs.nl, {nicola.beume, boris.naujoks}@cs.uni-dortmund.de
http://ls11-www.cs.uni-dortmund.de/
http://www.liacs.nl/

Abstract. The hypervolume measure is one of the most frequently applied measures for comparing the results of evolutionary multiobjective optimization algorithms (EMOA). The idea to use this measure for selection is self-evident. A steady-state EMOA will be devised, that combines concepts of non-dominated sorting with a selection operator based on the hypervolume measure. The algorithm computes a well distributed set of solutions with bounded size thereby focussing on interesting regions of the Pareto front(s). By means of standard benchmark problems the algorithm will be compared to other well established EMOA. The results show that our new algorithm achieves good convergence to the Pareto front and outperforms standard methods in the hypervolume covered.
We also studied the applicability of the new approach in the important field of design optimization. In order to reduce the number of time consuming precise function evaluations, the algorithm will be supported by approximate function evaluations based on Kriging metamodels. First results on an airfoil redesign problem indicate a good performance of this approach, especially if the computation of a small, bounded number of well-distributed solutions is desired.

1 Introduction

Pareto optimization [1, 2] has become a well established technique for detecting interesting solution candidates for multiobjective optimization problems. It enables the decision maker to filter efficient solutions and to discover trade-offs between opposing objectives among these solutions. Provided a set of objective functions $f_{1,...,n} : \mathbb{S} \to \mathbb{R}$ defined on some search space $\mathbb{S}$ to be minimized, in Pareto optimization the aim is to detect the *Pareto-optimal set* $M = \{\mathbf{x} \in \mathbb{S} | \nexists \mathbf{x}' \in \mathbb{S} : \mathbf{x}' \prec \mathbf{x}\}$, or at least a good approximation to this set.

In practice, the decision maker wishes to evaluate only a limited number of Pareto-optimal solutions. This is due to the limited amount of time for examining the applicability of the solutions to be realized in practice. Typically these solutions should include extremal solutions as well as solutions that are located in parts of the solution space, where balanced trade-offs can be found.

A measure for the quality of a non-dominated set is the hypervolume measure or $\mathcal{S}$ metric [3]. Until now, research mainly focussed on two approaches to utilize the $\mathcal{S}$ metric for multiobjective optimization: Fleischer [4] suggested to recast the multiobjective optimization problem to a single objective one by maximizing the $\mathcal{S}$ metric of a finite set of non-dominated points. Knowles et al. utilized the $\mathcal{S}$ metric within an archiving strategy for EMOA [5, 6].

Going one step further, our aim was to construct an algorithm in which the $\mathcal{S}$ metric governs the selection operator of an EMOA in order to find a set of solutions well distributed on the Pareto front. The basic idea of this EMOA is to integrate new points in the population, if replacing a member increases the hypervolume covered by the population. Moreover, we aimed at an algorithm that can easily be parallelized and is simple and transparent. It should be extendable by problem specific features, like approximate function evaluations. Thus, a steady-state $(\mu + 1)$-EMOA, the so-called $\mathcal{S}$ *metric selection EMOA (SMS-EMOA)*, is proposed.

Notice that in contrast to Knowles et al. [6], we do not evaluate an archiving operator solely, but the dynamics of a complete EMOA based on $\mathcal{S}$ metric selection. In our opinion, the design of an EA suitable for a given problem or a series of test problems is a multiobjective task again. This way we look at archiving strategies as only one component of the whole EMOA.

The article is structured as follows: The hypervolume or $\mathcal{S}$ metric that is used in the selection of our algorithm is discussed first (section 2). Afterwards, the integration in an EMOA as well as some features are described (section 3). Section 4 deals with the performance on several test problems whereas the results achieved on a real world design problem are the topic of section 5, including results with approximate function evaluations. In particular, the coupling of our method to a metamodel assisted fitness function approximation tool is presented here. We close with a summary and an outlook to implied future tasks (section 6).

2 The Hypervolume Measure

The hypervolume measure or S metric was originally proposed by Zitzler and Thiele [3], who called it the *size of the space covered* or *size of dominated space*. Coello Coello, Van Veldhuizen and Lamont [2] described it as the Lebesgue measure Λ of the union of hypercubes a_i defined by a non-dominated point m_i and a reference point x_{ref}:

$$\mathcal{S}(M) := \Lambda(\{\bigcup_i a_i | m_i \in M\}) = \Lambda(\bigcup_{m \in M} \{x | m \prec x \prec x_{ref}\}). \qquad (1)$$

Zitzler and Thiele note that this measure prefers convex regions to non-convex ones [3]. A major drawback was the computational time for recursively calculating the values of $\mathcal{S}$. Knowles and Corne [5] estimated $O(k^{n+1})$ with k being the number of solutions in the Pareto set and n being the number of objectives. Furthermore, an accurate calculation of the $\mathcal{S}$ metric requires a normalized and positive objective space and a careful choice of the reference point. In [5,7] Knowles and Corne gave an example with two Pareto fronts, A and B, in the two dimensional case. They showed either $\mathcal{S}(A) < \mathcal{S}(B)$ or $\mathcal{S}(B) < \mathcal{S}(A)$ depending on the choice of the reference point.

Despite these disadvantages, the $\mathcal{S}$ metric is currently the only unary quality measure that is complete with respect to weak out-performance, while also indicating with certainty that one set is not worse than another [6]. It was used in several comparative studies of EMOA, e.g. [8,9,10]. Quite recently, Fleischer [4] proved that the maximum of $\mathcal{S}$ is a necessary and sufficient condition for a finite true Pareto front ($|PF_{true}| < \infty$):

$$PF_{known} = PF_{true} \iff \mathcal{S}(PF_{known}) = max(\mathcal{S}(PF_{known})). \qquad (2)$$

Moreover, he developed a method for computing the $\mathcal{S}$ metric of a set in polynomial time: $O(k^3 n^2)$ [4]. This algorithm led to the efficient integration of the $\mathcal{S}$ metric in archiving strategies [6].

In addition, the $\mathcal{S}$ metric of a set of non-dominated solutions is suggested as a mapping to a scalar value. Fleischer proposed the use of metaheuristics to optimize this scalar. His idea was to try simulated annealing (SA) resulting in a provable global convergent algorithm towards the true Pareto front [4].

3 The Algorithm

Our aim was to design an EMOA that covers a maximal hypervolume with a limited number of points. Furthermore, we wanted to diminish the problem of choosing the right reference point. Our SMS-EMOA combines ideas borrowed from other EMOA, like the well established NSGA-II [11] and archiving strategies presented by Knowles, Corne, and Fleischer [5,6]. It is a steady-state evolutionary algorithm with constant population size that firstly uses non-dominated sorting as a ranking criterion. Secondly the hypervolume is applied as selection criterion to discard that individual, which contributes least hypervolume to the worst-ranked Pareto-optimal front.

3.1 Details of the SMS-EMOA

A basic feature of the SMS-EMOA is that it updates a population of individuals within a steady-state approach, i. e. by generating only one new individual in each iteration. The basic algorithm is described in algorithm 1. Starting with an initial population of μ individuals, a new individual is generated by means of random variation operators[1]. The individual enters the population, if replacing a member

---

[1] We employed the variation operators used by Deb et al. for their ϵ-MOEA algorithm [10]. These are the SBX recombination and a polynomial mutation operator, described in detail in [1]. We used the implementation available on the KanGAL home page http://www.iitk.ac.in/kangal/.

increases the hypervolume covered by the population. By this rule, individuals may always enter, if they replace dominated individuals and therefore contribute to a higher quality of the population. Apparently, the selection criterion assures that no non-dominated individual is replaced by a dominated one.

Before we will further explicate this selection strategy, we will spend a few more words on the steady-state approach. A steady-state scheme seems to be well suited for our approach, since it can be easily parallelized, enables the algorithm to keep a high diversity, and allows for an efficient implementation of the selection based on the hypervolume measure.

Algorithm 1 SMS-EMOA

1: $P_0 \leftarrow$ init() /* Initialize random start population of μ individuals */
2: $t \leftarrow 0$
3: **repeat**
4: $\quad q_{t+1} \leftarrow$ generate(P_t) /* Generate one offspring by variation operators */
5: $\quad P_{t+1} \leftarrow$ Reduce($P_t \cup \{q_{t+1}\}$) /* Select μ individuals for the new population */
6: $\quad t \leftarrow t+1$
7: **until** stop criterium reached

In contrast to other strategies that store non-dominated individuals in an archive, the SMS-EMOA keeps a population of non-dominated and dominated individuals at constant size. A variable population size might lead to single individual populations in the worst case and therefore to a crucial loss of diversity for succeeding populations. If the population size is kept constant, the population may also have to include dominated individuals. In order to decide, which individuals are eliminated in the selection, also preferences among the dominated solutions have to be established.

Algorithm 2 Reduce(Q)

1: $\{\mathcal{R}_1, \ldots, \mathcal{R}_I\} \leftarrow$ fast-nondominated-sort(Q)
2: $\quad$ /* all I non-dominated fronts of Q */
3: $r \leftarrow \text{argmin}_{s \in \mathcal{R}_I}[\Delta_\mathcal{S}(s, \mathcal{R}_I)]$ /* detect element of $\mathcal{R}_I$ with lowest $\Delta_\mathcal{S}(s, \mathcal{R}_I)$ */
4: $Q' \leftarrow Q \setminus \{r\}$ /* eliminate detected element */
5: **return** Q'

Algorithm 2 describes the replacement procedure Reduce employed. In order to decide, which individuals are kept in the population, the concept of Pareto front ranking from the well-known NSGA-II is be adopted. First, the Pareto fronts with respect to the non-domination level (or rank) are computed using the fast-nondominated-sort-algorithm [11]. Afterwards, one individual is discarded from the worst ranked front. If this front comprises $|\mathcal{R}_I| > 1$ individuals, the individual $s \in \mathcal{R}_I$ is eliminated that minimizes

$$\Delta_\mathcal{S}(s, \mathcal{R}_I) := \mathcal{S}(\mathcal{R}_I) - \mathcal{S}(\mathcal{R}_I \setminus \{s\}). \qquad (3)$$

figure). This indicates that good compromise solutions, which are located near knee-points of convex parts of the Pareto front are given better ranks in the SMS-EMOA than in the NSGA-II algorithm. Practically, solution x_5 is less interesting than solution x_4, since in the vicinity of x_5 little gains in objective f_2 can only be achieved at the price of large concession in objective f_1, which is not what is sought to be a well-balanced solution. Thus, the new method leads to more interesting solutions with fair trade-offs. It concentrates on knee-points without losing extremal points. This serves the practitioner who is mainly interested in a limited number of solutions on the Pareto front.

4 Test Problems

The SMS-EMOA from the last section was tested on several test problems from literature. We aimed at comparability to the papers of Deb and his coauthors presenting their ϵ-MOEA approach [9, 10]. That is why we also invoked the variation operators used for that approach. The test problems named ZDT1 to ZDT4 and ZDT6 from [10, 12] have been considered. For reasons of a clear overview, we copied the results for the hypervolume measure and the convergence achieved in [10] to table 1. This way, we compared our SMS-EMOA to NSGA-II, C-NSGA-II, SPEA2, and ϵ-MOEA.

4.1 Settings

We chose the parameters according to the ones given in [9, 10]. We set μ=100, calculated 20000 evaluations and used exactly the same variation operators as used for the ϵ-MOEA. The results of five runs are considered to create the values in table 1.

The hypervolume or $\mathcal{S}$ metric of the set of non-dominated points is calculated as described above, using the same reference point as in [9, 10]. The convergence measure is the average closest euclidean distance to a point of the true Pareto front as used in [10]. Note that the convergence measure is calculated concerning a set of 1000 equally distributed solution of the true Pareto front. Even an arbitrary point of the true Pareto front does not have a convergence value of 0, unless exactly equalling one of these 1000 points. Thus, the values are only comparable up to a certain degree of accuracy.

4.2 Results

The SMS-EMOA is ranked best concerning the $\mathcal{S}$ metric in all functions except for ZDT6. Concerning the convergence measure, it has two first, two second and one third rank. According to the sum of ranks of the two measures on each function, one can state that the SMS-EMOA provides best results on all considered functions, except for ZDT6, where it is outperformed by SPEA2. Building the sum of the achieved ranks of each measure shows that our algorithm obtains best results concerning both the convergence measure (with 9) and the

Table 1. Results

Test-function	Algorithm	Convergence measure			S metric		
		Average	Std. dev.	Rank	Average	Std. dev.	Rank
ZDT1	NSGA-II	0.00054898	6.62e-05	3	0.8701	3.85e-04	5
	C-NSGA-II	0.00061173	7.86e-05	4	0.8713	2.25e-04	2
	SPEA2	0.00100589	12.06e-05	5	0.8708	1.86e-04	3
	ϵ-MOEA	**0.00039545**	1.22e-05	1	0.8702	8.25e-05	4
	SMS-EMOA	0.00044394	2.88e-05	2	**0.8721**	2.26e-05	1
ZDT2	NSGA-II	**0.00037851**	1.88e-05	1	0.5372	3.01e-04	5
	C-NSGA-II	0.00040011	1.91e-05	2	0.5374	4.42e-04	3
	SPEA2	0.00082852	11.38e-05	5	0.5374	2.61e-04	3
	ϵ-MOEA	0.00046448	2.47e-05	4	0.5383	6.39e-05	2
	SMS-EMOA	0.00041004	2.34e-05	3	**0.5388**	3.60e-05	1
ZDT3	NSGA-II	0.00232321	13.95e-05	3	1.3285	1.72e-04	3
	C-NSGA-II	0.00239445	12.30e-05	4	1.3277	9.82e-04	5
	SPEA2	0.00260542	15.46e-05	5	1.3276	2.54e-04	4
	ϵ-MOEA	0.00175135	7.45e-05	2	1.3287	1.31e-04	2
	SMS-EMOA	**0.00057233**	5.81e-05	1	**1.3295**	2.11e-05	1
ZDT4	NSGA-II	0.00639002	0.0043	4	0.8613	0.00640	2
	C-NSGA-II	0.00618386	0.0744	3	0.8558	0.00301	4
	SPEA2	0.00769278	0.0043	5	0.8609	0.00536	3
	ϵ-MOEA	0.00259063	0.0006	2	0.8509	0.01537	5
	SMS-EMOA	**0.00251878**	0.0014	1	**0.8677**	0.00258	1
ZDT6	NSGA-II	0.07896111	0.0067	4	0.3959	0.00894	5
	C-NSGA-II	0.07940667	0.0110	5	0.3990	0.01154	4
	SPEA2	**0.00573584**	0.0009	1	**0.4968**	0.00117	1
	ϵ-MOEA	0.06792800	0.0118	3	0.4112	0.01573	3
	SMS-EMOA	0.05043192	0.0217	2	0.4354	0.02957	2

S metric (with 6). So in conjunction, concerning this bundle of test problems, the SMS-EMOA can be regarded as the best one.

ZDT1 has a smooth convex Pareto front where the SMS-EMOA is ranked best on the S metric and near to the best concerning the convergence measure. ZDT4 is a multi-modal function with multiple parallel Pareto fronts, whereas the best front is equivalent to that of ZDT1. On the basis of the given values from [9, 10], we assume that all algorithms achieved to jump above the second front with most solutions and aimed at the first front, like our SMS-EMOA. The worse values of the other algorithms seem to stem from disadvantageous distributions. ZDT2 has a smooth concave front and the SMS-EMOA covers most hypervolume, despite the criticism that the S metric favors convex regions. ZDT3 has a discontinuous Pareto front that consists of five slightly convex parts. Here, the SMS-EMOA is a little bit better concerning the S metric than the second ranked ϵ-MOEA and really better concerning the convergence. ZDT6 has a concave Pareto front that is equivalent to that of ZDT2, except for the differences that the front is truncated to a smaller range and that points are non-uniformly spaced. Here, the

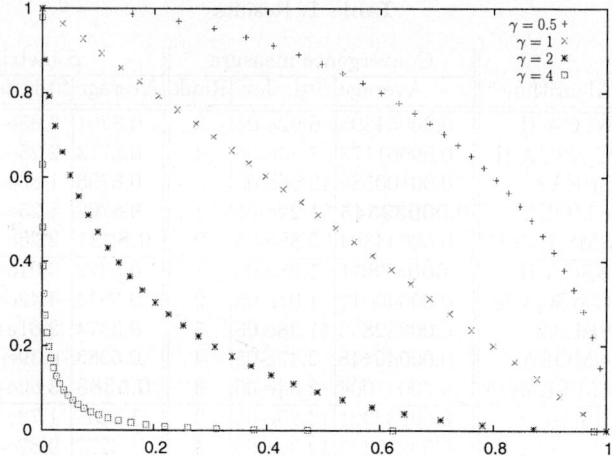

Fig. 2. This study visualizes results on the EBN problem family with Pareto fronts of different curvature computed by SMS-EMOA for a 20-dimensional search space

SMS-EMOA is ranked second on both measures, only outperformed by SPEA2, which shows apparently bad results on the other easier functions.

The outstanding performance concerning the $\mathcal{S}$ metric is a very encouraging result even though good results seem to be natural because of the use of the $\mathcal{S}$ metric as selection criterion. One should appreciate that our approach is a rather simple one with only one population and it is steady-state, resulting in a low selection pressure. Neither there are any special variation operators fitted to the selection strategy, nor it is tuned for performance in any way. All these facts would normally imply not that good results.

The good results in the convergence measure are maybe more surprising. Especially on the function that are supposed to be more difficult, the SMS-EMOA achieves very good results. A possible explanation might be that a population of well distributed points is able to sample individuals with larger improvement. Further investigations are required to clarify this topic.

4.3 Distribution of Solutions

In order to get an impression of how the SMS-EMOA distributes solutions on Pareto fronts of different curvature, we conducted a study on simple but high dimensional test functions. The aim is to observe the algorithms behavior on convex, concave and linear Pareto fronts. For the study, we devised the following family of simple generic functions:

$$f_1(\mathbf{x}) = (\sum_{i=1}^{d} |x_i|)^\gamma d^{-\gamma}, \quad f_2(\mathbf{x}) := (\sum_{i=1}^{d} |x_i - 1|)^\gamma d^{-\gamma}, \quad \mathbf{x} \in [0,1]^d, \quad (6)$$

with d being the number of object variables. The ideal criterion vectors for these bicriterial problems (which we will abbreviate EBN) are given by $\mathbf{x}_1^* = (0, \ldots, 0)$,

$f(x_1^*) = (0,1)^T$ and $x_2^* = (1,\ldots,1)$, $f(x_2^*) = (1,0)^T$. By the choice of the parameter γ the behavior of these functions can be adjusted. Parameter $\gamma = 1$ leads to a linear Pareto front, while $\gamma > 1$ yields convex fronts and $\gamma < 1$ concave ones.

Figure 2 shows that the solutions are not equally distributed on the Pareto front. The results demonstrate that the SMS-EMOA concentrates solutions in regions where the Pareto front has knee-points and captures the regions with fair trade-offs between different objectives. The regions with unbalanced trade-offs, located on the flanks of the Pareto front, are covered with less density, although extremal solutions are always maintained. On the linear Pareto front the points get uniformly distributed. In case of a concave Pareto front the regions with fair trade-offs are emphasized. These are located near the angular point of the Pareto front. The results can be explained by the way the contributing hypervolume is defined and is discussed in the previous sections.

5 Design Optimization

A frequently addressed multiobjective design problem is the two-dimensional NACA redesign of an airfoil [13, 14]. Here, two target airfoils are given, each almost optimal for predefined flow conditions. A computational fluid dynamics (CFD) tool based on the solutions of Navier-Stokes equations calculates the properties, e.g. the pressure distribution of airfoils proposed by the coupled optimization technique. From these results, the differences in pressure distribution to the target airfoils are calculated and serve as the two objectives to minimize. The computation of objective function values based on CFD calculations are usually very time consuming with one evaluation typically taking several minutes, hence only a limited number of evaluations can be afforded. Here, we allow 1000 evaluations to stay comparable to previous studies on this test problem.

5.1 Integration of Fitness Function Approximations

We use Kriging metamodels [15] as fitness function approximation tools to accelerate the SMS-EMOA. The Kriging methods allows for a prediction of the objective function values for new design points x' from previously evaluated points stored in a database. Basically, Kriging is a distance based interpolation method. In addition to the predicted value, Kriging also provides a confidence value for each prediction. Based on the statistical assumption of Kriging, the predicted result $y(x')$ and the confidence value $s(x')$ can be interpreted as the mean value and standard deviation of a one-dimensional gaussian distribution describing the probability for the 'true' outcome of the evaluation. We refer to [15] for technical details of this procedure and the statistical assumptions about the continuous random process that – as it is assumed – generated the landscape $y(x)$.

As Kriging itself tends to be time consuming for a large number of training points, Kriging models are only build from the $2d$ nearest neighbors of each point, where d denotes the dimension of the search space.

Algorithm 3 Metamodel-assisted SMS-EMOA

1: $P_0 \leftarrow$ init()   /* Initialize and evaluate start population of μ individuals */
2: $D \leftarrow P_0$ /* Initialize database of precisely evaluated solutions */
3: $t \leftarrow 0$
4: **repeat**
5:     Draw s_t randomly out of P_t
6:     $a_i \leftarrow$ mutate(s_t), $i = 1, \ldots, \lambda$ /* Generate λ solutions via mutation */
7:     approximate($D, a_1, \ldots, a_\lambda$)   /* Approximate results with local metamodels */
8:     $q_{t+1} \leftarrow$ filter($a_1, \ldots, a_\lambda$)   /* Detect 'best' approximate solution */
9:     evaluate q_{t+1} /* Evaluate selected solution precisely */
10:    $D \leftarrow D \cup \{q_{t+1}\}$
11:    $P_{t+1} \leftarrow$ Reduce($P_t \cup \{q_{t+1}\}$)   /* Select new population of μ individuals */
12:    $t \leftarrow t + 1$
13: **until** stop criterion reached

The new method is depicted in algorithm 3. In order to make extensive use of approximate evaluations, it proved to be a good strategy, to produce a surplus of λ individuals by mutation of the same parent individual. For these new individuals an approximation is computed by means of the local metamodel. The filter procedure selects the most promising solution then. The chosen solution gets evaluated precisely and is considered for the Reduce method in the SMS-EMOA. This ensures that only precisely evaluated solutions enter the population P and that the amount of approximations employed can be scaled by the user. All precisely evaluated solutions enter a database, so they can subsequently be considered for the metamodeling procedure.

The basic idea of the filter algorithm is to devise a criterion based on the approximate evaluation of a search point. Criteria for the integration of approximations in EMOA have already been suggested in [14]. Here, confidence interval boxes in the solution space were calculated as $l_i = \hat{y}_i - \omega \hat{s}_i$ and $u_i = \hat{y}_i + \omega \hat{s}_i$, $i = 1, \ldots, n$, where n is the number of objectives and ω is a confidence factor that can be used to scale the confidence level. An illustrative example for approximations with Kriging and confidence interval boxes in a 2-D solution space is given in figure 3.

Among the criteria introduced in [14], two criteria seemed to be of special interest: First, the predicted result from the Kriging method, the mean value of the confidence interval box, is considered as a surrogate for the objective function value. This corresponds to the frequently employed approach to use merely the estimated function values as surrogates for the true objective functions and thus ignore the degree of uncertainty for these approximations. The second criterion goes one step further and upvalues those points with a high degree of uncertainty, by using the lower bound edge $\hat{\mathbf{y}} - \omega \hat{\mathbf{s}}$ of the interval boxes instead of its center $\hat{\mathbf{y}}$ for the prediction. This offers us a best case estimation for the solution.

Both surrogate points are employed to evaluate a criterion based on the $\mathcal{S}$ metric that is used for sorting the candidate solutions. For the mean value surrogate this is the most likely improvement (MLI) in hypervolume for population P when selecting $\mathbf{x}$:

$$\text{MLI}(\mathbf{x}) = \mathcal{S}(P_t \cup \{\hat{\mathbf{y}}(\mathbf{x})\}) - \mathcal{S}(P_t) \tag{7}$$

and for the lower bound edge this is the potential improvement in hypervolume (LBI), that reads:

$$\text{LBI}(\mathbf{x}) = \mathcal{S}(P_t \cup \{\hat{\mathbf{y}} - \omega \hat{\mathbf{s}}(\mathbf{x})\}) - \mathcal{S}(P_t). \tag{8}$$

It may occur that all values of the criterion are zero, if all surrogate points are dominated by the old population. In that case, the Pareto fronts of lower dominance level are considered for computing the values of MLI or LBI, respectively.

For the metamodel-assisted SMS-EMOA the user has to choose the parameters ω and λ. If the lower bound criterion is used, the choice of ω determines the degree of global search by the metamodel. For high values of ω the search focuses more on the unexplored regions of the search space.

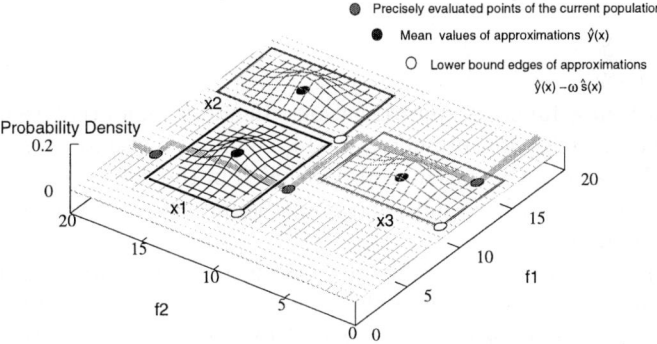

Fig. 3. Filtering of approximate solutions: Within the mean value criterion only x_3 is pre-selected while within the lower bound criterion the contributing hypervolume values of x_1 and x_3 are computed

5.2 Results

Like on the test problems, the SMS-EMOA provided very good and encouraging results on the design optimization problem. For this test series, we collected five runs for each setting again. We considered SMS-EMOA without fitness function approximation as well as the metamodel-assisted SMS-EMOA with mean value and lower bound criterion as described above.

For reasons of comparability, we utilized a method to average Pareto fronts from [13]. In short, parallel lines are drawn through the corresponding region of the search space. From the Pareto front of each run, the points with the shortest distance to these lines are considered for the calculation of the averaged front.

In the left hand part of figure 4 the different dotted sets describe three of the five Pareto fronts received from the different runs utilizing SMS-EMOA without Kriging. The line represents the received averaged Pareto front. This front is additionally copied to the right hand side figure for the reason of easier comparability. That figure compares the averaged fronts received using SMS-EMOA with

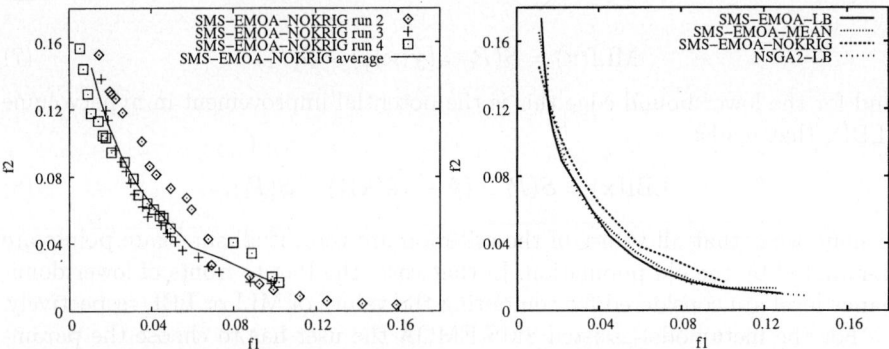

Fig. 4. The left hand side shows three of five runs used for averaging and the corresponding averaged front. The right hand side part compares SMS-EMOA without Kriging, using Kriging with lower bound (LB) and mean value (MEAN) criterion, next to NSGA-II using Kriging with lower bound criterion

and without fitness function approximations. In addition, a prior result, the best one from the investigation presented in [14] coupling metamodeling techniques with multiobjective optimization is also included in the figure. This result stems from NSGA-II runs with Kriging and lower bound criterion within 1000 exact evaluations as well.

The points on the Pareto fronts achieved using fitness function approximations are much better distributed than the ones obtained without. In the left figure, each received Pareto front is biased towards one special region. In the runs utilizing fitness function approximations no focuses can be recognized. The solutions are more equally distributed all over the Pareto front, with the aspired higher density in regions with fair trade-offs as discussed above. The reason are the thousands of preevaluations that are used to find promising regions of the search space to place exact evaluations. Compared to the results with Kriging the runs without Kriging seem not to tap their full potential due to the too small amount of evaluations.

A clear superiority of the algorithms utilizing metamodels can be recognized. The averaged front without metamodel integration is the worst front all over the search space except for the upper left corner, the extreme f_2 flank of the front. In most other regions the SMS-EMOA with lower bound criterion seems to be better than the other algorithms shortly followed by the old results from NSGA-II with lower bound criterion. The SMS-EMOA with mean value criterion yielded the worst front with metamodel integration.

In the extreme f_2 flank of the front the results seem to be turned upside down. Here, the averaged front from runs without model integration achieved the best results. The left hand side of the figure, however suggests that this might be an effect of the averaging technique. It seems to be that one run achieves outstanding results here, which leads to an unbalanced average point that is better than the averaged points of the other algorithms. This extreme effect could be avoided by averaging over more than five runs which is a small and statistically not significant number of course.

Notice, that the lower bound approximation technique yielded better results than the mean value approximation again. This was also observed in [14] and seems to be a general achievement, where more attention should be drawn to.

6 Summary and Outlook

The SMS-EMOA has been devised in this work, which is a promising algorithm for Pareto optimization, especially if a small, limited number of solutions is desired and areas with balanced trade-offs shall be emphasized. The results on academic test problems show that the algorithm is rather competitive to established EMO algorithms like SPEA2 and NSGA-II regarding the convergence measure. It clearly outperforms these methods, if the $\mathcal{S}$ metric is considered as performance measure.

Compared to many other EMOA the new approach is simple and efficient for the two objective case. The selection and variation procedures do not interfere with an extra archive and the number of strategy parameters is very low (population size and reference point). Instead of specifying a reference point the SMS-EMOA can also work with an infinite reference point.

The focus of the performance assessment was on the two objective case. We demonstrated for this case that the approach is of special elegance, since its implementation is quite simple and the update of the population can be computed efficiently. Future research will have to deal with the performance assessment for three and more objectives and for constraint problems.

For a real world airfoil design problem Kriging metamodels have been employed to save time consuming precise function evaluations. The results indicate that these techniques can be used to further enhance the performance of the SMS-EMOA.

Acknowledgements

This work was supported by the Deutsche Forschungsgemeinschaft (DFG) as part of the *Collaborative Research Center* 'Computational Intelligence' (SFB 531).

References

1. Deb, K.: Multi-Objective Optimization using Evolutionary Algorithms. John Wiley & Sons, Chichester, UK (2001)
2. Coello Coello, C.A., Van Veldhuizen, D.A., Lamont, G.B.: Evolutionary Algorithms for Solving Multi-Objective Problems. Kluwer Academic Publishers, New York (2002)

3. Zitzler, E., Thiele, L.: Multiobjective Optimization Using Evolutionary Algorithms—A Comparative Study. In Eiben, A.E., ed.: Parallel Problem Solving from Nature V, Amsterdam, Springer-Verlag (1998) 292–301
4. Fleischer, M.: The Measure of Pareto Optima. Applications to Multi-objective Metaheuristics. In Fonseca, C.M., Fleming, P.J., Zitzler, E., Deb, K., Thiele, L., eds.: Evolutionary Multi-Criterion Optimization. Second International Conference, EMO 2003, Faro, Portugal, Springer. Lecture Notes in Computer Science. Volume 2632 (2003) 519–533
5. Knowles, J., Corne, D.: Properties of an Adaptive Archiving Algorithm for Storing Nondominated Vectors. IEEE Transactions on Evolutionary Computation **7** (2003) 100–116
6. Knowles, J.D., Corne, D.W., Fleischer, M.: Bounded Archiving using the Lebesgue Measure. In: Proceedings of the 2003 Congress on Evolutionary Computation (CEC'2003). Volume 4., Canberra, Australia, IEEE Press (2003) 2490–2497
7. Knowles, J., Corne, D.: On Metrics for Comparing Nondominated Sets. In: Congress on Evolutionary Computation (CEC'2002). Volume 1., Piscataway, New Jersey, IEEE Service Center (2002) 711–716
8. Zitzler, E.: Evolutionary Algorithms for Multiobjective Optimization: Methods and Applications. PhD thesis, Swiss Federal Institute of Technology (ETH), Zurich, Switzerland (1999)
9. Deb, K., Mohan, M., Mishra, S.: Towards a Quick Computation of Well-Spread Pareto-Optimal Solutions. In Fonseca, C.M., Fleming, P.J., Zitzler, E., Deb, K., Thiele, L., eds.: Evolutionary Multi-Criterion Optimization. Second International Conference, EMO 2003, Faro, Portugal, Springer. Lecture Notes in Computer Science. Volume 2632 (2003) 222–236
10. Deb, K., Mohan, M., Mishra, S.: A Fast Multi-objective Evolutionary Algorithm for Finding Well-Spread Pareto-Optimal Solutions. KanGAL report 2003002, Indian Institute of Technology, Kanpur, India (2003)
11. Deb, K., Pratap, A., Agarwal, S., Meyarivan, T.: A Fast and Elitist Multiobjective Genetic Algorithm: NSGA–II. IEEE Transactions on Evolutionary Computation **6** (2002) 182–197
12. Zitzler, E., Deb, K., Thiele, L.: Comparison of Multiobjective Evolutionary Algorithms: Empirical Results. Evolutionary Computation **8** (2000) 173–195
13. Naujoks, B., Willmes, L., T.Bäck, W.Haase: Evaluating multi-criteria evolutionary algorithms for airfoil optimisation. In Guervós, J.J.M., Adamidis, P., Beyer, H.G., Fernández-Villacañas, J.L., Schwefel, H.P., eds.: Parallel Problem Solving from Nature – PPSN VII, Proc. Seventh Int'l Conf., Granada, Berlin, Springer (2002) 841–850
14. Emmerich, M., Naujoks, B.: Metamodel assisted multiobjective optimisation strategies and their application in airfoil design. In Parmee, I.C., ed.: Adaptive Computing in Design and Manufacture VI, London, Springer (2004) 249–260
15. Sacks, J., Welch, W.J., Mitchell, W.J., Wynn, H.P.: Design and analysis of computer experiments. Statistical Science **4** (1989) 409–435

The Combative Accretion Model – Multiobjective Optimisation Without Explicit Pareto Ranking

Adam Berry and Peter Vamplew

School of Computing, University of Tasmania, Private Bag 100,
Hobart, Tasmania, Australia
{Adam.Berry, Peter.Vamplew}@utas.edu.au

Abstract. Contemporary evolutionary multiobjective optimisation techniques are becoming increasingly focussed on the notions of archiving, explicit diversity maintenance and population-based Pareto ranking to achieve good approximations of the Pareto front. While it is certainly true that these techniques have been effective, they come at a significant complexity cost that ultimately limits their application to complex problems. This paper proposes a new model that moves away from explicit population-wide Pareto ranking, abandons both complex archiving and diversity measures and incorporates a continuous accretion-based approach that is divergent from the discretely generational nature of traditional evolutionary algorithms. Results indicate that the new approach, the Combative Accretion Model (CAM), achieves markedly better approximations than NSGA across a range of well-recognised test functions. Moreover, CAM is more efficient than NSGAII with respect to the number of comparisons (by an order of magnitude), while achieving comparable, and generally preferable, fronts.

1 Introduction

As the artificial intelligence community realises the importance of multiobjective optimisation in real-world problem domains, research attention continues to grow, with a majority of the effort being focussed on the development and investigation of Multi-Objective Evolutionary Algorithms (MOEA) [1]. At the core of much of this research rests Pareto-ranking – a concept that has been prevalent since Goldberg's early work [2] and features in a host of techniques (such as NSGA [3], MOGA [4], NSGAII [5], SPEA [6] and SPEAII [7]). Such popularity is grounded on the assumption that "Pareto ranking is the most appropriate way to generate an entire Pareto front" [1], and results investigating its use certainly support such a theory. However, despite garnering both popularity and legitimately impressive results, Pareto-ranking is not without considerable limitations. The approach carries a significant complexity cost due to its reliance on population-wide comparisons and is typically accompanied by a diversity controlling parameter that is both difficult to tune and generally expensive to use. Such is the level of computational burden that populations are fundamentally limited in size and the potential for MOEA use in high-dimensional or difficult real-world problem domains is restricted.

While a minority of contemporary algorithms endeavour to address the complexities introduced by the Pareto-ranking approach (see section 2.1), most second generation MOEA techniques extend the procedure through the inclusion of archiving, elitism and minor variations in the selection procedure (such as SPEA, SPEAII, PAES [8] and, to a lesser extent, PESA [9] and PESAII [10]). The inference that can be drawn from such a trend is that archiving and elitism are beneficial inclusions for general MOEA design, though results supporting such a claim are limited and lacking theoretical rigour. Moreover, given that the inclusion of an active secondary population generally incurs increased complexity, it is worth considering that such archiving need not be a pre-requisite for contemporary MOEA systems at all.

Consequently, this paper presents a model that moves away from active-archiving, while adopting an adaptable, inexpensive and implicit Pareto-ranking scheme that is grounded in pair-wise comparisons and simple diversity control. Furthermore, the population life-cycle is continuous rather than discrete (akin to Artificial Life systems) and agent generation is largely accretion – based on a consolidation of genes from an adaptive gene pool. Thus, the Combative Accretion Model (CAM) proposed herein represents a particularly novel approach to MOEA design that focuses on reducing complexity whilst maintaining high levels of performance.

2 Background

2.1 Pareto Ranking

Since the aim of all multiobjective optimisers is to develop an approximation of the Pareto optimal front, it is not particularly surprising that both contemporary and traditional efforts largely favour population-based Pareto dominance as a measure of fitness. By promoting those solutions that are non-dominated with respect to the current population, selection pressure favours exploration of potentially promising areas of the search space and focuses investigation on the current non-dominated front.

While it is apparent that measuring Pareto dominance is valuable in determining the direction of search, it is also significantly expensive. Even in the simplest case, where the population is divided into just two classes, the complexity[1] is $O(n^2)$ and infers a limiting bound on feasible population sizes. Such expense is only exacerbated as ranking becomes more fine-grained and continuous subdivision of dominated fronts is required (as in NSGA and NSGAII).

The complexities inherent in population-wide ranking have led to a number of algorithmic alternatives in the literature. Perhaps the most obvious approach, and the one adopted by Horn and Nafpliotis [11], is to reduce the percentage of the population under consideration when assessing dominance. In the Niched-Pareto Genetic Algorithm (NPGA), a tournament selection procedure is used, where the victor is determined by a single layer ranking process based on only ten percent of the total population (with ties broken through diversity estimation). By limiting the size of the

---

[1] This paper measures complexity in terms of the number of solution comparisons per evaluation (as per [1]) – objective comparisons are an equally valid measure and can be obtained by increasing comparison complexity by a factor of k, where k is the number of objectives.

population used, NPGA can gain increases in efficiency of up to an order of magnitude, while still capitalising on the general features of simple ranking. However, since the ratio of the selected population is statically defined, the process is inflexible – unable to adapt when more fine-grained analysis is required and less likely to encourage the exploration of poorly populated, but highly beneficial fronts. Moreover, results indicate that the overall quality of NPGA produced fronts is considerably worse than those utilising complete population sets [12].

A differing, though similarly motivated, approach is offered in the Pareto Archived Evolution Strategy (PAES), which considers only pair-wise dominance between parent and offspring until incomparability forces single-layered ranking against an archived set. While certainly promising, PAES suffers from its hill climbing characteristics – with the potential for significant performance degradation in problems featuring large local optima and disconnected fronts.

More recently, work has commenced on improving the efficiency of Pareto ranking by analysing the naïve list-based storage and linear search methods used in conventional MOEA. By incorporating modified versions of pre-existing efficient data structures and search algorithms, Jensen [13] outlines improvements for a host of MOEA and focuses particularly on the popular NSGAII. Although promising, and certainly worthy of continued research focus, results are minimal and suggest that tangible improvements are most noticeable with a reasonably small number of objectives. Moreover, irrespective of results, the development of more efficient structural representations and sorting methodologies should not preclude the refinement or extension of MOEA algorithms – efficiency gains in either area are likely to be of a complimentary nature and can only benefit the applicability of multiobjective optimisers in real world problem domains.

Given that approaches which endeavour to reduce the impact of ranking have only met with limited success, it is surprising that more MOEA research has not endeavoured to abandon its use altogether. The Artificial Life community has placed some focus on this concept, limiting comparisons in predator-prey systems to strictly pair-wise procedures [14, 15] and abandoning the use of dominance entirely in plant-based algorithms [16]. While such approaches typically induce significant reductions in complexity, results are of a strictly preliminary nature and require further investigation before gaining widespread acceptance.

Thus, Pareto ranking simultaneously represents the impetus for both performance efficacy and for efficiency degradation – it is the double-edged sword of multiobjective optimisation. Until the corresponding performance issues are addressed and effectively dealt with – be it through structural representation, improved search techniques or algorithmic refinement – the cost of high complexity will inevitably loom large over multiobjective optimisers in practical domains. It is not enough for researchers to focus simply on end results any longer – the utility of Pareto ranking has long since been known – the key now is to achieve those end results *efficiently*.

2.2 Complex Diversity Preservation

Where Pareto ranking explicitly guides the population towards the Pareto front, diversity preservation techniques are charged with ensuring that solutions remain well distributed along that front. By capitalising on techniques such as fitness sharing, diversity preservation reduces the likelihood of genetic drift and aids in developing a better picture of the true shape of the Pareto optimal region.

Although diversity preservation is generally a secondary operation, the complexity costs incurred through its inclusion can be as high as fitness assignment [13] and must therefore be considered prohibitively expensive. Such efficiency degradation is particularly evident in those algorithms that are reliant on niching, where the use of nearest-neighbour and clustering style techniques can yield $O(n^2)$ processing times [13] (as in SPEA and SPEAII). Moreover, in the general case, no optimisation of these niching procedures exists [13] and thus alternatives must be sought.

Beyond run-time efficiency, the performance of traditional multiobjective optimisers (such as MOGA, NSGA and NPGA) is tightly bound and extremely sensitive to the bias assigned to diversity preservation [1]. A failure to correctly specify the weighting of diversity in fitness assignment (typically referred to as the sharing factor) can lead to systems that prematurely diverge or converge. Thus, while guidelines exist for approximating appropriate sharing factors (see [4]), most practical systems will require significant tuning of this parameter to achieve optimal results.

Partially to address the inherent complexities associated with existing techniques, NSGAII introduces a more cost-effective approach to diversity maintenance that avoids $O(n^2)$ processing time and excessive parameter tuning. By utilising a simple crowding-distance metric, which exploits $O(\log n)$ objective-value sorting to enable low-cost nearest-neighbour measurements, NSGAII reduces diversity preservation complexity to $O(n \log n)$. Moreover, since the crowding-distance is only considered when breaking fitness-ties during tournament selection, no fitness sharing parameter is required. While the improvements made over existing techniques are impressive, complexity remains non-linear and evidence suggests ([1] citing [7]) that a notable search bias inhibits performance on higher objective problems.

In contrast, both PAES and PESA employ diversity-maintenance strategies that incur only linear complexity. These approaches divide the objective space into a hyper-grid and use the number of solutions occupying each cell to determine the relative crowding of that area. Although the move towards linear complexity is an important practical improvement, both approaches require the definition of cell-sizes, which will inevitably lead to additional parameter tuning. Furthermore, because the nature of grids is coarse, there is potential for the approach to miss or de-emphasise narrow regions of unexplored space.

Irrespective of chosen approach though, diversity preservation remains a complex, and largely unsolved, problem. Although existing techniques effectively distribute solutions across the objective-space, virtually all lead to optimisers that are susceptible to front deterioration due to the successive replacement of non-dominated solutions [17]. Such inability to maintain important solutions is significant and illustrates the complexities associated with balancing MOEA design – overly elitist approaches will lack the diversity to derive a successful spread along the Pareto optimal front, while diversity-preservation can slow and even prevent convergence onto the front. The negotiation of such issues, in addition to a continued focus on technique development, is of significant importance and warrants further attention in the MOEA research community.

2.3 Archive-Based Elitism

The stochastic nature of multiobjective optimisers – where the final set of solutions may not be representative of the best set of solutions found – requires that most, if not all ([18] citing [19]), practical installations of MOEAs capitalise on some form of solution repository. While traditional techniques use archiving purely as a background storage device, more contemporary approaches have included archival solutions as part of the selection process (SPEA, SPEAII and PAES) or as active members of the core population (PESA and NSGAII). By incorporating the archive into the evolutionary process, contemporary algorithms employ explicit elitism to bias the search around areas that have previously yielded the best results.

While empirical outcomes suggest that the incorporation of elitism into existing multiobjective algorithms can yield significant benefits [12], the use of archiving is not without limitations. Since the archive is now an active participant in the evolutionary process, careful bounds must be placed on its size to limit the negative impact that population growth will have on run-time complexity. Furthermore, since archives are typically composed of non-dominated solutions, archive maintenance can be complex and is largely based on diversity preservation principles – which, as seen earlier (Section 2.2), are both difficult to balance and potentially costly to execute. Moreover, as with any elitist approach, active-archiving infers a marked increase in selection pressure around promising solutions that can potentially lead to stagnation and premature convergence [1]. Thus, as with most core-concepts in MOEA design, archive-based elitism is as affected by complexity and balance limitations, as it is effective in achieving more rapid optimal convergence.

3 The Combative Accretion Model

The observations made in previous sections are not designed to cast popular pre-existing methods under a negative light, but to illustrate that while core MOEA techniques have achieved impressive results, they are not without their flaws. With this in mind, the development of unique approaches that aim to address existing problems can only aid in the continuing refinement and growth of multiobjective optimisers as a whole. It is such motivation that has led to the burgeoning growth of multiobjective research in areas as diverse as Artificial Life, Ant Colony simulation, Simulated Annealing and Messy Genetic Algorithms. It is also the primary motivation for this work: to develop a disparate, novel approach to multiobjective optimisation that explores new avenues for MOEA design while addressing the problems inherent in existing approaches.

In particular, the Combative Accretion Model is focussed on the reduction of complexity through the incorporation of implicit ranking, the removal of expensive diversity measures and a departure from conventional elitist archive design. Moreover, this goal is achieved in a unique system that is grounded in pair-wise dominance-based confrontation and accretion agent generation.

3.1 Agent Interaction

Central to CAM is the notion of agents and agent interaction. Borrowing terminology from the Artificial Life community, an agent is representative of a complete solution to the given multiobjective problem and carries an explicit mutable size that is representative of performance in the population[2].

Agent interaction is strictly pair-wise and the results are dictated by the dominance relationship between the two individuals (see Section 3.5). Thus, combat in CAM is derivative of binary tournaments, though selection probabilities are based around agent size rather than an expensive Pareto rank. Furthermore, the result of an interaction does not necessarily infer agent reproduction, as in conventional tournaments, but rather dictates agent survival and changes in agent size (see Section 3.5).

Since the size of an agent is the basis for selection and is subject to performance-based change, it represents endogenous fitness and results in an implicit and adaptable ranking of importance within the population – the greater the size of an agent, the more influence it will have. Such ordering is particularly significant when considering the efficiency of locating dominated solutions. In a traditional Pareto-ranking scheme, dominance is determined through an inherently expensive linear search of the current population – it does not fully capitalise on the performance of previously ranked solutions. Contrastingly, in CAM, a solution will be biased towards comparisons against the more successful agents – facilitating more rapid determination of dominated solutions (since these will generally perform poorly against the large, non-dominated agents).

3.2 The Gene Pool

To drive agents towards the current Pareto front, CAM makes use of a unique elitist concept based around the temporary storage of genes sourced from successful agents. This gene accumulation, which is ultimately used in agent creation (see Section 3.3), is referred to as the gene pool and is a finitely sized collection of alleles for each gene position. The pool is updated by any agent that passes a pre-specified size threshold (and is thus considered suitably fit) – with a random member of each gene-position collection (GPC) replaced by the corresponding agent allele (as outlined in the following equation and in Figure 1):

$$\forall i \in \{0, 1, .., g-1\}, \; GPC_{i_{|R|A|}} = x_i \quad (1)$$

where g is the number of genes per solution; x is the collection of agent genes; R is a random number between 0 and 1; GPC_i is the gene-position collection for the ith gene position; and A is the array of alleles in the given GPC.

It is tempting to find a correlation between the gene pool concept and the notion of building blocks seen in messy genetic algorithms (see [20, 21]) – however, the two approaches are quite significantly different. Messy Genetic Algorithms are charged

---

[2] Solution, agent and chromosome are considered synonymous in this work. Alleles and genes will be used to refer to specific components of a solution.

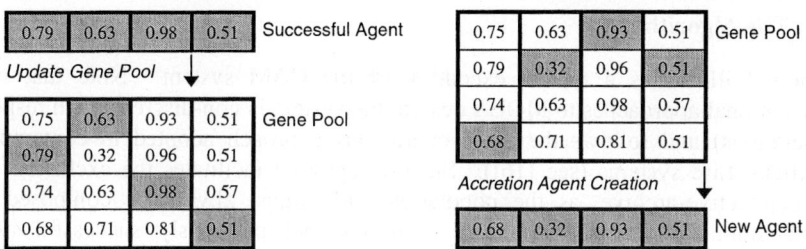

Fig. 1. Using the Gene Pool

with establishing the best possible linkage of building blocks in order to generate good solutions. In contrast, the gene pool concept utilises good solutions to identify important genes. Thus, the two approaches both aim to build on apparently successful components to drive evolution towards the Pareto optimal front, but the way in which those components are identified is diametrically opposed.

3.3 Recombination

Though reproduction can be of an asexual nature, most agent generation is performed via accretion creation – whereby the gene pool is randomly harvested to create a new individual (see Figure 1 & Equation 2). Specifically, for each GPC in the gene pool, a random member is selected and included as part of the new chromosome (with potential mutation dictated by mutation rate μ, and the new agent represented by x):

$$\forall i \in \{0,1,...,g-1\}, \; x_i = GPC_{i_{|R|A|}} \; ; \text{if} \; (R < \mu) \text{ then mutate}(x_i) \qquad (2)$$

To ensure a static population size, agent creation only ever occurs upon the death of another agent. Thus, the system is essentially a continuous poor-performer replacement scheme, whereby successful agents are retained simply by surviving.

3.4 Agent Death and Elitism

In CAM, agent death occurs under special conditions of domination or when an agent passes a pre-determined exhaustion threshold (see Section 3.5). The exhaustion threshold, which dictates the maximum number of times a single agent can contribute to the gene pool, in association with the maximum size threshold, determines the level of system elitism. Increasing the size and exhaustion thresholds infers greater pressure on high performance, while decreases improve the likelihood of diversity in the system by reducing the influence of dominant agents. The relationship between these two parameters (in addition to gene pool size) and the development of corresponding heuristics and automated tuning techniques are important areas of future work that will maximise the simplicity and practical applicability of CAM.

3.5 The Algorithm

Figure 2 illustrates a typical execution of the CAM system. Note that unlike conventional approaches to MOEA design, the system is non-discrete (with respect to generations) and follows the more continuous approach adopted in contemporary Artificial Life systems (see [16]). Such a departure facilitates the exclusion of an explicit active-archive, as the population will almost always be composed of a combination of recently generated solutions and previously successful agents. Furthermore, note that beyond the influence of thresholds and simple checks for equality[3] and incomparability, there is no explicit diversity operator charged with keeping a well-spread distribution of solutions. Results will indicate the effect of excluding such expensive techniques under the current CAM implementation, though the base model itself does not preclude their use.

Also significant is the initialisation procedure – where both the agent set and gene pool are randomly filled. While the effect is likely to be minimal and should only impede early recombinations, it is perhaps preferable to initialise the gene pool to the

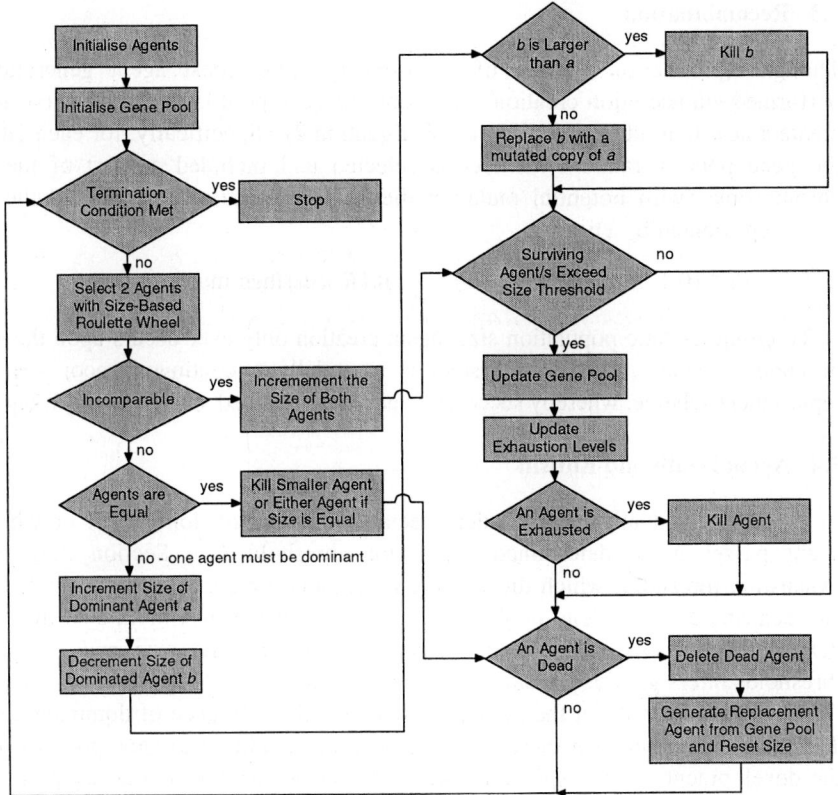

Fig. 2. The Execution Cycle of a CAM System

---

[3] Currently, agent equality is assessed in objective-space, though this is not a requirement of the system.

empty set and prevent agent generation until the pool is at least partially filled with potentially successful genes. Future work should consider this trivial amendment, though there does exist some potential for over-elitism if the pool is particularly sparse across a significant number of accretion creations.

Of the remaining processes, most are self-explanatory or have been discussed in previous sections, though both selection and combat may require further elucidation. The selection of agents for combat is as-per conventional binary roulette-wheel selection, though traditional fitness measures are replaced by a simple normalisation of agent size (using the corresponding size threshold). In the case of combat, to maximise diversity of the population, equality always leads to the death of one agent and incomparability results in growth of both agents. When an agent is dominated in combat it always dies, being replaced by a mutated clone of the dominant agent (when the size of the dominating agent is large and is thus likely to constitute a good solution) or an accretion creation (when the size of the dominating agent is small).

4 Results

Within the Multiobjective research community much debate exists as to which of a diverse set of performance metrics provide the most accurate representation of optimiser performance (see [22, 23] for a sample). This paper does not seek to settle the debate, but uses a broad range of both complexity and front analysis metrics that provide a detailed picture of CAM efficiency and effectiveness (see Table 1). To further elucidate CAM performance, and place it in a contemporary context, results are compared with both NSGA and NSGAII. The choice of systems here is important: both are popular techniques, with the original NSGA providing comparison with a non-elitist approach and NSGAII illustrating performance against an elitist system that has yielded impressive results [12] and is explicitly charged with reducing complexity.

All systems are tested on a broad range of well-recognised problems (see [12] for details) that emphasise the characteristics typically found in real world multiobjective optimisation – namely: convex (T1), concave (T2) and discontinuous fronts (T3); multi-modality (T4); and non-uniformly distributed fronts (T6). Note that since this CAM implementation is designed for real-value use only, T5 is excluded. The extension of CAM into binary problem domains lies as an important area of future work.

The presented results are representative of runs using test parameters specified in Table 2, with duplicate objective-space values removed to negate unreasonable biasing of the distribution metric. Parameter settings for NSGA and NSGAII are derived from system-defined defaults included in pre-existing implementations, or otherwise according to [5].

CAM parameter values underwent only limited tuning and thus further refinement and the associated development of corresponding heuristics can only improve overall performance. Note also that mutation in CAM is non-gaussian[4] – this is largely an

---

[4] Strictly, a distribution (with probabilities of 0.25, 0.375 and 0.375 respectively) between random ($v' = R(max\text{-}min)+min$), geometric ($v' = v \pm 0.075v$) and incremental ($v' = v \pm 0.075(max\text{-}min)$) mutation, where v is the initial value; v' is the new value; and max and min represent the range of allowed values.

arbitrary choice, though it is inspired by its recent use in Artificial Life systems. It may be beneficial to investigate the relative utility of this choice over more conventional operators in subsequent studies.

Table 1. Description of metrics used for system analysis

Metric	Motivation	Methodology
Front Graph	Establishes a visual hierarchy between competing systems and illustrates the proximity of produced solutions to the Pareto optimal front.	For each test display all non-dominated points from the amalgamation of solutions produced across three runs.
Coverage [6]	Provides a relative comparison between two non-dominated sets – determining, for one system, the ratio of solutions that are non-dominated with respect to a front produced by another system. Specifies which front is preferable – but not by how much.	For each test and system, compare each of the non-dominated sets produced across fifty runs against those sets produced by a competitor. Display the average.
Avg. Distance [12]	Specifies the average distance from a front produced by a given system to the Pareto optimal front. Thus, this measure encapsulates the accuracy of a system.	For each test and system, calculate the average Euclidean distance from the non-dominated solutions found in a given run to the Pareto optimal front. Average across fifty runs.
Extent [12]	Describes how widely spread solutions are by measuring the Euclidean distance between extreme points in a given front. The metric thus illustrates a system's ability to find boundary solutions.	For each test and system, calculate the average objective-space extent across the fifty non-dominated sets produced.
Delta Distribution [5]	Illustrates how well spread a front is by using the distance between consecutive solutions (d_i) to give: $$\Delta = 1/n \left(\sum_{i=1}^{n} |d_i - \bar{d}| \right); \text{ where } \bar{d} = 1/n \left(\sum_{i=1}^{n} d_i \right)$$ Note the large influence of front size in this function.	For each test and system, calculate the average objective-space distribution across the fifty non-dominated sets produced.
Population	Specifies the number of non-dominated solutions found across a given run. In combination with distribution, the population dictates the richness of the solution set.	For each test and system, find the number of unique objective-space solutions produced in each of the fifty runs. Display the average.
Complexity Graph	Establishes a visual hierarchy between competing systems by illustrating the average number of comparisons required per evaluation. A single comparison is considered to be any form of dominance check between two solutions.	For each test display the comparison averages per evaluation sourced from fifty distinct runs.
Big O Notation	Provides a theoretical assessment of the computational complexity of a given system.	Deduce the run-time complexity for the provided systems.

Table 2. System Settings

NSGAII and NSGA		CAM	
Uniform Crossover Rate	0.9	Gene Pool Collection Size	$n/10$
Gaussian Mutation Rate	$1/g$	Max Agent Size (s)	5
Crossover/Mutation Distribution Index	20	Initial/Reset Agent Size	$\lfloor s/2 \rfloor$
NSGA		Exhaustion Threshold (e)	30
Param. Space Sigma Share	$0.5*0.1^{1/g}$	Non-Gaussian Mutation Rate	$1/g$
Common			
Population Size (n)	50	Termination Condition	Number of Evals.

4.1 Front Quality

Figures 3-7 illustrate that for a broad range of problems, CAM is capable of finding both highly accurate and well-distributed approximations of the true Pareto optimal front. Furthermore, the arrival and development of these fronts is particularly efficient, requiring at most 5000 evaluations for all but the complex and multi-modal T4 problem. While there is degradation in overall efficiency for CAM in this instance, it still displays a marked improvement over NSGAII, which fails to converge to the true Pareto optimal front and stagnates in local optima. Moreover, T4 is generally recognised as the most difficult problem in the test-suite and existing studies have shown that systems such as SPEA and NPGA fail to locate the Pareto front even with larger populations [12]. Thus, results indicate that CAM can perform well on complex problem domains and is capable of moving through false local fronts with only minimal population sizes.

Furthermore, CAM clearly outperforms NSGAII on T6 – a non-uniform problem that has again been shown to cause difficulties for existing techniques such as SPEA [12]. Such robustness of performance irrespective of domain characteristics is an important feature of the CAM system and suggests broad practical applicability.

In addition to the superiority shown in the T4 and T6 graphs, coverage measurements (Figure 8) illustrate that CAM fronts are preferable to those produced by NSGAII on all remaining problems (excluding T3, where the systems are approximately equivalent). Moreover, while not displayed, CAM fronts completely dominate those generated by NSGA on every test function excluding T6 (where CAM achieves 94% coverage). Given that NSGA is a popular early system that forgoes archiving, such a comprehensive improvement by CAM is particularly significant.

To further clarify the contributing factors that define a given front, Figure 9 illustrates the average Euclidean distance, extent, distribution and non-dominated set size for each of the systems across the given test functions. In all cases, CAM has improved accuracy compared to both NSGA and NSGAII, finding solutions that are more closely positioned to the Pareto optimal front. Such a feature, in league with good extent values, illustrates the ability of CAM to rapidly develop highly accurate solutions, without converging onto a small region of objective-space. Indeed, across the entirety of the five fifty-run tests, the resultant CAM fronts never converged onto a single non-dominated point. Such avoidance of solution homogeny is particularly significant given the propensity for NSGA and NSGAII to become fixated on narrow areas of the objective-space (as in T4, where the resultant fronts of NSGAII and NSGA converged to a single non-dominated solution in 28% and 92% of runs respectively; and in T2, where NSGAII had such convergence in 42% of the runs).

However, it is important to note that while CAM features acceptable distribution levels which exceed those of NSGA (excluding T3), it is comparatively worse than NSGAII (excluding T6 and single-member populations). Such a difference is likely caused, and at least heavily influenced, by the increased frontal occupation of NSGAII (when it avoids single-point convergence), which reflects a more richly populated, though lower quality, approximation set than CAM. Thus, future work should examine the use of diversity-guided reproduction (via the manipulation of gene pool mechanics) to aid in the development of higher-cardinality high-quality sets in CAM.

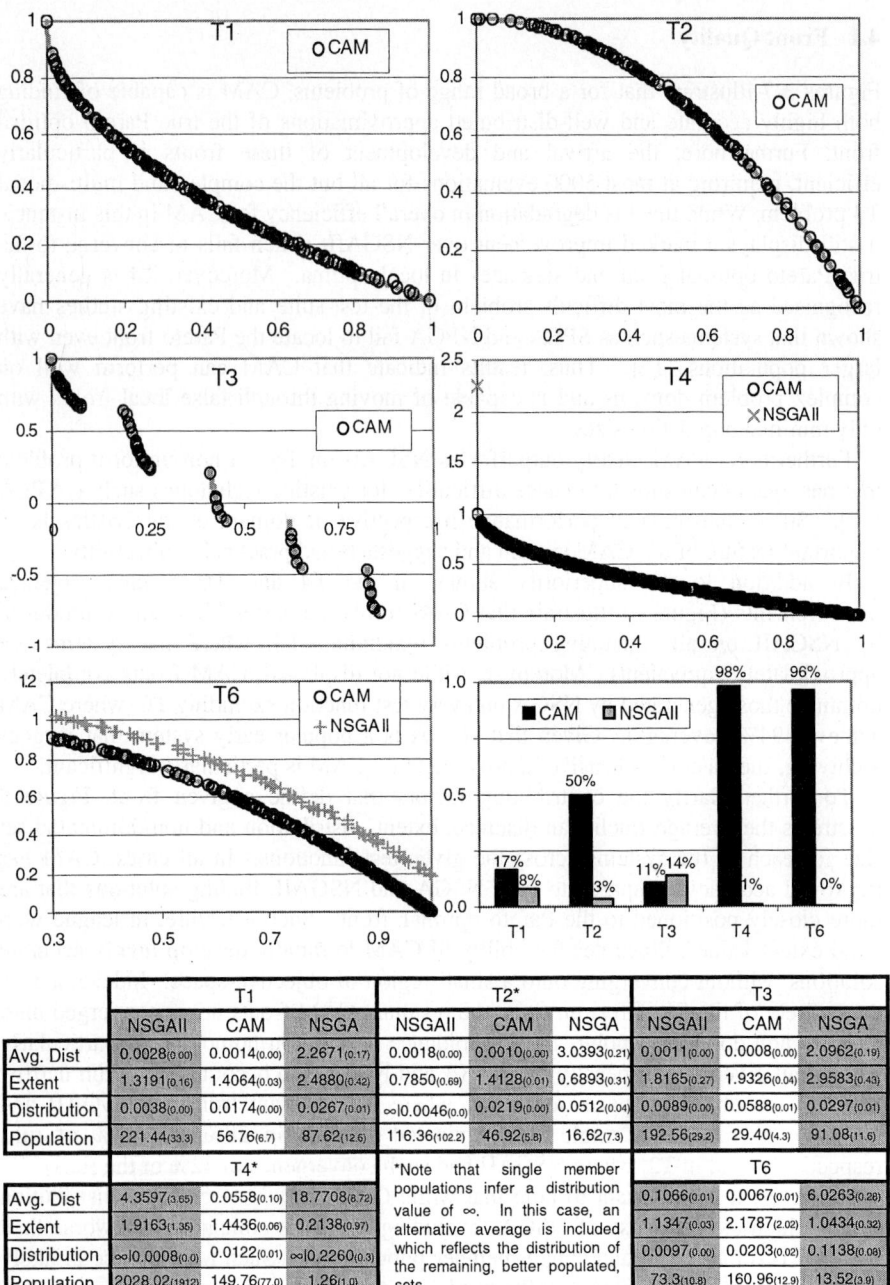

Figs. 3-9. Offline front analysis for runs of 5000 (T1-T3, T6) and 40000 evaluations (T4)
Figs. 3-7. Non-dominated solutions produced during three distinct runs – NSGAII excluded when clarity of graph is lost; *reference line* indicates the optimal front; *x-axis* is objective one; *y*-axis is objective two **Fig. 8.** The average relative coverage of non-dominated sets across fifty distinct runs **Fig. 9.** Average performances on given metrics (standard deviations provided in brackets)

4.2 Complexity Analysis

While the quality of fronts produced by CAM is impressive, the performance of any system cannot stand on quality alone. Indeed, for contemporary multiobjective optimisation, the utility of an algorithm is also contingent on the corresponding run-time complexity. With this in mind, the following section addresses the issue of complexity, utilising NSGAII to highlight relative performance improvements.

Empirical evidence illustrates that CAM is consistently and significantly more efficient than NSGAII across all of the tested areas for a population size of fifty (Figures 10-13). The poor performance of NSGAII can be attributed to the increased computational burden accrued from archiving, which essentially doubles the population size, sorting for fitness sharing and an explicit ranking scheme. By avoiding these techniques, CAM achieves an average run-time complexity that is faster than NSGAII by an order of magnitude and tends towards $O(ns)$.

The shape of the complexity graphs is also significant and warrants some discussion. CAM is most efficient early in the run where it can quickly identify dominated solutions via implicit size ordering (as discussed in Section 3.1). In sharp contrast, NSGAII is generally extremely inefficient during this period, requiring numerous passes through the population to determine explicit ranking, whilst lacking any existing Pareto ordering of solutions to maximise efficiency. Furthermore, the increase in the number of comparisons towards the completion of a CAM run indicates that dominance determination is flexible and adaptive, unlike the complexity reduction methods employed by NPGA. As the number of good solutions grow, so too does the breadth of search to ensure accurate non-dominated front representation – a capability that is beyond statically defined sub-population methods.

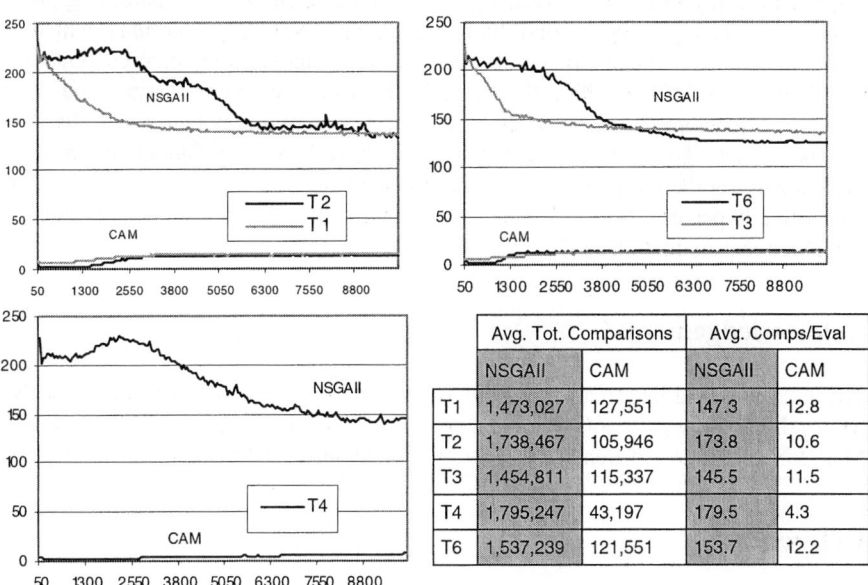

Figs. 10-13. Complexity Analysis **Figs. 10-12.** Average number of comparisons (*y-axis*) per evaluation (*x-axis*) over fifty runs. **Fig. 13.** Summaries of the total number of comparisons and the overall average number of comparisons per evaluation across fifty runs of 10000 evaluations

In terms of complexity, CAM performs worst when the population is highly non-dominated and incomparable – where solutions must generally wait until exhaustion to be removed. In this case, CAM has $O(nes)$ performance. Since complexity becomes $O(n^2)$ when $es ~ n$, it is theoretically possible to have a worst case complexity equivalent to NSGAII. Since such complexity occurs only on near-complete and persistent stagnation of a well distributed optimal or pseudo-optimal front, it is unlikely for such complexity to occur prior to completion of a typical run (as evidenced in Figures 10-13). Moreover, since the complexity rise is directly tied to stagnation of well distributed fronts, at worst the performance of CAM degenerates to NSGAII levels when on potentially good fronts, and at best, this peak in complexity can be used as an additional termination condition (since it indicates either the generation of a good Pareto front approximation or premature front convergence).

5 Conclusions and Future Work

This paper has presented a novel approach to multiobjective optimisation that is driven by agent-based pair-wise dominance interactions and the development of an elitist gene pool for accretion agent creation. By avoiding explicit ranking, complex diversity preservation and expensive active-archiving procedures, results have shown that the Combative Accretion Model demonstrates an order-of-magnitude improvement over the run-time complexity of NSGAII across a wide range of functions. Moreover, CAM consistently produces good approximations of the Pareto optimal front irrespective of diverse problem characteristics, while achieving frontal quality that is typically beyond both NSGA and NSGAII – notions which are substantiated both graphically and through coverage and accuracy metrics. Given the promising nature of the achieved results, future work is certainly merited and should focus on extensions to the CAM system, the refinement and potential automation of parameters, and the application of the approach to real-world problems. In particular, work regarding the integration of diversity into the accretion process, or the use of a PAES-like hyper-grid for biasing pair-wise results, may aid in achieving more richly populated approximation sets.

Acknowledgements

The authors would like to acknowledge the following people and institutions, without whom this work would not have been possible: Trixie Berry, Michael Berry, Pauline Mak, Ian Lewis, David Benda, the School of Computing and the GAI research group.

References

1. Coello Coello, C.A., Lamont, G.B., and Van Veldhuizen, D.A., 'Evolutionary Algorithms for Solving Multi-Objective Problems'. 2002, New York: Kluwer Academic.
2. Goldberg, D.E., 'Genetic Algorithms in Search, Optimization, and Machine Learning' 1989, Reading, Massachusetts: Addison-Wesley.

3. Srinivas, N. and Deb, K., *Multiobjective Optimization using Non-dominated Sorting in Genetic Algorithms*. Evolutionary Computation, 1994. 2(3): p. 221 - 248.
4. Fonesca, C.M. and Fleming, P.J. *Genetic Algorithms for Multiobjective Optimization: Formulation, Discussion and Generalization*. in Genetic Algorithms: Proceedings of the Fifth International Conference. 1993. San Mateo, CA.
5. Deb, K., Agrawal, S., Pratab, A., and Meyarivan, T., *A Fast Elitist Multi-Objective Genetic Algorithm for Multi-Objective Optimization: NSGA-II*. Proceedings of the Parallel Problem Solving from Nature Conference VI, 2000: p. 849-858.
6. Zitzler, E. and Thiele, L., *Multiobjective Evolutionary Algorithms: A Comparative Case Study and the Strength Pareto Approach*. Lecture Notes in Computer Science, 1999.
7. Zitzler, E., Laumanns, M., and Thiele, L., *SPEA2: Improving the Strength Pareto Evolutionary Algorithm For Multiobjective Optimization*. Evolutionary Methods for Design, Optimization and Control, 2002.
8. Knowles, J. and Corne, D. *The Pareto Archived Evolution Strategy: A New Baseline Algorithm for Pareto Multiobjective Optimisation*. in Proceedings of the Congress on Evolutionary Computation. 1999. Washington, D.C.: IEEE Press.
9. Corne, D.W., Knowles, J.D., and Oates, M.J., *The Pareto-Envelope based Selection Algorithm for Multiobjective Optimisation*. PPSN VI, 2000: p. 869-878.
10. Corne, D.W., Jerram, N.R., Knowles, J.D., and Oates, M.J. *PESA-II: Region-Based Selection in Evolutionary Multiobjective Optimization*. in Proceedings of GECCO. 2001.
11. Horn, J., Nafpliotis, N., and Goldberg, D.E. *A Niched Pareto Genetic Algorithm for Multiobjective Optimization*. in Proceedings of the First CEC. 1994.
12. Zitzler, E., Thiele, L., and Deb, K. *Comparison of Multiobjective Evolutionary Algorithms: Empirical Results*. in 1999 Genetic and Evolutionary Computation Conference. 1999.
13. Jensen, M.T., *Reducing the Run-time Complexity of Multi-Objective EAs: The NSGA-II and Other Algorithms*. IEEE Transactions on Evolutionary Computation, 2003. 7(5).
14. Socha, K. and Kisiel-Dorohinicki, M. *Agent-Based Evolutionary Multiobjective Optimisation*. in Proceedings of CEC 2002 - Congress on Evolutionary Computation. 2002.
15. Laumanns, M., Rudolph, G., and Schwefel, H.-P. *A Spatial Predator-Prey Approach to Multi-Objective Optimization: a Preliminary Study*. in PPSN V. 1998.
16. Berry, A. and Vamplew, P. *A Simplified Artificial Life Model for Multiobjective Optimisation: A Preliminary Report*. in The Congress on Evolutionary Computation (CEC). 2003. Canberra, ACT.
17. Laumanns, M., Thiele, L., Deb, K., and Zitzler, E., *Combing Convergence and Diversity in Evolutionary Multi-Objective Optimization*. Evolutionary Computation, 2002. 10(3).
18. Knowles, J.D. and Corne, D.W., *Properties of an Adaptive Archiving Algorithm for Storing Nondominated Vectors*. IEEE Transactions on Evolutionary Computation, 2003. 7(2).
19. Horn, J., *Multicriterion Decision Making*. Handbook of Evolutionary Computation, 1997.
20. Van Veldhuizen, D.A. and Lamont, G.B., *Multiobjective Optimization with Messy Genetic Algorithms*. Symposium on Applied Computing, 2000: p. 470-476.
21. Zydallis, J.B., Van Veldhuizen, D.A., and Lamont, G., *A Statistical Comparison of Multiobjective Evolutionary Algorithms Including the MOMGA-II*. EMO 2001, 2001.
22. Okabe, T., Jin, Y., and Sendhoff, B. *A Critical Survey of Performance Indices for Multi-Objective Optimisation*. in Congress on Evolutionary Computation. 2003. Canberra.
23. Zitzler, E., Thiele, L., Laumanns, M., Fonesca, C.M., and Grunert de Fonesca, V., *Performance Assessment of Multiobjective Optimizers: An Analysis and Review*. IEEE Transactions on Evolutionary Computation (Accepted for Publication), 2002.

Parallelization of Multi-objective Evolutionary Algorithms Using Clustering Algorithms

Felix Streichert, Holger Ulmer, and Andreas Zell

Center for Bioinformatics Tübingen (ZBIT), University of Tübingen,
Sand 1, 72076 Tübingen, Germany
streiche@informatik.uni-tuebingen.de

Abstract. While single-objective Evolutionary Algorithms (EAs) parallelization schemes are both well established and easy to implement, this is not the case for Multi-Objective Evolutionary Algorithms (MOEAs). Nevertheless, the need for parallelizing MOEAs arises in many real-world applications, where fitness evaluations and the optimization process can be very time consuming. In this paper, we test the 'divide and conquer' approach to parallelize MOEAs, aimed at improving the speed of convergence beyond a parallel island MOEA with migration. We also suggest a clustering based parallelization scheme for MOEAs and compare it to several alternative MOEA parallelization schemes on multiple standard multi-objective test functions.

1 Introduction

For some time, Evolutionary Algorithms (EAs) have developed into a feasible optimization approach for many industrial applications. Especially, Multi-Objective EAs (MOEAs) receive a lot of attention, because many practical optimization problems are often multi-objective. Unfortunately, EA approaches often require huge amounts of fitness evaluations to solve a given optimization problem. This holds true especially in case of multi-objective optimization problems, where the search space may be of same size, but a significantly larger portion of it needs to be explored to obtain the whole Pareto front.

There are two kinds of limitations on fitness evaluations that make EAs and MOEAs infeasible for many real world applications. First, a fitness evaluation requires real-world experiments, which can be both costly and time consuming. Or second, the fitness evaluation is computationally too expensive to allow optimization through EAs and MOEAs in reasonable time.

The first kind of limitation can only be solved by making EAs and MOEAs more efficient by means of improved evolutionary operators, supporting the EA with problem specific knowledge or local search heuristics. Alternatively, a surrogate model of the fitness function can be used instead of true fitness function evaluations. This approach can be applied both to EAs [10] and MOEAs [8].

The second problem is all about speed and can be resolved by using parallelization schemes for EAs and MOEAs on multiple processors. Due to the population based approach of EAs, single-objective EAs are very easy to parallelize. There is a large amount of literature on parallel EAs [3] and we will outline three major parallelization schemes for Single-Objective EAs in Sec. 2.1. But parallelization schemes for MOEAs are not as frequent, a good overview is given in [18], although the need for efficient parallelization schemes is even greater.

The fact that MOEAs search for a whole set of solutions (the Pareto front) instead of a single solution, suggests dividing the optimization problem into multiple subproblems, which are hopefully more efficient to solve. This approach is called the 'divide and conquer' approach and offers the prospect of parallelization schemes, which are even more efficient than non-parallelized MOEAs. The problem is, how to find a suitable partitioning of a given optimization problem without additional *a priori* knowledge about the topology of the search space? And second, is the 'divide and conquer' approach actually feasible? We suggest to use clustering algorithms to analyze the current population to find a suitable search space partitioning to aid the 'divide and conquer' approach and test whether or not an advantage of the 'divide and conquer' idea can be observed on standard benchmark functions.

The remaining publication is structured as follows: First, we outline the current state of the art regarding parallelization schemes for single and multi-objective EAs in Sec. 2 and we give details on the new clustering based parallelization scheme for MOEAs in Sec. 3. Then, the new algorithm is compared to other MOEA parallelization schemes on multiple test functions in Sec. 4. Finally, we discuss whether or not the 'divide and conquer' approach is actually feasible and how we intend to proceed with our future research on parallelizing MOEAs given in Sec. 5.

2 Related Work

As outlined before, EAs are well suited for parallelization, due to the population based search strategy. First, because individual alternative solutions of a population can be evaluated in parallel, same holds true for mutation and crossover. And second, although selection typically occurs on the whole population in a standard EA, selection could also act on a local subset of a population for several generations instead, without suffering major losses in performance. These two properties are usually used to parallelize EAs.

2.1 Parallel Evolutionary Algorithms

The most common parallel EAs can be grouped into one of three categories [3]:
The Island Model. utilizes the second property by running independent subpopulations on multiple processors. To prevent premature convergence and to increase convergence speed, every m_{rate} generations the w best individuals are migrated from one subpopulation to another. This approach is well suited for computer clusters, due to the limited amount of communication.

The Master-Slave Model. concentrates on the first property: a central master process stores a global population and distributes only fitness evaluations and eventually mutation and crossover over multiple slave processors. This approach requires the fitness evaluation to be significantly more time-consuming than the necessary communication. Apart from the additional communication overhead this parallelization scheme behaves like a standard EA.

The Diffusion Model. was developed for massively parallel computers, which provide numerous processors with a local fast communication network. It uses both properties, an individual is evaluated and mutated on a single processor and selection and crossover is limited to a few neighbors directly reachable by the network topology.

2.2 Parallel Multi-objective Evolutionary Algorithms

The Island Model. can also be used for parallelizing MOEAs [14, 9]. The most straightforward implementation of island MOEAs runs a number of MOEA populations independently, each trying to obtain the complete Pareto front and every m_{rate} generations migration takes place. Beyond this simple strategy many researchers believe that a 'divide and conquer' approach on multi-objective optimization problem could be more successful, because individual subpopulations could specialize on certain areas of the Pareto front and thus be more efficient.

One approach in this direction was proposed by Miki et al. [13]. That paper also applied an island MOEA, but at regular intervals, the subpopulations are gathered, sorted according to an alternating objective, and then redistributed onto the different processors. This approach allows the subpopulation to specialize on given objectives during the optimization, and was also used in [4, 5].

Another approach by Deb et al. uses the dominance principle [1] to guide the individual subpopulation to different sections of the global Pareto front [7]. Unfortunately, the individual search directions have to be set beforehand, which requires 'a priori' knowledge of the shape of the Pareto front, and moreover this approach cannot be applied to concave Pareto fronts.

The cone separation technique uses a geometrical approach to subdivide a given Pareto front [2], see Fig. 1 for an example. A reference point R is given by the extreme values of the current Pareto front and R is the origin of k subdividing demarcation lines. The authors point out, that in order to have each subpopulation focusing on a specific region in objective space, the demarcation lines for each region have to be treated as zone constraints using the constrained dominance principle [6]. Again, this approach has several drawbacks. In case of discontinuous or non evenly distributed Pareto fronts small or empty subpopulations can be generated, which do not reflect any problem inherent structure. And finally, the geometrical subdivision scheme of cone separation becomes rather complicated in case of more than two objectives.

The Master-Slave Model. for MOEA is simple but efficient and has been implemented several times, see for example [11]. But since it behaves exactly like a standard non-parallelized MOEA, but requires heavy communication.

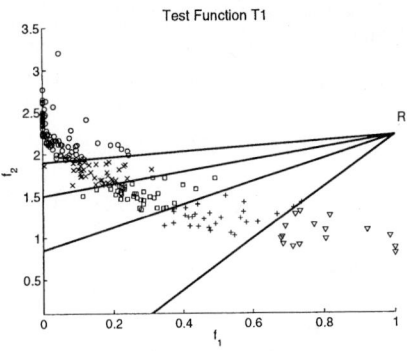 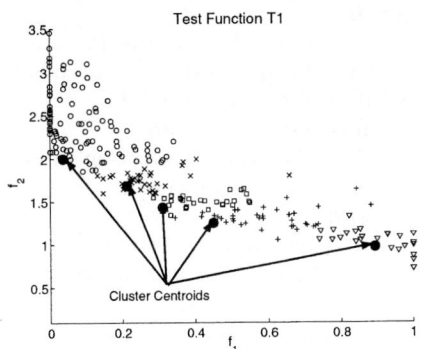

Fig. 1. Exemplary partitioning using the cone separation approach [2]

Fig. 2. Exemplary partitioning using k-Means ($k=5$) on the Pareto front

The Diffusion Model. To our best knowledge there is no publication on a MOEA implementation of the diffusion model.

3 Clustering Based Parallelization Scheme

We consider the island model together with the 'divide and conquer' principle of the cone separation approach to be an intuitive and promising approach to parallelize MOEAs. But we believe that the geometric subdivision scheme of cone separation not likely to be able to identify any problem inherent structures nor local search spaces, which would allow a successful 'divide and conquer' approach. Therefore, we decided to use clustering algorithms instead, to search for an suitable partitioning for each individual problem instance. We hope the additional overhead caused by the clustering algorithm is either compensated by an increased quality of final results or is negligible compared to a fitness evaluation.

Clustering algorithms have been used together with EAs searching for niches in multi-modal search spaces [19, 16]. Their application for parallelizing MOEAs in arbitrary in- or output dimensions is therefore most straightforward.

Following the paper on cone separation, we utilize an island MOEA for parallelizing an NSGA-II implementation [6], see Alg. 1. Each m_{rate} generations, all subpopulations $P_{i,remote}$ are gathered, clustered and redistributed onto the available processors. For clustering, we decided to use k-Means clustering on the current Pareto front, because k-Means allows use to choose the number of clusters according to the number of available processors k. In case the size of the current Pareto front is smaller than k, next level Pareto fronts are also used for clustering. We further distinguish between two variants for clustering, first a search space based clustering and second an objective space clustering. Fig. 2 gives an impression of an objective space based clustering.

To limit subpopulations to their specific region, we implement zone constraints based on the constrained dominance principle [6] using the cluster cen-

```
g = 0;
for (i = 0; i < k; i++) do  P_{i,remote}.initialize();
foreach P_{i,remote} do P_{i,remote}.evaluate();
while isNotTerminated() do
    foreach P_{i,remote} do
        P_{i,remote}.evolveOneGeneration();
        P_{i,remote}.evaluate();
    end
    if (g%m_{rate} == 0) then
        /*Migration and/or partitioning scheme */
        P_{local}.initialize();
        foreach P_{i,remote} do P_{local}.addPopulation(P_{i,remote});
        foreach P_{local}.cluster(k) do P_{i,remote} = P_{local}.cluster(k).getCluster(i);
        if useConstraints then foreach P_{i,remote} do P_{i,remote}.addConstraints();
    end
    g = g +1;
end
```

Algorithm 1: General scheme of the clustering based parallelization scheme for MOEA, with k number of processors used, m_{rate} the migration rate, $P_{i,remote}$ a remote population and P_{local} the local population.

troids. In case an individual is assigned to a different cluster centroid than the current subpopulation belongs to, the individual is marked as invalid. This interpretation complies with the implementation of cone separation, but we also tried the parallel MOEAs without zone constraints.

4 Experimental Results

We compare the new clustering based parallelization scheme with both objective space based K-Means (kosMOEA) and search space based K-Means (kssMOEA) on four different test function, see appendix, to three other approaches. First, an island model MOEA implementation without migration (pMOEA). We expect this approach to serve as worst case scenario, because this approach is only able to limit the chance of premature convergence to local optima and suffers significantly from decreasing population size with increased number of processors, see Tab. 1. Second, an island model MOEA with migration (iMOEA) where the subpopulations can profit from each others achievements. And finally, the cone separation MOEA (csMOEA). We further compare the pMOEA and the iMOEA

Table 1. Population and archive size of $P_{i,remote}$ per number of processors

Processors	1	2	3	4	5	6
Population size per processor	600	300	200	150	120	100
Archive size per processor	300	150	100	75	60	50
Accumulated population size	600	600	600	600	600	600

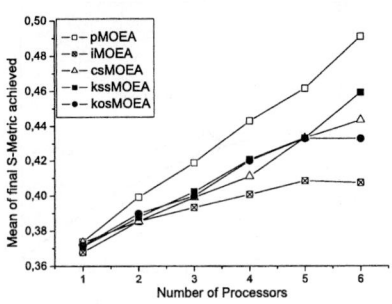 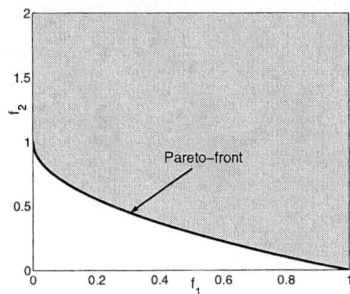

Fig. 3. Performance on T1 depending on the number of processors used

Fig. 4. The T1 test function and its search space

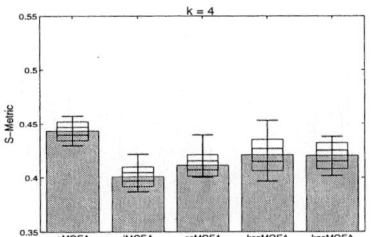 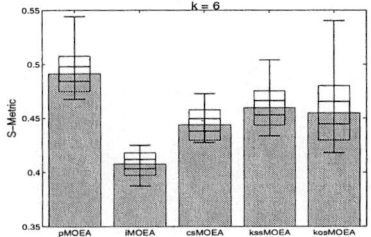

Fig. 5. Comparison of the parallelization schemes for k=4 (6) on T1, with mean of S-Metric, extreme values, standard deviation and 95% confidence interval

to both the cluster based and the cone separation MOEAs with zone constraints in Sec. 4.1 and without zone constraint in Sec. 4.2.

Each MOEA implementation uses NSGA-II [6] and every $m_{rate} = 2$ migration or search space partitioning is allowed. In case of migration (iMOEA) we migrate $w = 5$ individuals between each subpopulation. We apply the MOEAs with real-valued genotype and use self-adaptive local mutation [15] and discrete 1-point-crossover. For mutation/crossover operators we use a mutation probability of $p_m = 1.0$, a crossover probability of $p_c = 0.5$ and two parents for crossover. We compare the different parallel MOEA implementations on four test functions, three of them given from literature, see appendix. To see how each implementation scales with increased number of processors, we use up to six processors on each test function. To allow comparison we decrease the size of the subpopulations $P_{i,remote}$ with increased number of processors, see Tab. 1, while allowing 25,000 overall fitness evaluations for each optimization run.

For comparison, we use the hyper-volume under the accumulated population P_{local} (S-Metric) of each parallel MOEA implementation averaged over 25 multi-runs for each problem instance and processor configuration. The resulting metric is to be minimized, but typically converges to nonzero values, see Fig. 3.

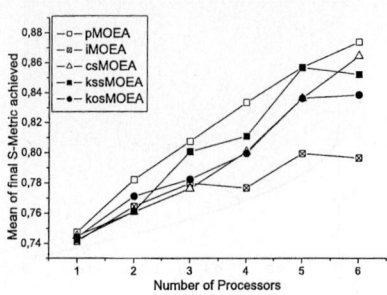 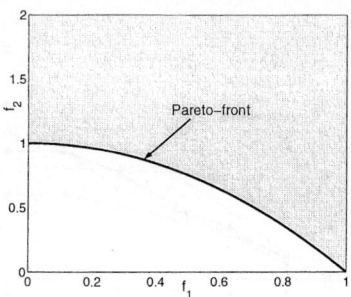

Fig. 6. Performance on T2 depending on the number of processors used

Fig. 7. The T2 test function and its search space

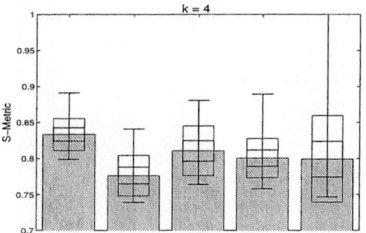 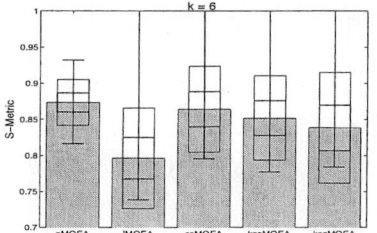

Fig. 8. Comparison of the parallelization schemes for k=4 (6) on T2, with mean of S-Metric, extreme values, standard deviation and 95% confidence interval

4.1 Results with Zone Constraints on T1-T4

Comparing pMOEA, iMOEA to csMOEA, kosMOEA and kssMOEA with zone constraints on T1 is rather disappointing, see Fig. 3. As expected, the pMOEA without migration performs worst and it suffers from a nearly linear decline due to reduced population size and fitness evaluations per subpopulation. With migration the iMOEA still suffers from a significant decline from increased number of processors, but the iMOEA performs significantly better than the pMOEA. Surprisingly, the 'divide and conquer' approaches csMOEA, kosMOEA and kssMOEA with zone constraints only outperform the simple pMOEA without communication, but all perform worse than the iMOEA with communication.

The differences between the 'divide and conquer' approaches are not really significant, as shown by the overlapping confidence intervals in Fig. 5.

The same result can be seen on the T2 test function, see Fig. 6. The pMOEA without migration suffers from a linear decline in performance while the iMOEA with migration performs best, despite the significant loss in performance with increasing number of processors. But on the T2 test function the 'divide and conquer' approaches with zone constraints do not even perform significantly better than the pMOEA without migration, see Fig. 8.

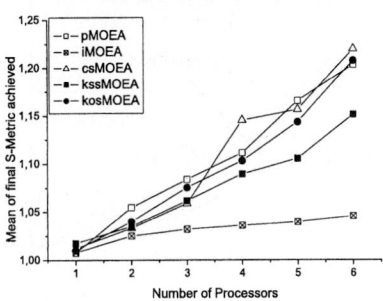
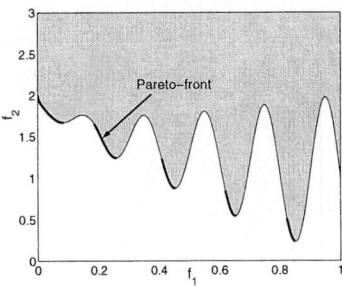

Fig. 9. Performance on T3 depending on the number of processors used

Fig. 10. The T3 test function and its search space

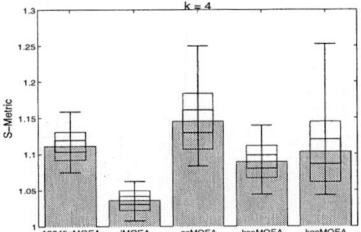
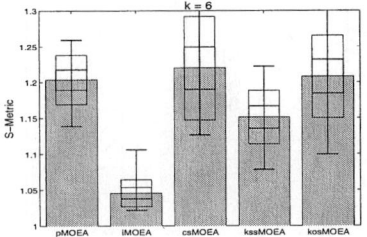

Fig. 11. Comparison of the parallelization schemes for k=4 (6) on T3, with mean of S-Metric, extreme values, standard deviation and 95% confidence interval

The disappointing performance of the 'divide and conquer' approaches can been linked to the evident contiguousness of the Pareto front in case of T1 and T2. Therefore, we foster more hopes while approaching the T3 test function, see Fig. 10. But again the 'divide and conquer' approaches with zone constraints performs very disappointing, see Fig. 9. While the iMOEA with migration performs slightly more stable than on T1 and T2 with increasing number of processors, only the kosMOEA is able to outperform the simple pMOEA without migration significantly, but is still much worse than the iMOEA.

Having a closer look at the T1-T3 test functions, see appendix, reveals that the Pareto fronts are not only contiguous in objective space, but also in search space. The Pareto-optimal solutions consist of vectors where $x_{i \neq 1} = 0$. A single solution on the true Pareto front could explore the whole Pareto front simply by mutating x_1. Therefore, it remains doubtful, and the experimental evidence does not support the claim, whether the 'divide and conquer' approach can be applied successfully on these test functions.

On the contrary frequent exchange of information, e.g. individuals, seems to be essential to explore these types of contiguous Pareto fronts efficiently, as can be concluded from the good performance of the iMOEA with migration. From this perspective zone constraints could be considered as hindering, as in case a subpopulation succeeds in reaching the true Pareto front any lateral exploit is

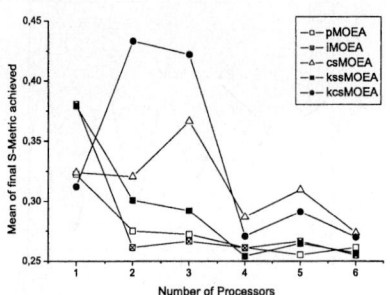 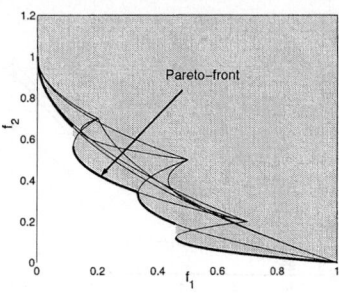

Fig. 12. Performance on T4 depending on the number of processors used

Fig. 13. The T4 test function and its search space

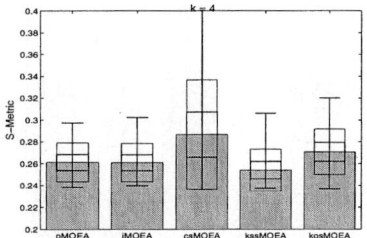 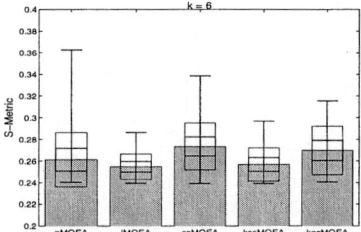

Fig. 14. Comparison of the parallelization schemes for k=4 (6) on T4, with mean of S-Metric, extreme values, standard deviation and 95% confidence interval

punished and further exploitation prevented if it leaves the local zone. Therefore, the local success is difficult to communicate to neighboring zones, although the individuals would be redirected into the neighboring zone during migration.

To investigate this problem further, we first introduce a new non-contiguous test function T4. Secondly, we disable the zone constraints on T1-T4 in Sec. 4.2.

The non-contiguous test function T4 is a simplified version of the real world multi-objective portfolio selection problem [12], for a detailed definition and parameters for T4 see appendix. For the T4 problem the distribution of $N = 5$ assets is to be optimized, minimizing risk (f_1) and loss (f_2). The other $n - N$ variables act as local penalties and need to be set to zero. To make this problem non-contiguous we add a cardinality constraint limiting the number of assets with non-zero weights to $\sum_{i=1}^{N} |sign(x_i)| = 2$. This results in a Pareto front with multiple local fronts, but only four global fronts, see Fig. 13. We adjust the parameters of this problem in such a way, that the number of global Pareto fronts is within the number of processors used in this experiment. The portfolio selection problem and the way to resolve the cardinality constraints using repair together with Lamarckism is described in [17].

Due to the multi-modal characteristics of the T4 test function, even the mean results for $k = 1$ are subject to considerable noise. For the same reason that both the pMOEA and the iMOEA are able to improve with increasing number

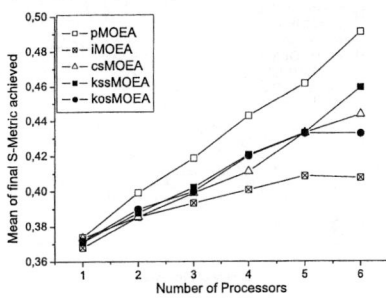

Fig. 15. 'Divide and conquer' approaches with zone constraints on T1

Fig. 16. 'Divide and conquer' approaches without zone constraints on T1

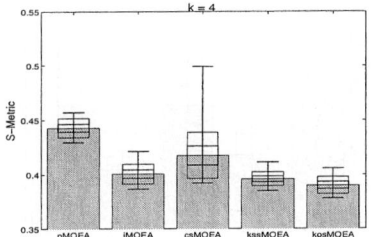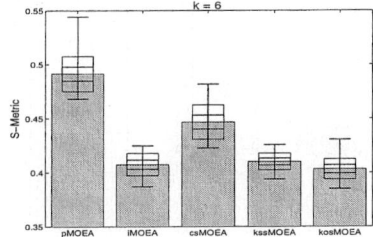

Fig. 17. 'Divide and conquer' approaches without zone constraints on T1 for k=4(6)

of processors due to the reduced chance of premature convergence, see Fig. 13. Regarding the clustering based approaches there are two interesting elements. On the one hand the kosMOEA and the kssMOEA improve significantly for $k \geq 4$, which complies with the number of global Pareto fronts. And on the other hand that the kssMOEA performs significantly better than the kosMOEA, which is confirmed by the clearly separated confidence intervals for $k = 4(6)$ in Fig. 14. This is related to the fact that most local Pareto fronts can only be separated in search space and not in objective space.

4.2 Results Without Zone Constraints on T1-T4

As discussed in Sec. 4.2 we believe that the zone constraints may have a negative effect on the performance of the 'divide and conquer' approaches. Therefore, we deactivated the zone constraints for csMOEA, kssMOEA and kosMOEA in the following experiments and compare them to the previous results.

On the T1 test function the performance of kssMOEA and kosMOEA is significantly improved by deactivating the zone constraints, compare Fig. 15 and Fig. 16. Now they equal the performance of the iMOEA with migration or even perform slightly better, see Fig. 17. It is interesting to note that the objective space based kosMOEA outperforms the kssMOEA. Unfortunately, the

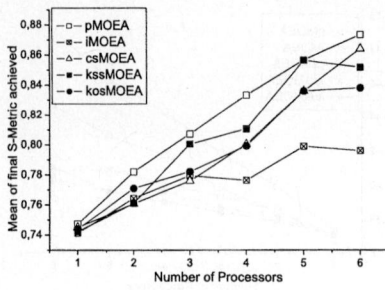 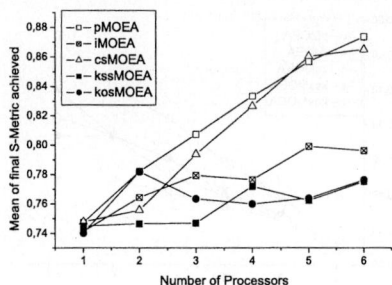

Fig. 18. 'Divide and conquer' approaches with zone constraints on T2

Fig. 19. 'Divide and conquer' approaches without zone constraints on T2

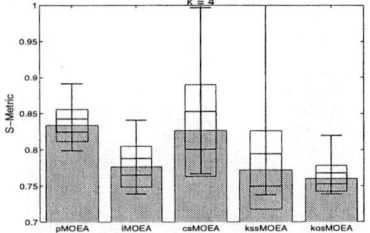 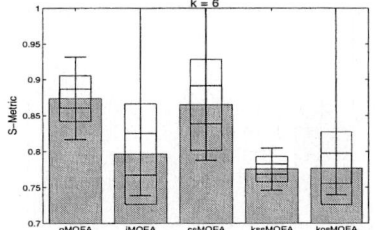

Fig. 20. 'Divide and conquer' approaches without zone constraints on T2 for k=4(6)

csMOEA is not able to improve at a similar rate, but seems to perform slightly worse than with activated zone constraints.

On the T2 test function again kssMOEA and kosMOEA are significantly better without zone constraints, while the csMOEA again performs worse than with active zone constraints, compare Fig. 18 and Fig. 19. Unfortunately, few outliers on random experimental runs cause the confidence intervals to overlap. Therefore, interpretation is limited but we tend to conclude, that the cluster based partitioning schemes for parallel MOEA without zone constraints perform better or at least as well on the T2 test function as the iMOEA with migration.

The results on the T3 test function also confirm this conclusion, compare Fig. 21 and Fig. 22. Again both cluster based partitioning schemes for parallel MOEA perform better without zone constraint and equal the performance of the iMOEA with migration and clearly outperform the csMOEA. On the T3 test function we can further show, that the K-Mean clustering based on objective space (kosMOEA) outperforms the K-Mean clustering based on search space (kssMOEA) for $k > 2$, see Fig. 20. This can be accounted to the shape of the T3 function. Here the Pareto front is non-contiguous in objective space. The fact that the kosMOEA outperforms the kssMOEA on T3 gives another example for successful 'divide and conquer'.

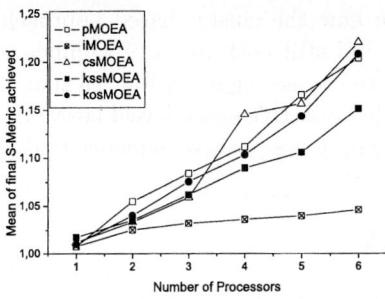

Fig. 21. 'Divide and conquer' approaches with zone constraints on T3

Fig. 22. 'Divide and conquer' approaches without zone constraints on T3

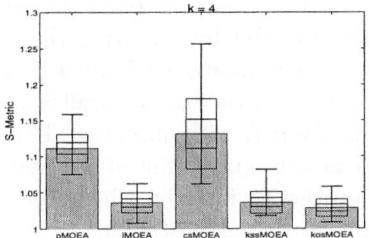 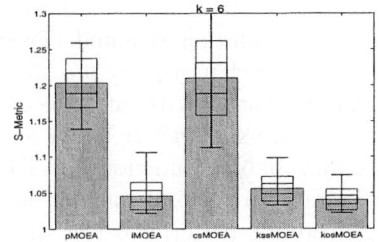

Fig. 23. 'Divide and conquer' approaches without zone constraints on T3 for k=4(6)

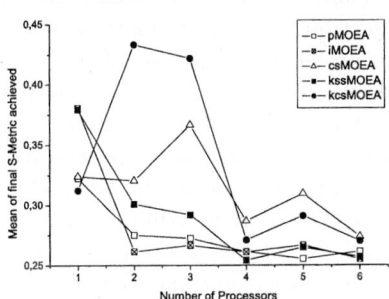 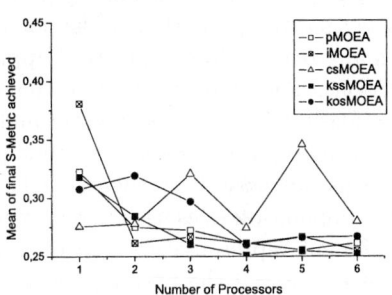

Fig. 24. 'Divide and conquer' approaches with zone constraints on T4

Fig. 25. 'Divide and conquer' approaches without zone constraints on T4

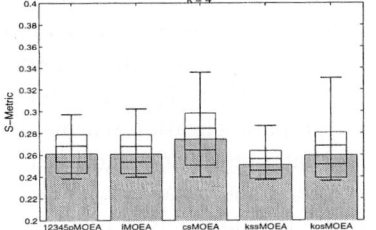 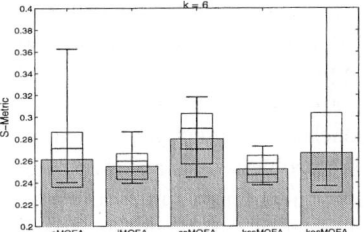

Fig. 26. 'Divide and conquer' approaches without zone constraints on for T4 k=4(6)

On the T4 test function we can still see how the cluster based approaches improve with deactivated zone constraints, and still both improve significantly for $k \geq 4$, see Fig. 24 and 12. Additionally, we can see again in Fig. 26 that the search space based kssMOEA outperforms the objective space based kosMOEA, contrary to the T1-T3 test functions. This again gives positive evidence that the 'divide and conquer' appraoch is actually feasible.

5 Conclusions and Future Work

We were able to show that the superiority of the 'divide and conquer' approach to parallelizing MOEAs, although charming and intuitive, is not easy to prove on the typical MOEA test functions. In fact the standard island MOEA with migration proves to be quite robust and hard to beat. We found that the simple and contiguous structure of the standard test functions allows lateral exploration once a good solution is found. This has been verified by removing the zone constraints, which may prevent a 'divide and conquer' approach to exploit a good solution efficiently. Without zone constraints the cluster based parallelization scheme performs as well or better as the island MOEA with migration. Despite the unfavorable test functions T1-T3 we were able to give a hint of the positive effect from a 'divide and conquer' approach on the T3 function when objective space clustering performs better than search space based clustering.

On the T4 test function on the other hand we were able to prove more clearly that the 'divide and conquer' approach is actually a suitable approach. And we believe that the properties of the portfolio selection problem are more exemplary for many real world or industrial application problems than those of the T1-T3 functions. Therefore, we believe the 'divide and conquer' approach is actually more feasible for many real world multi-objective optimization problems.

Comparing our approach to cone separation, we found that cone separation performs not as well as suggested in [2]. This may be due to the significantly larger population sizes used in our experiments, the different termination criteria or different representation and operators used. Nevertheless the cluster based approach showed the better performance in this paper, and has the advantage over cone separation, that it is easy to extend to multi-dimensional objective spaces. It proved to be able to identify problem specific structures, which allow the 'divide and conquer' approach to become suitable.

Finally, comparing our cluster based approach to the island MOEA allows us to give an outlook on future work. On contiguous test functions the island MOEA performs quite well, but alternative clustering techniques would enable our approach to distinguish between problem instances where a 'divide and conquer' approach can be useful and problem instances where it cannot. Such an approach could react more flexibly and adopt the parallelization scheme to the specific problem instance. For example, in case that a clustering algorithm identifies a single cluster in a contiguous problem space, the parallelization scheme could proceed with a simple island MOEA with migration. But in case multiple clusters are encountered a 'divide and conquer' approach could become more efficient.

Further, we would also like to implement persistent clusters (subpopulations) for the clustering based MOEA. This would enable us to monitor convergence states and focus on unexplored or more promising areas of the search space. We would also like to investigate more active migration strategies for 'divide and conquer' approaches instead of implicit migration.

Acknowledgements. This research has been funded by ALTANA Pharma AG, Konstanz, Germany.

References

1. J. Branke, T. Kauler, and H. Schmeck. Guidance in evolutionary multi-objective optimization. *Advances in Engineering Software*, 32:499–507, 2001.
2. J. Branke, H. Schmeck, K. Deb, and R. S. Maheshwar. Parallelizing multi-objective evolutionary algorithms: Cone separation. In *Congress on Evolutionary Computation (CEC 2004)*, pages 1952–1957, Portland, Oregon, USA, 2004. IEEE Press.
3. E. Cantu-Paz. A survey of parallel genetic algorithms. *Calculateurs Paralleles, Reseaux et Systems Repartis*, 10(2):141–171, 1998.
4. F. de Toro Negro, J. Ortega, J. Fernandez, and A. Diaz. PSFGA: A parallel genetic algorithm for multi-objective optimization. In F. Vajda and N. Podhorszki, editors, *Euromicro Workshop on Parallel Distributed and Network-Based Processing*, pages 849–858. IEEE, 2002.
5. F. de Toro Negro, J. Ortega, E. Ros, S. Mota, B. Paechter, and J. Martin. PSFGA: Parallel processing and evolutionary computation for multi-objective optimization. *Parallel Computing*, 30:721–739, 2004.
6. K. Deb, S. Agrawal, A. Pratab, and T. Meyarivan. A Fast Elitist Non-Dominated Sorting Genetic Algorithm for Multi-Objective Optimization: NSGA-II. In M. Schoenauer, K. Deb, G. Rudolph, X. Yao, E. Lutton, J. J. Merelo, and H.-P. Schwefel, editors, *Proceedings of the Parallel Problem Solving from Nature VI Conference*, pages 849–858, Paris, France, 2000. Springer. Lecture Notes in Computer Science No. 1917.
7. K. Deb, P. Zope, and A. Jain. Distributed computing of pareto-optimal solutions with evolutionary algorithms. In C. Fonseca, P. Fleming, E. Zitzler, K. Deb, and L. Thiele, editors, *Evolutionary Multi-Criterion Optimization*, volume 2632 of *LNCS*, pages 534–549. Springer-Verlag, 2003.
8. M. Emmerich, A. Giotis, M. zdemir, K. Giannakoglou, and T. Bck. Metamodel assisted evolution strategies. In *Parallel Problem Solving from Nature VII*, pages 362–370. Springer, 2002.
9. H. Horii, M. Miki, T. Koizumi, and N. Tsujiuchi. Asynchronous migration of island parallel GA for multi-objective optimization problems. In L. Wang, K. Tan, T. Furuhashi, J.-H. Kim, and X. Yao, editors, *Asia-Pacific Conference on Simulated Evolution and Learning*, pages 86–90, Nanyang University, Singapore, 2002.
10. Y. Jin. A comprehensive survey of fitness approximation in evolutionary computation. *Soft Computing Journal in press*, 2004.
11. R. Mäkinen, P. Neittaanmäki, J. Periaux, M. Sefrioui, and J. Toivanen. Parallel genetic solution for multiobjective MDO. In A. Schiane, A. Ecer, J. Periaux, and N. Satofuka, editors, *Parallel CFD'96 Conference*, pages 352–359. Elsevier, 1996.
12. H. M. Markowitz. *Portfolio Selection: efficient diversification of investments*. John Wiley & Sons, 1959.

13. M. Miki, T. Hiroyasu, and S. Watanabe. The new model of parallel genetic algorithm in multiobjective genetic algorithms. In *Congress on Evolutionary Computation CEC 2000*, volume 1, pages 333–340, 2000.
14. D. Quagliarella and A. Vicini. Subpopulation policies for a parallel multi-objective genetic algorithm with applications to wing design. *IEEE International Conference on Systems, Man and Cybernetics*, 4:3142–3147, 1998.
15. H.-P. Schwefel. *Numerical Optimization of Computer Models*. John Wiley & Sons, Chichester, U.K., 1977.
16. F. Streichert, G. Stein, H. Ulmer, and A. Zell. A clustering based niching ea for multimodal search spaces. In *6th International Conference on Artificial Evolution*, volume 2723 of *LNCS*, pages 293–304, Marseille, 27-30 October 2003.
17. F. Streichert, H. Ulmer, and A. Zell. Evaluating a hybrid encoding and three crossover operators on the constrained portfolio selection problem. In *Congress on Evolutionary Computation (CEC 2004)*, pages 932–939, Portland, Oregon, USA, 2004. IEEE Press.
18. D. A. Van Veldhuizen. Considerations in engineering parallel multiobjective evolutionary algorithms. *IEEE Transactions on Evolutionary Computation*, 7(2):144–173, 2003.
19. X. Yin and N. Germany. A fast genetic algorithm with sharing using cluster analysis methods in multimodal function optimization. In *Proceedings of the International Conference on Artificial Neural Nets and Genetic Algorithms*, pages 450–457, Innsbruck, Austria, 1993.
20. E. Zitzler, K. Deb, and L. Thiele. Comparison of multiobjective evolutionary algorithms: Empirical results. *Evolutionary Computation*, 8(2):173–195, 2000.

Appendix: Test Functions

The test functions T1-T3 are from [20] and have the basic structure of:

$$f_1(\bar{x}) = x_1 \tag{1}$$
$$f_2(\bar{x}) = g(\bar{x})h\left(f_1(\bar{x}), g(\bar{x})\right)$$

with $\bar{x} \in [0,1]^n$ and $n = 30$. They differ only in the definition of $g(x), h(x)$.
Test Function T1 has a convex Pareto front, see Fig. 4:

$$g(\bar{x}) = 1 + \frac{9}{n-1} \sum_{i=2}^{n} x_i \tag{2}$$
$$h(f_1, g) = 1 - \sqrt{f_1/g}$$

Test Function T2 has a concave Pareto front, see Fig. 7:

$$g(\bar{x}) = 1 + \frac{9}{n-1} \sum_{i=2}^{n} x_i \tag{3}$$
$$h(f_1, g) = 1 - (f_1/g)^2$$

Test Function T3 has a discontinuous Pareto front, see Fig. 10:

$$g(\bar{x}) = 1 + \frac{9}{n-1}\sum_{i=2}^{n} x_i \qquad (4)$$

$$h(f_1, g) = 2 - \sqrt{f_1/g} - (f_1/g)sin(10\pi f_i)$$

Test Function T4 resembles the constrained portfolio selection problem minimizing risk (f_1) and loss (f_2) of N assets. We limit to $N = 5$ assets such that the number of local Pareto fronts is well in the number of processors used. The remaining $n - N$ span the search space, see Fig. 13:

$$f_1(\bar{x}) = \sum_{i=1}^{N}\sum_{j=1}^{N} x_i x_j \sigma_{ij} \qquad (5)$$

$$f_2(\bar{x}) = \sum_{i=1}^{N} x_i \cdot \mu_i + \sum_{i=N+1}^{n} x_i^2 \cdot x_i \bmod N$$

μ_i	σ_{ij}				
0	1.0	0	0.1	0	0.3
1.0	0	0	0	0	0
0.2	0.1	0	0.7	0.3	-0.1
0.5	0	0	0.3	0.5	0
0.7	0.3	0	-0.1	0	0.2

with $n = 30$, $N = 5$, $x_i \in [0, 1]$, $\sum_{i=1}^{N} x_i = 1$ and $\sum_{i=1}^{N} |sign(x_i)| = 2$.

An Efficient Multi-objective Evolutionary Algorithm: OMOEA-II

Sanyou Zeng[1,2], Shuzhen Yao[1], Lishan Kang[3], and Yong Liu[1]

[1] Dept. of Computer Science, China University of GeoSciences,
430074 Wuhan, Hubei, P.R.China
yliu@u-aizu.ac.jp, sanyou-zeng@263.net
[2] Dept. of Computer Science, Zhuzhou Institute of Technology,
412008 Zhuzhou, Hunan, P.R.China
[3] State Key Laboratory of Software Engineering, Wuhan University
430072 Wuhan, Hubei, P.R.China

Abstract. An improved orthogonal multi-objective evolutionary algorithm (OMOEA), called OMOEA-II, is proposed in this paper. Two new crossovers used in OMOEA-II are orthogonal crossover and linear crossover. By using these two crossover operators, only small orthogonal array rather than large orthogonal array is needed for exploiting optimal in the global space. Such reduction in orthogonal array can avoid exponential creation of solutions of OMOEA and improve the performance in robustness without degrading precision and distribution of solutions. Experimental results show that OMOEA-II can solve problems with high dimensions and large number of local Pareto-optimal fronts better than some existing algorithms recently reported in the literatures.

Keywords: evolutionary algorithms; multi-objective optimization; Pareto optimal set.

1 Introduction

Almost every real-world problem involves simultaneous optimization of several incommensurable and often competing objectives. Evolutionary algorithms have the ability to find multiple Pareto-optimal solutions in one single simulation run. They have often been used to solve multi-objective problems. Such as vector evaluated genetic algorithm (VEGA)[1], Hajela and Lins genetic algorithm(HLGA) [2], pareto-based ranking procedure(FFGA) [3], niched Pareto genetic algorithm (NPGA) [4], pareto archived evolution strategy (PAES)[5], nondominated sorting genetic algorithm (NSGA-II)[6], strength pareto evolutionary algorithm (SPEA2) [7], rMOGAxs [8], and generalized regression GA (GRGA) [9].

Orthogonal design method [10] is developed to sample a small, but representative set of combinations for experimentation to obtain good combination. Leung and Zhang incorporated orthogonal design in genetic algorithm for single objective problems[11][12], found such method was more robust and statistically sound. In [13], orthogonal design method is used in multi-objective evolutionary algorithm and developed algorithm was called OMOEA. It was showed that OMOEA could find good solutions. But OMOEA degraded its performance on both precision and distribution of the yielded solutions for problems with strong interaction between variables, and when the number of objectives increases, the solutions yielded by OMOEA increased exponentially.

In this paper, an improved version of OMOEA (OMOEA-II) is proposed. Orthogonal design method is nested in crossover operator to select better genes as offsprings, and consequently, enhances the performance of OMOEA. Both orthogonal crossover and linear crossover are used in OMOEA-II. By combining two crossover operators, faster convergence and better solutions are obtained.

The rest of this paper is organized as follows. Section 2 briefly describes multi-objective optimization problem and orthogonal design method. Section 3 presents the proposed OMOEA-II. Section 4 shows experiment results and discussions. Finally, Section 5 concludes with a summary of the paper.

2 Preliminary

2.1 Problem Definition

Definition 1. *(Multi-objective Optimization Problem(MOP))* *A general MOP includes a set of N parameters (decision variables), a set of K objective functions, and a set of L constraints. Objective functions and constraints are functions of the decision variables. The optimization goal is to*

$$\begin{aligned}
&minimize\ \boldsymbol{y} = \boldsymbol{f}(x) = (f_1(\boldsymbol{x}), f_2(\boldsymbol{x}), ..., f_K(\boldsymbol{x})) \\
&subject\ to\ \boldsymbol{e}(\boldsymbol{x}) = (e_1(\boldsymbol{x}), e_2(\boldsymbol{x}), ..., e_L(\boldsymbol{x})) \leq \boldsymbol{0} \\
&where\quad \boldsymbol{x} = (x_1, x_2, ..., x_N) \in \mathcal{X} \\
&\qquad \mathcal{X} = \{(x_1, x_2, ..., x_N) | l_i \leq x_i \leq u_i, i = 1, 2, ..., N\} \\
&\qquad \boldsymbol{z} = (z_1, z_2, ..., z_N) \\
&\qquad \boldsymbol{l} = (l_1, l_2, ..., l_N) \\
&\qquad \boldsymbol{u} = (u_1, u_2, ..., u_N) \\
&\qquad \boldsymbol{y} = (y_1, y_2, ..., y_K) \in \mathcal{Y}
\end{aligned} \quad (1)$$

where $\boldsymbol{x}$ is the **decision vector**, *$\boldsymbol{y}$ is the* **objective vector**, *$\mathcal{X}$ denotes the decision space, $\boldsymbol{z}$ is the center of the decision space, $\boldsymbol{l}$ and $\boldsymbol{u}$ are the upper bound and lower bound of the decision space, and $\mathcal{Y}$ is called the objective space.*

2.2 Orthogonal Design Methods

An example was introduced in [14] to explain the basic concept of experimental design methods. The yield of a vegetable depends on: 1) the temperature, 2) the amount of fertilizer, and 3) the pH value of the soil. These three quantities are called the factors of the experiment. Each factor has three possible levels shown in Table 1. To find the best combination of levels for a maximum yield, we can do one experiment for each combination, and then select the best one. In the above example, there are $3 \times 3 \times 3 = 27$ combinations. and hence there are 27 experiments needed. In general, when there are N factors and Q levels, there are Q^N combinations. When N and Q are large, it may not be possible to do all Q^N experiments. Therefore, it is desirable to sample a small, but representative set of combinations for experimentation. The orthogonal design was developed for such purpose [14]. Let $L_M(Q^N)$ be an orthogonal array for N factors and Q levels, where "L" denotes a Latin square and M the number of combination of levels. It has M rows ,where every row represents a combination of levels. By applying orthogonal array $L_M(Q^N)$, we only select M combinations to be tested, where M may be much smaller than Q^N, For convenience, we denote $L_M(Q^N) = [a_{i,j}]_{M \times N}$ where the jth factor in the ith combination has level $a_{i,j}$ and $a_{i,j} \in \{1, 2, ..., Q\}$, and the corresponding yields of the M combinations by $[y_i]_{M \times 1}$, where the ith combination (experiment) has yield y_i. The following is an orthogonal array:

$$L_9(3^3) = \begin{bmatrix} 1 & 1 & 1 \\ 1 & 2 & 2 \\ 1 & 3 & 3 \\ 2 & 1 & 2 \\ 2 & 2 & 3 \\ 2 & 3 & 1 \\ 3 & 1 & 3 \\ 3 & 2 & 1 \\ 3 & 3 & 2 \end{bmatrix} \qquad (2)$$

In $L_9(3^3)$, there are three factors, three levels per factor, and nine combination of levels. The three factors have respective levels $1, 1, 1$ in the first combination, $1, 2, 2$ in the second combination, etc. We apply orthogonal array $L_9(3^3)$ to select nine combinations to be tested. The nine combination and their yields are shown in Table 2. From the yields of the selected combinations, a promising solution can be obtained by statistical methods. Firstly, the mean value of the yield for each factor at each level is calculated, where each factor has a level with best mean value. Secondly, the combination of the best levels is chosen as promising solution. For example, the mean yields for temperature at levels 1, 2 and 3 can be calculated by averaging yields for the experiments $1-2-3$, $4-5-6$ and $7-8-9$, respectively. The mean yields at different levels for other factors can be computed in a similar manner. The mean yields are shown in Table 3. From Table 3, we can see the best levels of temperature, amount of fertilizers and pH values are 25, 150 and 8 respectively. Therefore, we regard $(25°C, 150g/m^2, 8)$ as a promising solution. Such solution may not really be optimal. However, orthogonal design has been proven to be optimal for additive and quadratic models.

A special class of orthogonal arrays $L_M(Q^P)$, which we shall use a simple permutation method to construct, will be used in this paper, where Q is prime and $M = Q^J$, where J is a positive integer satisfying

$$P = \frac{Q^J - 1}{Q - 1} \qquad (3)$$

Table 1. Experimental design problem with three factors and three levels per factor

	Factors		
Levels	Temperature	Amount of fertilizers	pH value
Level 1	$20°C$	$100g/m^2$	6
Level 2	$25°C$	$150g/m^2$	7
Level 3	$30°C$	$200g/m^2$	8

Table 2. Nine representative combinations for experimentation and their yields based on the orthogonal array $L_9(3^3)$

	Factors			
combination	Temperature(°C)	Amount of fertilizers(g/m^2)	pH value	yield
1	1(20)	1(100)	1(6)	2.75
2	1(20)	2(150)	2(7)	4.52
3	1(20)	3(200)	3(8)	4.65
4	2(25)	1(100)	2(7)	4.60
5	2(25)	2(150)	3(8)	5.58
6	2(25)	3(200)	1(6)	4.10
7	3(30)	1(100)	3(8)	5.32
8	3(30)	2(150)	1(6)	4.10
9	3(30)	3(200)	2(7)	4.37

Table 3. The mean yields for each factor at different levels

	Mean yield		
Level	Temperature	Amount of fertilizers	pH value
Level1	3.97	4.22	3.65
Level2	4.76	4.73	4.50
Level3	4.60	4.37	5.18

Denote the jth column of the orthogonal array $[a_{i,j}]_{M \times P}$ by $\mathbf{a}_j$. Column $\mathbf{a}_j$ for $j = 1, 2, (Q^2-1)/(Q-1)+1, (Q^3-1)/(Q-1)+1, ..., (Q^{J-1}-1)/(Q-1)+1$ are called *basic columns*. and the others are called *nonbasic columns*. The algorithm first constructs the basic columns, and then generates the nonbasic columns. The details are as follows.

Algorithm 1 *Construction of orthogonal array $L_M(Q^P)$*
//Construct the basic columns as follows:
  FOR $k = 1$ TO J
    $j = \frac{Q^{k-1}-1}{Q-1} + 1$;
    FOR $i = 1$ TO Q^J

$a_{i,j} = \lfloor \frac{i-1}{Q^{j-k}} \rfloor \mod Q;$
ENDFOR
ENDFOR
//Construct the nonbasic columns as follows:
FOR $k = 2$ TO J
$j = \frac{Q^{k-1}-1}{Q-1} + 1;$
FOR $s = 1$ TO $j - 1$, $t = 1$ TO $Q - 1$
$\mathbf{a}_{j+(s-1)(Q-1)+t} = (\mathbf{a}_s \times t + \mathbf{a}_j) \mod Q;$
ENDFOR
ENDFOR
Increment $a_{i,j}$ by one for $-1 \leq i \leq M$ and $1 \leq j \leq P$; ♯

The used orthogonal array in this paper is required to satisfy $P \geq N$ and M is as small as possible. That is the columns P of $L_M(Q^P)$ must be larger than the number of factors (or decision variables) in the hope of sampling small number of points (combinations) for obtaining better solution. It only needs to determine Q and J for determining $L_M(Q^P)$. $L_M(Q^P)$ is determined by solving the following minimization problem:

$$Minimize\ M = Q^J$$
$$Subject\ to\ P = \frac{Q^J-1}{Q-1} > N \quad (4)$$

where Q is a prime and $Q \geq 3$, J is a positive integer.

$L_M(Q^P)$ is the full size of the orthogonal array, which has P columns. For a problem with N decision variables, we discard the last $P - N$ columns of $L_M(Q^P)$ and get an orthogonal array $L_M(Q^N)$.

The proposed algorithm will require the mean value of the objective at each level of each factor. Denote the objective values of the orthogonal experiments by $[y_i]_{M \times 1}$ where the objective has the value y_i at the ith combination, the mean values by $[\Delta_{k,j}]_{Q \times N}$ where the objective has the mean value $\Delta_{k,j}$ at the kth level of the jth factor, and

$$\Delta_{k,j} = \frac{Q}{M} \sum_{a_{i,j}=k} y_i \quad (5)$$

where the orthogonal array $L_M(Q^N)$ has the value $a_{i,j}$ at ith row and jth column. That is, the jth factor has level $a_{i,j}$ in the ith combination(experiment). The objective has value y_i at the ith combination, and $\sum_{a_{i,j}=k} y_i$ implies the sum of y_i where $\forall i$ satisfy $a_{i,j} = k$. The details of the algorithm are as follows

Algorithm 2 Calculation of mean value $[\Delta_{k,j}]_{Q \times N}$
$[\Delta_{k,j}]_{Q \times N} = [0]_{Q \times N};$
//Add up objective result for each factor at each level
FOR $i = 1$ TO M, $j = 1$ TO N
$q = a_{i,j}; \Delta_{q,j} = \Delta_{q,j} + y_i;$

ENDFOR
//Average results for each factor at each level
$[\Delta_{k,j}]_{Q\times N} = [\Delta_{k,j}]_{Q\times N} \times Q/M$ ♯

Each factor has its best level according to the mean value matrix $[\Delta_{k,j}]_{Q\times N}$, The combination **s** of the best levels is potentially a good solution. For minimization problems, it is calculated by

$$\begin{aligned}\Delta_{k_j,j} &= min\{\Delta_{1,j}, \Delta_{2,j}, ..., \Delta_{Q,j}\}, j = 1, 2, ..., N \\ \mathbf{s} &= (k_1, k_2, ..., k_N)\end{aligned} \quad (6)$$

3 OMOEA-II

OMOEA, proposde by the author [13], is an evolutionary algorithm by introducing orthogonal design for multi-objective optimization. It is good at finding good solutions for problems with decision variables relatively independent. However, it was found there were two problems.

1. Strong interaction between variables degrades the performance of OMOEA including both precision and distribution of the yielded solutions.
2. As number of objectives increase, the yielded solutions will increase exponentially.

The idea of OMOEA-II is to replace a large orthogonal array by a small orthogonal array in order to exploit optimal efficiently relatively small space which is generated by randomly choosing individuals in the population.

3.1 Framework of OMOEA-II

Algorithm 3 *Framework of OMOEA-II*

1. *Randomly create population $\mathbf{P}_0$ with size N_p. Set counter $t = 0$*
2. *Execute crossover operator (Algorithm 6) on $\mathbf{P}_t$ which yields offsprings $\mathbf{Q}_t$ with size N_p*
3. $\mathbf{R}_t = \mathbf{P}_t \cup \mathbf{Q}_t$
4. *Execute Selection operator (Algorithm 7) on $\mathbf{R}_t$ whiche yields next population $\mathbf{P}_{t+1}, t = t + 1$*
5. *If stopping criterion satisfied goto Step 6, else goto Step 2*
6. *Output $\mathbf{P}_t$* ♯

3.2 Crossover Operator

OMOEA-II uses two kind of multi-parents crossover operators. Parents are randomly selected from current population. Denote parents by $\mathbf{m}_1, \mathbf{m}_2$ and $\mathbf{m}_j = (m_{j,1}, m_{j,2}, ..., m_{j,N}), j = 1, 2$.

1 Orthogonal Crossover

Orthogonal design method is used on the subspace extended by the randomly chosen parents.

$$\mathbf{H} = \{(x_1, x_2, ..., x_N) | l'_i \leq x_i \leq u'_i, i = 1, 2, ..., N\} \quad (7)$$
$$l'_i = min\{m_{1,i}, m_{2,i}\} \quad u'_i = max\{m_{1,i}, m_{2,i}\}$$

For $(x_1, x_2, ..., x_N)$ in $\mathbf{H}$, x_i is regarded as the ith factor. Orthogonal array is selected by Equation (4). Each factor i is parted into $Q-1$ equal portions and yields Q levels $x_{1,i}, x_{2,i}, ..., x_{Q,i}$, where the design parameter Q must be prime and $x_{q,i}$ is given by

$$x_{q,i} = \begin{cases} l'_i & q = 1 \\ l'_i + (q-1)\delta_i & 2 \leq q \leq Q-1 \\ u'_i & q = Q \end{cases} \quad (8)$$

$$where \ \delta_i = \frac{u'_i - l'_i}{Q-1}$$

In other words, the difference between two successive levels is the same. For convenience, denote $x_j = \{x_{1,j}, x_{2,j}, ..., x_{Q,j}\}$, and call $x_{q,j}$ the qth level of the jth factor.

Algorithm 4 *Orthogonal crossover operator*

1. *Randomly choose parents m_1 and m_2, Construct subspace $\mathbf{H}$ according to Equation (7)*
2. *Choose an objective k from the K objectives as optimizing objective*
3. *Employ orthogonal design method on $\mathbf{H}$, where orthogonal array is determined by Equation (4)*
4. *s_1 is the combination with best objective value among the $M = Q^J$ combinations in the orthogonal experiment. s_2 is the potentially good solutions from Equation (6)*
5. *Output s_1, s_2 as offsprings* ♯

2 Linear Crossover

Algorithm 5 *Linear crossover operator*

1. *Randomly choose parents m_1 and m_2*
2. *For each decision variable x_i, $i = 1, 2, ..., N$, choose one of the two sub-steps to execute with probability p_c for Sub-step (a) and $1 - p_c$ for Sub-step (b)*
   (a) *Randomly create a_1 and a_2 with $a_1 + a_2 = 1$ in the range $-0.5 \leq a_1, a_2 \leq 1.5$. Let $s_i = a_1 m_{1,i} + a_1 m_{2,i}$*
   (b) *Randomly create a_1 and a_2 in the range $-0.5 \leq a_1, a_2 \leq 1.5$. Let $s_i = a_1(m_{1,i} - z_i) + a_2(m_{2,i} - z_i) + z_i$, where $z = (z_1, z_2, ..., z_N)$ is the center of the decision space (See Equation (1))*
3. *Output $s = (s_1, s_2, ..., s_N)$ as offspring* ♯

Note that sub-step (a) is helpful for the population to converge faster to good solutions while sub-step (b) helps to diverge the population.

Algorithm 6 *Crossover operator*

1. *Empty* Q_t.
2. *choose one of the two sub-steps to execute with probability p_o for Sub-step (a) and $1 - p_o$ for Sub-step (b)*
   (a) *Execute orthogonal crossover (Algorithm 4) which yields offsprings s_1, s_2, $Q_t = Q_t \cup \{s_1, s_2\}$.*
   (b) *Execute linear crossover (Algorithm 5) which yields offsprings s, $Q_t = Q_t \cup \{s\}$.*
3. *If the size of Q_t, i.e., $|Q_t|$, is smaller than N_p, then goto Step 2, else goto Step 4.*
4. *Output Q_t.*   ♯

3.3 Selection Operator

Deb etc employed non-dominated sort technology in NSGA [6]. In OMOEA-II, non-dominated sort is executed by the selection operator on R_t where $R_t = P_t \cup Q_t$. The details of the operator are as follows

Algorithm 7 *Selection operator*

1. *Empty P_{t+1}.*
2. *Find the non-dominated set B of R_t. If $|B| = N_p$ then $P_{t+1} \Leftarrow B$; if $|B| > N_p$ then execute cutoff operator (Algorithm 8) which eliminate $|B| - N_p$ elements from B and assigned the reduced B to P_{t+1}; if $|B| < N_p$ then move B from R_t to P_{t+1}, i.e., $R_t \Leftarrow R_t \setminus B$ and $P_{t+1} \Leftarrow P_{t+1} \cup B$, repeat the process of finding the non-dominated set of reduced R_t and moving the non-dominated set from R_t to P_{t+1} till $|P_{t+1}| = N_p$.*
3. *Output P_{t+1}.*   ♯

The goal of the cutoff operator on B is to enhance distribution uniformity of the reduced B after the cutoff. Zitzler etc. used cluster analysis to serve this goal in SPEA [7]. The cluster analysis is also employed in this paper. The details of the cutoff operator are as follows.

Algorithm 8 *Cutoff operator*

1. *Initialize cluster set Ψ: $\Psi = \cup_{i \in B} \{\{i\}\}$ where each individual $i \in B$ constitute a distinct cluster.*
2. *If $|\Psi| \leq N_p$, go to Step 5, else go to Step 3.*
3. *Calculate the distance of all possible pairs of clusters. The distance d_c of two cluster $C_1, C_2 \in \Psi$ is given as the average distance between pairs of individuals across the two clusters*

$$d_c = \frac{1}{|C_1|.|C_2|} \sum_{i_1 \in C_1, i_2 \in C_2} d(i_1, i_2)$$

where $d(i_1, i_2)$ is the distance between two individuals i_1 and i_2 (here the distance in objective space is used).

4. Determine two clusters C_1 and C_2 with minimal distance d_c; the chosen clusters are amalgamated into a large cluster: $\Psi = \Psi \setminus \{C_1, C_2\} \cup \{C_1 \cup C_2\}$. Go to Step 2
5. For each cluster, select a representative individual and remove all other individuals from the cluster. We consider the centroid (the point with minimal average distance to all other points in the cluster) as the representative individual. Compute the reduced non-dominated set by uniting the representative of the clusters: $\boldsymbol{P}_{t+1} = \cup_{C \in \Psi} C$. ♯

4 Numerical Experiments and Discussion

4.1 Test Functions

Some benchmark problems are taken to test OMOEA-II. They are: FON [15], $ZDT_1, ZDT_2, ZDT_3, ZDT_4, ZDT_6$ [16], $DTLZ_1, DTLZ_3, DTLZ_4, DTLZ_6$ [17]. The number of decision variables N: $N_{FON} = 3$, $N_{ZDT_1} = N_{ZDT_2} = N_{ZDT_3} = 30$, $N_{ZDT_4} = N_{ZDT_6} = 10$, $N_{DTLZ_1} = N_{DTLZ_3} = N_{DTLZ_4} = N_{DTLZ_6} = 12$. FON, $ZDT_1, ZDT_2, ZDT_3, ZDT_4, ZDT_6$ have 2 objectives and $DTLZ_1$, $DTLZ_3, DTLZ_4, DTLZ_6$ have 3 ones. ZDT_1 has a convex Pareto-optimal front while ZDT_2 has a nonconvex one. ZDT_3 represents the discreteness feature, its Pareto-optimal front consists of several noncontiguous convex parts. ZDT_4 contains 21^9 local Pareto-optimal fronts and $DTLZ_3$ contains 3^9 local ones. ZDT_6 has non-uniformity of objective space. Some decision variables of $DTLZ_1$, $DTLZ_3, DTLZ_4$ and $DTLZ_6$ are of linkage.

The parameter settings for OMOEA-II are as follows: function evaluations 25000 (about generations 250), population size 100, orthogonal crossover probability $p_o = 0.1$, linear crossover probability $1 - p_o = 0.9$ where convergence operator probability $p_c = 0.8$ and diversity operator probability $1 - p_c = 0.2$. 10 runs are repeated for each problem. These parameters were chosen after initial experiments. They are not meant to be optimal.

4.2 Results and Discussion

Metric Υ [6] was to measure the extent of convergence of known set of Pareto-optimal set. Metric Δ [6] measured the extent of spread achieved among the obtained solutions, Δ is only able to measure problems with two objectives. The smaller both the metric Υ and Δ, the better the obtained solutions. For comparison, the metric Υ and Δ of the obtained solutions on the test problems with two objectives by SPEA, NSGA-II and OMOEA-II are shown in Tables 4 and 5, where the data for NSGA-II and SPEA were from literature [6] and both of them also had population size 100 and evolution generations 250. For problems FON, $ZDT_1, ZDT_2, ZDT_3, ZDT_4, ZDT_6$ with two objectives, OMOEA-II found more accurate solutions than those of SPEA and NSGA-II except the results FON. Meanwhile, OMOEA-II could keep rather smaller or comparable Δ for diversity. For the test problems with three objectives, only the simulation results of OMOEA-II are shown; the metric Υ are shown in Table 4 while the spread of the

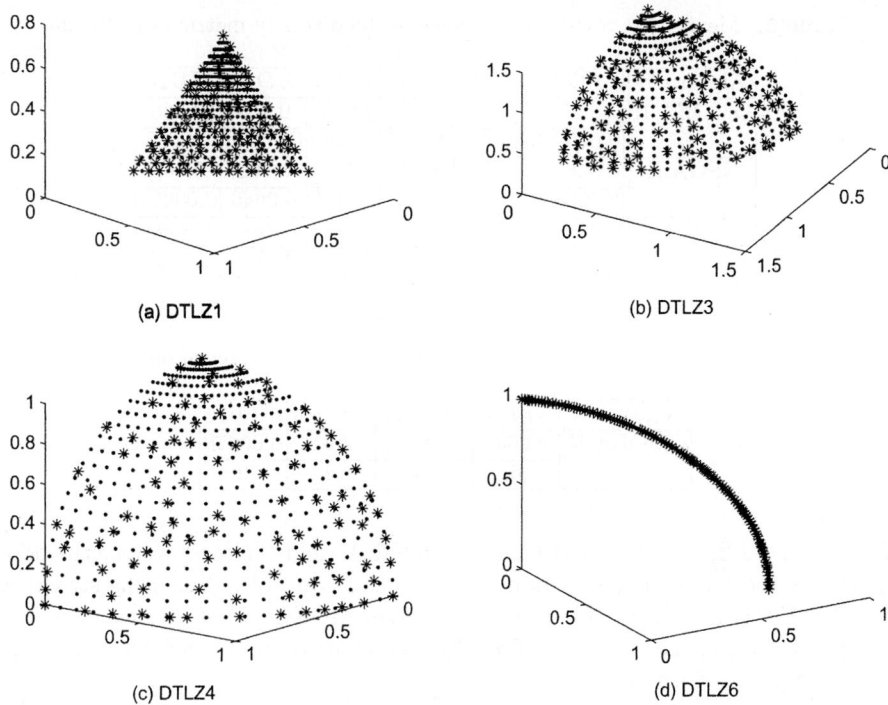

Fig. 1. Distribution of obtained solutions for $DTLZ1$, $DTLZ3$, $DTLZ4$, $DTLZ6$, where '.' denote true Pareto-optimal fronts and '*' close-to-Pareto-optimal fronts

Table 4. Mean (first rows) and variance of the convergence metric Υ in 10 runs

Algorithm	FON	ZDT_1	ZDT_2	ZDT_3	ZDT_4	ZDT_6
OMOEA-II	0.0032	0.00072	0.00057	0.0021	0.0032	0.000032
	0	0	0	0	0	0
NSGA-II	0.0019	0.03348	0.0724	0.1145	0.513	0.2965
	0	0.00475	0.0317	0.00794	0.118	0.0131
SPEA	0.1256	0.0018	0.00134	0.04751	7.34	0.0011
	0.00003	0.000001	0	0.00005	6.5725	0.00045

solutions are illustrated on Figure 1. As what Deb in [17] described that both NSGA-II and SPEA2 could not quit converge to true Pareto optimal on $DTLZ_3$ and $DTLZ_6$ where they evolved 500 generations, for $DTLZ_4$, they could not yield well distributed set of solutions, while OMOEA-II found solutions with both good spread and precision on these four problems (See Table 6, Figures 1) without changing parameter settings.

The population size of OMOEA in [13] is variable, and, for many problems, increases exponentially as the number of objectives increase. For problems FON, ZDT_1, ZDT_2, ZDT_3, ZDT_4, ZDT_6 with two objectives, the precision and distribution of solutions of OMOEA-II are comparable with OMOEA. But, for

Table 5. Mean (first rows) and variance of the diversity metric Δ in 10 runs

Algorithm	FON	ZDT_1	ZDT_2	ZDT_3	ZDT_4	ZDT_6
OMOEA-II	0.235	0.439	0.55	0.8159	0.250	0.235
	0.000283	0.0088	0.0088	0.0099	0.00044	0.0003
NSGA-II	0.378	0.39	0.43	0.7385	0.702	0.668
	0.000639	0.0019	0.0047	0.0197	0.0646	0.00992
SPEA	0.792	0.7845	0.7551	0.6729	0.7984	0.8493
	0.0055	0.0044	0.0045	0.0036	0.0146	0.0027

Table 6. Mean (first rows) and variance of the convergence metric Υ

Algorithm	$DTLZ_1$	$DTLZ_3$	$DTLZ_4$	$DTLZ_6$
OMOEA-II	0.0043	0.0127	0.01	0.00078
	0	0.000001	0	0

$DTLZ_1$, $DTLZ_3$, $DTLZ_4$ and $DTLZ_6$, OMOEA yields no results since the population is too large, that results from the linkage variables. Since the structure of OMOEA is not much the same as common MOEAs/MOGAs, usual metric comparison is much convincible. Therefore, comparison between OMOEA-II and OMOEA was not listed.

5 Conclusion

OMOEA-II overcomes the difficulties existed in its previous version (OMOEA). Orthogonal design method is nested in crossover operator to select better genes as offsprings, and consequently, enhances result precision. The other crossover called linear crossover is used, in order to converge fast to better solutions and diverge the population for global optima by changing parameters. Experimental results show that OMOEA-II can solve problems with high dimensions, large number of local Pareto-optimal fronts, non-uniformity or discontinuity or convex or nonconvex of global Pareto-optimal front and yield better solutions than some reported results.

Acknowledgment. This work was supported by The National Natural Science Foundation of China (No.s: 60473037, 60483081, 40275034, 60204001, 60133010) and by China Postdoctoral Science Foundation (No. 2003034505).

References

1. Schaffer, J. D. (1984).*Multiple Objective Optimization with Vector Evaluated Genetic Algorithms*. Ph. D. thesis, Vanderbilt University. Unpublished.
2. Hajela, P. and Lin, C. Y. (1992). Genetic search strategies in multicriterion optimal design. Structural Optimization 4, 99C107.

3. Fonseca, C. M. and Fleming, P. J. (1993). Genetic algorithms for multiobjective optimization: Formulation, discussion and generalization. In S. Forrest (Ed.), *Proceedings of the Fifth International Conference on Genetic Algorithms,* San Mateo, California, pp. 416C423. Morgan Kaufmann.
4. Horn, J. and Nafpliotis, N. (1993). Multiobjective optimization using the niched pareto genetic algorithm. IlliGAL Report 93005, Illinois Genetic Algorithms Laboratory, University of Illinois, Urbana, Champaign.
5. Knowles, J.D., Corne, D.W.(2000): Approximating the Nondominated Front Using the Pareto Archived Evolution Strategy. Evolutionary Computation 8(2), 149C172.
6. Kalyanmoy Deb, Amrit Pratap, Sameer Agarwal, and T. Meyarivan (2002). A Fast and Elitist Multiobjective Genetic Algorithm: NSGA-II. IEEE Transactions on Evolutionary Computation, 6(2):182-197.
7. Eckart Zitzler, Marco Laumanns, and Lothar Thiele(2002). SPEA2: Improving the Strength. Pareto Evolutionary Algorithm for Multiobjective Optimization. In K.C. Giannakoglou et al., editor, Proceedings of the EUROGEN2001 Conference, pages 95-100, Barcelona, Spain, CIMNE
8. Purshouse, R. C., Fleming, P. J. (2001). The Multi-objective Genetic Algorithm Applied to Benchmark Problems—an Analysis. Research Report No. 796. Department of Automatic Control and Systems Engineering University of Sheffield, Sheffield, S1 3JD, UK.
9. Tiwari, A., Roy, R.(2002). Generalised Regression GA for Handling Inseparable Function Interaction: Algorithm and Applications. Proceedings of the seventh international conference on parallel problem solving from nature. (PPSN VII). Granada, Spain
10. Wu, Q.(1978). On the optimality of orthogonal experimental design. Acta Math. Appl Sinica, 1(4), pp. 283-299.
11. Leung, Y. W. and Zhang, Q. (1997). Evolutionary algorithms + experimenta design methods: A hybrid apprpach for hard optimization and search problems, Res. Grant Proposal, Hong Kong Baptist Univ.
12. Jinn-Tsong Tsai, Tung-Kuan Liu, and Jyh-Horng Chou (2004). Hybrid Taguchi-Genetic Algorithm for Global Numerical Optimization, IEEE Transactions on Evolutionary Computation, 8(4),pp365-377.
13. Sanyou Y.Zeng, Lishan S.Kang, Lixing X.Ding (2004). An Orthogonal Multi-objective Evolutionary Algorithm for Multi-objective Optimization Problems with Constraints. Evolutionary Computation, 12(1), pp77-98.
14. Montgomery, D. C.(1991). Design and Analysis of Experiments. 3rd ed. New York: Wiley.
15. Carlos M. Fonseca and Peter J. Fleming(1998). Multiobjective Optimization and Multiple Constraint Handling with Evolutionary Algorithms–Part I: A Unified Formulation. IEEE Transactions on Systems, Man, and Cybernetics, Part A: Systems and Humans, 28(1):26-37.
16. Zitzler, E., Deb, K. and Thiele, L. (2000). Comparison of multiobjective evolutionary algorithms: empirical results. Evolutionary Computation 8(2): 173-195.
17. Deb. K, Lothar Thiele, Marco Laumanns and Eckart Zitzler (2001). Scalable Test Problems for Evolutionary Multi-Objective Optimization. KanGAL Report Number 2001001.

Path Relinking in Pareto Multi-objective Genetic Algorithms

Matthieu Basseur, Franck Seynhaeve, and El-Ghazali Talbi

Laboratoire d'Informatique Fondamentale de Lille (LIFL),
UMR CNRS 8022, University of Lille,
59655 Villeneuve d'Ascq Cedex, France
{basseur, seynhaev, talbi}@lifl.fr

Abstract. Path relinking algorithms have proved their efficiency in single objective optimization. Here we propose to adapt this concept to Pareto optimization. We combine this original approach to a genetic algorithm. By applying this hybrid approach to a bi-objective permutation flow-shop problem, we show the interest of this approach.

In this paper, we present first an Adaptive Genetic Algorithm dedicated to obtain a first well diversified approximation of the Pareto set. Then, we present an original hybridization with Path Relinking algorithm, in order to intensify the search between solutions obtained by the first approach. Results obtained are promising and show that cooperation between these optimization methods could be efficient for Pareto optimization.

1 Introduction

In solving Multi-objective Optimization Problems (MOPs), many methods scalarize the objective vector into a single objective. However, since several years, interest concerning MOPs using Pareto approaches always grows. Many of these studies use Evolutionary Algorithms (EAs) to solve MOPs [1, 2, 3].

The evolutionary approach called scatter search, and its generalized form called Path Relinking (PR), contrast with other evolutionary procedures, such as genetic algorithms, by providing unifying principles for joining solutions based on generalized path constructions. Joining solutions can be realized in both decisional and the objective spaces. Path relinking algorithms have recently been investigated in a number of studies for single-objective optimization, and especially in [4], where the Flow-shop problem is solved, in its single objective form.

In this paper, we propose a multi-objective approach to integrate Path relinking algorithms into EAs. We have to take into account several classical questions to implement a PR algorithm, and we propose some solutions for Pareto optimization. We have to define which distance operator has to be to used to join solutions. We propose a distance measure to compute distance in respect to an efficient neighborhood operator, the *Shift* operator. Then we define techniques to have an initial population (with EA), neighborhood generation to approach goal solutions from initial solutions, and path selection between solutions. Then, we propose to integrate path relinking into Pareto evolutionary algorithms to solve MOPs. We combine an Adaptive Genetic Algorithm (AGA) with Path relinking technique. In order to evaluate the effectiveness of this hybridization, we apply it to solve a Bi-Objective Flow-shop Scheduling Problem (BOFSP).

This paper is organized as follows. In section 2, we present the BOFSP. In section 3, we present a Pareto EA (AGA) developed to find an initial Pareto population. In section 4, we present cooperation between AGA and multi-objective Path relinking. Section 5 presents results on a large class of instances, which are non-exactly solved with exact approaches. In the last section, we discuss the effectiveness of this approach and perspectives of this work.

2 A Bi-objective Flow Shop Problem (BOFSP)

The Flow-shop Scheduling Problem (FSP) is one of the numerous scheduling problems. The FSP has been widely studied in the literature. Proposed methods for its resolution vary between exact methods such as the branch & bound algorithm [5], specific heuristics [6] and meta-heuristics [7]. However, the majority of works on flow-shop problem studies the problem in its single criterion form and aims mainly to minimize makespan, which is the total completion time. Some bi-objective approaches exist in the literature. Sayin et al. proposed a branch and bound strategy to solve the two-machine flow-shop scheduling problem, minimizing the makespan and the sum of completion times [5]. Sivrikaya-Serifoglu et al. proposed a comparison of branch & bound approaches for minimizing the makespan and a weighted combination of the average flowtime, applied to the two-machine flow-shop problem [8]. Rajendran proposed a specific heuristic to minimize the makespan and the total flowtime [6]. Nagar et al. proposed a survey of the existing multi-criteria approaches of scheduling problems [7].

FSP can be presented as a set of N jobs $J_1, J_2, \ldots, J_N$ to be scheduled on M machines. Machines are critical resources: one machine cannot be assigned to two jobs simultaneously. Each job J_i is composed of M consecutive tasks $t_{i1}, \ldots, t_{iM}$, where t_{ij} represents the jth task of the job J_i requiring the machine m_j. To each task t_{ij} is associated a processing time p_{ij}. Each job J_i must be achieved before its due date d_i. In our study, we are interested in permutation FSP where jobs must be scheduled in the same order on all the machines.

In this work, we minimize two objectives: C_{max}, the makespan (total completion time), and T, the total tardiness. Each task t_{ij} being scheduled at the time s_{ij}, the two objectives can be computed as follows:

$$C_{max} = Max\{s_{iM} + p_{iM} | i \in [1 \ldots N]\}$$

$$T = \sum_{i=1}^{N} [max(0, s_{iM} + p_{iM} - d_i)]$$

In the Graham et al. notation [9], this problem is denoted: F/perm, $d_i/(C_{max}, T)$.

In [10] C_{max} minimization has been proved to be NP-hard for more than two machines. The total tardiness objective T has been studied only a few times for M machines [11], but total tardiness minimization for one machine has been proved to be NP-hard [12]. The evaluation of the performances of our algorithm has been realized on some Taillard benchmarks for the FSP [13], extended to the bi-objective case [14][1].

3 An Adaptive Genetic Algorithm (AGA)

In this part, we present AGA which is applied to MOFSP. AGA's objective is to explore a large and diversified part of the landscape, to offer good solutions to the hybrid part of the algorithm (i.e. PR in our case). AGA proposes an adaptive selection between different mutation operators. In a first time, let us present classical operators of AGA:

- Initialization: Individuals are generated randomly.
- Selection: Elitist selection with NSGA ranking [15].
- Crossover Operator: 2-point crossover (See Fig. 1).
- Mutation Operators: Adaptive selection between 4 mutation operators: insertion, reciprocal exchange, random and inversion operator (See Fig. 2,3,4,5). Adaptive selection is described below.
- Diversification: Combined Sharing (sharing is realized in the decision space and in the objective space). Niche sizes are defined adaptively after each GA generation [16].
- Replacement: Generational (the population is automatically replaced by the new individuals).
- Stopping criterion: Fixed time.

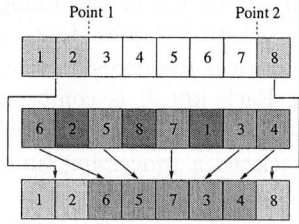

Fig. 1. 2-point crossover operator

An Adaptive Mutation Selection
Whatever the problem, we can choose between many mutation and crossover operators. Many evaluations must be done on each operator in order to know its effectiveness. Moreover, the efficiency of an operator may change during the

---

[1] bi-objective benchmarks and results obtained are available on the web at http://www.lifl.fr/~basseur

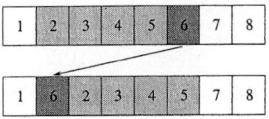

Fig. 2. Shift operator

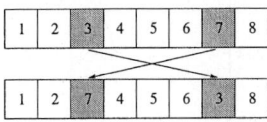

Fig. 3. Swap operator

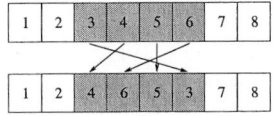

Fig. 4. Random operator

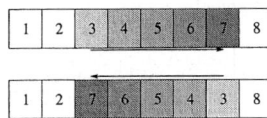

Fig. 5. Inversion operator

evolution process. An operator may offer a better convergence at the beginning of the GA, but this convergence may stop earlier than with another operator. The success of an operator may also depends on the instance of the problem.

Therefore, we have proposed an adaptive Pareto GA, in which the choice of the operator is done dynamically during the search. The purpose is to change the probability selection of each operator according to its efficiency.

This adaptive mechanism has the interest to diversify the Pareto population by the use of several operators. Moreover, it allows us to define the best operator to use for the PR algorithm presented in the next section.

To adapt mutation rate during the GA, each mutation operator M_i applied to the individual I was associated with a progress value (I_{M_i} is the individual I modified by the mutation M_i). This progress value allows to refine mutation rates after each GA generation. This approach has been initially proposed by Hong et al. [17] for the single-objective case. To have a good evaluation of each operator in the multi-objective case, we use the elitist and ranking approach. As a consequence, the progress value is expressed as follows:

$$\Pi(I_{M_i}) = C_{I_{M_i}} * \left(\frac{R_{I_{M_i}}}{R_I}\right)^k \qquad (1)$$

with $C_{I_{M_i}} = \frac{1}{R_{I_{M_i}}}$ ($C_{I_{M_i}}$ is an elitist coefficient), where $R_{I_{M_i}}$ is the rank of the solution after mutation, R_I is the rank of the solution before mutation, and k is how much we encourage the progress performed by a mutation operator. For our application, we set k to 2.

Then, the global progress of a mutation M_i is defined as follows:

$$Progress(M_i) = \frac{\sum \Pi(I_{M_i})}{\sum C_{I_{M_i}}} \qquad (2)$$

The new selection probabilities are computed proportionally to these values, with a minimum selection probability of δ:

$$P_{M_i} = \frac{Progress(M_i)}{\sum_{j=1}^{n} Progress(M_j)} * (1 - n*\delta) + \delta \qquad (3)$$

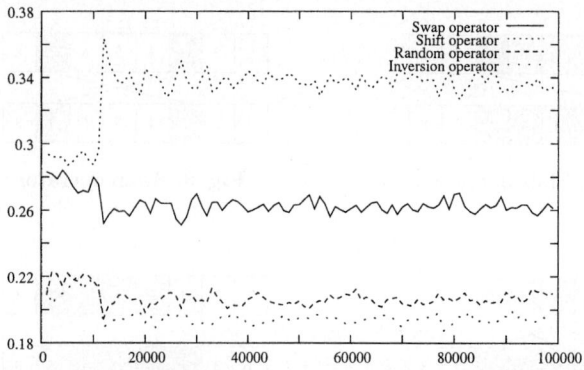

Fig. 6. Mutation rate evolution during GA run: ta_100_5_01 instance

In figure 6, we show an example of mutation rate on a problem with 100 jobs and 5 machines. Global results (detailed in [18]) show that *shift* operator is the most efficient neighborhood operator for this instance, and for the other instances.

Computational results presented in the last section show that AGA obtain a good initial approximation of the Pareto front. In order to improve these results by intensify the search within the set of solutions founded by AGA, we propose a cooperative scheme between AGA and Path relinking.

4 A Multi-objective Path Relinking Algorithm

Path Relinking (PR) was originally proposed by Glover [19] as an intensification strategy exploring trajectories connecting elite solutions obtained by tabu search or scatter search. [20] offers a good overview of this technique.

In this section we propose a cooperation scheme between PR and AGA in the multi-objective case. The goal is to show the interest of this type of approach, to improve solutions obtained by AGA. In our knowledge, a PR algorithm has already been proposed to solve Flow-shop problem, but only in its single objective optimization form [4]. The concept of PR algorithm is schematized in figure 7. Two solutions are chosen in an initial population (if possible, good solutions). Then by iterative applications of neighborhood operator, join the first solution to the second. These two solutions are called *initiating* and *guiding* solutions. During the algorithm, solutions are evaluated, and in order to improve efficiency of PR algorithms, some of them are selected to be improved by a local search algorithm.

Path relinking is naturally an extension of scatter search algorithms. Only a few studies propose scatter search algorithms to solve MOPs [21]. A study by Gandibleux et al. [22] propose to incorporate PR principles in a multiobjective heuristic using genetic heritage. In a general way, many questions have to be considered before designing a PR algorithm. Several answers exist for single

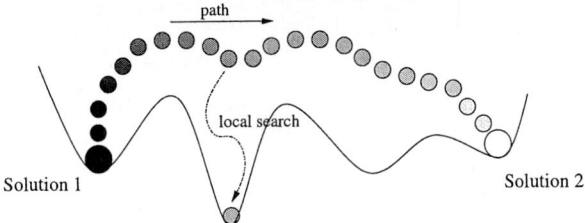

Fig. 7. Path Relinking: the concept

objective optimization, but they have to be reconsidered for the multi-objective case. We have to establish these following mechanisms:
- Neighborhood structure.
- Distance measure (correlated, or not, with the neighborhood).
- Selection criteria, to choose solutions to link.
- Path selection, to establish which path(s) has to be generated.
- Improvement of solutions. A local search algorithm may be applied on solutions belonging to the path. A mechanism of solutions selection to apply the local search algorithm has to be proposed.

Here we propose a first initiative to answer these difficulties, but there are still some open questions.

4.1 Neighborhood Operator

In order to link solutions, we need to select a neighborhood operator which allows us to generate the intermediate solutions. In [4], Reeves and Yamada propose a distance measure for the *swap* operator, but not for the *shift* operator. Then they propose other distance measures which are not correlated with these operators. Here, we choose the most powerful operator of AGA algorithm applied on BOFSP: the *shift* operator. We propose next a distance measure in respect to this operator.

4.2 Distance in Decision Space

Genotypic distance between two solutions S_1 and S_2, in respect to the *shift* operator, is: $d_{perm} = N - s_{max}$, where N is the number of jobs, and s_{max} is the size of the greatest shared substring between S_1 and S_2. In the appendix, we prove that $d_{perm}(S_1, S_2)$ is a distance, and that it corresponds to the minimum number of permutations required to join S_1 with S_2.

Computing a greatest shared substring between two solutions is a well-known problem in the genomic community. Computing a greatest shared substring between two solutions is a classical application of dynamic programming [23]. We implement this algorithm to compute distance between solutions in the decision space, its complexity is in $\mathcal{O}(N^2)$.

An example of distance between two solutions x and y obtained by computing a greatest shared substring is shown in figure 8. In this example, three inversions

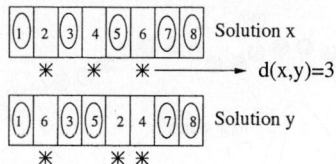

Fig. 8. Distance between two solutions x and y

are necessary to link x from y - the jobs 2, 4 and 6 have to be moved to link these solutions. They correspond to the set of jobs which are not in the greatest shared substring between the two solutions (let remark us that the greatest shared substring is not necessary unique: the substrings '13478' and '12478' can be viewed as largest shared substring too). As in many Path relinking implementations, we choose to generate only minimum path between solutions.

4.3 Initial Solutions for Path Relinking

In many single objective optimization path relinking algorithms, good local optima solutions are chosen to be linked. We can favor distant solutions to favor the exploration of the search space, or adjacent solutions to favor intensification of the search around good solutions. Then, in the multi-objective case, it may be interesting to link solutions which are closed to each other in the objective space in order to intensify the search around solutions with similar quality on the different objectives.

For this study, we randomly choose among Pareto solutions obtained by the GA. After linking the two first solutions, we compute the new Pareto set. The selection of the two next solutions to link is realized on this set.

4.4 Neighborhood Generation

We use the insertion neighborhood operator to generate the path from a solution x to a solution y. The goal is to explore only solutions which reduce distance between the generated solution and y.

So, after defining the set of "well placed" jobs, by computing the largest shared substring between x and y, we have to move misplaced jobs in order to increase the size of this largest shared substring.

Then, we compute the available positions for misplaced jobs, as shown in figure 9. On this example, a greatest shared substring is first computed and represented in this figure by encircled jobs (i. e. the string 13578). Possible moves to apply to x to approach y are also computed for jobs 2, 4 and 6. These available positions are chosen to increase the size of the largest shared substring, and so to decrease the distance between the current solution and the guiding solution.

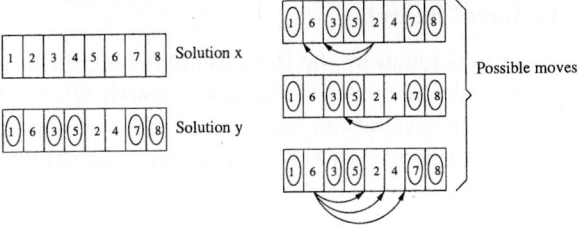

Fig. 9. Neighbors explored (greatest shared substring represented by encircled solutions)

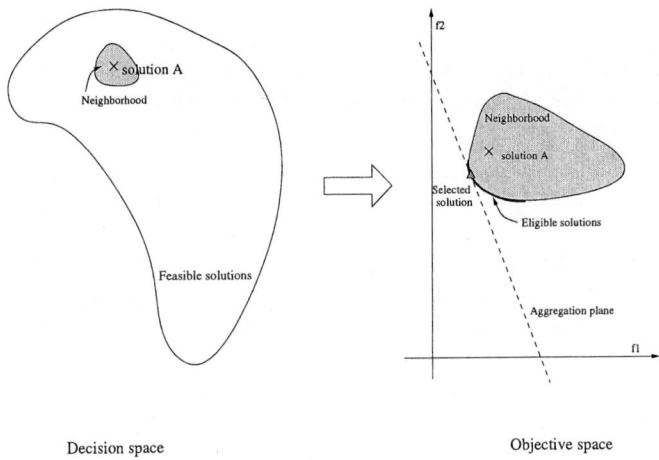

Fig. 10. Path Relinking algorithm: neighborhood exploration

4.5 Generation of the Path

We can generate from x a set of neighbors which reduce the distance between this solution x and the goal solution y. Now, we have to choose which path we will generate to iterate this mechanism and explore the largest landscape possible.

A first possible approach is to generate all the possible paths. But with this approach, the number of explored solutions grows exponentially with the distance between the two considered solutions. So, experimentally, we have to explore only a subset of the possible paths. Many PR algorithms which select a subset of the possible paths propose to explore only the best solutions generated. In the multi-objective case, the best solutions are those which are non-dominated by all the neighborhood. This set is represented in figure 10 (eligible solutions). In order to reduce the size of the exploration, we apply a random aggregation of the objectives to select only one solution in the set of eligible solutions (fig. 10). In future works, it is conceivable to select the totality of the set of eligible solutions, with a mechanism which limits the size of the exploration.

4.6 A Pareto Local Search

To affine Pareto solutions found by the Path Relinking algorithm, we implement a population-based local search or Pareto Local Search (PLS). Local searches are realized after each PR generation. The local search technique is described in algorithm 1. The neighborhood structure used for this algorithm is the insertion operator.

Algorithm 1 PLS algorithm

Generate an initial Pareto set PO (in our case, with PR algorithm)
do
$S' \leftarrow PO$.
Generate the neighborhood PN_x for each solution x of S'.
Let PO be the set of non-dominated solutions of $S' \cup_x PN_x$.
Until $PO=S'$ (the population has reached the local optima).
Let PO^* be the Pareto optimal set of $PO^* \cup S'$.
Return the Pareto set PO^*

To introduce PLS in PR algorithm, we select the set of non-dominated solutions discovered during the PR algorithm. Then we apply the PLS algorithm with this set of solutions as initial population. This mechanism is shown in figure 11. Let remark us that PLS algorithm may comsume larger computation time if the problem size or the number of objective increase. In this case, PLS algorithm may be implemented with other mecanisms as clustering, neighborhood restriction, etc. to be effective.

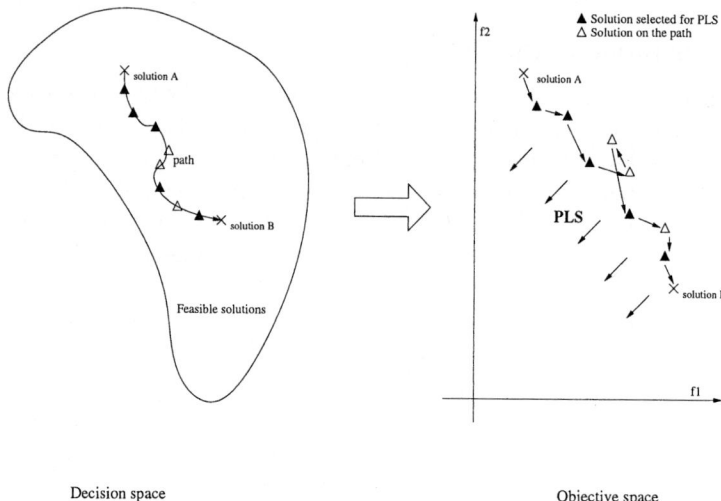

Fig. 11. Path relinking algorithm with local search

5 Experimental Results

5.1 Quality Assessment of Pareto Set Approximation

Solutions' quality can be assessed in different ways. Here, we use the contribution metric [24] to evaluate the proportion of Pareto solutions given by each front, and the S metric [25] which evaluate the dominated area with a reference point. This metric is advised in [26].

Contribution Metric: The contribution of a set of solutions PO_1 relatively to a set of solutions PO_2 is the ratio of non-dominated solutions produced by PO_1 in PO^*, where PO^* is the set of Pareto solutions of $PO_1 \cup PO_2$.

- Let PO be the set of solutions in $PO_1 \cap PO_2$.
- Let W_1 (resp. W_2) be the set of solutions in PO_1 (resp. PO_2) that dominate some solutions of PO_2 (resp. PO_1).
- Let L_1 (resp. L_2) be the set of solutions in PO_1 (resp. PO_2) that are dominated by some solutions of PO_2 (resp. PO_1).
- Let N_1 (resp. N_2) be the other solutions of PO_1 (resp. PO_2): $N_i = PO_i \setminus (PO \cup W_i \cup L_i)$.

$$Cont(PO_1/PO_2) = \frac{\frac{\|PO\|}{2} + \|W_1\| + \|N_1\|}{\|PO^*\|}$$

S Metric: A definition of the S metric is given in [25]. Let PO be a non-dominated set of solutions. S metric calculates the hyper-volume of the multi-dimensional region enclosed by PO and a reference point Z_{ref}.

Let PO_1 and PO_2 be two sets of solutions. To evaluate quality of PO_1 against PO_2, we compute the ratio $(S(PO_1) - S(PO_2))/S(PO_2)$. For the evaluation, the reference point is the one with the worst value on each objective among all the Pareto solutions found over the runs.

5.2 Computational Results

Evaluations were realized for 10 runs per instance, on a 1.6Ghz machine. We test the different algorithms of several instances proposed in [14]. The smallest instances are exactly solve in a bi-objective exact approach proposed by Lemesre et al. [27]. We test the different approaches on instances with 50 and 100 jobs.

We compare performance of AGA with PR against results obtained only with AGA. In the experiments, runs are realized with a specified time limit (1000 minutes in our experiments). For all tests, 10 runs have been done on each benchmark. For the cooperative approach $(AGA + PR)$, we choose to run AGA for 10% of the total run time. Results of tested instances are listed in tables 1, 2 and 3, with:

- S_{Min}, respectively C_{Min}, is the minimum value of S, respectively of the contribution.
- S_{Max}, respectively C_{Max}, is the maximum value of S, respectively of the contribution.

Table 1. Quality assessment (S metric): S(AGA)

Benchmark	\multicolumn{4}{c}{S(AGA)}			
	S_{Min}	S_{Max}	Average	Std dev
ta_50_10_01	516521	828610	709023.6	93279.2
ta_50_20_01	1841592	2476210	2170418.6	205943.7
ta_100_5_01	431760	446264	439799.2	5526.3
ta_100_10_01	5414203	6956336	6218845.8	464814.8
ta_100_20_01	17167424	20139584	19117894.4	866011.0

Table 2. Quality assessment (S metric): S(AGA+PR)

Benchmark	S(AGA+PR)				
	S_{Min}	S_{Max}	Average	Std dev	Avg Imp
ta_50_10_01	870448	1065523	988079.6	67443.2	39.4%
ta_50_20_01	2853166	3113992	3020664.0	93279.6	39.2%
ta_100_5_01	445695	446878	446323.0	286.2	1.5%
ta_100_10_01	7813940	8148625	7978420.0	93335.8	28.3%
ta_100_20_01	26209816	28745238	27752835.2	690416.0	45.2%

Table 3. Quality assessment (Contribution metric): Cont(AGA/AGA+PR) - 1000 minutes runs

Benchmark	Cont(AGA+PR/AGA)			
	C_{Min}	C_{Max}	Average	Std dev
ta_50_10_01	1.00	1.00	1.000	0.000
ta_50_20_01	1.00	1.00	1.000	0.000
ta_100_5_01	0.71	1.00	0.836	0.085
ta_100_10_01	1.00	1.00	1.000	0.000
ta_100_20_01	1.00	1.00	1.000	0.000

- **Average** is the average value.
- **Std dev** is the average deviation between the different measures.
- **Avg Imp** is the average improvement realized by AGA+PR against AGA in terms of dominance area.

The instance $ta\_N\_M\_n$, correspond to the bi-criteria version of the number n problem with N jobs and M machines in Taillard's benchmarks. Results show a great improvement for all the instances excepting $ta\_100\_5\_01$. For this problem, we can expect that we are really near the optimal Pareto front. In fact, the best value for C_{max} objective given by the hybrid algorithm is the optimal value (found by single objectives algorithms). For the other instances evaluated, with 10 and 20 machines, improvement, in term of dominance area, varies from 27.7% and 45.2%. Table 3 shows that for these problems, all the Pareto set generated by $AGA+PR$ algorithm dominate those generated by AGA ($C(AGA+PR/AGA) = 1.00$). Figure 12 shows an example of Pareto set generated. The first Pareto set

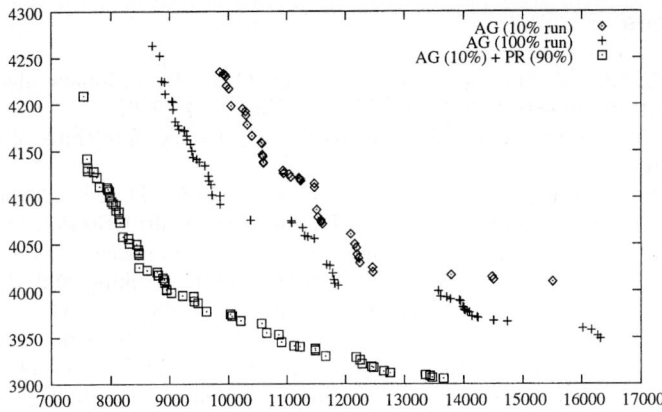

Fig. 12. Result on *ta_50_20_01* instance

is generated by *AGA*. The second corresponds to the same algorithm in a longer run. The third corresponds to the path relinking algorithm applied on the first Pareto set.

6 Conclusion and Perspectives

In this paper, we have first presented an Adaptive Genetic Algorithm to solve MOPs. This approach has been applied on a BOFSP. Then we have proposed a multi-objective path relinking algorithm, with a distance measure corresponding to the best neighborhood operator we select during *AGA* experiments. We propose several mechanisms to answer some difficulties of multi-objective path relinking implementation. Then we have proposed an original hybridization between *AGA* and path relinking to improve results. This approach has been tested, and its effectivenesses has been shown in comparison with Pareto fronts obtained by AGA algorithm. These results show the interest of cooperation with path relinking, which can be applied after many meta-heuristics, and especially evolutionary algorithms.

These results could be improved by adding other mechanisms of Path Relinking algorithms, such as replacing a genetic operator of *AGA* with the PR algorithm, or using a more evolved path selection. Another interesting approach could be to change the transition rules between *AGA* and *PR*. We can adapt the transition in respect to the evolution rate of the Pareto set.

Moreover, the hybrid approach presented in this article can be easily parallelized, by running several path generations and local searches simultaneously. This work will be integrated in the framework ParaDisEO (Parallel and Distributed Evolutionary Objects) [28].

References

1. Coello, C.A.C., Veldhuizen, D.A.V., Lamont, G.B.: Evolutionary algorithms for solving Multi-Objective Problems. Kluwer, New York (2002)
2. Deb, K.: Multi-objective optimization using evolutionary algorithms. Wiley, Chichester, UK (2001)
3. Zitzler, E., Deb, K., Thiele, L., Coello, C.A.C., Corne, D., eds.: Proceedings of the First International Conference on Evolutionary Multi-Criterion Optimization (EMO 2001). Volume 1993 of Lecture Notes in Computer Science. (2001)
4. Reeves, C., Yamada, T.: Genetic algorithms, path relinking and the flowshop sequencing problem. Evolutionary Computation **6** (1998) 230–234
5. Sayin, S., Karabati, S.: A bicriteria approach to the two-machine flow shop scheduling problem. European journal of operational research (1999) 435–449
6. Rajendran, C.: Heuristics for scheduling in flowshop with multiple objectives. European journal of operational research (1995) 540–555
7. Nagar, A., Haddock, J., Heragu, S.: Multiple and bicriteria scheduling: A litterature survey. European journal of operational research (1995) 88–104
8. Sivrikaya, F., Ulusoy, G.: A bicriteria two-machine permutation flowshop problem. European journal of operational research (1998) 414–430
9. Graham, R.L., Lawler, E.L., Lenstra, J.K., Kan, A.H.G.R.: Optimization and approximation in deterministic sequencing and scheduling: a survey. In: Annals of Discrete Mathematics. Volume 5. (1979) 287–326
10. Lenstra, J.K., Kan, A.H.G.R., Brucker, P.: Complexity of machine scheduling problems. Annals of Discrete Mathematics **1** (1977) 343–362
11. Kim, Y.D.: Minimizing total tardiness in permutation flowshops. European Journal of Operational Research **33** (1995) 541–551
12. Du, J., Leung, J.Y.T.: Minimizing total tardiness on one machine is NP-hard. Mathematics of operations research **15** (1990) 483–495
13. Taillard, E.: Benchmarks for basic scheduling problems. European Journal of Operations Research **64** (1993) 278–285
14. Talbi, E.G., Rahoual, M., Mabed, M.H., Dhaenens, C.: A hybrid evolutionary approach for multicriteria optimization problems : Application to the flow shop. In: Evolutionary Multi-Criterion Optimization (EMO'01). Volume 1993 of Lecture Notes in Computer Science. (2001) 416–428
15. Srinivas, N., Deb, K.: Multiobjective optimisation using non-dominated sorting in genetics algorithms. In: Evolutionary Computation. Volume 2. (1994) 221
16. Basseur, M., Seynhaeve, F., Talbi, E.G.: Design of multi-objective evolutionary algorithms: Application to the flow-shop scheduling problem. In: Congress on Evolutionary Computation (CEC'02), Honolulu, Hawaii, USA (2002) 1151–1156
17. Hong, T.P., Wang, H.S., Chen, W.C.: Simultaneous applying multiple mutation operators in genetic algorithm. Journal of Heuristics **6** (2000) 439–455
18. Basseur, M., Seynhaeve, F., Talbi, E.G.: Adaptive mechanisms for multi-objective evolutionary algorithms. In: Congress on Engineering in System Application (CESA'03), Lille, France (2003) 72–86
19. Glover, F.: Tabu search and adaptive memory programing advances, applications and challenges. In: Interfaces in Computer Science and Operations Research, Kluwer Academic Publishers, Boston (96) 1–75
20. Glover, F., Laguna, M.: Fundamentals of scatter search and path relinking. Control and Cybernetics **29** (1999) 653–684
21. Beausoleil, R.P.: Multiple Criteria Scatter Search. In: 4th Metaheuritics International Congress, Porto, Portugal (2001) 539–543

22. Gandibleux, X., Morita, H., Katoh, N.: Impact of clusters, path-relinking and mutation operators on the heuristic using a genetic heritage for solving assignment problems with two objectives. In: Metaheuristics International Conference (MIC'03), Kyoto, Japan (2003) 23(1-6)
23. Cormen, T.H., Leiserson, C.E., Rivest, R.L.: 15. In: Introduction to algorithms. The MIT Press, Cambridge, Massachuchetts (1990) 350–355
24. Meunier, H., Talbi, E.G., Reininger, P.: A multiobjective genetic algorithm for radio network optimisation. In: CEC. Volume 1., Piscataway, New Jersey, IEEE Service Center (2000) 317–324
25. Zitzler, E.: Evolutionary algorithms for multiobjective optimization: Methods and applications. Master's thesis, Swiss federal Institute of technology (ETH), Zurich, Switzerland (1999)
26. Knowles, J.D., Corne, D.W.: On metrics for comparing non-dominated sets. In Center, I.S., ed.: Congress on Evolutionary Computation (CEC'2002). Volume 1., Piscataway, New Jersey (2002) 711–716
27. Lemesre, J., Dhaenens, C., Talbi, E.: A parallel exact method for a bicriteria permutation flow-shop problem. In: Project Management and Scheduling (PMS'04), Nancy, France (2004) 359–362
28. Cahon, S., Melab, N., Talbi, E.G.: Paradiseo: a framework for the flexible design of parallel and distributed hybrid metaheuristics. Journal of Heuristics **10** (2004) 357–380
29. Basseur, M.: Cooperative models for multi-objective optimization. PhD thesis, University of Lille ((to appear))

Appendix

We want to prove that d_{perm} is a distance measure and that only minimum paths are generated. Let us introduce several notations:

- $Shift(i,j,X)$ corresponds to the permutation generated by moving the ith job of a sequence to a new position j.
- $pos_X(i)$ corresponds to the position of the job i in a permutation X.
- $GSS(X,Y)$ corresponds to the greatest shared substring between two permutations X and Y.

First, let us introduce a trivial lemma (proved in [29]):

Lemma 1 *For all the permutations X and Y, $d_{perm}(X,Y)-1 \leq d(X, Shift(i,j,Y)) \leq d(X,Y)+1$*

In fact, the largest shared substring between X and Y can not vary more than 1 by deleting and adding a 'letter' (job).

Now, let introduce another lemma:

Lemma 2 *For all the permutations X and Y of a size N, with $d_{perm}(X,Y) \neq 0$, $\exists\, i,j \in [0..N] \mid d_{perm}(Shift(i,j,X),Y) = d_{perm}(X,Y) - 1$.*

2. **Pareto optimal:** A solution x^0 is said to be *nondominated* (*Pareto optimal*) if and only if: $\neg \exists x^1 \in X : x^1 \succ x^0$.
3. **Pareto optimal set:** The set P_S of all Pareto optimal solutions:
$P_S = \{x^0 \mid \neg \exists x^1 \in X : x^1 \succ x^0\}$.
4. **Pareto optimal front:** The set P_F of all objective function values corresponding to the solutions in P_S: $P_F = \{f(x) = (f_1(x), \cdots, f_m(x)) \mid x \in P_S\}$.

The optimal result for such multiobjective optimization is no other than the Pareto optimal set P_S. However, the size of this set may be infinite, and it is impossible to find this set by using a finite number of solutions. In this case, a representative subset of P_S is desired. Generally, the characteristic of multiobjective evolutionary algorithms (MOEAs) is to search the decision space by maintaining a finite population of individuals (corresponding to the points in the decision space), which work according to the procedures that resemble the principles of natural selection and evolution. Because we only consider the subset of all the final nondominated individuals resulted from a MOEA, we call such subset an *approximation set* and denote it by S, and we call the corresponding objective set a *resulting final Pareto optimal front* and denote it by PF_{final}. Ideally; we are interested in finding an S of finite size, which contains a selection of individuals from such that the individuals in PF_{final} are diversified as possible. Unfortunately, we usually have no access to P_F on beforehand. We have to get close to P_F but in such a way that PF_{final} we found is as diversified as possible without compromising as much as possible the proximity of PF_{final} with respect to P_F. Thus, the concept of *proximity* and *diversity* should be outlined. Regarding this diversity, it is of importance to note that it depends on the mapping function whether a good diversity of the individuals in the decision space is also a good diversity of the individuals in the objective space correspondingly. However, it is common practice to search for a good diversity of the individuals in the objective space because decision makers will ultimately have to pick a single individual as final solution according to its objective vector values. Therefore, it is often best to present a wide variety of tradeoff individuals for the specified goals in constructing MOEAs.

During the past decade, various MOEAs have been proposed and applied [1]. A representative collection of these influential algorithms includes the Nondominated Sorting Genetic Algorithm (NSGA) and NSGA2 by Srinivas and Deb et al [2] [3], the Strength Pareto Evolutionary Algorithm (SPEA) and SPEA2 by Zitzler et al [4] [5], the Pareto Archived Evolution Strategy (PAES) and the memetic PAES (M-PAES)by Knowles and Corne [6] [7] etc. Although these MOEAs differ from each other, they share the common purpose — searching for a near-optimal, well-extended and uniformly diversified PF_{final} for a given multiobjective optimization. However, this ultimate goal is far from being accomplished by the existing MOEAs as documented in the literature, e.g., [1],[5]. In one respect, most of multiobjective optimizations are very complicated and have their own inherent characters and variabilities, so computational resources are required to be homogenously distributed in a high-dimensional decision space. On the other hand, those fitter individuals generally have strong tendencies to restrict searching efforts within local areas because of the genetic drift phenomenon [8], which results in the loss of diversity. This highlighted the hot issue how to improve the algorithm's robustness and how to balance proximity and diversity during searching process. However, the latter issue we considered in this paper has not been attended yet. Although in some elitism-based MOEAs, several techniques have been adopted such as crowded comparison [3], amalgamation strategy [4], archive truncation [5] and preselection scheme [9] etc, the manipulation of balance choice remains manual and presetting. Therefore, either the pressure for proximity or for diversity should be autonomously regulated by the current and historical search.

The present work proposed a novel dynamic archive evolution strategy (DAES) inherited from evolution strategy (ES) [10] to investigate the adaptive balance between the proximity and diversity. A special aspect of importance in which ES differs from most other EAs is that it has a long self-adaptive mechanism by usually including strategy parameters that can adaptively guide the search to explore and exploit the local and global topology of the decision space. DAES hold such mechanism too and have developed some new features. The prominent innovation of DAES is the dynamic external archive with its form, purpose and managing scheme. Other techniques served for exploration and exploitation have also been designed and work cooperatively to guarantee satisfactory results.

The remainder of this paper is organized as follows. In section 2, we introduce some basic rationale used to balance proximity and diversity in exploration stages and exploitation stages respectively by summarizing some prominent state-of-the-art MOEAs. Section 3 describes DAES arithmetic in detail. The empirical results and comparisons between DAES and SPEA2 [5], NSGA2 [3], M-PAES [7] on three prominent benchmark functions are presented in Section 4. Eventually, we conclude the paper with some remarks and future researches in the last section.

2 How to Balance Proximity and Diversity in DAES

2.1 Role of the External Archive

In order to obtain new and diverse nondominated individuals, especially when the set of nondominated individuals have approached P_F, the concept of elitism that the best individuals of the current population are copied into the next population is accepted to be a very important role for improving the results obtained by some MOEAs [4]. Alternatively, an external archive may be more commonly used that contains nondominated individuals, and the current population and the external archive are separated and managed by exchanging individuals between them. It also helps allow preserving the good individuals that are hard to be generated in the exploration stage.

In the evolutionary process, there are such individuals that they are not ensured nondominated responding to the current population, but are relatively better to their parents. Since these individuals may still hold some useful information, they should be wisely kept for a certain generation for further exploration and exploitation. However, if under the complete nondominance selection we intend to apply in the exploitation stage, they will definitely lose opportunity to survive into the next population due to their absence of qualification to be stored in the archive. So we need changing the role of the archive by some means. Furthermore, the archive size is highly sensitive to manage selected individuals. If the size is too small, there will not be enough schemas to exploit, resulting in a premature or non-uniformly PF_{final}. Otherwise, an excessive size will not be desirable since it may require unnecessarily large computation resources and even confuse search guide into stagnancy. Therefore, a dynamic size adjusted autonomously by the online characteristics of the proximity and diversity information of the archive will be more efficient and effective than a constant size. As a result, the archive should be endued with the responsibility of saving elitist individuals as well as relatively better individuals and a dynamic size of it is preferable. The archive increase scheme and the archive decrease scheme are designed concurrently in order to manage the archive in the probability sense.

2.2 Exploration of Proximity and Diversity

It is of importance to have an exploration operator that is capable of producing new nondominated individuals and diversifying them to guide the search towards P_F. Based on an appropriate selection of parents, a competent exploration operator is expected to be able to produce offspring in which good features of the selected parents are inherited, which does not differ much from the rationale in single objective EAs essentially. Since excessive proximity without adequate diversity will inevitably lead to premature convergence or local optimum while excessive diversity without adequate proximity will always confuse the search guide into stagnancy, it is necessary to deal with them equally and adaptively.

An interesting and relatively new field is the combinatorial operator, which attempts to model the regularities of parent structure by means of combination of several operators. In this way different operators are cooperated with to develop each merit and eliminate each weakness, where heuristic method is also applied. An example is taken in the fastEP for single optimization in which Gaussian and Cauchy mutations are combined to produce offspring from the same parent and the better ones are chosen into the next population [11]. Algorithms that use the similar approaches have obtained an increasing amount of attention over the last few years, and certainly have obtained promising results on a large variety of problems [12],[13]. These approaches are also constructive and beneficial for multiobjective optimization as well [14]. By doing that, the exploration has been shown to be more effectively stimulated [15]. Inspired by this idea, DAES appropriately combines the following three operators: discrete recombination, Gaussian mutation and Cauchy mutation. Therein, Gaussian mutation mainly severs for the proximity while Cauchy mutation for the diversity according to status of the parents offered to them and the offspring generated by them.

2.3 Exploitation of Proximity and Diversity

It is also essential to have a competent exploitation scheme that is capable of selecting a diverse set of individuals close to the set of nondominated individuals as possible. Since we can only indicate how diverse we are but cannot indicate how close to P_F we are to, the best we can do is to find individuals that are not dominated by any other individuals and are widely diversified in order to ensure pressure on both proximity selection and diversity preservation. Therefore, the exploitation stage is usually split into two steps: proximity selection first and diversity preservation later, and the later should never precede the former. So a straightforward way to obtain strong pressure toward the proximity is a *complete nondominance selection* in which all dominated individuals are neglected and only nondominated individuals have the opportunity to survive and reproduce in each generation [16]. This approach is considered cautious because it easily leads to premature convergence or local optimum unless other strong diversity preservation is accompanied with. Another class such as the concept of *domination count* [14] and *domination rank* [3] are introduced for proximity selection while *crowded comparison* [3] serves for diversity preservation. Although these approaches performed successfully in some situation, the problem caused by inappropriate exploitation scheme was still not completely resolved. Here in DAES, we proposed an exploitation scheme that is also separated into two sides respectively, but these incompatible sides are designed to work in mutual benefit. A complete non-dominance selection derived from [16] ensures maximal pressure for the proximity selection while a fitness assignment determined by dominance and population information ensures maximal diversity preservation. Particulars will be discussed in the following Section.

3 Arithmetic of DAES

A few items that are of major importance to implement an adaptive balance between the proximity and diversity can be outlined when constructing our DAES.

- The external archive should be used to save elitist individuals as well as relatively better individuals, and a dynamic size of it is preferable.
- The update scheme of the archive should be devised to accompany dynamic size including the archive increase scheme and the archive decrease scheme.
- The exploration operator should produce better offspring than either of its parents.
- Survive of the better individuals should be automatically done based on both the proximity selection and the diversity preservation.

The global pseudocode of DAES is shown in Fig.1 based on these considerations.

---

function *result* = DAES (μ, λ, *age_min*, *size_min*)
   Generate initial population P of μ random individuals and evaluate;
   Copy nondominated member of P in the external archive G;
   Initialize generation counter t zero;
   do //exploration stage
     Reset the current intermediate population H empty;
     Set offspring counter r zero;
     do //generate λ offspring
       Pick individuals a and b (not dominated by a) from $P \cup G$ randomly;
       c = ProduceOffspring (a, b, P, G); //via our exploration operator
       IncreaseArchive (c, a, G, *size_min*); Add c into H; $r = r + 1$;
     while ($r < \lambda$)
     DecreaseArchive (G, *age_min*, G, λ); //below is exploitation stage
     AssignFitness (H); //via our fitness assignment scheme
     Sort H in descending order of fitness;
     Pick the first μ members into the next population P(t+1); $t = t + 1$;
   while (terminal criterion is not satisfied)
   Set *result* the unique nondominated subset of $P(t) \cup G$; return *result*;
end function.

Fig. 1. Global pseudocode of DAES

It should be noted that the heart of DAES involves two populations: the current population P and the external archive G. In detail, P works similar to the single objective ES; but G serves a binary purpose — storage of elitist individuals as well as relatively better individuals found during the run, and the cooperative role of one-side parent in producing offspring. In the beginning of each generation, parents are randomly selected from P and G (the later selected parent side should not be dominated by the former in order to accelerate prepotency), and λ offspring will be produced within a loop in the exploration stage while the archive increase scheme is performed simultaneously. After the current intermediate population H has been filled with the λ offspring, the archive decrease scheme starts, and subsequently, the individuals of H are sorted in descending order of the assigned fitness to select the first μ individuals with higher fitness into the next population. These sequences of instructions are repeated until terminal criterion is satisfied.

3.1 Code Representation

In DAES, the real number representation is used instead of the binary string implementation. Since many problems in the real-world are expressed in real variables, faster computation can be obtain without conversions between different representations. It is characterized that DAES borrow a ternary group representation from [17], where an individual was denoted by (x, σ, θ) group. That is, $x = (x_1, x_2, ..., x_n)$ is the original decision vector, corresponding to a point in the decision space, and $\sigma =(\sigma_1, \sigma_2,..., \sigma_n)$ is the standard deviation used to instruct mutation, and $\theta =(\theta_{1,2}, \theta_{1,3},...,\theta_{1,n}, \theta_{2,3},..., \theta_{(n-1), n})$ is the rotation angle used to change orientation of the mutation associated with all possible pairs of the decision vectors. Both σ and θ are called strategy parameters. It is very similar to a hillclimber algorithm with a self-adaptive step σ and angle θ. In other words, all components are submitted to the evolutionary process by applying the exploration operator and the exploitation scheme on them. Thus, an appropriate adjustment and diversity of parameters can be automatically modified in demand, and this modification just corresponds to the local regulation.

3.2 Archive Increase Scheme

Generally, if an MOEA has an external archive with fixed size, a replacement scheme is always applied. In this scheme, in order to keep archive size unchanged, a new-added individual will replace one members of archive if it is considered to be better than the other individual. However, this scheme brings up a problem that some of the replaced individuals may still be very valuable and have not been well explored or exploited yet before they are replaced. Although some approaches such as amalgamation strategy in [4] and preselection scheme in [9] have been introduced, the problem caused by the replacement scheme is still not completely resolved. Therefore, DAES adopts two independent schemes—the increase scheme and the decrease scheme. The first scheme only focuses on pure population increment and ensures that each individual survives enough generations so that it can contribute its valuable schemas. Meanwhile an archive decrease scheme is also enforced to prevent the population size from growing excessively. The second scheme will be discussed in the next subsection.

Because we used the archive to store elitist individuals as well as relatively better individuals that obtain progress responding to their parents, the dominated individuals also have the opportunity to survive in the archive. This eliminates the disadvantage in [16] where dominated individuals are completely discarded and none of their useful information is under consideration. Therefore, we adopt the *diffusion scheme* in particle swarm optimization (PSO) [18] to guide archive increase scheme— an individual shares its information with the leading individuals in order to locate its moving direction. This idea is inspired by its significant performance. In DAES, if a newborn offspring is better than either of its parents in the proximity or the diversity (indicated by dominance or location of less crowded region [6]), it has the priority to be added into the archive; otherwise, it will be drastically discarded. Since the property of *age* is a crucial factor for the archive decrease scheme, we initialize age of new individual to be one when it is added for the first time. As a result, this scheme will guarantee that a new-added individual in the archive will have higher proximity or diversity than at least one of its parents, which helps DAES provide both elitist individuals and relatively better individuals participating in the exploration and exploitation stages so as to cover all the unexplored regions in the objective space. Fig. 2 describes the pseudocode of the archive increase scheme.

---

subprocedure IncreaseArchive (*offspring, par*1, *par*2, *ex_ar, size_min*)
  if (size of *ex_ar* is less than *size_min*) {
    Add *offspring* into *ex_ar*; Set age of *offspring* one; //initialize age }
  else
    if (*offspring* dominates *par*1 or *par*2); { //apply dominance
      Add *offspring* into *ex_ar*; Set age of *offspring* one; }
    else if (*offspring* in a less crowded region of *ex_ar* than *par*1 or *par*2) {
      Add *offspring* into *ex_ar*; Set age of offspring one; }
end subprocedure.

Fig. 2. Pseudocode of the archive increase scheme

3.3 Archive Decrease Scheme

An archive decrease scheme is necessary to prevent the archive from growing without bound. In DAES, whether an individual will be removed from the archive or not depends on its *age* and *fitness*. The initial age is one, and it will grow generation by generation as long as it survives in the archive. To ensure that an removed individual has a lowest fitness value and has been adequately explored and exploited, the scheme removes individuals in each generation according to the following principles:

- In order to keep balance of the archive in the probability sense, there are λ trials to remove relatively worse individuals in each decrease scheme because λ offspring have been produced and correspondingly there are λ trials to add relatively better individuals in each increase scheme. This means that there are at most λ individuals can be removed if we perform the decrease scheme once. Fig. 3 shows an example.
- We only remove the individuals with lowest fitness (either positive value or zero value) and whose ages are larger than the prespecified age threshold *age_min*. From the example in Fig. 3, the archive is different after decrease due to their different age, fitness and age threshold when λ is 4. Fig. 4 describes this pseudocode.

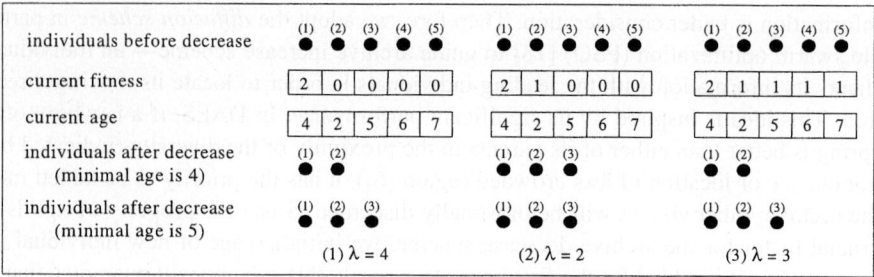

Fig. 3. Illustration of the decrease scheme: (1) There are 4 trials to remove the worse individuals, so the last three individuals (all zero fitness) are removed where minimal age is 4; (2) Only two worst individuals (all zero fitness) are removed and an individual whose age is larger than age threshold exist. (3) The worst individuals (positive fitness) are removed

```
subprocedure DecreaseArchive ( ex_ar, size_min, age_min, offspring_num )
  if ( size of ex_ar is larger than size_min ) {
    for ( each member m of ex_ar ) Increase age of m by one;
    Initialize remover counter r zero;
    AssignFitness ( ex_ar ); //assign fitness for direct remove
    do //remove dominated members from ex_ar within offspring_num trials
      if ( find a member m1 with fitness 0 and age > age_min ) {
        Remove m1 from ex_ar; Decrease size of ex_ar by one; }
      else { if ( find a member m2 whose age > age_min ) {
        Remove m2 from ex_ar; Decrease size of ex_ar by one; }} r = r + 1;
    while ( r < offspring_num ∧ size of ex_ar is larger than size_min ); }
end subprocedure.
```

Fig. 4. Pseudocode of the archive decrease scheme

3.4 Exploration Operator

In DAES, exploration includes recombination and mutation. Basically, recombination works choosing parents with uniform probability, and characteristics of parents are mixed to create one offspring. DAES employs the discrete recombination [19] that is commonly used in ES, and has also produced good results with real-coded MOEAs. Each component of offspring inherits from one of the parents randomly. If parent-1 is $(x^{(1)}, \sigma^{(1)}, \theta^{(1)})$ and parent-2 $(x^{(2)}, \sigma^{(2)}, \theta^{(2)})$, then we produce one offspring as

$$(x',\sigma',\theta') = ((x_1^{q_1},\cdots,x_n^{q_n}),(\sigma_1^{q_1},\cdots,\sigma_n^{q_n}),(\theta_{1,2}^{q_1},\theta_{1,3}^{q_1},\cdots,\theta_{1,n}^{q_1},\theta_{2,3}^{q_1},\cdots,\theta_{(n-1),n}^{q_1})) \quad (2)$$

where, q_i is chosen between 1 and 2 with probability 0.5. Notice that recombination is performed independently on the decision vectors as well as on the strategy parameters.

Generally, mutation is more emphasized than recombination in ES. It is typically implemented as Gaussian distribution around the generated individual being mutated. A new individual is produced via Gaussian mutation as

$$\begin{aligned}\sigma_i'' &= \sigma_i' \exp(\tau' N(0,1) + \tau N_i(0,1)); \\ \theta_{i,j}'' &= \theta_{i,j}' + \gamma N_{i,j}(0,1); \quad \forall i,j \in \{1,\cdots,n\}, j > i; \\ x'' &= x' + N(0, cov(\sigma'', \theta''))\end{aligned} \quad (3)$$

where, τ', τ and γ are parameters. $N(0, 1)$ denotes to generate a random scalar characterized by Gaussian distribution with zero mean and deviation one, and additional subscripts denote independent regeneration for each element respectively. $N(0, cov)$ denotes a vector function which returns a random vector distribution that is Gaussian distributed with zero mean and covariance matrix cov^{-1}. The rotations along with the variances are used to fill the covariance matrix and the variances form diagonal of the covariance matrix as

$$c_{i,i} = \sigma'_i; \quad c_{i,j} = \theta'_{i,j}, \forall i,j \in \{1,\cdots,n\}, j > i \quad (4)$$

Recently, a new type of mutation using Cauchy distribution was inspired in [11] and [12], which all claimed that Cauchy mutation outperformed Gaussian one on diversity preservation and search efficiency. Cauchy probability density function is

$$f(x) = \beta/\pi(\beta^2 + (x-\alpha)^2) \quad \alpha > 0, \beta > 0, -\infty < x < \infty \quad (5)$$

represented as $C(\alpha, \beta)$, where α and β are two parameters. We may produce a new individual via Cauchy mutation as

$$\begin{aligned}
\sigma''_i &= \sigma'_i \exp(\tau' C(0,1) + \tau C_i(0,1)); \\
\theta''_{i,j} &= \theta'_{i,j} + \gamma C_{i,j}(0,1); \quad \forall i,j \in \{1,\cdots,n\}, j > i; \\
x'' &= x' + C(0, cov(\sigma'', \theta''))
\end{aligned} \quad (6)$$

where $C(0,\cdot)$ denotes to yield random scalar or vector submitted to Cauchy distribution with respective parameters. The similar denotations are omitted here. We plot two distributions in Fig.5 by applying the same parameters. We can investigate that Cauchy distribution is symmetrical and long-tailed, and in the other word, it has a lower extremum in middle and a slower horizontal decline than Gaussian one, and the horizontal decline is getting smaller as it departs from the middle.

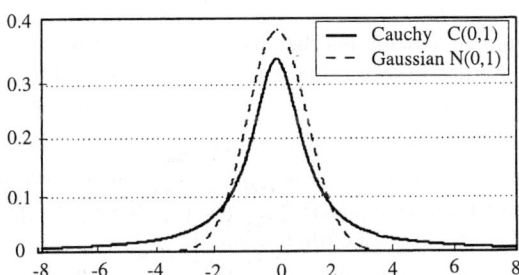

Fig. 5. Cauchy and Gaussian Distributions (with the same parameters)

Therefore probabilistically speaking, Cauchy distribution is more expanded, and it allows larger mutations and in this way producing more diversified individuals and covering more major space. The opposite, Gaussian mutation is accomplished in accurate search of its nearest space for the proximity exploration. If we take measures to develop the respective merits from two mutations, it sounds possible to provide higher performance on both proximity and diversity, Fig. 6 shows the measures that will be taken to realize our idea. Coupling the recombination and mutation, DAES produces an offspring completely as pseudocode in Fig. 7 describes.

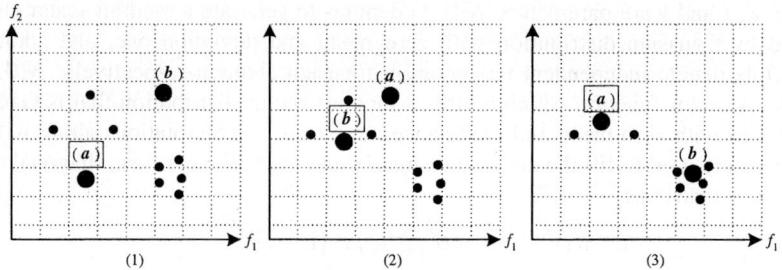

Fig. 6. Illustration of measures for exploration operator: (1) *a* dominates *b*, so *a* is chosen; (2) *b* dominates *a*, so *b* is chosen; (3) incomparable, *b* is located in a more crowded region, *a* is chosen

---
function *offspring* = ProduceOffspring (*par*1, *par*2, *cur_pop*, *ex_ar*)
    Produce *G_offspring* with Gaussian mutation; // via Eq.(2) & Eq.(3)
    Produce *C_offspring* with Cauchy mutation; //via Eq.(2) & Eq.(6)
    // compare two offspring and decide which one is accepted
    if (*G_offspring* dominates *C_offspring*) *offspring* = *G_offspring*;
    else if (*C_offspring* dominates *G_offspring*) *offspring* = *C_offspring*;
    else //incomparable, so the crowded location is applied
        if (*G_offspring* in a less crowded region of *cur_pop*∪ *ex_ar* than *C_offspring*)
            offspring = *G_offspring*;
        else *offspring* = *C_offspring*;
    return *offspring*;
end function.

Fig. 7. Pseudocode of exploration operator

3.5 Fitness Assignment Scheme

Since fitness assignment scheme is crucially accompanied with the complete nondominance selection, we use fitness value as anz indicator to distinguish dominated and nondominated individuals. It should be devised to meet the following requirements.

- The dominated individuals must share the lowest fitness than nondominated ones due to their absence of qualification for survival into next generation.
- The fitness of nondominated individuals must indicate their properties; so much better individuals of them can be filtered into the next step after competition.
- The individuals far away from the center of the current nondominated set must be assigned a higher fitness so as to share a higher probability of survival, which contributes to the diversity preservation and reducing the risk of premature convergence.

Therefore, for all the dominated individual of the intermediate population or the archive, we assigned their fitness the same zero; but for any nondominated individual, the average Euclid distance in the objective space between it and other *m* nearest individuals is equal to evaluate its contribution to the diversity preservation, defined as

$$F(j) = \min \sum_{l=1}^{m} \{\|x - x_l\|_{obj}\} / m, \, x, x_l \in H \qquad (7)$$

where, *j* denotes the nondominated individual, *x* is the decision vector of *j* and function $\|\cdot\|_{obj}$ calculates the Euclid distance in the objective space between two individuals.

4 Comparison Study

In order to validate the proposed DAES and quantitatively compare its performance with other advanced MOEAs, three recently designed benchmark functions [20] described in Table.1 are tested by three existed algorithms — SPEA2 [5], NSGA2 [3], M-PAES [7] and the proposed DAES in this comparison study, and each algorithm runs 50 times independently for each function to obtain statistical results. Since the comparison focuses on minimizing the proximity of PF_{final} as well as on maximizing

Table 1. Three benchmark functions. All P_F is formed with $g = 1$. n is the number of decision vector

No	Function	n	Boundary	Characters
T_1	$f_1(x) = x_1; f_2(x) = g \cdot (1 - \sqrt{f_1/g})$ $g = 1 + 9 \cdot (\sum_{i=2}^{n} x_i)/(n-1)$	30	$[0,1]^n$	continuous convex
T_2	$f_1(x) = x_1; f_2(x) = g \cdot (1 - f_1/g)^2$ $g = 1 + 9 \cdot (\sum_{i=2}^{n} x_i)/(n-1)$	30	$[0,1]^n$	continuous concave
T_3	$f_1(x) = x_1; f_2(x) = g \cdot (1 - \sqrt{f_1/g} - (f_1/g)\sin(10\pi f_1))$ $g = 1 + 9 \cdot (\sum_{i=2}^{n} x_i)/(n-1)$	30	$[0,1]^n$	discrete

Table 2. General parameters setting of four algorithms

Common Parameters	DAES	SPEA2	NSGA2	M-PAES
Chromosome length	-	15×dec_num	15×dec_num[3]	15×dec_num
Population size	20	80	100	1
Archive size	60[1]	20	0[2]	100
Offspring per generation	60	10	10	10
Crossover rate	-	0.7	0.7	-
Individuals perform mutation	1	1	1	10
Number of binary bit flipped	-	1	1	1
Maximum generation	10,000	10,000	10,000	10,000

(1) It is the minimal size of the archive.
(2) Instead of using archive, NSGA2 combines the populations from two consecutive generations.
(3) *dec_num* represents the number of decision variables.

the diversity of PF_{final}, we have to consider both the online performance and offline performance of each algorithm, and the nondominated set in the final population and the archive was taken as the output of an optimization run. Table 2 lists the general parameters setting of four algorithms for all runs, referred to literatures, e.g., [4], [5]. Then we use two methods: (1) graphical presentation for visual inspection and (2) the performance metric to show quantitative inspection.

4.1 Graphical Presentation

The first method is the graphical presentation. We unify the outcomes of each benchmark function from each algorithm, and all the nondominated individuals are plotted as shown in Fig. 8 and Fig. 9 in company with its *true P_F* (called PF_{true} in this section).

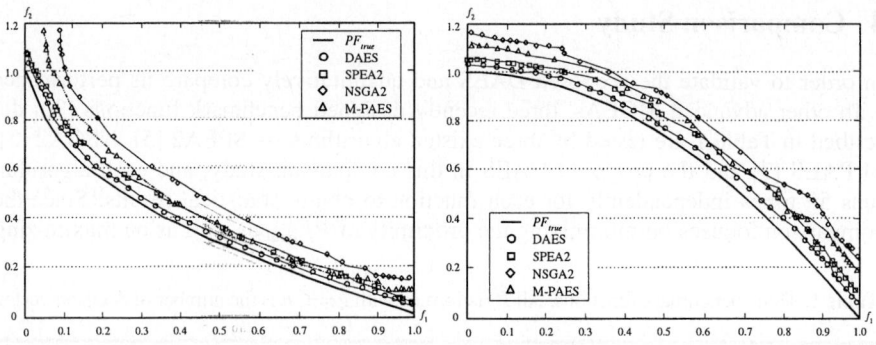

Fig. 8. Graphical presentation based on T_1 (left) and T_2 (right)

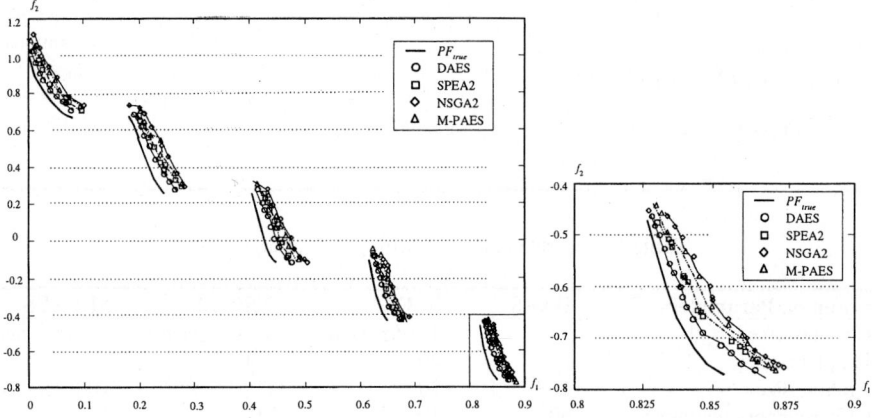

Fig. 9. Graphical presentation based on T_3 (left) and enlarged rectangle of T_3 from left (right)

For continuous function T_1 and T_2, the lines plotted by the solutions of DAES are much closer to the front than all its competitors. Besides, the solution distribution also varied. DAES distributes its solutions symmetrically in both middle and tail without obvious difference, which more than seventy percent of solutions of the latter three algorithms are crowded within thirty percent of region in the tail. This asymmetric distribution makes it difficult for decision making on middle compromise solutions. The results of function T_3 from either of the algorithms are a bit close. Both show less perfect than what has achieved in continuous functions. To show the discrepancy, we enlarged the region of the closest proximity between 0.8 and 0.9 of f_1 and displayed it on the right. It also demonstrates that DAES distributes their solutions closer to PF_{true} and more diversely than the others do. As a result, DAES is superior and the predominance decreases from continuous functions to discrete function to some extent.

4.2 Performance Metrics

We use three indicators to benchmark the comparison. The first indicator is the general distance (GD) [21] that can show how far PF_{final} are away from PF_{true}, calculated by

$$GD \triangleq (\sum_{i=1}^{N} d_i^p)^{1/p} / N, d_i = \min_j (|f_1^i(x) - f_1^j(x)| + |f_2^i(x) - f_2^j(x)|) \quad (8)$$

where N is the number of nondominated individuals in PF_{final}, $p = 2$, and d_i is the Euclidean distance in the objective space between each of these individuals and a closest point on PF_{true}. A smaller value of GD is preferable.

Because if another desire to measure diversity of solutions. We adopt the spacing (SP) to measure the range variance of neighboring individuals, defined as

$$SP \triangleq \sqrt{\sum_{i=1}^{N} (\bar{d} - d_i)^2 / (N-1)}, \bar{d} = (\sum_{i=1}^{N} d_i) / N \quad (9)$$

A zero value of SP indicates the ideal diversity that all nondominated individuals are equidistantly and uniformly spaced, and a smaller value of SP is preferable.

Moreover, the C value [20] is also included to compare the dominance relationship between two algorithms. It maps the ordered pair (X_i, X_j) to interval [0, 1], defined as

$$C(X_i, X_j) \triangleq |\{y \in X_j \wedge (\exists x \in X_i : x \succ y \vee x = y)\}| / |X_j| \quad (10)$$

where X_i and X_j denote PF_{final} from algorithm i and algorithm j respectively. The value $C(X_i, X_j) = 1$ means that all individuals in X_j are dominated by, or equal to, individuals in X_i. The opposite, $C(X_i, X_j) = 0$, represents the situation when none of the points in X_j are covered by the set X_i. Both should be considered independently.

Here, *box plots* [22] are used to visualize the distribution of these indicator samples. The box plots concerning the three indicators are shown in Fig. 10-12 respectively.

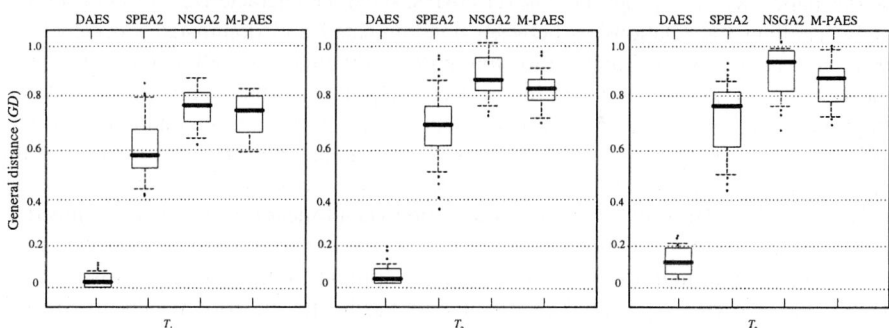

Fig. 10. Box plots based on metrics of the general distance (GD) of three functions

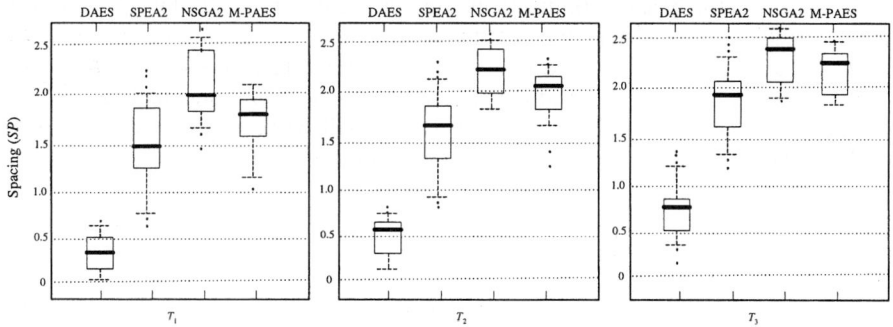

Fig. 11. Box plots based on metrics of the spacing (SP) of three functions

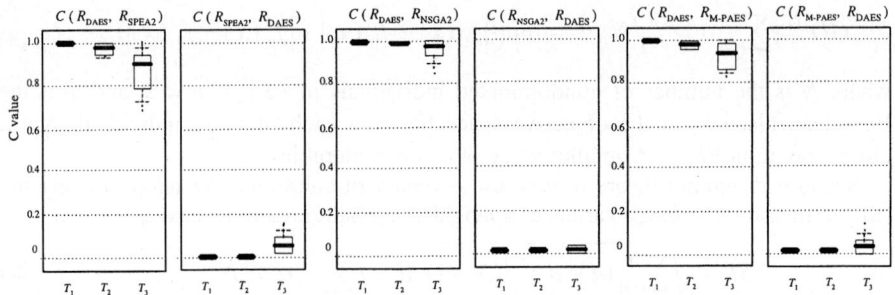

Fig. 12. Box plots based on the metric of C value of three functions

It is revealed that DAES is completely superior to all other algorithms on metrics of GD and SP. As far as C values is concerned, DAES covers all its competitors except for a few occasions, however, none of the other algorithms covers any of DAES' solutions. Consequently, the inspection of performance metrics shows the similar trend as the graphical presentation has discovered.

5 Conclusions

In this paper, we have proposed a novel DAES. It can be characterized as: (1) appropriately handling the relationship between nondominated and dominated individuals by using a dynamic external archive to store elitist individuals as well as relatively better individuals; (2) adaptively increasing or decreasing the external archive; (3) effectively improving the competence of exploration operator to provide new nondominated individuals and diversify them well by means of a combinatorial operator; (4) powerfully ensuring maximal pressure on proximity selection via a complete nondominance selection and ensuring maximal diversity preservation via a fitness assignment determined by dominance and population diversity information; and (5) converging to a near-optimal, well-extended and uniformly diversified Pareto optimal front. From comparison study, DAES has shown its potential in producing statistically superior results to SPEA2, NSGA2 and M-PAES on three prominent benchmark functions. So we suggest that DAES be a potential candidate in solving complicated problems.

However, as the benchmark functions used in this paper are still far from covering all the challenging characteristics of multiobjective optimization, a more profound study by applying DAES in dealing with other real-world problems is absolutely necessary in the future work. Additionally, there is no special design built in DAES to handle strong constraints. In near future, DAES will be revised to deal with those multiobjective optimization with strong and complicated constraints.

Acknowledgements

The authors gratefully acknowledge the support of the National Grand Fundamental Research 973 Program of China under Grant No.2003CB415206.

References

1. Fonseca, C.M., Fleming, P.J.: An Overview of Evolutionary Algorithms in Multiobjective Optimization. Evolutionary Computation. 1 (1995) 1–16
2. Srinivas, N., Deb, K.: Multiobjective Optimization Using Nondominated Sorting in Genetic Algorithms. Evolutionary Computation. 2 (1994) 221–248
3. Deb, K., Agrawal, S., Pratap, A., et al.: A Fast Elitist Nondominated Sorting Genetic Algorithm for Multiobjective Optimization: NSGA-II. Evolutionary Computation. 2 (2002) 182–197
4. Zitzler, E., Thiele, L.: Multiobjective Evolutionary Algorithms: A Comparative Case Study and the Strength Pareto Approach. Evolutionary Computation. 1 (1999) 257–271
5. Zitzler, E., Laumanns, M., Thiele, L.: SPEA2: Improving the Strength Pareto Evolutionary Algorithm for Multiobjective Optimization. In: EUROGEN 2001, Athens, Greece (2001)
6. Knowles, J.D., Corne, D.W.: Approximating the Nondominated Front Using the Pareto Archived Evolution Strategy. Evolutionary Computation. 2 (2000) 149-172
7. Knowles, J.D., Corne, D.W.: M-PAES: A Memetic Algorithm for Multiobjective Optimization. In: Proceedings of the 2000 congress on evolutionary computation. Piscataway, NJ: IEEE Press (2000) 325–332
8. Mahfoud, S.W.: Genetic Drift in Sharing Methods. In: Grefenstette, J.J. (eds.): Proceedings of the 1st IEEE conference on evolutionary computation. Piscataway, NJ: IEEE Press (1994) 67–72
9. Schwefel, H.P, Back, T.: A Survey of Evolutionary Strategies, In: Belew, R. (eds.): Proceedings of the 4th international conference on genetic algorithms. Morgan Kaufmann publishers, San Mateo, CA (1991) 92-99
10. Bosman, P.A.N., Thierens, D.: The Balance Between Proximity and Diversity in Multiobjective Evolutionary Algorithms. Evolutionary Computation. 7 (2003) 174-188
11. Yao, X., Liu, Y.: Fast Evolutionary Programming. In: Proceedings of the 5th annual conference on evolutionary programming, The MIT Press, Cambridge, MA (1996) 451-460
12. Gomes, J.R., Saavedra, O.R.: A Cauchy-based Evolution Strategy for Solving the Reactive Power Dispatch Problem. Electrical Power and Energy Systems. 24 (2002) 277-283
13. Shi, L.B., Xu, G.Y.: Self-adaptive Evolutionary Programming and its Application to Multiobjective Optimal Operation of Power Systems. Electric Power Systems Research. 57 (2001) 181-187
14. Bosman, P.A.N., Thierens, D.: Multiobjective Optimization with Diversity Preserving Mixture-based Iterated Density Estimation Evolutionary Algorithms. Int. J. Approx. Reasoning. 31 (2002) 259–289
15. Tan, K.C., Lee, T., Khor, E.: Evolutionary Algorithms with Dynamic Population Size and Local Exploration for Multiobjective Optimization. Evolutionary Computation. 12 (2001) 565–588
16. Sarker, R., Liang, K.H., Newtom, C.: A New Multiobjective Evolutionary Algorithm. European Journal of Operational Research. 140 (2002) 12-23
17. Schwefel, H.P.: Numerical Optimization for Computer Models. John Wiley. Chichester, UK (1981) 129-132
18. Hu, X., Eberhart, R.C.: Multiobjective Optimization Using Dynamic Neighborhood Particle Swarm Optimization. In: Proceedings of the 2002 congress on evolutionary computation. Piscataway, NJ: IEEE Press (2002) 1677–1681
19. Back, T., Schwefel, H.P.: An Overview of Evolutionary Algorithm for Parameter Optimization. Evolutionary Computation. 2 (1993) 1-23
20. Zitzler, E., Deb, K., Thiele, L.: Comparison of Multiobjective Evolutionary Algorithms: Empirical Results. Evolutionary Computation. 2 (2000) 173-195
21. van Veldhuizen, D.A., Lamont, G.B.: On Measuring Multiobjective Evolutionary Algorithm Performance. In: Proceedings of the 2000 congress on evolutionary computation. Piscataway, NJ: IEEE Press (2000) 204-211
22. Chambers, J.M., Cleveland, W.S., Kleiner, B., et al.: Graphical Methods for Data Analysis. Pacific Grove, CA: Wadsworth & Brooks/Cole (1983)

Searching for Robust Pareto-Optimal Solutions in Multi-objective Optimization

Kalyanmoy Deb and Himanshu Gupta

Kanpur Genetic Algorithms Laboratory (KanGAL),
Indian Institute of Technology Kanpur,
Kanpur, PIN 208016, India
{deb, himg}@iitk.ac.in

Abstract. In optimization studies including multi-objective optimization, the main focus is usually placed in finding the global optimum or global Pareto-optimal frontier, representing the best possible objective values. However, in practice, users may not always be interested in finding the global best solutions, particularly if these solutions are quite sensitive to the variable perturbations which cannot be avoided in practice. In such cases, practitioners are interested in finding the so-called *robust* solutions which are less sensitive to small changes in variables. Although robust optimization has been dealt in detail in single-objective optimization studies, in this paper, we present two different robust multi-objective optimization procedures, where the emphasis is to find the robust optimal frontier, instead of the global Pareto-optimal front. The first procedure is a straightforward extension of a technique used for single-objective robust optimization and the second procedure is a more practical approach enabling a user to control the extent of robustness desired in a problem. To demonstrate the subtle differences between global and robust multi-objective optimization and the differences between the two robust optimization procedures, we define four test problems and show simulation results using NSGA-II. The results are useful and should encourage further studies considering robustness in multi-objective optimization.

1 Introduction

For the past decade or more, the primary focus of the research and application in the area of evolutionary multi-criterion optimization (EMO) has been placed in finding the globally best Pareto-optimal solutions. Such solutions are non-dominated to each other and there exists no other solution in the entire search space which dominates any of these solutions. From a theoretical point of view, such solutions are of utmost importance in a multi-objective optimization problem. However, in practice, often a solution cannot be implemented with arbitrary precision for various reasons and the implemented solution may be slightly different from the desired solution. If a global optimal solution is *sensitive* to variable perturbation in its vicinity, the implemented solution may correspond to different objective values than that of the theoretical optimal solution. Thus, from a practical standpoint, such solutions are of not much importance and the emphasis must be made in finding *robust* solutions, which are less sensitive to variable perturbations in their vicinity.

In single-objective optimization, a number of studies have been devoted for finding such robust solutions. Branke [1] suggested a number of heuristics for searching robust solutions. In another study, Branke [2] suggested a number of methods for alternate fitness estimation. Later, Branke [1] also pointed out key differences between searching optimal solutions in a noisy environment and searching for robust solutions. Jin and Sendhof [3] posed the issue of finding robust solutions in single-objective optimization problem as a multi-objective optimization problem with the objectives being maximizing robustness and performance. Tsutsui and Ghosh [4] presented a mathematical model for obtaining robust solutions using schema theorem for single-objective genetic algorithms. Parmee [5] suggested a hierarchical strategy of searching several high performance regions in a fitness landscape simultaneously. Teich [6] extended Pareto-dominance for handling uncertain objectives and Hughes [7] computed the error estimate for using deterministic Pareto-dominance in noisy functions. However, to our knowledge, there does not exist a systematic study introducing robustness in evolutionary multi-objective optimization.

In this paper, we make an effort to extend an existing approach for finding robust solutions in single-objective optimization for multi-objective optimization. Essentially, in this approach, instead of optimizing the original objective functions, we optimize the mean effective objective values computed at a point by averaging the function values of a few solutions in its vicinity. The solutions which are less sensitive to local perturbations will fair well in terms of the mean effective objective values and the resulting Pareto-optimal front will be the robust frontier. To illustrate the working of this approach, we first suggest four different controllable test problems and then employ NSGA-II. We also present a new definition of robustness in which original objectives are optimized but a constraint limiting the change in function values due to local perturbations is added. The latter approach is more pragmatic and a user has a control on the desired level of robustness on the obtained solutions. The differences between these two robust procedures and fundamental differences between global and robust optimization in the context of multi-objective optimization are clearly demonstrated.

Rest of the paper is designed as followed. Section 2 introduces the concept of robustness in multi-objective optimization and stresses its importance. Sections 3 and 4 discuss the two robust optimization schemes and results obtained using NSGA-II. Finally, a conclusion of this study is presented in Section 5.

2 Robustness in Optimization

We consider a multi-objective optimization problem of the following type:

$$\left. \begin{array}{l} \text{Minimize } (f_1(\mathbf{x}), f_2(\mathbf{x}), \ldots, f_M(\mathbf{x})), \\ \text{subject to } \mathbf{x} \in \mathcal{S}, \end{array} \right\} \quad (1)$$

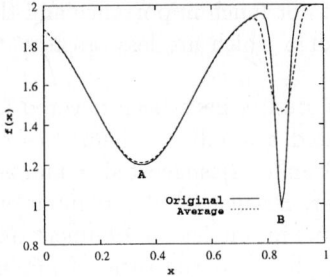
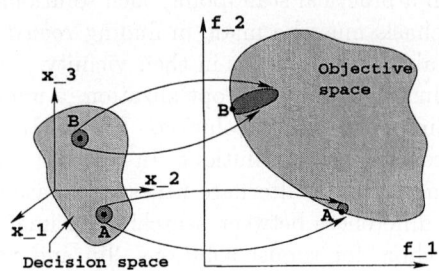

Fig. 1. Illustration of global versus robust solutions in a single-objective optimization problem

Fig. 2. Point A is less sensitive to variable perturbation than point B

where S is the feasible search space. A robust solution is defined as the one which is less sensitive to the perturbation of the decision variables in its neighborhood. Let us consider the single-objective function shown in Figure 1. Of the two optimal solutions, solution A is considered robust as a small variation in the decision variables does not alter the objective function value of the solution. On the other hand, solution B is quite sensitive to the variable perturbation and often cannot be recommended in practice, despite having a better function value than solution A. Several EA researchers suggested different procedures of defining and finding such robust solutions in a single-objective optimization problem [2, 8, 3, 4, 5].

For solving multi-objective optimization problems, an EMO procedure attempts to find a finite number of Pareto-optimal solutions, instead of a single optimum. Since Pareto-optimal solutions collectively *dominate* any other feasible solution in the search space, they all are considered to be better than any other solution [9]. The concept of robustness discussed above for single-objective optimization can be extended for multi-objective optimization as well and is worth from a practical standpoint. In Figure 2, two Pareto-optimal solutions (A and B) are checked for their sensitivity in the decision variable space. Since the local perturbation of point B causes a large change in objective values, this solution may not be a robust solution, whereas solution A which does not cause a large change in objective values due to a local perturbation in its vicinity, is a robust solution. To qualify as a robust solution, each Pareto-optimal solution now has to demonstrate its insensitivity towards small perturbations in its decision variable values. The main differences with a single-objective robust solution is that (i) the sensitivity now has to be established with respect to all M objectives. That is, a combined effect of variations in all M objectives has to be used as a *measure* of sensitivity to variable perturbation, and (ii) there are many solutions to be checked for robustness.

2.1 Robust Optimization Approaches

One of the main approaches portrayed in the single-objective literature is to use a *mean effective* objective function ($f^{\text{eff}}(\mathbf{x})$) for optimization, instead of the original objective function ($f(\mathbf{x})$) itself. Here, we give a definition for a generic M-objective optimization problem:

Definition 1. Multi-objective Robust Solution of Type I: *A solution $\mathbf{x}^*$ is called a multi-objective robust solution of type I, if it is the Pareto-optimal solution to the following multi-objective minimization problem defined with respect to a δ-neighborhood ($\mathcal{B}_\delta$):*

$$\left. \begin{array}{l} \text{Minimize } (f_1^{\text{eff}}(\mathbf{x}), f_2^{\text{eff}}(\mathbf{x}), \ldots, f_M^{\text{eff}}(\mathbf{x})), \\ \text{subject to } \mathbf{x} \in \mathcal{S}, \end{array} \right\} \quad (2)$$

where $f_j^{\text{eff}}(\mathbf{x})$ is defined as follows:

$$f_j^{\text{eff}}(\mathbf{x}) = \frac{1}{|\mathcal{B}_\delta|} \int_{\mathbf{y} \in \mathbf{x} + \mathcal{B}_\delta} f_j(\mathbf{y}) d\mathbf{y}. \quad (3)$$

where $|\mathcal{B}_\delta|$ is the hypervolume of the chosen neighborhood.

To use it in practice, a finite set of H solutions ($\mathbf{y}$) can be randomly (or in some structured manner) chosen around a δ-neighborhood ($\mathbf{y} \in \mathbf{x} + \mathcal{B}_\delta$, where $\mathcal{B}_\delta = \{\mathbf{z} | z_i \in [-\delta_i, \delta_i]\}$) of a solution $\mathbf{x}$ in the variable space and the mean effective objectives (f_j^{eff}) are optimized by an EMO procedure. This way, instead of an individual's own function value (f_j), an agglomerate objective value in its vicinity is used as the objective for optimization.

Another approach would be to restrict a normalized change in perturbed objective vector from its original objective vector by a user-specified limit η:

Definition 2. Robust Solution of Type II: *For the minimization of a multi-objective problem, a solution $\mathbf{x}^*$ is called a robust solution of type II, if it is the Pareto-optimal solution to the following problem:*

$$\left. \begin{array}{l} \text{Minimize } (f_1(\mathbf{x}), f_2(\mathbf{x}), \ldots, f_M(\mathbf{x})), \\ \text{subject to } \frac{\|\mathbf{f}^p(\mathbf{x}) - \mathbf{f}(\mathbf{x})\|}{\|\mathbf{f}(\mathbf{x})\|} \leq \eta, \\ \mathbf{x} \in \mathcal{S}. \end{array} \right\} \quad (4)$$

The perturbed objective vector $\mathbf{f}^p$ can be chosen as the mean effective function value ($\mathbf{f}^{\text{eff}}$) or the worst function value (among H chosen solutions) in the neighborhood. The operator $\|\cdot\|$ can be any norm measure.

Both definitions for robustness in multi-objective optimization raise some interesting issues. For example, due to the variable sensitivities, a part of the original global Pareto-optimal front may not qualify as a robust front. In another scenario, the original global Pareto-optimal front (given in Equation 1) may be completely non-robust and a original local Pareto-optimal front may become robust. Depending on how robust the original global Pareto-optimal front is with respect to the above definition, there can be the four different scenarios, which we discuss next.

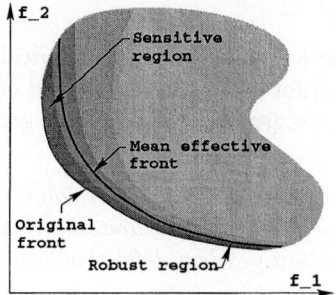
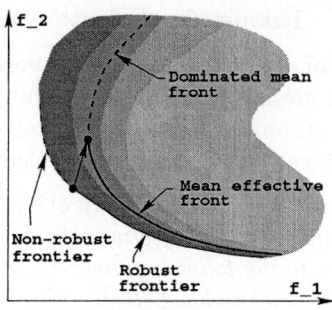

Fig. 3. Case 1: Complete Pareto-optimal front is robust

Fig. 4. Case 2: A part of the Pareto-optimal front is robust

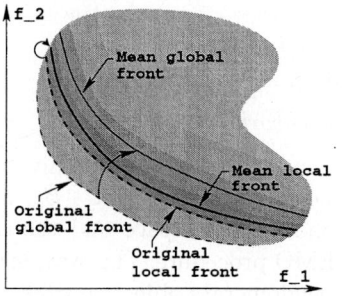
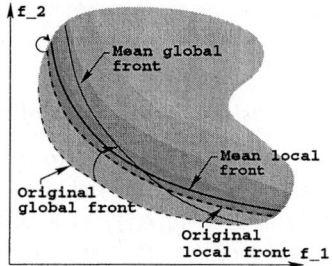

Fig. 5. Case 3: A local Pareto-optimal front is now robust

Fig. 6. Case 4: A part of the global Pareto-optimal front is not robust

2.2 Case 1: Original Pareto-optimal Front Remains Robust

This is the simplest case in which the original Pareto-optimal front remains Pareto-optimal with respect to the mean effective function values. Figure 3 illustrates such a problem. The complete original front remains robust in this type of problems.

2.3 Case 2: Only a Part of Original Front Remains Robust

Here, the complete original Pareto-optimal front is not robust with respect to the above definition of robustness of type I. In most real-world scenarios such a problem is expected, as some portion of the Pareto-optimal front may lie in a sensitive region in the decision variable space. In such a problem, the task of a multi-objective robust optimizer would be to find only that part of the Pareto-optimal front which is robust. Figure 4 shows that the Pareto-optimal front corresponding to the mean effective objectives does not span over the entire original Pareto-optimal region.

2.4 Case 3: Original Local Front Is Robust

Cases 3 and 4 correspond to more difficult problems in which the original problem may have more than one Pareto-optimal fronts (global and local) [9]. In Case 3, the mean effective front constructed using the original global Pareto-optimal solutions is completely dominated by that constructed using the local Pareto-optimal front, thereby meaning that the original global Pareto-optimal front is not a robust one. Figure 5 demonstrates such a problem. This type of problems, if encountered, must be solved for finding the robust Pareto-optimal front, instead of the sensitive global Pareto-optimal front.

2.5 Case 4: Only a Part of Original Global Front Is Robust

Instead of the complete original global Pareto-optimal front being sensitive to the variable perturbation, Case 4 problems cause a part of it to be adequately robust. In the remaining part, a new front appears to be robust. Figure 6 illustrates this problem.

Certainly, other scenarios are possible, where instead of an original local Pareto-optimal front becoming robust, a completely new frontier emerges to be robust. However, we argue that the above four scenarios most likely cover different types of robust multi-objective optimization problems which can be encountered in practice and an algorithm capable of solving these scenarios would be adequate to solve other simpler kinds.

2.6 Test Problems

In this section, we now construct a mathematical two-objective test problem for each of the above four cases.

Test Problem 1. This problem is an illustration to Case 1 discussed above:

$$\text{Minimize } (f_1(\mathbf{x}), f_2(\mathbf{x})) = (x_1, h(x_1) + g(\mathbf{x})S(x_1)),$$
$$\text{Subject to } 0 \le x_1 \le 1, -1 \le x_i \le 1, \quad i = 2, 3, \ldots, n,$$
$$\text{where } h(x_1) = 1 - x_1^2,$$
$$g(\mathbf{x}) = \sum_{i=2}^{n} \left(10 + x_i^2 - 10\cos(4\pi x_i)\right), \quad S(x_1) = \frac{\alpha}{0.2 + x_1} + \beta x_1^2. \quad (5)$$

Here, we use $\alpha = 1$ and $\beta = 1$. The Pareto-optimal front corresponds to $x_i^* = 0$ for $i = 2, 3, \ldots, n$ and for any value of x_1 in the prescribed domain $[0, 1]$. At these solutions, $g(\mathbf{x}) = 0$, thereby making the following relationship between original objectives:

$$f_2^* = 1 - f_1^{*2}. \quad (6)$$

The mean effective objectives in a δ-neighborhood for a Pareto-optimal solution $\mathbf{x}$ (for $x_1 \in [0, 1]$) are given as follows:

$$f_1^{\text{eff}}(\mathbf{x}) = x_1,$$
$$f_2^{\text{eff}}(\mathbf{x}) = (1 - x_1^2) - \tfrac{1}{3}\delta_1^2 + \left[\alpha \tfrac{1}{2\delta_1} \log\left(\tfrac{0.2 + x_1 + \delta_1}{0.2 + x_1 - \delta_1}\right)\right.$$
$$\left. + \beta\left(x_1^2 + \tfrac{1}{3}\delta_1^2\right)\right] \sum_{i=2}^{n} \left(10 + \tfrac{1}{3}\delta_i^2 - \tfrac{10}{4\pi\delta_i} \sin 4\pi\delta_i\right). \quad (7)$$

Test Problem 2. This problem is an illustration of Case 2. The mathematical formulation of this problem is identical to that in test problem 1, except that here we use $\alpha = 1$ and $\beta = 10$. The corresponding Pareto-optimal frontier for the original problem and for the mean effective objectives can be obtained from Equation 6 and Equation 7, respectively, by substituting the above parameter values.

Test Problem 3. This problem is an instantiation of Case 3. Since, this problem requires the concept of local and global Pareto-optimal front, we construct a multi-modal multi-objective optimization problem:

$$\text{Minimize } (f_1(\mathbf{x}), f_2(\mathbf{x})) = (x_1, h(x_2)(g(\mathbf{x}) + S(x_1))),$$
$$\text{Subject to } 0 \leq x_1, x_2 \leq 1, -1 \leq x_i \leq 1, \quad i = 3, 4, \ldots, n,$$
$$\text{where } h(x_2) = 2 - 0.8 \exp\left(-\left(\tfrac{x_2 - 0.35}{0.25}\right)^2\right) - \exp\left(-\left(\tfrac{x_2 - 0.85}{0.03}\right)^2\right), \quad (8)$$
$$g(\mathbf{x}) = \sum_{i=3}^{n} 50 x_i^2, \quad S(x_1) = 1 - \sqrt{x_1}.$$

Once again, the Pareto-optimal front corresponds to $x_i = 0$ for $i = 3, 4, \ldots, n$. Thus, at this front, $f_2(x_1, x_2) = h(x_2) S(x_1)$. Since, $f_1(\mathbf{x}) = x_1$, the local and global Pareto-optimal frontiers will correspond to the local and global minima of $h(x_2)$, respectively. A careful look at $h()$ function (shown in Figure 1) will reveal that there are two minima, of which the global minimum is at $x_2^* = 0.85$ ($h(x_2^*) \approx 1.0$). Similarly, the local Pareto-optimal front corresponds to $x_2^* = 0.35$ (with $h(x_2^* \approx 1.2)$. Approximate relationships between f_1 and f_2 at these two fronts are as follows:

$$f_2 = 1 - \sqrt{f_1} \quad \text{(global)}, \qquad f_2 = 1.2(1 - \sqrt{f_1}) \quad \text{(local)}.$$

The mean effective objective values at these two fronts are given as follows:

$$f_1^{\text{eff}}(\mathbf{x}) = x_1, \qquad (9)$$

$$f_2^{\text{eff}}(\mathbf{x}) = H(x_2^*, \delta_2) \sum_{i=3}^{n} \left[\frac{50}{3} \delta_i^2 + \left(1 - \frac{1}{3\delta_1}\left((x_1 + \delta_i)^{1.5} - (x_1 - \delta_i)^{1.5}\right)\right)\right], \qquad (10)$$

where $H(x_2^*, \delta_2) = \frac{1}{2\delta_2} \int_{x_2^* - \delta_2}^{x_2^* + \delta_2} h(y) dy$ and is equal to 1.154 and 1.237 for local and global Pareto-optimal solutions, respectively, with $\delta_2 = 0.03$.

Test Problem 4. To represent Case 4, we construct a problem which is the same as test problem 3, with a couple of modifications: (i) the function is $h()$ is dependent on two variables:

$$h(x_1, x_2) = 2 - x_1 - 0.8 \exp\left(-\left(\frac{x_1 + x_2 - 0.35}{0.25}\right)^2\right) - \exp\left(-\left(\frac{x_1 + x_2 - 0.85}{0.03}\right)^2\right).$$

and (ii) the variable bound on x_2 is different: $-0.15 < x_2 < 1$. The problem has its global Pareto-optimal front somewhere near $x_1 + x_2 = 0.85$ and the local Pareto-optimal front near $x_1 + x_2 = 0.35$, as before. However, the global Pareto-optimal front for the mean effective objectives corresponds to a mix of these two sets of values of x_1 and x_2. For f_1 smaller than about 0.5, the front corresponds to $x_1 + x_2 \approx 0.35$ and for $f_1 \geq 0.6$ values the front corresponds to $x_1 + x_2 \approx 0.85$.

3 Simulation Results

Here, we use NSGA-II [10] procedure to obtain the robust Pareto-optimal front, although any other EMO algorithm can also be used. Various parameters which would determine the extent and nature of shift of the mean effective front from the original front are as follows:

- The extent of the neighborhood (δ vector) considered to each variable.
- Number of neighboring points (H) used to compute the mean effective objectives.

We discuss the effect of these two parameters in detail for the first two test problems. However, before we discuss the results, there is an important matter which we discuss next.

There can be a number of ways of generating H neighboring points in the vicinity of a solution to compute the mean effective objective values [8]. The simplest strategy can be to randomly create H points in the neighborhood of every solution. However, this introduces additional randomness in evaluating the same solution more than once and it was suggested that a random pattern of points around a solution be created in the beginning of a generation and the same pattern be used for evaluating all population members. To create a pattern systematically, we divide the perturbation domain of each variable (around $[-\delta, \delta]$) into exactly H equal grids, thereby dividing the δ-neighborhood into n^H small hyperboxes. Thereafter, we pick exactly H hyperboxes randomly from n^H hyperboxes so that in each variable exactly one hyperbox for each of H grids is picked. Once the hyperboxes are identified, a random point within each hyperbox is chosen and is used for the computation of the mean effective objective values.

In all simulations, we have used the simulated binary crossover (SBX) and the polynomial mutation operator with distribution indices of 10 and 50, respectively. A population size of 100 is run for 10,000 generations to have confidence in the location of the robust optimal front, although the final effective frontier appears well within 1,000 generations.

3.1 Test Problems 1 and 2

Effect of Neighborhood Size, δ: To not have a significant effect due to finite neighboring points and variation in problem size, we use $H = 50$ and $n = 5$. To have an identical normalized neighborhood size for each variable, we use $\delta_1 = \delta$ and $\delta_i = 2\delta$ for $i \geq 2$. Figure 7 shows the theoretical mean effective front obtained using Equation 7 for four different values of neighborhood size, δ. It is clear from the figure that as δ increases, the mean effective front moves away from the original Pareto-optimal front (marked as the 'original front'). Although for this test problem, all solutions corresponding to the mean effective front are identical to those lying on the original Pareto-optimal front for any neighborhood size, the change in shape of the front is interesting. For the four δ used here, the mean effective front is non-convex, whereas the original front was convex. It

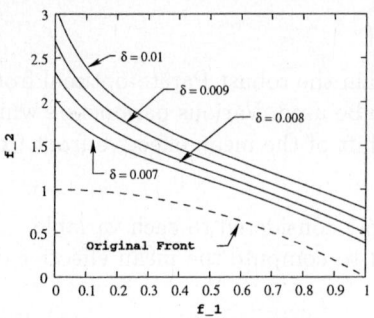
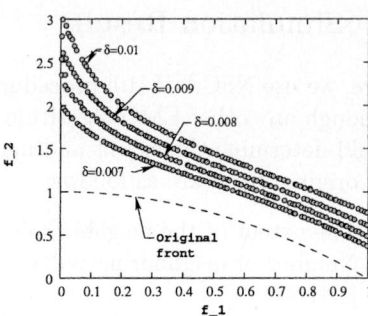

Fig. 7. Theoretical mean effective fronts showing the effect of δ on test problem 1

Fig. 8. Robust NSGA-II solutions show the effect of δ on test problem 1

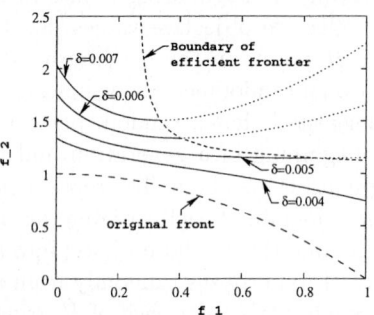
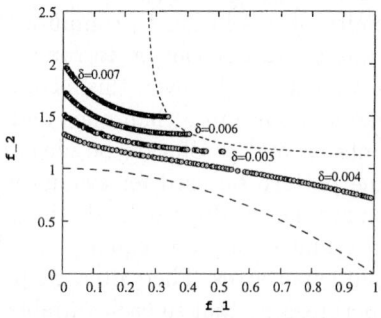

Fig. 9. Theoretical mean effective fronts showing the effect of δ on test problem 2

Fig. 10. Robust NSGA-II solutions show the effect of δ on test problem 2

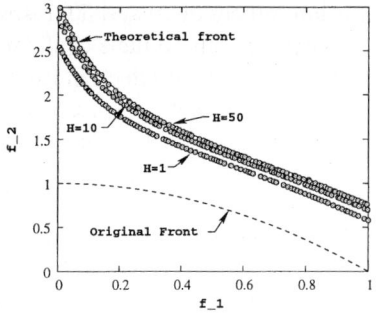
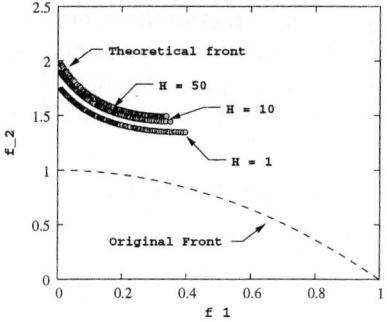

Fig. 11. Effect of H (theoretical and NSGA-II) on test problem 1

Fig. 12. Effect of H (theoretical and NSGA-II) on test problem 2

is important to highlight here that for robust optimization an EMO algorithm works on the mean effective objectives and thus may have difficulty in solving the robust optimization problem of handling a non-convex problem compared to the original convex problem. Figure 8 shows the obtained NSGA-II solutions for the same four δ values. A close investigation will reveal that the obtained front is exactly the same as that obtained using the exact mathematical analysis (Figure 7).

Figures 9 and 10 show theoretical and NSGA-II results on test problem 2. In this problem, not only the shape of the mean effective front is different from the original one, some original Pareto-optimal solutions are no more robust. It is clear from Figure 9 that for $\delta = 0.006$, original Pareto-optimal solutions having x_1^* greater than about 0.4 now get dominated. This simply means that these Pareto-optimal solutions are quite sensitive to variable perturbation and are not robust. When performing a robust multi-objective optimization, an algorithm should then find only those Pareto-optimal solutions which are robust. Figure 10 shows that NSGA-II finds only the robust portion of the original Pareto-optimal front.

Effect of Neighboring Points, H: It is intuitive that if more neighboring points are chosen for computing the mean effective objectives, the objective values will be closer to the theoretical average values; however, the computation time will be more. Figure 11 shows the effect of using different values of H on test problem 1. Here, we use $\delta = 0.01$ and $n = 5$. The theoretical mean effective front (ideally for $H = \infty$) is also shown with a solid line in the figure. It is clear that as H is increased, the mean effective front shifts away from the original front and asymptotically approaches the theoretical front. Figure 12 shows the effect of H on test problem 2 (with $n = 5$ and $\delta = 0.007$). The front obtained using a small H overestimates the true robust front, but at a much smaller computational time.

3.2 Test Problems 3 and 4

For problems 3 and 4, we show the effect of local and global fronts of the original problem in deciding on the true robust front. For both problems, we use $\delta = 0.03$, $H = 50$, and $n = 5$. Figure 13 shows the theoretical results obtained using Equations 9 and 10. The original local and global fronts are shown in dashed lines. The mean effective local and global fronts are also shown in the figure with solid lines. It is clear that the mean effective local front is the robust frontier of this problem, meaning that the original local Pareto-optimal solutions are robust solutions and original global Pareto-optimal solutions are sensitive to the variable perturbation and are not robust solutions. Figure 14 shows NSGA-II solutions applied to mean objective values obtained by averaging H function values in the δ-neighborhood of a solution. The NSGA-II front is very close to the theoretical local mean effective front.

To show the difference between original Pareto-optimal front and robust front, we show all 100 obtained NSGA-II solutions for two cases. In Figure 15, we show the solutions obtained for optimizing the original problem (without robustness consideration). It is clear that for all solutions, x_2 is close to 0.85. Variables x_3 to x_5 are all settled to a value zero and the variation in the front appears due to the variation in x_1. Figure 16 shows all solutions for the robust optimization. Here, all solutions take a value close to $x_2 = 0.35$.

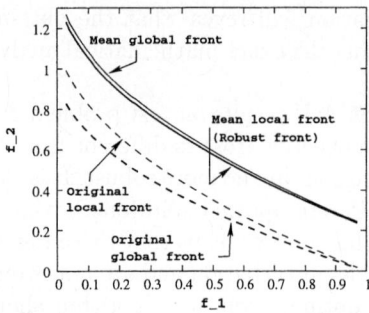

Fig. 13. Theoretical robust front for test problem 3

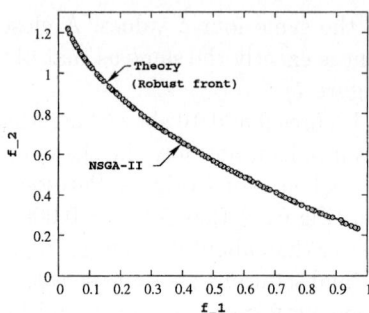

Fig. 14. NSGA-II robust front for test problem 3

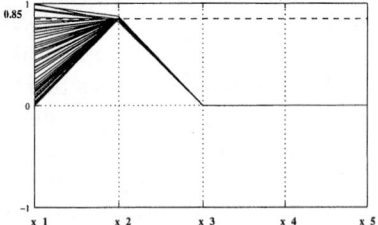

Fig. 15. NSGA-II solutions of the original test problem 3

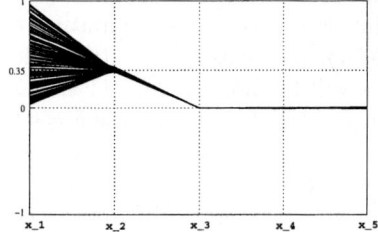

Fig. 16. NSGA-II robust solutions for test problem 3

Next, we consider test problem 4. The theoretical fronts for the original problem are shown in Figure 17 in dashed lines and corresponding mean effective fronts are shown in solid lines. It is clear from the figure that the robust frontier is constituted with a part of the local Pareto-optimal solutions and a part of the global Pareto-optimal solutions. Figure 18 shows the robust solutions obtained using NSGA-II. The deviation in the global part of the robust frontier from theory is due to the choice of a finite H (50 here). The original function landscape at the global frontier is quite sensitive to parameter changes, and it becomes difficult for an optimization algorithm to converge to the exact global frontier. When we rerun the problem with $H = 500$, the obtained NSGA-II solutions lie on the theoretical frontier. Figures 19 and 20 show the relationship between x_1 and x_2 in the solutions obtained for the original problem and that obtained for the mean effective objectives, respectively. It is clear that for solutions $f_1 \leq 0.5$ the relationship more or less follows $x_1 + x_2 = 0.35$ and for $f_1 > 0.6$ the relationship is $x_1 + x_2 = 0.85$. The latter condition corresponds to the original global Pareto-optimal front, as shown in Figure 19.

The above discussion on simulation results amply demonstrates that by optimizing the mean effective objectives (instead of the original objective functions) computed by averaging a few neighboring solutions, the robust frontier of type I can be found by using an EMO procedure. In a problem, the computation of

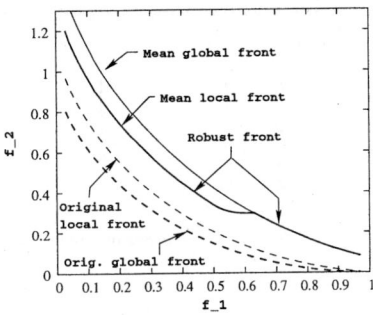

Fig. 17. Theoretical robust front for test problem 4

Fig. 18. NSGA-II robust front for test problem 4

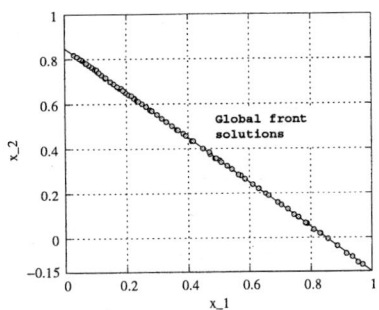

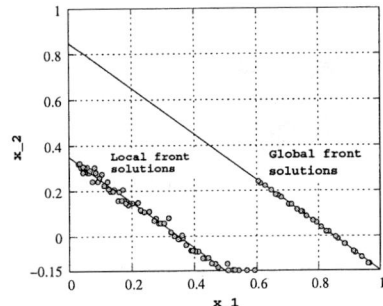

Fig. 19. NSGA-II solutions of the original test problem 4

Fig. 20. NSGA-II robust solutions for test problem 4

the robust front is more useful and provides a user with the information about robust solutions directly. It has been also found that the neighborhood size and the number of neighboring points used to compute the mean objective values are important parameters in obtaining the true robust frontier. However, the type I definition of robustness is somewhat less practical and yields in a robust frontier which cannot be controlled. For a given problem, the above definition constitutes a particular front as a robust front, mainly from the consideration of mean objective values. However, a user may like a preferred limiting change in function values for defining robustness and would be interested in knowing the corresponding robust frontier. For this purpose, we have defined the robust solutions of type II earlier and discuss it in the next section.

4 Multi-objective Robust Solutions of Type II

The robust solution of type II were defined earlier (Definition 2). Here, we use f_j^{eff} for f_j^p and the Euclidean norm for $\|\cdot\|$ operator. The limiting parameter η is considered constant in a simulation run and is a user-defined parameter. We

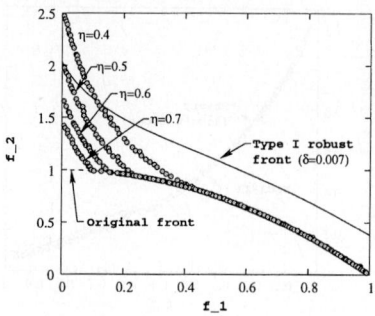
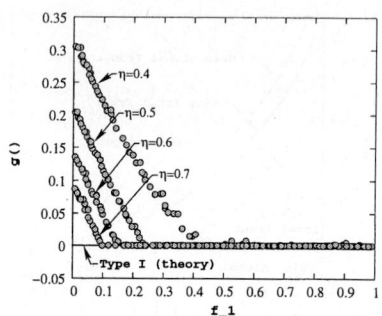

Fig. 21. Robust fronts with different η obtained using exact f^{eff} for problem 1

Fig. 22. Function $g()$ of the robust solutions shown in Figure 21 for problem 1

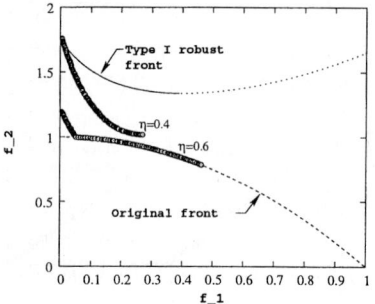
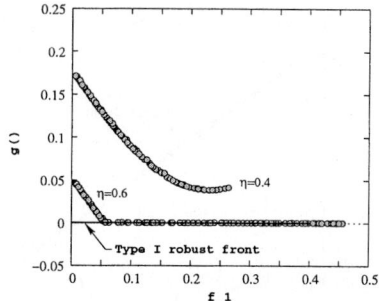

Fig. 23. Robust fronts with different η obtained using exact f^{eff} for problem 2

Fig. 24. Function $g()$ of the robust solutions shown in Figure 23 for problem 2

simply employ NSGA-II to solve the corresponding constrained optimization problem by using the constrained-domination principle [9].

To demonstrate the nature of robust solutions of type II, here we consider test problems 1 and 2, for brevity. We use $\delta = 0.007$ and 0.006 for problems 1 and 2, respectively. All other parameters are the same as before.

4.1 Test Problem 1

Figure 21 shows NSGA-II solutions obtained for different pre-defined η values on test problem 1. Here, the mathematical mean effective objective functions (Equation 7) are optimized with the additional η constraint by using NSGA-II. On separate NSGA-II runs, similar fronts are obtained when the mean effective objective values are computed using $H = 50$ neighboring solutions. The figure demonstrates that the sensitive region of the original Pareto-optimal front is vulnerable to the chosen value of η. For a more tight (smaller) limiting η, the corresponding front is further away from the original front in the sensitive region. As η is increased, the robust frontier gets closer to the original front. However, on the less sensitive portion of the original frontier, there is no change.

For comparison, the robust front obtained with type I robustness is also shown for identical δ and H parameter values in Figure 21. Recall that in the case of type I robustness, the robust solutions for this problem corresponds to $x_i = 0$ for $i > 1$ (thereby making the $g(\mathbf{x}^*) = 0$). However, with type II robustness, different solutions appear in the sensitive portion of the robust frontier having $g(\mathbf{x}^*) \geq 0$. To demonstrate this aspect, we plot $g(\mathbf{x}^*)$ values in Figure 22 for two cases: type I robust frontier (theoretical) and type II robust frontier. Although solutions having $g(\mathbf{x}^*) \geq 0$ were not the Pareto-optimal solutions of the original problem, the definition of robustness of type II causes them to be robust optimal solutions with respect to a particular η.

4.2 Test Problem 2

Figure 23 shows the NSGA-II solutions obtained using $H = 100$ neighboring points and with $\eta = 0.4$ and $\eta = 0.6$. As discussed earlier and as shown in the figure, the complete Pareto-optimal front was not robust of type I in this problem. For both η values, the robust frontiers of type II also do not cover the entire range of the original Pareto-optimal front. However, as η is increased the robust frontier comes closer to the original front. Figure 24 compares the $g(\mathbf{x}^*)$ values for all robust solutions of type I (theoretical) and type II ($\eta = 0.4$ and $\eta = 0.6$). The theoretical type I robust solutions correspond to $f_1 \leq 0.4$ and the corresponding $g()$ value for all solutions is zero. However, for the robust solutions of type II, we observe that the $g()$ values are nonzero in the most sensitive region. The NSGA-II procedure finds solutions which were non-optimal before but are robust with respect to the chosen η parameter.

5 Conclusions

This paper takes the first step towards defining robust multi-objective solutions. First, a straightforward extension of a mean effective objective approach suggested for single-objective optimization is defined for multiple objectives. In this approach (we redefined it as a robust optimization of type I), an EMO methodology can be applied to the mean effective objective values obtained by averaging a finite set of neighboring solutions. Second, we have suggested robust optimization of type II, in which the original objectives are optimized, but an additional constraint restricting a pre-defined limiting change in objective values is considered. We have argued that such a procedure is more practical, as it allows a user to find robust solutions with a user-defined limit to the extent of change in objective values with respect to local perturbations.

Additionally, we have identified four different scenarios which can happen to a robust frontier in real-world problems and suggested variable-wise scalable two-objective test problems. Simulation results of NSGA-II on these test problems have been illustrated and explained to understand the differences between two robust optimization procedures.

A number of salient issues remains. In this research, we have considered $H = 50$ neighboring solutions to compute the mean effective objectives. Thus, in principle, this method is 50 times computationally more expensive than the regular non-robust optimization methods. This issue needs an immediate attention before such a method becomes really practical. We are currently pursuing the use of an updatable archive to store a large number of previously-computed solutions as a reservoir for neighboring solutions. Such a technique has been successfully tried for single-objective robust optimization [8], but, new insertion and deletion rules honoring the two distinct goals of multi-objective optimization – convergence and distribution – may have to be considered. Other efficient statistical techniques involving evaluations of a fewer number of neighboring solutions may also be tried. It is also important to extend the study for handling active constraints while finding robust Pareto-optimal solutions. Nevertheless, this initial study should motivate more detailed studies in the future and may encourage interested readers to understand and apply robust optimization procedures to real-world multi-objective optimization problems.

Acknowledgements

The first author wishes to acknowledge and thank Juergen Branke, Hartmut Schmeck, and Marco Farina, for some initial discussions.

References

1. Jurgen Branke. Creating robust solutions by means of an evolutionary algorithm. *Parallel Problem Solving from Nature*, pages 119–128, 1998.
2. Jurgen Branke and C. Schmidt. Faster convergence by means of fitness estimation. In *Soft Computing*, 2000.
3. Yaochu Jin and Bernhard Sendhoff. Trade-off between performance and robustness: An evolutionary multiobjective approach. In *EMO2003*, pages 237–251, 2003.
4. S. Tsutsui and A. Ghosh. Genetic algorithms with a robust solution searching scheme. *IEEE transactions on Evolutionary Computation*, pages 201–219, 1997.
5. I.C. Parmee. The maintenance of search diversity for effective design space decomposition using cluster-oriented genetic algorithms(cogas) and multi-agent strategies(gaant). In *ACEDC*, 1996.
6. J. Teich. Pareto-front exploration with uncertain objectives. In *Proceedings of the First International Conference on Evolutionary Multi-Criterion Optimization (EMO-01)*, pages 314–328, 2001.
7. E. J. Hughes. Evolutionary multi-objective ranking with uncertainty and noise. In *Proceedings of the First International Conference on Evolutionary Multi-Criterion Optimization (EMO-01)*, pages 329–343, 2001.
8. Jurgen Branke. Efficient evolutionary algorithms for searching robust solutions. *ACDM*, pages 275–286, 2000.
9. K. Deb. *Multi-objective optimization using evolutionary algorithms*. Chichester, UK: Wiley, 2001.
10. K. Deb, S. Agrawal, A. Pratap, and T. Meyarivan. A fast and elitist multi-objective genetic algorithm: NSGA-II. *IEEE Transactions on Evolutionary Computation*, 6(2):182–197, 2002.

Multi-objective MaxiMin Sorting Scheme

E. J. Solteiro Pires[1], P. B. de Moura Oliveira[2], and J. A. Tenreiro Machado[3]

[1] Universidade de Trás-os-Montes e Alto Douro, Dep. de Engenharia Electrotécnica,
Quinta de Prados, 5000–911 Vila Real, Portugal
http://www.utad.pt/~epires

[2] Universidade de Trás-os-Montes e Alto Douro, CETAV,
Quinta de Prados, 5000–911 Vila Real, Portugal
{epires,oliveira}@utad.pt
http://www.utad.pt/~oliveira

[3] Instituto Superior de Engenharia do Porto, Dep. de Engenharia Electrotécnica,
Rua Dr. António Bernadino de Almeida, 4200-072 Porto, Portugal
jtm@dee.isep.ipp.pt
http://www.dee.isep.ipp.pt/~jtm

Abstract. Obtaining a well distributed non-dominated Pareto front is one of the key issues in multi-objective optimization algorithms. This paper proposes a new variant for the elitist selection operator to the NSGA-II algorithm, which promotes well distributed non-dominated fronts. The basic idea is to replace the crowding distance method by a maximin technique. The proposed technique is deployed in well known test functions and compared with the crowding distance method used in the NSGA-II algorithm. This comparison is performed in terms of achieved front solutions distribution by using distance performance indices.

1 Introduction

Multi-objective techniques using genetic algorithms (GAs) have been increasing in relevance as a research area. In 1989, Goldberg [1] suggested the use of a GA to solve multi-objective problems and since then other investigators have been developing new methods, such as multi-objective genetic algorithm (MOGA) [2], non-dominated sorted genetic algorithm (NSGA) [3] and niched Pareto genetic algorithm (NPGA) [4], among many other variants [5].

Achieving a well-spread and well-diverse Pareto solution front can be a time consuming computational problem, associated with multi-objective evolutionary algorithms (MOEAs). A good background review about the use of bounded archive population in MOEAs can be found in [6]. The computational complexity is directly related with the level of diversity and distribution the MOEAs aims to obtain. The higher this level, the larger computational power will be required. Indeed, as it was stated in [7] "For example, NSGA-II uses a crowding approach which has a computational complexity of $O(N \log N)$, where N is the population size. On the other hand, SPEA uses a clustering approach which has computational complexity of $O(N^3)$". Also it was found that while for two objectives problems the difference in terms of the achieved solution diversity with NSGA-II and SPEA is not significant, for three objectives problem the SPEA proved to be clearly better, but at the expensive of a higher computational load.

Thus, new computational schemes which can find good distributed Pareto fronts with reasonable computational effort are a highly desired feature.

Maximin is a well known method used in classic multi-attribute problems [8] and in the game theory [9, 10]. Recently Balling [11] proposed a multi-objective optimization technique based on a fitness function derived from using the maximin strategy [9] and Li [12] used the maximin fitness in a particle swarm multi-objective optimizer.

Bearing these ideas in mind, this paper, proposes a sorting scheme to select the best solutions in order to promote its diversity within MOEAs. In each generation, the achieved set is initially formed by the best solutions for each objective. Then, the achieved population is completed, one solution at a time, by the maximum of the minimal norm between a solution and the set of solutions already selected.

The article is organized as follows: section 2 describes the proposed method. Section 3 presents the MOEA settings, test functions and performance indices used for performance comparison. Section 4 shows the results and analysis of experiments carried out with maximin sorting scheme. Finally, section 5 outlines the main conclusions.

2 MaxiMin Sorting Scheme

This section presents the maximin sorting algorithm to render the following generation, having a good solution distribution.

The problem addressed by the proposed sorting scheme (maximin) is the selection of the best distributed N solutions from an original population with size M ($M > N$). As it is well known, this is a very useful feature in elitism MOEAs when it is necessary to choose the solutions which passes to the following generation. Similarly to the NSGA-II, the proposed algorithm is based on non-dominated fronts [13]. However, when the last allowed front is being considered, and there are more solutions in the last front than the remaining slots in the population, a maximin function is called to select the last solutions, in spite of using the crowding distance.

The main idea behind the maximin sorting scheme is to select the solutions in order to decrease the large gap areas existing in the already selected population. For example, let us consider the non-dominated solutions in figure 1. Initially the extreme solutions are selected $S \equiv \{a, b\}$. Then, solution c is selected because it has the greater distance to the set S. After that, solutions d and e are selected into the set $S \equiv \{a, b, c\}$, for the same reason. The process is repeated until the S set is completed.

The maximin sorting scheme is depicted in algorithm 1, assuming a minimization problem. Table 1 presents the notation used in algorithm 1. In each generation the new population is merged with the archive population, resulting in set T, from with the new archive is obtained by applying the proposed algorithm (lines 0–1). After that, the algorithm selects the best solutions for each objective (lines 2-4) into the S set. Then the non-dominated front is removed from the non-selected population T into S set until the last allowed front being considered does not fit into the S set (lines 5-9). Therefore, the square of the distance, c_i (1a), between each non-dominated solution and the set of solutions already selected, S, is evaluated, selecting the solution whose distance to the set is greater (1b). Every time a solution enter to the set S the cost c_i, of non-dominated solutions, is reevaluated (lines 10-19). This process ends when the S set is completed.

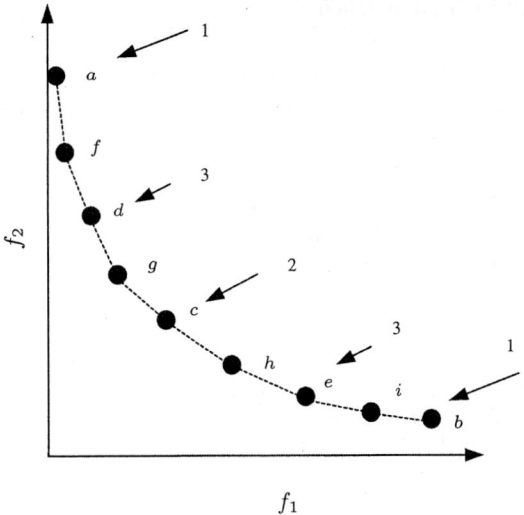

Fig. 1. Non-dominated front solutions

Table 1. Description of variables and functions used in maximin algorithm

	Description
D	Offspring population
A, T	Auxiliary populations
S	Archive or Parent population
k	Actual maximum
a_{c_i}	Minimum euclidian norm between solution i and set S
getMin(X,i)	Remove from set X the solution whose objective i is minimal
getMaxCi(X)	remove the solution whose norm value c_i is maximum
getND(X)	Remove all non-dominated solutions from set X

$$c_{a_j} = \min_{s_i \in S} \| f_{a_j} - f_{s_i} \| \tag{1a}$$

$$S = S \cup \max_{a_j \in A} a_j \tag{1b}$$

We verify that the different bulk of this algorithm, in relation to the crowding method (lines 10–19), requires at most $O(N^2)$ computations, which is worst than the $O(N \log N)$ computations required by the crowding distance scheme.

3 Test Problems and Performance Indices

In this section the maximin sorting scheme performance is compared to the crowding distance performance used in the NSGA-II. In order to study the diversity obtained by the two algorithms, they will be studied using some functions and performance indices.

Algorithm 1 MaxiMin Sorting Scheme

```
 0:  T = S ∪ D
 1:  S = ∅
 2:  for i = 1 to N_obj do
 3:      S = S∪ getMin(T, i)
 4:  end for
 5:  A = getND(T)
 6:  while ( #S + #A ≤ pop_size )
 7:      S = S ∪ A
 8:      A = getND(T)
 9:  end while
10:  for j = 1 to #A do
11:      c_{a_j} = min_{s_i ∈ S} {||f_{a_j} − f_{s_i}||}
12:  end for
13:  while ( #S < pop_size )
14:      k = getMaxCi(A)
15:      S = S ∪ k
16:      for l = 1 to #A do
17:          c_{a_l} = min{||f_{a_l} − f_k||, c_{a_l}}
18:      end for
19:  end while
```

3.1 Genetic Settings

To compare the results a gray binary MOEA is used, with 24 bits by parameter, crossover probability $p_c = 0.8$, mutation provability $p_m = 1/l$ (l is the string length), niching parameter $\sigma_{\text{share}} = 0.5$, dominance pressure of 100%, population size of 100 strings. Moreover, the algorithm is executed in 1000 generations per test.

3.2 Functions Used

The test bed is formed by a total of five functions. Three functions $\{F_1, F_2, F_3\}$, proposed by Zitzler *et al.* [14] as ZDT1 (2), ZDT2 (3) and ZDT3 (4), each with two objectives $\{f_1, f_2\}$. Two other functions $\{F_4, F_5\}$, used by Deb *et al.* in [15] as DTLZ2 (5) and DTLZ4, each with three objectives $\{f_1, f_2, f_3\}$.

$$F_1 = \begin{cases} f_1(X) = x_1 \\ g(X) = 1 + 9 \sum_{i=2}^{m} \frac{x_i}{m-1} \\ h(f_1, g) = 1 - \sqrt{\frac{f_1}{g}} \\ f_2(X) = g(X) h(f_1, g) \end{cases} \quad (2)$$

$$F_2 = \begin{cases} f_1(X) = x_1 \\ g(X) = 1 + 9 \sum_{i=2}^{m} \frac{x_i}{m-1} \\ h(f_1, g) = 1 - \left(\frac{f_1}{g}\right)^2 \\ f_2(X) = g(X) h(f_1, g) \end{cases} \quad (3)$$

$$F_3 = \begin{cases} f_1(X) = x_1 \\ g(X) = 1 + 9 \sum_{i=2}^{m} \frac{x_i}{m-1} \\ h(f_1, g) = 1 - \sqrt{\frac{f_1}{g}} - \frac{f_1}{g} \sin(10\pi f_1) \\ f_2(X) = g(X) h(f_1, g) \end{cases} \quad (4)$$

$$F_4 = \begin{cases} f_1(X) = [1 + g(X)] \cos(x_1 \pi/2) \cos(x_2 \pi/2) \\ f_2(X) = [1 + g(X)] \cos(x_1 \pi/2) \sin(x_2 \pi/2) \\ f_3(X) = [1 + g(X)] \sin(x_1 \pi/2) \\ g(X) = 1 + 9 \sum_{i=3}^{m} (x_i - 0.5)^2 \end{cases} \quad (5)$$

Function F_5 is similar to F_4 but it has a meta-variable mapping $x_i \to x_i^\alpha$ with $\alpha = 100$. The vector X is formed by m parameters, $m_j = \{30, 30, 30, 12, 12\}, j = 1, \ldots, 5$, for $F_j = \{F_1, F_2, F_3, F_4, F_5\}$, with each value $x_i \in [0, 1], i = 1, \ldots, m_j$.

3.3 Performance Indices

To study the solution diversity the following indices are used: the spacing index (SP) [16] (6), the distance-based distribution index (Δ') [17] (7) and the minimal distance graph (MDG) (8). In all indices $\bar{d}$ is the average of d_i distances.

$$SP(S) = \sqrt{\frac{1}{\#S - 1} \sum_{i=1}^{\#S} (d_i - \bar{d})^2} \quad (6a)$$

$$d_i = \min_{s_k \in S \wedge s_k \neq s_i} \sum_{m=1}^{M} |f_m(s_i) - f_m(s_k)| \quad (6b)$$

The distance d_i of Δ' index is evaluated by euclidian distance between consecutive solutions in S. Therefore, this index can only be used in two objective problems.

$$\Delta'(S) = \sum_{i=1}^{\#S-1} \frac{|d_i - \bar{d}|}{\#S - 1} \quad (7)$$

The MDG performance index is calculated based on the minimal distances, d_i, that links all the non-dominated solutions of the population. For example, considering the solutions $\{s_1 = (3, 5, 7), s_2 = (2, 6, 5), s_3 = (7, 7, 2), s_4 = (5, 4, 8), s_5 = (4, 5, 6), s_6 = (1, 6, 5)\}$ the minimal distances that link all the solutions are: $\{\overline{s_1 s_2}, \overline{s_1 s_4}, \overline{s_1 s_5}, \overline{s_2 s_6}, \overline{s_3 s_5}\}$, as it is illustrated in figure 2(a). The MDG index has some advantage over the SP index because it never uses the same distance more than once and relates all solutions together (see figure 2). Moreover, the MDG index can be used in any dimensional space in spite of the Δ' index.

$$MDG(S) = \sqrt{\frac{1}{\#S - 2} \sum_{i=1}^{\#S-1} (d_i - \bar{d})^2} \quad (8)$$

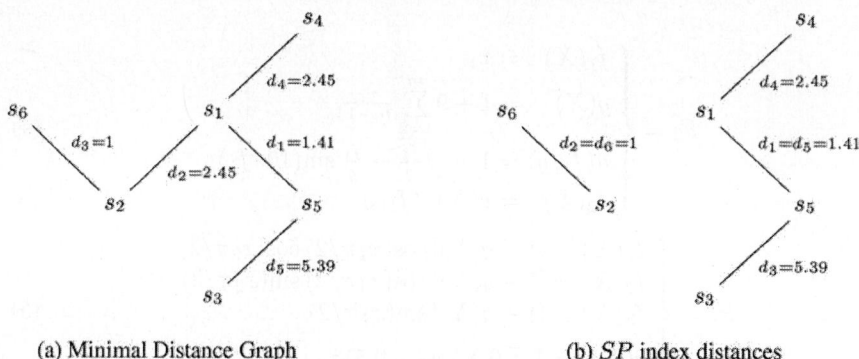

(a) Minimal Distance Graph (b) SP index distances

Fig. 2. Distances used by MDG and SP indices

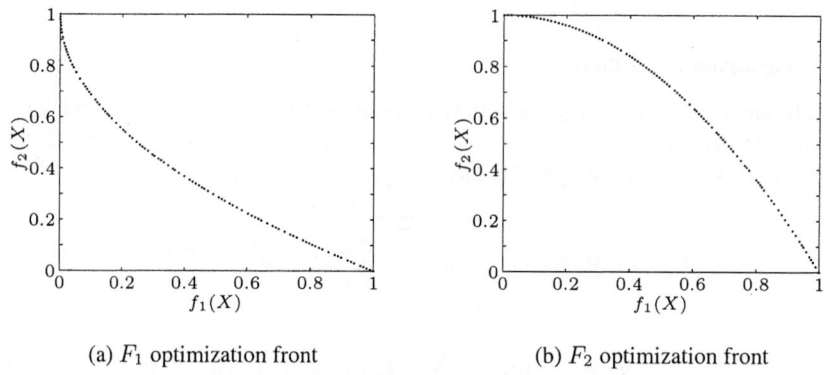

(a) F_1 optimization front (b) F_2 optimization front

Fig. 3. F_1 and F_2 optimization with maximin sorting scheme

4 Simulation Results

To compare the maximin sorting scheme and the crowding distance technique several simulations were conducted involving $n = 101$ runs each. For the results analysis are considered the median, average, standard deviation, and the minimum and the maximal solutions obtained.

Table 2. F_1 test results

	Maximin sorting scheme			Crowding distance		
	SP	Δ'	MDG	SP	Δ'	MDG
Median	**0.005**	**0.067**	**0.005**	0.008	0.075	0.007
Average	**0.005**	**0.067**	**0.005**	0.008	0.075	0.007
Std dev	0.000	0.001	0.000	0.001	0.003	0.001
Min	**0.004**	**0.064**	0.005	0.006	0.068	0.006
Max	0.006	0.069	0.006	0.009	0.083	0.009

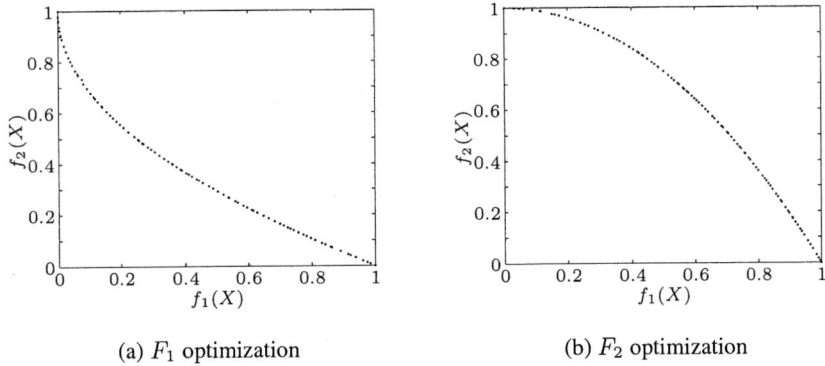

(a) F_1 optimization (b) F_2 optimization

Fig. 4. F_1 and F_2 optimization with crowding distance

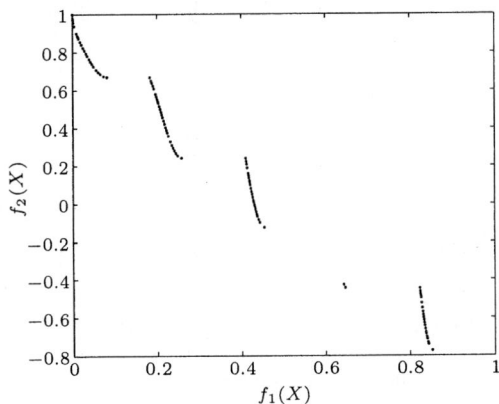

Fig. 5. A F_3 optimization front with maximin sorting scheme

Table 3. F_2 test results

	Maximin sorting scheme			Crowding distance		
	SP	Δ'	MDG	SP	Δ'	MDG
Median	0.005	0.067	0.005	0.008	0.075	0.007
Average	0.005	0.067	0.005	0.008	0.075	0.007
Std dev	0.000	0.001	0.000	0.001	0.002	0.001
Min	0.004	0.064	0.005	0.006	0.066	0.006
Max	0.006	0.070	0.006	0.010	0.086	0.009

For the F_1 and F_2 optimizations both schemes obtain good diversity (figures 3 and 4). Nevertheless, the maximin sorting scheme achieved better performance indices (tables 2 and 3), reaching better results, even though, the worst results obtained with the maximin sorting scheme are in the same range of the best results with the crowding distance. Due to the proximity of achieved indices values, it is difficult to conclude about the

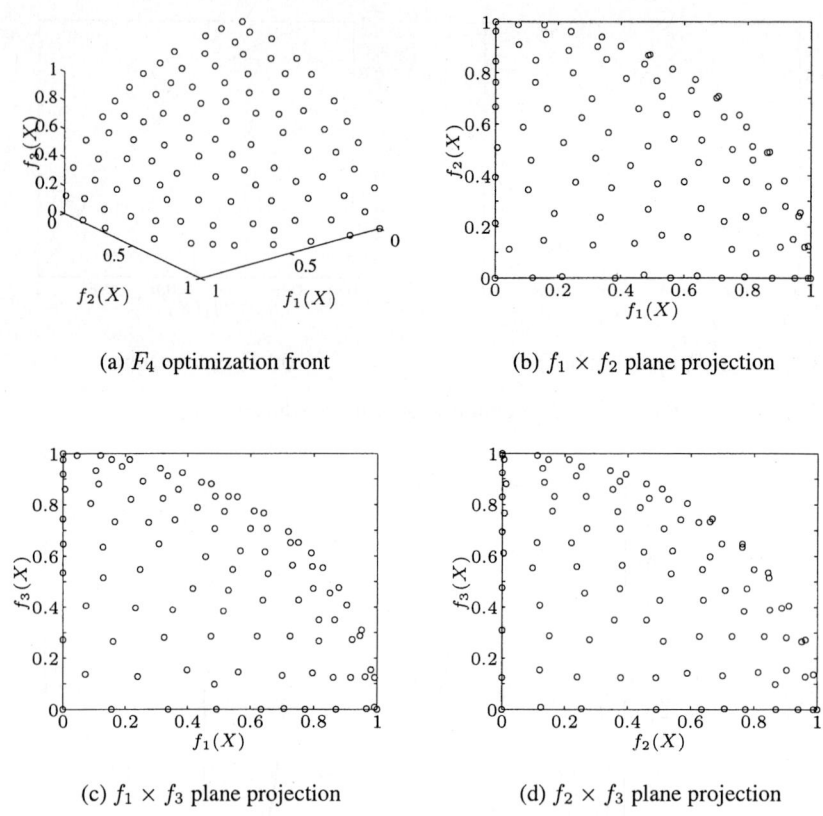

Fig. 6. F_4 optimization with maximin sorting scheme

Table 4. F_3 test results

	Maximin sorting scheme			Crowding distance		
	SP	Δ'	MDG	SP	Δ'	MDG
Median	0.014	0.122	0.039	**0.008**	0.121	**0.034**
Average	0.015	**0.121**	0.037	**0.011**	0.121	0.035
Std dev	0.011	0.006	0.006	0.007	0.005	0.006
Min	**0.004**	**0.109**	**0.027**	0.006	0.111	**0.027**
Max	0.052	0.138	0.049	0.052	0.138	0.054

superiority of any algorithm. Thus, this results were subjected to the Mann-Whitney test which shows that the proposed method is significantly better for $p > 0.05$.

For F_3 the crowding distance method presents better performance indices (see table 4). However, the maximin sorting scheme reaches the best results in terms of the achieved distribution. The reason for this is due to the maximin algorithm has difficulty in converging for some tests to the entire non-dominate fronts (see figure 5).

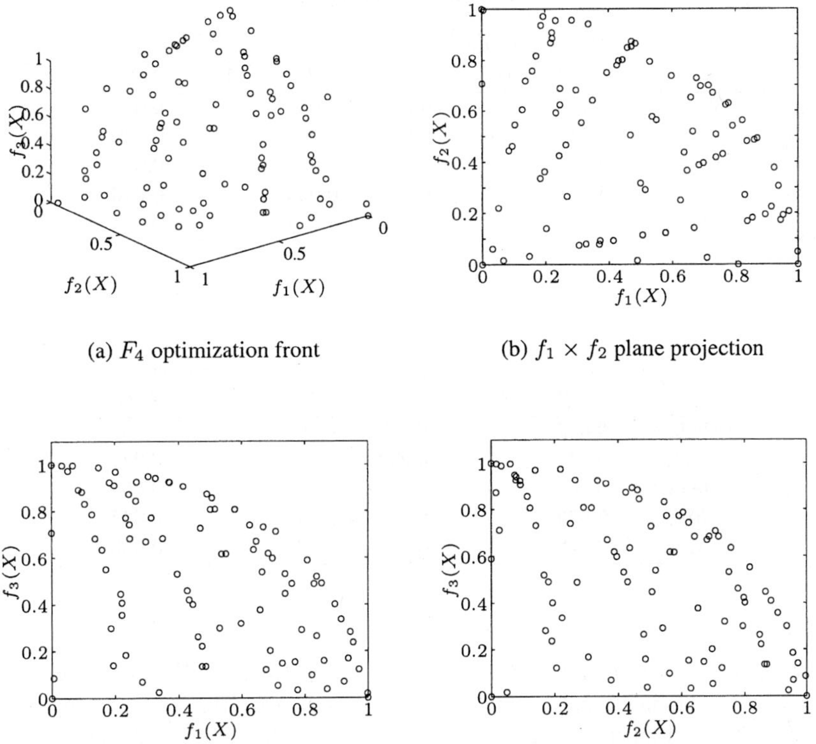

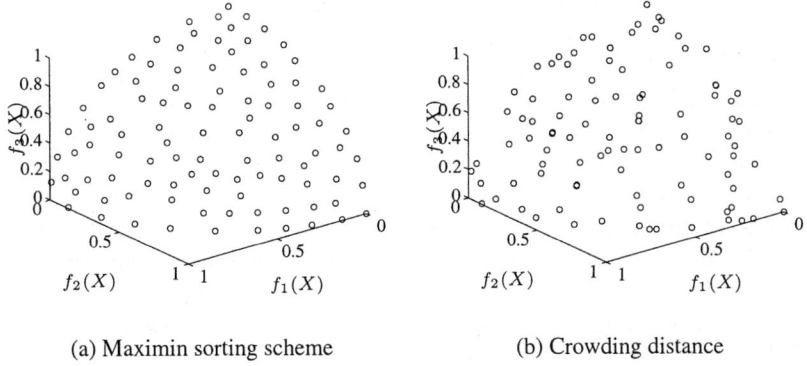

(a) F_4 optimization front

(b) $f_1 \times f_2$ plane projection

(c) $f_1 \times f_3$ plane projection

(d) $f_2 \times f_3$ plane projection

Fig. 7. F_4 optimization with crowding distance

(a) Maximin sorting scheme

(b) Crowding distance

Fig. 8. F_5 optimization

Table 5. F_4 test results

	Maximin sorting scheme			Crowding distance		
	SP	Δ'	MDG	SP	Δ'	MDG
Median	**0.028**	–	**0.017**	0.060	–	0.048
Average	**0.028**	–	**0.017**	0.060	–	0.047
Std dev	0.003	–	0.002	0.005	–	0.003
Min	**0.021**	–	**0.013**	0.044	–	0.040
Max	0.038	–	0.025	0.073	–	0.055

Table 6. F_5 test results

	Maximin sorting scheme			Crowding distance		
	SP	Δ'	MDG	SP	Δ'	MDG
Median	**0.029**	–	**0.019**	0.060	–	0.048
Average	**0.029**	–	**0.020**	0.061	–	0.049
Std dev	0.004	–	0.004	0.006	–	0.003
Min	**0.020**	–	**0.015**	0.049	–	0.043
Max	0.048	–	0.043	0.078	–	0.060

For F_4 and F_5 the maximin scheme leads to much better results, as can be seen by figures 6 to 8 and tables 5 and 6. In these simulations the worst test case obtained by the maximin scheme is superior to the best case obtained by the crowding distance method.

5 Conclusions and Further Work

A maximin sorting algorithm to select non-dominated solutions which complete the new population was proposed. The new algorithm was integrated in a MOEA in order to test its capacity to generate well distributed non-dominated Pareto fronts. This technique was deployed in well known test functions and the results compared with the ones obtained by using the NSGA-II crowding distance method. The analysis was made considering some distance performance indices and points out that the algorithm reaches a set of non-dominated solutions, along the Pareto front, with a good spread, outperforming the crowding method in most of the test functions, particularly when involving three objectives. The proposed algorithm is more demanding in terms of computational load. However, the improvement gained in terms of diversity compensates this cost.

In the nearby future the proposed method will be compared with other techniques such as the C-NSGAII, ϵ-MOEA, and SPEA. These tests will incorporate more performance criteria and access the execution computational time.

Acknowledgment

This work is partially supported by the grant Prodep III (2/5.3/2001) from FSE.

References

1. Goldberg, D.E.: Genetic Algorithms in Search, Optimization, and Machine Learning. Addison – Wesley (1989)
2. Fonseca, C.M., Fleming, P.J.: An overview of evolutionary algorithms in multi-objective optimization. Evolutionary Computation Journal **3** (1995) 1–16
3. Deb, K.: Multi-Objective Optimization using Evolutionary Algorithms. Wiley-Interscience Series in Systems and Optimization. (2001)
4. Horn, J., Nafploitis, N., Goldberg, D.: A niched pareto genetic algorithm for multi-objective optimization, Proceedings of the First IEEE Conference on Evolutionary Computation (1994) 82–87
5. Coello, C., Carlos, A.: A comprehensive survey of evolutionary-based multiobjective optimization techniques. Knowledge and Information Systems **1** (1999) 269–308
6. Knowles, J.D., Corne, D.W., Fleischer, M.: Bounded archiving using the lebesgue measure. In: CEC – Congress on Evolutionary Computation, Canberra, Australia (2003)
7. Kalyanmoy Deb, Manikanth Mohan, S.M.: Towards a quick computation of well-spread pareto-optimal solutions. In Fonseca, C.M., Fleming, P.J., Zitzler, E., Deb, K., Thiele, L., eds.: Evolutionary Multi-Criterion Optimization, Second International Conference, EMO 2003, Faro, Portugal, April 8-11, 2003, Proceedings. Volume 2632 of Lecture Notes in Computer Science., Springer (2003) 222–236
8. K. Paul Yoon, C.L.H.: Multiple Attribute Decision Making : An Introduction (Quantitative Applications in the Social Sciences). SAGE Publications (1995)
9. Luce, R.D., Raiffa, H.: Introduction and Critical Survey. John Wiley & Sons, Inc, New York (1957)
10. Rawls, J.: A Theory of Justice. Oxford University Press, London (1971)
11. Balling, R.: The maximin fitness function; multi-objective city and regional planning. In Fonseca, C.M., Fleming, P.J., Zitzler, E., Deb, K., Thiele, L., eds.: Evolutionary Multi-Criterion Optimization, Second International Conference, EMO 2003, Faro, Portugal, April 8-11, 2003, Proceedings. Volume 2632 of Lecture Notes in Computer Science., Springer (2003) 1–15
12. Li, X.: Better spread and convergence: Particle swarm multiobjective optimization using the maximin fitness function. In Deb, K.e.a., ed.: Proceeding of Genetic and Evolutionary Computation Conference 2004 (GECCO'04). Volume LNCS 3102 of Lecture Notes in Computer Science, Springer-Verlag., Seattle, USA (2004) 117–128
13. Deb, K.: Multi-Objective Optimization Using Evolutionary Algorithms. John Wiley & Sons, LTD (2001)
14. Zitzler, E., D.K., Thiele, L.: Comparison of multiobjective evolutionary algorithms: Empirical results. Evolutionary Computation **8** (2000) 173–195
15. Deb, K., Thiele, L., Laumanns, M., Zitzler, E.: Scalable multi-objective optimization test problems. In Fogel, D.B., El-Sharkawi, M.A., Yao, X., Greenwood, G., Iba, H., Marrow, P., Shackleton, M., eds.: Proceedings of the 2002 Congress on Evolutionary Computation CEC2002, IEEE Press (2002) 825–830
16. Schott, J.R.: Fault Tolerant Design Using Single and Multicriteria Genetic Algorithm Optimization. Master's thesis, Massachusetts Institute of Technology, Department of Aeronautics and Astronautics, Cambridge, Massachusetts (1995)
17. Deb, K., Agrawal, S., Pratap, A., Meyarivan, T.: A fast elitist non-dominated sorting genetic algorithm for multi-objective optimization: NSGA-II. In Schoenauer, M., Deb, K., Rudolph, G., Yao, X., Lutton, E., Merelo, J.J., Schwefel, H.P., eds.: Parallel Problem Solving from Nature – PPSN VI. Volume 1917 of LNCS., Berlin, Springer (2000) 849–858

Multiobjective Optimization on a Budget of 250 Evaluations

Joshua Knowles[1] and Evan J. Hughes[2]

[1] School of Chemistry, University of Manchester, Faraday Building, Sackville Street, PO Box 88, Manchester M60 1QD, UK
j.knowles@manchester.ac.uk
[2] Cranfield University, Shrivenham, Swindon SN6 8LA, UK

Abstract. In engineering and other 'real-world' applications, multiobjective optimization problems must frequently be tackled on a tight evaluation budget — tens or hundreds of function evaluations, rather than thousands. In this paper, we investigate two algorithms that use advanced initialization and search strategies to operate better under these conditions. The first algorithm, Bin_MSOPS, uses a binary search tree to divide up the decision space, and tries to sample from the largest empty regions near 'fit' solutions. The second algorithm, ParEGO, begins with solutions in a latin hypercube and updates a *Gaussian processes* surrogate model of the search landscape after every function evaluation, which it uses to estimate the solution of largest expected improvement. The two algorithms are tested using a benchmark suite of nine functions of two and three objectives — on a budget of only 250 function evaluations each, in total. Results indicate that the two algorithms search the space in very different ways and this can be used to understand performance differences. Both algorithms perform well but ParEGO comes out on top in seven of the nine test cases after 100 function evaluations, and on six after the first 250 evaluations.

Keywords: multiobjective optimization, expensive black-box functions, ParEGO, DACE, Bin_MSOPS, landscape approximation, response surfaces, test suites.

1 Introduction

The vast majority of research effort in developing modern multiobjective evolutionary algorithms (MOEAs) has concentrated on improving algorithm performance and efficiency on runs, typically, of ten thousand function evaluations or more. In this paper, we consider multiobjective problems where a 'budget' of at most 250 evaluations is imposed because of the expensive nature of evaluating candidate solutions. More specifically, we are interested in problems where most or all of the features described in Fig.1 are true.

Features 1–4 limit the numbers of function evaluations possible, while features 5–8 make it reasonable to apply global search techniques rather than either random search or hillclimbing. Problems exhibiting these features include various combinatorial biochemistry and materials science applications [6, 26], as well as instrument set-up optimization in analytical chemistry [20, 24]. In [20], a standard MOEA, PESA-II, was successfully used to substantially improve the settings of a GC-MS spectrometer, using

1. the time taken to perform one evaluation is of the order of minutes or hours,
2. only one evaluation can be performed at one time (no parallelism is possible),
3. the total number of evaluations to be performed is limited by financial considerations,
4. no realistic simulator or other method of approximating the full evaluation is readily available,
5. noise is low (repeated evaluations yield very similar results),
6. the overall gains in quality (or reductions in cost) that can be achieved are high,
7. the search landscape is multimodal but not highly rugged,
8. the dimensionality of the search space is low-to-medium,
9. the problem has multiple, possibly incommensurable, objectives.

Fig. 1. Features exhibited by problems of interest

just 180 evaluations. However, it is clear that given such a restricted number of evaluations, and no particular restriction on computational overhead (since each experiment requires 20 minutes), a search strategy that more carefully considers each evaluation would be more appropriate.

Scanning the optimization literature reveals that a sparse but varied array of different techniques (that were proposed or could be used) for economizing on evaluations in multiobjective optimization has already been examined. One strand in this focuses on the use of *neural networks* for modeling the search landscape during optimization, in order to replace some real function evaluations with approximated ones [19, 8, 9], or to replace standard variation operators with adaptive ones [1]. The simpler concept of *fitness inheritance* has also been investigated in multiobjective optimization to economize on function evaluations [2, 5]. And a third strand is to use Bayesian network and/or other probabilistic model-building algorithms in a multiobjective scenario, e.g. [17].

However, while the above methods may offer some performance gains over standard MOEAs when function evaluations are expensive, not one of the studies above has demonstrated a significant performance advantage within the challenging evaluation budget we are interested in here. In this paper, we present and compare two recently proposed algorithms that take very different approaches to this challenge. The first algorithm, Binary-MSOPS, which is summarized below, is based on two separate pieces of work previously published by the second author [11, 12]. The second algorithm, ParEGO, was first described in a recent technical report [16], and is described here again in some detail. We evaluate these algorithms over a range of problems and, unlike in other studies, we focus explicitly on the first 250 evaluations only.

The rest of the paper is organized as follows. Sections 2 and 3 describe the two algorithms, while 4, 5 and 6 detail the test functions, performance assessment methods and parameter settings of the algorithms, respectively. Section 7 presents results and section 8 discusses findings and concludes.

2 Binary-MSOPS

The Binary-MSOPS algorithm is based primarily on the Binary Search Algorithm [11], summarized below. This method, which can be combined with almost any fitness assignment scheme, attempts to improve decision space sampling to ensure that promising

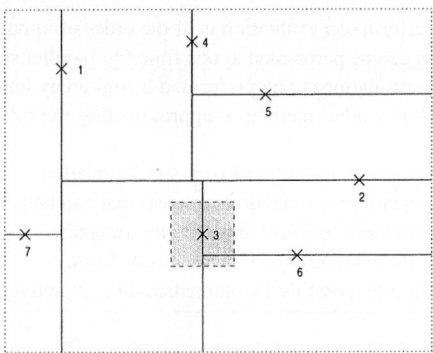

Fig. 2. The Binary Search process illustrated in a two dimensional decision space, with the first seven search points shown. The box around the third point indicates a small distance around a 'fit' point, and how this intersects with several empty regions

regions are not missed or over-sampled in the early stages of the search, and to explicitly balance exploitation and exploration. Combining this with the MSOPS ranking method [12] — a computationally efficient means of assigning fitness in multiobjective optimization, based on target vectors —, Binary-MSOPS is both efficient and frugal with evaluations.

2.1 Binary Search Algorithm

The overall strategy of Bin_MSOPS uses a binary search tree [11] to divide the decision space into empty regions, allowing the largest empty region to be approximated. The search tree is constructed as shown in Fig. 2 by generating a point at random within a chosen hypercube, then dividing the hypercube along the dimension that yields the most 'cube-like' subspaces.

The basic algorithm for constructing the binary search tree (and generating new solutions) works by repeatedly choosing an exploration or exploitation step:

Exploration: Next point is generated at random within the largest empty region (i.e. global search),
Exploitation: Next point is generated within the largest empty region that is within a small distance of a selected *good* point (i.e. local search),

where the choice is random but biased by a parameter specifying the exploration/exploitation ratio.

The identification of a local region for exploitation is illustrated in Fig. 2. A small offset distance d_p is used to generate a hypercube of interest about the chosen point (chosen with tournament selection). The small hypercube is placed around the point of interest simply to provide an efficient means of identifying large neighbouring regions. A new point is then generated at random using a normal distribution in the largest region that intersects the hypercube.

At each iteration, the tree search to find the largest empty region is at worst $O(d\,P)$, where P is the number of evaluation points so far and d is the number of dimensions. Tree pruning can lead to $O(d\,log(P))$ performance for exploitation, and at worst $O(d\,P)$. Thus a computational explosion is avoided.

2.2 Population Ranking by MSOPS

In order to decide which are 'good' points, the entire population is ranked, and a tournament between a random subset of the population, based on rank value, decides on the next solution to update. In order to control the computational complexity, a non-Pareto ranking approach has been applied, that like ParEGO (see next section), is also capable of handling many-objective problems. For this, the Multiple Single Objective Sampling (MSOPS) [12], with $O(P\,log(P))$ time complexity, has been used.

The concept of MSOPS is to generate a set of T target vectors, and evaluate the performance of every individual in the population, of size P, for every target vector, based on a conventional aggregation method. As aggregation methods (e.g. weighted min-max, ϵ-constraint, goal attainment etc.) are very simple to process, the calculation of each of the performance metrics is fast.

Thus each of the P members of the population has a set of T scores that indicate how well the population member satisfied the range of target conditions. The scores are held in a score matrix, S, which has dimensions $P \times T$. Each *column* of the matrix S corresponds to one target vector (each column containing P entries) and is ranked, with the best performing population member on the corresponding target vector being given a rank of 1, and the worst a rank of P. The rank values are stored in a matrix R. Each *row* of the rank matrix R may now be sorted, with the ranks for each population member placed in ascending order. The R matrix now holds in the first column the highest rank achieved for each population member across the set of target vectors. The second column will hold the second highest rank achieved etc. Thus the matrix R may be used to rank the population, with the most fit being the solution that achieved the most scores that were ranked 1, etc.

The flexibility of the approach is such that the target vectors can be arbitrary, either generated using some structure, or generated at random within certain limits. As the ranking method employed is based on the number of target vectors that are satisfied the best, a solution at the edge of the objective space will often satisfy vectors that cannot be attained. Thus the focus of the optimization is naturally drawn to interesting regions of surface such as the boundary of the optimization surface and discontinuities.

3 ParEGO: Landscape Modeling Using Gaussian Processes

Learning a cost landscape from a set of solution/cost pairs is variously called surrogate, approximate or meta- modeling in the literature [13]. In design engineering, meta-modeling is usually known as the response surface method [18], and involves fitting a low order polynomial via some form of least squares regression. A closely related approach, deriving from geology, is Kriging, whereby *Gaussian process* models are parameterized by maximum likelihood estimation. A particular example of this is known as the Design and Analysis of Computer Experiments (DACE) model [22], which forms the basis of the EGO search algorithm [14]. EGO has been designed specifically for optimization on a very restricted evaluation budget: e.g. in [14], four low-dimensional multimodal test functions are optimized to within 1% of optimal in the order of 100 function evaluations.

The EGO algorithm begins by first generating a number of solutions in a latin hypercube, and by then finding the maximum likelihood DACE model that best explains these solutions (making use of some suitable optimization algorithm). To generate a new solution to evaluate, EGO searches for the solution that maximizes what Jones et al [14] call "the expected improvement" — the expected value of that part of the standard error curve that lies below the best cost sampled so far. This effectively means that EGO weighs up both the predicted value of solutions, *and the error in this prediction*, in order to find the one that has the greatest *potential* to improve the minimum cost. EGO does *not* just choose the solution that the model predicts would minimize the cost. Rather, it *automatically* balances exploitation and exploration: where a solution has low predicted cost and low error, it may not be as desirable as a solution whose predicted cost is higher but whose associated error of prediction is also higher. Once a new solution has been chosen and evaluated (using the true, expensive cost function), the DACE model is updated with this new information, and the next solution is chosen using this updated model.

The EGO algorithm could be extended for use with multiobjective optimization problems in a number of different ways. One simple approach recently proposed by the first author in [16] (and that has the advantage of scaling to many objectives), converts the k different cost values of a solution into a single cost via a parameterized scalarizing weight vector. By choosing a different (parameterization of the) weight vector at each iteration of the search, an approximation to the whole Pareto front can be gradually built up. This multiobjective extension of EGO is called ParEGO.

ParEGO begins by normalizing the k cost functions with respect to the known (or estimated) limits of the cost space, so that each cost function lies in the range [0,1]. Then, at each iteration of the algorithm, a weight vector λ is drawn uniformly at random from the set of evenly distributed vectors defined by:

$$\Lambda = \left\{ \lambda = (\lambda_1, \lambda_2, \ldots, \lambda_k) \mid \sum_{j=1}^{k} \lambda_j = 1 \wedge \forall j, \lambda_j = l/s, l \in 0..s \right\}, \quad (1)$$

with $|\Lambda| = \binom{s+k-1}{k-1}$, so that the choice of s determines how many vectors there are in total [10]. The scalar cost of a solution $f_\lambda(x)$ is then computed using the augmented Tchebycheff function [23]:

$$f_\lambda(x) = \max_j (\lambda_j \cdot f_j(x)) + \rho \sum_j \lambda_j \cdot f_j(x), \quad j \in 1..k \quad (2)$$

where f_j is the raw cost value on objective j and ρ is a small positive value, which ensures all minima are proper Pareto optima, and which we set to 0.05. The scalar costs of all previously visited solutions are computed and, using all or a selection of these, a DACE model of the landscape is constructed by maximum likelihood. The solution that maximizes the expected improvement with respect to this DACE model is determined. This becomes the next point, and is evaluated on the real, expensive cost function, completing one iteration of ParEGO.

Algorithm 1 ParEGO pseudocode

```
 1: procedure PAREGO(f, d, k, s)
 2:     xpop[] ← LATINHYPERCUBE(d)           /* Initialize using procedure: line 15 */
 3:     for each i in 1 to 11d − 1 do
 4:         ypop[i] ← EVALUATE(xpop[i], f)    /* See line 36 */
 5:     end for
 6:     while not finished do
 7:         λ ← NEWLAMBDA(k, s)               /* See line 19 */
 8:         model ← DACE(xpop[], ypop[], λ)   /* See line 22 */
 9:         xnew ← EVOLALG(model, xpop[])     /* See line 28 */
10:         xpop[] ← xpop[] ∪ {xnew}
11:         ynew ← EVALUATE(xnew, f)
12:         ypop[] ← ypop[] ∪ {ynew}
13:     end while
14: end procedure

15: procedure LATINHYPERCUBE(d)
16:     divide each dimension of search space into 11d − 1 'rows' of equal width
17:     return 11d − 1 vectors x such that no two share the same 'row' in any dimension
18: end procedure

19: procedure NEWLAMBDA(k, s)
20:     return a k-dimensional scalarizing weight vector chosen uniformly at random from
            amongst all those defined by equation 1
21: end procedure

22: procedure DACE(xpop[], ypop[], λ)
23:     compute the scalar fitness f_λ of every cost vector in ypop[], using equation 2
24:     choose a subset of the population based on the computed scalar fitness values
25:     maximize the likelihood of the DACE model for the chosen population subset
26:     return the parameters of the maximum likelihood DACE model
27: end procedure

28: procedure EVOLALG(model, xpop[])
29:     initialize a temporary population of solution vectors, some as mutants of xpop[] and
            others purely randomly
30:     while set number of evaluations not exceeded do
31:         evaluate the expected improvement of solutions using the model
32:         select, recombine and mutate to form new population
33:     end while
34:     return best evolved solution
35: end procedure

36: procedure EVALUATE(x, f)
37:     call the expensive evaluation function f with the solution vector x
38:     return true cost vector y of solution x
39: end procedure
```

> **Population size:** 20 solutions
> **Population update:** steady state (one offspring produced per generation, from either a crossover or cloning event, followed by a mutation)
> **Generations/evaluations:** 10,000 evaluations
> **Reproductive selection:** binary tournament without replacement
> **Crossover:** simulated binary crossover [3] with probability 0.2, producing one offspring
> **Mutation:** decision value shifted by $\pm 1/100.\mu.\rho$, where μ is drawn uniformly at random from $(0.0001, 1)$, ρ is the range of the decision variable, and p_m, the per-gene mutation probability, is $1/d$.
> **Replacement:** offspring replaces (first) parent if it is better, else it is discarded
> **Initialization:** 5 solutions are mutants[a] of the 5 best solutions evaluated on the real fitness function under the prevailing λ vector; the remaining 15 solutions are generated in a latin hypercube in decision space
>
> ---
> [a] The mutation is carried out as described above except that mutants are checked to ensure they are different than parents.

Fig. 3. The EA used in ParEGO to search for the 'best' next solution

Pseudocode for the entire ParEGO algorithm is given in Algorithm 1. The Nelder and Meads downhill simplex algorithm is used (with 20 restarts) to maximize the likelihood of the DACE model (line 25 of Algorithm 1). The evolutionary algorithm used within ParEGO to search for the solution that maximizes the expected improvement (line 28) is implemented as detailed in Fig. 3.

In practice, on a very expensive cost function, all solutions previously evaluated should be used to update the DACE model, at every iteration. However, to save computational overhead in our experiments (because of the need to do 21 runs on a large number of functions to collect performance data), we used a simple, heuristic method of choosing a subset of the solutions evaluated to update the model, as follows: At each iteration: (i) if the iteration number $iter$ is less than 25, all $11d - 1 + iter$ solutions evaluated so far, are used to update the model; and (ii) if $iter \geq 25$ a subset of $11d - 1 + 25$ solutions is used, where the first half of them are the best j solutions under the prevailing scalarizing vector λ and the other half are selected at random without replacement. Further details of the parameter settings used in ParEGO are given in Section 6.

4 Test Function Suite

4.1 Notes on the Selection of Functions

A number of good attempts at designing test function suites and/or general schemes for test function generation have been proposed in the multiobjective optimization literature, of which those described in [4, 21, 25] are some of the best. We make a selection of nine test functions, borrowing from these, and adapting some of them slightly for our purposes. Overall, our suite contains functions from two to eight decision variables; functions with a very low density of solutions at the Pareto front; functions with locally optimal Pareto fronts; functions where the Pareto set follows a complicated curve in the decision space; functions where the Pareto front is disconnected in objective space;

> **KNO1 [16]** Features: Two decision variables; two objectives; Fifteen locally optimal Pareto fronts.
> **OKA1 [21]** Features: Two decision variables; two objectives; Pareto optima lie on curve; density of solutions low at PF.
> **OKA2 [21]** Features: Three decision variables; two objectives; Pareto optima lie on spiral-shaped curve; density of solutions very low at PF.
> **VLMOP2 [25]** Features: Two decision variables; two objectives; concave PF.
> **VLMOP3 [25]** Features: Two decision variables; three objectives; disconnected Pareto optimal set and PF is a curve 'following a convoluted path through objective space'.
> **DTLZ1a, adapted from [4]** Features: Six decision variables; two objectives; local optima on the way to the PF.
> **DTLZ2a and DTLZ4a, adapted from [4]** Features: Eight decision variables; three objectives; DTLZ4a biases the density distribution of solutions toward the $f_3 - f_1$ and $f_2 - f_1$ planes.
> **DTLZ7a, adapted from [4]** Features: Eight decision variables, three objectives; four disconnected regions in the Pareto front (in objective space).

Fig. 4. Summary of the nine test functions

and functions where the density of points parallel to the Pareto front is non-uniformly distributed. There is thus a good deal of variety in the difficulties that they pose. We have nonetheless been restrictive in some particular aspects: all functions are unconstrained and while difficult, are not overly high-dimensional (in decision space), and have a reasonable, rather than pathological degree of ruggedness. And, we have kept to functions of two and three objectives only. These restrictions accord with our description (in Section 1) of certain kinds of expensive engineering/scientific problem, where we hope to obtain good results in a very small number of function evaluations. We do not reproduce the equations of all functions here but they can be found in [16] and are summarized in Fig. 4.

5 Selected Performance Analysis Techniques

In accordance with the analyses presented in [27], we choose the hypervolume indicator to assess the approximation sets obtained by Bin_MSOPS and ParEGO. We supplement these tabulated values and significance levels with a visual representation based on *summary attainment surfaces*, for some of the 2-objective functions.

5.1 Hypervolume Indicator

The hypervolume indicator assesses the size (hypervolume or Lebesgue integral) of the region weakly dominated by an approximations set, thus larger values indicate better nondominated sets. It is "the only unary indicator we are aware of that is capable of detecting that A is not worse than B for all pairs $A \triangleright$ [better than] B" [27], where A and B are two approximation sets.

The weakly dominated region being measured must be bounded from above in some way, and for this some point b is chosen, which must be itself dominated by every point

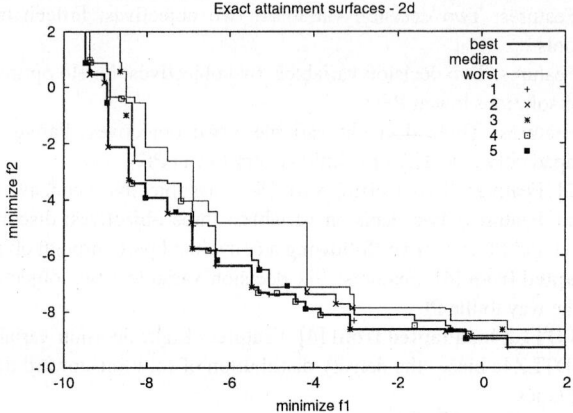

Fig. 5. Five sets of nondominated points and the best, median and worst attainment surfaces that they define. The interpretation of the median attainment surface is that, for every point on it (independently), a point (weakly) dominating this was obtained in at least 50% of the nondominated sets. Similarly, the worst attainment surface indicates the level achieved in 100% of the sets. The best attainment surface indicates the level achieved by the aggregation of all sets. *In this study, the best attainment surface is irrelevant and is* not *included in plotted results*

in the sample set. In order to choose a bounding point for application of the hypervolume indicator, we use the following method. First, the collection of nondominated point sets from all runs of both algorithms (on the relevant function) are aggregated into a single superset. Then, the ideal and the anti-ideal point of this superset are found. The bounding point is then the anti-ideal point shifted by δ times the range, in each objective:

$$\mathbf{b} = (b_1, b_2, \ldots, b_k), \quad \text{with}$$
$$b_j = \max_j + \delta(max_j - min_j), \quad j \in 1..k,$$

where max_j and min_j are the maximum and minimum value, respectively, on the jth objective, found within the superset. We use $\delta = 0.01$ here.

For the analysis of multiple runs, we compute the hypervolume indicator of each individual run, and report the mean and the standard deviation of these. Since the distribution of Bin_MSOPS' and ParEGO's results are not necessarily normal, we use the Mann-Whitney rank-sum test to indicate if there is a statistically significant difference in the position of the two distributions.

5.2 Median and Worst Summary Attainment Surface Plots

A *summary attainment surface* is a visual way of summarizing a number of runs of a multiobjective optimizer, based on the notion of an attainment surface [7]. For illustration, we plot five sets of nondominated points and their exact sample median, best and worst, summary attainment surfaces in Fig. 5.

For the two-objective problems in this paper, we give the median and worst attainment surfaces *only* of ParEGO and of Bin_MSOPS on the same plot, with ParEGO's two

Table 1. ParEGO parameter settings, where d is the number of decision variables

Parameter	setting
Initial population in latin hypercube	$11d - 1$
Total maximum evaluations	250
Number of scalarizing vectors	11 (for 2 objectives), 15 (for 3 objectives)
Scalarizing function	augmented Tchebycheff
Internal GA evals per iteration	200 000
Crossover probability	0.2
Real-value mutation probability	$1/d$
Real-value SBX parameter	10
Real-value mutation parameter	50

surfaces shown in solid and Bin_MSOPS's two surfaces shown with dashed lines. We do not give plots for the three-objective problems here, because of space restrictions.

6 Experimental Details

To evaluate Bin_MSOPS and ParEGO on the test suite, each algorithm is run 21 times, and all solutions visited are stored. The nondominated sets achieved after a particular number of function evaluations can then be determined and used to estimate performance.

6.1 Bin_MSOPS Parameter Settings

Weighted Min-Max was used as the aggregation method within the MSOPS ranking algorithm. The weighted min-max score s of k objectives is calculated using (3), where λ_i is the weight for the ith objective value, f_i.

$$s = \max_{i=1}^{k}(\lambda_i f_i), \tag{3}$$

A set of objective weights constitutes a single target vector.

Thirty target vectors were used, spaced so that the angle to their nearest neighbour was constant across the set of 30. Thus in trials with 3 objectives, the set of weight vectors, although evenly spaced, was non-unique.

To choose a 'good' point, a tournament size with a maximum of 20, without replacement, was used throughout all the experiments.

A search interval (see figure 2) of $d_p = 0.02$ was used and a lower limit was set on the coverage area of allowable cells of 0.005^2 (in normalized decision space with all variables in the range [0,1]). If the cells near to the chosen 'good' point were below the cell area limit, a search was performed to find the nearest cell that is large enough to split. The lower limit promotes a wider search around interesting points, but does prevent a tight-formation search occurring, which may be detrimental in problems with a very low density at the Pareto set (e.g. test function OKA1).

Initially, the algorithm performed a global exploration search for approximately the first 20 points, then global exploration was performed 6% of the time.

6.2 ParEGO Parameter Settings

The full set of parameter settings used in *all* runs of ParEGO are given in Table 1. These were determined empirically from a few exploratory trials.

7 Results

Tables 2 and 3 present the results of applying the hypervolume indicator to the 21 runs of Bin_MSOPS and ParEGO after, respectively, 100 and 250 function evaluations. Because it is not possible to use these values in comparisons with other algorithms, we make available the raw results at [15]. Note also that the appearance of the hypervolume

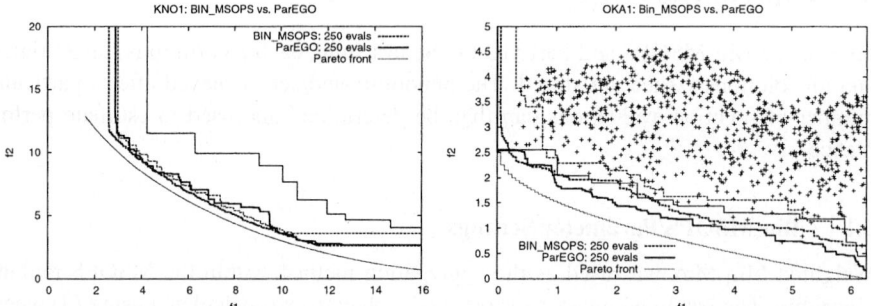

Fig. 6. Attainment surface plot on the KNO1 function after 250 function evaluations (left), and attainment surface plot with PF and 1000 random search points also shown on the OKA1 function after 250 function evaluations (right)

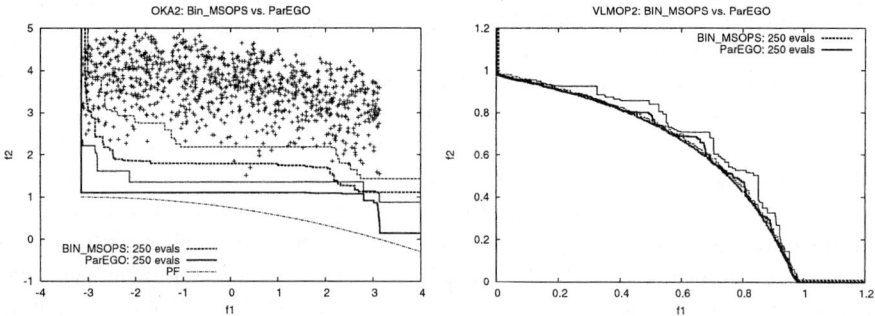

Fig. 7. Attainment surface plot on OKA2 after 250 function evaluations, with the true PF and 1000 random search points also shown (left), and attainment surface plot on VLMOP2 after 250 function evaluations (right)

Table 2. Mean and SD values of the hypervolume indicator after 100 evaluations of Bin_MSOPS / ParEGO from 21 runs of each. Larger values indicate better performance. The distributions of the values are tested using the Mann-Whitney rank sum test. The z values and significance level are indicated. ParEGO is significantly better than Bin_MSOPS unless stated

Function	Bin_MSOPS mean (SD)	ParEGO mean (SD)	z-value	significance	
OKA1	14.8496 (0.571733)	17.9532 (0.372966)	-5.546841	> 99%	
OKA2	13.6773 (1.53528)	19.7183 (0.887686)	-5.546841	> 99%	
KNO1	86.9879 (5.9303)	78.4856 (7.24012)	-3.735627	> 99%	Bin_MSOPS wins
VLMOP2	0.317661 (0.00440428)	0.307027 (0.00564198)	-4.792169	> 99%	Bin_MSOPS wins
VLMOP3	7.22865 (0.681038)	7.61652 (0.157667)	-1.547078	> 90%	
DTLZ1a	185317 (2573.43)	189262 (209.001)	-5.521685	> 99%	
DTLZ2a	3.85168 (0.139089)	3.97869 (0.0890044)	-3.358291	> 99%	
DTLZ4a	0.533989 (0.0394657)	0.673329 (0.263526)	-2.477840	> 99%	
DTLZ7a	10.5544 (1.7988)	12.9893 (0.655795)	-4.490300	> 99%	

Table 3. Mean and SD values of the hypervolume indicator after 250 evaluations of Bin_MSOPS / ParEGO from 21 runs of each. Larger values indicate better performance. The distributions of the values are tested using the Mann-Whitney rank sum test. The z values and significance level are indicated. ParEGO is significantly better than Bin_MSOPSunless stated

Function	Bin_MSOPS mean (SD)	ParEGO mean (SD)	z-value	significance	
OKA1	14.8169 (0.446187)	15.9849 (0.372917)	-5.219816	> 99%	
OKA2	13.9093 (1.13887)	18.4416 (0.467239)	-5.546841	> 99%	
KNO1	94.972 (1.70367)	84.2247 (5.5432)	-5.521685	> 99%	Bin_MSOPS wins
VLMOP2	0.324363 (0.000664073)	0.310789 (0.00398408)	-5.546841	> 99%	Bin_MSOPS wins
VLMOP3	4.94459 (0.0599138)	4.75726 (0.113527)	-4.892791	> 99%	Bin_MSOPS wins
DTLZ1a	63980.6 (602.738)	64869.6 (6.98978)	-5.546841	> 99%	
DTLZ2a	2.79582 (0.0960357)	3.08818 (0.0276524)	-5.546841	> 99%	
DTLZ4a	0.111108 (0.124952)	1.67242 (0.478067)	-5.521685	> 99%	
DTLZ7a	10.6171 (1.26015)	18.1657 (0.431554)	-5.546841	> 99%	

decreasing from 100 to 250 evaluations on some problems is only due to the choice of a different bound point (see above).

Fig. 5 and 6 visualize the median and worst summary attainment surfaces of the 21 runs of both algorithms on selected 2-objective problems. Fig. 7 and 8 show the decision space points visited by the first run of the two algorithms on KNO1 and OKA1, respectively.

From these results a number of observations can be made:

- ParEGO is statistically significantly better than Bin_MSOPS on seven of the nine functions at 100 evaluations, and on six of the nine functions after 250 evaluations, under the hypervolume indicator. (cf. [16], where ParEGO was better than a standard setup of NSGA-II with population size 20 on all test functions under two different indicators).
- The standard deviations of the two algorithms are generally comparable, with a large difference evident on only one problem, DTLZ1a, at 250 evaluations (2 orders of magnitude less deviation for ParEGO). Results reported in [16] indicated

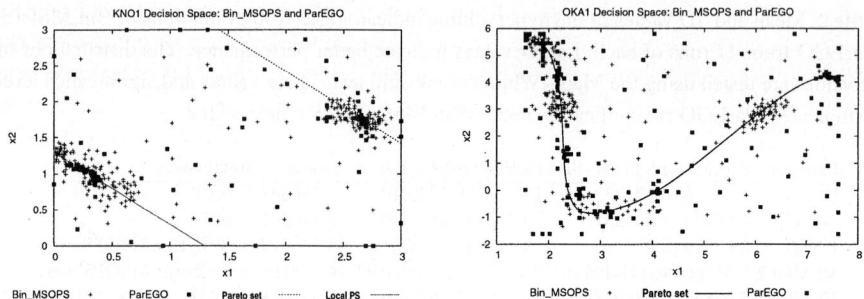

Fig. 8. Decision space points visited by Bin_MSOPS and ParEGO on KNO1 and OKA1. ParEGO fares worse on the former because it focuses too much on getting exactly on the PF instead of spreading out along it, while Bin_MSOPS extends further along it. But ParEGO's strategy does better than Bin_MSOPS's on OKA1 because points a small way off the Pareto set in decision space are far from the PF in objective space

that ParEGO's standard deviations were frequently one or two orders of magnitude lower than NSGA-II's on these problems, so the results here show that Bin_MSOPS is performing comparatively robustly — an important feature on problems where only one run may be possible.

- Fig. 5 and 6 indicate that ParEGO run for 250 evaluations is superior to a random search of 1000 evaluations on the difficult OKA1 and OKA2 functions, where there is a low density of solutions near the Pareto front. On OKA1, ParEGO's median attainment surface dominates all 1000 randomly generated points and on OKA2, even ParEGO's worst attainment surface does.
- Fig. 7 and 8 demonstrate that the search patterns of Bin_MSOPS and ParEGO are very different, and generally complementary. It is clear that while Bin_MSOPS attempts to get an even coverage of the search space, both globally and locally, ParEGO operates by combining a well-spread global search with full exploitation of local 'niches' (highly fit regions). This explains the far superior performance of ParEGO on the test functions with either low density at the Pareto front, local Pareto fronts or severe discontinuities. Fig. 7 (right) is a good example, however, of a smoother, more dense function where Bin_MSOPS has provided good even coverage, but ParEGO has focused too much on local niches. Fig. 8 really shows how ParEGO homes in on parts of the true Pareto set more aggressively, but sometimes fails to spread across it.

8 Summary and Conclusion

In many optimization scenarios, the number of fitness evaluations that can be performed is severely limited by cost or other constraints. In this study, the performance of two advanced multiobjective optimization algorithms, Bin_MSOPS and ParEGO, was measured on much shorter runs than used in most previous MOEA studies. A suite of nine difficult, but low-dimensional, multiobjective test functions of limited ruggedness were used to evaluate and compare the algorithms. The results of the comparison indicated that ParEGO's use of a surrogate model to establish both the expected multiobjective cost of candidate solutions *and* the uncertainty in these predictions, enables rapid advances in the early phase of the optimization process. The less directed search performed by Bin_MSOPS is good, but can fail to capitalize effectively on previously gathered information when the solution density at the Pareto front is low.

Overall, the experiments reported here can serve as a benchmark for other algorithms aimed at these type of expensive, low-dimensional multiobjective problems. To facilitate the comparison of Bin_MSOPS and ParEGO with other methods, raw results are available at [15]. Comparisons of ParEGO with NSGA-II can already be found in [16].

Future work will focus on three possible extensions. 1. an adaptive update of the scalarizing vectors to get a better distribution on the Pareto front; 2. constraint handling mechanisms; and 3. investigation of a hybrid between Bin_MSOPS and ParEGO that might offer a good combination of computational efficiency and high performance in short-to-medium run lengths (say up to 500 evaluations) for problems where evaluations are faster but still otherwise limited.

Acknowledgments. JK is supported by a David Phillips fellowship from the Biotechnology and Biological Sciences Research Council (BBSRC), UK. Thanks to J. Handl for proof-reading and ParEGO implementation advice.

References

1. D. Büche, G. Guidati, P. Stoll, and P. Kourmoursakos. Self-organizing maps for Pareto optimization of airfoils. In *Parallel Problem Solving from Nature—PPSN VII*, pages 122–131, September 2002. Springer-Verlag.
2. J.-J. Chen, D. E. Goldberg, S.-Y. Ho, and K. Sastry. Fitness inheritance in multi-objective optimization. In *Proceedings of the Genetic and Evolutionary Computation Conference (GECCO'2002)*, pages 319–326, July 2002. Morgan Kaufmann Publishers.
3. K. Deb and H. Beyer. Self-adaptive genetic algorithms with simulated binary crossover. *Evolutionary Computation*, 9(2):197–221, 2001.
4. K. Deb, L. Thiele, M. Laumanns, and E. Zitzler. Scalable test problems for evolutionary multi-objective optimization. Technical Report 112, Computer Engineering and Networks Laboratory (TIK), Swiss Federal Institute of Technology (ETH), Zurich, Switzerland, 2001.
5. E. I. Ducheyne, B. De Baets, and R. De Wulf. Is fitness inheritance useful for real-world applications? In *Evolutionary Multi-Criterion Optimization. Second International Conference, EMO 2003*, pages 31–42, April 2003. Springer.
6. J. R. G. Evans, M. J. Edirisinghe, and P. V. C. J. Eames. Combinatorial searches of inorganic materials using the inkjet printer: science philosophy and technology. *Journal of the European Ceramic Society*, 21:2291–2299, 2001.
7. C. M. Fonseca and P. J. Fleming. On the performance assessment and comparison of stochastic multiobjective optimizers. In *Parallel Problem Solving from Nature—PPSN IV*, pages 584–593, 1996. Springer-Verlag.
8. A. Gaspar-Cunha and A. Vieira. A multi-objective evolutionary algorithm using neural networks to approximate fitness evaluations. *International Journal of Computers, Systems, and Signals*, 2004 (in press).

9. A. Gaspar-Cunha and A. S. Vieira. A hybrid multi-objective evolutionary algorithm using an inverse neural network. Hybrid Metaheuristics (HM 2004) Workshop at ECAI 2004, pages 25-30, 2004. http://iridia.ulb.ac.be/ hm2004/proceedings/
10. M. P. Hansen and A. Jaszkiewicz. Evaluating the quality of approximations of the nondominated set. Technical Report IMM-REP-1998-7, Technical University of Denmark, 1998.
11. E. J. Hughes. Multi-objective binary search optimisation. In *Second International Conference on Evolutionary Multi-Criterion Optimisation, EMO'03*, pages 102–117, April 2003. Springer.
12. E. J. Hughes. Multiple single objective Pareto sampling. In *Congress on Evolutionary Computation 2003*, pages 2678–2684, December 2003. IEEE.
13. Y. Jin, M. Olhofer, and B. Sendhoff. A framework for evolutionary optimization with approximate fitness functions. *IEEE Transactions on Evolutionary Computation*, 6(5):481–494, 2002.
14. D. Jones, M. Schonlau, and W. Welch. Efficient global optimization of expensive black-box functions. *Journal of Global Optimization*, 13:455–492, 1998.
15. Knowles' webpage. http://dbk.ch.umist.ac.uk/knowles/
16. J. Knowles. ParEGO: A hybrid algorithm with on-line landscape approximation for expensive multiobjective optimization problems. Technical Report TR-COMPSYSBIO-2004-01, University of Manchester, UK, 2004. Available from http://dbk.ch.umist.ac.uk/knowles/pubs.html
17. M. Laumanns and J. Ocenasek. Bayesian optimization algorithms for multi-objective optimization. In *Parallel Problem Solving from Nature—PPSN VII*, pages 298–307, September 2002. Springer-Verlag.
18. R. Myers and D. Montgomery. *Response Surface Methodology*. Wiley, New York, 1995.
19. P. K. S. Nain and K. Deb. A computationally effective multi-objective search and optimization technique using coarse-to-fine grain modeling. Technical Report Kangal Report No. 2002005, IITK, Kanpur, India, 2002.
20. S. O'Hagan, W. Dunn, M. Brown, J. Knowles, and D. Kell. Closed-loop, multiobjective optimization of analytical instrumentation: gas chromatography/time-of-flight mass spectrometry of the metabolomes of human serum and of yeast fermentations. *Analytical Chemistry*, 2004 (in press). http://pubs.acs.org/cgi-bin/asap.cgi/ancham/asap/html/ac049146x.html
21. T. Okabe, Y. Jin, M. Olhofer, and B. Sendhoff. On test functions for evolutionary multi-objective optimization. In *Parallel Problem Solving from Nature VIII*, pages 792–802, September 2004, Springer.
22. J. Sacks, W. Welch, T. Mitchell, and H. Wynn. Design and analysis of computer experiments (with discussion). *Statistical Science*, 4:409–435, 1989.
23. R. E. Steuer and E.-U. Choo. An interactive weighted Tchebycheff procedure for multiple objective programming. *Mathematical Programming*, 25:326–344, 1983.
24. S. Vaidyanathan, D. I. Broadhurst, D. B. Kell, and R. Goodacre. Explanatory optimization of protein mass spectrometry via genetic search. *Analytical Chemistry*, 75(23):6679–6686, 2003.
25. D. A. V. Veldhuizen and G. B. Lamont. Multiobjective evolutionary algorithm test suites. In *Proceedings of the 1999 ACM Symposium on Applied Computing*, pages 351–357, 1999. ACM.
26. D. Weuster-Botz and C. Wandrey. Medium optimization by genetic algorithm for continuous production of formate dehydrogenase. *Process Biochemistry*, 30:563–571, 1995.
27. E. Zitzler, L. Thiele, M. Laumanns, C. M. Fonseca, and V. G. da Fonseca. Performance assessment of multiobjective optimizers: An analysis and review. *IEEE Transactions on Evolutionary Computation*, 7(2):117–132, April 2003.

Initial Population Construction for Convergence Improvement of MOEAs

Christian Haubelt, Jürgen Gamenik, and Jürgen Teich

Hardware-Software-Co-Design
Department of Computer Science 12
University of Erlangen-Nuremberg, Germany
{haubelt, teich}@cs.fau.de

Abstract. Nearly all Multi-Objective Evolutionary Algorithms (MOEA) rely on random generation of initial population. In large and complex search spaces, this random method often leads to an initial population composed of infeasible solutions only. Hence, the task of a MOEA is not only to converge towards the Pareto-optimal front but also to guide the search towards the feasible region. This paper proposes the incorporation of a novel method for constructing initial populations into existing MOEAs based on so-called Pareto-Front-Arithmetics (PFA). We will provide experimental results from the field of embedded system synthesis that show the effectiveness of our proposed methodology.

1 Introduction and Related Work

Many optimization techniques have been proposed in the literature to solve global optimization problems [1]. A special class of stochastic optimization methods that can be applied to Multi-objective Optimization (MOP) problems is called *Multi-Objective Evolutionary Algorithms (MOEA)* [2]. MOEAs are *iteratively improving* optimization techniques, i.e., starting from a set of initial solutions, the so-called *initial population*, a MOEA tries to improve this set of solutions. Due to complexity reasons, nearly all MOEAs use a simple random sampling from search space to construct the initial population. In the presence of search spaces containing only a few feasible solutions, these random sampling methods are expected to produce only infeasible solutions. Hence, it is the task of the MOEA to guide the search not only towards the *Pareto-optimal front* but also towards the *feasible region*. To find the feasible region can be as complicated as improving a feasible solution to find the Pareto-optimal front.

This paper proposes the incorporation of constructive methods into existing MOEAs to create the initial set of solutions. Therefore, a novel approach called *Pareto-Front-Arithmetics (PFA)* is proposed which allows a fast approximation of the Pareto-optimal front [3]. Although this method is expected to generate infeasible and suboptimal solutions, we will show by experiment that this approach is indeed useful when applied to the task of initial population construction. The key idea is as follows: First the MOP is separated into several subproblems, these subproblems are optimized independently using any standard optimization strategy. Afterwards, the results of the suboptimizations are combined in the fast PFA step. The obtained non-dominated solutions are used as initial solutions to the overall optimization problem. An advantage of the proposed methodology is that this technique can be integrated into any existing MOEA.

In [4], Gandibleux et al. compare population-based optimization runs which use different seeding solutions. Their basic idea is, that some solutions can be computed efficiently by constructive of heuristic methods. Their idea is is based on the fact that solutions of each combinatorial optimization problem are composed of so-called *supported efficient solutions*, i.e., solutions which can be computed by a weighted sum approach suggesting a convex Pareto front, and so-called *non-supported efficient solutions*. In their test cases, Gandibleux et al. use either constructed single-objective solutions, the supported efficient solutions, or an approximation of the supported efficient solutions for seeding the population-based optimization strategy. The same authors present in [5] a multi-objective optimization approach that incorporates knowledge of supported efficient solutions in the crossover operator. In their experiments it can be seen that using this information during crossover improves the convergence of the optimization of their particular problem enormously. Hence, the motivation is very similar to the one presented in this paper.

However, the proposed PFA methodology is more comparable to subdivision techniques. In subdivision approaches, the optimization complexity is reduced by separating the MOP into several subproblems. By solving these subproblems, the solutions to the original problem may be found. These techniques have some limitations regarding the optimization problem [6, 7, 8]. This will be discussed in detail in Section 3. However, these methods are proposed as stand alone approaches, whereas our idea is the use of subdivision techniques for the initialization.

The rest of the paper is organized as follows: Section 2 provides the necessary mathematical background and the problem formulation to this paper. In Section 3, a method called *Pareto-Front-Arithmetics*, for initial population construction is discussed in detail. Experimental results showing the effectiveness of our approach are presented in Section 4. In all test cases, the method using Pareto-Front-Arithmetics on average outperforms the random-based traditional method. Finally, Section 5 concludes the paper.

2 MOPs and MOEAs

This section will provide the formal background and the problem description this paper is dedicated to. We will start with a formal notation of multi-objective optimization problems.

Definition 1 (Multi-objective Optimization Problem). *A* multi-objective optimization problem *(*MOP*) is given by:*

minimize $f(x)$,
subject to $c(x) \leq 0$

where $x = (x_1, x_2, \ldots, x_m) \in X$ is the *decision vector and X is called the decision space*. Furthermore, the constraints $c(x) \leq 0$ determine the set of feasible solutions, where c is k-dimensional, i.e., $c(x) = (c_1(x), c_2(x), \ldots, c_k(x))$.

The *objective function* f is n-dimensional, i.e., n objectives are optimized simultaneously. Only those *decision vectors* $x \in X$ that satisfy all constraints c_i are in the set of feasible solutions, or for short in the *feasible set* called $X_f \subseteq X$. The *objective space* Y is the image of X under f and is defined as $Y = f(X) \subset \mathbb{R}^n$, where the objective function f on the set X is given by $f(X) = \{f(x) \mid x \in X\}$. Analogously, the *feasible region* of the objective space is denoted by $Y_f = f(X_f) = \{f(x) \mid x \in X_f\}$.

Without loss of generality, only minimization problems are considered. In contrast to single-objective optimization problems, a MOP may have not just one, but many optimal solutions. Due to the many, and often competing, objectives in a MOP, there are several tradeoff solutions which are optimal in a sense that there is no solution better in all objectives simultaneously. These optimal solutions are called *Pareto-optimal solutions*.

Definition 2 (Pareto dominance). *For any two decision vectors a and b,*

$a \succ b$ (*a dominates b*) *iff* $f(a) \leq f(b) \wedge f(a) \neq f(b)$
$a \succeq b$ (*a weakly dominates b*) *iff* $f(a) \leq f(b)$
$a \sim b$ (*a is incomparable to b*) *iff* $f(a) \not\leq f(b) \wedge f(a) \not\geq f(b)$.

where the relations $\circ \in \{=, \leq, <, \geq, >\}$ *are defined as:* $f(a) \circ f(b)$ *iff* $\forall j = 1, \ldots, n : f_j(a) \circ f_j(b)$.

In multi-objective optimization problems, there is generally not only one global optimum, but a set of so-called *Pareto-optimal solutions*. A Pareto-optimal solution x_p is a decision vector which is not worse than any other decision vector $\tilde{x} \in X$ in all objectives. The set of all Pareto-optimal solutions is called the *Pareto-optimal set*, or the *Pareto set* X_p for short.

Definition 3 (Pareto optimality). *A decision vector* $x \in X_f$ *is said to be non-dominated regarding a set* $A \subseteq X_f$ *iff*

$$\nexists a \in A : a \succ x.$$

A decision vector x is said to be Pareto-optimal iff x is non-dominated regarding $X_f \setminus \{x\}$. *The* Pareto-optimal front *is given by* $Y_p = f(X_p)$.

An approximation of the Pareto-set X_p will be termed *approximation set* X_a. All elements in X_a are incomparable to each other. In order to approximately solve a MOP, *Multi-Objective Evolutionary Algorithms (MOEAs)* are particularly well suited. This is because (i) they improve a set of solutions, (ii) they do not explicitly assume any properties of the objective functions, and (iii) they work well in large and non-convex search spaces [9].

Alg. 1 Optimization procedure

```
OPTIMIZE
IN: MOP multi-objective optimization problem
OUT: P' archive containing best solutions
BEGIN
    t ← 0
    P_t ← initialize(MOP)
    P'_t ← update(P_t)
    WHILE (!terminate(P'_t, t)) DO
        t = t + 1
        P_t ← generate(P_{t-1}, P'_{t-1})
        P'_t ← update(P_t, P'_{t-1})
    ENDWHILE
    RETURN P'_t
END
```

Alg. 1 outlines a generic optimization strategy as proposed by nearly all MOEA applications. In a first step, the population P of solutions is initialized. This is usually done using some random sampling from search space. Next, the archive P' is updated. Then, in a loop, new solutions are constructed from the solutions in the population P and the archive P', until some termination criterion is fulfilled.

Using a random sampling method for the initial population often results in many infeasible solutions. Alg. 2 shows an improved version of the algorithm as proposed in this paper. Firstly, the multi-objective optimization problem MOP is separated into l disjunctive subproblems $\{MOP_1, MOP_2, \ldots, MOP_l\}$. Although not in the scope of this paper, this separation can be simply done by partitioning the decision space or by any other and more sophisticated technique. Secondly, a novel method, called *Pareto-Front-Arithmetics* (*PFA*) [3], constructs the initial population. Hopefully, this initial population contains better solutions to the overall MOP than a random generated initial population. Finally, the optimization is performed as already outlined in Alg. 1.

Of course, there are several limitations to this approach as already discussed in literature (cf. [10]), but as will be shown by experiments in this paper, a constructive approach can outperform the random-based approaches in many practical problems. Next, the proposed construction of the initial population and its limitations will be discussed in detail.

3 Pareto-Front-Arithmetics

An interesting approach of reducing the complexity in solving MOPs was proposed by Abraham et al. [7] where *feasibility filters* and *optimality filters* were used to find the Pareto-optimal front. After separating the multi-objective optimization problem MOP in subproblems $\Theta(MOP) = \{MOP_1, \ldots, MOP_l\}$, in a first step, the partial decision spaces X_i corresponding to the subproblems

Alg. 2 Improved optimization procedure

```
OPTIMIZE
IN: MOP multi-objective optimization problem
OUT: P' archive containing best solutions
BEGIN
    t ← 0
    // Partition MOP into l subproblems
    Θ(MOP) = {MOP₁, MOP₂, ..., MOPₗ}
    // Use Pareto-Front-Arithmetics to construct initial population
    Pₜ ← pfa(Θ(MOP))
    P'ₜ ← update(Pₜ)
    WHILE (!terminate(P'ₜ, t)) DO
        t = t + 1
        Pₜ ← generate(Pₜ₋₁, P'ₜ₋₁)
        P'ₜ ← update(Pₜ, P'ₜ₋₁)
    ENDWHILE
    RETURN P'ₜ
END
```

MOP_i are filtered regarding feasibility and optimality. Afterwards in a second step, the remaining solutions are combined to form a new decision space at the next level of hierarchy. The resultant decision space is again filtered. A global feasibility test is needed since not all constraints can be separated and shifted to lower levels of hierarchy. Abraham et al. [7] proved that they can construct the true Pareto-optimal set if the decomposition of the design space is *monotonic*, i.e., the MOP is *separable*.

In this section, an approach called *Pareto-Front-Arithmetics (PFA)* will be proposed that is somehow similar to the approach proposed by Abraham et al. but does not assume any monotonicity property of the objective function. PFA will be used later on for the construction of the initial population of a MOEA. For the PFA method, the MOP is decomposed into l subproblems $\Theta(MOP) = \{MOP_1, MOP_2, \ldots, MOP_l\}$ [3]. Next, for each partial optimization problem MOP_i the optimization is performed. This can be done by using a MOEA as outlined in Alg. 1, i.e., a MOEA using randomly generated initial population. The results are combined in a special combination step, the PFA. This is outlined in Figure 1.

The inputs to Pareto-Front-Arithmetics are the approximation sets $X_{i,a}$ resulting from the individual optimization of the partial problems $MOP_i \in \Theta(MOP)$ which are filtered regarding local feasibility (f) and Pareto optimality (P). These approximation sets are combined at higher levels of abstraction according to the structure of the overall MOP. This combination step is discussed in detail below. The combined set X' is again filtered regarding Pareto optimality and feasibility leading to an approximation set X_a of the overall MOP. Of course, this approximation set may contain infeasible and suboptimal solutions.

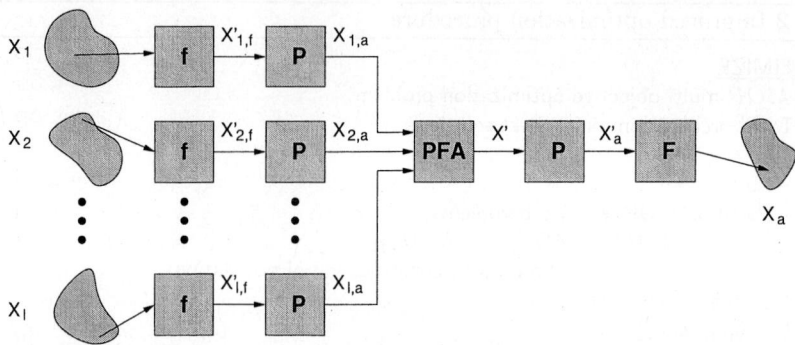

Fig. 1. Concept of Pareto-Front-Arithmetics. In a first step, the approximation sets of the subproblems named $X_1, X_2, \ldots, X_l$ are combined at a higher level of hierarchy to X'. In a second step, these results are filtered regarding Pareto optimality and feasibility leading to an approximation set X_a

The motivation for Pareto-Front-Arithmetics is the same as given by Abraham et al. who name three advantages of hierarchical decomposition [7]:

1. The size of each subproblem's decision space is smaller than the top-level decision space, i.e.,

$$\forall_{MOP_i \in \Theta(MOP)} : |X(MOP_i)| \leq |X(MOP)|.$$

2. The evaluation effort for each subproblem MOP_i is low because of the smaller complexity of the subproblem, i.e., the evaluation of the objective functions $f(x)$ which can be substantial in practical problems is reduced.
3. The number of top-level solutions to be evaluated is a small fraction of the size of the original decision space.

$$|X_f(MOP)| \leq \prod_{MOP_i \in \Theta(MOP)} |X_f(MOP_i)|.$$

The last and most important advantage states that a feasible solution at the top-level must be composed of feasible partial solutions. Thus, the search space can be reduced dramatically.

Abraham et al. define necessary and sufficient conditions of the decomposition function of the objectives which guarantee Pareto optimality for the top-level solution depending on the Pareto optimality of solutions of the subproblems [7]. Pareto-optimal solutions of a top-level MOP are composable of Pareto-optimal solutions of the partial MOPs iff the composition function of each objective function is a *monotonic function*, i.e., the top-level MOP is *separable*. Although this observation is important and interesting, many practical objective functions unfortunately do not possess these monotonicity properties.

Despite these results, we will use PFA also in non-separable problems to construct an initial population to a MOEA optimization run. The key idea of PFA is to do the necessary combinations in the *objective space* only. This substantially reduces the computation time. Moreover, optimality filtering is performed as soon as possible in order to restrict the search space. From the previous discussions, it should be obvious that these combinations may lead to suboptimal optimization results. This is due to the non-monotonic property of the decomposition operator for many objective functions. Especially, in the application of the technique to design space exploration of embedded systems, as used in this paper as case study later on, all properties are non-monotonically decomposable. On the other hand, the combined results may be infeasible as well. That results from the omitted feasibility check at deeper levels of hierarchy. Of course, each partial solution is tested for feasibility, but only independent from the top-level problem, i.e., it cannot be guaranteed that a feasible partial solution may contribute to a feasible overall solution. Nevertheless, as will be shown by experiments later on, the PFA method may contribute to a fast convergence of the top-level problem by using this method for construction of an initial population. But first, the operations performed by the box named PFA in Figure 1 will be discussed.

In general, a PFA operation can be defined as an n-dimensional function:

$$f = \mathrm{h}(y_1, y_2, \ldots, y_l), \text{ where } y_j \in Y_{j,\mathrm{a}} \text{ with } 1 \leq j \leq l$$

Note, the combination is only done in objective space. The objective values of a solution are determined by the objective values of the partial solutions Y_j obtained at lower levels of hierarchy. If this function is monotonically increasing or monotonically non-decreasing in a minimization problem, the top-level design is indeed Pareto-optimal as discussed above. Such multi-level optimization problems are termed *separable optimization problems* (see, e.g., [10]). Unfortunately, in many MOPs most objectives functions cannot be formulated as monotonic functions in a hierarchical context.

Although these drawbacks are crucial, monotonic functions will be used in Pareto-Front-Arithmetics in order to approximate the true Pareto front. It is also a common technique in other technical fields to determine worst case and best case approximations by using over-simplified assumptions. Hence, one has to understand that PFA is able to rapidly construct an approximation of the overall problem, but this approximation may contain suboptimal and infeasible solutions. On the other hand, as mentioned above, by only considering local feasible partial solutions in the combination step, there might be a high probability to construct a feasible implementation as well. Moreover, the constructed top-level solution, even if infeasible is composed of many feasible parts. An interesting fact which can be seen by the experiments is that these feasible parts introduce genetic information into the initial population which definitely contributes to the convergence speed of a MOEA.

Here, only the worst case approximation represented by an addition will be discussed. The best case approximation can be performed similarly by applying the maximum operator. Figure 2 outlines the addition of the objective of two or more Pareto points: Figure 2(a) shows two partial objective spaces. Each

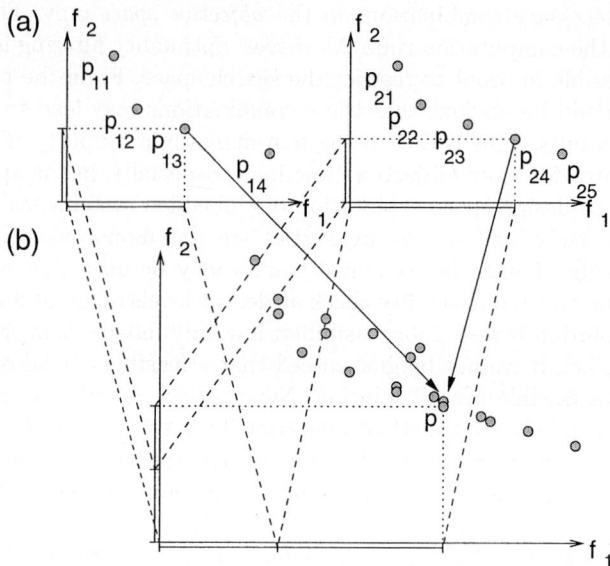

Fig. 2. Pareto-Front Arithmetics addition operation. (a) Two approximation sets. (b) The two approximation sets are combined using the addition Pareto-Front Arithmetics operation, e.g., p_{13} and p_{24} result in $p = (f_1(p_{13}) + f_1(p_{24}), f_2(p_{13}) + f_2(p_{24}))$

Pareto-optimal solution p_{1i} is combined with each point p_{2j}. Here, the resulting objectives are calculated as the sum of the objectives of the subsystems, i.e., $f_k(p) = f_k(p_{1i}) + f_k(p_{2j})$ for $k = 1, 2$. The results are shown in Figure 2(b).

To get a better approximation of the Pareto-front, the algorithm described above has to prevent the rejection of good points. The PFA approach can be improved by considering both, the best and worst case approximation, simultaneously (cf. [3]), leading to the notion of *property intervals*. A property interval is represented by the combination of best case and worst case methods, leading to a lower and an upper bound of the objectives. For example, in the design of embedded systems, the implementation cost of a system that is composed of two subsystems can be restricted by the maximum implementation cost of each subsystem and the sum of those cost. The maximum of the cost of the two subsystems corresponds to the case where both subsystems share the same resource (best case), while the sum of the cost model the fact that both subsystems are implemented using dedicated resources (worst case). Unfortunately, by using property intervals, the definition of dominance as introduced so far becomes meaningless. To solve this problem, Teich [11] proposed the notion of *probabilistic dominance* for Pareto optimality. Due to space limitations, the discussion of probabilistic dominance will be omitted here.

4 Experimental Results

In this section, we will provide some experimental results from using Pareto-Front-Arithmetics for initial population construction of population-based multiobjective optimization methods, e.g., MOEAs.. To analyze the performance of our proposed strategy, we have chosen a MOP from the area of embedded system synthesis [12, 13]. The actual optimization problem is a combinational selection and graph partitioning MOP: Starting from a mathematical problem formulation called *specification graph*, a set of applications given as task graphs (hierarchically organized) as well as resources from a target architecture must be selected. Later the set of applications has to be mapped onto the selected architecture. Each solution to the MOP represents an *implementation* of the embedded system. An implementation may be feasible or infeasible due to constraints imposed on the partitioning of tasks (for a detailed discussion cf. [12]). The three objectives used during the optimization are technical properties of embedded systems, namely *implementation cost*, *power dissipation*, and *latency*. All these properties are to be minimized and they are non-monotonic due to resource sharing, power consumption being dependent on the binding, etc. Hence, PFA may fail to construct optimal and feasible implementations. But note: Even if not the best solutions are constructed, *good* implementations can be generated in less time by using PFA. In order to apply the PFA approach, the partitioning of the MOP is naturally given by the hierarchical structure of the task graph, i.e., a subproblem is defined by a leaf task graph, the target architecture and the mapping relation corresponding to the selected subgraph.

The following subsection provides quantitative results from the comparison between the two optimization strategies proposed in Alg. 1 and Alg. 2, i.e., a MOEA with a randomly generated initial population and a MOEA with a PFA generated initial population. The PISA (Platform independent Interface for Search Algorithms) [14] framework was chosen for optimization purposes. In the present work, the SPEA2 selection procedure [15] was applied.

4.1 Comparison Methodology

The experiments are performed as follows: A generator program is used to randomly construct MOP instances (specification graphs), where several parameters determine the architecture template, the application, possible mappings, and the attributes used to compute the objective values. Due to different random values the generated problem instances are similar in structure, but not equal. Each MOP instance is optimized by both methods (with and without PFA). It is noteworthy that the optimization of the subproblems terminates if no improvement in terms of coverage and distance thresholds between two consecutive generations is obtained (in the forthcoming test cases the coverage threshold was chosen to be 90% and the distance threshold to be 0.05) or a maximum number of generations is reached (140 in forthcoming test cases). After the optimization of each problem instance, the non-dominated solutions found by both methods are combined in a single *reference set*. This reference set is Pareto-filtered and is used to quantitatively assess the performance of both methods.

The MOP instances (specification graphs) can be generated from a few parameters. The most important ones are: (i) The number of resources r in the architecture template. From these resources a subset must be chosen during optimization. Hence, this number affects the problem size. (ii) the number of hierarchical tasks t_h and non-hierarchical tasks t_{nh} in the task graph. The application selection is done from a hierarchical task graph. Using the hierarchical structure the problem separation can be easily done by selecting the leaf graphs. Leaf graphs only contain non-hierarchical tasks. Again, these numbers affect the MOP size. (iii) The number of hierarchical levels l and number of refining subgraphs s per hierarchical task in the task graph. Obviously, these numbers also directly affect the problem size. From all subgraphs refining a hierarchical task exactly one has to be selected during optimization (algorithm selection). (iv) The number of mapping relations m per non-hierarchical task. A task can be only partitioned into a cluster corresponding to a resource which is selected and a mapping relation exists between the task and the resource. Hence, the number of mapping relations has a large influence of the MOP size. (v) The number of edges in the tasks graph is given by a probability value p. This value determines the probability that two tasks are connected by an edge. An edge represents a data dependency among these two tasks. Due to the feasibility requirement, the number of data dependencies affects the complexity of the optimization problem. The complexity increases with the number of data dependencies.

The performance indicators used in the present work are: the *coverage* [16] and the *normalized distance* [10] to assess the convergence, and the entropy indicator with 100 grid points [17] to assess the diversity. A detailed discussion on performance indicators can be found in [18]. The approximation sets obtained from both optimization methods are compared to the reference set by using the coverage and normalized distance indicators. Since the entropy indicator is a unary indicator, no reference set is needed. Moreover, the average time needed for a fix number of generations is calculated as well. The estimation of the consumed processor time is realized by the "clock"-function of the Linux operating system.

4.2 Quantitative Results

In this section, the four most interesting test cases are discussed. In both methods, with and without PFA, the same kind of selection, crossover, mutation, parameters for the SPEA2 algorithm, and random seed is used to have a clear comparison. All results are averaged over 20 MOP instances, where MOP instances were omitted if both methods were not able to find any feasible solution. Infeasibilities are treated as objective functions, i.e., the number of errors in an implementation is counted and used for optimization. In an infeasible solution, all other objectives are set to infinity. The number of generations is set to 400. For the first two cases the worst case approximation, in cases three and four the property intervals are used for PFA. Table 1 shows the parameter values chosen for the four test cases:

Table 1. The four test cases for performance assessment

Test Case	r	t_h	t_{nh}	l	s	m	p
TC1	49	6	10	2	2	4	0.3
TC2	49	6	20	2	2	4	0.3
TC3	49	2	10	3	2	4	0.3
TC4	81	6	10	2	2	4	0.6

TC1. In this test case, there exist $2 \cdot 6 = 12$ subgraphs and $12 \cdot 10 = 120$ leaf tasks, each having four different mappings to one of the 49 resources. Thus, there are 4^{120} possible mappings. The average number of non-dominated solutions found was 232 (minimum 6, maximum 656).

The averaged optimization results are illustrated in Figure 3. Figure 3(a) shows the coverage of the reference set by both methods. Figure 3(b) shows the normalized distance of both methods to the reference set. One can see that the method using PFA for initial population constructions starts on average 180 seconds later than the method using random initial population. This is exactly the time needed for the construction of the initial population. The second result is, that after this initialization phase, the method using PFA produces better results than the method without PFA, i.e., it covers a higher percentage of the reference set and it is closer to the points in the reference set. With respect to the time, the exploration with PFA is always quantitatively better. After about 700 seconds both methods converge in both, coverage and distance.

TC2. Taking the same parameters as in TC1, only the number of non-hierarchical tasks is increased to be 20 instead of 10, i.e., the subproblem size is increased. Again, there are 12 refinements and 49 resources, but now, the number of leaf tasks increases to be 240. Hence, there are 4^{240} possible mappings. On average, 491 non-dominated solutions were found (minimum 47, maximum 760).

The results are nearly analogous to TC1 and are shown in Figure 4. However, the differences between both methods in coverage and distance are smaller in this test case. This results, from the fact that the search space contains more feasible solutions, and, hence, both methods are more likely to find these solutions.

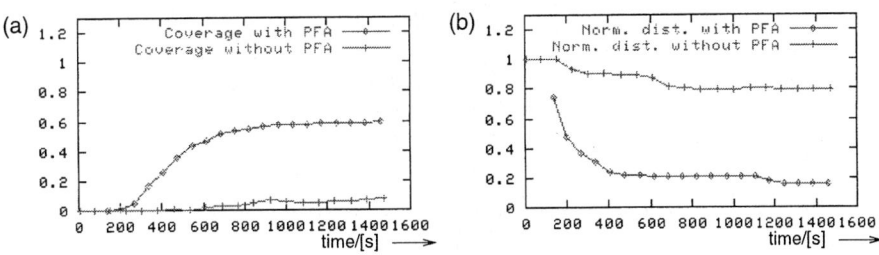

Fig. 3. (a) Coverage of reference set by both methods in case TC1. (b) Distance of both methods to reference set in case TC1

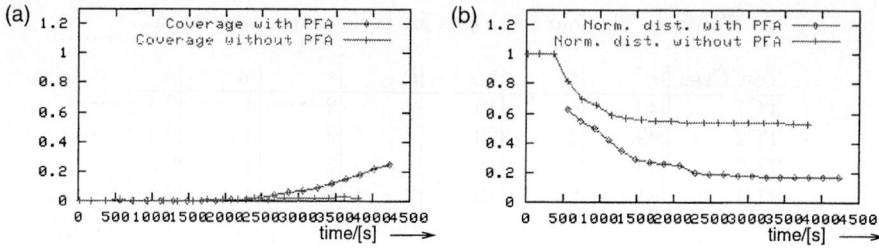

Fig. 4. (a) Coverage of reference set by both methods in case TC2. (b) Distance of both methods to reference set in case TC2

TC3. Similar parameters to TC1 were chosen. However, an additional level of hierarchy was introduced, i.e., the number of subproblems was increased. The depth of the task graph is three. As a consequence, more PFA steps are necessary than in TC1 or TC2. Moreover, PFA is performed with property intervals. To limit the complexity of the search space, the task graph only contains two hierarchical tasks at each level. There are 16 leaf subgraphs and 160 leaf tasks with 4^{160} mapping relations. On average, 257 non-dominated solutions have been found (minimum 18, maximum 594).

The results of TC3 are shown in Figure 5. Again, the construction of the initial population consumes a lot of time using PFA. But, this method clearly outperforms the method using random-based population generation.

TC4. Here, the probability for data dependencies between processes has been doubled (60%) in comparison to TC1-TC3. Furthermore, there are additional resources (81) and the PFA is performed with property intervals. The number of subgraphs, leaf tasks, and mapping relations is the same as in TC1. Due to the increased number of resources and data dependencies, the search space is expected to contain less feasible solutions than in TC1-TC3. This could be shown by the average number of non-dominated solutions to be 7 (minimum 2, maximum 29).

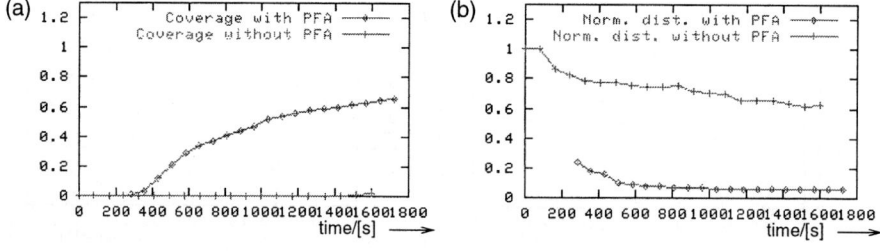

Fig. 5. (a) Coverage of reference set by both methods in case TC3. (b) Distance of both methods to reference set in case TC3

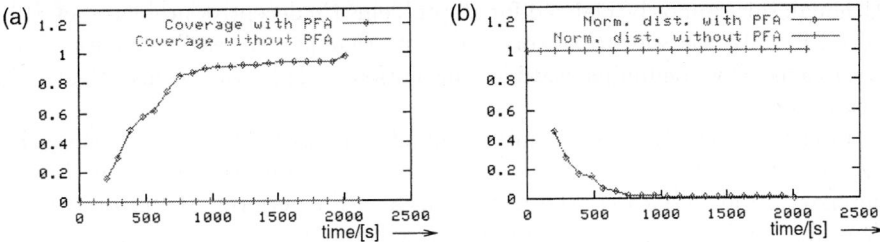

Fig. 6. (a) Coverage of reference set by both methods in case TC4. (b) Distance of both methods to reference set in case TC4

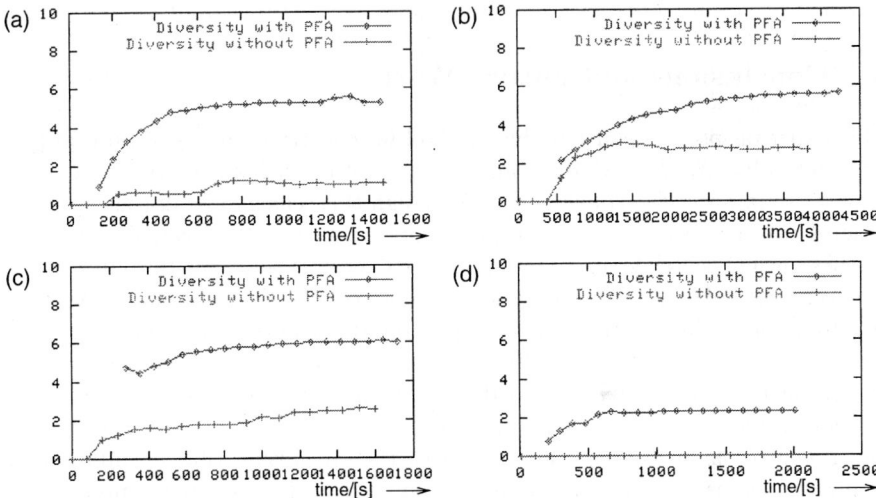

Fig. 7. (a) Entropy of both methods in case TC1, (b) TC2, (c) TC3, and (d) TC4

The results are presented in Figure 6. Again, the method with PFA performed better than the method without PFA. The increasing number of data dependencies and the additional resources result in a search space with only a small fraction of feasible solutions. Only the method using initial population construction by PFA was able to find feasible solutions at all. 61 MOP instances have been tested, until 20 of them led to feasible solutions, indicating the large number of infeasible solutions in the corresponding search spaces. In all 20 MOP instances, the randomly generated initial population did not contain any feasible solution and the succeeding MOEA run could not construct feasible solutions. An interesting fact is that in 4 of the 20 MOP instances, the initial population constructed from Pareto-Front-Arithmetics did not contain any feasible solution. Despite this fact, the succeeding MOEA run found feasible solutions! Obviously, using PFA introduces better genetic information into the chromosomes.

Although this cannot be claimed for all problems (in deceptive problems, cf. [19], PFA will fail), the way PFA works suggests that partial feasible solutions are constructed (by combining feasible subsolutions). This will be investigated in future work.

Finally, Figure 7 shows the entropy of all test cases. The method using PFA show a better diversity than the method using random-based construction. In summary, in all test cases, the method using PFA performs better on average than the MOEA using just a randomly generated initial population. As expected, the initial populations constructed by PFA were superior to the random generated initial populations on average (we also compared the initial population directly). These better populations were the basis of better overall solutions in the later optimization run. Moreover, in complicated cases, PFA has led the search towards the feasible region.

5 Conclusions and Future Work

This paper proposes a constructive method for construction of good initial population in MOEAs. Any MOEA can be enhanced by such a constructive method in order to improve the convergence. Although this seems to be not useful in the general case (especially deceptive problems), such a method can be helpful in the presence of search spaces containing some feasible solutions (also in the case of deceptive problems). The proposed method, called Pareto-Front-Arithmetics, which performs the construction by approximating the solutions in the objective space only, was applied to the MOP of optimizing resource allocation and binding problems encountered in embedded system design. In all experiments, the method using PFA for initial population construction outperformed a random-based approach on average. This leads to the conclusion that spending some computational effort in constructing initial populations helps to improve the convergence in optimization. In future work, we will evaluate more test cases to study the effect of initial populations on the optimization results even more. Here, also different separation techniques will be of concern. Moreover, we will apply archive reduction techniques known from, e.g., SPEA 2, NSGA-II, etc., to reduce the intermediate results in the Pareto-Front-Arithmetics step.

References

1. Coello Coello, C.A., Van Veldhuizen, D.A., Lamont, G.B.: Evolutionary Algorithms for Solving Multi-Objective Problems. Kluwer Academic Publishers (2002)
2. Deb, K.: Multi-Objective Optimization using Evolutionary Algorithms. John Wiley & Sons, Ltd. (2001)
3. Haubelt, C., Teich, J.: Accelerating Design Space Exploration Using Pareto-Front Arithmetics. In: Proceedings of Asia and South Pacific Design, Automation and Conference, Kitakyushu, Japan (2003) 525–531

4. Gandibleux, X., Morita, H., Katoh, N.: The Supported Solutions Used as a Genetic Information in a Population Heuristic. In: Proceedings of the First International Conference on Evolutionary Multi-Criterion Optimization, Lecture Notes in Computer Science (LNCS). Volume 1993., Zurich, Switzerland (2001) 429–442
5. Gandibleux, X., Morita, H., Katoh, N.: Use of a Genetic Heritage for Solving the Assignment Problem with Two Objectives. In: Proceedings of the Second International Conference on Evolutionary Multi-Criterion Optimization, Lecture Notes in Computer Science (LNCS). Volume 2632., Faro, Portugal (2003) 43–57
6. Josephson, J.R., Chandrasekaran, B., Carroll, M., Iyer, N., Wasacz, B., Rizzoni, G., Li, Q., Erb, D.A.: An Architecture for Exploring Large Design Spaces. In: Proceedings of the fifteenth national Conference on Artificial intelligence/Innovative applications of artificial intelligence (AI), Madison, USA (1998) 143–150
7. Abraham, S.G., Rau, B.R., Schreiber, R.: Fast Design Space Exploration Through Validity and Quality Filtering of Subsystem Designs. Technical report, Hewlett Packard, Compiler and Architecture Research, HP Laboratories Palo Alto (2000)
8. Szymanek, R., Catthoor, F., Kuchcinski, K.: Time-Energy Design Space Exploration for Multi-Layer Memory Architectures. In: Proceedings of the Design, Automation and Test in Europe Conference, Paris, France (2004) 10318–10323
9. Blickle, T.: Theory of Evolutionary Algorithms and Application to System Synthesis. PhD thesis, Swiss Federal Institute of Technology Zurich (1996)
10. Veldhuizen, D.A.V.: Multiobjective Evolutionary Algorithms: Classifications, Analyses, and New Innovations. PhD thesis, Graduate School of Engineering, Air Force Institute of Technology (1999)
11. Teich, J.: Pareto-Front Exploration with Uncertain Objectives. In: Proc. of the First Int. Conf. on Evolutionary Multi-Criterion Optimization, Lecture Notes in Computer Science (LNCS). Volume 1993., Zurich, Switzerland (2001) 314–328
12. Blickle, T., Teich, J., Thiele, L.: System-Level Synthesis Using Evolutionary Algorithms. In Gupta, R., ed.: Design Automation for Embedded Systems. 3. Kluwer Academic Publishers, Boston (1998) 23–62
13. Zitzler, E.: Evolutionary Algorithms for Multiobjective Optimization: Methods and Applications. PhD thesis, Eidgenössische Technische Hochschule Zürich (1999)
14. Bleuler, S., Laumanns, M., Thiele, L., Zitzler, E.: PISA - A Platform and Programming Language Independent Interface for Search Algorithms. In: Proceedings of the Conference on Evolutionary Multi-Criterion Optimization (EMO 2003), Faro, Protugal (2003) 494–508
15. Zitzler, E., Laumanns, M., Thiele, L.: SPEA2: Improving the Strength Pareto Evolutionary Algorithm for Multiobjective Optimization. In: Evolutionary Methods for Design, Optimisation, and Control, Barcelona, Spain (2002) 19–26
16. Zitzler, E., Thiele, L.: Multiobjective Optimization Using Evolutionary Algorithms – A Comparative Case Study. In: Proceedings of Parallel Problem Solving from Nature – PPSN-V, Amsterdam, The Netherlands (1998) 292–301
17. Gunawan, S., Farhang-Mehr, A., Azarm, S.: Multi-Level Multi-Objective Genetic Algorithm Using Entropy to Preserve Diversity. In: Evolutionary Multi-Criterion Optimization. Volume 2632 of Lecture Notes in Computer Science (LNCS). (2003) 148–161
18. Zitzler, E., Thiele, L., Laumanns, M., Fonseca, C.M., Grunert da Fonseca, V.: Performance Assessment of Multiobjective Optimizers: An Analysis and Review. IEEE Transactions on Evolutionary Computation **7** (2003) 117—132
19. Deb, K.: Multi-objective Genetic Algorithms: Problem Difficulties and Construction of Test Problems. Evolutionary Computation **7** (1999) 205–230

Multi-objective Go with the Winners Algorithm: A Preliminary Study

Carlos A. Brizuela and Everardo Gutiérrez

Computer Science Department,
CICESE Research Center,
Km 107 Carr. Tijuana-Ensenada, Ensenada, B.C., México
{cbrizuel, egutierr}@cicese.mx
+52-646-175-0500

Abstract. This paper introduces a new algorithm to deal with multi-objective combinatorial and continuous problems. The algorithm is an extension of a previous one designed to deal with single objective combinatorial problems. The original purpose of the single objective version was to study in a rigorous way the properties the search graph of a particular problem needs to hold so that a randomized local search heuristic can find the optimum with high probability. The extension of these results to better understand multi-objective combinatorial problems seems to be a promising line of research. The work presented here is a first small step in this direction. A detailed description of the multi-objective version is presented along with preliminary experimental results on a well known combinatorial problem. The results show that the algorithm has the desired characteristics.

1 Introduction

Real world optimization problems are usually multi-objective (MO) in nature. The lack of exact methodologies to tackle these problems makes them attractive theoretically, and practically.

In the last few years many population-based strategies have been proposed to deal with these problems. These strategies seem especially well suited for MO problems where a set of solutions, instead of a single solution, is needed. Among them Genetic Algorithms [1, 9] have been the most widely used. Other approaches were proposed as an extension of their mono-objective counter part, some of these are: MO Simulated Annealing [14, 15], MO Ant Colony Optimization [16, 17], MO Particle Swarm Optimization [18], MO Evolutionary Strategy [21], MO Tabu Search [19, 20], among others.

The mainstream of research with these methods has been directed to experimental rather than theoretical issues. The reasons for this are simple: first, the analysis of these algorithms are extremely complex and the insight gained with the results are usually poor; second, the urgent need for methodologies to solve real problems is high.

On the other hand the research community in analysis of algorithms has focused in trying to find a model for the random local search heuristics. This with the objective of giving a rigorous explanation of these algorithms behavior. One such attempt has been the model of heuristics where good solutions are preferred over bad solutions in what was called the "Go with the Winner" (GWW) scheme.

The GWW algorithm have been introduced by Aldous and Vazirani [2] as an attempt to give a rigorous explanation of the behavior of some algorithms based on a non crossover "survival of the fittest" paradigm. This algorithm resembles the well known simulated annealing strategy and is restricted to work with search graph with a tree structure [2]. Dimitriou and Impagliazzo [5,6] generalize the algorithm to deal with general search graphs. In this paper an extension of the generalized version of GWW algorithm to deal with MO optimization problems is presented.

In order to extend all theoretical results of the single objective case to the MO case, a great deal of work will be required. Our first step is to propose an extension of the algorithm and to study its performance on a NP-$hard$ scheduling problem.

Scheduling appears to be one of the most challenging combinatorial MO problems. In a real scheduling problem we are interested not only in minimizing the latest completion time (makespan) but also in minimizing the total time all jobs exceed their respective due dates. These objectives are usually in conflict with each other.

The remainder of the paper is organized as follows. Section 2 presents the GWW algorithm and extends it to deal with MO problems. Section 3 states the problem to be used as a study case. Section 4 shows the experimental setup and results. Finally, section 5 presents the summary of this work.

2   The Go with the Winners Algorithm

Originally, the GWW algorithm was designed for search graphs with a tree structure and the main result is as follows [2]. If we want to increase the success probability in a randomized optimization heuristic from p (generally small) to 0.99, one way is to re-run the heuristic $O(1/p)$ times and select the best answer. In case of the GWW algorithm this number can be reduced to $log(1/p)$ runs of the algorithm by introducing some interactions between the runs. Dimitriou and Impagliazzo [5] introduced a variant of this algorithm that can be used for search graphs that are not trees. They show that for any one of a large class of distributions on search graphs their version of the GWW finds, in polynomial time, the optimum with high probability, for almost all search graphs drawn from the distribution. The most important result is that they also found a sufficient condition on the search graph structure (a combinatorial property) for the algorithm to reach the optimum with high probability. This property is known as the "local expansion property." Carson [3] uses the GWW algorithm to analyze, experimentally as well as analytically the search graph for the bisection problem. The analysis performed can give information regarding which instances will be easy to solve by a given local search algorithm. Basically, search graphs having the local expansion property will be easy to solve by the GWW algorithm. The pseudocode for this algorithm is given below.

Algorithm 1. GWW for integer minimization [3].

Parameters. P: number of particles in the population. L: number of steps in the random walk.
Step 1. Initialization. Generate P random solutions. Place particles on each of these solutions. Initialize i to be the maximum objective function value in the starting population.
Step 2. Until all particles are at local minima do:
Step 3. Restriction. $i = i - 1$. /*decrease the threshold*/
Step 4. Redistribution. Let **M** be the set of particles at solutions of cost exactly $i+1$. If $|\mathbf{M}| = P$, then exit. Otherwise, delete the particles in **M** and make $|\mathbf{M}|$ copies of randomly selected members of the remaining population.
Step 5. Randomization. For every duplicate made in the previous step, execute the following random walk for L steps; select a random neighbor; if the neighbor is of cost at most i, then move to that solution, otherwise stay at the current solution. (Note that selecting a neighbor but not moving to it still counts as one of the L steps in the random walk).
Step 6. Continue to stage $i = i - 1$.

The main idea with this algorithm is to improve the iterative local search procedure by the introduction of interactions between different runs of the algorithm. The interactions are given in the restriction step since i determines which solutions continue (go with the winners) and which are deleted.

The **initialization** step consists on the generation of P initial solutions. The assumption in this step is that the particles are uniformly distributed in the search graph. The **restriction** step forces the search graph to be partitioned into components with different levels, where each level represents a value of the objective function and an element of a component with no down connections represents one local optimum for the instance. In the **redistribution** stage all solutions over the current threshold level are deleted and replaced by copies of solutions under the threshold level (minimization problem). From each of these solutions, in the **randomization** step, a length L random walk is performed. This is a way to uniformly explore the subset of solutions at the current threshold level. The algorithm works on the hypothesis that the following loop invariant holds "at each iteration of the algorithm steps 3, 4 and 5 the search graph is uniformly sampled."

2.1 The Multi-objective GWW

With the same philosophy we apply these steps to the MO version of the algorithm whose pseudocode is described as follows.

Algorithm 2. MOGWW.

Parameters. P: number of particles in the population. L: number of steps in the random walk.
Step 1. Initialization. Generate P random solutions. Place particles on each of these solutions. Classify the solutions according to their dominance relations. Initialize i to be the worst nondominated front rank.
Step 2. Until all particles are nondominated do:
Step 3. Restriction. $i = i - 1$. /*decrease the threshold*/
Step 4. Redistribution. Let **M** be the set of particles at solutions in a nondominated front of rank $i+1$. If $|\mathbf{M}| = P$, then exit. Otherwise, delete the particles in **M** and make $|\mathbf{M}|$ copies of randomly selected members of the remaining population in higher remain fronts.
Step 5. Randomization. For every duplicate made in the previous step, execute the following multi-objective random walk for L steps; select a random neighbor; if the neighbor is not dominated by any solution in the front with rank i then move to it, otherwise stay at the current solution. (Note that selecting a neighbor but not moving to it still counts as one of the L steps in the random walk).
Step 6. Reclassify the solutions according to their dominance relations. Initialize i to be the worst nondominated front rank. Continue to the next stage.

This algorithm works with the same principles as Algorithm 1 does. The extension is performed by introducing a proper threshold condition and a corresponding multi-objective random walk.

The main idea regarding the threshold is as follows. At each stage the solutions are classified in nondominated fronts, here the poorest front is assigned the highest index and it is selected as the threshold, i.e. all solutions which are in that front are deleted and replaced with solutions in fronts with lower indices. An option to generalize this step can be to do the replacement with solutions coming only from a particular set of fronts (*e.g.* coming only from the best front).

Each one of the duplicates is the starting point of a length L random walk, at each step in this walk the selected neighbor is compared to solutions originally in the front adjacent to the previously deleted one. In this way it is ensured that the new solutions will not be worse than the front which is going to be the threshold at the next stage. The function of this step is to uniformly sample the search graph at a given level of solution quality. This mechanism allows the algorithm to keep a diverse set of solutions, something that is not easy to achieve, for instance in Genetic Algorithms.

An advantage for this algorithm is that it needs only two parameters L and P, where L can be determined previous to the running of the algorithm. The appropriate number of particles will depend on the search graph structure. If this is composed of many disconnected components then the required P will be large. When the graph is composed of a single component the P does not need to be large. Another point in favor of MOGWW is that it presents a 100%

4 Experimental Setup and Results

The experiments have two main objectives. The first one is related to the performance study of the algorithm when different random walk lengths are used. The second one has to do with the study of the relative performance when comparing the MOGWW with a well known genetic algorithm.

The neighborhood for this particular problem is defined by the insert operator described in [7]. The distance measure used in the decorrelation test is based on the minimum number of times this operator needs to be applied to one solution to reach the other.

Before any experiment, aimed at measuring the algorithm performance, is carried out the decorrelation test is applied for the analyzed instance. Two different lengths were tried $L = 35$ and $L = 75$, the obtained results for the mean decorrelation probability were a 0.678 (2.26% standard deviation) and 0.522 (3.08% standard deviation) respectively. As the reference value is 0.5 we can see that $L = 35$ fails to pass the test and $L = 75$ success in the test.

Once we know that the length to use is $L = 75$, we need to study the performance of this algorithm. Previous to this, some performance measures are defined.

4.1 Performance Measures

The performance measures are related to the dominance relations between nondominated fronts generated by the studied algorithms. The measures can be defined as follows [7]: m^1 counts the number of times a dominance relation can not be establish between a solution given by algorithm **A** and the set of solutions given by algorithm **B**, m^2 counts the number of solutions given by algorithm **B** dominate solutions generated by algorithm **A**, m^3 counts the number of times solutions given by algorithm **A** dominate solutions given by algorithm **B**, finally m^4 counts the number of times solution has copies of itself in the other nondominated set.

4.2 Random Walk Length and Performance Comparison

The set of instances used here are the same as those described in a recent work on the same problem [7], and which are available on the web [1]. Ten instances of 75-jobs and 20-machines were used to perform the experiments. For each instance the number of particles is set to $P = 100$ and the decorrelation length L is fixed to 75. The length was set by using the decorrelation test proposed by Carson [3].

For each studied instance 50 runs are performed taking the union of all the resulting nondominated solutions. Figures 1, 2 and 3 show the projections of the solutions obtained by two different decorrelation lengths, one which does not pass the decorrelation test ($L = 35$), and another which does ($L = 75$). Table 1 shows the dominance relations between these two sets of solutions. The

---

[1] http://www.cicese.mx/~cbrizuel/MOGWW/instances.html

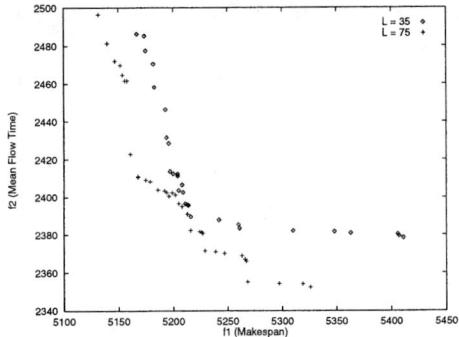

Fig. 1. Projection of Makespan and Mean Flow Time for the union of all nondominated solutions of 50 runs with $L = 35$ and $L = 75$ (Instance 6, $P = 100$)

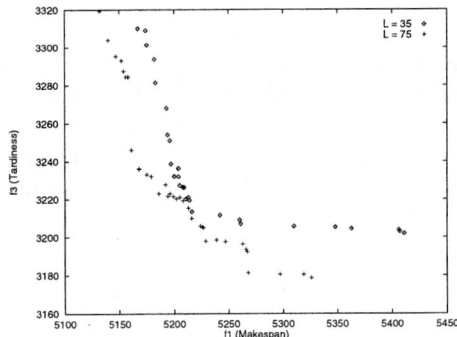

Fig. 2. Projection of Makespan and Tardiness for the union of all nondominated solutions of 50 runs with $L = 35$ and $L = 75$ (Instance 6, $P = 100$)

Table 1. Dominance relations between the union of all nondominated solutions of 50 runs with $L = 35$ and $L = 75$ (Instance 6, $P = 100$)

L	m^4	m^3	m^2	m^1
35	0.00%	0.00%	20.00%	80.00%
75	0.00%	20.00%	0.00%	80.00%

obtained results with the length which pass the decorrelation test dominate 20% of the solutions obtained with the length which does not pass the test ($L = 35$). Equivalent results were obtained for the other instances (not shown here). This result tell us about the importance of using the appropriate length L in order to have a good uniformity, and therefore a good sampling performance.

4.3 GA Versus MOGWW

The set of nondominated solutions obtained with length ($L = 75$) which pass the decorrelation test for each analyzed instance is compared to the results ob-

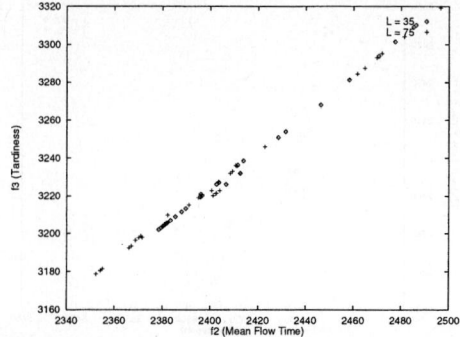

Fig. 3. Projection of Mean Flow Time and Tardiness for the union of all nondominated solutions of 50 runs with $L = 35$ and $L = 75$ (Instance 6, $P = 100$)

tained in a previous work [7]. Figures 4, 5 and 6 show the projections of the nondominated fronts obtained by the MOGWW and the one reported in [7]. The parameters used by the GA are as follows: population size is 100, number of generations is 2000, crossover rate is 1.0 and mutation rate is 0.1. Table 2 shows the nondominance relations between these two sets of solutions. The results obtained by the MOGWW algorithm clearly dominate to the solutions obtained by the GA algorithm.

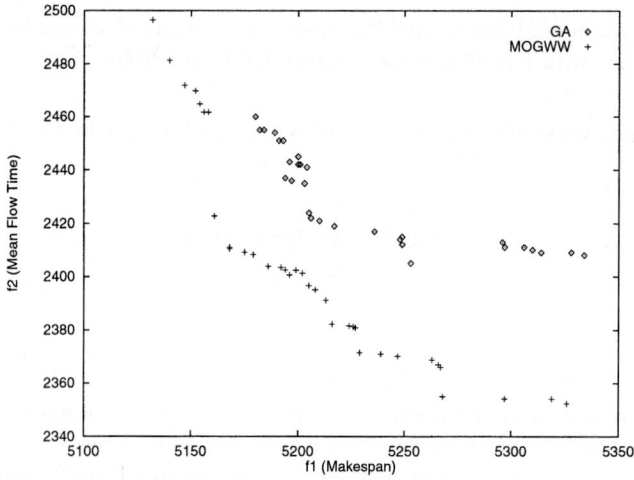

Fig. 4. Projection of Makespan and Mean Flow Time for the union of all nondominated solutions of 50 runs of GA and MOGWW (Instance 6)

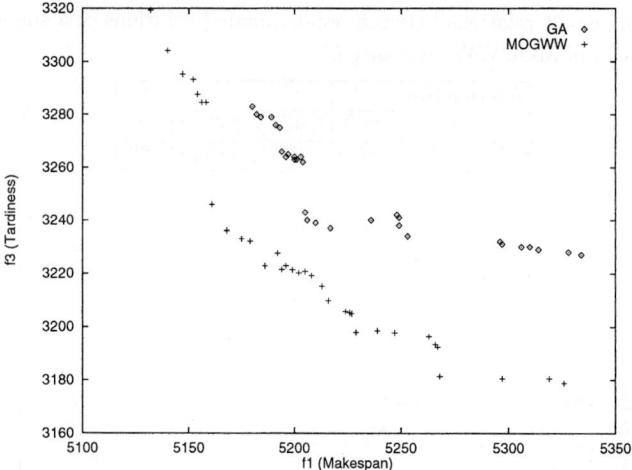

Fig. 5. Projection of Makespan and Tardiness for the union of all nondominated solutions of 50 runs of GA and MOGWW (Instance 6)

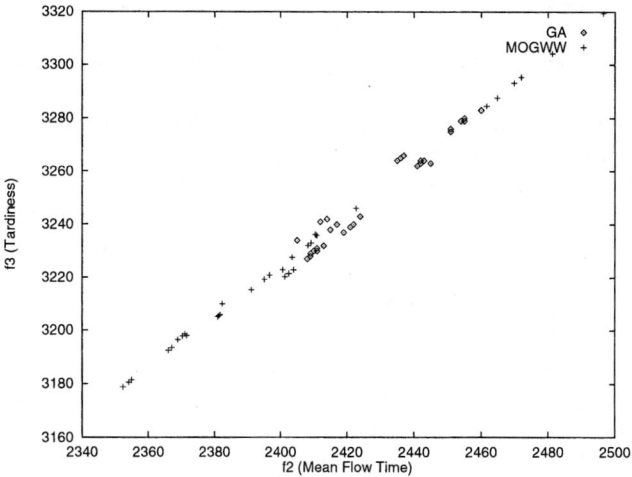

Fig. 6. Projection of Mean Flow Time and Tardiness for the union of all nondominated solutions of 50 runs of GA and MOGWW (Instance 6)

Table 2. Dominance relations between the union of all nondominated solutions of 50 runs of GA and MOGWW (Instance 6)

Algorithm	m^4	m^3	m^2	m^1
GA	0.00%	0.00%	36.57%	63.43%
MOGWW	0.00%	36.57%	0.00%	63.43%

Table 3. Dominance relations between nondominated solutions of a single run of GA and a single run of MOGWW (Instance 6)

Algorithm	m^4	m^3	m^2	m^1
GA	0.00%	2.36%	5.26%	92.38%
MOGWW	0.00%	5.26%	2.36%	92.38%

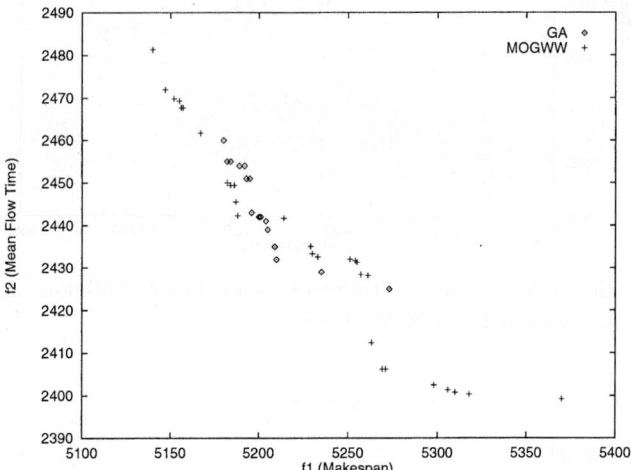

Fig. 7. Projection of Makespan and Mean Flow Time for the nondominated solutions of one run of GA and MOGWW (Instance 6)

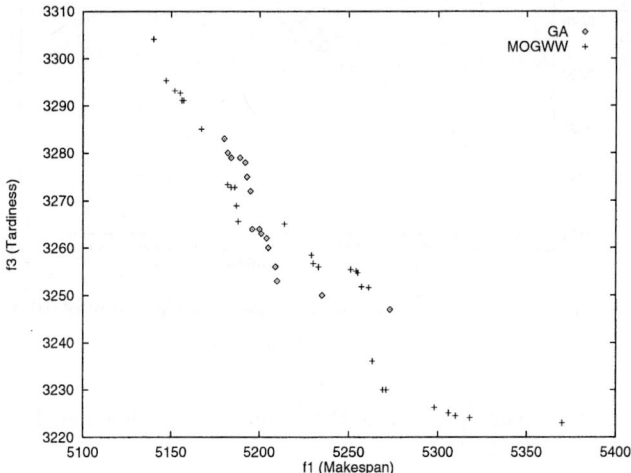

Fig. 8. Projection of Makespan and Tardiness for the nondominated solutions of one run of GA and MOGWW (Instance 6)

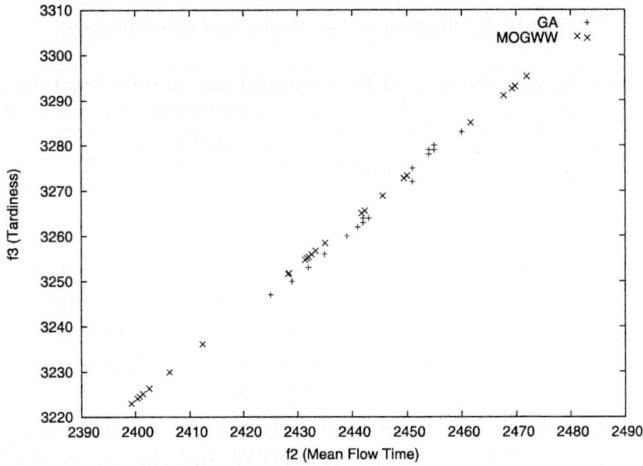

Fig. 9. Projection of Mean Flow Time and Tardiness for the nondominated solutions of one run of GA and MOGWW (Instance 6)

Table 4. Dominance relations between the union of all nondominated solutions of 50 runs of GA and MOGWW (10 instances)

	Algorithm	m^4	m^3	m^2	m^1
Instance 0	GA	0.00%	0.00%	19.23%	80.77%
	MOGWW	0.00%	19.23%	0.00%	80.77%
Instance 1	GA	0.00%	0.00%	43.69%	56.31%
	MOGWW	0.00%	43.69%	0.00%	56.31%
Instance 2	GA	0.00%	0.00%	30.62%	69.38%
	MOGWW	0.00%	30.62%	0.00%	69.38%
Instance 3	GA	0.00%	0.00%	16.74%	83.26%
	MOGWW	0.00%	16.74%	0.00%	83.26%
Instance 4	GA	0.00%	0.00%	23.12%	76.88%
	MOGWW	0.00%	23.12%	0.00%	76.88%
Instance 5	GA	0.00%	0.00%	32.27%	67.73%
	MOGWW	0.00%	32.27%	0.00%	67.73%
Instance 7	GA	0.00%	0.00%	32.80%	67.20%
	MOGWW	0.00%	32.80%	0.00%	67.20%
Instance 8	GA	0.00%	0.00%	24.34%	75.66%
	MOGWW	0.00%	24.34%	0.00%	75.66%
Instance 9	GA	0.00%	0.00%	59.10%	40.90%
	MOGWW	0.00%	59.10%	0.00%	40.90%

Figures 7, 8 and 9 show the projections of the solutions obtained by one run of the MOGWW and one of the GA [7], both selected randomly. Table 3 shows the nondominance relations between these two sets of solutions. The result shows

Table 5. Number of iterations and evaluations

Algorithm Name	Mean Number of Iterations	Standard Deviation (Iterations)	Mean Number of Evaluations	Standard Deviation (Evaluations)
GA	2000.00	0.00%	200000.00	0.00%
MOGWW	1028.26	10.19%	715776.00	10.83%

a very tight case where more than 90% of the solutions are non comparable. Even though the fronts are similar with respect to the nondominance relations, they are different in the front dispersion given by the algorithms. In this sense MOGWW has a bigger front dispersion than the GA. This is a result of the good diversity maintenance mechanism of the MOGWW.

Table 4 shows the dominance relations between the union of all nondominated solutions obtained by 50 runs of the MOGWW and the union of all nondominated solutions obtained by 50 runs reported in [7]. The results for all instances including those in Table 2 show a clear superiority of the proposed MOGWW over the GA. The algorithm used in [7] is probably not the best up to date GA for the problem, but the results are competitive. Similar results, not shown here, were obtained by comparing the MOGWW with the well-known NSGA-II [4] for the same set of instances [8].

Table 5 shows the mean number of iterations and evaluations and their standard deviations for the MOGWW and the GA. This table shows that the proposed MOGWW needs more computing time than the GA, therefore we may increase the population size of the GA and perhaps obtain better results. However, the main point here is to show that the algorithm works similar to what its single objective version does. On the other side an interesting characteristic is that the algorithm needs only two parameters, one of which can be easily tuned previous to the run of the algorithm.

5 Conclusions

A new algorithm for continuous and combinatorial multi-objective optimization problems have been proposed. The algorithm is based on the "Go with the winner" paradigm introduced to model randomized local search. A detailed description of the algorithm was presented along with the ideas behind each step. The new algorithm is tested with a classical scheduling problem, the permutation flow shop. Preliminary results show that the algorithm works following the principles of its single objective version.

Future research is planned to extend the motivating theoretical results obtained with the algorithm in single objective problems to the multi-objective cases. Additionally future research comparing the performance of the proposed algorithm against a multi-objective multi-start local search with a replacement criteria based on pareto-dominance relations would be interesting.

Acknowledgments

The authors would like to the thank the anonymous referees for their valuable comments and interesting ideas for future research.

References

1. Coello Coello, C. A., Van Veldhuizen, D. A., and Lamont, G. B. *Evolutionary Algorithms for Solving Multi-objective Problems.* Kluwer Academic Publishers, New York (2002)
2. Aldous, D., Vazirani, U. "Go with the winners algorithms". In Proceedings of the *35th IEEE Symposium on Foundations of Computer Science* pp. 492-501 (1994)
3. Carson, T. Empirical and analitic approaches to understanding local search heuristics. PhD Thesis, University of California. San Diego, California (2001)
4. Deb, K., Agrawal, S., Pratap, A., and Meyarivan, T. "A fast elitist non-dominated sorting algorithm for multi-objective optimization: NSGA-II". In *Parallel Probem Solving from Nature – PPSN VI*, Berlin. Springer (Marc Shoenauer, et al., ed.), pp. 849–858 (2000).
5. Dimitriou, A, Impagliazzo, R. "Towards a rigorous analysis of local optimization algorithms". In Proceedings of the *25th ACM Symposium on the Theory of Computing* (1996)
6. Dimitriou, A, Impagliazzo, R. "Go-with-the-winner algorithms for graph bisection". In Proceedings of the *9th Annual SIAM Symposium on Discrete Algorithms* pp. 510-520 (1998)
7. C. A. Brizuela and R. Aceves, "Experimental genetic operators analysis for the multi-objective permutation flowshop." In Fonseca C., Fleming P., Zitzler E., Deb K. and Thiele L. (Editors) *Evolutionary Multi-criterion Optimization*, LNCS Vol. 2632, pp. 578 - 592, (2003).
8. R. Aceves and C. A. Brizuela, "Análisis Experimental de Operadores Genéticos en NSGA-II para un Problema de Calendarización Multi-objetivo." In Botello S., Hernández A. and Coello C. (Eds.) *Congreso Mexicano de Computación Evolutiva*, pp. 55 - 66, (2003).
9. Deb, K. *Multi-Objective Optimization using Evolutionary Algorithms* John Wiley & Sons (2001)
10. Isibuchi, H. and Murata, T. "Multi-objective Genetic Local Search Algorithm". Proceedings of the 1996 *International Conference on Evolutionary Computation*, pp:119-124, (1996).
11. Tamaki, H., and Nishino, E. A "Genetic Algorithm approach to multi-objective scheduling problems with regular and non-regular objective functions". Proceedings of *IFAC LSS'98*, pp:289-294 (1998).
12. Bagchi, T. P. *Multiobjective Scheduling by Genetic Algorithms.* Kluwer Academic Publishers (1999).
13. Srinivas, N. and Deb, K. Multi-Objective function optimization using non-dominated sorting genetic algorithms. *Evolutionary Computation*, 2(3), pp:221-248 (1995).
14. Serafini, P. "Simulated Annealing for Multiple Objective Optimization Problems". In Tzeng, G., Wang, H., Wen, U., and Yu, P. (editors) *Proceedings of the Tenth International Conference on Multiple Criteria Decision Making: Expand and Enrich the Domains of Thinking and Application*, volume 1 pp. 283-292 (1994)

15. Thompson, M. "Application of Multi Objective Evolutionary Algorithms to Analogue Filter Tuning". In Zitzler, E., Deb, K., Thiele, L., Coello Coello, C. A., and Corne, D., (editors) *First International Conference on Evolutionary Multi-Criterion Optimization.* LNCS No. 1993 pp. 546-559 (2001)
16. Mariano, C. E., Morales, E. "MOAQ an Ant-Q Algorithm for Multiple Objective Optimization Problems". In Banzhaf, W., Daida, J., Eiben, A. E., Garzon, M. H., Honavar, V., Jakiela, M., and Smith, R. E., (editors) *Genetic and Evolutionary Computing COnference (GECCO 99)*, volume 1, pp. 894-901 (1999)
17. Gambardella, L. M., Taillard, E., Agazzi, G. "MACS-VRPTW: A Multiple Ant Colony System for Vehicle Routing Problems with Time Windows". In Corne, D., Dorigo, M., and Glover, F., (editors) *New Ideas in Optimization.* McGraw Hill pp. 63-76 (1999).
18. Kennedy, J., Eberhart, R. C. "Particle Swarm Optimization". In Proceedings of the 1995 *IEEE International Conference on Neural Networks*, pp. 1942-1948 (1995)
19. Hansen, M. P. "Tabu Search in Multiobjective Optimisation: MOTS". In Proceedings of the *13th International Conference on Multiple Criteria Decision Making (MCDM'97)* (1997)
20. Gandibleux, X., Mezdaoui, N., and Fréville, A. "A Tabu Search Procedure to Solve Combinatorial Optimisation Problems". In Caballero, R., Ruiz, F., and Steuer, R. E., (editors) *Advances in Multiple Objectives and Goal Programming*, volume 455 of Lecture Notes in Economics and Mathematical Systems, pp. 291-300 (1997)
21. Binh, T. T., and Korn, U. "An evolution strategy for the multiobjective optimization". In The *Second International Conference on Genetic Algorithms* (Mendel 96) pp. 23-28 (1996)

Exploiting Comparative Studies Using Criteria: Generating Knowledge from an Analyst's Perspective

Daniel Salazar[1,*], Néstor Carrasquero[2], and Blas Galván[3]

[1] Postgrado de Investigación de Operaciones. Facultad de Ingeniería,
Universidad Central de Venezuela. Caracas, 1041-A, Venezuela
daniel_salazar@cantv.net

[2] Grupo de Optimización Combinatoria Emergente (GOCE). Facultad de Ingeniería,
Universidad Central de Venezuela. Caracas, 1041-A, Venezuela
nestor_carrasquero@cantv.net

[3] Instituto de Sistemas Inteligentes y Aplicaciones Numéricas en Ingeniería (IUSIANI)
División de Computación Evolutiva y Aplicaciones (CEANI)
Universidad de Las Palmas de Gran Canaria, Islas Canarias, España
bgalvan@step.es

Abstract. In this work the use of qualitative preferences for classifying and selecting MOEAs is introduced. The classical notions of the Analyst and the so called Prescriptive Analysis are introduced explicitly in EMO, identifying some difficulties in exploiting the results of the comparative studies performed by the current fashion. A methodology is developed that allows the analyst to translate DM's general preferences as well as quantitative benchmarking results into a practical tool for the comparison of MOEAs, facilitating the selection of the proper method and/or parameters for the MCDM problem at hand. A comparative experimentation is performed using well known state of the art functions, allowing drawing clear conclusions about the utility of the proposed methodology. The results are useful for research, practitioners and analysts involved in benchmarking, comparative studies and prescriptive analysis for EMO.

1 Introduction

When Multiple Criteria Decision Making (MCDM) is modelled, different stages and actors can appear as a part of the whole process. An actor is defined as any individual, group of individuals or entity, playing any role during the decision making process [1][2]. In this sense, besides the Decision Maker (DM), it is useful to identify another actor called the *analyst*. Arsham [3] lists a sequence -which allows feedback loops- of tasks accomplished by the analyst:

1. Understanding the Problem
2. Constructing an Analytical Model
3. Finding a Good Solution
4. Communicating the Results with the Decision-Maker

---

* The author is now also with 3.

technical aspects of EMO, and becomes a very relevant point when there is any matter that makes exploring the search space a difficult task. But let us focus now on the particular features an analyst must take into account when he works with a posteriori method, and the difficulties which take place for an analyst for the current EMO state of art.

Once the analyst has determined to use a posteriori method, the second step is to choose a particular MOEA, setting the proper parameters for the selected algorithm. The main tools to tackle this task are the published comparative studies or the studies that the analyst could make by his own. However, when analyzing the literature, is not easy to make solid conclusions about which method is better. Consider the problem of the parameters in the classification results. In [11] the authors found a "hierarchy of method" between eight different MOEAs for a determined group of parameters, but admitted that the situation may be different for other parameters settings and other test problems [11, page 193]. Such situations clearly generate doubts in the analyst. Furthermore, when introduced three well known methods (NSGA-II [12], SPEA2 [13], PESA-II [14]), each author reports their results using different parameters settings (crossover and mutation rates), and even different representation techniques (binary and real) with crossover operators distinct from each other. A brief review of more recent works, including new methods, applications and comparisons studies, ([15]-[21]) shows that the diversity in parameters settings, genetic representations and genetic operators still remains in EMO, not allowing the analyst to make direct comparisons and results extrapolation between researches (in fact, only [15] performed tune before comparing). Finally, even the traditional heuristic of mutating one bit per chromosome may prevent an algorithm to yield maximum efficiency in certain conditions [22][23].

The analyst may consider using auto adaptative or parameter-less algorithms ([24]-[27]) to avoid the problem of setting parameters, but these work schemata may present drawbacks [28] and employing them does not solve the problem completely but the selection of the proper algorithm remains open and features like genetic operators or internal parameters may be changed, possibly affecting the overall performance.

On the other hand, there is the issue of the performance metrics. When solving a MOP, most *a posteriori* MOEAs produce as outcome a set of non dominated solutions. These outcomes can be compared using unary or binary metrics. Some outperformance relationships have been introduced to analyze the quality of unary metrics [29] [30]. Latter in [31] a general framework for comparing outcomes is presented, demonstrating that unary metrics have theoretical limitations when no preferences information is used, and providing some indicators for binary metrics based comparisons. However, even when binary metrics do not have such theoretical limitations of unary metrics [31, page 127], in practice they present some difficulties, like they are not conceived for multiple runs, and some could lead to conclusions different from those which one could come by expressing some preference (e.g. see figures 2 and 3 in [32]).

In summary, selecting a particular solving technique is a decision making problem itself, where the analyst is the actor in charge of making the selection based on the desires, aspirations and general preferences of the DM. Some difficulties to accomplish this task are:

- The published comparative studies were performed employing different parameters settings, and analyzed with different metrics. Such a variety makes difficult to exploit the information presented in those works.
- Even when, for the sake of objectivity, researchers tend to work with metrics not based on DM's preferences, in practice at least some general information could be useful for working with unary and binary metrics ([29] and [33] are based on the assumption that some information is available)

In the following sections we present a methodology for the incorporation of DM's general preferences providing an aid to the analyst in the model building process. The methodology, as is presented, is based on a complete experimentation, which among other things, clarify some questions formulated before.

3 Exploiting Information Using Criteria

The following subsections present the development of the methodology proposed from the generation of experimental data to their interpretation using DM general preferences.

3.1 Experiment Design

In order to generate knowledge about the behaviour of the MOEAs when changing parameters settings, the experimental design was formulated as follows:

- SPEA2, PESA-II and NSGA-II were selected to be studied using functions ZDT2 to ZDT4 and ZDT6 [12] (see table 1).

Table 1. Multiobjective test problems selected for experimentation

Name	Formulation:	Domain	Optimal set
ZDT2	$f_1(x) = x_1$ $f_2(x) = g(x)\left(1 - (x_1/g(x))^2\right)$ $g(x) = 1 + 9\left(\sum_{i=2}^{9} x_i\right)/(n-1)$	$x_i \in [0,1]$ $n = 30$	$x_1 \in [0,1]$ $x_i = 0$ $i \in \{2,...,n\}$
ZDT3	$f_1(x) = x_1$ $f_2(x) = g(x)\left(1 - \sqrt{x_1/g(x)} - \frac{x_1}{g(x)}\sin(10\pi x_1)\right)$ $g(x) = 1 + 9\left(\sum_{i=2}^{9} x_i\right)/(n-1)$	$x_i \in [0,1]$ $n = 30$	$x_1 \in [0,1]$ $x_i = 0$ $i \in \{2,...,n\}$
ZDT4	$f_1(x) = x_1$ $f_2(x) = g(x)\left(1 - \sqrt{x_1/g(x)}\right)$ $g(x) = 1 + 10(n-1) + \sum_{i=2}^{9}\left(x_i^2 - 10\cos(4\pi x_i)\right)$	$x_1 \in [0,1]$ $x_i \in [-5,5]$; $i \in \{2,...,n\}$ $n = 10$	$x_1 \in [0,1]$ $x_i = 0$ $i \in \{2,...,n\}$
ZDT6	$f_1(x) = 1 - \exp(-4x_1)\sin^6(4\pi x_1)$ $f_2(x) = g(x)\left(1 - (x_1/g(x))^2\right)$ $g(x) = 1 + 9\left[\left(\sum_{i=2}^{9} x_i\right)/(n-1)\right]^{1/4}$	$x_i \in [0,1]$ $n = 10$	$x_1 \in [0,1]$ $x_i = 0$ $i \in \{2,...,n\}$

- The chromosomes were represented with binary strings with one-point crossover and bitwise mutation (for the number of bits see table 1).
- Parameters were fixed as follows: six values of crossover rate (pc) and three for mutation (pm) rates were employed (pc = {0.5, 0.6, 0.7, 0.8, 0.9, 1}, pm = {0.001, 0.010, 0.100}). Population size – archive size ratios (M:N) were fixed to 1:1 and 2:1. The maximum number of generation was set to 250.
- For each function and parameters settings (pc, pm, M:N) ten runs were performed. The outcomes were analyzed comparing the mean value of metric S (for hypervolume covered) [30].
- Computational complexity was not considered.

3.2 Building Preferences Maps

To aid the analyst during the formulation of the whole decision making model, preferences maps were conceived as the main tool to translate the quantitative information obtained from the metrics into qualitative information. Naturally, the selection of the metrics is related with the features considered important by the DM. In our case, we chose unary metrics because they allow statistic treatment. The selection of S is recommended in [30] for few objectives. Any Analyst/DM may base his maps in other metrics.

Figure 2 present the results of metric S for function ZDT4. The matrix layout allows comparisons between methods and between parameters for each method. The reader can easily note some variations in means values. At this stage, it is necessary to express some general criteria for classify the vectors of means (the curves). Hence, four preference relationships were proposed to build the classifications, they express: ▷ *Strictly preference*, ⊵ *Preference*, ‖ *Indifference*, ⇎ *No preference*. They were inspired in the relationships formulated in [31], but not for express dominance but preference.

Finally, very simple and intuitive statements were formulated here as classification criteria (lexicographically):

- *if* one curve is closer the optimal than another, *then* that curve is strictly preferable, *else*
- *if* one curve is closer to the optimal or at the same level than another but it has lower variance, *then* that curve is preferable, *else*
- *if* one curve intersects another, *then* that curve is indifferent, *else*
- the curve is no preferable.

Substituting the if-then-else sentences any analyst/DM can build an appropriate classification for his/her criteria.

For the construction of the map, few topics were considered. First, the crossover rate was divided in two groups, low crossover (LC) with comprises pc from 0.5 to 0.7 and high crossover for pc from 0.8 to 1.0. Then, for each group and populations ratio, an assessment mutation rates were accomplished, based on the rules described above.

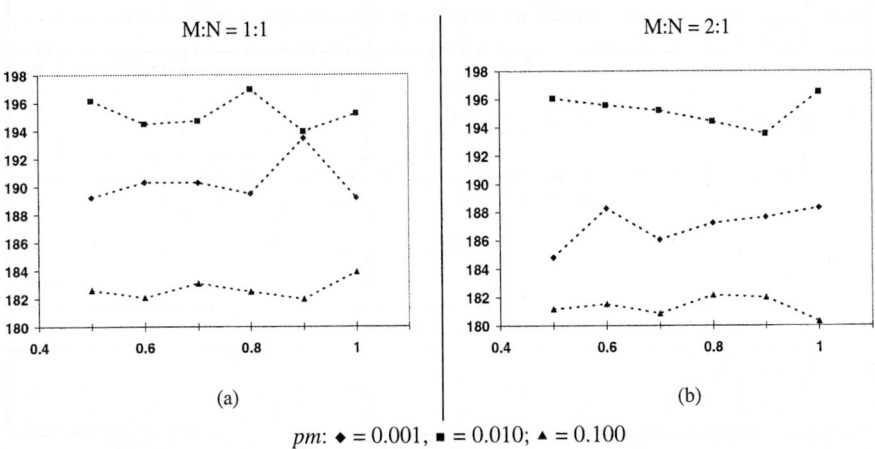

Fig. 2. Mean values of metric S vs. pc for SPEA2 over function ZDT4

M:N	Metric S					
	1:1			2:1		
LC	10^{-3}	▷	⋫	10^{-3}	▷	⋫
	⋫	10^{-2}	⋫	⋫	10^{-2}	⋫
	▷	▷	10^{-1}	▷	▷	10^{-1}
HC	10^{-3}	▷	⋫	10^{-3}	▷	⋫
	⋫	10^{-2}	⋫	⋫	10^{-2}	⋫
	▷	▷	10^{-1}	▷	▷	10^{-1}
M:N	Global population-size ratio					
	1:1			2:1		
1:1	---			⋫		
2:1	▷			---		

Fig. 3. SPEA2 preference map for metric S and function ZDT4. For reading always start vertically!

Figure 3 shows an excerpt of the whole map for function ZDT6. Note that the numbers in the diagonals indicate the different mutation rates. The table is reading as follows, for assess the effect of pc and pm, the analyst choose one crossover rates group (LC or HC) and one populations ratio (1:1 or 2:1), then starting *vertically* for one pm value and finding the intersection with another pm value, he/she has the

MOEA		ZDT2						ZDT3					
	M:N	1:1			2:1			1:1			2:1		
SPEA2	LC	10^{-3}	▶̸	▶̸	10^{-3}	▶̸	▶̸	10^{-3}	∥	▶̸	10^{-3}	∥	▶̸
		▷	10^{-2}	▶̸	▷	10^{-2}	▶̸	∥	10^{-2}	▶̸	∥	10^{-2}	▶̸
		▷	▷	10^{-1}	▷	▷	10^{-1}	▷	▷	10^{-1}	▷	▷	10^{-1}
	HC	10^{-3}	▶̸	▶̸	10^{-3}	▶̸	▶̸	10^{-3}	▶̸	▶̸	10^{-3}	▶̸	▶̸
		▷	10^{-2}	▶̸	▷	10^{-2}	▶̸	▷	10^{-2}	▶̸	▷	10^{-2}	▶̸
		▷	▷	10^{-1}	▷	▷	10^{-1}	▷	▷	10^{-1}	▷	▷	10^{-1}
		Global population-size ratio											
	1:1	---			∥			---			∥		
	2:1	∥			---			∥			---		
NSGA-II	LC	10^{-3}	▷	▶̸	10^{-3}	▷	▶̸	10^{-3}	▶̸	▶̸	10^{-3}	▶̸	▷
		▶̸	10^{-2}	▶̸	▶̸	10^{-2}	▶̸	▷	10^{-2}	▶̸	▷	10^{-2}	▷
		▷	▷	10^{-1}	▷	▷	10^{-1}	▷	▷	10^{-1}	▶̸	▶̸	10^{-1}
	HC	10^{-3}	▷	▶̸	10^{-3}	∥	▶̸	10^{-3}	▶̸	▶̸	10^{-3}	∥	▷
		▶̸	10^{-2}	▶̸	∥	10^{-2}	▶̸	▷	10^{-2}	▶̸	∥	10^{-2}	▷
		▷	▷	10^{-1}	▷	▷	10^{-1}	▷	▷	10^{-1}	▶̸	▶̸	10^{-1}
		Global population-size ratio											
	1:1	---			▶̸			---			▶̸		
	2:1	▷			---			▷			---		
PESA-II	LC	10^{-3}	∥	▶̸	10^{-3}	∥	▶̸	10^{-3}	∥	▶̸	10^{-3}	∥	▶̸
		∥	10^{-2}	▶̸	∥	10^{-2}	▶̸	∥	10^{-2}	▶̸	∥	10^{-2}	▶̸
		▷	▷	10^{-1}	▷	▷	10^{-1}	▷	▷	10^{-1}	▷	▷	10^{-1}
	HC	10^{-3}	∥	▶̸	10^{-3}	▶̸	▶̸	10^{-3}	∥	▶̸	10^{-3}	⊵	▶̸
		∥	10^{-2}	▶̸	▷	10^{-2}	▶̸	∥	10^{-2}	▶̸	▶̸	10^{-2}	▶̸
		▷	▷	10^{-1}	▷	▷	10^{-1}	▷	▷	10^{-1}	▷	▷	10^{-1}
		Global population-size ratio											
	1:1	---			▶̸			---			⊵		
	2:1	⊵			---			▶̸			---		

Fig. 4. Preference map for metric S and function ZDT2 and ZDT3

MOEA	M:N	ZDT4						ZDT6					
		1:1			2:1			1:1			2:1		
SPEA2	LC	10^{-3}	▷	⊯	10^{-3}	▷	⊯	10^{-3}	⊯	⊯	10^{-3}	∥	⊯
		⊯	10^{-2}	⊯	⊯	10^{-2}	⊯	▷	10^{-2}	⊯	∥	10^{-2}	⊯
		▷	▷	10^{-1}	▷	▷	10^{-1}	▷	▷	10^{-1}	▷	▷	10^{-1}
	HC	10^{-3}	▷	⊯	10^{-3}	▷	⊯	10^{-3}	⊯	⊯	10^{-3}	∥	⊯
		⊯	10^{-2}	⊯	⊯	10^{-2}	⊯	▷	10^{-2}	⊯	∥	10^{-2}	⊯
		▷	▷	10^{-1}	▷	▷	10^{-1}	▷	▷	10^{-1}	▷	▷	10^{-1}
		Global population-size ratio											
	1:1	---			⊯			---			⊯		
	2:1	▷			---			▷			---		
NSGA-II	LC	10^{-3}	▷	▷	10^{-3}	▷	▷	10^{-3}	▷	⊯	10^{-3}	▷	▷
		⊯	10^{-2}	∥	⊯	10^{-2}	▷	⊯	10^{-2}	⊯	⊯	10^{-2}	⊯
		⊯	∥	10^{-1}	⊯	⊯	10^{-1}	▷	▷	10^{-1}	⊯	▷	10^{-1}
	HC	10^{-3}	∥	▷	10^{-3}	▷	▷	10^{-3}	▷	⊯	10^{-3}	▷	▷
		∥	10^{-2}	▷	⊯	10^{-2}	▷	⊯	10^{-2}	⊯	⊯	10^{-2}	⊯
		⊯	⊯	10^{-1}	⊯	⊯	10^{-1}	▷	▷	10^{-1}	⊯	▷	10^{-1}
		Global population-size ratio											
	1:1	---			⊯			---			⊯		
	2:1	▷			---			▷			---		
PESA-II	LC	10^{-3}	▷	▷	10^{-3}	▷	∥	10^{-3}	▷	⊯	10^{-3}	▷	⊯
		⊯	10^{-2}	⊯	⊯	10^{-2}	⊯	⊯	10^{-2}	⊯	⊯	10^{-2}	⊯
		⊯	▷	10^{-1}	∥	▷	10^{-1}	▷	▷	10^{-1}	▷	▷	10^{-1}
	HC	10^{-3}	▷	⊯	10^{-3}	▷	⊯	10^{-3}	▷	⊯	10^{-3}	▷	⊯
		⊯	10^{-2}	⊯	⊯	10^{-2}	⊯	⊯	10^{-2}	⊯	⊯	10^{-2}	⊯
		▷	▷	10^{-1}	▷	▷	10^{-1}	▷	▷	10^{-1}	▷	▷	10^{-1}
		Global population-size ratio											
	1:1	---			∥			---			▷		
	2:1	∥			---			⊯			---		

Fig. 5. Preference map for metric S and function ZDT4 and ZDT6

relationship that stands for this case. For instance, for LC and 1:1, we have $10^{-3} \not\triangleright 10^{-2}$ and $10^{-3} \triangleright 10^{-1}$, and in the same way $10^{-1} \not\triangleright 10^{-2}$ and $10^{-1} \not\triangleright 10^{-3}$. The same procedure is applied for the global M:N, but starting always *vertically*. In this case, $1:1 \trianglerighteq 2:1$ because in most of the cases the curves for 1:1 are above or at the same level of the curves for 2:1 for each *pm*.

So far, we have not defined what *at the same level* means. It is clear that is not possible to expect numerical equality in the means, but there is space for statistical equality or intervals. Visual inspection of the results is possible in some cases, but impractical if a huge number of functions and metrics are considered during the experimentation and the analysis. Therefore, the preferences maps can be generated automatically employing the lexicographical rules and studying the statistical equivalence of the means for assure the conclusions are according with the preferences statements. Other options is to calculate the difference between means labeling the means as equal if the value belongs to an interval previously defined (by the analyst/DM), or if the amount of the difference is lower than a certain percentage of value between the upper and the lower means of the curves under comparison.

3.3 Extracting Information from Preferences Maps

Figures 4 and 5 report the analysis of the metric *S* for the outcomes produced by SPEA2, NSGA-II and PESA-II over functions ZDT2, ZDT3, ZDT4 and ZDT6.

Note that the maps may help in selecting the proper parameters for a particular method, providing information about what dimension in *pm* and which population size ratio are preferable. However, more information can be obtained from the qualitative classification introducing weights as numerical equivalence of the relationships. For instance, the sensitivity to population size ratio is assessed assigning $\triangleright = 3$, $\trianglerighteq = 2$, and $\| = 1$ and counting the occurrences of each preference relationship in both maps (in *global population-size ratio* section), then a weighted sum is calculated. Figure 6 present the results. Notice that SPEA2 is less sensible to population size ratio than the other methods in terms of hypervolume covered (*S*), while NSGA-II is the most sensitive.

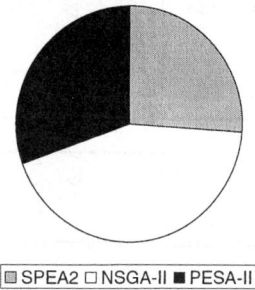

Preference Relationship	Weight	SPEA2	NSGA-II	PESA-II
$\triangleright$	3	0	2	0
$\trianglerighteq$	2	2	2	3
$\|$	1	2	0	1
Total (mean)		1.5	2.5	1.75

(a)   (b)

Fig. 6. Sensitivity of metric *S* to M:N, for SPEA2, NSGA-II and PESA-II

An assessment of the sensitivity to the crossover rate is also possible. In this case, the number of changes (jumps) in the preference relationships when *pc* changes from LC to HC is tabulated in figure 7b. The jumps are labeled as short jumps when the preference relationship changes to an immediate one (e.g. from ▷ to ⊵), otherwise they are labeled as long jumps. Short jumps are weighted by 1 and long jumps by 2. The weighted sum is then presented graphically in Fig. 7a and numerically in Fig 7b. Note that all MOEAs are less or equal sensitive to *pc* when M:N ratio is 1:1.

Notice in Fig 7 that there is a hierarchy of sensitivity to *pc*, which could lead to conclude SPEA2 ▷ PESA-II ▷ NSGA-II for the studied feature. In general, for any analyst/DM interested in selecting an algorithm, starting from the outcomes and calculating the metrics values with their statistics, preferences maps may be build according to the DM's general preferences. Then, by analyzing the maps, sensitivity information can be extracted. Finally, for making a decision about the algorithm to choose, a new structure of preferences must be expressed, indicating if-then-else sentences (if lexicographically) or weights (for weighted sum), that will identify the proper method. For example, if the DM's needs point to a search performed with small populations and as less sensitive as possible to *pc*, by examining the figures 6 and 7 the analyst can conclude that SPEA2 is the more appropriate to satisfy DM's desires. Nevertheless, the conclusions depend on the needs, preferences, aspiration levels, etc, expressed by the DM.

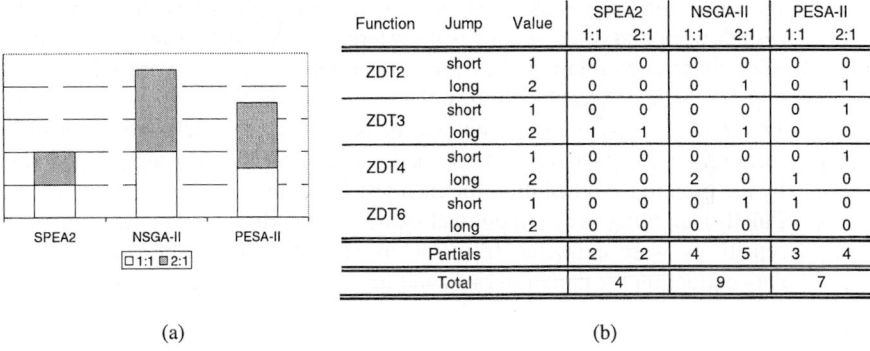

(a) (b)

Fig. 7. Sensitivity of metric *S* to *pc* for SPEA2, NSGA-II and PESA-II

In addition, consider that the maps allow identifying bounds for parameters settings. For instance, by examining figures 4 and 5 and making a census of strictly preference and preference relationships when *pm* changes from one value to another, an analyst can easily conclude that $[10^{-3}, 10^{-2}]$ is the best interval for fixing *pm* for ZDT2-ZDT4 and $[10^{-2}, 10^{-1}]$ for ZDT6, when working with SPEA2 and PESA-II. It is worthy to note these intervals enclose the one-bit mutation rate value.

In summary, the methodology presented allows the construction of a simple and a useful tool that helps the analyst during the decision making modeling. With a small number of general preferences, the methodology guides to the classification of the numerical results from a DM/Analyst perspective (translation the data into preferences). Including more metrics in the preferences maps, it is possible to extract more information, like convergence or diversity/distribution, which can enrich the number of options to consider during the decision making model building.

4 Conclusions

In the presented work we have introduced explicitly the classical notion of the analyst into EMO, remarking the decision making modeling as one of the main steps or stages of a real MDCM problem. Then we have identified some difficulties which take place due to the diversity of values in parameters settings and metrics that can be found when analyzing the comparative studies already published. Finally, we have presented and described a methodology for the incorporation of DM's general preferences during the MOEA selection and the parameters settings. This methodology showed the ability to help the analyst in extracting and exploit the information obtained from comparative studies and translate them into preferences relationships, which can help to choose a particular technique and set the parameters, in a very easy way.

Some remarks of the proposed methodology are:

- An appropriate experimentation was carried out to assess the influence of the crossover and mutation rates, and the population sizes ratio in the behavior of the selected MOEAs.
- Employing a metric for measure covered hypervolume, the means values of several experiments were calculated.
- By means of simple and very general lexicographic rules, the quantitative results were classified in terms of preferences (translating into preferences). Four preference relationships for comparing the results were employed to express the lexicographical rules, then
- Preferences maps were introduced for representing the classification of the results in a compact way. These maps contain all the information generated by the preference translation for each test function.
- Finally, applying very simple techniques of weighting, some information of sensitivity were extracted from the maps. Additionally, brief examples of data extraction for mutation rates and for MOEA selection were presented.

References

1. Roy, B.: A French-English Decision Aiding glossary. Newsletter of the European Working Group "Multicriteria Aid for Decisions". Series 3, n°1, Spring, 2000.
2. Valls, A.: ClusDM: A Multiple Criteria Decision Making Method for Heterogeneous Data Sets. PhD. Thesis. Universitat Politècnica de Catalunya. September 2002.
3. Arsham, H.: Applied Management Science: Making Good Strategic Decisions. 1994. In http://home.ubalt.edu/ntsbarsh/Business-stat/opre/opre640.htm (last visited in October 2004).
4. Powell, D.: Multiobjective Optimization with Genetic Algorithms is Now Considered Mainstream. Evolutionary Computation in Industry. Workshop Proceedings, Tutorials, Late Breaking Papers, and Evolutionary Computation in Industry Track Presentations. Genetic and Evolutionary Computation Conference (GECCO-2004) (CD-ROM) X-CD Technologies. (2004).

5. Van Veldhuizen, D., and Lamont, G.: Multiobjective Evolutionary Algorithms: Analyzing the State-of-the-Art. Evolutionary Computation 8(2) (2000) 125-147.
6. Corne, D.W., Knowles, J.D.: No Free Lunch and Free Leftovers Theorems for Multiobjective Optimization Problems. Evolutionary Multi-Criterion Optimization (EMO 2003), Proceedings of the Second International Conference, Portugal (2003) 327-341.
7. Corne, D.W., Knowles, J.D.: Some Multiobjective Optimizers are Better than Others. Proceedings of the IEEE Congress on Evolutionary Computation. (2003) 2506-2512.
8. Horn, J.: F1.12: Multicriteria Decision Making and Evolutionary Computation. IlliGAL Report No. 9600X. University of Illinois. (1996).
9. Barba-Romero, S., Pomerol, J-C.: Decisiones Multicriterio. Fundamentos Teóricos y Utilización Práctica. Universidad de Alcalá. (1997).
10. Zeleny, M.: Multiple Criteria Decision Making. McGraw-Hill. (1982).
11. Zitzler, E., Deb, K., Thiele, L.: Comparison of Multiobjective Evolutionary Algorithms: Empirical Results. Evolutionary Computation **8**(2) (2000) 173-195.
12. Deb, K., Pratap, A., Agarwal, S., Meyarivan, T.: A Fast and Elitist Multi-Objective Genetic Algorithm: NSGA-II. KanGAL Report No. 200001. Kanpur Genetic Algorithms Laboratory (KanGAL). Indian Institute of Technology (2001).
13. Zitzler, E., Laummans, M., Thiele, L.: SPEA2: Improving the Strength Pareto Evolutionary Algorithm. TIK Report No. 103. Swiss Federal Institute of Technology (ETH). Computer Engineering and Networks Laboratory (TIK). (2001).
14. Corne, D.W., Jerram, N.R., Knowles, J.D., and Oates, M.J.: PESA-II: Region-based Selection in Evolutionary Multiobjective Optimization. Proceedings of the Genetic and Evolutionary Computation Conference (GECCO-2001), Morgan Kaufmann Publishers (2001) 283-290.
15. Khare, V., Yao, X., Deb, K.: Performance Scaling of Multi-objective Evolutionary Algorithms. Evolutionary Multi-Criterion Optimization (EMO 2003), Proceedings of the Second International Conference, Portugal (2003) 376-390.
16. Koch, T.E., and Zell, A.: Multi-Objective Clustering Selection Evolutionary Algorithm. Proceedings of Genetic and Evolutionary Computation Conference (GECCO-2002), Morgan Kaufmann, San Francisco, CA. (2002) 423-430.
17. Ishibuchi, H., and Shibata, Y.: An Empirical Study on the Effect of Matting Restriction on the Search Ability of EMO Algorithm. Evolutionary Multi-Criterion Optimization (EMO 2003), Proceedings of the Second International Conference, Portugal (2003) 433-447.
18. Wanatabe, S., Hiroyasu, and T. Miki, M.: Multi-objective Rectangular Packing Problem and Its Applications. Evolutionary Multi-Criterion Optimization (EMO 2003), Proceedings of the Second International Conference, Portugal (2003) 565-577.
19. Greiner, D., Galván, B., and Winter, G.: Safety Systems Optimum Design by Multicriteria Evolutionary Algorithms. Evolutionary Multi-Criterion Optimization (EMO 2003), Proceedings of the Second International Conference, Portugal (2003) 722-736.
20. Shu, L-S., Ho, S-J, Ho, S-Y., Chen, J-H, and Hung, M-H.: A Novel Multi-objective Orthogonal Simulated Annealing Algorithm for Solving Multi-objective Optimization Problems with a Large Number of Parameters. Proceedings of Genetic and Evolutionary Computation Conference (GECCO-2004), Springer-Verlag, Germany. (2004) 737-747.
21. Deb, K., and Gupta, N. K.: Optimal Operating Conditions for Overhead Crane Maneuvering Using Multi-objective Evolutionary Algorithms. Proceedings of Genetic and Evolutionary Computation Conference (GECCO-2004), Springer-Verlag, Germany. (2004) 1042-1053.

22. Laumanns, M., Zitzler, E., and Thiele, L.: On The Effects of Archiving, Elitism, and Density Based Selection in Evolutionary Multi-Objective Optimization. En E. Zitzler et al. (eds.): Evolutionary Multi-criterion Optimization (EMO 2001), First International Conference, EMO 2001, Zurich, Switzerland, March 7-9 2001, Proceedings. Lecture Notes in Computer Science Vol. 1993, Springer. (2001) 181-196.
23. Ochoa, G.: Setting the Mutation Rate: Scope and Limitations of the 1/L Heuristics. Proceedings of Genetic and Evolutionary Computation Conference (GECCO-2002), Morgan Kaufmann, San Francisco, CA. (2002) 495-502.
24. Toscano, G., Coello, C.: The Micro Genetic Algorithm 2: Towards Online Adaptation in Evolutionary Multiobjective Optimization. Evolutionary Multi-Criterion Optimization (EMO 2003), Proceedings of the Second International Conference, Portugal (2003) 252-266.
25. Büche, D., Müller, S., Koumoutsatkos, P.: Self-Adaptation for Multi-objective Evolutionary Algorithms. Evolutionary Multi-Criterion Optimization (EMO 2003), Proceedings of the Second International Conference, Portugal (2003) 267-281.
26. Groşan, C.: An Evolutionary Approach for Multiobjective Optimization using Adaptive Representation of Solutions. Late Breaking Papers. Workshop Proceedings, Tutorials, Late Breaking Papers, and Evolutionary Computation in Industry Track Presentations. Genetic and Evolutionary Computation Conference (GECCO-2004) (CD-ROM) X-CD Technologies. (2004).
27. Salazar, D., Galván, B., and Winter, G.: Enhancing A Multiobjective Evolutionary Algorithm Through Flexible Evolution. Late Breaking Papers. Workshop Proceedings, Tutorials, Late Breaking Papers, and Evolutionary Computation in Industry Track Presentations. Genetic and Evolutionary Computation Conference (GECCO-2004) (CD-ROM) X-CD Technologies. (2004).
28. Laumanns, M., Rudolph, G., and Schwefel, H.-P.: Mutation Control and Convergence in Evolutionary Multi-Objective Optimization. In Matousek and Osmera (eds.): Proceedings of the 7th International Mendel Conference on Soft Computing (MENDEL 2001), Czech Republic, 2001. (2001) 24-29.
29. Hansen, P., Jaszkiewicz, A.: Evaluating the quality of approximations of the non-dominated set. Technical Report IMM-Rep-1998-7. Technical University of Denmark, Lyngby, Denmark (1998)
30. Knowles, J.D., Corne, D.W.: On Metrics for Comparing Non-Dominated Sets. In Proceedings of the 2002 Congress on Evolutionary Computation Conference (CEC02), IEEE Press (2002) 711-716
31. Zitzler, E., Laummans, M., Thiele, L. Fonseca, C. M., Grunert da Fonseca, V.: Performance Assessment of Multiobjective Optimizers: An Analysis and Review. IEEE Transactions on Evolutionary Computation 7(2) (2003) 117-132.
32. Bosman, P., and Thierens, D.: The Balance Between Proximity and Diversity in Multiobjective Evolutionary Algorithms. IEEE Transactions on Evolutionary Computation 7(2) (2003) 174-188.
33. Farhang-Mehr, A., and Azarm, S.: Minimal Sets of Quality Metrics. Evolutionary Multi-Criterion Optimization (EMO 2003), Proceedings of the Second International Conference, Portugal (2003) 405-417.

A Multiobjective Evolutionary Algorithm for Deriving Final Ranking from a Fuzzy Outranking Relation

Juan Carlos Leyva-Lopez[1,2] and Miguel Angel Aguilera-Contreras[2]

[1] Universidad de Occidente,
Carr. a Culiacancito Km. 1.5 Culiacan, Sinaloa, Mexico, 80020
jleyva@culiacan.udo.mx
[2] Universidad Autonoma de Sinaloa,
Prol. Josefa O. de Dominguez s/n, Ciudad Universitaria
Culiacan, Sinaloa, Mexico, 80040
{jleyva, aguilera}@uas.uasnet.mx

Abstract. The multiple criteria aggregation methods allow us to construct a recommendation from a set of alternatives based on the preferences of a decision maker. In some approaches, the recommendation is immediately deduced from the preferences aggregation process. When the aggregation model of preferences is based on the outranking approach, a special treatment is required, but some non-rational violations of the explicit global model of preferences could happen. In this case, the exploitation phase could then be treated as a multiobjective optimization problem. In this paper a new multiobjective evolutionary algorithm, which allows exploiting a known fuzzy outranking relation, is introduced with the purpose of constructing a recommendation for ranking problems. The performance of our algorithm is evaluated on a set of test problems. Computational results show that the multiobjective genetic algorithm-based heuristic is capable of producing high-quality recommendations.

1 Introduction

Multiple Criteria Decision Analysis provides two major approaches of constructing a global preference model from an actor involved in the decision process. The first one is the functional model, which has been widely used within the framework of multi-attribute utility theory (e.g. [10, 17, 30]). The second one is the relational model, which has its most known representation in the form of a fuzzy or crisp outranking relation (e.g. [26]). This paper is concerned with the outranking approach to Multiple Criteria Decision Aid. Methods related to this approach, including the well-known family of ELECTRE methods, are often presented as the combination of two phases: aggregation (or construction) and exploitation. The aggregation process corresponds to the operation, which transforms the marginal evaluations of separate criteria into a global outranking relation between every pair of alternatives, which is generally neither transitive nor complete. Outranking relations, in most methods, are built using a concordance-discordance principle.

It is well known that this principle does not, in general, lead to binary relations possessing "remarkable properties" such as transitivity and completeness [2]. The exploitation process deals with the outranking relation in order to clarify the decision through a partial or total preordering reflecting some of the irreducible indifferences and incomparabilities [8]. ELECTRE-III, PROMETHEE and other methods for decision aid (e.g. [25, 1, 8]) build and exploit a fuzzy outranking relation.

Let A be the set of decision alternatives or potential actions and let us consider a fuzzy outranking relation S_A^σ defined on AXA; this means that we associate with each ordered pair $(a,b) \in AXA$ a real number $\sigma(a,b)$ $(0 \leq \sigma(a,b) \leq 1)$ reflecting the degree of strength of the arguments favoring the crisp outranking aSb. The exploitation phase transforms the global information included in S_A^σ into a global ranking of the elements of A. Usually; three different ways are used [8]:

1: transform S_A^σ into another valued relation R that presents some interesting property needed for ranking purposes, i.e. transitivity,
2: determine a crisp binary relation, close to S_A^σ which presents crisp properties needed for ordering,
3: use a ranking method to obtain a score function.

Way 1 includes the process of finding the transitive closure or the intersection of traces. Way 3 is most commonly used in classical procedures like ELECTRE-III and PROMETHEE. But the main difficulty consists in finding reasonable ways of dealing with the intransitivities without losing too much of the contents of the outranking relation. In this sense, the methods included in ways 1 and 2 lose information coming from S_A^σ when exploiting a not so close transitive valued relation R, or a crisp binary relation with desirable properties for ranking purposes. On the other hand, the methods based in score functions do not perform well in presence of irrelevant alternatives or in case of complex graphs with many circuits. Nonrational situations could happen when the prescription is constructed. Most significant is the following: Suppose that a_i and a_j are two actions such that $\sigma(a_i,a_j) \geq \lambda$ and $\sigma(a_j,a_i) \leq \lambda - \beta$, $(\beta > 0)$; if $\lambda \geq c$ and $\beta \geq t$ (c and t representing consensus and threshold levels respectively), we should accept that "a_i outranks a_j" $(a_i S^\lambda a_j)$ and "a_j does not outrank a_i" $(a_j n S^\lambda a_i)$; in this case the global preference model captured in outranking relation is giving a presumed preference favoring a_i. However, a score function or any other similar method could lead to a final ordering in which a_j is ranked better. ELECTRE and PROMETHEE methods do not have a way to minimize this kind of irregularity. In any case, the exploitation phase could then be treated as a multiobjective optimization problem [19]. In this way, a number of solutions can be found which provide the decision maker with insight into the characteristics of the problem before a final solution is chosen.

Evolutionary Multiobjective Optimization (EMOO) seeks to optimize the components of a vector-valued cost function. Unlike single objective optimization, the solution to this problem is not a single point, but a family of points known as the Pareto-optimal set. Each point in this surface is optimal in the sense that no improvement can be achieved in one cost vector component that does not lead to degradation in at least one of the remaining components. Assuming, without loss of generality, a minimization problem, each element in the Pareto-optimal set constitutes a non-inferior solution to the EMOO problem. Non-inferior solutions have been obtained by solving appropriately formulated ranking problems. Methods used include the contained in the ways (1), (2) and (3) and recently a method based on a genetic algorithm [19, 7].

By maintaining a population of solutions, multiobjective evolutionary algorithms (MOEAs) can search for many non-inferior solutions in parallel. This characteristic makes MOEAs very attractive for solving EMOO problems.

In this paper, we propose a multiobjective evolutionary algorithm for improving the quality of recommendation when a fuzzy outranking relation is exploited, which is of particular interest for solving the Multiple Criteria Ranking Problem. This approach rests on the main idea of reducing differences between the global model of preferences and final ranking. In the next section the exploitation of a fuzzy outranking relation formulated as a multiobjective optimization problem is described, and on this background we present our proposal in section 3. A test problem and computational result is given in section 4, and finally, in section 5 some conclusions are discussed.

2 The Exploitation of a Fuzzy Outranking Relation as a Multi-objective Combinatorial Optimization Problem

Let A be a finite set of decision alternatives, which is the object of the decision process. This set is not the universe of the potentially feasible alternatives; it is only the set under consideration in a specific decision problem. Let $\sigma(a,b)$ be a valued binary relation defined on AXA with image in [0,1]. Note that the fuzzy outranking relation S_A^σ of the past section is a particular case of $\sigma(a,b)$. $\sigma(a,b)$ can be interpreted as the credibility degree of the predicate "a is at least as good as b". Let λ be a cut level such that if $\sigma(a,b) \geq \lambda$, we say that a outranks b with credibility λ, denoted by $aS^\lambda b$. Otherwise, the outranking is rejected $anS^\lambda b$.

We assume the existence of a threshold $\beta > 0$ such that if $aS^\lambda b$ and also $\sigma(b,a) \leq (\lambda - \beta)$, then there is an asymmetric preference relation favoring a that will be denoted by $aP^{\lambda,\beta}b$. One can agree that for some values of λ and β, the conditions defining $P^{\lambda,\beta}$ are good arguments for justifying a strict preference relation in the sense proposed by Roy (cf. [24]).

Let E be a way of exploiting σ and R_A the complete ranking derived from applying E to σ. E is a function assigning a ranking R_A to each σ defined on AXA. R_A defines a weak order R on A. $\forall (a,b) \in AXA$, aRb if and only if b is not ranked before a in R_A. We think that the quality of a final ranking should be judged according to the number of its discrepancies and concordances with σ and the crisp relations S^λ, and $P^{\lambda,\beta}$. Let V be the set of strong discrepancies (violations) defined as $V = \{(a,b) \in AXA: aP^{\lambda,\beta}b, bRa\}$ and $n_V = \text{cardinality of }(V)$. Note that n_V is a function of R, λ, and β.

We propose to consider the best ordering as the best compromise solution of the following multiple objective optimization problem:

$$Min(n_V), \quad Min(f), \quad Max(\lambda) \tag{1}$$

Subject to

$$R \subset AXA, \quad \lambda, \beta \in [0,1], \quad \lambda \geq \lambda_0$$

(λ_0 is a minimum level of credibility, usually greater than 0.5)

Where f is a measure counting the number of incomparable pairs i.e. counting all the pairs $(a,b) \in AXA$ such that $anS^\lambda b$ and $bnS^\lambda a$.

The minimization of n_V can be seen as a process of reducing the magnitude of the arguments against R. The minimization of f can be seen as a process of increasing the number of comparable pair of alternatives in S^λ. Increasing λ improves the credibility of $P^{\lambda,\beta}$ and S^λ, relations on which the ranking is based. The most important objective is n_V. A value of λ between 0.65 and 0.75 is often considered good and no further increments are necessary. The structure of (1) strongly suggests the use of evolutionary algorithms.

3 Multiobjective Evolutionary Algorithm for Exploiting a Fuzzy Outranking Relation

This section presents an evolutionary algorithm that solves (1) overcoming the limitations of the genetic algorithm presented by [19] by taking advantage of the structure of the objective space.

3.1 Multiobjective Evolutionary Optimization

Evolutionary algorithms are stochastic search techniques that mimic the natural selection process. The goal of evolutionary algorithms is to obtain better individuals (i.e., solutions) as the algorithm progresses. In any given generation (i.e., iteration), the population of individuals is combined and altered to obtain a children population. The parents and children undergo an evaluation and selection process, where the better individuals have a higher chance of survival. Algorithms sharing the same spirit and based on the same natural selection principle have been proposed in the fields of evolutionary strategies [28], evolutionary programming [9], and genetic algorithms [14].

Evolutionary algorithms have been applied to solve complex problems where traditional optimization methods have failed to provide a good solution. Solving optimization problems with multiple objectives is, generally, a very difficult task. Evolutionary algorithms are particularly well suited for multiobjective optimization due to their ability to explore a vast set of alternatives, partially because they are population-based and can evaluate several solutions in parallel [33]. Evolutionary algorithms can also be designed to search for the efficient frontier in a single run, without making assumptions about the shape and mathematical properties of the frontier [3]. Moreover, there are few competitive alternatives to multiobjective optimization with noise and uncertain objective functions [15].

Multiobjective evolutionary algorithms have become an active line of research since the first algorithm was proposed in the mid eighties [27]. Taxonomy of multiobjective evolutionary algorithms is possible based on the decision maker's preferences being made before, during, or after the optimization process. [32], present such a classification of multiobjective evolutionary algorithms based on prior, progressive, and posterior articulation of preferences. In the latter years, the algorithms based on posterior articulation of preferences have received the most attention. These evolutionary algorithms move the population of solutions toward and efficient frontier. Among these algorithms, the most prominent are the Multiobjective Genetic Algorithm (MOGA) [11], the Niched Pareto Genetic Algorithm (NPGA) [16], the Nondominated Sorting Genetic Algorithm (NSGA) [29, 6], the Strength Pareto Evolutionary Algorithm (SPEA2) [33], and the Pareto Archived Evolution Strategy (PAES) [18]. For a thorough exposition of multiobjective evolutionary algorithms the reader is referred to [3, 4, 32, 5].

3.2 The Multiobjective Evolutionary Algorithm

In this section we present a multiobjective evolutionary algorithm based on posterior articulation of preferences, able to exploit a known fuzzy outranking relation with the purpose of constructing a recommendation for the multiple criteria ranking problem. The algorithm borrows fundamental elements from MOGA [11], which has become one of the leading multiobjective evolutionary algorithms.

In the following subsections we present further detail on the fundamental aspects of the Multiobjective evolutionary algorithm.

Encoding the Solutions. A potential solution of a ranking problem is represented as an ordinal representation. In general, a potential solution is a ranking of the set of decision alternatives or actions by decreasing order of preference. These actions (known as genes) are joined together forming a string of values (known as chromosome). Any symbol in this string is refereed to as an allele [13, 20]. The chromosome is represented as the string of m-ary alphabet where m is the number of actions into the decision problem. In such representation, each action is coded into m-ary form. Actions are then linked together to produce one long m-ary string or chromosome. An action coded with value a_{k_i} in the i-th entry of the string means that the action coded with value a_{k_i} is ranked in the i-th place of the ordering and a_{k_i} is preferred to a_{k_j} if $i<j$, where $a_{k_i} \in A = \{a_1, a_2, ..., a_m\}$, $i = 1,2,...,m$, and $[k_1, k_2, ..., k_m]$ is a permutation of $[1,2,...,m]$.

Objective functions f, u and λ. Each potential solution or individual in the population is associated with a number λ ($0 \leq \lambda \leq 1$), which will be connected with the credibility level of a crisp outranking relation defined on the set of genes. The fitness of an individual with credibility level λ is calculated according to a given *fitness procedure*. The approach for defining individual's fitness involves the non dominated solutions in a similar form of MOGA [11]. In accordance with (1), we define the objective function f of an individual $\tilde{p}$ with credibility level λ as follows: Let $\tilde{p} = a_{k_1} a_{k_2} \ldots a_{k_m}$ be the schematic representation of an individual's chromosome and suppose that given a_{k_i} and a_{k_j}, two actions such that $\sigma(a_{k_i}, a_{k_j}) \geq \lambda$ and $\sigma(a_{k_j}, a_{k_i}) \leq \lambda - \beta$ ($\beta > 0$, representing a threshold level), we accept that "a_{k_i} outranks a_{k_j}" ($a_{k_i} S^{\lambda} a_{k_j}$) and "$a_{k_j}$ does not outrank a_{k_i}" ($a_{k_j} nS^{\lambda} a_{k_i}$). In this case, into the crisp outranking relation generated by λ, S_A^{λ}, a presumed preference favoring a_{k_i}, holds. Then:

$$f(\tilde{p}) = \left| \{ (a_{k_i}, a_{k_j}) : a_{k_i} nSa_{k_j} \text{ and } a_{k_j} nSa_{k_i} ; \; i = 1,2,\ldots,m-1, \; j = 2,3,\ldots,m, \; i < j \} \right| \quad (2)$$

where $[k_1, k_2, \ldots, k_m]$ is a permutation of $[1, 2, \ldots, m]$

$f(\tilde{p})$ is the number of incomparabilities between pairs of actions (a_{k_i}, a_{k_j}) into the individual $\tilde{p} = a_{k_1} a_{k_2} \ldots a_{k_m}$ in the sense of the crisp relation S_A^{λ}. Note that the quality of solution increases with decreasing f score.

The objective function u of an individual $\tilde{p}$ measures the amount of unfeasibility (in relative terms) and we chose to define it as:

$$u(\tilde{p}) = \left| \{ (a_{k_i}, a_{k_j}) : a_{k_i} Sa_{k_j} \text{ and } a_{k_j} nSa_{k_i} ; \; i = 1,2,\ldots,m, \; j = 1,2,\ldots,m, \; i > j \} \right| \quad (3)$$

$u(\tilde{p})$ is the number of preferences between actions into the individual $\tilde{p}$ which are not "well-ordered" in the sense of S_A^{λ}.

An individual $\tilde{p}$ is feasible if $u(\tilde{p}) = 0$ and infeasible if $u(\tilde{p}) > 0$. Defining the objective function u taking the zero minimum value if and only if the solution is feasible seems a natural approach. Each individual $\tilde{p}$ can then be represented by a triad of values f, u, and λ.

We are interested in:

i) Individuals whose objective function u value is equal to zero. This assures us that the ordering represented by the individual is transitive; this is one of two characteristics that should be exhibited by all recommendation (solution) of ranking problems [31].

ii) Individuals whose objective function f value is equal (or near) to zero. This objective improves the comparability of S on A.

iii) Individuals whose credibility level λ is near to 1. This indicates us that the ordering represented by the individual with credibility level λ is trustier whenever the objective functions u and f values are zero or near to zero. In practice, the requirement connected to function f does not permit that λ values approach to 1 because in this case we could have many incomparable genes.

Then, we use an evolutionary search for solving the multiobjective problem:

$$Min(u), \quad Min(f), \quad Max(\lambda) \tag{4}$$

Subject to

$R_S, \quad \lambda \in [0,1], \quad \lambda \geq \lambda_0$

(where R_S is a strict total order of A and λ_0 is a minimum level of credibility)

We can see that the objective function u coincides with n_V, (see (1) of section 2).

Fitness Assignment Procedure. Most of the approaches of the multiobjective decision making seek elements of *all* the *Pareto optimal set* P^* which in the jargon of Multiobjective Evolutionary Algorithms (MOEA) it is often denoted as P_{true} [4]. During MOEA execution, a "current" set of Pareto optimal solutions is determined at each EA generation and termed $P_{current}(t)$, where t represents the generation number. Many MOEA implementations also use a secondary population storing nondominated solutions found through the generations. This secondary population is named $P_{known}(t)$. This term is also annotated with t to reflect its possible changes in membership during MOEA execution. $P_{known}(0)$ is defined as the empty set (ϕ) and P_{known} only as the final set of solutions returned by the MOEA at termination. $P_{current}(t)$, P_{known}, and P_{true} are sets of MOEA genotypes; each set's corresponding phenotypes form an approximated Pareto front. The associated approximated Pareto front for each of these solution sets is called $PF_{current}(t)$, PF_{known}, and PF_{true}. Most of the methods based on MOEA attempt to evolve a population toward the *true Pareto frontier* PF_{true}. The hope is that by the end of the run, $P_{known} = P_{true}$, $P_{known} \subset P_{true}$, or $\{\bar{u}_i \in PF_{known}, \bar{u}_j \in PF_{true} : \forall i, \forall j \quad min[distance(\bar{u}_i, \bar{u}_j)] < \varepsilon\}$, where distance is defined over some norm (of course in an open problem we generally have no way of knowing P_{true}).

For solving the multicriteria ranking problem using a MOEA it is not necessary seek all the *Pareto optimal set* P_{true} or the associated Pareto front PF_{true} because of the fact that a lot of nondominated solutions are not of interest for the decision maker, we will use the strategy of attempting to find in each EA generation the most promising and attractive solutions for the decision maker which in our case are those individuals of which u, f score are near to value zero and has a sufficiently high value of λ. Is sufficient to seek a *restricted Pareto optimal set*, which for our purpose it is defined as following.

$$P_{true}^{restricted} = \left\{\bar{p} \in P_{true} : \left\| (u(\bar{p}), f(\bar{p})) \right\|_\infty \leq \varepsilon, \text{ where } \varepsilon \text{ is a small no-negative number} \right\} \tag{5}$$

After that the child has been produced through the MOEA operators, it will replace the "less fitting" member of the population. The average of the population will improve if the child solution has lower scores than those of the solutions being replaced. In this algorithm, every new offspring is replacing the worst chromosome in the population. Each time that we replace a new offspring by the worst individual, the new population is sorted with the same criterion.

4 Computational Example

In order to benchmark the algorithm performance, some test problems were selected for numerical experiments. Here, only one test problem is shown. The proposed algorithm was coded in C++ and tested on a PC with processor Intel Pentium IV (1.5 GHz). In our computational study, 1 test problem instance of the MOEA heuristic was generated for solving the following test problem. The algorithm stopped when 10,000 populations had been generated. The population size was set to 40. The crossover probability was chosen 0.85 and the mutation probability was 0.30.

4.1 Test Problem 1

The fuzzy outranking relation is given by the following credibility matrix(10x10) between actions $A_0, A_1, A_2, A_3, A_4, A_5, A_6, A_7, A_8, A_9$ (Table 1).

Table 1. Credibility Matrix Between Actions $A_0, ..., A_9$

	A0	A1	A2	A3	A4	A5	A6	A7	A8	A9
A0	1.0	0.6	0.3	0.5	0.2	0.7	0.9	0.3	0.7	0.8
A1	0.2	1.0	0.4	0.3	0.4	0.5	0.5	0.5	0.6	0.3
A2	0.7	0.9	1.0	0.6	0.4	0.7	0.8	0.9	0.7	0.7
A3	0.6	0.8	0.3	1.0	0.4	0.7	0.8	0.4	0.3	0.8
A4	0.6	0.8	0.7	0.9	1.0	0.7	0.8	0.5	0.7	0.9
A5	0.4	0.65	0.4	0.4	0.2	1.0	0.4	0.3	0.5	0.5
A6	0.5	0.4	0.3	0.2	0.5	0.6	1.0	0.4	0.3	0.2
A7	0.6	0.8	0.3	0.8	0.2	0.8	0.9	1.0	0.8	0.7
A8	0.4	0.6	0.5	0.4	0.4	0.3	0.9	0.3	1.0	0.5
A9	0.5	0.8	0.3	0.6	0.3	0.5	0.4	0.3	0.7	1.0

Table 1 was processed with the proposal of section 3. The Restricted Pareto front $PF_{known}^{restricted}$ found and the associated final set of solutions returned by the MOEA at termination $P_{known}^{restricted}$ are presented in Table 2.

Table 2. Restricted Pareto Front Found and the Associated Individual of the Solutions Space

Individual: Final set of solutions returned by the MOEA	$\hat{p}_1$	$\hat{p}_2$	$\hat{p}_3$	$\hat{p}_4$	$\hat{p}_5$	$\hat{p}_6$	$\hat{p}_7$	$\hat{p}_8$	$\hat{p}_9$
	4	2	4	4	4	4	4	4	4
	2	7	2	7	2	7	2	0	2
	7	4	3	5	0	2	0	2	7
	3	3	7	2	7	0	7	7	0
	0	0	0	0	3	3	3	3	3
	9	9	9	3	5	9	9	9	9
	8	5	8	9	9	8	5	8	8
	1	1	1	8	8	1	1	1	1
	6	8	5	6	1	5	8	5	5
	5	6	6	1	6	6	6	6	6

Function u	01	02	03	07	00	02	00	01	00
Function f	06	06	06	06	13	13	14	14	26
Lambda	0.5992	0.5998	0.5999	0.6000	0.6497	0.6500	0.6997	0.7000	0.8000
Fitness value	41.44	41.44	41.44	41.44	31.30	31.30	29.06	29.06	15.65

The number $T(i, j)$, $(1 \leq i, j \leq m)$, of times that an alternative was found at a certain place in the ranking of the individual associated to the members of the final restricted Pareto front is given in table 3. Based on the table 3 we found a compromise solution following the next procedure: as the ranking of the alternatives is of significant importance, the number of times that an alternative is found at a certain place in the ranking is weighted according to the importance of the alternatives to be ranked. It is reasonable to conclude that in certain cases, the rank of the alternatives would not be of equal importance.

Relative importance could be reflected in a weighting $w_i = m - i + 1$ of each rank i, where m is the length of an individual. After that, we calculate the weighted sum $\sum_{i=1}^{m} w_i T(i, j)$, $j=1,2, ..., m$. Finally, we obtaining a succession in decreasing order of preference, generating of this manner, a recommendation for the decision maker.

Table 3 suggests the following final ranking

$$A_4 \succ A_2 \succ A_7 \succ A_0 \succ A_3 \succ A_9 \succ A_8 \succ A_5 \succ A_1 \succ A_6 \qquad (10)$$

where $A \succ B$ means that "alternative "A" is preferred to alternative "B"".

The Genetic algorithm of [19] obtains the following results: The number of times that an alternative was found at a certain place in the ranking is given in table 4 with respect to 100 variations in the seed parameter.

Table 3. The Number of Times that an Alternative was Found at a Certain Place in the Ranking

Weight w_i	Rank	A_4	A_2	A_7	A_0	A_3	A_9	A_8	A_5	A_1	A_6
10	1	8	1	0	0	0	0	0	0	0	0
9	2	0	5	3	1	0	0	0	0	0	0
8	3	1	2	2	2	1	0	0	1	0	0
7	4	0	1	4	2	2	0	0	0	0	0
6	5	0	0	0	4	5	0	0	0	0	0
5	6	0	0	0	0	1	7	0	1	0	0
4	7	0	0	0	0	0	2	5	2	0	0
3	8	0	0	0	0	0	0	2	0	7	0
2	9	0	0	0	0	0	0	2	4	1	2
1	10	0	0	0	0	0	0	0	1	1	7
$\sum_{i=1}^{m} w_i T(i,j)$		88	78	71	63	57	43	30	30	24	11
Minimum Lamda= 0.5992											

Table 4. Results of the Genetic Algorithm of [19]

	A_4	A_2	A_7	A_0	A_3	A_8	A_5	A_9	A_6	A_1
1	81	12	0	4	0	2	0	0	0	0
2	4	86	6	2	0	2	0	0	0	0
3	8	0	54	36	0	2	0	0	0	0
4	4	0	34	28	18	6	4	6	0	0
5	2	0	2	18	46	20	4	8	0	0
6	0	2	2	10	20	22	22	20	0	2
7	0	0	2	2	14	20	30	22	8	2
8	0	0	0	0	0	16	12	34	14	24
9	0	0	0	0	0	10	20	8	36	26
10	0	0	0	0	2	0	8	2	42	46

The best compromise solutions were individuals in which $u=0$, $f=13$, $\lambda = 0.69$ mostly corresponding to rankings $A_4 A_2 A_7 A_0 A_3 A_9 A_8 A_5 A_1 A_6$ and $A_4 A_2 A_0 A_7 A_3 A_9 A_8 A_1 A_5 A_6$. These results, combined with Table 4, suggest the final ranking:

$$A_4 \succ A_2 \succ A_7 \succ A_0 \succ A_3 \succ A_9 \succ A_8 \succ A_5 \succ A_1 \succ A_6 \qquad (11)$$

Although without a clear preference between A_5 and A_1.

ELECTRE-III suggest the following ranking:

$$A_4 \succ (A_2, A_7) \succ A_0 \succ A_3 \succ (A_9, A_8) \succ A_5 \succ A_1 \succ A_6 \qquad (12)$$

In this ordering, we have $u = 0$ with $\lambda = 0.69$, but we can hardly agree with the fact that A_2 and A_7 are posed in the same position. It is clear that $A_2 S^\lambda A_7$ and the contrary is false. The net flow score associated to A_2 is higher, and also, if we consider the set $A - \{A_4\}$, A_2 is the best alternative in the sense of [22].

5 Conclusions

This paper presents a Multiobjective Evolutionary Algorithm suitable for exploiting fuzzy outranking relations. The basic concepts and formulations of the new method were given with an improved technique to handle the Pareto front. The intrinsic advantages of the method are that it is not sensitive to the shape of the Pareto front and that it is partially adaptive to inadequate input from a user as a complex graph with several cycles. The most similar method, the genetic algorithm of [19], was benchmarked against the new MOEA. The comparison study contrasted both methods in generation of efficient solutions and in the final recommendation. The MOEA present a wide area of efficient solutions whilst the other generates only one. Moreover, the recommendation for decision maker is obtained with only one run whilst the genetic algorithm procedure should be run n times for obtaining a final recommendation. In the numerical experiments carried out, our proposal performed very well, in the sense of quality of solution as well in the sense of computational effort. The Multiobjective Evolutionary Algorithm procedure also looks more effective than other approaches that exploit a fuzzy outranking relation; it could work easily with preference graphs of considerable size and it looks more robust than other dealing with irrelevant alternatives. Approaches based on a function score do not have a way to reduce nonrational violations of explicit preferences, while the evolutionary one may identify this kind of irregularity and try to control and minimize it.

References

1. Brans, J. P., Vincke, Ph.:A Preference Ranking Organization Method. Management Science 31 (1985) 647-656
2. Bouyssou, D., Vincke, Ph.: Ranking Alternatives on the Basis of Preference Relations: A Progress Report with Special Emphasis on Outranking Relations. Technical report *IS-MG 95/03*. Institut de Statistique et de Recherche Opérationnelle, Universite Libre de Bruxelles. Serie: Mathématiques de la Gestion (1995)
3. Coello, C.A.: A Comprehensive Survey of Evolutionary-Based Multiobjective Optimization Techniques. Knowledge and Information Systems 1 (1999) 269-308
4. Coello, C.A., Van Veldhuizen, D.A., and Lamont, G.B.: Evolutionary Algorithms for Solving Multi-Objective Problems. Kluwer Academic Publishers New York (2002)
5. Deb, K.: Multi-Objective Optimization Using Evolutionary Algorithms. John Wiley and Son, Chichester, UK (2001)
6. Deb, K., Pratap, A., Agarwal, S., Meyarivan, T.: A Fast and Elitist Multi-Objective Genetic Algorithm: NSGA-II. IEEE Transactions on Evolutionary Computation. 6 (2002) 182-197

7. Fernandez E., Leyva López J. C.: A method based on multiobjective optimization for deriving a ranking from a fuzzy preference relation. European Journal of Operational Research 154 (2004) 110-124
8. Fodor J., Roubens M: Fuzzy Preference Modeling and Multicriteria Decision Support. Kluwer, Dordrecht (1994)
9. Fogel, L.J., Owens, A.J., Walsh, M.J.: Artificial Intelligence Through Simulated Evolution. John Wiley, New York (1966)
10. French S.: Decision Theory: An Introduction to the Mathematics of Rationality, Halsted Press, New York Chichester Brisbane Toronto (1986)
11. Fonseca, C.M., Fleming, P.J.: Genetic Algorithms for Multiobjective Optimization: Formulation, Discussion and Generalization. In: Genetic Algorithms: Proceedings of the Fifth International Conference. Morgan Kaufmann (1993) 416-423
12. Fonseca, C.M., Fleming, P.J.: An Overview of Evolutionary Algorithm in Multiobjective Optimization. Evolutionary Computation. 3 (1995) 1-16
13. Goldberg, D.: Genetic algorithms in search, optimization, and machine learning. Addison-Wesley (1989)
14. Holland, J.H.: Adaptation in Natural and Artificial Systems. University of Michigan Press, Ann Arbor Michigan (1975)
15. Horn, J.: Multicriterion Decision Making. In: Back, T., Fogel, D. B., Michalewicz, Z. (eds.): Handbook of Evolutionary Computation. IOP Publishing Ltd and Oxford University Press, Bristol U.K. (1997)
16. Horn, J. Nafploitis, N., Goldberg, D.E.: A Niched Pareto Genetic Algorithm for Multiobjective Optimization". In: Michalewicz, Z. (ed.): Proceeding of the First IEEE Conference on Evolutionary Computation.: IEEE Service Center, Piscataway, New Jersey (1994) 82-87
17. Keeney R., Raiffa, H.: Decision with Multiple Objectives: Preferences and Value Tradeoffs. Wiley, New York (1976)
18. Knowles, J.D., Corne, D.W.: Approximating the Nondominated Front using the Pareto Archived Evolution Strategy. Evolutionary Computation. 8 (2000) 149-172
19. Leyva-López J.C., Fernández-González, E.: A Genetic Algorithm for Deriving Final Ranking from a Fuzzy Outranking Relation. Foundations of Computing and Decision Sciences. 24 (1999) 33-47
20. Michalewicz, Z.: Genetic Algorithms + Data Structures = Evolution Programs, Springer – Verlag, Berlin Heidelberg New York (1996)
21. Ordoñez Reinoso G., Valenzuela Rendón M.: Permutation Optimization with Genetic Algorithms: The Traveling Salesman Problem (in Spanish). Proc. of the 3th. Latin-American Congress of Artificial Intelligence. (1992) 271-282
22. Orlovski, S.A.: Decision-Making with a Fuzzy Preference Relation. Fuzzy Sets and Systems 1 (1978) 155-167
23. Poon P.W., Carter J.N.: Genetic Algorithm Crossover Operators for Ordering Applications. Computers & Operations Research 22 (1995) 135-147
24. Roy B.: Multicriteria Methodology for Decision Aiding. Kluwer (1996)
25. Roy B.: The Outranking Approach and the Foundations of ELECTRE Methods. In Bana e Costa, C.A. (ed.): Reading in Multiple Criteria Decision Aid. Springer-Verlag, Berlin (1990) 155-183
26. Roy B.: Decision-Aid and Decision-Making. European Journal of Operational Research. 45 (1990) 324-331

27. Schaffer, J.D.: Multiple Objective Optimization with Vector Evaluated Genetic Algorithms. In: Grefenstette, J.J. (ed.): Genetic algorithms and their Applications: Proceedings of the First International Conference on Genetic Algorithms. Hillsdale, New Jersey: (1985) 93-100
28. Schwefel, H.P.: Numerical Optimization of Computer models. Wiley, Chichester UK (1981)
29. Srinivas, N., Deb, K.: Multiobjective Function Optimization using Nondominated Sorting Genetic Algorithms. Evolutionary Computation. 2 (1995) 221-248
30. Triantaphyllou, E.: Multicriteria Decision Making Methods: A Comparative Study. Kluwer Academic Publishers, Boston MA (2000)
31. Vanderpooten, D.: The Construction of Prescriptions in Outranking Methods. In: Bana e Costa, C.A. (ed.): Reading in Multiple Criteria Decision Aid, Springer-Verlag, Berlin (1990) 184-215
32. Van Veldhuizen, D.A., Lamont, G.A.: Multiobjective Evolutionary Algorithms: Analyzing the State-of-the-Art. Evolutionary Computation. 8 (2000) 1-26
33. Zitzler, E., Laumanns, M., Thiele, L.: SPEA2: Improving the Strength Pareto Evolutionary Algorithm. Technical Report 103, Computer Engineering and Networks Laboratory (TIK), Swiss Federal Institute of Technology (ETH) Zurich, Gloriastrasse 35, CH-8092 Zurich, Switzerland, (2001)

Exploring the Performance of Stochastic Multiobjective Optimisers with the Second-Order Attainment Function

Carlos M. Fonseca[1], Viviane Grunert da Fonseca[1,2], and Luís Paquete[3,4]

[1] CSI – Centro de Sistemas Inteligentes
Faculdade de Ciências e Tecnologia
Universidade do Algarve
Faro, Portugal
cmfonsec@ualg.pt
vgrunert@csi.fct.ualg.pt

[2] INUAF – Instituto Superior D. Afonso III
Loulé, Portugal

[3] Fachgebiet Intellektik, Fachbereich Informatik
Technische Universität Darmstadt
Darmstadt, Germany
lpaquete@intellektik.informatik.tu-darmstadt.de

[4] Faculdade de Economia
Universidade do Algarve
Faro, Portugal

Abstract. The attainment function has been proposed as a measure of the statistical performance of stochastic multiobjective optimisers which encompasses both the quality of individual non-dominated solutions in objective space and their spread along the trade-off surface. It has also been related to results from random closed-set theory, and cast as a mean-like, first-order moment measure of the outcomes of multiobjective optimisers. In this work, the use of more informative, second-order moment measures for the evaluation and comparison of multiobjective optimiser performance is explored experimentally, with emphasis on the interpretability of the results.

1 Introduction

Stochastic multiobjective optimisers, such as evolutionary algorithms and other metaheuristics, produce Pareto-set approximations which consist of sets of non-dominated points in objective space. Given the stochastic nature of the optimisers, such non-dominated point sets may be seen as realisations of corresponding random non-dominated point sets, the stochastic behaviour of which is tied both to the problem and to the optimiser considered.

The performance of a multiobjective optimiser is intimately related to the quality of the Pareto-set approximations it produces. In the literature, several attempts have been made to quantify the quality of (deterministic) Pareto-set approximations through so-called unary quality indicators, which are functions which assign real values to each Pareto-set approximation. In the face of *random* Pareto-set approximations, unary quality indicators provide a convenient transformation from random sets to random variables. However, this approach suffers from inherent limitations of unary quality indicators. In fact, even a finite number of unary quality indicators which, in combination with each other, would completely describe a deterministic set of non-dominated points in objective space cannot exist in practice [1]. As a result, information is irremediably lost by finite-dimensional unary quality indicators even before any statistical analysis takes place.

To be able to retain all of the information available in the original non-dominated sets, a quality indicator must be infinite-dimensional, such as the binary field [2] derived from the set of goals attained by a Pareto-set approximation (attained set or attained region). Describing a Pareto-set approximation by a (real-valued) function defined over the whole of the objective space may not seem to be very sensible in the deterministic case. However, in the random case, it provides a useful link to existing random set theory, where distributions of random sets are studied directly (possibly, up to a complete distributional description) and not indirectly through distributions of summary measures (indicators) of the sets. The attainment function of a random Pareto-set approximation, for example, has been identified as the first-order moment measure of the binary random field derived from the corresponding random attained set [3], and, as such, is a concept perfectly integrated in random set theory.

By combining fundamentally different features of the quality of Pareto-set approximations, such as the quality of individual solutions and their spread along the trade-off surface, into a real-valued function of the goals, the attainment function can already effectively describe an important aspect of the distribution of random Pareto-set approximations, namely its location. To address the dependence structure of individual solutions within these approximation sets, two additional measures of performance are considered in this work, both of which are of second-order moment type: the second-order attainment function and its centred version, the covariance function.

In section 2, background is given on the attainment function. In section 3, the second-order attainment function and the covariance function are introduced. Empirical estimates of the two functions, obtained from experimental data, are presented graphically, and their interpretation is discussed. Section 4 is devoted to the comparison of optimiser performance using statistical hypothesis tests based on first-order or second-order attainment functions. To illustrate the application of the attainment function approach, experimental results obtained on two different optimisation problems are presented and discussed in section 5. The paper concludes with some remarks and directions for further work in section 6.

2 Background

The outcome of a multiobjective optimiser is considered to be the set of non-dominated objective vectors evaluated during one optimisation run. If the optimiser is stochastic, such Pareto-set approximations are random, and their distribution becomes of interest when assessing the optimiser's performance. Consider the following definitions (assuming minimisation of all objective functions without loss of generality):

Definition 1 (Random non-dominated point set). *A random point-set*

$$\mathcal{X} = \{X_1, \ldots, X_M \in \mathbb{R}^d : P(X_i \leq X_j) = 0, \ i \neq j\},$$

where both the number of elements M and the elements X_j themselves are random and $P(0 \leq M < \infty) = 1$, is called a random non-dominated point set (RNP-set).

Random Pareto-set approximations produced by stochastic multiobjective optimisers on d-objective problems are RNP-sets in $\mathbb{R}^d$.

Definition 2 (Attained set). *The random set*

$$\mathcal{Y} = \{y \in \mathbb{R}^d \mid X_1 \leq y \ \vee \ X_2 \leq y \ \vee \ldots \vee \ X_M \leq y\}$$
$$= \{y \in \mathbb{R}^d \mid \mathcal{X} \trianglelefteq y\}$$

is the set of all goals $y \in \mathbb{R}^d$ attained by the RNP-set $\mathcal{X}$ (see Figure 1).

The distributions of both random sets, $\mathcal{X}$ and $\mathcal{Y}$, are equivalent, i.e. a characterisation of the distribution of $\mathcal{X}$ automatically provides a characterisation of the distribution of $\mathcal{Y}$, and vice versa.

Definition 3 (Attainment indicator). *Let $\mathbf{I}\{\cdot\} = \mathbf{I}_{\{\cdot\}}(z)$ denote the indicator function. Then, the random variable $b_\mathcal{X}(z) = \mathbf{I}\{\mathcal{X} \trianglelefteq z\}$ is called the attainment indicator of $\mathcal{X}$ at goal $z \in \mathbb{R}^d$.*

The set of all attainment indicators indexed by $z \in \mathbb{R}^d$ is the binary random field $\{b_\mathcal{X}(z), z \in \mathbb{R}^d\}$. For the deterministic case, this binary field fully characterises a single Pareto-set approximation, as one can always be obtained from

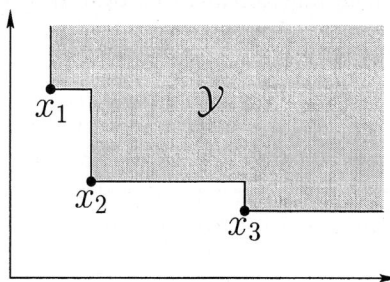

Fig. 1. *RNP-set $\mathcal{X}$ with non-dominated realisations x_1, x_2, and x_3 and the attained set $\mathcal{Y}$ (here as a realisation); compare with Grunert da Fonseca et al. [3]*

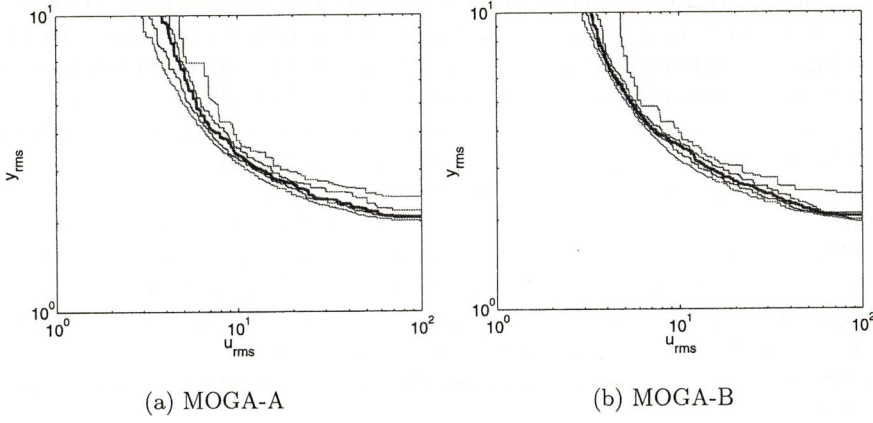

(a) MOGA-A (b) MOGA-B

Fig. 2. EAF contour plots (first example)

the other. As an infinite-dimensional quality indicator, it can be used to construct a comparison method which is complete and compatible with respect to weak-dominance [1]. Although this may not seem to be very useful in the deterministic case, such a quality indicator provides an interesting assessment tool in the random case:

Definition 4 (Attainment function). *The function* $\alpha_\mathcal{X} : \mathbb{R}^d \longmapsto [0,1]$ *with*

$$\alpha_\mathcal{X}(z) = P\bigl(b_\mathcal{X}(z) = 1\bigr)$$

is called the attainment function of $\mathcal{X}$.

As identified in Grunert da Fonseca et al. [3], the attainment function is the first-order moment measure of the binary random field $\{b_\mathcal{X}(z), z \in \mathbb{R}^d\}$ derived from $\mathcal{Y}$ (the set attained by $\mathcal{X}$) and, as such, it offers a useful description of the *location* of the distribution of $\mathcal{Y}$ (and also of $\mathcal{X}$). Note that for $M = 1$, the optimiser produces a single random objective vector X per run, and the attainment function reduces to the usual multivariate distribution function $F_X(z) = P(X \leq z)$. A natural empirical counterpart of the (theoretical) attainment function $\alpha_\mathcal{X}(\cdot)$ may be defined as follows:

Definition 5 (Empirical attainment function). *Let* $b_1(z), \ldots, b_n(z)$ *be* n *realizations of the attainment indicator* $b_\mathcal{X}(z)$, $z \in \mathbb{R}^d$. *Then, the function defined as* $\alpha_n : \mathbb{R}^d \longmapsto [0,1]$ *with*

$$\alpha_n(z) = \frac{1}{n} \cdot \sum_{i=1}^{n} b_i(z)$$

is called the empirical attainment function of $\mathcal{X}$ *(EAF).*

The realizations $b_1(z), \ldots, b_n(z)$ correspond to n runs of the optimiser under study. In Figure 2, contour plots of the EAFs obtained from 21 independent

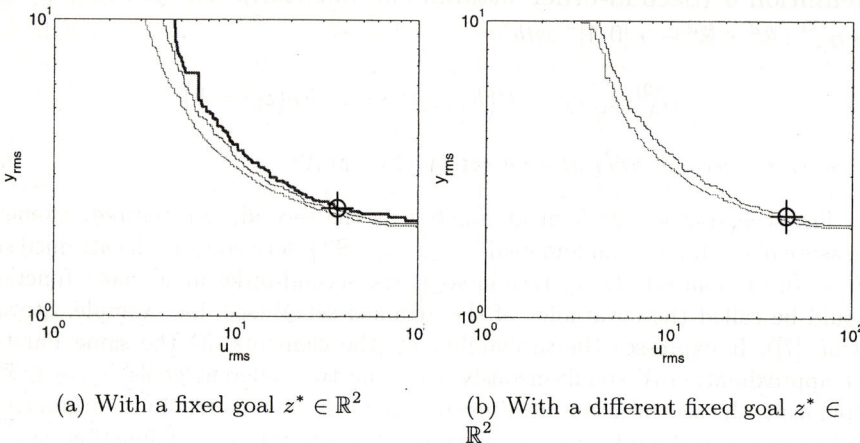

(a) With a fixed goal $z^* \in \mathbb{R}^2$

(b) With a different fixed goal $z^* \in \mathbb{R}^2$

Fig. 3. Contour plot of the marginal second-order EAF (MOGA-A)

function $\alpha_n^{(2)}(z, z^*)$, it is possible to explore how the marginal second-order EAF changes as a function of the goal z^*. By pulling the goal z^* further downwards, Figure 3(b) is obtained. Note how another contour disappears, and how the ones that remain tend to move away from the best Pareto-set approximation known. Continuing to pull the goal z^* downwards until it cannot be attained by any of the runs, all contours would disappear as the marginal second-order EAF became equal to zero over the whole objective space. On the other hand, moving the goal z^* upwards until it can be attained by all optimisation runs will lead to the contours of the (first-order) EAF.

An alternative way of studying the optimiser's second-order behaviour is given by the second, *centred*, moment measure of the binary random field $\{b_\mathcal{X}(z), z \in \mathbb{R}^d\}$. In line with the random set theory literature [7], this will be referred to as the covariance function.

Definition 8 (Covariance function). *The function* $\mathrm{cov}_\mathcal{X} : \mathbb{R}^d \times \mathbb{R}^d \longmapsto [-0.25, 0.25]$ *with*

$$\mathrm{cov}_\mathcal{X}(z_1, z_2) = \alpha_\mathcal{X}^{(2)}(z_1, z_2) - \alpha_\mathcal{X}(z_1) \cdot \alpha_\mathcal{X}(z_2)$$

is called the covariance function of $\mathcal{X}$.

For each pair of goals $(z_1, z_2) \in \mathbb{R}^d \times \mathbb{R}^d$, the value of the covariance function indicates the direction and strength of the relationship between the two (random) attainment indicators $b_\mathcal{X}(z_1)$ and $b_\mathcal{X}(z_2)$. If the value $\mathrm{cov}_\mathcal{X}(z_1, z_2)$ equals zero the two random variables are uncorrelated, i.e. there exists no linear relationship between them. On the other hand, if all elements of $\mathcal{X}$ were independent $\mathrm{cov}_\mathcal{X}(z_1, z_2)$ would equal zero for all $z_1, z_2 \in \mathbb{R}^d$, $z_1 \neq z_2$. Moreover, if the value $\mathrm{cov}_\mathcal{X}(z_1, z_2)$ is positive, there is a positive correlation between $b_\mathcal{X}(z_1)$ and $b_\mathcal{X}(z_2)$, in the sense that the differences $b_\mathcal{X}(z_1) - \alpha_\mathcal{X}(z_1)$ and $b_\mathcal{X}(z_2) - \alpha_\mathcal{X}(z_2)$ will be

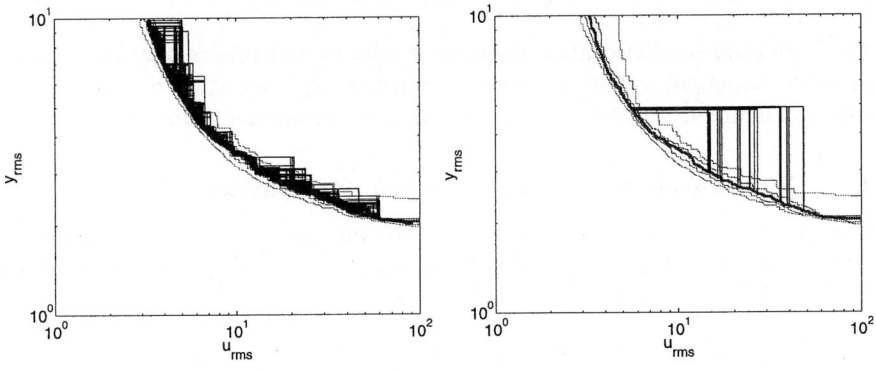

(a) Covariance value greater than 0.21   (b) Covariance value less than -0.21

Fig. 4. Pairs (z_1, z_2) showing a covariance below or above a threshold (MOGA-B)

likely to show the same sign. In other words, the attainment of goal z_1 tends to coincide with the attainment of goal z_2. The tendency *not* to attain two particular goals simultaneously is reflected by a negative covariance function value. The maximum of $\text{cov}_\mathcal{X}(z_1, z_2)$ equals 0.25 and is reached by the variance function

$$\text{var}_\mathcal{X}(z) = \text{cov}_\mathcal{X}(z, z)$$

at all $z \in \mathbb{R}^d$ where $\alpha_\mathcal{X}(z) = 0.5$. The minimum possible value of $\text{cov}_\mathcal{X}(z_1, z_2)$ is -0.25 and will be reached at any pairs of goals which *cannot* be attained together, but where each goal has probability of being attained individually equal to 0.5.

An empirical counterpart of the covariance function may be defined as follows:

Definition 9 (Empirical covariance function). *The function* $\text{cov}_n : \mathbb{R}^d \times \mathbb{R}^d \longmapsto [-0.25, 0.25]$ *with*

$$\text{cov}_n(z_1, z_2) = \alpha_n^{(2)}(z_1, z_2) - \alpha_n(z_1) \cdot \alpha_n(z_2)$$

is called the empirical covariance function of $\mathcal{X}$ *(ECF).*

As with the second-order EAF, visualisation of the ECF requires a workaround. In Figure 4, the pairs of goals $(z_1, z_2) \in \mathbb{R}^2 \times \mathbb{R}^2$ which exhibit a covariance above or below a certain threshold are indicated in objective space by a solid bracket beginning at one goal and ending at the other, while the contours of first-order EAF are plotted as a reference in the background. Figure 4(a) shows that goals which are likely to be attained simultaneously by MOGA-B are generally located close to each other in objective space, whereas goals which are likely to be attained in alternative to each other are located farther apart, as Figure 4(b) indicates.

4 Comparison of Optimiser Performance

The performance of two multiobjective stochastic optimisers may be compared through statistical hypothesis tests of Smirnov-type based on either the (first-order) attainment function or the second-order attainment function.

4.1 First-Order Attainment Function Comparison

Given two optimisers A and B applied to the same optimisation problem independently from each other, the following two-sample test problem can be formulated for first-order comparison:

$$H_0 : \alpha_{\mathcal{X}_A}(z) = \alpha_{\mathcal{X}_B}(z) \quad \text{for all } z \in \mathbb{R}^d$$

v.s.

$$H_1 : \alpha_{\mathcal{X}_A}(z) \neq \alpha_{\mathcal{X}_B}(z) \quad \text{for at least one } z \in \mathbb{R}^d,$$

where $\alpha_{\mathcal{X}_A}(\cdot)$ and $\alpha_{\mathcal{X}_B}(\cdot)$ represent the (first-order) attainment function for optimiser A and optimiser B, respectively. Note that the equality of the two attainment functions stated in the null hypothesis H_0 is *not* equivalent to the equality of the performance of the two optimisers, since attainment functions solely address the *location* of the RNP-set distributions in objective space.

Let $\alpha_n^A(\cdot)$ and $\alpha_m^B(\cdot)$ be the EAF of optimiser A and B, respectively, after n and m optimisation runs. The above null hypothesis H_0 is rejected when the observed value of the test statistic

$$D_{n,m} = \sup_{z \in \mathbb{R}^d} |\alpha_n^A(z) - \alpha_m^B(z)|$$

is large. The test is a generalisation of the multivariate two-sided Kolmogorov-Smirnov test for two independent samples, and like the latter it is not distribution-free under H_0. However, critical values can be obtained by using the permutation argument [8], i.e. reject H_0 if $D_{n,m}$ is greater than the $(1 - \alpha)$-quantile of the resulting permutation distribution of the test statistic under H_0. Note that, the permutation approach has also been used to formulate a distribution-free version for the multivariate two-sample Kolmogorov-Smirnov test [9].

4.2 Second-Order Attainment Function Comparison

A similar test problem can be formulated based on the second-order attainment function:

$$H_0 : \alpha_{\mathcal{X}_A}^{(2)}(z_1, z_2) = \alpha_{\mathcal{X}_B}^{(2)}(z_1, z_2) \quad \text{for all } z_1, z_2 \in \mathbb{R}^d$$

v.s.

$$H_1 : \alpha_{\mathcal{X}_A}^{(2)}(z_1, z_2) \neq \alpha_{\mathcal{X}_B}^{(2)}(z_1, z_2) \quad \text{for at least one pair } (z_1, z_2) \in \mathbb{R}^d \times \mathbb{R}^d,$$

where $\alpha_{\mathcal{X}_A}^{(2)}(\cdot, \cdot)$ and $\alpha_{\mathcal{X}_B}^{(2)}(\cdot, \cdot)$ represent the second-order attainment function for optimiser A and optimiser B, respectively. Again, the equality of the two second-order attainment functions stated in H_0 is *not* equivalent with the equality of the performance of the two optimisers. However, the Smirnov-like test formulated in the following should already inherit more statistical power than the respective test based on the (first-order) attainment function as it uses more information from the data. In other words, not rejecting H_0 is generally less likely to be a wrong decision.

As before, construct a permutation test which rejects H_0 if

$$D_{n,m}^{(2)} = \sup_{z_1, z_2 \in \mathbb{R}^d} |\alpha_n^{A(2)}(z_1, z_2) - \alpha_m^{B(2)}(z_1, z_2)|$$

is greater than the $(1-\alpha)$-quantile of the resulting permutation distribution of the test statistic under H_0. Here, $\alpha_n^{A(2)}(\cdot, \cdot)$ and $\alpha_n^{B(2)}(\cdot, \cdot)$ are the second-order EAF of optimiser A and optimiser B, respectively, after n and m optimisation runs.

5 Experimental Results

To illustrate how the first-order and the second-order attainment functions may be used to study the performance of stochastic multiobjective optimisers, two application examples are considered here.

The first example consists of the optimisation of a multiobjective Linear-Quadratic-Gaussian controller design problem proposed by Barratt and Boyd [10] under controller complexity constraints, as formulated in [11], with multiobjective genetic algorithms (MOGA). Two MOGAs were applied to the problem, one without sharing or mating restriction (MOGA-A) and another one with sharing and mating restriction in the decision variable domain (MOGA-B), as described in [12]. Each algorithm was run 21 times for 100 generations, and the cumulative set of non-dominated objective vectors found in each run was taken as the outcome of that run.

The second example consists of the optimisation of a multiobjective Travelling Salesman Problem (TSP) by stochastic local search techniques. The problem itself is a benchmark instance based on a pair of 100-city TSP instances, kroA100.tsp and kroB100.tsp, available at TSPLIB[1]. The optimiser used was the Pareto local search (PLS) algorithm proposed in Paquete et al. [13], which uses an archive of non-dominated solutions and an acceptance criterion which takes into account the concept of Pareto optimality. Two different neighbourhoods were considered, the standard 2-opt neighbourhood and a 2-opt extension proposed by Bentley [14] and known as 2H-opt, leading to two variants of the algorithm, respectively PLS-A and PLS-B. Each variant was run 25 times until all solutions in the archive were locally Pareto-optimal, and the corresponding non-dominated objective vectors were taken as the outcome of each run. Table 1 gives some additional information about the Pareto-set approximations obtained in the two examples.

---

[1] http://www.iwr.uni-heidelberg.de/groups/comopt/software/TSPLIB95/

Table 1. Pareto-set approximation statistics

Optimiser	No. of runs	No. of elements		
		min	average	max
MOGA-A	21	48	120.38	191
MOGA-B	21	87	170.95	259
PLS-A	25	1973	2386.1	2891
PLS-B	25	2052	2541.5	3032

Table 2. Hypothesis test results ($\alpha = .05$)

Optimiser	Hypothesis test	Test statistic	Critical value	p-value	decision
MOGA	1st-order EAF	0.571	0.571	0.091	do not reject H_0
MOGA	2nd-order EAF	0.762	0.714	0.016	reject H_0
PLS	1st-order EAF	0.680	0.560	0.004	reject H_0
PLS	2nd-order EAF	0.840	0.720	0.002	reject H_0

For each set of runs, the (first-order) EAF was computed at all relevant goals in objective space (where the EAF exhibits transitions) and a record was kept of which runs attained which goals. This record was used to compute the second-order EAF and the corresponding empirical covariance function values. The observed values of the test statistic of each of the two Smirnov-like tests were determined by pooling the outcomes of the two optimisers involved in the test, and computing the first-order EAF of the resulting pooled sample. Then, the maximum absolute difference between the two individual EAFs to be compared (whether first or second-order) was determined over the the set of goals (or pairs of goals) at which the corresponding EAF of the pooled sample exhibited transitions. Finally, the permutation distribution of the two test statistics was simulated by considering 10000 random permutations of the runs in the corresponding pooled samples, assigning half of the runs to each algorithm variant, and recomputing the test statistic for each permutation. The results of the tests are summarised in Table 2, and will be discussed separately for each example.

5.1 First Example

First-order EAF plots for the outcomes of MOGA-A and MOGA-B have already been presented in Figure 2. Visual inspection suggests that MOGA-B may be more likely to attain certain goals than MOGA-A, as the weight of the EAF of MOGA-B seems to be more concentrated downwards and to the left than in the case of MOGA-A. However, the hypothesis test based on the first-order EAF does not lead to the rejection of the null hypothesis at the 0.05-significance level, indicating that any differences between the first-order attainment functions of the two algorithms are not statistically significant.

With respect to second-order behaviour, MOGA-B was shown in Figure 4 to exhibit strong negative covariance at certain pairs of goals. In comparison, MOGA-A does not exhibit as strong a negative covariance anywhere in objective

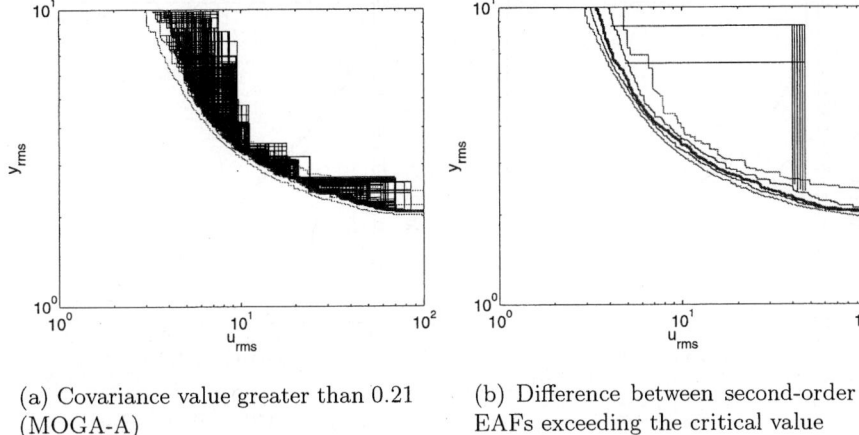

(a) Covariance value greater than 0.21 (MOGA-A)

(b) Difference between second-order EAFs exceeding the critical value

Fig. 5. Additional results (first example)

space. Thus, the effect of the niche induction techniques used in MOGA-B seems to be the imposition of stronger dependencies between the elements of the Pareto-set approximations. As for strong positive covariance, the analogue to Figure 4(a) for MOGA-A is presented in Figure 5(a). MOGA-A seems to exhibit stronger positive covariance between more distant goals in objective space, which may simply be due to the fact that the corresponding Pareto-set approximations generally contain fewer points than those produced by MOGA-B.

Finally, the hypothesis test based on the second-order EAF does lead to rejection of the null hypothesis at the 0.05-significance level. Hence, there is evidence for a statistically significant difference between the second-order attainment functions and, consequently, in the performance of the two algorithms. Figure 5(b) represents the pairs of goals where the absolute difference between second-order EAFs exceeded the critical value, with the contours of the first-order EAF of the pooled sample plotted in the background. Closer inspection of the individual second-order EAFs indicates that MOGA-B has a greater probability of attaining these pairs of goals than MOGA-A, which supports the idea that niche induction techniques had a positive effect on the performance of MOGA-B.

5.2 Second Example

First-order EAF contour plots for the outcomes of PLS-A and PLS-B on the multiobjective TSP example are presented in Figure 6. Due to the characteristics of both EAFs, only the ϵ-, 0.5- and $(1-\epsilon)$-levels are displayed. The hypothesis test based on the first-order EAF leads to a clear rejection of the null hypothesis, even at significance level 0.01. In Figure 7(a), the goals at which the critical value of the test statistic is exceeded are marked with black dots. Inspection of the

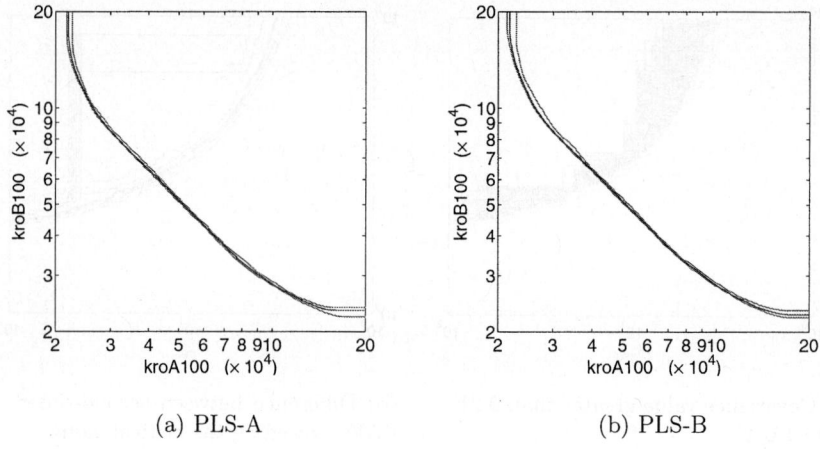

Fig. 6. EAF contour plots (second example)

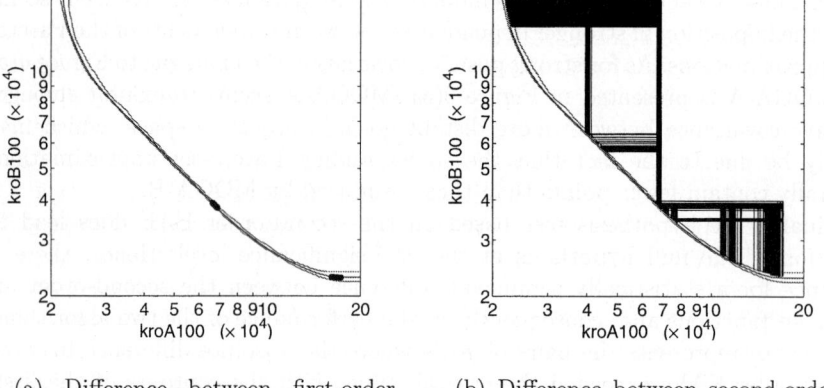

(a) Difference between first-order EAFs exceeding the critical value

(b) Difference between second-order EAFs exceeding the critical value

Fig. 7. Hypothesis test results (second example)

individual EAFs shows that PLS-B is more likely to attain those goals than PLS-A, which is hardly surprising since the 2H-opt neighbourhood contains the 2-opt neighbourhood and both algorithms were run until the archive contained exclusively local optima.

Also not surprisingly, the hypothesis test based on the second-order EAF leads to an even clearer rejection of the null hypothesis, with absolute differences in second-order EAF values greater than the critical value being observed at many pairs of goals (see Figure 7(b)). Again, the probability of PLS-B attaining these pairs of goals is greater than that of PLS-A, confirming the benefits of a larger neighbourhood under the experimental conditions considered.

6   Concluding Remarks

In this paper, it was shown how the performance of multiobjective optimisers may be studied using the attainment function approach. Whereas the first-order attainment function expresses the probability of given goals being attained independently from each other in one optimisation run, the second-order attainment function and the covariance function consider the probability of pairs of goals being attained simultaneously and, thus, take the dependence between the non-dominated elements of individual Pareto-set approximations into account.

By relating the probability of simultaneously attaining given pairs of goals with the probability of attaining them independently from each other, the covariance function provides information about the behaviour of the optimiser. In the first example, the use of niche induction techniques in a MOGA lead to a different covariance structure, suggesting stronger dependencies between the points in each random Pareto-set approximation when those techniques were used.

The second-order attainment function, on the other hand, may be said to extend the first-order attainment function in the task of assessing the ability of an optimiser to consistently produce good solutions, not only in isolation, but also in combination with each other. This is particularly relevant when *comparing* the performance of two optimisers, as differences in performance may not be related only to the quality of the non-dominated solutions produced, but also to whether or not they are likely to occur together. In the first example, only the hypothesis test based on the second-order attainment function was able to detect significant performance differences between the two algorithms. In the second example, the first-order attainment function could already distinguish the performance of the two optimisers in a statistically significant way. In that case, it is reasonable to expect even larger differences between second-order attainment functions to arise, as was observed in this example.

The two examples presented here show that the attainment function approach to optimiser assessment is currently applicable to bi-objective problems, using realistic sample sizes (numbers of runs) and large non-dominated point sets. Determining critical values for the second-order EAF hypothesis tests is by far the most computationally demanding aspect of the approach (6.5 hours and 40 days on a single Athlon MP 1900+ processor for the first and second examples, respectively, using 10000 permutations). Clearly, there is still much scope for work on the computational aspects of the approach, especially as the number of objectives and/or the size of the data sets grow.

References

1. Zitzler, E., Thiele, L., Laumanns, M., Fonseca, C.M., Grunert da Fonseca, V.: Performance assessment of multiobjective optimizers: An analysis and review. IEEE Transactions on Evolutionary Computation **7** (2003) 117–132

2. Goutsias, J.: Modeling random shapes: An introduction to random closed set theory. In Haralick, R.M., ed.: Mathematical Morphology: Theory and Hardware. Oxford Series in Optical & Imaging Sciences. Oxford University Press, New York (1996)
3. Grunert da Fonseca, V., Fonseca, C.M., Hall, A.O.: Inferential performance assessment of stochastic optimisers and the attainment function. In Zitzler, E., Deb, K., Thiele, L., Coello Coello, C.A., Corne, D., eds.: Evolutionary Multi-Criterion Optimization. First International Conference, EMO 2001, Zürich, Switzerland, March 2001, Proceedings. Number 1993 in Lecture Notes in Computer Science. Springer Verlag, Berlin (2001)
4. Justel, A., Peña, D., Zamar, R.: A multivariate Kolmogorov-Smirnov test of goodness of fit. Statistics and Probability Letters **35** (1997) 251–259
5. Fonseca, C.M., Fleming, P.J.: On the performance assessment and comparison of stochastic multiobjective optimizers. In Voigt, H.M., Ebeling, W., Rechenberg, I., Schwefel, H.P., eds.: Parallel Problem Solving from Nature – PPSN IV. Number 1141 in Lecture Notes in Computer Science. Springer Verlag, Berlin, Germany (1996) 584–593
6. Shaw, K.J., Nortcliffe, A.L., Thompson, M., Love, J., Fleming, P.: Assessing the performance of multiobjective genetic algorithms for optimization of a batch process scheduling problem. In: Proceedings of the Congress on Evolutionary Computation (CEC99). Volume 1., Washington DC (1999) 37–45
7. Stoyan, D., Kendall, W.S., Mecke, J.: Stochastic Geometry and its Applications. 2nd edn. Wiley Series in Probability and Statistics. Wiley & Sons, Chichester (1995)
8. Good, P.I.: Permutation Tests: A Practical Guide to Resampling Methods for Testing Hypotheses. 2nd edn. Springer Series in Statistics. Springer Verlag, New York (2000)
9. Bickel, P.J.: A distribution free version of the Smirnov two sample test in the p-variate case. The Annals of Mathematical Statistics **40** (1969) 1–23
10. Barratt, C., Boyd, S.: Example of exact trade-offs in linear control design. IEEE Control Systems Magazine **9** (1989) 46–52
11. Fonseca, C.M., Fleming, P.J.: Multiobjective optimal controller design with genetic algorithms. In: Proc. IEE Control'94 International Conference. Volume 1., Warwick, U.K. (1994) 745–749
12. Fonseca, C.M., Fleming, P.J.: Multiobjective genetic algorithms made easy: selection, sharing and mating restriction. In: First IEE/IEEE International Conference on Genetic Algorithms in Engineering Systems: Innovations and Applications, Sheffield, U.K. (1995) 45–52
13. Paquete, L., Chiarandini, M., Stützle, T.: Pareto local optimum sets in the biobjective traveling salesman problem: An experimental study. In Gandibleux, X., Sevaux, M., Sörensen, K., T'kindt, V., eds.: Metaheuristics for Multiobjective Optimisation. Volume 535 of Lecture Notes in Economics and Mathematical Systems. Springer Verlag (2004) 177–200
14. Bentley, J.: Fast algorithms for geometric traveling salesman problems. ORSA Journal on Computing **4** (1992) 387–411

Recombination of Similar Parents in EMO Algorithms

Hisao Ishibuchi and Kaname Narukawa

Department of Industrial Engineering, Osaka Prefecture University,
1-1 Gakuen-cho, Sakai, Osaka, 599-8531, Japan
{hisaoi, kaname}@ie.osakafu-u.ac.jp

Abstract. This paper examines the effect of crossover operations on the performance of EMO algorithms through computational experiments on knapsack problems and flowshop scheduling problems using the NSGA-II algorithm. We focus on the relation between the performance of the NSGA-II algorithm and the similarity of recombined parent solutions. First we show the necessity of crossover operations through computational experiments with various specifications of crossover and mutation probabilities. Next we examine the relation between the performance of the NSGA-II algorithm and the similarity of recombined parent solutions. It is shown that the quality of obtained solution sets is improved by recombining similar parents. Then we examine the effect of increasing the selection pressure (i.e., increasing the tournament size) on the similarity of recombined parent solutions. An interesting observation is that the increase in the tournament size leads to the recombination of dissimilar parents, improves the diversity of solutions, and degrades the convergence performance of the NSGA-II algorithm.

1 Introduction

Since Schaffer's study [21], various evolutionary multiobjective optimization (EMO) algorithms have been proposed to find well-distributed Pareto-optimal or near Pareto-optimal solutions of multiobjective optimization problems (Coello et al. [2] and Deb [4]). Recent EMO algorithms usually share some common ideas such as elitism, fitness sharing and Pareto ranking. While mating restriction has been often discussed in the literature, it has not been used in many EMO algorithms as pointed out in some reviews on EMO algorithms [7, 23, 26]. In this paper, we examine the effect of recombining similar parents on the performance of EMO algorithms to find well-distributed Pareto-optimal or near Pareto-optimal solutions.

Mating restriction was suggested by Goldberg [8] for single-objective genetic algorithms. Hajela & Lin [9] and Fonseca & Fleming [6] used it in their EMO algorithms. The basic idea of mating restriction is to ban the recombination of dissimilar parents from which good offspring are not likely to be generated. In the implementation of mating restriction, a user-definable parameter σ_{mating} called the mating radius is usually used for banning the recombination of two parents whose distance is larger than σ_{mating}. The distance between two parents is measured in the decision space or the objective space. The necessity of mating restriction in EMO algorithms was also stressed by Jaszkiewicz [17] and Kim et al. [18]. In the parallelization of EMO algorithms, mating restriction is implicitly realized since similar individuals are likely to be assigned to the same processor or the same island (e.g., see Branke et al. [1]). On the other hand, Zitzler & Thiele [25] reported that no improvement was achieved by mating restriction in their computational experiments. Van Veldhuizen & Lamont [23] mentioned that the empirical evidence presented in the literature could be interpreted as an argument either for or against the use of mating restriction. Moreover, there was also an argument for the selection of dissimilar parents. Horn et al. [10] argued that information from very different types of tradeoffs could be combined to yield other kinds of good tradeoffs. Schaffer [21] examined the selection of dissimilar parents but observed no improvement.

A similarity-based mating scheme was proposed in Ishibuchi & Shibata [13] to examine positive and negative effects of mating restriction on the performance of EMO algorithms. In their mating scheme, one parent (say Parent A) was chosen by the standard fitness-based binary tournament scheme while its mate (say Parent B) was chosen among a pre-specified number of candidates (say β candidates) based on their similarity or dissimilarity to Parent A. To find β candidates, the standard fitness-based binary tournament selection was iterated β times. Almost the same idea was independently proposed in Huang [11] where Parent B was chosen from two candidates (i.e., the value of β was fixed as $\beta = 2$). Ishibuchi & Shibata [14] extended their similarity-based mating scheme as shown in Fig. 1. That is, first a pre-specified number of candidates (say α candidates) were selected by iterating the standard fitness-based binary tournament selection α times. Next the average vector of those candidates was calculated in the objective space. The most dissimilar candidate to the average vector was chosen as Parent A. On the other hand, the most similar one to Parent A among β candidates was chosen as Parent B. Furthermore, it was demonstrated in [15] that the diversity-convergence balance can be dynamically adjusted by controlling the values of the two parameters α and β.

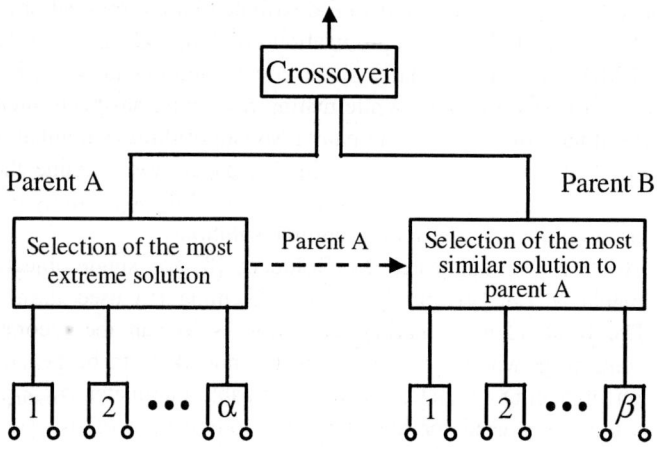

Fig. 1. Mating scheme in Ishibuchi & Shibata [14]

In this paper, we examine the effect of crossover operations on the performance of EMO algorithms through computational experiments on knapsack problems and flowshop scheduling problems using the NSGA-II algorithm of Deb et al. [5]. We focus on the relation between the performance of the NSGA-II algorithm and the similarity of recombined parents. First we show the necessity of crossover operations through computational experiments with various specifications of crossover and mutation probabilities. It is shown that crossover operations play an important role in the NSGA-II algorithm while its performance is not very sensitive to the crossover probability if compared with the mutation probability. Next we examine the relation between the performance of the NSGA-II algorithm and the similarity of recombined parents. We use the similarity-based mating scheme in Fig. 1 to choose dissimilar parents as well as similar parents. That is, the most similar or dissimilar solution to Parent A among β candidates is chosen as Parent B in our computational experiments where the value of α is fixed as $\alpha = 1$ (i.e., Parent A is selected by the standard fitness-based binary tournament selection). The similarity is measured in the decision space and the objective space. It is shown that the performance of the NSGA-II algorithm can be improved by recombining similar parents. Then we examine the effect of increasing the selection pressure (i.e., increasing the tournament size) on the similarity of recombined parents. An interesting observation is that the increase in the tournament size leads to the recombination of dissimilar parents, improves the diversity of solutions, and degrades the convergence performance of the NSGA-II algorithm on some test problems.

2 Test Problems

Multiobjective 0/1 knapsack problems with k knapsacks (i.e., k objectives and k constraints) and n items in Zitzler & Thiele [26] can be written as follows:

$$\text{Maximize } \mathbf{f}(\mathbf{x}) = (f_1(\mathbf{x}), f_2(\mathbf{x}), ..., f_k(\mathbf{x})), \tag{1}$$

$$\text{subject to } \sum_{j=1}^{n} w_{ij} x_j \leq c_i, \quad i = 1, 2, ..., k, \tag{2}$$

$$\text{where } f_i(\mathbf{x}) = \sum_{j=1}^{n} p_{ij} x_j, \quad i = 1, 2, ..., k. \tag{3}$$

In this formulation, $\mathbf{x}$ is an n-dimensional binary vector (i.e., $(x_1, x_2, ..., x_n) \in \{0, 1\}^n$), p_{ij} is the profit of item j according to knapsack i, w_{ij} is the weight of item j according to knapsack i, and c_i is the capacity of knapsack i. Each solution $\mathbf{x}$ is handled as a binary string of length n in EMO algorithms. The k-objective n-item knapsack problem is referred to as a k-n knapsack problem in this paper.

Zitzler & Thiele [26] examined the performance of several EMO algorithms using nine test problems with two, three, four objectives and 250, 500, 750 items. In this paper, we use the three 500-item test problems (i.e., 2-500, 3-500, and 4-500 knapsack problems) while we can only report a part of experimental results due to the page limitation.

We also generate two-objective and three-objective 20-machine 80-job flowshop scheduling problems in the same manner as Ishibuchi et al. [12, 16]. These test problems are denoted as 2-20-80 and 3-20-80 scheduling problems, respectively. The makespan and the maximum tardiness are considered as two objectives to be minimized in the 2-20-80 scheduling problem. In addition to these two objectives, the minimization of the total flowtime is considered in the 3-20-80 scheduling problem. Each solution of these two test problems is represented as a permutation of 80 jobs.

In the case of the knapsack problems, the distance between two solutions (i.e., two binary strings) is measured by the Hamming distance in the decision space. On the other hand, the distance of two solutions (i.e., two permutations of 80 jobs) is calculated as the sum of the distance between the positions of each job. The calculation of the distance between two permutation-type strings is illustrated in Fig. 2. The distance between the positions of Job 1 (denoted by J1 in Fig. 2) is 4 since it is placed in the first position of String 1 and the fifth position of String 2. The distance between the positions of the other jobs is calculated in the same manner (i.e., 1 for Job 2, 0 for Job 3 and Job 4, and 3 for Job 5). Thus the distance between the two strings in Fig. 2 is calculated as 8 (i.e., 4+1+0+0+3).

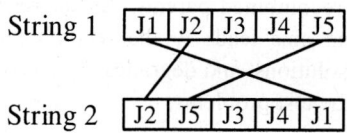

Fig. 2. Distance between two strings for five-job flowshop scheduling problems

The distance between two solutions in the objective space is calculated by the Euclidean distance in both the knapsack problems and the scheduling problems. That is, the distance between two solutions **x** and **y** is calculated in the objective space as

$$|\mathbf{f}(\mathbf{x}) - \mathbf{f}(\mathbf{y})| = \sqrt{|f_1(\mathbf{x}) - f_1(\mathbf{y})|^2 + \cdots + |f_k(\mathbf{x}) - f_k(\mathbf{y})|^2}, \quad (4)$$

where $\mathbf{f}(\mathbf{x}) = (f_1(\mathbf{x}), ..., f_k(\mathbf{x}))$ is the k-dimensional objective vector corresponding to the solution **x**.

3 Performance Measures

Various performance measures have been proposed in the literature to evaluate a non-dominated solution set. As explained in Knowles & Corne [19], Okabe et al. [20], and Zitzler et al. [27], no single performance measure can simultaneously evaluate various aspects of a non-dominated solution set (e.g., convergence and diversity). Moreover, some performance measures are not designed to simultaneously compare many solution sets but to compare only two solution sets with each other. For various performance measures, see [2, 4, 19, 20, 27].

In this paper, we use the following performance measures to simultaneously compare a number of solution sets:

1. Generational distance (GD)
2. $D1_R$ measure ($D1_R$)
3. Spread measure (Spread)
4. Hypervolume measure (Hypervolume)
5. Ratio of non-dominated solutions (Ratio)

Let S and S^* be a set of non-dominated solutions and the set of all Pareto-optimal solutions. The generational distance [22] is the average distance from each solution in S to its nearest Pareto-optimal solution in S^*. The distance is measured in the objective space using the Euclidean distance. On the other hand, the $D1_R$ measure is the average distance from each Pareto-optimal solution in S^* to its nearest solution in S. This measure was used in Czyzak & Jaszkiewicz [3]. The generational distance and the $D1_R$ measure need all Pareto-optimal solutions of each test problem. Since true Pareto-optimal solutions are not available for each test problem except for the 2-500 knapsack problem, we use as S^* a set of near Pareto-optimal solutions obtained for each test problem using much longer CPU time and much larger memory storage than computational experiments reported in this paper.

The spread measure is calculated for the solution set S as follows:

$$Spread = \sum_{i=1}^{k} [\max_{\mathbf{x} \in S} \{f_i(\mathbf{x})\} - \min_{\mathbf{x} \in S} \{f_i(\mathbf{x})\}]. \quad (5)$$

This measure is similar to the maximum spread of Zitzler [24]. The hypervolume measure [25] calculates the volume of the dominated region by the solution set S in the objective space.

The ratio of non-dominated solutions is calculated for a solution set with respect to other solution sets. We used this measure in our former studies to compare multiple non-dominate rule sets [12, 16]. This measure is similar to the coverage measure of Zitzler & Thiele [25], which was proposed to compare two non-dominated rule sets. Let us assume that we have m solution sets $S_1, S_2, ..., S_m$. By merging these solution sets, we construct another solution set S as $S = S_1 \cup S_2 \cup ... \cup S_m$. Let S_{ND} be the set of non-dominated solutions in S. The ratio of non-dominated solutions is calculated for each solution set S_i as $|S_i \cap S_{ND}|/|S_i|$ where $|S_i|$ denotes the cardinality of S_i (i.e., the number of solutions in S_i).

4 Conditions of Computational Experiments

Our computational experiments on the knapsack problems are performed using the NSGA-II algorithm under the following parameter specifications:

Crossover probability (one-point crossover): 0.8,
Mutation probability (bit-flip mutation): 1/500 (per bit),
Population size: 200 (2-500 problem), 250 (3-500 problem), 300 (4-500 problem),
Stopping condition: 500 generations.

The average value of each performance measure is calculated over 50 runs from different initial populations for each knapsack problem.

On the other hand, we use the following parameter specifications for the scheduling problems:

Crossover probability (two-point order crossover): 0.8,
Mutation probability (shift mutation): 0.5 (per string),
Population size: 200,
Stopping condition: 500 generations.

The average value of each performance measure is calculated over 20 runs with different initial populations for each scheduling problem.

It should be noted that the above-mentioned parameter values are basic settings in our computational experiments. Various specifications of the crossover and mutation probabilities are examined in the next section.

5 Effects of Crossover Operations

For examining the necessity of crossover operations in the NSGA-II algorithm, we apply it to each knapsack problem using the following specifications of the crossover and mutation probabilities:

Crossover probability (P_C): 0.0, 0.1, 0.2, 0.3, 0.4, 0.5, 0.6, 0.7, 0.8, 0.9, 1.0,
Mutation probability (P_M): 0.0001, 0.0002, 0.0005, 0.001, 0.002, ..., 0.1 (per bit).

In the application of the NSGA-II algorithm to each scheduling problem, the following parameter specifications are examined:

Crossover probability (P_C): 0.0, 0.1, 0.2, 0.3, 0.4, 0.5, 0.6, 0.7, 0.8, 0.9, 1.0,
Mutation probability (P_M): 0.1, 0.2, 0.3, 0.4, 0.5, 0.6, 0.7, 0.8, 0.9, 1.0 (per string).

Experimental results on the 3-500 knapsack problem are shown in Fig. 3. In these experimental results, we can not observe the necessity of the crossover operation. That is, good results are obtained even in the case of no crossover (i.e., $P_C = 0$). The performance of the NSGA-II algorithm in Fig. 3 is sensitive to the mutation probability (P_M) and insensitive to the crossover probability (P_C).

While the crossover operation seems to be unnecessary in Fig. 3, its necessity is clearly shown by the average ratio of non-dominated solutions in Fig. 4 (a) where the 10×11 combinations of the crossover and mutation probabilities are compared. That is, 10×11 solution sets obtained from these combinations are compared to calculate the ratio of non-dominated solutions. From Fig. 4 (a), we can see that all solutions obtained from the NSGA-II with no crossover (i.e., $P_C = 0$) are dominated by other solutions obtained from that with crossover (i.e., $P_C > 0$). The same observation is obtained from computational experiments on the 3-20-80 scheduling problem in Fig. 4 (b). The performance of the NSGA-II algorithm for the 3-20-80 scheduling problem is not sensitive to the crossover probability as shown in Fig. 5. Similar results are obtained for all the five test problems in this paper.

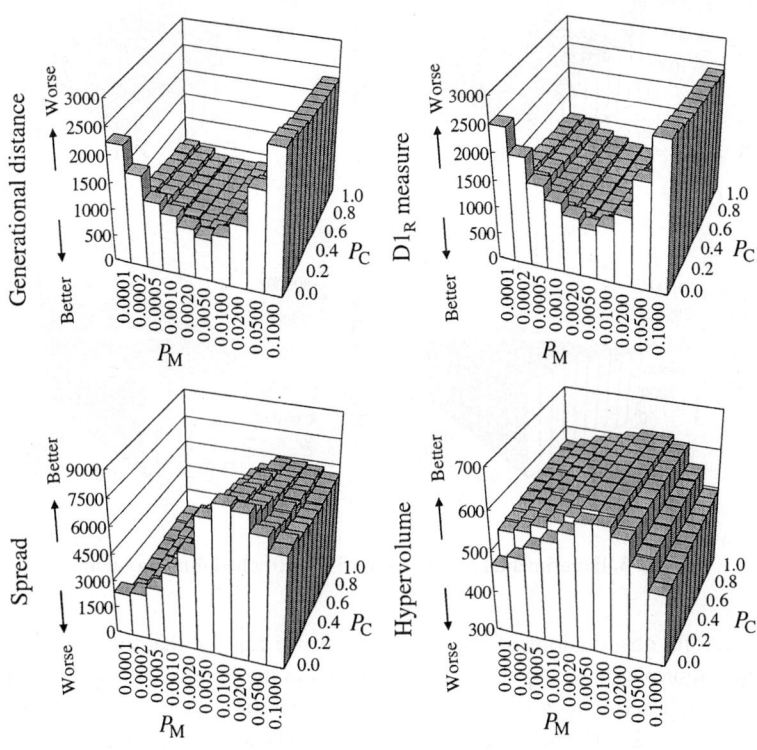

Fig. 3. Experimental results on the 3-500 knapsack problem

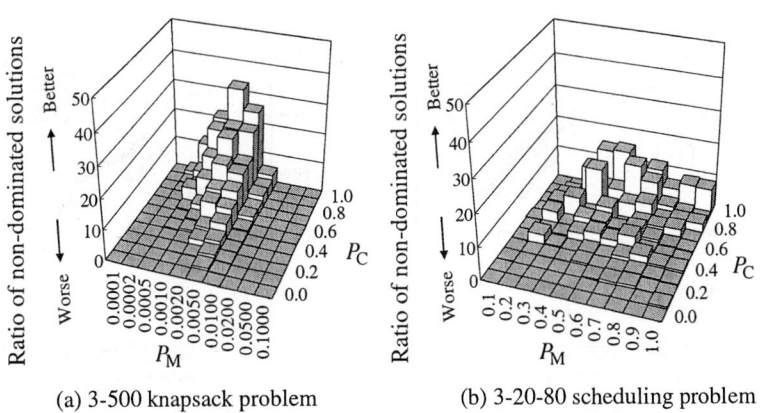

(a) 3-500 knapsack problem (b) 3-20-80 scheduling problem

Fig. 4. Average ratio of non-dominated solutions

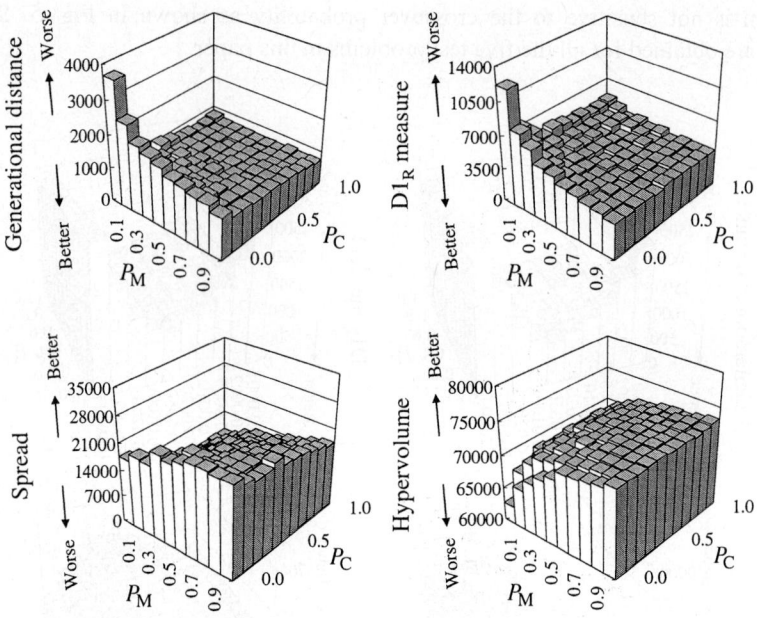

Fig. 5. Experimental results on the 3-20-80 scheduling problem

From careful observations of Fig. 3 and Fig. 5, we can see that the use of crossover operations improved the convergence of solutions to the Pareto front and degraded the diversity of solutions. Such effects of crossover operations are visually demonstrated in Fig. 6 for the 2-500 knapsack problem. Fig. 6 shows non-dominated solutions at each generation of a single run of the NSGA-II algorithm with crossover (Fig. 6 (a)) and without crossover (Fig. 6 (b)). The mutation rate was specified as 1/500 (i.e., its basic setting) in Fig. 6.

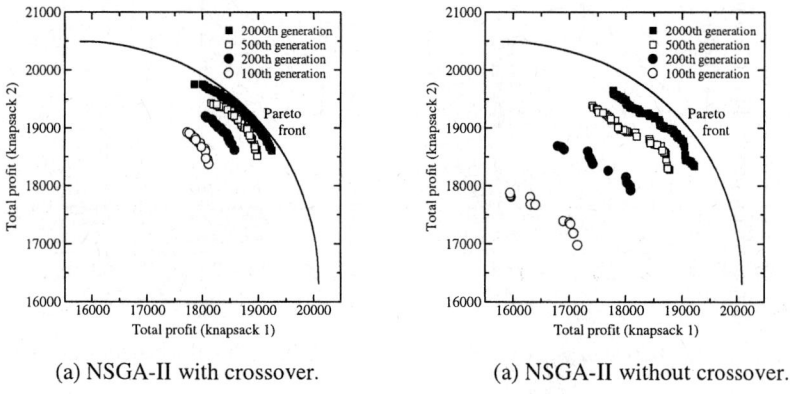

(a) NSGA-II with crossover.   (a) NSGA-II without crossover.

Fig. 6. Non-dominated solutions at each generation for the 2-500 knapsack problem

6 Effects of Recombination of Similar Parents

We examine the effect of recombining similar parents using the mating scheme in Fig. 1. When a pair of parents is to be chosen, one parent (say Parent A) is selected by the standard fitness-based binary tournament selection in the same manner as the NSGA-II algorithm. That is, the value of α in Fig. 1 is fixed as $\alpha = 1$ in order to focus on the effect of recombining similar parents. Next we iterate the standard fitness-based binary tournament selection β times to find β candidates for the selection of the other parent (say Parent B). The most similar candidate to Parent A is chosen as Parent B (i.e., the mate of Parent A). For comparison, we also examine the choice of the most dissimilar parent to Parent A. This mating scheme is exactly the same as the standard fitness-based binary tournament selection in the NSGA-II algorithm when the values of α and β are specified as $\alpha = 1$ and $\beta = 1$. A large value of β means a strong bias toward the choice of similar (or dissimilar) parents.

We apply the NSGA-II algorithm with the mating scheme in Fig. 1 to each test problem using various values of β. Experimental results on the 3-500 knapsack problem are summarized in Fig. 7. We can see from Fig. 7 that the performance of the NSGA-II algorithm is improved by recombining similar parents. We can also see that similar results are obtained independent of the choice between the objective space and the decision space where the similarity of solutions is measured. The average distance between recombined parents is shown in Fig. 8. The improvement in the performance of the NSGA-II algorithm is also observed in computational experiments on the 3-20-80 scheduling problem in Fig. 9 where we show average results over 50 runs (Fig. 4 (b) and Fig. 5 were average results over 20 runs on the 3-20-80 scheduling problem).

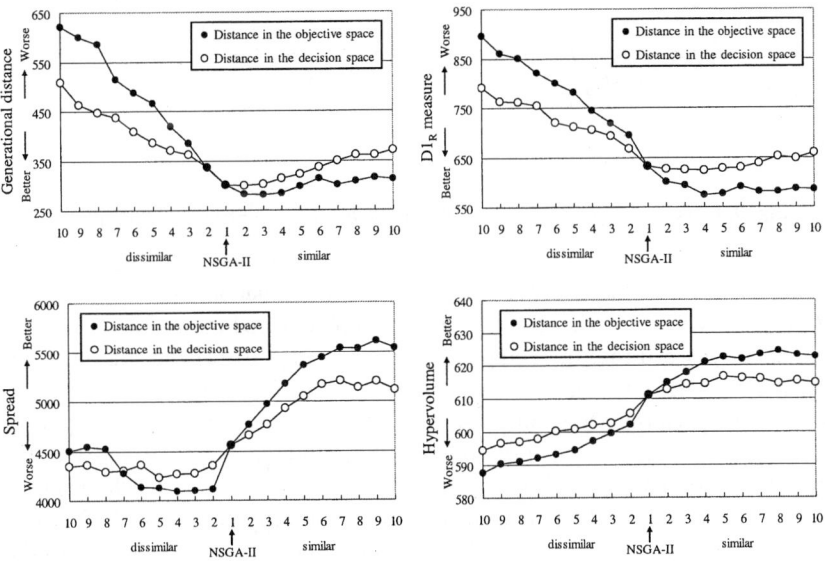

Fig. 7. Effects of recombining similar parents for the 3-500 knapsack problem

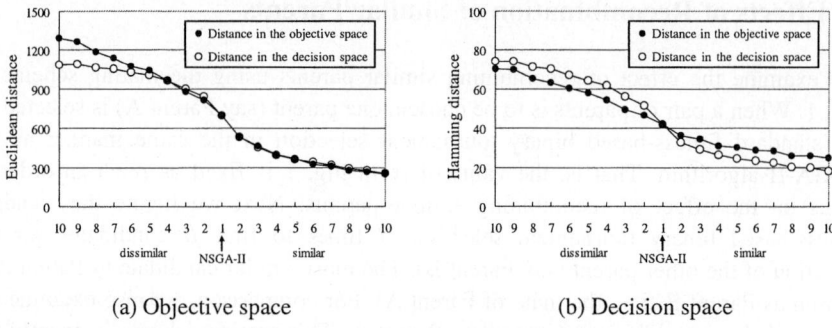

Fig. 8. Average distance between recombined parents for the 3-500 knapsack problem

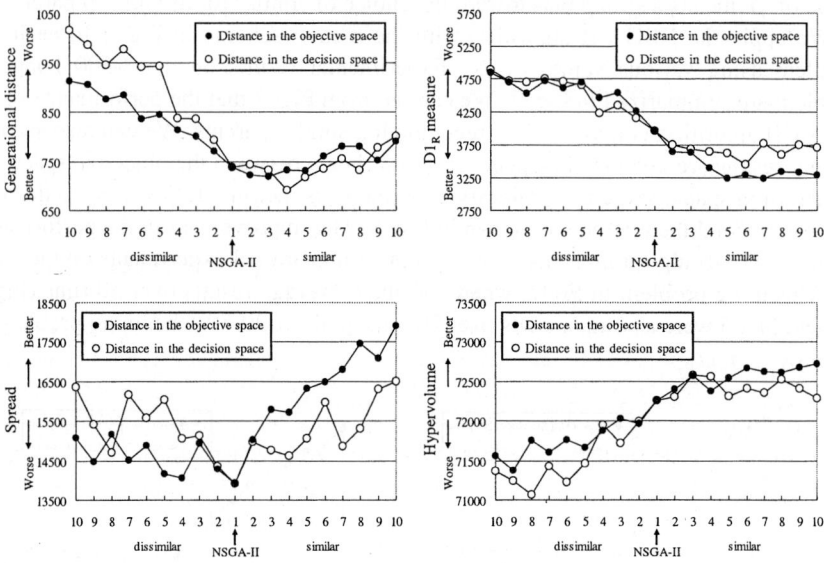

Fig. 9. Effects of recombining similar parents for the 3-20-80 scheduling problem

We visually show the effect of choosing similar parents on the performance of the NSGA-II algorithm for the 2-500 knapsack problem in Fig. 10 (a) where β was specified as $\beta = 5$. From the comparison between Fig. 10 (a) and Fig. 6 (a) by the original NSGA-II algorithm, we can see that the recombination of similar parents improved the performance of the NSGA-II algorithm (especially with respect to the diversity of solutions). Such improvement was observed in Fig. 7 and Fig. 9. The effect of our similarity-based mating scheme in Fig. 1 becomes much clearer if we choose extreme and similar parents using the two parameters α and β. We visually show the effect of choosing extreme and similar parents in Fig. 10 (b) where the values of the two parameters α and β were specified as $\alpha = 5$ and $\beta = 5$. We can clearly see that the diversity of solutions was improved in Fig. 10 (b) from Fig. 6 (a).

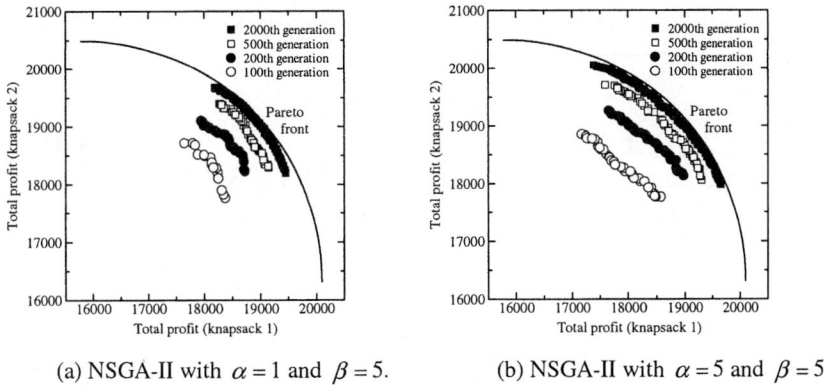

Fig. 10. Effects of our meting scheme on the performance of the NSGA-II algorithm

7 Effects of High Selection Pressure

We examine the performance of the NSGA-II algorithm on each test problem using various specifications of the tournament size. Experimental results on the 3-500 knapsack problem are shown in Fig. 11. Each closed circle in Fig. 11 shows the average result over 50 runs for each specification of the tournament size. The standard deviation is also shown in Fig. 11 as the radius of each interval (i.e., the distance between the closed circle and each edge of the interval). It should be noted in Fig. 11 that the tournament size 1 means the random selection from the current population. Even in this case, the NSGA-II algorithm still has a somewhat strong selection pressure since the best individuals are chosen from the parent population and the offspring population in the generation update phase. We can see from Fig. 11 that the increase in the tournament size improves the diversity of solutions (i.e., the spread and hypervolume measures) and degrades the convergence of solutions to the Pareto front (i.e., the generational distance and the $D1_R$ measure). This observation may be counterintuitive for some readers because the strong selection pressure does not lead to the improvement of the convergence but the improvement of the diversity in Fig. 11. During the computational experiments in Fig. 11, we also measure the distance between recombined parents. The average distance between recombined parents is shown in Fig. 12. From this figure, we can see that the increase in the tournament size leads to the recombination of dissimilar parents. Similar results are also obtained for the 3-20-80 scheduling problem in Fig. 13 and Fig. 14. Average results over 50 runs are shown in these figures.

In Fig. 15, we visually show the effect of increasing the tournament size on the performance of the NSGA-II algorithm for the 2-500 knapsack problem. From the comparison of Fig. 15 with Fig. 6 (a) by the binary tournament selection, we can see that the increase in the tournament size (i.e., the use of a higher selection pressure) improved the diversity of solutions while it degraded the convergence of solutions to the Pareto front. Such observations were also obtained from Fig. 11 and Fig. 14.

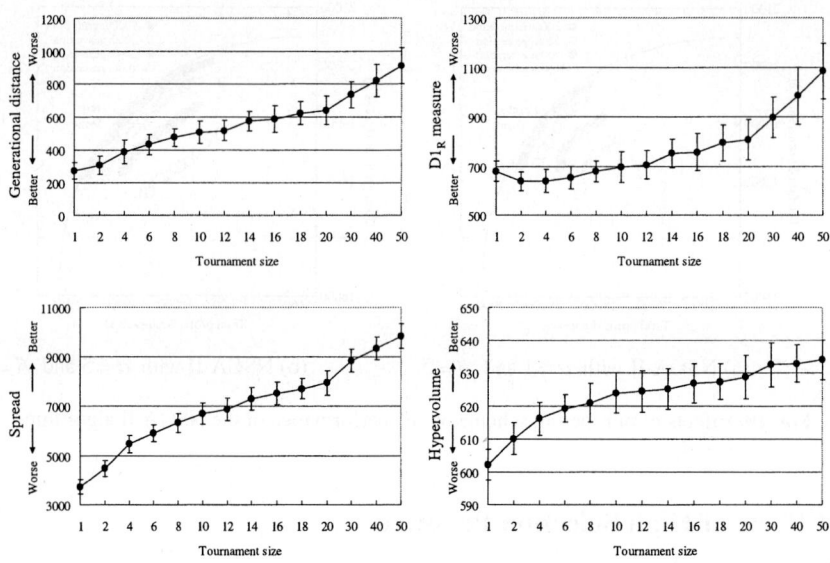

Fig. 11. Effects of the tournament size for the 3-500 knapsack problem

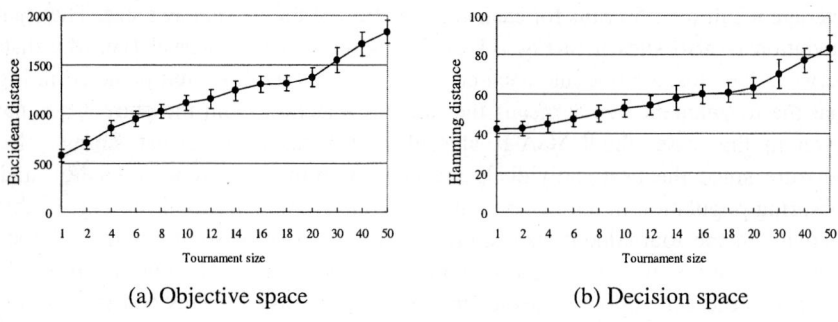

(a) Objective space (b) Decision space

Fig. 12. Average distance between recombined parents for the 3-500 knapsack problem

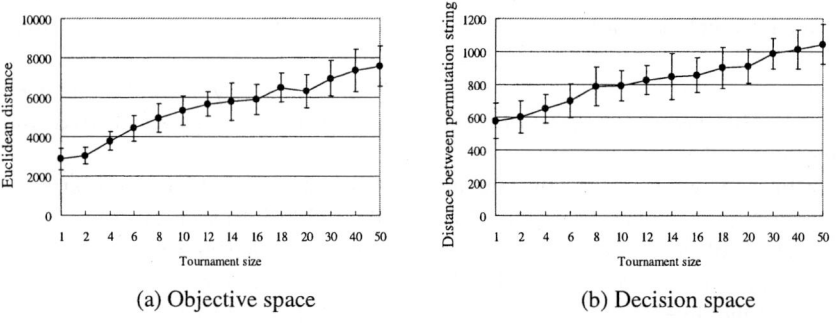

(a) Objective space (b) Decision space

Fig. 13. Average distance between recombined parents for the 3-20-80 scheduling problem

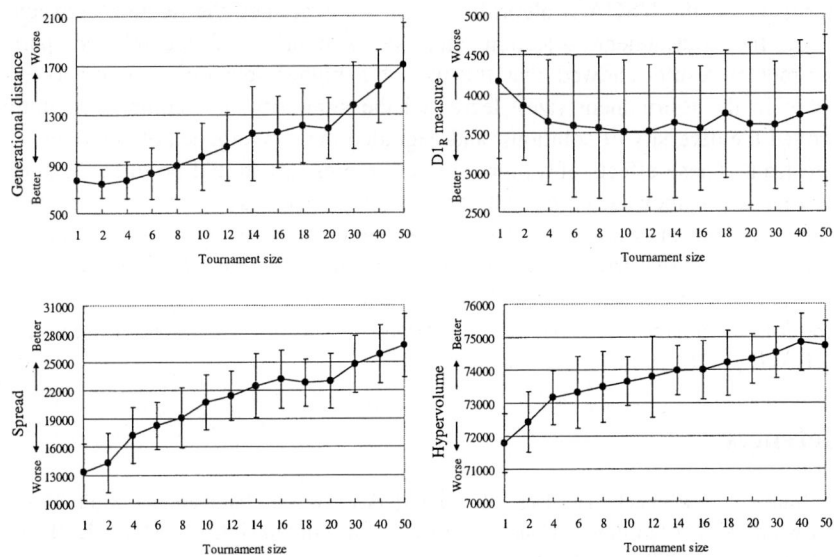

Fig. 14. Effects of the tournament size for the 3-20-80 scheduling problem

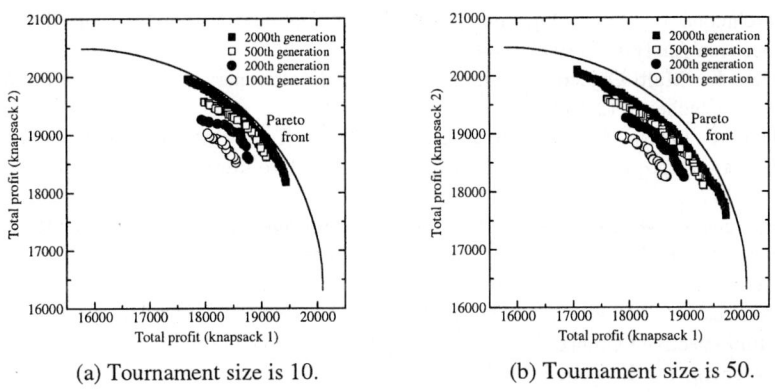

Fig. 15. Effects of the tournament size on the performance of the NSGA-II algorithm

8 Concluding Remarks

Through computational experiments on multiobjective 0/1 knapsack problems and multiobjective flowshop scheduling problems, we examined the effect of crossover operations on the performance of the NSGA-II algorithm. First we showed that crossover operations played an important role while mutation operations seemed to have a larger effect on the performance of the NSGA-II algorithm. Next we empirically demonstrated that the recombination of similar parents improved the performance of the NSGA-II algorithm. Then we examined the relation between the size of the tournament selection and the similarity of recombined parents. Experimental results showed that the use of a higher selection pressure (i.e., the increase in the tournament size) decreased the similarity of recombined solutions, improved the diversity of solutions, and degraded the convergence of solutions to the Pareto front. These effects of a higher selection pressure should be further examined since they are somewhat counterintuitive. That is, one may think that the use of a higher selection pressure may lead to the improvement in the convergence of solutions to the Pareto front and the deterioration in the diversity of solutions.

The authors would like to thank the financial support from Japan Society for the Promotion of Science (JSPS) through Grand-in-Aid for Scientific Research (B): KAKENHI (14380194).

References

1. Branke, J., Schmeck, H., Deb, K., and Reddy, S. M.: Parallelizing Multi-Objective Evolutionary Algorithms: Cone Separation. Proc. of 2004 Congress on Evolutionary Computation (2004) 1952-1957.
2. Coello Coello, C. A., Van Veldhuizen, D. A., and Lamont, G. B.: Evolutionary Algorithms for Solving Multi-Objective Problems. Kluwer Academic Publishers, Boston (2002).
3. Czyzak, P. and Jaszkiewicz, A.: Pareto-Simulated Annealing – A Metaheuristic Technique for Multi-Objective Combinatorial Optimization. Journal of Multi-Criteria Decision Analysis 7 (1998) 34-47.
4. Deb, K.: Multi-Objective Optimization Using Evolutionary Algorithms. John Wiley & Sons, Chichester (2001).
5. Deb, K., Pratap, A., Agarwal, S., and Meyarivan, T.: A Fast and Elitist Multiobjective Genetic Algorithm: NSGA-II. IEEE Trans. on Evolutionary Computation 6 (2002) 182-197.
6. Fonseca, C. M. and Fleming, P. J.: Genetic Algorithms for Multiobjective Optimization: Formulation, Discussion and Generalization. Proc. of 5th International Conference on Genetic Algorithms (1993) 416-423.
7. Fonseca, C. M. and Fleming, P. J.: An Overview of Evolutionary Algorithms in Multiobjective Optimization. Evolutionary Computation 3 (1995) 1-16.
8. Goldberg, D. E.: Genetic Algorithms in Search, Optimization, and Machine Learning. Addison-Wesley, Reading (1989).
9. Hajela, P. and Lin, C. Y.: Genetic Search Strategies in Multicriterion Optimal Design. Structural Optimization 4 (1992) 99-107.
10. Horn, J., Nafpliotis, N., and Goldberg, D. E.: A Niched Pareto Genetic Algorithm for Multi-Objective Optimization. Proc. of 1st IEEE International Conference on Evolutionary Computation (1994) 82-87.
11. Huang, C. F.: Using an Immune System Model to Explore Mate Selection in Genetic Algorithms. Lecture Notes in Computer Science, Vol. 2723 (Proc. of GECCO 2003), Springer, Berlin (2003) 1041-1052.
12. Ishibuchi, H. and Murata, T.: A Multi-Objective Genetic Local Search Algorithm and Its Application to Flowshop Scheduling. IEEE Trans. on Systems, Man, and Cybernetics - Part C: Applications and Reviews 28 (1998) 392-403.

13. Ishibuchi, H. and Shibata, Y.: An Empirical Study on the Effect of Mating Restriction on the Search Ability of EMO Algorithms. Lecture Notes in Computer Science, Vol. 2632 (Proc. of EMO 2003), Springer, Berlin (2003) 433-447.
14. Ishibuchi, H. and Shibata, Y.: A Similarity-based Mating Scheme for Evolutionary Multiobjective Optimization. Lecture Notes in Computer Science, Vol. 2723 (Proc. of GECCO 2003), Springer, Berlin (2003) 1065-1076.
15. Ishibuchi, H. and Shibata, Y.: Mating Scheme for Controlling the Diversity-Convergence Balance for Multiobjective Optimization. Lecture Notes in Computer Science, Vol. 3102 (Proc. of GECCO 2004), Springer, Berlin (2004) 1259-1271.
16. Ishibuchi, H., Yoshida, T., and Murata, T.: Balance between Genetic Search and Local Search in Memetic Algorithms for Multiobjective Permutation Flowshop Scheduling. IEEE Trans. on Evolutionary Computation **7** (2003) 204-223.
17. Jaszkiewicz, A.: On the Performance of Multiple-Objective Genetic Local Search on the 0/1 Knapsack Problem - A Comparative Experiment. IEEE Trans. on Evolutionary Computation **6** (2002) 402-412.
18. Kim M., Hiroyasu, T., Miki, M., and Watanabe, S.: SPEA2+: Improving the Performance of the Strength Pareto Evolutionary Algorithm 2. Lecture Notes in Computer Science, Vol. 3242 (Proc. of PPSN VIII), Springer, Berlin (2004) 742-751.
19. Knowles, J. D. and Corne, D. W.: On Metrics for Comparing Non-dominated Sets. Proc. of 2002 Congress on Evolutionary Computation (2002) 711-716.
20. Okabe, T., Jin, Y., and Sendhoff, B.: A Critical Survey of Performance Indices for Multi-Objective Optimization. Proc. of 2003 Congress on Evolutionary Computation (2003) 878-885.
21. Schaffer, J. D.: Multiple Objective Optimization with Vector Evaluated Genetic Algorithms. Proc. of 1st International Conference on Genetic Algorithms and Their Applications (1985) 93-100.
22. Van Veldhuizen, D. A.: Multiobjective Evolutionary Algorithms: Classifications, Analyses, and New Innovations. Ph. D dissertation, Air Force Institute of Technology (1999).
23. Van Veldhuizen, D. A. and Lamont, G. B.: Multiobjective Evolutionary Algorithms: Analyzing the State-of-the-Art. Evolutionary Computation **8** (2000) 125-147.
24. Zitzler, E.: Evolutionary Algorithms for Multiobjective Optimization: Methods and Applications. Ph. D dissertation, Shaker Verlag, Aachen (1999).
25. Zitzler, E. and Thiele, L.: Multiobjective Optimization using Evolutionary Algorithms – A Comparative Case Study. Proc. of 5th International Conference on Parallel Problem Solving from Nature (1998) 292-301.
26. Zitzler, E. and Thiele, L.: Multiobjective Evolutionary Algorithms: A Comparative Case Study and the Strength Pareto Approach, IEEE Transactions on Evolutionary Computation **3** (1999) 257-271.
27. Zitzler, E., Thiele, L., Laumanns, M., Fonseca, C. M., and da Fonseca, V. G.: Performance Assessment of Multiobjective Optimizers: An Analysis and Review. IEEE Trans. on Evolutionary Computation **7** (2003) 117-132.

A Scalable Multi-objective Test Problem Toolkit

Simon Huband[1], Luigi Barone[2], Lyndon While[2], and Phil Hingston[1]

[1] Edith Cowan University, Mount Lawley WA 6050, Australia
{s.huband, p.hingston}@ecu.edu.au
[2] The University of Western Australia, Crawley WA 6009, Australia
{luigi, lyndon}@csse.uwa.edu.au

Abstract. This paper presents a new toolkit for creating scalable multi-objective test problems. The WFG Toolkit is flexible, allowing characteristics such as bias, multi-modality, and non-separability to be incorporated and combined as desired. A wide variety of Pareto optimal geometries are also supported, including convex, concave, mixed convex/concave, linear, degenerate, and disconnected geometries.

All problems created by the WFG Toolkit are well defined, are scalable with respect to both the number of objectives and the number of parameters, and have known Pareto optimal sets. Nine benchmark multi-objective problems are suggested, including one that is both multi-modal and non-separable, an important combination of characteristics that is lacking among existing (scalable) multi-objective problems.

1 Introduction

There have been several attempts to define test suites and toolkits for testing multi-objective evolutionary algorithms (MOEAs) [1, 2, 3, 4]. However, existing multi-objective test problems do not test a wide range of characteristics, and are often poorly designed. Typical defects include not being scalable and being susceptible to simple search strategies. Moreover, many problems are poorly constructed, with unknown Pareto optimal sets, or featuring parameters with poorly located optima.

As suggested for single-objective problems by Whitley et al. [5] and Bäck and Michalewicz [6], test suites should include scalable problems that are resistant to hill climbing strategies, are non-linear, non-separable[1], and multi-modal. Such requirements are also a good start for multi-objective test suites, but unfortunately are poorly represented in the literature.

Addressing this problem, this paper presents the Walking Fish Group (WFG) Toolkit, which places an emphasis on allowing test problem designers to construct scalable test problems with any number of objectives, where features such

---

[1] Separable problems can be optimised by considering each parameter in turn, independently of one another. A non-separable problem is thus characterised by parameter dependencies, is more difficult, and is more representative of real world problems.

as modality and separability can be customised as required. Test problems in the WFG Toolkit are defined in terms of a simple underlying problem that defines the fitness space and a series of composable, configurable transformations that allow the test problem designer to add arbitrarily levels of complexity to the test problem. Problems created by the WFG Toolkit are well defined, are scalable with respect to both the number of objectives and the number of parameters, and have known Pareto optimal sets.

The next section of the paper introduces the multi-objective terminology used throughout. Section 3 briefly examines previous multi-objective test suites, highlighting the deficiencies with them. Section 4 specifies our new WFG Toolkit, generalising the concepts introduced in these previous test suites to produce a configurable toolkit that allows for the construction of scalable, well-behaved test problems. Section 5 then describes how the WFG Toolkit can be used to construct an example test problem. Some experimental results are presented in Section 6. A suite of nine test problems are proposed in Section 7 that exceeds the functionality of previous test suites. Section 8 concludes the paper.

2 Terminology

Consider a multi-objective optimisation problem given in terms of a search space of allowed values of n parameters $x_1, \ldots, x_n$, and a vector of M objective functions $\{f_1, \ldots, f_M\}$ mapping parameter vectors into fitness space. The mapping from the search space to fitness space defines the *fitness landscape*.

In multi-objective optimisation, we aim to find the set of optimal trade-off solutions known as the *Pareto optimal set*. The Pareto optimal set is the set of all Pareto optimal parameter vectors, and the corresponding set of objective vectors is the *Pareto optimal front*. The Pareto optimal set is a subset of the search space, whereas the Pareto optimal front is a subset of the fitness space.

The following types of relationships are useful because they allow us to separate the convergence and spread aspects of sets of solutions for a problem. A *distance parameter* is one that when modified only ever results in a dominated, dominating, or equivalent parameter vector. A *position parameter* is one that when modified only ever results in an incomparable or equivalent parameter vector. All other parameters are *mixed parameters*.

When the projection of the Pareto optimal set onto the domain of a single parameter, the parameter optima, is a single value at the edge of the domain, then we call the parameter an *extremal parameter*. If instead the parameter optima cluster around the middle of the domain, then it is a *medial parameter*. Extremal parameters can be unduly favoured by truncation based mutation correction strategies, whereas medial parameters can be favoured by EAs that employ intermediate recombination [7].

3 Previous Multi-objective Test Problems

Deb's toolkit [1] for constructing two-objective problems is the only toolkit for multi-objective problems of which we are aware. Deb's toolkit segregates parameters into distance and position parameters — mixed parameters are atypical. Three functionals are used that control the shape of and position on the trade-off surface, and the distance to the Pareto optimal front. Deb's toolkit provides a number of functionals, including multi-modal and biased functions, most of which are scalable parameter-wise.

Deb's toolkit has various limitations: it was designed for two-objective problems, no real-valued deceptive functions are suggested, the suggested functions do not facilitate the construction of problems with degenerate Pareto optimal front geometries[2], only one non-separable function is suggested (but it scales poorly and has but weak parameter dependencies), and position and distance parameters are always independent of one another[3].

Related by authorship to Deb's toolkit are the DTLZ test problems [4, 8], which, unlike the majority of multi-objective problems, are scalable objective-wise. This important characteristic has facilitated several recent investigations into what are commonly called "many" objective problems. Like Deb's toolkit, the DTLZ problems have distinct distance and position components, have known Pareto optimal fronts, and are simple to employ. The DTLZ problems also address a variety of problem characteristics, including multi-modality, bias, and several Pareto optimal front geometries.

However, the DTLZ test suite has serious limitations: none of its problems is deceptive, none of its problems is non-separable[4], and the number of position parameters is always fixed relative to the number of objectives. DTLZ5 and DTLZ6 also deserve special mention, as they are both meant to be problems with degenerate Pareto optimal fronts. However, we have found that this is untrue for instances with four or more objectives (due to space limitations, we omit the proof). As DTLZ5 and DTLZ6 do not behave as expected, their Pareto optimal fronts are unclear beyond three-objectives.

Whilst other test problems exist, including those employed by Van Veldhuizen [9], Zitzler et al. [10], and others [11, 12], they tend to be of limited scope, and are often ad hoc and consequently difficult to analyse (but by the same token, some also have unusual Pareto optimal geometries). Many are restricted to three or fewer parameters or objectives, some have poorly located parameter optima, few are non-separable (and even fewer are both non-separable and

---

[2] A degenerate front is a front that is of lower dimension than the objective space in which it is embedded, less one.

[3] Deb does suggest a way of making position and distance parameters mutually non-separable. However, the suggested approach can lead to cyclical dependencies, potentially causing unwanted side effects on the fitness landscape.

[4] Technically speaking, the majority of the DTLZ problems are non-separable, but only marginally so.

multi-modal), and those that are non-separable are either not scalable, or have unknown Pareto optimal fronts.

Despite the variety of existing test problems, there is clear need for additional work. At present there is no toolkit for creating problems with an arbitrary number of objectives, where desirable features can easily be incorporated or omitted as desired. We remedy this problem with our WFG Toolkit.

4   The WFG Toolkit

The WFG Toolkit defines a problem in terms of an underlying vector of parameters **x**. The vector **x** is always associated with a simple underlying problem that defines the fitness space. The vector **x** is derived, via a series of transition vectors, from a vector of working parameters **z**. Each transition vector adds complexity to the underlying problem, such as multi-modality and non-separability. The EA directly manipulates **z**, through which **x** is indirectly manipulated.

Unlike previous test suites in which complexity is "hard-wired" in an ad-hoc manner, the WFG Toolkit allows a test problem designer to control, via a series of composable transformations, which features will be present in the test problem. To create a problem, the test problem designer selects several shape functions to determine the geometry of the fitness space, and employs a number of transformation functions that facilitate the creation of transition vectors. Transformation functions must be designed carefully such that the underlying fitness space (and Pareto optimal front) remains intact with a relatively easy to determine Pareto optimal set. The WFG Toolkit provides a variety of predefined shape and transformation functions to help ensure this is the case.

For convenience, working parameters are labelled as either distance- or position-related parameters (even if they are actually mixed parameters), depending on the type of the underlying parameter being mapped to.

All problems created by the WFG Toolkit conform to the following format:

Given $\quad \mathbf{z} = \{z_1, \ldots, z_k, z_{k+1}, \ldots, z_n\}$

Minimise $\quad f_{m=1:M}(\mathbf{x}) = x_M + S_m h_m(x_1, \ldots, x_{M-1})$

where $\quad \mathbf{x} = \{x_1, \ldots, x_M\} = \{\max(t_M^p, A_1)(t_1^p - 0.5) + 0.5, \ldots,$
$\quad\quad\quad\quad\quad\quad\quad\quad\quad\quad \max(t_M^p, A_{M-1})(t_{M-1}^p - 0.5) + 0.5, t_M^p\}$

$\quad\quad\quad \mathbf{t}^p = \{t_1^p, \ldots, t_M^p\} \leftarrow [\,\mathbf{t}^{p-1} \leftarrow [\,\ldots \leftarrow [\,\mathbf{t}^1 \leftarrow [\,\mathbf{z}_{[0,1]}\,]$

$\quad\quad\quad \mathbf{z}_{[0,1]} = \{z_{1,[0,1]}, \ldots, z_{n,[0,1]}\} = \{z_1/z_{1,\max}, \ldots, z_n/z_{n,\max}\}$

where M is the number of objectives, **x** is a set of M underlying parameters (where x_M is an underlying distance parameter, and $x_{1:M-1}$ are underlying position parameters), **z** is a set of $k + l = n \geq M$ working parameters (the first k and the last l working parameters are position- and distance-related parameters respectively), $A_{1:M-1} \in \{0, 1\}$ are degeneracy constants (for each $A_i = 0$, the dimensionality of the Pareto optimal front is reduced by one), $h_{1:M}$ are shape functions, $S_{1:M} > 0$ are scaling constants, and $\mathbf{t}^{1:p}$ are transition vectors, where

"↩" indicates that each transition vector is created from another vector via transformation functions. The domain of all $z_i \in \mathbf{z}$ is $[0, z_{i,\max}]$ (the lower bound is always zero for convenience), where all $z_{i,\max} > 0$. Note that all $x_i \in \mathbf{x}$ will have domain $[0, 1]$.

Some observations can be made about the above formalism: substituting in $x_M = 0$ and disregarding all transition vectors provides a parametric equation that covers and is covered by the Pareto optimal front of the actual problem, working parameters can have dissimilar domains (which would encourage EAs to normalise parameter domains), and employing dissimilar scaling constants results in dissimilar Pareto optimal front tradeoff ranges (this is more representative of real world problems, and encourages EAs to normalise fitness values).

4.1 Shape Functions

Shape functions determine the nature of the Pareto optimal front, and map parameters with domain $[0, 1]$ onto the range $[0, 1]$. Each of $h_{1:M}$ must be associated with a shape function. For example, letting $h_1 = \text{linear}_1$, $h_{m=2:M-1} = \text{convex}_m$, and $h_M = \text{mixed}_M$ indicates that h_1 uses the linear shape function, h_M uses the mixed shape function, and all of $h_{2:M-1}$ use convex shape functions.

Table 1 presents five different types of shape functions. Example Pareto optimal fronts constructed using these shape functions are given in Fig. 1.

Table 1. Shape functions. In all cases, $x_1, \ldots, x_{M-1} \in [0, 1]$. A, α, and β are constants

---

Linear

$$\text{linear}_1(x_1, \ldots, x_{M-1}) = \prod_{i=1}^{M-1} x_i$$
$$\text{linear}_{m=2:M-1}(x_1, \ldots, x_{M-1}) = \left(\prod_{i=1}^{M-m} x_i\right)(1 - x_{M-m+1})$$
$$\text{linear}_M(x_1, \ldots, x_{M-1}) = 1 - x_1$$

When $h_{m=1:M} = \text{linear}_m$, the Pareto optimal front is a linear hyperplane, where $\sum_{m=1}^{M} h_m = 1$.

Convex

$$\text{convex}_1(x_1, \ldots, x_{M-1}) = \prod_{i=1}^{M-1}(1 - \cos(x_i \pi/2))$$
$$\text{convex}_{m=2:M-1}(x_1, \ldots, x_{M-1}) = \left(\prod_{i=1}^{M-m}(1 - \cos(x_i \pi/2))\right)(1 - \sin(x_{M-m+1}\pi/2))$$
$$\text{convex}_M(x_1, \ldots, x_{M-1}) = 1 - \sin(x_1 \pi/2)$$

When $h_{m=1:M} = \text{convex}_m$, the Pareto optimal front is purely convex.

Concave

$$\text{concave}_1(x_1, \ldots, x_{M-1}) = \prod_{i=1}^{M-1} \sin(x_i \pi/2)$$
$$\text{concave}_{m=2:M-1}(x_1, \ldots, x_{M-1}) = \left(\prod_{i=1}^{M-m} \sin(x_i \pi/2)\right) \cos(x_{M-m+1}\pi/2)$$
$$\text{concave}_M(x_1, \ldots, x_{M-1}) = \cos(x_1 \pi/2)$$

When $h_{m=1:M} = \text{concave}_m$, the Pareto optimal front is purely concave, and a region of the hyper-sphere of radius one centred at the origin, where $\sum_{m=1}^{M} h_m^2 = 1$.

Mixed convex/concave ($\alpha > 0$, $A \in \{1, 2, \ldots\}$)

$$\text{mixed}_M(x_1, \ldots, x_{M-1}) = \left(1 - x_1 - \frac{\cos(2A\pi x_1 + \pi/2)}{2A\pi}\right)^\alpha$$

Causes the Pareto optimal front to contain both convex and concave segments, the number of which is controlled by A. The overall shape is controlled by α: when $\alpha > 1$ or when $\alpha < 1$, the overall shape is convex or concave respectively. When $\alpha = 1$, the overall shape is linear.

Disconnected ($\alpha, \beta > 0$, $A \in \{1, 2, \ldots\}$)

$$\text{disc}_M(x_1, \ldots, x_{M-1}) = 1 - (x_1)^\alpha \cos^2(A(x_1)^\beta \pi)$$

Causes the Pareto optimal front to have disconnected regions, the number of which is controlled by A. The overall shape is controlled by α (when $\alpha > 1$ or when $\alpha < 1$, the overall shape is concave or convex respectively, and when $\alpha = 1$, the overall shape is linear), and β influences the location of the disconnected regions (larger values push the location of disconnected regions towards larger values of x_1, and vice versa).

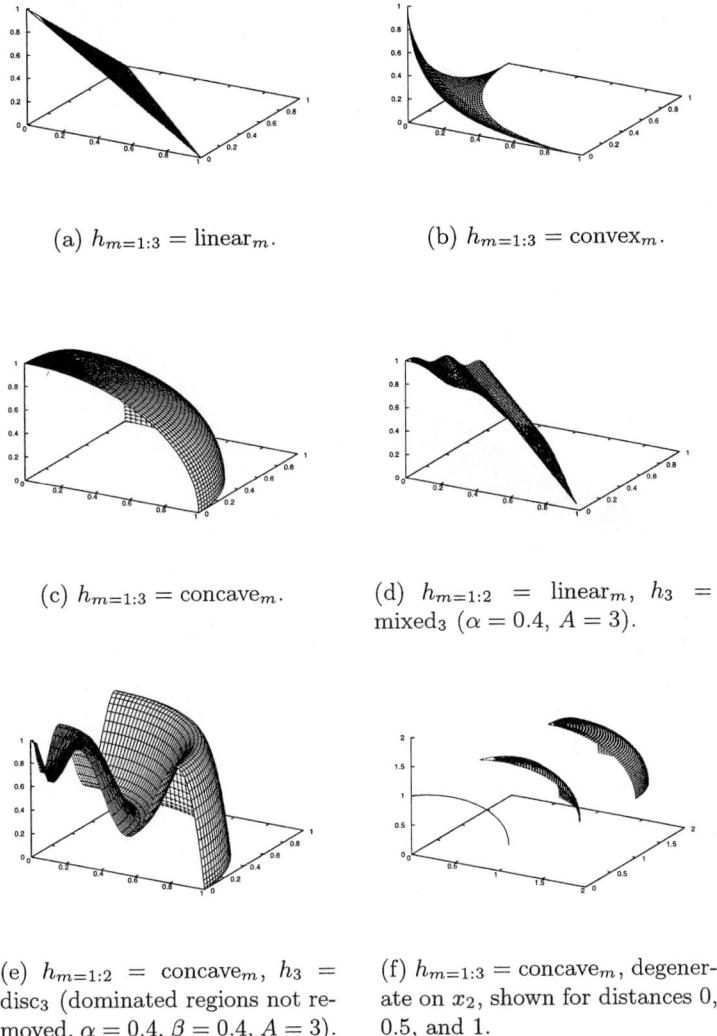

(a) $h_{m=1:3} = \text{linear}_m$.

(b) $h_{m=1:3} = \text{convex}_m$.

(c) $h_{m=1:3} = \text{concave}_m$.

(d) $h_{m=1:2} = \text{linear}_m$, $h_3 = \text{mixed}_3$ ($\alpha = 0.4$, $A = 3$).

(e) $h_{m=1:2} = \text{concave}_m$, $h_3 = \text{disc}_3$ (dominated regions not removed, $\alpha = 0.4$, $\beta = 0.4$, $A = 3$).

(f) $h_{m=1:3} = \text{concave}_m$, degenerate on x_2, shown for distances 0, 0.5, and 1.

Fig. 1. Example three-objective Pareto optimal fronts

4.2 Transformation Functions

Transformation functions map input parameters with domain $[0, 1]$ onto the range $[0, 1]$. All transformation functions take a vector of parameters (called the primary parameters) and map them to a single value. Transformation functions may also employ constants and secondary parameters that further influence the

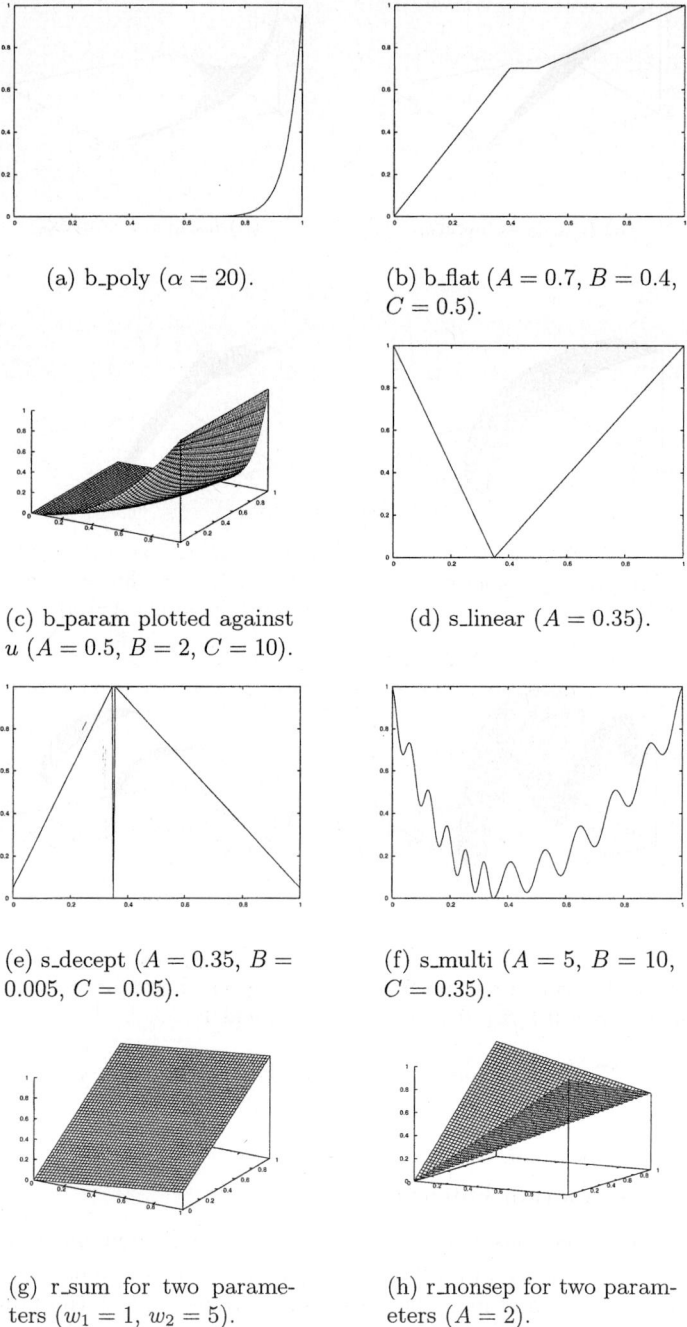

Fig. 2. Example transformations. Each example plots the value of the input primary parameter(s) versus the result of the transformation

Table 2. Transformation functions. The primary parameters y and $y_1, \ldots, y_{|\mathbf{y}|}$ always have domain $[0, 1]$. A, B, C, α, and β are constants. For b_param, $\mathbf{y}'$ is a vector of secondary parameters (of domain $[0, 1]$), and u is a reduction function

---

Bias: Polynomial ($\alpha > 0$, $\alpha \neq 1$)
$$\text{b\_poly}(y, \alpha) = y^\alpha$$
When $\alpha > 1$ or when $\alpha < 1$, y is biased towards zero or towards one respectively.

Bias: Flat Region ($A, B, C \in [0, 1]$, $B < C$, $B = 0 \Rightarrow A = 0 \wedge C \neq 1$, $C = 1 \Rightarrow A = 1 \wedge B \neq 0$)
$$\text{b\_flat}(y, A, B, C) = A + \min(0, \lfloor y - B \rfloor) \frac{A(B-y)}{B} - \min(0, \lfloor C - y \rfloor) \frac{(1-A)(y-C)}{1-C}$$
Values of y between B and C (the area of the flat region) are all mapped to the value A.

Bias: Parameter Dependent ($A \in (0, 1)$, $0 < B < C$)
$$\text{b\_param}(y, \mathbf{y}', A, B, C) = y^{B + (C-B)v(u(\mathbf{y}'))}$$
$$v(u(\mathbf{y}')) = A - (1 - 2u(\mathbf{y}')) \left| \lfloor 0.5 - u(\mathbf{y}') \rfloor + A \right|$$
A, B, C, and the secondary parameter vector $\mathbf{y}'$ together determine the degree to which y is biased by being raised to an associated power: values of $u(\mathbf{y}') \in [0, 0.5]$ are mapped linearly onto $[B, B + (C - B)A]$, and values of $u(\mathbf{y}') \in [0.5, 1]$ are mapped linearly onto $[B + (C - B)A, C]$.

Shift: Linear ($A \in (0, 1)$)
$$\text{s\_linear}(y, A) = \frac{|y - A|}{||A - y| + A|}$$
A is the value for which y is mapped to zero.

Shift: Deceptive ($A \in (0, 1)$, $0 < B \ll 1$, $0 < C \ll 1$, $A - B > 0$, $A + B < 1$)
$$\text{s\_decept}(y, A, B, C) = 1 + (|y - A| - B) \times$$
$$\left(\frac{\lfloor y - A + B \rfloor (1 - C + \frac{A - B}{B})}{A - B} + \frac{\lfloor A + B - y \rfloor (1 - C + \frac{1 - A - B}{B})}{1 - A - B} + \frac{1}{B} \right)$$
A is the value at which y is mapped to zero, and the global minimum of the transformation. B is the "aperture" size of the well/basin leading to the global minimum at A, and C is the value of the deceptive minima (there are always two deceptive minima).

Shift: Multi-modal ($A \in \{1, 2, \ldots\}$, $B \geq 0$, $(4A + 2)\pi \geq 4B$, $C \in (0, 1)$)
$$\text{s\_multi}(y, A, B, C) = \frac{1 + \cos\left[(4A+2)\pi \left(0.5 - \frac{|y - C|}{2(\lfloor C - y \rfloor + C)}\right)\right] + 4B\left(\frac{|y - C|}{2(\lfloor C - y \rfloor + C)}\right)^2}{B + 2}$$
A controls the number of minima, B controls the magnitude of the "hill sizes" of the multi-modality, and C is the value for which y is mapped to zero. When $B = 0$, $2A + 1$ values of y (one at C) are mapped to zero, and when $B \neq 0$, there are $2A$ local minima, and one global minimum at C. Larger values of A and smaller values of B create more difficult problems.

Reduction: Weighted Sum ($|\mathbf{w}| = |\mathbf{y}|$, $w_1, \ldots, w_{|\mathbf{y}|} > 0$)
$$\text{r\_sum}(\mathbf{y}, \mathbf{w}) = \left(\sum_{i=1}^{|\mathbf{y}|} w_i y_i \right) / \sum_{i=1}^{|\mathbf{y}|} w_i$$
By varying the constants of the weight vector $\mathbf{w}$, EAs can be forced to treat parameters differently.

Reduction: Non-separable ($A \in \{1, \ldots, |\mathbf{y}|\}$, $|\mathbf{y}| \bmod A = 0$)
$$\text{r\_nonsep}(\mathbf{y}, A) = \frac{\sum_{j=1}^{|\mathbf{y}|} \left(y_j + \sum_{k=0}^{A-2} |y_j - y_{1 + (j+k) \bmod |\mathbf{y}|}| \right)}{\frac{|\mathbf{y}|}{A} \lceil A/2 \rceil (1 + 2A - 2\lceil A/2 \rceil)}$$
A controls the degree of non-separability (noting that $\text{r\_nonsep}(\mathbf{y}, 1) = \text{r\_sum}(\mathbf{y}, \{1, \ldots, 1\})$).

---

mapping. Primary parameters allow us to qualify working parameters as being position- and distance-related.

There are three types of transformation functions: bias, shift, and reduction functions. Bias and shift functions only ever employ one primary parameter, whereas reduction functions can employ many.

Bias transformations have a natural impact on the search process by biasing the fitness landscape. Shift transformations move the location of optima. In the absence of any shift, all distance-related parameters would be extremal parameters, with optimal value at zero. Shift transformations can be used to set the location of parameter optima (subject to skewing by bias transformations), which is useful if medial and extremal parameters are to be avoided. We

Table 3. Transformation function restrictions

Restriction	Comment
Constants	Must be fixed (not tied to the value of any parameters).
Primary parameters	For any given transition vector, all parameters of the originating transition vector must be employed exactly once as a primary parameter (counting parameters that appear independently as primary parameters), and in the same order in which they appear in the originating transition vector.
Secondary parameters	Care must be taken to avoid cyclical dependencies in b_param. Consider the following terminology: if a is a primary parameter of b_param, and b is one of the secondary parameters, then we say that a *depends* on b. If b likewise depends on c, then a also (indirectly) depends on c. When a depends on some parameter b, then there is an associated dependency between the corresponding working parameters. To prevent cyclical dependencies, no two working parameters should be dependent on one another. In addition, a parameter should not depend on itself.
Shifts	Parameters should not be subjected to more than one shift transformation.
Reductions	Reduction transformations should belong to transition vectors that are closer to the underlying parameter vector than any shift transformation.
b_flat	When $A = 0$, b_flat should only belong to transition vectors that are further away from the underlying parameter vector than any shift or reduction transformation.

recommend that all distance-related parameters be subjected to at least one shift transformation.

The transformation functions are specified in Table 2 and plotted in Fig. 2. To ensure problems are well designed, some restrictions apply as given in Table 3. For brevity, we have omitted a weighted product reduction function (analogous to the weighted sum reduction function).

By incorporating secondary parameters via a reduction function, b_param can create dependencies between distinct parameters, including position- and distance-related parameters. Moreover, when employed before any shift transformation, b_param can create objectives that are effectively non-separable — a separable optimisation approach would fail unless given multiple iterations, or a specific order of parameters to optimise.

The deceptive and multi-modal shift transformations make the corresponding problem deceptive and multi-modal respectively[5]. When applied to position-related parameters, some regions of the Pareto optimal set can become difficult to find, and the mapping from the Pareto optimal set to the Pareto optimal front will be many-to-one (even when $k = M - 1$)[6]. When applied to distance-related parameters, finding any Pareto optimal solution becomes more difficult.

The flat region transformation can have a significant impact on the fitness landscape[7], and can also be used to create a stark many-to-one mapping from the Pareto optimal front to the Pareto optimal set.

---

[5] Multi-modal problems are difficult because an optimiser can become stuck in local optima. Deceptive problems (as defined by Deb [1]) exacerbate this difficulty by placing the global optimum in an unlikely place.

[6] Many-to-one mappings from the Pareto optimal set to the Pareto optimal front present difficulties to the optimiser, as choices must be made between two otherwise equivalent parameter vectors.

[7] Optimisers can have difficulty with flat regions due to a lack of gradient information.

5 Building an Example Test Problem

Creating problems with the WFG Toolkit involves three main steps: specifying values for the underlying formalism (including scaling constants and parameter domains), specifying the shape functions, and specifying transition vectors. To aid in construction, a computer-aided design tool or meta-language could be used to help select and connect together the different components making up the test problem. With the use of sensible default values, the test problem designer then need only specify which features of interest they desire in the test problem. An example scalable test problem is specified in Table 4 and expanded in Fig. 3.

This example problem is scalable both objective- and parameter-wise, where the number of distance- and position-related parameters can be scaled independently. For a solution to be Pareto optimal, it is required that all of:

$$z_{i=k+1:n} = \begin{cases} 0.35^{(0.02+1.96\text{r\_sum}(\{z_{i+1},\ldots,z_n\},\{1,\ldots,1\}))^{0.5}}, & i \neq n \\ 0.35, & i = n \end{cases}$$

which can be found by first determining z_n, then z_{n-1}, and so on, until the required value for z_{k+1} is determined. Once the optimal values for $z_{k+1:n}$ are

Table 4. An example test problem. The number of position-related parameters, k, must be divisible by the number of underlying position parameters, $M-1$ (this simplifies $\mathbf{t}^3$). The number of distance-related parameters, l, can be set to any positive integer. To enhance readability, for any transition vector $\mathbf{t}^i$, we let $\mathbf{y} = \mathbf{t}^{i-1}$. For $\mathbf{t}^1$, let $\mathbf{y} = \mathbf{z}_{[0,1]} = \{z_1/2, \ldots, z_n/(2n)\}$

Type	Setting
Constants	$S_{m=1:M} = 2m$ $A_{1:M-1} = 1$ The settings for $S_{1:M}$ ensures the Pareto optimal front will have dissimilar trade-off magnitudes, and the settings for $A_{1:M-1}$ ensures the Pareto optimal front is not degenerate.
Domains	$z_{i=1:n,\max} = 2i$ The working parameters have domains of dissimilar magnitude.
Shape	$h_{m=1:M} = \text{concave}_m$ The purely concave Pareto optimal front facilitates the use of some performance metrics, where the distance of a solution to the nearest point on the Pareto optimal front must be determined.
$\mathbf{t}^1$	$t^1_{i=1:n-1} = \text{b\_param}(y_i, \text{r\_sum}(\{y_{i+1}, \ldots, y_n\}, \{1, \ldots, 1\}), \frac{0.98}{49.98}, 0.02, 50)$ $t^1_n = y_n$ By employing the parameter dependent bias transformation, this transition vector ensures that distance- and position-related working parameters are inter-dependent and somewhat non-separable.
$\mathbf{t}^2$	$t^2_{i=1:k} = \text{s\_decept}(y_i, 0.35, 0.001, 0.05)$ $t^2_{i=k+1:n} = \text{s\_multi}(y_i, 30, 95, 0.35)$ This transition vector makes some parts of the Pareto optimal front more difficult to determine (due to the deceptive transformation), and also makes it more difficult to converge to the Pareto optimal front (due to the multi-modal transformation). The multi-modality is similar to Rastrigin's function, with many local optima ($61^l - 1$), and one global optimum, where the "hill size" between adjacent local optima is relatively small.
$\mathbf{t}^3$	$t^3_{i=1:M-1} = \text{r\_nonsep}(\{y_{(i-1)k/(M-1)}, \ldots, y_{ik/(M-1)}\}, k/(M-1))$ $t^3_M = \text{r\_nonsep}(\{y_{k+1}, \ldots, y_n\}, l)$ This transition vector ensures that all objectives are non-separable, and also reduces the number of parameters down to M, as required by the framework.

Given $\mathbf{z} = \{z_1, \ldots, z_k, z_{k+1}, \ldots, z_n\}$
Minimise $f_1(\mathbf{x}) = x_M + 2 \prod_{i=1}^{M-1} \sin(x_i \pi/2)$
$f_{m=2:M-1}(\mathbf{x}) = x_M + 2m \left(\prod_{i=1}^{M-m} \sin(x_i \pi/2) \right) \cos(x_{M-m+1} \pi/2)$
$f_M(\mathbf{x}) = x_M + 2M \cos(x_1 \pi/2)$
where $x_{i=1:M-1} = \text{r\_nonsep}(\{y_{(i-1)k/(M-1)}, \ldots, y_{ik/(M-1)}\}, k/(M-1))$
$x_M = \text{r\_nonsep}(\{y_{k+1}, \ldots, y_n\}, l)$
$y_{i=1:k} = \text{s\_decept}(y'_i, 0.35, 0.001, 0.05)$
$y_{i=k+1:n} = \text{s\_multi}(y'_i, 30, 95, 0.35)$
$y'_{i=1:k} = \text{b\_param}\left(z_i/(2i), \sum_{j=i+1}^{n} z_j/(2n-2i), 0.5, 0.02, 50 \right)$
$y'_{i=k+1:n} = \text{b\_param}\left(z_i/(2i), \sum_{j=1}^{i-1} z_j/(2i-2), 0.5, 0.02, 50 \right)$

Fig. 3. The expanded form of the problem defined in Table 4. $|\mathbf{z}| = n = k + l$, $k \in \{M-1, 2(M-1), 3(M-1), \ldots\}$, $l \in \{1, 2, \ldots\}$, and the domain of all $z_i \in \mathbf{z}$ is $[0, 2i]$

determined, the position-related parameters can be varied arbitrarily to obtain different Pareto optimal solutions.

The example problem has a distinct many-to-one mapping from the Pareto optimal set to the Pareto optimal front due to the deceptive transformation of the position-related parameters. All objectives are non-separable, deceptive, and multi-modal, the latter with respect to the distance component. The problem is also biased in a parameter dependent manner.

This example constitutes a well designed scalable problem that is both non-separable and multi-modal — we are not aware of any problem in the literature with comparable characteristics.

6 Implications

In this section, we consider the performance of NSGA-II [13] on five related problem instances with varying levels of complexity (see Table 5). I1 is separable and hence the simplest problem. I2 adds complexity by making position-related parameters depend on distance-related parameters (and other position-related parameters), whereas I3 makes distance-related parameters depend on position-related parameters (and other distance-related parameters). I4 instead employs a non-separable reduction. I5 combines the difficulties of both I3 and I4. I1–I5 all have concave Pareto optimal front, and are all uni-modal.

To facilitate analysis, I1–I5 are tested only for two objectives, with $k = 1$ and $l = 10$. The following (un-optimised) NSGA-II settings were used: population size 100, 250 generations, crossover probability 0.9, real parameter mutation probability $\frac{1}{n}$, SBX parameter 10, and mutation parameter 50. NSGA-II was run 35 times on each of I1–I5. The 50% attainment surfaces [14] (the "median" non-dominated fronts) are plotted in Fig. 4.

These are the first results we know of for NSGA-II on scalable, non-separable problems with known Pareto optimal sets, and clearly show the effects different types of problem complexity can have. For I1, NSGA-II effectively finds the

Table 5. Test problems I1–I5. To enhance readability, for any transition vector $\mathbf{t}^i$, we let $\mathbf{y} = \mathbf{t}^{i-1}$. For $\mathbf{t}^1$, let $\mathbf{y} = \mathbf{z}_{[0,1]} = \{z_1/2, \ldots, z_n/(2n)\}$

Problem	Type	Setting
I1	Constants	$S_{1:M} = 1$
		$A_{1:M-1} = 1$
	Domains	$z_{i=1:n,\max} = 1$
	Shape	$h_{m=1:M} = \text{concave}_m$
	$\mathbf{t}^1$	$t^1_{i=1:n} = y_i$
	$\mathbf{t}^2$	$t^2_{i=1:k} = y_i$
		$t^2_{i=k+1:n} = \text{s\_linear}(y_i, 0.35)$
	$\mathbf{t}^3$	$t^3_{i=1:M-1} = \text{r\_sum}(\{y_{(i-1)k/(M-1)+1}, \ldots, y_{ik/(M-1)}\}, \{1, \ldots, 1\})$
		$t^3_M = \text{r\_sum}(\{y_{k+1}, \ldots, y_n\}, \{1, \ldots, 1\})$
I2	As I1, except the following replaces $\mathbf{t}^1$:	
	$\mathbf{t}^1$	$t^1_{i=1:n-1} = \text{b\_param}(y_i, \text{r\_sum}(\{y_{i+1}, \ldots, y_n\}, \{1, \ldots, 1\}), \frac{0.98}{49.98}, 0.02, 50)$
		$t^1_n = y_n$
I3	As I1, except the following replaces $\mathbf{t}^1$:	
	$\mathbf{t}^1$	$t^1_1 = y_1$
		$t^1_{i=2:n} = \text{b\_param}(y_i, \text{r\_sum}(\{y_1, \ldots, y_{i-1}\}, \{1, \ldots, 1\}), \frac{0.98}{49.98}, 0.02, 50)$
I4	As I1, except the following replaces $\mathbf{t}^3$:	
	$\mathbf{t}^3$	$t^3_{i=1:M-1} = \text{r\_nonsep}(\{y_{(i-1)k/(M-1)+1}, \ldots, y_{ik/(M-1)}\}, k/(M-1))$
		$t^3_M = \text{r\_nonsep}(\{y_{k+1}, \ldots, y_n\}, l)$
I5	As I1, except use $\mathbf{t}^1$ from I3, and $\mathbf{t}^3$ from I4.	

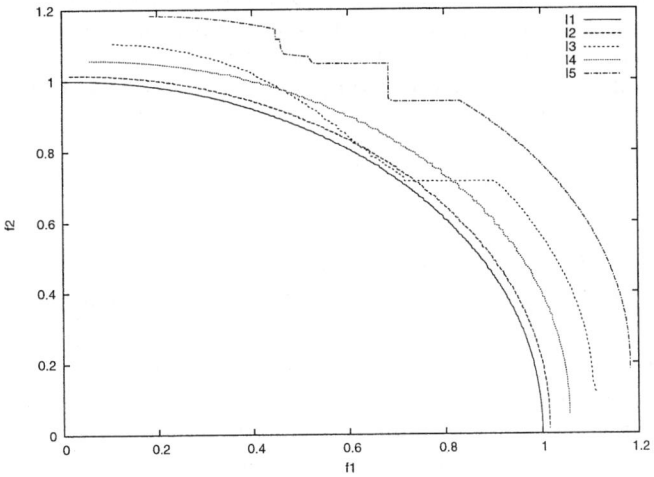

Fig. 4. The 50% attainment surfaces obtained by NSGA-II on I1–I5

Pareto optimal front, whereas having distance parameters depend on position parameters, as with I3 and I5, causes much difficulty. Moreover, I1–I5 are all uni-modal, and as the tests are only in two-objectives, more challenging problem instances are easily envisaged — NSGA-II can clearly be challenged by a variety of problem characteristics.

7 A Suggested Test Suite

In this section, we propose a test suite that consists of nine scalable, multi-objective test problems (WFG1–WFG9) that focuses on some of the more pertinent problem characteristics. Table 6 specifies WFG1–WFG9, the properties of which are summarised in Table 7.

We make the following additional observations: WFG1 skews the relative significance of different parameters by employing dissimilar weights in its weighted sum reduction, only WFG1 and WFG7 are both separable and uni-modal, the non-separable reduction of WFG6 and WFG9 is more difficult than that of WFG2 and WFG3, the multi-modality of WFG4 has larger "hill sizes" (and is thus more difficult) than that of WFG9, the deceptiveness of WFG5 is more difficult than that of WFG9 (WFG9 is only deceptive on its position parameters), the position-related parameters of WFG7 are dependent on its distance-related parameters (and other position-related parameters) — WFG9 employs a similar type of dependency, but distance-related parameters also depend on other distance-related parameters, the distance-related parameters of WFG8 are dependent on its position-related parameters (and other distance-related parameters) and as a consequence the problem is non-separable, and the predominance of concave Pareto optimal fronts facilitates the use of performance metrics that require knowledge of the distance to the Pareto optimal front.

For WFG1–WFG7, a solution is Pareto optimal iff $z_{k+1} = \ldots = z_n = 0.35$, noting WFG2 is disconnected. For WFG8, it is required that all of:

$$z_{i=k+1:n} = 0.35^{\left(0.02+49.98\left(\frac{0.98}{49.98} - (1-2u)\left|\lfloor 0.5-u \rfloor + \frac{0.98}{49.98}\right|\right)\right)^{0.5}}$$
$$u = \text{r\_sum}(\{z_1, \ldots, z_{i-1}\}, \{1, \ldots, 1\})$$

To obtain a Pareto optimal solution, the position should first be determined by setting $z_{1:k}$ appropriately. The required distance-related parameter values can then be calculated by first determining z_{k+1} (which is trivial given $z_{1:k}$ have been set), then z_{k+2}, and so on, until z_n has been calculated. Unlike the other WFG problems, different Pareto optimal solutions will have different distance-related parameter values, making WFG8 a difficult problem.

The WFG test suite exceeds the functionality of previous existing test suites. In particular, it includes a number of problems that exhibit properties not evident in the commonly-used DTLZ test suite. These include: non-separable problems, deceptive problems, a truly degenerative problem, a mixed shape Pareto front problem, problems scalable in the number of position-related parameters[8], and problems with dependencies between position- and distance-related parameters. The WFG test suite provides a fairer means of assessing the true performance of optimisation algorithms on a wider range of different problems.

---

[8] The DTLZ test suite uses a fixed (relative to the number of objectives) number of position parameters.

A Scalable Multi-objective Test Problem Toolkit   293

Table 6. The WFG test suite. The number of position-related parameters, k, must be divisible by the number of underlying position parameters, $M-1$ (this simplifies reductions). The number of distance-related parameters, l, can be set to any positive integer, except for WFG2 and WFG3, for which l must be a multiple of two (due to the nature of their non-separable reductions). To enhance readability, for any transition vector $\mathbf{t}^i$, we let $\mathbf{y} = \mathbf{t}^{i-1}$. For $\mathbf{t}^1$, let $\mathbf{y} = \mathbf{z}_{[0,1]} = \{z_1/2, \ldots, z_n/(2n)\}$

Problem	Type	Setting
All	Constants	$S_{m=1:M} = 2m$ $A_1 = 1$ $A_{2:M-1} = \begin{cases} 0, \text{ for WFG3} \\ 1, \text{ otherwise} \end{cases}$ The settings for $S_{1:M}$ ensures the Pareto optimal fronts have dissimilar trade-off magnitudes, and the settings for $A_{1:M-1}$ ensures the Pareto optimal fronts are not degenerate, except in the case of WFG3, which has a one dimensional Pareto optimal front.
All	Domains	$z_{i=1:n,\max} = 2i$ The working parameters have domains of dissimilar magnitude.
WFG1	Shape $\mathbf{t}^1$ $\mathbf{t}^2$ $\mathbf{t}^3$ $\mathbf{t}^4$	$h_{m=1:M-1} = \text{convex}_m$ $\quad h_M = \text{mixed}_M$ (with $\alpha = 1$ and $A = 5$) $t^1_{i=1:k} = y_i$ $t^1_{i=k+1:n} = \text{s\_linear}(y_i, 0.35)$ $t^2_{i=1:k} = y_i$ $t^2_{i=k+1:n} = \text{b\_flat}(y_i, 0.8, 0.75, 0.85)$ $t^3_{i=1:n} = \text{b\_poly}(y_i, 0.02)$ $t^4_{i=1:M-1} = \text{r\_sum}(\{y_{(i-1)k/(M-1)+1}, \ldots, y_{ik/(M-1)}\},$ $\quad\quad\quad \{2(i-1)k/(M-1)+1, \ldots, 2ik/(M-1)\})$ $t^4_M = \text{r\_sum}(\{y_{k+1}, \ldots, y_n\}, \{2(k+1), \ldots, 2n\})$
WFG2	Shape $\mathbf{t}^1$ $\mathbf{t}^2$ $\mathbf{t}^3$	$h_{m=1:M-1} = \text{convex}_m$ $\quad h_M = \text{disc}_M$ (with $\alpha = \beta = 1$ and $A = 5$) As $\mathbf{t}^1$ from WFG1. (Linear shift.) $t^2_{i=1:k} = y_i$ $t^2_{i=k+1:k+l/2} = \text{r\_nonsep}(\{y_{k+2(i-k)-1}, y_{k+2(i-k)}\}, 2)$ $t^3_{i=1:M-1} = \text{r\_sum}(\{y_{(i-1)k/(M-1)+1}, \ldots, y_{ik/(M-1)}\}, \{1, \ldots, 1\})$ $t^3_M = \text{r\_sum}(\{y_{k+1}, \ldots, y_{k+l/2}\}, \{1, \ldots, 1\})$
WFG3	Shape $\mathbf{t}^{1:3}$	$h_{m=1:M} = \text{linear}_m$ (degenerate) As $\mathbf{t}^{1:3}$ from WFG2. (Linear shift, non-separable reduction, and weighted sum reduction.)
WFG4	Shape $\mathbf{t}^1$ $\mathbf{t}^2$	$h_{m=1:M} = \text{concave}_m$ $t^1_{i=1:n} = \text{s\_multi}(y_i, 30, 10, 0.35)$ $t^2_{i=1:M-1} = \text{r\_sum}(\{y_{(i-1)k/(M-1)+1}, \ldots, y_{ik/(M-1)}\}, \{1, \ldots, 1\})$ $t^2_M = \text{r\_sum}(\{y_{k+1}, \ldots, y_n\}, \{1, \ldots, 1\})$
WFG5	Shape $\mathbf{t}^1$ $\mathbf{t}^2$	$h_{m=1:M} = \text{concave}_m$ $t^1_{i=1:n} = \text{s\_decept}(y_i, 0.35, 0.001, 0.05)$ As $\mathbf{t}^2$ from WFG4. (Weighted sum reduction.)
WFG6	Shape $\mathbf{t}^1$ $\mathbf{t}^2$	$h_{m=1:M} = \text{concave}_m$ As $\mathbf{t}^1$ from WFG1. (Linear shift.) $t^2_{i=1:M-1} = \text{r\_nonsep}(\{y_{(i-1)k/(M-1)+1}, \ldots, y_{ik/(M-1)}\}, k/(M-1))$ $t^2_M = \text{r\_nonsep}(\{y_{k+1}, \ldots, y_n\}, l)$
WFG7	Shape $\mathbf{t}^1$ $\mathbf{t}^2$ $\mathbf{t}^3$	$h_{m=1:M} = \text{concave}_m$ $t^1_{i=1:k} = \text{b\_param}(y_i, \text{r\_sum}(\{y_{i+1}, \ldots, y_n\}, \{1, \ldots, 1\}), \frac{0.98}{49.98}, 0.02, 50)$ $t^1_{i=k+1:n} = y_i$ As $\mathbf{t}^1$ from WFG1. (Linear shift.) As $\mathbf{t}^2$ from WFG4. (Weighted sum reduction.)
WFG8	Shape $\mathbf{t}^1$ $\mathbf{t}^2$ $\mathbf{t}^3$	$h_{m=1:M} = \text{concave}_m$ $t^1_{i=1:k} = y_i$ $t^1_{i=k+1:n} = \text{b\_param}(y_i, \text{r\_sum}(\{y_1, \ldots, y_{i-1}\}, \{1, \ldots, 1\}), \frac{0.98}{49.98}, 0.02, 50)$ As $\mathbf{t}^1$ from WFG1. (Linear shift.) As $\mathbf{t}^2$ from WFG4. (Weighted sum reduction.)
WFG9	As the example in Section 5.	

Table 7. Properties of the WFG problems. All WFG problems are scalable, have no extremal nor medial parameters, have dissimilar parameter domains and Pareto optimal tradeoff magnitudes, have known Pareto optimal sets, and can be made to have a distinct many-to-one mapping from the Pareto optimal set to the Pareto optimal front by scaling the number of position parameters

Problem	Obj.	Separability	Modality	Bias	Geometry
WFG1	$f_{1:M}$	separable	uni	polynomial,flat	convex, mixed
WFG2	$f_{1:M-1}$	non-separable	uni	–	convex, disconnected
	f_M	non-separable	multi		
WFG3	$f_{1:M}$	non-separable	uni	–	linear, degenerate
WFG4	$f_{1:M}$	separable	multi	–	concave
WFG5	$f_{1:M}$	separable	deceptive	–	concave
WFG6	$f_{1:M}$	non-separable	uni	–	concave
WFG7	$f_{1:M}$	separable	uni	parameter dependent	concave
WFG8	$f_{1:M}$	non-separable	uni	parameter dependent	concave
WFG9	$f_{1:M}$	non-separable	multi,deceptive	parameter dependent	concave

8 Conclusions

The WFG Toolkit offers a substantial range of features. Test problem designers can construct problems with a diverse range of Pareto optimal geometries and can incorporate a variety of important features in the manner of their choosing. A suite of nine test problems is presented that exceeds the functionality of existing test suites. Significantly, the WFG Toolkit allows for the construction of scalable problems that are both non-separable and multi-modal. Given the relevance of both characteristics to real world problems and the corresponding lack of such problems in the literature, the WFG Toolkit offers an important contribution in assessing the quality of optimisation algorithms on these types of problems.

Acknowledgments

This work was partly supported by an Australian Research Council linkage grant.

References

1. Deb, K.: Multi-objective genetic algorithms: Problem difficulties and construction of test problems. Evolutionary Computation **7** (1999) 205–230
2. Van Veldhuizen, D.A.: Multiobjective Evolutionary Algorithms: Classifications, Analyses, and New Innovations. PhD thesis, Air Force Institute of Technology, Wright-Patterson AFB, Ohio (1999)
3. Zitzler, E., Deb, K., Thiele, L.: Comparison of multiobjective evolutionary algorithms: Empirical results. Evolutionary Computation **8** (2000) 173–195
4. Deb, K., Thiele, L., Laumanns, M., Zitzler, E.: Scalable test problems for evolutionary multi-objective optimization. KanGAL Report 2001001, Kanpur Genetic Algorithms Laboratory, Indian Institute of Technology, Kanpur, India (2001)
5. Whitley, D., Mathias, K., Rana, S., Dzubera, J.: Building better test functions. 6th International Conference on Genetic Algorithms, Morgan Kaufmann Publishers (1995) 239–246

6. Bäck, T., Michalewicz, Z.: Test landscapes. Handbook of Evolutionary Computation. Institute of Physics Publishing (1997) B2.7 14–20
7. Fogel, D.B., Beyer, H.G.: A note on the empirical evaluation of intermediate recombination. Evolutionary Computation **3** (1995) 491–495
8. Deb, K., Thiele, L., Laumanns, M., Zitzler, E.: Scalable multi-objective optimization test problems. CEC'02. Volume 1., IEEE (2002) 825–830
9. Van Veldhuizen, D.A., Lamont, G.B.: Multiobjective evolutionary algorithm test suites. 1999 ACM Symposium on Applied Computing, ACM (1999) 351–357
10. Zitzler, E., Laumanns, M., Thiele, L.: SPEA2: Improving the strength Pareto evolutionary algorithm for multiobjective optimization. EUROGEN 2001, CIMNE, Barcelona, Spain (2001) 95–100
11. Bentley, P.J., Wakefield, J.P.: Finding acceptable solutions in the pareto-optimal range using multiobjective genetic algorithms. Soft Computing in Engineering Design and Manufacturing, Springer-Verlag (1998) 231–240
12. Knowles, J.D., Corne, D.W.: Approximating the nondominated front using the Pareto archived evolution strategy. Evolutionary Computation **8** (2000) 149–172
13. Deb, K., Pratap, A., Agarwal, S., Meyarivan, T.: A fast and elitist multiobjective genetic algorithm: NSGA-II. IEEE Transactions on Evolutionary Computation **6** (2002) 182–197
14. Fonseca, C.M., Fleming, P.J.: On the performance assessment and comparison of stochastic multiobjective optimizers. Parallel Problem Solving from Nature — PPSN IV, Springer-Verlag (1996) 584–593

Extended Multi-objective fast messy Genetic Algorithm Solving Deception Problems

Richard O. Day and Gary B. Lamont

*Air Force Institute of Technology,
Dept of Electrical and Computer Engineering,
Graduate School of Engineering & Management,
Wright-Patterson AFB (Dayton) OH, 45433, USA
{Richard.Day, Gary.Lamont}@afit.edu

Abstract. Deception problems are among the hardest problems to solve using ordinary genetic algorithms. Designed to simulate a high degree of epistasis, these deception problems imitate extremely difficult real world problems. [1]. Studies show that Bayesian optimization and explicit building block manipulation algorithms, like the fast messy genetic algorithm (fmGA), can help in solving these problems. This paper compares the results acquired from an *extended* multiobjective fast messy genetic algorithm (MOMGA-IIa), ordinary multiobjective fast messy genetic algorithm (MOMGA-II), multiobjective Bayesian optimization algorithm (mBOA), and the non-dominated sorting genetic algorithm-II (NSGA-II) when applied to three different deception problems. The *extended* MOMGA-II is enhanced with a new technique exploiting the fmGA's basis function to improve partitioned searching in both the genotype and phenotype domain. The three deceptive problems studied are: interleaved minimal deceptive problem, interleaved 5-bit trap function, and interleaved 6-bit bipolar function. The *unmodified* MOMGA-II, by design, explicitly learns building block linkages, a requirement if an algorithm is to solve these hard deception problems. Results using the MOMGA-IIa are excellent when compared to the non-explicit building block algorithm results of both the mBOA and NSGA-II.

1 Introduction

Algorithms that solve problems by realizing good building blocks (BBs) are useful in solving extremely difficult problems: Protein Structure Prediction [2], 0/1 Modified Knapsack [3], Multiple Objective Quadratic Assignment Problem [4,5], Digital Amplitude-Phase Keying Signal Sets with M-ary Alphabets and many academic problems [6,7]. MOMGA-IIa originated as a single objective

---

* The views expressed in this article are those of the authors and do not reflect the official policy or position of the United States Air Force, Department of Defense, or the United States Government.

messy GA (mGA). It evolved from being a single objective mGA into a multi-objective mGA called the MOMGA [8]. Many different MultiObjective Evolutionary Algorithms (MOEAs) were produced during this time period; however, the MOMGA is the only MOEA explicitly using good BBs to solve problems – even the Bayesian optimization algorithm (BOA) uses a probabilistic model to find good building blocks. The MOMGA has a population size limitation: as the BB size increases so does the population size during the Partially Enumerative Initialization (PEI) phase. This renders the MOMGA less useful on large problems. To overcome this problem, the MOMGA-II, based on the single objective fmGA, is designed. The fmGA is similar to the mGA in that it specifically uses BBs to find solutions; however, it has a reasonable population size and lower run time complexity (See Table 1) when compared to the mGA. MOMGA-II includes many different repair, selection, and crowding mechanisms. Unfortunately, the MOMGA-II is found to be limited when solving large deception problems [6]. This called for the development of basis function diversity measures in the MOMGA-IIa which are designed for smart BB searching in both the geno- and pheno-type domains. Also discussed in this investigation is the mBOA and the NSGA-II [7] neither of which compares well to MOMGA-IIa results.

The next section discusses in detail the MOMGA-II and MOMGA-IIa algorithm domains. In addition, a short description of the mBOA and the NSGA-II is provided. The mBOA and NSGA-II have been used to solve these three multiobjective problems (MOPs) in previous research [10,6]. The three MOPs are then described in detail in Section 3. Next, experimental design, resources, parameter settings, and algorithm efficiency are discussed briefly in Section 4. Finally, in the results section, the mBOA, NSGA-II, MOMGA-II and MOMGA-IIa results are compared and analyzed.

Table 1. Complexity Estimates for serial GAs

Phase	Single Objective Algorithm				Multiple Objective Algorithm		
	sGA^a	$ssGA^b$	mGA	$fmGA$	$NSGA\text{-}II$	$mo\text{-}BOA$	$MOMGA\text{-}IIa$
Initialization	$O(l^n)$	$O(l^n)$	$O(l^k)$	$O(l)$			
Recombination	$O(gnq)$	$O(g)$					
Primordial	$O(\emptyset)$		$O(\emptyset)$	$O(l^2)^c$			
Juxtapositional			$O(l \log l)$	$O(l \log l)$			
Overall	$O(l^n)$	$O(l^n)$	$O(l^k)$	$O(l^2)$	$O(mn^3)^d$	$O(n^{3.5})^e$	$O(megn^2)^f$

[a] l is the length of chromosome, n is the size of population, q is group size for tournament selection, g is the number of generations.
[b] l is the length of chromosome, n is the size of population, g is the number of generations of reproduction.
[c] Building Block Filtering
[d] m is the number of objectives
[e] This complexity is problem specific and in this case has been taken from the spin glass problem.[9]
[f] e = number of eras, g = max number of generations

2 Extended Multiobjective fast messy GA (MOMGA-II)

The MOMGA-II(a)[1] is a multiobjective version of the fmGA that has the ability to achieve a near-partitioned search in both the genotype and phenotype domains during execution. It is an algorithm that exploits "good" building blocks (BBs) in solving optimization problems. These BBs represent "good" information in the form of partial strings that can be combined to obtain even better solutions. The BB approach is used in the fmGA to increase the number of "good" BBs that are present in each subsequent generation of the algorithm. The fmGA algorithm executes in three phases: Initialization, Building Block Filtering, and Juxtapositional Phase. See Figure 1 for diagram of the program flow for MOMGA-IIa.

The algorithm begins with the Probabilistically Complete Initialization Phase. This phase randomly generates a user specified number of population members. These population members are of the *a priori* specified chromosome length and each is evaluated to determine its respective fitness values. Our implementation utilizes a binary scheme in which each bit is represented with either a 0 or 1.

The fitness functions used to calculate each string's merit are described in Section 3. In MOMGA-II each string has m fitness values, while in MOMGA-IIa each string has $f = (c * m + i + o) * m$ fitness values associated with it – corresponding to the m objective functions to optimize, c competitive templates, i inverse templates (equal to $c * m$), and o orthogonal templates.

The Building Block Filtering (BBF) Phase follows by randomly deleting locus points and their corresponding allele values in each of the population member's chromosomes. This process completes once the length of the population member's chromosomes have been reduced to a predetermined BB size. In order to evaluate these population members a competitive template (CT) is utilized to fill in the missing allele values. The competitive template is a fully specified chromosome and evolves, by allowing the best member found to replace the old competitive template(s), after each BB generation.

In the MOMGA-II, the next competitive template(s) is(are) randomly chosen from among the non-dominated points within the population. This is the difference between the MOMGA-II and MOMGA-IIa.

Where the MOMGA-II's competitive template selection mechanism selects the next competitive template randomly from the non-dominated points without regard to pareto front (PF) point placement, the MOMGA-IIa competitive template generation, replacement, and evolution is engaged in making sure competitive template selection partitions the search space in both the phenotype and genotype domain.

The innovative balance is achieved through the use of two mechanisms: Orthogonal competitive template generation and Target Vector (TV) guidance. Orthogonal competitive template generation is directed to partition the geno-

---

[1] When the reader sees MOMGA-II(a) in this report, the sentence is referring to both the MOMGA-II and the MOMGA-IIa.

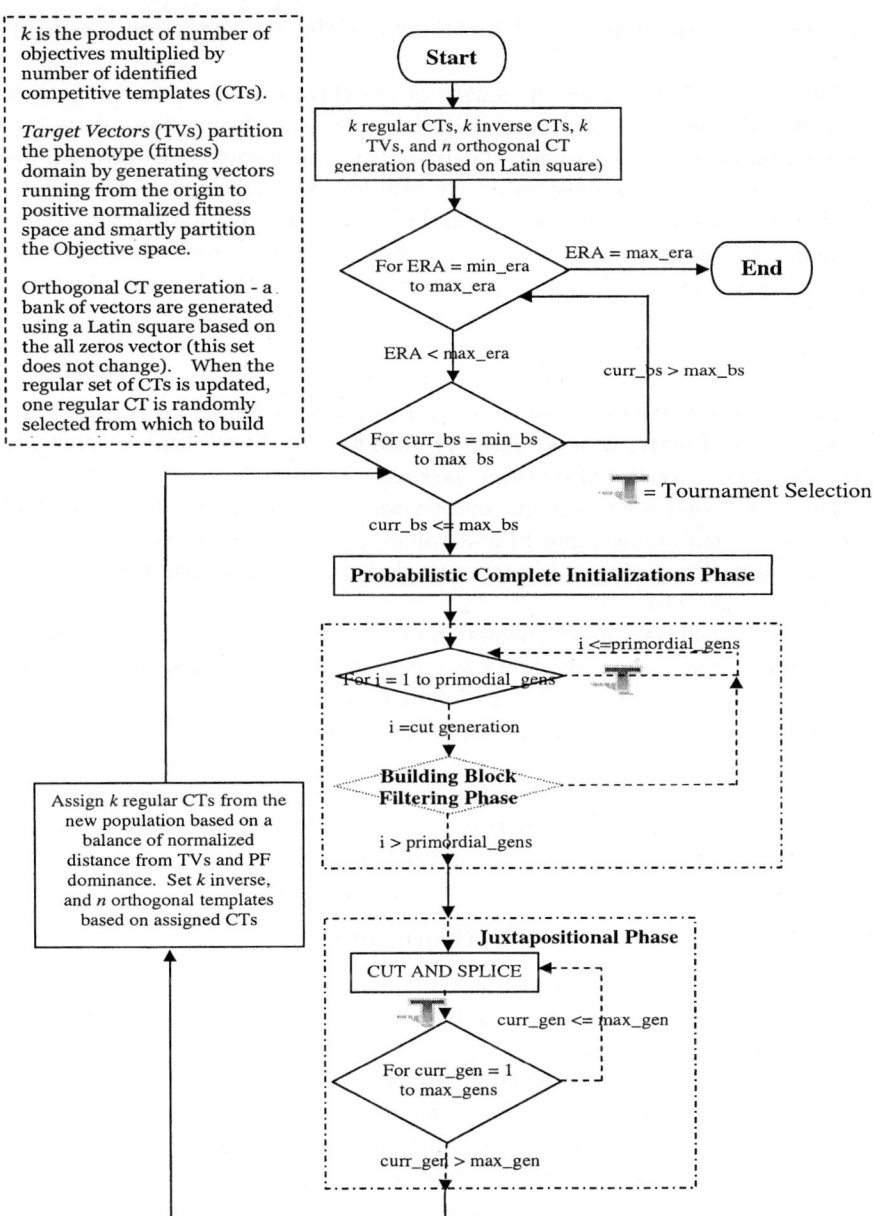

Fig. 1. Illustrated in this figure is the program flow of the MOMGA-IIa. Note the placement of each phase and where tournament selection is performed. Additionally, the MOMGA-IIa exploits and partitions in both the phenotype and genotype domains by updating and generating regular, inverse, and orthogonal competitive templates. See Section 2 for a detailed description of the algorithm

type space while keeping a good partitioning of the phenotype space using TV guidance.

Through the BBF phase the length of the chromosome decreases, yet each chromosome continues to be evaluated for selection. During this phase these chromosomes are referred to as "underspecified" since each locus position does not have an associated allele value. To evaluate an underspecified population member, the member is overlayed upon the competitive template to fully specify the member. This process uses the allele values from the template to fill in any missing allele values in the population member to allow the fitness evaluation to take place and is repeated any time an underspecified population member needs to be evaluated. The BBF process is alternated with a selection mechanism to keep only the strings with the "best" BBs found, or those with the best number of fitness values. In the case of a tie, where two strings each have an equal number of better fitness values (ie. each have $\frac{m}{2}$ "best" fitness values), the string is randomly selected between the two.

The MOMGA-IIa has a more complex selection mechanism than MOMGA-II because it maintains more fitness values per solution; however, ultimately it is the same comparing only each string's best objective fitness value for all templates considered.

The juxtapositional phase follows and uses the BBs found through the BBF phase and recombination operators to create chromosomes that are fully specified. A chromosome is referred to as fully specified if it is not missing any locus positions, or in other words does not need to use the competitive template for evaluation.

Furthermore, the algorithms have an outer and inner loop and must completely iterate through each BB size a number of epochs before it terminates. For more information on the fmGA and BB theory, see [3, 11].

2.1 Non-dominated Sorting Algorithm-II

The non-dominated sorting algorithm-II (NSGA-II) is a generic non-explicit BB MOEA applied to multiobjective problems (MOPs) – based on the original design of NSGA. It builds a population of compete individuals, ranks and sorts each individual according to non-domination level, applies Evolutionary Operations (EVOPs) to create new pool of offspring, and then combines the parents and offspring before partitioning the new combined pool into fronts. The NSGA-II then conducts niching by adding a crowding distance to each member. It uses this crowding distance in its selection operator to keep a diverse front by making sure each member stays a crowding distance apart. This keeps the population diverse and helps the algorithm to explore the fitness landscape.

2.2 Multiobjective Bayesian Optimization Algorithm

The multiobjective Bayesian optimization algorithm (mBOA) was also used to solve these MOPs in previous research [12]. mBOA is identical to the single objective Bayesian Optimization Algorithm (BOA) [12] minus the selection procedure. The mBOA's selection procedure is replaced by the non-dominated sorting and selection mechanism of NSGA-II. The BOA generates a child population of size n from a parent population. The child and parent population is then merged and the combined population is pareto ranked. Based on the pareto ranking and crowding distance function a new population is created from which BOA builds a new probabilistic model to generate children again.

3 Deception Problems

In 1987, Goldberg's research group introduced deception to test the abilities of current genetic algorithms [13]. They designed problems having specific difficulties which genetic algorithms (GAs) might face in problem solving. These *deception* problems are often challenging to optimize and involve some degree of deception – resulting in conflicting objectives (e.g. the k-arm bandit competitions between hyperplanes [14]). Later, Whitley [14] proved deceptive attractors must have complementary bit patterns to the global optimum pattern in order to be either fully deceptive or consistently deceptive problems. He then defines a deceptive problem at least one more relevant lower order hyperplane competitor that guides a genetic search away from the global winner. Imagine a hill climbing search algorithm starting anywhere except with the bit configuration *111x*. The hill climbing algorithm always finds the suboptimal fitness of 9 as a solution. This example illustrates how a competitor hyperplane might guide a GA away from the optimal solution. Furthermore, it is every GA engineer's desire to build an algorithm that finds proper linkages within a problem, overcoming this type of deception.

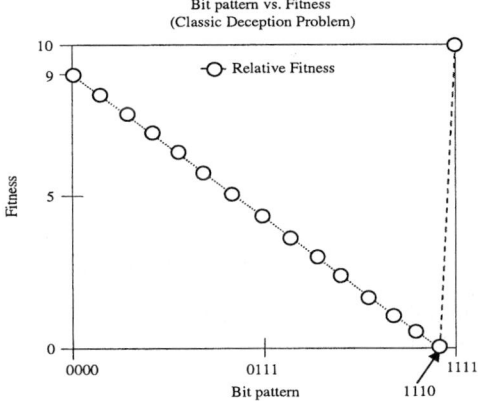

Fig. 2. This figure illustrates a classical deception problem

We evaluate the following five test functions in this investigation:

1. T1 - Interleaved minimal deception problem
2. T2 - Complement of T1

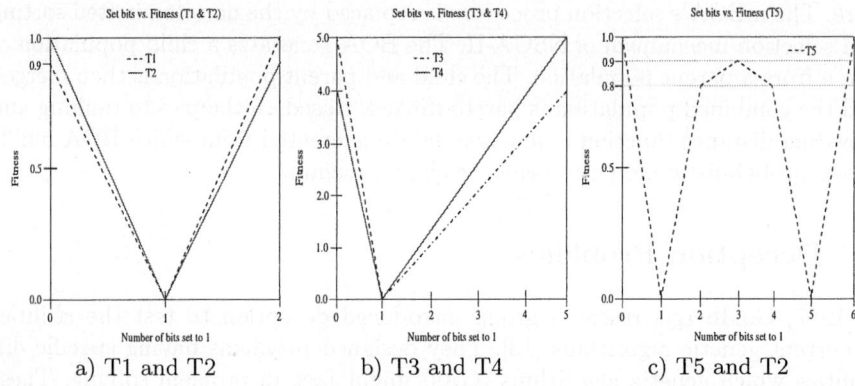

Fig. 3. These figures illustrate the fitness landscape for each function. Subfigure *a* illustrates deception problems T1 and T2, subfigure *b* illustrates deception problems T3 and T4, and subfigure *c* illustrates deception problems T5 and T2.

3. T3 - Interleaved 5-bit trap function
4. T4 - Complement of T3
5. T5 - Interleaved 6-bit bipolar deception function

These test functions are claimed to be difficult in four respects: deception, loose linkage, mutlimodality, and combinatorially - having a large search space [15, 16, 17]. Sections 3.1 through 3.3 included a detailed discussion of these deceptive problems.

In addition to solving these five test functions, difficulty is added by combining these functions together to make three multiobjective problems. By aggregating these test functions together, the order of deception is increased because the functions are paired in a manner that adds a relevant lower order hyperplane competitor to guide a genetic search away from the global winner. The following is the list of the MOPs investigated in this paper:

1. MOP 1: T1 and T2
2. MOP 2: T3 and T4
3. MOP 3: T2 and T5

3.1 Interleaved Minimal Deceptive Problem (T1 & T2)

The interleaved minimal deceptive problems are designed to test an algorithm's ability to discover loosely linked bits by dividing the string into two halves and coupling one bit from each half. Figure 4 illustrates how the bits are correlated. Bits having the same pattern are rewarded, while alternating couplets are not. Additionally, Figure 3.a illustrates the bit couplet fitness for T1 and T2.

3.2 Interleaved 5-Bit Trap Function (T3 & T4)

The interleaved 5-bit trap function is devised to test an algorithm's ability to find loose linkages having non-consecutive bits. Bits in problems T3 and T4 both

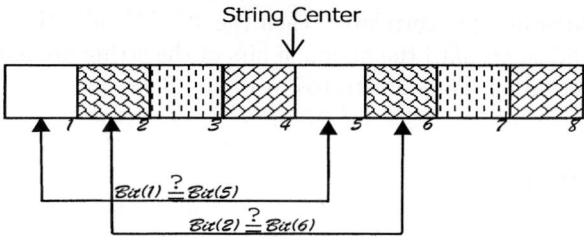

Fig. 4. This figure illustrates bit linkage in an eight bit solution. Notice that every i^{th} and $(\frac{l}{2} + i)^{th}$ bit is linked such that $i \leq \frac{l}{2}$, l=length of string, and the 1st bit is bit 1

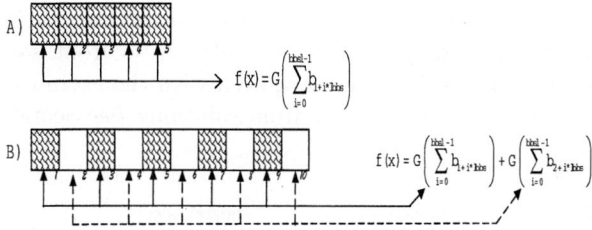

Fig. 5. This figure illustrates bit linkage in an 5 and 10 bit solution. In the figure, the fitness function for each set of bits is shown to the right with arrows indicating which bits contribute to each term of the fitness summation

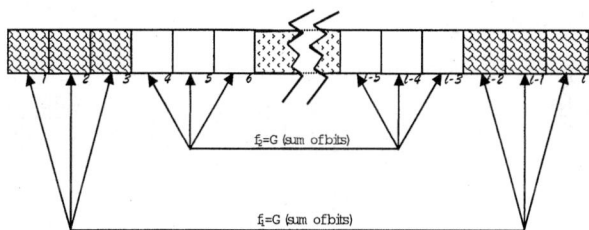

Fig. 6. This figure illustrates 6-bit bipolar linkage in an l bit solution. The bit string is broken in the middle to enhance the idea that string can be of any size as long it is a multiple of 6

have correlated bits with a distance of $\frac{l}{5}$ from one another. Figure 5 illustrates how the bits in groups of 5 are coupled. Additionally, Figure 3.b graphically illustrates how the fitness behavior varies according to the number of bits that are set in the described 5-bit linkage pattern. Notice that T3 in Figure 3.b is similar to the classical deception problem example illustrated in Figure 2.

3.3 Interleaved 6-Bit Bipolar Function (T5)

The interleaved 6-bit Bipolar function is constructed as a loose linkage problem having correlated bits in variable placement in the string. The first three bits and the last three bits are correlated, then the 4th, 5th, 6th, $(l-4)^{th}$, $(l-5)^{th}$, and $(l-6)^{th}$ and so on until the middle 6 bits of the string are left. Graphically, the fitness function for T5 is illustrated in Figure 3.c.

4 Experiments

The experiment for all MOMGA-II MOPs were run simultaneously on the two computational clusters (ASPEN and Polywells) listed in Table 2. The MOMGA-IIa ran in serial on one computational cluster (TAHOE). The MOMGA-II is given 30 to 50 experiments to solve the three MOPs while the MOMGA-IIa is run for 10 experiments or less. Statistically, 10 runs are required to be able to compare MOEAs mathematically; however, the MOMGA-II is given more experiments in attempt to allow it to find all pareto front solutions for the larger deception problems. Unfortunately, even with the extra experiments, the MOMGA-II still did not find all pareto front solutions. See Section 5 for results.

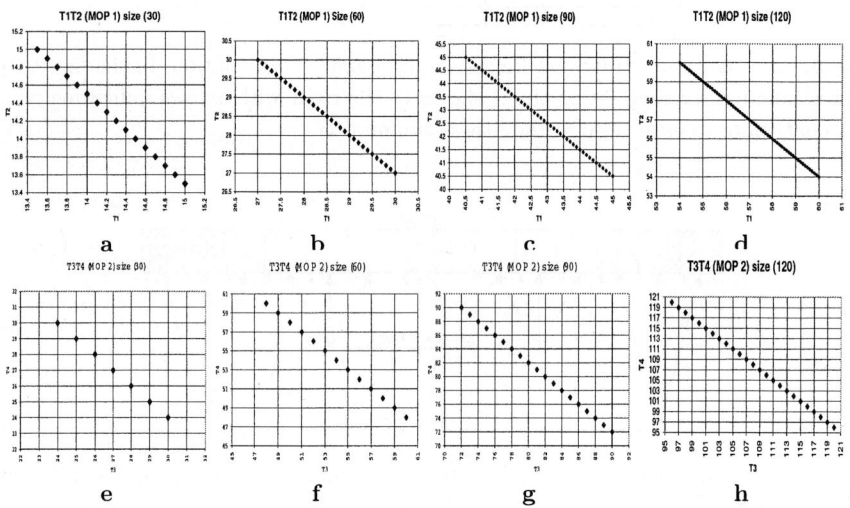

Fig. 7. This figure illustrates the pareto front findings for the T1 vs. T2 and T3 vs. T4 experiment using a string length of 30, 60, 90, and 120 bits

4.1 Resources

Table 2 lists the resources used for these experiments. Each MOMGA-II experiment took approximately 1 week to complete including the time to process all data for presentation. Each MOMGA-IIa experiment took approximately 2 days. This includes the time jobs sat idle in the scheduler queue and the post mortem collection of data for analysis. NSGA-II and mBOA run times were not recorded in this investigation and are unknown [10].

Table 2. System Configuration

Cluster 1 (TAHOE)	Cluster 2 (ASPEN)	Cluster 3 (Polywells)
Fedora Core 2	Redhat Linux 9.0	Redhat Linux 7.3
Dual Opteron 2.2 ghz	Ath XP 3000+ 2.1ghz	Ath XP 2800+ 2.0ghz
Cache(L1 I 64,D 64/L2 1024)KB	(64,64/512)KB	(64,64/512)KB
Gb Ethernet	Fast Ethernet	Gb Ethernet
RAM 4 GByte	1 GByte	1 GByte
Crossbar Switch	Crossbar Switch	Crossbar Switch
RAID 5	RAID 5	RAID 5
48 node,2 CPUS/node	48 node,2 CPUS/node	16 node,1 CPU/node

4.2 Parameter Settings

The MOMGA-II(a) has many parameters for proper program execution. BB sizes must be determined, elitism percentages, cut probability, splice probability, mutate probability, population sizing variable (n_a), era generations, and BB filtering schedule. The program is run x times and pareto front members are collected from each solution set. Table 4 indicates the number of times the program is run on each MOP. In some cases the experiments were terminated early if all pareto front solutions were found. All MOMGA-II(a) experiments are executed using a multiple objective fmGA. Tables 3 and 4 list all the parameters used for the experiments conducted in this study. n_a values for the MOMGA-II can be found in [6] while n_a values were set to 500 for the MOMGA-IIa. Table 3 lists the constant parameter settings while Table 4 identifies parameters that were varied for each MOP. In the cases where the programs were run for less than 30 times, this is because the MOMGA-II(a) found the optimal pareto front before each run completed.

Table 3. Summary of Static parameters set for each experiment regardless of MOP and problem size. MOMGA-II(MOMGA-IIa)

Parameter	Setting	Parameter	Setting
Maximization	1(n/a)	Thresholding	0(n/a)
mGA(0)/fmGA(1)	1(n/a)	Shuffle Number	2(2)
		Tiebreaking	0(n/a)
Overflow	2.0(2.0)	Reduced initial pop	0(n/a)
Elitism %	25(0)	Extra pop members	0(n/a)
Prob cut	0.02(0.02)	Stop criteria factor	1.00(n/a)
Prob splice	1.0(1.0)	Partition file	0(n/a)
Prob mutation		Plotting file	0(n/a)
allelic	0.0(n/a)	Pop record file	0(n/a)
genic	0.0(n/a)	Copies	5 1 1(n/a)
Inverse Template	n/a(Y)	CT Guesses	n/a(1)

BB size selection for these problems is tricky because the identification of the length of one particular linkage (say 5 bits long) may not be enough to transform solutions out of the search basis provided by the resident competitive templates. This is prominent for MOMGA-II when solving the MOP 2 of size 90 where the competitive templates used constrict the search into a space where the small BB sizes cannot overcome the competitive template basis. Normally, a BB size can be selected upon knowing the length of these linkages; however, in MOP 2 there are multiple linkages of five bits long each magnifying the difficulty and requiring a larger BB size to allow the MOMGA-II to find more PF_{true} points. It should be noted here that when the BB sizes increase, population size is also increased as well as run time for the algorithm to complete. MOMGA-II's limitation was found in MOP 2 with a size greater than or equal to 90. This limitation is overcome by MOMGA-IIa by using specially chosen competitive templates for search after each BB generation.

Table 4. Summary of Era parameters settings for each experiment using MOMGA-II(MOMGA-IIa)

Experiment	Start ERA	End ERA	Runs	CTs	Inverse CTs	Orthogonal CTs
MOP 1(30)	1	10(4)	30(3)	4(18)	0(18)	0(41)
MOP 1(60)	1	10(4)	30(9)	4(18)	0(18)	0(45)
MOP 1(90)	1	10(4)	30(10)	4(18)	0(14)	0(9)
MOP 1(120)	1	10(4)	30(10)	4(18)	0(14)	0(9)
MOP 2(30)	1	10(4)	30(10)	4(18)	0(18)	0(41)
MOP 2(60)	1	8(4)	50(10)	4(18)	0(18)	0(45)
MOP 2(90)	1	6(4)	50(10)	4(18)	0(18)	0(10)
MOP 2(120)	1	4(4)	30(8)	4(14)	0(14)	0(49)
MOP 3(30)	1	10(4)	10(1)	4(14)	0(14)	0(41)
MOP 3(60)	1	12(4)	10(1)	4(14)	0(14)	0(41)

4.3 Efficiency Finding Pareto-Front Points

Reduction of relative execution time for MOMGA-IIa is achieved by keeping the pareto front points (including duplicates) in memory. The MOMGA-IIa benefits from a creatively designed structure maintained as a dynamic linked list object which holds all pareto front members. Dominated solutions are deleted from the structure and from memory including all duplicates for that particular point. As the program runs, the pareto front point listings can become long. This is a disadvantage; however, elitism selection is $O(p)$ run time as all solutions, p, are readily listed. In addition to this enhancement, epsilon-difference dominance, crowding techniques and dominance linkage can be instantiated. The MOMGA-IIa also has the ability to trace evolutionary solutions throughout the search process. This enhanced feature comes at a cost in space (memory or disk) and efficiency. All experiments in this paper are run with an active trace feature to allow for post mortem analysis of pareto front point conception.

5 Results

The MOMGA-IIa results are superior. In every case the MOMGA-IIa has either found more pareto front solutions or more unique solutions than all other algorithms tested on these three MOPs. Along with the enhanced algorithm, deception problems of larger sizes are added into this investigation to allow for even more difficult problem resolution. As expected from the last investigation [6], this investigation found MOP 2 to be the most difficult to solve. Difficulty comes in the form of time to find all pareto front members.

The following sections describe, in detail, the experiment results for each MOP. Notice that the pareto front for MOP 1 and MOP 2 is linear. This is due to the linear slopes of the individual objective functions making up each of these MOPs. It is also worth noting that each point on the MOP 1 and MOP 2 pareto front may have many unique strings (solutions) equating to that

Table 5. Summary of Results for all experiments. Included in this table are the number of optimal pareto front points, number of unique strings making up these Optimal points, and the number of points each algorithm (MOMGA-IIa, MOMGA-II, mBOA, and NSGA-II) have found in each category

MOP 1								
T1 vs. T2 - Sizes: (30/60/90/120)								
Unique Strings: $\{(2^{15}/2^{30}/2^{45}/2^{60})\}$ Pareto Front Strings: (16/31/46/61)								
Algorithm	Unique Strings Found				Pareto Front pts Found			
	30	60	90	120	30	60	90	120
MOMGA-IIa	**32768**	**300776**	**57661**	**32876**	**16**	**31**	**46**	**61**
MOMGA-II	596	21364	28	138	**16**	**31**	16	56
mBOA	224		591		**16**		**46**	
NSGA-II	7		5		6		3	
MOP 2								
T3 vs. T4 - Sizes: (30/60/90/120)								
Unique strings: $\{(2^6/2^{12}/2^{18}/2^{24})\}$ Pareto Front Strings: (7/13/19/25)								
Algorithm	Unique Strings Found				Pareto Front pts Found			
	30	60	90	120	30	60	90	120
MOMGA-IIa	**64**	**565**	**1280**	**3594**	**7**	**13**	**19**	**25**
MOMGA-II	32	98	54	31	**7**	12	6	14
mBOA	30	102	327		**7**	**13**	**19**	
NSGA-II	0	1	0		0	1	0	
MOP 3								
T5 vs. T2 - Sizes: (30/90)								
Unique strings: (1/1) Pareto Front Strings: (1/1)								
Algorithm	Unique Strings Found				Pareto Front pts Found			
	30		90		30		90	
MOMGA-IIa	1		1		1		1	
MOMGA-II	1		1		1		1	
mBOA	1		1		1		1	
NSGA-II	0		0		0		0	

particular pareto front point. This phenomena is illustrated in Table 5 where both the number of pareto front points and unique strings are listed.

5.1 T1 Versus T2 (MOP 1)

Figures 7a~d reflect the results of the MOP 1 of sizes 30, 60, 90, and 120. Diamonds indicate the pareto front solutions found. Notice that these points are linear and discrete. For problem sizes 30, 60, 90, and 120 there are 16, 31, 46, and 61 total pareto front points. Not only does MOMGA-IIa find the same or more true pareto front points for MOP 1 than every other algorithm tested, it also finds more unique strings making up each of these pareto front points. Table 5 numerically shows in bold that MOMGA-IIa is the front runner in this experiment. In the problem size 30, the MOMGA-IIa has actually found all unique strings corresponding to each of the 16 pareto front solutions.

5.2 T3 Versus T4 (MOP 2)

Figures 7e~h illustrate the results of the MOP 2 experiment. Similarly, diamonds indicate the pareto front solutions found. Notice that these points are linear and discrete. For the problem sizes 30, 60, 90, and 120 there are 7, 13, 19, and 25 total pareto front points in the entire search space. Not only does MOMGA-IIa find the same or more true pareto front points for MOP 1 than every other algorithm tested, it also finds more unique strings making up these pareto front points. Specifically, MOMGA-IIa overcomes the problem size issue and finds more pareto front points than its predecessor, MOMGA-II. Table 5 numerically shows in bold that MOMGA-IIa is the front runner in this experiment. In addition to finding all the pareto front solutions for MOP 2 size 30, the MOMGA-IIa finds all corresponding unique strings.

5.3 T5 Versus T2 (MOP 3)

No figures in this paper reflects the results of MOP 3 for it is rather uninteresting. See [6] for an example. There is only one point on the pareto front. All algorithms but the NSGA-II found the one and only pareto front point and the only unique string representing this solution.

5.4 Comparison

MOMGA-IIa results are excellent. The MOMGA-IIa finds all pareto front members in every MOP and at every size. Additionally, the MOMGA-IIa also finds more unique strings than any other algorithm tested in this investigation. Clearly, the MOMGA-IIa outperforms the MOMGA-II and is able to scale up to solve larger deception problems by having a pool of better competitive templates.

6 Conclusion

In conclusion, this experiment illustrates an explicit building block genetic algorithm's ability to solve deception problems. MOMGA-IIa's capabilities to explicitly find and use good multiobjective Building Blocks (MOBBs) is illustrated. MOMGA-IIa can find the loosely linked bits in deceptive problems using its manipulation of BBs and it has been shown to scale when "good" competitive templates are selected. The pedagogical problems solved in this report are limited in testing the capabilities of MOMGA-IIa; however, they do show that the MOMGA-IIa can solve difficult problems with ease. The MOMGA-IIa implicitly solves problems by identifying good MOBBs when iteratively selecting "good" competitive templates that partition both the geno- and pheno-type domains. Important aspects of this algorithm are the BB size, BB schedule and, competitive template numbers (regular, inverse, and orthogonal). An important conjecture learned from this investigation is the fact that as the search space increases it is important to increase the number of competitive templates - thus limiting the largest required BB size and ultimately limiting the population size generated by the MOMGA-IIa. In addition, a second long standing conjecture that finding larger BBs allows for more pareto front points along the extremes of the front to be found [18]. The new innovative tracing technique has identified interesting results concerning this second conjecture.

7 Future Work

Future work includes the results of algorithm runs on larger deception problems found in Table 5. Additionally, a detailed description of design on the MOMGA-IIa is in order along with the correction of the count of unique strings found. Finally, an examination of the evolutionary trace function that is embedded within the MOMGA-IIa is needed to illustrate where solutions are drawn from with respect to operators and BB sizes.

References

1. Colin Reeves and Christine Wright. An experimental design perspective on genetic algorithms. In L. Darrell Whitley and Michael D. Vose, editors, *Foundations of Genetic Algorithms 3*, pages 7–22. Morgan Kaufmann, San Francisco, CA, 1995.
2. Richard O. Day. A multiobjective approach applied to the protein structure prediction problem. Ms thesis, Air Force Institute of Technology, March 2002. Sponsor: AFRL/Material Directorate.
3. Jesse B. Zydallis. *Explicit Building-Block Multiobjective Genetic Algorithms: Theory, Analysis, and Development*. Dissertation, Air Force Institute of Technology, AFIT/ENG, BLDG 642, 2950 HOBSON WAY, WPAFB (Dayton) OH 45433-7765, Feb 2002.
4. Richard O. Day, Mark P. Kleeman, and Gary B. Lamont. Solving the Multiobjective Quadratic Assignment Problem Using a fast messy Genetic Algorithm. In *Congress on Evolutionary Computation (CEC'2003)*, volume 1, pages 2277–2283, Piscataway, New Jersey, December 2003. IEEE Service Center.
5. Mark P. Kleeman. Optimization of heterogeneous uav communications using the multiobjective quadratic assignment problem. Ms thesis, Air Force Institute of Technology, March 2004. Sponsor AFRL.
6. Richard O. Day and Gary B. Lamont. Multi-objective fast messy genetic algorithm solving deception problems. *Congress on Evolutionary Computation; Portland, Oregon*, 4:1502–1509, June 19 - 23 2004.
7. Carlos A. Coello Coello, David A. Van Veldhuizen, and Gary B. Lamont. *Evolutionary Algorithms for Solving Multi-Objective Problems*. Kluwer Academic Publishers, New York, May 2002.
8. Carlos M. Fonseca and Peter J. Fleming. Genetic Algorithms for Multiobjective Optimization: Formulation, Discussion and Generalization. In Stephanie Forrest, editor, *Proceedings of the Fifth International Conference on Genetic Algorithms*, pages 416–423, San Mateo, California, 1993. University of Illinois at Urbana-Champaign, Morgan Kauffman Publishers.
9. Martin Pelikan. *Bayesian Optimization Algorithm: From Single Level to Hierarchy*. Thesis, University of Illinois at Urbana-Champaign, 117 Transportation Building, 104 S. Mathews Avenue Urbana, IL 61801, July 2002. IlliGAL Report No. 2002023.

10. Nazan Khan, David E. Goldberg, and Martin Pelikan. Multi-objective byesian optimization algorithm. ILLiGAL Report 2002009, University of Illinois at Urbana-Champaign, 117 Transportation Building, 104 S. Mathews Avenue Urbana, IL 61801, March 2002.
11. David E. Goldberg, Kalyanmoy Deb, Hillol Kargupta, and Georges Harik. Rapid, accurate optimization of difficult problems using fast messy genetic algorithms. *Proceedings of the Fifth International Conference on Genetic Algorithms*, pages 56–64, July 1993.
12. Nazan Khan. Bayesian optimization algorithms for multiobjective and heierarchically difficult problems. Master thesis, University of Illinois at Urbana-Champaign, 117 Transportation Building, July 2003.
13. David Goldberg. *Simple Genetic Algofithms and the Minimal, Decptive Problem*, chapter Genetic Algorithms and Simulated Annealing, pages pp: 74–88. Morgan Kaufmann, Pubs, 1987.
14. L. Darrell Whitley. Fundamental principles of deception in genetic search. In Gregory J. Rawlins, editor, *Foundations of genetic algorithms*, pages 221–241. Morgan Kaufmann, San Mateo, CA, 1991.
15. Kalyanmoy Deb, Jeffrey Horn, and David E. Goldberg. Multimodal deceptive functions. Technical Report IlliGAL Report No 92003, University of Illinois, Urbana, 1992.
16. Kalyanmoy Deb and David E. Goldberg. Analyzing deception in trap functions. In L. Darrell Whitley, editor, *Foundations of Genetic Algorithms 2*, pages 93–108. Morgan Kaufmann, San Mateo, CA, 1993.
17. David E Goldberg, Kalyanmoy Deb, and Jeffrey Horn. Massive multimodality, deception and genetic algorithms. In R. Männer and B. Manderick, editors, *Parallel Problem Solving from Nature 2*, Amsterdam, 1992. Elsevier Science Publishers B.V.
18. Jesse B. Zydallis and Gary B. Lamont. Explicit Building-Block Multiobjective Evolutionary Algorithms for NPC Problems. In *Congress on Evolutionary Computation (CEC'2003)*, volume 4, pages 2685–2695, Piscataway, New Jersey, December 2003. IEEE Service Center.

Comparing Classical Generating Methods with an Evolutionary Multi-objective Optimization Method

Pradyumn Kumar Shukla, Kalyanmoy Deb, and Santosh Tiwari

Kanpur Genetic Algorithms Laboratory (KanGAL),
Indian Institute of Technology Kanpur,
Kanpur, PIN 208016, India
{shukla, deb, tiwaris}@iitk.ac.in

Abstract. For the past decade, many evolutionary multi-objective optimization (EMO) methodologies have been developed and applied to find multiple Pareto-optimal solutions in a single simulation run. In this paper, we discuss three different classical generating methods, some of which were suggested even before the inception of EMO methodologies. These methods specialize in finding multiple Pareto-optimal solutions in a single simulation run. On visual comparisons of the efficient frontiers obtained for a number of two and three-objective test problems, these algorithms are evaluated with an EMO methodology. The results bring out interesting insights about the strengths and weaknesses of these approaches. Further investigations of such classical generating methodologies and their evaluation should enable researchers to design a hybrid multi-objective optimization algorithm which may be better than each individual method.

1 Introduction

Multi-objective optimization has been a rapidly growing area in modern optimization. There exist a plethora of methods and algorithms for solving multi-objective optimization problems. The methods can be divided in two categories: (i) classical methods which use direct or gradient-based methods following some mathematical principles and (ii) non-traditional methods which follow some natural or physical principles. Of them, the evolutionary multi-objective optimization (EMO) has been getting growing attention over the past decade. The classification is also appropriate from two other perspectives. The classical approaches usually use deterministic transition rules, whereas non-traditional approaches use stochastic rules. They are also different from each other from another vital consideration. Classical methods mostly attempt to scalarize multiple objectives and perform repeated applications to find a set of Pareto-optimal solutions. On the other hand, EMO methods attempt to find multiple Pareto-optimal solutions in a single simulation run.

However, there exist a few classical generating methods (stochastic and deterministic) which attempt to find multiple Pareto-optimal solutions in a single simulation run, very much similar to the way EMO methods work. In this paper, we present three such algorithms and provide simulation results on a number of two and three-objective optimization problems. We also compare their performance with an EMO methodology and unveil the problem classes where the classical generating methods are better and the problem classes where the EMO methods have their niche. The study reveals important insights about the working of the algorithms, which can be combined together in a hybrid manner to develop an algorithm even better than individual algorithms.

2 Classical Generating Methods

Although most classical generating multi-objective optimization methods use an iterative scalarization scheme of standard procedures such as weighted-sum or epsilon-constraint methods [8], we have found at least three generating methods which attempt to find multiple Pareto-optimal solutions in a single simulation run. In the following subsections, we describe these methods.

2.1 Schäffler's Stochastic Method (SSM)

A stochastic method for the solution of unconstrained multi-objective optimization problems was proposed by Schäffler et. al. [9] in 2002. The method is based on the solution of a set of stochastic differential equations. This method requires the objective functions to be twice continuously-differentiable. It may be used for the computation of all or a large number of Pareto-optimal solutions. In each iteration, a trace of non-dominated points is constructed by calculating at each point $\mathbf{x}$, a direction $(-q(\mathbf{x}))$ in the decision space which is a direction of descent for *all* objective functions. The direction of descent is obtained by solving a quadratic subproblem. The following initial value problem (IVP) for a multi-objective optimization problem is then set up:

$$\dot{\mathbf{x}} = -q(\mathbf{x}(t)), \quad \mathbf{x}(0) = \mathbf{x}_0,$$

where x_0 is a starting point. The numerical solution of the above IVP gives a single point where the first-order weak Pareto-optimality conditions are fulfilled. After such a solution is obtained, a set of non-dominated solutions is obtained by perturbing it using a Brownian motion concept. The following stochastic differential equation is employed for this purpose:

$$d\mathbf{X}_t = -q(\mathbf{X}_t)d(t) + \varepsilon dB_t, \quad \mathbf{X}_0 = \mathbf{x}_0, \tag{1}$$

where $\varepsilon > 0$ and B_t is a n-dimensional Brownian motion having the following properties:

1. The expected value is zero,
2. The increments B_0, $(B_{t_1} - B_{t_0})$, $(B_{t_2} - B_{t_1})$ for every $t_0(=0) < t_1 < t_2 < \ldots$ are stochastically independent, and
3. For every $s < t$, the increment $(B_s - B_t)$ is normally distributed with mean equal to zero and a variance equal to $(s-t)I_n$, where I_n is a n-dimensional identity matrix.

Thus, starting from an initial solution, a number of solutions converging to the efficient frontier are expected to be generated by this procedure. The $-q(\mathbf{X}_t)d(t)$ term in Equation 1 is the deterministic descent part, while the Brownian motion is the local random search term. In all simulations here, to solve the above equation numerically, we employ the Euler's method. The approach needs two parameters to be set properly: (i) the parameter ε which controls the amount of local search and (ii) the step size σ used in the Euler's approach which controls the accuracy of the integration procedure. At the end of a pre-specified number of iterations, a non-domination check of the obtained solutions is performed and the resulting solutions are declared as the obtained Pareto-optimal solutions. For more information on this algorithm, interested readers may refer to the original study [9].

2.2 Timmel's Population Based Method (TPM)

As early as in 1980, Timmel [10] proposed a population-based stochastic approach for finding multiple Pareto-optimal solutions of a differentiable multi-objective optimization problem. In this method, first a feasible solution set (we call it a population) is randomly created. The non-dominated solutions $(\mathbf{X}_0 = \{\mathbf{x}_1^0, \mathbf{x}_2^0, \ldots, \mathbf{x}_s^0\})$ are identified and they serve as the first approximation to the Pareto-optimal set. Thereafter, from each solution $\mathbf{x}_k^0$, a child solution is created in the following manner:

$$\mathbf{x}_k^1 = -\sum_{i=1}^{M} t_i u_i \nabla f_i(\mathbf{x}_k^0),$$

where u_i is a uniformly distributed random number (between 0 and 1) and t_i is step-length in the i-th objective. It is a simple exercise to show that the above formulation ensures that not all functions can be worsened simultaneously. Thus, the child solution is either non-dominated to the parent solution $\mathbf{x}_k^0$, or it dominates the parent. However, the variation of the step-length over iterations must be made carefully to ensure convergence to the efficient frontier. The original study suggested the following sequence for updating t_i:

$$\lim_{i \to +\infty} t_i = 0, \quad \sum_{i=1}^{\infty} t_i = \infty, \quad \sum_{i=1}^{\infty} t_i^2 < \infty.$$

After the child population is created, it is combined with the parent population and only the non-dominated solutions are retained. This set then becomes the second approximation to the Pareto-optimal set. This procedure is continued for a pre-specified number of iterations. Note that the population size can vary with iterations. In fact, in most problems, an increase in the population size is expected.

The step-length variation mentioned above ensures the following aspects:

1. The step size should slowly decrease to zero as solutions closer to the Pareto-optimal set are found and
2. The decrease of the step size must not be slow enough so that the algorithm gets caught in sub-optimal points.

Thus, it is clear that the update of the step length is a crucial part of the working of the algorithm and a tuning of the update strategy may have to be done for every problem. Here, we use the following strategy: $t_i = c/i$ (where c is a positive constant), which satisfies all the above-mentioned conditions. For the interested readers, we refer to the original study [10, 11] for further details. It is interesting to note that this algorithm uses an elitist strategy, in which best of parent and offspring populations is retained.

2.3 Normal Boundary Intersection Method (NBI)

The NBI method was developed by Das et. al. [1] for finding a uniform spread of Pareto-optimal solutions for a general nonlinear multi-objective optimization problem. The weighted-sum scalarization approach has a fundamental drawback of not being able to find a uniform spread of Pareto-optimal solutions given a uniform spread of weights. The NBI approach uses a scalarization scheme with a property that a uniform spread in parameters will give rise to a uniform spread in points on the efficient frontier. Also, the method is independent of the relative scales of different objective functions. The scalarization scheme is briefly described below.

Let us consider a multi-objective problem as $\min_{\mathbf{x} \in S} F(\mathbf{x})$, where $S = \{\mathbf{x} \mid h(\mathbf{x}) = 0; g(\mathbf{x}) \leq 0, a \leq \mathbf{x} \leq b\}$ be the constraint set. Let $F^* = (f_1^*, f_2^*, \ldots, f_M^*)^T$ be the utopia point of the multi-objective optimization problem with M objective functions and n variables. Let the individual minima of the functions be attained at $\mathbf{x}_i^*$ for each $i = 1, 2, \ldots, M$. The convex hull of the individual minima is then obtained. The simplex obtained by the convex hull of the individual minima can be expressed as $\Phi\beta$, where $\Phi = (F(\mathbf{x}_1^*), F(\mathbf{x}_2^*), \ldots F(\mathbf{x}_M^*))$ is a $M \times M$ matrix and $\beta = \{(b_1, b_2, \ldots, b_M)^T \mid \sum_{i=1}^{M} b_i = 1\}$. The original study suggested a systematic method of setting β vectors in order to find a uniformly distributed set of efficient points. The NBI scalarization scheme takes a point on the simplex and then searches for the maximum distance along the normal pointing towards the origin. This chosen point may or may not be a Pareto-optimal point. In non-convex situations, even the Pareto-optimal points which cannot be obtained by the usual weighted-sum schemes, are possible to be obtained by this method. The NBI subproblem (NBI$_\beta$) for a given vector β is as follows:

$$\max_{(\mathbf{x},t)} \quad t,$$
$$\text{subject to } \Phi\beta + t\hat{n} = F(\mathbf{x}), \qquad (2)$$
$$\mathbf{x} \in S,$$

where $\hat{n}$ is the normal direction at the point $\Phi\beta$ pointing towards the origin. The solution of the above problem gives the maximum t and also the corresponding Pareto-optimal solution, $\mathbf{x}$. The method works even when the normal direction is not an exact one, but a quasi-normal direction. The following quasi-normal direction vector is suggested in Das et al. [1]: $\hat{n} = -\Phi e$, where $e = (1, 1, \ldots, 1)^T$ is a $M \times 1$ vector. The above quasi-normal direction has the property that NBI$_\beta$ is independent of the relative scales of the objective functions. A modified version of the NBI approach (called the recursive knee approach) was developed elsewhere

[2] for convex problems. Another study extended the approach by using a suitable inequality constraint to define a subproblem [7].

3 Comparison with NSGA-II

In this section, we compare the above three classical generating methods with NSGA-II on a number of two and three-objective test problems. The test problems are chosen in such a way so as to systematically investigate various aspects of an algorithm. In the test problems, the exact knowledge of the Pareto-optimal front is available. For classical methods, a limited parametric study is performed for each test problem and results from the best parameter setting are presented. For NSGA-II, we use a standard real-parameter SBX and polynomial mutation operator with $\eta_c = 10$ and $\eta_m = 10$, respectively [3]. For all problems solved using NSGA-II, we use a population of size 100.

3.1 Two-Objective Test Problems

First, we consider two-objective ZDT test problems [3, 4]. The test problems are slightly modified so that they become unconstrained multi-objective optimization problems, as the SSM method is only able to tackle unconstrained problems. A constrained version of SSM algorithm is currently being investigated by the authors.

Modified ZDT1 Test Problem: The modified ZDT1 test problem can be stated as follows:

$$\text{Minimize } f_1(\mathbf{x}) = x_1,$$
$$\text{Minimize } f_2(\mathbf{x}) = g(\mathbf{x})\left(1 - \sqrt{\frac{x_1}{g(\mathbf{X})}}\right), \quad (3)$$
$$\text{where } g(x) = 1 + \frac{9}{n-1}\sum_{i=2}^{n} x_i^2,$$

where the box constraints are $x_1 \in [0, 1]$, and $x_i \in [-1, 1]$ for $i = 2, 3, \ldots, n$. Here, we choose $n = 30$. This modified ZDT1 problem has a convex Pareto-optimal front. The Pareto-optimal solutions correspond to $0 \leq x_1^* \leq 1$ and $x_i^* = 0$ for $i = 2, 3, \ldots, n$. This problem offers a difficulty in handling a large number of variables.

The Euler's method with a step size of $\sigma = 0.8$ along with $\epsilon = 0.05$ is used in SSM. An initial starting point is randomly created using the box constraints. It is to be noted that the SSM method requires gradient information. To make a fair comparison with an EMO methodology, gradients are calculated numerically here and the overall function evaluations is recorded. Figure 1 shows the obtained distribution of efficient solutions after 20,000 (inset plot) and 100,000 function evaluations. Due to the use of a descent direction, the SSM method quickly converges near to the efficient frontier in this problem. However, the spread of solutions along the efficient frontier is very slow. Notice that from 20,000 function evaluations till 100,000 function evaluations, the procedure finds a spread from $x_1 = 0.4$ to $x_1 = 0.7$. The use of a Brownian motion for spread seems to be too generic to get a faster spread along the efficient frontier. After even 100,000 function evaluations, the solutions are not quite on to the efficient frontier. The nu-

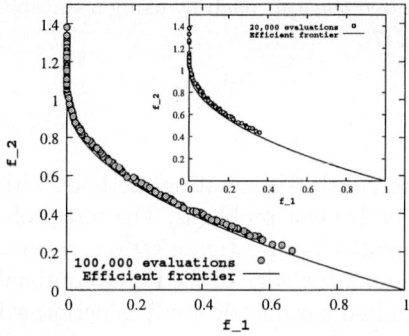

Fig. 1. Performance of SSM method on ZDT1

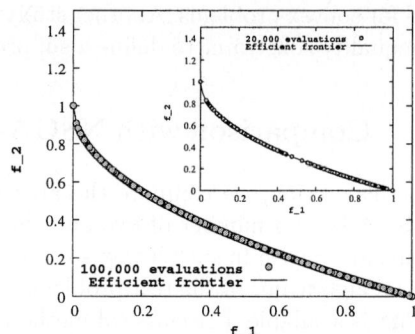

Fig. 2. Performance of TPM method on ZDT1

merical gradient evaluation is costly, requiring $2n$ function evaluations for each gradient. With a large number of variables, such methods may become computationally expensive. However, the simulation results show that this test problem does not offer too much of a difficulty to the SSM method in quickly converging near to the efficient frontier.

Next, we apply the TPM method. We begin the search with a single solution ($s = 1$), randomly created satisfying the box constraints. Figure 2 shows the obtained front after 20,000 (inset plot) and 100,000 function evaluations. It is clear that the TPM method performs extremely well on ZDT1 both in terms of convergence and maintenance of diversity.

The NBI method needs the computation of the utopia point. This requirement causes an added difficulty for the NBI method. Here, the subproblems are solved using the sequential quadratic programming (SQP) method. Figure 3 (inset) shows that the NBI method is capable of finding a good spread of Pareto-optimal solutions even with 20,000 function evaluations. Since, a systematic initial points are considered in this approach, a good spread is obtained. If more β-vectors are used, a more dense set of solutions can be found.

Finally, we apply NSGA-II for a total of 20,000 (inset plot) and 100,000 function evaluations. Figure 4 shows that even with 20,000 evaluations, a good distribution is achieved. Based on all these simulations, it can be concluded that the ZDT1 problem is best solved by using a systematic procedure such as the NBI method, whereas the TPM or NSGA-II also performs well on this problem.

Modified ZDT2 Test Problem: The modified ZDT2 test problem can be stated as follows:

$$\text{Minimize } f_1(\mathbf{x}) = x_1,$$
$$\text{Minimize } f_2(\mathbf{x}) = g(\mathbf{x})\left(1 - \left(\frac{x_1}{g(\mathbf{X})}\right)^2\right), \tag{4}$$
$$\text{where } g(x) = 1 + \frac{9}{n-1}\sum_{i=2}^{n} x_i^2,$$

where the box constraints are $x_1 \in [0,1]$ and $x_i \in [-1,1]$ for $i = 2, 3, \ldots, n$. Here again we use $n = 30$. This problem has a non-convex efficient frontier. The Pareto-

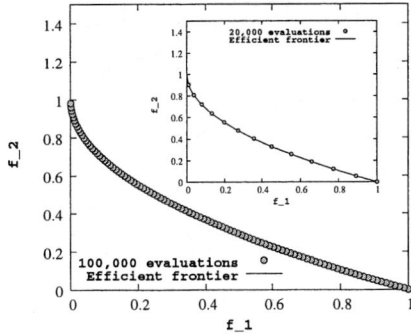
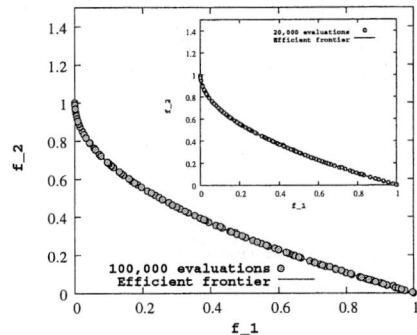

Fig. 3. Performance of NBI method on ZDT1

Fig. 4. Performance of NSGA-II method on ZDT1

optimal solutions correspond to $0 \leq x_1^* \leq 1$ and $x_i^* = 0$ for $i = 2, 3, \ldots, n$. This problem provides two difficulties to an optimization algorithm: (i) large number of variables and (ii) a non-convex efficient frontier.

The Euler's method with a step size of $\sigma = 0.1$ along with $\epsilon = 0.01$ is used in the SSM algorithm. Figure 5 shows the obtained distribution of solutions after 20,000 and 100,000 function evaluations. Although the convergence near the efficient front is quick similar to that in ZDT1, the distribution is poor. In the TPM method, we use a population of size 100 randomly created satisfying the box constraints. Solutions after 20,000 and 100,000 evaluations are shown in Figure 6. A good convergence and diversity of solutions is observed. Figure 7 shows the solutions obtained using the NBI method. A good set of solutions even with 20,000 function evaluations is apparent from the figure. Figure 8 shows the NSGA-II solutions for 20,000 function evaluations. It is clear that the non-convexity of the efficient frontier did not provide any problem to NBI and TPM approaches, while the performance of the SSM approach is poor.

Modified ZDT3 Test Problem: The modified ZDT3 test problem can be stated as follows:

$$\text{Minimize } f_1(\mathbf{x}) = x_1,$$
$$\text{Minimize } f_2(\mathbf{x}) = g(x)\left(1 - \sqrt{\tfrac{x_1}{g(\mathbf{X})}} - \tfrac{x_1}{g(\mathbf{X})}\sin(10\pi x_1)\right), \quad (5)$$
$$\text{where } g(x) = 1 + \tfrac{9}{n-1}\sum_{i=2}^{n} x_i^2,$$

where the box constraints are $x_1 \in [0, 1]$, and $x_i \in [-1, 1]$ for $i = 2, 3, \ldots, n$. We use $n = 30$. This problem has a convex discontinuous efficient frontier. The Pareto optimal solutions correspond to $0 \leq x_1^* \leq 1$ and $x_i^* = 0$ for $i = 2, 3, \ldots, n$. The Euler's method with a step size $\sigma = 0.5$ along with an $\epsilon = 0.01$ is used in SSM. Figure 9 shows the obtained distribution after 100,000 functions evaluations. Only a portion of the efficient frontier is discovered by this method. The TPM method is applied with an initial population of size 5,000, randomly created satisfying the box constraints. Figure 10 shows the obtained solutions after 100,000 eval-

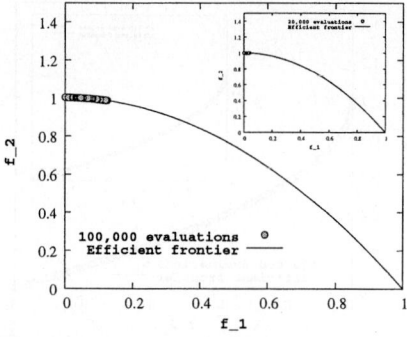

Fig. 5. Performance of SSM method on ZDT2

Fig. 6. Performance of TPM method on ZDT2

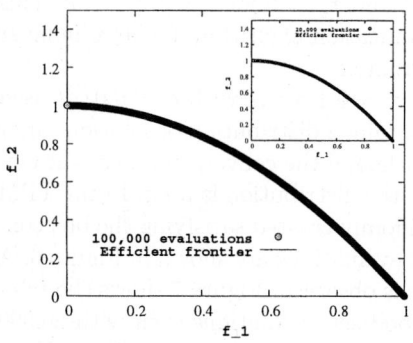
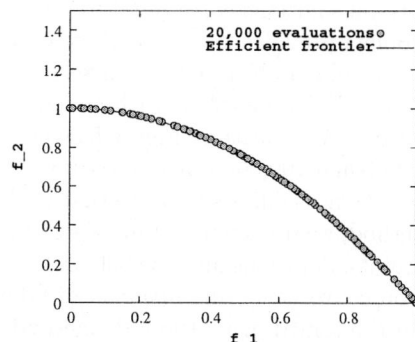

Fig. 7. Performance of NBI method on ZDT2

Fig. 8. Performance of NSGA-II method on ZDT2

uations. The figure shows that all disconnected efficient fronts are discovered by this method. It is noteworthy that with 20,000 evaluations the complete front was not fully discovered by the TPM method.

Figure 11 shows the distribution of the NBI method after 100,000 evaluations. It is apparent that not all disconnected efficient fronts are discovered by this approach. Although the NBI method performed very well on ZDT1 and ZDT2 problems having a continuous efficient front, the disconnectedness of the efficient frontier seems to have provided difficulty to this approach. Since not all fronts are discovered, this method ends up finding a dominated portion of the true efficient frontier. Since the idea of non-domination is not built in the NBI approach, it ends up finding some non-efficient points. A comparison with the NSGA-II results (Figure 12) indicates the NSGA-II with only 20,000 evaluations is able to find all disconnected efficient fronts.

ZDT4 Test Problem: Next, we use the 10-variable ZDT4 test problem [3]. This problem has a total of 100 distinct local efficient fronts in the objective space.

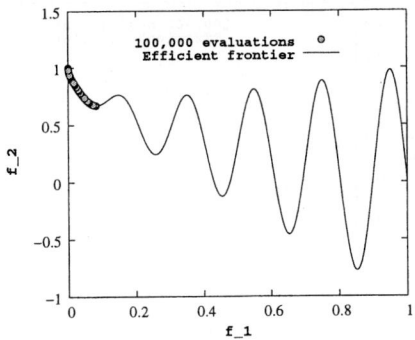

Fig. 9. Performance of SSM method on ZDT3

Fig. 10. Performance of TPM method on ZDT3

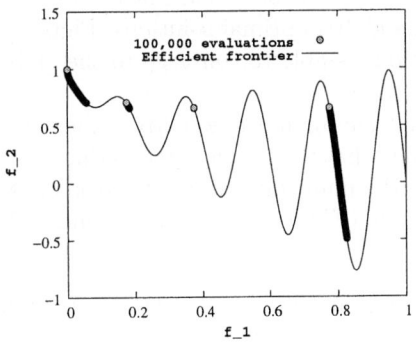
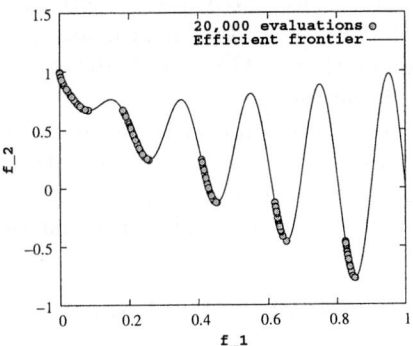

Fig. 11. Performance of NBI method on ZDT3

Fig. 12. Performance of NSGA-II method on ZDT3

The global Pareto-optimal solutions correspond to $0 \leq x_1^* \leq 1$ and $x_i^* = 0$ for $i = 2, 3, \ldots, n$. The algorithms face a difficulty in overcoming a large number of local fronts and converging to the global front.

The Euler's method with a step size of $\sigma = 0.1$ along with $\epsilon = 0.001$ is used in SSM. Only a few weak Pareto-optimal solutions ($f_1 = 0$ and $f_2 = 70$ to 70.4) are found after 20,000 evaluations. Since the SSM method requires functions to be twice continuously-differentiable and since ZDT4 is not twice differentiable precisely at $x_1 = 0$, the gradient computation is erroneous at $x_1 = 0$, resulting in a failure of the method.

The TPM method is applied with 2,000 initial solutions randomly created satisfying the box constraints. Figure 13 shows that a set of dominated local-Pareto-optimal solutions is discovered after 100,000 evaluations. The optimization algorithm used in the TPM method can get stuck to a local-optimal solution and the ZDT4 problem with many local efficient frontier provides enough difficulty to this approach for finding the true global efficient frontier. The multi-modality of the

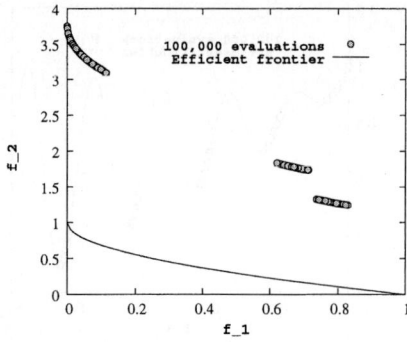 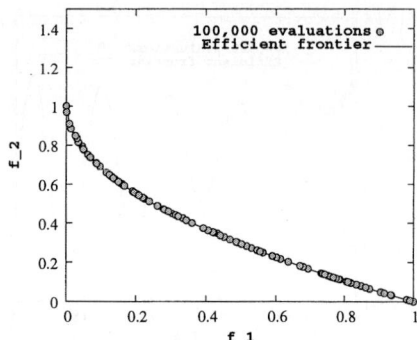

Fig. 13. Performance of TPM method on ZDT4

Fig. 14. Performance of NSGA-II method on ZDT4

search space also causes the NBI method to not find the global efficient frontier. The SQP method is inadequate to find the global optimal solutions. Figure 14 shows that NSGA-II with 100,000 evaluations is able to converge to the global efficient frontier.

The problem ZDT4 provides difficulty in terms of multi-modality of the search space. It is evident from the simulation results that the classical generating methods face enormous problems in overcoming the multi-modalities, whereas in this type of problems evolutionary multi-objective (EMO) methods are found to be useful.

Modified ZDT6 Test Problem: The $n = 10$ variable modified ZDT6 test problem is as follows:

$$\text{Minimize } f_1(\mathbf{x}) = 1 - \exp(-4x_1)\sin^6(4\pi x_1),$$
$$\text{Minimize } f_2(\mathbf{x}) = g(\mathbf{x})\left(1 - \left(\frac{f_1(\mathbf{X})}{g(\mathbf{X})}\right)^2\right), \tag{6}$$
$$\text{where } g(x) = 1 + 9\left(\sum_{i=2}^{n} x_i^2/(n-1)\right)^{0.25},$$

where the box constraints are $x_1 \in [0,1]$ and $x_i \in [-1,1]$ for $i = 2, 3, \ldots, n$. This problem has a non-convex and non-uniformly spaced Pareto-optimal solutions. The Pareto-optimal solutions correspond to $0 \leq x_1^* \leq 1$ and $x_i^* = 0$ for $i = 2, 3, \ldots, n$. The Euler's method with a step size of $\sigma = 0.15$ along with $\epsilon = 0.001$ is used in SSM. Figure 15 shows the distribution of obtained solution after 100,000 function evaluations. The algorithm is not able to find a well-converged set of solutions. Although there is no local efficient frontier at the location where the algorithm gets stuck, parameters play an important role in the success of SSM and in this problem it is seen that for small values of parameters there is an ascent in functions, instead of a descent in them. A theoretical analysis suggests that at each point $\mathbf{x}$ the direction $(-q(x))$ is a descent direction for all functions, however with a finite step size this result does not hold.

The TPM method with 1,000 initial random solutions produces a set of solutions closer to the efficient frontier, but there are only a few solutions found even

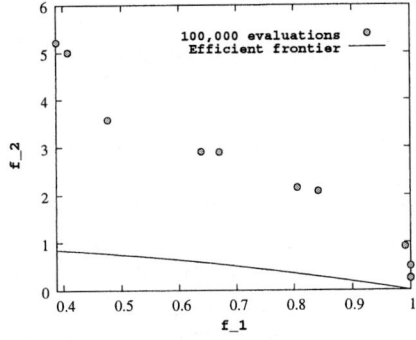

Fig. 15. Performance of SSM method on ZDT6

Fig. 16. Performance of TPM method on ZDT6

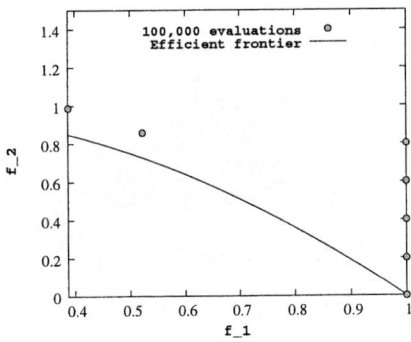
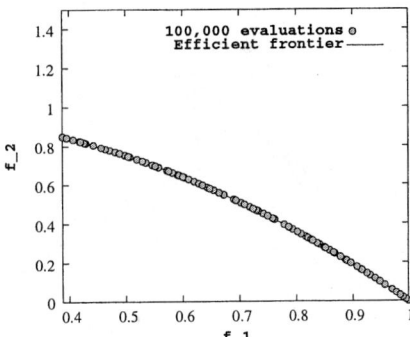

Fig. 17. Performance of NBI method on ZDT6

Fig. 18. Performance of NSGA-II method on ZDT6

after 100,000 function evaluations (Figure 16). For this problem, there is a slow improvement in each iteration and by the time the solution reaches near the efficient frontier, the step size t_i becomes very small and it would take a long time before the solutions fall on the efficient frontier.

The NBI method (Figure 17) performs poorly on this problem. Since the density of solutions along the frontier is non-uniform, the SQP method along with the NBI strategy is unable to find a good distribution. On the other hand, NSGA-II is able to find a good convergence and distribution with 100,000 function evaluations (Figure 18).

Based on these simulations, we infer that a non-uniform density of solutions in the objective space (which occurs in many real-world problems [3]) provides enough difficulty to the classical generating methods. These are another class of optimization problems in which EMO methodology performs comparatively better than the classical methods.

3.2 Three-Objective Test problems

Now, we consider a couple of three-objective test problems developed elsewhere [6] to study the behavior of all four algorithms.

DTLZ2 Test Problem: First, we consider the 12-variable DTLZ2 test problem having a spherical efficient front satisfying $f_1^2 + f_2^2 + f_3^2 = 1$ in the range $f_1, f_2 \in [0, 1]$. The Euler's method with a step size of $\sigma = 0.1$ and $\epsilon = 0.01$ is used in SSM. Figure 19 shows all obtained solutions after 100,000 evaluations. It is clear that the SSM approach is able to get the solutions on the frontier, but the distribution of solutions (obtained mainly by the Brownian approach) is not adequate. It will take enormous number of evaluations for the algorithm to find a distribution across the complete efficient frontier. However as apparent from the figure, due to the descent direction it needs only few iteration to reach the efficient frontier.

The Timmel's method is applied next with 1,000 initial random solutions. After 100,000 evaluations, the approach is able to find a good coverage of the en-

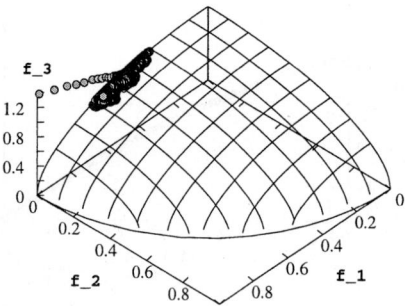

Fig. 19. Performance of SSM method on DTLZ2

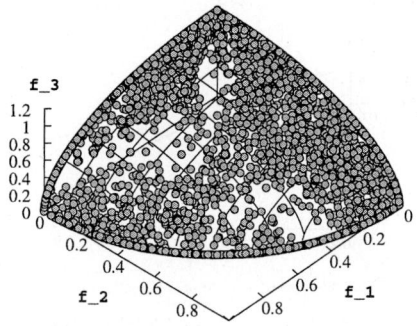

Fig. 20. Performance of TPM method on DTLZ2

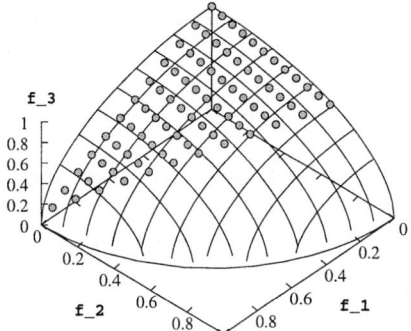

Fig. 21. Performance of NBI method on DTLZ2

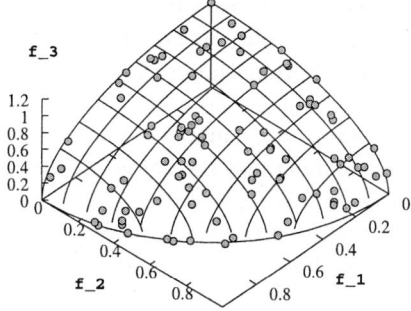

Fig. 22. Performance of NSGA-II method on DTLZ2

tire efficient frontier (Figure 20). It is interesting that the boundary solutions are adequately discovered by this approach. The NBI approach, after 100,000 evaluations, finds a few well-distributed solutions (Figure 21). If more evaluations are allowed, the remaining portion of the efficient frontier may also be discovered by this method, however the requirement of a large number of evaluations for high-dimensional objective space is a drawback of this algorithm.

The spread of solutions using NSGA-II (with 20,000 evaluations) is shown in Figure 22. Although the distribution is not as regular as in the NBI approach, the obtained solutions spread across the entire front. As pointed elsewhere, a better niching operator than the crowding-distance operator, such as a clustered NSGA-II [5] or another EMO such as SPEA2 can employ a better distribution of solutions in problems having more than two objectives.

If these algorithms are applied on DTLZ3 which has a number of local efficient frontiers as in ZDT4, the classical algorithms will have similar difficulties in converging to the true efficient frontier. Thus, we do not show the results on DTLZ3.

DTLZ5 Test Problem: The DTLZ5 is a 12-variable problem having a Pareto-optimal curve: $f_3^2 = 1 - f_1^2 - f_2^2$ with $f_1 = f_2 \in [0, 1]$. This problem, although a

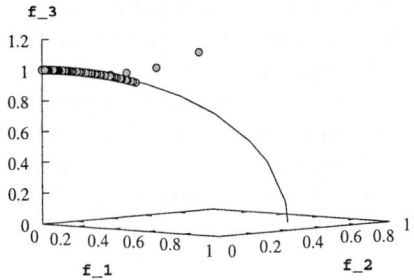

Fig. 23. Performance of SSM method on DTLZ5

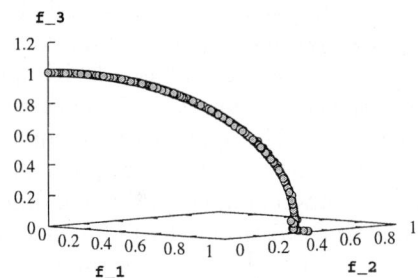

Fig. 24. Performance of TPM method on DTLZ5

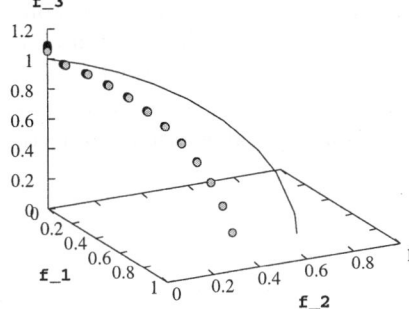

Fig. 25. Performance of NBI method on DTLZ5

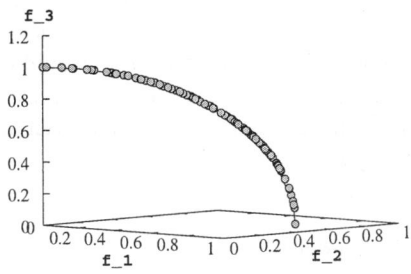

Fig. 26. Performance of NSGA-II method on DTLZ5

three-objective one, has a one-dimensional efficient frontier. The SSM, using Euler's method with a step size of $\sigma = 0.5$ and $\epsilon = 0.01$, finds the partial front after 100,000 evaluations, as shown in Figure 23. The TPM approach (with 500 initial random solutions) finds the complete front, as shown in Figure 24. On the other hand, the NBI approach finds a different one-dimensional curve as the efficient frontier (Figure 25).

On the other hand, like TPM, NSGA-II (with 20,000 function evaluations) does not have any problem in finding a good distribution on the true frontier, as shown in Figure 26.

4 Conclusions

This study brings into light three different classical generating methods which can be used to find a set of Pareto-optimal solutions in a single simulation run. The comparison of these methods with NSGA-II on a number of test problems have adequately demonstrated that these methods perform very well when the problem size and search space complexity is small. Among the three methods, the SSM approach seems to find only a part of the entire efficient front. However, due to the use of a direction of descent on all objective functions simultaneously, it usually reaches a local efficient front quickly. The TPM approach is similar to an elite-preserving population-based EMO approach with an exception that with iterations the population size can increase indefinitely, thereby making the latter iterations slow. The approach also requires fixing a step-size update scheme, which requires fine-tuning for every problem. The NBI approach is a systematic mathematical programming approach in which a number of searches are performed from a uniformly-distributed set of points in the objective space.

On a number of two and three-objective test problems, it has been observed that the TPM and NBI are better than the SSM approach. However, for problems having multi-modal efficient fronts or non-uniform density of points in the objective space, all three methods do not perform well. They either get stuck to a local efficient frontier or to suboptimal solutions. On the other hand, on all problems considered here, NSGA-II with an identical parameter setting, has performed well. One way to extend the study would be to replace the SQP or classical optimization approach embedded to these classical algorithms with an evolutionary algorithm. Another approach would be to use some of the classical principles as an additional operator in an EMO methodology. Some such extensions would be an immediate focus for useful research and application in the area of multi-objective optimization.

References

1. I. Das and J.E. Dennis. Normal-boundary intersection: A new method for generating the Pareto surface in nonlinear multicriteria optimization problems. *SIAM Journal of Optimization*, 8(3):631–657, 1998.

2. I. Das and J.E. Dennis. An improved technique for choosing parameters for Pareto surface generation using normal-boundary intersection. In *Proceedings of the Third World Congress on Structutal and Multidisciplinary Optimization*, 1999.
3. K. Deb. *Multi-objective optimization using evolutionary algorithms*. Chichester, UK: Wiley, 2001.
4. K. Deb, S. Agrawal, A. Pratap, and T. Meyarivan. A fast and elitist multi-objective genetic algorithm: NSGA-II. *IEEE Transactions on Evolutionary Computation*, 6(2):182–197, 2002.
5. K. Deb, M. Mohan, and S. Mishra. Towards a quick computation of well-spread pareto-optimal solutions. In *Proceedings of the Second Evolutionary Multi-Criterion Optimization (EMO-03) Conference (LNCS 2632)*, pages 222–236, 2003.
6. K. Deb, L. Thiele, M. Laumanns, and E. Zitzler. Scalable multi-objective optimization test problems. In *Proceedings of the Congress on Evolutionary Computation (CEC-2002)*, pages 825–830, 2002.
7. A. Messac and C. A. Mattson. Normal constraint method with guarantee of even representation of complete pareto frontier. *AIAA Journal*, in press.
8. K. Miettinen. *Nonlinear Multiobjective Optimization*. Kluwer, Boston, 1999.
9. S. Schäffler, R. Schultz, and K. Weinzierl. Stochastic method for the solution of unconstrained vector optimization problems. *Journal of Optimization Theory and Applications*, 114(1):209–222, 2002.
10. G. Timmel. Ein stochastisches suchverrahren zur bestimmung der optimalen kompromilsungen bei statischen polzkriteriellen optimierungsaufgaben. *Wiss. Z. TH Ilmenau*, 6:159–174, 1980.
11. G. Timmel. Modifikation eines statistischen suchverfahrens der vektoroptimierung. *Wiss. Z. TH Ilmenau*, 5:139–148, 1982.

A New Analysis of the LebMeasure Algorithm for Calculating Hypervolume

Lyndon While

The University of Western Australia, WA 6009, Australia
lyndon@csse.uwa.edu.au

Abstract. We present a new analysis of the LebMeasure algorithm for calculating hypervolume. We prove that although it is polynomial in the number of points, LebMeasure is exponential in the number of objectives in the worst case, not polynomial as has been claimed previously. This result has important implications for anyone planning to use hypervolume, either as a metric to compare optimisation algorithms, or as part of a diversity mechanism in an evolutionary algorithm.

Keywords: Multi-objective optimisation, evolutionary computation, performance metrics, hypervolume.

1 Introduction

Multi-objective optimisation problems abound, and many evolutionary algorithms have been proposed to derive good solutions for such problems, for example [4,11,1,5]. However, the question of what metrics to use in comparing the performance of these algorithms remains difficult[7,9,4]. One metric that has been favoured by many people is *hypervolume*[4,8], also known as the S-metric[10] or the Lebesgue measure[3]. The hypervolume of a set of solutions measures the size of the portion of objective space that is dominated by those solutions collectively. Generally, hypervolume is favoured because it captures in a single scalar both the closeness of the solutions to the optimal set and, to some extent, the spread of the solutions across objective space. It also has nicer mathematical properties than many other metrics[2,12], although it can be sensitive to scaling and to the presence or absence of extremal values.

Unfortunately, hypervolume is very expensive to calculate for problems with more than 2–3 objectives. A recently published algorithm that seemed to solve this problem is the LebMeasure algorithm[2,3]. The original analysis of LebMeasure claimed that this algorithm has complexity $O(m^3n^2)$ for m points in n objectives, thus allowing the efficient calculation of hypervolume for large sets of points in large numbers of objectives. However, we prove in this paper that although LebMeasure is polynomial in the number of points, it is exponential in the number of objectives in the worst-case, and thus its performance deteriorates sharply as the number of objectives increases. We present three strands of evidence.

Empirical Data. We give some pathological patterns of sets of points that clearly exhibit exponential slowdown under LebMeasure as the number of objectives increases.

A Lower-Bound on the Complexity. For one pattern of sets of two points, we describe their slowdown with a recurrence relation in the number of objectives n, and we prove that this recurrence relation is equal to 2^{n-1}, thus proving that LebMeasure is exponential in the number of objectives in the worst case.

An Upper-Bound on the Complexity. It is hard to establish an exact complexity for LebMeasure in the general case, because it is difficult to be certain which sets of points will suffer the biggest slowdown. However, we can establish a likely upper-bound on the worst-case complexity by considering illegal sets of points that are likely to perform worse than any legal sets of points[1]. We do this for one pattern of sets of points, defining a recurrence relation for their slowdown in m and n, and proving that this recurrence relation is $O(m^n)$.

Taken together, this evidence demonstrates that in the worst case, LebMeasure is exponential in the number of objectives, and thus that it cannot provide a general solution to the performance problem (currently) inherent in calculating hypervolume.

The rest of this paper is structured as follows. Section 2 describes LebMeasure and its behaviour. Section 3 gives an empirical analysis of some patterns of sets of points which exhibit exponential growth (in terms of the number of objectives) under LebMeasure. Section 4 gives a recurrence relation for one pattern of sets of points, and proves that the time to process this pattern grows exponentially with the number of objectives. Section 5 discusses the possibility of placing an upper-bound on the complexity of LebMeasure, and Section 6 concludes the paper and outlines some future work.

2 The LebMeasure Algorithm

2.1 The Definition of Hypervolume

Given a set S containing m points in n objectives, the hypervolume of S is the size of the portion of objective space that is dominated by at least one point in S. We say that a point $\overline{x}$ dominates a point $\overline{y}$ iff $\overline{x}$ is at least as good as $\overline{y}$ in every objective *and* $\overline{x}$ is strictly better than $\overline{y}$ in at least one objective. The hypervolume of S is calculated relative to a reference point which is worse than (or equal to) every point in S in every objective. Given two sets of points S and S' containing solutions to a given problem, if S occupies a greater hypervolume than S', then S is taken to be a better solution to the problem.

---

[1] A set of points is legal only if its members are mutually non-dominating.

2.2 The Behaviour of LebMeasure

LebMeasure works by processing the points in a set one at a time. It calculates the volume that is dominated exclusively by the first point, then it discards that point and moves on to the subsequent points, until all points have been processed and all volumes have been summed. This is particularly efficient when the volume dominated exclusively by a point $\bar{x}$ is "hyper-cuboid", but where this is not the case, LebMeasure lops-off the largest hyper-cuboid volume that is dominated exclusively by $\bar{x}$, and replaces $\bar{x}$ with a set of up to n "spawned" points that dominate the remainder of $\bar{x}$'s exclusive hypervolume. A spawn is discarded if it dominates no exclusive hypervolume, either because it has one or more objective values equal to the reference point, or because it is dominated by an unprocessed point.

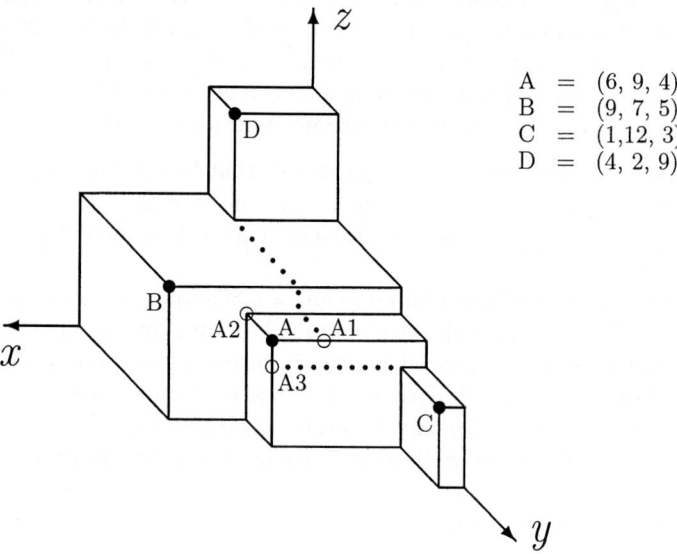

Fig. 1. Spawning in LebMeasure. The blocks denote the hypervolume dominated by {A, B, C, D} in a maximisation problem relative to the origin. The filled circles represent the four points, and the empty circles represent the three potential spawns of A. Each spawn is generated by reducing one objective to the largest smaller value from the other points. It is clear that A2 is dominated by B, so A is replaced by A1 and A3

Consider as an example the set of points shown in Fig. 1, with Point A to be processed first. The largest hyper-cuboid dominated exclusively by A is bounded at the opposite corner by (4,7,3). Thus the three potential spawns of A are

A1 = (4, 9, 4)   A2 = (6, 7, 4)   A3 = (6, 9, 3)

However, A2 is dominated by B (from the main list of points), so only A1 and A3 dominate exclusive hypervolume of their own, and only those two are added to the main list to be processed.

LebMeasure continues lopping-off volume, and replacing points with their spawns, until all points (and spawns, and spawns of spawns, etc.) have been processed.

Fig. 2 gives pseudo-code for LebMeasure.

```
hypervolume (ps):
    pl = a stack containing the points from ps in some order,
         each paired with n, the number of objectives
    hypervolume = 0
    while pl /= emptylist
      (p, z) = head (pl)
      pl = tail (pl)
      a = oppositeCorner (p, pl)
      hypervolume = hypervolume + volBetween (p, a)
      ql = spawns (p, z, a, pl)
      prepend ql to pl
    return hypervolume

spawns (p, z, a, pl):
    ql = emptylist
    for k = 1 to z
      if a[k] /= referencePoint[k]
        q = p
        q[k] = a[k]
        if not (dominated (q, pl))
          append (q, k) to ql
    return ql

oppositeCorner (p, pl) returns the point that bounds the
largest hyper-cuboid that is dominated exclusively by the
point p relative to the points on pl

volBetween (p, q) returns the hypervolume between the points p and q

dominated (p, pl) returns True iff the point p is dominated by
at least one of the points on pl
```

Fig. 2. Pseudo-code for LebMeasure. Each point on the main stack is paired with the highest index that it can use to generate a non-dominated spawn

2.3 The Complexity of LebMeasure

[2, 3] claim that the worst-case complexity of LebMeasure is $O(m^3 n^2)$. This claim is based largely on an observation about the spawning behaviour of the algorithm. Consider the situation in the above example, after A has been processed:

A1 = (4, 9, 4)
A3 = (6, 9, 3)

```
B = (9,  7,  5)
C = (1, 12,  3)
D = (4,  2,  9)
```

The three potential spawns of A1 are

```
A11 = (1, 9, 4)    A12 = (4, 7, 4)    A13 = (4, 9, 3)
```

Both A12 and A13 are dominated (respectively by B and A3). Moreover, *this is guaranteed to be the case*. For example, A13 is generated by reducing the first and third objectives from A, *but* A3 is still unprocessed, and that point is generated by reducing only the third objective from A. Similarly, A12 is guaranteed to be dominated by A2 (or in this example, by the point that dominated A2: dominance is transitive).

Thus these two points are guaranteed to be dominated, and LebMeasure need never generate them in the first place. This is a major optimisation for the algorithm. However, we shall see that this optimisation is not enough to change the time complexity of LebMeasure: it provides only a constant-factor speed-up.

The optimisation does however change the space complexity of LebMeasure. As described in [2, 3], the size of the stack of points is now bounded by $m+n-1$, because the number of potential spawns generated from a point P is bounded by the lowest-numbered objective that was reduced during the formation of P.

3 Empirical Evidence of the Complexity of LebMeasure

It is clear that the complexity of LebMeasure is at least equal to the number of points processed *that actually contribute to the result* (including spawns, spawns of spawns, etc.): this is the number of times that the while loop in Figure 2 is traversed. We can thus estimate a lower bound on the complexity by counting the numbers of contributing points for some concrete sets of points. We consider two pathological sets of points that together suggest strongly that LebMeasure is exponential in the number of objectives.

Consider first the sets of points described by the pattern shown in Fig. 3. The numbers of points processed for this pattern are as given in Table 1. The

1	5	⋯	5
2	4	⋯	4
3	3	⋯	3
4	2	⋯	2
5	1	⋯	1

Fig. 3. A pathological example for LebMeasure. This pattern describes sets of five points in n objectives, $n \geq 2$. All columns except the first are identical. The pattern can be generalised for other numbers of points

Table 1. The numbers of points processed for the pattern from Fig. 3, equal to m^{n-1}

n	m = 2	m = 5	m = 8	m = 10
2	2	5	8	10
3	4	25	64	100
4	8	125	512	1,000
5	16	625	4,096	10,000
6	32	3,125	32,768	100,000
7	64	15,625	262,144	1,000,000
8	128	78,125	2,097,152	10,000,000
9	256	390,625	16,777,216	100,000,000

numbers clearly show growth that is exponential in the number of objectives, indicating that this pattern generates m^{n-1} points.

However, this pattern raises the question of the order in which the points are processed. If the points from Fig. 3 are reversed before being presented to LebMeasure, then no non-dominated spawns are generated and the number of points is m, for all values of n. The order shown in Fig. 3 is in fact the worst-case ordering for this pattern.

Thus one way to improve the overall performance of LebMeasure is to develop heuristics to select the best order in which to present the points to the algorithm. However, we illustrate now that there are patterns of points that exhibit exponential complexity under LebMeasure even in their best-case ordering. Consider the sets of points described by the pattern shown in Fig. 4.

By evaluating all $m!$ permutations of each set of points, we can see that the numbers of points processed for this pattern when the points are presented in their optimal ordering are as given in Table 2. Again, we see exponential growth: this pattern generates approximately $m(m!)^{(n-2)div\ m}((n-2)mod\ m)!$ points, even for the best-case ordering.

Finally, [6] proposes the use of LebMeasure in an evolutionary algorithm that calculates hypervolume incrementally to promote diversity. Structurally, LebMeasure is well-suited to this task, because it calculates explicitly the hypervolume dominated exclusively by its first point. However, even evaluating the hypervolume dominated by a single point can be exponential in LebMeasure, as

1	1	2	3	4	5	1 ⋯	5
2	2	3	4	5	1	2 ⋯	4
3	3	4	5	1	2	3 ⋯	3
4	4	5	1	2	3	4 ⋯	2
5	5	1	2	3	4	5 ⋯	1

Fig. 4. A second pathological example for LebMeasure. This pattern describes sets of five points in n objectives, $n \geq 2$. The first and last columns are fixed, and if $n > 2$, the second column is fixed and each other column is a simple rotation of the previous column. The pattern can be generalised for other numbers of points

Table 2. The best-case numbers of points processed for the pattern from Fig. 4, approximately equal to (but not bounded by) $m(m!)^{(n-2) \, div \, m}((n-2) \, mod \, m)!$. The values of m reported were limited by the numbers of permutations that must be tested

n	$m=2$	$m=3$	$m=4$	$m=5$
2	2	3	4	5
3	2	3	4	5
4	4	6	8	10
5	4	17	23	29
6	8	17	88	112
7	8	35	88	549
8	16	105	180	549
9	16	105	558	1,115
10	32	213	2,248	3,421
11	32	641	2,248	14,083
12	64	641	4,528	70,889
13	64	1,289	13,708	70,889
14	128	3,873	54,976	142,309
15	128	3,873	54,976	428,449
16	256	7,761	110,160	1,721,605
17	256	23,297	331,128	8,618,577

shown by the numbers in Table 3. Note that in this context, where we need to evaluate the hypervolume contribution of a single point, changing the order of the other points makes no difference to the algorithm's performance.

Thus a simple empirical study of LebMeasure suggests that it is exponential in the number of objectives.

Table 3. The numbers of points processed for the first point of the pattern from Fig. 3, equal to $m^{n-1} - (m-1)^{n-1}$, i.e. $O(m^{n-2})$

n	$m=2$	$m=5$	$m=8$	$m=10$
2	1	1	1	1
3	3	9	15	19
4	7	61	169	271
5	15	369	1,695	3,439
6	31	2,101	15,961	40,951
7	63	11,529	144,495	468,559
8	127	61,741	1,273,609	5,217,031
9	255	325,089	11,012,415	56,953,279

4 A Lower Bound on the Complexity of LebMeasure

To prove a lower bound on the worst-case complexity of LebMeasure, we give a recurrence *hcs* for the number of points processed for the two-point pathological example in Fig. 5.

$$\begin{array}{|cccc|}\hline 1 & 2 & \cdots & 2 \\ 2 & 1 & \cdots & 1 \\ \hline\end{array}$$

Fig. 5. A third pathological example for LebMeasure. This is the pattern from Fig. 3, with two points in n objectives

It is useful to introduce some terminology at this point. Given a point $\overline{x}$: the *descendants* of $\overline{x}$ are the contributing spawns of $\overline{x}$, and their contributing spawns, and so on; and the *relatives* of $\overline{x}$ are the points derived from the same original point as $\overline{x}$.

$$h(1,k) = 1 \tag{1}$$

$$h(n,k) = 1 + \sum_{i=0}^{k-1} h(n-1,i) \tag{2}$$

$$hcs(n) = h(n-1, n-1) + 1 \tag{3}$$

$h(n,k)$ returns the number of points processed for a point (or spawn) which has n 2s, of which we can reduce k and still generate spawns that aren't dominated by their relatives. The value of k in the call for a point $\overline{x}$ depends on the ancestry of $\overline{x}$, in particular on which 2s were reduced during the formation of $\overline{x}$: basically k is the number of contiguous 2s in the lower indices of $\overline{x}$'s objective values.

(1) says that if a point has only one 2, it can't generate any non-dominated spawns: they would be dominated by the second original point (but count one for itself). (2) says that if a point can reduce k 2s, it can generate k spawns by reducing any one of them (and count one for itself). Each spawn will have $n-1$ 2s, and each will be able to reduce the 2s with lower indices than the one just reduced.

For example, the point $(2,2,2,1,2,1)$ corresponds to $h(4,3)$, and it generates three spawns: $(1,2,2,1,2,1)$ corresponds to $h(3,0)$; $(2,1,2,1,2,1)$ corresponds to $h(3,1)$; and $(2,2,1,1,2,1)$ corresponds to $h(3,2)$.

$hcs(n)$ returns the number of points processed for the points from Fig. 5 in n objectives. (3) calls h for the first point, but obviously the second point generates no spawns.

It is easy to see by expansion that

$$hcs(n) = 2^{n-1}$$

We now give an inductive proof of this equality.

Lemma 1:
$$h(k,k) = \sum_{i=1}^{k} h(i, i-1) \tag{4}$$

Proof: The proof is by induction on k. The base case is $k=1$:

$$h(1,1) = h(1,0) = \sum_{i=1}^{1} h(i, i-1), \text{ by (1)}$$

Inductive case: Suppose the formula holds for $k = j$, i.e.

$$h(j,j) = \sum_{i=1}^{j} h(i, i-1) \qquad (5)$$

$$h(j+1, j+1) = 1 + \sum_{i=0}^{j} h(j, i), \text{ by (2)}$$

$$= 1 + h(j,j) + \sum_{i=0}^{j-1} h(j, i)$$

$$= 1 + \sum_{i=1}^{j} h(i, i-1) + \sum_{i=0}^{j-1} h(j, i), \text{ by (5)}$$

$$= h(j+1, j) + \sum_{i=1}^{j} h(i, i-1), \text{ by (2)}$$

$$= \sum_{i=1}^{j+1} h(i, i-1)$$

Hence the formula holds for $k = j+1$, and by induction, it holds for all k.

Lemma 2:

$$h(n, k) = 2^k, \text{ if } n > k \qquad (6)$$

Proof: The proof is by induction on n. The base case is $n = 1$:

$$h(1, 0) = 1 = 2^0, \text{ by (1)}$$

Inductive case: Suppose the formula holds for $n = j$, i.e.

$$h(j, k) = 2^k, \text{ if } j > k \qquad (7)$$

$$h(j+1, k) = 1 + \sum_{i=0}^{k-1} h(j, i), \text{ by (2)}$$

$$= 1 + \sum_{i=0}^{k-1} 2^i, \text{ by (7), given } j > k$$

$$= 1 + 2^k - 1, \text{ sum of a geometric series}$$

$$= 2^k$$

Hence the formula holds for $n = j+1$, and by induction, it holds for all n.

Theorem 1:

$$hcs(n) = 2^{n-1}$$

Proof:

$$hcs(n) = h(n-1, n-1) + 1, \text{ by (3)}$$
$$= \sum_{i=1}^{n-1} h(i, i-1) + 1, \text{ by (4)}$$
$$= \sum_{i=1}^{n-1} 2^{i-1} + 1, \text{ by (6)}$$
$$= \frac{1}{2} \sum_{i=1}^{n-1} 2^i + 1$$
$$= \frac{1}{2}(2^n - 2) + 1, \text{ sum of a geometric series}$$
$$= 2^{n-1}$$

Thus the two-point pathological example in Fig. 5 has complexity that is exponential in the number of objectives n, and we can say immediately that LebMeasure is exponential in the number of objectives in the worst case.

5 An Upper-Bound on the Complexity of LebMeasure

To develop an upper-bound on the general worst-case complexity, consider the "spawning tree" for three objectives, illustrated in Fig. 6. The root of the tree is an original point, and where a line connects two points, the lower point is a spawn of the higher point: thus the tree contains all of the descendants of the point at the root. The points are processed in depth-first order, left-to-right across the tree. The key feature of the tree is that if Objective i is reduced at any time during the formation of a point $\bar{x}$, then any potential spawn of $\bar{x}$ where Objective j, $j > i$, is reduced is guaranteed to be dominated by an unprocessed relative of $\bar{x}$. This limits the size of the tree to some extent.

We can derive a recurrence q that counts the points spawned in this tree.

$$q(0, y, z) = 1 \tag{8}$$

$$q(x, y, z) = 1 + q(x-1, y, z) + \sum_{i=1}^{y-1} q(z, i, z) \tag{9}$$

$$p(m, n) = \sum_{i=0}^{m-1} q(i, n, i) \tag{10}$$

$q(x, y, z)$ returns the number of descendants of a point, where

- Objective y is the highest-indexed objective that can be changed,
- there are up to x points smaller than the current one in Objective y, and
- there are up to z points smaller than the current one in Objectives $1 \ldots y-1$.

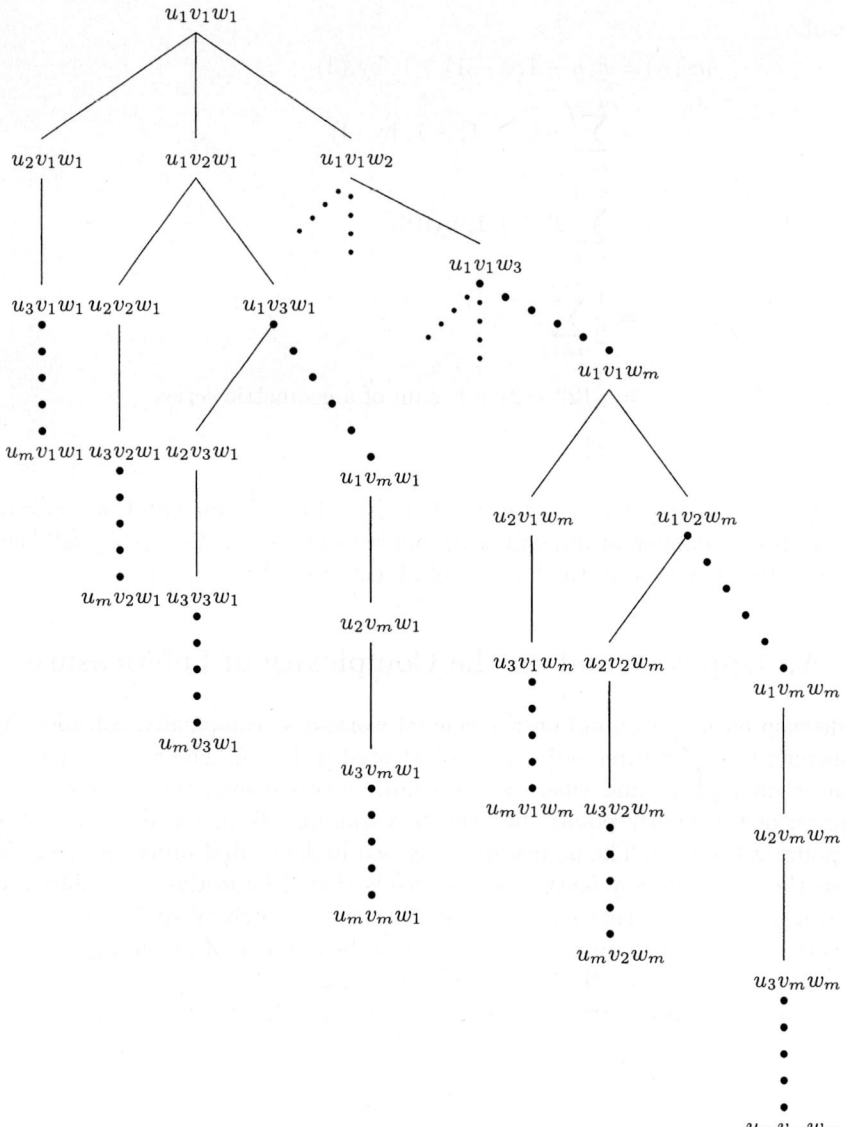

Fig. 6. The "spawning tree" for LebMeasure for m points in three objectives. The original point at the root of the tree generates all of the points in the tree, in the worst case. In each objective x, the values occur in the worsening order $x_1, x_2, \ldots, x_m$. The key feature is that when the ith objective has been reduced, no "higher" objective can subsequently be reduced, because the resulting spawn would certainly be dominated (explicitly: if v has been reduced, w can't be reduced; and if u has been reduced, neither v nor w can be reduced). The tree generalises naturally to other numbers of objectives

(8) says that if there are no points smaller in the current objective (i.e. the current objective equals the reference point), just count one for the point itself. (9) says that the current point can generate y spawns, either by:

- reducing Objective y, so the first argument is decremented, or
- reducing any one of Objectives $1 \ldots y-1$, so the first argument is restored,

and count one for the point itself.

$p(m, n)$ returns the worst-case number of points processed for m points in n objectives. (10) calls q separately for each original point: in each call, i is the number of original points yet to be processed.

Note that this recurrence models patterns of sets of the form shown in Fig. 7, which are illegal, because the points are not mutually-non-dominating. Hence the

m	m	$\cdots$	m
m-1	m-1	$\cdots$	m-1
$\vdots$	$\vdots$		$\vdots$
1	1	$\cdots$	1

Fig. 7. A (theoretical but illegal) pathological example for LebMeasure. This pattern describes sets of m points in n objectives, $n \geq 2$. All columns are identical

solution to p will be only an upper-bound on the general complexity. However, it is difficult to derive an exact general complexity, because it is difficult to know which sets of points will suffer the worst performance[2].

It is easy to see by expansion that

$$p(m, n) = \sum_{i=1}^{m} i^n$$

We now give an inductive proof of this equality.

Lemma 3:

$$q(x, y, z) = 1 + x(1 + \sum_{i=1}^{y-1} q(z, i, z)) \qquad (11)$$

Proof: The proof is by induction on x. The base case is $x = 0$:

$$q(0, y, z) = 1 = 1 + 0(1 + \sum_{i=1}^{y-1} q(z, i, z)), \text{ by (8)}$$

---

[2] It is also unnecessary to some extent: the lower-bound is the more significant result.

Inductive case: Suppose the formula holds for $x = j$, i.e.

$$q(j, y, z) = 1 + j(1 + \sum_{i=1}^{y-1} q(z, i, z)) \tag{12}$$

$$q(j+1, y, z) = 1 + q(j, y, z) + \sum_{i=1}^{y-1} q(z, i, z), \text{ by (9)}$$

$$= 1 + 1 + j(1 + \sum_{i=1}^{y-1} q(z, i, z)) + \sum_{i=1}^{y-1} q(z, i, z), \text{ by (12)}$$

$$= 1 + (j+1)(1 + \sum_{i=1}^{y-1} q(z, i, z))$$

Hence the formula holds for $x = j+1$, and by induction, it holds for all n.

Lemma 4:

$$q(x, n, x) = (x+1)^n \tag{13}$$

Proof: The proof is by strong induction on n. The base case is $n = 1$:

$$q(x, 1, x) = 1 + q(x-1, 1, x) + \sum_{i=1}^{0} q(x, i, x), \text{ by (9)}$$

$$= 1 + q(x-1, 1, x)$$

$$= 1 + 1 + (x-1)(1 + \sum_{i=1}^{0} q(x, i, x)), \text{ by (11)}$$

$$= (x+1)^1$$

Inductive case: Suppose the formula holds for all $1 \leq n \leq j$, i.e.

$$q(x, n, x) = (x+1)^n, \text{ if } n \leq j \tag{14}$$

$$q(x, j+1, x) = 1 + q(x-1, j+1, x) + \sum_{i=1}^{j} q(x, i, x), \text{ by (9)}$$

$$= 1 + 1 + (x-1)(1 + \sum_{i=1}^{j} q(x, i, x)) + \sum_{i=1}^{j} q(x, i, x), \text{ by (11)}$$

$$= x + 1 + x \sum_{i=1}^{j} q(x, i, x)$$

$$= x + 1 + x \sum_{i=1}^{j} (x+1)^i, \text{ by (14), given } i \leq j$$

$$= x + 1 + x\frac{(x+1)^{j+1} - (x+1)}{x}, \text{ sum of a geometric series}$$
$$= (x+1)^{j+1}$$

Hence the formula holds for $n = j + 1$, and by induction, it holds for all n.

Theorem 2:
$$p(m,n) = \sum_{i=1}^{m} i^n$$

Proof:
$$p(m,n) = \sum_{i=0}^{m-1} q(i,n,i), \text{ by (10)}$$
$$= \sum_{i=0}^{m-1} (i+1)^n, \text{ by (13)}$$
$$= \sum_{i=1}^{m} i^n$$

Thus the illegal example in Fig. 7 has complexity that is $O(m^n)$. We expect that this would be worse than any legal set of points.

6 Conclusions and Future Work

We have proved that, for a set of m points in n objectives, while LebMeasure is polynomial in m, it is exponential in n:

- we have given data that exhibits exponential slowdown as n increases;
- we have proved that the pattern of sets of points in Fig. 5 generates 2^{n-1} contributing points, giving a lower-bound on the worst-case complexity;
- we have proved that the pattern of illegal sets of points in Fig. 7 generates $O(m^n)$ contributing points, which is likely to be worse than any legal set.

Thus LebMeasure (as it stands) cannot provide a general solution to the performance problem (currently) inherent in calculating hypervolume.

As discussed in Section 3, the performance of LebMeasure can be improved (sometimes dramatically) by presenting the points to the algorithm in the best order (although the worst-case complexity is still exponential). Points with many poor objective values generate few non-dominated spawns, and should be processed first. We have developed some initial heuristics to exploit this.

Another promising line of research is in re-ordering the objective values in the points before calculating their hypervolume. This too can deliver significant performance improvement for some sets of points.

Finally, we speculate that no polynomial-time algorithm exists that can calculate hypervolume exactly, due to the irregularity of the shapes in the worst case. We would love to be able to prove this!

Acknowledgments

We would like to thank Phil Hingston, Mark Fleischer, and Simon Huband for helpful discussions on LebMeasure and its behaviour.

References

1. K. Deb, A. Pratap, S. Agarwal, and T. Meyarivan. A fast and elitist multi-objective genetic algorithm: NSGA-II. *IEEE TEC*, 6(2):182–97, 2002.
2. M. Fleischer. The measure of Pareto optima: Applications to multi-objective metaheuristics. Technical Report ISR TR 2002-32, University of Maryland, 2002.
3. M. Fleischer. The measure of Pareto optima: Applications to multi-objective metaheuristics. *EMO 2003*, vol 2632 of *LNCS*, pp 519–33. Springer-Verlag, 2003.
4. S. Huband, P. Hingston, L. While, and L. Barone. An evolution strategy with probabilistic mutation for multi-objective optimization. *CEC'03*, vol 4, pp 2284–91. IEEE, 2003.
5. J. Knowles and D. Corne. M-PAES: A memetic algorithm for multi-objective optimization. In *CEC'00*, vol 1, pp 325–32. IEEE, 2000.
6. J. Knowles, D. Corne, and M. Fleischer. Bounded archiving using the Lebesgue measure. *CEC'03*, vol 4, pp 2490–7. IEEE, 2003.
7. T. Okabe, Y. Jin, and B. Sendhoff. A critical survey of performance indices for multi-objective optimisation. *CEC'03*, vol 2, pp 878–85. IEEE, 2003.
8. R. Purshouse. *On the evolutionary optimisation of many objectives*. PhD thesis, The University of Sheffield, 2003.
9. J. Wu and S. Azarm. Metrics for quality assessment of a multi-objective design optimization solution set. *Journal of Mechanical Design*, 123:18–25, 2001.
10. E. Zitzler. *Evolutionary algorithms for multi-objective optimization: Methods and applications*. PhD thesis, Swiss Federal Inst of Technology (ETH) Zurich, 1999.
11. E. Zitzler, M. Laumanns, and L. Thiele. SPEA2: Improving the strength Pareto evolutionary algorithm for multi-objective optimization. *EUROGEN 2001*, pp 95–100. Int Center for Numerical Methods in Engineering, Barcelona, 2001.
12. E. Zitzler, L. Thiele, M. Laumanns, C. M. Fonseca, and V. Grunert da Fonseca. Performance assessment of multi-objective optimizers: An analysis and review. *IEEE TEC*, 7(2):117–32, 2003.

Effects of Removing Overlapping Solutions on the Performance of the NSGA-II Algorithm

Yusuke Nojima, Kaname Narukawa, Shiori Kaige, and Hisao Ishibuchi

Department of Industrial Engineering, Osaka Prefecture University,
1-1 Gakuen-cho, Sakai, Osaka, 599-8531, Japan
{nojima, kaname, shiori, hisaoi}@ie.osakafu-u.ac.jp

Abstract. The focus of this paper is the handling of overlapping solutions in evolutionary multiobjective optimization (EMO) algorithms. In the application of EMO algorithms to some multiobjective combinatorial optimization problems, there exit a large number of overlapping solutions in each generation. We examine the effect of removing overlapping solutions on the performance of EMO algorithms. In this paper, overlapping solutions are removed from the current population except for a single solution. We implement two removal strategies of overlapping solutions. One is the removal of overlapping solutions in the objective space. In this strategy, one solution is randomly chosen among the overlapping solutions with the same objective vector and left in the current population. The other overlapping solutions with the same objective vector are removed from the current population. As a result, each solution in the current population has a different location in the objective space. It should be noted that the overlapping solutions in the objective space are not necessary the same solution in the decision space. Thus we also examine the other strategy where the overlapping solutions in the decision space are removed from the current population except for a single solution. As a result, each solution in the current population has a different location in the decision space. The effect of removing overlapping solutions is examined through computational experiments where each removal strategy is combined into the NSGA-II algorithm.

1 Introduction

The design of evolutionary multiobjective optimization (EMO) algorithms has been discussed in the literature to find well-distributed Pareto-optimal or near Pareto-optimal solutions as many as possible (e.g., see Coello et al. [1] and Deb [2]). The handling of overlapping solutions, however, has not been discussed explicitly in many studies. This is mainly because the performance evaluation of EMO algorithms has been performed through computational experiments on multiobjective optimization problems with a large number of Pareto-optimal solutions. Since EMO algorithms usually have diversity-preserving mechanisms, many overlapping solutions are not likely to exist in each generation when they are applied to multiobjective optimization problems with continuous decision variables and/or many objective functions. On the other hand, the handling of overlapping solutions becomes an important issue in the application of EMO algorithms to multiobjective combinatorial optimization problems with only a few objective functions. In such an application, there may exist a large number of overlapping solutions in each generation as we will show in this paper through computational experiments on some test problems.

In this paper, we examine the effect of removing overlapping solutions on the performance of EMO algorithms. Overlapping solutions are removed from the current population. We examine two removal strategies of overlapping solutions. One removal strategy is performed in the objective space. Only a single solution among the overlapping solutions with the same objective vector is left in the current population. That is, overlapping solutions are removed so that each solution in the current population has a different location in the objective space. It should be noted that the overlapping solutions with the same objective vector are not necessary the same solution in the decision space. Thus we also examine the other strategy where the removal of overlapping solutions is performed in the decision space. Only a single solution among the overlapping solutions with the same decision vector is left in the current population. That is, overlapping solutions are removed so that each solution in the current population has a different location in the decision space. In this strategy, multiple solutions with the same objective vector can exist in the current population if they are not the same solution in the decision space.

The effect of removing overlapping solutions on the performance of EMO algorithms is examined through computational experiments on multiobjective 0/1 knapsack problems where each removal strategy is combined into the NSGA-II algorithm of Deb et al. [3]. First we show that there actually exist a large number of overlapping solutions in each generation in the application of the NSGA-II algorithm to two-objective 0/1 knapsack problems. Next we show that the removal of overlapping solutions improves the performance of the NSGA-II algorithm on those test problems especially in terms of the diversity of obtained non-dominated solutions. No clear differences are observed in the performance between the two removal strategies. Finally we show that these two removal strategies have a large effect on the performance of the NSGA-II algorithm when they are used together with a weighted sum-based tournament selection scheme of parent solutions with a large tournament size. This may be because a good balance is realized between the diversity-preserving effect of the removal strategies and the high selection pressure toward the Pareto front by the weighted sum-based parent selection scheme.

2 Handling of Overlapping Solutions

The original NSGA-II algorithm of Deb et al. [3] has no explicit mechanism to remove overlapping solutions while smaller fitness values are likely to be assigned to overlapping solutions than non-overlapping ones with the same non-dominated rank due to its diversity-preserving mechanism. In this section, we first briefly explain the NSGA-II algorithm. Then we explain two strategies for removing overlapping solutions, each of which is combined into the NSGA-II algorithm in computational experiments on multiobjective 0/1 knapsack problems in the next section.

2.1 The NSGA-II Algorithm

As in many evolutionary algorithms, first an initial population P_0 of size N is randomly generated in the NSGA-II algorithm [3]. That is, N initial solutions are randomly generated. It is possible in this initialization phase that some initial solutions are the same in the objective space. Those overlapping solutions can be the same solutions or different ones in the decision space. Next an offspring population Q_0 of size N is generated from the initial population P_0 by genetic operations (i.e., selection, recombination and mutation operations). It is possible in this genetic search phase that some of newly generated offspring solutions are overlapping in the objective space. It is also possible that some parent and offspring solutions are overlapping in the objective space. Then the initial population P_0 and its offspring population Q_0 are combined to construct a merged population R_0. The next population P_1 is constructed by choosing the best N solutions from the merged population R_0. The genetic operations for generating an offspring population and the generation update are iterated until a pre-specified stopping condition is satisfied (see Fig. 1).

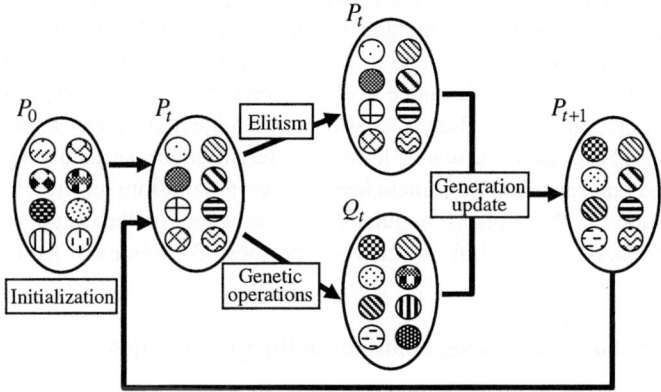

Fig. 1. Outline of the NSGA-II algorithm

In the generation update phase (also in the parent selection phase), each solution is evaluated based on a non-dominated sorting scheme and a crowding measure. Lower rank solutions are viewed as being better than higher rank solutions in the NSGA-II algorithm according to the Pareto dominance relation. Among solutions with the same rank, solutions in less crowded regions are viewed as being better than those in more crowded regions according to the crowding measure in the NSGA-II algorithm. Thus overlapping solutions have lower fitness among solutions with the same rank. Low rank solutions in the merged population, however, are likely to be included in the next population even if they are overlapping solutions. When the number of non-dominated solutions in the merged population is smaller than N, all non-dominated solutions (i.e., all solutions with rank 1) are always included in the next population independent of overlapping or non-overlapping. Only when the number of non-dominated solutions in the merged population is larger than N, the diversity-preserving mechanism based on the crowding measure removes overlapping non-dominated solutions in the generation update phase of the NSGA-II algorithm.

2.2 Removal of Overlapping Solutions in the Objective Space

This removal strategy does not permit more than one solution with the same objective vector in each generation. Thus each solution in the current population has a different location in the objective space. In the generation phase of an initial population, we should generate N different solutions in the objective space. That is, we iterate the random generation procedure of initial solutions until N different solutions in the objective space are generated.

We do not have to modify the generation mechanism of an offspring population. That is, overlapping solutions can be generated as offspring solutions as in the original NSGA-II algorithm. This is because at least N different solutions are always included in the merged population in the case of this removal strategy (it should be noted that the current population always includes N different solutions in the objective space). In the generation update phase, overlapping solutions with the same objective vector are removed from the merged population except for a single solution. The single solution left in the merged population is randomly chosen among the overlapping solutions with the same objective vector. As a result, each solution in the merged population has a different location in the objective space. It should be noted that the merged population has at least N different solutions in the objective space because the current population includes N different solutions in the objective space. Each solution in the merged population is evaluated in the same manner as the original NSGA-II algorithm to choose the best N solutions from the merged population.

2.3 Removal of Overlapping Solutions in the Decision Space

The removal strategy in Subsection 2.2 can be performed in the decision space. That is, the existence of overlapping solutions in the objective space is permitted if they are not the same solutions in the decision space. Only when overlapping solutions have the same decision vector, those solutions are removed except for a single solution. As a result, each solution in the current population has a different location in the decision space while some solutions may have the same location in the objective space.

3 Computational Experiments

3.1 Conditions of Computational Experiments

As test problems, we use multiobjective 0/1 knapsack problems with k objectives, k constraints (i.e., k knapsacks) and n items in Zitzler & Thiele [13]:

$$\text{Maximize } \mathbf{f}(\mathbf{x}) = (f_1(\mathbf{x}), f_2(\mathbf{x}), ..., f_k(\mathbf{x})), \tag{1}$$

$$\text{subject to } \sum_{j=1}^{n} w_{ij}x_j \leq c_i, \quad i=1,2,...,k, \tag{2}$$

where

$$f_i(\mathbf{x}) = \sum_{j=1}^{n} p_{ij}x_j, \quad i=1,2,...,k. \tag{3}$$

In this formulation, $\mathbf{x}$ is an n-dimensional binary vector (i.e., $(x_1, x_2, ..., x_n) \in \{0, 1\}^n$), p_{ij} is the profit of item j according to knapsack i, w_{ij} is the weight of item j according to knapsack i, and c_i is the capacity of knapsack i. Each solution $\mathbf{x}$ is handled as a binary string of length n in EMO algorithms. The k-objective n-item knapsack problem is referred to as a k-n knapsack problem in this paper. We examine 2-250, 2-500, 2-750, 3-500, and 4-500 knapsack problems. We employ the following parameter specifications:

Crossover probability (one-point crossover): 0.8,
Mutation probability (bit-flip mutation): $4/n$ (n: the number of items),
Population size: 150 (2-250), 200 (2-500), 250 (2-750), 250 (3-500), 300 (4-500),
Tournament size: 1, 2, 5, 10, 20,
Stopping condition: 500 generations.

The average performance is calculated for each test problem over 50 runs with different initial populations for each parameter specification (i.e., for each value of the tournament size).

Various performance measures have been proposed in the literature for evaluating a set of non-dominated solutions. As explained in Knowles & Corne [11], Okabe et al. [12], and Zitzler et al. [14], no performance measures can simultaneously evaluate various aspects of a solution set. Thus we visually show a solution set by a single run and the 50% attainment surface over 50 runs for each of the two-objective test problems. We also use four performance measures that are applicable to simultaneous comparison of many solution sets.

Let S be a solution set obtained by an EMO algorithm. The proximity of the solution set S to the Pareto front is evaluated by the generational distance (GD) defined as follows:

$$GD(S) = \frac{1}{|S|} \sum_{\mathbf{x} \in S} \min\{d_{\mathbf{xy}} \mid \mathbf{y} \in S^*\}, \tag{4}$$

where S^* is a reference solution set (i.e., the set of all Pareto-optimal solutions) and $d_{\mathbf{xy}}$ is the distance between a solution $\mathbf{x}$ and a reference solution $\mathbf{y}$ in the k-dimensional objective space:

$$d_{\mathbf{xy}} = \sqrt{(f_1(\mathbf{x}) - f_1(\mathbf{y}))^2 + \cdots + (f_k(\mathbf{x}) - f_k(\mathbf{y}))^2}. \tag{5}$$

For evaluating both the diversity of solutions in the solution set S and their convergence to the Pareto front, we calculate the $D1_R$ measure defined as follows:

$$D1_R(S) = \frac{1}{|S^*|} \sum_{y \in S^*} \min\{d_{xy} \mid x \in S\}. \tag{6}$$

It should be noted that $D1_R(S)$ is the average distance from each reference solution y in S^* to its nearest solution in S while $GD(S)$ in (4) is the average distance from each solution x in S to its nearest reference solution in S^*. The generational distance evaluates the proximity of the solution set S to the reference solution set S^*. On the other hand, the $D1_R$ measure evaluates how well the solution set S approximates the reference solution set S^*. Since all the Pareto-optimal solutions are known for the 2-250 and 2-500 test problems, we can use them as the reference solution set S^* for each test problem. Since true Pareto-optimal solutions are not available for the other test problems, we use as S^* a set of near Pareto-optimal solutions obtained for each test problem using much longer CPU time and much larger memory storage than computational experiments reported in this paper.

The spread measure is calculated as follows:

$$Spread = \sum_{i=1}^{k} [\max_{x \in S_i}\{f_i(x)\} - \min_{x \in S_i}\{f_i(x)\}]. \tag{7}$$

The hypervolume measure calculates the volume of the dominated region by the solution set S_i in the objective space. The ratio of non-dominated solutions is calculated for a solution set with respect to other solution sets. Let us assume that we have m solution sets $S_1, S_2, ..., S_m$. By merging these solution sets, we construct another solution set S as $S = S_1 \cup S_2 \cup ... \cup S_m$. Let S_{ND} be the set of non-dominated solutions in S. The ratio of non-dominated solutions is calculated for the solution set S_i as $|S_i \cap S_{ND}|/|S_i|$ where $|S_i|$ denotes the cardinality of S_i (i.e., the number of solutions in S_i).

3.2 Number of Overlapping Solutions in the Original NSGA-II Algorithm

The average number of overlapping solutions in each generation of the original NSGA-II algorithm is shown in Fig. 2 (a) for the two-objective test problems and Fig. 2 (b) for the 500-item test problems. We can see from Fig. 2 (a) that the average number of overlapping solutions monotonically increases during the multiobjective evolution by the NSGA-II algorithm for all the two-objective test problems except for the final stage of the evolution for the 2-250 test problem. The final population includes about 135 overlapping solutions on the average. On the other hand, we can see from Fig. 2 (b) that the number of overlapping solutions is not large in the case of the three-objective and four-objective test problems.

In the same manner as Fig. 2, we show the average number of different solutions in the objective space in Fig. 3. From Fig. 3 (a), we can see that the number of different solutions monotonically decreases during the multiobjective evolution by the NSGA-II algorithm except for the final stage of the evolution for the 2-250 test problem. On the other hand, we can see from Fig. 3 (b) that the number of different solutions does not decrease during the computational experiments for the 3-500 and 4-500 test problems. Experimental results in Fig. 2 and Fig. 3 suggest that the issue of overlapping solutions is very important for multiobjective combinatorial optimization problems with two objective functions. When the number of objective functions is large, many overlapping solutions do not exist in each generation.

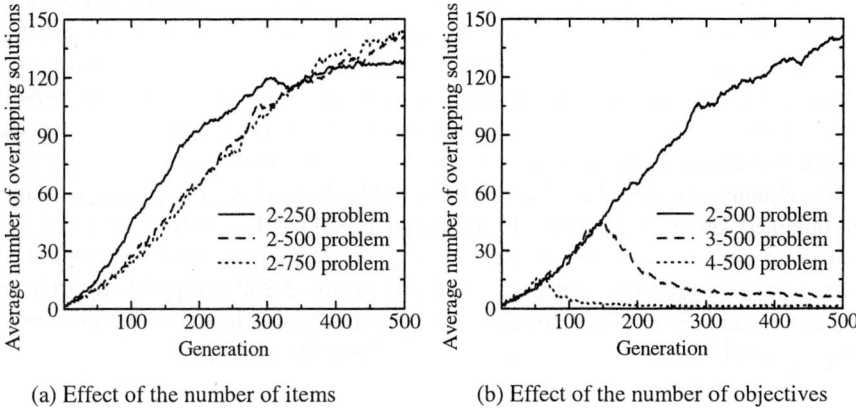

Fig. 2. Average number of overlapping solutions in each generation

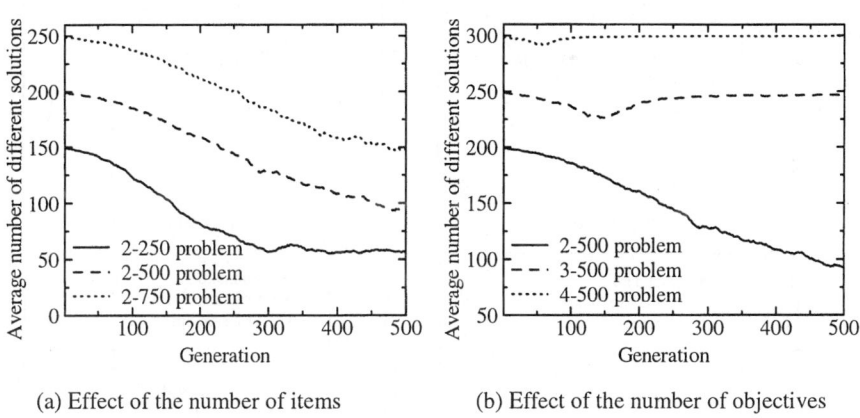

Fig. 3. Average number of different solutions in the objective space

3.3 Performance Evaluation of Modified NSGA-II Algorithms

Experimental results by the original NSGA-II algorithm and its two variants are summarized in Figs. 4-6. In these figures, the two removal strategies are labeled as

"Objective space" and "Decision space" depending on the space where the removal of overlapping solutions is performed. The original NSGA-II algorithm is labeled as "Standard". Each dashed line in these figures shows the average result by the original NSGA-II with the tournament size 2. From these figures, we can see that all of the four performance measures are improved by removing overlapping solutions when the tournament size is 2. On the other hand, the removal of overlapping solutions degrades some performance measures when the tournament size is 20. It should be noted in Fig. 4 that the $D1_R$ measure, the spread measure, and the hypervolume measure are improved by increasing the tournament size while the GD measure is degraded. That is, the increase in the tournament size improves the diversity of solutions and degrades the convergence to the Pareto front. As a result, the positive effect of removing overlapping solutions on the diversity of solutions is not clear when the tournament size is large.

We visually examine the validity of the above discussion by depicting a solution set obtained by a single run of the original NSGA-II algorithm and the 50% attainment surface over its 50 runs in Fig. 7 where two specifications of the tournament size are examined (i.e., 2 and 20) for the 2-250 test problem. From this figure, we can see that the increase in the tournament size increases the diversity of solutions and degrades the convergence to the Pareto front.

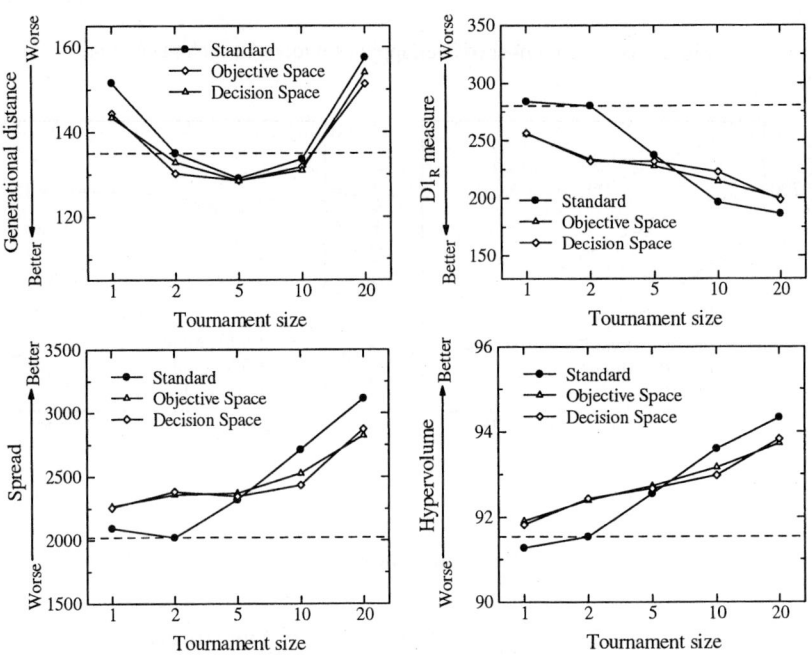

Fig. 4. Average results over 50 runs on 2-250 problem

Effects of Removing Overlapping Solutions 349

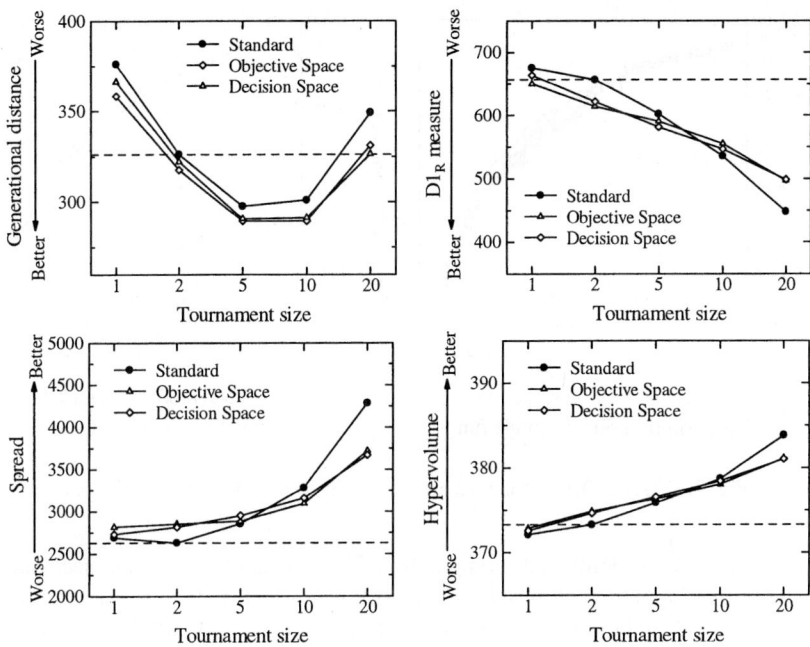

Fig. 5. Average results over 50 runs on 2-500 problem

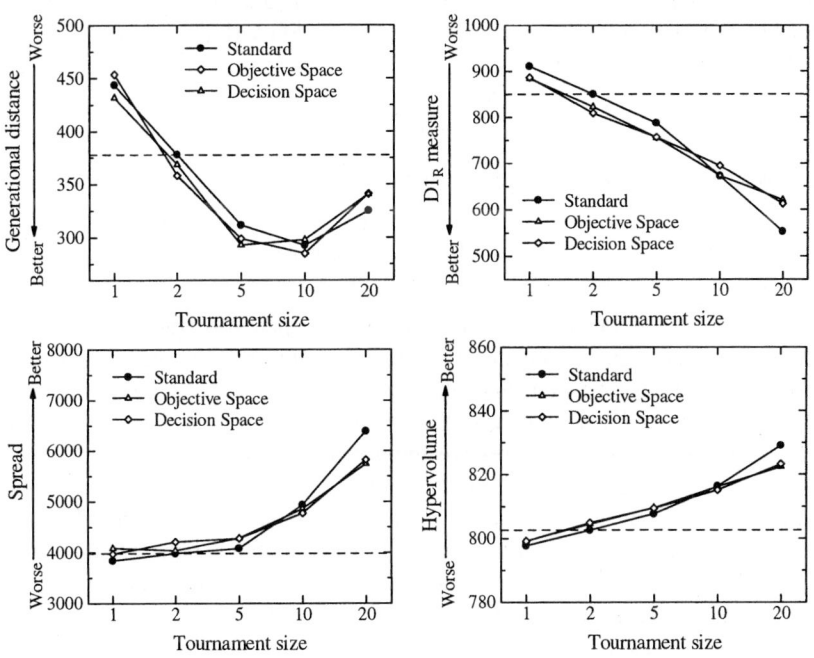

Fig. 6. Average results over 50 runs on 2-750 problem

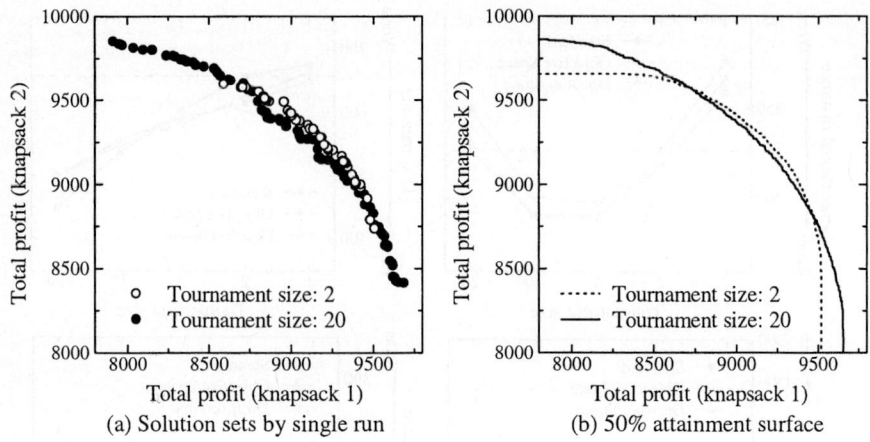

Fig. 7. Effect of the tournament size on obtained non-dominated solutions

From the above-mentioned computational experiments, we can see that the removal of overlapping solutions does not work well together with a large tournament size. This is because both the removal of overlapping solutions and the increase in the tournament size have the same effect on the multiobjective evolution by the NSGA-II algorithm (i.e., they both increase the diversity of solutions). In order to achieve a good balance between the diversity of solutions and the convergence to the Pareto front, we examine the use of the following weighted sum of multiple objectives as a scalar fitness function in the selection phase of the NSGA-II algorithm:

$$f(\mathbf{x}, \lambda) = \sum_{i=1}^{k} \lambda_i f_i(\mathbf{x}), \qquad (8)$$

where $\lambda = (\lambda_1, \lambda_2, ..., \lambda_k)$ is a weight vector:

$$\forall i \ \lambda_i \geq 0 \ \text{ and } \ \sum_{i=1}^{k} \lambda_i = 1. \qquad (9)$$

The weight vector is randomly updated whenever a pair of parent solutions is to be chosen. This scalar fitness function was used in multiobjective genetic local search algorithms (MOGLS) of Ishibuchi et al. [5, 8] and Jaszkiewicz [9, 10].

Experimental results are shown in Figs. 8-10. From these figures, we can see that all the four performance measures are improved by increasing the tournament size and removing overlapping solutions. It should be noted that the effect of removing overlapping solutions is much larger in Figs. 8-10 with the weighted sum-based fitness function than in Figs. 4-6 with the fitness function of the original NSGA-II algorithm. This may be because the lack of a diversity-preserving mechanism in the weighted sum-based fitness function is compensated by the removal of overlapping solutions. This discussion is supported by obtained non-dominated solutions in Fig. 11 with the removal strategy in the objective space.

Effects of Removing Overlapping Solutions   351

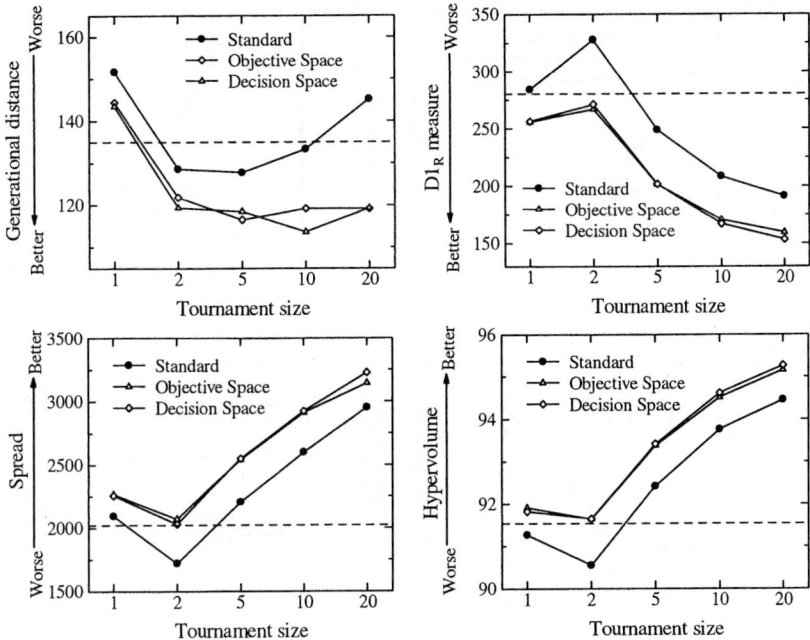

Fig. 8. Average results by the weighted sum-based fitness function for the 2-250 problem

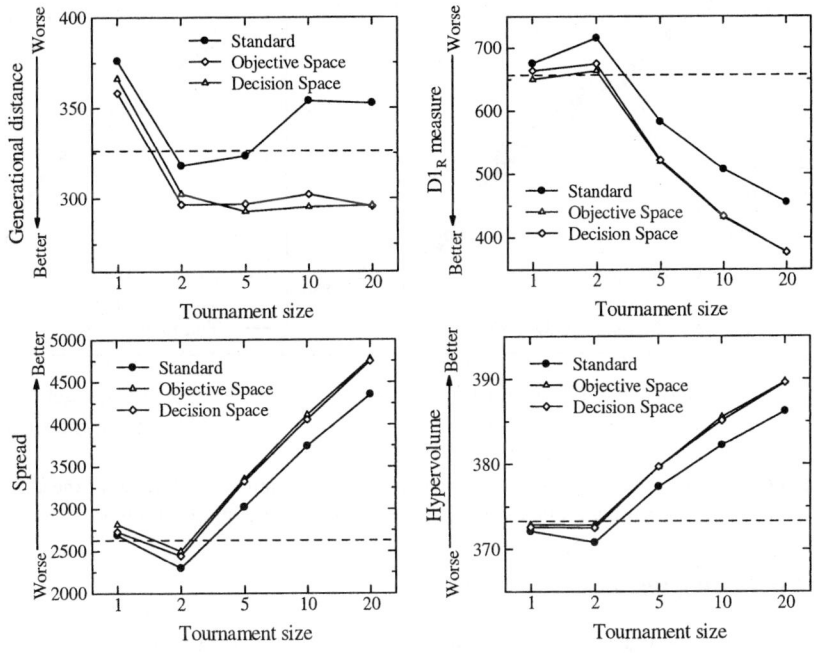

Fig. 9. Average results by the weighted sum-based fitness function for the 2-500 problem

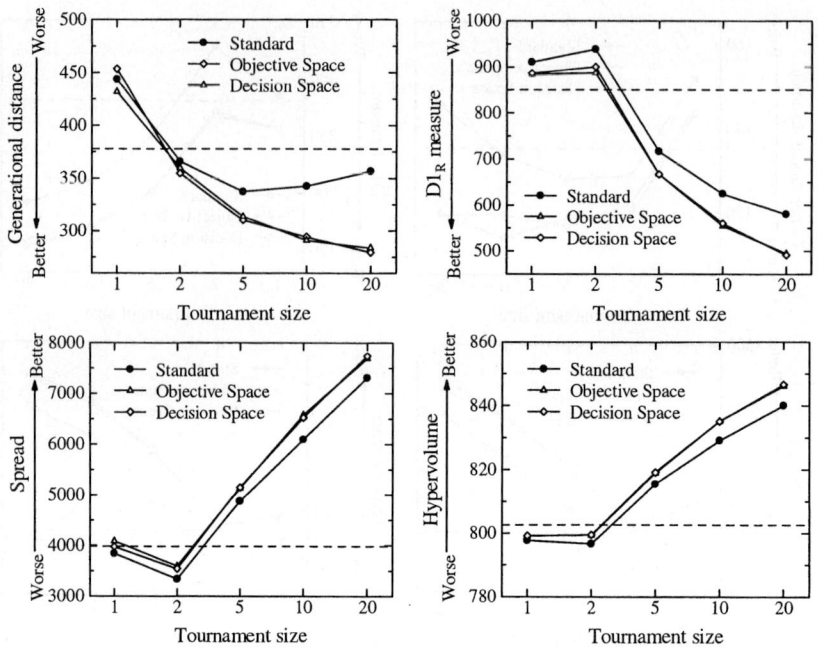

Fig. 10. Average results by the weighted sum-based fitness function for the 2-750 problem

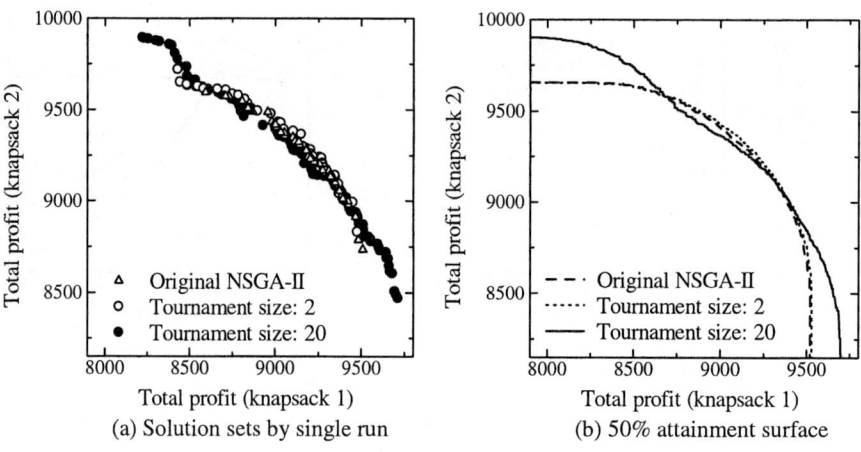

Fig. 11. Effect of the tournament size on obtained non-dominated solutions in the case of the weighted sum-based fitness function with the removal of overlapping solutions

4 Concluding Remarks

In this paper, we examined the handling of overlapping solutions in EMO algorithms through computational experiments on multiobjective 0/1 knapsack problems using the NSGA-II algorithm. First we showed that the number of overlapping solutions increased during the multiobjective evolution by the NSGA-II algorithm for two-objective knapsack problems. We also showed that the number of overlapping solutions was very small in the case of three-objective and four-objective knapsack problems. These results suggest that the issue of overlapping solutions is very important only for multiobjective combinatorial optimization problems with two objective functions. Next we showed that the performance of the NSGA-II algorithm with binary tournament selection was improved by removing overlapping solutions for two-objective knapsack problems. When we increased the tournament size, the diversity of non-dominated solutions obtained by the NSGA-II algorithm increased and the positive effect of removing overlapping solutions disappeared. This may be because the increase in the tournament size and the removal of overlapping solutions have the same effect on the multiobjective evolution by the NSGA-II algorithm. That is, they both increase the diversity of solutions. Finally we suggested the use of the weighted sum-based fitness function together with the removal of overlapping solutions. In this case, the increase in the tournament size improved the convergence to the Pareto front while the removal of overlapping solutions improved the diversity of solutions. As a result, a good balance between the convergence and the diversity was achieved by increasing the tournament size. Experimental results on two-objective knapsack problems showed that the performance of the NSGA-II algorithm was improved by the weighted sum-based tournament selection with a large tournament size and the removal of overlapping solutions.

In computational experiments in this paper, we only used multiobjective 0/1 knapsack problems as test problems. As future studies, we are planning to examine the effect of removing overlapping solutions through computational experiments on other multiobjective combinatorial optimization problems (e.g., multiobjective rule selection problems [7, 6] and multiobjective flowshop scheduling problems [5, 8]) as well as multiobjective function optimization problems (e.g., standard multiobjective test problems in Deb et al. [4]). Such computational experiments will clarify how the handling of overlapping solutions is important in the application of EMO algorithms to different types of multiobjective optimization problems. Computational experiments were performed using only the NSGA-II algorithm in this paper. We will also examine the effect of removing overlapping solutions on the performance of other EMO algorithms in future studies.

References

1. Coello Coello, C. A., Van Veldhuizen, D. A., and Lamont, G. B.: Evolutionary Algorithms for Solving Multi-Objective Problems, Kluwer Academic Publishers, Boston (2002).
2. Deb, K.: Multi-Objective Optimization Using Evolutionary Algorithms, John Wiley & Sons, Chichester (2001).
3. Deb, K., Pratap, A., Agarwal, S., and Meyarivan, T.: A Fast and Elitist Multiobjective Genetic Algorithm: NSGA-II. IEEE Trans. on Evolutionary Computation **6** (2002) 182-197.
4. Deb, K., Thiele, L., Laumanns, M., and Zitzler, E.: Scalable Multiobjective Optimization Test Problems. Proc. of 2002 Congress on Evolutionary Computation (2002) 825-830.

5. Ishibuchi, H. and Murata, T.: A Multi-Objective Genetic Local Search Algorithm and Its Application to Flowshop Scheduling. IEEE Trans. on Systems, Man, and Cybernetics - Part C: Applications and Reviews **28** (1998) 392-403.
6. Ishibuchi, H., and Yamamoto, T.: Effects of Three-Objective Genetic Rule Selection on the Generalization Ability of Fuzzy Rule-Based Systems. Lecture Notes in Computer Science **2632**: Evolutionary Multi-Criterion Optimization, Springer, Berlin (2003) 608-622.
7. Ishibuchi, H., and Yamamoto, T.: Fuzzy Rule Selection by Multi-Objective Genetic Local Search Algorithms and Rule Evaluation Measures in Data Mining. Fuzzy Sets and Systems 141 (2004) 59-88.
8. Ishibuchi, H., Yoshida, T., and Murata, T.: Balance between Genetic Search and Local Search in Memetic Algorithms for Multiobjective Permutation Flowshop Scheduling. IEEE Trans. on Evolutionary Computation **7** (2003) 204-223.
9. Jaszkiewicz, A.: Comparison of Local Search-based Metaheuristics on the Multiple Objective Knapsack Problem. Foundations of Computing and Decision Sciences **26** (2001) 99-120.
10. Jaszkiewicz, A.: On the Performance of Multiple-Objective Genetic Local Search on the 0/1 Knapsack Problem - A Comparative Experiment. IEEE Trans. on Evolutionary Computation **6** (2002b) 402-412.
11. Knowles, J. D. and Corne, D. W.: On Metrics for Comparing Non-dominated Sets. Proc. of 2002 Congress on Evolutionary Computation (2002) 711-716.
12. Okabe, T., Jin, Y., and Sendhoff, B.: A Critical Survey of Performance Indices for Multi-Objective Optimization. Proc. of 2003 Congress on Evolutionary Computation (2003) 878-885.
13. Zitzler, E. and Thiele, L.: Multiobjective Evolutionary Algorithms: A Comparative Case Study and the Strength Pareto Approach. IEEE Trans. on Evolutionary Computation 3 (1999) 257-271.
14. Zitzler, E., Thiele, L., Laumanns, M., Fonseca, C. M., and da Fonseca, V. G.: Performance Assessment of Multiobjective Optimizers: An Analysis and Review. IEEE Trans. on Evolutionary Computation **7** (2003) 117- 132.

Selection, Drift, Recombination, and Mutation in Multiobjective Evolutionary Algorithms on Scalable MNK-Landscapes

Hernán E. Aguirre and Kiyoshi Tanaka

Faculty of Engineering, Shinshu University,
4-17-1 Wakasato, Nagano, 380-8553 Japan
{ahernan, ktanaka}@gipwc.shinshu-u.ac.jp

Abstract. This work focuses on the working principles, behavior, and performance of state of the art multiobjective evolutionary algorithms (MOEAs) on discrete search spaces by using MNK-Landscapes. Its motivation comes from the performance shown by NSGA-II and SPEA2 on epistatic problems, which suggest that simpler population-based multiobjective random one-bit climbers are by far superior. Adaptive evolution is a search process driven by selection, drift, mutation, and recombination over fitness landscapes. We group MOEAs features and organize our study around these four important and intertwined processes in order to understand better their effects and clarify the reasons to the poor performance shown by NSGA-II and SPEA2. This work also constitutes a valuable guide for the practitioner on how to set up its algorithm and gives useful insights on how to design more robust and efficient MOEAs.

1 Introduction

Epistasis in the context of evolutionary algorithms (EAs) describes nonlinearities in fitness functions due to changes in the values of interacting bits. Epistasis is recognized as an important factor that makes a problem difficult for optimization algorithms and its influence on the performance of single objective EAs is being increasingly investigated. Particularly, Kauffman's NK-Landscapes model of epistatic interactions [1] has been the center of several studies, both for the statistical properties of the generated landscapes and for their *EA-hardness*. See for example [2, 3, 4, 5] and there in. Studies on the behavior of single objective EAs on NK-Landscapes have proved useful to advance our understanding of EA's working principles and served to design robust and better algorithms [5].

Contrary to single objective EAs, studies concerning epistasis within the context of multiobjective evolutionary algorithms (MOEAs) are few and its effects still not well understood. Recently, Aguirre and Tanaka [6] have extended Kauffman's NK-Landscapes model of epistatic interactions to multiobjective MNK-Landscapes, giving insights into their properties in order to understand how the parameters of the landscapes relate to multiobjective concepts such as shape of the fronts, number of non-dominated fronts, number of non-dominated solutions, accessibility to the true Pareto front, correlation between and within fronts, and metrics. From a multiobjective random test problem generator standpoint [7], desirable features of MNK-Landscapes are that the problems are easy to construct and can scale to any number of objectives M, number of bits N, and number of epistatic interactions K, allowing the creation of sub-classes of combinatorial non-linear problems for discrete search spaces in which we can test the working principles of MOEAs in order to design better and more robust algorithms. Aguirre and Tanaka have also studied the behavior of multiobjective random one-bit climbers (moRBCs) [8] on MNK-Landscapes and have provided initial results on the performance of two well known representatives of the latest generation of elitist MOEAs [9], namely NSGA-II [10] and SPEA2 [11].

This work focuses on the working principles, behavior, and performance of state of the art MOEAs on discrete search spaces by using MNK-Landscapes. Its motivation comes from the performance shown by NSGA-II and SPEA2 on epistatic problems [9, 8], which suggest that simpler population-based moRBCs are by far superior. Adaptive evolution is a search process driven by selection, drift, mutation, and recombination over fitness landscapes [1]. We group MOEAs features and organize our study around these main processes. In most of the latest generation MOEAs [10, 12] selection incorporates elitism and it is biased by Pareto dominance and a diversity preserving strategy in objective space. Genetic operators vary according to whether the search space is continuous or discrete. In discrete search spaces, like MNK-Landscapes, recombination is usually implemented as one-point or two-point crossover and mutation as the standard bit flipping method. Some approaches also include specialized mutation operators to perform local search. In addition to these features explicit to the algorithm design, drift is also an important process that drives evolution and it is implicit to all stochastic algorithms working on finite small populations, although sometimes highly overlooked. In this paper we study the effects of these important and intertwined processes in order to understand them better, clarifying the reasons to the poor performance shown by NSGA-II and SPEA2. This work also constitutes a valuable guide for the practitioner on how to set up its algorithm and gives useful insights on how to design more robust and efficient MOEAs.

2 Multiobjective MNK-Landscapes

A multiobjective MNK-Landscape is defined as a vector function mapping binary strings into real numbers $\boldsymbol{f}(\cdot) = (f_1(\cdot), f_2(\cdot), \cdots, f_M(\cdot)) : \mathcal{B}^N \rightarrow \Re^M$, where M is the number of objectives, $f_i(\cdot)$ is the i-th objective function, $\mathcal{B} = \{0, 1\}$, and N is the bit string length. $\boldsymbol{K} = \{K_1, \cdots, K_M\}$ is a set of integers where K_i ($i = 1, 2, \cdots, M$) is the number of bits in the string that epistatically interact with each bit in the i-th landscape. Each $f_i(\cdot)$ can be expressed as an average of N functions as follows

$$f_i(\boldsymbol{x}) = \frac{1}{N} \sum_{j=1}^{N} f_{i,j}(x_j, z_1^{(i,j)}, z_2^{(i,j)}, \cdots, z_{K_i}^{(i,j)}) \qquad (1)$$

where $f_{i,j} : \mathcal{B}^{K_i+1} \to \Re$ gives the fitness contribution of bit x_j to $f_i(\cdot)$, and $z_1^{(i,j)}, z_2^{(i,j)}, \cdots, z_{K_i}^{(i,j)}$ are the K_i bits interacting with bit x_j in the string $\boldsymbol{x}$. The fitness contribution $f_{i,j}$ of bit x_j is a number between [0.0, 1.0] drawn from a uniform distribution. Thus, each $f_i(\cdot)$ is a non-linear function of $\boldsymbol{x}$ expressed by a Kauffman's NK-Landscape model of epistatic interactions [1].

For a given N, we can tune the ruggedness of the fitness function $f_i(\cdot)$ of the i-th objective by varying K_i. In the limits, $K_i = 0$ corresponds to a model in which there are no epistatic interactions and the fitness contribution from each bit value is simply additive, which yields a single peaked smooth i-th fitness landscape. On the opposite extreme, $K_i = N - 1$ corresponds to a model in which each bit value is epistatically affected by all the remaining bit values yielding a maximally rugged fully random i-th fitness landscape. Varying K_i from 0 to $N - 1$ gives a family of increasingly rugged multi-peaked landscapes.

Besides defining N and K_i for each $f_i(\cdot)$, it is also possible to arrange the epistatic pattern between bit x_j and the K_i other interacting bits. That is, the distribution $D_i = \{random, nearest\ neighbor\}$ of K_i bits among N. Thus, M, N, $\boldsymbol{K} = \{K_1, K_2, \cdots, K_M\}$, and $\boldsymbol{D} = \{D_1, D_2, \cdots, D_M\}$, completely specify a multiobjective MNK-Landscape. By varying these parameters we can analyze the properties of the multiobjective landscapes and study the effects of the number of objectives, size of the search space, intensity of epistatic interactions, and epistatic pattern on the performance of multiobjective optimization algorithms on combinatorial discrete search spaces.

3 The Algorithms

In this work we present results by NSGA-II [10], SPEA2 [11], and moRBC(δ : 1+1) [8], a multiobjective random one-bit climber using a population for restarts. Since we refer extensively to NSGA-II and moRBC(δ : 1 + 1) is an important reference for comparison we include a brief description of both algorithms.

3.1 NSGA-II

NSGA-II keeps at the t-th generation a parent population $\mathcal{P}_t$ and an offspring population $\mathcal{Q}_t$, both of same size μ. The parent population $\mathcal{P}_{t+1}$ at the $t + 1$-th generation is a subset of the best individuals obtained by truncating the combined population of parents and offspring $\mathcal{R}_t = \mathcal{P}_t \cup \mathcal{Q}_t$. That is, $\mathcal{P}_{t+1} \subset \mathcal{R}_t$, where $|\mathcal{R}_t| = 2\mu$ and $|\mathcal{P}_{t+1}| = \mu$. To obtain $\mathcal{P}_{t+1}$, $\mathcal{R}_t$ is first classified into non-dominated fronts. The first front $\mathcal{F}_1$ contains the best non-dominated solutions $\mathcal{S}_1$. The subsequent fronts $\mathcal{F}_j, j > 1$, contain lower level non-dominated solutions and are obtained by disregarding solutions corresponding to the previous higher non-dominated fronts, i.e. $\mathcal{F}_j, j > 1$, is obtained from the set $\mathcal{R}_t - \bigcup_{k=1}^{j-1} \mathcal{S}_k$. Once the classification of non-dominated fronts is over, the parent population $\mathcal{P}_{t+1}$ is filled with solutions belonging to the higher fronts, starting with front $\mathcal{F}_1$. If the whole front $\mathcal{F}_i$ does not fit, the required number of individuals with best crowding distance are selected to fill the parent population. Each solution in $\mathcal{P}_t$

is assigned a rank (fitness) equal to its nondomination level (1 is the best level). Binary tournament selection with crowded tournament operator, recombination, and mutation operators are used to create the offspring population $\mathcal{Q}_{t+1}$ from $\mathcal{P}_{t+1}$. During selection, solution x wins a tournament if it has a better rank than y. If x and y have the same rank, the solution with best crowding distance wins.

3.2 moRBC($\delta : 1 + 1$)

moRBC($\delta : 1+1$) is a random one-bit climber that at all times keeps one parent individual from which it creates one offspring. It begins with a randomly created parent string of length N. Then, a random permutation π of the string positions is generated. A child is created by cloning the parent and flipping the bit at position π_i, the child is evaluated and replaces the parent if it *dominates* the parent. Child creation, evaluation, and (possibly) parent replacement are repeated for all π_i, $1 \leq i \leq N$. If no parent replacements were detected a *dominance local optimum* has been found and moRBC($\delta : 1 + 1$) RESTARTS the search. Testing continues by going back to create a new permutation π. This process ends once a given number of evaluations has been expended. A *Population* of up to δ solutions non-dominated by the parent and amongst themselves are kept during the process. moRBC($\delta : 1 + 1$) RESTARTS the search by replacing the parent with one individual chosen from the collected *Population*. If *Population* is empty, the parent is replaced with a random string created anew. Additionally, the non-dominated solutions found throughout the search are kept in an *Archive* of limited capacity. The procedures that update the *Population* and the *Archive* use NSGA-II's diversity preserving mechanism in objective space, where non-dominated individuals with better crowding distance [10](p.236) are preferred in case the *Population/Archive* has reached its capacity. Duplicate solutions are not allowed in the *Population* or in the *Archive*.

4 Metric, Test Problems, and Parameters

In this work we use the hypervolume metric $\mathcal{H}$ proposed by Zitzler [13] to evaluate and compare the performance of the algorithms. Let $\mathcal{A}$ be a set of non-dominated solutions. The metric $\mathcal{H}$ calculates the volume of the M-dimensional region in objective space enclosed by the elements of $\mathcal{A}$ and a dominated reference point, hence computing the size of the region $\mathcal{A}$ dominates. The hypervolume can be expressed as

$$\mathcal{H}(\mathcal{A}) = \cup_{i=1}^{|\mathcal{A}|}(\mathcal{V}_i - \cap_{j=1}^{i-1}\mathcal{V}_i\mathcal{V}_j) \qquad (2)$$

where $\mathcal{V}_i$ is the hypervolume rendered by the point $x_i \in \mathcal{A}$ and the reference point. The hypervolume is among the few recommended metrics for comparing non-dominated sets [14] and there is some theoretical evidence [15] that the maximization of the hypervolume constitutes the necessary and sufficient condition for the solutions in objective space to be maximally diverse Pareto optimal solutions of a discrete, multiobjective, optimization problem. The reference point to calculate the hypervolume is set to $[0.0, \cdots, 0.0]$.

In our study we use MNK-Landscapes with $M = \{2, 3, 5\}$ objectives, $N = \{20, 50, 100\}$ bits, vary the number of epistatic interactions from %0 to %50 of N simultaneously in all objectives $(K_1, \cdots, K_M = K)$, and set *random* epistatic patterns among bits for all objectives $(D_1, \cdots, D_M = random)$. For each combination of M, N and K, 50 different problems randomly generated are employed.

NSGA-II and SPEA2 use a population size of 100 individuals, two point crossover for recombination with probability $p_c = 0.6$, and bit flipping mutation with probability $p_m = 1/N$ per bit. moRBC($\delta : 1+1$) also uses a population $\delta = 100$ individuals. For all algorithms, the number of evaluations is set to 3×10^5 and the *Archive* size is set to 100.

5 Performance by Conventional NSGA-II and SPEA2

First, we present results by conventional NSGA-II and SPEA2 on scalable MNK-Landscapes for various values of M, N, and K, in order to have a broad view of the performance of these algorithms on combinatorial multiobjetive epistatic problems. **Fig. 1** plots the Archive's average hypervolume over the number of epistatic interactions K for $N = \{20, 50, 100\}$ bits landscapes. The average hypervolume of the true Pareto front obtained by enumeration is also included for $N = 20$ bits landscapes. Vertical bars overlaying the mean hypervolume curves represent 95% confidence intervals.

From **Fig. 1 (a)**, note that on $N = 20$ bits landscapes the trend of the hypervolume of the true Pareto front for any value of M is to rapidly increase with K, from $K = 0$ to small values of K, and to remain high for medium and large K. A similar trend is expected for the hypervolume of the true Pareto front on landscapes with higher values of N. Looking at results by NSGA-II and SPEA2, in **Fig. 1 (a)** we can see that the hypervolume of the solutions found by these algorithms approach the hypervolume of true Pareto front on $N = 20$ bits landscapes only for $K \leq 15\%N$. Increasing the number of bits N, we see that the value of the hypervolume of the non-dominated solutions found by the algorithms decreases continuously from $K \geq 8\%N$ for $N = 50$ and from $K \geq 5\%N$ for $N = 100$ bits. See **Fig. 1 (b)** and **Fig. 1 (c)**, respectively. These decreasing values are against the expected trend of the hypervolume of the true Pareto front and indicate that the search performance of the algorithms is worsening significantly as K increases.

In the following we focus on NSGA-II and especially look into the effects of selection, drift, recombination, and mutation.

6 Selection and Drift

The main processes that drive evolution are selection, drift, mutation, and recombination. In this section we observe the effects of selection and drift, which decrease genetic variation being the homogenization of the population an extreme consequence of it. Selection features are made explicit during the design

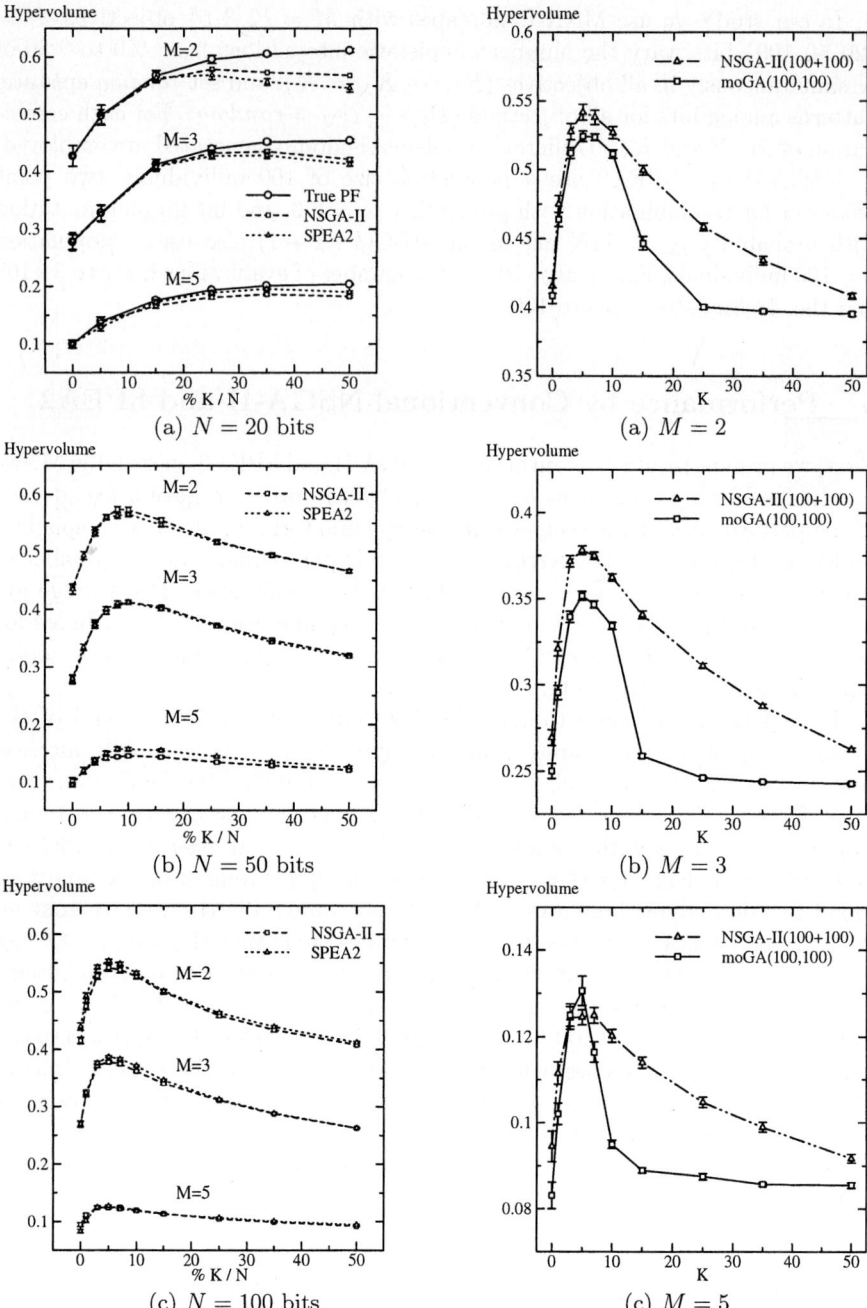

Fig. 1. Average hypervolume $\bar{\mathcal{H}}$ over number of epistatic interactions K by conventional NSGA-II and SPEA2

Fig. 2. Effect of elitism. $\bar{\mathcal{H}}$ over K by conventional NSGA-II($\mu + \lambda$) using elitism and moGA(μ, λ) without elitism. N=100 bits landscapes

of the algorithms. In most of the latest generation MOEAs selection incorporates elitism and it is biased by Pareto dominance and a diversity preserving strategy in objective space. On the other hand, drift is a process implicit to all stochastic algorithms working on finite small populations.

6.1 Elitism

Elitism is considered an important component of the selection process in state of the art evolutionary multiobjective optimizers. In order to have a clear idea of its contribution to the performance of MOEAs on epistatic problems this section compare results by NSGA-II($\mu+\lambda$) and moGA(μ,λ). NSGA-II implements elitism by keeping for the next generation the best (μ) individuals from the joined population ($\mu+\lambda$) of parents and offspring. On the other hand, moGA(μ,λ) replaces the parent population (μ) by its offspring population (λ) at each generation. moGA(μ,λ)'s other features are the same used by NSGA-II.

Fig. 2 shows results by NSGA-II(100+100) and moGA(100,100) for $M = \{2,3,5\}$ objectives on $N = 100$ bits landscapes. From this figure we can see that if elitism is not included there is a severe deterioration in performance for all values of K and M, except for $M = 5$ and $K = 5$. Note that the performance by moGA(100,100) falls sharply from $K = 10\%N$ to $K = 15\%N$ in $M = 2$ objectives. The fall in performance is even more pronounced for $M = 3$ and $M = 5$ objectives. These results are in accordance with the expectation that elitism is a very important feature for multiobjective combinatorial optimization. However, elitism can also bring about undesired side effects that could severely affect the efficacy and efficiency of the algorithms. Throughout the following sections we discuss some of them.

6.2 Genetic Drift

Genetic drift is a phenomenon that emerges from the stochastic operators of selection, recombination, and selection. It refers to the change on bit (allele) frequencies due to chance alone especially in small populations. In single objective EAs it is well known that genetic drift is one important factor that affects negatively the performance of EAs especially if a strong selection pressure is used, such as truncated selection (μ, λ) without elitism where $\mu < \lambda$. See [5], for example. The presence of elitism, for instance in the form of truncated selection ($\mu + \lambda$) used by NSGA-II, would increase selection pressure making elitist algorithms even more prone to the effects of drift.

In this section we enhance NSGA-II by preventing fitness duplicates from the population in order to observe the effect of genetic drift on the performance of the algorithm. In the enhanced algorithm, called NSGA-II^{ed}, if several individuals have exactly the same fitness in all objectives then one is chosen at random and kept. The other equal fitness individuals are eliminated from the population. Fitness duplicates elimination is carried out before truncating the population from ($\mu + \lambda$) to (μ) individuals. This process aims to effectively eliminate clones without the need to compare hamming distances, postpone genetic drift, and remove an unwanted source of selective bias.

To explain and understand better the effects of duplicates on the performance of multiobjective algorithms, **Fig. 3** shows cumulative bar diagrams of the average number of individuals per non-dominated front over the number of epistatic interactions K. Results by the conventional NSGA-II and by the enhanced NSGA-II^{ed} that eliminates duplicates are presented for the top five fronts in $M = \{2, 3, 5\}$ objectives. In the case of conventional NSGA-II, the figure also shows with lines the number of duplicate individuals δ in the whole population $(\mu + \lambda)$ before truncation, in the first non-dominated front (F_1) also before truncation, and in the truncated population (μ). For NSGA-II^{ed} results are presented after elimination of duplicates. Horizontal lines indicate the truncation site.

Looking at results by NSGA-II in **Fig. 3**, the following observations are relevant. (i) The number of duplicates increases as we increase the epistatic interactions K. (ii) The presence of duplicates reduces increasing the number of objectives M. (iii) Most duplicates belong to the first non-dominated front and a large number remain after truncation, especially for large K. Conversely, looking at the size of the cumulative bars by NSGA-II^{ed} in **Fig. 3**, we can deduce that the average number of duplicates eliminated at each generation by NSGA-II^{ed} is only a small fraction of the whole $(\mu + \lambda)$ population and it is similar for all K and M. For example, for $M = 2$ note that in NSGA-II the number of duplicates augment from 8% to 90% of the truncated population (μ) and from 18% to 60% of the whole population $(\mu + \lambda)$ increasing K from 0 to 50, respectively. By contrast, the average number of duplicates in NSGA-II^{ed} is around 9% for all K. The number of duplicates observed in NSGA-II^{ed} could be taken as the homogenization effect of drift and selection at each generation, whereas the number of duplicates in NSGA-II should be taken as the amplified effect of drift and selection throughout the generations.

Duplicates hinder exploration and selection as well. If duplicates are not eliminated at each generation they accumulate rapidly decreasing the likelihood that the algorithm will explore a larger number of different candidate solutions during a run. Also, since the chances of selecting a given genotype are multiplied by the number of clones of that genotype present in the population, duplicate genotypes end up with higher selective advantage than unique genotypes. This unwanted selective bias is not based in actual fitness and cannot be avoided by ranking procedures, scaling mechanism, or even truncated deterministic mechanisms. A reduced explorative capability combined with an unwanted selective bias can considerably affect the possibility of finding better non-dominated solutions.

Fig. 4 shows the hypervolume by NSGA-II^{ed}(100+100) that eliminates duplicates and by the conventional NSGA-II(100+100) to illustrate the effect of duplicates on the performance of the algorithms. From this figure, we can see that elimination of duplicates improves the performance of NSGA-II in two and five objective landscapes for $K_i \geq 5$ and $3 \leq K_i \leq 35$, respectively. In three objectives landscapes we see almost no improvement by eliminating fitness duplicates. Note that the largest overall performance difference between NSGA-II^{ed} and NSGA-II is for $M = 2$ objective landscapes, where precisely the accumulation of duplicates in the conventional NSGA-II is the highest as shown in **Fig. 3**.

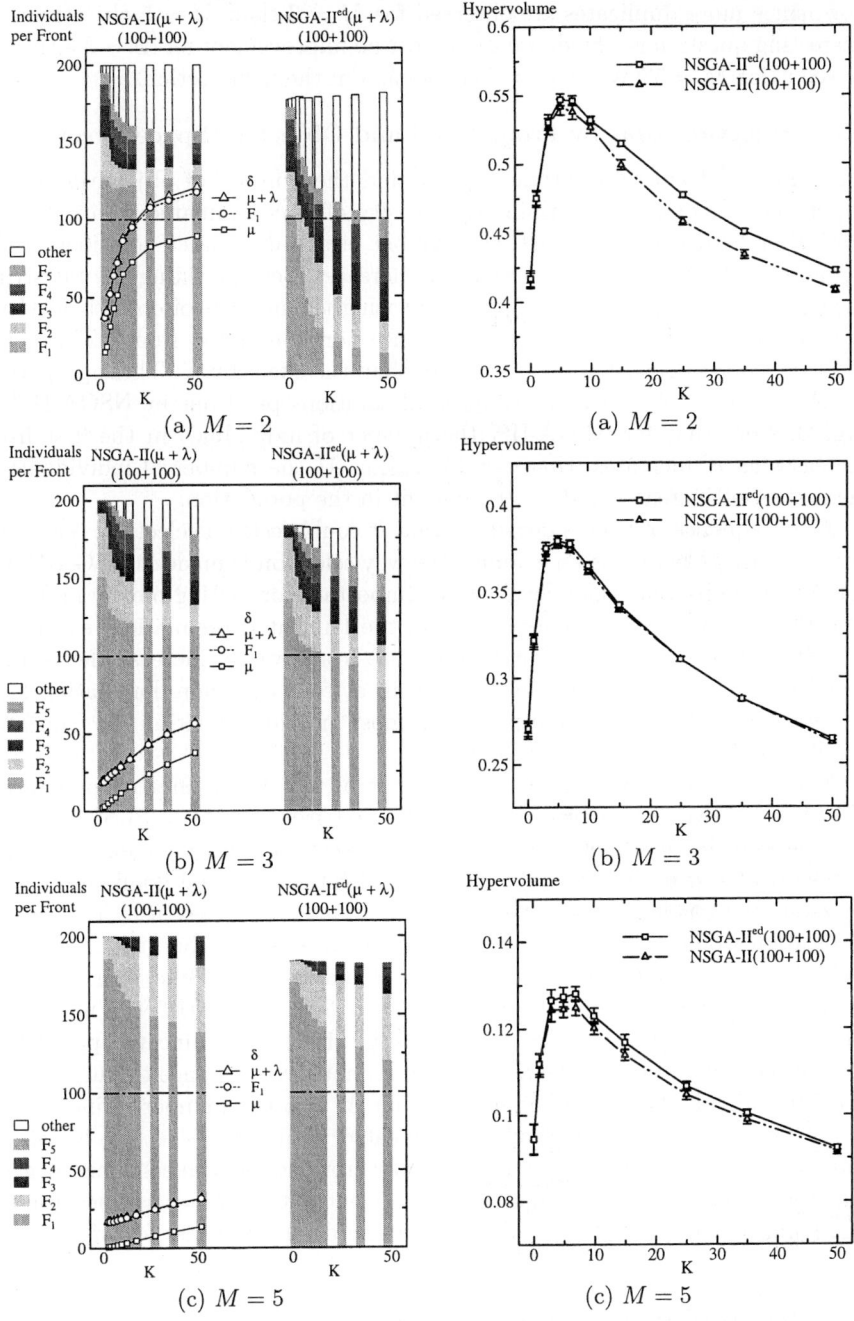

Fig. 3. Average number of individuals per front and average number of duplicates by conventional NSGA-II and enhanced NSGA-II^{ed}. $N = 100$ bits

Fig. 4. Effect of drift on performance. $\bar{\mathcal{H}}$ over K by conventional NSGA-II and NSGA-II^{ed} that eliminates fitness duplicates. $N = 100$ bits

Now, since more duplicates are observed for $M=3$ than $M=5$ objectives, an interesting question is why do we see almost no improvement for $M=3$ although we see it for $M=5$? We answer this question in the next section.

6.3 Selection Bias by Front Level and Objective Space Diversity

The number of epistatic interactions K and the number of objectives M are important factors that determine the density of non-dominated fronts in multiobjective landscapes [6]. First, the number of non-dominated solutions in the top non-dominated fronts reduces as K increases. Second, fixing the bit string length (size of the search space) and increasing the number of objectives M we have landscapes with fewer but more dense non-dominated fronts. Third, the effect on front's density by M is stronger than the effect by K. These properties are clearly reflected in the distribution of solutions per front by NSGA-II^{ed} in **Fig. 3**. Note that for NSGA-II^{ed} the number of individuals in the first front decreases as K increases. However, increasing M the number of individuals in the first front increases and fewer fronts fit in the population.

A consequence of front's density is that it could restrain selection, especially during mating. Taking as an example the way selection is made in NSGA-II, see **3.1**, Pareto non-domination level will be important for mating mostly in $M=2$ objectives where several fronts fit within the truncated population, except for $K=0$. Increasing M, fewer but more dense fronts would increase the relative importance of crowding of solutions within a front over non-domination level as criterion to bias selection, especially if most individuals within the truncated population belong to the same front.

From the same **Fig. 3**, we can see that in fact the truncated population of NSGA-II^{ed} would mostly come from the first front for $M=\{3,5\}$. Thus, in both cases during mating the criterion to bias selection would be mainly crowding factor since most solutions would be ranked with the same non-domination level. However, there is a difference between $M=3$ and $M=5$ given by truncation. Truncation would reinforce diversity in objective space by purging individuals with high crowded factor if the number of individuals in the front is larger than the population size. Note that for $M=3$ in NSGA-II^{ed} the number of individuals in the first front F_1 is close to the size of the truncated population (μ), for most K, and thus the algorithm does not have a chance to purge highly crowded individuals. On the other hand, for $M=5$ the number of individuals in the first front is greater than the truncated population size (μ), for all K, and truncation can contribute purging highly crowded individuals. This suggests that the increase in performance for $M=5$ but not for $M=3$ is because the algorithm by means of truncation is preserving diversity better for $M=5$.

7 Recombination and Mutation

EAs to function effectively must balance the processes of evolution that decrease genetic variation with those that increase it. In previous sections we have restricted our discussion to selection and drift, which decrease genetic variation.

In the following we focus on the effectiveness of recombination and mutation, mechanisms that increase genetic variation. We also discuss issues that hinder exploration under elitist selection and try ways to make mutation more effective.

7.1 Recombination

In this section we observe the effect on performance of (not) using recombination. **Fig. 5** shows results by NSGA-II^{ed}(100+100) and M^{ed}(100+100). M^{ed} is an NSGA-II^{ed} algorithm with recombination turned off using mutation as the sole variation operator, i.e. $p_c = 0.0$ and $p_m = 1/N$ per bit. From **Fig. 5** note that NSGA-II^{ed} that includes recombination and mutation performs better than M^{ed} that uses only mutation for $K \leq 1$, $K \leq 3$, and $K \leq 7$ for $M = \{2, 3, 5\}$ objectives, respectively. For other values of K we do not see any contribution to performance by including recombination. In fact, we can see that mutation alone performs better for some values of K, especially in landscapes with $M = 2$.

Results by recombination are in accordance with the effects of epistasis on multiobjective landscapes. In [6] it is shown that for small values of K non-dominated solutions of top fronts are highly correlated in decision space (genotype), in objective space (phenotype), and between spaces. However, this correlation decreases rapidly by increasing K, being its fall faster for smaller M. For small K, recombination of high fitness individuals would likely produce high fitness offspring. However, as the number of epistatic interactions K increases the likelihood that offspring would be far from the parents in objective space also increases considerably. The properties of MNK-Landscapes offer no much hope for blind mating and recombination, i.e. just taking any two individuals from the best non-dominated front in the population and recombining them. In the literature there are some reports suggesting that mating based on proximity in decision or objective space could help recombination in MOEAs. It will be interesting to asses in the future the benefit of these approaches on scalable espistatic landscapes. How helpful are them as we increase K?

7.2 Elite's Age and Mutation Explorative Range

An undesired side effect of elitism combined with a short explorative range by recombination and/or mutation is cyclically exploring same points. To explain this it is useful to see the probability of recombination p_c governing the application of two operators. One is recombination followed by mutation (p_c) and the other one is mutation alone ($1 - p_c$). In this section we focus on elitism and mutation and do not consider the case of recombination followed by mutation.

Conventional NSGA-II, for example, uses a $(\mu + \lambda)$ selection where elite solutions could remain in the population indefinitely. Additionally, mutation rate is often set to $p_m = 1/N$, which means that mutation will explore solutions in average one bit away from the parent in decision space. In this case, eventually after some generations offspring created from elite solutions would likely not be different from offspring created before, even in the case of perfect sampling (no drift). The expected time for mutation to start sampling again same points from

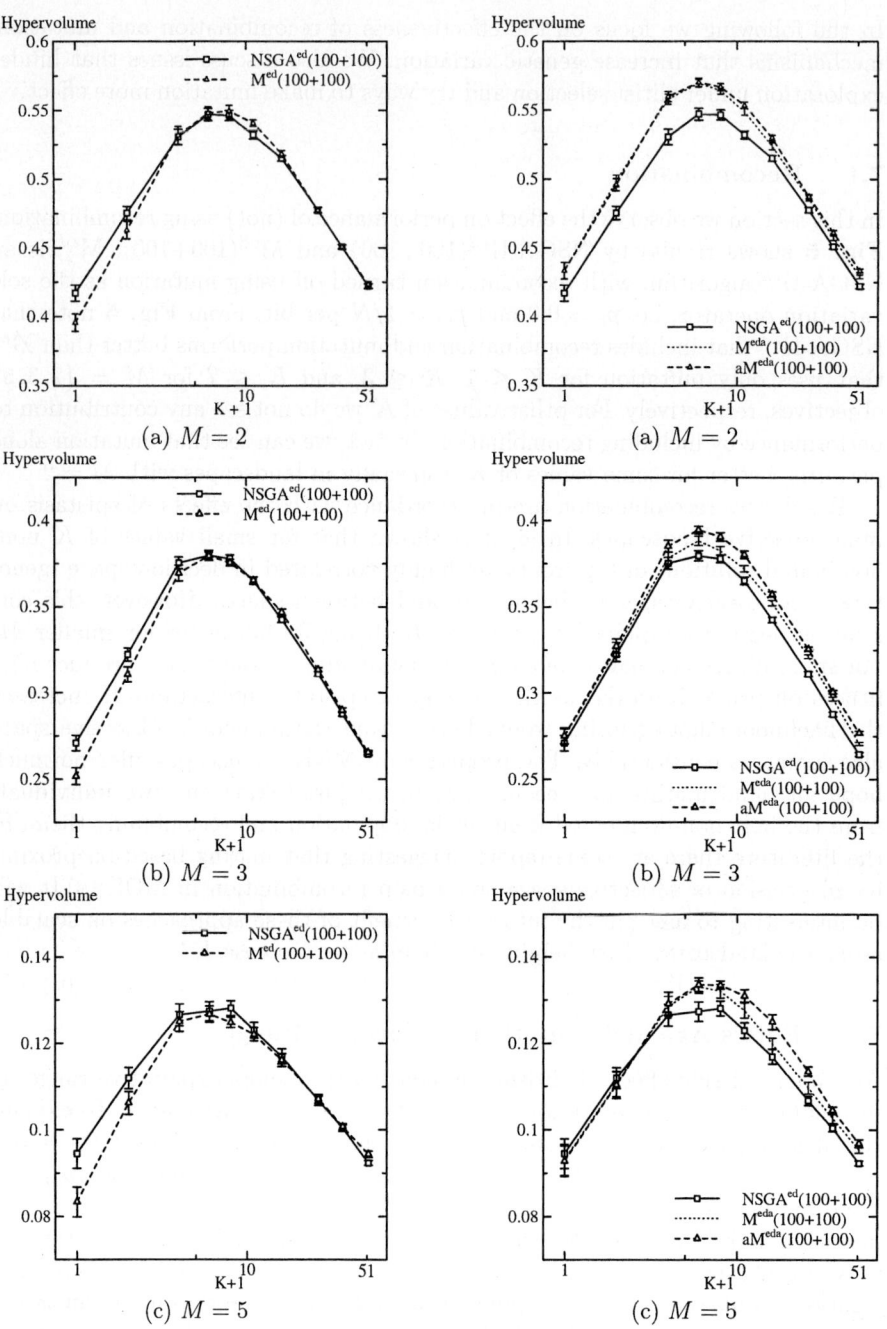

Fig. 5. Effect of recombination on performance. $\bar{\mathcal{H}}$ over K by NSGA-II^{ed} using crossover followed by mutation and M^{ed} using mutation alone. $N = 100$ bits landscapes

Fig. 6. Elite's age and mutation explorative range. $\bar{\mathcal{H}}$ over K by NSGA-II^{ed}, M^{eda} that eliminate old elite individuals, and aM^{eda} that also uses age to guide mutation. $N = 100$ bits landscapes

an elite individual would be a function of the number of occasions the same elite individual has been selected for reproduction, the rate at which mutation is applied alone given by the probability $1 - p_c$, and the bit string length N.

A way to avoid this undesired cycles and enhance exploration is to put an age limit to elite solutions and bias selection accordingly. To observe the effects of age of elite solutions we create M^{eda} from NSGA-II^{ed}. M^{eda} increases by one the age of an elite solution each time it is selected for reproduction. Age is also incremented by one at each generation. Before truncation, fitness duplicates and individuals with age greater than N are eliminated from the population. It uses mutation with probability $p_m = 1/N$ per bit but does not use crossover ($p_c = 0.0$) for simplicity. The rationale for this is that selecting N times the elite individual for reproduction will suffice to sample a good number of solutions that lay within the average explorative range set by the mutation rate, i.e. one-bit neighbors. Of course, mutation with $p_m = 1/N$ will sometimes flip more than one bit, none at all, or sample the same bit more than once during the N trials.

We also verify whether a local search-like strategy would be more effective than the conventional bit flipping mutation strategy. To do that we create aM^{eda} from M^{eda}. aM^{eda}, in addition to eliminating duplicates and very old elite individuals, it also uses the age to guide mutation. The bit string of length N is subdivided in S segments of length L, $N = S \times L$. For elite individuals, age greater than one, mutation flips one bit at the position indicated by $j + i$, $j = S \times [rand() \bmod S]$ and $i = [age \bmod N] \bmod L$, i.e. the mutation segment is chosen at random and the bit within the segments is given by the age of the individual. This kind of mutation makes sure that only one bit will be flipped and increases the chances of exploring most one-bit neighbors of an elite individual as its age approaches N. In our experiments $N = 100$ and $S = L = 10$. For individuals whose age is one standard flipping mutation is applied ($p_c = 0.0$).

Fig. 6 shows results by the mutation-only algorithms M^{eda}(100+100) and aM^{eda}(100+100) together with results by NSGA-II^{ed}. Looking at results by M^{eda} we can see that preventing old elite individuals increases substantially the performance of NSGA-II^{ed} for all number of objectives M and most values of epistatic interactions K. Note that in this case there is no more a performance advantage offered by recombination in small K landscapes, except for $M = 5$ and $K \leq 1$. Looking at results by aM^{eda}(100+100) also note that eliminating old elite individuals combined with local search-like mutation strategy informed by age further improves performance, especially for medium and high K.

8 Comparison with moRBC

Finally, we compare the performance of conventional NSGA-II, the enhanced algorithm aM^{aed}, and the population-based multiobjective random one-bit climber moRBC($\delta : 1+1$). Results are shown in **Fig. 7**. From this figure we can see that the performance of conventional NSGA-II is worse by several standard deviations than the performance of moRBC($\delta : 1+1$), for all values of K and M. In contrast, note that the performance of aM^{eda} approaches the performance

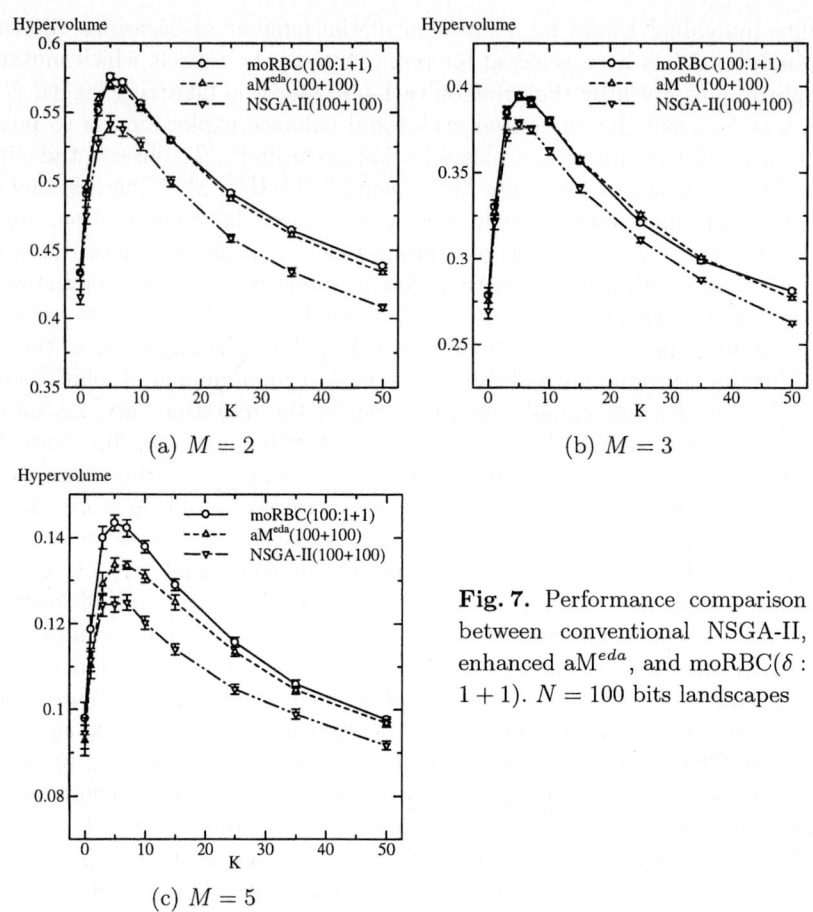

Fig. 7. Performance comparison between conventional NSGA-II, enhanced aM^{eda}, and moRBC($\delta : 1+1$). $N = 100$ bits landscapes

of moRBC($\delta : 1 + 1$) in $M = 2$ and $M = 3$ objectives. However, for $M = 5$ objectives moRBC($\delta : 1 + 1$) still performs better.

9 Conclusions

In this work we have studied the effects of selection, drift, recombination, and mutation in MOEAs on discrete search spaces by using MNK-Landscapes. We have shown that enhancing selection and postponing drift by eliminating fitness duplicates and removing old elite individuals help to increase substantially the performance of MOEAs. We also observed that recombination adds to the performance of standard bit flipping mutation only for small values of epistatic interactions. However, any gain by recombination is largely surpassed by doing a more effective exploration with short-ranged mutation alone. Enhancements in selection, postponing drift, and explorative efficiency have considerably increased the robustness of MOEAs across several classes of epistatic problems and number of objectives. Yet, these enhancements are not enough to surpass the performance of simpler population-based multiobjective random one-bit climbers and we should look for ways to design better MOEAs. Results in this work strongly suggest that elitism combined with an efficient short-range explorative capability by mutation is highly effective and likely to be a required feature of MOEAs. The advantages, if any, of elitism combined with mutation using larger explorative ranges should be investigated in the future. In addition, it would be interesting to look into special mating strategies for recombination to further clarify its role in multiobjective discrete search spaces.

References

1. S. A. Kauffman, *The Origins of Order: Self-Organization and Selection in Evolution*, Oxford University Press, 1993.
2. R. Heckendorn, S. Rana, and D. Whitley, "Test Function Generators as Embedded Landscapes", *Foundations of Genetic Algorithms 5*, pp.183–198, Morgan Kaufmann, 1999.
3. R. Smith and J. Smith, "An Examination of Tunable, Random Search Landscapes", *Foundations of Genetic Algorithms 5*, pp.165–182, Morgan Kaufmann, 1999.
4. K. E Mathias, L. J. Eshelman, and D. Schaffer, "Niches in NK-landscapes", *Foundations of Genetic Algorithms 6*, pp.27–46, Morgan Kaufmann, 2001.
5. H. Aguirre and K. Tanaka, "A Study on the Behavior of Genetic Algorithms on NK-Landscapes: Effects of Selection, Drift, Mutation, and Recombination", *IEICE Trans. Fundamentals*, (9):2270–2279, 2003.
6. H. Aguirre and K. Tanaka, "Insights on Properties of Multiobjective MNK-Landscapes", *Proc. 2004 IEEE Congress on Evolutionary Computation*, pp.196–203, IEEE Press, 2004.
7. K. Deb, L. Thiele, M. Laumanns, and E. Zitzler, "Scalable Multi-Objective Optimization Test Problems". *Proc. 2002 Congress on Evolutionary Computation*, pp.825–830, IEEE Press, 2002.
8. H. Aguirre and K. Tanaka, "Effects of Elitism and Population Climbing on Multiobjective MNK-Landscapes", *Proc. 2004 IEEE Congress on Evolutionary Computation*, pp.449-456, IEEE Press, 2004.
9. H. Aguirre, M. Sato, and K. Tanaka, "Preliminary Study on the Performance of Multiobjective Evolutionary Algorithms with MNK-Landscapes", *Proc. RISP Intl. Workshop on Nonlinear Circuits and Signal Processing*, pp.315–318, 2004.
10. K. Deb, *Multi-Objective Optimization using Evolutionary Algorithms*, John Wiley & Sons, Chichester, West Sussex, England, 2001.
11. E. Zitzler, M. Laumanns, and L. Thiele, "SPEA2: Improving the Strength Pareto Evolutionary Algorithm", Technical Report 103, TIK-Report, 2001.
12. C. Coello, D. Van Veldhuizen, and G. Lamont, *Evolutionary Algorithms for Solving Multi-Objective Problems*. Kluwer Academic Publishers, Boston, 2002.
13. E. Zitzler, *Evolutionary Algorithms for Multiobjective Optimization: Methods and Applications*, PhD thesis, Swiss Federal Institute of Technology, Zurich, 1999.
14. J. Knowles and D. Corne, "On Metrics for Comparing Non-dominated Sets", *Proc. 2002 Congress on Evolutionary Computation*, pp.711–716. IEEE Press, 2002.
15. M. Fleischer, "The Measure of Pareto Optima: Applications to Multi-objective Metaheuristics" *Second Intl. Conf. on Evolutionary Multi-Criterion Optimization*, Lecture Notes in Computer Science, vol.2632, pp.519–533, Springer, 2003.

Comparison Between Lamarckian and Baldwinian Repair on Multiobjective 0/1 Knapsack Problems

Hisao Ishibuchi, Shiori Kaige, and Kaname Narukawa

Department of Industrial Engineering, Osaka Prefecture University,
1-1 Gakuen-cho, Sakai, Osaka, 599-8531, Japan
{hisaoi, shiori, kaname}@ie.osakafu-u.ac.jp

Abstract. This paper examines two repair schemes (i.e., Lamarckian and Baldwinian) through computational experiments on multiobjective 0/1 knapsack problems. First we compare Lamarckian and Baldwinian with each other. Experimental results show that the Baldwinian repair outperforms the Lamarckian repair. It is also shown that these repair schemes outperform a penalty function approach. Then we examine partial Lamarckianism where the Lamarckian repair is applied to each individual with a prespecified probability. Experimental results show that a so-called 5% rule works well. Finally partial Lamarckianism is compared with an island model with two subpopulations where each island has a different repair scheme. Experimental results show that the island model slightly outperforms the standard single-population model with the 50% partial Lamarckian repair in terms of the diversity of solutions.

1 Introduction

Since 1990s, multiobjective 0/1 knapsack problems have been frequently used to evaluate the performance of various multiobjective metaheuristics including evolutionary multiobjective optimization (EMO) algorithms [4-7, 11, 16, 18]. When EMO algorithms are applied to multiobjective 0/1 knapsack problems, unfeasible solutions are often generated by genetic operations. That is, generated solutions do not always satisfy the constraint conditions. Thus several constraint handling methods have been examined in the application of EMO algorithms to multiobjective 0/1 knapsack problems (e.g., Ishibuchi & Kaige [4], Mumford [11], and Zydallis & Lamont [18]). Constraint handling methods for multiobjective 0/1 knapsack problems can be roughly classified into the following three categories:

Greedy Repair: An unfeasible solution is repaired by removing items until all the constraint conditions are satisfied. The order in which items are removed is pre-specified based on a heuristic measure for evaluating each item.

Penalty Function: Objective functions are penalized when constraint conditions are violated.

Permutation Coding: Each solution is not represented by a binary string but a permutation of items. That is, the order of items is used as a string to represent each solution. A feasible solution is obtained from each permutation-type string by adding items to the knapsacks in the order specified by that string.

In this paper, we concentrate on the comparison between two implementation schemes of greedy repair: Lamarckian and Baldwinian. In the Lamarckian implementation, a feasible solution is generated from an unfeasible one by removing items until all the constraint conditions are satisfied. That is, the genetic information of the unfeasible solution is modified by greedy repair as shown in Fig. 1. As a result, each population includes no unfeasible solutions. Since Zitzler & Thiele [16], the Lamarckian implementation has been implicitly used in almost all computational experiments of EMO algorithms with greedy repair on multiobjective 0/1 knapsack problems [4-7, 11, 18].

On the other hand, the genetic information of an unfeasible solution is not changed in the Baldwinian implementation where greedy repair is used only to evaluate the fitness value of each solution. As shown in Fig. 2, the same feasible solution as in Fig. 1 is generated from the unfeasible solution by greedy repair. This feasible solution is used only to assign the fitness value to the unfeasible solution. As a result, each population becomes a mixture of feasible and unfeasible solutions.

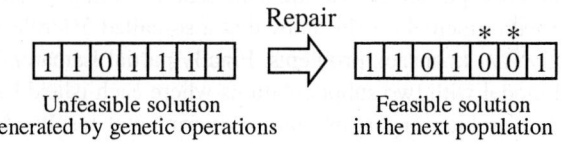

Fig. 1. Illustration of the Lamarckian implementation of greedy repair

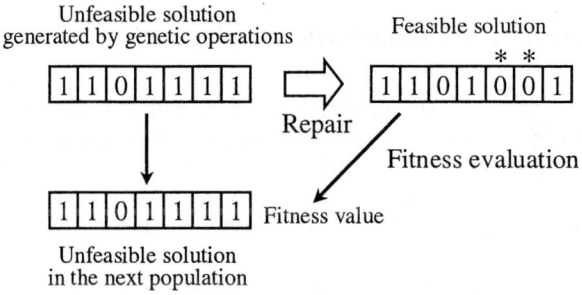

Fig. 2. Illustration of the Baldwinian implementation of greedy repair

In this paper, we first briefly explain multiobjective 0/1 knapsack problems and two repair methods examined in Ishibuchi & Kaige [4] and Zydallis & Lamont [18]. The two repair methods are the maximum ratio repair and the weighted scalar repair. Next we examine the two implementation schemes (i.e., Lamarckian and Baldwinian) of these repair methods. The two implementation schemes are compared with each other through computational experiments on multiobjective 0/1 knapsack problems in Zitzler & Thiele [16] using the NSGA-II algorithm of Deb et al. [2]. While better results can be obtained by memetic EMO algorithms (e.g., MOGLS [6]) for multiobjective 0/1 knapsack problems, we use the NSGA-II algorithm because it seems to be the most frequently used EMO algorithm in the literature. Experimental results show that the Baldwinian repair outperforms the Lamarckian repair as in many other studies on single-objective combinatorial optimization problems (e.g., Liu et al. [9] and Orvosh & Davis [12,13]). We also evaluate the performance of the repair methods in comparison with a penalty function approach where objective functions are penalized when constraint conditions are violated. Then we examine the performance of partial Lamarckianism where the Lamarckian repair is applied to each unfeasible solution with a prespecified probability. When the Lamarckian repair is not applied, the unfeasible solution is handled by the Baldwinian repair.

Partial Lamarckianism has been examined in some studies on single-objective optimization problems. For example, Orvosh & Davis [12,13] found that good results were obtained for combinatorial optimization problems by the application of the Lamarckian repair to each unfeasible solution with a 5% probability while Michalewicz & Nazhiyath [10] used a 20% probability for continuous optimization problems. On the other hand, Houck et al. [3] showed that good results were obtained from 20% and 40% partial Lamarckianism search strategies on a number of test problems. Our experimental results show that a so-called 5% rule [12,13] works well on multiobjective 0/1 knapsack problems. Finally partial Lamarckianism is compared with an island model with two subpopulations where each island has a different repair scheme (i.e., Lamarckian or Baldwinian). Experimental results show that the island model slightly outperforms the standard single-population model with the 50% partial Lamarckian repair in terms of the diversity of solutions.

2 Multiobjective 0/1 Knapsack Problems

The following k-objective 0/1 knapsack problem with k knapsacks and n items was used in Zitzler & Thiele [16] where an objective function as well as a constraint condition was related to each knapsack:

$$\text{Maximize } \mathbf{f}(\mathbf{x}) = (f_1(\mathbf{x}), f_2(\mathbf{x}), ..., f_k(\mathbf{x})), \tag{1}$$

$$\text{subject to } \sum_{j=1}^{n} w_{ij} x_j \leq c_i, \quad i = 1, 2, ..., k, \tag{2}$$

where

$$f_i(\mathbf{x}) = \sum_{j=1}^{n} p_{ij} x_j, \quad i = 1, 2, ..., k. \tag{3}$$

In this formulation, $\mathbf{x}$ is an n-dimensional binary vector (i.e., $(x_1, x_2, ..., x_n) \in \{0, 1\}^n$), p_{ij} is the profit of item j according to knapsack i, w_{ij} is the weight of item j according to knapsack i, and c_i is the capacity of knapsack i. Each solution $\mathbf{x}$ is handled by a binary string of length n in our computational experiments. As a test problem, we use a two-objective 500-item (i.e., 2-500) knapsack problem in [16].

3 Repair Methods

Zitzler & Thiele [16] used a greedy repair method where items were removed in the ascending order of the maximum profit/weight ratio q_j over all knapsacks:

$$q_j = \max\{p_{ij}/w_{ij} \mid i=1,2,...,k\}, \quad j=1,2,...,n. \quad (4)$$

The maximum profit/weight ratio q_j in (4) has been used in many studies on EMO algorithms (e.g., Knowles & Corne [7]). In this paper, we refer to this repair method as ***maximum ratio repair***.

While Pareto ranking was used to evaluate each solution in many EMO algorithms (e.g., Deb et al. [2] and Zitzler & Thiele [16]), the following weighted scalar fitness function was used in some EMO algorithms (e.g., Jaszkiewicz [5, 6]):

$$f(\mathbf{x},\lambda) = \sum_{i=1}^{k} \lambda_i f_i(\mathbf{x}), \quad (5)$$

where

$$\forall i \; \lambda_i \geq 0 \; \text{ and } \; \sum_{i=1}^{k} \lambda_i = 1. \quad (6)$$

In a multiobjective genetic local search (MOGLS) algorithm of Jaszkiewicz [5, 6], the weighted scalar fitness function in (5) was used in the following manner. When a pair of parent solutions is to be selected, first the weight vector $\lambda = (\lambda_1, ..., \lambda_k)$ is randomly specified. Next the best K solutions are selected from the current population using the weighted scalar fitness function with the current weight vector. Then a pair of parent solutions is randomly chosen from those K solutions in order to generate an offspring by genetic operations from the selected pair. The same weighted scalar fitness function with the current weight vector is used in the repair for the generated offspring where items are removed in the ascending order of the following ratio:

$$q_j = \sum_{i=1}^{k} \lambda_i p_{ij} \bigg/ \sum_{i=1}^{k} \lambda_i w_{ij}, \quad j=1,2,...,n. \quad (7)$$

We refer to this repair method as ***weighted scalar repair***. A local search procedure is applied to the repaired offspring using the same scalar fitness function with the current weight vector. The weighted scalar repair is also used in the local search phase.

It should be noted that the weighted scalar repair is directly applicable only to the MOGLS with the weighted scalar fitness function. In the application to the NSGA-II algorithm [2] in this paper, we use the weighted scalar repair by randomly updating the weight vector $\lambda = (\lambda_1, ..., \lambda_k)$ for each unfeasible solution. That is, a different weight vector is assigned to each unfeasible solution.

For illustrating each repair method, we randomly generate an n-dimensional binary vector $\mathbf{x} = (x_1, ..., x_n)$ by assigning 0 with the probability 0.4 and 1 with the probability 0.6 to each x_j. Then we generate a feasible solution using one of the two repair methods if the randomly generated binary vector is unfeasible. When the randomly generated binary vector is feasible, we return to the first step in order to generate another binary vector. The random generation of an unfeasible solution and the application of a repair method to the generated unfeasible solution are iterated to obtain a prespecified number of feasible solutions. We can draw the trajectory from each unfeasible solution to its repaired one in the objective space.

In Fig. 3 (a), we show the trajectories from 10 unfeasible solutions by the maximum ratio repair for the 2-500 test problem. In this figure, unfeasible and feasible solutions are denoted by open circles and closed circles, respectively. It should be noted that the same order of items specified by (4) is always used for all the 10 unfeasible solutions in the case of the maximum ration repair. Thus the directions of the trajectories are similar to each other in Fig. 3 (a). On the other hand, Fig. 3 (b) shows the trajectories from the same 10 unfeasible solutions by the weighted scalar repair. In the weighted scalar repair, a different order of items is used for each unfeasible solution because the weight vector $\lambda = (\lambda_1, ..., \lambda_k)$ in (7) is randomly updated for each unfeasible solution. As a result, we observe various directions of the trajectories in Fig. 3 (b).

From the comparison between Fig. 3 (a) and Fig. 3 (b), one may think that the weighted scalar repair has a positive effect on the diversity of obtained solution sets by EMO algorithms. Our computational experiments in the next section show that better solution sets with larger diversity are actually obtained from the weighted scalar repair than the maximum ratio repair. Similar results have been reported with respect to the performance of the two repair methods in the literature [4, 18].

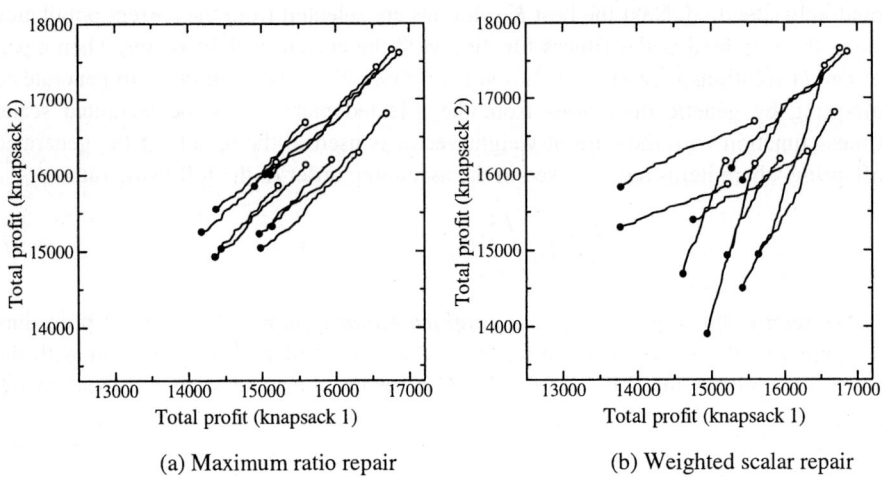

Fig. 3. Unfeasible solutions (open circles) and repaired solutions (closed circles)

Each repair method is implemented in the two implementation schemes: Lamarckian and Baldwinian. This means that we examine the four combinations of the two repair methods and the two implementation schemes. We also implement a partial Lamarckian repair strategy, which can be viewed as a hybrid version of the Lamarckian and Baldwinian implementation schemes.

Just for comparison, we also examine the performance of a penalty function approach. Using a positive constant α representing a unit penalty with respect to the violation of each constraint condition, we formulate the following k-objective optimization problem with no constraint conditions from the original k-objective 0/1 knapsack problem in (1)-(3):

$$\text{Maximize } \mathbf{g}(\mathbf{x}) = (g_1(\mathbf{x}), g_2(\mathbf{x}), ..., g_k(\mathbf{x})), \tag{8}$$

where

$$g_i(\mathbf{x}) = f_i(\mathbf{x}) - \alpha \cdot \max\left(0, \sum_{j=1}^{n} w_{ij}x_j - c_i\right), \quad i = 1, 2, ..., k. \tag{9}$$

In this formulation, each objective $f_i(\mathbf{x})$ is penalized when the corresponding constraint condition is violated. The application of EMO algorithms to the k-objective optimization problem in (8)-(9) is straightforward because no constraint conditions are involved. When we evaluate the performance of an EMO algorithm with the penalty function approach, we only examine feasible solutions in the final solution set obtained by each run of the EMO algorithm (i.e., unfeasible solutions of the original knapsack problem are not taken into account in the performance evaluation).

4 Computational Experiments

4.1 Conditions of Computational Experiments

We incorporate each of the four combinations of the two repair methods and the two implementation schemes into the NSGA-II algorithm [2]. As a test problem, we use the two-objective 500-item knapsack problem (i.e., 2-500 test problem) in Zitzler & Thiele [16]. Each solution is coded as a binary string of length 500. Each of the four variants of the NSGA-II algorithm is applied to the test problem under the following parameter specifications:

Crossover probability (one-point crossover): 0.8,
Mutation probability (bit-flip mutation): 4/500,
Population size: 200,
Stopping condition: 500 generations.

The average performance of each variant was calculated over 30 runs with different initial populations.

A number of performance measures have been proposed for evaluating a set of non-dominated solutions in the literature. As explained in Knowles & Corne [8] and Zitzler et al. [17], no performance measure can simultaneously evaluate various aspects of a solution set. In this paper, we visually compare each variant of the NSGA-II algorithm by drawing the 50% attainment surface in the two-dimensional objective space over 30 runs. We also use two performance measures that are applicable to simultaneous comparison of many solution sets. Let S be a solution set obtained by an EMO algorithm. The proximity of the solution set S to the Pareto front is evaluated by the generational distance (GD) as follows [15]:

$$GD(S) = \frac{1}{|S|} \sum_{\mathbf{x} \in S} \min\{d_{\mathbf{xy}} \mid \mathbf{y} \in S^*\}, \qquad (10)$$

where S^* is a reference solution set (i.e., the set of all Pareto-optimal solutions) and $d_{\mathbf{xy}}$ is the distance between a solution $\mathbf{x}$ and a reference solution $\mathbf{y}$ in the k-dimensional objective space ($k = 2$ in our computational experiments):

$$d_{\mathbf{xy}} = \sqrt{(f_1(\mathbf{x}) - f_1(\mathbf{y}))^2 + \cdots + (f_k(\mathbf{x}) - f_k(\mathbf{y}))^2} . \qquad (11)$$

For evaluating both the diversity of solutions in the solution set S and the convergence to the Pareto front, we calculate the $D1_R$ measure as follows [1]:

$$D1_R(S) = \frac{1}{|S^*|} \sum_{\mathbf{y} \in S^*} \min\{d_{\mathbf{xy}} \mid \mathbf{x} \in S\} . \qquad (12)$$

It should be noted that $D1_R(S)$ is the average distance from each reference solution $\mathbf{y}$ in S^* to its nearest solution in S while $GD(S)$ in (10) is the average distance from each solution $\mathbf{x}$ in S to its nearest reference solution in S^*. The generational distance evaluates the proximity of the solution set S to the reference solution set S^*. On the other hand, the $D1_R$ measure evaluates how well the solution set S approximates the reference solution set S^*. Since all the Pareto-optimal solutions are known for the 2-500 test problem, we can use them as the reference solution set S^*.

4.2 Comparison of Two Implementation Schemes: Lamarckian and Baldwinian

Experimental results are summarized in Table 1 where the average value of each performance measure over 30 runs is shown together with the corresponding standard deviation in parentheses. It should be noted that smaller values of each performance measure mean better results. In Table 1, the better result between the two implementation schemes (i.e., Lamarckian and Baldwinian) is indicated by boldface for each combination of the two performance measures and the two repair methods. From Table 1, we can see that the Baldwinian implementation consistently outperforms the Lamarckian implementation for all the four combinations of the two performance measures and the two repair methods. We can also see from Table 1 that much better results are obtained from the weighted scalar repair than the maximum ratio repair as we expected from the trajectories by each repair method in Fig. 3. For visually demonstrating the above-mentioned observations, we show in Fig. 4 the 50% attainment surface over 30 runs for each combination of the two implementation schemes and the two repair methods. We can see from Fig. 4 that better results are obtained from the Baldwinian implementation scheme and the weighted scalar repair.

Table 1. Average value of each performance measure over 30 runs and the corresponding standard deviation in parentheses

Repair method	Generational distance (GD)		$D1_R$ measure	
	Lamarckian	Baldwinian	Lamarckian	Baldwinian
Maximum ratio	324 (32)	**239** (24)	632 (42)	**502** (47)
Weighted scalar	155 (17)	**100** (19)	269 (26)	**103** (18)

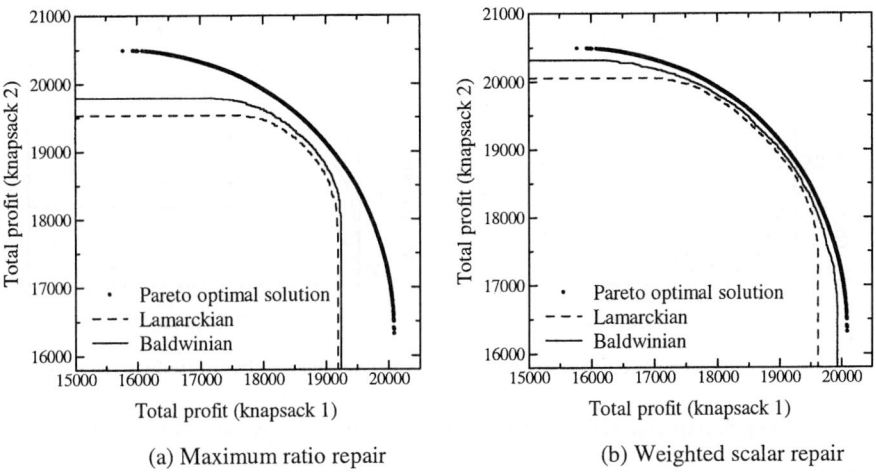

Fig. 4. Pareto front of the 2-500 test problem and the 50% attainment surface obtained from 30 runs using each combination of the two implementation schemes and the two repair methods

In order to further examine the two implementation schemes, we monitor the number of feasible solutions among 200 individuals in each generation. The number of feasible solutions in each generation is shown in Fig. 5 (a) for a single run with the maximum ratio repair. As we have already explained, all the 200 individuals at each generation are always feasible in the case of the Lamarckian implementation scheme. On the contrary, the number of feasible solutions rapidly decreases to zero in the early stage of evolution in the case of the Baldwinian implementation scheme in Fig. 5 (a).

We also monitor the average number of items in each solution during the same computational experiments as Fig. 5 (a). Experimental results are summarized in Fig. 5 (b). From Fig. 5 (b), we can see that much more items are included in each solution in the case of the Baldwinian repair than the Lamarckian repair. We can also see that the increase in the average number of items is very slow after the 100th generation even in the case of the Baldwinian repair where all solutions are infeasible after the 3rd generation as shown in Fig. 5 (a).

Furthermore, we monitor the number of removed items from each individual in each generation during the same computational experiments as Fig. 5. It should be noted that excess items are not actually removed from individuals in the case of the Baldwinian repair. We monitor the number of excess items that are tentatively removed to evaluate each individual in the case of the Baldwinian repair. The maximum number and the average number of removed items over 200 individuals in each generation are shown in Fig. 6 (a) and Fig. 6 (b), respectively. From Fig. 6 (b), we can see that only a few items are removed in the case of the Lamarckian repair on average while each individual includes about 60 excess items in the case of the Baldwinian repair.

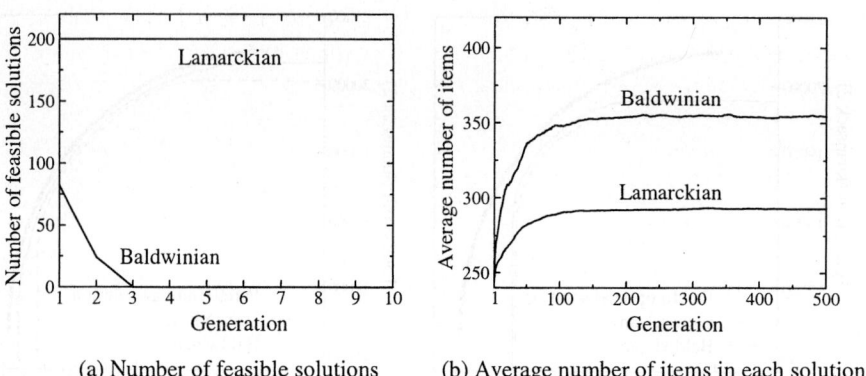

Fig. 5. Experimental results of a single run with the maximum ratio repair

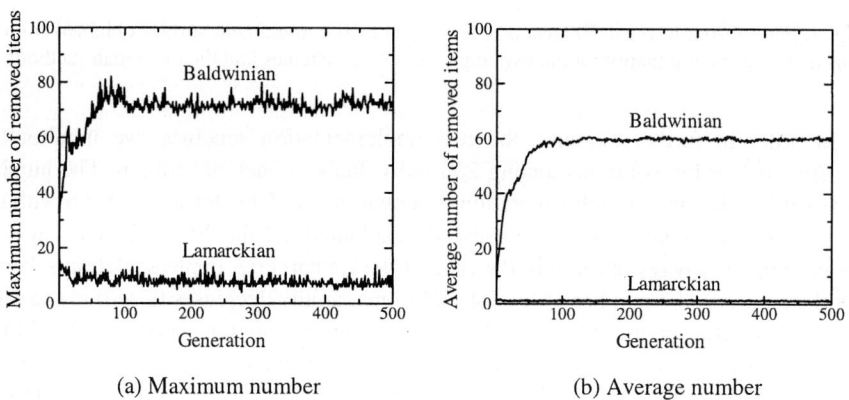

Fig. 6. Maximum and average numbers of excess items removed by each implementation scheme of the maximum ratio repair

In the same manner as Fig. 5 and Fig. 6, experimental results by the weighted scalar repair are shown in Fig. 7 and Fig. 8. From Figs. 5-8, we can see that the number of excess items consistently increases over 500 generations only when we use the Baldwinian implementation of the weighted scalar repair (see Fig. 8 (b)). As a result, much more items are included in each individual in this case. Even when the weighted scalar repair is used, only a few excess items are included in each individual in the case of the Lamarckian implementation (see Fig. 8 (b)). The difference in the average number of excess items between the two repair methods can be explained by the increase in the diversity of solutions by the weighted scalar repair. As shown in Fig. 3 (b), the weighted scalar repair spreads each individual over a wide range of the objective space. Thus the increase in the number of excess items contributes to the increase in the diversity of solutions in the objective space when we use the Baldwinian implementation of the weighted scalar repair. This explanation is consistent with the good result of the 50% attainment surface by the Baldwinian implementation of the weighted scalar repair in Fig. 4 (b).

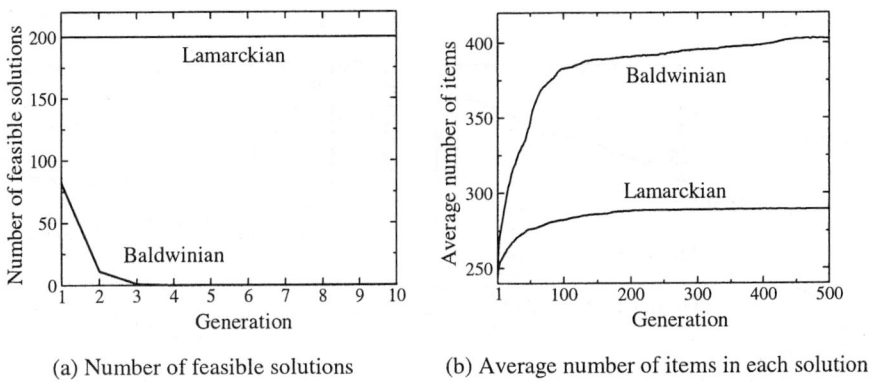

Fig. 7. Experimental results of a single run with the weighted scalar repair

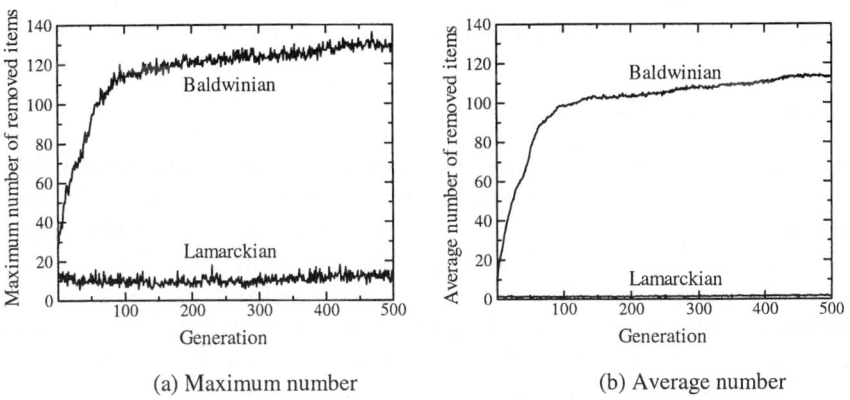

Fig. 8. Maximum and average numbers of excess items removed by each implementation scheme of the weighted scalar repair

4.3 Comparison with Penalty Function Approach

In the same manner as in Subsection 4.2, we apply the NSGA-II algorithm to the 2-500 test problem 30 times using the penalty function approach in (8)-(9). We examine the various specifications of the unit penalty α : $\alpha = 1.0, 1.2, 1.4, ..., 3.0$. Average results over 30 runs are depicted by open circles in Fig. 9 where we also show the average results by the Baldwinian implementation of the two repair methods in Subsection 4.2. It should be noted that smaller values mean better results in Fig. 9. From Fig. 9, we can see that good results are not obtained by the penalty function approach while we examine a wide range of parameter values. More specifically, we can see from Fig. 9 (a) that the convergence to the Pareto front degrades as the unit penalty α increases. On the other hand, we can observe in Fig. 9 (b) that the diversity of obtained solutions degrades as the unit penalty α decreases.

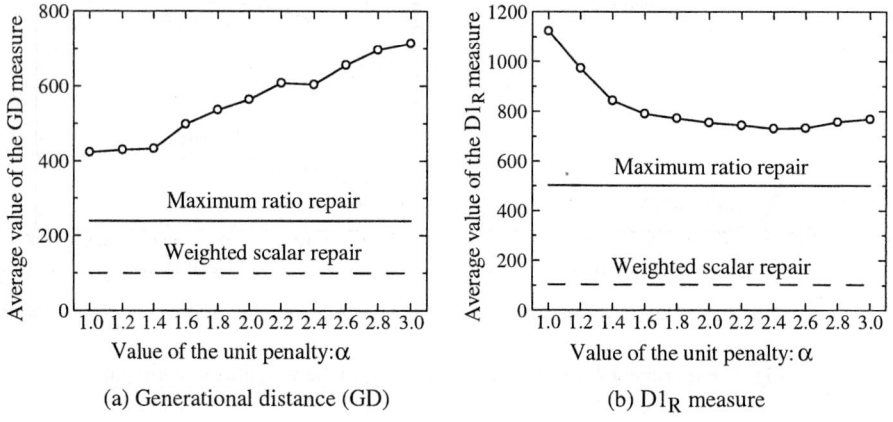

Fig. 9. Average results by the penalty function approach with various parameter values

4.4 Partial Lamarckianism

It has been demonstrated in the literature [3, 10, 12, 13] that a partial Lamarckianism search strategy outperforms both pure Lamarckian and pure Baldwinian. We examine the effect of the partial Lamarckian implementation where the Lamarckian repair is applied to each individual with a prespecified probability (say P_L %). When the Lamarckian repair is not applied, the individual is handled in the framework of the Baldwinian repair. We examine such a partial Lamarckian implementation for several specifications of the probability P_L of the Lamarckian repair: $P_L = 0, 20, 40, 60, 80, 100$ (%). It should be noted that $P_L = 0$ and $P_L = 100$ mean the pure Baldwinian repair and the pure Lamarckian repair, respectively. Experimental results are shown in Fig. 10 for the case of the maximum ratio repair. Fig. 10 (a) and Fig. 10 (b) show the average values of the generational distance (GD) and the $D1_R$ measure over 30 runs, respectively. From Fig. 10, we can see that the quality of obtained solution sets is degraded by increasing the probability of the Lamarckian repair.

In order to examine the validity of a so-called 5% rule [12, 13], we perform exhaustive computational experiments using various values of the probability of the Lamarckian repair between 0% and 20%. Average results over 500 runs are shown in Fig. 11. From this figure, we can see that the partial Lamarckian repair improves the $D1_R$ measure of the pure Baldwinian repair when the probability of the Lamarckian repair is specified around 5%. That is, the diversity of obtained solutions is improved by invoking the Lamarckian repair with a low probability. This observation supports the validity of the 5% rule in the implementation of a partial Lamarckianism repair for multiobjective 0/1 knapsack problems.

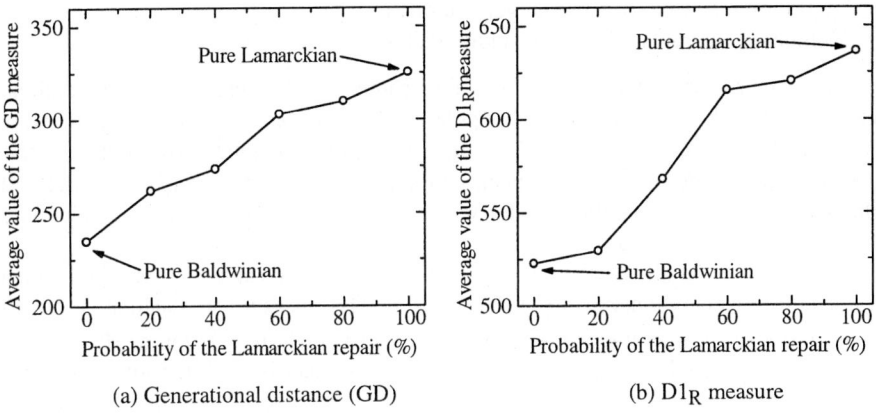

Fig. 10. Average results by the partial Lamarckian implementation of the maximum ratio repair over 30 runs

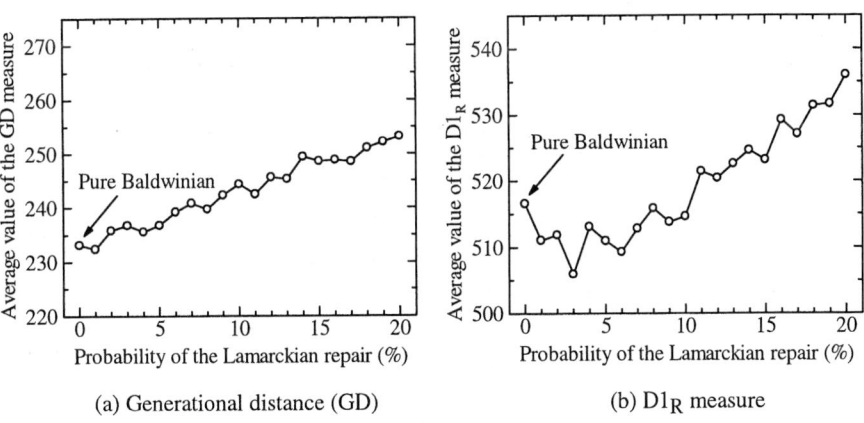

Fig. 11. Further examination of the partial Lamarckian implementation of the maximum ratio repair through computational experiments with 500 runs

4.5 Comparison Between Partial Lamarckianism and Island Model

Recently Skolicki & De Jong [14] clearly demonstrated high search ability of multi-representation island models for single-objective continuous optimization problems where each island used a different coding method. We examine a similar idea using the two implementation schemes of greedy repair (i.e., Lamarckian and Baldwinian). That is, we examine an island model with two subpopulations where one island uses the Lamarckian repair and the other island uses the Baldwinian repair. A prespecified number of individuals are randomly chosen from non-dominated solutions in each island (i.e., each subpopulation) at intervals of a prespecified number of generations. Then their copies are inserted to the other island. This island model is executed using the following parameter specifications:

Subpopulation size: 100 (individuals),
Migration interval: 10, 50, 100 (generations),
Number of migrants: 10, 20, 50 (individuals).

The other parameter values are the same as the previous computational experiments. The size of each subpopulation is specified as 100 so that the total number of individuals (i.e., 2×100) is the same as the population size in the previous computational experiments. This specification is for comparing the island model with the standard single-population model with the 50% partial Lamarckian repair under the same computation load.

Experimental results by the island model are shown in Table 2 where the average values of the generational distance (GD) and the $D1_R$ measure are calculated over 30 runs. We also show average results by the standard single-population model with the 50% partial Lamarckian repair. The maximum ratio repair is used in Table 2. The best result for each performance measure is indicated by boldface. From this table, we can see that the island model improves the diversity of obtained solutions (see the fourth column labeled as $D1_R$) while it degrades the convergence to the Pareto front (see the third column labeled as GD).

Table 2. Average results over 30 runs by the island model with various parameter specifications. The maximum ratio repair is used in computational experiments in this table

Migration interval	Number of migrants	GD	$D1_R$
10	10	296	556
10	20	290	579
10	50	313	594
50	10	288	566
50	20	285	568
50	50	300	560
100	10	295	570
100	20	296	**549**
100	50	291	561
50% Lamarckian		**277**	586

5 Concluding Remarks

We examined two implementation schemes (i.e., Lamarckian and Baldwinian) of greedy repair through computational experiments on multiobjective 0/1 knapsack problems using the NSGA-II algorithm. While the Lamarckian implementation has been almost always used in the application of EMO algorithms to multiobjective 0/1 knapsack problems in the literature, better results were obtained by the Baldwinian implementation in this paper. The main contribution of this paper is that the superiority of the Baldwinian implementation over the Lamarckian implementation was clearly demonstrated through computational experiments on multiobjective 0/1 knapsack problems. This observation is consistent with some reported results on single-objective optimization problems [9, 10, 12, 13].

We also compared two greedy repair methods (i.e., maximum ratio repair and weighted scalar repair) with each other using the two implementation schemes. Independent of the choice of an implementation scheme, better results were obtained by the weighted scalar repair than the maximum ratio repair. This observation is consistent with some reported results [4, 5, 6, 18]. For comparison, we also examined the performance of the penalty function approach. In our computational experiments, better results could not be obtained from the penalty function approach in comparison with greedy repair.

In addition to pure Lamarckian and pure Baldwinian, we also examined their hybrid version (i.e., partial Lamarckianism search strategy) where the Lamarckian repair was applied to each individual with a prespecified probability. Some of our experimental results supported a so-called 5% rule where the Lamarckian repair and the Baldwinian repair were applied to each individual with 5% and 95% probabilities, respectively. That is, only 5% unfeasible solutions were actually repaired. Finally, we examined the performance of an island model with two subpopulations where each island used a different implementation scheme (i.e., Lamarckian or Baldwinian). In our computational experiments, we did not observe clear improvement by such an island model from the standard single-population model with the 50% partial Lamarckian repair. The use of the island model slightly improved the diversity of solutions.

While we empirically showed the superiority of the Baldwinian implementation over the Lamarckian implementation through computational experiments on multiobjective 0/1 knapsack problems, we could not clearly explain why the Baldwinian implementation outperformed the Lamarckian implementation in the application of EMO algorithms to multiobjective 0/1 knapsack problems. We did not examine the performance of the two implementation schemes for other test problems, either. Further empirical studies as well as theoretical studies are left for future research with respect to the comparison between the two implementation schemes for multiobjective optimization problems.

The authors would like to thank the financial support from Japan Society for the Promotion of Science (JSPS) through Grand-in-Aid for Scientific Research (B): KAKENHI (14380194).

References

1. Czyzak, P. and Jaszkiewicz, A.: Pareto-Simulated Annealing – A Metaheuristic Technique for Multi-Objective Combinatorial Optimization. Journal of Multi-Criteria Decision Analysis **7** (1998) 34-47.
2. Deb, K., Pratap, A., Agarwal, S., and Meyarivan, T.: A Fast and Elitist Multiobjective Genetic Algorithm: NSGA-II. IEEE Trans. on Evolutionary Computation **6** (2002) 182-197.
3. Houck, C. R., Joines, J. A., Kay, M. G., and Wilson, J. R.: Empirical Investigations of the Benefits of Partial Lamarckianism. Evolutionary Computation **5** (1997) 31-60.
4. Ishibuchi, H. and Kaige, S.: Effects of Repair Procedures on the Performance of EMO Algorithms for Multiobjective 0/1 Knapsack Problems. Proc. of 2003 Congress on Evolutionary Computation (2003) 2254-2261.
5. Jaszkiewicz, A.: Comparison of Local Search-based Metaheuristics on the Multiple Objective Knapsack Problem. Foundations of Computing and Decision Sciences **26** (2001) 99-120.
6. Jaszkiewicz, A.: On the Performance of Multiple-Objective Genetic Local Search on the 0/1 Knapsack Problem - A Comparative Experiment. IEEE Trans. on Evolutionary Computation **6** (2002) 402-412.
7. Knowles, J. D. and Corne, D. W.: A Comparison of Diverse Approaches to Memetic Multiobjective Combinatorial Optimization. Proc. of 2000 Genetic and Evolutionary Computation Conference Workshop Program (2000) 103-108.
8. Knowles, J. D. and Corne, D. W.: On Metrics for Comparing Non-dominated Sets. Proc. of 2002 Congress on Evolutionary Computation (2002) 711-716.
9. Liu, B., Haftka, R. T., Akgun, M. A., and Todoroki, A.: Permutation Genetic Algorithm for Stacking Sequence Design of Composite Laminates. Computer Methods in Applied Mechanics and Engineering **186** (2000) 357-372.
10. Michalewicz, Z. and Nazhiyath, G.: Genocop III: A Co-evolutionary Algorithm for Numerical Optimization Problems with Nonlinear Constraints. Proc. of 2nd IEEE International Conference on Evolutionary Computation **2** (1995) 647-651.
11. Mumford, C. L.: Comparing Representations and Recombination Operators for the Multi-Objective 0/1 Knapsack Problem. Proc. of 2003 Congress on Evolutionary Computation (2003) 854-861.
12. Orvosh, D. and Davis, L.: Shall We Repair? Genetic Algorithms, Combinatorial Optimization, and Feasibility Constraints. Proc. of 5th International Conference on Genetic Algorithms (1993) 650.
13. Orvosh, D. and Davis, L.: Using Genetic Algorithms to Optimize Problems with Feasibility Constraints. Proc. of 1st IEEE Conference on Evolutionary Computation (1994) 548-553.
14. Skolicki, Z. and De Jong, K.: Improving Evolutionary Algorithms with Multi-representation Island Models. Lecture Notes in Computer Science, Vol. 3242 (Proc. of PPSN VIII), Springer, Berlin (2004) 420-429.
15. Van Veldhuizen, D. A.: Multiobjective Evolutionary Algorithms: Classifications, Analyses, and New Innovations. Ph. D dissertation, Air Force Institute of Technology (1999).
16. Zitzler, E. and Thiele, L.: Multiobjective Evolutionary Algorithms: A Comparative Case Study and the Strength Pareto Approach. IEEE Trans. on Evolutionary Computation **3** (1999) 257-271.

17. Zitzler, E., Thiele, L., Laumanns, M., Fonseca, C. M., and da Fonseca, V. G.: Performance Assessment of Multiobjective Optimizers: An Analysis and Review. IEEE Trans. on Evolutionary Computation **7** (2003) 117- 132.
18. Zydallis, J. B. and Lamont, G. B.: Explicit Building-Block Multiobjective Evolutionary Algorithms for NPC Problems. Proc. of 2003 Congress on Evolutionary Computation (2003) 2685-2695.

The Value of Online Adaptive Search: A Performance Comparison of NSGAII, ε-NSGAII and εMOEA

Joshua B. Kollat[1] and Patrick M. Reed[2]

[1] Department of Civil and Environmental Engineering,
The Pennsylvania State University,
406B Sackett Building, University Park,
PA 16802-1408
juk124@psu.edu

[2] Department of Civil and Environmental Engineering,
The Pennsylvania State University,
212 Sackett Building, University Park,
PA 16802-1408
preed@engr.psu.edu

Abstract. This paper demonstrates how adaptive population-sizing and epsilon-dominance archiving can be combined with the Nondominated Sorted Genetic Algorithm-II (NSGAII) to enhance the algorithm's efficiency, reliability, and ease-of-use. Four versions of the enhanced Epsilon Dominance NSGA-II (ε-NSGAII) are tested on a standard suite of evolutionary multiobjective optimization test problems. Comparative results for the four variants of the ε-NSGAII demonstrate that adapting population size based on online changes in the epsilon dominance archive size can enhance performance. The best performing version of the ε-NSGAII is also compared to the original NSGAII and the εMOEA on the same suite of test problems. The performance of each algorithm is measured using three running performance metrics, two of which have been previously published, and one new metric proposed by the authors. Results of the study indicate that the new version of the NSGAII proposed in this paper demonstrates improved performance on the majority of two-objective test problems studied.

1 Introduction

Deb *et al.* [1] identified three primary goals in multiobjective (MO) optimization using evolutionary algorithms (EAs): (1) to obtain good convergence toward the Pareto-optimal solution set, (2) to develop a diverse, or evenly distributed set of non-dominated solutions and to maintain this diversity throughout the entire run of the algorithm, and (3) to achieve the first two goals at the lowest computational cost and in the most efficient manner possible. The third goal can be realized through the development of new techniques to achieve convergence and diversity at the lowest possible computational cost and to develop online adaptive EAs that can assess on-line performance and modify key algorithm parameters throughout a run. The ultimate goal of online adaptation is to enhance algorithmic efficiency, reliability, and ease-of-use.

This study presents alternative techniques by which epsilon-dominance archiving [2], the Nondominated Sorted Genetic Algorithm-II (NSGAII) [3], and parameter adaptation [4] can be combined. These techniques use online performance assessment to adapt population size and to automatically terminate search based on minimal user input and can potentially be integrated into any multiobjective evolutionary algorithm (MOEA) to improve algorithm efficiency, reliability, and ease-of-use. In this paper, section 2 provides a description of the algorithms being compared as well as justification for their selection. Sections 3 and 4 discuss the performance metrics and test problems used in the study. Section 5 presents the results of the simulation in two parts: (1) a comparison of four versions of the ε-NSGAII and (2) a comparison of the best version of the ε-NSGAII identified in the first half of the study to the NSGAII [3] and the εMOEA [1]. Conclusions and potential future research is provided in section 6.

2 Tested Algorithms

The current study is conducted in two parts. The first compares the performance of four versions of the ε-NSGAII, and the second compares the best version of the ε-NSGAII with the NSGAII and the εMOEA.

2.1 Overview of ε-NSGAII

The primary objective of this study is to demonstrate the ε-NSGAII's efficacy at solving multiobjective optimization problems quickly, efficiently, and reliably. The ε-NSGAII is based on the NSGAII, which uses a fast non-dominated sorting approach to classify solutions according to level of non-domination and a crowding distance operator to preserve solution diversity [3]. The ε-NSGAII extends these concepts by adding ε-dominance [2], adaptive population sizing, and self termination to minimize the need for parameter calibration as demonstrated by Reed *et al.* [4].

ε-dominance is a concept whereby the user is able to specify the precision with which they want to obtain the Pareto-optimal solutions to a multiobjective problem, in essence giving them the ability to assign a relative importance to each objective. This is accomplished by applying a grid (sized by user specified ε values) to the search space of the problem. Larger ε values result in a courser grid (and ultimately fewer solutions) while smaller ε values produce a finer grid. The fitness of each solution is then mapped to a box fitness based on the specified ε values. Non-domination sorting is then conducted using each solution's box fitness, and solutions with identical box fitness (i.e., solutions that occur in the same grid block) are compared and those that are dominated within the grid block are eliminated. This results in no more than one non-dominated solution existing in any one grid block, preventing clustering of solutions and promoting a more even search of the objective space. The interested reader can refer to prior work by Laumanns *et al.* [2] and Deb *et al.* [1] for a more detailed description of ε-dominance.

The adaptive population sizing scheme used in the original form of the ε-NSGAII is based on the population sizing theory of Harik *et al.* [5] and the automatic parameterization methodology proposed by Reed *et al.* [4]. The ε-NSGAII uses a series of "connected runs" where small populations are exploited to pre-condition search with successively doubled population sizes. Pre-conditioning occurs by injecting current solutions within the epsilon-dominance archive into the initial generations of larger population runs. For example, the initial population (usually five individuals) is evolved until it is no longer making significant progress. When this occurs, the population size is increased, a subset of archived solutions are injected into the next population, and the search continues. Under the current design of the algorithm, two injection scenarios exist. If the archive size is smaller then the population into which it will be injected, the remaining individuals needed to fill the population are randomly generated. However, if the archive is larger than the subsequent population, then individuals are randomly selected from the archive to fill the population.

The search is terminated using two user-specified criteria: (1) the *intra-run* criterion and (2) the *inter-run* criterion. The intra-run criterion defines the two cases when the current population N will be doubled: (1) if search within w generations (termed the lag window) fails to yield a specified percentage increase in the number of archived solutions or (2) the maximum run duration has been reached. The termination of search across all runs (i.e., across all populations used) compares how the archive size changes at the end of two successive runs of the ε-NSGAII. For example, a run that uses a population of N to evolve an ε-nondominated set composed of A individuals will be compared to a second run that used a population of 2N to evolve an ε-nondominated set of K individuals. The results of these runs are used in equation (1), to define which of the two following courses of action will be taken: (1) population size is again doubled, resulting in 4N individuals to be used in an additional run of the ε-NSGA-II or (2) the algorithm stops to allow the user to assess if the ε-nondominated set has been quantified to sufficient accuracy. Δ was set to 10-percent for this study as recommended by Reed et al. [4].

$$if\ \Delta < \left(\frac{|K-A|}{A}\right)100\ then\ double\ N\ and\ continue\ search \qquad (1)$$

else stop search

The solutions obtained in the archive at the end of the final run represent a sufficient approximation of the true Pareto front based on user defined accuracy goals.

2.1.1 Improving Termination

This study investigates improvements in the way the ε-NSGAII adaptively sizes its population and self-terminates. The initial version of the algorithm based its inter- and intra-run termination strictly on changes in the number of solutions stored in the epsilon dominance archive. However, a shortcoming to this method exists when the algorithm finds a number of new solutions dominating the same number of solutions in the archive. The algorithm could terminate in this case because the quantity of solutions in the archive has remained constant. By this scenario the algorithm may be making significant progress while the archive size remains constant.

To ameliorate this issue, a second version of the algorithm has been developed which accounts for solution quality in the termination criteria. This is accomplished by monitoring both the archive size and the number of solutions that have been replaced when the archive is updated. The solutions that differ will be improved solutions since a new solution will only be accepted into the archive if it ε-dominates one or more existing solutions. In addition, since the computational cost of a quality comparison is higher than simply checking for a change in solution quantity, the quality comparison is only conducted if the quantity comparison results in insufficient improvement. With the improved termination criteria, the ε-NSGAII will continue the search and seek better convergence to the true Pareto set.

2.1.2 Improving Population Sizing and Injection

The current population doubling scheme used by the ε-NSGAII has a possible flaw in that it lacks a bound on population growth. This issue is particularly important for high-dimensional, difficult problems with large Pareto optimal sets. This study investigates this issue by analyzing two alternative population sizing schemes and by comparing their performance to the prior population doubling methodology.

The first scheme bases the size of the population on the size of the epsilon dominance archive by maintaining a 25% injection rate. For example, if 25 ε-nondominated solutions exist in the archive at the end of a run, the subsequent population size will be four times the archive size, or 100 individuals. This scheme bounds the maximum size of the population to four times the number of solutions that exist at the user specified ε resolution. Theoretically, this approach allows population sizes to increase or decrease, and in the limit when the epsilon dominance archive size stabilizes, the ε-NSGAII's "connected runs" are equivalent to time continuation [6]. Since this method guarantees that new individuals will be introduced at the end of each run by 25% injection, the chances of escaping local nondominated fronts are greatly improved.

The second proposed population sizing scheme attempts to merge both the old doubling scheme and the new scheme based on 25% injection. This scheme, hereafter referred to as the adaptive scheme, bases the subsequent population size on the smaller of either population doubling, or 25% injection. This method typically provides a slower population growth rate while also bounding the size of the population using the 25% injection scheme.

2.2 Other Algorithms

The ε-NSGAII's performance has been tested in this study relative to the NSGAII [3] and the εMOEA [1]. The NSGAII was chosen for comparison since it is the original algorithm from which the ε-NSGAII was derived. The εMOEA is a steady state MOEA that co-evolves both an evolutionary algorithm population and an archive population by randomly mating individuals from the population and the archive to generate new solutions [1]. The εMOEA was chosen for comparison because it uses the concept of ε-dominance to preserve solution diversity and its effectiveness on the test problems examined in this study has been demonstrated previously [1].

3 Metrics Used to Assess Performance

The performance of the algorithms tested in this study is assessed using a convergence metric and a diversity metric proposed by Deb and Jain [7] as well as a combined convergence/diversity metric proposed by the authors. The convergence metric proposed by Deb and Jain [7] measures the average Euclidean distance between the non-dominated solutions generated by an algorithm and the set of Pareto-optimal solutions. Deb and Jain's diversity metric attempts to alleviate many of the shortcomings associated with previously proposed diversity metrics. This metric measures diversity by projecting algorithm-based solutions and reference solutions onto an M-1 dimensional plane where M is the number of objectives. The projected solutions are then compared to the projected reference solutions in terms of their distribution across the objective space. The metric value is determined by favoring well distributed solutions with a high weight and assigning low weights to clustered solutions. In this study, a new metric is proposed that effectively measures both convergence and diversity. This metric uses the concept of ε-dominance to measure the percentage of solutions that have been found within a user specified ε distance of a reference set. Steps in calculating the metric are as follows:

1. Apply ε-dominance to a reference solution set according to user specified ε values.
2. At each generation, match each solution generated by the algorithm to its corresponding ε-nondominated reference set solution. Each reference solution can only have one algorithm solution associated with it. If more than one solution exists within ε of the reference solution, then the closest solution in terms of Euclidean distance is chosen. This accounts for overlapping ε regions in the reference set and frees up the additional solutions to be matched with other ε-nondominated reference solutions.
3. Each ε-nondominated reference solution that has a corresponding algorithm solution receives a score of one, while each reference solution that has no corresponding algorithm solution receives a score of zero. The metric is then calculated according to equation (2).

$$\varepsilon(P^{(t)}) = \sum_{i=1}^{n} h_i / n \qquad (2)$$

where h_i is one or zero for each solution i in the ε-nondominated reference set and n is the total number of solutions in the reference set.

This metric measures convergence by accounting for solutions that have converged to within ε of a reference set. Diversity is accounted for by including only one solution for each ε-nondominated reference solution, regardless of the existence of additional solutions in that ε-block. This ensures that clustered solutions do not contribute to the calculation of the metric.

4 Test Problem Suite

The relative performance of the algorithms' was assessed using a suite of two-objective test problems that have been commonly employed in prior literature. Each of the test problems are discussed briefly below. The naming convention used in referring to the test problems is adopted from [8]. A more detailed description of the construction of these test problems can be found in [8] and [9].

- T_1: a convex, thirty variable test problem ($m = 30$ and $x_i \in [0, 1]$) with the Pareto front formed at $g(\mathbf{x}) = 1$.
- T_2: a 30 variable test problem ($m = 30$ and $x_i \in [0, 1]$) with a concave Pareto front formed at $g(\mathbf{x}) = 1$.
- T_3: a 30 variable test problem ($m = 30$ and $x_i \in [0, 1]$) with a discontinuous Pareto front formed at $g(\mathbf{x}) = 1$.
- T_4: a multi-modal (21^9 local fronts), convex, ten variable test problem ($m = 10$, $x_1 \in [0, 1]$, and $x_2,...,x_m \in [-5, 5]$) with the Pareto front formed at $g(\mathbf{x}) = 1$.
- T_6: a non-uniform, non-convex, ten variable test problem ($m = 10$ and $x_i \in [0, 1]$) with the Pareto front formed at $g(\mathbf{x}) = 1$.

5 Simulation Results

5.1 ε-NSGAII Version Comparison

The first portion of the study compared four versions of the ε-NSGAII to identify the impacts of population sizing and termination criteria on solving the two-objective test problem suite. The versions of the ε-NSGAII compared are identified as the ε-NSGAIIv1 (the original version of the algorithm) and the ε-NSGAIIv2 (a version in which the termination criteria have been changed to account for solution quality). The ε-NSGAIIv2 was further tested using population doubling, 25% injection, and the adaptive population sizing scheme identified previously in section 2.1.2. Each version of the algorithm used an initial population size of n = 5. Each population was run for a maximum of 250 generations. Based on Deb's recommendations, the re-combination scaling factor η_c was set equal to 15 and the probability of crossover was set equal to 1.0. The polynomial mutation operator used a scaling factor $\eta_m = 20$ and the probability of mutation was set equal to $1/i$, where i is the number of real-coded decision variables. The inter- and intra-run termination criteria were each set to 10%. The ε-NSGAIIv1 and the ε-NSGAIIv2 using population doubling and adaptive population sizing used an initial lag window of 50 generations for the first three runs and a final lag window of 25 thereafter (lag window is the number of generations across which the algorithm is checked for termination conditions). The ε-NSGAIIv2 that used 25% injection used an initial lag window of 50 for the first run and a lag window of 10 thereafter since the archive tends to grow more quickly by this approach. Each of the performance metrics used a reference set of solutions containing approximately 1000 points. ε values were chosen for each test problem to result in approximately 100 solutions, and the grid sizes, G, used to compute the diversity metric, were obtained using 1/ε (Table 1). The analysis was conducted using 50 random seeds and all results reflect all random seeds.

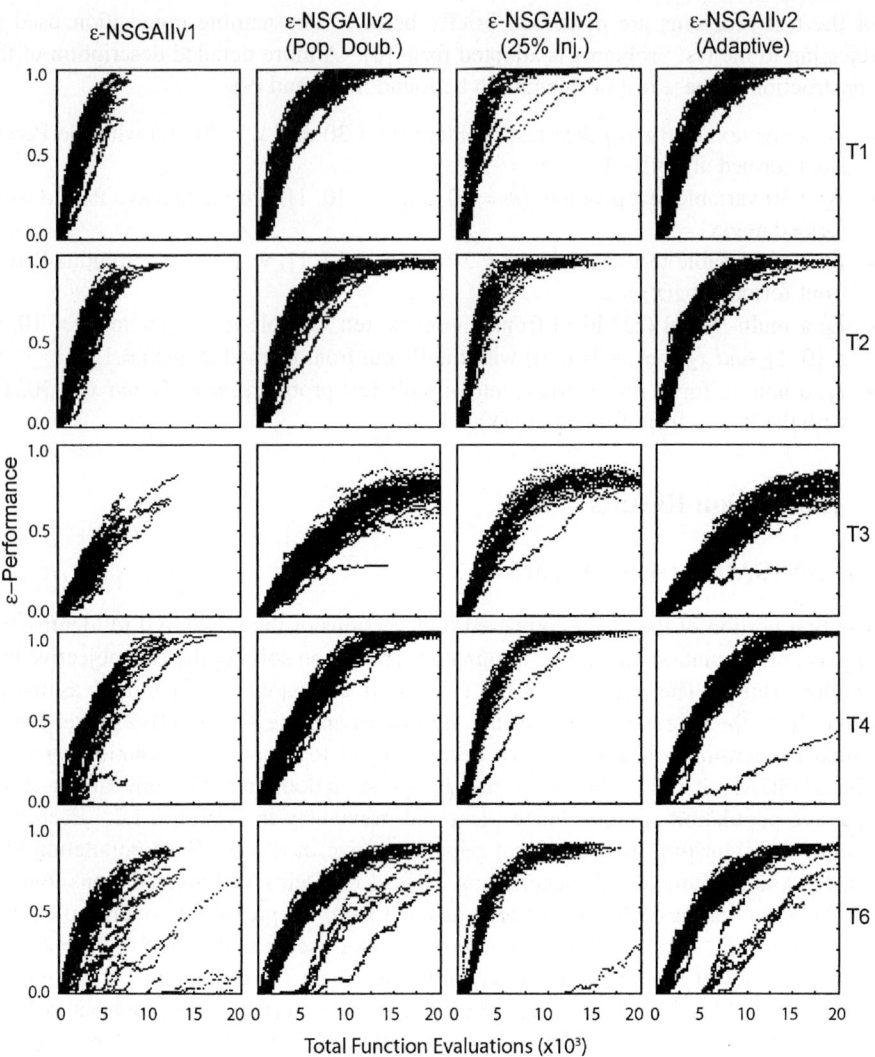

Fig. 1. Running ε-performance metric results for each version of the ε-NSGAII examined on test problems T_1, T_2, T_3, T_4, and T_6. Results represent 50 random seeds plotted as ε-performance versus total function evaluations in thousands. This figure can be cross-referenced with Table 2 to determine final mean ε-performance

The results of the ε-performance metric versus total function evaluations in thousands for 50 random seeds and each test problem are shown in Figure 1. These plots reflect the spread of results for all random seeds across all generations. Based on the scatter plots in Figure 1, the ε-NSGAIIv1 performs the poorest in terms of ε-performance as is evidenced by its premature termination for many test problems and

The Value of Online Adaptive Search   393

Table 1. ε values and G (the number of grid blocks used in the diversity metric) for each of the two-objective test problems used in the study. ε values were chosen to generate approximately 100 solutions for each test problem

Test Problem	ε_1	ε_2	G
T1	0.0075	0.0075	133
T2	0.0076	0.0076	131
T3	0.00261	0.00261	383
T4	0.0074	0.0074	135
T6	0.0067	0.0067	149

Table 2. Final results comparing four versions of the ε-NSGAII (ε-NSGAIIv1, ε-NSGAIIv2 using population doubling, ε-NSGAIIv2 using 25% injection, and ε-NSGAIIv2 using adaptive population sizing). Included are the means and standard deviations for total solutions found, total function evaluations, ε-performance, convergence, diversity, and CPU time in seconds for 50 random seeds. The best result in each category is indicated by bold-underlined text

	MOEA	Sol.		NFE		ε-Perf ($\times 10^{-4}$)		Conv. ($\times 10^{-6}$)		Diver. ($\times 10^{-4}$)		Time (s) ($\times 10^{-3}$)	
		Avg.	σ	Avg.	σ	Avg.	σ	Avg.	σ	Avg.	σ	Avg.	σ
T1	v1	90	15	**5550**	**1416**	8212	1536	401	65	7581	1193	**213**	**64**
	v2 - Doub.	96	3	10540	3270	9272	501	**378**	**24**	8030	202	500	220
	v2 - 25% Inj.	**99**	**0**	12738	2081	**9855**	**71**	389	25	**8226**	**31**	1137	183
	v2 - Adapt.	96	3	10540	3270	9273	501	**378**	**24**	8030	202	492	224
T2	v1	88	20	**6085**	**2183**	7648	1937	397	75	6927	1524	**233**	**117**
	v2 - Doub.	98	2	14425	5819	9147	514	402	**21**	7606	208	859	665
	v2 - 25% Inj.	**99**	**0**	12585	1400	**9451**	**142**	406	22	**7683**	**88**	1153	163
	v2 - Adapt.	97	3	12538	5415	8986	578	406	25	7517	235	667	458
T3	v1	80	18	**6067**	**2227**	4612	1342	608	1393	6783	1502	**209**	**112**
	v2 - Doub.	96	11	55010	27249	7392	740	**269**	23	8170	943	10714	12166
	v2 - 25% Inj.	**98**	**5**	19259	3894	**7832**	**377**	271	**18**	**8390**	**405**	1809	449
	v2 - Adapt.	97	11	39677	9084	7598	769	270	23	8308	902	3364	1043
T4	v1	70	38	**6813**	**4181**	5758	3735	11998	49703	5884	3126	**221**	**182**
	v2 - Doub.	97	4	18270	9266	9266	952	**383**	**24**	8042	246	1086	865
	v2 - 25% Inj.	**100**	**1**	14392	9796	**9796**	**97**	384	24	**8175**	**43**	1214	225
	v2 - Adapt.	97	4	18878	9246	9246	963	399	77	8035	268	1099	869
T6	v1	93	6	**9926**	**4720**	5102	2744	189	250	6385	342	**394**	340
	v2 - Doub.	96	5	18225	11707	7752	794	454	**112**	6620	280	1293	2430
	v2 - 25% Inj.	**100**	**1**	14295	3501	**8363**	**258**	447	133	**6809**	**46**	1175	**284**
	v2 - Adapt.	97	4	16532	7069	7704	845	471	160	6612	275	797	671

random seeds. In addition, there is a large spread in ε-performance across random seeds for many of the test problems. Examining the scatter plots of the three versions of the ε-NSGAIIv2 for test problem T_1, the version using 25% injection converges the fastest with high reliability across random seeds. The reliability of the algorithm can be assessed by examining the spread of the scatter plot (e.g., a tighter plot indicates high reliability). This is also the case for test problems T_2, T_3, and T_4. For test problem T_6, the ε-NSGAIIv2 using 25% injection clearly outperforms the other three versions of the algorithm.

Table 2 presents a summary of the mean results and standard deviations for the performance of each version of the ε-NSGAII on each test problem using 50 random seeds. Results displayed include total solutions found, total function evaluations, ε-performance, convergence, diversity, and the CPU time in seconds (i.e. wall clock time). The best performance in each category is indicated by bold-underlined text. The ε-NSGAIIv2 using 25% injection found the most solutions, achieved the highest ε-performance and the highest diversity with the lowest standard deviations for all test problems. In terms of total function evaluations, the ε-NSGAIIv1 outperformed the other versions on all test problems, but as a result of premature termination. When examining the convergence metric, the ε-NSGAIIv2 using 25% injection outperformed the ε-NSGAIIv1 in all test problems except T_2. In terms of CPU time, but neglecting results from the ε-NSGAIIv1 due to premature termination, CPU time was sacrificed for improved performance in test problems T_1, T_2, T_4, and T_6. However, there is a large improvement in CPU time for test problem T_3. On average, the ε-NSGAIIv2 using 25% injection was the most robust at solving the two-objective test problems examined in this study. Based on these findings, the performance of the ε-NSGAII using 25% injection was subsequently compared to NSGAII and εMOEA.

5.2 Comparing NSGAII, ε-NSGAII, and εMOEA

For the second half of the study, the second version of ε-NSGAII that used 25% injection (hereafter referred to as ε-NSGAII for simplicity) as its population sizing scheme was chosen for comparison to the NSGAII and the εMOEA, as it was found to be the most robust version of the algorithm. The same two-objective test problems (T_1, T_2, T_3, T_4, and T_6) are again used for comparison. Again, the analysis was performed using 50 random seeds and all results shown reflect all 50 random seeds. However, since the ε-NSGAII self-terminates, in order to make a fair algorithmic comparison, the NSGAII and the εMOEA were set to run for the same mean number of function evaluations that the ε-NSGAII required to completely solve each test problem and self terminate. This methodology provides a snapshot in time of the performance of the other algorithms at the time that the ε-NSGAII terminates on average. The NSGAII was parameterized using a population size of 100 individuals, uniform crossover with probability $P_c = 1.0$, and probability of mutation, $P_m = 1/i$ where i is the number of real coded variables in the test problem. The εMOEA was parameterized using the same settings as previously mentioned for NSGAII and the ε settings and grid resolution, G, were set identically to those values contained in Table 1 for each test problem. However, as noted above, the number of generations that each algorithm was set to run differed depending on the results of the ε-NSGAII. For the εMOEA, the total iterations was determined by dividing the average total function evaluations required by the ε-NSGAII by two, since each iteration results in two function evaluations. Table 3 provides a summary the generational settings of each algorithm based on the results of the ε-NSGAII from part one of the study.

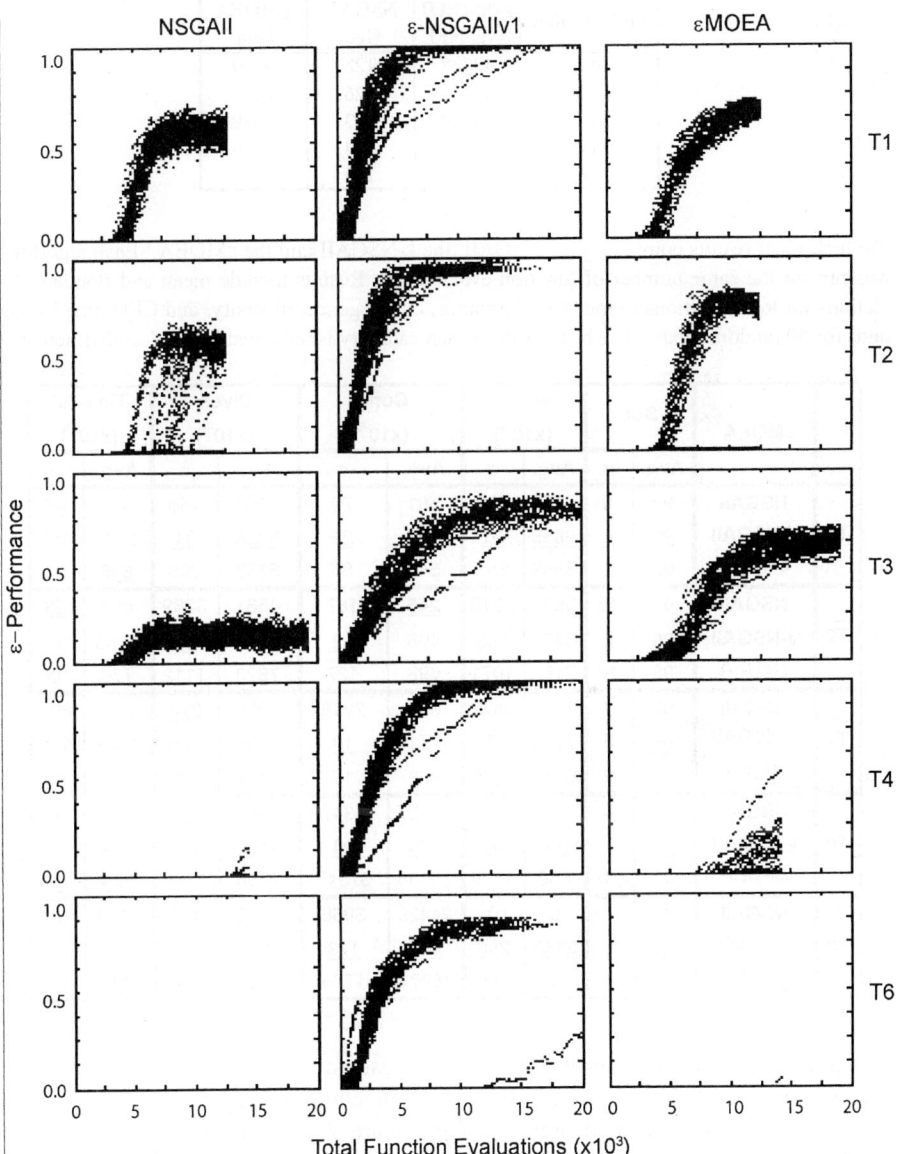

Fig. 2. Running ε-performance metric results for the NSGAII, the ε-NSGAII, and the εMOEA examined on test problems T_1, T_2, T_3, T_4, and T_6. Results represent 50 random seeds plotted as ε-performance versus total function evaluations in thousands. This figure can be cross-referenced with Table 4 to determine final mean ε-performance

Table 3. Average total function evaluations required by the ε-NSGAII to solve each test problem, and the corresponding number of generations that the NSGAII and the εMOEA were set to run for each test problem

Test Problem	ε-NSGAII Avg. NFE	NSGAII Gen.	εMOEA Gen.
T1	12738	128	6369
T2	12585	126	6293
T3	19259	193	9630
T4	14392	144	7196
T6	14295	143	7148

Table 4. Final results comparing the NSGAII, the ε-NSGAII and the εMOEA. Each algorithm was run for the same number of function evaluations. Results include mean and standard deviations for total solutions found, ε-performance, convergence, diversity, and CPU time in seconds for 50 random seeds. The best result in each category is indicated by bold-underlined text

	MOEA	Sol.		ε-Perf ($\times 10^{-4}$)		Conv. ($\times 10^{-6}$)		Diver. ($\times 10^{-4}$)		Time (s) ($\times 10^{-3}$)	
		Avg.	σ	Avg.	σ	Avg.	σ	Avg.	σ	Avg.	σ
T1	NSGAII	98	0	5455	462	561	79	7153	146	920	**<u>46</u>**
	ε-NSGAII	**<u>99</u>**	**<u>0</u>**	**<u>9855</u>**	**<u>71</u>**	**<u>389</u>**	**<u>25</u>**	**<u>8226</u>**	**<u>31</u>**	1137	183
	εMOEA	99	2	6659	251	543	65	8132	235	**<u>846</u>**	53
T2	NSGAII	62	45	3531	2618	**<u>246</u>**	187	4581	3389	982	**<u>56</u>**
	ε-NSGAII	**<u>99</u>**	**<u>0</u>**	**<u>9451</u>**	**<u>142</u>**	406	**<u>22</u>**	7683	**<u>88</u>**	1153	163
	εMOEA	98	14	7190	1071	698	127	**<u>7820</u>**	1112	**<u>776</u>**	100
T3	NSGAII	92	**<u>2</u>**	1301	365	659	2218	6936	**<u>270</u>**	1538	697
	ε-NSGAII	**<u>98</u>**	5	**<u>7832</u>**	**<u>377</u>**	271	**<u>18</u>**	**<u>8390</u>**	405	1809	449
	εMOEA	95	8	6172	363	**<u>268</u>**	19	8101	605	**<u>1198</u>**	**<u>84</u>**
T4	NSGAII	51	44	34	178	92637	113472	3086	273	1102	57
	ε-NSGAII	**<u>100</u>**	**<u>1</u>**	**<u>9796</u>**	**<u>97</u>**	**<u>384</u>**	**<u>24</u>**	**<u>8175</u>**	**<u>43</u>**	1214	225
	εMOEA	45	20	680	1135	20158	37345	3500	1665	**<u>359</u>**	**<u>33</u>**
T6	NSGAII	96	1	0	0	24429	3966	5703	130	1004	**<u>54</u>**
	ε-NSGAII	**<u>100</u>**	**<u>1</u>**	**<u>8363</u>**	**<u>258</u>**	**<u>447</u>**	**<u>133</u>**	**<u>6809</u>**	**<u>46</u>**	1175	284
	εMOEA	95	4	9	63	10074	1775	6497	192	**<u>862</u>**	76

The results of the ε-performance metric (measured as the percentage of solutions that fall within ε of each objective) versus total function evaluations for all 50 random seeds and each test problem are shown in Figure 2. When the NSGAII and the εMOEA are given the same opportunity to solve each test problem (i.e., they are each permitted the same number of function evaluations), the performance of the ε-NSGAII is superior for each test problem. In fact, for test problems T_4, and T_6, NSGAII and εMOEA achieve little to no measure of ε-performance. This is most likely the result of an insufficient number of function evaluations (based on the performance of the ε-NSGAII) for solving these particular test problems.

Table 4 shows the final mean results for the NSGAII, the ε-NSGAII, and the εMOEA. The ε-NSGAII finds the most solutions in all test problems except T_1. The ε-NSGAII achieves the highest ε-performance and the best convergence for all test problems except T_2, and T_3 where the NSGAII and the εMOEA are superior. The ε-NSGAII achieves the highest diversity in all test problems except T_2 where it is slightly outperformed by the εMOEA. Average CPU time is again sacrificed to achieve higher performance in all other measures by the ε-NSGAII. In summary, the ε-NSGAII showed improved performance above the NSGAII and the εMOEA in all respects except CPU time for test problems T_1, T_4, and T_6, and achieved comparable performance on test problems T_2 and T_3.

6 Conclusions and Future Work

Adaptive population sizing based on epsilon archive size produces the most robust and reliable performance for the ε-NSGAII algorithm in solving the two-objective test problem suite used in this study. In addition, the ε-NSGAII outperforms its predecessor, the NSGAII, and the steady state algorithm, the εMOEA, on test problems T_1, T_4, and T_6 and performed comparably to the NSGAII and the εMOEA on test problems T_2 and T_3 in terms of solution quality and reliability.

Future research in this area will include a comparative study using higher dimensional test problems. In addition, the affects of integrating adaptive population sizing and termination into other MOEAs will be explored. The authors are currently working on two MOEA comparative studies involving real-world applications to long-term groundwater monitoring and integrated hydrologic model calibration. The performance of the NSGAII, the ε-NSGAII, the εMOEA, the SPEA2, and the MOSCEM applied to these real-world problems is currently being explored. Reed and Devireddy [10, 11] have attained up to a 90% reduction in computational costs over the NSGAII when applying ε-dominance archiving and automatic parameterization to long-term groundwater monitoring problems. Preliminary results of our studies indicate that the ε-NSGAII's ability to adaptively size population and self-terminate while allowing the user to specify the precision of the resulting Pareto-optimal solutions eliminates much of the trial and error analysis associated with traditional MOEA parameterization.

References

1. K. Deb, M. Mohan, S. Mishra. A Fast Multi-objective Evolutionary Algorithm for Finding Well-Spread Pareto-Optimal Solutions. *KenGAL*, Report No. 2003002. Indian Institute of Technology, Kanpur, India, 2003.
2. M. Laumanns, L. Thiele, K. Deb, E. Zitzler, Combining Convergence and Diversity in Evolutionary Multiobjective Optimization. *Evolutionary Computation*, 10(3):263-282, 2002.
3. K. Deb, A. Pratap, S. Agarwal, T. Meyarivan, A Fast and Elitist Multiobjective Genetic Algorithm: NSGA-II. *IEEE Trans. Evol. Computation*, 6(2):182-197, 2002.

4. P. Reed, B.S. Minsker, D.E. Goldberg, Simplifying Multiobjective Optimization: An Automated Design Methodology for the Nondominated Sorted Genetic Algorithm-II. *Water Resources Research*, 39(7):1196-1201, 2003.
5. G.R. Harik, E. Cuantu-Paz, D.E. Goldberg, B.L. Miller. The Gambler's Ruin Problem, Genetic Algorithms, and the Sizing of Populations. in *Proceedings of the 1997 IEEE Conference on Evolutionary Computation*, Piscataway, NJ, IEEE Press. pages 7-12, 1997.
6. D.E. Goldberg, *The Design of Innovation: Lessons from and for Competent Genetic Algorithms*. Kluwer Academic Publishers, Norwell, MA, 2002.
7. K. Deb, S. Jain. Running Performance Metrics for Evolutionary Multi-Objective Optimization. *KanGAL*, Report No. 2002004. Indian Institute of Technology, Kanpur, India, 2002.
8. E. Zitzler, K. Deb, L. Thiele, Comparison of multiobjective evolutionary algorithms: Empirical results. *Evolutionary Computation*, 8(2):125-148, 2000.
9. K. Deb, Multi-objective Genetic Algorithms: Problem Difficulties and Construction of Test Problems. *Evolutionary Computation*, 7(3):205-230, 1999.
10. P. Reed, V. Devireddy, *Groundwater Monitoring Design: A Case Study Combining epsilon-Dominance Archiving and Automatic Parameterization for the NSGA-II*, in *Applications of Multi-Objective Evolutionary Algorithms*, C. Coello-Coello, Editor. World Scientific, New York. 2004 (In Press).
11. P. Reed, V. Devireddy. Using Interactive Archives in Evolutionary Multiobjective Optimization: Case Studies for Long-Term Groundwater Monitoring Design. in *The International Environmental Modeling and Software Society Conference*, Osnabruck, Germany. 2004 (In Press).

Fuzzy-Pareto-Dominance and Its Application in Evolutionary Multi-objective Optimization

Mario Köppen, Raul Vicente-Garcia, and Bertram Nickolay

Fraunhofer IPK,
Pascalstr. 8-9, 10587 Berlin, Germany
{mario.koeppen, raul, nickolay}@ipk.fraunhofer.de

Abstract. This paper studies the fuzzification of the Pareto dominance relation and its application to the design of Evolutionary Multi-Objective Optimization algorithms. A generic ranking scheme is presented that assigns dominance degrees to any set of vectors in a scale-independent, non-symmetric and set-dependent manner. Based on such a ranking scheme, the vector fitness values of a population can be replaced by the computed ranking values (representing the "dominating strength" of an individual against all other individuals in the population) and used to perform standard single-objective genetic operators. The corresponding extension of the Standard Genetic Algorithm, so-called Fuzzy-Dominance-Driven GA (FDD-GA), will be presented as well. To verify the usefulness of such an approach, an analytic study of the Pareto-Box problem is provided, showing the characteristic parameters of a random search for the Pareto front in a unit hypercube in arbitrary dimension. The basic problem here is the loss of dominated points with increasing problem dimension, which can be successfully resolved by basing the search procedure on the fuzzy dominance degrees.

1 Introduction

In multiobjective optimization, the optimization goal is given by more than one objective to be extreme. Formally, given a domain as subset of R^n, there are assigned m functions $f_1(x_1, \ldots, x_n), \ldots, f_m(x_1, \ldots, x_n)$. Usually, there is not a single optimum but rather the so-called PARETO set of *non-dominated* solutions.

Evolutionary Computation (EC) has been shown to be a powerful technique for multi-objective optimization [1][2][3] (EMO - Evolutionary Multi-Objective Optimization). This biologically inspired methodology offers both flexibility in goal specification and good performance in multimodal, nonlinear search spaces.

If we want to solve a highly complex multi-objective optimization problem, we might select one of the best ranked evolutionary approaches reviewed in the literature, like NSGA-II [4] or SPEA2 [5] and hopefully start reaching good results quickly. However, all these algorithms need dominated individuals in the population, to perform the corresponding genetic operators. For a higher number of objectives, this might become a problem, since the probability of having a dominated individual in the population will rapidly go to zero.

The need for a revision of the Pareto dominance relation for also handling a larger number of objectives was already pointed out in a few studies, esp. given by Farina and Amato [6]. There, we also find the suggestion to use fuzzy membership degrees for the degree of a point belonging to the Pareto set (so-called fuzzy optimality). Authors design their revised dominance measure in a way that the approach to the Pareto front can be registered more early in the search. The approach was shown to work successfully in the domain of more than three objectives. It came out that the use of fuzzy concepts is fruitful in this regard. However, the approach did not provide a direct means to formulate corresponding EMO algorithms. So, the limitation is that still only the relation between two points is considered.

In this paper, we are going to use concepts from fuzzy fusion theory to achieve a more far reaching goal. Instead of introducing only a degree of dominance for two points, we are going to fuse the mutual degrees of dominance within a set of points and assign a *ranking* value to each point within this set of points. The circumstances now allow for using these ranking values in the same fashion as single objectives, e.g. to rank individuals within a population and thus easily expanding the application field of a Standard Genetic Algorithm to the multi-objective domain. Note that this is not the same as reducing multiple objectives into a single objective (as it is done e.g. with the weighted sum aproach): the ranking values are only fixed for a point within a given set of points. Once the set changes, the ranking values can vary as well.

In [7] this approach was already shown to handle search problems with separated Pareto fronts. However, for a better understanding of such an approach, suitable test problems are also needed, what has been a long-termed research issue in EMO as well. Usually, the different algorithms are compared by measuring their performance indices in difficult test searches [1][2][3][8]. However, these problems are hard to analyze, and the characteristics of a random or genetic search can not be provided. This prevents us from keeping track of the populations dynamics unambiguously (as already stated by Coello in [2]). Thus, we also introduce an "easy" multi-objective test function that allows us to observe the search progress and that is yet easily scalable to higher number of objectives as well. The PARETO-Box Problem, which will be presented and studied in this paper as well, unifies these crucial properties. It will help us to know more about how the PARETO-front is searched for in EMO, and to measure the progress of the novel Fuzzy PARETO Dominance-Driven Genetic Algorithm (FDD-GA) approach in search problems with higher number of objectives.

In the following section, we will consider the Fuzzy-Pareto Dominance relation, and base an EMO algorithm on it in section 3. Then, section 4 defines the PARETO-Box problem and its analysis for random search. These results will be used in an exemplary manner to study the dynamics of EMOs in section 5. The paper ends with the conclusion and the reference.

2 Fuzzification of Pareto Dominance Relation

In this section, we are going to study the fuzzification of the Pareto dominance relation. The goal is to yield a "softer" and practically usable numerical representation of the dominance relation between two vectors that can be employed in EMO. The issue was studied in more detail in [9]. This work showed the principal problems related to the specification of such a degree of dominance. Fuzzy dominance degrees can be computed once the following two conditions are taken into account:

1. The measure is not symmetric, and between two vectors a and b the two measures "a dominates b by degree α" and "a is dominated by b to degree α" have to be distinguished. Moreover, if a dominates b, either one measure is numerically 0 and the other lower-or-equal to 1, or one is greater-or-equal to 0 and the other 1^1.
2. The dominance degrees are set-dependent and can not be assigned in an absolute manner to single vectors alone.

A generic fuzzy ranking scheme for a set S of multivariate data (vectors) a_i with real-valued components a_{ij} and $1 \leq i \leq N$ can be based on the provision of a *comparison function* $f_x(y) : R \times R \to [0, 1]$ and a T-norm. Then, the following two steps are performed:

1. We compute the *comparison values* for any two vectors $a_i = (a_{ik})$ and $a_j = (a_{jk})$ by $c_{a_i}(a_j) = T(f_{a_{ik}}(a_{jk}) \,|\, k = 1, \ldots, N)$ with N the number of components of each vector.
2. We compute the *ranking values* for any element a_i of S by

$$r_S(a_i) = \max[c_{a_i}(a_j) | j \neq i].$$

Then, we consider vectors with lower numerical ranking values to be on a higher ranking position. For step 2, instead of max the min operator can be used as well, depending on the ranking to be favoured in increasing or decreasing order.

When using the comparison function bounded division and the algebraic (or product) norm as T-norm, the ranking scheme fulfills several useful properties like scale-independency in the data. The fuzzification of PARETO dominance relation can be written then as follows: It is said that vector a dominates vector b by degree μ_a with

$$\mu_a(a, b) = \frac{\prod_i \min(a_i, b_i)}{\prod_i a_i} \tag{1}$$

and that vector a is dominated by vector b at degree μ_p with

$$\mu_p(a, b) = \frac{\prod_i \min(a_i, b_i)}{\prod_i b_i} \tag{2}$$

---

[1] In [9] it was demonstrated that otherwise the complexity of the corresponding function specification would grow exponentially, as well as the number of discontinuities.

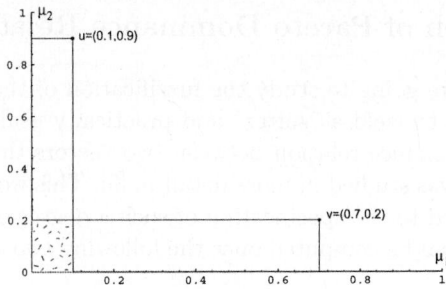

Fig. 1. Definition of Fuzzy-PARETO-Dominance. Here, u dominates v by degree $0.1 \cdot 0.2/0.1 \cdot 0.9 = 0.\bar{2}$ and is dominated by v by degree $0.1 \cdot 0.2/0.7 \cdot 0.2 \approx 0.143$

For a PARETO-dominating b, $\mu_a(a, b) = 1$ and $\mu_p(b, a) = 1$, but $\mu_p(a, b) < 1$ and $\mu_a(b, a) < 1$. Figure 1 gives a numerical example for the fuzzy PARETO dominance considered here. Note that the case of having an a_i or b_i equal to 0 is handled by the exclusion of the corresponding index in the products in the numerator and denominator.

3 Fuzzy-Pareto Dominance Driven Genetic Algorithm

In this section, we will show how the fuzzy ranking scheme easily extends a standard genetic algorithm to the multi-objective case.

We may use the dominance degrees of eq. (2) to rank the set M of multivariate data (vectors) given by the fitness values of a multiobjective optimization problem. Each element of M is assigned the maximum degree of being dominated by any other element of M, and the elements of M are sorted according to the ranking values in increasing order:

$$r_M(a) = \max_{b \in M \setminus \{a\}} \mu_p(a, b) \qquad (3)$$

Note again that this definiton is related to a set. A *ranking value* of a within M can only be assigned with reference to a set M containing a.

By sorting the elements of M according to the ranking values in increasing order (FPD ranking, FPD for Fuzzy-Pareto-Dominance), we obtain a partial ranking of the elements of M.

From the definition of the ranking scheme, it can be seen that an individual has two ways to reduce its comparison values: by increasing the objectives (thus increasing the denominator in the comparison values), or/and by being larger in some components than other vectors, i.e. being diverse from other vectors. Thus, both goals of evolutionary multi-objective optimization can be met by using such a measure: to approach the Pareto front, and to maintain a diverse population.

The foregoing discussion leads to the (Fuzzy-Dominance-Driven) FDD-GA algorithm, a Genetic Algorithm (GA) variant that employs the fuzzy ranking values of the fitness values (represented as vectors in case of multiobjective op-

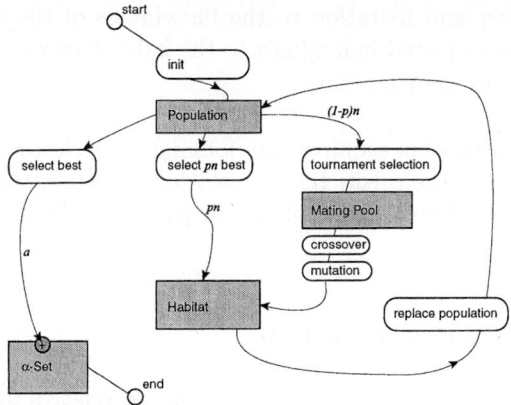

Fig. 2. Schematic view of FDD-GA algorithm

timization) for defining selection operators. The algorithm and its components can be seen in fig. 2.

FDD-GA maintains four pools of individuals:

- Population: contains n individuals as in standard GA.
- Mating Pool: Contains individual pairs that were selected for crossover operation.
- Habitat: This pool is composed of individuals from other pools and used to replace the population of generation n by generation $n+1$.
- α-Set (or *archive*): In this pool, all non-dominated individuals are collected. This pool also gives the output of the FDD-GA algorithm.

After random initialization of the population, the FDD-GA algorithm iteratively repeats the following steps until a stopping criteria (number of generations, size of α-Set) is met:

1. Rank population by FPD ordering of fitness vectors of the individuals in the population (see section 2).
2. Select *best individual a* from the ranked population (one individual with lowest ranking value) and conditionally add it to the α-set. Adding a to the α-set is only possible, when fitness of a is not dominated by the fitness of any individual already in the α-set, and if fitness of a is not equal to any individual's fitness there. In case a is added, all individuals in the α-set with fitness values dominated by fitness of a are removed from the α-set.
3. Add best pn of population individuals, according to FPD ordering ranking values, to the habitat ($0 \leq p \leq 1$).
4. Select $(1-p)n$ pairs from population by roulette-wheel selection, using the negated logarithms of the ranking values of the ranked population for selection (lower ranking value counts better), and put these pairs into the mating pool.

5. Apply crossover and mutation to the individuals of the mating pool, and add these newly created individuals to the habitat as well.
6. Replace population by habitat.

The FDD-GA algorithm acquires non-dominated (with respect to their fitness values) individuals in the α-set. In an evolutionary sense, those "FDD Pareto Set" approaches the Pareto front of the multiobjective optimization problem under study.

4   The Pareto-Box Problem

The advantages of basing the genetic operators on relative dominance degrees can be seen if the number of objectives is increasing. The following discussion will show how the chances of having a true dominance case in the population rapidly goes to 0. This problem can be clearly circumvented by the use of dominace degrees.

We are going to study a multi-objective optimization problem for arbitrary number of objectives where one can still provide a complete numerical analysis. The problem is referred to as Pareto-Box problem.

Given are m uniformly randomly selected n-dimensional points P_i in the n-dimensional unit hypercube ($1 \leq i \leq m$), with coordinates P_{ij} ($1 \leq j \leq n$). Thus, for each P_{ij} we have $0 \leq P_{ij} \leq 1$. The problem we state is:

PARETO-Box Problem: *What is the expectation value for the size of the* PARETO *set of these points?*

Here, we use the minimum version of PARETO dominance, so for two n-dimensional vectors $a = (a_i)$ and $b = (b_i)$ it is said that a dominates b (written as $a \prec b$) if and only if

$$\forall i:\ a_i \leq b_i,\ \wedge\ \exists j:\ a_j < b_j\ (1 \leq i, j \leq n) \tag{4}$$

For a set M of points, its PARETO set $P(M)$ is the subset for which none of its elements is dominated by any element of M. The PARETO set of the complete unit hypercube contains only one element, the point 0. The random sampling represents a random search in the unit hypercube, thus we are also going to answer the question if random search can find the PARETO set of the unit hypercube.

Obviously, the PARETO set of this problem is not hard to find, and there is also no conflict in the objectives. However, the following analysis will show that it is a hard problem for multi-objective optimization, once the dimension n of the problem is increased. Moreover, this problems allows for a precise analysis of the *progress* of algorithmic search, including the approach to the PARETO front and the entering of concave regions of the PARETO front.

In the following, $e_m(n)$ denotes the expectation value for the size of the PARETO set of m randomly selected points in the n-dimensional unit hypercube. Then, the following theorems hold:

Theorem 1. *Given are m randomly selected points in the n-dimensional hypercube. For the expectation value of the size of the PARETO set of these m points we have the recursive relation:*

$$e_1(n) = 1 \tag{5}$$
$$e_m(1) = 1$$
$$e_m(n) = e_{m-1}(n) + \frac{1}{m} e_m(n-1) \quad (n, m \geq 2)$$

Theorem 2. *The expectation value for the size of the PARETO set of $m \geq 1$ randomly selected points in the n-dimensional hypercube ($n \geq 1$) is*

$$e_m(n) = \sum_{k=1}^{m} \frac{(-1)^{k+1}}{k^{n-1}} \binom{m}{k} \tag{6}$$

Due to space limitations, the proofs of these theorems can not be given here. The appendix will give a sketch on how to derive these expressions.

Theorems 1 and 2 allow for the specification of the limiting behaviour of the expectation values for increasing number of points and increasing dimensions. This is stated in the following central theorem.

Theorem 3. *For fixed dimension $n > 1$ and the number of points $m \to \infty$, the expectation value $e_m(n) \to \infty$, the ratio of the non-dominated points $e_m(n)/m \to 0$ and for fixed $m > 1$ and dimension $n \to \infty$ it holds $e_m(n) \to m$.*

Proof. We see that

$$e_m(2) = \sum_{k=1}^{m} \frac{1}{k} = 1 + \frac{1}{2} + \frac{1}{3} + \frac{1}{4} + \ldots + \frac{1}{m} \tag{7}$$

which is the harmonic series and known to be divergent. Now, eq. (6) shows that for $n > 2$ always $e_m(n) \geq e_m(n-1) \geq \ldots \geq e_m(2)$, so for $m \to \infty$ $e_m(n) \to \infty$ as well. From the corresponding property of the harmonic series, $e_m(n)/m \to 0$ for $m \to \infty$ can be seen in a similar manner.

On the other hand, if $m > 1$ is fixed, all terms in eq. (6) but the one for $k = 1$ will go to 0 for $n \to \infty$, and the term for $k = 1$ itself computes to m. So, it is easy to see that $e_m(n) \to m$ for $n \to \infty$.

We can express this result as follows: for increasing number of sample points in the hypercube, the number of non-dominated points will also increase, and never "shrink" to the PARETO set of the hypercube, which only contains the point 0. So, random search will not solve the problem to find the PARETO set of the hypercube in any dimension.

For increasing dimension, it will become more and more unlikely to find *any* dominated point in a population of random sampling points. In fact, the probability falls exponentially. The PARETO set of m points will contain nearly all m points.

handling of randomly selected points by the FPD ranking scheme (as it happens when applied to the PARETO-Box problem). The distribution has a tail at the side of smaller ranking values, so roulette-wheel selection will acknowledge the fact that such individuals gradually perform better (with respect to PARETO-dominance). Such a behaviour can not be achieved when an EMO is relying on the presence of dominated individuals alone.

It has to be noted (but will not be detailed here) that nevertheless NSGA-II, in this set-up, is also finding the optimum up to a problem dimension of 8. In low dimensions (2-3) the FDD-GA is also outperformed by NSGA-II.

6 Conclusions

In this paper, issues related to the fuzzification of Pareto dominance were considered. A ranking scheme was presented that assigns dominance degrees to any set of vectors in a scale-independent, non-symmetric and set-dependent manner. Based on such a ranking scheme, the vector fitness values of a population can be replaced by the computed ranking values (representing the "dominating strength" of an individual against all other individuals in the population) and used to perform standard single-objective genetic operators. The corresponding extension of the Standard Genetic Algorithm, the FDD-GA, was presented as well. To verify the usefulness of such an approach, an analytic study of the Pareto-Box problem was provided, showing the characteristical parameters of a random search for the Pareto front in a unit hypercube in arbitrary dimension. The basic problem here is the loss of dominated points with increasing problem dimension, which can be successfully resolved by founding the search procedure on the fuzzy dominance degrees.

Acknowledgment

The research presented in this paper was supported by German research project "VisionIC," funded by German Ministry of Research and Education (BMBF) as handled by "Deutsche Gesellschaft für Luft- und Raumfahrt" (DLR). The authors also would like to express their thanks to Frank Hoffmann from University of Dortmund and Anna Ukovich from University of Trieste for their valuable comments and inspiring discussions about the topics of this work.

References

1. Coello Coello, C.A.: A short tutorial on evolutionary multiobjective optimization. In Zitzler, E., Deb, K., Thiele, L., Coello, C.A.C., Corne, D., eds.: First International Conference on Evolutionary Multi-Criterion Optimization. Volume 1993 of Lecture Notes in Computer Science. Springer Verlag (2001) 21–40
2. Coello Coello, C.A.: An updated survey of GA-based multiobjective optimization techniques. Technical Report RD-98-08, Laboratorio Nacional de Informática Avanzada (LANIA), Xalapa, Veracruz, México (1998)

3. Fonseca, C.M., Fleming, P.J.: An overview of evolutionary algorithms in multiobjective optimization. Evolutionary Computation **3(1)** (1995) 1–16
4. Deb, K., Agrawal, S., Pratap, A., Meyarivan, T.: A fast elitist non-dominated sorting genetic algorithm for multi-objective optimization: NSGA-II. In Schoenauer, M., Deb, K., Rudolph, G., Yao, X., Lutton, E., Merelo, J.J., Schwefel, H.P., eds.: Proceedings of the Parallel Problem Solving from Nature VI Conference. Volume 1917 of Lecture Notes in Computer Science., Paris, France, Springer (2000) 849–858
5. Zitzler, E., Laumanns, M., Thiele, L.: SPEA2: Improving the strength pareto evolutionary algorithm. Technical Report 103, Computer Engineering and Networks Laboratory (TIK), Swiss Federal Institute of Technology, Zurich, ETH Zentrum, Gloriastr 35, CH-8092 Zurich, Switzerland (2001)
6. Farina, M., Amato, P.: Fuzzy optimality and evolutionary multiobjective optimization. In C.M. Fonseca et al., ed.: EMO 2003, LNCS 2632. (2003) 58–72
7. Köppen, M., Franke, K., Nickolay, B.: Fuzzy-pareto-dominance driven multiobjective genetic algorithm. In: Proceedings of the 10th IFSA World Congress. (2003) 450–453
8. Okabe, T., Jin, Y., Sendhoff, B.: A critical survey of performance indices for multiobjective optimization. In: Proceedings of the 2003 Congress on Evolutionary Computation (CEC'2003), Canberra, Australia, IEEE Press (2003) 878–885
9. Köppen, M., Vicente Garcia, R.: A fuzzy scheme for the ranking of multivariate data and its application. In: Proceedings of the 2004 Annual Meeting of the NAFIPS (CD-ROM), Banff, Alberta, Canada, NAFIPS (2004) 140–145
10. Corne, D.W., Knowles, J.D., Oates, M.J.: The pareto envelope-based selection algorithm for multiobjective optimization. In: Proceedings of the Parallel Problem Solving from Nature VI Conference. Volume 1917 of Lecture Notes in Computer Science., Springer (2000) 839–848

Appendix

Derivation of Theorems 1 and 2

Here, we shortly sketch the derivation of Theorems 1 and 2. For general dimensions n we assign ranking vectors to all m point. We indicate the coordinate directions with $x_1, x_2, \ldots, x_n$. If a point P_i gets assigned the ranking vector $(i_1, i_2, \ldots, i_n)$ with $i_k \in \{1, 2, \ldots, m\}$ this means that the x_1 coordinate of this point is the i_1-th smallest among all the m x_1-coordinates of the m points, the x_2-coordinate is the i_2-th smallest among all x_2-coordinates of the m points and so forth. For all m points and all $1 \leq k \leq n$, the ranking vectors at position k form a permutation of the set $\{1, 2, \ldots, m\}$. So, a set of n permutations of the set $\{1, 2, \ldots, m\}$ may be derived from any selection of m points. However, a ranking scheme is already sorted into one dimension, e.g. the x_1-dimension, so only $(n-1)$ permutations are independent. As a result, there are $m!^{n-1}$ different ranking schemes for m points in n dimensions.

Among the m points there is one point P_l with the largest x_1-coordinate value. We consider the *projection* of all m points into the $(n-1)$-dimensional ranking scheme, spanned by the coordinates $x_2, x_3, \ldots, x_n$ and $x_1 = n$. There are

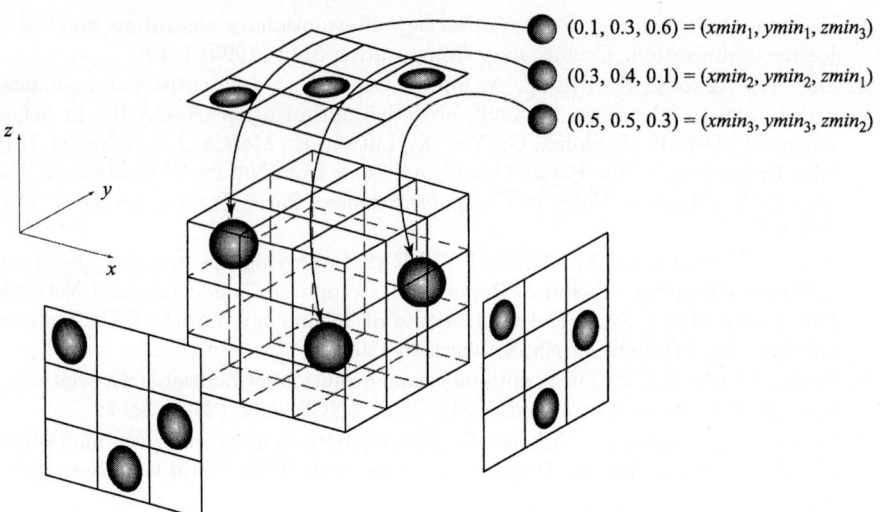

Fig. 5. Entering the three points $(0.1, 0.3, 0.6), (0.3, 0.4, 0.1)$ and $(0.5, 0.5, 0.3)$ into one of the $3!^2 = 36$ ranking schemes for 3 points in 3 dimensions, together with the projections into 2-dimensional ranking schemes of 3 points. Two of the projections specify the ranking scheme completely, since the points are already sorted in the third dimension. The analysis is based on the fact that the PARETO set size of this ranking scheme is the PARETO set size of the two points with the lowest x-coordinate plus 1 in case the point with the largest x-coordinate is projected onto a point of the PARETO set of the projected ranking scheme, which is parallel to the (y, z)-plane

$m!^{n-2}$ such ranking schemes, and each of them will be obtained by the projection of $m!$ different n-dimensional ranking schemes (since $m!^{n-1}/m!^{m-2} = m!$).

Now, a moment of reasoning gives that the point P_l will belong to the PARETO set of the n-dimensional ranking scheme if and only if its projection belongs to the PARETO set of the $(n-1)$-dimensional projected ranking scheme: We are considering whether the point P, onto which P_l is projected, belongs to the PARETO set of the $(n-1)$-dimensional projected ranking scheme or not. If it belongs to the PARETO set of the $(n-1)$-dimensional ranking scheme, this means that there is no other point having lower ranking positions than P in all the dimensions $x_2, x_3, \ldots, x_n$ simultaneously. But this means that there is also no point in the n-dimensional ranking scheme having lower ranking positions than P_l in all the dimensions $x_1, x_2, x_3, \ldots, x_n$ simultaneously as well. Thus, if P_l is projected onto an element of the PARETO set of the $(n-1)$-dimensional ranking scheme, it belongs to the PARETO set of the n-dimensional ranking scheme.

If P does not belong to the PARETO set of the $(n-1)$-dimensional projected ranking scheme, there is another point P_1 in the projected $(n-1)$-dimensional ranking scheme having lower ranking positions than P in all dimensions $x_2, x_3, \ldots, x_n$ simultaneously. Once P_l is projected onto this point, it has to be taken into account that P_l is the point with the highest x_1-coordinate value, so *each* other point in the n-dimensional ranking scheme will have a lower x_1-coordinate value, including the point from which P_1 originated. So, there is a point in the n-dimensional ranking scheme dominating P_l and P_l will not belong to the PARETO set of the n-dimensional ranking scheme.

The PARETO set of the $(m-1)$ points different from P_l is not influenced by P_l, since P_l can never dominate any of these points (it fails already to dominate in the x_1-coordinate). So, the size of the PARETO set of the m points is either P_s or $P_s + 1$, with P_s being the size of the PARETO set of the $(m-1)$ points that are different from P_l. It is $P_s + 1$ if P_l belongs to the PARETO set of the n-dimensional ranking scheme, P_s otherwise. But we have just seen that P_l belongs to the PARETO set of the n-dimensional ranking scheme if and only if it belongs to the PARETO set of the $(n-1)$-dimensional projected ranking scheme. Putting it all together:

- There are $m!^{n-1}$ different ranking schemes for m points in the n-dimensional hypercube. We denote them by $R_1, R_2, \ldots, R_N$.
- Each of these ranking schemes R_i can be related to a ranking scheme S_j of $(m-1)$ points in n dimensions by removing the point P_l with the largest x_1-coordinate ($j = 1, \ldots, (m-1)!^{n-1}$). Given any ranking scheme S_j for $(m-1)$ points in the n-dimensional hypercube, it can be made a ranking scheme of m points in n dimensions by adding a "last" point P_l to it. The number of ways to add the m-th point to a ranking scheme S_j does not depend on the ranking scheme itself, so any S_j can generate the same number of R_i, say l. It follows $l = m!^{n-1}/(m-1)!^{n-1} = m^{n-1}$.
- Each R_i can be projected into a ranking scheme s_i of m points in the $(n-1)$-dimensional hypercube by removing the ranking according to the x_1-coordinate ($i = 1, \ldots, m!^{n-2}$). There are always $m!$ ranking schemes R_i that are projected into the same s_j.
- The size of the PARETO set of a ranking scheme R_i will be denoted by r_i, the size of the PARETO set of a ranking scheme S_i by p_i and the size of the PARETO set of a ranking scheme s_i by q_i.
- From the $m!$ cases that a ranking scheme R_i is projected onto a ranking scheme s_j with PARETO set size q_j, in exactly $m! \cdot q_j/m = (m-1)! \cdot q_j$ cases its point P_l with the largest x_1-coordinate will be projected into an element of the PARETO set of s_j, thus belong to the PARETO set of R_i in addition to the r_i points that comprise the PARETO set of the $(m-1)$ points different from P_l.

Now we sum the PARETO set sizes over all R_i and divide by the number of R_i to get the expectation value $e_m(n)$. We can decompose this sum into two contributions: the contributions coming from the reduced ranking schemes S_i with PARETO set sizes p_i and the contribution coming from the PARETO sets of the projected ranking schemes s_i with PARETO set sizes q_i:

$$e_m(n) = \frac{1}{m!^{n-1}} \sum_{k=1}^{m!^{n-1}} r_k \tag{8}$$

$$= \frac{1}{m!^{n-1}} \left[m^{n-1} \sum_{k=1}^{(m-1)!^{n-1}} p_k + (m-1)! \sum_{k=1}^{m!^{n-2}} q_k \right]$$

$$= \frac{1}{m!^{n-1}} \left[m^{n-1} \cdot e_{m-1}(n) \cdot (m-1)!^{n-1} + (m-1)! \cdot e_m(n-1) \cdot m!^{n-2} \right]$$

$$= e_{m-1}(n) + \frac{1}{m} e_m(n-1)$$

by using that the sum of all p_i equals the expectation value for $(m-1)$ points in the n-dimensional hypercube times the number $(m-1)!^{n-1}$ of their ranking schemes S_i, and the sum of all q_i equals the expectation value for m points in $(n-1)$ dimensions times the number $m!^{n-2}$ of their ranking schemes s_i.

When adding the obvious relations $e_1(n) = e_m(1) = 1$ this will give Theorem 1. By showing that the expression in eq. (6) fulfills the recursive equation in Theorem 1, Theorem 2 can be established as well. The proof goes via complete induction over n and m.

Multi-objective Optimization of Problems with Epistemic Uncertainty

Philipp Limbourg

Institute of Information Technology, Department of Engineering,
University of Duisburg-Essen, Bismarckstr. 90,
47057 Duisburg, Germany
limbourg@uni-duisburg.de
http://iit.uni-duisburg.de

Abstract. Multi-objective evolutionary algorithms (MOEAs) have proven to be a powerful tool for global optimization purposes of deterministic problem functions. Yet, in many real-world problems, uncertainty about the correctness of the system model and environmental factors does not allow to determine clear objective values. Stochastic sampling as applied in noisy EAs neglects that this so-called epistemic uncertainty is not an inherent property of the system and cannot be reduced by sampling methods. Therefore, some extensions for MOEAs to handle epistemic uncertainty in objective functions are proposed. The extensions are generic and applicable to most common MOEAs. A density measure for uncertain objectives is proposed to maintain diversity in the nondominated set. The approach is demonstrated to the reliability optimization problem, where uncertain component failure rates are usual and exhaustive tests are often not possible due to time and budget reasons.

1 Introduction

The traditional way to define optimization problems is to create a model of the system and state it to be exact and deterministic. Clearly defined decision values are mapped to likewise clearly defined, non-varying objective values.

Respecting the fact that nature doesn't adhere to determinism, stochastic optimization problems and their evolutionary solution methods emerged and gained importance [1]. Yet, the main part of this approaches still abide to the certainty of observed objectives. High sampling rates of a given decision value could simply reveal the underlying distribution of the random processes modelled by the system [2,3]. MOEA approaches dealing with aleatory uncertainty are presented in [4] and [5].

Models of real systems are built without perfect knowledge of the system simulated. Often the objective values stay highly uncertain even if the real (aleatory) variance is minimal because of a fundamental lack of information about environmental factors or the system itself. In this case even infinitive sampling rates won't help as we simply don't know the distributions to sample from. This so-called epistemic uncertainty must not be ignored in the optimization process.

Indeed, there is a trend in reliability science and other application areas to formulate models that incorporate and propagate epistemic uncertainties to the simulation outputs rather than generating sharp values. The results are very often only given as intervals or belief functions and thus a need for algorithms capable to handle this types of data is needed. Some approaches towards this issue can be found in [6] and [7].

This work is structured as followed. Section 2 introduces epistemic uncertainty modelling and its mathematical and computational representation. Section 3 discusses different possible extensions of common decision criteria for single and multi-objective evolutionary algorithms (MOEAs). Two ways of redefining the Pareto order over objective vectors and the order over one-dimensional objective functions are introduced. Section 4 shows a way to extend standard MOEAs to handle uncertain objectives in both selection and repository processes. The extension is generic and thus could be applied to most of the commonly used MOEAs. Section 5 proposes a niching strategy to prevent diversity among uncertain solutions. Section 6 shows the application of the proposed approach to the reliability design problem. Finally, some outlines and further research directions are proposed.

2 Belief, Plausibility and the Representation of Epistemic Uncertainty

2.1 Aleatory and Epistemic Uncertainty

There are at least two types of uncertainty that have to be distinguished because of their difference in origin, modelling and effects: Aleatory and epistemic uncertainty. Oberkampf et al. [8] defines aleatory uncertainty as the "inherent variation associated with the physical system or the environment under consideration". Aleatory uncertainty of a quantity can often be distinguished from other types of uncertainty by its characterization as a random value with known distribution. The exact value will change but is expected to follow the distribution. A simple example for aleatory uncertainty is the uncertainty about the outcome of a dice toss $X \in \{1,2,3,4,5,6\}$. We are uncertain about the number we will receive, but we are sure that each of the numbers will occur with a probability $p(X = 1) \cdots p(X = 6) = 1/6$.

On the contrary, epistemic uncertainty describes not uncertainty about the outcome of some random event due to system variance but the uncertainty of the outcome due to "any lack of knowledge or information in any phase or activity of the modelling process"[8]. This shows the important difference between this two types of uncertainty. Epistemic uncertainty is not an inherent property of the system. A gain of information about the system or environmental factors can lead to a reduction of epistemic uncertainty. We now focus again on the dice example. Somebody told us that the dice is pronged and so we expect that the probability is limited as $p(X = 1) \ldots p(X = 6) \in [1/12, 7/12]$. Of course, the dice would follow a distribution and if we would carry out an infinite number of experiments, we would find out that it is $p(X = 1)\ldots p(X = 5) = 1/9$, $p(X = 6) = 4/9$. Before we do this, we don't have enough information to assume any possible distribution without neglecting that reality may be anywhere else. Hence, epistemic uncertainty is our inability to model reality.

Epistemic uncertainty is often ignored and some arbitrary distribution over the uncertain value is stated as "the best/most realistic/most intuitive". Alternative approaches [9, 10] include epistemic uncertainty in the modelling process and apply frameworks of probability calculus that allow arithmetic with uncertain probability values.

2.2 The Dempster-Shafer-Framework of Evidence

The probabilistic calculus used in this work is the Dempster-Shafer-Framework of evidence first described by Dempster [11] and extended by Shafer [12]. It has proven to be a well-suited framework for representing both epistemic and aleatory uncertainty and has found application in various fields [13, 14]. Thus, a short introduction to the general concepts of this theory is given here. A more detailed overview can be found in [15].

In the classical discrete probability calculus, a probability mass $m(a)$ is defined for each possible value of X and $p(X = a) = m(a)$. Dempster-Shafer-Structures on the real line are similar to discrete distributions with one important difference. The probability mass function is not a mapping $\mathbb{R} \to [0,1]$ but instead a mapping from $2^{\mathbb{R}} \to [0,1]$, where probability masses are assigned to sets instead of discrete values. A Dempster-Shafer-Structure can be described by its basic probability assignment (bpa) or by its focal elements. For the problems modelled, it is adequate (but not necessary) to restrict the focal elements to intervals rather than more complicated sets.

Definition 1. *A basic probability assignment m over the real line is a mapping $m : 2^{\mathbb{R}} \to [0,1]$ provided:*

$$m(\emptyset) = 0 \tag{1}$$

$$\sum_{B \subseteq \mathbb{R}} m(B) = 1 \tag{2}$$

Definition 2. *A focal element $A = [\underline{a}, \overline{a}] \subseteq \mathbb{R}$ is an interval with a nonzero mass $m(A) > 0$.*

Because of the uncertainty modelled it is not possible to give an exact probability $p(X \in B)$ for a value or interval B, yet upper and lower bounds can be calculated. Associated with each bpa are two functions $Bel, Pl : 2^{\mathbb{R}} \to [0,1]$ which are referred to as belief and plausibility of an event.

Definition 3. *The belief and plausibility of an interval $B \subseteq \mathbb{R}$ are given by*

$$Bel(B) = \sum_{A \subseteq B} m(A) \tag{3}$$

$$Pl(B) = \sum_{A \cap B \neq \emptyset} m(A) \tag{4}$$

It is obvious that $Bel(B) \leq Pl(B)$ because $A \subseteq B \Rightarrow A \cap B \neq \emptyset$. In fact $Bel(B)$ and $Pl(B)$ can be interpreted as bounds on the probability $p(X \in B)$.

Informally the belief function represents the maximal value that we despite all epistemic uncertainty "believe" to be smaller than $p(X \in B)$, the plausibility function represents the highest "plausible" value of $p(X \in B)$. Belief and plausibility values are used in this work to describe the output of an objective function.

3 Decision Criteria Using Belief Functions

Classical optimization problem formulations are mappings from the decision space X which could be of any type to an objective space $Y \subseteq \mathbb{R}^n$ where the goal is to find a vector $\mathbf{x}_{opt} \in X$ which maximizes the objective function

$$f : X \to Y$$

$$f(\mathbf{x}) = \mathbf{y} = \begin{pmatrix} y_1 \\ \vdots \\ y_n \end{pmatrix} \tag{5}$$

respective to a partial order relation $\succ$ over Y. If $n = 1$, then we deal with a single-objective optimization problem and $\succ$ is a total order relation (the "bigger than" relation $>$). If $n > 1$, then there is a need to define an order relation $\succ$ over vectors $\mathbf{y} \in \mathbb{R}^n$. In the following work, the order relation inside vector dimension $i \in 1...n$ of the objective vector are denoted as $\succ^i$ while relations between objective vectors are described as $\succ$. Two common approaches should be named here: The aggregation approach and the Pareto approach. The aggregation approach uses an aggregation function u which maps the objective space to the real line (or an arbitrary other space providing a total order).

$$\begin{aligned} u &: Y \to \mathbb{R} \\ u(\mathbf{y}) &= z \in \mathbb{R} \end{aligned} \tag{6}$$

The resulting order relation is backpropagated to Y. The second approach that has gained much attention in evolutionary optimization during the last years is the Pareto approach using the Pareto dominance criterion [16].

Definition 4. *A vector* $\mathbf{y} \in \mathbb{R}^n$ *Pareto dominates another vector* $\mathbf{y}' \in \mathbb{R}^n$ *(*$\mathbf{y} \succ_p \mathbf{y}'$*) if:*

$$\forall i \in 1...n : y_i \geq y'_i \tag{7}$$
$$\exists i \in 1...n : y_i > y'_i \tag{8}$$

The Pareto relation relies on the total order inside the dimensions of the element vector. But what if the results of f are uncertain values, represented as intervals of belief and plausibility values. Formally the objective function changes to:

$$f : X \to Y \subseteq (\mathbb{R} \times \mathbb{R})^n$$
$$f(\mathbf{x}) = \mathbf{y} = \begin{pmatrix} [\underline{y}_1, \overline{y}_1] \\ \vdots \\ [\underline{y}_n, \overline{y}_n] \end{pmatrix} \tag{9}$$

There are several ways to define partial or total order relations over intervals that can make sense depending on the application.

Definition 5. *An interval $y = [\underline{y}, \overline{y}] \subseteq \mathbb{R}$ dominates another interval $y' = [\underline{y}', \overline{y}'] \subseteq \mathbb{R}$ certain ($y \succ_c y'$), if*
$$\underline{y} > \overline{y}' \tag{10}$$

Definition 6. *An interval $y = [\underline{y}, \overline{y}] \subseteq \mathbb{R}$ dominates another interval $y' = [\underline{y}', \overline{y}'] \subseteq \mathbb{R}$ uncertain ($y \succ_{uc} y'$), if*
$$\underline{y} \geq \underline{y}' \wedge \overline{y} \geq \overline{y}' \wedge y \neq y' \tag{11}$$

The certain and uncertain domination criterion is only a partial order. If e.g. $y' \subset y$, then neither $y \succ y'$ nor $y' \succ y$ holds. $y\|y'$ will be used to denote this indifference. Certain dominance is a stronger criterion than uncertain dominance ($y \succ_c y' \Rightarrow y \succ_{uc} y'$). If the certain dominance criterion holds, we can be sure that an uncertain value is better than another. This cannot be inferred by uncertain dominance. Nevertheless it could be argued that uncertain dominance is also a reasonable relation to speed up the optimization process because it does not stay indifferent when there is high uncertainty.

Both certain and uncertain dominance can't help if $y \subset y'$. There is no straightforward relation without background knowledge that can give us a hint which value is superior. Many different interval aggregation functions of the form

$$\begin{aligned}\mathbb{R} \times \mathbb{R} &\to \mathbb{R} \\ f_{agg}(y) &= z \in \mathbb{R}\end{aligned} \tag{12}$$

have been proposed that map intervals to a total ordered space. We will denote this relation as $\succ_{agg}$. A good survey is given in [17]. Their big drawback is that they explicitly or implicitly assume a distribution on y like averaging or the maximum entropy approach [18].

In case of multi-objective problems we have to redefine the Pareto dominance relation if we deal with partial orders in the vector dimensions. Definition 4 can be extended for partial orders as follows.

Definition 7. *A vector $\mathbf{y}$ Pareto dominates another vector $\mathbf{y}'$ weak ($\mathbf{y} >_w \mathbf{y}'$), if*
$$\begin{aligned}\forall i \in 1...n &: y_i \succ^i y_i' \vee y_i \| y_i' \vee y_i = y_i' \\ \exists i \in 1...n &: y_i \succ^i y_i'\end{aligned} \tag{13}$$

Definition 8. *A vector* y *Pareto dominates another vector* y′ *strong* (y $>_s$ y′), *if*

$$\forall i \in 1...n : y_i \succ^i y'_i \vee y_i = y'_i$$
$$\exists i \in 1...n : y_i \succ^i y'_i \quad (14)$$

Both relations reduce to $\succ_p$ when $i = 1...n : y_i$ and y'_i are degenerated intervals ($\underline{y}_i = \overline{y}_i$) or $i = 1...n :\succ^i$ are total orders. The one to be applied depends again on the aims of the user. Strong dominance will only hold if y is at least equal to y′ in all objectives and better in one while weak dominance holds if y is at least indifferent y′ in all objectives and better in one.

4 Algorithmic Approach

In this section, an approach to integrate the introduced relations in a Pareto-based multi-objective evolutionary algorithm (PMOEA) is proposed. Various different PMOEAs are used amongst practitioners, possibly the most popular are NSGA2 [19] and SPEA2 [20]. Almost all PMOEAs follow an algorithmic scheme as given in Fig. 1. Binary tournament selection will be used because of its common use in PMOEAs and for the same reason one-point crossover. In two different parts, a relation over Y is used:

Selection relation R_{sel}. This relation is used by the tournament selection operator. It is applied each time two individuals are compared in the selection process. The winner survives and can generate offsprings.

Repository relation R_{rep}. This relation defines the nondominated solutions stored in the repository. Is is used only in PMOEAs which maintain a set of currently nondominated solutions found during the optimization process. Its task is to determine which solutions from both population and repository are nondominated and thus have to be kept.

If we deal with certain objective values, both R_{sel} and R_{rep} are standard Pareto relations $\succ_p$ while the relations inside the objective functions $\succ^{1...n}$ are "bigger than" operators over the real line. We thus denote $R_{sel} = R_{rep} = (\succ, \{i = 1...n :>^i\})$. Which types of relations can be meaningful for R_{sel} and R_{rep} in case

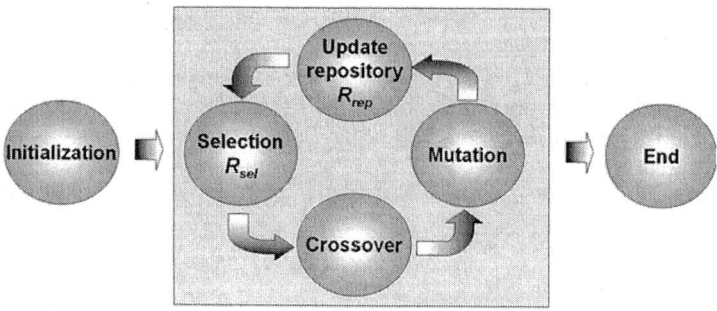

Fig. 1. Standard PMOEA algorithm

of uncertain values? In [5], an aggregation function for intervals is used. A known distribution over the intervals $[\underline{y},\overline{y}]$ and $[\underline{y}',\overline{y}']$ is assumed and the expected values $E([\underline{y},\overline{y}])$ and $E([\underline{y}',\overline{y}'])$ are compared mapping the problem to a standard multi-objective form. The relations used are $R_{sel} = R_{rep} = (\succ, \{i=1...n : \succ^i_{agg}\})$. This approach is not feasible in our context, where the distribution could not be assumed without risk. If the model designers and parameter estimators are not able to come up with a distribution inside their uncertain results, how could the optimizer be able to assume the correct one. Before selecting among the proposed relations, desired characteristics of R_{sel} and R_{rep} will be outlined which lead to a choice use in the algorithm.

Selection Relation R_{sel}

High sensitivity. The selection relation should be able to decide between different solutions as often as possible. Therefore, a relation with a high degree of indifference could degrade the evolution process to a randomized search.

Lower accuracy. Too many "wrong" decisions between individuals have to be prevented. Yet, when dealing with uncertainty we have to make a compromise between the incommensurable goals sensitivity and accuracy. If we want a sensitive decision criterion, some decisions that would have been wrong knowing the certain values can't be prevented. As EAs have proven to be robust optimization methods, sensitivity is more important.

Repository Relation R_{rep}

High accuracy. In the repository, candidate solutions for a posteriori selection of the user are stored. Using uncertain objective functions, we cannot simply leave out a solution that occupies an interesting region of the objective space only because it is eventually worse than another solution. We must be certain that this solution is dominated.

Low sensitivity. The impact of solutions from the repository on the optimization process is low. The repository is often passive and there is no feedback to the current population. The optimization speed therefore can't be slowed down by a high amount of indifference. Yet if the repository is bounded in size, high indifference can rapidly lead to an overflow of nondominated solutions. This problem is well-known and many different methods to restrain the size while preserving diversity of the nondominated set have been proposed [21, 19, 20, 22]. A method handling uncertainty is shown in section 5. Therefore relations should be chosen that guarantee high accuracy.

Taking into account this demands, the following choice of relations is proposed:

$$R_{rep} = (\succ_s, \{i=1...n : \succ^i_{ic}\}) \quad (15)$$

$$R_{sel} = (\succ_w, \{i=1...n : \succ^i_{iuc}\}) \quad (16)$$

Table 1. Selection relations $TS1$ and $TS2$: Comparison results

	Bel	Pl
$m(TS1)$	1.520	3.415
$m(TS2)$	1.363	3.051
$v(TS1)$	0.021	0.097
$v(TS2)$	0.026	0.101
$P(H_0)$	$2.31E-8$	$2.33E-9$

The strong Pareto dominance $\succ_s$ guarantees that only solutions are thrown out of the archive, if at least one solution is equal or better in all objectives. Indifference in only one objective prevents the solution from being eliminated. This fulfils the need for accuracy. For all objective dimensions, the certain domination $\succ_{ic}^i$ should be used as it is the only presented relation that provides maximal accuracy on this level.

The selection relation gains sensitivity through the uncertain dominance relation $\succ_{iuc}$ that is much less indifferent than $\succ_{ic}$. The vector relation is also relaxed to the weak Pareto dominance $\succ_w$, which allows us to make a decision between two alternatives even if there is indifference in some objectives. To prove this arguments, 100 runs on each of the test sets $TS1 : R_{sel} = (\succ_w, \{i = 1...n :\succ_{iuc}^i\})$ and $TS2 : R_{sel} = (\succ_s, \{i = 1...n :\succ_{ic}^i\})$ were carried out. Test problem and parameter settings are presented in detail in 6. The results were evaluated with a normalized hypervolume metric. Table 1 shows the median m and variance v of the hypervolume values (belief and plausibility). Furthermore both belief and plausibility values were tested on the equality of medians $P(H_0)$ by a Wilcoxon signed rank test. The highly significant results show that $TS1$ performs better and the selection relation choice is meaningful.

5 A Niching Strategy for Uncertain Objectives

In this section, a straightforward extension of the nearest neighbor method as described in [23] is introduced. This density estimate is used e.g. in SPEA2 [20] and has proven to be effective in both conserving diversity and promote variability among solutions in the repository. [5] extends the method to uncertain solutions with known distributions using expectation values. Each nondominated solution with uncertain objective vector y is assigned a fitness value $f_{del}(y)$ which depends on the distance of the k-nearest neighbor. As the scales of the objective dimensions may differ in orders of magnitude, the Euclidean distance is normalized by the maximal extension $\hat{d}$ of the nondominated repository Rep = $\{y', y'', ...\}$:

$$\hat{d}_i = \frac{1}{\max_{y' \in \text{Rep}}(y'_i) - \min_{y'' \in \text{Rep}}(y''_i)} \qquad (17)$$

The normalized distance between two objective vectors $\mathbf{y}, \mathbf{y}'$ is then given as:

$$d(\mathbf{y}, \mathbf{y}') = \sqrt{\sum_{i=1\ldots n} \hat{d}_i^{\,2} \cdot (y_i - y_i')} \tag{18}$$

Due to uncertainty of $\mathbf{y}$ and $\mathbf{y}'$, $d(\mathbf{y}, \mathbf{y}')$ is also an uncertain value with belief and plausibility. Interval calculus defines that

$$Bel(d(\mathbf{y}, \mathbf{y}')) = \min_{\mathbf{a} \in \mathbf{y}, \mathbf{a}' \in \mathbf{y}'} (d(\mathbf{a}, \mathbf{a}')) \tag{19}$$

$$Pl(d(\mathbf{y}, \mathbf{y}')) = \max_{\mathbf{a} \in \mathbf{y}, \mathbf{a}' \in \mathbf{y}'} (d(\mathbf{a}, \mathbf{a}')) \tag{20}$$

and thus require twice the solution of an optimization problem. As $\sqrt{\ }$ and $\sum$ are monotonically growing functions, the problems reduce to:

$$\min_{a_i \in y_i,\, a_i' \in y_i'} (a_i - a_i')^2 \tag{21}$$

$$\max_{b_i \in y_i,\, b_i' \in y_i'} (b_i - b_i')^2 \tag{22}$$

Both problems can be analytically solved as:

$$\min_{a_i \in y_i,\, a_i' \in y_i'} (a_i - a_i')^2 = \begin{cases} 0, & \text{if } bel(y_i) - pl(y_i') < 0 \vee bel(y_i') - pl(y_i) < 0 \\ \min((bel(y_i) - pl(y_i'))^2, (bel(y_i') - pl(y_i))^2) & \text{else} \end{cases} \tag{23}$$

$$\max_{a_i \in y_i,\, a_i' \in y_i'} (a_i - a_i')^2 = \max((bel(y_i) - pl(y_i'))^2, (bel(y_i') - pl(y_i))^2) \tag{24}$$

The distance of the k-nearest neighbour $d_k(\mathbf{y})$ is then defined as the kth normalized distance sorted in ascending order:

$$d_1(\mathbf{y}) \leq d_2(\mathbf{y}) \ldots \leq d_n(\mathbf{y}) \tag{25}$$

$Bel(d_k(\mathbf{y}))$ and $Pl(d_k(\mathbf{y}))$ are given as the kth distance of the sorted Belief/Plausibility values $Bel(d(\mathbf{y}, \mathbf{y}'))$, $Pl(d(\mathbf{y}, \mathbf{y}'))$. The deletion fitness function $f_{del}(\mathbf{y})$ is then defined as:

$$f_{del}(\mathbf{y}) = -d_k(\mathbf{y}) \tag{26}$$

If the repository grows above the constraining size, individuals are selected by binary tournament selection and their deletion fitness f_{del} is compared by the $\succ_{uc}$ relation. The dominated individual is deleted from the repository. If the comparison stays indifferent, one of the individuals is chosen at random. This approach differs from [5] where a distribution inside the intervals is known or at least assumed. To preserve solutions at the edges of the nondominated set, the solutions with best plausibility or belief values in one objective obtain $f_{del}(\mathbf{y}) = -\infty$. The right choice of the parameter k is also a critical problem. Small values prevent small clusters while large values lead to a more global diversity but do not prevent small clusters. [23] suggests to the square root of the considered point number, so $k = \sqrt{|Rep|}$ is chosen in the examples.

6 Application - The Optimal Reliability Design Problem

A practical application of uncertain data is the reliability design problem as recently presented e.g. in [24, 25], and in a comprehensive overview of older approaches in [26]. In the time where a wide range of "Commercial-of-the-Shelf" (COTS) components with different characteristics is available, it becomes a big problem to choose the best component combination for a technical system that is both reliable and cheap in production. Zafiropoulos and Dialynas [27] describe this task as a multi-objective optimization problem where they regard only components with constant failure rates. Their chosen objectives were cost and system failure rate (95% confidence interval) that was obtained through Monte Carlo simulation of the system. In this work, a different formulation which aims to find an optimal design regarding cost and system mean time to failure (MTTF) is defined:

1. Maximize the system mean time to failure $MTTF(\mathbf{x})$. $MTTF(\mathbf{x})$ is defined as the expected time until the system consisting of components $x_1...x_n$ fails.
2. Minimize the system costs $CS(\mathbf{x})$.

The system (Fig. 2) is given as a reliability block diagram $G = (\{C_{1...n}\}, E)$ [28], a special case of a stochastic flow network, where each block symbolizes a component placeholder that can be replaced by one of several component alternatives (Table 2). The system is functional if and only if there is a path of working components from the source node A to the sink node B. The calculation of system reliability from component reliability is performed by the minimal cut set method [28] that can be extended to handle uncertain probabilities. The system reliability was obtained as a Dempster-Shafer-Structure and the MTTF given as the uncertain expected value. Component failure probability were defined in one of three possible ways.

1. Exponential failure distribution $f_{\exp}(t) = \frac{1}{\lambda}e^{-\frac{t}{\lambda}}$ with uncertain parameter λ (Exp).
2. Weibull failure distribution $f_{weib}(t) = \alpha\beta t^{\beta-1}e^{-at^\beta}$ with uncertain parameters α, β (Weib).
3. Estimates of an arbitrary failure distribution defined by a bpa (Raw).

In system reliability analysis, two distribution functions have gained high popularity for modelling component failure functions. Electronic components neither

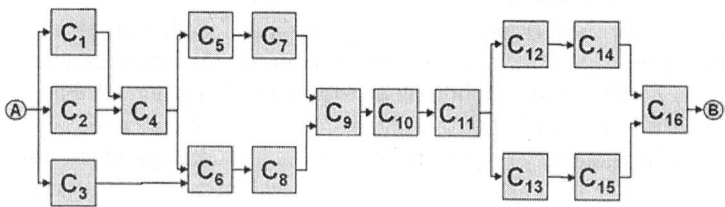

Fig. 2. Example system in block diagram structure

Table 2. Exemplary component failure and cost data

Components	Choice	Type	Parameters	Costs
1	1	Exp.	λ=[33000,43000]	[1000,1010]
	2	Exp.	λ=[51000,61000]	[1100,1105]
	3	Exp.	λ=[75000,82000]	[1200,1210]
	4	Exp.	λ=[88000,96000]	[1280,1290]
2	1	Weib.	α=[15000,20000], β=[0.75,0.85]	[970,1000]
	2	Weib.	α=[29000,35000], β=[0.7,0.8]	[1210,1220]
	3	Weib.	α=[45000,54000], β=[0.8,0.9]	[1120,1140]
	4	Weib.	α=[70000,90000], β=[0.7,0.85]	[1150,1160]
3	1	Raw	[0,4000,m=0.1], [3000,15000,m=0.1], [12000,19000,m=0.1], [18000,30000,m=0.1], [25000,50000,m=0.15], [45000,80000,m=0.15], [60000,90000,m=0.15], [80000,128000,m=0.15]	[635,640]
	2	Raw	[0,8000,m=0.1], [5000,12000,m=0.1], [12000,25000,m=0.1], [22000,40000,m=0.1], [35000,60000,m=0.15], [40000,90000,m=0.15], [60000,100000,m=0.15], [80000,180000,m=0.15]	[710,725]
	3	Raw	[0,9000,m=0.05], [6000,15000,m=0.1], [10000,30000,m=0.1], [28000,49000,m=0.05], [45000,87000,m=0.25], [70000,120000,m=0.25]	[780,785]
...	...	...	...	...

suffer heavily from wear out nor from teething problems if tested before. Because of its property of being memoryless (constant failure rate) the exponential distribution has proven to be adequate for modelling such parts.

Mechanical systems normally tend to degrade over time and therefore require more complex distribution types. This is modelled through a two-parameter Weibull distribution which is often used to estimate component failures from field failure and accelerated lifetime test data [29].

A third possibility is the specification of an arbitrary failure function by expert estimates which is represented through focal elements of a Dempster-Shafer-Structure.

The component costs are considered with low uncertainty as they are normally available at a stipulated price. Yet, there are sources of uncertainty, e.g. unknown integration costs. This costs can be especially high for mechanical parts where interchanging a component can lead to substantial design changes. In a commonly used cost function [30], $CS(x)$ is calculated as the sum of all component costs:

$$CS(\mathbf{x}) = - \sum_{i=1...n} CS(c_i) \qquad (27)$$

Using this system, exemplary runs with repository size 25 and 100 were carried out. Mutation probability was set to 0.1, crossover probability to 0.9. The runs over 100 generations were carried out with a population size of 20. The results (Fig. 3-5) show a sample front of nondominated solutions regarding the proposed repository relations $R_{rep} = (\succ_s, \{i = 1...n :\succ_{ic}^i\})$ after 100 generations. The nondominated solutions of repository size 25 (shown as rectangles to reflect uncertain values in Fig. 3) are forming a diverse front. Repository size 100 spreads even better and achieves better results. Due to the clarity of visualization, only belief and plausibility were plotted (Fig. 5). The overlapping intervals could be seen in a contour plot of the

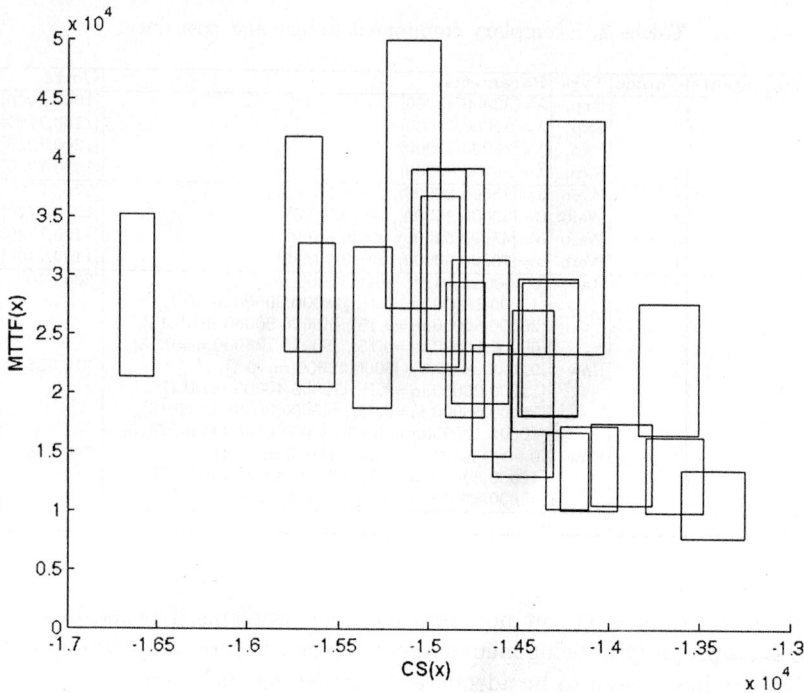

Fig. 3. Nondominated repository of size 25 after 100 generations

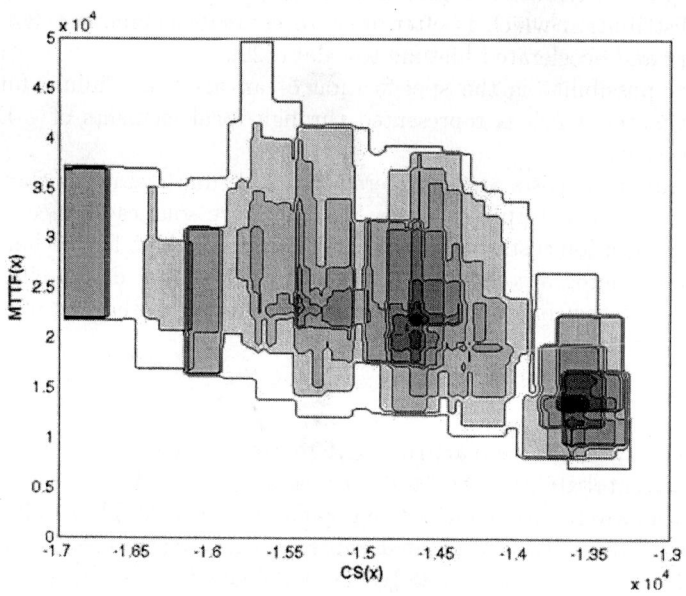

Fig. 4. Nondominated repository of size 100 after 100 generations, density plot

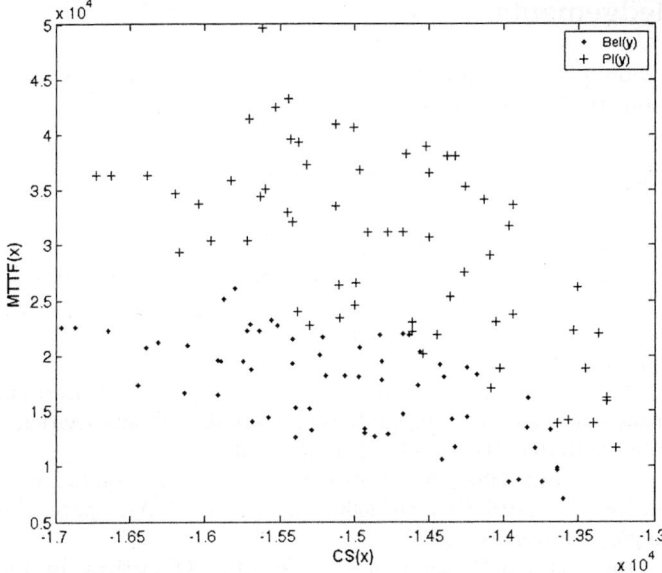

Fig. 5. Nondominated repository of size 100 after 100 generations (belief and plausibility values)

repository assuming uniform distribution inside the uncertain values (Fig. 4). As said before, this distribution is not used as density estimation and serves only for visualization purposes. From this nondominated set of uncertain results, the decision maker could a posteriori choose a solution to realize.

7 Conclusions and Further Research Directions

Problems incorporating epistemic uncertainty are of great practical importance as parametric and modelling knowledge is never perfect. The initial approach proposed shows that multi-objective evolutionary optimization is possible even if the objective function is disrupted by uncertainty resulting from a lack of knowledge. Some general and easy-to-implement extensions to the repository, selection and diversity measure are necessary to enable standard MOEAs to handle such problems. The reliability design problem well-known in the field of reliability analysis and engineering as introduced. In early development stages, when design changes are still inexpensive, parametric uncertainties are sometimes in the order of magnitudes. In an example it was shown how MOEAs can be applied even in this cloudy and blurred modelling phase.

Much work can be done in optimizing and specializing MOEAs towards this problem field. A first step to this empirical approach must be the extension of multi-objective performance metrics to uncertain solutions which leads to even more difficulties in assessing the quality of an algorithm than currently exist. Then different density estimates, selection methods and other algorithmic features can be analyzed and compared to the proposed approach.

Acknowledgements

The author would like to thank Dr. Jörg Petersen and the anonymous reviewers for their inspirational comments.

References

1. Arnold, D.V.: Noisy Optimization with Evolution Strategies. Volume 8 of Genetic Algorithms and Evolutionary Computation. Kluwer, Boston (2002)
2. Fetz, T., Oberguggenberger, M.: Propagation of uncertainty through multivariate functions in the framework of sets of probability measures. Reliability Engineering & System Safety **85** (2004) 73–87
3. Agarwal, H., Renaud, J.E., Preston, E.L., Padmanabhan, D.: Uncertainty quantification using evidence theory in multidisciplinary design optimization. Reliability Engineering & System Safety **85** (2004) 281–294
4. Hughes, E.J.: Multi-Objective Probabilistic Selection Evolutionary Algorithm. Technical Report DAPS/EJH/56/2000, Department of Aerospace, Power & Sensors, Cranfield University (2000)
5. Teich, J.: Pareto-Front Exploration with Uncertain Objectives. In: First International Conf. on Evolutionary Multi-Criterion Optimization. Volume 1993 of Lecture Notes in Computer Science., Springer-Verlag (2001) 314–328
6. Rudolph, G.: A partial order approach to noisy fitness functions. In: Proc. 2001 Congress on Evolutionary Computation (CEC'01), Seoul, Piscataway NJ, IEEE Press (2001) 318–325
7. Rudolph, G.: Evolutionary search under partially ordered fitness sets. In: Proc. Int'l NAISO Congress on Information Science Innovations (ISI 2001), Millet, Kanada, ICSC Academic Press (2001) 818–822
8. Oberkampf, W.L., Helton, J.C., Joslyn, C.A., Wojtkiewicz, S.F., Ferson, S.: Challenge problems: uncertainty in system response given uncertain parameters. Reliability Eng. & System Safety **85** (2004) 11–19
9. Walley, P.: Statistical Reasoning with Imprecise Probabilities. Chapman and Hall, London (1991)
10. Yager, R.R.: Uncertainty modeling and decision support. Reliability Eng. & System Safety **85** (2004) 341–354
11. Dempster, A.: Upper and lower probabilities induced by a multivalued mapping. Annals of Math. Statistics **38** (1967) 325–339
12. Shafer, G.: A Mathematical Theory of Evidence. Princeton University Press, Princeton, NJ, USA (1976)
13. Hégarat-Mascle, S.L., Richard, D., Ottlé, C.: Multiscale data fusion using demspter-shafer evidence theory. Integrated Comp. Aided Eng. **10** (2003) 9–22
14. Tonon, F.: Using random set theory to propagate epistemic uncertainty through a mechanical system. Reliability Eng. & System Safety **85** (2004) 169–181
15. Yager, R.R., Kacprzyk, J., Fedrizzi, M.: Advances in the Dempster-Shafer theory of evidence. Wiley, New York (1994)
16. Pareto, V.: Manual of political Economy. A.M. Kelley, New York (1971)
17. Yager, R.R., Kreinovich, V.: Decision Making under Interval Probabilities. International Journal of Approximate Reasoning **22** (1999) 195–215
18. Kreinovich, V.: Maximum entropy and interval computations. Reliable Computing **2** (1996) 63–79

19. Deb, K., Agrawal, S., Pratab, A., Meyarivan, T.: A Fast Elitist Non-Dominated Sorting Genetic Algorithm for Multi-Objective Optimization: NSGA-II. In: Proceedings of the Parallel Problem Solving from Nature VI Conf., Paris, France (2000) 849–858
20. Zitzler, E., Laumanns, M., Thiele, L.: SPEA2: Improving the Strength Pareto Evolutionary Algorithm. In: Evolutionary Methods for Design, Optimisation, and Control, Barcelona, Spain (2001)
21. Knowles, J.D., Corne, D.W.: The pareto archived evolution strategy: A new baseline algorithm for pareto multiobjective optimisation. In: Congress on Evolutionary Computation. Volume 1., IEEE Press (1999) 98–105
22. Bartz-Beielstein, T., Limbourg, P., Mehnen, J., Schmitt, K., Parsopoulos, K.E., Vrahatis, M.N.: Particle swarm optimizers for Pareto optimization with enhanced archiving techniques. In: Congress on Evolutionary Computation (CEC'03). Volume 3., Canberra, IEEE Press (2003) 1780–1787
23. Silverman, B.W.: Density estimation for statistics and data analysis. Chapman and Hall, London (1986)
24. Coit, D., Jin, T.: Multi-criteria optimization: Maximization of a system reliability estimate and minimization of the estimate variance. In: Proceedings of the 2001 European Safety & Reliability International Conf. (ESREL), Turin, Italy, (2001)
25. Elegbede, C., Adjallah, K.: Availability allocation to repairable systems with genetic algorithms: a multi-objective formulation. Reliability Eng. & System Safety **82** (2003) 319–330
26. Kuo, W., Prasad, V.: An annotated overview of system-reliability optimization. IEEE Transaction on Reliability **49** (2000) 176–187
27. Zafiropoulos, E.P., Dialynas, E.N.: Reliability and cost optimization of electronic devices considering the component failure rate uncertainty. Reliability Eng. & System Safety **84** (2004) 271 – 284
28. Billinton, R., Allan, R.N.: Reliability evaluation of engineering systems: concepts and techniques. Plenum, New York (1992)
29. Pozsgai, P., Krolo, A., Bertsche, B., Fritz, A.: SYSLEB-a tool for the calculation of the system reliability from raw failure data. In: Reliability and Maintainability Symposium, 2002. (2002) 542–549
30. Coit, D.W., Smith, A.E.: Reliability optimization of series-parallel systems using a genetic algorithm. IEEE Transaction on Reliability **45** (1996) 254–260

The Naive MIDEA: A Baseline Multi–objective EA

Peter A.N. Bosman and Dirk Thierens

Institute of Information and Computing Sciences,
Utrecht University, The Netherlands
{Peter.Bosman, Dirk.Thierens}@cs.uu.nl

Abstract. Estimation of distribution algorithms have been shown to perform well on a wide variety of single–objective optimization problems. Here, we look at a simple - yet effective - extension of this paradigm for multi–objective optimization, called the naive MIDEA. The probabilistic model in this specific algorithm is a mixture distribution, and each component in the mixture is a univariate factorization. Mixture distributions allow for wide–spread exploration of the Pareto front thus aiding the important preservation of diversity in multi–objective optimization. Due to its simplicity, speed, and effectiveness the naive MIDEA can well serve as a baseline algorithm for multi–objective evolutionary algorithms.

1 Introduction

Estimation of distribution algorithms (EDAs) are a class of evolutionary algorithms that build at each generation a probabilistic model from the selected solutions. EDAs are mainly characterized by the kind of probabilistic model they learn. Somewhat surprisingly, it has become clear that even the use of a simple univariate factorization as probabilistic model often leads to good performance for discrete, single objective optimization problems. Considering their ease of implementation, univariate EDAs have become a baseline algorithm that can be used to generate reasonable good solutions quickly, thus setting a performance level that more elaborated algorithms need to surpass to justify their use. The goal of this paper is to show that a similar baseline algorithm can be constructed for multi–objective optimization problems. The resulting algorithm - called the naive MIDEA - is a simple, fast, and efficient multi–objective evolutionary algorithm (MOEA) based on the concept of Pareto dominance, and can thus be used as a baseline algorithm for more elaborate MOEAs.

The remainder of this paper is organized as follows. In Section 2 we specify the naive MIDEA algorithm as an instance of the more general MIDEA (mixture–based, multi–objective iterated density–estimation evolutionary algorithm) framework. In Section 3 we test the performance of the algorithm on four multi–objective, combinatorial optimization problems, compare the results with two state–of–the–art MOEAs, and discuss our findings. We present our conclusions in Section 4.

2   The Naive MIDEA

2.1   Mixture Probability Distributions

A mixture probability distribution is a weighted sum of k probability distributions. Each probability distribution in the mixture distribution is called a mixture component. Let $\mathcal{Z} = (Z_0, Z_1, \ldots, Z_{l-1})$ be a vector of random variables Z_i associated with the i-th problem variable. A mixture probability distribution is then defined as:

$$P^{mixture}(\mathcal{Z}) = \sum_{i=0}^{k-1} \beta_i P^i(\mathcal{Z}) \qquad (1)$$

where $\beta_i > 0$, $i \in \{0, 1, \ldots, k-1\}$, and $\sum_{i=0}^{k-1} \beta_i = 1$. The β_i with which the mixture components are weighted in the sum are called mixing coefficients.

The general advantage of mixture probability distributions is that a larger class of dependency relations between the random variables can be expressed than when using non–mixture probability distributions since a mixture probability distribution makes a combination of multiple probability distributions. In many cases, simple probability distributions can be used as mixture components to get accurate descriptions of the data in different parts of the sample space. By using mixture probability distributions, a powerful, yet computationally tractable type of probability distribution can be used within EDAs, that provides for processing complicated interactions between a problem's variables.

For multi–objective optimization, mixture distributions have an additional advantage that renders them particularly useful. The specific advantage is geometrical in nature. If we, for instance, cluster the solutions as observed in the objective space and then estimate a simpler probability distribution in each cluster, the probability distributions in these clusters can portray specific information about the different regions along the Pareto optimal front that we are ultimately interested in. Drawing new solutions from the resulting mixture probability distribution gives solutions that are more likely to be well spread along the front as each mixture component delivers a subset of new solutions. The use of such a mixture distribution thus results in a parallel exploration along the current Pareto front. This parallel exploration may very well provide a better spread of new solutions along the Pareto front than when a single non–mixture distribution is used to capture information about the complete Pareto front.

To complete the construction of the mixture distribution we also need to determine the mixing coefficients β_i. This can be done in various ways. Here we set β_i proportional to the size of the i-th cluster with respect to the sum of the sizes of all clusters.

2.2   Selection Operator

In multi-objective optimization we want the solutions to be as close to the Pareto optimal front as possible, and we want a good diverse representation of the Pareto optimal front. In a practical application, we have no indication of how close we are to the Pareto optimal front. To ensure selection pressure toward the

Pareto optimal front in the absence of such information, the best we can do is to find solutions that are dominated as little as possible by any other solution. A straightforward way to obtain selection pressure toward non–dominated solutions is therefore to count for each solution in the population the number of times it is dominated by another solution in the population, which is called the domination count of a solution [1, 5].

In the MIDEA selection we discern two cases. Call n the population size and $\tau \in [0...1]$ the selection threshold. If the number of non–dominated solutions - those with a domination count of 0 - is less than or equal to $\lfloor \tau n \rfloor$ we simply select the $\lfloor \tau n \rfloor$ solutions with lowest domination count (ties are broken at random). However, if the number of non–dominated solutions is larger than $\lfloor \tau n \rfloor$ we first collect all non–dominated solutions in a set $\mathcal{S}^P$. Next, the final selection $\mathcal{S}$ is obtained from $\mathcal{S}^P$ using a nearest neighbor heuristic to enforce diversity. We start by picking a solution from $\mathcal{S}^P$ with an optimal value for a randomly chosen objective and move it to the set $\mathcal{S}$ which has now a single element. Note that the choice of objective is arbitrary as the goal is to find a diverse selection of solutions. For all solutions in $\mathcal{S}^P$, the nearest neighbor distance is computed to the solution in $\mathcal{S}$. The distance that we use is the Euclidean distance scaled to the sample range in each objective. The solution in $\mathcal{S}^P$ with the *largest* distance is then deleted from $\mathcal{S}^P$ and added to $\mathcal{S}$. The distances in $\mathcal{S}^P$ are updated by investigating whether the distance to the newly added point in $\mathcal{S}$ is smaller than the currently stored distance. These last two steps are repeated until $\lfloor \tau n \rfloor$ solutions are in the final selection $\mathcal{S}$. This selection operator has a running time complexity of $\mathcal{O}(n^2)$ which is equal to the complexity for computing the domination counts.

2.3 The Naive MIDEA

The naive MIDEA is an instance of the MIDEA framework for multi–objective optimization using EDAs [1, 9]. The MIDEA can be specified in pseudo–code as follows:

MIDEA
1 Initialize a population of n random solutions and evaluate their objectives
2 Iterate until termination
2.1 Compute the domination counts
2.2 Select $\lfloor \tau n \rfloor$ solutions with the diversity preserving selection operator
2.3 Estimate a mixture probability distribution $P^{mixture}(\mathcal{Z})$
2.4 Replace the non–selected solutions with new solutions drawn from $P^{mixture}(\mathcal{Z})$
2.5 Evaluate the objectives of the new solutions

The naive MIDEA uses a simple univariate factorized probability distributions in each cluster: $P^{univariate}(\mathcal{Z}) = \prod_{i=0}^{l-1} P(Z_i)$. For discrete random variables, this amounts to repeatedly counting frequencies and computing proportions for a single random variable. Since in each cluster we thus disregard all dependencies between random variables, we call this specific MIDEA instance naive in analogy with the well–known naive Bayes classifier. However, the clusters are expected to already provide a large benefit for multi-objective optimization. Moreover, algorithms such as UMDA [8] and the compact GA [6] that also use the univariate marginal probability distribution (without clustering) have proved to be reasonably effective for single-objective optimization problems.

For computational efficiency reasons we apply a fast clustering algorithm. Possibly this adds to the naiveness of our naive MIDEA instance, but other clustering algorithms are easily implemented if required. The algorithm that we use here is the leader algorithm, which is one of the fastest partitioning algorithms. The use of it can thus be beneficial if the amount of overhead that is introduced by factorization mixture selection methods is desired to remain small. There is no need to specify in advance how many partitions there should be. The first solution to make a new partition is appointed to be its leader. The leader algorithm goes over the solutions exactly once. For each solution it encounters, it finds the first partition that has a leader being closer to the solution than a given threshold $\mathfrak{T}_d$. If no such partition can be found, a new partition is created containing only this single solution. To prevent the first partitions from becoming quite a lot larger than the later ones, we randomize the order in which the partitions are inspected. One of the drawbacks of the randomized leader algorithm is that it is not invariant given the sequence of the input solutions. Therefore, to be sure that the ordering of the solutions is not subject to large repeating sequences of solutions, we randomize the ordering of the solutions each time the leader algorithm is applied.

Summarizing, the pseudo-code of the naive MIDEA is as follows:

---

naive MIDEA
(*instantiation of steps 2.3 and 2.4 of the general* MIDEA *framework*)

---

1 $(c^0, c^1, \ldots, c^{k-1}) \leftarrow$ LeaderAlgorithm($\mathfrak{T}_d$)
2 for $i \leftarrow 0$ to $k - 1$ do
  2.1 $\beta_i \leftarrow |c^i|/\lfloor \tau n \rfloor$
  2.2 for $j \leftarrow 0$ to $l - 1$ do
    2.2.1 Estimate the univariate marginal probability distribution $P^{i,j}(Z_j)$ for random variable Z_j from the solutions in the i-th cluster (i.e. c^i)
3 for $i \leftarrow \lfloor \tau n \rfloor$ to $n - 1$ do
  3.1 Initialize a new solution z
  3.2 Choose an index $q \in \{0, 1, \ldots, k-1\}$ with probability β_q
  3.3 for $j \leftarrow 0$ to $l - 1$ do
    3.3.1 Draw a value for Z_j from the univariate marginal probability distribution $P^{q,j}(Z_j)$ associated with the q-th cluster
  3.4 Add z to the set of new offspring.

---

3 Experiments

In this section we compare the naive MIDEA to two state-of-the-art MOEAs that aim at obtaining a diverse set of solutions along the Pareto front. The SPEA algorithm by Zitzler and Thiele [13] and the NSGA-II algorithm by Deb et al. [3] showed superior performance compared to most other MOEAs [3, 11].

The multi-objective optimization problems are described in Section 3.1. The performance measures we use to score the results of the algorithms with are described in Section 3.2. In Section 3.3 we present our experimental setup. Finally, in Section 3.4 we discuss the obtained results.

3.1 Multi-objective Optimization Problems

The test–suite we have used consists of four multi–objective optimization problems with two different dimensionalities resulting in a total test size of 8 problems. Figure 1 specifies the four problems. Next to being binary, these problems are also multi–objective variants of well–known combinatorial optimization problems. The number of objectives for these problems is not restricted to two and is denoted by m.

Maximum satisfiability. In the maximum satisfiability problem, we are given a propositional formula in conjunctive normal form. The goal is to satisfy as many clauses as possible. The solution string is a truth assignment to the involved literals. These formulas can be represented by a matrix in which row i specifies

Name	Definition
MS (Maximum Satisfiability)	**Maximize** $(f_0(x), f_1(x), \ldots, f_{m-1}(x))$ **Where** • $\forall i : f_i(x) = \sum_{j=0}^{c_i-1} \text{sgn}\left(\left[\sum_{k=0}^{l-1}(C_i)_{jk} \otimes x_k\right]\right)$ • $\text{sgn}(x) = \begin{cases} 1 & \text{if } x > 0 \\ 0 & \text{if } x = 0 \\ -1 & \text{if } x < 0 \end{cases}$ • $\begin{array}{c\|cc} \otimes & 0 & 1 \\ \hline -1 & 1 & 0 \end{array}$ $\begin{array}{c\|cc} \otimes & 0 & 1 \\ \hline 0 & 0 & 0 \end{array}$ $\begin{array}{c\|cc} \otimes & 0 & 1 \\ \hline 1 & 0 & 1 \end{array}$
KN (Knapsack)	**Maximize** $(f_0(x), f_1(x), \ldots, f_{m-1}(x))$ **Where** • $\forall i : f_i(x) = \sum_{j=0}^{l-1} P_{ij} x_j$ **Such that** • $\forall i : \sum_{j=0}^{l-1} W_{ij} x_j \leq c_i$
SC (Set Covering)	**Minimize** $(f_0(x), f_1(x), \ldots, f_{m-1}(x))$ **Where** • $\forall i : f_i(x) = \sum_{j=0}^{l-1} C_{ij} x_j$ **Such that** • $\forall i : \forall_{0 \leq j < r} : \sum_{k=0}^{l-1}(A_i)_{jk} x_k \geq 1$
MST (Minimal Spanning Tree)	**Minimize** $(f_0(x), f_1(x), \ldots, f_{m-1}(x))$ **Where** • $\forall i : f_i(x) = \sum_{j=0}^{l-1} W_{ij} x_j$ **Such that** • $\forall_{S \subseteq V} : \sum_{x_j \in (S \times (V-S))} x_j \geq 1$ • $\forall_{S \subseteq V} : \sum_{x_j \in (S \times S)} x_j \leq \|S\| - 1$

Fig. 1. Binary multi–objective combinatorial optimization test problems

what literals appear either positive (1) or negative (−1) in clause i. In the multi–objective variant of this problem, we have m of such matrices and only a single solution to satisfy as many clauses as possible in each objective at the same time.

Knapsack. The multi–objective knapsack problem was first used by Zitzler and Thiele [13] to test MOEAs. We are given m knapsacks with a specified capacity and n items. Each item can have a different weight and profit in every knapsack. Selecting item i in a solution implies placing it in every knapsack. A solution may not cause exceeding the capacity of any knapsack.

Set covering. In the set covering problem, we are given l locations at which we can place some service at a specified cost. Furthermore, associated with each location is a set of regions $\subseteq \{0, 1, \ldots r-1\}$ that can be serviced from that location. The goal is to select locations such that *all* regions are serviced against minimal costs. In the multi–objective variant of set covering, m services are placed at a location. Each service however covers its own set of regions when placed at a certain location and has its own cost associated with a certain location. A binary solution indicates at which locations the services are placed.

Minimal spanning tree. In the minimal spanning tree problem we are given an undirected graph (V, E) such that each edge has a certain weight. We are interested in selecting edges $E_T \subseteq E$ such that (V, E_T) is a spanning tree. The objective is to find a spanning tree such that the weight of all its edges is minimal. In the multi–objective variant of this problem, each edge can have a different weight in each objective.

3.2 Performance Indicators

To measure the performance of a MOEA we only consider the subset of all non–dominated solutions that is contained in the final population that results from running the MOEA. We call such a subset an approximation set and denote it by $\mathcal{S}$. The size of the approximation set depends on the settings used to run the MOEA with. To actually measure performance, performance indicators are used. A performance indicator is a function that, given an approximation set $\mathcal{S}$, returns a real value that indicates how good $\mathcal{S}$ is with respect to a certain feature that is measured by the performance indicator. More detailed information regarding the importance of using good performance indicators to evaluate MOEAs may be found in dedicated literature [2, 7, 12]. Here we will use three performance indicators:

1. The *Front Spread* (**FS**) indicator measures the size of the objective space covered by an approximation set [13]. A larger **FS** indicator value is preferable given equal values for the other indicators. The **FS** indicator for an approximation set $\mathcal{S}$ is defined to be the maximum Euclidean distance inside the smallest m–dimensional bounding–box that contains $\mathcal{S}$:

$$\mathbf{FS}(\mathcal{S}) = \sqrt{\sum_{i=0}^{m-1} \max_{(z^0,z^1) \in \mathcal{S} \times \mathcal{S}} \{(f_i(z^0) - f_i(z^1))^2\}} \qquad (2)$$

2. The *Front Occupation* (**FO**) indicator measures the size of the set of non–dominated solutions [10]. Since a larger set of trade–off points is more desirable, a larger **FO** indicator value is preferable given equal values for the other indicators.

$$\mathbf{FO}(\mathcal{S}) = |\mathcal{S}| \qquad (3)$$

3. The *Front to Set Distance* indicator ($\boldsymbol{D_{\mathcal{P}_F \to \mathcal{S}}}$) computes for each solution in the discrete Pareto optimal set the distance to the closest solution in an approximation set $\mathcal{S}$ and takes the average as the indicator value:

$$\boldsymbol{D_{\mathcal{P}_F \to \mathcal{S}}}(\mathcal{S}) = \frac{1}{|\mathcal{P}_\mathcal{S}|} \sum_{z^1 \in \mathcal{P}_\mathcal{S}} \min_{z^0 \in \mathcal{S}} \{d(z^0, z^1)\} \qquad (4)$$

Since we are interested in performance as measured in the objective space, the distance between two multi–objective solutions z^0 and z^1 is the Euclidean distance between their objective values $f(z^0)$ and $f(z^1)$.

The $\boldsymbol{D_{\mathcal{P}_F \to \mathcal{S}}}$ indicator represents both the goal of getting close to the Pareto optimal front as well as the goal of getting a diverse, wide–spread front of solutions. A smaller value for this performance indicator is preferable.

A performance indicator that is closely related to the $\boldsymbol{D_{\mathcal{P}_F \to \mathcal{S}}}$ indicator, is the hypervolume indicator by Knowles and Corne [7]. In the hypervolume indicator, a point in the objective space is picked such that it is dominated by all points in the approximation sets that need to be evaluated. The indicator value is then equal to the hypervolume of the multi–dimensional region enclosed by the approximation set and the picked reference point. This value is an indicator of the region in the objective space that is dominated by the approximation set. The main difference between the hypervolume indicator and the $\boldsymbol{D_{\mathcal{P}_F \to \mathcal{S}}}$ indicator is that for the hypervolume indicator a reference point has to be chosen. Different reference points lead to different indicator values. Moreover, different reference points can lead to indicator values that indicate a preference for different approximation sets. Since in the $\boldsymbol{D_{\mathcal{P}_F \to \mathcal{S}}}$ indicator the true Pareto optimal front is used, the $\boldsymbol{D_{\mathcal{P}_F \to \mathcal{S}}}$ indicator does not suffer from this drawback. Of course, a major drawback of the $\boldsymbol{D_{\mathcal{P}_F \to \mathcal{S}}}$ indicator is that in a real application the true Pareto optimal front is not known beforehand. In that case, the Pareto front of all approximation sets could be used as a substitute for the actual Pareto optimal front.

3.3 Experimental Setup

Optimization Problem Dimensionalities. We used test instances with dimensionality $l = 100$ and $l = 1000$. For the maximum satisfiability problem, we generated the test instances by generating 2500 clauses for $l = 100$ and 12500

EA	$D_{\mathcal{P}_F \to \mathcal{S}}$							
	MS^{100}	KN^{100}	SC^{100}	MST^{105}	MS^{1000}	KN^{1000}	SC^{1000}	MST^{1035}
SPEA^{UX}	12.5	9.59	2.92	1.43	183	67.3	518	6.10
SPEA^{1X}	11.6	8.50	2.99	1.50	277	83.1	452	5.75
NSGA-II^{UX}	11.4	7.75	2.61	1.21	185	84.1	260	6.45
NSGA-II^{1X}	11.5	8.87	2.63	1.49	289	121.0	329	5.95
naive MIDEA	**7.95**	**4.13**	**1.52**	**1.19**	**37.2**	**30.4**	**117**	**3.39**

Fig. 2. Average of the $D_{\mathcal{P}_F \to \mathcal{S}}$ performance indicator on all combinatorial problems

EA	Front Spread **FS**							
	MS^{100}	KN^{100}	SC^{100}	MST^{105}	MS^{1000}	KN^{1000}	SC^{1000}	MST^{1035}
SPEA^{UX}	116	69.5	**64.6**	30.6	288	254	631	52.1
SPEA^{1X}	126	82.6	51.0	32.5	399	308	**636**	50.8
NSGA-II^{UX}	120	78.3	14.8	26.3	370	288	144	33.7
NSGA-II^{1X}	129	76.6	12.8	23.9	364	291	107	36.1
naive MIDEA	**172**	**115**	24.7	**34.3**	**538**	**453**	204	**57.0**

Fig. 3. Average of the **FS** performance indicator on all combinatorial problems

clauses for $l = 1000$ with a random number of literals between 1 and 5. For the knapsack problem, we generated instances by generating random weights in $[1; 10]$ and random profits in $[1; 10]$. The capacity of a knapsack was set at half of the total weight of all the items, weighted according to that knapsack objective. For set covering, the costs were generated at random in $[1; 10]$. We used 250 regions and 2500 regions to be serviced for $l = 100$ and $l = 1000$ respectively. We varied the problem difficulty through the region–location adjacency relation. This relation was generated by making each location adjacent to 70 and 50 randomly selected regions for $l = 100$ and $l = 1000$ respectively. Finally, for the minimum spanning tree problem, we used full graphs with 105 edges (15 vertices) and 1035 edges (46 vertices). The dimensionality of these problems is therefore not precisely 100 and 1000. The weights of the edges were generated randomly in $[1; 10]$.

EA	Front Occupation **FO**							
	MS^{100}	KN^{100}	SC^{100}	MST^{105}	MS^{1000}	KN^{1000}	SC^{1000}	MST^{1035}
SPEA^{UX}	46.8	46.5	**25.0**	42.8	49.4	49.5	26.2	48.8
SPEA^{1X}	46.1	**77.6**	24.3	**80.1**	49.9	49.7	**26.5**	**95.0**
NSGA-II^{UX}	33.5	35.5	7.80	32.3	35.4	33.1	7.50	64.7
NSGA-II^{1X}	41.1	37.5	6.80	24.5	42.0	36.4	7.20	64.8
naive MIDEA	**52.6**	57.8	10.4	23.7	**116**	**104**	6.27	60.0

Fig. 4. Average of the **FO** performance indicator on all combinatorial problems

	Population Size n							
EA	MS^{100}	KN^{100}	SC^{100}	MST^{105}	MS^{1000}	KN^{1000}	SC^{1000}	MST^{1035}
SPEA^{UX}	25	25	25	25	25	25	25	25
SPEA^{1X}	25	50	25	100	25	25	25	50
NSGA-II^{UX}	350	325	300	200	200	250	200	250
NSGA-II^{1X}	100	175	250	200	150	200	150	200
naive MIDEA	750	625	1300	4700	1100	1175	1625	1875

Fig. 5. Population sizes used for the combinatorial problems

Optimization Problem Constraints. The set covering, knapsack and minimal spanning tree problems have constraints. To deal with them, we can use a repair mechanism to transform infeasible solutions into feasible solutions. Another approach is based on the notion of constraint–domination introduced by Deb et al. [4]. This notion allows to deal with constrained multi–objective problems in a general fashion. A solution z^0 is said to constraint-dominate solution z^1 if any of the following is true:

1. Solution z^0 is feasible and solution z^1 is infeasible
2. Solutions z^0 and z^1 are both infeasible, but z^0 has a smaller overall constraint violation
3. Solutions z^0 and z^1 are both feasible and $z^0 \succ z^1$

The overall constraint violation is the amount by which a constraint is violated, summed over all constraints. We have used this principle for the set covering problem. For the knapsack problem, an elegant repair mechanism was proposed earlier by Zitzler and Thiele [13]. For the minimal spanning tree problem, the

Statistically Significant Improvement Matrices	$D_{\mathcal{P}_F \to \mathcal{S}}$						Front Spread **FS**						Front Occ. **FO**					
	SPEA^{UX}	SPEA^{1X}	NSGA-II^{UX}	NSGA-II^{1X}	naive MIDEA	Sum	SPEA^{UX}	SPEA^{1X}	NSGA-II^{UX}	NSGA-II^{1X}	naive MIDEA	Sum	SPEA^{UX}	SPEA^{1X}	NSGA-II^{UX}	NSGA-II^{1X}	naive MIDEA	Sum
SPEA^{UX}	0	-1	-2	0	-8	-11	0	-4	1	0	-4	-7	0	-4	6	6	-2	6
SPEA^{1X}	1	0	-4	0	-8	-11	4	0	8	7	-3	16	4	0	8	8	2	22
NSGA-II^{UX}	2	4	0	4	-7	3	-1	-8	0	1	-8	-16	-6	-8	0	-1	-3	-18
NSGA-II^{1X}	0	0	-4	0	-8	-12	0	-7	-1	0	-8	-16	-6	-8	1	0	-5	-18
naive MIDEA	8	8	7	8	0	31	4	3	8	8	0	23	2	-2	3	5	0	8

Fig. 6. Number of times an improvement was found to be statistically significant in the $D_{\mathcal{P}_F \to \mathcal{S}}$, FS and FO performance indicators, summed over all tested problems. The numbers in a single row indicate the summed number of significantly better or worse results compared to the algorithms in the different columns

number of constraints grows exponentially with the problem size l. We therefore propose to use repair mechanisms for these latter two problems.

Knapsack repair mechanism. If a solution violates a constraint, the repair mechanism iteratively removes items until all constraints are satisfied. The order in which the items are investigated, is determined by the maximum profit/weight ratio. The items with the lowest profit/weight ratio are removed first.

Minimal spanning tree repair mechanism. First the edges are removed from the currently constructed graph and they are sorted according to their weight. Next, they are added to the graph such that no cycles are introduced. This is done by only allowing edges to be introduced between the connected components in the graph. If after this phase, the number of connected components has not been reduced to 1, all edges between the connected components are regarded in increasing weight and again the connected components are merged until a single component is left.

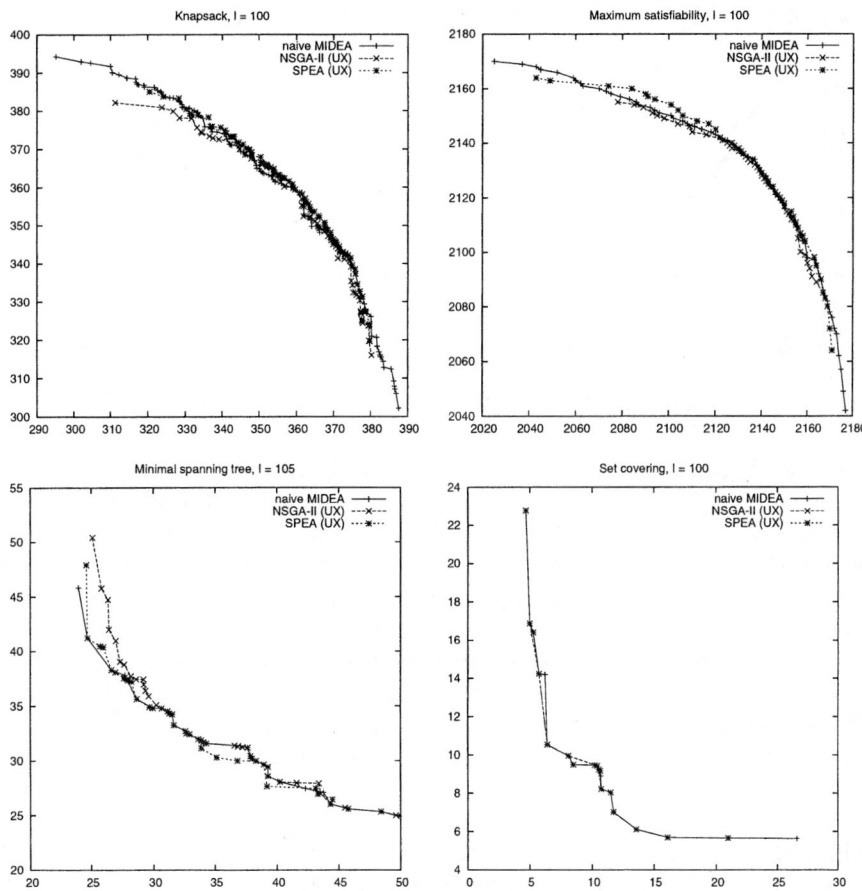

Fig. 7. Pareto fronts over 50 runs on all tested problems with dimensionality $l = 100$

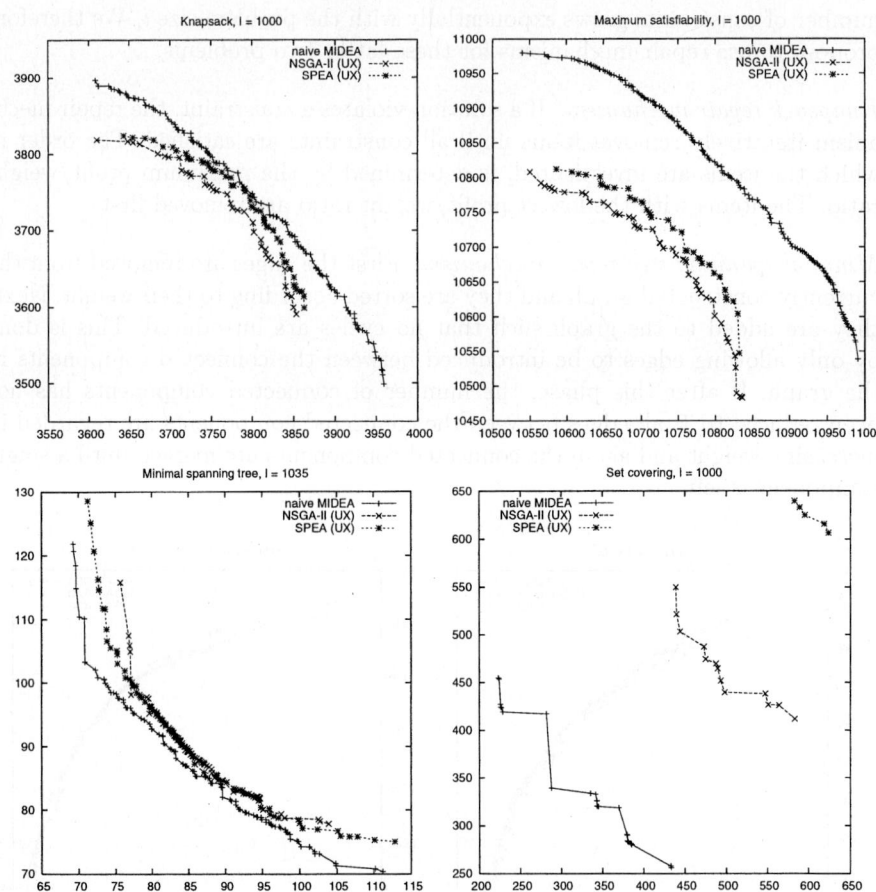

Fig. 8. Pareto fronts over 50 runs on all tested problems with dimensionality $l = 1000$

General Algorithmic Setup. We ran every algorithm 50 times on each problem. In any single run we chose to allow a maximum of $20 \cdot 10^3$ evaluations for the problems of dimensionality $l = 100$ and a maximum of $100 \cdot 10^3$ evaluations for the problems of dimensionality $l = 1000$. As a result of imposing the restriction of a maximum of evaluations, a value for the population size n exists for each MOEA such that the MOEA will perform best. For too large population sizes, the search will become a random search and for too small population sizes, there is not enough information to perform adequate model selection and induction. We therefore increased the population size in steps of 25 to find the best results. To actually select the best population size, we selected the result with the lowest value for the $D_{\mathcal{P}_F \to \mathcal{S}}$ indicator.

Algorithms. We tested three MOEAs. In the following we will describe the details that are required in addition to the details given in earlier sections for constructing the actual MOEAs that we will use for testing.

1. For SPEA, we used uniform crossover and one–point crossover with a probability of 0.8. Bit–flipping mutation was used in combination with either of these recombination operators with a probability of 0.01. These settings were used previously by the SPEA authors [11]. We allowed the size of the external storage in SPEA to become as large as the population size.
2. For NSGA–II, we used the same crossover and mutation operators as above.
3. For the naive MIDEA, we used the leader clustering algorithm in the objective space such that four clusters were constructed on average. If the number of clusters becomes too large, the requirements for the population size increases in order to facilitate proper factorization selection in each cluster. We do not suggest that the number of clusters we use is optimal, but it will serve to indicate the effectiveness of parallel exploration along the Pareto front as well as diversity preservation. For the truncation percentile, we used the often used value $\tau = 0.3$.

3.4 Results

To compare the MOEAs, we investigated their average performance with respect to performance indicators introduced in Section 3.2. For the $D_{\mathcal{P}_F \to \mathcal{S}}$ performance indicator, we used the Pareto front over all results obtained by all MOEAs.

For each of the performance indicators, we computed their average and standard deviation over the 50 runs to get an assessment of their performance. The averages are tabulated in Figures 2 through 4 (standard deviations can be found in the technical report). The best results are written in boldface. The population sizes that led to the best performance, are tabulated in Figure 5. Although the average behavior is the most interesting, the standard deviations are vital to determine whether the differences in the average behavior of the different algorithms are significant. To investigate these significances, we have performed Aspin–Welch–Satterthwaite (AWS) statistical hypothesis T–tests at a significance level of $\alpha = 0.05$. The AWS T–test is a statistical hypothesis test for the equality of means in which the equality of variances is not assumed. For each problem, we verified for each pair of algorithms whether the average obtained performance indicator values differ significantly. We assigned a value of 1 if an algorithm scored significantly better and a value of -1 if an algorithm scored significantly worse. We summed the so obtained matrices over all problems to get the statistically significant improvement matrices that are shown in Figure 6. We also computed the sum for each algorithm of its significant improvement values over all other algorithms to indicate the summed relative statistically significant performance of the algorithms. A less detailed summary of the statistical significance tests is shown in Figure 9. In this figure histograms are used to indicate the sum of the results of the statistical significance tests for each algorithm compared with all other algorithms. The histogram represents the sums for the different tested dimensionalities and their average.

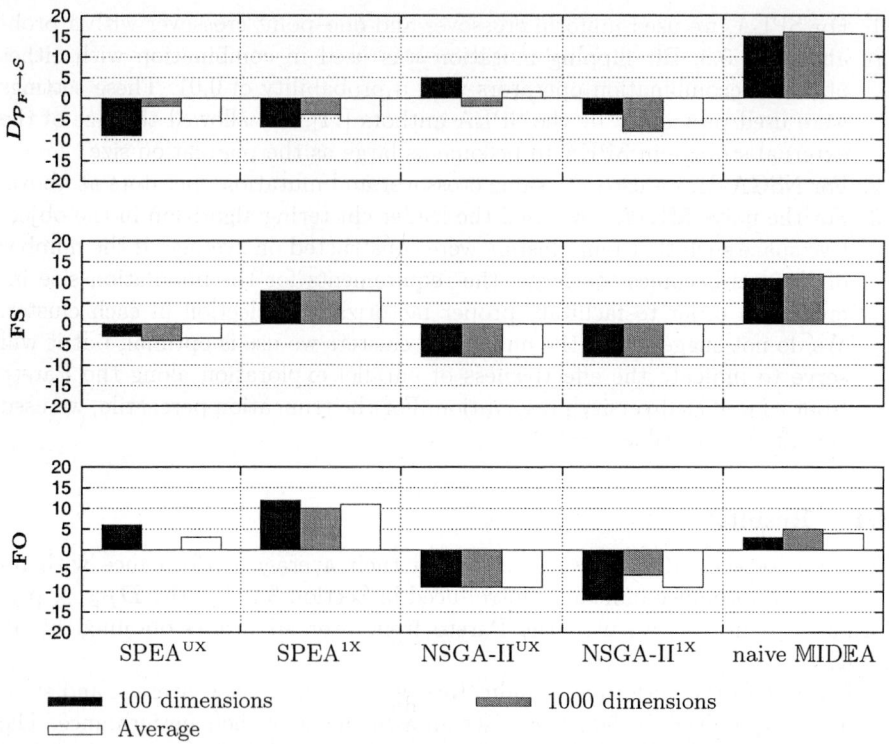

Fig. 9. A summary of the results of the statistical hypothesis tests performed for each pair of algorithms. For each algorithm, the sum of the outcome of the statistical hypothesis tests is shown for the combinatorial problems for each dimensionality separately. Furthermore, the average of these values is also shown, which serves as a global indicator of the performance of an algorithm relative to the other tested algorithms

3.5 Discussion

The naive MIDEA performs obviously better when the dimensionality of the problem becomes larger. This is most likely due to the efficient diversity exploration and preservation in MIDEA. As the dimensionality of the problem goes up, the parameter search space becomes larger and the number of solutions in the objective space becomes larger as well. In Figures 7 and 8 the Pareto fronts over 50 runs for all algorithms are plotted on one problem from each problem class and dimensionality. The better diversity preservation and proper distribution of the points along the front can be seen clearly for the problems of larger dimensionality. For the lower dimensionality problems, better diversity preservation can also be observed, which is most exemplified by the fact that the naive MIDEA obtains non–dominated solutions at the outer ends of the front for the knapsack problem with $l = 100$.

The naive MIDEA is arguably a very effective algorithm on the test suite used. Moreover, the naive MIDEA runs quickly, even for problems with many variables. The experimental results indicate that clustering the objective space to construct mixture probability distributions in MIDEAs leads to efficient MOEAs, even for simple univariate probabilistic models. It can be expected that the use of clustering is also an appealing technique for more traditional MOEAs that do not use probabilistic models. Actually, a MOEA that applies uniform crossover and restricted mating within the clusters should behave rather similar as the naive MIDEA. Whether one is 'more baseline' as the other seems to be a matter of personal taste.

4 Conclusions

In this paper we have presented the naive MIDEA for multi–objective optimization. The naive MIDEA clusters the selected solutions in the objective space, after which it estimates a univariate factorization in each cluster separately. New solutions are then drawn from the so–obtained mixture probability distribution. The naive MIDEA is a specific instance of the algorithmic framework MIDEA which is a general form of an EDA for multi–objective optimization in which a probabilistic model is learned. For the specific task of multi–objective optimization, the use of mixture distributions obtained by clustering the objective space has been observed to stimulate the desirable parallel exploration along the Pareto front. The naive MIDEA has only little computational overhead since clustering in the objective space can be done very fast as can the estimation of a univariate factorization. Furthermore, although no further exploitation of dependencies between a problem's variables is used in the naive MIDEA, the results obtained compare favorably to results obtained with algorithms in which clustering the objective space is not used. Concluding, the naive MIDEA has been found to be a fast, easy–to–use and effective algorithm for multi–objective optimization. Considering its simplicity, speed, and effectiveness the algorithm might play a role as baseline algorithm for MOEAs.

References

1. P. A. N. Bosman and D. Thierens. Multi–objective optimization with diversity preserving mixture–based iterated density estimation evolutionary algorithms. *International Journal of Approximate Reasoning*, 31:259–289, 2002.
2. P. A. N. Bosman and D. Thierens. The balance between proximity and diversity in multi–objective evolutionary algorithms. *IEEE Transactions on Evolutionary Computation*, 7:174–188, 2003.
3. K. Deb, S. Agrawal, A. Pratab, and T. Meyarivan. A fast elitist non-dominated sorting genetic algorithm for multi-objective optimization: NSGA-II. In M. Schoenauer et al., editor, *Parallel Problem Solving from Nature – PPSN VI*, pages 849–858. Springer, 2000.
4. K. Deb, A. Pratap, and T. Meyarivan. Constrained test problems for multi–objective evolutionary optimization. In E. Zitzler, K. Deb, L. Thiele, C. A. Coello Coello, and D. Corne, editors, *First International Conference on Evolutionary Multi-Criterion Optimization*, pages 284–298, Berlin, 2001. Springer–Verlag.

5. C. M. Fonseca and P. J. Fleming. An overview of evolutionary algorithms in multiobjective optimization. *Evolutionary Computation*, 3(1):1–16, 1995.
6. G. Harik, F. Lobo, and D. E. Goldberg. The compact genetic algorithm. In *Proceedings of the 1998 IEEE International Conference on Evolutionary Computation*, pages 523–528. IEEE Press, 1998.
7. J. Knowles and D. Corne. On metrics for comparing non–dominated sets. In *Proceedings of the 2002 Congress on Evolutionary Computation CEC 2002*, pages 666–674, Piscataway, New Jersey, 2002. IEEE Press.
8. H. Mühlenbein and G. Paaß. From recombination of genes to the estimation of distributions I. binary parameters. In A. E. Eiben et al., editor, *Parallel Problem Solving from Nature – PPSN V*, pages 178–187. Springer, 1998.
9. D. Thierens and P. A. N. Bosman. Multi–objective mixture–based iterated density estimation evolutionary algorithms. In L. Spector et al., editor, *Proceedings of the GECCO-2001 Genetic and Evolutionary Computation Conference*, pages 663–670, San Francisco, California, 2001. Morgan Kaufmann.
10. D. A. Van Veldhuizen. *Multiobjective Evolutionary Algorithms: Classifications, Analyses, and New Innovations*. PhD thesis, Graduate School of Engineering of the Air Force Institute of Technology, WPAFB, Ohio, 1999.
11. E. Zitzler, K. Deb, and L. Thiele. Comparison of multiobjective evolutionary algorithms: Empirical results. *Evolutionary Computation*, 8(2):173–195, 2000.
12. E. Zitzler, M. Laumanns, L. Thiele, C. M. Fonseca, and V. Grunert da Fonseca. Why quality assessment of multiobjective optimizers is difficult. In W. B. Langdon et al., editor, *Proceedings of the 2002 Genetic and Evolutionary Computation Conference*, pages 666–674, San Francisco, California, 2002. Morgan Kaufmann.
13. E. Zitzler and L. Thiele. Multiobjective evolutionary algorithms: A comparative case study and the strength pareto approach. *IEEE Transactions on Evolutionary Computation*, 3(4):257–271, 1999.

New Ideas in Applying Scatter Search to Multiobjective Optimization

Antonio J. Nebro, Francisco Luna, and Enrique Alba

Departamento de Lenguajes y Ciencias de la Computación,
E.T.S. Ingeniería Informática
Campus de Teatinos, 29071 Málaga (Spain)
{antonio, flv, eat}@lcc.uma.es

Abstract. This paper elaborates on new ideas of a scatter search algorithm for solving multiobjective problems. Our approach adapts the well-known scatter search template for single objective optimization to the multiobjective field. The result is a simple and new metaheuristic called SSMO, which incorporates typical concepts from the multiobjective optimization domain such as Pareto dominance, crowding, and Pareto ranking. We evaluate SSMO with both constrained and unconstrained problems and compare it against NSGA-II. Preliminary results indicate that scatter search is a promising approach for multiobjective optimization.

1 Introduction

Most optimization problems in the real world involve the optimization of more than one function, which in turn can require a significant computational time to be evaluated. This feature and the fact that the search space tends to be very large in multiobjective problems (MOPs) make deterministic techniques difficult to apply in order to obtain the Pareto-optimal solutions of MOPs. As a consequence, stochastic techniques have been widely proposed and applied in this domain. Among them, evolutionary algorithms have been investigated by many authors, and some of the most well-known algorithms for solving MOPs belong to this class (e.g. NSGA-II [1], PAES [2], SPEA-2 [3], and micro-GA [4]).

Many evolutionary algorithms for solving MOPs are some kind of genetic algorithm. This implies they use the concepts of population, crossover, mutation, and similar genetic operators (an exception is PAES, which is an (1+1) evolution strategy). We are interested in studying the application of scatter search, another kind of population-based evolutionary algorithm, to solve MOPs. Scatter search has proved to be very effective for solving a diverse set of single objective optimization problems from both classical and real world settings [5], but little attention has been paid to its use in multiobjective optimization (existing works almost reduce to [6,7,8]).

Scatter search is based on using a small population known as the reference set, whose individuals are combined to construct new solutions which, in contrast to other evolutionary algorithms, are obtained in a systematic way (i.e., stochastic procedures such as crossover and mutation are not used). Furthermore, these solutions can be improved by applying a local search method. The reference set is initialized from an initial population composed of dispersed solutions, and it is updated taking into account the results of the local search improvement.

The *scatter search template* presented in [9] has served as the main reference for most of the scatter search implementations to date. The template consists of five methods: diversification generation, improvement, reference set update, subset generation, and solution combination. This template is used in [10] to design a scatter search procedure for single objective optimization problems with continuous bounded variables. In this paper, we have taken this implementation as the basis of a scatter search algorithm for multiobjective optimization, trying to modify it as little as possible with the idea of getting a simple algorithm. We have named this algorithm SSMO (Scatter Search for Multiobjective Optimization). Our main goal is to identify and study new issues that can affect the performance of the algorithm for MOPs.

The contributions of our work can be summarized as follows:

- We propose a scatter search algorithm for solving constrained as well as unconstrained MOPs. The algorithm is based on incorporating the concepts of Pareto dominance, ranking, and crowding, and they are applied to define the improvement and reference set update methods of the scatter search algorithm.
- Two strategies for building the reference set are studied. The first one uses ranking and crowding to carry out a sorting of the population to obtain the best individuals, while the second strategy is based on applying a clustering technique to get a set of centroids of the individuals with best rank.
- The algorithm is evaluated using a benchmark of constrained plus unconstrained MOPs, and it is compared against the NSGA-II algorithm.

The remaining of the paper is organized as follows. In Section 2, we discuss related works concerning multiobjective optimization and scatter search. In Section 3, we describe our proposal. Experimental results are presented in Section 4. Finally, in Section 5 we give some conclusions and lines for future research.

2 Related Work

The application of scatter search to multiobjective optimization has received little attention until recently. We analyze here the proposals presented in [6], [7], and [8]. We use the following terminology: P is the initial set, k is the number of objective functions, and the reference set is composed of $p + q$ individuals, which are obtained by selecting the best p solutions of P, while the remaining q individuals are selected from both P and the current reference set by using a mechanism promoting diversity.

MOSS [6] is an algorithm that proposes a tabu/scatter search hybrid method for solving nonlinear multiobjective optimization problems. Tabu search is used in the diversification generation method to obtain a diverse approximation to the Pareto-optimal set of solutions; it is also applied to rebuild the reference set after each iteration of the scatter search algorithm. To measure the quality of the solutions, MOSS uses a weighted sum approach. This algorithm is compared against NSGA-II, SPEA-2, and PESA on a set of unconstrained test functions.

Similarly to MOSS, SSPMO [7] is a scatter search algorithm which includes tabu search, although they differ in the use of different tabu search algorithms. SSPMO obtains a part of the reference set by selecting the best solutions of the initial set P for each of the k objective functions. The rest of the reference set is obtained by using the usual approach of selecting the remaining solutions in P that maximize the distance to the solutions already in the reference set. In contrast to MOSS, the set P is updated with solutions generated in the scatter search main loop. SSPMO is evaluated by using a benchmark of unconstrained test functions.

Compared to MOSS and SSPMO, our proposal is also applied for solving MOPs with continuous bounded variables, but we additionally consider constrained MOPs. We use a non-dominating sorting procedure to build the reference set from the initial set P, and a local search based on a mutation operator is used instead of a tabu search to improve the solutions obtained from the reference set. MOSS and SSPMO do not seem to search a Pareto front with a bounded number of solutions, as it is usual in many evolutionary algorithms for solving MOPS, but they search as many solutions as possible; we consider the former goal, and it is achieved by using the set P as a population where all the non-dominated solutions found in the scatter search loop are stored.

In [8] a scatter search algorithm for solving the bi-criteria multi-dimensional knapsack problems is proposed. This algorithm is tailored to solve a specific problem, so the scatter search methods differ significantly of those used in this work.

Concerning evolutionary algorithms, the micro-GA [4] is similar to our proposal in the sense that they two use a small population and a reinitialization process. However, in the micro-GA this population is very small (typically four members), it is obtained by randomly choosing individuals of another population composed of non-variable and variable parts, and crossover and mutation operators are used to generate new individuals. In contrast, in SSMO the size of the reference set ranges between ten to twenty solutions, which are selected from an initial population choosing the best p individuals according to two different strategies, and new individuals are obtained from the reference set by applying a systematic combination procedure. Finally, the micro-GA uses an external archive to store the non-dominated solutions found, while SSMO uses the initial set P already included in the standard scatter search template.

3 Non-dominated Sorting Scatter Search Algorithm

SSMO is based on the scatter search template proposed in [9] and its application to solve bounded continuous single objective optimization problems [10]. The template consists of the definition of five methods, as depicted in Fig. 1. We

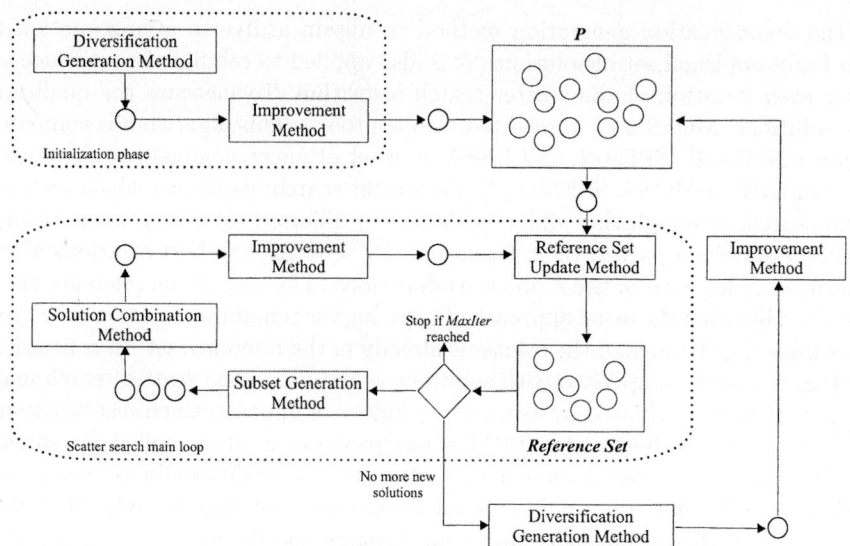

Fig. 1. Outline of the standard scatter search algorithm

first describe these methods, focusing mainly in the improvement and reference set update procedures, which constitute the basis of our proposal. Then, we detail how the initial population P is managed. Finally, we outline the overall algorithm.

3.1 Scatter Search Methods

Diversification Generation Method. This method is basically the same one proposed in [10]. The goal is to obtain an initial set P of diverse solutions. The method consists in dividing the range of each variable in a number of sub-ranges of equal size; then, each solution is obtained in two steps. First, a sub-range is randomly selected, with the probability of selecting a sub-range being inversely proportional to its frequency count (the number of times the sub-range has been selected); second, a value is randomly generated within the selected range.

Improvement Method. The idea behind this method is to use a local search algorithm to improve the solutions in the initial set P. In contrast to [10], where a simplex method is used, we have to deal with MOPs which have constraints, so simplex seems not adequate. Instead, we propose an improvement method based on a mutation operator and a Pareto dominance test. We describe the method in Fig. 2.

The improvement method is simple. Taking as an argument an individual, this is repeatedly mutated with the aim of obtaining a better individual. The term "better" is defined here in a similar way as the constrained-dominance approach used in NSGA-II [1]. The constraint violation test checks whether two individuals are feasible or not. If one of them is feasible and the other one is

```
Individual improvement(Individual originalIndividual, int iter) {
  Individual improvedIndividual
  repeat iter times {
    mutatedIndividual = mutation(originalIndividual)
    if (the problem has constraints) {
      evaluateConstraints(mutatedIndividual)
      best = constraintTest(mutatedIndividual, originalIndividual)
      if (none of them is better than the other one) {
        evaluate(mutatedIndividual)
        best = dominanceTest(mutatedIndividual, originalIndividual)
      } // if
      else if (mutatedIndividual is best)
        evaluate(mutatedIndividual)
    } // if
    else { // the problem has no constraints
      evaluate(mutatedIndividual)
      best = dominanceTest(mutatedIndividual, originalIndividual)
    } // else
    if (mutatedIndividual is best)
      originalIndividual = mutatedIndividual
    else if (originalIndividual is best)
      delete(mutatedIndividual)
    else { // both individuals are non-dominated
      add originalIndividual to P
      originalIndividual = mutatedIndividual
    } // else
  } // repeat
  return originalIndividual
} // improvement
```

Fig. 2. Pseudocode describing the improvement method

not, or both are infeasible but one of them has an smaller overall constraint violation, the test returns the winner. Otherwise, a dominance test is performed to decide whether one of the individuals dominates the other one. If the original individual wins, the mutated one is deleted; if the mutated individual wins, it replaces the original one; finally, if they are both non-dominated, the original individual is moved into the initial set P and the mutated individual becomes the new original one.

We can point out several features of the proposed improvement method. First, mutated individuals are only evaluated if they are going to replace the original individual. Second, in the case of finding several non-dominated solutions in the procedure, they are inserted into P, which could eventually fill. The strategy we propose to deal with this issue is explained in Section 3.2. Finally, we can adjust the improvement effort by tuning the parameter *iter*.

Reference Set Update Method. The reference set, *RefSet*, is a collection of both high quality solutions and diverse solutions that are used to generate new

```
referenceSetUpdate(bool build) {
  if (build) { // build a new reference set
    select the p best individuals of P
    build the RefSet1 with these p individuals
    compute Euclidean distances in P to obtain q individuals
    build the RefSet2 with these q individuals
  } // if
  else { // update the reference set
    for (each new solution s) {
      test to insert s in RefSet1
      if (test fails)
        test to insert s en RefSet2
          if (test fails)
            delete s
    } // for
  } // else
} // referenceSetUpdate
```

Fig. 3. Pseudocode describing the reference set update method

individuals by applying the solution combination method. The set itself is composed of two subsets $RefSet_1$ and $RefSet_2$ of size p and q, respectively. The first subset contains the best quality solutions in P, while the second subset should be filled with solutions promoting diversity. In [7] the $RefSet_2$ is constructed by selecting from P those individuals whose minimum Euclidean distance to the $RefSet_1$ is the highest. We keep the same strategy for building the $RefSet_2$, but, as is usual in the multiobjective optimization domain, we have to define the concept of "best individual" to build the $RefSet_1$. On the other hand, the reference set update method is used to generate the reference set, but also to update it with the new solutions obtained in the scatter search main loop (see Fig. 1). A scheme of this method is included in Fig. 3.

To select the best p individuals in P we propose the following two strategies:

1. The first approach is to carry out a non-dominated sorting of P. However, as there will typically be several individuals per rank, some kind of niching value can be assigned to them to decide which are the most promising solutions. We have used the crowding distance used in NSGA-II, but other kind of niching measurement are valid (e.g., the density applied in SPEA-2). Thus, the $RefSet_1$ is composed of the best ranked individuals, being the individuals with the same rank ordered by its crowding distance.

2. The second strategy consists in ranking the population P and then applying a clustering algorithm, such as k-means or a minimum spanning tree method, to obtain p centroids of the set composed of individuals with best rank. The centroids are individuals which are representative of the set of solutions they derive, so they can be promising elements to compose the $RefSet_1$. We use the Euclidean distance as the metric to assess the similarity among the individuals.

```
// Test to update the RefSet1 with individual s
bool dominated = false
for (each solution r in RefSet1)
   if (s dominates r)
      remove r from RefSet1
   else (if r dominates s)
      dominated = true
if (not dominated)
   if (RefSet1 not full)
      add s to RefSet1
   else
      add s to P
else // the individual s is dominated
   // test to update the RefSet2 with individual s
   ...
```

Fig. 4. Pseudocode describing the test to add new individuals to $RefSet_1$

Once the reference set is completed, its solutions are combined to obtain new solutions which, after applying the improvement method to them, are checked against those belonging to the reference set. According to the scatter search template, a new solution can become a member of the reference set if either one of the following conditions is satisfied:

- The new individual has better objective function value than the individual with the worst objective value in $RefSet_1$.
- The new individual has a better distance value to the reference set than the individual with the worst distance value in $RefSet_2$.

While the second condition holds in the case of multiobjective optimization, we have again the decide about the concept of best individual concerning the first condition. To determine whether a new solution is better than another one in $RefSet_1$ (i.e., the test to insert a new individual s in $RefSet_1$, as it appears in Fig. 3) we cannot use a ranking procedure because the size of this population usually is small (typically the size of the whole reference set is 20 or less). Our approach is to compare each new solution i to the individuals in $RefSet_1$ using a dominance test. This test is included in Fig. 4. (For the sake of simplicity, we do not consider here constraints in the MOP. The procedure to deal with constraints is as explained in the improvement method in Fig. 2.)

Let us note that a new individual does not replace another one in $RefSet_1$. Instead, it is inserted into that set if it is non-dominated by $RefSet_1$ and this is not full; otherwise, it is sent to the set P. This way, we try to keep all non-dominated solutions found by using P as a kind of archive of non-dominated solutions. As we mentioned in the improvement method subsection, this can lead the set P to fill. This issue is considered in Section 3.2.

Subset Generation Method. This method generates subsets of individuals, which will be used for creating new solutions with the solution combination method. Several kinds of subsets are possible [10]. We restrict this method to generate all pairwise combinations of solutions in the reference set.

Solution Combination Method. The idea of this method is to find linear combinations of reference solutions. Again, we use the same method proposed in [10], where each pair of solutions x_1 and x_2 can lead to two, three, or four new solutions, depending on whether x_1 and x_2 belong to $RefSet_1$ or $RefSet_2$.

3.2 Managing the Initial Population

Prior to describing the full algorithm of SSMO, we still need to define a procedure to manage the set P. In particular, when new non-dominated solutions are found by the improvement and reference set update methods, they can be inserted in P, which can eventually fill. This issue is also important because diversity can be improved depending on the applied strategy.

Our approach is to allow the set P to grow until a certain limit. Therefore, if P is intended to store up to t individuals, we extend this limit to, for example, $2t$. When a new individual is going to be added to P, a test checking whether there is already an individual with the same objective function values is executed. If the test is successful, the individual is deleted; otherwise, it is inserted into P. If P has reached to its limit, a cutoff procedure is invoked.

The cutoff procedure performs a ranking of P and then it removes all individuals except those with the best rank (i.e., the individuals with rank equal to 0). If after the removing the size is greater than t, the crowding distance of the individuals is calculated, P is ordered according to this value, and the solutions falling into positions beyond t are removed.

3.3 Outline of SSMO

Once the five methods of the scatter search have been proposed and a procedure to manage the population P has been defined, we are now ready to give an overall view of the technique. The outline depicted in Fig. 5 shows that the SSMO algorithm is simple.

Initially, the diversification generation method is invoked to generate s initial solutions, and each of them is passed to the improvement method. The result is the initial set P. Then, a number of iterations is performed (the outer loop in Fig. 5). In each iteration, the reference set is built, the subset generation method is invoked, and the main loop of the scatter search algorithm is executed until there are no new solutions. Then, the individuals in $RefSet_1$ are inserted into P. The number of iterations can be fixed, or it can depend on other conditions; here, we have used as stop condition the computation of a preprogrammed number of fitness evaluations (see next section). Finally, the cutoff procedure is invoked to remove dominated solutions from P.

```
construct the initial set P
// outer loop
until (stop condition) {
  referenceSetUpdate(build=true)
  subsetGeneration()
  // scatter search main loop
  while (new subsets are generated) {
    combination()
    for (each combinated individual) {
      improvement() ;
      referenceSetUpdate(build=false)
    } // for
    subsetGeneration()
  } // while
  add RefSet1 to P
} // until
cutoff()
```

Fig. 5. Outline of the SSMO algorithm

4 Computational Results

This section is devoted to the evaluation of SSMO. We have chosen several test problems taken from the specialized literature, and we have analyzed the results taking as a reference those obtained with NSGA-II.

Given that SSMO is a real-coded evolutionary algorithm, we have used the real-coded NSGA-II with the parameter settings suggested in [1]. A crossover probability of $p_c = 0.9$ and a mutation probability $p_m = 1/n$ (where n is the number of decision variables) are used. The operators for crossover and mutation are simulated binary crossover (SBX) and polynomial mutation, with distribution indexes of $\eta_c = 20$ and $\eta_m = 20$, respectively. The population size is 100 individuals, and the algorithm is run for 250 iterations.

For SSMO we have chosen a reasonable set of values, and we have not made any effort to find the best parameter settings. The size of P is 100, and it can grow up to 200 individuals. The mutation operator used in the improvement method is the same as in NSGA-II, polynomial mutation, with the same value of η_m. The size of the *RefSet*$_1$ and *RefSet*$_2$ is 10 in both sets. The number of iterations in the improvement method has a value of $iter = 10$. The algorithm is run until 25000 function evaluations are computed.

SSMO is written in C++. We have compiled the software with GCC V3.2 and optimization level -O3, and the experiments have been executed in a Pentium 4 at 2.8GHz with 512 MB of RAM, running Suse Linux 8.1 (kernel 2.4.19).

4.1 Test Problems

We have selected both constrained and unconstrained problems that have been used in studies in this area. Given that they are widely known, we do not include

Table 1. Unconstrained test functions

Problem	Objective functions	Variable bounds	n
Schaffer	$f_1(x) = x^2$ $f_2(x) = (x-2)^2$	$-10^5 \leq x \leq 10^5$	1
Fonseca	$f_1(x) = 1 - e^{-\sum_{i=1}^{n}(x_i - \frac{1}{\sqrt{n}})^2}$ $f_2(x) = 1 - e^{-\sum_{i=1}^{n}(x_i + \frac{1}{\sqrt{n}})^2}$	$-4 \leq x_i \leq 4$	3
Kursawe	$f_1(x) = \sum_{i=1}^{n-1}(-10e^{(-0.2*\sqrt{x_i^2 + x_{i+1}^2})})$ $f_2(x) = \sum_{i=1}^{n}(\lvert x_i \rvert^a + 5\sin(x_i)^b)$	$-5 \leq x_i \leq 5$	3
ZDT1	$f_1(x) = x_1$ $f_2(x) = g(x)[1 - \sqrt{x_1/g(x)}]$ $g(x) = 1 + 9(\sum_{i=2}^{n} x_i)/(n-1)$	$0 \leq x_i \leq 1$	30
ZDT2	$f_1(x) = x_1$ $f_2(x) = g(x)[1 - (x_1/g(x))^2]$ $g(x) = 1 + 9(\sum_{i=2}^{n} x_i)/(n-1)$	$0 \leq x_i \leq 1$	30
ZDT3	$f_1(x) = x_1$ $f_2(x) = g(x)[1 - \sqrt{\frac{x_1}{g(x)}} - \frac{x_1}{g(x)}\sin(10\pi x_1)]$ $g(x) = 1 + 9(\sum_{i=2}^{n} x_i)/(n-1)$	$0 \leq x_i \leq 1$	30
ZDT4	$f_1(x) = x_1$ $f_2(x) = g(x)[1 - (x_1/g(x))^2]$ $g(x) = 1 + 10(n-1) + \sum_{i=2}^{n}[x_i^2 - 10\cos(4\pi x_i)]$	$0 \leq x_1 \leq 1$ $-5 \leq x_i \leq 5$ $i = 2, ..., n$	10
ZDT6	$f_1(x) = 1 - e^{-4x_1}\sin^6(6\pi x_1)$ $f_2(x) = g(x)[1 - (f_1(x)/g(x))^2]$ $g(x) = 1 + 9[(\sum_{i=2}^{n} x_i)/(n-1)]^{0.25}$	$0 \leq x_i \leq 1$	10

full details of them here for space constraints. They can be found in the cited references and also in books such as [11] and [12].

The selected unconstrained problems include the studies of Schaffer [13], Fonseca [14], and Kursawe [15], as well as the problems ZDT1, ZDT2, ZDT3, ZDT4, and ZDT6, which are defined in [16]. Their formulation is provided in Table 1. The constrained problems are Osyczka2 [17], Tanaka [18] (which are respectively known as MOP-C2 and and MOP-C4 in [11]), Srinivas [19], Constr_Ex [1], and Golinski [20]. They are described in Table 2.

4.2 Performance Metrics

Several metrics have been proposed for measuring the results of Pareto-based multi-objective optimization algorithms. In this work we use the metrics M_1^* [16] and Δ [1]. The former gives the average distance to the Pareto optimal set and the latter is a diversity metric that measures the extent of spread achieved among the obtained solutions. Their formulations are:

$$M_1^* = \frac{1}{|Y'|} \sum_{d' \in Y'} \min\{\|d' - \bar{d}\|^*; \bar{d} \in \bar{Y}\} \tag{1}$$

where Y is the set of all possible objective vectors, $Y' \subseteq Y$ is the set of objective vectors found, and $\bar{Y} \subseteq Y$ is the set of solutions of the Pareto optimal set. This metric ideally should be zero. Δ is defined as:

$$\Delta = \frac{d_f + d_l + \sum_{i=1}^{N-1}|d_i - \bar{d}|}{d_f + d_l + (N-1)\bar{d}} \tag{2}$$

Table 2. Constrained test functions

Problem	Objective functions	Constraints	Variable bounds	n
Osyczka2	$f_1(x) = -(25(x_1-2)^2 +$ $(x_2-2)^2 +$ $(x_3-1)^2(x_4-4)^2 +$ $(x_5-1)^2)$ $f_2(x) = x_1^2 + x_2^2 +$ $x_3^2 + x_4^2 + x_5^2 + x_6^2$	$g_1(x) = 0 \le x_1 + x_2 - 2$ $g_2(x) = 0 \le 6 - x_1 - x_2$ $g_3(x) = 0 \le 2 - x_2 + x_1$ $g_4(x) = 0 \le 2 - x_1 + 3x_2$ $g_5(x) = 0 \le 4 - (x_3-3)^2 - x_4$ $g_6(x) = 0 \le (x_5-3)^3 + x_6 - 4$	$0 \le x_1, x_2 \le 10$ $1 \le x_3, x_5 \le 5$ $0 \le x_4 \le 6$ $0 \le x_6 \le 10$	6
Tanaka	$f_1(x) = x_1$ $f_2(x) = x_2$	$g_1(x) = -x_1^2 - x_2^2 + 1+$ $0.1\cos(16\arctan(x_1/x_2)) \le 0$ $g_2(x) = (x_1 - 0.5)^2 +$ $(x_2 - 0.5)^2 \le 0.5$	$-\pi \le x_i \le \pi$	2
Constr_Ex	$f_1(x) = x_1$ $f_2(x) = (1+x_2)/x_1$	$g_1(x) = x_2 + 9x_1 \ge 6$ $g_2(x) = -x_2 + 9x_1 \ge 1$	$0.1 \le x_1 \le 1.0$ $0 \le x_2 \le 5$	2
Srinivas	$f_1(x) = (x_1-2)^2 +$ $(x_2-1)^2 + 2$ $f_2(x) = 9x_1 - (x_2-1)^2$	$g_1(x) = x_1^2 + x_2^2 \le 225$ $g_2(x) = x_1 - 3x_2 \le -10$	$-20 \le x_i \le 20$	2
Golinski	$f_1(x) = 0.7854x_1x_2^2(10x_3^2/3 +$ $14.933x_3 - 43.0934)$ $-1.508x_1(x_6^2 + x_7^2) +$ $7.477(x_6^3 + x_7^3)$ $+0.7854(x_4x_6^2 + x_5x_7^2)$ $f_2(x) = \frac{\sqrt{(\frac{745.0x_4}{x_2x_3})^2 + 1.69*10^7}}{0.1x_6^3}$	$g_1(x) = \frac{1.0}{x_1x_2^2x_3} - \frac{1.0}{27.0} \le 0$ $g_2(x) = \frac{1.0}{x_1x_2^2x_3} - \frac{1.0}{27.0} \le 0$ $g_3(x) = \frac{x_4^3}{x_2x_3^2x_6^4} - \frac{1.0}{1.93} \le 0$ $g_4(x) = \frac{x_5^3}{x_2x_3x_7^4} - \frac{1.0}{1.93} \le 0$ $g_5(x) = x_2x_3 - 40 \le 0$ $g_6(x) = x_1/x_2 - 12 \le 0$ $g_7(x) = 5 - x_1/x_2 \le 0$ $g_8(x) = 1.9 - x_4 + 1.5x_6 \le 0$ $g_9(x) = 1.9 - x_5 + 1.1x_7 \le 0$ $g_{10}(x) = f_2(x) \le 1300$ $a = 745.0x_5/x_2x_3$ $b = 1.575*10^8$ $g_{11}(x) = \frac{\sqrt{(a)^2+b)}}{0.1x_7^3} \le 1100$	$2.6 \le x_1 \le 3.6$ $0.7 \le x_2 \le 0.8$ $17.0 \le x_3 \le 28.0$ $7.3 \le x_4 \le 8.3$ $7.3 \le x_5 \le 8.3$ $2.9 \le x_6 \le 3.9$ $5.0 \le x_7 \le 5.5$	7

where d_i is the Euclidean distance between consecutive solutions, $\bar{d}$ is the mean of these distances, and d_f and d_l are the Euclidean distances to the *extreme* solutions of the exact Pareto front in the objective space (see [1] for the details). The ideal value of Δ is also 0, indicating a perfect spreadout of the solutions along the Pareto front. To calculate these metrics we have used the true Pareto fronts obtained by enumeration of the test problems (excepting the ZDTx family) publicly available at http://neo.lcc.uma.es/software/esam.

4.3 Discussion of the Results

The results are summarized in Table 3 (M_1^*) and Table 4 (Δ). For each problem, we have carried out 30 independent runs, and the tables include the mean $\bar{x}$ and standard deviation σ_n, as well as the result of an ANOVA (Analysis of Variance) test with a 5% of significance level (marked as "+" in tables). An ANOVA [21] tests the difference between the means of two or more sets of numeric values.

We have tested two versions of our algorithm. The first version, SSMOv1, selects the best individuals from the reference set using ranking and crowding,

Table 3. Mean and standard deviation of the convergence metric $M1^*$

	NSGA-II		SSMOv1		SSMOv2		
Problem	$\bar{x}$	σ_n	$\bar{x}$	σ_n	$\bar{x}$	σ_n	A
Schaffer	0.0223	0.0011	0.0225	0.0011	0.0225	0.0012	-
Fonseca	0.0025	0.0002	0.0021	0.0002	0.0019	0.0002	+
Kursawe	0.0134	0.0027	0.0531	0.0663	0.0185	0.0036	+
ZDT1	0.0005	6.3e-5	0.0004	5.7e-5	0.0004	4.2e-5	+
ZDT2	0.0004	3.2e-5	0.0004	2.9e-5	0.0004	2.8e-5	+
ZDT3	0.0025	0.0002	0.0016	9.3e-5	0.0020	0.0002	+
ZDT4	0.0044	0.0029	25.6412	12.6266	38.4334	11.1194	+
ZDT6	0.0764	0.0076	0.5954	0.3170	0.2604	0.1773	+
Tanaka	0.0044	0.0003	0.0095	0.0010	0.0096	0.0048	+
Osyczka2	5.4688	8.6247	16.8277	15.2031	13.8621	11.9258	+
Srinivas	0.2570	0.0421	0.2839	0.0489	0.2274	0.0558	+
Constr_Ex	0.0053	0.0003	3.5802	0.4226	3.0383	0.2215	+
Golinski	4.0790	0.7839	35.2460	24.9212	10.9907	11.8089	+

Table 4. Mean and standard deviation of the diversity metric Δ

	NSGA-II		SSMOv1		SSMOv2		
Problem	$\bar{x}$	σ_n	$\bar{x}$	σ_n	$\bar{x}$	σ_n	A
Schaffer	0.4202	0.0264	0.3984	0.0273	0.4323	0.0268	+
Fonseca	0.3756	0.0259	0.3448	0.0324	0.3695	0.0359	+
Kursawe	0.5380	0.0285	0.6720	0.0785	0.5043	0.0542	+
ZDT1	0.5252	0.0333	0.4855	0.0770	0.4871	0.0455	+
ZDT2	0.5149	0.0385	0.5981	0.1660	0.4450	0.0371	+
ZDT3	0.6278	0.0288	1.1508	0.0685	0.7004	0.0582	+
ZDT4	0.4843	0.1750	0.9219	0.0544	0.9766	0.0886	+
ZDT6	0.6078	0.0404	1.3229	0.0604	1.2989	0.1379	+
Tanaka	0.6427	0.0256	0.8694	0.0463	1.0619	0.0931	+
Osyczka2	0.7884	0.0958	1.0817	0.1451	0.9398	0.2197	+
Srinivas	0.3843	0.0353	0.3515	0.0347	0.3964	0.0378	+
Constr_Ex	0.7655	0.0371	0.9070	0.0468	0.8057	0.0228	+
Golinski	0.7516	0.0289	0.9908	0.1682	0.6840	0.1137	+

while the second version, SSMOv2, uses a clustering procedure to obtain the centroids that will compose the *RefSet1* (see Section 3.1).

We analyze first the two versions of SSMO. Considering the metric $M1^*$, we can observe that SSMOv2 converges better in 7 out of the 13 problems, while SSMOv1 provides better results only in 3 problems. If we take into account the metric Δ, SSMOv1 and SSMOv2 behave better in seven and six problems, respectively. These observations indicate that the use of centroids for building the reference set is a promising approach for improving the accuracy of the algorithm that can be used as the basis of future developments.

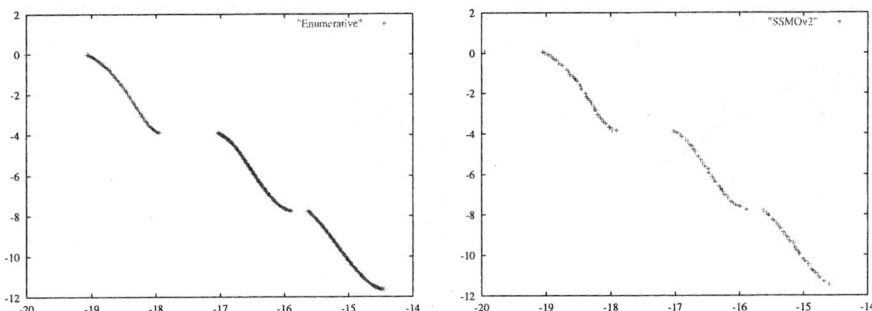

Fig. 6. Exact Pareto front for the problem Kursawe (left) and nondominated solutions obtained with SSMOv2 (right)

We now turn to compare our algorithm with NSGA-II. If we analyze the unconstrained problems, we observe that, with the exception of the problems ZDT4 and ZDT6, the two versions of SSMO obtain competitive performance. If we consider that this is just a first approach of using scatter search for MOPs, this sounds quite promising.

Concerning the constrained problems, the two metrics indicate that the results of SSMO are comparable to NSGA-II in terms of diversity, but they are slightly worse in terms of convergence. With respect to the problems Tanaka and Srinivas our two proposals show comparable performance to NSGA-II.

In order to better illustrate the working principles of SSMO, we show a typical simulation result with the problem Kursawe. This problem has three discontinuous regions in the Pareto-optimal front. Fig 6 shows the true Pareto front obtained by enumeration (left) and the solutions obtained by SSMOv2. Next, we show the nondominated solutions of the problem Srinivas. The true Pareto front yielded by the enumerative search appears in Fig 7 (left), while the right part of that figure shows the resulting front from SSMOv2. Finally, in Fig. 8 we show the results of solving the problem ZDT3 with NSGA-II (left) and SSMOv2 (right). This problem has a number of disconnected Pareto-optimal fronts.

We conclude that a more comprehensive study is still necessary to understand the behavior of SSMO. The choice of a size of 10 for both the $RefSet_1$ and $RefSet_2$ intends to keep a balance between intensification and diversification, but most probably other values would enhance the search. On the other hand, our experiments reveal that using a different strategy to update the $RefSet_1$ with new improved individuals allows better performance to be obtained with some difficult problems; for example, in the case of the problem ZDT4, the mean value of the metric $M1^*$ reduces from 38.4334 to 10.4382, and in 15 of the 30 experiments, the value of $M1^*$ is below 0.0038.

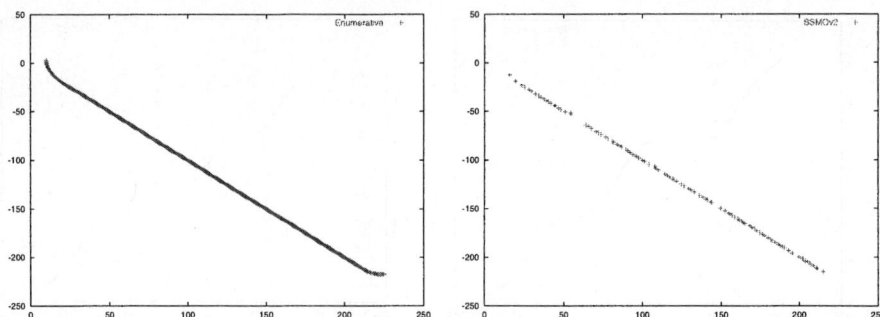

Fig. 7. Exact Pareto front for the problem Srinivas (left) and nondominated solutions obtained with SSMOv2 (right)

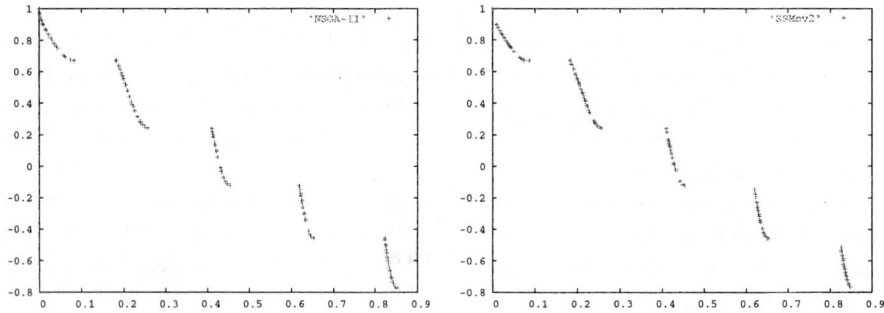

Fig. 8. Nondominated solutions obtained for the problem ZDT3 with NSGA-II (left) and SSMOv2 (right)

5 Conclusions and Future Work

We have proposed a first approximation to the utilization of a scatter search method to solve multiobjective optimization problems. Two variants of our approach have been compared against NSGA-II in thirteen different difficult problems taken from the literature. In the unconstrained test functions used, our algorithm has obtained comparable performance to NSGA-II in six of the eight selected problems. In the unconstrained problems, a more obscure scenario appears, being acccurate for Tanaka and Srinivas, but showing difficulties to solve the problems Constr_Ex, Osyczka2, and Golinski.

A deep study to find the best parameters defining the behavior of SSMO is a matter of future work. In this sense, we also plan to study new strategies for the improvement and reference set update methods, as well as other approaches to store and manage the non-dominated solutions encountered during the execution of the algorithm, such as using an external archive of nondominated solutions.

References

1. Deb, K., Pratap, A., Agarwal, S., Meyarivan, T.: A Fast and Elitist Multiobjective Genetic Algorithm: NSGA-II. IEEE Transactions on Evolutionary Computation **6** (2002) 182–197
2. Knowles, J., Corne, D.: The Pareto Archived Evolution Strategy: A New Baseline Algorithm for Multiobjective Optimization. In: Proceedings of the 1999 Congress on Evolutionary Computation, Piscataway, NJ, IEEE Press (1999) 9–105
3. Zitzler, E., Laumanns, M., Thiele, L.: SPEA2: Improving the Strength Pareto Evolutionary Algorithm. Technical Report 103, Swiss Federal Institute of Technology (ETH), Zurich, Switzerland (2001)
4. Coello, C.A., Toscano, G.: Multiobjective Optimization Using a Micro-Genetic Algorithm. In: GECCO-2001. (2001) 274–282
5. Glover, F., Laguna, M., Martí, R.: Fundamentals of Scatter Search and Path Relinking. Control and Cybernetics **29** (2000) 653–684
6. Beausoleil, R.P.: MOSS: Multiobjective Scatter Search Applied to Nonlinear Multiple Criteria Optimization. To appear in the European Journal of Operational Research (2004)
7. Caballero, R., Laguna, M., Molina, J., Martí, R.: SSPMO: A Scatter Search Procedure for Non-Linear Multiobjective Optimization. Submitted to INFORMS Journal on Computing (2004)
8. da Silva, C.G., Clímaco, J., Figueira, J.: A Scatter Search Method for the Bi-Criteria Multi-Dimensional {0,1}-Knapsack Problem using Surrogate Relaxation. Journal of Mathematical Modelling and Algorithms **3** (2004) 183–208
9. Glover, F.: A Template for Scatter Search and Path Relinking. In J. K. Hao and E. Lutton and E. Ronald and M. Shoenauer and D. Snyers, ed.: Artificial Evolution. Lecture Notes in Computer Science, Springer Verlag (1997)
10. Glover, F., Laguna, M., Martí, R.: Scatter Search. In Ghosh, A., Tsutsui, S., eds.: Advances in Evolutionary Computing: Theory and Applications. Springer (2003)
11. Coello, C.A., Van Veldhuizen, D.A., Lamont, G.B.: Evolutionary Algorithms for Solving Multi-Objective Problems. Kluwer Academic Publishers (2002)
12. Deb, K.: Multi-Objective Optimization Using Evolutionary Algorithms. John Wiley & Sons (2001)
13. Schaffer, J.D.: Multiple Objective Optimization with Vector Evaluated Genetic Algorithms. In Grefensttete, J., ed.: First International Conference on Genetic Algorithms, Hillsdale, NJ (1987) 93–100
14. Fonseca, C.M., Flemming, P.J.: Multiobjective Optimization and Multiple Constraint Handling with Evolutionary Algorithms - Part II: Application Example. IEEE Transactions on System, Man, and Cybernetics **28** (1998) 38–47
15. Kursawe, F.: A Variant of Evolution Strategies for Vector Optimization. In Schwefel, H., Männer, R., eds.: Parallel Problem Solving for Nature, Berlin, Germany, Springer-Verlag (1990) 193–197
16. Zitzler, E., Deb, K., Thiele, L.: Comparison of Multiobjective Evolutionary Algorithms: Empirical Results. IEEE Transactions on Evolutionary Computation **8** (2000) 173–195
17. Osyczka, A., Kundo, S.: A New Method to Solve Generalized Multicriteria Optimization Problems Using a Simple Genetic Algorithm. Structural Optimization **10** (1995) 94–99
18. Tanaka, M., Watanabe, H., Furukawa, Y., Tanino, T.: GA-Based Decision Support System for Multicriteria Optimization. In: Proceedings of the IEEE International Conference on Systems, Man, and Cybernetics. Volume 2. (1995) 1556–1561

19. Srinivas, N., Deb, K.: Multiobjective Function Optimization Using Nondominated Sorting Genetic Algorithms. Evolutionary Computation **2** (1995) 221–248
20. Kurpati, A., Azarm, S., Wu, J.: Constraint Handling Improvements for Multi-Objective Genetic Algorithms. Structural and Multidisciplinary Optimization **23** (2002) 204–213
21. Montgomery, D.C.: Design and Analysis of Experiments. 3 edn. John Wiley, New York (1991)

A MOPSO Algorithm Based Exclusively on Pareto Dominance Concepts

Julio E. Alvarez-Benitez*, Richard M. Everson, and Jonathan E. Fieldsend**

Department of Computer Science, University of Exeter, UK

Abstract. In extending the Particle Swarm Optimisation methodology to multi-objective problems it is unclear how global guides for particles should be selected. Previous work has relied on metric information in objective space, although this is at variance with the notion of dominance which is used to assess the quality of solutions. Here we propose methods based exclusively on dominance for selecting guides from a nondominated archive. The methods are evaluated on standard test problems and we find that probabilistic selection favouring archival particles that dominate few particles provides good convergence towards and coverage of the Pareto front. We demonstrate that the scheme is robust to changes in objective scaling. We propose and evaluate methods for confining particles to the feasible region, and find that allowing particles to explore regions close to the constraint boundaries is important to ensure convergence to the Pareto front.

1 Introduction

Evolutionary algorithms (EA) have been used since the mid-eighties to solve complex single and multi-objective optimisation problems (see, for example, [1, 2, 3]). More recently the Particle Swarm Optimisation (PSO) heuristic, inspired by the flocking and swarm behaviour of birds, insects, and fish schools has been successfully used for single objective optimisation, such as neural network training and non-linear function optimisation [4]. Briefly, PSO maintains a balance between exploration and exploitation in a population (swarm) of solutions by moving each solution (particle) towards both the global best solution located by the swarm so far and towards the best solution that the particular particle has so far located. The global best and personal best solutions are often called *guides*.

Since PSO and EA algorithms have structural similarities (such as the presence of a population searching for optima and information sharing between population members) it seems a natural progression to extend PSO to multi-objective problems (MOPSO). Some attempts in this direction have been made with promising results such as [5, 6, 7, 8, 9]. In the most recent heuristics the

---

* Supported by and currently with Banco de la República, Colombia.
** Supported by EPSRC, grant GR/R24357/01.

guides are selected from the set of non-dominated solutions found so far. However, in a multi-objective problem each of non-dominated solutions is a potential global guide and there are many ways of selecting a guide from among them for each particle in the swarm. Heuristics to date have relied on proximity in objective space to determine this selection, however the relative weightings of the objectives are *a priori* unknown and the use of metric information in objective space is at variance with the notion of dominance that is central to the definition of Pareto optimality. In this paper we propose and examine MOPSO heuristics based entirely on Pareto dominance concepts. The manner in which particles are constrained to lie within the search space can have a marked effect on the optimisation efficiency: the other central purpose of this paper is to propose and compare constraint methods.

We start by briefly reviewing basic definitions of multi-objective problems and Pareto concepts (section 2), after which we describe the single objective PSO methodology in section 3. The multi-objective PSO algorithm is presented in section 4, and we present and evaluate methods for selecting guides here. Techniques for confining particles to the feasible region are described and evaluated in section 5. Finally, conclusions are drawn in section 6.

2   Dominance and Pareto Optimality

In a multi-objective optimisation problem we seek to simultaneously extremise D objectives: $y_i = f_i(\mathbf{x})$, where $i = 1, \ldots, D$ and where each objective depends upon a vector $\mathbf{x}$ of K parameters or decision variables. The parameters may also be subject to the J constraints: $e_j(\mathbf{x}) \geq 0$ for $j = 1, \ldots, J$.

Without loss of generality it is assumed that these objectives are to be minimised, as such the problem can be stated as:

$$\text{minimise } \mathbf{y} = \mathbf{f}(\mathbf{x}) \equiv (f_1(\mathbf{x}), f_2(\mathbf{x}), \ldots, f_D(\mathbf{x})) \tag{1}$$
$$\text{subject to } \mathbf{e}(\mathbf{x}) \equiv (e_1(\mathbf{x}), e_2(\mathbf{x}), \ldots, e_J(\mathbf{x})) \geq \mathbf{0}. \tag{2}$$

A decision vector $\mathbf{u}$ is said to *strictly dominate* another $\mathbf{v}$ (denoted $\mathbf{u} \prec \mathbf{v}$) if $f_i(\mathbf{u}) \leq f_i(\mathbf{v}) \; \forall i = 1, \ldots, D$ and $f_i(\mathbf{u}) < f_i(\mathbf{v})$ for some i; less stringently $\mathbf{u}$ *weakly dominates* $\mathbf{v}$ (denoted $\mathbf{u} \preceq \mathbf{v}$) if $f_i(\mathbf{u}) \leq f_i(\mathbf{v})$ for all i. A set of decision vectors is said to be a *non-dominated set* if no member of the set is dominated by any other member. The *true* Pareto front, $\mathcal{P}$, is the non-dominated set of solutions which are not dominated by any feasible solution.

3   Particle Swarm Optimisation – PSO

The particle swarm optimisation method evolved from a simple simulation model of the movement of social groups such as birds and fish [4], in which it was observed that local interactions underlie the group behaviour and individual members of the group can profit from the discoveries and experiences of other members. In PSO each solution (particle) $\mathbf{x}_n$ in the swarm of N particles is endowed with a velocity which determines its location at the next time step:

$$\mathbf{x}_n^{(t+1)} = \mathbf{x}_n^{(t)} + \chi \mathbf{v}_n^{(t)} + \boldsymbol{\epsilon}^{(t)} \qquad (3)$$

where $\chi \in [0,1]$ is a constriction factor which controls the velocity's magnitude; in the work reported here $\chi = 1$. The final term in (3) is a small stochastic perturbation, known as the *turbulence factor*, added to the position to help prevent the particle becoming stuck in local minima and to promote wide exploration of the decision space. Although originally introduced as a normal perturbation [8], here a perturbation to each dimension was added with probability 0.01 and ϵ_k itself was a perturbation from a Laplacian density $p(\epsilon_k) \propto e^{-|\epsilon_k|/\beta}$ with $\beta = 0.1$. The Laplacian distribution yields occasional large perturbations thus enabling wider exploration.

The velocities of each particle are modified to fly towards two different guides: their personal best, $\mathbf{P}_n$, for exploiting the best results found so far by each of the particles, and the global best, $\mathbf{G}$, the best solution found so far by the whole swarm for encouraging further exploration and information sharing between the particles. This is achieved by updating the K components of each particle's velocity as follows:

$$v_{nk}^{(t+1)} = w v_{nk}^{(t)} + c_1 r_1 (P_{nk} - x_{nk}^{(t)}) + c_2 r_2 (G_{nk} - x_{nk}^{(t)}) \qquad (4)$$

r_1 and r_2 are two uniformly distributed random numbers in the range $[0,1]$. The constants c_1 and c_2 control the effect of the personal and global guides, and the parameter w, known as the *inertia*, controls the trade-off between global and local experience; large w motivates global exploration by giving large weight to the current velocity. In the work reported here $c_1 = c_2 = 1$ and $w = 0.5$. The global guide carries a subscript n because for multi-objective PSO a (possibly different) global guide is associated with each particle; this is in contrast to uni-objective PSO in which there is a single global guide, namely the best solution located so far.

4 Multi-objective PSO

The main difficulty in extending PSO to multi-objective problems is to find the best way of selecting the guides for each particle in the swarm; the difficulty is manifest as there are no clear concepts of personal and global bests that can be clearly identified when dealing with D objectives rather than a single objective. Previous MOPSO implementations [5, 6, 7, 8, 9, 10] have all used metrics in objective space (either explicitly or implicitly) in the selection of guides – thus making them susceptible to different scalings in objective space.

The algorithms we propose here are similar to recent MOPSO algorithms [7, 8, 9, 10] in that they use an archive or repository, A, which contains the non-dominated solutions found by the algorithm so far. We emphasise that we do not restrict the size of A by gridding, clustering or niching (as done in, for example, [10]) as that may lead to oscillation or shrinking of the Pareto front [11, 12].

Algorithm 1 Multi-objective PSO

1 : $A := \emptyset$ *Initially empty archive*
2 : $\{\mathbf{x}_n, \mathbf{v}_n, \mathbf{G}_n, \mathbf{P}_n\}_{n=1}^{N} := \texttt{initialise}()$ *Random locations and velocities*
3 : for $t := 1 : G$ *G generations*
4 :   for $n := 1 : N$
5 :   for $k := 1 : K$ *Update velocities and positions*
6 : $v_{nk} := wv_{nk} + r_1(P_{nk} - x_{nk}) + r_2(G_{nk} - x_{nk})$
7 : $x_{nk} := x_{nk} + v_{nk} + \epsilon$
8 :   end
9 : $\mathbf{x}_n := \texttt{enforceConstraints}(\mathbf{x}_n)$
10 : $\mathbf{y}_n := \mathbf{f}(\mathbf{x}_n)$ *Evaluate objectives*
11 :   if $\mathbf{x}_n \not\prec \mathbf{u} \; \forall \; \mathbf{u} \in A$ *Add non-dominated $\mathbf{x}_n$ to A*
12 : $A := \{\mathbf{u} \in A \,|\, \mathbf{u} \not\prec \mathbf{x}_n\}$ *Remove points dominated by $\mathbf{x}_n$*
13 : $A := A \cup \mathbf{x}_n$ *Add $\mathbf{x}_n$ to A*
14 :   end
15 :   end
16 :   if $\mathbf{x}_n \preceq \mathbf{P}_n \vee (\mathbf{x}_n \not\prec \mathbf{P}_n \wedge \mathbf{P}_n \not\prec \mathbf{x}_n)$ *Update personal best*
17 : $\mathbf{P}_n := \mathbf{x}_n$
18 :   end
19 : $\mathbf{G}_n := \texttt{selectGuide}(\mathbf{x}_n, A)$
20 : end

At the start of the optimisation, which is outlined in Algorithm 1, A is empty and the locations and velocities of the N particles are initialised randomly. The personal bests for each particle are initialised to be the starting location, $\mathbf{P}_n = \mathbf{x}_n$; likewise the global guide for each particle is initialised to be its initial location: $\mathbf{G}_n = \mathbf{x}_n$.

At each generation t the velocities $\mathbf{v}_n$ and locations $\mathbf{x}_n$ of each particle are updated according to (4) and (3) (lines 5–8 of Algorithm 1). Following updating, it is possible that the particle positions lie outside the region of feasible solutions. In this case it must be constrained to the feasible region; this is indicated in the Algorithm 1 by the function enforceConstraints, and we discuss methods for enforcing the constraints in section 5. With $\mathbf{x}_n$ in the feasible region the objectives may be evaluated (line 10), and any solutions which are not weakly dominated by any member of the archive are added to A (line 13) and any elements of A which are dominated by $\mathbf{x}_n$ are deleted from A, thus ensuring that A is a non-dominating set.

The crucial parts of the MOPSO algorithm are selecting the personal and global guides. Selection of $\mathbf{P}_n$ is straightforward: if the current position of the n-th particle, $\mathbf{x}_n$, weakly dominates $\mathbf{P}_n$ or $\mathbf{x}_n$ and $\mathbf{P}_n$ are mutually non-dominating, then $\mathbf{P}_n$ is set to the current position (lines 16–18). Since members of A are mutually non-dominating and no member of the archive is dominated by any $\mathbf{x}_n$, so that in some senses the archive is globally 'better' than each member of the swarm, all the members of A are candidates for the global guide and we now present alternative ways of selecting a global guide *for each* particle in the swarm from A.

Algorithm 2 ROUNDS selection of global guides

1 :	$X' := X$	*Swarm*
2 :	$A' := \emptyset$	*Candidate guides*
3 :	while $|X'| > 0$	
4 :	if $|A'| = 0$, then $A' = A$	*New round: all A are candidates*
5 :	for $\mathbf{a} \in A'$	
6 :	$X_{\mathbf{a}} := \{\mathbf{x} \in X' \,|\, \mathbf{a} \prec \mathbf{x}\}$	*Swarm members dominated by* $\mathbf{a}$
7 :	end	
8 :	$\mathbf{a}^\star := \arg\min_{\mathbf{a} \in A' \wedge |X_{\mathbf{a}}| > 0}(|X_{\mathbf{a}}|)$	$\mathbf{a}^\star$ *dominates fewest particles*
9 :	$\mathbf{x}_n := \text{choose}(X_{\mathbf{a}^\star})$	*Random selection from* $X_{\mathbf{a}^\star}$
10 :	$\mathbf{G}_n := \mathbf{a}^\star$	
11 :	$X' := X' \setminus \mathbf{x}_n$	*Guide selected for* $\mathbf{x}_n$
12 :	$A' := A' \setminus \mathbf{a}^\star$	*Assigned, so delete from candidates*
11 :	end	

4.1 Selecting Global Guides

Here we focus on methods of selecting global guides which are based solely on Pareto dominance and do not attempt to use metric information in objective space. Three alternatives are examined: ROUNDS, which is most complex and explicitly promotes diversity in the population; RANDOM, which is simple and promotes convergence; and PROB, which is a weighted probabilistic method and forms a compromise between RANDOM and ROUNDS. It may be supposed that the archive members which dominate particle $\mathbf{x}_n$ would be better global guides than those archive members which do not, and each of these schemes is based on the idea of selecting a guide for a particle from the members of the archive which dominate the particle.

ROUNDS. The idea underlying this method is that in order to promote diversity in the population by attracting the swarm towards sparsely populated regions, members of the archive that dominate the fewest $\mathbf{x}_n$ should be preferentially assigned as global guides. As shown in Algorithm 2, this is achieved by first locating the member of the archive $\mathbf{a}^\star$ which dominates the fewest particles (but at least one), which is then assigned to be the guide of one of the particles in $X_{\mathbf{a}^\star}$, the set of particles which it dominates. Having assigned $\mathbf{a}^\star$ as a guide, it is removed from consideration as a possible guide until all the other archive members have been assigned a particle to guide and a new round begins (line 4). Clearly, the algorithm can be coded more efficiently than the outlined in Algorithm 2, however, the procedure can be computationally expensive when the archive is large.

RANDOM. While the ROUNDS methods associates a member of the archive with one of the particles in the swarm that it dominates, the RANDOM selection methods focuses on the particle $\mathbf{x}_n$ and selects a guide from among the archive members that dominate $\mathbf{x}_n$. If $A_{\mathbf{x}} = \{\mathbf{a} \in A \,|\, \mathbf{a} \prec \mathbf{x}\}$ is the set of archived points that dominate $\mathbf{x}$, then the RANDOM selection method simply chooses an element

Table 1. Test problems DTLZ1, DTLZ2 & DTLZ3 of [13] for 3 objectives

DTLZ1	$f_1(\mathbf{x}) = \frac{1}{2} x_1 x_2 (1 + g(\mathbf{x}))$ $f_2(\mathbf{x}) = \frac{1}{2} x_1 (1 - x_2)(1 + g(\mathbf{x}))$ $f_3(\mathbf{x}) = \frac{1}{2} (1 - x_1)(1 + g(\mathbf{x}))$ $g(\mathbf{x}) = 100[|\mathbf{x}| - 2 + \sum_{k=3}^{K} (x_k - 0.5)^2 - \cos(20\pi(x_k - 0.5))]$ $0 \leq x_k \leq 1$, for $k = 1, 2, \ldots, K$, $K = 7$
DTLZ2	$f_1(\mathbf{x}) = \cos(x_1 \pi/2) \cos(x_2 \pi/2)(1 + g(\mathbf{x}))$ $f_2(\mathbf{x}) = \cos(x_1 \pi/2) \sin(x_2 \pi/2)(1 + g(\mathbf{x}))$ $f_3(\mathbf{x}) = \sin(x_1 \pi/2)(1 + g(\mathbf{x}))$ $g(\mathbf{x}) = \sum_{k=3}^{K} (x_k - 0.5)^2$ $0 \leq x_k \leq 1$, for $k = 1, 2, \ldots, K$, $K = 12$
DTLZ3	$f_1(\mathbf{x}) = \cos(x_1 \pi/2) \cos(x_2 \pi/2)(1 + g(\mathbf{x}))$ $f_2(\mathbf{x}) = \cos(x_1 \pi/2) \sin(x_2 \pi/2)(1 + g(\mathbf{x}))$ $f_3(\mathbf{x}) = \sin(x_1 \pi/2)(1 + g(\mathbf{x}))$ $g(\mathbf{x}) = 100[|\mathbf{x}| - 2 + \sum_{k=3}^{K} (x_k - 0.5)^2 - \cos(20\pi(x_k - 0.5))]$ $0 \leq x_k \leq 1$, for $k = 1, 2, \ldots, K$, $K = 7$

of $A_{\mathbf{x}_n}$ with equal probability to be the guide for $\mathbf{x}_n$. If $\mathbf{x}_n \in A$ then, clearly, $A_{\mathbf{x}_n}$ is empty, so in this case a guide is selected from the entire archive. Thus

$$\mathbf{G}_n = \begin{cases} \mathbf{a} \in A \text{ with probability } |A|^{-1} & \text{if } \mathbf{x}_n \in A \\ \mathbf{a} \in A_{\mathbf{x}_n} \text{ with probability } |A_{\mathbf{x}_n}|^{-1} & \text{otherwise.} \end{cases} \quad (5)$$

PROB. The RANDOM selection method gives equal probability of being chosen as the guide to all archive members dominating a particle. However, archive members in sparsely populated regions of the front and towards the 'edges' of the front are likely to dominate fewer particles than those in well populated regions or close to the centre of the front. To guide the search towards the sparse regions and edges, we adapt the RANDOM method to favour archive members that dominate the least points. Let $X_\mathbf{a} = \{\mathbf{x} \in X \mid \mathbf{a} \prec \mathbf{x}\}$ be the set of particles dominated by $\mathbf{a}$. Then guides are chosen as:

$$\mathbf{G}_n = \begin{cases} \mathbf{a} \in A \text{ with probability } \propto |X_\mathbf{a}|^{-1} & \text{if } \mathbf{x}_n \in A \\ \mathbf{a} \in A_{\mathbf{x}_n} \text{ with probability } \propto |X_\mathbf{a}|^{-1} & \text{otherwise.} \end{cases} \quad (6)$$

The PROB selection method thus combines the intention behind ROUNDS with the simplicity of RANDOM. With efficient data structures [12] or relatively small populations the computational expense in calculating $|X_\mathbf{a}|$ and $A_\mathbf{x}$ is not exorbitant and can be efficiently incorporated into the updating of A (lines 11–14 of Algorithm 1).

4.2 Experiments

We compared the efficiency of the global guide selection methods on standard test problems DTLZ1–DTLZ3 [13], whose definitions for three objectives are provided in Table 1.

Table 2. GD(A) and $\mathcal{V}_\mathcal{P}(A)$ measures for the methods proposed to select guides. The best value across methods is highlighted in bold

	GD(A)				$\mathcal{V}_\mathcal{P}(A)$			
DTLZ1	MT	ROUNDS	RANDOM	PROB	MT	ROUNDS	RANDOM	PROB
Best	0.0002	0.0048	3.77×10^{-5}	$\mathbf{3.15 \times 10^{-5}}$	0.9992	0.9823	**0.9997**	**0.9997**
Worst	0.7481	0.1761	**0.031**	0.0349	0.9270	0	0.9744	**0.9796**
Average	0.1016	0.0696	$\mathbf{3.10 \times 10^{-3}}$	5.55×10^{-3}	0.9824	0.5201	0.9965	**0.9974**
Median	0.0303	0.0656	3.23×10^{-4}	$\mathbf{1.41 \times 10^{-4}}$	0.9947	0.4952	0.9979	**0.9992**
S. dev.	0.2068	0.0572	7.30×10^{-3}	0.0114	0.0231	0.3413	0.0057	0.0047
DTLZ3								
Best	0.003	0.001	4.81×10^{-5}	5.77×10^{-5}	0.9921	0.9972	0.9979	**0.9981**
Worst	0.2195	1.0244	**0.1446**	0.2621	0	0	**0.7798**	0.704
Average	0.0413	0.2044	**0.0217**	0.0305	0.8743	0.1602	0.9352	**0.945**
Median	0.0134	0.1343	$\mathbf{1.44 \times 10^{-3}}$	1.52×10^{-3}	0.9635	0	0.9486	**0.9946**
S. dev.	0.0627	0.2451	0.0429	0.0687	0.2290	0.316	0.0674	0.086

Unlike single objective problems, solutions to multi-objective optimisation problems can be assessed in several different ways. Here we use the Generational Distance (*GD*) introduced in [14] and used by others (e.g., [10]) as a measure of the mean distance between elements of the archive and the true Pareto front:

$$GD(A) = \left[\frac{1}{|A|} \sum_{\mathbf{a} \in A} d(\mathbf{a})^2 \right]^{\frac{1}{2}} \quad (7)$$

where $d(\mathbf{a})$ is the shortest Euclidean distance between $\mathbf{a}$ and the front $\mathcal{P}$. Clearly, this measure depends on the relative scaling of the objective functions, however, it yields a fair comparison here because the objectives for the DTLZ test functions have similar ranges.

An alternative measure which also measures the spread of the solutions found across the front is the volume measure, $\mathcal{V}_\mathcal{P}(A)$, which is defined as the fraction of the minimum axis-parallel hyper-rectangle containing $\mathcal{P}$ which is dominated by both $\mathcal{P}$ and A. It may be straightforwardly calculated by Monte Carlo sampling; see [12] for details.

We present results of two sets of experiments performed, firstly, in order to evaluate the selection methods and, secondly, illustrate the robustness of the selected method to rescaling of the objectives.

To evaluate the selection methods proposed, we assessed the fronts generated by the ROUNDS, RANDOM and PROB methods together with the fronts generated by an implementation of the Mostaghim & Teich's MOSPO (designated MT in this paper) [9]. In order to permit fair comparisons we did not limit the archive size in the MT algorithm. During early experimentation it was observed that more rapid convergence may be achieved (for all algorithms, including MT) by initially promoting more aggressive search (wider exploration); in all the work reported here this was done by ignoring the contribution from global guides

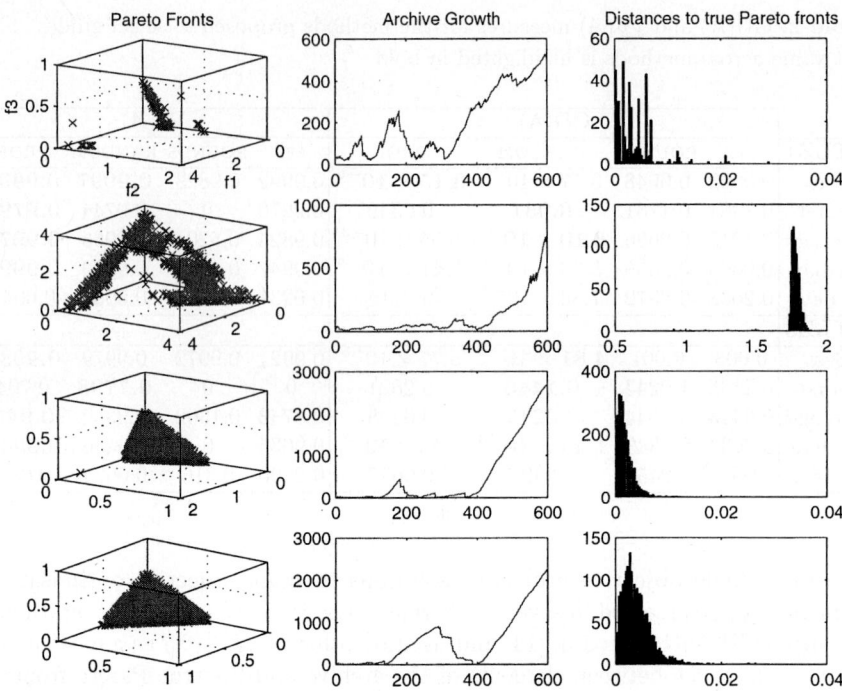

Fig. 1. Archives, archive growth and histograms of distances to the DTLZ1 Pareto fronts corresponding to the median result of the GD metric for (top to bottom) MT, ROUNDS, RANDOM, PROB selection methods. (The more distant clusters of particles were cut from the MT histogram for visualisation purposes)

($c_2 = 0$ in (4)) when $|A| < 100$. For all methods $N = 100$ particles comprised the swarm and the algorithms were run for 600 generations.

Table 2 shows the mean, standard deviation, median, worst and best values of the $GD(A)$ and $\mathcal{V}_\mathcal{P}(A)$ measures over 20 different random initialisations of each method. On the basis of these results it is difficult to distinguish between the MT and ROUNDS methods, but it is clear that the RANDOM and PROB methods are generally superior to both of them. In terms of the GD measure the RANDOM selection scheme appears to be slightly superior to the PROB method, but the $\mathcal{V}_\mathcal{P}(A)$ measure favours the PROB method. This reflects the explicit promotion of search towards edges and sparsely populated regions by PROB, resulting in better coverage of the front, which is measured by $\mathcal{V}_\mathcal{P}(A)$, rather than merely distance from the front which is quantified by $GD(A)$. The fronts from the 20 different runs of RANDOM and PROB methods were compared pairwise by calculating the volume in objective space dominated by one front but not by the other [12]: in over 60% of the comparisons PROB outperformed RANDOM.

Figures 1 and 2 show for DTLZ1 and DTLZ3 respectively the archives, archive growth and histograms of the distances from $\mathcal{P}$ for the median run according to the GD measure. For both problems it is apparent that RANDOM selection

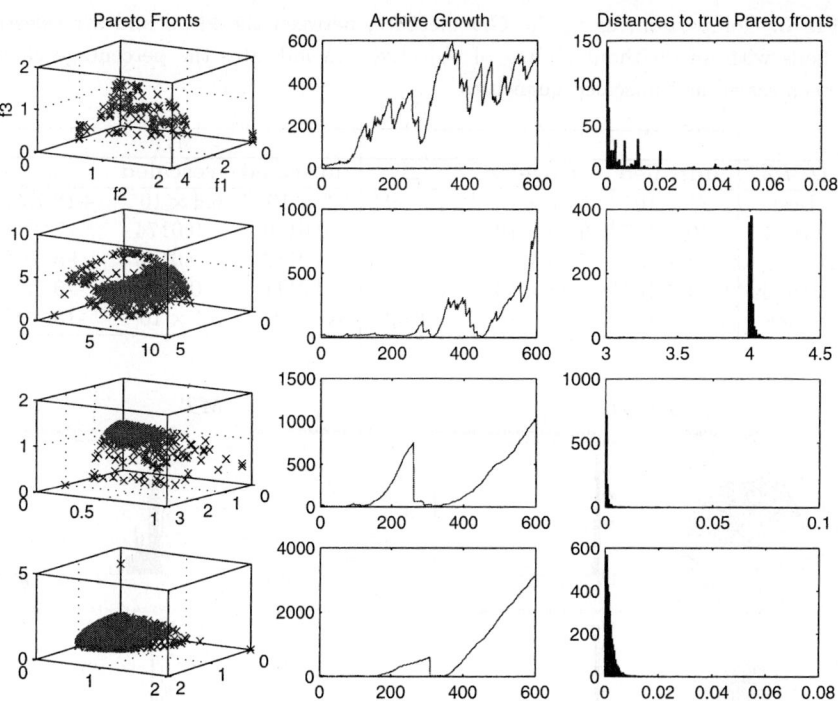

Fig. 2. Archives, archive growth and histograms of distances to the DTLZ3 Pareto fronts corresponding to the median result of the GD measure for (top to bottom) MT, ROUNDS, RANDOM, PROB selection methods. (The more distant clusters of particles were cut from the MT histogram for visualisation purposes)

achieves tightly grouped solutions close to the true front, but the PROB scheme yields a better coverage of particles. These figures also show that the PROB and RANDOM schemes both result in significantly larger archives than MT and ROUNDS. It is also interesting to note that although ROUNDS often fails to converge well it does provide good coverage; note that the ROUNDS front shown in Figure 1 is distant from the true front giving a false impression of its coverage.

These results, along with the preference for an algorithm promoting diversity, lead us to choose PROB selection as the best alternative and from now on we concentrate on this method.

As mentioned previously, the selection methods introduced here do not depend upon metric information in objective space and thus may be expected to be unaffected by the scales on which the objectives are measured. To illustrate the robustness of the method we compared 20 optimisations of the DTLZ2 test problem in which one of the objectives was rescaled with 20 optimisations in which there was no rescaling of objectives. (All optimisations started from different random initial particle locations.) On the i-th optimisation one of the objectives (chosen cyclically) for the rescaled run was multiplied by $(i + 1)$. The fronts

Table 3. Comparison, using the GD measure, between the PROB and MT selection methods with and without scaling of objectives. Δ indicates the percentage change between scaled and unscaled quantities

$DTLZ2$	PROB			MT		
	unscaled	rescaled	Δ	unscaled	rescaled	Δ
Best	5.79×10^{-4}	5.87×10^{-4}	+1.38%	5.2×10^{-3}	6.4×10^{-3}	+18.75%
Worst	1.10×10^{-3}	9.95×10^{-4}	-9.54%	0.0189	0.0174	-7.91%
Average	7.18×10^{-4}	7.06×10^{-4}	-1.67%	0.0113	0.0120	+5.83%
Median	6.64×10^{-4}	6.84×10^{-4}	+2.92%	0.0111	0.0119	+6.72%
S. dev.	1.44×10^{-4}	1.03×10^{-4}	-28.4%	3.8×10^{-3}	3.1×10^{-3}	-18.42%

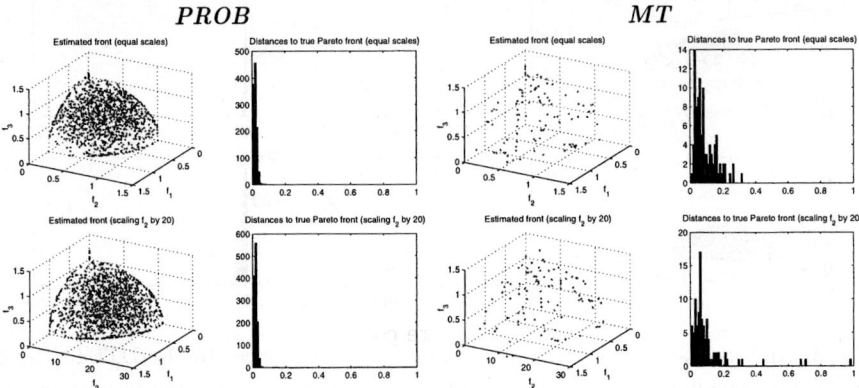

Fig. 3. Pareto fronts and histograms of distances to the true Pareto fronts corresponding for unscaled (top) and with f_2 rescaled by 20 (bottom) using the PROB (left) and MT (right) selection rules

obtained after 45 generations were assessed using the $GD(A)$ measure, but to facilitate comparison the relevant objective was rescaled back to the usual scale.

Average results are shown in Table 3 and Figure 3 compares the estimated Pareto fronts for runs in which f_2 was multiplied by 20 with fronts from unscaled runs. As the table shows, the optimisations using PROB selection are unaffected by the rescaling, in contrast to the MT method which relies on objective space distances for its selection of guides. We emphasise again that in terms of the GD metric the performance of the PROB method is an order of magnitude better than the MT on both the scaled and unscaled problems

5   Keeping Particles Within the Search Space

The velocity and position updates (4) and (3) are liable to cause particles to exceed the boundaries of the feasible regions and both single and multi-objective PSO algorithms must be modified to keep the particles within the constraints. The manner in which this is done may have a great impact on the performance of the algorithm as it affects the way in which particles move around the search space and it is particularly important when the optimum decision variables values lie on or near to the boundaries. In Algorithm 1 this is delegated to the `enforceConstraints` function and in this section we discuss methods for ensuring that particles remain in the feasible region.

A number of alternatives for this have been proposed: A straightforward method [10] is to truncate the location at the exceeded boundary at this generation and reflect the velocity in the boundary so that the particle moves away at the next generation. An alternative [15] is to resample the stochastic terms in the velocity update formula (4) until a feasible position is achieved. Other schemes rely on limiting the magnitude of the velocities, either explicitly [16] or by modifying the constriction factor χ and the other 'constants' w, c_2 and c_2 appearing in the update equations [17].

Other methods may involve using *a priori* knowledge about the particular problem being optimised. For example, in [9] the trespass rule for the first $D-1$ parameters was different from the remainder [18], exploiting the knowledge that in the DTLZ test functions the first $D-1$ parameters determine the coverage of the front while the remainder determine the distance from the front. This approach does improve the quality of solutions but is not used here because we are interested in examining generic methods and not those dependant on prior knowledge about the functions to be optimised.

Here we examine four methods of constraining the particles. In describing these we assume that the constraints are constraints on individual parameters (i.e., constraints of the form $L \leq x_k \leq U$ for some upper and lower limits, L and U), however, they are easily generalised to oblique or curved feasible regions.

TRC. Particles exceeding a boundary are truncated at the boundary for this generation and the velocity is reflected in the boundary so that they tend to move away on the next update [10].

SHR. In reflecting the particle at the boundary the TRC method endows the particle at the next generation with a velocity *away* from the boundary, which can be detrimental to finding optima if the optimal decision parameters lie on the boundary. To combat this the SHR method shrinks the magnitude of the velocity vector of the particle so that it arrives exactly at the boundary, but does not alter its direction, permitting the particle to stay in the vicinity of the boundary. Suppose that the k-th component of the particle's position exceeds a boundary at U, then the SHR scheme sets

$$\mathbf{x}_n^{(t+1)} = \mathbf{x}_n^{(t)} + \sigma(\chi \mathbf{v}_n^{(t)} + \boldsymbol{\epsilon}) \tag{8}$$

with

$$\sigma = \frac{x_{nk}^{(t)} - U}{\chi v_{nk}^{(t)} + \epsilon_k} \tag{9}$$

Note that, in contrast to the other methods discussed here, the SHR scheme affects all components of the particle's position, rather than just the component that has exceeded a constraint.

RES. The resampling method merely resamples the stochastic variables r_1 and r_2 in (4) and ϵ_k in (3) for each velocity component until the particle location is in the feasible region [15].

EXP. The final method we examine updates the position component with a random draw when that particular component, say $x_k^{(t)}$, would have been updated to a position beyond a boundary at, say, b. For convenience, suppose $x_k^{(t)} < U$. In this case we sample from a truncated exponential distribution oriented so that there is a high probability of samples close to the boundary and a lower probability of samples at the current position $x_k^{(t)}$. More precisely a new location $x_k^{(t+1)}$ is drawn with probability:

$$p(x_k^{(t+1)}) \propto \begin{cases} \exp\left\{-\frac{|U-x_k^{(t+1)}|}{|U-x_k^{(t)}|}\right\} & \text{if } x_k^{(t)} \leq x_k^{(t+1)} \leq U \\ 0 & \text{otherwise} \end{cases} \qquad (10)$$

with obvious modifications if $U < x_k^{(t)}$. In a similar manner to the SHR method this scheme tends to allow particles that would have exceeded the boundaries to remain close to the boundaries.

5.1 Experiments

To determine the impact of each of the four methods, we compared the fronts located for the DTLZ1 and DTLZ3 problems using each of them in conjunction with the PROB guide selection scheme. The fronts were all assessed against the true Pareto front using the $GD(A)$ and $\mathcal{V}_\mathcal{P}(A)$ measures. Each version was run 20 times (using the same parameters as described above) and the results are presented in Table 4.

It is clear that the SHR method, which shrinks the velocity vector so that the particle arrives exactly at the boundary, yields superior results on both test problems according to both the generational distance and volume measures. The EXP method, which resamples giving preference to locations close to the boundary is the next best, while the two methods that tend to move a particle away from the boundary, TRC and RES, give the poorest results. Indeed RES and TRC occasionally prevent convergence.

Further insight into the way in which the RES and SHR methods behave may be gained by examining the trajectory of a single particle, as shown in Figure 4. The figure shows 7 coordinates of a single particle during an optimisation of the DTLZ1 problem. As remarked previously, in this problem the optimum value for variables x_3 to x_7 is 0.5, while $0 \leq x_1, x_2 \leq 1$ provide coverage of the front when x_3 to x_7 are at their optimum value. It is clear from Figure 4 that the resampling method RES promotes greater movements across the space which may be beneficial for exploration. However during the resampling the particle is pushed away from the optimal locations. In contrast the SHR scheme permits the particle to remain close to the boundaries during the search process.

Table 4. Generational distance and $\mathcal{V}_\mathcal{P}(A)$ measures for constraint handling methods compared on DTLZ1 & DTLZ3

DTLZ1	GD(A)				$\mathcal{V}_\mathcal{P}(A)$			
	SHR	RES	EXP	TRC	SHR	RES	EXP	TRC
Best	3.77×10^{-5}	1.3588	6.36×10^{-5}	7×10^{-3}	**0.9997**	0.1289	0.9996	0.9957
Worst	**0.0349**	11.8362	0.1996	0.4706	**0.9796**	0	0.6958	0
Average	5.55×10^{-3}	8.2132	0.0336	0.1747	**0.9974**	0.0084	0.9645	0.6983
Median	1.41×10^{-4}	8.6158	0.0178	0.2064	**0.9992**	0	0.992	0.8221
S. dev.	0.011426	2.3872	0.05	0.147	0.0047	0.0297	0.0699	0.3035
DTLZ3								
Best	5.77×10^{-5}	21.75	2.03×10^{-4}	0.0295	**0.9981**	0	0.9965	0.9964
Worst	**0.2621**	41.08	1.5826	4.09	**0.704**	0	0	0
Average	**0.0305**	31.74	0.1539	1.61	**0.945**	0	0.8145	0.2189
Median	1.52×10^{-3}	32.29	0.0381	1.54	**0.9946**	0	0.9367	0.0955
S. dev.	0.068707	4.25	0.3501	1.21	0.086	0	0.2892	0.2945

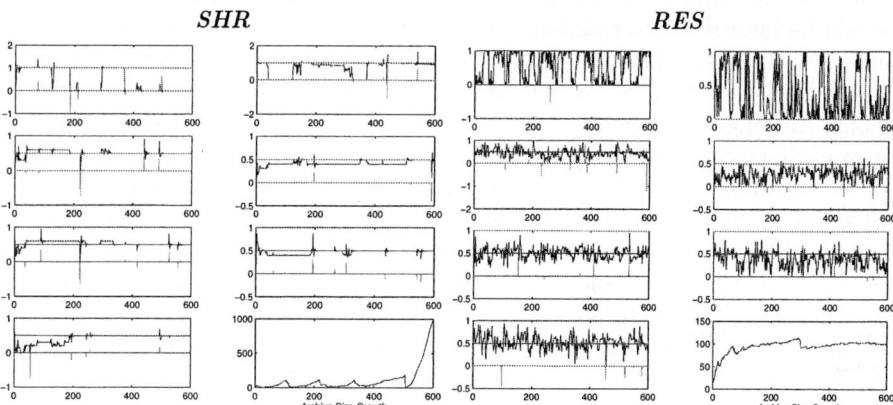

Fig. 4. Movements of a particle and growth of the archive during the searching process when using SHR and RES methods on DTLZ1. (x_1 to x_7 are shown top-left, top-right, ..., to bottom-right)

The DTLZ test problems which we analyse here are special in that the roles of the decision variables may be clearly distinguished. However, it is likely that in real problems optima may lie close to or on the constraint boundaries or these regions will be visited *en route* to the optima and it will be important to permit particles to properly explore these regions.

6 Conclusions

We have examined several methods of choosing global guides in multi-objective extensions of particle swarm optimisers. Unlike previous work, guides are selected without reference to distance information in the objective space, which renders them robust to the relative scalings of the objectives. Indeed, if the relative importance or scales of the objectives were known in advance it might be more straightforward to optimise a single, appropriately weighted, sum of the objectives. Notions of dominance and Pareto optimality are well suited to handling competing objectives whose relative importance is *a priori* unknown and it is therefore natural to eschew metric information in favour of dominance concepts when choosing guides. We find that selecting guides probabilistically from the archive of non-dominated solutions, giving more weight to solutions that dominate few particles, provides both good convergence and widespread coverage. This method yields superior performance to an existing MOPSO technique and is robust to changes of scale in the objective functions.

The computation involved in the selection is a more extensive than other recently proposed schemes (e.g., [9, 10]) but is more than compensated for by improved convergence and coverage.

The PROB selection method selects guides with probability inversely proportional to the number of particles the potential guide dominates (c.f., (6)). It would be interesting to examine the performance of an algorithm which selects guides with probability proportional to $|X_a|^{-q}$; as $q \to 0$ the method becomes the RANDOM method, but as q increases additional weight is given to sparse regions. Although this might enable finer control of the convergence and coverage it would introduce an additional parameter to be 'tweaked'.

It was found that the manner in which particles are constrained to the feasible region can vastly affect the performance of a MOPSO. Four methods of constraining particles were examined and it was found that a method which permits particles to remain close to the boundaries enables more rapid location of the Pareto front. We anticipate that careful handling of solutions to allow exploration close to the boundaries will be important not only in MOPSO, but also in other approaches to multi-objective optimisation.

References

1. Srinivas, N., Deb, K.: Multiobjective Optimization Using Nondominated Sorting in Genetic Algorithms. IEEE Transactions on Evolutionary Computation **2(3)** (1995) 221–248
2. Zitzler, E., Thiele, L.: Multiobjective Evolutionary Algorithms: A Comparative Case Study and the Strength Pareto Approach. IEEE Transactions on Evolutionary Computation **3(4)** (1999) 257–271
3. Laumanns, M., Zitzler, E., Thiele, L.: A Unified Model for Multi-Objective Evolutionary Algorithms with Elitism. In: Proceedings of the 2000 Congress on Evolutionary Computation. (2000) 46–53
4. Kennedy, J., Eberhart, R.: Particle Swarm Optimization. In: Proceedings of the Fourth IEEE International Conference on Neural Networks. (1995) 1942–1948
5. Hu, X., Eberhart, R.: Multiobjective Optimization Using Dynamic Neighborhood Particle Swarm Optimization. In: Proceedings of the 2002 Congess on Evolutionary Computation, IEEE Press (2002)

6. Parsopoulos, K., Vrahatis, M.: Particle Swarm Optimization Method in Multi-objective Problems. In: Proceedings of the 2002 ACM Symposium on Applied Computing (SAC 2002). (2002) 603–607
7. Coello, C., Lechunga, M.: MOPSO: A Proposal for Multiple Objective Particle Swarm Optimization. In: Proceedings of the 2002 Congress on Evolutionary Computation, IEEE Press (2002) 1051–1056
8. Fieldsend, J., Singh, S.: A Multi-Objective Algorithm based upon Particle Swarm Optimisation, an Efficient Data Structure and Turbulence. In: Proceedings of UK Workshop on Computational Intelligence (UKCI 02). (2002) 37–44
9. Mostaghim, S., Teich, J.: Strategies for Finding Good Local Guides in Multi-Objective Particle Swarm Optimization (MOPSO). In: IEEE 2003 Swarm Intelligence Symposium. (2003) 26–33
10. Coello, C., Pulido, G., Lechunga, M.: Handling Multiple Objectives with Particle Swarm Optimization. IEEE Transactions on Evolutionary Computation **8(3)** (2004) 256–279
11. Hanne, T.: On the convergence of multiobjective evolutionary algorithms. European Journal of Operational Research **117** (1999) 553–564
12. Fieldsend, J., Everson, R., Singh, S.: Using Unconstrained Elite Archives for Multi–Objective Optimisation. IEEE Transactions on Evolutionary Computation **7** (2003) 305–323
13. Deb, K., Thiele, L., Laumanns, M., Zitzler, E.: Scalable Multi–Objective Optimization Test Problems. In: Congress on Evolutionary Computation (CEC'2002). Volume 1. (2002) 825–830
14. Veldhuizen, D.V., Lamont, G.: Multiobjective Evolutionary Algorithms Research: A History and Analysis. Technical Report TR-98-03, Dept. Elect. Comput. Eng., Graduate School of Eng., Air Force Institute Technol., Wright-Patterson AFB, OH (1998)
15. Fieldsend, J.: Multi–Objective Particle Swarm Optimization Methods. Technical Report No. 419, Department of Computer Science, University of Exeter (2004)
16. van den Bergh, F.: An Analysis of Particle Swarm Optimizers. PhD thesis, Faculty of Natural and Agricultural Science, University of Pretoria (2001)
17. Clerc, M.: The Swarm and the Queen: Towards a Deterministic and Adaptative Particle Swarm Optimization. In: Proceedings of the Congress on Evolutionary Computation, IEEE Press (1999) 1951–1957
18. Mostaghim, S., Teich, J.: Personal communication. (2004)

Clonal Selection with Immune Dominance and Anergy Based Multiobjective Optimization

Licheng Jiao[1], Maoguo Gong[1], Ronghua Shang[1],
Haifeng Du[1, 2], and Bin Lu[1]

[1] Institute of Intelligent Information Processing, P.O. Box 224,
Xidian University, Xi'an, 710071, P.R. China
lchjiao@mail.xidian.edu.cn, maoguo_gong@hotmail.com
[2] School of Mechanical Engineering, Xi'an Jiaotong University,
Xi'an, 710049, P.R. China
Haifengdu72@163.com

Abstract. Based on the concept of Immunodominance and Antibody Clonal Selection Theory, we propose a new artificial immune system algorithm, Immune Dominance Clonal Multiobjective Algorithm (IDCMA). The influences of main parameters are analyzed empirically. The simulation comparisons among IDCMA, the Random-Weight Genetic Algorithm and the Strength Pareto Evolutionary Algorithm show that when low-dimensional multiobjective problems are concerned, IDCMA has the best performance in metrics such as Spacing and Coverage of Two Sets.

1 Introduction

In 1984, Schaffer put forward a vector evaluated genetic algorithm (VEGA) by modifying the fitness assignment and the individual selection strategy[1]. His work is regarded as the beginning of solving multiobjective optimization problems by genetic algorithm. Until the middle 1990s, the number of literatures on multiobjective evolutionary algorithms (MOEAs) increased greatly. Among them, Fonseca et al's Multiobjective Genetic Algorithm, Horn et al's Niched Pareto Genetic Algorithms and Srinivas et al's Nondominated Sorting in Genetic Algorithm attracted more attention[2]. These evolutionary algorithms show better performance in solving multiobjective problems than traditional algorithms. However, they didn't adopt elitism preserving strategy definitely, which was recognized and supported by experiments in the following years. In recent years, a lot of newly improved algorithms were proposed, such as Deb et al's A Fast Elitist Non-dominated Sorting Genetic Algorithm, Corne et al's Pareto Envelope based Selection Algorithm and Zitzler's the strength Pareto evolutionary algorithm (SPEA and SPEA2). In particular, Zitzler et al's SPEA and SPEA2 have shown many good performances[3,4]. At the same time, Coello Coello et al presented their own multiobjective evolutionary algorithms and proposed an multiobjective algorithm named by the Multiobjective Immune System Algorithm (MISA)[5] using the clonal selection principle. They set out that MISA was very promising based on a few simulations.

Artificial immune system (AIS) makes use of the mechanism of vertebrate immune system, and constructs new intelligent algorithms with immunology terms and fundamental. Artificial immune system provides the evolutionary learning mechanism like noise enduring, non-teacher learning, self-organization, and memory, thus it has the potential for providing novel method for solving problems, and its research production refers to many fields like control, data processing, optimization learning and trouble diagnosing, and it has been a research hot spot after the neural network, fuzzy logic and evolutionary computation.[6]

After defining several basic concepts of artificial immune system in Section 2, a novel multiobjective optimization algorithm, Immune Dominance Clonal Multiobjective Algorithm (IDCMA), is put forward in Section 3. The influences of main parameters are analyzed empirically, then five representative low-dimensional multiobjective problems and three famous multiobjective algorithms, the Random-Weight Approach proposed by Ishibuchi[7], and the Strength Pareto Evolutionary Algorithm proposed by Zitzler[3], and Multiobjective Immune System Algorithm proposed by Coello Coello [5] are selected for simulation tests in Section 4.

2 Basic Definitions

Although an antigen has many epitopes (antigenic determinants), only one works to induce a special immune response for the host cells. The phenomenon is called immunodominance, the epitope is called a dominant epitope, immunodominance is produced by the action of antibody and antigen[8]. The clonal selection theory (F. M. Burnet, 1959) is used in the immune system to describe the basic features of an immune response. Its main idea lies in that the antigens can selectively react to the antibodies, which are the native production and spread on the cell surface in the form of peptides. The reaction leads to cell proliferating clonally and the colony has the same antibodies. Some clonal cells divide into antibodies that produce cells, and others become immune memory cells to boost the second immune response. The clonal selection is a dynamic process of the immune system self-adapting antigen stimulation. From the viewpoint of the Artificial Intelligence, some biologic characters such as learning, memory and antibody diversity can be used in artificial immune system.

In order to describe the algorithm for multiobjective optimization problems well, we just define the glossary as follows.

Definition 1. Antigen
In AIS, antigen usually means the problem and its constraints. Especially, for multiobjective optimization problems, we have

$$(P) \begin{cases} \min F(x) = (f_1(x), f_2(x), ... f_p(x))^T \\ S.T. \quad g_i(x) \leq 0 \quad i = 1, 2, \cdots m \end{cases} \quad (1)$$

where, $x = (x_1, x_2, ... x_n)$, $p \geq 2$, the antigen is defined as a function of objective function $F(x)$, namely, $G(x) = g(F(x))$. Similar to the function of antigen in immunology, it is the initial factor for the artificial immune system algorithm. Usually, we let $G(x) = F(x)$ when not mentioned especially.

Definition 2. Antibody

Antibodies represent candidates of the problem. The limited-length character string $a = a_1 a_2 \cdots a_l$ is the antibody coding of variable x, denoted by $a = e(x)$, and x is called the decoding of antibody a, expressed as $x = e^{-1}(a)$. In practice, binary coding and decimal coding are often used. For example, an antibody of binary coding whose length is 8 can be written as '0-1-1-1-0-1-0-0'. Set I is called antibody space, namely $a \in I$. The antibody population $A = \{a_1, a_2, \cdots, a_n\} \in I^n$ is an n-dimensional group of antibody a, namely,

$$I^n = \{A : A = (a_1, a_2, \cdots, a_n), \ a_k \in I, \ 1 \leq k \leq n\} \quad (2)$$

where the positive integer n is the antibody population size.

Definition 3. Antibody-Antigen Affinity

Antibody-Antigen Affinity is the reflection of the total combination power locates between antigen and antibodies. In AIS, it generally indicates values of objective functions or fitness measurement of the problem.

Definition 4. Antibody-Antibody Affinity

Antibody-Antibody Affinity is the reflection of the total combine power locates between two antibodies. In this paper, we compute the antibody-antibody affinity as reference [5]. Namely, if the coding of an antibody a_i is '1 1 0 0 0 0 1 0', and the coding of another antibody a_{di} is '1 1 0 1 0 1 1 0', then the number of genes matched between the two antibodies is 6, the matched gene strings whose length are greater than 2 are '110' and '10', and the corresponding lengths are 3 and 2, so the antibody-antibody affinity between a_i and a_{di} is $6 + 3^2 + 2^2 = 19$. Of course, other distance measures are possible.

Definition 5. Immune Dominance

For problem (P), the antibody a_i is an immune dominance antibody in antibody population $A = \{a_1, a_2, \cdots, a_n\}$, iff there is no antibody a_j $(j = 1, 2, \cdots n \wedge j \neq i)$ in population A satisfied the formula (3)

$$(\forall k \in \{1, 2, \cdots p\} \ f_k(e^{-1}(a_j)) \geq f_k(e^{-1}(a_i))) \wedge (\exists l \in \{1, 2, \cdots p\} \ f_l(e^{-1}(a_j)) > f_l(e^{-1}(a_i))) \quad (3)$$

So the immune dominance antibodies are the Pareto-optimal individuals in the current population.

Definition 6. Clonal Operation

In the immunology, clone is the process of antibody proliferation. In AIS, the clonal operation to the antibody population is defined as:

$$Y(k) = T_c^C(A(k)) = [T_c^C(a_1(k)) \quad T_c^C(a_2(k)), \quad \cdots \quad ,T_c^C(a_n(k))]^T \tag{4}$$

where $T_c^C(a_{ci}(k)) = I_{ci} \times a_{ci}(k)$ $i = 1, 2 \cdots n$, I_{ci} is a q_{ci}-dimensional identity row vector. The process is called the q_{ci} clone of antibody a_i, namely $q_{ci}(k) = \hbar(n_c, \Theta_i)$, where Θ_i stands for the affinity function of antibody a_i and other antibody, and n_c is the clonal scale.

Definition 7. Immune Differential Degree
In this paper, the Immune Differential Degree denotes the relative distribution of an immune dominance antibody. Namely, assuming that there are N immune dominance antibodies (Pareto-optimal solutions) in current population, f_{kl} is the value of the K-th objective function of the l-th antibody. The Immune Differential Degree of the l-th antibody a_l can be calculated as follow,

$$d_l^* = \min\left\{d_l(m) = \sqrt{\sum_{k=1}^{q}\left(\frac{\phi(f_{kl}) - \phi(f_{km})}{\phi(f_{kl})}\right)^2} \,\middle|\, l = 1, 2, \cdots N; m = 1, 2, \cdots N \wedge m \neq l\right\} \tag{5}$$

where $\phi(\bullet)$ is an incremental function without the value of zero.

3 Algorithm Description

Inspired from the immuodominace of the biology immune system and the clonal selection mechanism, we designed a novel artificial immune system algorithm based on clonal selection with immune dominance and clone anergy for multiobjective optimization problems which can be implemented as follow:

Step 1: Give the termination generation $G_{\max}$, the size of Immune Dominance Antibody population n_d, the size of Generic Antibody population n_b, the size of Dominance Clonal Antibody population n_t, and clonal scale n_c. Set the mutation probability p_m, recombination probability p_c and coding length c. Randomly generate the original antibody population $A(0) = \{a_1(0), a_2(0), \cdots a_{n_b}(0)\} \in I^{n_b}$, $k := 0$;

Step 2: Compute the antibody-antigen affinities of all the antibodies in $A(k)$;

Step 3: According to the affinities, select all the immune dominance antibodies to constitute the population $DT(k)$, if the number of antibodies in $DT(k)$ is no larger than n_d, let Immune Dominance Antibody population $D(k)=DT(k)$, go to Step6; otherwise go to Step4;

Step 4: Compute the Immune Differential Degrees of all the antibodies in population $DT(k)$;

Step 5: Sort all the antibodies in $DT(k)$ by descending of their Immune Differential Degrees, and select the first n_d antibodies to constitute the current Immune Dominance Antibody population $D(k)$;

Step 6: If $k=G_{max}$, export $D(k)$ as the output of the algorithm, Stop. Otherwise, replace the immune dominance antibodies in $A(k)$ by new antibodies generated randomly. Then marked the antibody population as $B(k)$;

Step 7: Select an immune dominance antibody a_{di} randomly from $D(k)$. Compute the antibody-antibody affinities between the antibodies in $B(k)$ and the antibody a_{di}.

Step 8: Sort all the antibodies in $B(k)$ by descending of their antibody-antibody affinities, select the first n_t antibodies to constitute the Dominance Clonal Antibody population $TC(k)$, and other antibodies to constitute the Immune Anergy Antibody population $NR(k)$.

Step 9: Compute the clonal proportion $q_{ci}(k)$ of each antibody a_{ci} in $TC(k)$ according to antibody-antibody affinity and the clonal scale.

Step 10: Implement the Antibody Clonal Operation T_c^C at $TC(k)$ and get the antibody population $CO(k)$ after clonal operation.

Step 11: Implement the recombination operation at $CO(k)$ with the probability p_c and get the antibody population $CO'(k)$, $CO'(k) = T_r^C(CO(k))$; Namely, for the antibody $y_i(k)$ in $CO(k)$, implement the following operation:

$$y'_i(k) = T_r^C(y_i(k), a_{di}(k)), \quad y_i(k) \in CO(k), \quad a_{di}(k) \in D(k) \tag{6}$$

and $CO'(k) = \{y'_i(k)\}\ i = 1, 2, \cdots n_c$.

For binary coding, the recombination operation in this paper is as follow:

$$y'_i(k) = T_r^C(y_i(k), a_{di}(k)) = y_i(k)_{a \sim b \to 0} + a_{di}(k)_{1 \sim a | b \sim c \to 0}, \quad y_i(k) \in CO(k), \quad a_{di}(k) \in D(k) \tag{7}$$

where c is the coding length, a, b are random integers between 1 and c. $a \sim b \to 0$ means the bits from a to b set zeros, $1 \sim a | b \sim c \to 0$ means the bits from 1 to a and the bits from b to c set zeros, + means 'or' operation by bits.

Step 12: Implement the mutation operation at $CO'(k)$ with the probability p_m generating the antibody population $COT(k)$, $COT(k) = T_m^C(CO'(k))$:

For binary coding, the clonal antibody population is mutated as follow:

$$COT(k) = (-1)^{random \leq p_m} CO'(k) \tag{8}$$

$(-1)^{random \leq p_m} CO'(k)$ means each element of $CO'(k)$ multiplies -1 with probability of p_m.

Step 13: Combine the populations $COT(k)$, $D(k)$ and $NR(k)$ to form the antibody population $A(k+1)$, $k:=k+1$, go to Step 2.

From the description above we can see that the new algorithm divides all the antibodies into three sorts, and stores them in three populations. Different evolutionary strategies are adopted at different populations, but they are not isolated. The combination of the three populations helps to increase the global search ability. The operation of immune dominance antibody population based on the Immune Differential Degree retains the diversity of the population. The operation of dominance clonal antibody population based on the antibody-antibody affinity can select the effective local in antibody space and assure the validity of the search in next generation. The existence of the immune anergy antibody population assures the diversity of populations and simulates the immune response process more meticulously. In additional, the worst time complexity of one generation for immune dominance clonal multiobjective optimization algorithm is $O(n_d + n_b)^2$, where n_d is the size of Immune Dominance Antibody population, and n_b is the size of Generic Antibody population.

4 Simulation Analyses

In this section, we adopt two popular metrics, Coverage of Two Sets and Spcing, which are defined as follows:

Coverage of Two Sets[9]: Let $A', A'' \subseteq X$ be two sets of decision vectors. The finction ς maps the ordered pair (A', A'') to the interval [0, 1]:

$$\varsigma(A', A'') \triangleq \frac{|\{a'' \in A''; \exists a' \in A' : a' \triangleright a''\}|}{|A''|} \quad (9)$$

Where $\triangleright$ means Pareto dominate or equal. The value $\varsigma(A', A'') = 1$ means that all decision vectors in A'' are weakly dominated by A'. $\varsigma(A', A'') = 0$ implies the opposite. Note that always both directions have to be considered because $\varsigma(A', A'')$ is not necessarily equal to $1 - \varsigma(A'', A')$.

A metric called Spacing was proposed by Schott[10] as a way of measuring the range variance of neighboring vectors in the Pareto front known. This metric is defined as:

Spacing[10]: Let $A' \subseteq X$ be a set of decision vectors. The function S

$$S \triangleq \sqrt{\frac{1}{|A'|-1} \sum_{i=1}^{|A'|} (\bar{d} - d_i)^2} \quad (10)$$

Where $d_i = \min_j \left\{ \sum_{k=1}^{p} |f_k(x_i) - f_k(x_j)| \right\}$ $x_i, x_j \in A'$ $i, j = 1, \cdots |A'| \ \Box \ \bar{d}$ is the mean of all d_i, and p is the number of objective functions.

4.1 Analysis of the Influences of Main Parameters

In IDCMA, the main parameters are $G_{\max}$, n_d, n_b, n_t, n_c, p_m, p_c and the coding length c. The influences of the parameters $G_{\max}$, n_d, n_b, n_t and c to the performance are obvious, if not take the complexity into account, the larger their values, the better the results. The influences of n_c, p_m and p_c to the algorithm performance is more complex. The following is the empirical analysis results.

IDCMA can get different $\varsigma(P_{ag}, P_{ture}^3)$ values with different parameter settings, where P_{ture}^3 is a set of solutions equidistantly spaced at the Pareto-optimal fronts and P_{ag} is a set of decision vectors. We take the following test function for example:

$$\min F(x,y) = (f_1(x,y), f_2(x,y)),$$
$$f_1(x,y) = \frac{1}{x^2+y^2+1}, \quad f_2(x,y) = x^2 + 3y^2 + 1, \quad (11)$$
$$S.T. -3 \le x, y \le 3$$

We code x and y with binary string of 10 bits long. The main parameters are as follows: $G_{max}=150$, $n_d=100$, $n_b=100$, $n_t=50$. We analyze the parameters n_c, p_m and p_c one after one by sampling a parameter with the same interval while fixing the other parameters. Choose 5000 solutions equidistantly spaced at the ideal Pareto-optimal front to compose P_{ture}^3. The data are the statistical results obtained from 10 times of random running.

i   Influence of Clonal Scale

Let $p_m=1/c$ and $p_c=1$. The clonal scale n_c is sampled by the same interval of 50 between 100 and 500 and runs 10 times to get the maximum, the minimum, the

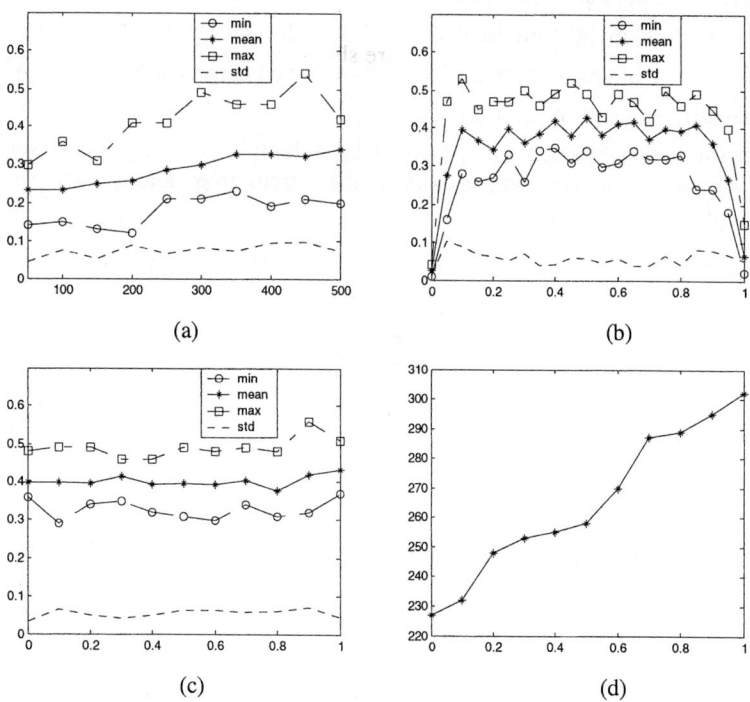

Fig. 1. The influence of the main parameters to the performance of the algorithm

average values and the deviation of $\varsigma(\boldsymbol{P}_{ag}, \boldsymbol{P}_{ture}^3)$ which are shown in Figure 1(a). It can be seen from the results that the influence of clonal scale to the algorithm is notable and the average value of $\varsigma(\boldsymbol{P}_{ag}, \boldsymbol{P}_{ture}^3)$ increases by approximate linearity with the increasing of n_c. Practically, under the given experiment condition, the value of ς will increase approximately by 0.0122 when n_c increases by 50, but the computational complexity will increase $o((50-n_t)^2 + 50 \times c)$ accordingly. Similar results are obtained by a great deal of experiments to other test problems.

ii Influence of Mutation Probability

Let $n_c = 300$, $p_c=1$. Sample the mutation probability by the interval of 0.05 from 0 to 1 and run 10 times to get the maximum, the minimum, the average values and the deviation of $\varsigma(\boldsymbol{P}_{ag}, \boldsymbol{P}_{ture}^3)$ which are shown in Figure 1(b). It can be seen that the change of ς is not obviously when the immune probability gets values from 0.1 to 0.8, but ς will decrease obviously when the mutation probability doesn't get the values from 0.1 to 0.8, which is obviously different from the influence of the mutation operation in genetic algorithm.

iii Influence of Recombination Probability

Let $n_c = 300$, $p_m=0.3$. Sample the recombination probability by the interval of 0.1 from 0 to 1 and run 10 times to get the maximum, the minimum, the average values and the deviation of $\varsigma(\boldsymbol{P}_{ag}, \boldsymbol{P}_{ture}^3)$ which are shown in Figure 1(c). It can be seen that the influence of recombination probability to ς is not obviously. Actually, p_c influences mainly the convergent speed. In order to explain the influence of p_c quantitatively, we adopt the following estimate manner:

Let the terminal generation be 10 and compare the influences of p_c to the amount of the nondominated solutions obtained from IDCMA. In order to eliminate the influence of other parameters, we set the parameters as follows:

$G_{max}=10$, the immune dominance antibody population size is not confined, $n_b = 100$, $n_t = 50$, $n_c = 300$, and $p_m=0.3$. The changes of the amount of nondominated solutions obtained are shown in Figure 1(d). It can be seen that the influence of p_c to the amount of nondominated solutions is obviously.

4.2 Test and Results Analysis

In order to validate the algorithm, we compare the algorithm with another three algorithms. They are Ishibuchi's Random-Weight genetic algorithm (RWGA)[7], Zitzler's Strength Pareto Evolutionary Algorithm (SPEA)[3] and Coello Coello's Multiobjective Immune System Algorithm (MISA)[5]. We design the software emulator of IDCMA using Matlab 6.1, and simulate RWGA and SPEA as exactly as we can under the same conditions. It is necessary to note that the performance of a MOEA in tackling multiobjective constrained optimization problems maybe largely depend on the constraint-handling technique used, so we don't consider side-constrained problems in this Paper. The parameters setting are as follows.

Spacing. It can be seen from Figure 3 that the distribution of the solutions gained by IDCMA is better than those gained by RWGA, SPEA and MISA.

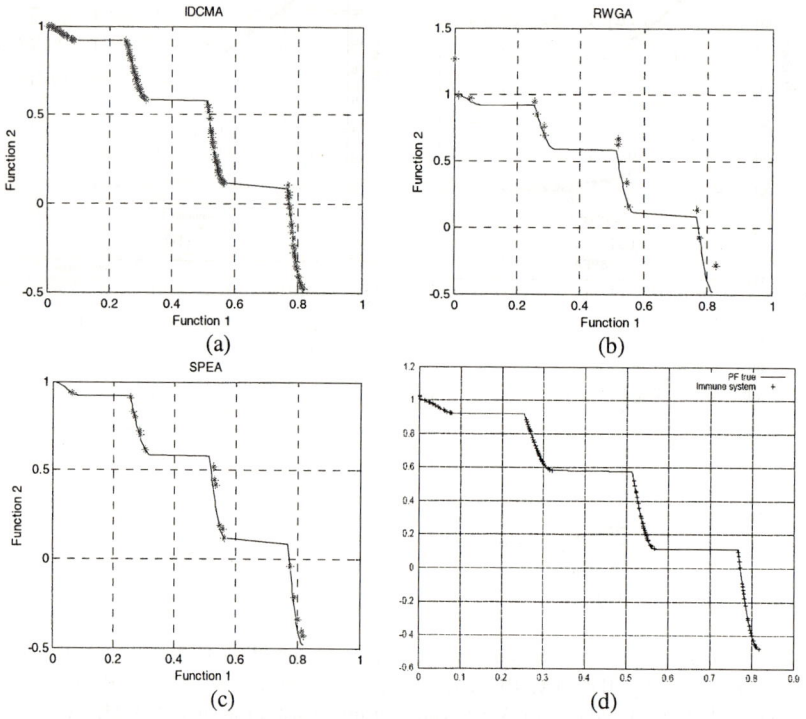

Fig. 3. The Pareto-optimal solution distributions corresponding to test 2

Test 3
Next, we consider another multiobjective problem[11].

$$\min F(x, y) = \left(f_1(x, y), f_2(x, y)\right),$$
$$f_1(x,y) = \sqrt[8]{x^2 + y^2}, \quad f_2(x,y) = \sqrt[4]{(x-0.5)^2 + (y-0.5)^2}, \quad (14)$$
$$S.T. -5 \leq x, y \leq 10$$

This problem is a typical reflection of many to one. Figure 4(a) shows the global Pareto-optimal front in the $f_1 - f_2$ space. Figures 4(b) (c) and (d) show the Pareto-optimal solution distributions solved by IDCMA, RWGA and SPEA separately.

For this problem, IDCMA can obtain 100 Pareto-optimal solutions per time, while SPEA can get 26 Pareto-optimal solutions and RWGA can get 10 Pareto-optimal solutions per time on average. Because this is a reflection of many to one, one dot in the objective space correspond to many dots in the domain, so the number of the dots for 100 optimal solutions gained by algorithm IDCMA displayed in the objective space is much less than 100. Figure 4 (a) shows that its Pareto-optimal front is very complicated, especially for the solution of $f_1 = 0$, SPEA and RWGA are difficult to find it, while IDCMA always find it in 30 times running. Figure 4 shows that the Pareto-optimal solution distribution solved by IDCMA is much better than those solved by SPEA and RWGA. The values of Spacing also show that quantitatively. The values in table 1 show that IDCMA produced solutions that clearly dominated or equals those generated by SPEA and RWGA.

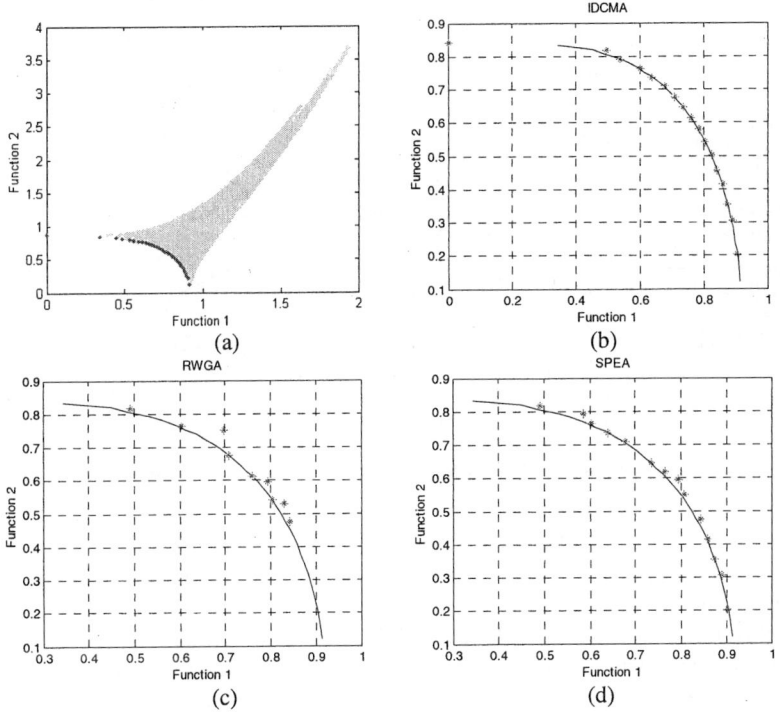

Fig. 4. The Pareto-optimal solution distributions corresponding to test 3. (a) The global Pareto-optimal front; (b) The solution distribution solved by IDCMA; (c) The solution distribution solved by RWGA; (d) The solution distribution solved by SPEA

Test 4
Let us consider a three-objective optimization problem having two variables[11]:

$$\min F(x,y) = (f_1(x,y), f_2(x,y), f_3(x,y))$$

$$f_1(x,y) = \frac{(x-2)^2}{2} + \frac{(y+1)^2}{13} + 3, \quad f_2(x,y) = \frac{(x+y-3)^2}{36} + \frac{(-x+y+2)^2}{8} - 17,$$

$$f_3(x,y) = \frac{(x+2y-1)^2}{175} + \frac{(2y-x)^2}{17} - 13,$$

$$S.T. -4 \le x, y \le 4$$

(15)

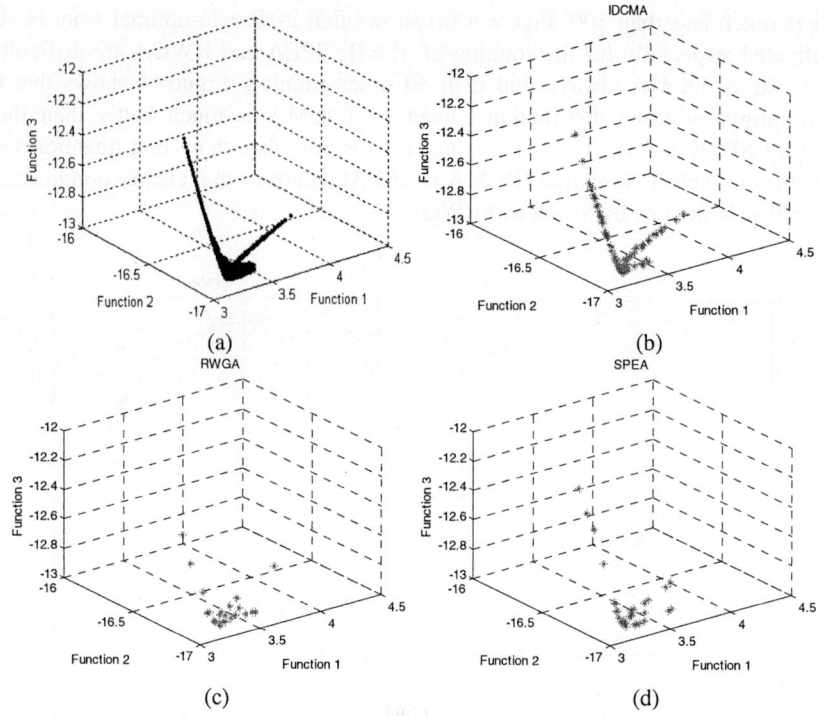

Fig. 5. The Pareto-optimal solution distributions corresponding to test 4

Figure 5(a) shows the global Pareto-optimal front in function space. Figures 5(b) (c) and (d) show the Pareto-optimal solution distributions solved by IDCMA, RWGA and SPEA separately.

For this three-objective problem, IDCMA can also obtain 100 Pareto-optimal solutions per time, while SPEA can only get 40 Pareto-optimal solutions and RWGA can get 27 Pareto-optimal solutions per time on average. For this problem, IDCMA still has the obvious dominance. Figure 5(b) is most close to Figure 5(a), which reflects directly that the distribution of the solutions gained by IDCMA is the most ideal. It can be seen from table 1 that $\varsigma(X^1, X^S)$ and $\varsigma(X^1, X^R)$ are much greater than $\varsigma(X^S, X^1)$ and $\varsigma(X^R, X^1)$, which reflects objectively that the optimal solutions gained by IDCMA dominate those gained by the other two algorithms. It can be seen from metric Spacing that the solutions gained by IDCMA is the most uniform.

Test 5
At last, we consider a two-objective problem having three variables [11]:

$$\min F(\bar{x}) = \left(f_1(\bar{x}), f_2(\bar{x}) \right),$$

$$f_1(\bar{x}) = \sum_{i=1}^{n-1} \left(-10 e^{(-0.2)\sqrt{x_i^2 + x_{i+1}^2}} \right), \quad f_2(\bar{x}) = \sum_{i=1}^{n} \left(|x_i|^{0.8} + 5\sin(x_i)^3 \right) \quad (16)$$

$$S.T. -5 \le x_i \le 5, i = 1, 2, 3, n = 3$$

Figure 6(a) shows the global Pareto-optimal front in function space. Figures 6(b) (c) and (d) show the Pareto-optimal solution distributions solved by IDCMA, RWGA and SPEA separately.

For this problem, IDCMA can obtain 100 Pareto-optimal solutions per time, while SPEA can get 24 Pareto-optimal solutions and RWGA can get 12 Pareto-optimal solutions per time on average. It can be seen from table 1 that $\varsigma(X^1, X^S)$ and $\varsigma(X^1, X^R)$ are much greater than $\varsigma(X^S, X^1)$ and $\varsigma(X^R, X^1)$. The values of metric Spacing show that the solutions gained by IDCMA are the most uniform. In addition, especially for the isolated optimal point $f_1 = 0$ in the objective space, SPEA and RWGA can not find it in 30 times runs, but IDCMA can find this point very well, which can show adequately that IDCMA have a stronger ability for the global search.

The statistical results for data of the two metrics of these nondominated solutions are shown in table 1 and table 2. In which X^1 denotes the solutions solved by IDCMA, X^S denotes the solutions solved by SPEA, and X^R denotes the solutions solved by RWGA. '/' means no correlative data.

Table 1. Average results of the metric Coverage of Two Sets

No.	$\varsigma(X^1, X^S)$	$\varsigma(X^S, X^1)$	$\varsigma(X^1, X^R)$	$\varsigma(X^R, X^1)$	$\varsigma(X^S, X^R)$	$\varsigma(X^R, X^S)$
Test 1	0.727525	0.466000	0.722667	0.298667	0.516000	0.342929
Test 2	0.833333	0.014333	0.792857	0.014667	0.238095	0.206667
Test 3	0.994872	0.548667	0.996667	0.277333	0.626667	0.319231
Test 4	0.428333	0.037000	0.388889	0.038000	0.161728	0.163333
Test 5	0.788889	0.016000	0.905556	0.004333	0.441667	0.101389

Table 2. Average results of the metrics Spacing

	Algorithm	IDCMA	MISA	SPEA	RWGA
Spacing S	Test 1	0.057842	0.107427	0.051856	0.127075
	Test 2	0.047654	0.114692	0.077124	0.438289
	Test 3	0.033705	/	0.045433	0.059720
	Test 4	0.032992	/	0.076490	0.127397
	Test 5	0.134716	/	0.479929	0.738698

We adopted two popular numerical metrics, Convergence of Two Sets and Spacing, selected five typical multiobjective problems, and compared with the other three advanced multiobjective algorithms. The simulation results show that Immune Dominance Clone Multiobjective algorithm proposed in this paper can solve the low-dimensional multiobjective problems very well. Especially for the discontinuous Pareto-optimal fronts or the isolated optimal solutions, IDCMA can also construct and find them while the other algorithms seem incapable sometimes. In addition, Figure 2 to Figure 6 testifies the rationality of the conclusion above intuitively.

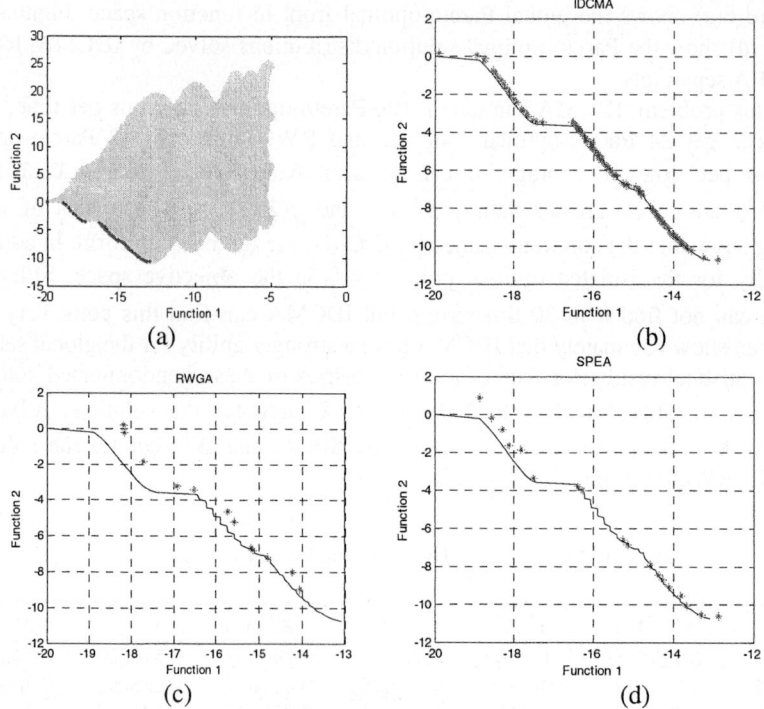

Fig. 6. The Pareto-optimal solution distributions corresponding to test 5. (a) The global Pareto-optimal front; (b) The solution distribution solved by IDCMA; (c) The solution distribution solved by RWGA; (d) The solution distribution solved by SPEA

5 Conclusion and Prospective

In this paper, the basic concepts of artificial immune system are presented and a novel algorithm, Immune Dominance Clonal Multiobjective Algorithm, inspired by the concept of immunodominance and the clonal selection theory, is proposed. When compared with RWGA, SPEA and MISA, IDCMA is more effective for low-dimensional multiobjective optimization problems in the two popular metrics, Spacing and Coverage of Two Sets.

Although IDCMA can solve some low-dimensional multiobjective problems preferably, it adopts binary coding, so it can not solve high-dimensional problems with low computational complexity. To design a suitable antibody coding mode and computing method of antibody-antibody affinity is our next work.

References

1. Schaffer, J.D.: Multiple objective optimization with vector ecaluated genetic algorithms. PhD thesis, Vanderbilt University (1984)
2. Abido, M.A.: Environmental economic power dispatch using multiobjective evolutionary algorithms. IEEE Trans. Power Systems, Vol.18, No. 4, November (2003)

3. Zitzler, E., Thiele, L.: Multiobjective Evolutionary Algorithms: A Comparative Case Study and the Strength Pareto Approach. IEEE Trans. Evolutionary Computation. Vol. 3, No. 4, November (1999)
4. Zitzler, E., Laumanns, M., Thiele, L.: SPEA2: Improving the Strength Pareto Evolutionary Algorithm. In: Giannakoglou, K., Tsahalis, D., Periaux, J., Papailou, P., Fogarty, T. (eds.): EUROGEN 2001, Evolutionary Methods for Design, Optimization and Control with Applications to Industrial Problems. Athens, Greece (2002)95–100
5. Coello Coello, C.A., Nareli, C.C.: An Approach to Solve Multiobjective Optimization Problems Based on an Artificial Immune System. In: Jonathan, T., Peter, J.B. (eds.): Proceedings of the First International Conference on Artificial Immune Systems. Canterbury, UK (2002)212–221
6. Du, H.F., Jiao, L.C., Gong, M.G., Liu, R.C.: Adaptive Dynamic Clone Selection Algorithms. In: Zdzislaw, P., Lotfi, Z. (eds.): Proceedings of the Fourth International Conference on Rough Sets and Current Trends in Computing. Uppsala, Sweden (2004)
7. Ishibuchi, H., Murata, T.: A multiobjective genetic local search algorithm and its application to flowshop scheduling. IEEE Trans. System, Man and Cybernetics. Vol. 28, No. 3, (1998) 392–403
8. Abbas, A.K., Lichtman, A.H., Pober, J.S.: Cellular and Molecular Immunology. 3rd edn. W. B. Saunders Company, New York (1998)
9. Zitzler, E.: Evolutionary Algorithms for Multiobjective Optimization: Methods and Applications. A dissertation submitted to the Swiss Federal Institute of Technology Zurich for the degree of Doctor of Technical Sciences. Diss. Eth No. 13398 (1999)
10. Schott, J.R.: Fault Tolerant Design Using Single and Multictiteria Genetic Algorithm Optimization. Master's thesis, Massachusetts Institute of Technology,Cambridge, Massachusetts, May (1995)
11. David, A.V.: Multiobjective Evolutionary Algorithms: Classification, Analyses, and New Innovations. PhD thesis. Presented to the Faculty of the Graduate School of Engineering of he Air Force Institute of Technology. Air University. USA. AFIT/DS/ENG (1999)

A Multi-objective Tabu Search Algorithm for Constrained Optimisation Problems

Daniel Jaeggi, Geoff Parks, Timoleon Kipouros, and John Clarkson

Engineering Design Centre, Department of Engineering,
University of Cambridge, Trumpington Street,
Cambridge CB2 1PZ, United Kingdom

Abstract. Real-world engineering optimisation problems are typically multi-objective and highly constrained, and constraints may be both costly to evaluate and binary in nature. In addition, objective functions may be computationally expensive and, in the commercial design cycle, there is a premium placed on rapid initial progress in the optimisation run. In these circumstances, evolutionary algorithms may not be the best choice; we have developed a multi-objective Tabu Search algorithm, designed to perform well under these conditions. Here we present the algorithm along with the constraint handling approach, and test it on a number of benchmark constrained test problems. In addition, we perform a parametric study on a variety of unconstrained test problems in order to determine the optimal parameter settings. Our algorithm performs well compared to a leading multi-objective Genetic Algorithm, and we find that its performance is robust to parameter settings.

1 Introduction

Real-world optimisation problems have a number of characteristics which must be taken into account when developing optimisation algorithms. Real-world problems are typically multi-objective; trade-offs between risk and reward, and cost and benefit exist at a fundamental level throughout the natural world and are a deep-seated part of human consciousness. These trade-offs carry over directly to the business world, and thus into any form of design activity. Any optimisation method which is to have any serious benefit to the design process must be able to handle multiple objectives.

Real-world problems also tend to be highly constrained. The nature of these constraints and their effect on the optimisation landscape varies from problem to problem. However, optimisation problems in a number of fields have constraints with similar characteristics, and this is discussed further in Section 1.1 below. The optimisation landscape – regions of feasible, highly constrained design space and the variations of objective function values within that space – is strongly influenced by the parameterisation scheme, for any one given problem. Good parameterisation schemes for aerodynamic shape optimisation problems – the particular focus of our work – as shown by Harvey [1], Kellar [2] and Gaiddon *et al.* [3], tend to produce optimisation landscapes that are highly constrained, have many variables, and many local minima. Thus, the optimisation algorithm must be chosen to perform well in these circumstances [4].

These characteristics quickly rule out the use of traditional gradient-based optimisation methods: notwithstanding their requirement of gradient information (which may be difficult, expensive, or impossible to obtain), these algorithms perform poorly in problems which are highly constrained and contain local minima. Harvey [1] tested a number of meta-heuristic methods on a representative aerodynamic design optimisation problem and found Tabu Search (TS) to be superior to the Genetic Algorithm (GA) and Simulated Annealing (SA) methods.

Numerous of multi-objective GAs exist [5]. Similarly, multi-objective SA methods have been developed [6]. However, despite its popularity in single-objective optimisation problems, very few attempts have been made at developing a multi-objective version of TS. Jones [7] reviewed the literature on multi-objective meta-heuristics and found only 6% of 124 papers concerned with TS.

Given that it may well perform better than a GA or SA method (assuming Harvey's results carry over into multi-objective optimisation) on aerodynamic design optimisation problems, and there is a strong real-world requirement to perform multi-objective optimisation, there appears to be both a need and an opportunity to develop a new multi-objective TS algorithm.

1.1 Constraint Handling

It is important that the constraint handling method of an optimisation algorithm is able to deal with constraints which are binary, as happens in a number of real-world engineering problems. The constraint handling in many multi-objective evolutionary algorithms requires the ability to assign some kind of constraint violation distance to points in design space which violate constraints, and points are then ranked accordingly [5]. In the presence of binary constraints, such an approach cannot be used.

Such constraints occur typically in shape optimisation problems, especially when the cost function is evaluated using a finite difference type of method (including finite element and finite volume methods) which solves a system of equations over a finite mesh. The constraints for these problems arise from three sources, amongst others:

1. *Geometric considerations.* The parameterisation scheme may give rise to shapes which are physically impossible (*i.e.* negative volumes). Conceptually, a distance measure in design space may be formulated by considering an offset vector $\Delta \bar{x}$ which can be added to the design vector $\bar{x}$ to make the design feasible; in practice this may be too costly due to the interdependence between design variables.
2. *Mesh considerations.* Given a valid geometry, it may be impossible to fit a mesh that satisfies certain criteria relevant to the numerical solution of the problem (*i.e.* skewed cells). Again, finding an offset vector $\Delta \bar{x}$ is conceptually possible, but is in practice even more costly than finding a $\Delta \bar{x}$ that satisfies just the geometric considerations.

3. *Numerical solution considerations.* Given a valid mesh and geometry, it may still be impossible to reach a numerical solution of the system of equations for that geometry (*i.e.* lack of convergence in a finite difference method). Yet again, finding a $\Delta \bar{x}$ that makes the solution converge is possible but totally unrealistic, given the massive computational cost of a solution.

The extent to which such constraint violations are encountered is strongly dependent on the parameterisation scheme used. For any problem, it is probably possible to employ a parameterisation scheme that gives a continuous design space and constraint violations for which $\Delta \bar{x}$ may be easily calculated. However, it is the authors' belief that the parameterisation scheme should be chosen first and then the optimisation algorithm, not the other way round. Of course, there is a trade-off: a parameterisation scheme that produces a design space that is almost impossible to search is of virtually no use. Yet we can clearly enhance our choice of parameterisation scheme by easing the restrictions placed on it by the optimiser's constraint handling method.

1.2 Existing Multi-objective Tabu Search Algorithms

There are two approaches to solving a multi-objective optimisation problem. The first reduces the multiple objectives to a single objective by generating a composite objective function, usually from a weighted sum of the objectives. This composite objective function can be optimised using existing single-objective optimisers. However, the weights must be pre-set, and the solution to this problem will be a single vector of design variables rather than the entire Pareto-optimal (PO) set. This can have undesirable consequences: setting the weights implicitly introduces the designer's preconceptions about the relative trade-off between objectives. Real-world problems can produce surprising PO sets which may profoundly affect design decisions, and the potential to generate novel designs is a key benefit of optimisation [6].

The second approach to solving the multi-objective problem is to search directly for the entire PO set. This can be achieved in a number of ways and requires modification to existing single-objective algorithms.

The authors know of only two attempts to produce a multi-objective TS algorithm which finds multiple PO solutions in a single run. Hansen's algorithm [8] is an extension of the composite objective approach: his algorithm performs a number of composite objective Tabu searches in parallel. Each search has a different and dynamically updated set of weights, and in this way the search can be driven to explore the entire Pareto front. This algorithm, although a good implementation of TS, suffers the problems common to all weighted-sum approaches: for problems with concave Pareto fronts, there may be regions of the front that are not defined by a combination of weights and conversely certain combinations of weights represent two points on the front. Thus, this algorithm may not adequately locate the entire PO set.

Baykasoglu *et al.* [9] developed a TS algorithm combining a downhill local search with an *intensification memory* (IM) to store non-dominated points that were not selected in the search. When the search fails to find a downhill move, a point from the IM is selected instead. When the IM is empty and all search paths exhausted, the algorithm stops. This cannot be considered a true TS algorithm: in restricting the search to only downhill moves its originators reject one of the basic tenets of TS, that "a bad strategic choice can yield more information than a good random choice" [10]. Also, the lack of any diversification strategy renders the algorithm incomplete and merely an elaborate local search algorithm.

The other TS algorithms reviewed by Jones [7] either use a composite objective function or are little more than local search algorithms similar to the algorithm of Baykasoglu *et al.*

2 Multi-objective Tabu Search Adaptation

The single-objective TS implementation of Connor and Tilley [11] is used as a starting point for our multi-objective variant. This uses a Hooke and Jeeves (H&J) local search algorithm (designed for continuous optimisation problems) [12] coupled with short, medium and long term memories to implement search intensification and diversification as prescribed by Glover and Laguna [10].

TS operates in a sequential, iterative manner: the search starts at a given point and the algorithm selects a new point in the search space to be the next current point. The basic search pattern is the H&J search.

Recently visited points are stored in the *short term memory* (STM) and are tabu – the search is not allowed to revisit these points. Optimal or near-optimal points are stored in the *medium term memory* (MTM) and are used for intensification, focusing the search on areas of the search space with good objective function values. The *long term memory* (LTM) records the regions of the search space which have been explored, and is used on diversification, directing the search to regions which are under-explored. This is achieved by dividing each control variable into a certain number of regions and counting the number of solutions evaluated in those regions. A local iteration counter i_local is used and reset upon a successful addition to the MTM. When i_local reaches user-specified values, the algorithm will diversify or intensify the search, or reduce the search step size and restart the search from the best solution found.

Thus, TS combines an exhaustive, systematic local search with a stochastic element and an intelligent coverage of the entire search space. Our multi-objective TS implementation of [11] is modified in the following areas: search point comparison; the H&J move; optimal point archiving and the MTM; search intensification and restart strategy. These modifications are described briefly below, along with some further improvements.

2.1 Search Point Comparison

In a single-objective optimisation problem, points may be compared using the operators $==$, $>$ and $<$ acting on the objective function values for those points. Similarly, points in a multi-objective problem can be compared in the same way

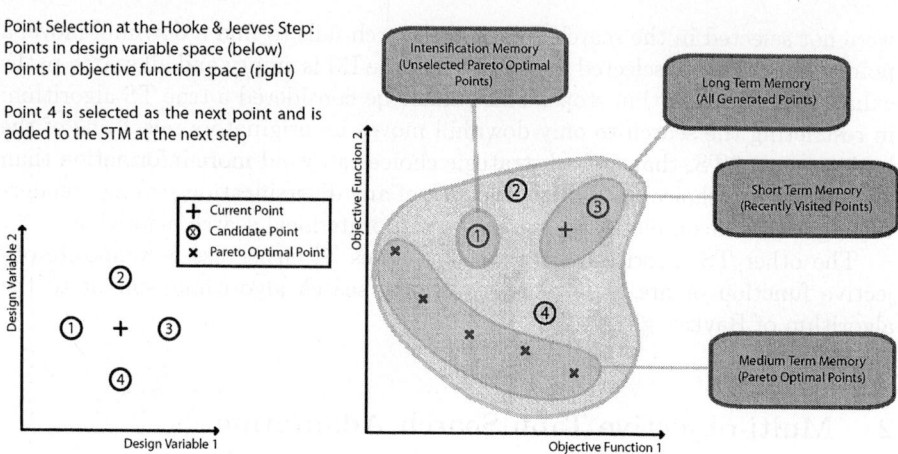

Fig. 1. Point Selection for the Hooke & Jeeves move and Tabu Search Memories

(thus preserving the logic of the single-objective algorithm) by using the concepts of Pareto equivalence (==) and dominance (> and <).

2.2 The Hooke and Jeeves Move

At each iteration, a H&J move is made. $2n\_var$ new points are generated by incrementing and decrementing each design variable by a given step around the current point. The objective functions for each new point are evaluated and, as long as the point is neither tabu (*i.e.* not a member of the STM) nor violates any constraints, it is considered as a candidate for the next point in the search.

In the single-objective TS algorithm, these candidates are sorted and the point with the lowest objective is chosen as the next point. A similar logic can be applied to the multi-objective case: however, the possibility of multiple points being Pareto equivalent (PE) and optimal must be allowed for.

This is achieved by classifying each candidate point according to its domination or Pareto equivalence to the current point. If there is a single dominating point, it is automatically accepted as the next point. If there are multiple dominating points, the dominated points within that group are removed and one is selected at random from those remaining. The other points become candidates for intensification (discussed below). If there are no dominating points, the same procedure is applied to those candidate points which are PE to the current point. If there are no PE points, a dominated point is selected in the same fashion. Thus, our strategy accepts both downhill and uphill moves – the next point is simply the "best" point (or one of the PE best points) selected from the candidate solutions. This logic is shown in Fig. 1 for clarity.

In addition, a *pattern move* strategy is implemented in the same way as Connor and Tilley [11]. Before every second H&J move, the previous move is repeated. This new point is compared to the current point, and, if it dominates

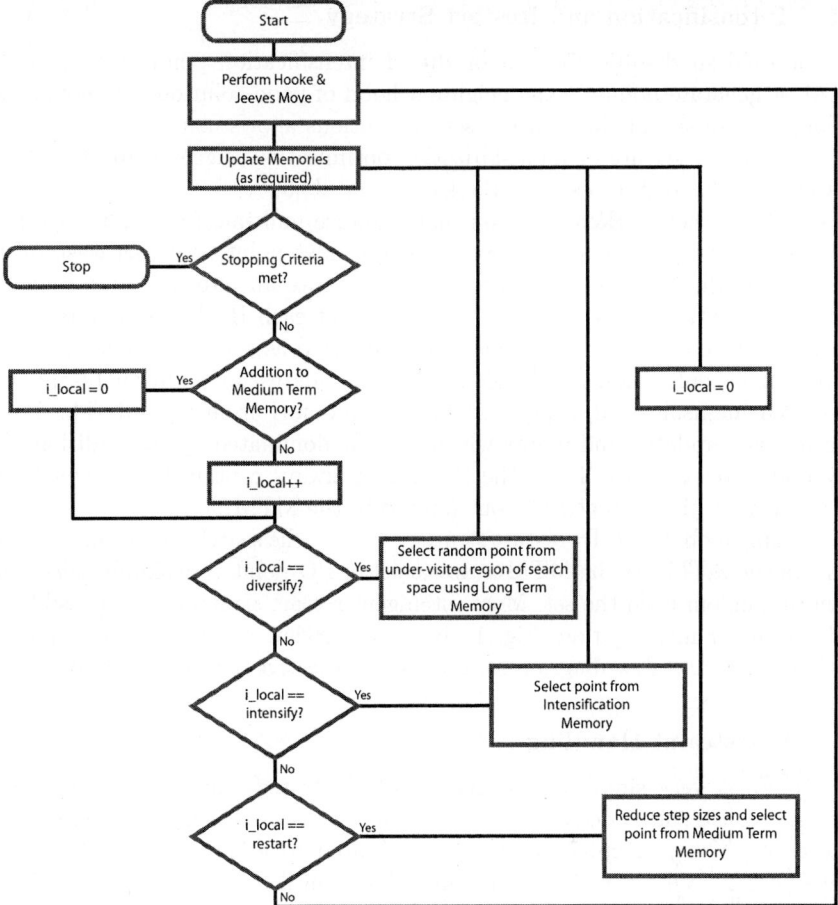

Fig. 2. Flow Diagram of the Multi-objective Tabu Search Algorithm

it, is accepted as the next point; if not, the standard H&J move is made. In this way, the search may be accelerated along known downhill directions.

2.3 Optimal Point Archiving and the Medium Term Memory

In Connor's single-objective TS [11], the MTM is a bounded, sorted list of near-optimal solutions. As the concept of a single optimal point does not exist in multi-objective optimisation (see Section 1.2), we replace the MTM in our multi-objective TS variant by an unbounded set of non-dominated solutions produced by the search. As new points are evaluated, they become candidates for addition to this set. Thus, the MTM represents the PO set for the problem at that stage in the search.

2.4 Intensification and Restart Strategy

The original single-objective TS produced intensification points by using the MTM to generate points in the neighbourhood of good solutions. Although the replacement of the MTM by a PO set of solutions allows us to use a variant of this strategy, a feature of multi-objective optimisation suggests an alternative strategy, similar to that used by Baykasoglu et al. [9].

A multi-objective H&J iteration may produce multiple PO points (see Fig. 1). As only one point may be selected as the next point, it seems wasteful to discard the other points. Therefore, we incorporate an intensification memory into our algorithm. This is a set of PE points; at each H&J step, points which dominate the current solution, but are not selected as the next point (of which there can be only one), are considered as candidates for addition to the set. At search intensification, a point is chosen randomly from the IM. The IM is continuously updated and points which become dominated by the addition of a new point are removed. Thus, the IM should always contain points which are on, or near to, the current PO front (stored in the MTM).

The single-objective TS restart strategy returns the search to the current best point in the MTM. As the MTM is now a set of PO points, we simply select one point at random from the set. More intelligent restart strategies are possible [6] and are under investigation. Fig. 1 gives an overview of the various memories used, and Fig. 2 a flow diagram for our multi-objective TS implementation.

2.5 Constraint Handling

We employ a very simple constraint handling strategy: any point which violates any constraint is deemed to be tabu and the search is not allowed to visit that point. Thus, accepted solutions are limited to feasible space. On search diversification, we allow the algorithm to loop until a feasible point has been found. Depending on the problem, it would also be possible to introduce penalty functions to handle certain constraints.

2.6 Improving Local Search Efficiency and Parallelisation Strategy

The H&J local search strategy requires roughly $2n\_var$ solution evaluations (allowing for points that are tabu or violate constraints) at each step, where n_var is the number of design variables. A real-world problem may contain a large number of variables (the shape optimisation of a Boeing 747 wing required 90 variables [13]) and this strategy could become prohibitively expensive. One solution to this is to incorporate an element of random sampling in the H&J step.

We generate the $2n\_var$ new points, remove those that are tabu, and only evaluate $n\_sample \leq 2n\_var$ points from those that remain, selecting randomly to avoid introducing any directional bias. If one of these points dominates the current point, it is automatically accepted as the next point. If more than one point dominates the current point, a non-dominated point from these is randomly selected. If no points dominate the current point, a further n_sample points are sampled and the comparison is repeated. If all the feasible, non-tabu points have

Table 1. Test Functions

Function Name	Number of Variables	Number of Objectives	Constraint Types
SCH	1	2	None
FON	3	2	None
POL	2	2	None
KUR	3	2	None
ZDT1	30	2	None
ZDT2	30	2	None
ZDT3	30	2	None
ZDT4	10	2	None
ZDT6	10	2	None
CONSTR	2	2	2 Inequality
SRN	2	2	2 Inequality
TNK	2	2	2 Inequality
WATER	3	5	7 Inequality

been sampled without finding a point that dominates the current solution, the standard selection procedure is employed.

Any optimisation procedure that forms part of a real-world design cycle must be able to complete in a reasonable time-frame. Parallel processing offers a large potential speed-up; any serious optimisation algorithm should be designed with this in mind. Our multi-objective Tabu Search algorithm is parallelised by means of functional decomposition. At each H&J move, the required objective function evaluations are computed in parallel.

3 Tabu Search Parameter Investigation

The performance of this algorithm has already been tested on nine standard unconstrained test problems and compared to the performance of NSGA-II [14]. Over these nine problems, the algorithm performed comparably with NSGA-II. In that initial study, no attempt was made to find optimal TS parameter settings; they were set to reasonable values, based on experience. Here, we conduct a more systematic parameter setting investigation. The parameters that may be set in the algorithm are shown in Table 2. The parameter settings used in the benchmarking in [14] are also given.

In the studies that follow, the same test conditions as in [14] were used. Parameters not being varied were kept fixed at the values given in Table 2. Performance was assessed using the convergence metric Υ, described by Deb et al. [15]; the mean and standard deviation for the results of 45 runs were calculated, and the same set of random number generator seeds used. Each run was stopped and the results calculated after 25000 function evaluations.

Table 2. Tabu Search parameters

Parameter	Initial Value	Description
$diversify$	10	Diversify search when $i\_local == diversify$
$intensify$	20	Intensify search when $i\_local == intensify$
$reduce$	50	Reduce step sizes and restart when $i\_local == reduce$
n_stm	20	STM size – the last n_stm visited points are tabu
$n\_regions$	2	In the LTM each variable is divided into $n\_regions$ regions
SS	8%	Initial H&J step size as percentage of variable range
$SSRF$	0.5	Factor by which step sizes are reduced on restart
n_sample	6	Number of points randomly sampled at each H&J move

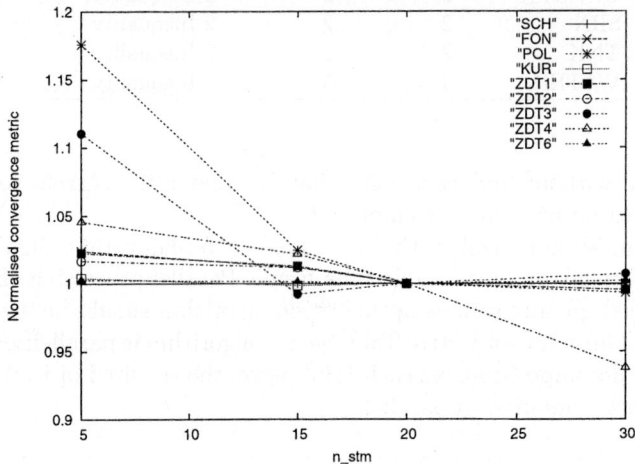

Fig. 3. Normalised convergence metric $\hat{\Upsilon}$ vs n_stm over 9 test problems

3.1 Variation in Short Term Memory Size

One of the distinguishing features of TS algorithms in general is the use of a short term memory to define points that are tabu and may not be revisited. This gives the search algorithm a means by which it can climb out of local minima. We might expect that the size of the STM may affect the performance of the algorithm: a STM of zero size would reduce the algorithm to a mere local search algorithm coupled with a global random search; a STM of infinite size would prevent the search from refining solutions in known good regions of the search space. There are also algorithm run time considerations: the computational cost of the algorithm increases with STM size, but for most real-world problems this cost is negligible compared to the cost of function evaluations.

We consider variations in the STM size in the range $5 \leq n\_stm \leq 30$ for the nine test problems used in [14], keeping all other parameter values constant as given in Table 2. The results are shown in Fig. 3; Υ is normalised with respect

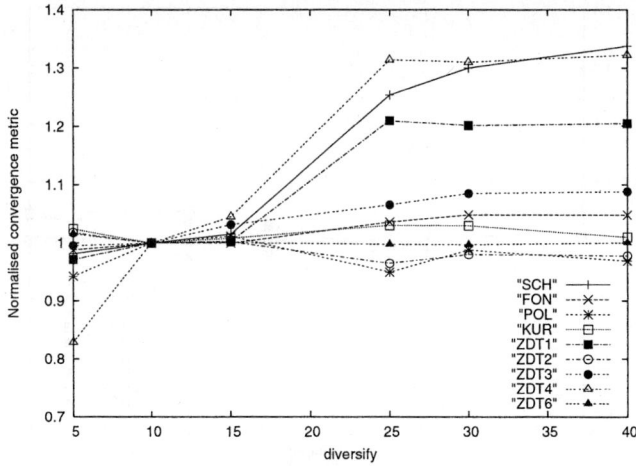

Fig. 4. Normalised convergence metric $\hat{\Upsilon}$ vs *diversify* over 9 test problems

to its value for $n\_stm = 20$, the value used in [14], and is plotted against n_stm. Performance is improved by increasing n_stm from 5 up to around 15 on most test problems. Increases beyond that improve performance but only marginally, and there does not appear to be much benefit in this. On problems with low numbers of design variables, there is hardly any variation in performance; this is most likely due to the lower potential for "cycling" (the search repeating a recent search path) as there are fewer paths that can be taken. The exception is Poloni's problem which shows quite a large improvement in performance between $n\_stm = 5$ and $n\_stm = 15$ despite having only two design variables. Problem ZDT3 also displays a large improvement in performance over this range of n_stm; on this problem, it appears that a larger STM is required to prevent cycling.

The other test problem that exhibits sensitivity to changes in n_stm is problem ZDT4; this shows continually improving performance with increases in n_stm. However, because the presented results have been normalised to show relative performance the incredibly poor absolute performance of our algorithm on ZDT4 has been masked. This is commented on in [14].

3.2 Variations in *Intensify* and *Diversify*

The parameters *intensify* and *diversify* control the balance between a local search of the design space and a global one. Therefore, they are critical in governing performance: certain problems, particularly multi-modal ones, will benefit strongly from a strategy which favours diversification; other problems, such as those with clearly defined local regions containing many near PO points, will be better searched by a strategy favouring intensification.

Fig. 4 shows the effect of varying *diversify* on the normalised convergence metric $\hat{\Upsilon}$; as in Section 3.1, Υ is normalised against its value for *diversify* = 10, used in the original study [14]. There appear to be two trends: for one group of

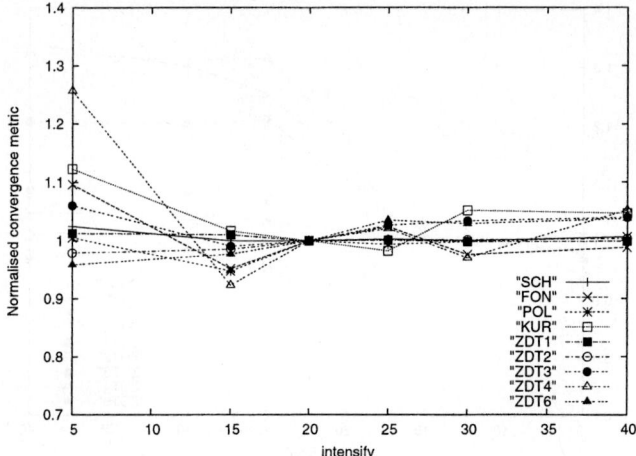

Fig. 5. Normalised convergence metric $\hat{\Upsilon}$ vs *intensify* over 9 test problems

problems, the point at which diversification takes place has little effect on overall performance; another group definitely favours early diversification. In general, early diversification brings performance benefits for these problems.

It also is worth noting two points. First, *intensify* was fixed at 20: thus, over the range of values for *diversify* used in this study, the order in which intensification and diversification takes place changes. Some of the performance variation is attributable to this. In particular, on problem ZDT4, which contains a large number of false Pareto fronts, intensifying before diversifying traps the optimiser in these false fronts and hinders the location of the true Pareto front.

Second, early diversification tends to increase the total number of diversification moves performed during an optimisation run. This behaviour is beneficial on problems such as SCH; good performance on this problem depends on how fast the region of design space in which the Pareto front is located is found. Thus, more random behaviour in the search speeds its discovery; once this region is found, the local search component in TS effectively finds the rest of the front.

Similarly, Fig. 5 shows the effect on $\hat{\Upsilon}$ of varying *intensify*; Υ is normalised against its value for *intensify* = 20.

For the majority of problems, the absolute value of *intensify* appears relatively unimportant; of more importance is whether *intensify* is less or greater than *diversify*. The results suggest that for good performance on these test functions in general, diversification must occur first and its value primarily governs performance. However, the results also show that there is some benefit in reducing the gap between diversification and intensification; a value of *intensify* = 15 gives, on average, better performance. This would prevent the algorithm from needlessly searching poor areas of the design space.

These results raise a concern about the use of test functions in determining algorithm performance on real-world problems. Kipouros et al. [16] used a variant of this TS algorithm to perform a multi-objective optimisation of a gas-

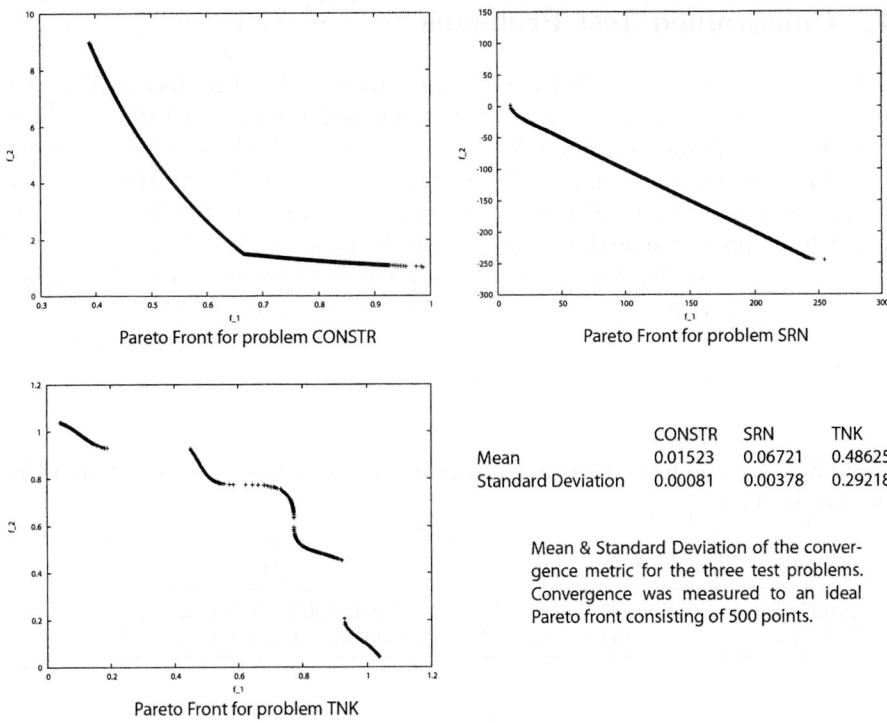

Fig. 6. Pareto fronts for constrained test problems CONSTR, SRN & TNK and convergence metric results

turbine compressor blade. For that particular application, intensification was found to be particularly beneficial to overall performance; the optimiser was able to make steady progress (indicated by rate of addition to the MTM) by performing many intensification steps. This suggests that the problem contains regions with many locally PO points and TS is able to effectively search all these through its intensification strategy. In contrast, this characteristic does not appear to be present in this set of test functions. There appears to be an urgent need to devise test problems that accurately reflect characteristics of real-world problems.

For both these sets of results, error bars based on the standard deviation of $\hat{\Upsilon}$ were not plotted, for reasons of clarity. Although there was some variation in the standard deviation with parameter setting, for the most part, these variations were small and the values remained close to the nominal values reported in [14]. There are two main exceptions to this: with $diversify = 5$ seven problems showed an increase in the standard deviation; the standard deviations for ZDT4 varied greatly, which is probably due to the algorithm's poor performance on this problem.

4 Constrained Test Problems

Deb et al. [15] also tested NSGA-II on four constrained test problems. Although no quantitative performance results were published, plots showed that NSGA-II was capable of finding a good spread of results along the Pareto front for each. We have tested our algorithm on those four constrained problems: the results for the problems CONSTR, SRN and TNK are shown in Fig. 6; results for problem WATER are presented as the range of values found for all five objectives and are given in Table 3. We used the lower limit of 20000 function evaluations prescribed by Deb et al. The parameter settings used were the same as given in Table 2, with the exception that *intensify* was reduced to 15 as a result of the parameter study presented above.

Table 3. Lower and upper bounds of objective function values on the Pareto front for problem WATER

	f_1	f_2	f_3	f_4	f_5
Multi-objective TS	0.804-0.918	0.022-0.857	0.104-0.962	0.056-1.320	0.129-3.121
NSGA-II	0.798-0.920	0.027-0.900	0.095-0.951	0.031-1.110	0.001-3.124

On problem CONSTR, coverage of the Pareto front is excellent, except in a small region near the tail where f_1 approaches 1.0. The range of solutions found is comparable to NSGA-II and far superior to Ray-Tai-Seow's algorithm presented in [15]. Similar performance is achieved on problem SRN – convergence to and coverage of the Pareto front are good and comparable to NSGA-II.

Problem TNK is slightly harder – the Pareto front is discontinuous and, as shown by Deb et al., some algorithms have difficulty finding the entire central continuous region. Although the spread of solutions that our algorithm finds in this region is slightly worse than on the rest of the Pareto front, it succeeds in locating the continuous region correctly.

Problem WATER is a five-objective problem; due to the difficulty of visualising the Pareto front, figures are presented in Table 3 for the minimum and maximum values of the objective functions found on the Pareto front. Deb et al. presented similar figures for NSGA-II on this problem. Of the 10 minimum/maximum values, NSGA-II finds better values on 7, although the differences in most cases are small.

5 Conclusions

In this paper, we have presented a multi-objective TS algorithm with features that make it particularly attractive to real-world optimisation problems, in particular with regard to its constraint handling. In previous work, we benchmarked this algorithm against NSGA-II on a number of test functions and found that it performed comparably. Here, we performed a study on the effect of varying the TS parameters on the algorithm performance on the same nine test functions.

As regards the STM size, the results suggest that a value of $n\_stm \geq 15$ gives good performance on a range of test functions. Increasing n_stm beyond 25 does not appear to give much performance benefit on the functions tested. Although the computational cost of the algorithm increases with n_stm, this cost is usually negligible for real-world problems where function evaluations are expensive.

Results for varying *intensify* and *diversify* show that algorithm performance is, in general, dominated by diversification; performance is improved by increasing the diversification element, and intensification has relatively little effect. This is slightly at odds with experience of using this algorithm on aerodynamic shape optimisation problems, where intensification is a more effective means of advancing the search. This raises a concern about the use of test functions to derive algorithm performance information for use in real-world problems.

Over the majority of test functions, performance is reasonably independent of the TS parameter settings; this suggests that the power of TS comes from its fundamental elements – the combination of a local search algorithm with a variety of search memories – rather than particular, carefully chosen parameter settings. This should ease its application to new problems.

Finally, our TS algorithm was tested on four constrained test problems. On all the problems, the algorithm was able to find a good spread of solutions along the Pareto front, despite our strict constraint handling approach. We believe our multi-objective TS algorithm is well suited to use on real-world optimisation problems where constraints can be binary in nature and such an approach is required to allow the optimiser to run. Indeed the only way to show this is to actually apply it to real-world problems (as we have done in a companion paper [17]); test problems that exist in the literature do not share these characteristics, and it is hard to draw meaningful conclusions from tests using these problems.

However, future work should include a more rigorous performance comparison with other leading MO optimisation algorithms. The performance metrics used in this study are not optimal [18] but were chosen to allow comparison with previously published data [15]. Finally, determining optimal settings for this algorithm requires further more detailed work.

Acknowledgements

This research is supported by the UK Engineering and Physical Sciences Research Council (EPSRC) under grant number GR/R64100/01. The authors would also like to thank Prof. Bill Dawes for his support and encouragement.

References

1. Harvey, S.: The Design Optimisation of Turbomachinery Blade Rows. PhD thesis, Cambridge University Engineering Department (2002)

2. Kellar, W.: Geometry Modelling in Compuational Fluid Dynamics and Design Optimisation. PhD thesis, Cambridge University Engineering Department (2002)
3. Gaiddon, A., Greard, J., Pagan, D., Knight, D.: Automated optimization of supersonic missile performance taking into account design uncertainties. Technical Report AIAA-2003-3879, AIAA (2003)
4. Duvigneau, R., Visonneau, M.: Hybrid genetic algorithms and artificial neural networks for compex design optimization in CFD. International Journal for Numerical Methods in Fluids **44** (2004) 1257–1278
5. Deb, K.: Multi-Objective Optimization using Evolutionary Algorithms. John Wiley & Sons, Ltd., Winchester, UK (2001)
6. Suppapitnarm, A., Seffen, K., Parks, G., Clarkson, J.: A simulated annealing algorithm for multiobjective optimization. Engineering Optimization **33** (2000) 59–85
7. Jones, D., Mirrazavi, S., Tamiz, M.: Multi-objective meta-heuristics: An overview of the current state-of-the-art. European Journal of Operational Research **137** (2002) 1–9
8. Hansen, M.: Tabu search for multiobjective optimization: MOTS. In: MCDM, Cape Town, South Africa. (1997)
9. Baykasoglu, A., Owen, S., Gindy, N.: A taboo search based approach to find the pareto optimal set in multiple objective optimization. Engineering Optimization **31** (1999) 731–748
10. Glover, F., Laguna, M.: Tabu Search. Kluwer Academic Publishers, Boston, MA (1997)
11. Connor, A., Tilley, D.: A tabu search method for the optimisation of fluid power circuits. IMechE Journal of Systems and Control **212** (1998) 373–381
12. Hooke, R., Jeeves, T.: Direct search solution of numerical and statistical problems. Journal of the ACM **8** (1961) 212–229
13. Jameson, A.: A perspective on computational algorithms for aerodynamic analysis and design. Progress in Aerospace Sciences **37** (2001) 197–243
14. Jaeggi, D., Asselin-Miller, C., Parks, G., Kipouros, T., Bell, T., Clarkson, P.: Multi-objective parallel tabu search. In Yao, X., Burke, E., Lozano, J.A., Smith, J., Merelo-Guervos, J., Bullinaria, J., Rowe, J., Tino, P., Kaban, A., Schwefel, H.P., eds.: Parallel Problem Solving from Nature – PPSN VIII. Lecture Notes in Computer Science, Vol. 3242, Springer-Verlag, Berlin (2004) 732–741
15. Deb, K., Pratap, A., Agarwal, S., Meyarivan, T.: A fast and elitist multiobjective genetic algorithm: NSGA–II. IEEE Transactions on Evolutionary Computation **6** (2002) 182–197
16. Kipouros, T., Parks, G., Savill, A., Jaeggi, D.: Multi-objective aerodynamic design optimisation. In: ERCOFTAC Design Optimisation: Methods and Applications Conference Proceedings, On CDRom. Paper ERCODO2004_239. (2004)
17. Kipouros, T., Jaeggi, D., Dawes, W., Parks, G., Savill, A.: Multi-objective optimisation of turbomachinery blades using tabu search. In: 3rd Int. Conf. Evolutionary Multi-Criterion Optimization. Lecture Notes in Computer Science. (2005)
18. Zitzler, E., Thiele, L., Laumanns, M., Fonseca, C., Grunert da Fonseca, V.: Performance assessment of multiobjective optimizers: an analysis and review. IEEE Transactions on Evolutionary Computation **7** (2003) 117–132

Improving PSO-Based Multi-objective Optimization Using Crowding, Mutation and ϵ-Dominance

Margarita Reyes Sierra and Carlos A. Coello Coello

CINVESTAV-IPN (Evolutionary Computation Group),
Electrical Eng. Department, Computer Science Dept.,
Av. IPN No. 2508, Col. San Pedro Zacatenco, México D.F. 07300, México
mreyes@computacion.cs.cinvestav.mx
ccoello@cs.cinvestav.mx

Abstract. In this paper, we propose a new Multi-Objective Particle Swarm Optimizer, which is based on Pareto dominance and the use of a crowding factor to filter out the list of available leaders. We also propose the use of different mutation (or *turbulence*) operators which act on different subdivisions of the swarm. Finally, the proposed approach also incorporates the ϵ-dominance concept to fix the size of the set of final solutions produced by the algorithm. Our approach is compared against five state-of-the-art algorithms, including three PSO-based approaches recently proposed. The results indicate that the proposed approach is highly competitive, being able to approximate the front even in cases where all the other PSO-based approaches fail.

1 Introduction

Kennedy and Eberhart [1] initially proposed the swarm strategy for optimization. The particle swarm optimization (PSO) algorithm is a population-based search algorithm based on the simulation of the social behavior of birds within a flock. In PSO, individuals, referred to as particles, are "flown" through hyperdimensional search space. Changes to the position of the particles within the search space are based on the social-psychological tendency of individuals to emulate the success of other individuals. A swarm consists of a set of particles, where each particle represents a potential solution. The position of each particle is changed according to its own experience and that of its neighbors. Let $x_i(t)$ denote the position of particle p_i, at time step t. The position of p_i is then changed by adding a velocity $v_i(t)$ to its current position, i.e.: $x_i(t) = x_i(t-1) + v_i(t)$. The velocity vector drives the optimization process and reflects the socially exchanged information. In the global best version (used here) of PSO, the social knowledge used to drive the movement of particles includes the position of the best particle from the entire swarm (*gbest*) and its history of experiences in terms of its own best solution thus far (*pbest*). In this case, the velocity vector changes in the following way: $v_i(t) = Wv_i(t-1) + C_1 r_1 (x_{pbest_i} - x_i(t)) + C_2 r_2 (x_{gbest} - x_i(t))$, where W is the inertia weight, C_1 and C_2 are the learning factors (usually defined as constants), and $r_1, r_2 \in [0,1]$ are random values. The successful application of PSO in many single objective optimization problems reflects the effectiveness of PSO. However, in order to handle multiple objectives, PSO must be obviously modified. In most approaches (which will be generically called MOPSOs, for Multiple-Objective Particle Swarm Optimizers), the major modifications to the basic PSO algorithm are the selection process of *pbest* and *gbest* [2, 3]. In this paper, we present a new proposal which is based on Pareto dominance and the use of a crowding factor for the selection of leaders. We also incorporate mutation operators (taken from the evolutionary algorithms literature) and the concept of ϵ-dominance. This paper is organized as follows. The previous related work is reviewed in Section 2. In Section 3, we describe our proposed approach. The obtained results and discussion are presented in Sections 4 and 5, respectively. Finally, the conclusions and future work are described in Section 6.

2 Related Work

There have been several proposals to extend PSO to handle multiple objectives. We will review next the most representative of them:

Ray and Liew [4]: This algorithm uses Pareto dominance and combines concepts of evolutionary techniques with the particle swarm. The approach uses crowding to maintain diversity and a multilevel sieve to handle constraints.

Hu and Eberhart [5]: In this algorithm, only one objective is optimized at a time using a scheme similar to lexicographic ordering. In further work, Hu et al. [6] adopted a secondary population (called "extended memory") and introduced some further improvements to their dynamic neighborhood PSO approach.

Fieldsend and Singh [7]: This approach uses an unconstrained elite archive (in which a special data structure called "dominated tree" is adopted) to store the nondominated individuals found along the search process. The archive interacts with the primary population in order to define local guides. This approach also uses a "turbulence" (or mutation) operator.

Coello et al. [2]: This approach uses a global repository in which every particle deposits its flight experiences. Additionally, the updates to the repository are performed considering a geographically-based system defined in terms of the objective function values of each individual; this repository is used by the particles to identify a leader that will guide the search. It also uses a mutation operator that acts both on the particles of the swarm, and on the range of each design variable of the problem to be solved. In more recent work, Toscano and Coello [8] adopted clustering techniques in order to divide the population of particles into several swarms in order to have a better distribution of solutions in decision variable space. In each sub-swarm, a PSO algorithm is executed and, at some point, the different sub-swarms exchange information: the leaders of each swarm are migrated to a different swarm to variate the selection pressure.

Mostaghim and Teich [3]: They proposed a sigma method in which the best local guides for each particle are adopted to improve the convergence and diversity of a PSO approach used for multiobjective optimization. They also use a "turbulence" operator. In further work, the authors [9] studied the influence of ϵ-dominance [10] on MOPSO methods. ϵ-dominance is compared with existing clustering techniques for fixing the archive size and the solutions are compared in terms of computational time, convergence and diversity. In more recent work, the authors [11] proposed a new method called *covering*MOPSO (cvMOPSO). This method works in two phases. In phase 1, a MOPSO' algorithm is run with a restricted archive size and the goal is to obtain a good approximation of the Pareto-front. In phase 2, the non-dominated solutions obtained from phase 1 are considered as the input archive of the cvMOPSO. The particles in the population of the cvMOPSO are divided into subswarms around each non-dominated solution after the first generation. The task of the subswarms is to cover the gaps between the non-dominated solutions obtained from phase 1.

Li [12]: This approach incorporates the main mechanisms of the NSGA-II [13] into PSO. It combines the population of particles and all the *pbest* positions of each particle, and selects the best from them to conform the next population. It also selects the leaders randomly from the leaders set based on both a niche count and a crowding distance. In more recent work, Li [14] proposed the *maximinPSO*, which uses a fitness function that requires no additional clustering or niching procedure to maintain diversity.

3 Description of Our Approach

It should be obvious that the main issue when extending PSO to deal with multiple objectives is how to generalize the concept of leader in the presence of several (equally good) solutions. The most straightforward approach is simply to consider every non-dominated solution as a new leader. This approach has, however, the drawback of increasing the size of the set of leaders very quickly. In our approach, we use a crowding factor [13] in order to establish a second discrimination criterion (additional to Pareto dominance). This criterion is also adopted to decide what leaders to keep over generations when the maximum list size has been exceeded. For each particle, we select the leader by means of a binary tournament based on the crowding value of the leaders. The maximum size of the set of leaders is fixed equal to the size of the swarm (or population). After each generation, the set of leaders is updated, and so are the corresponding crowding values. If the size of the set of leaders is greater than the maximum allowable size, only the best leaders are retained based on their crowding value. The rest of the leaders are eliminated. Although there are previous approaches that use the crowding factor to select the leaders (see for example [4, 12]), our approach is the first to adopt this information to fix the size of the set of leaders. This feature of our algorithm considerably simplifies the mechanism to control the set of leaders without requiring any additional parameter or selection criterion. We also propose the use of two mutation operators that are well-known in the EA literature: uniform mutation (i.e., the variability range allowed for each decision variable is kept constant over generations) and non-uniform mutation (i.e., the variability range allowed for each decision variable decreases over time). These operators modify the values of the decision variables of a particle with a certain probability. This makes a significant difference with respect to the previous proposals in which all the decision variables are modified when the turbulence (or mutation) operator is applied. Additionally, we considered the possibility of not using mutation at all. Given the uncertainty regarding the type of mutation to apply, we proposed a scheme by which the swarm is subdivided in three parts (of equal size). Each sub-part of the swarm will adopt a different mutation scheme: the first sub-part will have no mutation at all, the second sub-part will have uniform mutation and the third sub-part will have non-uniform mutation. With the use of these different operators we are aiming to have the ability of exploring (uniform mutation) and exploiting (non-uniform mutation) the search space as the process progresses. The available set of leaders is the same for each of these sub-parts. Additionally, each particle can use as a leader a particle produced by a different sub-part of the swarm. In this way, the three different sub-parts of the swarm will share their particular success and the final results will be a combination of using different behaviors inside the same swarm. In order to avoid the definition of extra parameters for the mutation operators, we adopt a rule of thumb normally used in the EA literature [15]: the mutation rate is defined as $1/codesize$, where $codesize$ refers to the total length of the string that encodes all the decision variables of the problem (the number of variables in our case). Finally, we adopt the concept of ϵ-dominance [10] in order to fix the size of the external archive that contains the (non-dominated) solutions that will be reported by the algorithm. A decision vector x_1 is said to ϵ-dominate a decision vector x_2 for some $\epsilon > 0$ iff: $f_i(x_1)/(1+\epsilon) \leq f_i(x_2), \forall i = 1, ..., m$ and $f_i(x_1)/(1+\epsilon) < f_i(x_2)$, for at least one $i = 1, ..., m$. It is worth noting that, when using ϵ-dominance, the size of the final external archive depends on the ϵ-value, which is normally a user-defined parameter [10]. For the sake of simplicity, in this paper, we consider the same value of ϵ for all the objective functions of a given problem. For each problem, the value of ϵ was tuned based on the desired amount of points in the final Pareto-front. Figure 1 shows the way in which our algorithm works. First, we initialize the swarm. The non-dominated particles found in the swarm will be introduced into the set of leaders. Later on, the crowding factor of each leader is calculated. At each generation, for each particle, we perform the flight and apply the corresponding mutation operator based on the subdivision of the swarm previously described. In order to perform the flight of each particle, the changes to the velocity vector are done in the following way: $v_i(t) = Wv_i(t-1) + C_1 r_1 (x_{pbest_i} - x_i(t)) + C_2 r_2 (x_{gbest} - x_i(t))$, where $W = random(0.1, 0.5)$, $C_1, C_2 = random(1.5, 2.0)$, and $r_1, r_2 = random(0.0, 1.0)$. Note that most of the previous PSO proposals fix the values of W, C_1 and C_2 instead of using random values as in our case. The only exception that we know (in the specific case of MOPSOs) is some of our own previous work [8]. We adopted this scheme since we found it as a more convenient way of dealing with the difficulties of fine tuning the parameters W, C_1 and C_2 for each specific test function. Then, we proceed to evaluate the particle and update its personal best value (*pbest*). A new particle replaces its *pbest* value if such value is dominated by the new particle or if both are non-dominated with respect to each other. After all the particles have been updated, the set of leaders is updated, too. Obviously, only the particles that outperform their *pbest* value will try to enter the leaders set. Once the leaders set has been updated, the ϵ-archive is updated. Finally, we proceed to update the crowding values of the set of leaders and we eliminate as many leaders as necessary in order avoid exceeding the allowable size of the leaders set. This process is repeated a fixed number (*gmax*) of iterations. The parameters needed by our approach are: (1) *swarmsize* (size of the swarm), (2) *gmax* (number of iterations), (3) *pm* (mutation rate—automatically computed), and (4) ϵ (value for bounding the size of the ϵ-archive).

```
Begin
    Initialize swarm. Initialize leaders. Send leaders to ϵ-archive
    crowding(leaders), g = 0
    While g < gmax
        For each particle
            Select leader. Flight. Mutation. Evaluation. Update pbest.
        EndFor
        Update leaders, Send leaders to ϵ-archive
        crowding(leaders), g++
    EndWhile
    Report results in ϵ-archive
End
```

Fig. 1. Pseudocode of our algorithm

4 Comparison of Results

To validate our approach, we performed both quantitative (adopting four performance measures) and qualitative comparisons (plotting the Pareto fronts produced) with respect to two MOEAs that are representative of the state-of-the-art in the area: the SPEA2 [16] and the NSGA-II [13]. We also compared our approach against three PSO-based approaches recently proposed: MOPSO [2], Sigma-MOPSO [3] and Cluster-MOPSO [8]. For our comparative study, we implemented two unary and two binary measures of performance. The following are the unary measures:

Success Counting (SCC): This measure counts the number of vectors (in the current set of nondominated vectors available) that are members of the Pareto optimal set: $SCC = \sum_{i=1}^{n} s_i$, where n is the number of vectors in the current set of nondominated vectors available; $s_i = 1$ if vector i is a member of the Pareto optimal set, and $s_i = 0$ otherwise. It should then be clear that $SCC = n$ indicates an ideal behavior. For a fair comparison, when we use this measure, all the algorithms should limit their final number of non-dominated solutions to the same value.

Inverted Generational Distance (IGD): In this measure, we use as a reference the true Pareto front, and we compare each of its elements with respect to the front produced by an algorithm. This measure is defined as: $IGD = \frac{\sqrt{\sum_{i=1}^{n} d_i^2}}{n}$ where n is the number of elements in the true Pareto front and d_i is the Euclidean distance (measured in objective space) between each of these and the nearest member of the set of nondominated vectors found by the algorithm. It should be clear that a value of $IGD = 0$ indicates that all the elements generated are in the true Pareto front of the problem.

The binary measures adopted are the following:

Two Set Coverage (SC): This measure was proposed in [17]. Consider X', X'' as two sets of phenotype decision vectors. SC is defined as the mapping of the order pair (X', X'') to the interval $[0,1]$: $SC(X', X'') \triangleq |\{a'' \epsilon X''; \exists a' \epsilon X' : a' \preceq a''\}|/|X''|$. If all points in X' dominate or are equal to all points in X'', then by definition $SC = 1$. $SC = 0$ implies the opposite. In general, $SC(X', X'')$ and $SC(X'', X')$ both have to be considered due to set intersections not being empty. If $SC(X', X'') = 0$ and $SC(X'', X') = 1$, we say that X'' is better than X'.

Two Set Difference Hypervolume (HV) This measure was proposed in [18]. Consider X', X'' as two sets of phenotype decision vectors. HV is defined by: $HV(X', X'') = \delta(X' + X'') - \delta(X'')$, where the set $X' + X''$ is defined as the nondominated vectors obtained from the union of X' and X'', and δ is the unary hypervolume measure. $\delta(X)$ is defined as the hypervolume of the portion of the objective space that is dominated by X. In this way, $HV(X', X'')$ gives the hypervolume of the portion of the objective space that is dominated by X' but not for X''. In this paper, we use the origin as a reference point to compute the hypervolume. So, since all the test functions have to be minimized, with this measure we obtain a difference between the areas that *dominate* the analyzed Pareto fronts. In this way, if $HV(X', X'') = 0$ and $HV(X'', X') < 0$, we say that X'' is better than X'.

For each of the test functions shown below, we performed 20 runs per algorithm. The parameters of each approach were set such that they all performed 20000 objective function evaluations. The codes of NSGA-II and SPEA2 were obtained from PISA.[1] The code of MOPSO was obtained from the EMOO repository.[2] The codes of Sigma-MOPSO and Cluster-MOPSO were provided by their authors. The code of our approach is available via email request to the first author. We adopted several test functions [19], however, given the available space we only present results corresponding to the following four: ZDT1, ZDT2, ZDT4 [20] and DTLZ6 [21]. All the algorithms compared adopted real-numbers encoding. The parameters for SPEA2 were $\alpha = \mu = \lambda = 100$ and 200 generations, and for the NSGA-II we used *popsize*=100 and 200 generations. As recommended in [21], for the NSGA-II and SPEA2, the crossover probability p_c was set to 1.0 and the mutation probability p_m was set to $1/codesize$. For our proposed approach and the MOPSO algorithm the parameters were: swarm size of 100 particles and 200 iterations. Cluster-MOPSO used 40 particles, 4 swarms, 5 iterations per swarm and a total number of iterations of 100. In the case of Sigma-MOPSO, 200 particles were used through 100 iterations (author suggestion). As recommended in [3], the Sigma-MOPSO used a turbulence probability of 0.01 for all functions, except for ZDT4 in which the turbulence probability used was 0.05. As recommended by its authors, the MOPSO used a mutation probability of 0.5. Our proposed approach used a probability mutation of $1/codesize$. The Pareto fronts that we will show correspond to the nondominated vectors obtained from the union of the 20 Pareto fronts produced by each approach. It should be noted that the Pareto fronts shown were also used to apply the binary measures of performance. All the algorithms, except for cMOPSO, were set such that they provided Pareto fronts with 100 points. cMOPSO does not have a scheme to fix the size of its final archive. Thus, in order to allow a fair comparison the values of the SCC measures were scaled (in the case of cMOPSO) to the interval [0,100]. From Table 1 to Table 6 we show the values of the performance measures obtained for each of the algorithms compared.

---

[1] http://www.tik.ee.ethz.ch/pisa/
[2] http://delta.cs.cinvestav.mx/~ccoello/EMOO

5 Discussion of Results

Through the use of binary measures of performance, and under certain conditions, we can conclude that an algorithm is better than another [22]. In this work, since we use two different binary measures, we will conclude that an algorithm is better than another when at least one of the measures indicates so, according to the definitions given in Section 4. Since the conditions to conclude that an algorithm is better that another one using the binary measures are very difficult to satisfy in most cases, we will use the values obtained by the SC binary measure in order to conclude *partial* results: We will say that an algorithm A is *relatively* better than algorithm B when $SC(A,B) > SC(B,A)$, and *almost* better than B when $SC(B,A)=0$ and $SC(A,B)>0.9$. The values of the HV binary measure will be used to make only conclusions of the type: algorithm A is better than algorithm B, just like it was defined in Section 4.

ZDT1. From Table 1, we can conclude that the best results with respect to the SCC measure were obtained by the sMOPSO algorithm. In this case, our proposed approach

Table 1. Comparison of results between our approach (denoted by OMOPSO), NSGA-II [13], SPEA2 [16], MOPSO [2], sMOPSO [3] and cMOPSO [8], for **ZDT1**

		\multicolumn{6}{c}{Test Function ZDT1}
		OMOPSO	NSGA-II	SPEA2	MOPSO	sMOPSO	cMOPSO
SCC	best	82	38	47	0	93	37
	median	43	20	26	0	58	7
	worst	5	9	15	0	23	1
	average	40	21	27	0	**59**	8
	std. dev.	21.5	7.5	8.1	0	24.2	7.9
IGD	best	0.0009	0.0008	0.0006	0.0240	0.0031	0.0016
	median	0.0010	0.0008	0.0007	0.0276	0.0260	0.0029
	worst	0.0013	0.0011	0.0008	0.0385	0.0448	0.0041
	average	0.0010	0.0009	**0.0007**	0.0286	0.0269	0.0030
	std. dev.	0.00008	0.00009	0.00006	0.0040	0.0095	0.0007
\multicolumn{8}{c}{Test Function ZDT1 - Two Set Coverage Measure SC}
$SC(X,$		OMOPSO)	NSGA-II)	SPEA2)	MOPSO)	sMOPSO)	cMOPSO)
OMOPSO		0.00	0.64	0.55	0.96	0.03	0.95
NSGA-II		0.05	0.00	0.22	1.00	0.01	0.99
SPEA2		0.11	0.49	0.00	1.00	0.01	1.00
MOPSO		0.00	0.00	0.00	0.00	0.00	0.00
sMOPSO		0.50	0.69	0.66	0.99	0.00	0.90
cMOPSO		0.00	0.01	0.00	1.00	0.00	0.00
\multicolumn{8}{c}{Test Function ZDT1 - Two Set Hypervolume Measure HV}
$HV(X,$		OMOPSO)	NSGA-II)	SPEA2)	MOPSO)	sMOPSO)	cMOPSO)
OMOPSO		0.000000	-0.001834	-0.001339	-0.323693	0.006828	-0.019389
NSGA-II		0.001304	0.000000	-0.000037	-0.322274	0.007647	-0.016607
SPEA2		0.001302	-0.000534	0.000000	-0.322771	0.007733	-0.017104
MOPSO		0.001719	0.000000	0.000000	0.000000	0.000000	0.000000
sMOPSO		-0.000922	-0.003241	-0.002658	-0.333162	0.000000	-0.020032
cMOPSO		0.000356	0.000000	0.000000	-0.305667	0.007463	0.000000

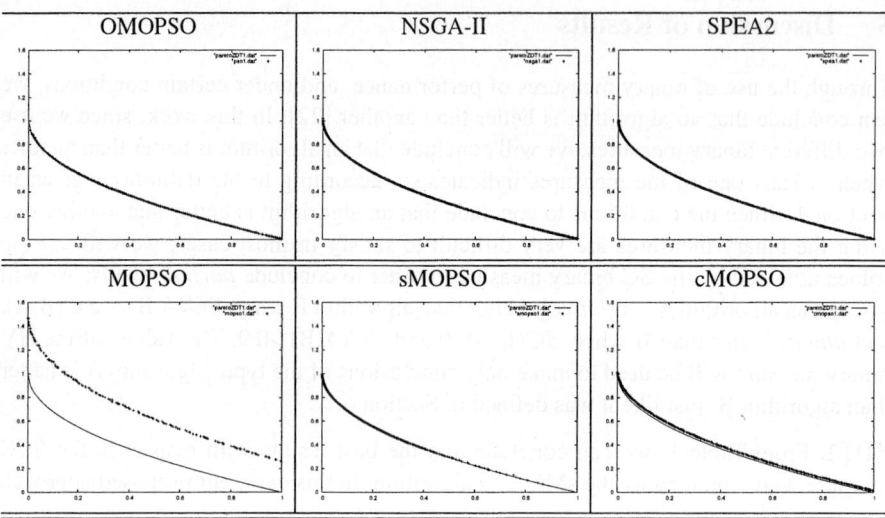

Fig. 2. Pareto fronts obtained by all the approaches for **ZDT1**. Our algorithm is denoted by OMOPSO and, in this case, it used $\epsilon=0.0075$

is the second best. However, the sMOPSO algorithm was unable to generate the complete Pareto front, as we can appreciate in Figure 2. This fact is reflected in the values of the IGD measure in Table 1. With respect to the IGD measure, our approach obtained results as good as those obtained by all the other MOEAs compared, improving the results obtained by the other three PSO-based approaches. Regarding the binary measures (Table 1) and considering both of them, we can conclude that the NSGA-II and SPEA2 are better than cMOPSO and MOPSO. Also, we can conclude that sMOPSO and cMOPSO are better than MOPSO. On the other hand, we can conclude that OMOPSO and sMOPSO are *relatively* better than the rest of the algorithms, in particular, OMOPSO is *almost* better than MOPSO and cMOPSO. Finally, sMOPSO is *relatively* better than OMOPSO. We will now analyze in more detail the results obtained by our algorithm in these measures. We can't conclude that OMOPSO is better than the NSGA-II (for example) since SC(OMOPSO,NSGA-II)$\neq$ 1 and SC(NSGA-II,OMOPSO)$\neq$ 0, but, since SC(OMOPSO,NSGA-II)> SC(NSGA-II,OMOPSO), OMOPSO is *relatively* better than NSGA-II. On the other hand, we have SC(MOPSO , OMOPSO)= 0 and SC(OMOPSO,MOPSO)=0.95, so OMOPSO is *almost* better than MOPSO. Although it should be clear that OMOPSO is better than MOPSO, the results obtained do not allow to reach this conclusion since OMOPSO lost the extreme superior point of the front, as we can see in Figure 2. This is due to the use of the ϵ-dominance scheme to fix the number of solutions in the external archive. This also explains the positive values obtained for the binary hypervolume measure in the column of OMOPSO in Table 1, since the hypervolume corresponding to the front obtained from the union of the MOPSO and OMOPSO fronts is marginally bigger than the hypervolume corresponding to the front of OMOPSO, giving a positive value to the difference in the binary measure. This exemplifies the sort of anomalous behavior that can go undetected even when using binary performance measures.

Table 2. Comparison of results between our approach (denoted by OMOPSO), NSGA-II [13], SPEA2 [16], MOPSO [2], sMOPSO [3] and cMOPSO [8], for **ZDT2**, with respect to the unary measures

		Test Function ZDT2					
		OMOPSO	NSGA-II	SPEA2	MOPSO	sMOPSO	cMOPSO
SCC	best	99	30	34	0	1	94
	median	49	0	0	0	1	0
	worst	0	0	0	0	1	0
	average	**43**	6	7	0	1	29
	std. dev.	34.1	9.8	10.4	0	0	38.9
IGD	best	0.0006	0.0008	0.0007	0.0271	0.0723	0.0030
	median	0.0009	0.0724	0.0723	0.1098	0.0723	0.0723
	worst	0.0303	0.0737	0.0736	0.3525	0.0723	0.0852
	average	**0.0034**	0.0512	0.0404	0.1561	0.0723	0.0680
	std. dev.	0.0078	0.0337	0.0367	0.0952	0.0000	0.0152

Table 3. Comparison of results between our approach (denoted by OMOPSO), NSGA-II [13], SPEA2 [16], MOPSO [2], sMOPSO [3] and cMOPSO [8], for **ZDT2**, with respect to the binary measures

	Test Function ZDT2 - Two Set Coverage Measure SC					
$SC(X,$	OMOPSO)	NSGA-II)	SPEA2)	MOPSO)	sMOPSO)	cMOPSO)
OMOPSO	0.00	0.93	0.94	1.00	0.00	0.21
NSGA-II	0.01	0.00	0.34	1.00	0.00	0.21
SPEA2	0.01	0.21	0.00	1.00	0.00	0.21
MOPSO	0.00	0.00	0.00	0.00	0.00	0.00
sMOPSO	0.01	0.01	0.01	0.44	0.00	0.00
cMOPSO	0.01	0.02	0.02	0.99	0.00	0.00
	Test Function ZDT2 - Two Set Hypervolume Measure HV					
$HV(X,$	OMOPSO)	NSGA-II)	SPEA2)	MOPSO)	sMOPSO)	cMOPSO)
OMOPSO	0.000000	-0.004947	-0.005765	-0.342087	-0.666684 *	-0.036710
NSGA-II	0.000547	0.000000	-0.000493	-0.336593	-0.672178 *	-0.031559
SPEA2	0.000708	0.000486	0.000000	-0.335614	-0.673157 *	-0.030560
MOPSO	0.000000	0.000000	0.000000	0.000000	-0.897843 *	0.000000
sMOPSO	0.000000	0.000000	0.000000	-0.110928	0.000000	0.000000
cMOPSO	0.000126	-0.000217	-0.000197	-0.305251	-0.703520 *	0.000000

ZDT2. From Table 2, we can conclude that our algorithm (OMOPSO) obtained the best results in both unary measures, with the largest number of points (on average) belonging to the true Pareto front and the minimum IGD (on average). Regarding the binary measures (considering both of them) (Table 3), we can conclude that OMOPSO, NSGA-II, SPEA2 and cMOPSO are better than MOPSO and sMOPSO. Also, we can say that OMOPSO is *almost* better than NSGA-II, SPEA2 and cMOPSO. We can see in the SC binary measure values that almost 80% of the points of the cMOPSO algorithm are concentrated on the top part of the true Pareto front. Thus, although the major part of the observed front of the cMOPSO algorithm (see Figure 3) is not on the true Pareto front, the

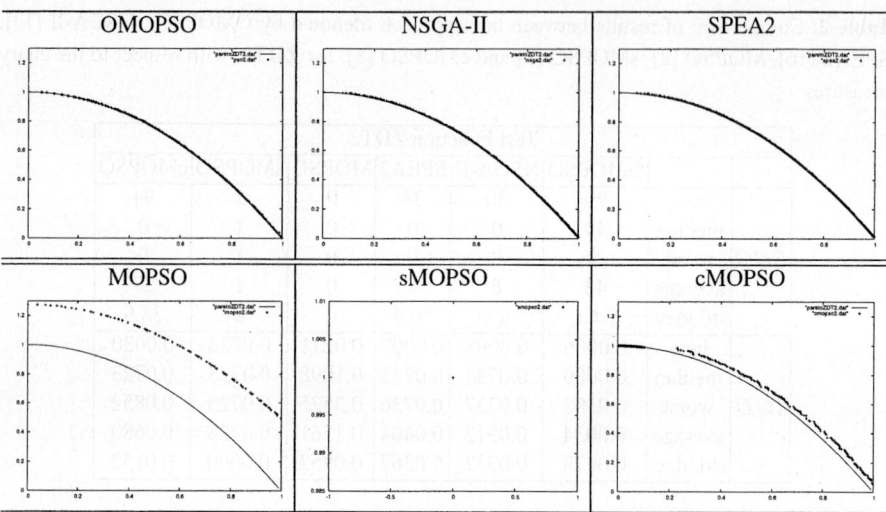

Fig. 3. Pareto fronts obtained by all the approaches for **ZDT2**. Our algorithm is denoted by OMOPSO and, in this case, it used $\epsilon=0.0075$

corresponding results on the SC measure are not as expected. Additionally, the sMOPSO algorithm obtained just one point: (0.0,1.0). None of the other algorithms were able to generate this point, as we can see in the SC measure values from Table 3. For example, in this case, our algorithm (OMOPSO) preserved two extreme points: (0.0, 1.0005) and $(5 \times 10^{-10}, 1.0)$ (although they are not visible in Figure 3). These two points let the

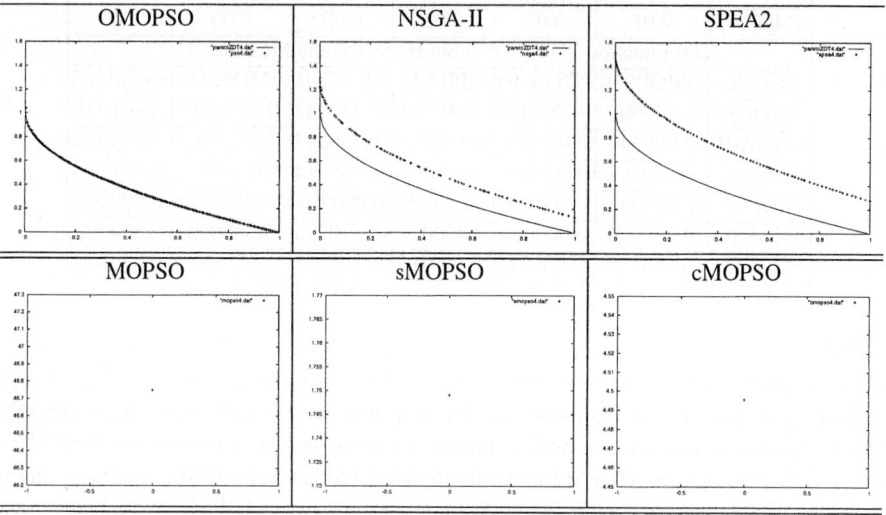

Fig. 4. Pareto fronts obtained by all the approaches for **ZDT4**. Our algorithm is denoted by OMOPSO and, in this case, it used $\epsilon=0.0075$

Table 4. Comparison of results between our approach (denoted by OMOPSO), NSGA-II [13], SPEA2 [16], MOPSO [2], sMOPSO [3] and cMOPSO [8], for **ZDT4**

		Test Function ZDT4					
		OMOPSO	NSGA-II	SPEA2	MOPSO	sMOPSO	cMOPSO
SCC	best	98	0	0	0	0	0
	median	88	0	0	0	0	0
	worst	0	0	0	0	0	0
	average	**77**	0	0	0	0	0
	std. dev.	26.2	0	0	0	0	0
IGD	best	0.0009	0.0126	0.0256	4.6415	0.1541	0.4203
	median	0.0009	0.1317	0.0811	12.407	0.7393	1.6404
	worst	0.0432	0.3219	0.3464	15.250	1.2865	4.1864
	average	**0.0030**	0.1508	0.1224	9.9195	0.7591	1.8621
	std. dev.	0.0095	0.0973	0.0943	4.0106	0.3147	0.9357
		Test Function ZDT4 - Two Set Coverage Measure SC					
$SC(X,$		OMOPSO)	NSGA-II)	SPEA2)	MOPSO)	sMOPSO)	cMOPSO)
OMOPSO		0.00	0.92	0.94	0.00	0.00	0.00
NSGA-II		0.00	0.00	1.00	1.00	1.00	1.00
SPEA2		0.00	0.00	0.00	1.00	1.00	1.00
MOPSO		0.00	0.00	0.00	0.00	0.00	0.00
sMOPSO		0.00	0.00	0.00	1.00	0.00	1.00
cMOPSO		0.00	0.00	0.00	1.00	0.00	0.00
		Test Function ZDT4 - Two Set Hypervolume Measure HV					
$HV(X,$		OMOPSO)	NSGA-II)	SPEA2)	MOPSO)	sMOPSO)	cMOPSO)
OMOPSO		0.000000	-0.164026	-0.343057	-0.333325 *	-0.333325 *	-0.333325 *
NSGA-II		0.000574	0.000000	-0.179995	-0.497925 *	-0.497925 *	-0.497925 *
SPEA2		0.001538	0.000000	0.000000	-0.677920 *	-0.677920 *	-0.677920 *
MOPSO		0.000000	0.000000	0.000000	0.000000	0.000000	0.000000
sMOPSO		0.000000	0.000000	0.000000	0.000000	0.000000	0.000000
cMOPSO		0.000000	0.000000	0.000000	0.000000	0.000000	0.000000

front obtained by OMOPSO to completely dominate the front obtained by MOPSO, but not the front obtained by sMOPSO. Fortunately, these problems with the SC measure are overcome by the HV measure with a small modification: the values that we have marked with an asterisk (*) in Table 3 were originally positive. However, we changed them to correspond more closely with reality, since the hypervolume corresponding to the front of sMOPSO is zero.

ZDT4. Based on the results shown in Table 4, we can conclude that our OMOPSO obtained the best results with respect to the two unary measures adopted, with the largest number of points (on average) belonging to the true Pareto front and the minimum IGD (on average). Regarding the binary measures and considering both of them (see Table 4), we can conclude that OMOPSO, NSGA-II and SPEA2 are better than the other three PSO-based approaches. Also, we can say that the NSGA-II is better than SPEA2, and that OMOPSO is *almost* better than NSGA-II and SPEA2. In this case, our approach is only *almost* better than the NSGA-II and SPEA2 for the same reason that we discussed in the case of function ZDT1. As we can see in Figure 4, OMOPSO lost the top extreme

Table 5. Comparison of results between our approach (denoted by OMOPSO), NSGA-II [13], SPEA2 [16], MOPSO [2], sMOPSO [3] and cMOPSO [8], for **DTLZ6**, with respect to the unary measures

		Test Function DTLZ6					
		OMOPSO	NSGA-II	SPEA2	MOPSO	sMOPSO	cMOPSO
SCC	best	93	1	2	0	1	0
	median	23	0	0	0	1	0
	worst	0	0	0	0	1	0
	average	**32**	0.1	0.6	0	1	0
	std. dev.	30.9	0.3	0.9	0	0	0
IGD	best	0.0024	0.0064	0.0037	0.0375	0.0673	0.0110
	median	0.0029	0.0088	0.0045	0.0583	0.0673	0.0345
	worst	0.0213	0.0314	0.0214	0.1185	0.0673	0.0742
	average	**0.0065**	0.0132	0.0067	0.0658	0.0673	0.0373
	std. dev.	0.0060	0.0083	0.0051	0.0205	0.0000	0.0172

point of the Pareto front due to the use of the ϵ-dominance scheme. For this reason, OMOPSO can't dominate completely the fronts produced by NSGA-II and SPEA2. In fact, it can't even dominate the isolated points obtained by the other PSO-based approaches. Additionally, for this same reason we find positive values in the column of OMOPSO for the binary hypervolume measure. However, the binary hypervolume measure lead us to conclude the superiority of OMOPSO compared with the other PSO-based approaches. It is very important to note that our algorithm was the only PSO-based approach that was able to generate the entire Pareto front of this function. This illustrates

Table 6. Comparison of results between our approach (denoted by OMOPSO), NSGA-II [13], SPEA2 [16], MOPSO [2], sMOPSO [3] and cMOPSO [8], for **DTLZ6**, with respect to the binary measures

Test Function DTLZ6 - Two Set Coverage Measure SC						
$SC(X,$	OMOPSO)	NSGA-II)	SPEA2)	MOPSO)	sMOPSO)	cMOPSO)
OMOPSO	0.00	0.68	0.64	0.92	0.00	0.57
NSGA-II	0.01	0.00	0.31	1.00	0.00	0.80
SPEA2	0.01	0.30	0.00	0.98	0.00	0.57
MOPSO	0.00	0.00	0.00	0.00	0.00	0.00
sMOPSO	0.00	0.07	0.12	0.45	0.00	0.24
cMOPSO	0.00	0.04	0.12	1.00	0.00	0.00
Test Function DTLZ6 - Two Set Hypervolume Measure HV						
$HV(X,$	OMOPSO)	NSGA-II)	SPEA2)	MOPSO)	sMOPSO)	cMOPSO)
OMOPSO	0.000000	-0.109957	-0.095828	-0.333723	-3.606059*	-0.629977
NSGA-II	0.109963	0.000000	0.001996	-0.113803	-3.825408*	-0.410346
SPEA2	0.141673	0.019577	0.000000	-0.096222	-3.843163*	-0.392616
MOPSO	0.000000	0.000000	0.000000	0.000000	-3.880365*	0.000000
sMOPSO	0.000000	0.000572*	0.000398*	0.059418*	0.000000	0.001163*
cMOPSO	0.000180	-0.000109	0.000040	0.296434	-4.235054*	0.000000

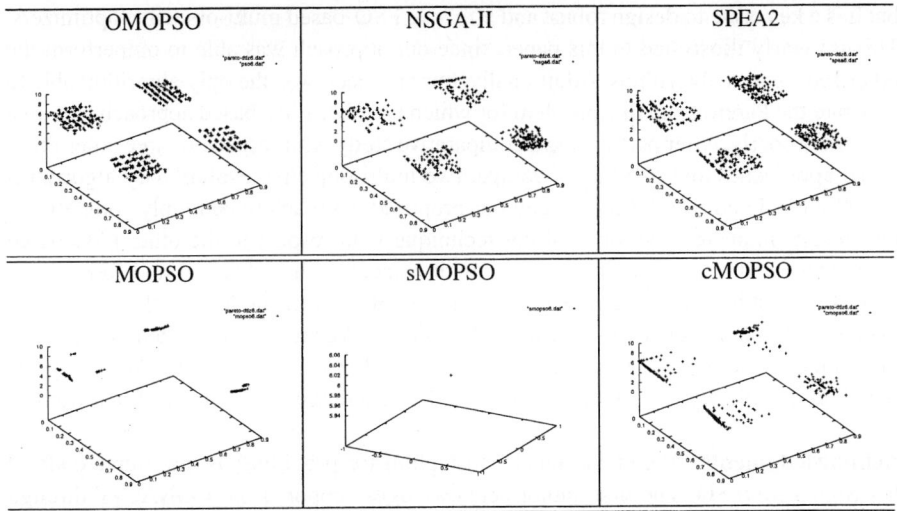

Fig. 5. Pareto fronts obtained by all the approaches for **DTLZ6**. Our algorithm is denoted by OMOPSO and, in this case, it used $\epsilon=0.05$

the effectiveness of the mechanisms adopted in our approach to maintain diversity and to select and filter out leaders.

DTLZ6. From Table 5, we can conclude that our algorithm (OMOPSO) obtained the best results with respect to the two unary measures adopted, with the largest number of points (on average) belonging to the true Pareto front and the minimum IGD (on average). Regarding the binary measures and considering both of them (see Table 6), we can conclude that OMOPSO, NSGA-II, SPEA2 and cMOPSO are better than MOPSO, and that OMOPSO is better than sMOPSO. Also, OMOPSO is *relatively* better than NSGA-II, SPEA2 and cMOPSO. We can see the Pareto fronts obtained for this function in Figure 5.

Overall Discussion. With respect to the unary performance measures, our approach obtained the best results in all functions, except for ZDT1. Thus, this indicates that OMOPSO was able to obtain a good approximation and a good number of points of the true Pareto of all the test functions used in this paper. Regarding the binary measures, although in function ZDT1 our approach was relatively outperformed by sMOPSO, in general OMOPSO was clearly superior compared with the other PSO-based approaches adopted in our comparative study. Also, the results obtained by OMOPSO showed that it is highly competitive with respect to both NSGA-II and SPEA2.

6 Conclusions and Future Work

We have proposed a new multi-objective particle swarm optimizer which uses Pareto dominance and a crowding-based selection mechanism to identify the leaders to be removed when there are too many of them. The selection of such in-excess leaders has been a topic often disregarded in the literature of multi-objective particle swarm optimizers, but it is a key issue to design robust and effective PSO-based multi-objective optimizers. This is clearly illustrated in this paper, since our approach was able to outperform the other PSO-based algorithms. Additionally, our approach was the only algorithm able to generate the Pareto front of a problem for which no other PSO-based approach was able to work properly. After performing a comparative study with respect to three other PSO-based approaches and two highly competitive multi-objective evolutionary algorithms (the NSGA-II and SPEA2), we found our proposed approach to be highly competitive. Our results indicate superiority of our technique with respect to the other PSO-based approaches and a very similar behavior with respect to the NSGA-II and SPEA2. As part of our future work, we intend to fix the problem with the loss of the extrema of the Pareto fronts caused by the use of $\epsilon-$dominance. We are also interested in exploring mechanisms that can accelerate convergence and that allow our approach to stop the search automatically (i.e., without having to define a maximum number of iterations).

Acknowledgments. We thank Sanaz Mostaghim for providing us the source code of her Sigma-MOPSO. The first author acknowledges support from CONACyT through a scholarship to pursue graduate studies at CINVESTAV-IPN. The second author acknowledges support from CONACyT project number 42435-Y.

References

1. Kennedy, J., Eberhart, R.C.: Particle Swarm Optimization. In: Proceedings of the 1995 IEEE International Conference on Neural Networks, Piscataway, New Jersey, IEEE Service Center (1995) 1942–1948
2. Coello Coello, C.A., Toscano Pulido, G., Salazar Lechuga, M.: Handling Multiple Objectives With Particle Swarm Optimization. IEEE Transactions on Evolutionary Computation **8** (2004) 256–279
3. Mostaghim, S., Teich, J.: Strategies for Finding Good Local Guides in Multi-objective Particle Swarm Optimization (MOPSO). In: 2003 IEEE Swarm Intelligence Symposium Proceedings, Indianapolis, Indiana, USA, IEEE Service Center (2003) 26–33
4. Ray, T., Liew, K.: A Swarm Metaphor for Multiobjective Design Optimization. Engineering Optimization **34** (2002) 141–153
5. Hu, X., Eberhart, R.: Multiobjective Optimization Using Dynamic Neighborhood Particle Swarm Optimization. In: Congress on Evolutionary Computation (CEC'2002). Volume 2., Piscataway, New Jersey, IEEE Service Center (2002) 1677–1681
6. Hu, X., Eberhart, R.C., Shi, Y.: Particle Swarm with Extended Memory for Multiobjective Optimization. In: 2003 IEEE Swarm Intelligence Symposium Proceedings, Indianapolis, Indiana, USA, IEEE Service Center (2003) 193–197
7. Fieldsend, J.E., Singh, S.: A Multi-Objective Algorithm based upon Particle Swarm Optimisation, an Efficient Data Structure and Turbulence. In: Proceedings of the 2002 U.K. Workshop on Computational Intelligence, Birmingham, UK (2002) 37–44
8. Toscano Pulido, G., Coello Coello, C.A.: Using Clustering Techniques to Improve the Performance of a Particle Swarm Optimizer. In et al., K.D., ed.: Proceedings of the 2004 Genetic and Evolutionary Computation Conference. Part I, Seattle, Washington, USA, Springer-Verlag, Lecture Notes in Computer Science Vol. 3102 (2004) 225–237
9. Mostaghim, S., Teich, J.: The role of ε-dominance in multi objective particle swarm optimization methods. In: Proceedings of the 2003 Congress on Evolutionary Computation (CEC'2003). Volume 3., Canberra, Australia, IEEE Press (2003) 1764–1771

10. Laumanns, M., Thiele, L., Deb, K., Zitzler, E.: Combining Convergence and Diversity in Evolutionary Multi-objective Optimization. Evolutionary Computation **10** (2002) 263–282
11. Mostaghim, S., Teich, J.: Covering pareto-optimal fronts by subswarms in multi-objective particle swarm optimization. In: 2004 Congress on Evolutionary Computation (CEC'2004). Volume 2., Portland, Oregon, USA, IEEE Service Center (2004) 1404–1411
12. Li, X.: A Non-dominated Sorting Particle Swarm Optimizer for Multiobjective Optimization. In et al., E.C.P., ed.: Genetic and Evolutionary Computation—GECCO 2003. Proceedings, Part I, Springer. Lecture Notes in Computer Science Vol. 2723 (2003) 37–48
13. Deb, K., Pratap, A., Agarwal, S., Meyarivan, T.: A Fast and Elitist Multiobjective Genetic Algorithm: NSGA–II. IEEE Transactions on Evolutionary Computation **6** (2002) 182–197
14. Li, X.: Better Spread and Convergence: Particle Swarm Multiobjective Optimization Using the Maximin Fitness Function. In et al., K.D., ed.: Proceedings of the 2004 Genetic and Evolutionary Computation Conference. Part I, Seattle, Washington, USA, Springer-Verlag, Lecture Notes in Computer Science Vol. 3102 (2004) 117–128
15. Bäck, T., ed.: Evolutionary Algorithms in Theory and Practice. Oxford University Press, New York (1996)
16. Zitzler, E., Laumanns, M., Thiele, L.: SPEA2: Improving the Strength Pareto Evolutionary Algorithm for Multiobjective Optimization. In et al., K.G., ed.: Proceedings of the EUROGEN2001 Conference, Barcelona, Spain, CIMNE (2002) 95–100
17. Zitzler, E., Thiele, L.: Multiobjective Optimization Using Evolutionary Algorithms—A Comparative Study. In Eiben, A.E., ed.: Parallel Problem Solving from Nature V, Amsterdam, Springer-Verlag (1998) 292–301
18. Zitzler, E.: Evolutionary Algorithms for Multiobjective Optimization: Methods and Applications. PhD thesis, Swiss Federal Institute of Technology (ETH), Zurich, Switzerland (1999)
19. Reyes-Sierra, M., Coello, C.A.C.: A New Multi-Objective Particle Swarm Optimizer with Improved Selection and Diversity Mechanisms. Technical Report EVOCINV-05-2004, Sección de Computación, Depto. de Ingeniería Eléctrica, CINVESTAV-IPN, México (2004)
20. Zitzler, E., Deb, K., Thiele, L.: Comparison of Multiobjective Evolutionary Algorithms: Empirical Results. Evolutionary Computation **8** (2000) 173–195
21. Deb, K., Thiele, L., Laumanns, M., Zitzler, E.: Scalable Multi-Objective Optimization Test Problems. In: Congress on Evolutionary Computation (CEC'2002). Volume 1., Piscataway, New Jersey, IEEE Service Center (2002) 825–830
22. Zitzler, E., Thiele, L., Laumanns, M., Fonseca, C.M., da Fonseca, V.G.: Performance Assessment of Multiobjective Optimizers: An Analysis and Review. IEEE Transactions on Evolutionary Computation **7** (2003) 117–132

DEMO: Differential Evolution for Multiobjective Optimization

Tea Robič and Bogdan Filipič

Department of Intelligent Systems, Jožef Stefan Institute,
Jamova 39, SI-1000 Ljubljana, Slovenia
tea.robic@ijs.si
bogdan.filipic@ijs.si

Abstract. Differential Evolution (DE) is a simple but powerful evolutionary optimization algorithm with many successful applications. In this paper we propose Differential Evolution for Multiobjective Optimization (DEMO) – a new approach to multiobjective optimization based on DE. DEMO combines the advantages of DE with the mechanisms of Pareto-based ranking and crowding distance sorting, used by state-of-the-art evolutionary algorithms for multiobjective optimization. DEMO is implemented in three variants that achieve competitive results on five ZDT test problems.

1 Introduction

Many real-world optimization problems involve optimization of several (conflicting) criteria. Since multiobjective optimization searches for an optimal vector, not just a single value, one solution often cannot be said to be better than another and there exists not only a single optimal solution, but a set of optimal solutions, called the *Pareto front*. Consequently, there are two goals in multiobjective optimization: (i) to discover solutions as close to the Pareto front as possible, and (ii) to find solutions as diverse as possible in the obtained nondominated front. Satisfying these two goals is a challenging task for any algorithm for multiobjective optimization.

In recent years, many algorithms for multiobjective optimization have been introduced. Most originate in the field of Evolutionary Algorithms (EAs) – the so-called Multiobjective Optimization EAs (MOEAs). Among these, the NSGA-II by Deb et al. [1] and SPEA2 by Zitzler et al. [2] are the most popular. MOEAs take the strong points of EAs and apply them to Multiobjective Optimization Problems (MOPs). A particular EA that has been used for multiobjective optimization is Differential Evolution (DE). DE is a simple yet powerful evolutionary algorithm by Price and Storn [3] that has been successfully used in solving single-objective optimization problems [4]. Hence, several researchers have tried to extend it to handle MOPs.

Abbass [5, 6] was the first to apply DE to MOPs in the so-called Pareto Differential Evolution (PDE) algorithm. This approach employs DE to create new individuals and then keeps only the nondominated ones as the basis for the next generation. PDE was compared to SPEA [7] (the predecessor of SPEA2) on two test problems and found to outperform it.

Madavan [8] achieved good results with the Pareto Differential Evolution Approach (PDEA[1]). Like PDE, PDEA applies DE to create new individuals. It then combines both populations and calculates the nondominated rank (with Pareto-based ranking assignment) and diversity rank (with the crowding distance metric) for each individual. Two variants of PDEA were investigated. The first compares each child with its parent. The child replaced the parent if it had a higher nondominated rank or, if it had the same nondominated rank and a higher diversity rank. Otherwise the child is discarded. This variant was found inefficient – the diversity was good but the convergence slow. The other variant simply takes the best individuals according to the nondominated rank and diversity rank (like in NSGA-II). The latter variant has proved to be very efficient and was applied to several MOPs, where it produced favorable results.

Xue [9] introduced Multiobjective Differential Evolution (MODE). This algorithm also uses the Pareto-based ranking assignment and the crowding distance metric, but in a different manner than PDEA. In MODE the fitness of an individual is first calculated using Pareto-based ranking and then reduced with respect to the individual's crowding distance value. This single fitness value is then used to select the best individuals for the new population. MODE was tested on five benchmark problems where it produced better results than SPEA.

In this paper, we propose a new way of extending DE to be suitable for solving MOPs. We call it DEMO (Differential Evolution for Multiobjective Optimization). Although similar to the existing algorithms (especially PDEA), our implementation differs from others and represents a novel approach to multiobjective optimization. DEMO is implemented in three variants (DEMO/parent, DEMO/closest/dec and DEMO/closest/obj). Because of diverse recommendations for the crossover probability, three different values for this parameter are investigated. From the simulation results on five test problems we find that DEMO efficiently achieves the two goals of multiobjective optimization, i.e. the convergence to the true Pareto front and uniform spread of individuals along the front. Moreover, DEMO achieves very good results on the test problem ZDT4 that poses many difficulties to state-of-the-art algorithms for multiobjective optimization.

The rest of the paper is organized as follows. In Section 2 we describe the DE scheme that was used as a base for DEMO. Thereafter, in Section 3, we present DEMO in its three variants. Section 4 outlines the applied test problems and performance measures, and states the results. Further comparison and discussion of the results are provided in Section 5. The paper concludes with Section 6.

---

[1] This acronym was not used by Madavan. We introduce it to make clear distinction between his approach and other implementations of DE for multiobjective optimization.

Differential Evolution
1. Evaluate the initial population $\mathcal{P}$ of random individuals.
2. While stopping criterion not met, do:
   2.1. For each individual P_i ($i = 1, \ldots, popSize$) from $\mathcal{P}$ repeat:
       (a) Create candidate C from parent P_i.
       (b) Evaluate the candidate.
       (c) If the candidate is better than the parent, the candidate replaces the parent. Otherwise, the candidate is discarded.
   2.2. Randomly enumerate the individuals in $\mathcal{P}$.

Fig. 1. Outline of DE's main procedure

Candidate creation
Input: Parent P_i
1. Randomly select three individuals $P_{i_1}, P_{i_2}, P_{i_3}$ from $\mathcal{P}$, where i, i_1, i_2 and i_3 are pairwise different.
2. Calculate candidate C as $C = P_{i_1} + F \cdot (P_{i_2} - P_{i_3})$, where F is a scaling factor.
3. Modify the candidate by binary crossover with the parent using crossover probability $crossProb$.
Output: Candidate C

Fig. 2. Outline of the candidate creation in scheme *DE/rand/1/bin*

2 Differential Evolution

DE is a simple evolutionary algorithm that creates new candidate solutions by combining the parent individual and several other individuals of the same population. A candidate replaces the parent only if it has better fitness. This is a rather greedy selection scheme that often outperforms traditional EAs.

The DE algorithm in pseudo-code is shown in Fig. 1. Many variants of creation of a candidate are possible. We use the DE scheme DE/rand/1/bin described in Fig. 2 (more details on this and other DE schemes can be found in [10]). Sometimes, the newly created candidate falls out of bounds of the variable space. In such cases, many approaches of constraint handling are possible. We address this problem by simply replacing the candidate value violating the boundary constraints with the closest boundary value. In this way, the candidate becomes feasible by making as few alterations to it as possible. Moreover, this approach does not require the construction of a new candidate.

3 Differential Evolution for Multiobjective Optimization

When applying DE to MOPs, we face many difficulties. Besides preserving a uniformly spread front of nondominated solutions, which is a challenging task for any MOEA, we have to deal with another question, that is, when to replace the parent with the candidate solution. In single-objective optimization, the decision

Differential Evolution for Multiobjective Optimization
1. Evaluate the initial population $\mathcal{P}$ of random individuals.
2. While stopping criterion not met, do:
   2.1. For each individual P_i ($i = 1, \ldots, popSize$) from $\mathcal{P}$ repeat:
      (a) Create candidate C from parent P_i.
      (b) Evaluate the candidate.
      (c) If the candidate dominates the parent, the candidate replaces the parent.
          If the parent dominates the candidate, the candidate is discarded.
          Otherwise, the candidate is added in the population.
   2.2. If the population has more than $popSize$ individuals, truncate it.
   2.3. Randomly enumerate the individuals in $\mathcal{P}$.

Fig. 3. Outline of DEMO/parent

is easy: the candidate replaces the parent only when the candidate is better than the parent. In MOPs, on the other hand, the decision is not so straightforward. We could use the concept of dominance (the candidate replaces the parent only if it dominates it), but this would make the greedy selection scheme of DE even greedier. Therefore, DEMO applies the following principle (see Fig. 3). The candidate replaces the parent if it dominates it. If the parent dominates the candidate, the candidate is discarded. Otherwise (when the candidate and parent are nondominated with regard to each other), the candidate is added to the population. This step is repeated until $popSize$ number of candidates are created. After that, we get a population of the size between $popSize$ and $2 \cdot popSize$. If the population has enlarged, we have to truncate it to prepare it for the next step of the algorithm.

The truncation consists of sorting the individuals with nondominated sorting and then evaluating the individuals of the same front with the crowding distance metric. The truncation procedure keeps in the population only the best $popSize$ individuals (with regard to these two metrics). The described truncation is derived from NSGA-II and is also used in PDEA's second variant.

DEMO incorporates two crucial mechanisms. The immediate replacement of the parent individual with the candidate that dominates it, is the core of DEMO. The newly created candidates that enter the population (either by replacement or by addition) instantly take part in the creation of the following candidates. This emphasizes elitism within reproduction, which helps achieving the first goal of multiobjective optimization – convergence to the true Pareto front. The second mechanism is the use of nondominated sorting and crowding distance metric in truncation of the extended population. Besides preserving elitism, this mechanism stimulates the uniform spread of solutions. This is needed to achieve the second goal – finding as diverse nondominated solutions as possible. DEMO's selection scheme thus efficiently pursues both goals of multiobjective optimization.

The described DEMO's procedure (outlined in Fig. 3) is the most elementary of the three variants presented in this paper. It is called DEMO/parent.

The other two variants were inspired by the concept of Crowding DE, recently introduced by Thomsen in [11].

Thomsen applied crowding-based DE in optimization of multimodal functions. When optimizing functions with many optima, we would sometimes like not only to find one optimal point, but also discover and maintain multiple optima in a single algorithm run. For this purpose, Crowding DE can be used. Crowding DE is basically conventional DE with one important difference. Usually, the candidate is compared to its parent. In Crowding DE, the candidate is compared to the most similar individual in the population. The applied similarity measure is the Euclidean distance between two solutions.

Crowding DE was tested on numerous benchmark problems and its performance was impressive. Because the goal of maintaining multiple optima is similar to the second goal of multiobjective optimization (maintaining diverse solutions in the front), we implemented the idea of Crowding DE in two additional variants of the DEMO algorithm. The second, DEMO/closest/dec, works in the same way as DEMO/parent, with the exception that the candidate solution is compared to the most similar individual in decision space. If it dominates it, the candidate replaces this individual, otherwise it is treated in the same way as in DEMO/parent. The applied similarity measure is the Euclidean distance between the two solutions in decision space. In the third variant, DEMO/closest/obj, the candidate is compared to the most similar individual in objective space.

DEMO/closest/dec and DEMO/closest/obj need more time for one step of the procedure than DEMO/parent. This is because at every step they have to search for the most similar individual in the decision and objective space, respectively. Although the additional computational complexity is notable when operating on high-dimensional spaces, it is negligible in real-world problems where the solution evaluation is the most time consuming task.

4 Evaluation and Results

4.1 Test Problems

We analyze the performance of the three variants of DEMO on five ZDT test problems (introduced in [12]) that were frequently used as benchmark problems in the literature [1, 8, 9]. These problems are described in detail in Tables 1 and 2.

4.2 Performance Measures

Different performance measures for evaluating the efficiency of MOEAs have been suggested in the literature. Because we wanted to compare DEMO to other MOEAs (especially the ones that use DE) on their published results, we use three metrics that have been used in these studies.

For all three metrics, we need to know the true Pareto front for a problem. Since we are dealing with artificial test problems, the true Pareto front is not

Table 1. Description of the test problems ZDT1, ZDT2 and ZDT3

ZDT1	
Decision space	$\mathbf{x} \in [0,1]^{30}$
Objective functions	$f_1(\mathbf{x}) = x_1$
	$f_2(\mathbf{x}) = g(\mathbf{x}) \left(1 - \sqrt{x_1/g(\mathbf{x})}\right)$
	$g(\mathbf{x}) = 1 + \frac{9}{n-1} \sum_{i=2}^{n} x_i$
Optimal solutions	$0 \le x_1^* \le 1$ and $x_i^* = 0$ for $i = 2, \ldots, 30$
Characteristics	convex Pareto front

ZDT2	
Decision space	$\mathbf{x} \in [0,1]^{30}$
Objective functions	$f_1(\mathbf{x}) = x_1$
	$f_2(\mathbf{x}) = g(\mathbf{x}) \left(1 - (x_1/g(\mathbf{x}))^2\right)$
	$g(\mathbf{x}) = 1 + \frac{9}{n-1} \sum_{i=2}^{n} x_i$
Optimal solutions	$0 \le x_1^* \le 1$ and $x_i^* = 0$ for $i = 2, \ldots, 30$
Characteristics	nonconvex Pareto front

ZDT3	
Decision space	$\mathbf{x} \in [0,1]^{30}$
Objective functions	$f_1(\mathbf{x}) = x_1$
	$f_2(\mathbf{x}) = g(\mathbf{x}) \left(1 - \sqrt{x_1/g(\mathbf{x})} - \frac{x_1}{g(\mathbf{x})} \sin(10\pi x_1)\right)$
	$g(\mathbf{x}) = 1 + \frac{9}{n-1} \sum_{i=2}^{n} x_i$
Optimal solutions	$0 \le x_1^* \le 1$ and $x_i^* = 0$ for $i = 2, \ldots, 30$
Characteristics	discontinuous Pareto front

hard to obtain. In our experiments we use 500 uniformly spaced Pareto-optimal solutions as the approximation of the true Pareto front[2].

The first metric is *Convergence metric* Υ. It measures the distance between the obtained nondominated front Q and the set P^* of Pareto-optimal solutions:

$$\Upsilon = \frac{\sum_{i=1}^{|Q|} d_i}{|Q|},$$

where d_i is the Euclidean distance (in the objective space) between the solution $i \in Q$ and the nearest member of P^*.

Instead of the convergence metric, some researchers use a very similar metric, called *Generational Distance GD*. This metric measures the distance between

---

[2] These solutions are uniformly spread in the decision space. They were made available online by Simon Huband at http://www.scis.ecu.edu.au/research/wfg/datafiles.html.

Table 2. Description of the test problems ZDT4 and ZDT6

ZDT4	
Decision space	$\mathbf{x} \in [0,1] \times [-5,5]^9$
Objective functions	$f_1(\mathbf{x}) = x_1$
	$f_2(\mathbf{x}) = g(\mathbf{x})\left(1 - (x_i/g(\mathbf{x}))^2\right)$
	$g(\mathbf{x}) = 1 + 10(n-1) + \sum_{i=2}^{n}\left(x_i^2 - 10\cos(4\pi x_i)\right)$
Optimal solutions	$0 \leq \mathbf{x}_1^* \leq 1$ and $\mathbf{x}_i^* = 0$ for $i = 2, \ldots, 10$
Characteristics	many local Pareto fronts

ZDT6	
Decision space	$\mathbf{x} \in [0,1]^{10}$
Objective functions	$f_1(\mathbf{x}) = 1 - \exp^{-4x_1} \sin(6\pi x_1)^6$
	$f_2(\mathbf{x}) = g(\mathbf{x})\left(1 - (f_1(\mathbf{x})/g(\mathbf{x}))^2\right)$
	$g(\mathbf{x}) = 1 + \frac{9}{n-1}\sum_{i=2}^{n} x_i$
Optimal solutions	$0 \leq \mathbf{x}_1^* \leq 1$ and $\mathbf{x}_i^* = 0$ for $i = 2, \ldots, 10$
Characteristics	low density of solutions near Pareto front

the obtained nondominated front Q and the set P^* of Pareto-optimal solutions as

$$GD = \frac{\sqrt{\sum_{i=1}^{|Q|} d_i^2}}{|Q|},$$

where d_i is again the Euclidean distance (in the objective space) between the solution $i \in Q$ and the nearest member of P^*.

The third metric is *Diversity metric* Δ. This metric measures the extent of spread achieved among the nondominated solutions:

$$\Delta = \frac{d_f + d_l + \sum_{i=1}^{|Q|-1} |d_i - \overline{d}|}{d_f + d_l + (|Q|-1)\overline{d}},$$

where d_i is the Euclidean distance (in the objective space) between consecutive solutions in the obtained nondominated front Q, and $\overline{d}$ is the average of these distances. The parameters d_f and d_l represent the Euclidean distances between the extreme solutions of the Pareto front P^* and the boundary solutions of the obtained front Q.

4.3 Experiments

In addition to investigating the performance of the three DEMO variants, we were also interested in observing the effect of the crossover probability (*crossProb* in Fig. 2) on the efficiency of DEMO. Therefore, we made the following experiments: for every DEMO variant, we run the respective DEMO algorithm with crossover probabilities set to 30%, 60% and 90%. We repeated all tests 10 times

Table 3. Statistics of the results on test problems ZDT1, ZDT2 and ZDT3

Algorithm	ZDT1	
	Convergence metric	Diversity metric
NSGA-II (real-coded)	0.033482±0.004750	0.390307±0.001876
NSGA-II (binary-coded)	0.000894±0.000000	0.463292±0.041622
SPEA	0.001799±0.000001	0.784525±0.004440
PAES	0.082085±0.008679	1.229794±0.004839
PDEA	N/A	0.298567±0.000742
MODE	0.005800±0.000000	N/A
DEMO/parent	0.001083±0.000113	0.325237±0.030249
DEMO/closest/dec	0.001113±0.000134	0.319230±0.031350
DEMO/closest/obj	0.001132±0.000136	0.306770±0.025465

Algorithm	ZDT2	
	Convergence metric	Diversity metric
NSGA-II (real-coded)	0.072391±0.031689	0.430776±0.004721
NSGA-II (binary-coded)	0.000824±0.000000	0.435112±0.024607
SPEA	0.001339±0.000000	0.755148±0.004521
PAES	0.126276±0.036877	1.165942±0.007682
PDEA	N/A	0.317958±0.001389
MODE	0.005500±0.000000	N/A
DEMO/parent	0.000755±0.000045	0.329151±0.032408
DEMO/closest/dec	0.000820±0.000042	0.335178±0.016985
DEMO/closest/obj	0.000780±0.000035	0.326821±0.021083

Algorithm	ZDT3	
	Convergence metric	Diversity metric
NSGA-II (real-coded)	0.114500±0.007940	0.738540±0.019706
NSGA-II (binary-coded)	0.043411±0.000042	0.575606±0.005078
SPEA	0.047517±0.000047	0.672938±0.003587
PAES	0.023872±0.000010	0.789920±0.001653
PDEA	N/A	0.623812±0.000225
MODE	0.021560±0.000000	N/A
DEMO/parent	0.001178±0.000059	0.309436±0.018603
DEMO/closest/dec	0.001197±0.000091	0.324934±0.029648
DEMO/closest/obj	0.001236±0.000091	0.328873±0.019142

with different initial populations. The scaling factor F was set to 0.5 and was not tuned. To match the settings of the algorithms used for comparison, the population size was set 100 and the algorithm was run for 250 generations.

4.4 Results

Tables 3 and 4 present the mean and variance of the values of the convergence and diversity metric, averaged over 10 runs. We provide the results for all three DEMO variants. Results of other algorithms are taken from the literature (see

Table 4. Statistics of the results on test problems ZDT4 and ZDT6

	ZDT4	
Algorithm	Convergence metric	Diversity metric
NSGA-II (real-coded)	0.513053±0.118460	0.702612±0.064648
NSGA-II (binary-coded)	3.227636±7.307630	0.479475±0.009841
SPEA	7.340299±6.572516	0.798463±0.014616
PAES	0.854816±0.527238	0.870458±0.101399
PDEA	N/A	0.840852±0.035741
MODE	0.638950±0.500200	N/A
DEMO/parent	0.001037±0.000134	0.359905±0.037672
DEMO/closest/dec	0.001016±0.000091	0.359600±0.026977
DEMO/closest/obj	0.041012±0.063920	0.407225±0.094851

	ZDT6	
Algorithm	Convergence metric	Diversity metric
NSGA-II (real-coded)	0.296564±0.013135	0.668025±0.009923
NSGA-II (binary-coded)	7.806798±0.001667	0.644477±0.035042
SPEA	0.221138±0.000449	0.849389±0.002713
PAES	0.085469±0.006664	1.153052±0.003916
PDEA	N/A	0.473074±0.021721
MODE	0.026230±0.000861	N/A
DEMO/parent	0.000629±0.000044	0.442308±0.039255
DEMO/closest/dec	0.000630±0.000021	0.461174±0.035289
DEMO/closest/obj	0.000642±0.000029	0.458641±0.031362

[1] for the results and parameter settings of both versions of NSGA-II, SPEA and PAES, [8] for PDEA, and [9] for MODE[3]).

Results for PDEA in [8] were evaluated with generational distance instead of the convergence metric. Because PDEA is the approach that is the most similar to DEMO, we present the additional comparison of their results in Tables 5 and 6. Once more, we present the mean and variance of the values of generational distance, averaged over 10 runs.

Table 5. Generational distance achieved by PDEA and DEMO on the problems ZDT1, ZDT2 and ZDT3

	Generational distance		
Algorithm	ZDT1	ZDT2	ZDT3
PDEA	0.000615±0.000000	0.000652±0.000000	0.000563±0.000000
DEMO/parent	0.000230±0.000048	0.000091±0.000004	0.000156±0.000007
DEMO/closest/dec	0.000242±0.000028	0.000097±0.000004	0.000162±0.000013
DEMO/closest/obj	0.000243±0.000050	0.000092±0.000004	0.000169±0.000017

---

[3] The results for MODE are the average of 30 instead of 10 runs. In [9] no diversity metric was calculated.

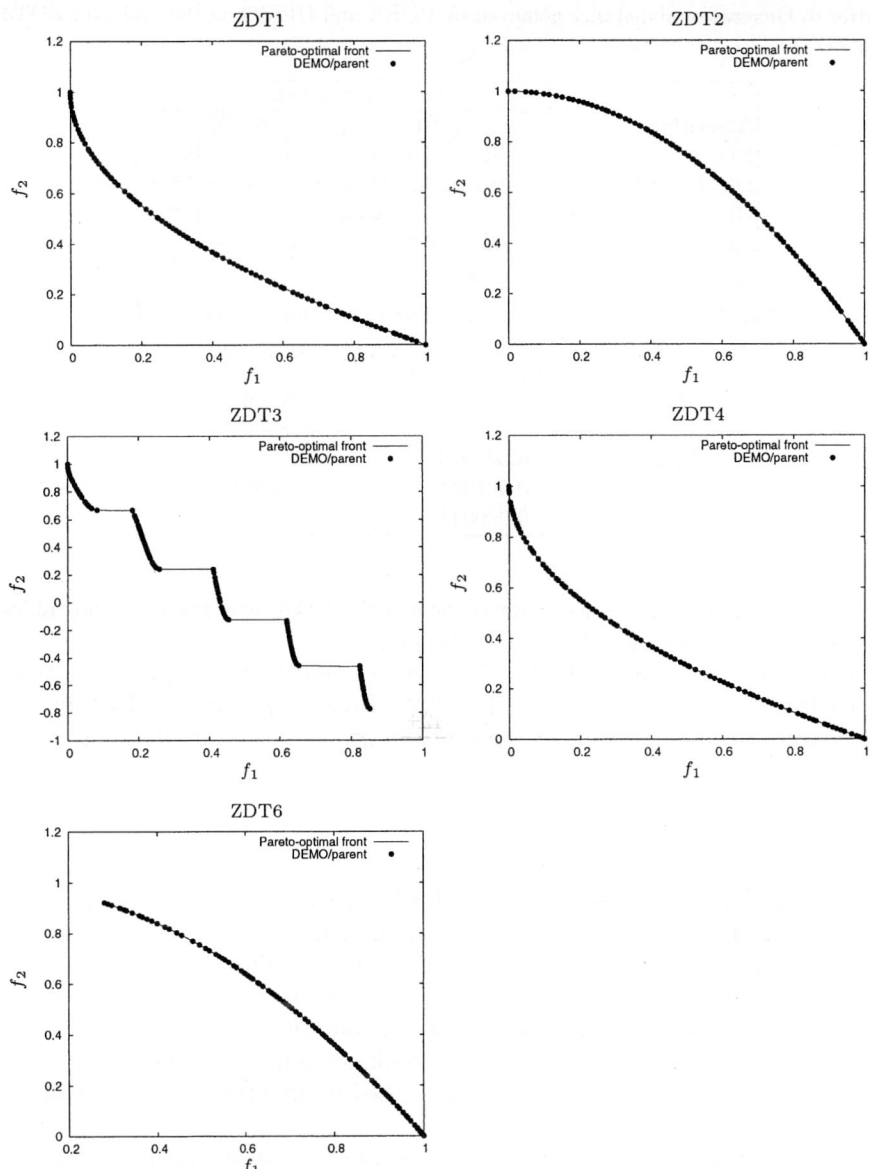

Fig. 4. Nondominated solutions of the final population obtained by DEMO on five ZDT test problems (see Table 7 for more details on these fronts). The presented fronts are the outcome of a single run of DEMO/parent

In Tables 3, 4, 5 and 6, only the results of DEMO with crossover probability 30% are shown. The results obtained with crossover probabilities set to 60% and

Table 6. Generational distance achieved by PDEA and DEMO on the problems ZDT4 and ZDT6

	Generational distance	
Algorithm	ZDT4	ZDT6
PDEA	0.618258±0.826881	0.023886±0.003294
DEMO/parent	0.000202±0.000053	0.000074±0.000004
DEMO/closest/dec	0.000179±0.000048	0.000075±0.000002
DEMO/closest/obj	0.004262±0.006545	0.000076±0.000003

Table 7. Metric values for the nondominated fronts shown in Fig. 4

Problem	Convergence metric	Diversity metric
ZDT1	0.001220	0.297066
ZDT2	0.000772	0.281994
ZDT3	0.001220	0.274098
ZDT4	0.001294	0.318805
ZDT6	0.000648	0.385088

90% for all three variants were always worse than the ones given in the tables and are not presented for the sake of clarity[4].

Figure 4 shows the nondominated fronts obtained by a single run of DEMO/parent with crossover probability 30%. Table 7 summarizes the values of the convergence and diversity metrics for the nondominated fronts from Fig. 4.

5 Discussion

As mentioned in the previous section, DEMO with crossover probability of 30% achieved the best results. This is in contradiction with the recommendations by the authors of DE [10] and confirms Madavan's findings in [8]. However, we have to take these results with caution. Crossover probability is highly related to the dimensionality of the decision space. In our study, only high-dimensional functions were used. When operating on low-dimensional decision spaces, higher values for crossover probabilities should be used to preserve the diversity in the population.

The challenge for MOEAs in the first three test problems (ZDT1, ZDT2 and ZDT3) lies in the high-dimensionality of these problems. Many MOEAs have achieved very good results on these problems in both goals of multiobjective optimization (convergence to the true Pareto front and uniform spread of solutions along the front). The results for the problems ZDT1 and ZDT2 (Tables 3 and 5) show that DEMO achieves good results, which are comparable to the

---

[4] The interested reader may find all nondominated fronts obtained by the three versions of DEMO with the three different crossover probabilities on the internet site http://dis.ijs.si/tea/demo.htm.

results of the algorithms NSGA-II and PDEA. On ZDT3, the three DEMO variants outperform all other algorithms used in comparison. On the first three test problems we cannot see a meaningful difference in performance of the three DEMO variants.

ZDT4 is a hard optimization problem with many (21^9) local Pareto fronts that tend to mislead the optimization algorithm. In Tables 4 and 6 we can see that all algorithms (with the exception of DEMO) have difficulties in converging to the true Pareto front. Here, we can see for the first time that there is a notable difference between the DEMO variants. The third variant, DEMO/closest/obj, performs poorly compared to the first two variants, although still better than other algorithms. While the first two variants of DEMO converge to the true Pareto optimal front in all of the 10 runs, the third variant remains blocked in a local Pareto front 3 times out of 10. This might be caused by DEMO/closest/obj putting too much effort in finding well spaced solutions and thus falling behind in the goal of convergence to the true Pareto optimal front. In this problem, there is also a big difference in results produced by the DEMO variants with different crossover probabilities. When using crossover probability 60% or 90%, no variant of DEMO ever converged to the true Pareto optimal front.

With the test problem ZDT6, there are two major difficulties. The first one is thin density of solutions towards the Pareto front and the second one non-uniform spread of solutions along the front. On this problem, all three DEMO variants outperform all other algorithms (see Tables 4 and 6). The results of all DEMO variants are also very similar and almost no difference is noted when using different crossover probabilities. Here, we note that the diversity metric value is worse than on all other problems. This is because of the non-uniform spread of solutions that causes difficulties although the convergence is good.

6 Conclusion

DEMO is a new DE implementation dealing with multiple objectives. The biggest difference between DEMO and other MOEAs that also use DE for reproduction is that in DEMO the newly created good candidates immediately take part in the creation of the subsequent candidates. This enables fast convergence to the true Pareto front, while the use of nondominated sorting and crowding distance metric in truncation of the extended population promotes the uniform spread of solutions.

In this paper, three variants of DEMO were introduced. The detailed analysis of the results brings us to the following conclusions. The three DEMO variants are as effective as the algorithms NSGA-II and PDEA on the problems ZDT1 and ZDT2. On the problems ZDT3, ZDT4 and ZDT6, DEMO achieves better results than any other algorithm used for comparison. As for the DEMO variants, we have not found any variant to be significantly better than another. Crowding DE thus showed not to bring the expected advantage over standard DE. Because DEMO/closest/dec and DEMO/closest/obj are computationally more expensive than DEMO/parent and do not bring any important advantage, we recommend the variant DEMO/parent be used in future experimentation.

In this study, we also investigated the influence of three different settings of the crossover probability. We found that DEMO in all three variants worked best when the crossover probability was low (30%). These findings, of course, cannot be generalized because our test problems were high-dimensional (10- or 30-dimensional). In low-dimensional problems, higher values of crossover probability should be used to preserve the diversity in the population. Seeing how crossover probability affects DEMO's performance, we are now interested if the other parameter used in candidate creation (the scaling factor F) also influences DEMO's performance. This investigation is left for further work. In the near future, we also plan to evaluate DEMO on additional test problems.

Acknowledgment

The work presented in the paper was supported by the Slovenian Ministry of Education, Science and Sport (Research Programme P2-0209 *Artificial Intelligence and Intelligent Systems*). The authors wish to thank the anonymous reviewers for their comments and Simon Huband for making the Pareto-optimal solutions for the ZDT problems available online.

References

1. Deb, K., Pratap, A., Agarwal, S., Meyarivan, T.: A fast and elitist multiobjective genetic algorithm: NSGA–II. IEEE Transactions on Evolutionary Computation **6** (2002) 182–197
2. Zitzler, E., Laumanns, M., Thiele, L.: SPEA2: Improving the strength pareto evolutionary algorithm. Technical Report 103, Computer Engineering and Networks Laboratory (TIK), Swiss Federal Institute of Technology (ETH) Zurich, Gloriastrasse 35, CH-8092 Zurich, Switzerland (2001)
3. Price, K.V., Storn, R.: Differential evolution – a simple evolution strategy for fast optimization. Dr. Dobb's Journal **22** (1997) 18–24
4. Lampinen, J.: (A bibliography of differential evolution algorithm) http://www2.lut.fi/~jlampine/debiblio.htm
5. Abbass, H.A., Sarker, R., Newton, C.: PDE: A pareto-frontier differential evolution approach for multi-objective optimization problems. In: Proceedings of the Congress on Evolutionary Computation 2001 (CEC'2001). Volume 2, Piscataway, New Jersey, IEEE Service Center (2001) 971–978
6. Abbass, H.A.: The self-adaptive pareto differential evolution algorithm. In: Congress on Evolutionary Computation (CEC'2002). Volume 1, Piscataway, New Jersey, IEEE Service Center (2002) 831–836
7. Zitzler, E., Thiele, L.: Multiobjective evolutionary algorithms: A comparative case study and the strength pareto approach. IEEE Transactions on Evolutionary Computation **3** (1999) 257–271
8. Madavan, N.K.: Multiobjective optimization using a pareto differential evolution approach. In: Congress on Evolutionary Computation (CEC'2002). Volume 2, Piscataway, New Jersey, IEEE Service Center (2002) 1145–1150

9. Xue, F., Sanderson, A.C., Graves, R.J.: Pareto-based multi-objective differential evolution. In: Proceedings of the 2003 Congress on Evolutionary Computation (CEC'2003). Volume 2, Canberra, Australia, IEEE Press (2003) 862–869
10. Storn, R.: (Differential evolution homepage) http://www.icsi.berkeley.edu/~storn/code.html
11. Thomsen, R.: Multimodal optimization using crowding-based differential evolution. In: 2004 Congress on Evolutionary Computation (CEC'2004). Volume 1, Portland, Oregon, USA, IEEE Service Center (2004) 1382–1389
12. Zitzler, E., Deb, K., Thiele, L.: Comparison of multiobjective evolutionary algorithms: Empirical results. Evolutionary Computation 8 (2000) 173–195

Multi-objective Model Selection for Support Vector Machines

Christian Igel

Institute for Neurocomputing,
Ruhr-University Bochum,
44780 Bochum, Germany
christian.igel@neuroinformatik.rub.de

Abstract. In this article, model selection for support vector machines is viewed as a multi-objective optimization problem, where model complexity and training accuracy define two conflicting objectives. Different optimization criteria are evaluated: Split modified radius margin bounds, which allow for comparing existing model selection criteria, and the training error in conjunction with the number of support vectors for designing sparse solutions.

1 Introduction

Model selection for supervised learning systems requires finding a suitable trade-off between at least two objectives, especially between model complexity and accuracy on a set of noisy training examples ($\to$ bias vs. variance, capacity vs. empirical risk). Usually, this multi-objective problem is tackled by aggregating the objectives into a scalar function and applying standard methods to the resulting single-objective task. However, this approach can only lead to satisfactory solutions if the aggregation (e.g., a linear weighting of empirical error and regularization term) matches the problem. Thus, choosing an appropriate aggregation itself is an optimization task. A better way is to apply multi-objective optimization (MOO) to approximate the set of Pareto-optimal trade-offs and to choose a final solution afterwards, as discussed in the context of neural networks in [1,2,3,4]. A solution is Pareto-optimal if it cannot be improved in any objective without getting worse in at least one other objective [5,6,7].

In the following, we consider MOO of the kernel and the regularization parameter of support vector machines (SVMs). We show how to reveal the trade-off between different objectives to guide the model selection process, e.g., for the design of sparse SVMs. One advantage of SVMs is that theoretically well founded bounds on the expected generalization performance exist, which can serve as model selection criteria.[1] However, in practice heuristic modifications of these bounds—e.g., corresponding to different weightings of capacity and empirical

---

[1] When used for model selection in the described way, the term "bound" is slightly misleading.

risk—can lead to better results [8]. The MOO approach enables us to compare model selection criteria proposed in the literature after optimization.

As we consider only kernels from a parameterized family of functions, our model selection problem reduces to multidimensional real-valued optimization. We present a multi-objective evolution strategy (ES) with self-adaptation for real-valued MOO. The basics of MOO using non-dominated sorting [6,9] and the ES are presented in the next section. Then SVMs and the model selection criteria we consider are briefly described in section 3.2 and the experiments in section 4.

2 Evolutionary Multi-objective Optimization

Consider an optimization problem with M objectives $f_1, \ldots, f_M : X \to \mathbb{R}$ to be minimized. The elements of X can be partially ordered using the concept of Pareto dominance. A solution $x \in X$ dominates a solution x' and we write $x \prec x'$ iff $\exists m \in \{1, \ldots, M\} : f_m(x) < f_m(x')$ and $\nexists m \in \{1, \ldots, M\} : f_m(x) > f_m(x')$. The elements of the (Pareto) set $\{x \mid \nexists x' \in X : x' \prec x\}$ are called Pareto-optimal. Without any further information, no Pareto-optimal solution can be said to be superior to another. The goal of multi-objective optimization (MOO) is to find in a single trial a diverse set of Pareto-optimal solutions, which provide insights into the trade-offs between the objectives. When approaching a MOO problem by linearly aggregating all objectives into a scalar function, each weighting of the objectives yields only a limited subset of Pareto-optimal solutions. That is, various trials with different aggregations become necessary—but when the Pareto front (see below) is not convex, even this inefficient procedure does not help (cf. [6,7]). Evolutionary multi-objective algorithms have become the method of choice for MOO [5,6]. The most popular variant is the non-dominated sorting genetic algorithm NSGA-II, which shows fast convergence to the Pareto-optimal set and a good spread of solutions [6,9]. On the other hand, evolution strategies (ES) are among the most elaborated and best analyzed evolutionary algorithms for real-valued optimization. Therefore, we propose a new method that combines ES with concepts from the NSGA-II.

2.1 Non-dominated Sorting

We give a concise description of the non-dominated sorting approach used in NSGA-II. For more details and efficient algorithms realizing this sorting we refer to [9]. First of all, the elements in a finite set $A \subseteq X$ of candidate solutions are ranked according to their level of non-dominance. Let the non-dominated solutions in A be denoted by $\mathrm{ndom}(A) = \{a \in A \mid \nexists a' \in A : a' \prec a\}$. The Pareto front of A is then given by $\{(f_1(a), \ldots, f_M(a)) \mid a \in \mathrm{ndom}(A)\}$. The elements in $\mathrm{ndom}(A)$ have rank 1. The other ranks are defined recursively by considering the set without the solutions with lower ranks. Formally, let $\mathrm{dom}_n(A) = \mathrm{dom}_{n-1}(A) \setminus \mathrm{ndom}_n(A)$ and $\mathrm{ndom}_n(A) = \mathrm{ndom}(\mathrm{dom}_{n-1}(A))$ for $n \in \{1, \ldots\}$ with $\mathrm{dom}_0 = A$. For $a \in A$ we define the level of non-dominance $\mathrm{r}(a, A)$ to be i iff $a \in \mathrm{ndom}_i(A)$.

As a second sorting criterion, non-dominated solutions A' are ranked according to how much they contribute to the spread (or diversity) of objective function values in A'. This can be measured by the crowding-distance. For M objectives, the crowding-distance of $a \in A'$ is given by $c(a, A') = \sum_{m=1}^{M} c_m(a, A')/(f_m^{\max} - f_m^{\min})$, where $f_m^{\max}$ and $f_m^{\min}$ are (estimates of) the minimum and maximum value of the mth objective and

$$c_m(a, A') = \begin{cases} \infty, \text{ if } f_m(a) = \min\{f_m(a') \mid a' \in A'\} \text{ or } f_m(a) = \max\{f_m(a') \mid a' \in A'\} \\ \min\{f_m(a'') - f_m(a') \mid \\ \qquad a', a'' \in A' : f_m(a') < f_m(a) < f_m(a'')\}, \text{ otherwise.} \end{cases}$$

Based on the level of non-dominance and the crowding distance we define the relation

$$a \prec_A a' \Leftrightarrow r(a, A) < r(a', A) \text{ or}$$
$$[(r(a, A) = r(a', A)) \land (c(a, \text{ndom}_{r(a', A)}(A)) > c(a', \text{ndom}_{r(a', A)}(A)))],$$

for $a, a' \in A$. That is, a is better than a' when compared using $\prec_A$ if either a has a better (lower) level of non-dominance or a and a' are on the same level but a is in a "lesser crowded region of the objective space" and therefore induces more diversity.

2.2 Multi-objective Evolution Strategy with Mutative Self-Adaptation

Evolution strategies (ES, cf. [10, 11]) are one of the main branches of evolutionary algorithms (EAs), i.e., a class of iterative, direct, randomized optimization methods mimicking principles of neo-Darwinian evolution theory. In EAs, a (multi-) set of μ individuals representing candidate solutions, the so called parent population, is maintained. In each iteration (generation) t, λ new individuals (the offspring) are generated based on the parent population. A selection procedure preferring individuals representing better solutions to the problem at hand determines the parent population of the next iteration.

In our ES, each individual $a_i^{(t)} \in \mathbb{R}^{2n}$ is divided into two parts, $a_i^{(t)} = (x_i^{(t)}, \sigma_i^{(t)})$, the object variables $x_i^{(t)} \in \mathbb{R}^n = X$ representing the corresponding candidate solution and the strategy parameters $\sigma_i^{(t)} \in \mathbb{R}^n$. For simplicity, we do not distinguish between $f_m(a_i^{(t)})$ and $f_m(x_i^{(t)})$. The strategy parameters are needed for self-adaptation, a key concept in EAs [12, 13] that allows for an online adaptation of the search strategy leading to improved search performance in terms of both accuracy and efficiency (similar to adaptive step-sizes / learning rates in gradient-based steepest-descent algorithms). The initial parent population consists of μ randomly created individuals. In each iteration t, new individuals $a_i^{(t+1)}$, $i = 1, \ldots, \lambda$ are generated. For each offspring $a_i^{(t+1)}$, two individuals, say $a_u^{(t)} = (\boldsymbol{x}_u^{(t)}, \boldsymbol{\sigma}_u^{(t)})$ and $a_v^{(t)} = (\boldsymbol{x}_v^{(t)}, \boldsymbol{\sigma}_v^{(t)})$, are chosen from the current parent population uniformly at random. The new strategy parameters $\boldsymbol{\sigma}_i^{(t+1)} = (\sigma_{i,1}^{(t+1)}, \ldots, \sigma_{i,n}^{(t+1)})$ of offspring i are given by

$$\sigma_{i,k}^{(t+1)} = \underbrace{\frac{1}{2}\left[\sigma_{u,k}^{(t)} + \sigma_{v,k}^{(t)}\right]}_{\text{intermediate recombination}} \cdot \underbrace{\exp\left(\tau' \cdot \zeta_i^{(t)} + \tau \cdot \zeta_{i,k}^{(t)}\right)}_{\text{log-normal mutation}}.$$

Here, the $\zeta_{i,k}^{(t)} \sim \mathcal{N}(0,1)$ are realizations of a normally distributed random variable with zero mean and unit variance that is sampled anew for each component k for each individual i, whereas the $\zeta_i^{(t)} \sim \mathcal{N}(0,1)$ are sampled once per individual and are identical for each component. The mutation strengths are set to $\tau = 1/\sqrt{2\sqrt{n}}$ and $\tau' = 1/\sqrt{2n}$ [12, 11]. Thereafter the objective parameters are altered using the new strategy parameters:

$$x_{i,k}^{(t+1)} = \underbrace{x_{c_{i,k}^{(t)},k}^{(t)}}_{\text{discrete recombination}} + \underbrace{\sigma_{i,k}^{(t+1)} z_{i,k}^{(t)}}_{\text{Gaussian mutation}},$$

where $z_{i,k}^{(t)} \sim \mathcal{N}(0,1)$. The $c_{i,k}^{(t)}$ are realizations of a random variable taking the values u and v with equal probability. After generating the offspring, (μ, λ)-selection is used, i.e., the μ best individuals of the offspring form the new parent population.

So far, we have described a canonical ES with mutative self-adaptation, for more details please see [11] and references therein. Now we turn this ES into a multi-objective algorithm by using the non-dominated sorting operator $\prec_{\{a_1^{(t+1)},\ldots,a_\lambda^{(t+1)}\}}$ for the ranking in the (μ, λ)-selection in iteration t. In addition, we keep an external archive $\mathcal{A}^{(t+1)} = \text{ndom}(\mathcal{A}^{(t)} \cup \{a_1^{(t+1)}, \ldots, a_\lambda^{(t+1)}\})$ of all non-dominated solutions discovered so far starting from the initial population. This extends the ideas in [14] and yields the first self-adaptive ES using non-dominated sorting, which we call NSES.

The NSES uses a non-elitist selection scheme (i.e., the best solutions found so far are not kept in the parent population), because self-adaption does not work well together with elitism. This is in contrast to NSGA-II [9] and the self-adaptive SPANN [2]. Of course, the NSES is elitist when looking at the concurrent archive $\mathcal{A}^{(t)}$.

3 Models Selection for SVMs

Support vector machines (SVMs, e.g., [15, 16, 17]) are learning machines based on two key elements: a general purpose linear learning algorithm and a problem specific kernel that computes the inner product of input data points in a feature space.

3.1 Support Vector Machines

We consider L_1-norm soft margin SVMs for the discrimination of two classes. Let $(\boldsymbol{x}_i, y_i), 1 \leq i \leq \ell$, be the training examples, where $y_i \in \{-1, 1\}$ is the label associated with input pattern $\boldsymbol{x}_i \in X$. The main idea of SVMs is to map the input vectors to a feature space F and to classify the transformed data by a linear function. The transformation $\phi : X \to F$ is implicitly done by a kernel $K : X \times X \to \mathbb{R}$, which computes an inner product in the feature space, i.e., $K(\boldsymbol{x}_i, \boldsymbol{x}_j) = \langle \phi(\boldsymbol{x}_i), \phi(\boldsymbol{x}_j) \rangle$. The linear function for classification in the feature space is chosen according to a generalization error bound considering a margin and the margin slack vector, i.e., the amounts by which individual training patterns fail to meet that margin (cf. [15, 16, 17]). This leads to the SVM decision function

$$f(\boldsymbol{x}) = \text{sign} \left(\sum_{i=1}^{\ell} y_i \alpha_i^* K(\boldsymbol{x}_i, \boldsymbol{x}) + b \right),$$

where the coefficients α_i^* are the solution of the following quadratic optimization problem:

$$\text{maximize} \quad W(\boldsymbol{\alpha}) = \sum_{i=1}^{\ell} \alpha_i - \frac{1}{2} \sum_{i,j=1}^{\ell} y_i y_j \alpha_i, \alpha_j K(\boldsymbol{x}_i, \boldsymbol{x}_j)$$

$$\text{subject to} \quad \sum_{i=1}^{\ell} \alpha_i y_i = 0$$

$$0 \leq \alpha_i \leq C, \quad i = 1, \ldots, \ell .$$

The optimal value for b can then be computed based on the solution $\boldsymbol{\alpha}^*$. The vectors $\boldsymbol{x}_i$ with $\alpha_i^* > 0$ are called support vectors. The number of support vectors is denoted by $\#\text{SV}$. The regularization parameter C controls the trade-off between maximizing the margin

$$\gamma = \left(\sum_{i,j=1}^{\ell} y_i y_j \alpha_i^* \alpha_j^* K(\boldsymbol{x}_i, \boldsymbol{x}_j) \right)^{-1/2}$$

and minimizing the L_1-norm of the final margin slack vector $\boldsymbol{\xi}^*$ of the training data, where

$$\xi_i^* = \max \left(0, 1 - y_i \left(\sum_{j=1}^{\ell} y_j \alpha_j^* K(\boldsymbol{x}_j, \boldsymbol{x}_j) + b \right) \right) .$$

3.2 Model Selection Criteria for L_1-SVMs

Choosing the right kernel for a SVM is important for its performance. When a parameterized family of kernel functions is considered, kernel adaptation reduces to finding an appropriate parameter vector. These parameters together with the regularization parameter are called hyperparameters of the SVM. In practice, the hyperparameters are usually determined by grid search. Because of the computational complexity, grid search is only suitable for the adjustment of very few parameters. Further, the choice of the discretization of the search space may be crucial. Perhaps the most elaborate techniques for choosing hyperparameters are gradient-based approaches [18, 8, 19, 20]. When applicable, these methods are highly efficient. However, they have some drawbacks and limitations: The kernel function has to be differentiable. The score function for assessing the performance of the hyperparameters (or at least an accurate approximation of this function) also has to be differentiable with respect to all hyperparameters, which excludes reasonable measures such as the number of support vectors. In some approaches, the computation of the gradient is only exact in the hard-margin case (i.e., for separable data / L_2-SVMs) when the model is consistent with the training data. Further, as the objective functions are indeed multi-modal, the performance of gradient-based heuristics depends on the initialization—the algorithms are prone to getting stuck in local optima. In [21, 22], single-objective evolution strategies were proposed for adapting SVM hyperparameters, which partly overcome these problems; in [23] a single-objective genetic algorithm was used for SVM feature selection (see also [24, 25, 26]) and adaptation of the (discretized) regularization parameter. Like gradient-based techniques, these methods are not suitable for MOO. Therefore we apply the NSES to directly address the multi-objective nature of model selection.

We optimize Gaussian kernels $k_{\boldsymbol{A}}(\boldsymbol{x}, \boldsymbol{z}) = \exp\left(-(\boldsymbol{x}-\boldsymbol{z})^T \boldsymbol{A}(\boldsymbol{x}-\boldsymbol{z})\right)$, $\boldsymbol{x}, \boldsymbol{z} \in \mathbb{R}^m$. We look at two parameterizations of the symmetric, positive definite matrix $\boldsymbol{A}$. In the standard scenario, we adapt $k_{\gamma \boldsymbol{I}}$, where $\boldsymbol{I}$ is the unit matrix and $\gamma > 0$ is the only adjustable parameter. In addition, we optimize m independent scaling factors weighting the input components and consider $k_{\boldsymbol{D}}$, where $\boldsymbol{D}$ is a diagonal matrix with arbitrary positive entries.

We want to design classifiers that generalize well. The selection criteria we consider are therefore (partly) based on bounds on the number of errors in the leave-one-out procedure, which gives an estimate of the expected generalization performance. However, most bounds were derived for the hard-margin case (i.e., for separable data / L_2-SVMs) and L_1-SVMs cannot be reduced to this scenario. Thus, we combine heuristics and results for the hard-margin case to selection criteria for L_1-SVMs.

3.3 Modified Radius Margin Bounds

Let R denote the radius of the smallest ball in feature space containing all ℓ training examples given by

$$R = \sqrt{\sum_{i=1}^{\ell} \beta_i^* K(\boldsymbol{x}_i, \boldsymbol{x}_i) - \sum_{i,j=1}^{\ell} \beta_i^* \beta_j^* K(\boldsymbol{x}_i, \boldsymbol{x}_j)} \;,$$

where $\boldsymbol{\beta}^*$ is the solution vector of the quadratic optimization problem

$$\underset{\boldsymbol{\beta}}{\text{maximize}} \ \sum_{i=1}^{\ell} \beta_i K(\boldsymbol{x}_i, \boldsymbol{x}_i) - \sum_{i,j=1}^{\ell} \beta_i \beta_j K(\boldsymbol{x}_i, \boldsymbol{x}_j)$$

$$\text{subject to} \ \sum_{i=1}^{\ell} \beta_i = 1$$

$$\beta_i \geq 0 \ , \quad i = 1, \ldots, \ell \ ,$$

see [27]. Following a suggestion by Olivier Chapelle, the modified radius margin bound

$$T_{\text{DM}} = (2R)^2 \sum_{i=1}^{\ell} \alpha_i^* + \sum_{i=1}^{\ell} \xi_i^* \ ,$$

was considered for model selection of L_1-SVMs in [28]. In practice, this expression did not lead to satisfactory results [8, 28]. Therefore, in [8] it was suggested to use

$$T_{\text{RM}} = R^2 \sum_{i=1}^{\ell} \alpha_i^* + \sum_{i=1}^{\ell} \xi_i^* \ ,$$

based on heuristic considerations and it was shown empirically that T_{RM} leads to better models than T_{DM}.[2] Both criteria are not differentiable [28]. They can be viewed as two different aggregations of the following two objectives

$$f_1 = R^2 \sum_{i=1}^{\ell} \alpha_i^* \ \text{and} \ f_2 = \sum_{i=1}^{\ell} \xi_i^* \tag{1}$$

penalizing model complexity and training errors, respectively. For example, a highly complex SVM classifier that very accurately fits the training data has high f_1 and small f_2.

3.4 Number of SVs and Training Error

There are good reasons to prefer SVMs with few support vectors: In the hard-margin case, the number of SVs is an upper bound on the expected number of errors made by the leave-one-out procedure (e.g., see [18, 17]). Further, the space and time complexity of the SVM classifier scales with the number of SVs. A natural measure for the performance of a classifier on a training set is the percentage of misclassified patterns of the training data $\text{CE}(D_{\text{train}})$. Hence, we consider

---

[2] Also for L_2-SVMs it was shown empirically that theoretically better founded weightings of such objectives (e.g., corresponding to tighter bounds) need not correspond to better model selection criteria [8].

$$f'_1 = \#\text{SV} + \theta\left(R^2 \sum_{i=1}^{\ell} \alpha_i^*\right) \text{ and } f'_2 = \ell \cdot \text{CE}(D_{\text{train}}) + \theta\left(\sum_{i=1}^{\ell} \xi_i^*\right), \quad (2)$$

where $\theta(x) = x/(1+x)$. For example, it is easy to achieve zero classification error when all training points become support vectors, but this solution is not likely to generalize well.

The optional $\theta(\dots)$ terms are used for smoothing the objective functions. In recent experiments, it turns out that they can be omitted without deterioration of performance.

4 Experiments

For the evaluation of our hyperparameter optimization method we used the common medical benchmark datasets *breast-cancer*, *diabetes*, *heart*, and *thyroid* with input dimensions m equal to 9, 8, 13, and 5, and ℓ equal to 200, 468, 170, and 140. The data originally from the UCI Benchmark Repository [29] are preprocessed and partitioned as in [30], where we consider the first of the splits into training and test set D_{train} and D_{test}. We applied the NSES to adapt C and the parameters of $k_{\gamma I}$ or k_D; the quality of a candidate solution was determined after SV learning. We set $\mu = 15$ and $\lambda = 75$. The strategy parameters $\boldsymbol{\sigma}$ were initialized to 1 when adapting $k_{\gamma I}$ and because of the increased number ($m+1$) of objective parameters to 0.1 when adapting k_D. The objective parameters of the initial individuals were chosen uniformly at random from $[0.1, 100]$ and $[0.01, 10]$ for C and the kernel parameters, respectively. The NSES parameters were not tuned, i.e, the efficiency of the ES could surely be improved. Figures 1, 2, and 3 depict the solutions in the final archives.[3]

4.1 Modified Radius Margin Bounds

Figure 1 shows the results of optimizing $k_{\gamma I}$ using the objectives (1); for each f_1 value of a solution in the final archive the corresponding f_2, T_{RM}, T_{DM}, and $100 \cdot \text{CE}(D_{\text{test}})$ are given. For *diabetes*, *heart*, and *thyroid*, the solutions lie on typical convex Pareto fronts; in the *breast-cancer* example the convex front looks piecewise linear. Assuming convergence to the Pareto-optimal set, the results of a single MOO trial are sufficient to determine the outcome of single-objective optimization of any (positive) linear weighting of the objectives. Thus, we can directly determine and compare the solutions that minimizing T_{RM} and T_{DM} would suggest. Our experiments substantiate the findings in [8] that the heuristic bound T_{RM} is better suited for model selection than T_{DM}: When looking at $\text{CE}(D_{\text{test}})$ and the minima of T_{RM} and T_{DM}, we can conclude that T_{DM} puts too

---

[3] Because it is difficult to present sets of sets, we discuss outcomes from typical (randomly chosen) trials. Additional results showing the robustness of our approach can be downloaded from http://www.neuroinformatik.rub.de/PEOPLE/igel/moo.

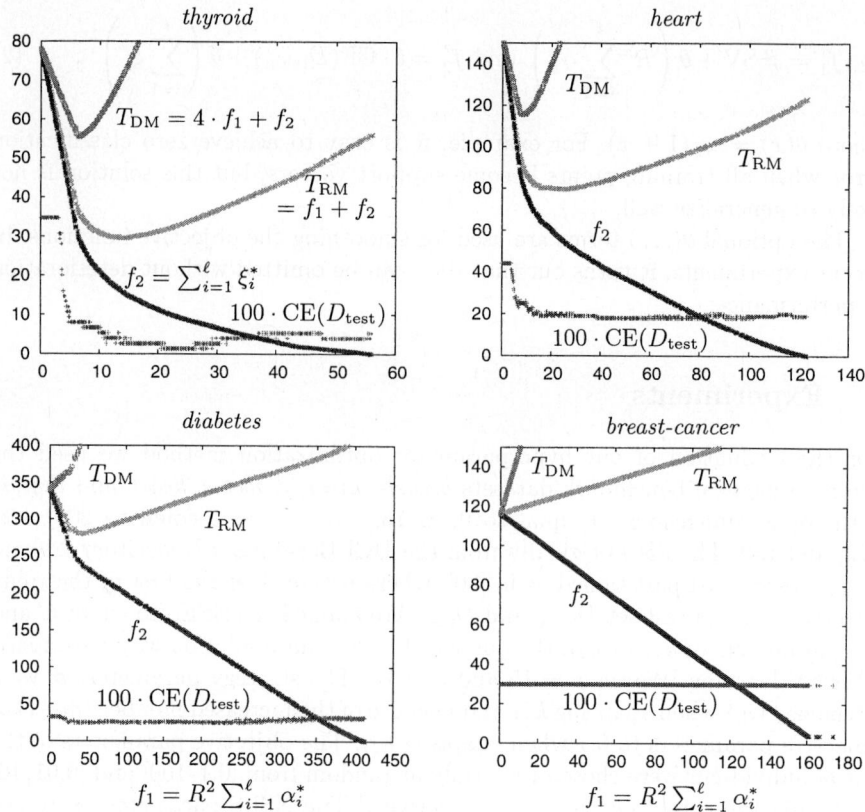

Fig. 1. Pareto fronts (i.e., f_2 vs. f_1) of $\mathcal{A}^{(100)}$—the outcome of the optimization after $100 \cdot \lambda$ evaluations—for $k_{\gamma I}$ and the four benchmark problems. Additionally, for every solution in $\mathcal{A}^{(100)}$ the values of T_{RM}, T_{DM}, and $100 \cdot \text{CE}(D_{\text{test}})$ are plotted against the corresponding f_1 value. Projecting the minimum of T_{RM} (for T_{DM} proceed analogously) along the y-axis on the Pareto front gives the (f_1, f_2) pair suggested by the model selection criterion T_{RM}—this would also be the outcome of single-objective optimization using T_{RM}. Projecting an (f_1, f_2) pair along the y-axis on $100 \cdot \text{CE}(D_{\text{test}})$ yields the corresponding error on an external test set (which is assumed to be not available for model selection)

much emphasis on the "radius margin part" yielding worse classification results on the external test set (except for *breast-cancer* where there is no difference on D_{test}). The *heart* and *thyroid* results suggest that even more weight should be given to the slack-variables (i.e., the performance on the training set) than in T_{RM}.

In the MOO approach, degenerated solutions resulting from a not appropriate weighting of objectives (which we indeed observed—without the chance to change the trade-off afterwards—in single-objective optimization of SVMs) become obvious and can be excluded. For example, one would probably not pick the

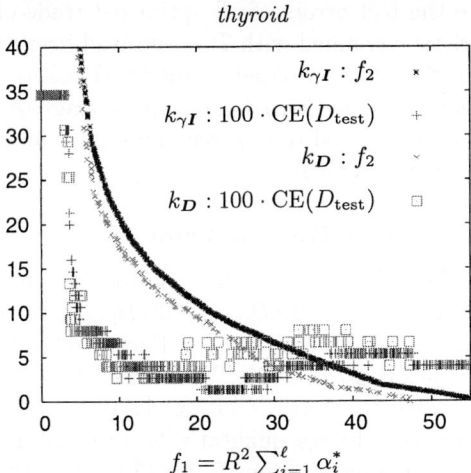

Fig. 2. Pareto fronts after optimizing $k_{\gamma I}$ and k_D for objectives (1) and *thyroid* data after 100 iterations. For both kernel parameterizations, f_2 and $100 \cdot \mathrm{CE}(D_{\text{test}})$ are plotted against f_1

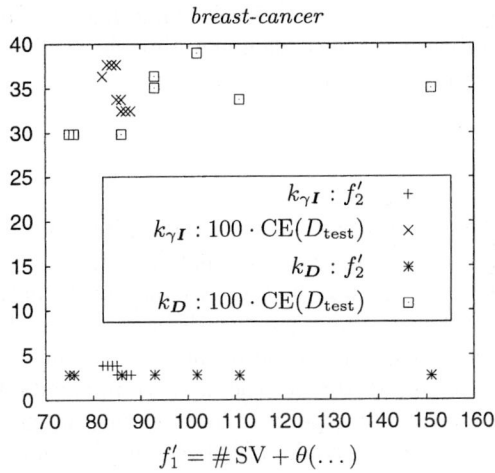

Fig. 3. Pareto fronts after optimizing $k_{\gamma I}$ and k_D for objectives (2) and *breast-cancer* data after 200 generations. For both kernel parameterizations, f'_2 and $100 \cdot \mathrm{CE}(D_{\text{test}})$ are plotted against f'_1

solution suggested by T_{DM} in the *diabetes* benchmark. A typical MOO heuristic is to choose a solution from the archive that corresponds to an "interesting" part of the Pareto front. In case of a typical convex front, this would be the area of highest "curvature" (the "knee", see figure 1). In our benchmark problems, this leads to results on a par with T_{RM} and much better than T_{DM} (except for *breast-cancer*, where the test errors of all optimized trade-offs were the same). Therefore, this heuristic combined with T_{RM} (derived from the MOO results) is an alternative for model selection based on modified radius margin bounds.

Adapting the scaling of the kernel (i.e., optimizing k_D) sometimes led to better objective values compared to $k_{\gamma I}$, see figure 2 for an example, but not to better generalization performance.

4.2 Number of SVs and Training Error

Now we describe the results achieved when optimizing the objectives (2). We write solutions as triples $[\# \text{SV}, \text{CE}(D_{\text{train}}), \text{CE}(D_{\text{test}})]$. In the *thyroid* example with two hyperparameters, the archive (the Pareto set) after 100 generations contained a single solution $[14, 0, 0.027]$. When additionally adjusting the scaling, the archive contained the single solution $[11, 0, 0.027]$. In case of the *heart* data, the archives did not change qualitatively after 200 generations. For $k_{\gamma I}$ we got a Pareto set containing solutions from $[49, 0.17, 0.2]$ to $[57, 0, 0.17]$. With scaling, the NSES converged to an archive containing only consistent solutions with 49 SVs. However, the classification error on the test set was either 0.23 or 0.24. In the *diabetes* task, the archives did not change qualitatively after 100 generations, where we got solutions from $[208, 0, 0.297]$ to $[212, 0, 0.287]$ for $k_{\gamma I}$ and an archive where all solutions corresponded to $[178, 0, 0.297]$ when adjusting k_D. Results for the *breast-cancer* data are shown in figure 3. In all four scenarios, we achieved lower objective values when adapting the scaling. These results did not necessarily correspond to lower $\text{CE}(D_{\text{test}})$.

5 Conclusion

Model selection is a multi-objective optimization (MOO) problem. We presented an evolution strategy combining non-dominated sorting [9] and self-adaptation [12, 14] for efficient MOO. It was successfully applied to optimize multiple hyperparameters of SVMs with Gaussian kernels based on conflicting, not differentiable criteria. In most experiments, better objective values were achieved when adapting individual scaling factors for the input components. However, these solutions did not necessarily correspond to lower errors on external test sets. The final Pareto fronts visualize the trade-off between model complexity and learning accuracy for guiding the model selection process. When looking at split modified radius margin bounds, standard MOO heuristics based on the curvature of the Pareto front led to comparable models as using the modified bound proposed in [8]. The latter puts more emphasis on minimizing the slack vector compared to the bound considered in [28], a strategy that is strongly supported by our results. In practice, the experiments involving minimization of the number of support vectors are of particular interest, because here the complexity objective is directly related to the speed of the classifier. Knowing the speed vs. accuracy trade-off is helpful when designing SVMs that have to obey real-time constraints.

References

1. Abbass, H.A.: An evolutionary artificial neural networks approach for breast cancer diagnosis. Artificial Intelligence in Medicine **25** (2002) 265–281
2. Abbass, H.A.: Speeding up backpropagation using multiobjective evolutionary algorithms. Neural Computation **15** (2003) 2705–2726
3. Jin, Y., Okabe, T., Sendhoff, B.: Neural network regularization and ensembling using multi-objective evolutionary algorithms. In: Congress on Evolutionary Computation (CEC'04), IEEE Press (2004) 1–8
4. Wiegand, S., Igel, C., Handmann, U.: Evolutionary multi-objective optimization of neural networks for face detection. International Journal of Computational Intelligence and Applications **4** (2004) 237–253 Special issue on Neurocomputing and Hybrid Methods for Evolving Intelligence.
5. Coello Coello, C.A., Van Veldhuizen, D.A., Lamont, G.B.: Evolutionary Algorithms for Solving Multi-Objective Problems. Kluwer Academic Publishers (2002)
6. Deb, K.: Multi-Objective Optimization Using Evolutionary Algorithms. Wiley (2001)
7. Sawaragi, Y., Nakayama, H., Tanino, T.: Theory of Multiobjective Optimization. Volume 176 of Mathematics in Science and Engineering. Academic Press (1985)
8. Chung, K.M., Kao, W.C., Sun, C.L., Lin, C.J.: Radius margin bounds for support vector machines with RBF kernel. Neural Computation **15** (2003) 2643–2681
9. Deb, K., Agrawal, S., Pratap, A., Meyarivan, T.: A fast and elitist multiobjective genetic algorithm: NSGA-II. IEEE Transactions on Evolutionary Computation **6** (2002) 182–197
10. Beyer, H.G.: The Theory of Evolution Strategies. Springer-Verlag (2001)
11. Beyer, H.G., Schwefel, H.P.: Evolution strategies: A comprehensive introduction. Natural Computing **1** (2002) 3–52
12. Bäck, T.: An overview of parameter control methods by self-adaptation in evolutionary algorithms. Fundamenta Informaticae **35** (1998) 51–66
13. Igel, C., Toussaint, M.: Neutrality and self-adaptation. Natural Computing **2** (2003) 117–132
14. Laumanns, M., Rudolph, G., Schwefel, H.P.: Mutation control and convergence in evolutionary multi-objective optimization. In Matousek, R., Osmera, P., eds.: Proceedings of the 7th International Mendel Conference on Soft Computing (MENDEL 2001), Brno, Czech Republic: University of Technology (2001) 24–29
15. Cristianini, N., Shawe-Taylor, J.: An Introduction to Support Vector Machines and other kernel-based learning methods. Cambridge University Press (2000)
16. Schölkopf, B., Smola, A.J.: Learning with Kernels: Support Vector Machines, Regularization, Optimization, and Beyond. MIT Press (2002)
17. Vapnik, V.N.: The Nature of Statistical Learning Theory. Springer-Verlag (1995)
18. Chapelle, O., Vapnik, V., Bousquet, O., Mukherjee, S.: Choosing multiple parameters for support vector machines. Machine Learning **46** (2002) 131–159
19. Gold, C., Sollich, P.: Model selection for support vector machine classification. Neurocomputing **55** (2003) 221–249
20. Keerthi, S.S.: Efficient tuning of SVM hyperparameters using radius/margin bound and iterative algorithms. IEEE Transactions on Neural Networks **13** (2002) 1225–1229
21. Friedrichs, F., Igel, C.: Evolutionary tuning of multiple SVM parameters. In Verleysen, M., ed.: 12th European Symposium on Artificial Neural Networks (ESANN 2004), Evere, Belgium: d-side publications (2004) 519–524

22. Runarsson, T.P., Sigurdsson, S.: Asynchronous parallel evolutionary model selection for support vector machines. Neural Information Processing – Letters and Reviews **3** (2004) 59–68
23. Fröhlich, H., Chapelle, O., Schölkopf, B.: Feature selection for support vector machines by means of genetic algorithms. In: 15th IEEE International Conference on Tools with AI (ICTAI 2003), IEEE Computer Society (2003) 142–148
24. Eads, D.R., Hill, D., Davis, S., Perkins, S.J., Ma, J., Porter, R.B., Theiler, J.P.: Genetic algorithms and support vector machines for time series classification. In Bosacchi, B., Fogel, D.B., Bezdek, J.C., eds.: Applications and Science of Neural Networks, Fuzzy Systems, and Evolutionary Computation V. Volume 4787 of Proceedings of the SPIE. (2002) 74–85
25. Jong, K., Marchiori, E., van der Vaart, A.: Analysis of proteomic pattern data for cancer detection. In Raidl, G.R., Cagnoni, S., Branke, J., Corne, D.W., Drechsler, R., Jin, Y., Johnson, C.G., Machado, P., Marchiori, E., Rothlauf, F., Smith, G.D., Squillero, G., eds.: Applications of Evolutionary Computing. Number 3005 in LNCS, Springer-Verlag (2004) 41–51
26. Miller, M.T., Jerebko, A.K., Malley, J.D., Summers, R.M.: Feature selection for computer-aided polyp detection using genetic algorithms. In Clough, A.V., Amini, A.A., eds.: Medical Imaging 2003: Physiology and Function: Methods, Systems, and Applications. Volume 5031 of Proceedings of the SPIE. (2003) 102–110
27. Schölkopf, B., Burges, C.J.C., Vapnik, V.: Extracting support data for a given task. In Fayyad, U.M., Uthurusamy, R., eds.: Proceedings of the First International Conference on Knowledge Discovery & Data Mining, AAAI Press (1995) 252–257
28. Duan, K., Keerthi, S.S., Poo, A.: Evaluation of simple performance measures for tuning SVM hyperparameters. Neurocomputing **51** (2003) 41–59
29. Blake, C., Merz, C.: UCI repository of machine learning databases (1998)
30. Rätsch, G., Onoda, T., Müller, K.R.: Soft margins for adaboost. Machine Learning **42** (2001) 287–32

Exploiting the Trade-Off — The Benefits of Multiple Objectives in Data Clustering

Julia Handl and Joshua Knowles

School of Chemistry, University of Manchester
Faraday Building, Sackville Street, PO Box 88, Manchester M60 1QD
http://dbk.ch.umist.ac.uk/handl/mock/

Abstract. In previous work, we have proposed a novel approach to data clustering based on the explicit optimization of a partitioning with respect to two complementary clustering objectives [6]. Here, we extend this idea by describing an advanced multiobjective clustering algorithm, MOCK, with the capacity to identify good solutions from the Pareto front, and to automatically determine the number of clusters in a data set. The algorithm has been subject to a thorough comparison with alternative clustering techniques and we briefly summarize these results. We then present investigations into the mechanisms at the heart of MOCK: we discuss a simple example demonstrating the synergistic effects at work in multiobjective clustering, which explain its superiority to single-objective clustering techniques, and we analyse how MOCK's Pareto fronts compare to the performance curves obtained by single-objective algorithms run with a range of different numbers of clusters specified.

Keywords: Clustering, multiobjective optimization, evolutionary algorithms, automatic determination of the number of clusters.

1 Introduction

Clustering is commonly defined as the task of finding natural groups within a data set such that data items within the same group are more similar than those within different groups. This is an intuitive but rather 'loose' concept, and it remains quite difficult to realize in general practice. Evidently, one reason for the difficulty is that, for many data sets, no unambiguous partitioning of the data exists, or can be established, even by humans. But even in cases where an unambiguous partitioning of the data *is* possible, clustering algorithms can drastically fail. This is because most existing clustering techniques rely on estimating the quality of a particular partitioning by means of just one *internal evaluation function*, an objective function that measures intrinsic properties of a partitioning, such as the spatial separation between clusters or the compactness of clusters. Hence, the internal evaluation function is assumed to reflect the quality of the partitioning reliably, an assumption that may be violated for certain data sets.

Given that many objective functions for clustering are complementary, the simultaneous optimization of several such objectives may help to overcome this weakness. In previous work [6], we have demonstrated this idea, showing that the simultaneous optimization of two clustering objectives results in clear performance gains with respect to single-objective clustering algorithms. However, the algorithm presented, VIENNA (Voronoi initialized evolutionary nearest neighbour algorithm), was limited in two respects: its application required knowledge of the correct number of clusters, and no mechanism was presented to select good solutions from the Pareto front obtained.

Our new algorithm, MOCK (multiobjective clustering with automatic determination of the number of clusters), overcomes these weaknesses. It uses a novel, flexible representation that permits us to efficiently generate clustering solutions that both correspond to different trade-offs between our two clustering objectives and that contain different numbers of clusters. An automated technique is employed to select high-quality solutions from the resulting Pareto front, and it thus simultaneously determines the number of clusters in a data set. We briefly present the algorithm in this paper and summarize analytical results demonstrating its robust performance across data sets that exhibit a wide range of different data properties. A more detailed description and additional analytical results are provided in [7]. Besides introducing MOCK, a second goal of this paper is to give more insight into the mechanisms underlying multiobjective clustering: in particular, we aim to show that MOCK's good performance arises as a direct consequence of the simultaneous optimization of several clustering objectives: an archetypal problem — serving to illustrate the synergistic effects at work in multiobjective clustering — is used for this purpose. We additionally underline the differences between single- and multiobjective clustering by comparing the shape of MOCK's Pareto fronts to the performance curves obtained for single-objective clustering methods run with a varying number of clusters specified.

The remainder of this paper is organized as follows. Section 2 briefly summarizes related work on clustering and evolutionary algorithms. This is followed by a description of our algorithm, MOCK (Section 2.1), and all other contestant methods used in this study (Section 3). Section 4 presents our experiments and discusses results, and Section 5 concludes.

2 Related Work

Clustering problems arise in a variety of different disciplines, ranging from biology to sociology to computer science. Consequently, they have been the subject of active research for several decades, and a multitude of clustering methods exist nowadays, which fundamentally differ in their basic principles and in the properties of the data they can tackle. For an extensive survey of clustering problems and algorithms the reader is referred to Jain et al. [8].

Evolutionary algorithms (EAs) have a history of being applied to clustering problems [4, 12, 13]. However, previous research in this respect has been limited to the single objective case: k-means' criterion of intra-cluster variance has been the objective most commonly employed, as this measure provides smooth incremental guidance in all parts of the search space. Due to the difficulty of deriving an encoding and operators that efficiently explore the very large clustering search space, actual hybridizations between EAs and k-means have also been particularly popular [12, 13].

Multiobjective evolutionary algorithms (MOEAs) have repeatedly been used to perform feature selection for clustering [5, 9], but have not previously been applied to the actual clustering task itself. This is despite the general agreement that clustering objectives can be conflicting or complementary, and that no single clustering objective can deal with every conceivable cluster structure [10]. To date, attempts to deal with this problem have focused on the *retrospective* combination of different clustering results by means of ensemble methods [11, 16, 18]. In order to construct clustering ensembles, different clustering results are first generated by repeatedly running the same algorithm (using different initializations, bootstrapping or a varying number of clusters) or several complementary methods (e.g. agglomerative algorithms based on diverse linkage criteria such as single-link and average-link). The resulting solutions are then combined into an ensemble clustering using graph-based approaches, expectation maximization or co-association methods.

Results reported in the literature demonstrate that clustering ensembles are often more robust and yield higher quality results than individual clustering methods, indicating that the combination of several clustering objectives is favourable. However, it is our contention that ensemble methods do not fully exploit the potential of using several objectives, as they are limited to the *a posteriori* integration of solutions rather than exploring trade-off solutions *during* the clustering process. We aim to overcome this limitation by tackling clustering as a truly multiobjective optimization problem.

2.1 Multiobjective Clustering

We based our multiobjective clustering algorithm on the elitist MOEA, PESA-II, described in detail in [3].

PESA-II. Briefly, PESA-II updates, at each generation, a current set of non-dominated solutions stored in an external population (of non-fixed but limited size), and uses this to build an internal population of fixed size to undergo reproduction and variation. PESA-II uses a selection policy designed to give equal reproduction opportunities to all regions of the current nondominated front; thus in the clustering application, it should provide a diverse set of solutions trading off different clustering measures. No critical parameters are associated with this 'niched' selection policy, as it uses an adaptive range equalization and normalization of the objectives. PESA-II may be used to optimize any number of objective functions, allowing us to simultaneously optimize several clustering measures, but in our algorithm MOCK we use just two (conceptually distant) measures as objectives, described below.

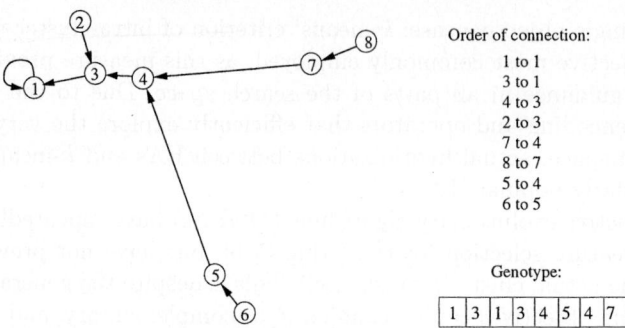

Fig. 1. Construction of the minimum spanning tree and its genotype coding. The data item with label 1 is first connected to itself, then Prim's algorithm is used to connect the other items. In the genotype, each gene (i.e. position in the string) represents the respective data item, and its allele value represents the item it points to (e.g. gene 2 has allele value 3 because data item 2 points to data item 3). The genotype coding for the full MST (as shown) is used as the first individual in the EA population

Genetic Representation and Operators. To apply PESA-II to the clustering problem, a suitable genetic encoding of a partitioning and one or more genetic variation operators (e.g. mutation and/or crossover) have to be chosen.

We employ the locus-based adjacency representation proposed in [14]. In this graph-based encoding (see Figure 1), each individual g consists of N genes $g_1, \ldots, g_N$, where N is the size of the data set given, and each gene g_i can take allele values j in the range $\{1, \ldots, N\}$. Thus, a value of j assigned to the ith gene, is then interpreted as a link between data items i and j: in the resulting clustering solution they will be in the same cluster. The decoding of this representation requires the identification of all subgraphs, which can be done in linear time. All data items belonging to the same subgraph are then assigned to one cluster.

This encoding scheme permits us to keep the number of clusters dynamic and is well-suited for use with standard crossover operators such as uniform, one-point or two-point crossover. We choose uniform crossover, and employ a specialized mutation operator that significantly reduces the size of the search space: each data item can only be linked to one of its L nearest neighbours. Hence, $g_i \in \{nn_{i1}, \ldots, nn_{iL}\}$, where nn_{il}, $l \in 1 \ldots L$, denotes the lth nearest neighbour of data item i.

Our initialization routine also exploits the link-based encoding and uses minimum spanning trees (MSTs). For a given data set, we first compute the complete MST using Prim's algorithm. The ith individual of the initial populations is then initialized by the MST with the $(i-1)$th largest links removed (see Figure 1).

Objective Functions. MOCK's clustering objectives have been chosen to reflect two fundamentally different aspects of a good clustering solution: the global concept of *compactness of clusters*, and the more local one of *connectedness of data points*.

In order to express cluster compactness we calculate the *overall deviation* of a partitioning. This is simply computed as the overall summed distances between data items and their corresponding cluster centre:

$$Dev(C) = \sum_{C_k \in C} \sum_{i \in C_k} \delta(i, \mu_k),$$

where C is the set of all clusters, μ_k is the centre of cluster C_k and $\delta(.,.)$ is the distance function chosen (Euclidean distance in this paper). As an objective, overall deviation should be minimized.

As an objective reflecting cluster connectedness, we use a measure, connectivity, which evaluates the degree to which neighbouring data-points have been placed in the same cluster. It is computed as

$$Conn(C) = \sum_{i=1}^{N} \left(\sum_{j=1}^{L} x_{i, nn_i(j)} \right), \quad x_{i, nn_i(j)} = \begin{cases} \frac{1}{j} & \text{if } \nexists C_k : i, nn_i(j) \in C_k \\ 0 & \text{otherwise}, \end{cases}$$

where $nn_i(j)$ is the jth nearest neighbour of datum i, and L is a parameter determining the number of neighbours that contribute to the connectivity measure. As an objective, connectivity should be minimized.

2.2 Automatic Solution Selection

The MOEA described above returns a set of nondominated solutions corresponding to different compromises of the two objectives, and containing different numbers of clusters. We next devise an automated method to select a single one of these solutions as the correct 'answer'. Unlike previous approaches to identifying promising solutions from the Pareto front (e.g. [2]), our method is application-specific. In particular, it makes use of several domain-specific considerations (inspired by Tibshirani et al.'s Gap statistic [17]) that enable us to derive a more effective technique for our particular purpose.

Intuitively, we expect the structure of the data to be reflected in the shape of the Pareto front. From the two objectives employed, overall deviation decreases with an increasing number of clusters, whereas connectivity increases. Hence, we can say that, incrementing the number of clusters k, we gain an improvement in overall deviation δD at the cost of a degradation in connectivity δC. For a number of clusters k smaller than the true number, we expect the ratio $R = \frac{\delta D}{\delta C}$ to be large: the separation of two clusters will trigger a great decrease in overall deviation, with only a small or no increase in connectivity. When we surpass the correct number of clusters this ratio will diminish: the decrease in overall deviation will be less significant but come at a high cost in terms of connectivity (because a true cluster is being split). Due to the natural bias of both measures, the solutions in the Pareto front are approximately ordered by the number of clusters they contain: plotting connectivity on the abscissa and overall deviation on the ordinate, k gradually increases from left to right. The distinct change in R occurring for the correct number of clusters can therefore be seen as a 'knee'. In order to help us correctly determine this knee, we use uniformly random reference data: clustering a number of such reference distributions using MOCK provides us with a set of 'reference fronts', which help us to abstract from k-specific biases in our clustering objectives (see Figure 2).

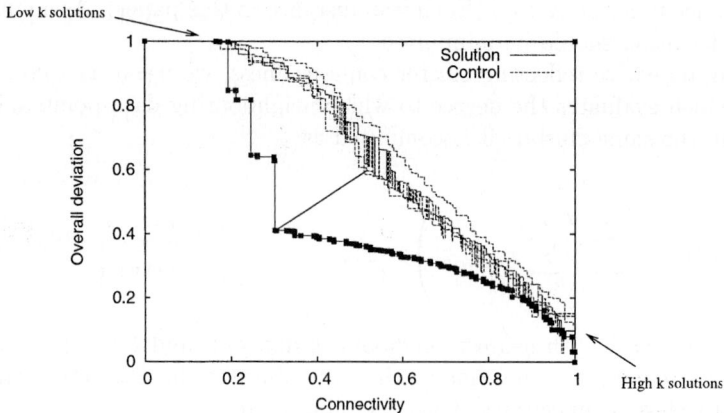

Fig. 2. Solution and control reference fronts for a run of MOCK on the *Square1* data set. The solution with the largest minimum distance to the reference fronts is indicated by the angled line, and corresponds to the correct $k = 4$ cluster solution

Briefly, we then determine good solutions as follows: after a normalization step, the distances between individual solutions and the attainment surfaces described by the reference fronts are computed. We plot the resulting *attainment scores* as a function of the number of clusters k. The maximum of the resulting curve provides us with the number of clusters k; also, the solution corresponding to the highest attainment score for this k is selected as the best solution. A more detailed motivation and description of this methodology (including pseudo-code) is provided in [7].

3 Contestant Methods

We evaluate MOCK by comparing it to four established single-objective clustering methods, whose implementations are described below. Three of these contestants are traditional and conceptually different clustering algorithms, namely k-means, single-link agglomerative clustering and average-link agglomerative clustering. Here, the algorithms k-means and single-link agglomerative clustering are of particular interest, as each of them uses a clustering objective that is conceptually quite similar to one of MOCK's. The fourth algorithm is an advanced clustering ensemble method by Strehl and Ghosh [16]. All four algorithms are

— differently to MOCK — given the same advantage of being provided with the correct number of clusters. A comparison of MOCK to a state-of-the-art method for the determination of the number of clusters, Tibshirani et al.'s Gap statistic [17], can be found in [7].

3.1 k-Means

Starting from a random partitioning, the k-means algorithm repeatedly (i) computes the current cluster centres (i.e. the average vector of each cluster in data space) and (ii) reassigns each data item to the cluster whose centre is closest to it. It terminates when no more reassignments take place. By this means, the intra-cluster variance, that is, the sum of squares of the differences between data items and their associated cluster centres, is locally minimized.

Our implementation of the k-means algorithm is based on the batch version of k-means, that is, cluster centres are only recomputed after the reassignment of all data items. As k-means can sometimes generate empty clusters, these are identified in each iteration and are randomly reinitialized. This enforcement of the correct number of clusters can prevent convergence, and we therefore set the maximum number of iterations to 100. To reduce suboptimal solutions k-means is run repeatedly (100 times) using random initialisation (which is known to be an effective initialization method [15]) and only the best result in terms of intra-cluster variance is returned.

3.2 Hierarchical Clustering

In general, agglomerative clustering algorithms start with the finest partitioning possible (i.e. singletons) and, in each iteration, merge the two least distant clusters. They terminate when the target number of clusters has been obtained. Single-link and average-link agglomerative clustering only differ in the linkage metric used. For the linkage metric of average-link, the distance between two clusters C_i and C_j is computed as the average dissimilarity between all possible pairs of data elements i and j with $i \in C_i$ and $j \in C_j$. For the linkage metric of single-link, the distance between two clusters C_i and C_j is computed as the smallest dissimilarity between all possible pairs of data elements i and j with $i \in C_i$ and $j \in C_j$.

3.3 Cluster Ensemble

Strehl and Ghosh's 'knowledge reuse framework' employs three conceptually different ensemble methods namely (1) CSPA (Cluster-based Similarity Partitioning Algorithm), (2) HGPA (Hyper-Graph Partitioning Algorithm) and (3) MCLA (Meta-Clustering Algorithm). The solutions returned by the individual combiners then serve as the input to a supra-consensus function, which selects the best solution in terms of average shared mutual information.

For the implementation of this cluster ensemble we use Strehl and Ghosh's original Matlab code with the correct number of clusters provided. In order to generate the input labels we use the algorithms described above, that is, k-means, average-link and single-link agglomerative clustering. As ensemble methods generally benefit from being provided partitionings of higher resolution (i.e. comprising more clusters), we run each algorithm for all $k \in \{2,\ldots,20\}$. The resulting 57 labelings then serve as the input to Strehl and Ghosh's method.

3.4 Parameter Settings for MOCK

Parameter settings for MOCK are given in Table 3.4 and are kept constant over all experiments.

Table 1. Parameter settings for MOCK, where N is data set size

Parameter	setting
Number of generations	200
External population size	1000
Internal population size	$\max(50, \frac{N}{20})$
Initialization	Minimum spanning tree
Mutation type	L nearest neighbours ($L = 20$)
Mutation rate p_m	$1/N$
Recombination	Uniform crossover
Recombination rate p_r	0.7
Objective functions	Overall deviation and connectivity ($L = 20$)
Constraints	$k \in \{1,\ldots,25\}$, cluster size > 2
Number of reference distributions	5

4 Experiments

The following experimental section is split into two major parts. We first provide a summary of our comparative study between MOCK and the clustering methods introduced above. MOCK's high performance leads us to to investigate the reasons for the robustness of multiobjective clustering; an archetypal example and the visualization of solutions in two-objective space are used for this purpose.

4.1 Analytical Evaluation

In our comparative study, MOCK has been evaluated using a range of synthetic and real data sets. In this paper, we present results on 19 two-dimensional data sets that are well-suited to demonstrate MOCK's performance for different data properties. The particular properties studied are overlap between clusters (on the *Square* and *Triangle* series), unequally sized clusters (in terms of the number of data points contained; on the *Sizes* series) and elongated cluster shapes (on the *Long*, *Spiral* and *Smile* series). Detailed descriptions and generators for the data sets are available at [1]. More results, including those for high-dimensional and real data, can be found in [7].

The clustering results of the five different algorithms are compared using an objective evaluation function, the *F-Measure* [19], which is an external evaluation

Table 2. Sample Median and Interquartile Range F-Measure [19] values for 50 runs of each algorithm on two-dimensional synthetic data sets exhibiting different data properties. The best value is shown in bold. MOCK is the only consistent method; all other algorithms have a 'nemesis' data set on which they are worst

Problem	single-link	average-link	k-means	clustering ensemble	MOCK
Long1	0.666444 (0.333556)	0.665104 (0.005896)	0.521989 (0.015211)	**1.0 (0.0)**	**1.0 (0.0)**
Long2	0.678444 (0.000178)	0.67714 (0.011155)	0.520026 (0.01683)	0.902 (0.0)	**1.0 (0.0)**
Long3	0.777895 (0.000556)	0.77514 (0.009774)	0.566661 (0.017321)	0.730591 (0.001802)	**1.0 (0.006)**
Sizes1	0.428323 (0.000242)	0.977935 (0.007071)	**0.989003 (0.005013)**	0.747616 (0.030904)	0.987 (0.006)
Sizes2	0.522742 (0.000477)	0.981947 (0.009885)	0.987051 (0.004999)	0.633283 (0.002187)	**0.988 (0.006)**
Sizes3	0.600841 (0.000782)	0.98502 (0.00905)	0.987114 (0.006899)	0.562078 (0.00984)	**0.99 (0.005)**
Sizes4	0.658308 (0.000676)	0.983953 (0.005826)	0.985274 (0.006851)	0.506595 (0.261149)	**0.989 (0.0041)**
Sizes5	0.702411 (0.001261)	0.986976 (0.007064)	0.984288 (0.005843)	0.487591 (0.311809)	**0.9909 (0.0079)**
Smile1	**1.0 (0.0)**	0.753036 (0.0)	0.665609 (0.009407)	**1.0 (0.0)**	**1.0 (0.0)**
Smile2	**1.0 (0.0)**	0.725156 (0.0)	0.586508 (0.009967)	0.91 (0.0)	**1.0 (0.0)**
Smile3	**1.0 (0.0)**	0.549761 (0.0)	0.505994 (0.007393)	0.776494 (0.001284)	**1.0 (0.0)**
Spiral	**1.0 (0.0)**	0.576 (0.0)	0.593 (0.002)	**1.0 (0.0)**	**1.0 (0.0)**
Square1	0.399759 (8e-05)	0.977997 (0.015005)	**0.987006 (0.004982)**	0.984 (0.008006)	0.985 (0.0051)
Square2	0.399759 (0.0)	0.961982 (0.009888)	**0.976019 (0.007988)**	0.97 (0.008002)	0.973 (0.009)
Square3	0.399759 (8e-05)	0.934935 (0.016238)	**0.956933 (0.00802)**	0.94599 (0.015982)	0.946 (0.0172)
Square4	0.399759 (8e-05)	0.883035 (0.02214)	**0.919999 (0.008024)**	0.908 (0.019006)	0.9041 (0.0184)
Square5	0.399759 (8e-05)	0.720672 (0.107357)	**0.86798 (0.014231)**	0.842965 (0.033088)	0.8361 (0.0324)
Triangle1	**1.0 (0.0)**	0.997 (0.004001)	0.98486 (0.00613)	0.999 (0.001)	**1.0 (0.0)**
Triangle2	0.45193 (0.116834)	0.986979 (0.013638)	0.957697 (0.011837)	0.810492 (0.068513)	**0.995 (0.004)**

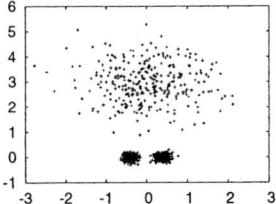

Fig. 3. A simple three-cluster data set posing difficulties to many clustering methods

function that requires knowledge of the correct class labels and compares a generated clustering solution to this 'gold standard'. The F-Measure can take on values in the interval $[0, 1]$ and should be maximized.

From the results presented in Table 2 it becomes clear that the single-objective algorithms all experience trouble for particular data properties. For k-means and average-link, the problematic data sets are those with arbitrary shaped or elongated clusters (in the *Long*, *Smile* and *Spiral* series) for single-link it is those that contain clusters with overlap or noise points (in the *Square*, *Sizes* and *Long* series). The clustering ensemble fares better, but severely breaks down for unequally sized clusters (in the *Sizes* series). Only MOCK shows a very strong performance for all types of different data properties and is best, or close to best, on almost all data sets.

4.2 A Simple Example Showing the Synergistic Effects at Work in MOCK

Given MOCK's performance, it is worth examining more closely the mechanisms underlying multiobjective clustering. In particular, we would like to show that MOCK's *access* to two conceptually different clustering objectives is *not sufficient* to explain its good performance, but that it is their *simultaneous optimization* that is the key to its success.

We first observe that, in an explicit single-objective optimization, it wouldn't be possible to keep the number of clusters dynamic (without the introduction of additional constraints) at all: due to the natural bias of both measures, an optimization method would, for *any data set*, necessarily converge to trivial solutions: these are singleton clusters for overall deviation and a single cluster for connectivity. Only the simultaneous optimization allows us to effectively explore interesting solutions, as we can exploit the oppositional trends exhibited by the two objectives with respect to changes in the number of clusters.

More importantly, even for a fixed number of clusters, there are cases where both single-objective versions will fail, but a multiobjective approach will work. In order to demonstrate this, let us consider the scenario shown in Figure 3. The data set consists of three clusters that contain the same number of data points and are easily discernible to the human eye. The difficulty of the data set arises from two facts: (1) the clusters are constituted from data points with very different densities, and (2) a couple of noise points at the border of the sparse cluster 'bridge the gap' to the smaller two clusters. As a direct result of these properties the assumptions made by overall deviation and connectivity are both violated:

- The correct three-cluster solution does not correspond to the minimum in overall deviation, as splitting of the large cluster results in a much greater reduction in overall deviation.
- The correct three-cluster solution does not correspond to the minimum in connectivity, as the separation of the large from the small clusters involves an increase in connectivity, and the assignment of outliers to their own cluster therefore becomes preferable.

Consequently, the optimization of any *one* of the two objectives, will not lead to the correct solution. Indeed, none of our contestant single-objective algorithms can solve this data set: they all fail to separate the two smaller clusters — instead, k-means splits the large cluster in half, and single-link and average-link separate outliers from this cluster.

MOCK in contrast, which explores the *trade-offs* between overall deviation and connectivity, easily manages to detect the correct solution. In particular, in many cases, MOCK can even *discard* those solutions that are optimal only under *overall deviation* or *connectivity*, as these evaluate particularly badly under the respective second objective, and are therefore (due to the natural biases in the measures) likely to be dominated by clustering solutions with a different number of clusters. Overall, the performance of MOCK seems to demonstrate that the quality of the trade-off between our two objectives is *in many cases* a much better indicator of clustering quality than the individual objectives themselves.

4.3 Performance Curves Versus Pareto Fronts

In this last section, we aim to visualize the different strategies pursued by single-objective clustering algorithms and our two-objective version. Towards this end, we compare the output of MOCK with the solutions obtained by k-means, average-link and single-link agglomerative clustering, when run for a range of different numbers of clusters $k \in 1, \ldots, 25$. We then evaluate all resulting solutions using overall deviation and connectivity, and visualize, in normalized two-objective space, the solutions obtained (see Figure 4). For MOCK, the resulting curve is simply its Pareto front and therefore monotonic. For the other three algorithms the resulting performance plots may contain dominated points, and may therefore be non-monotonic. This is because each of these algorithms only optimizes one of the two objectives: thus, solutions that are better in one objective are not necessarily worse in the other.

A combined analysis of these plots and the F-Measure values provided in Table 2 confirms that the trade-off between overall deviation and connectivity is a good indicator of quality on the data sets employed. There appears to be a distinct correlation between the quality of solutions and their distance to the 'knee' in MOCK's Pareto front. In particular, the algorithms that perform well on a given data set have at least one solution that comes close to or even dominates MOCK's solutions in the 'knee' region. This is generally the solution that best captures the structure of the data set, and also scores highest under the F-Measure. Straightforward examples of this are the k-means and average-link solutions on the *Square4* and *Sizes5* data sets, and the single-link solution on the *Spiral* data set. Results for the *Long1* data set are slightly more involved. Here, k-means and average-link are clearly both widely off track. The F-Measure value for single-link indicates that, due to noise points, the algorithm also has problems to find the correct clustering solution for $k = 2$. This is confirmed by the fact that, in Figure 4, the $k = 2$ solution for single-link is quite far off the 'knee'. However, the plot also reveals (and F-Measure values confirm) that there *are* single-link solutions closer to the optimum: these are the solutions for $k = 4$ or higher, for which single-link assigns noise points to their own clusters but *also* manages to correctly separate the two main clusters present in the data set.

In addition to the above, our visualization demonstrates some general aspects of the algorithms' clustering behaviour. On several data sets, we can observe single-link's tendency to gradually separate outliers, without any significant decrease in overall deviation (even for high numbers of clusters). Clearly, this property makes the algorithm very sensitive to noise, which is one of the reasons why it is rarely applied in practice. k-means, in contrast, quickly reduces overall deviation, but pays virtually no attention to the underlying local data distribution. This becomes evident in a rapid increase and apparent non-monotonicity (e.g. see *Long1*) in connectivity. Only average-link agglomerative clustering shows a more reasonable overall behaviour: its capabilities to satisfy both objectives seem

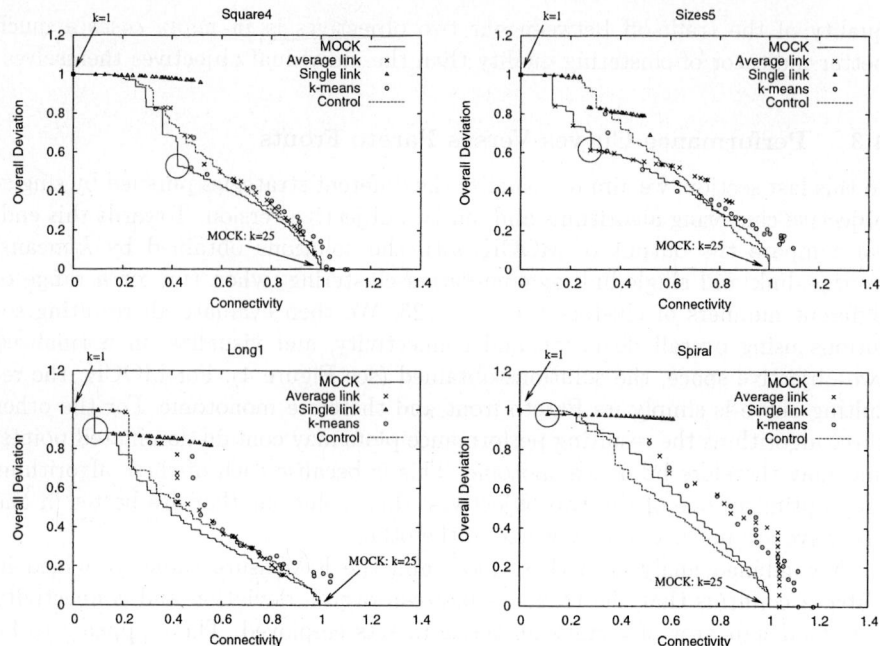

Fig. 4. Plots of all solutions, in normalized objective space, for $k \in [1, \ldots, 25]$ of k-means, single-link and average-link on the *Square4*, *Sizes5*, *Long1* and *Spiral* data sets. Both MOCK's Pareto front and one of its control fronts are visualized as attainment surfaces, and arrows indicate the position of the $k = 1$ solution (identical for all algorithms) and MOCK's $k = 25$ solution. Knees are indicated by circles centred around the solution identified by MOCK

somewhat superior to those of k-means and single-link. For increases in k, it is monotonic in both objectives and its solutions are distributed more uniformly along the Pareto front.

5 Conclusion

Existing clustering algorithms are limited to optimizing (explicitly or otherwise) one single clustering objective. This can lead to a lack of robustness with respect to different data properties, a limitation which we have suggested can be overcome by the use of several complementary objectives. In this paper we have introduced an advanced multiobjective clustering algorithm, MOCK. A comparative study has shown the robustness of the approach both at finding high quality solutions and determining the number of clusters. The origins of MOCK's good performance have then been investigated further. Using an archetypal problem we have demonstrated that the identification of trade-off solutions in clustering can be crucial: the simultaneous optimization of two complementary objectives may permit the solution of clustering problems that are not solvable by either of the two objectives individually. The differences between single-objective and multiobjective clustering have been further underlined by a visual comparison of the quality of the trade-off solutions generated by single- and multiobjective algorithms for a range of different numbers of clusters. The software for MOCK is available on request from the first author.

Acknowledgements. JH gratefully acknowledges support of a scholarship from the Gottlieb Daimler- and Karl Benz-Foundation, Germany. JK is supported by a David Phillips Fellowship from the Biotechnology and Biological Sciences Research Council (BBSRC), UK.

References

1. Supporting material for MOCK. http://dbk.ch.umist.ac.uk/handl/mock/
2. J. Branke, K. Deb, H. Dierolf, and M. Osswald. Finding knees in multi-objective optimization. In *Proceedings of the Eighth International Conference on Parallel Problem Solving from Nature*, pages 722–731. Springer-Verlag, 2004.
3. D. W. Corne, J. D. Knowles, and M. J. Oates. PESA-II: Region-based selection in evolutionary multiobjective optimization. In *Proceedings of the Genetic and Evolutionary Computation Conference*, pages 283–290. Morgan Kaufmann, 2001.
4. E. Falkenauer. *Genetic Algorithms and Grouping Problems*. John Wiley & Son Ltd, 1998.
5. G. Fleurya, A. Hero, S. Zareparsi, and A. Swaroop. Gene discovery using Pareto depth sampling distributions. *Special Number on Genomics, Signal Processing and Statistics, Journal of the Franklin Institute*, 341(1–2):55–75, 2004.
6. J. Handl and J. Knowles. Evolutionary multiobjective clustering. In *Proceedings of the Eighth International Conference on Parallel Problem Solving from Nature*, pages 1081–1091. Springer-Verlag, 2004.
7. J. Handl and J. Knowles. Multiobjective clustering with automatic determination of the number of clusters. Technical Report COMPYSYBIO-TR-2004-02, Department of Chemistry, UMIST, UK, August 2004.
8. A. K. Jain, M. N. Murty, and P. J. Flynn. Data clustering: A review. *ACM Computing Surveys*, 31(3):264–323, 1999.
9. Y. Kim, W. N. Street, and F. Menczer. Evolutionary model selection in unsupervised learning. *Intelligent Data Analysis*, 6:531–556, 2002.
10. J. Kleinberg. An impossibility theorem for clustering. In *Proceedings of the 15th Conference on Neural Information Processing Systems*, 2002. http://www.cs.cornell.edu/home/kleinber/nips15.ps.
11. M. H.C Law. Multiobjective data clustering. In *Proceedings of the IEEE Computer Society Conference on Computer Vision and Pattern Recognition*, pages 424–430. IEEE Press, 2004.
12. U. Maulik and S. Bandyopadhyay. Genetic algorithm-based clustering technique. *Pattern Recognition*, 33:1455–1465, 2000.
13. H. Pan, J. Zhu, and D. Han. Genetic algorithms applied to multi-class clustering for gene expression data. *Genomics, Proteomics & Bioinformatics*, 1(4), 2003.
14. Y.-J. Park and M.-S. Song. A genetic algorithm for clustering problems. In *Proceedings of the Third Annual Conference on Genetic Programming*, pages 568–575. Morgan Kaufmann, 1998.

15. J. M. Pena, J. A. Lozana, and P. Larranaga. An empirical comparison of four initialization methods for the k -means algorithm. *Pattern Recognition Letters*, 20(10):1027–1040, 1999.
16. A. Strehl and J. Ghosh. Cluster ensembles — a knowledge reuse framework for combining multiple partitions. *Journal on Machine Learning Research*, 3:583–617, 2002.
17. R. Tibshirani, G. Walther, and T. Hastie. Estimating the number of clusters in a dataset via the Gap statistic. Technical Report 208, Department of Statistics, Stanford University, USA, 2000.
18. A. Topchy, A. K. Jain, and W. Punch. Clustering ensembles: Models of consensus and weak partitions. Submitted to IEEE Transactions on Pattern Analysis and Machine Intelligence, 2004.
19. C. van Rijsbergen. *Information Retrieval, 2nd edition*. Butterworths, 1979.

Extraction of Design Characteristics of Multiobjective Optimization – Its Application to Design of Artificial Satellite Heat Pipe

Min Joong Jeong[1], Takashi Kobayashi[2], and Shinobu Yoshimura[3]

[1] Computational Biology Research Center, National Institute of Advanced Industrial Science and Technology, 2-43 Aomi, Koto-ku, Tokyo, 135-0064 Japan
jeong@cbrc.jp, http://www.cbrc.jp
[2] Design Systems Engineering Center, Mitsubishi Electric Corp., 5-1-1 Ofuna, Kamakura, Kanagawa, 247-8501 Japan
[3] Institute of Environmental Studies, The University of Tokyo, 7-3-1 Hongo, Bunkyo-ku, Tokyo, 113-8656 Japan

Abstract. An artificial satellite design requires severe design objectives such as performance, reliability, weight, robustness, cost, and so on. To solve the conflicted requirements at the same time, multiobjective optimization is getting more popular in the design. Using the optimization, it becomes ordinary to get many solutions, such as Pareto solutions, quasi-Pareto solutions, and feasible solutions. The alternative solutions, however, are very difficult to be adopted to practical engineering decision directly. Therefore, to make the decision, proper information about the solutions in a function, parameter and real design space should be provided. In this paper, a new approach for the interpretation of Pareto solutions is proposed based on multidimensional visualization and clustering. The proposed method is applied to a thermal robustness and mass optimization problem of heat pipe shape design for an artificial satellite. The information gleaned from the propose approach can support the engineering decision for the design of artificial satellite heat pipe.

1 Introduction

A multiobjective optimization yields ideally innumerable alternative solutions known as Pareto solutions. There is no any superior one in the solutions because of their definition. However it is a very difficult task to judge Decision-Making (DM) from Pareto solutions. Two kinds of methods are well known to overcome the difficulty. First, preference methods give fixed preferences to objectives before multiobjective optimization, and then find one solution for DM. Second, trade-off methods are used to make DM from Pareto solutions after optimization. However, attempts to support DM with the preference or trade-off may result in poor design characters [1, 2]. That is because the DM using one solution may be inadequate when we include all factors that influence the choice of a particular design, such as durability and manufacturability. Therefore, if the whole set of

Pareto solutions are provided to engineers with proper information, they can use this information to choose the best overall design.

Visualization has been one of the most useful tools to guide information of correlations and characters between parameters, functions, and actual design shapes [3]. However, Pareto solutions are normally in a multidimensional space, and deter from extracting the information. To overcome this difficulty, the authors propose a synchronous 3D visualization. In the visualization, a multidimensional parameter and function space is divided into several 2D or 3D subspaces, and visualized in parallel. Each datum corresponds to a line segment between the subspaces and its shape in the real design space at the same time. Therefore, engineers can understand the correlation and effect of each datum in optimized solutions.

Furthermore, it is introduced that a clustering approach which considers a set of Pareto solutions as a group of several distinct clusters. This approach is based on the concept that the solutions consist of obvious characters in their function and parameter space. To measure the similarity and dissimilarity of solutions more essentially, the Euclidean distance and a point symmetry distance[4, 5] are hybrided.

As a practical engineering application, the proposed approach for interpretation of multiobjective solutions is applied to a thermal robustness and mass optimization problem of heat pipe shape for an artificial satellite [6]. Two functions and five parameters are considered in the design optimization. Using the proposed approach, we search for the information that can support engineering decision for the design of artificial satellite heat pipe.

2 Multiobjective Optimization of Artificial Satellite Heat Pipe

For a cooling system of artificial satellite, heat pipes based isothermal radiator panels are generally employed. Fin efficiency is dramatically improved using orthogonal interconnected (matrix layout) heat pipes as shown in Fig. 1. To maximize the fin efficiency of isothermal panels, the minimization of the temperature gradient between the lateral and header heat pipes becomes a very important design object [7]. On the other hand, saving the total mass (weight) of a thermal control subsystem is highly important to reduce load (pay load cost) on a booster-rocket. The satellite panels contain many embedded aluminum heat pipes, which generally occupy over 50% of the total mass of the fundamental radiator panels. Thus the thermal design of artificial satellite requires both the fin efficiency and mass saving of the heat pipes at the same time. Additionally, the operating temperature of the heat pipes is very widely ranging from $-20.0°C$ to $60.0°C$ in orbit. The thermal performance of the heat pipes must be stable in the temperature change. Therefore, the temperature dependency must be also taken into account for the heat pipe design. In this study, a combination of Response Surface Methodology (RSM) and Monte Carlo simulation is applied at first to formulate functions of the mass and thermal performance of the pipe

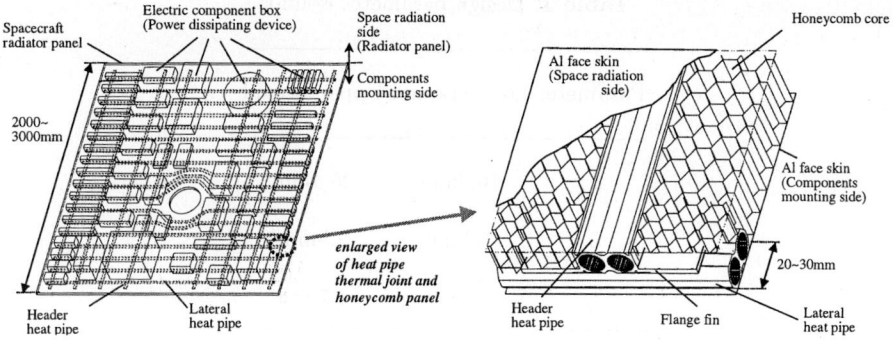

Fig. 1. Layout of heat pipes for satellite radiator

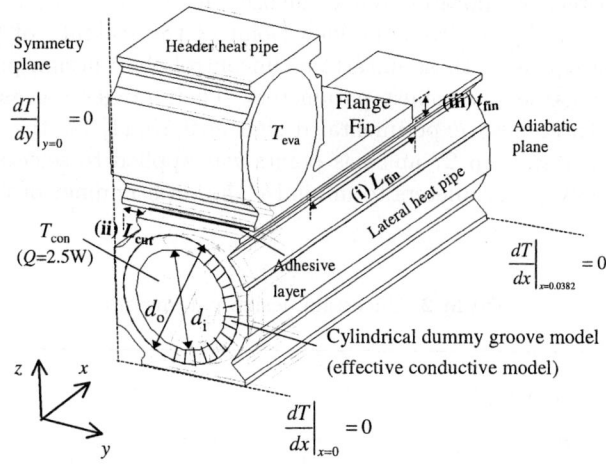

Fig. 2. Design parameters in 3D FE analysis model of heat pipes

structure [8, 10]. Here shape parameters of the heat pipes embedded within a satellite panel are design parameters, while the both of the mass and thermal robustness of the heat pipes are objective functions.

The design parameters determined by the mechanical designers are as follows: (i) length of fin (L_f), (ii) cutting length of adhesive attached area (L_c) and (iii) thickness of fin (t_f). These are illustrated in Fig. 2. Allowable ranges of the design parameters are given in Table 1. (iv) Adhesive thickness (t_b) and (v) operation temperature (T_{op}) are uncontrollable by the mechanical designers but affect the thermal performance of the heat pipes. The lateral and header heat pipes are bonded together by flange fin area with conductive epoxy. The adhesive thickness t_b has production tolerance, which influences the thermal performance of the heat pipes. To take account of this effect, it is assumed that the adhesive thickness has statistical normal distribution ranging from 0.12mm

Table 1. Design parameter bounds

Parameter	Lower Bound	Upper Bound
L_f	10.0mm	25.4mm
L_c	1.5mm	2.5mm
t_f	1.0mm	1.7mm

to 0.22mm. The operating temperature (T_{op}) for the heat pipes is required to range from $-20.0°C$ to $60.0°C$. The temperature dependency of heat transfer coefficient between evaporator and condenser at the inner wall of each heat pipe cannot be negligible from the view point of the stability of thermal performance. Consequently, these uncertain (uncontrollable) design parameters, t_b and T_{op}, are regarded as robust parameters. Their ranges are assumed as in Table 2. The three level experimental design [8, 9, 10, 11] for the 5 design parameters which results in 27 analysis points was applied to selecting combinations of the analysis parameters to minimize the total number of finite element analyses.

Table 2. Uncertain design parameters

Parameter	t_b	T_{op}
Lower Bound	0.12mm	$-20.0°C$
Upper Bound	0.22mm	$60.0°C$
Probability Distribution	Normal Distribution $\mu=0.17$, $\sigma=0.016$	Normal Distribution $\mu=20.0$, $\sigma=14.3$

2.1 Generation of Fitted Estimation Equations

27 Finite element analyses were performed to calculate the thermal performance of the parametric combinations which were given from the Taguchi orthogonal array - L27 [8, 9]. Those results were used to construct estimation equations for characteristic values G and M. Chebyshev's equation was considered to correlate

the regression coefficients in multiple linear regression models. The calculated value G is the thermal conductance across the thermal joint of the heat pipes, defined as:

$$G = \frac{Q}{T_{con} - T_{eva}} \quad (1)$$

where T_{con} is the condensing liquid temperature in the lateral heat pipe, T_{eva} is the evaporating vapor temperature in the header heat pipe, and Q is the assumed quantity of the transported heat of 2.5W per a thermal joint. The determined response surface equation of G is as follows:

$$\begin{aligned}\hat{G} = f(L_f, L_c, t_f, t_b, T_{op}) &= 0.3745378 - 0.9352909 t_b \\ &+ 1.01612 t_b^2 + 2.324128 e^{-2} L_c - 7.209993 e^{-3} L_c^2 \\ &+ 1.838379 e^{-3} L_f - 5.379707 e^{-5} L_f^2 + 2.447391 e^{-2} t_f \\ &+ 2.304583 e^{-3} t_f^2 - 6.483411 e^{-4} T_{op} - 9.232971 e^{-7} T_{op}^2 \\ &- 2.259702 e^{-2} t_b L_c - 4.735652 e^{-3} t_b L_c^2 + 0.1102442 t_b^2 L_c \\ &- 9.702533 e^{-3} t_b^2 L_c^2 + 5.382211 e^{-3} t_b L_f - 9.540484 e^{-5} t_b L_f^2 \\ &+ 5.15048 e^{-3} t_b^2 L_f - 1.232524 e^{-4} t_b^2 L_f^2 + 0.2972589 t_b t_f \\ &- 0.1052935 t_b t_f^2 - 0.5422262 t_b^2 t_f - 0.1829687 t_b^2 t_f^2\end{aligned} \quad (2)$$

The response surface equation for the total mass M also expressed in the following equation:

$$\begin{aligned}\hat{M} = f(L_f, L_c, t_f, t_b) &= (1313.877 - 75.5 * L_c + 11.0 L_c^2 \\ &+ 1.402597 L_f - 1.278314 e^{-15} L_f^2 + 62.38776 t_f \\ &- 6.122449 t_f^2 - 380.8 t_b + 1120 t_b^2) * 21;\end{aligned} \quad (3)$$

3 Multidimensional Pareto Solutions and Their Synchronous Visualization

To search the Pareto solutions of Eqs. 2 and 3, the Intermediate Tendency (IT) optimizer was used. Details of the optimizer described in Ref. [12, 13]. To put it briefly, the optimizer is a kind of genetic search algorithms, and consists of typical genetic operators such as fitness evaluation, selection, and mutation. To improve search efficiency, it adopts the IT recombination that is more robust in search ability than conventional intermediate recombination. In conventional recombination, such as the global intermediate recombination, any offspring individuals cannot deviate from d-dimensional search space covered by their parental individuals. In other word, if an optimum point is located out of the search space, any offspring cannot reach the optimum point by the recombination. The IT recombination, however, yields offsprings depending on a discrepancy between parental individuals and randomly selected ones. The discrepancy is considered as the *tendency* in an evolution process, and the offsprings are yielded by adding the tendency to their parental individuals. Therefore the individuals of subsequent

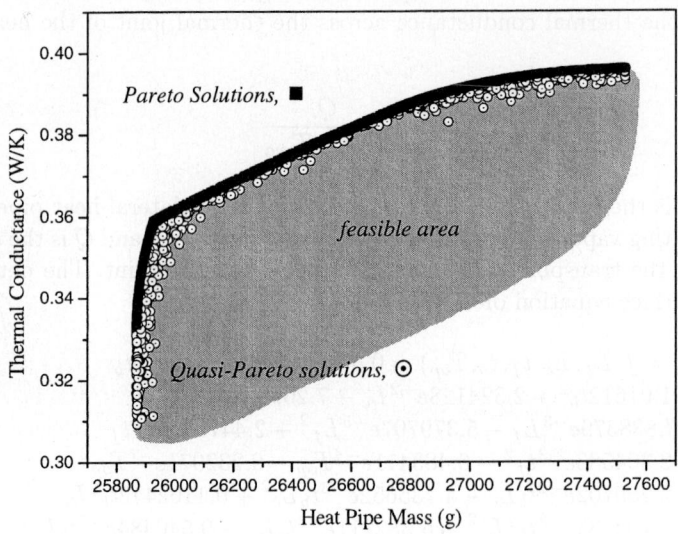

Fig. 3. Pareto solutions and quasi-Pareto solutions in function space

iteration are not bound to their parental search space. Superior performance of the IT recombination is shown in Ref. [12, 13].

For the multiobjective optimization, the optimizer is randomly changing preferences between the two objective functions and searches Pareto solutions in the function space. Figure 3 shows the Pareto solutions and also quasi-Pareto solutions, which were gathered during the optimization process. As shown in Fig. 3, there are many quasi-Pareto and Pareto solutions. Next it is explained how to search engineering information out of them.

3.1 Synchronous Visualization of Multidimensional Function and Parameter Spaces

Getting adequate information for determining a final solution is not an easy task especially if there are many Pareto solutions in a multidimensional space. Although parallel-coordinate methods [14, 15] handle multidimensional solutions, they have limitation in the number of solution sets and their amount of dimensions. To overcome the difficulty, a synchronous 3D visualization is proposed. Here, each of multidimensional parameter and function spaces is subdivided into several 2D or 3D subspaces, and visualized simultaneously. Each solution is visualized in all the subspaces, and those corresponding points in the subspaces are connected by line segments. The real world space is also visualized at the same time. Through the interactive operation of the present visualization system, engineers can explore and understand the correlation among multidimensional function and parameter spaces and the real world space.

In the present heat pipe optimization, the 5-dimensional design parameter space is split to 2D and 3D subspaces, i.e., the 2D space of t_b and T_{op}, and the 3D

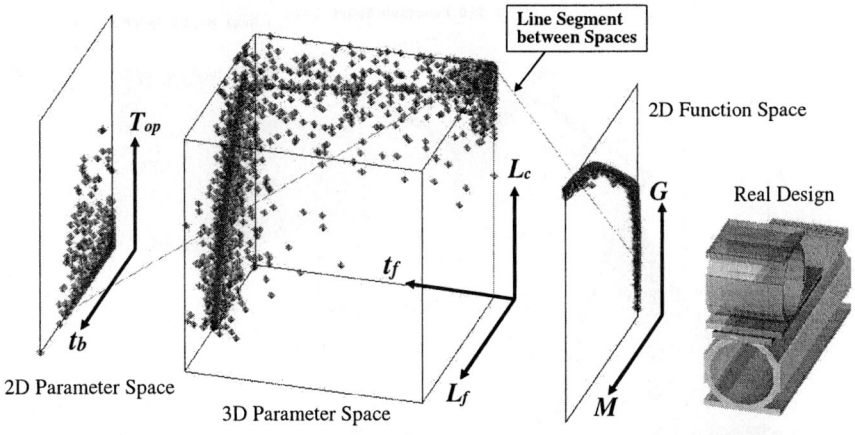

Fig. 4. Synchronous visualization of parameter subspaces, function space and real world design

space of L_f, L_c, and t_f, respectively. The 2D objective function space of $\hat{M}$ and $\hat{G}$ is remained as it is. The two parameter subspaces and the one function space are visualized simultaneously. Figure 4 shows the concept of the proposed visualization. The solutions in the original 7D space are separated into two parameter spaces and one function subspace. Moreover, a corresponding actual shape of the heat pipe is visualized with the subspaces. Corresponding points shown in different subspaces are connected by line segments. Therefore, it is easy to grasp the solution's correlation between the subspaces. The present visualization system was developed using the C language and the graphic-programming libraries named ADVENTURE AutoGL [16]. A programmable graphical user interface is also provided.

In Fig. 4, there are 3,552 quasi-Pareto solutions. It is clearly showed what kinds of parameter positions should make the optimum. For example, thin t_b and low T_{op} keep up the optimum. However, long L_f with small L_c and t_f does not correspond to the optimum. From this visualization, engineers can easily get ideas on parametric sensitivity of the present heat pipe shape design.

In addition, it can be interpreted the Pareto solutions. It is assessed the effects of t_b and T_{op} on the objective function space, i.e., of the mass and conductance in Fig. 5. In the figure, the line segments visualize the correlation among the subspaces. As shown in Fig. 5, the operating temperature is at its minimum, and only the variation of the adhesive thickness results in the changes of the objective functions $\hat{G}$ and $\hat{M}$. The three shape parameters L_f, L_c and t_f are almost tied up at the same point in their parameter subspace. Thus, without changing the shape parameter L_f, L_c and t_f, the design of heat pipes adopting thinner adhesive at lower operating temperature is expected to minimize the mass and to maximize the heat conductance. However, it is very difficult to

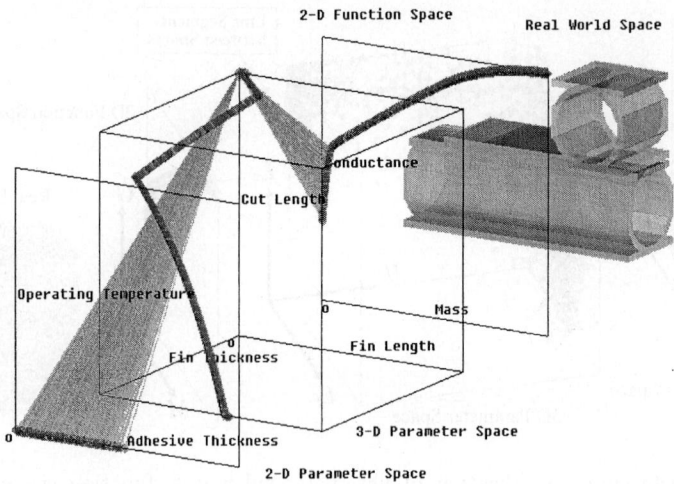

Fig. 5. Synchronous visualization of Pareto solutions of heat pipes

change the operating temperature because of the design limitation of an orbit and thermal control system. Moreover, controls over the thickness of adhesive involve great uncertainty in manufacturing. To improve heat pipe performance, therefore, it is required to focus on the correlation of the shape parameters and the objective functions. For this purpose, the equations of $\hat{G}$ and $\hat{M}$ are regenerated to consider uncertainty of t_b and T_{op}. That is, the equations consist of only the three shape parameters, while t_b and T_{op} are taken into account as probability distributions as in Table 2.

3.2 Estimation Equations Considering Uncertain Parameters

The adhesive thickness t_b is assumed to be composed of random values with the normal distribution of $\mu_{t_b} = 170.0\mu m$ and $\sigma_{t_b} = 16.7\mu m$. The operating temperature T_{op} is also assumed to be composed of random values with $\mu_{T_{op}} = 20.0K$ and $\sigma_{T_{op}} = 14.3K$. A direct sampling Monte Carlo simulation with the Box-Muller method is used to take into account random 2D parameters of the adhesive thickness t_b and the operating temperature T_{op}. The number of samples for the Monte Carlo simulation is 1,000,000. An average thermal conductance, $\bar{G}_R$, under consideration of the thermal robustness is defined as follows:

$$\bar{G}_R = \int_{-20}^{60} \int_{0.12}^{0.22} G(t_b, T_{op}) \cdot f_p(t_b, T_{op}) dt_b dT_{op} \qquad (4)$$

$$\cong \frac{1}{N} \sum_{m=1}^{1000} \sum_{n=1}^{1000} \hat{G}_R(t_{b,m}, T_{op,n}) \qquad (5)$$

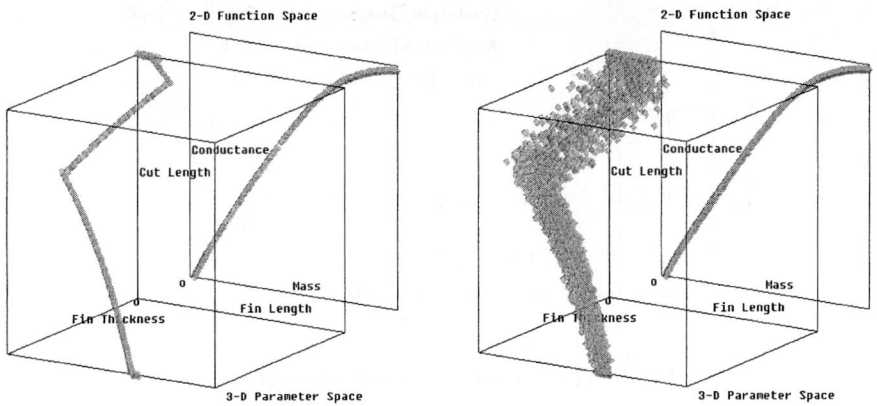

Fig. 6. Pareto (left) and quasi-Pareto (right) solutions of $\hat{\hat{G}}_R$ and $\hat{\hat{M}}_R$

where $f_p(x)$ is a probability density function, and N is the total number of samples for Monte Carlo Simulation ($N = m*n = 1,000,000$).

The fitted polynominal equations for G and M, which consider the probability density of the uncontrollable parameters t_b and T_{op}, are regenerated using a quadratic model as:

$$\hat{\hat{G}}_R = 2.261369e^{-2}L_c - 8.299937e^{-3}L_c{}^2 + 2.905449e^{-3}L_f \\ -7.364545e^{-5}L_f{}^2 + 5.925684e^{-2}t_f - 1.028177e^{-2}t_f{}^2 \\ +0.2312513 \tag{6}$$

$$\hat{\hat{M}}_R = (1283.375 + 1.402597L_f - 1.278314e^{-15}L_f{}^2 \\ -75.5L_c + 11.0L_c{}^2 + 62.38776t_f - 6.122449t_f{}^2)*21 \tag{7}$$

4 Clustering of Pareto Solutions

Figure 6 shows the Pareto and the quasi-Pareto solutions of $\hat{\hat{G}}_R$ and $\hat{\hat{M}}_R$. As shown in the figure, the Pareto solutions and the quasi-Pareto ones have almost the same function values. However, the quasi-Pareto solutions have much more variance in the parameter space than the Pareto ones. In the engineering sense, such quasi-Pareto solutions with parametric variance also seem beneficial because of their increasing design freedom such as manufacturability. Therefore, both Pareto and quasi-Pareto solutions should be examined in more detail before making the final decision of the heat pipe design.

To do so, one of clustering algorithms is applied to the solutions. Overviews of clustering algorithms can be found in Ref. [13, 17, 19]. If the solutions are appropriately classified into several clusters, it is expected that engineers can

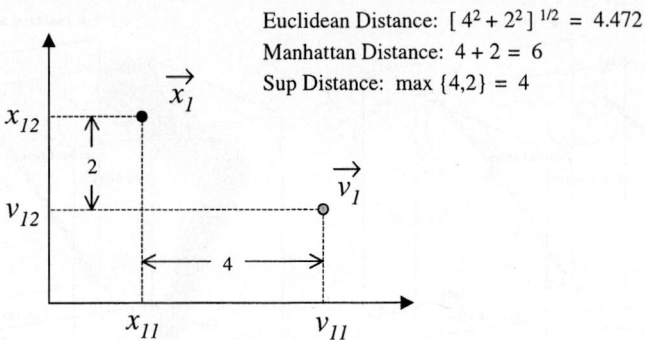

Fig. 7. Three common Minkowski metrics [17]

interpret mathematical as well as engineering characteristics of the solutions in a more abstract manner.

The clustering function without fuzziness is

$$C_F'(X, U, V) = \sum_{k=1}^{K} \sum_{i=1}^{n} u_{ik} \cdot dis(\boldsymbol{x}_i, \boldsymbol{v}_k) \qquad (8)$$

where $X = \{\boldsymbol{x}_1, \ldots, \boldsymbol{x}_n\} \subseteq \mathbb{R}^d$ is a set of n solutions in a d-dimensional real-valued space, n is the number of solutions to be clustered, K is the number of clusters, $u_{ik} \in \{0, 1\}$ is the membership of $\boldsymbol{x}_i$ belonging to the kth cluster, and $\boldsymbol{v}_k$ is the center of the kth cluster. $dis(\boldsymbol{x}_i, \boldsymbol{v}_k)$ means a distance between $\boldsymbol{x}_i$ and $\boldsymbol{v}_k$.

4.1 Point Symmetry Distance Measure

Many clustering algorithms are adapting the Minkowski [17] metric to measure dissimilarity, i.e., distance $dis(\boldsymbol{x}_i, \boldsymbol{v}_k)$, in the clustering function. The Minkowski metric for measuring the dissimilarity between a solution $\boldsymbol{x}_i = (x_{i1}, \cdots, x_{id})^T$ and a center (search vector) $\boldsymbol{v}_k = (v_{k1}, \cdots, v_{kd})^T$ is defined as:

$$d_{mink}(\boldsymbol{x}_i, \boldsymbol{v}_k) = [\sum_{j=1}^{d} |x_{ij} - v_{kj}|^r]^{1/r} \qquad (9)$$

where $r \geq 1$. The three common Minkowski metrics are illustrated in Fig. 7. The Euclidean distance ($r = 2$) is one of the most common Minkowski distance metrics. Conventional clustering algorithms with the Euclidean distance tend to detect hyperspherical-shaped clusters.

Since the distribution shapes of Pareto and quasi-Pareto solutions are much closer to combinations of hyperellipsoidal or hyperline shapes than hyperspherical shapes, the Euclidean distance measure may not be a good choice for searching the characteristics of the solutions. Instead, a point symmetry distance

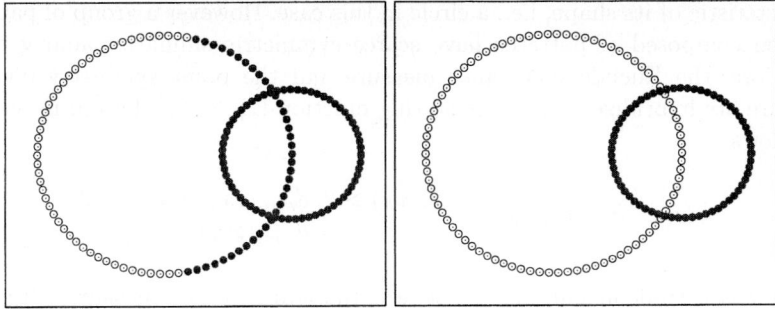

Fig. 8. Two clusters measured by the Euclidean distance (left) and point symmetry distance (right)

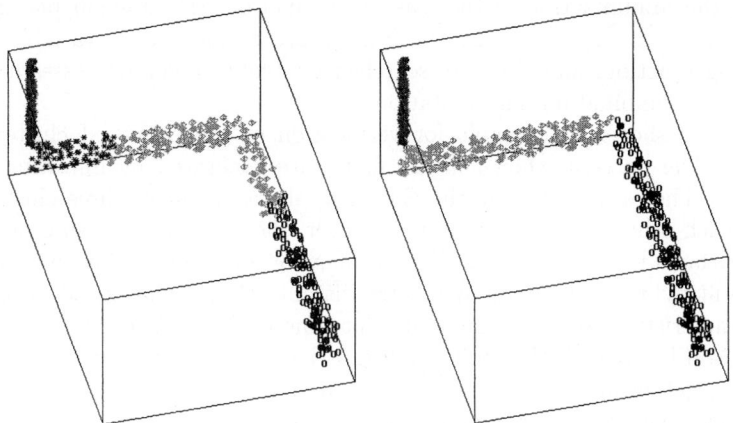

Fig. 9. Three clusters obtained by using the Euclidean distance (left) and point symmetry distance (right)

measure [18] is adopted. The distance measure is a more flexible to find clusters of hyperellipsoidal or hyperline shapes. Given n solutions, the point symmetry distance between a solution $\boldsymbol{x}_i$ and a cluster center $\boldsymbol{v}_k$ is defined as:

$$d_{sym}(\boldsymbol{x}_i, \boldsymbol{v}_k) = \min_{\substack{p=1,\cdots,n \\ and\ p \neq i}} \frac{||(\boldsymbol{x}_i - \boldsymbol{v}_k) + (\boldsymbol{x}_p - \boldsymbol{v}_k)||}{(||(\boldsymbol{x}_i - \boldsymbol{v}_k)|| + ||(\boldsymbol{x}_p - \boldsymbol{v}_k)||)} \quad (10)$$

where the denominator term is used to normalize the point symmetry distance. Due to such normalization, the point symmetry distance becomes insensitive to the Euclidean distances $||(\boldsymbol{x}_i - \boldsymbol{v}_k)||$ and $||(\boldsymbol{x}_p - \boldsymbol{v}_k)||$.

Figure 8 shows different clustering results of sample patterns obtained by the two different distance measures. The clustering result obtained by the Euclidean distance shows quantitatively well separated two clusters. The result obtained by the point symmetry distance shows that each clustered pattern possesses the characteristic of its shape, i.e., a circle in this case. However a group of patterns is often composed of patterns have scarce symmetric similarity among them. Therefore, the Euclidean distance measure and the point symmetry distance measure are hybridized for the clustering function Eq. 8. The hybrid measure is as follows:

$$dis(\boldsymbol{x}_i, \boldsymbol{v}_k) = \begin{cases} \text{if } d_{sym}(\boldsymbol{x}_i, \boldsymbol{v}_k) > \theta, \ d_{mink}(\boldsymbol{x}_i, \boldsymbol{v}_k), \ r = 2 \\ \text{else,} \qquad\qquad\qquad d_{sym}(\boldsymbol{x}_i, \boldsymbol{v}_k) \end{cases} \qquad (11)$$

where θ is a tradeoff parameter between the two distance measures. Iterative clustering algorithms such as the K-means algorithm [17, 19] depend their results on initial centers employed. In the proposed clustering function, the clustering results severely vary with θ. To overcome those problems simultaneously, the hybrid distance measure is adapted to the evolutionary clustering algorithm [12]. In the minimization of the clustering function, the tradeoff parameter is predefined as a constant. Only the centers are considered as variables of the clustering function, and they are searched by evolutionary processes including selection, recombination, and mutation.

Figure 9 shows an example for verification of the proposed clustering algorithm. The patterns shown in the figure are generated by imitating Pareto solutions. The result by using the Euclidean distance shows three clusters separated each other in geometry only. However, the proposed measure makes the break surfaces between clusters when the flow of solutions' variation is changed. The result by using the point symmetry distance shows what kinds of parametric characteristics are transformed in a parameter space. It is confirmed by the comparison between both results in Fig. 9 that the proposed distance is better suited for gathering engineering information. Therefore, the proposed clustering is employed with the hybrid distance in the following section.

4.2 Cluster Interpretation of Multiobjective Pareto and Quasi-Pareto Solutions

The proposed clustering method was applied to 4,061 Pareto and quasi-Pareto solutions of Eq. 6 and 7. The solutions were clearly classified into two clusters that have distinct parametric characteristics. One cluster #1 includes 1,993 solutions dominated mainly by the fin thickness and slightly by the cut length. The solutions in the cluster are not sensitive to the fin length. Cluster #2 has 2,068 solutions varying with the fin length and the cut length. In this case, the thickness of fins is fixed to its maximum value, 1.7mm. Figure 10 shows the projection of the two clusters in the parameter space onto the object function space. The figure clearly shows that the solutions in cluster #1 are in the range of increasing both thermal conductance and mass in a reasonable rate. On the other hand, the solutions in cluster #2 are in the range of increasing the mass exponentially. For example, the 63.6% possible increase of the mass raises the 84.8% of the possible thermal conductance in cluster #1. However, in cluster #2, the 46.9% of the mass increase makes only the 25.6% of the

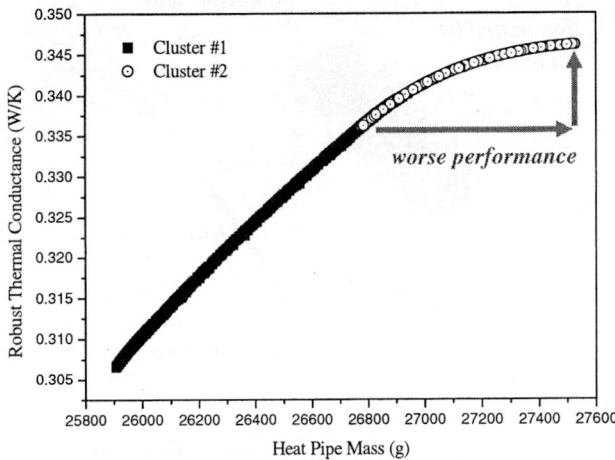

Fig. 10. Clustering in function space

conductance. Thus the solutions in cluster #1 must be candidates for the best overall design on the view point of minimizing the mass and maximizing the conductance.

Figure 11 shows the clustering result in both of the function and parameter spaces. Since thick fins reduce heat resistance, the solutions in cluster #1 having thick fins increase the thermal conductance. In case of cluster #2 which has the solutions with long fins, however, the fin parts away from the pipe junction have very small heat flux. Therefore the ends of the long fins just enlarge the mass without increasing the conductance. Through such interpretation based on the proposed visualization and clustering, thermal designers can obtain useful information for their decision of final design. Especially, the solutions on the border between the two clusters are the most possible for DM when we focus on the maximization of the thermal conductance.

4.3 Conclusions

In this paper, it was described a new procedure to interpret Pareto and quasi-Pareto solutions of multiobjective optimization. A synchronous 3D visualization was used to explore the structure and characteristics of multidimensional solutions. The visualization presents the correlations between a parameter, function, and actual design space in their subdivided spaces. Moreover, to make clusters of solutions in engineering meanings, we presented a clustering algorithm which has a hybrid distance measure of the Euclidean and point symmetry distance. The clustering algorithm shows similarity and dissimilarity among the solutions and beneficial information for designers.

As a practical engineering application, the proposed approach was applied to a multiobjective optimization of an artificial satellite. The design optimization has two functions, two uncontrollable parameters, and three shape parameters of

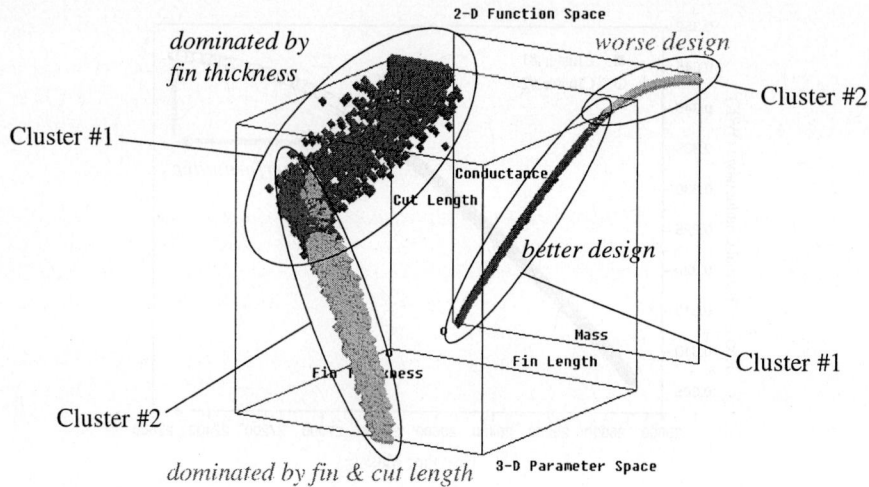

Fig. 11. Design parameters corresponding to optimum design

heat pipes of the satellite. The synchronous visualization helps one to understand the design effect of each solution from the Pareto and quasi-Pareto solutions. The clustering of the optimization solutions guides the shape parameters those corresponds to worse or better design clearly. Through such interpretation based on the proposed visualization and clustering, the thermal designers can obtain useful information for their decision of final design.

References

1. Tappeta, R. V., Renaud, J. E.: Iterative multiobjective optimization procedure. AIAA Journal **37**, 7 (1999) 881–889
2. Tappeta, R. V., Renaud, J. E.: Iterative multiobjective optimization design strategy for decision based design. Journal of Mechanical Design, **123**, June (2001) 205–215
3. Meng, Z., Pao, Y. H.: Visualization and selforganization of multidimensional data through equalized orthogonal mapping. IEEE Transactions on Neural Networks **11**, 4 (2000) 1031–1038
4. Miller,W.: Symmetry Groups and Their Applications. (1972) Academic Press, London
5. Weyl, H.: Symmetry. (1952) Princeton University Press, Princeton, NJ
6. Kobayashi, T., Nomura, T., Kamifuji, M., Yao, A., Ogushi, T.: Thermal robustness and mass optimization of heat pipe shape for spacecraft panel using a combination of responce surface methodology and monte carlo simulation. Proceedings of 28th Design Automation Conference, (2002) DETC2002/DAC-34055
7. Kelly, W. H., Reisenweber, J. H.: Thermal performance of embedded heat pipe spacecraft radiator panels. SAE Technical Paper, (1993) 932158
8. Taguchi, G.: Design of Experiment. Japanese Standards Association, (1979) in Japanese

9. Taguchi, G., Konishi, S.: Orthogonal Arrays and Linear Graphs. ASI press, (1987) Dearborn, MI
10. Myers, R. H., Montegomery, D. C.: Responce Surface Methodology: Process and Product Optimization Using Design Experiments. (1995) Wiley Inter-Science
11. Kashiwamura, T., Shiratori, M.: Structural optimization using the design of experiments and mathematical programming. Transactions of the JSME **62**, 601 (1996) 208–223, in Japanese
12. Jeong, M. J., Yoshimura, S.: An evolutionary clustering approach to pareto solutions in multiobjective optimization. Proceedings of 28th Design Automation Conference, (2002) DETC2002/DAC-34048
13. Jeong, M. J.: Integrated Support System for Decision-Making in Design Optimization. PhD Thesis, (2003) The University of Tokyo
14. Chen, J. X.: Data visualization: Parallel coordinates and dimension reduction. Computing in Science and Engineering **3**, 5 (2001) 110–113
15. Inselberg, A.: Visulaiztion and data mining of highdimensional data. Chemometrics and Intelligent Laboratory Systems **60**, (2002) 147–159
16. Kawai, H.: Development of a 3-d graphics and gui tookit for making pre- and post-processing tools. Proceedings of the Conference on Computational Engineering and Science **8**, 2 (2003) 889–892, in Japanese
17. Jain, A. K., Dubes, R. C.: Algorithms for Clustering Data. Prentice-Hall, (1989) Englewood Cliffs, New Jersey
18. Su, M., Chou, C.: A modified version of the k-means algorithm with a distance based on cluster symmety. IEEE Trans. on Pattern Analysis and Machine Intelligence **23**, 6 (2001) 674–680
19. Mirkin, B.: Mathematical Classification and Clustering. Kluwer Academic Publishers, (1996) New York

Gray Coding in Evolutionary Multicriteria Optimization: Application in Frame Structural Optimum Design

David Greiner, Gabriel Winter, José M. Emperador, and Blas Galván

Evolutionary Computation and Applications Division (CEANI),
Intelligent Systems and Numerical Applications in Engineering Institute (IUSIANI),
35017, Campus de Tafira Baja, Las Palmas de Gran Canaria University (ULPGC), Spain
dgreiner@iusiani.ulpgc.es

Abstract. A comparative study of the use of Gray coding in multicriteria evolutionary optimisation is performed using the SPEA2 and NSGAII algorithms and applied to a frame structural optimisation problem. A double minimization is handled: constrained mass and number of different cross-section types. Influence of various mutation rates is considered. The comparative statistical results of the test case cover a convergence study during evolution by means of certain metrics that measure front amplitude and distance to the optimal front. Results in a 55 bar-sized frame test case show that the use of the Standard Binary Reflected Gray code compared versus Binary code allows to obtain fast and more accurate solutions, more coverage of non-dominated fronts; both with improved robustness in frame structural multiobjective optimum design.

1 Introduction

Recently, interest in analysis and design of representations and operators for evolutionary computation has been liven up (e.g. a special Issue about this topic of the journal IEEE Transactions on Evolutionary Computation is coming). The motivation of this work is to analyse the influence of an adequate coding in multicriteria optimization, particularly we compare here the use of Gray coding versus binary coding. In multiobjective optimization the search has to deal with multiple requirements: the approximation to the optimum non-dominated front, the achievement of a smooth distribution along the front and also the completion of its maximum coverage [6][8]. So, the codification influence in the search towards the set of optimum solutions should be focused in such a plural way. The choice of the proper coding can have a drastic repercussion in the final results. The smoothness in the correspondence between the phenotypic and the genotypic space is the main claimed advantage of the Gray Code [29][30]. Guarantying this smoothness could be especially critical when the genotypic unit (represented with 0s and 1s) has its phenotypic correspondence in an ordered database, where each gene has a set of associated values, whose magnitudes can vary considerably even in consecutive genes. This is a frequent case when using discrete representation of the chromosome via 0s and 1s, for example in scheduling optimisation problems [10].

Here a discrete frame structural multicriteria optimization problem belonging to that archetype is handled. The first application of evolutionary algorithms to structural optimization is dated twenty-five years ago [13]: A ten-bar truss is optimized for the minimum constrained mass problem, with continuous variables for the section area of the bars. A pioneer article for frame structures optimization using evolutionary algorithms is [18], where a genetic algorithm is used for the optimal design of skeletal building structures considering discrete sizing, geometrical and topological variables in two design examples. A recent state of the art of structural optimization with special emphasis in evolutionary optimization is [1], where the recent developments in the field, for the period 1980 to 2000, are documented by the ASCE (American Society of Civil Engineering) Technical Committee on Optimal Structural Design of the Technical Administrative Committee on Analysis and Computation. Interesting reviews about multicriteria optimization in structural engineering are [4][5][23], and a set of applications of multicriteria evolutionary optimization in structural and civil engineering are summarized in [6].

The organization of this paper is described as follows: First, the multiobjective frame structural problem is described. Section 3 disserts about the evolutionary approach and the use of Gray Code in multiobjective frame optimization. Section 4 exposes the 55 bar-sized test case. After that, the experimental results are shown in section 5, ending with the conclusions section.

2 Frame Structural Optimum Design

2.1 Definition of the Problem

The frame structural design problem considered has two conflicting objectives: the minimization of the constrained mass and the minimization of the number of different cross-section types considered in the final design. This design problem was introduced in [11], being solved using a combination of weights. It has been solved with elitist multiobjective evolutionary algorithms in [14][15]. Both objectives are explained as follows.

The first objective, the *minimization of the constrained mass* is taken into account to minimize the raw material cost of the designed structure. The constraints consider those conditions that allow the designed frame to carry out its task without collapsing or deforming excessively. The constraints are the following, taking into account the Spanish design code (EA-95) guidelines:

a) Stresses of the bars: where the limit stress depends on the frame material and the comparing stress takes into account the axial and shearing stresses by means of the shear effort, and also the bending effort (a common value for steel is of 260 MPa - S275JR steel -), for each bar:

$$\sigma_{co} - \sigma_{\lim} \leq 0 \qquad (1)$$

b) Compressive slenderness limit: where the $\lambda_{\lim}$ value is 200 (to include the buckling effect the evaluation of the β factor, is based on Julian and Lawrence criteria). For each bar:

$$\lambda - \lambda_{\lim} \leq 0 \qquad (2)$$

c) Displacements of joints (in each of the three possible degrees of freedom) or middle points of bars. In the test cases, the maximum vertical displacement of each beam is limited (in the multiobjective test case the maximum vertical displacement of the beams is L / 500):

$$u_{co} - u_{\lim} \leq 0 \qquad (3)$$

The first objective function *constrained mass*, results:

$$ObjetiveFunction_1 = \left[\sum_{i=1}^{Nbars} A_i \cdot \rho_i \cdot l_i\right]\left[1+k \cdot \sum_{j=1}^{Nviols}(viol_j - 1)\right] \qquad (4)$$

where :

A_i = bar i cross-section area; ρ_i = bar i specific mass; l_i = length of bar i; k = constant that regulates the equivalence between mass and constraints; $viol_j$ = for each of the violated constraints (stress, displacement or slenderness), is the quotient between the value that violates the constraint limit: violated constraint value and its reference limit. The constraints reference limits are chosen according to the Spanish design codes. So, constraints if violated, are integrated into the mass of the whole structure as a penalty depending on the amount of the violation (for each constraint violation the total mass is incremented):

$$viol_j = \frac{Violated\ Constraint\ Value}{Constraint\ Limit} \qquad (5)$$

Moreover, the minimization of a second function is considered as a multicriteria optimization problem: the *number of different Cross-Section Types (CST)*, that supposes a condition of constructive order, and with special relevancy in structures with high number of bars [11][19]. It helps to a better quality control during the execution of the building site. It is a factor that has been also recently related with the life cost cycle optimization of steel structures, as claimed in [24][28], and evolutionary multicriteria optimization allows it to be integrated in the design optimization process.

2.2 The System Model

Numerous literature about evolutionary bar optimization problems is about trusses. They are characterized for its articulated nodes between bars, where no resistant moment is executed and the only required geometric magnitude of the bar section is its area. Nevertheless, here we handle with frames: the nodes between bars are rigid, and the moments have to be taken into account. More bar geometric magnitudes have to be considered: to evaluate the normal stresses, the area, the modulus of section and the relation of beam height are required; to evaluate the shearing stresses the web area is required; to evaluate the medium span displacement the moment of inertia is required. Moreover, considering the buckling effect implies to take into account the radius of gyration. Because of the design is performed using real cross-section types -

developing a discrete optimization problem with direct real application-, all these magnitudes are stored in a vector associated to each cross-section type, whose complete set constitute a database. The codification of the chromosome implies for each bar of the structure, a discrete value that is assigned to the order of cross-section types in the database.

The structural calculation implies the resolution of a finite element modelling -with Hermite approximation functions-, and its associated linear equation system. In the plane case, each node has three degrees of freedom: horizontal, vertical and gyration displacements (U), as well as three associated force (F) types: horizontal and vertical forces and momentum. Forces and displacements are related (F=KU) by the stiffness matrix (K), which depends on geometric and material properties, and is created with an assembling process from each element of the structure. A renumbering using the inverse Cuthill-McKee method [7] is used to reduce the matrix bandwidth, in order to reduce the calculation time of the system (essential when many evaluations are required), which is programmed in C++ language.

3 The Evolutionary Approach

The frame structural problem of determining the constrained minimum mass has many local minima [21], so a global optimization method is recommended. Moreover, we deal with a discrete search space. If the improvement in one criterion implies the worsening in another objective, as happens with the constrained mass and the number of different cross-section types, which are our two minimising objective functions, a multiobjective optimization is required. Because of the requirement of a global, discrete and multiobjective optimization method, the evolutionary multicriteria optimization methods are suitable. Among the most recent algorithms, those which include elitism and parameter independence are outstanding. For our study, the SPEA2 [34] and NSGA-II [9] have been selected. An improved adaptation of the truncation operator in SPEA2 specially suited for two dimensional multicriteria problems, proposed in [17] is implemented. It takes advantage of the linearity of the distribution of the non-dominated solutions in the bicriteria case.

Two metrics are considered, defined on objective space, concerning about accuracy and coverage of the front. They are averaged from thirty independent runs of each algorithm.

The first metric (approximation to the front) is the M1* metric of Zitzler, representative of the approximation to the optimal Pareto front. To evaluate this metric, belonging to the scalable metrics type, the best Pareto front should be known. Its expression is [33]:

$$M1*(U) = \frac{1}{|U|} \sum_{u \in U} \min\{\|u - y\| \mid y \in Y_p\} \quad (6)$$

where U = f(A) ⊆ Y (being A a set of decision vectors, Y the objective space and Yp referred to the Pareto set).

The second metric handles with the coverage of the front. The adopted criteria is using the number of solutions of the best non-dominated front achieved in each generation, because of the nature of our second fitness function: the number of different cross-section types, which produce a discrete non-dominated front with limited maximum number of solutions. For this reason, also there is no necessity for evaluating the smoothness of the spread of solutions along the front, because we have a discrete one, and apparently, there is no difficulty in obtaining one non-dominated solution for each number of different cross-section types between the extreme solutions of each generation.

Thirty independent runs have been considered for each algorithm (NSGAII and SPEA2) and codification case. A population size of 100 individuals and uniform crossover have been used. Three different values of the mutation rate: 0.4%, 0.8% and 1.5% are studied, comparing the binary and Gray coding.

3.1 Gray Coding for Multicriteria Optimization

The discrete nature of the search space of our problem, compound of the cross-section type of each bar, can be benefit from the discrete coding of binary coding, opposed to a less properly real coding in this particular case.

Traditionally, the more number of schemata per information unit among all the possible codifications has been claimed for binary coding [12], because of its low cardinality, being beneficial for the building blocks propagation. However, the implicit parallelism of genetic algorithms is not exclusive of the binary representation, existing for other alphabetic cardinalities of codification.

However, Binary coding suffers from not being homogeneous respect to its decimal numeric equivalent, which is normally used in its decoding. For example, the number 7 is followed by 8, but their binary representations are respectively 0111 and 1000, where every allele diverges from another. This is known as Hamming Cliff. In the phenotypic space both are consecutive values, but in the genotypic space both differ completely. It seems to be desirable a representation that maintains analogous smoothness in the phenotypic and genotypic spaces.

A Gray code is defined as a representation with 1s and 0s that permits a bijective equivalence between phenotype and genotype for consecutive integers differing only in one bit between them. It can be seen in table 1 a comparison between binary and Gray codes. It has been remarked in *italics* the values of the differing bits between two consecutive integers. It can be observed how in the case of Gray code, only one bit differ, but a more chaotic behaviour of the binary code.

Experimental studies over many kind of single objective test functions widely used in genetic algorithms show an improved behaviour of using Gray code versus the standard binary code [2][22][25].

There is an analogous theorem of the 'No Free Lunch' Theorem [32] of direct application in the comparison of binary / Gray representation [26]: 'All algorithms are equivalent compared over all possible representations'. Whitley [30] claims in this case as an example of contradiction between theory and practice, and shows that the coding influences the number of optimum that the phenotype generates for the same genotype. Gray coding can reduce this number of optima. However, it does not imply necessarily that with fewer optima the problem is solved easier, as is shown in [3]: a coding with greater optima can have less expected convergence time. There is also shown that the efficiency of the coding is dependent of the search operators. Gray coding is still an open question, as can be seen recently in [27].

Table 1. Gray Code versus Binary Code in a 4 bit string for structural optimization

Binary Code					Gray Code			
Area cm²	Moment of inertia cm⁴	IPE	String	Equivalent Integer	String	IPE	Area cm²	Moment of inertia cm⁴
7.64	80.1	80	0000	0	0000	80	7.64	80.1
10.3	171	100	0001	1	0001	100	10.3	171
16.4	541	140	0010	2	0011	120	13.2	318
13.2	318	120	0011	3	0010	140	16.4	541
33.4	2770	220	0100	4	0110	160	20.1	869
28.5	1940	200	0101	5	0111	180	23.9	1320
20.1	869	160	0110	6	0101	200	28.5	1940
23.9	1320	180	0111	7	0100	220	33.4	2770
116.0	48200	500	1000	8	1100	240	39.1	3890
98.8	33740	450	1001	9	1101	270	45.9	5790
72.7	16270	360	1010	10	1111	300	53.8	8360
84.5	23130	400	1011	11	1110	330	62.6	11770
39.1	3890	240	1100	12	1010	360	72.7	16270
45.9	5790	270	1101	13	1011	400	84.5	23130
62.6	11770	330	1110	14	1001	450	98.8	33740
53.8	8360	300	1111	15	1000	500	116.0	48200

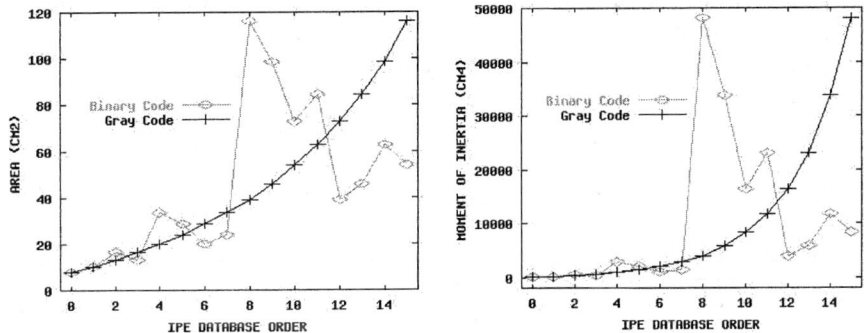

Fig. 1. Area and Moment of Inertia of IPE cross-section type database order using binary and Gray codes

Gray coding is not unique, and there are as many possibilities as combinations are allowed to establish a continuous mapping of the genotypic and phenotypic spaces. We have implemented the Standard Binary Reflected Gray Code [25]. The suggestion of [31] has been adopted, forming the Gray code from a right displacement of the binary vector and performing a XOR with the resulting vector and the original binary one. The resulting vector is the Gray code of the binary vector. It can be implemented in C++ language easily.

The influence of the coding in the database order and its correspondence with the included cross-section types can be viewed simultaneously in table 1 and figure 1. The above cited homogeneous correspondence between phenotype and genotype is shown, in terms of two of the geometric magnitudes of the cross-section types: area and moment of inertia of the first sixteen cross-section types of the IPE series (from IPE-80 until IPE-500). Analogous figures are obtainable for the other geometric magnitudes (modulus of section, web area, etc), and for other cross-section types series (HEB, etc).

The use of Gray coding has been proven advantageous in the single criteria optimization problem of constrained mass [16]. Here is proposed an analysis of Gray code versus binary code in a discrete frame structural multicriteria optimization problem, emphasizing its use in multiobjective optimisation.

4 Test Case

4.1 Description

The test case is represented in figure 2, based on a reference problem in [20]. The figure includes the numbering of the bars, and the precise loads in Tons. Moreover, in every beam there is a uniform load of 39945 N/m. The lengths of the beams are 5.6 m and the heights of the columns are 2.80 m. The columns belong to the HEB cross-section type series, and the beams belong to the IPE cross-section type series; being the admissible stresses of 200 MPa. and 220 MPa. respectively. The maximum vertical displacement in the middle point of each beam is established in $1/300 = 1.86 \cdot 10^{-2}$ m. The density and elasticity modulus are the typical values of steel: 7850 kg/m^3 and $2.1 \cdot 10^5$ MPa., respectively.

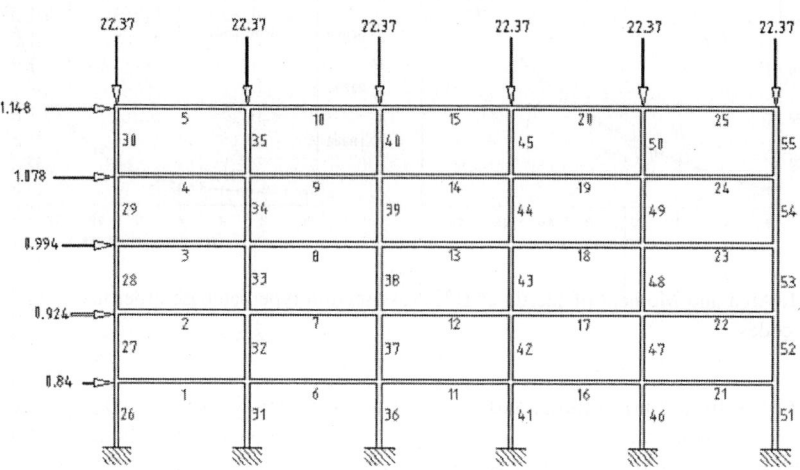

Fig. 2. Frame Test Case

4.2 Optimal Solutions

The cross-section type database is composed of the first sixteen IPE cross-section types (from IPE-80, 100, 120, 140, 160, 180, 200, 220, 240, 270, 300, 330, 360, 400, 450 to IPE-500) for the beams and the first sixteen HEB cross-section types (from

HEB-100, 120, 140, 160, 180, 200, 220, 240, 260, 280, 300, 320, 340, 360, 400 to HEB-450) for the columns. Extended to the 55 bar-sized test case (and four bits per bar), this implies a search space of $2^{55*4} = 2^{220} \approx 1.7 \cdot 10^{66}$. The optimum Pareto Front is not known for this problem, and we report here the best front we have found so far, which is the reference for the metrics comparison. It is represented in figure 3, and its solutions are detailed in table 2 (detailed numerical values of solutions), table 3 (cross-section types of columns) and table 4 (cross-section types of beams).

Table 2. Pareto Set Values of Test Case

Constrained Mass (kg) (F1)	10130.04	10212.07	10318.95	10517.65	10865.28	11394.45
Constraint (kg)	2.75	5.65	0.00	0.00	213.55	272.13
Mass (kg)	10127.29	10206.41	10318.95	10517.65	10651.73	11122.32
Number of different CST (F2)	8	7	6	5	4	3

Table 3. Pareto Set Detailed Cross-Section Types (CST) of Columns of Test Case

Number of different CST (F2)	8	7	6	5	4	3
Bar n° 26	HEB160	HEB160	HEB160	HEB160	HEB220	HEB220
Bar n° 27	HEB180	HEB180	HEB200	HEB200	HEB220	HEB220
Bar n° 28	HEB160	HEB160	HEB160	HEB160	HEB220	HEB220
Bar n° 29	HEB140	HEB140	HEB140	HEB160	HEB140	HEB180
Bar n° 30	HEB180	HEB180	HEB200	HEB200	HEB200	HEB220
Bar n° 31	HEB220	HEB220	HEB220	HEB220	HEB220	HEB220
Bar n° 32	HEB200	HEB200	HEB200	HEB200	HEB200	HEB220
Bar n° 33	HEB180	HEB180	HEB200	HEB200	HEB200	HEB180
Bar n° 34	HEB160	HEB160	HEB160	HEB160	HEB140	HEB180
Bar n° 35	HEB120	HEB140	HEB140	HEB160	HEB140	HEB180
Bar n° 36	HEB200	HEB200	HEB200	HEB200	HEB220	HEB220
Bar n° 37	HEB200	HEB200	HEB200	HEB200	HEB200	HEB220
Bar n° 38	HEB160	HEB160	HEB160	HEB160	HEB200	HEB180
Bar n° 39	HEB140	HEB140	HEB140	HEB160	HEB140	HEB180
Bar n° 40	HEB120	HEB140	HEB140	HEB160	HEB140	HEB180
Bar n° 41	HEB220	HEB220	HEB220	HEB220	HEB220	HEB220
Bar n° 42	HEB200	HEB200	HEB200	HEB200	HEB200	HEB220
Bar n° 43	HEB160	HEB160	HEB160	HEB160	HEB200	HEB180
Bar n° 44	HEB140	HEB140	HEB140	HEB160	HEB140	HEB180
Bar n° 45	HEB120	HEB140	HEB140	HEB160	HEB140	HEB180
Bar n° 46	HEB220	HEB220	HEB220	HEB220	HEB220	HEB220
Bar n° 47	HEB200	HEB200	HEB200	HEB200	HEB200	HEB220
Bar n° 48	HEB160	HEB160	HEB160	HEB160	HEB200	HEB220
Bar n° 49	HEB140	HEB140	HEB140	HEB160	HEB140	HEB180
Bar n° 50	HEB120	HEB140	HEB140	HEB160	HEB140	HEB180
Bar n° 51	HEB180	HEB180	HEB200	HEB200	HEB220	HEB220
Bar n° 52	HEB200	HEB200	HEB200	HEB200	HEB220	HEB220
Bar n° 53	HEB200	HEB200	HEB200	HEB200	HEB200	HEB180
Bar n° 54	HEB160	HEB160	HEB160	HEB160	HEB220	HEB180
Bar n° 55	HEB200	HEB200	HEB200	HEB200	HEB200	HEB220

Table 4. Pareto Set Detailed Cross-Section Types (CST) of Beams of Test Case

Number of different CST (F2)	8	7	6	5	4	3
Bar n° 1	IPE330	IPE330	IPE330	IPE330	IPE300	IPE300
Bar n° 2	IPE330	IPE330	IPE330	IPE330	IPE300	IPE300
Bar n° 3	IPE330	IPE330	IPE330	IPE330	IPE300	IPE300
Bar n° 4	IPE300	IPE300	IPE300	IPE300	IPE300	IPE300
Bar n° 5	IPE330	IPE330	IPE330	IPE330	IPE300	IPE300
Bar n° 6	IPE300	IPE300	IPE300	IPE300	IPE300	IPE300
Bar n° 7	IPE300	IPE300	IPE300	IPE300	IPE300	IPE300
Bar n° 8	IPE300	IPE300	IPE300	IPE300	IPE300	IPE300
Bar n° 9	IPE300	IPE300	IPE300	IPE300	IPE300	IPE300
Bar n° 10	IPE300	IPE300	IPE300	IPE300	IPE300	IPE300
Bar n° 11	IPE300	IPE300	IPE300	IPE300	IPE300	IPE300
Bar n° 12	IPE300	IPE300	IPE300	IPE300	IPE300	IPE300
Bar n° 13	IPE300	IPE300	IPE300	IPE300	IPE300	IPE300
Bar n° 14	IPE300	IPE300	IPE300	IPE300	IPE300	IPE300
Bar n° 15	IPE300	IPE300	IPE300	IPE300	IPE300	IPE300
Bar n° 16	IPE300	IPE300	IPE300	IPE300	IPE300	IPE300
Bar n° 17	IPE300	IPE300	IPE300	IPE300	IPE300	IPE300
Bar n° 18	IPE300	IPE300	IPE300	IPE300	IPE300	IPE300
Bar n° 19	IPE300	IPE300	IPE300	IPE300	IPE300	IPE300
Bar n° 20	IPE300	IPE300	IPE300	IPE300	IPE300	IPE300
Bar n° 21	IPE300	IPE300	IPE300	IPE300	IPE300	IPE300
Bar n° 22	IPE300	IPE300	IPE300	IPE300	IPE300	IPE300
Bar n° 23	IPE300	IPE300	IPE300	IPE300	IPE300	IPE300
Bar n° 24	IPE300	IPE300	IPE300	IPE300	IPE300	IPE300
Bar n° 25	IPE300	IPE300	IPE300	IPE300	IPE300	IPE300

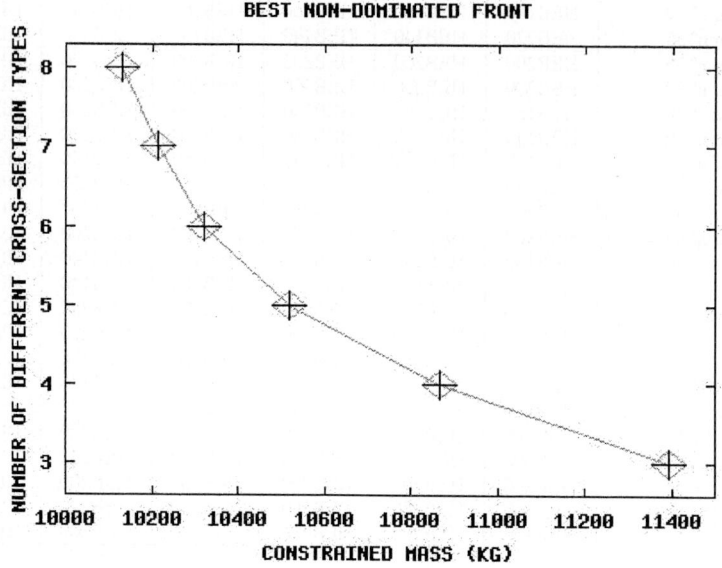

Fig. 3. Solution and Pareto Sets of Frame Structural Test Case

5 Results and Discussion

The obtained results for the SPEA2 are represented in figures from 4 to 9, where the black line corresponds to the Gray code and the grey line corresponds to the binary code. The left graph of each figure shows the mean over 30 independent runs and the right graph represents the typical deviation of the metric. Because in multiobjective optimization the search has to cope with multiple requirements, figures are organized in two groups, in order to analyse the effects of Gray code independently: approximation and coverage to the optimum front. Figures from 4 to 6 resume the results of the convergence to optimum front metric (varying mutation rate from 0.4% to 1.5%), whereas figures from 7 to 9 resume the results of the coverage front metric (varying mutation rate from 0.4% to 1.5%), as described in section 3. The x-axis corresponds to the number of evaluations of the fitness function. Also some results of the NSGAII have been included, for the mutation rate of 0.4% (figures 10 and 11).

5.1 Analysing the Approximation to the Optimum Front Metric

The nearest the value of this metric is to zero, the better. From the observation of figures 4 to 6, it can be seen a common behaviour for all the mutation rates tested. The Gray code mean metric achieves not only a better final value in all the cases (approximately half of the binary code metric value), but also during the whole convergence process it shows lower values and an initial steeper slope. It also outperforms the binary code with lower typical deviations in all cases. The variation of the mutation rate does not seem to alter the performance of this metric mean for both codings, whereas affects the performance of the typical deviation of the binary one, increasing it with the increase of the mutation rate. Similar qualitative results are also obtained using the NSGAII algorithm, as can be seen in figure 10.

5.2 Analysing the Coverage of the Front Metric

If we observe figure 3, where the optimum reference front is shown, it is noticeable that the best value of the coverage of the front using the metric described in section 3 is 6. So, the nearest the final value of the metric is to 6, the better. From the observation of figures 7 to 9, it can be seen a common behaviour for all the mutation rates tested. In the initial evaluations, high oscillations are present in this metric, which decrease significantly from 20000 evaluations. The Gray code metric achieves a final better value (around 4, versus binary, whose value is below 3.5) in all the cases, indicating a better coverage of the best non-dominated front (the most difficult solutions to reach are the right lower ones represented in figure 3). Its value is also greater during the whole convergence process, what means a wider front that increases the diversity of the non-dominated solution set. The typical deviation of Gray code is also smaller than the binary code metric, showing an enhanced robustness. The variation of the mutation rate does not seem to alter significantly the Gray code performance of this metric mean, whereas affects the binary one. It is noteworthy that similar qualitative results are also obtained using the NSGAII algorithm, as can be seen in figure 11.

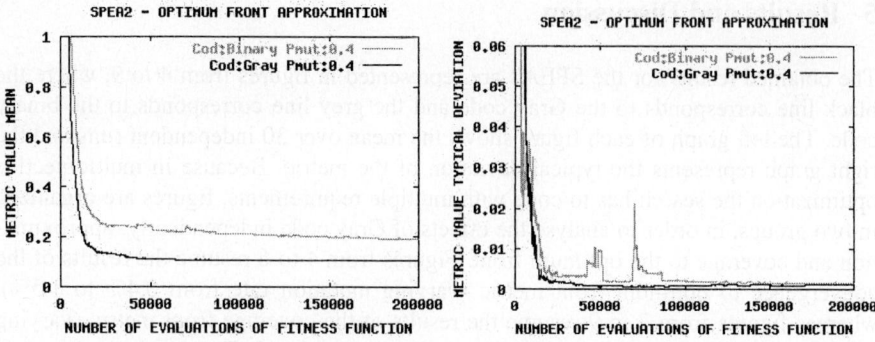

Fig. 4. Metric Approximation to Optimum Front: Mean and Typical Deviation over 30 independent runs of SPEA2 and Mutation rate 0.4%

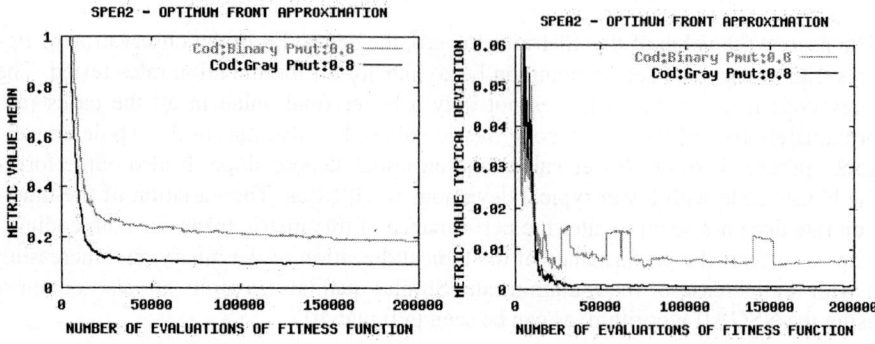

Fig. 5. Metric Approximation to Optimum Front: Mean and Typical Deviation over 30 independent runs of SPEA2 and Mutation rate 0.8%

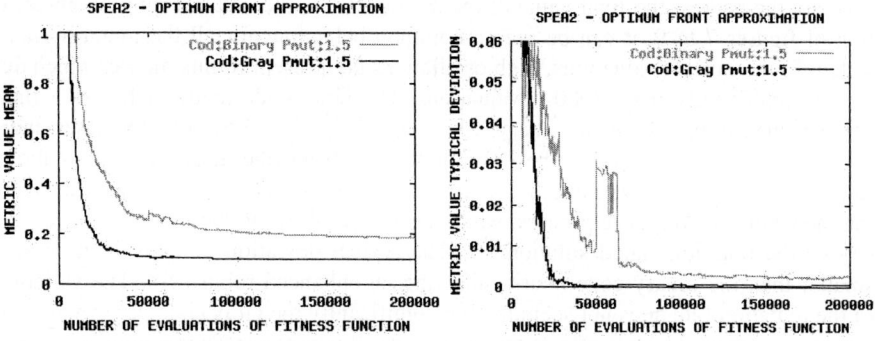

Fig. 6. Metric Approximation to Optimum Front: Mean and Typical Deviation over 30 independent runs of SPEA2 and Mutation rate 1.5%

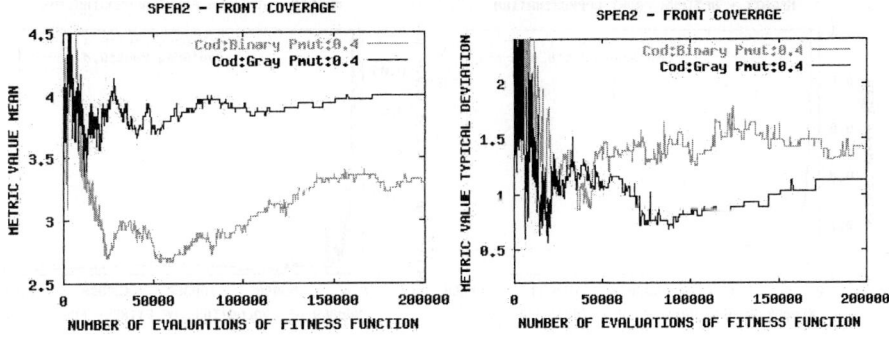

Fig. 7. Metric Front Coverage: Mean and Typical Deviation over 30 independent runs of SPEA2 and Mutation rate 0.4%

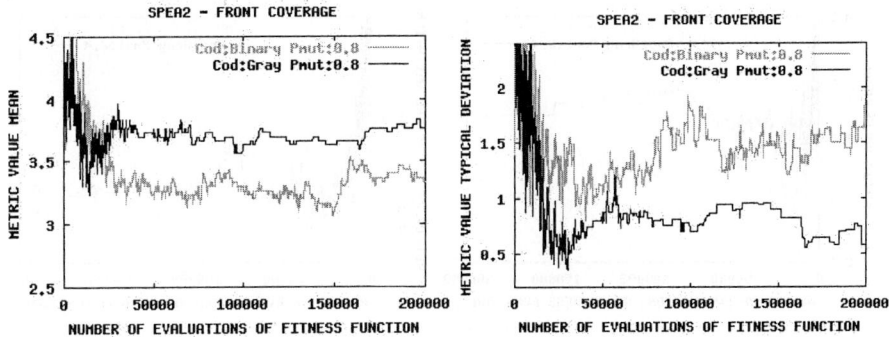

Fig. 8. Metric Front Coverage: Mean and Typical Deviation over 30 independent runs of SPEA2 and Mutation rate 0.8%

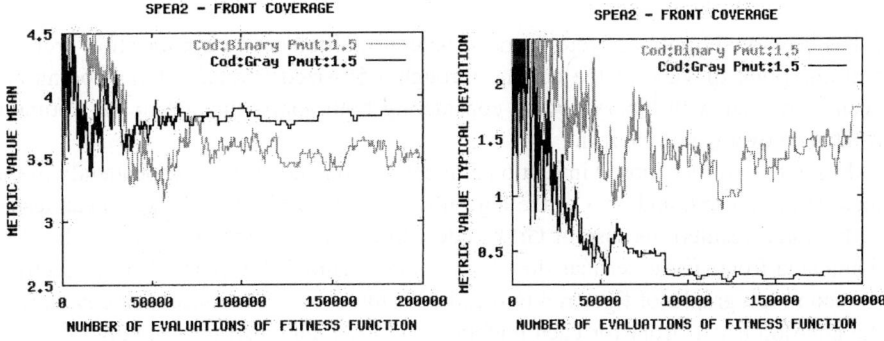

Fig. 9. Metric Front Coverage: Mean and Typical Deviation over 30 independent runs of SPEA2 and Mutation rate 1.5%

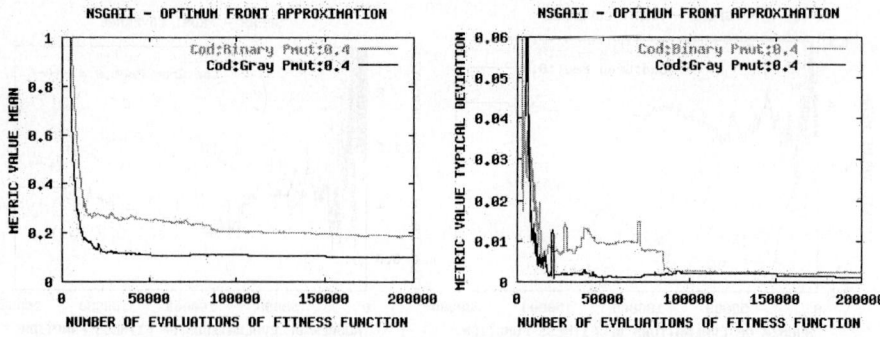

Fig. 10. Metric Approximation to Optimum Front: Mean and Typical Deviation over 30 independent runs of NSGAII and Mutation rate 0.4%

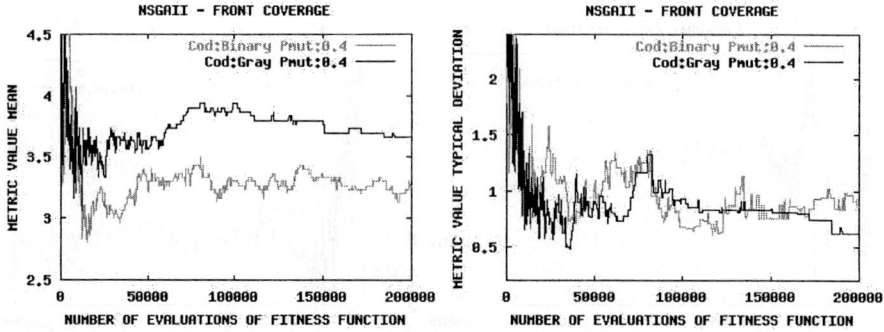

Fig. 11. Metric Front Coverage: Mean and Typical Deviation over 30 independent runs of NSGAII and Mutation rate 0.4%

6 Conclusions

It has been analysed a discrete frame structural multiobjective optimisation problem with the point of view of the coding, through a 55-sized test case. Gray coding has been compared with binary coding considering both approximation to the optimum front and coverage of the front.

From the results exposed in section 5, it can be concluded that independently of the mutation rate used, and also of the algorithm (SPEA2 and NSGAII have been tested with similar results), the use of Gray code allows a faster approximation to the non-dominated front (more vertical slope), and more accurate (lower value of the metric), as seen in left graphs of figures 4 to 6 and 10. Moreover, the amplitude of coverage of the non-dominated front for each mutation rate used, and also for each algorithm, as seen in left graphs of figures 7 to 9 and 11, is greater when using Gray code, indicating that a more complete front is obtained. Looking at the right graphs of figures 4 to

11, where the typical deviation of the metrics are displayed, for all cases it is lower with the Gray code, revealing a higher robustness of this codification in this test case.

So, the theoretical advantages that the Gray code has due to its greater homogeneity in the correspondence between the genotypic and phenotypic spaces, and that other application studies of single criteria optimisation claimed, are also corroborated in this work by means of the obtained experimental results, in a multiobjective optimisation problem. Results show that the use of Gray code allows to obtain fast and more accurate solutions, more coverage of non-dominated fronts; both with improved robustness in frame structural multiobjective optimum design.

A generalization of this study in other multiobjective design optimisation applications, where the coding implies a phenotypic correspondence with a database ordering resulting in a more homogeneous and easier resolution, could provide more light about the empirical performance of Gray coding in multicriteria optimization.

Acknowledgements. This investigation was funded partially by the Research Project DPI2001-3570 from the Ministry of Science and Technology of Spanish Government.

References

1. S.A. Burns, Ed. *Recent Advances in Optimal Structural Design*, Institute of American Society of ASCE-SEI (2002).
2. R. Caruana, J. Schaffer, Representation and Hidden Bias: Gray vs. Binary Coding for Genetic Algorithms. In *Proceedings of the Fifth International Conference on Machine Learning*. Morgan Kaufmann (1988) pp.153-161.
3. U.K. Chakraborty, C.Z. Janikow, "An analysis of Gray versus binary encoding in genetic search", *Information Sciences: an International Journal; Special Issue: Evolutionary Computation*, 156 (2003) pp. 253-269.
4. C.A. Coello Coello, A.D. Christiansen, "Multiobjective optimization of trusses using genetic algorithms" , *Computers & Structures* 75 (2000) 647-660.
5. C.A. Coello Coello, "An Empirical Study of Evolutionary Techniques for Multiobjective Optimization in Engineering Design", *PhD Thesis*, Tulane University, LA, U.S.A. (1996).
6. C. Coello, D. Van Veldhuizen, G. Lamont, *"Evolutionary Algorithms for solving multiobjective problems"*. Kluwer Academic Publishers - GENA Series (2002).
7. Cuthill, McKee, "Reducing the bandwidth of sparse symmetric matrixes", *Proceedings ACM National Conference*, New York (1969).
8. K. Deb, *Multiobjective Optimization using Evolutionary Algorithms*. John Wiley & Sons - Series in Systems and Optimization- (2001).
9. K. Deb, A. Pratap, S. Agrawal, T. Meyarivan, "A fast and elitist multiobjective genetic algorithm NSGAII", *IEEE Transactions on Evolutionary Computation* 6(2), (2002) pp. 182-197.
10. S. French, *"Sequencing and Scheduling: An Introduction to the Mathematics of the Job-Shop"*, Wiley, New York, 1982.
11. M. Galante, 'Genetic Algorithms as an approach to optimise real-world trusses' *International Journal for Numerical Methods in Engineering*, vol. 39, (1996) pp. 361-382.
12. D.E. Goldberg, *'Genetic algorithms for search, optimisation, and machine learning'* Reading, MA: Addison Wesley (1989).

13. D.E. Goldberg, M.P. Samtani, "Engineering Optimization via genetic algorithm". *Proceedings Ninth Conference on Electronic Computation*, ASCE, New York, NY, (1986), pp. 471-482.
14. D. Greiner, J.M. Emperador, G. Winter, "Single and Multiobjective Frame Optimization by Evolutionary Algorithms and the Auto-adaptive Rebirth Operator", in *Computer Methods in Applied Mechanics and Engineering,* Elsevier , 193 (2004) 3711-3743.
15. D. Greiner, J.M. Emperador, G. Winter, "Multiobjective Optimisation of Bar Structures by Pareto-GA", in *European Congress on Computational Methods in Applied Sciences and Engineering* (2000), CIMNE.
16. D. Greiner, G. Winter, J.M. Emperador, "Optimising Frame Structures by different strategies of GA", *Finite Elements in Analysis and Design*, Elsevier, 37 (5) pp.381-402, (2001).
17. D. Greiner, G. Winter, J.M. Emperador, B. Galván. An efficient adaptation of the truncation operator in SPEA2. In: *Proceedings of the First Spanish Congress on Evolutionary and Bioinspired Algorithms*. Eds: Herrera et al. Mérida, Spain, February 2002. (in Spanish)
18. D.E. Grierson, W.H. Pak, "Optimal sizing, geometrical and topological design using a genetic algorithm", *Structural Optimization*, 6-3 (1993) 151-159.
19. P. Hajela, C.Y. Lin, "Genetic search strategies in multicriterion optimal design", *Structural Optimization*, 4 (1992) 99-107.
20. S. Hernández Ibáñez, *'Structural Optimum Design Methods'*. Colección Seinor. Colegio de Ingenieros de Caminos, Canales y Puertos, Madrid, (1990).
21. S. Hernández Ibáñez, 'From Conventional Design to Optimum Design. Posibilities and Variants. Part I. Sensibility Analysis and local and global Optimisation', *International Journal of Numerical Methods for Calculus and Engineering Design*, (1993) CIMNE, pp. 91-110. (in Spanish)
22. R. Hinterding, H. Gielewski, T.C. Peachey, "The nature of mutation in genetic algorithms", *Proceedings of the Sixth International Conference on Genetic Algorithms* 1989, Morgan Kaufmann, pp. 70-79.
23. J. Koski, "Multicriteria optimization in structural design: state of the art", in: *Proceedings of the 19th Design Automation Conferences ASME*, (1993) pp. 621-629.
24. M. Liu, S.A. Burns, Y.K. Wen, Optimal seismic design of steel frame buildings based on life cycle cost considerations, *Earthquake Eng Struc*, 32-9 (2003) pp. 1313-1332.
25. K.E. Mathias, D. Whitley, Trasforming the Search Space with Gray Coding. *IEEE Int. Conference on Evolutionary Computation*, pp. 513-518. IEEE Service Center (1994).
26. N.J. Radcliffe, P.D. Surry, "Fundamental Limitations on Search Algorithms: Evolutionary Computing in Perspective", *Lecture Notes in Computer Science* 1000 (Computer Science Today: Recent Trends and Development), Springer, (1995) pp. 275-291.
27. J. Rowe, D. Whitley, L. Barbulescu, P. Watson, "Properties of Gray and Binary Representations", *Evolutionary Computation*, 12-1 (2004) pp. 46-76.
28. K. Sarma, H. Adeli, "Life-cycle cost optimization of steel structures", *International Journal for Numerical Methods in Engineering*, 55 (2002) 1451-1462.
29. C. Savage, "A Survey of Combinatorial Gray Codes", *SIAM Review* Vol. 39 (4) pp. 605-629 (1997).
30. D. Whitley, "A free lunch proof for Gray versus Binary Codings", *Proceedings of the Genetic and Evolutionary Computation Conference* 1999, Morgan Kaufmann pp. 726-733.
31. D. Whitley, S. Rana, R. Heckendorn, Representation Issues in Neighborhood Search and Evolutionary Algorithms, in Eds: D. Quagliarella, J. Périaux, C. Poloni, G. Winter; *Genetic Algorithms and Evolution Strategies in Engineering and Computer Science*. John Wiley & Sons (1997) pp. 39-57.

32. D.H. Wolpert, W.G. MacReady, "No Free Lunch Theorems for Search", *Technical Report* SFI-TR-95-02-010, Santa Fe Institute, July 1995.
33. E. Zitzler, Evolutionary Algorithms for Multiobjective Optimization : Methods and Applications. *PhD Thesis*. Swiss Federal Institute of Technology (ETH), Zurich 1999.
34. E. Zitzler, M. Laumanns, L. Thiele, "SPEA2: Improving the Strength Pareto Evolutionary Algorithm for Multiobjective Optimization", *Evolutionary Methods for Design, Optimization and Control with Applications to Industrial Problems*, CIMNE, pp. 95-100 (2002).

Multi-objective Genetic Algorithms to Create Ensemble of Classifiers

Luiz S. Oliveira, Marisa Morita, Robert Sabourin, and Flávio Bortolozzi

[1] Pontifícia Universidade Católica do Paraná, Curitiba, Brazil
[2] Universidade Tuiuti do Paraná, Curitiba, Brazil
[3] Ecole de Technologie Superiéure, Montreal, Canada
soares@ppgia.pucpr.br

Abstract. Feature selection for ensembles has shown to be an effective strategy for ensemble creation due to its ability of producing good subsets of features, which make the classifiers of the ensemble disagree on difficult cases. In this paper we present an ensemble feature selection approach based on a hierarchical multi-objective genetic algorithm. The algorithm operates in two levels. Firstly, it performs feature selection in order to generate a set of classifiers and then it chooses the best team of classifiers. In order to show its robustness, the method is evaluated in two different contexts: supervised and unsupervised feature selection. In the former, we have considered the problem of handwritten digit recognition while in the latter, we took into account the problem of handwritten month word recognition. Experiments and comparisons with classical methods, such as Bagging and Boosting, demonstrated that the proposed methodology brings compelling improvements when classifiers have to work with very low error rates.

1 Introduction

Ensemble of classifiers has been widely used to reduce model uncertainty and improve generalization performance. Developing techniques for generating candidate ensemble members is a very important direction of ensemble of classifiers research. It has been demonstrated that a good ensemble is one where the individual classifiers in the ensemble are both accurate and make their errors on different parts of the input space [7]. In other words, an ideal ensemble consists of good classifiers (not necessarily excellent) that disagree as much as possible on difficult cases.

The literature has shown that varying the feature subsets used by each member of the ensemble should help to promote this necessary diversity [6, 15, 18]. Traditional feature selection algorithms aim at finding the best trade-off between features and generalization. On the other hand, ensemble feature selection has the additional goal of finding a set of feature sets that will promote disagreement among the component members of the ensemble. The Random Subspace Method (RMS) proposed by Ho in [6] was one early algorithm that constructs an ensemble by varying the subset of features. Strategies based on genetic algorithms (GAs) also have been proposed [5, 15]. All these strategies claim better results than those produced by traditional methods for creating ensembles such as Bagging and Boosting. In spite of the good results brought by GA-based methods, they still can be improved in some aspects, e.g., avoiding classical methods such as the weighted sum to combine multiple objective functions. It is well known that when dealing with this kind of combination, one should deal with problems such as scaling and sensitivity towards the weights.

It has been demonstrated that feature selection through multi-objective genetic algorithm (MOGA) is a very powerful tool for finding a set of good classifiers [4, 14], since GA is quite effective in rapid global search of large, non-linear and poorly understood spaces [17]. Besides, it can overcome problems such as scaling and sensitivity towards the weights. Kudo and Sklansky [8] have compared several algorithms for feature selection and concluded that GAs are suitable when dealing with large-scale feature selection (number of features is over 50). This is the case of most of the problems in handwriting recognition, which is the test problem in this work.

In this light, we propose an ensemble feature selection approach based on a hierarchical MOGA. The underlying paradigm is the "overproduce and choose" [16]. The algorithm operates in two levels. The former is devoted to generate a set of good classifiers by minimizing two criteria: error rate and number of features. The latter combines these classifiers in order to find an ensemble by maximizing the following two criteria: accuracy of the ensemble and a measure of diversity. We demonstrated through experimentation that using diversity jointly with performance to guide selection can avoid overfitting during the search.

In order to show robustness of the proposed methodology, it was evaluated in two different contexts: supervised and unsupervised feature selection. In the former, we have considered the problem of handwritten digit recognition and used three different feature sets and multi-layer perceptron (MLP) neural networks as classifiers. In the latter, we took into account the problem of handwritten month word recognition and used three different feature sets and hidden Markov models (HMM) as classifiers. We demonstrate that it is feasible to find compact clusters and complementary high-level representations (codebooks) in subspaces without using the recognition results of the system. Experiments and comparisons with classical methods, such as Bagging and Boosting, demonstrated that the proposed methodology brings compelling improvements when classifiers have to work with very low error rates.

2 Methodology Overview

In this section we outline the hierarchical approach proposed. As stated before, it is based on an "overproduce and choose" paradigm where the first level generates several classifiers by conducting feature selection and the second one chooses the best ensemble among such classifiers. Figure 1 depicts the proposed methodology. Firstly, we carry out feature selection by using a MOGA. It gets as inputs a trained classifier and its respective data set. Since the algorithm aims at minimizing two criteria during the search[1], it will produce at the end a 2-dimensional Pareto-optimal front, which contains a set of classifiers (trade-offs between the criteria being optimized). The final step of this first level consists in training such classifiers.

---

[1] Error rate and number of features in the case of supervised feature selection and a clustering index and the number of features in the case of unsupervised feature selection.

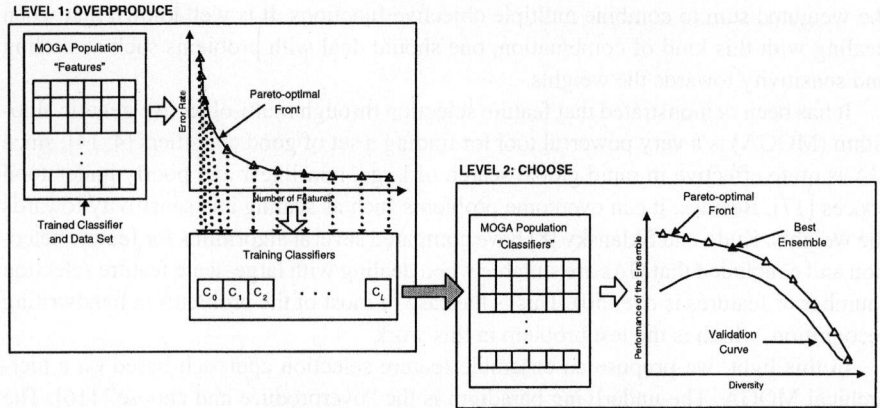

Fig. 1. An overview of the proposed methodology

Once the set of classifiers have been trained, the second level is suggested to pick the members of the team which are most diverse and accurate. Let $A = \{C_1, C_2, \ldots, C_L\}$ be a set of L classifiers extracted from the Pareto-optimal and B a chromosome of size L of the population. The relationship between A and B is straightforward, i.e., the gene i of the chromosome B is represented by the classifier C_i from A. Thus, if a chromosome has all bits selected, all classifiers of A will be included in the ensemble. Therefore, the algorithm will produce a 2-dimensional Pareto-optimal front which is composed of several ensembles (trade-offs between accuracy and diversity). In order to choose the best one, we use a validation set, which points out the most diverse and accurate team among all. Later in this paper, we will discuss the issue of using diversity to choose the best ensemble.

In both cases, MOGAs are based on bit representation, one-point crossover, and bit-flip mutation. In our experiments, MOGA used is a modified version of the Non-dominated Sorting Genetic Algorithm (NSGA) [2] with elitism.

3 Classifiers and Feature Sets

As stated before, we have carried out experiments in both supervised and unsupervised contexts. The remaining of this section describes the feature sets and classifiers we have used.

3.1 Supervised Context

To evaluate the proposed methodology in the supervised context, we have used three base classifiers trained to recognize handwritten digits of NIST SD19. Such classifiers were trained with three well-known feature sets: Concavities and Contour (CCsc) [13], Distances (DDDsc), and Edge Maps (EMsc). All classifiers here are MLPs trained with the gradient descent applied to a sum-of-squares error function.

The training (TRDBsc) and validation (VLDB1sc) sets are composed of 195,000 and 28,000 samples from hsf_0123 series respectively while the test set (TSDBsc) is

Table 1. Description and performance of the classifiers on TSDB (zero-rejection level)

Feature Set	Number of Features	Units in the Hidden Layer	Rec. Rate (%)	Rec. Rate Err=0.1%	Err=0.5%
CCsc	132	80	99.13	91.83	98.50
DDDsc	96	60	98.17	75.11	92.80
EMsc	125	70	97.04	60.11	85.10

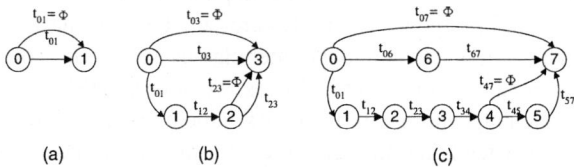

Fig. 2. Topologies of (a) space, (b), and (c) letter models

composed of 30,089 samples from the hsf_7. We consider also a second validation set (VLDB2sc), which is composed of 30,000 samples of hsf_7. This data is used to select the best ensemble of classifiers. Table 1 reports the performance of all classifiers at zero-rejection level and error rates fixed at low levels (0.10 and 0.50%). These numbers are much more meaningful when dealing with real applications since they describe the recognition rate in relation to a specific error rate, including implicitly a corresponding reject rate. They also corroborates that recognition of handwritten digits is still an open problem when very low error rates are required.

3.2 Unsupervised Context

To evaluate the proposed methodology in unsupervised context we have used three HMM-based classifiers trained to recognize handwritten Brazilian month words ("Janeiro", "Fevereiro", "Março", "Abril", "Maio", "Junho", "Julho", "Agosto", "Setembro", "Outubro", "Novembro", "Dezembro"). The training (TRDBuc), validation (VLDB1uc), and testing (TSDBuc) sets are composed of 1,200, 400, and 400 samples, respectively. In order to increase the training and validation sets, we have also considered 8,300 and 1,900 word images, respectively, extracted from the legal amount database. This is possible because we are considering character models. We consider also a second validation set (VLDB2uc) of 500 handwritten Brazilian month words. Such data is used to select the best ensemble of classifiers.

Given a discrete HMM-based approach, each word image is transformed as a whole into a sequence of observations by the successive application of preprocessing, segmentation, and feature extraction. Preprocessing consists of correcting the average character slant. The segmentation algorithm uses the upper contour minima and some heuristics to split the date image into a sequence of segments (graphemes), each of which consists of a correctly segmented, an under-segmented, or an over-segmented character. A detailed description of the preprocessing and segmentation stages is given in [12].

The word models are formed by the concatenation of appropriate elementary HMMs, which are built at letter and space levels. The topology of space model shown in Figure 2(a) consists of two states linked by two transitions that encode a space (transition t_{01}) or no space (transition $t_{01} = \Phi$).

Two topologies of letter models were chosen based on the output of our grapheme-based segmentation algorithm which may produce a correct segmentation of a letter, a letter under-segmentation or a letter over-segmentation into two, three, or four graphemes depending on each letter. In order to cope with these configurations of segmentations, we have designed topologies with three different paths leading from the initial state to the final state. Considering uppercase and lowercase letters, we need 42 models since the legal amount alphabet is reduced to 21 letter classes and we are not considering the unused ones. Thus, regarding the two topologies, we have 84 HMMs which are trained using the Baum-Welch algorithm with the Cross-Validation procedure.

The feature set that feeds the first classifier is a mixture of concavity and contour features (CCuc) [13]. In this case, each grapheme is divided into two equal zones (horizontal) where for each region a concavity and contour feature vector of 17 components is extracted. Therefore, the final feature vector has 34 components. The other two classifiers make use of a feature set based on distances. The former uses the same zoning discussed before (two equal zones), but in this case, for each region a vector of 16 components is extracted. This leads to a final feature vector of 32 components (DDD32$_u c$). For the latter we have tried a different zoning. Table 2 reports the performance of all classifiers on the test set at zero-rejection level and error rates fixed at 1 and 4%. We have chosen higher error rates in this case due to the size of the database we are dealing with.

Table 2. Performance of the classifiers on the test set

Feature Set	Number of Features	Codebook Size	Rec. Rate (%)	Rec. Rate Err=1%	Rec. Rate Err=4%
CCuc	34	80	86.1	61.0	79.2
DDD32uc	32	40	73.0	30.5	48.4
DDD64uc	64	60	64.5	24.9	37.0

It can be observed from Table 2 that the recognition rates with error fixed at 1 and 4% are very poor, hence, the number of rejected patterns is very high. We will see in the next sections that the proposed methodology can improve these results considerably.

4 Implementation

This section introduces how we have implemented both levels of the proposed methodology. First we discuss the supervised context and then the unsupervised.

4.1 Supervised Feature Subset Selection

The feature selection algorithm used in here was introduced in [14]. To make this paper self-contained, a brief description is included in this section.

As stated elsewhere, the idea of using feature selection is to promote diversity among the classifiers. To tackle such a task we have to optimize two objective functions: minimization of the number of features and minimization of the error rate of the classifier. Computing the first one is simple, i.e., the number of selected features. The problem lies in computing the second one, i.e., the error rate supplied by the classifier. Regarding a wrapper approach, in each generation, evaluation of a chromosome (a feature subset) requires training the corresponding neural network and computing its accuracy. This evaluation has to be performed for each of the chromosomes in the population. Since such a strategy is not feasible due to the limits imposed by the learning time of the huge training set considered in this work, we have adopted the strategy proposed by Moody and Utans in [9], who use the sensitivity of the network to estimate the relationship between the input features and the network performance.

Moody and Utans show that when variables with small sensitivity values with respect to the network outputs are removed, they do not influence the final classification. So, in order to evaluate a given feature subset we replace the unselected features by their averages. In this way, we avoid training the neural network and hence turn the wrapper approach feasible for our problem. Such a scheme makes it feasible to deal with huge databases in order to better represent the pattern recognition problem during the fitness evaluation. Moreover it can accommodate multiple criteria such as the number of features and the accuracy of the classifier, and generate the Pareto-optimal front in the first run of the algorithm.

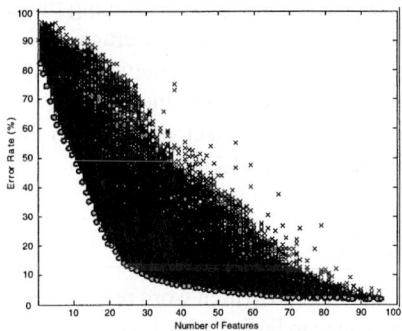

Fig. 3. Evolution of the population in the objective plane

It can be observed in Figure 3 that the Pareto-optimal front is composed of several different classifiers. To find out which classifiers of the Pareto-optimal front compose the best ensemble, we carried out a second level of search. Once we did not train the models during the search (the training step is replaced by the sensitivity analysis), the last step of feature selection consists of training the solutions provided by the Pareto-optimal front (1).

4.2 Choosing the Best Ensemble

As defined in Section 2 each gene of the chromosome is represented by a classifier produced in the previous level. Therefore, if a chromosome has all bits selected, all classifiers will compose the team. In order to find the best ensemble of classifiers, i.e., the most diverse set of classifiers that brings a good generalization, we have used two objective functions during this level of the search, namely, maximization of the recognition rate of the ensemble and maximization of a measure of diversity. We have tried different measures such as overlap, entropy, and ambiguity [7]. The results achieved with ambiguity and entropy were very similar. In this work we have used ambiguity as diversity measure. The ambiguity is defined as follows:

$$a_i(x_k) = [V_i(x_k) - \overline{V}(x_k)]^2 \qquad (1)$$

where a_i is the ambiguity of the ith classifier on the example x_k, randomly drawn from an unknown distribution, while V_i and $\overline{V}$ are the ith classifier and the ensemble predictions, respectively. In other words, it is simply the variance of ensemble around the mean, and it measures the disagreement among the classifiers on input x. Thus the ambiguity of an ensemble measured on a set of M samples is

$$\overline{A} = \frac{1}{N} \sum \frac{1}{M} \sum_{k=1}^{M} a_i(x_k) \qquad (2)$$

where N is the number of classifiers. So, if the classifiers implement the same functions, the ambiguity $\overline{A}$ will be low, otherwise it will be high.

At this level of the strategy we want to maximize the generalization of the ensemble, therefore, it will be necessary to use a way of combining the outputs of all classifiers to get a final decision. To do this, we have used the average, which is a simple and effective scheme of combining predictions of the neural networks. Other combination rules such as product, min, and max have been tested but the simple average has produced slightly better results. In order to evaluate the objective functions during the search described above we have used the validation set VLDB1sc.

4.3 Unsupervised Feature Subset Selection

A lot of work done in the field of handwritten word recognition take into account discrete HMMs as classifiers, which have to be fed with a sequence of discrete values (symbols). This means that before using a continuous feature vector, we must convert it to discrete values. A common way to do that is through clustering. The problem is that for the most of real-life situations we do not know the best number of clusters, what makes it necessary to explore different numbers of clusters using traditional clustering methods such as the K-means algorithm and its variants. In this light, clustering can become a trial-and-error work. Besides, its result may not be very promising especially when the number of clusters is large and not easy to estimate.

Unsupervised feature selection emerges as a clever solution to this problem. The literature contains several studies on feature selection for supervised learning, but only recently, the feature selection for unsupervised learning has been investigated [3]. The objective in unsupervised feature selection is to search for a subset of features that best uncovers "natural" groupings (clusters) from data according to some criterion. In this way, we can avoid the manual process of clustering and find the most discriminative features in the same time. Hence, we will have at the end a more compact and robust high-level representation (symbols).

In the above context, unsupervised feature selection also presents a multi-criterion optimization function, where the objective is to find compact and well separated hyperspherical clusters in the feature subspaces. Differently of the supervised feature selection, here the criteria optimized by the algorithm are a validity index and the number of features. [11].

In order to measure the quality of clusters during the clustering process, we have used the Davies-Bouldin (DB)-index [1] over 80,000 feature vectors extracted from the training set of 9,500 words. To make such an index suitable for our problem, it must be normalized by the number of selected features. This is due to the fact that it is based on geometric distance metrics and therefore, it is not directly applicable here because it is biased by the dimensionality of the space, which is variable in feature selection problems.

We have noticed that the value of DB index decreases as the number of features increases. We have correlated this effect with the normalization of DB-index by the number of features. In order to compensate this, we have considered as second objective the minimization of the number of features. In this case, one feature must be set at least. Figure 4 depicts the Pareto-optimal front found after the search, the relationship between the number of clusters and number of features and the relationship between the recognition rate on the validation set and the number of features.

Once we have a limited space here, we opted by not showing the Pareto-optimal front for unsupervised case. However, it is very similar to that presented in Figure 3. Figure 4 shows the relationship between the number of clusters and the number of features and the relationship between the recognition rate and the number of features. The way of choosing the best ensemble is exactly the same as introduced in Section 4.2.

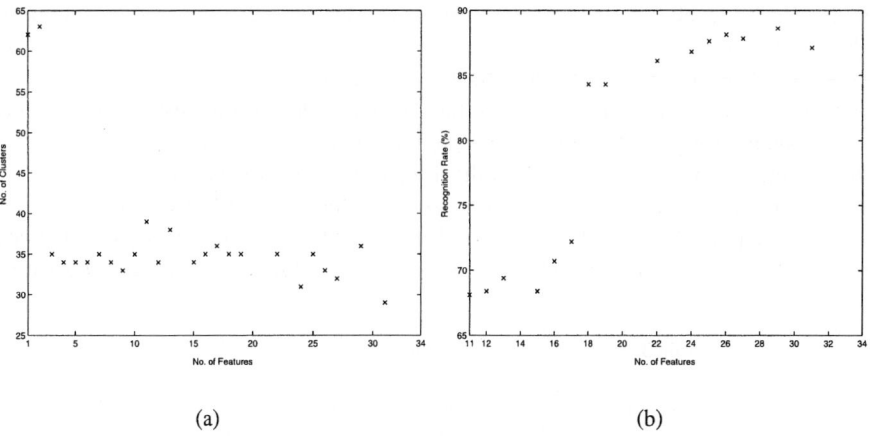

Fig. 4. (a)Relationship between the number of clusters and the number of features and (b) Relationship between the recognition rate and the number of features

5 Experimental Results

All experiments in this work were based on a single-population master-slave MOGA. In this strategy, one master node executes the genetic operators (selection, crossover and mutation), and the evaluation of fitness is distributed among several slave processors. We have used a Beowulf cluster with 17 (one master and 16 slaves) PCs (1.1Ghz CPU, 512Mb RAM) to execute our experiments.

The following parameter settings were employed in both levels: population size = 128, number of generations = 1000, probability of crossover = 0.8, probability of mutation = $1/L$ (where L is the length of the chromosome), and niche distance (σ_{share}) = [0.25,0.45]. The length of the chromosome in the first level is the number of components in the feature set (see Table 1), while in the second level is the number of classifiers picked from the Pareto-optimal front in the previous level.

In order to define the probabilities of crossover and mutation, we have used the one-max problem, which is probably the most frequently-used test function in research on genetic algorithms because of its simplicity. This function measures the fitness of an individual as the number of bits set to one on the chromosome. We have used a standard genetic algorithm with a single-point crossover and the maximum generations of 1000. The fixed crossover and mutation rates are used in a run, and the combination of the crossover rates 0.0, 0.4, 0.6, 0.8 and 1.0 and the mutation rates of $0.1/L$, $1/L$ and $10/L$, where L is the length of the chromosome. The best results were achieved with $P_c = 0.8$ and $P_m = 1/L$. The parameter σ_{share} was tuned empirically.

5.1 Experiments in the Supervised Context

Once all parameters have been defined, the first step, as described in Section 4.1, consists of performing feature selection for a given feature set. As depicted in Figure 3, this procedure produces quite a large number of classifiers, which should be trained for use in the second level. After some experiments, we found out that the second level never chooses those classifiers with poor performance (e.g., error > 60%) to compose the ensemble. Thus, in order to speed up the training process and the second level of search as well, we decide not to use them in the second level. To train such classifiers, the same databases reported in Section 3.1 were used. Table 3 summarizes the classifiers that undergoes to the second level for the three feature sets we have considered.

Considering for example the feature set CC_{sc}, the first level of the algorithm provided 81 classifiers which have the number of features ranging from 24 to 125 and recognition rates ranging from 90.5% to 99.1% on $TSDB_{sc}$. This shows the great di-

Table 3. Summary of the classifiers produced by the first level

Feature Set	No. of Classifiers	Range of Features	Range of Rec. Rates (%)
CCsc	81	24-125	90.5 - 99.1
DDDsc	54	30-84	90.6 - 98.1
EMsc	78	35-113	90.5 - 97.0

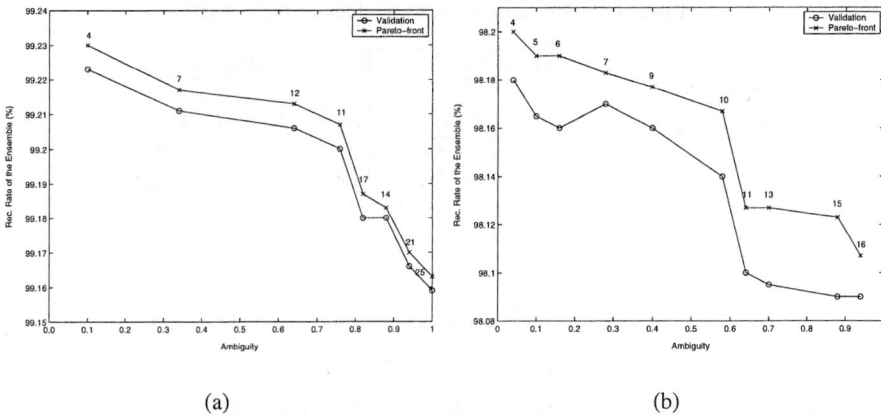

Fig. 5. The Pareto-optimal front produced by the second-level MOGA: (a) S_1 and (b) S_2

versity of the classifiers produced by the feature selection method. Based on Table 3 we define four sets of base classifiers as follows: $S_1 = \{CCsc_0, \ldots, CCsc_{80}\}$, $S_2 = \{DDDsc_0, \ldots, DDDsc_{53}\}$, $S_3 = \{EMsc_0, \ldots, EMsc_{77}\}$, and $S_4 = \{S_1 \bigcup S_2 \bigcup S_3\}$. All these sets could be seem as ensembles, but in this work we reserve the word ensemble to characterize the results yielded by the second-level of the algorithm. In order to assess the objective functions of the second-level of the algorithm (generalization of the ensemble and diversity) we have used the validation set (VLDB1sc).

Like the first level, the second one also generates a set of possible solutions which are the trade-offs between the generalization of the ensemble and its diversity. Thus the problem now lies in choosing the most accurate ensemble among all. Figure 5 depicts the variety of ensembles yielded by the second-level of the algorithm for S_1 and S_2. The number over each point stands for the number of classifiers in the ensemble. In order to decide which ensemble to choose we validate the Pareto-optimal front using VLDB2sc, which was not used so far. Since we are aiming at performance, the direct choice will be the ensemble that provides better generalization on VLDB2$_{sc}$. Table 4 summarizes the best ensembles produced for the four sets of base classifiers and their performance at zero-rejection level on the test set. For facility, we reproduce in this table the results of the original classifiers.

We can notice from Table 4 that the ensembles and base classifiers have very similar performance at zero-rejection level. On the other hand, it also shows that the ensembles respond better for error rates fixed at very low levels than single classifiers. The most expressive result was achieved for the ensemble S_3, which attains a reasonable performance at zero-rejection level but performs very poorly at low error rates. In such a case, the ensemble of classifiers brought an improvement of about 8%. We have noticed that the ensemble reduces the high outputs of some outliers so that the threshold used for rejection can be reduced and consequently the number of samples rejected is reduced. Thus, aiming for a small error rate we have to consider the important role of the ensemble. Another fact worth noting though, is the performance of S_4 at low error rates. For the error rate fixed at 1% it reached 95.0% against 93.5% of S_1. S_4 is composed of 14,

Table 4. Performance of the ensembles on the test set

Feature Set	No. Classif.	Ensembles Rec. Rate no Rej.	Err=0.1%	Err=0.5%	Original Rec. Rate no Rej.	Err=0.1%	Err=0.5%
S_1	4	99.22	93.49	98.86	99.13	91.83	98.50
S_2	4	98.18	79.22	95.28	98.17	75.11	92.80
S_3	7	97.10	68.50	89.00	97.04	60.11	85.10
S_4	24	99.25	95.03	98.94			

Table 5. Summary of the classifiers produced by the first level

Feature Set	Number of Classifiers	Range of Features	Range of Codebook	Range of Rec. Rates (%)
$CCuc$	15	10-32	29-39	68.1 - 88.6
$DDD32uc$	21	10-31	20-30	71.7 - 78.0
$DDD64uc$	50	10-64	52-80	60.6 - 78.2

6, and 4 classifiers from S_1, S_2, and S_3, respectively. This emphasizes the ability of the algorithm in finding good ensembles when more original classifiers are available.

5.2 Experiments in the Unsupervised Context

The experiments in the unsupervised context follow the same vein of the supervised one. As discussed in Section 4.3, the main difference lies in the way the feature selection is carried out. In spite of that, we can observe that the number of classifiers produced during unsupervised feature selection is quite large as well. To train the classifiers, the same databases reported in Section 3.2 were considered. Table 5 summarizes the classifiers (after training) produced by the first level for the three feature sets we have considered.

Considering for example the feature set $CCuc$, the first level of the algorithm provided 15 classifiers which have the number of features ranging from 10 to 32 and recognition rates ranging from 68.1% to 88.6% on VLDB1uc. This shows the great diversity of the classifiers produced by the feature selection method. Based on the classifiers reported in Table 5 we define four sets of base classifiers as follows: $F_1 = \{CCuc_0, \ldots, CCuc_{14}\}$, $F_2 = \{DDD32uc_0, \ldots, DDD32uc_{20}\}$, $F_3 = \{DDD64uc_0, \ldots, DDD64uc_{49}\}$, and $F_4 = \{F_1 \bigcup F_2 \bigcup F_3\}$.

Figure 6 depicts the variety of ensembles yielded by the second-level of the algorithm for F_1 and F_2. The number over each point stands for the number of classifiers in the ensemble. Like in the previous experiments, the second validation set (VLDB2uc) was used to select the best ensemble. After selecting the best ensemble the final step is to assess them on the test set. Table 6 summarizes the performance of the ensembles on the test set. For the sake of comparison, we reproduce in Table 6 the results presented in Table 2.

Like in the previous experiments (supervised context), the result achieved by the ensemble F_4 shows the ability of the algorithm in finding good ensembles when more

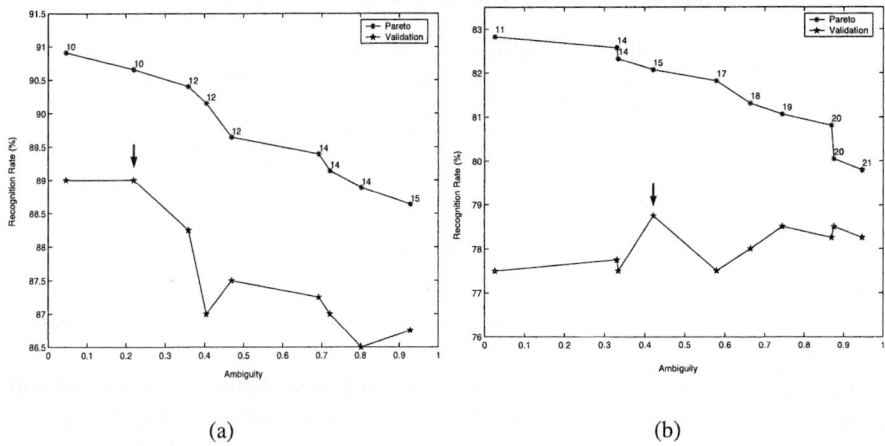

Fig. 6. The Pareto-optimal front (and validation curves where the best solutions are highlighted with an arrow) produced by the second-level MOGA: (a) F_1 and (b) F_2

Table 6. Comparison between ensembles and original classifiers

Feature Set	No. Classif.	Ensembles Rec. Rate no Rej.	Err=1%	Err=5%	Original Rec. Rate no Rej.	Err=1%	Err=4%
F_1	10	89.2	66.0	81.0	86.1	61.0	79.0
F_2	15	80.2	45.0	60.2	73.0	29.5	48.5
F_3	36	80.7	43.7	62.5	64.5	24.0	36.5
F_4	45	90.2	70.2	77.0			

base classifiers are considered. The ensemble F_4 is composed of 9, 11, and 25 classifiers from F_1, F_2, and F_3, respectively. In light of this, we decided to introduce a new feature set, which, based on our experience, has a good discrimination power when combined with other features such as concavities. This feature set, which we call "global features", is composed of primitives such as ascenders, descenders, and loops. The combination of these primitives plus a primitive that determines whether a grapheme does not contain ascender, descender, and loop produces a 20-symbol alphabet. For more details, see Ref. [10]. In order to train the classifier with this feature set, we have used the same databases described in Section 3.2. The recognition rates at zero-rejection level are 86.1% and 87.2% on validation and testing sets, respectively. This performance compares with the CCuc classifier.

Since we have a new base classifier, our sets of base classifiers must be modified to cope with it. Thus, $F_{1G} = \{F_1 \bigcup G\}$, $F_{2G} = \{F_2 \bigcup G\}$, $F_{3G} = \{F_3 \bigcup G\}$, and $F_{4G} = \{F_1 \bigcup F_2 \bigcup F_3 \bigcup G\}$. In such cases, G stands for the classifier trained with global features. Table 7 summarizes the ensembles found using these new sets of base classifiers. It is worthy of remark the reduction of the size of the teams and the improvement in the recognition rates. This shows the ability of the algorithm in finding not just diverse but

Table 7. Performance of the ensembles with global features

Base Classifiers	Number of Classifiers	Rec. Rate (%) no Rej.	Err=1%	Err=4%
F_{1G}	2	92.2	69.0	87.5
F_{2G}	2	89.7	53.2	80.2
F_{3G}	7	85.5	55.0	75.0
F_{4G}	23	92.0	75.0	88.7

also uncorrelated classifiers to compose the ensemble [19]. Besides, it corroborates to our claim that the classifier G when combined with other features bring an improvement to the performance.

Like the results at zero-rejection level, the improvement observed here also are quite impressive. Table 7 shows that F_{1G} and F_{4G} reach similar results on the test set at zero-rejection level, however, F_{1G} contains just two classifiers against 23 of F_{4G}. On the other hand, the latter features a slightly better error-reject trade-off in the long run.

Based on the experiments reported so far we can affirm that the unsupervised feature selection is a good strategy to generate diverse classifiers. This is made very clear in the experiments regarding the feature set DDD64. In such a case, the original classifier has a poor performance (about 65% on the test set), but when it is used to generate the set of base classifiers, the second-level MOGA was able to produce a good ensemble by maximizing the performance and the ambiguity measure. Such an ensemble of classifiers brought an improvement of about 15% in the recognition rate at zero-rejection level.

6 Discussion and Conclusion

The results obtained here attest that the proposed strategy is able to generate a set of good classifiers in both supervised and unsupervised contexts. To better evaluate our results, we have used two traditional ensemble methods (Bagging and Boosting) in the

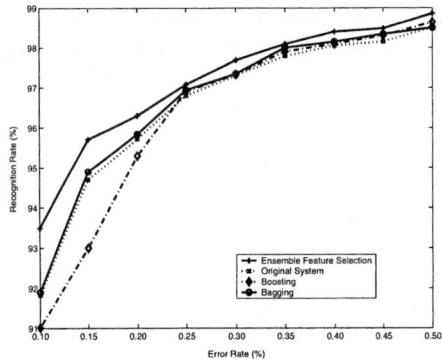

Fig. 7. Comparison among feature selection for ensembles, bagging, and boosting for CC_sc

supervised context. Figure 7 reports the results for CC_{sc}. As we can see, the proposed methodology achieved better results, especially when considering very low error rates.

Diversity is an issue that deserves some attention when discussing ensemble of classifiers. As we have mentioned before, some authors advocated that diversity does not help at all. In our experiments, most of the time, the best ensembles of the Pareto-optimal also were the best for the unseen data. This could lead one to agree that diversity is not important when building ensembles, since even using a validation set the selected team is always the most accurate and with less diversity.

However, if we look carefully the results, we will observe that there are cases where the validation curve does not have the same shape of the Pareto-optimal. In such cases diversity is very useful to avoid selecting overfitted solutions.

One can argue that using a single-objective GA and considering the entire final population, perhaps the similar solutions found in the Pareto-optimal produced by the MOGA will be there. To show that it does not happen, we have carried out some experiments with a single-objective GA where the fitness function was the maximization of the ensemble's accuracy. Since a single-objetive optimization algorithm searches for an optimum solution, it is natural to expect that it will converge towards the fittest solution, hence, the diversity of solutions presented in the Pareto-optimal is not present in the final population of the single-objective GA.

We have described a methodology for ensemble creation underpinned on the paradigm "overproduce and choose". It takes two levels of search where the first level overproduces a set of classifiers by performing feature selection while the second one chooses the best team of classifiers.

The feasibility of the strategy was demonstrated through comprehensive experiments carried out in the context of handwriting recognition. The idea of generating classifiers through feature selection was proved to be successful in both supervised and unsupervised contexts. The results attained in both situations and using different feature sets and base classifiers demonstrated the efficiency of the proposed strategy by finding powerful ensembles, which succeed in improving the recognition rates for classifiers working with a very low error rates. Such results compare favorably to traditional ensemble methods such as Bagging and Boosting.

Finally we have addressed the issue of using diversity to build ensembles. As we have seen, using diversity jointly with the accuracy of the ensemble as selection criterion might be very helpful to avoid choosing overfitted solutions. Our results certainly brings some contribution to the field, but this still is an open problem.

Acknowledgements

This research has been supported by The National Council for Scientific and Technological Development (CNPq) grant 150542/2003-8.

References

1. D. L. Davies and D. W. Bouldin. A cluster separation measure. *IEEE Trans. on Pattern Analysis and Machine Intelligence*, 1(224-227):550–554, 1979.

2. K. Deb. *Multi-Objective Optimization using Evolutionary Algorithms*. John Wiley and Sons Ltd, 2nd edition, April 2002.
3. J. G. Dy and C. E. Brodley. Feature subset selection and order identification for unsupervised learning. In *Proc. 17th International Conference on Machine Learning*, Stanford University-CA, July 2000.
4. C. Emmanouilidis, A. Hunter, and J. MacIntyre. A multiobjective evolutionary setting for feature selection and a commonality-based crossover operator. In *Proc. of Congress on Evolutionary Computation*, volume 1, pages 309–316, 2000.
5. C. Gerra-Salcedo and D. Whitley. Genetic approach to feature selection for ensemble creatin. In *Proc. of Genetic and Evolutionary Computation Conference*, pages 236–243, 1999.
6. T. K. Ho. The random subspace method for constructing decision forests. *IEEE Trans. on Pattern Analysis and Machine Intelligence*, 20(8):832–844, 1998.
7. A. Krogh and J. Vedelsby. Neural networks ensembles, cross validation, and active learning. In G.Tesauro et al, editor, *Advances in Neural Information Processing Systems 7*, pages 231–238. MIT Press, 1995.
8. M. Kudo and J. Sklansky. Comparision of algorithms that select features for pattern classifiers. *Pattern Recognition*, 33(1):25–41, 2000.
9. J. Moody and J. Utans. Principled architecture selection for neural networks: Application to corporate bond rating prediction. In J. Moody, S. J. Hanson, and R. P. Lippmann, editors, *Advances in Neural Information Processing Systems 4*. Morgan Kaufmann, 1991.
10. M. Morita, L. S. Oliveira, R. Sabourin, F. Bortolozzi, and C. Y. Suen. An HMM-MLP hybrid system to recognize handwritten dates. In *Proc. of International Joint Conference on Neural Networks*, pages 867–872, Honolulu-USA, 2002. IEEE Press.
11. M. Morita, R. Sabourin, F. Bortolozzi, and C. Y. Suen. Unsupervised feature selection using multi-objective genetic algorithms for handwritten word recognition. In *Procs of the 7th ICDAR*, pages 666–670. IEEE Computer Society, 2003.
12. M. Morita, A. El Yacoubi, R. Sabourin, F. Bortolozzi, and C. Y. Suen. Handwritten month word recognition on Brazilian bank cheques. In *Proc. 6th ICDAR*, pages 972–976, 2001.
13. L. S. Oliveira, R. Sabourin, F. Bortolozzi, and C. Y. Suen. Automatic recognition of handwritten numerical strings: A recognition and verification strategy. *IEEE Trans. on Pattern Analysis and Machine Intelligence*, 24(11):1438–1454, 2002.
14. L. S. Oliveira, R. Sabourin, F. Bortolozzi, and C. Y. Suen. A methodology for feature selection using multi-objective genetic algorithms for handwritten digit string recognition. *International Journal of Pattern Recognition and Artificial Intelligence*, 17(6):903–930, 2003.
15. D. W. Optiz. Feature selection for ensembles. In *Proc. of 16th International Conference on Artificial Intelligence*, pages 379–384, 1999.
16. D. Partridge and W. B. Yates. Engineering multiversion neural-net systems. *Neural Computation*, 8(4):869–893, 1996.
17. W. Siedlecki and J. Sklansky. A note on genetic algorithms for large scale on feature selection. *Pattern Recognition Letters*, 10:335–347, 1989.
18. A. Tsymbal, S. Puuronen, and D. W. Patterson. Ensemble feature selection with the simple Bayesian classification. *Information Fusion*, 4:87–100, 2003.
19. K. Tumer and J. Ghosh. Error correlation and error reduction in ensemble classifiers. *Connection Science*, 8(3-4):385–404, 1996.

Multi-objective Model Optimization for Inferring Gene Regulatory Networks

Christian Spieth, Felix Streichert, Nora Speer, and Andreas Zell

Centre for Bioinformatics Tübingen (ZBIT), University of Tübingen,
Sand 1, D-72076 Tübingen, Germany
spieth@informatik.uni-tuebingen.de
http://www-ra.informatik.uni-tuebingen.de

Abstract. With the invention of microarray technology, researchers are able to measure the expression levels of ten thousands of genes in parallel at various time points of a biological process. The investigation of gene regulatory networks has become one of the major topics in Systems Biology. In this paper we address the problem of finding gene regulatory networks from experimental DNA microarray data. We suggest to use a multi-objective evolutionary algorithm to identify the parameters of a non-linear system given by the observed data. Currently, only limited information on gene regulatory pathways is available in Systems Biology. Not only the actual parameters of the examined system are unknown, also the connectivity of the components is a priori not known. However, this number is crucial for the inference process. Therefore, we propose a method, which uses the connectivity as an optimization objective in addition to the data dissimilarity (relative standard error - RSE) between experimental and simulated data.

1 Introduction

Gene regulatory networks (GRNs) represent the dependencies of the different actors in a cell operating at the genetic level. They dynamically determine the level of gene expression for each gene in the genome by controlling whether a gene will be transcribed into RNA or not. A simple GRN consists of one or more input signalling pathways, several target genes, and the RNA and proteins produced from those target genes. In addition, such networks often include dynamic feedback loops that provide further network regulation activities and output. In order to understand the underlying structures of activities and interactions of intra-cellular processes one has to understand the dependencies of gene products and their impact on the expression of other genes. Therefore, finding a GRN for a specific biological process would explain this process from a logical point of view, thus explaining many diseases.

Therefore, the model reconstruction of gene regulatory networks has become one of the major topics in bioinformatics. However, the huge number of system components requires a large amount of experimental data to infer genome-wide networks. Recently, DNA microarrays have become one of the major tools in the research area of microbiology. This technology enables researchers to monitor the activities of thousands of genes in parallel and can therefore be used as a powerful tool to understand the regulatory mechanisms of gene expression in a cell. With this technique, cells can be studied under several conditions such as medical treatment or different environmental influences.

Microarray experiments often result in time series of measured values indicating the activation level of each tested gene in a genome. These data series can then be used to examine the reactions of the cell to external stimuli. A model would enable biologists to predict the reactions of intracellular signalling processes. To re-engineer or infer the regulatory processes computationally from these experimental data sets, one has to find a model that is able to produce the same time series data as the experiments. The idea is then that the model reflects the true system dependencies, i.e. the dependencies of the components of the regulatory system.

Several approaches have been made to address this problem. Many of them are only relying on the distance between the experimental data and the simulated data coming from the mathematical model, but the biological plausibility of the system is almost always neglected. And although Biologist know, that regulatory systems are sparse, i.e. one gene relies on average on a small number of other genes, this fact can be found only in some publications.

In this paper we propose a methodology for reverse engineering sets of time series data obtained by artificial expression analysis by combining two objectives into a multi-objective optimization problem. The first objective is the dissimilarity between the experimental and the simulated data (RSE). The second objective is the connectivity of the system. Both objectives are to be minimized to gain a system, which fits the data and at the same time is only sparsely connected and therefore biological plausible. With this approach, we systematically examine the impact of the connectivity of the regulatory network on the overall inference process.

The remainder of this paper is structured as follows. Section 2 of this paper presents an overview over related work and lists associated publications. Detailed description of our proposed method will be given in section 3 and example applications will be shown in section 4. Finally, conclusions and an outlook on future research will be covered by section 5.

2 Related Work

Researchers are interested in understanding the mechanisms of gene regulatory processes and therefore in inferring the underlying networks. The following section briefly describes the work that has been done in this area.

The earliest models to simulate regulatory systems found in the literature are boolean or random boolean networks (RBN) [8]. In boolean networks gene expression levels can be in one of two states: either 1 (on) or 0 (off). The quantitative level of expression is not considered. Two examples of algorithms for inferring GRNs with boolean networks are given by Akutsu [1] and the RE-

3.2 Multi-objective Algorithm

For examining the connectivity and the RSE in parallel, we used a multi-objective EA, which optimizes the parameters of $\mathcal{G}$, $\mathcal{H}$, α_i and β_i in respect to the following two optimization objectives:

I) For evaluating the RSE fitness of the individuals we used the following equation for calculation of the fitness values:

$$f_1 = \sum_{i=1}^{N}\sum_{k=1}^{T}\left\{\left(\frac{\hat{x}_i(t_k) - x_i(t_k)}{x_i(t_k)}\right)^2\right\} \quad (2)$$

where N is the total number of genes in the system, T is the number of sampling points taken from the experimental time series and $\hat{x}$ and x distinguish between estimated data of the simulated model and data sampled in the experiment. The problem is to minimize the fitness value f_1.

II) The second optimization objective is to minimize the connectivity of the system, as biologically the gene regulatory network is known to be sparse. The connectivity is defined in two different ways: first, the maximum connectivity of the genes, i.e. the total number of interactions of the system:

$$f_2^a = \sum_{i=1}^{N}(|\text{sign}(\alpha_i)| + |\text{sign}(\beta_i)|) + \sum_{i=1}^{N}\sum_{j=1}^{N}(|\text{sign}(\mathcal{G}_{i,j})| + |\text{sign}(\mathcal{H}_{i,j})|) \quad (3)$$

And secondly, the median average connectivity of all genes, i.e. the median average number of interactions of each gene:

$$f_2^b = \text{median}(\frac{f_2^a}{N}) \quad (4)$$

3.3 Test Cases

The optimization experiments were performed with different configurations. In the first test class (**Test class I: Hybrid**), an EA with hybrid encoding individuals was used to minimize the connectivity of the regulatory model, which was recently developed by the authors [19]. This individual combines a binary and a real valued genotype that are evolved in parallel. The binary variables are used to determine a topology or structure of the network and the double encoded optimization variables represent the corresponding model parameter. The individuals always encode all possible model parameters but only some of them are used for simulation according to the binary representation of the topology. Nevertheless, the unused variables are continuously evolved and subject to random walk and might be incorporated in the simulation if the bitset changes. This enables the optimizing algorithm to escape local optima. This algorithm has to optimize $2(N + N^2)$ real valued parameters for the

S-System plus $2N^2$ bits for each entry $\mathcal{G}_{i,j}$ and $\mathcal{H}_{i,j}$, thus in total $2N + 4N^2$ variables.

Test Class II (MA). is a memetic algorithm, where an individual of the MOEA represents the topology together with an evolution strategy, which searches for the best parameters for the given topology. The memetic algorithm uses a MOEA population to evolve populations of topologies of possible networks. These topologies are encoded as bitsets, where each bit represents the existence or absence of an interaction between genes and therefore of non-zero parameters in the mathematical model. The evaluation of the fitness of each individual within the MOEA population applies a local search to find suitable parameters. For evaluation of each structure suggested by the MOEA population an evolution strategy is used, which is suited for the parameter optimizing problem, since it is based on real values. The ES optimizes the parameters of the mathematical model used for representation of the regulatory network. This algorithm was introduced by the authors in [17]. The current implementation is working on both, $\mathcal{G}$ and $\mathcal{H}$, thus having the same number of bits as the algorithm in the previous case ($2N^2$). For future implementations it would be interesting to encode only the logical dependency between genes and set the corresponding bits in $\mathcal{G}$ and $\mathcal{H}$ at the same time. The difference between the hybrid encoding of I and the II is that the latter is optimizing only the S-System parameters for those values of the bitset that are true. Therefore, this algorithm has a dynamic range of the total number of model parameters between 0 (no connectivity at all) and $2(N + N^2)$ for the complete S-System.

Furthermore, a third test class (**Test class III: Skeletalizing**) was implemented, where a technique called *skeletalizing* was used [21]. This is an extension to a standard real-coded GA that introduces a threshold value t_{skel}, which represents a lower boundary for the parameters $\mathcal{G}_{i,j}$ and $\mathcal{H}_{i,j}$ in the mathematical model. If the absolute value of a decoded decision variable of the GA drops below this threshold during optimization the corresponding phenotype value is forced to 0.0 indicating no relationship between the components. Thus, $|\mathcal{G}_{i,j}| < t_{skel} \rightarrow \mathcal{G}_{i,j} = 0.0$. This algorithm has the same total number of parameters to optimize as the MA described above.

Overall, a total number of three different algorithms was tested on two problem instances (f_1, f_2^a) and (f_1, f_2^b) with artificial experimental data to examine the impact of the connectivity on the overall optimization process.

3.4 Algorithm Settings

The settings and chosen parameter for each of the three algorithm test classes will be given in the following sections.

Test Class I: Hybrid. The multi-objective runs were performed using a real-valued NSGA-II algorithm with a population size of 500 individuals, crowded tournament selection with a tournament group size of $t_{group} = 8$, Uniform-crossover recombination with $p_c = 1.0$ and a mutation probability $p_m = 0.1$. (Details on the implementation is given in Deb *et al.* [4]).

Test Class II: MA. The memetic algorithm used also an NSGA-II to evolve a population of possible topologies with crowded tournament selection with a tournament group size of $t_{group} = 8$, Uniform-crossover with $p_c = 1.0$ and a mutation probability $p_m = 0.1$. The local optimization was performed using a (μ,λ)-ES with $\mu = 10$ parents and $\lambda = 20$ offsprings together with a Covariance Matrix Adaptation (CMA) (see Hansen and Ostermeier [5]) mutation operator without recombination.

Test Class III: Skeletalizing. This test class was a standard real-coded NSGA-II using a population of 500 individuals, crowded tournament selection with a tournament group size of $t_{group} = 8$, Uniform-crossover with $p_c = 1.0$ and a mutation probability of $p_m = 0.1$ together with the threshold value $t_{skel} = 0.05$.

To keep track of the Pareto-front the multi-objective algorithms maintained an archive of (population size/2) individuals and used this archive as elite to achieve a faster convergence. Each test class experiment terminated after 100,000 fitness evaluations. Each example setting was repeated 20 times.

We further compared the MOEAs to a standard (μ,λ)-ES with $\mu = 10$ parents and $\lambda = 20$ offsprings together with CMA and no recombination. We tested this ES on the artificial systems with 20 multiruns and evaluated the overall best individual found in the runs.

4 Applications

To illustrate our method, we established two regulatory network systems, which were simulated to gain sets of expression data. After creating the data sets, we used our proposed algorithm to reverse engineer the correct model parameters. The following section show this for a 5-dimensional and a 10-dimensional example.

4.1 Artificial Gene Regulatory Network

Due to the fact that GRNs in nature are sparse systems, we created regulatory networks randomly with a maximum cardinality of $k <= 3$, i.e. each of the N genes depends on three or less other genes within the network. The total connectivity of the 5-dimensional example was 36, i.e. the number of non-zero parameters was 36. As a second and, due to the increased number of participating genes, more difficult test case, we created another regulatory network randomly with a maximum cardinality of $k <= 3$. The total connectivity of the 10-dimensional example was 78.

Target. The time dynamics of the systems can be seen in Fig. 1 and 2, respectively. In the figures, each x_i represents the RNA level of a certain gene. At this point, we do not differentiate between closely related molecules like mRNA and distantly related like proteins.

Obviously, the information about the true connectivity was not used in the optimization process it served only for validation purposes.

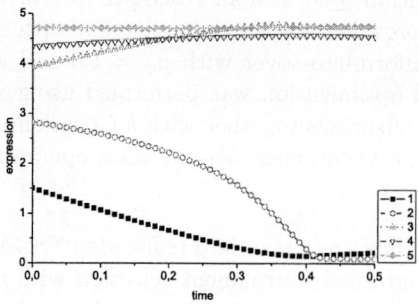

Fig. 1. 5-dimensional GRN. The numbers indicate the component genes

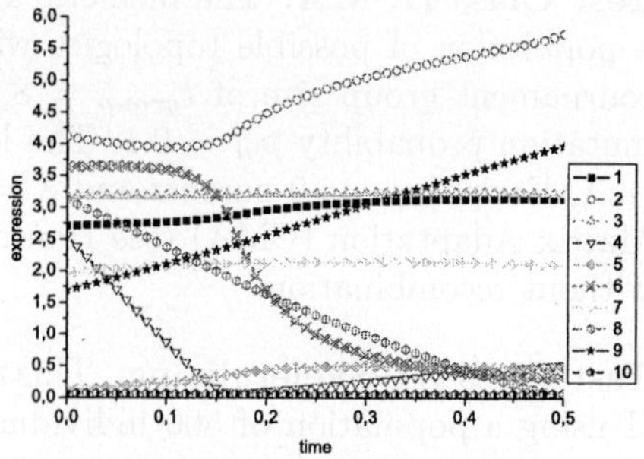

Fig. 2. 10-dimensional GRN. The numbers indicate the component genes

4.2 Initial Results

In the initial implementation, the MOEA found the only one Pareto-optimal solution with a connectivity of 0. This is caused by the trivial solution with zero interactions. With a connectivity of 0, the system is static and does not change over time. Because the expression levels of the system are in the range of 0.0 and 5.0, the data distance of the static case is very small compared to many of the systems with a larger connectivity and often unstable dynamics. To bypass this issue, we multiplied the data sets by a factor $f_{scale} = 1,000,000$ and introduced therefore a penalty for being static. During evaluation and postprocessing of the results, we reversed this scaling to gain fitness values for the original data range.

4.3 5-Dimensional Example

The actual results of the three test classes are shown for the 5-dimensional example in the next figures. Fig. 3 shows the Pareto-front results of the three test classes accumulated over all 20 multiruns with the RSE as ordinate and the total number of interactions in the inferred model as abscissa.

As one can see, with increasing number of interactions, the RSE is getting better due to the ambiguity in the data sets. Because the optimizer has more parameters and thus more model variables to use for fitting the data, it becomes more easy to actually reach the desired time series course. But as can be seen in the figure to the right (Fig. 4), the euclidian distance between the true system and the model found by the EA is getting worse, due to the sparseness of the true system. This validates the results of previous work [17].

The next two figures show the Pareto-front and the parameter distance of the 5-dimensional example in the case that the median average number of interactions of each gene was taken into account as the second optimization objective.

There is almost no difference between the two definitions of connectivity as can be see in the figure. All algorithms yield similar results as in the test case of the total connectivity. And as in the previous results shown, there is no significant difference between the two multi-objective implementations (MA and

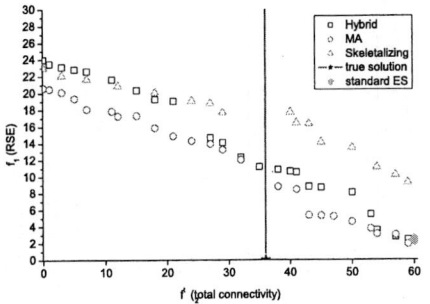

Fig. 3. Pareto-front of the 5-dim example (total connectivity)

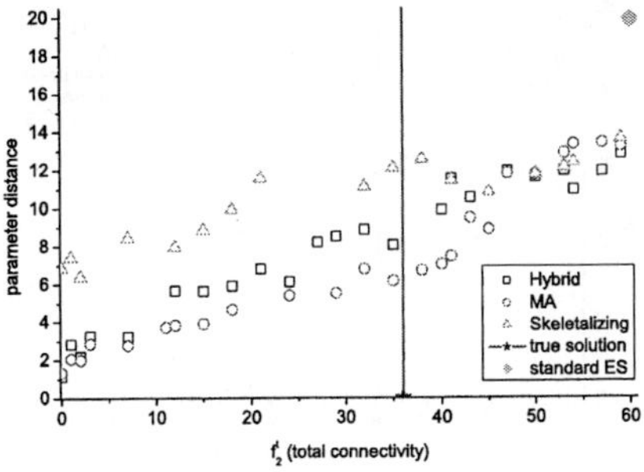

Fig. 4. Distances of the 5-dim example (total connectivity)

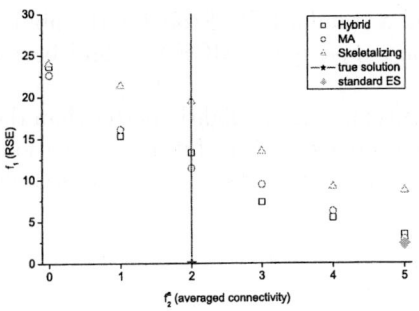
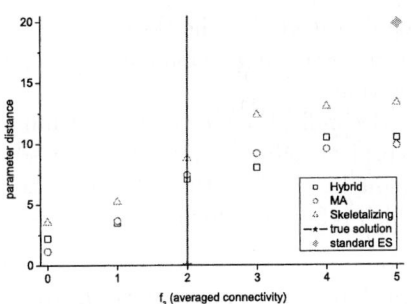

Fig. 5. Pareto-front of the 5-dim example (average connectivity)

Fig. 6. Distances of the 5-dim example (average connectivity)

hybrid). Both, the MA and the hybrid algorithm, outperform the skeletalizing GA.

The star in the figures show the true solution with a total connectivity of 36. None of the tested algorithm is able to reach the true solution. This might be partly due to the limited number of total fitness evaluations. But the major problem for all EAs is that the solution space is highly multi-modal and in case of the S-System very large.

For comparison, we tested additionally a standard ES on the artificial systems. These results are indicated in the figures with the rhombus symbol. As one can see, the standard ES prefers fully connected networks. The averaged best RSE fitness values in case of the 5-dimensional example are as good as those of the multi-objective EAs. But the second objective, the distance between the optimized parameters and the true system, is much better for the MOEAs. The ES found solutions, which fit the data comparably good, but show less resemblance with the original system.

Overall, two conclusions can be drawn from the figures: first, the MA and the hybrid algorithm perform significantly better than the algorithm taken from

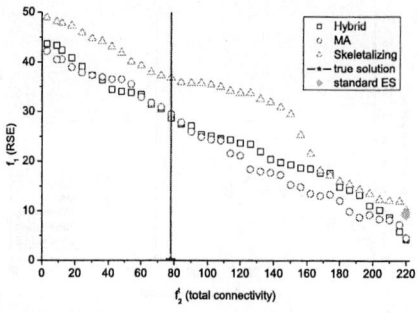 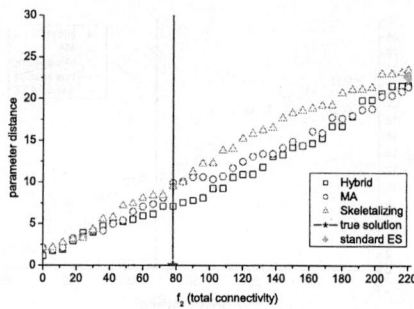

Fig. 7. Pareto-front of the 10-dim example (total connectivity)

Fig. 8. Distances of the 10-dim example (total connectivity)

the literature, i.e. the skeletalizing GA and a standard ES. Probably the ability to preserve the diversity in the population supports the MOEA to find better solutions.

Secondly, the memetic algorithm performs on average slightly better than the hybrid encoding individual. One advantage for the MA in future applications might be the fact that EAs tend to yield better solutions with a local search, which refines the found parameters.

4.4 10-Dimensional Example

The conclusions from the previous section are validated by the results of the 10-dimensional example as given in Fig. 7 and 8.

Again, the fitness values for fitting the data (RSE) are getting better for increasing numbers of interactions in the regulatory system. The distance to the true model parameters is getting worse. And as in the 5-dimensional example, the solutions were not even close to the true solution, indicated with the star. For the true connectivity, no algorithm was able to find the correct parameters. The reason for this can be understood, if one looks closer at the topologies for this connectivity. The algorithms examined the correct degree of dependencies but with wrong interactions partners. In this case, it is obviously not possible to find the correct values for the model parameters. The conclusion from this fact is that more topologies have to be examined, which is easy to perform with the MA or the hybrid algorithm and a higher number of total fitness evaluations.

The standard ES performed better than in the previous example. The RSE fitness values were comparably good as those of the totaly connected MOEA results. And also the parameter distances of the ES results corresponded with those of the MOEA solutions. But again, the ES was not able to find any solutions that were not fully connected. Considering that the ES had as many fitness evaluations as the MOEA, the MOEA performed much better. This supports the claim of the usefulness of the increased diversity in a MOEA population.

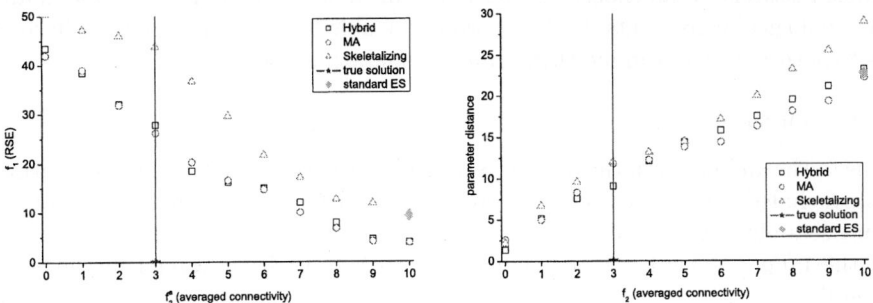

Fig. 9. Pareto-front of the 10-dim example (average connectivity)

Fig. 10. Distances of the 10-dim example (average connectivity)

Again, the choice of the connectivity definition did not influence the overall results as can be seen in figures 9 and 10. Both implementations resulted in similar solutions in respect to the parameter distance and the RSE fitness.

5 Discussion

5.1 Conclusions

In this paper we compared different multi-objective strategies to infer gene regulatory networks from time-series microarray data. The results of the test examples showed that the number of interactions has a crucial impact on the ability of an EA to fit a mathematical model to a given time series. High numbers of connections between system components yield in better results in respect to the data fitness. This is due to the large number of model parameters and the small number of data sets available, the system of equations is highly underdetermined. Therefore, multiple solutions exist, which fit the given data, but show only little resemblance with the original target system. This problem is known in literature but there are currently only few publications reflecting on this issue. The multi-objective strategy showed promising performance in comparison to standard single-objective algorithms. MOEA are better at preserving the diversity of the solutions in the population as standard algorithms do and are able to better cope with the ambiguity issues than single-objective optimization algorithms as we have shown in this paper. This is especially important for the highly multi-modal solutions space in the case of inference problems.

On the other side, with increasing number of interactions, it becomes more and more difficult to find the true system in respect to correct parameter values. Because biological systems are sparse, algorithm have to take this fact into account to find better solutions. In this example, we knew that the true solutions existed with the known connectivity. In real world applications, researches do not know, which connectivity is the correct one. MOEAs are again a promising approach to this problem as they result in a set of Pareto-optimal solutions from which researchers can choose the most suitable model complying with, for example, biological constraints. Biologist would be able to take one or more solutions as hypotheses and examine them in additional experiments.

5.2 Outlook

Due to the ambiguity in the data, it is difficult for EAs to find the correct solution as concluded above. Recently, the authors published a method to incorporate data sets obtained by additional experiments [16]. As one future enhancement of the proposed methods, we plan to incorporate such additional methods to identify the correct network.

As the initial results showed, precautions have to be taken to prevent the MOEA from finding the trivial solution of no connectivity. This can be done for example by scaling the experiment data as we did in this publication or one could introduce a penalty term to scale the objective function.

In future work we also plan to include a-priori information into the inference process of real microarray data like partially known pathways or information about co-regulated genes, which can be found in literature or in public databases. Here, the dissimilarity to the known pathway could be included as a third objective, which is also to be minimized. This would enable the MOEA to search for models consistent with current biological knowledge, but would also allow for alternative solutions where biological information is missing or faulty.

Furthermore, additional models for gene regulatory networks will be examined for simulation of the non-linear interaction system as listed in Sect. 2 to overcome the problems with a quadratic number of model parameters of the S-System.

Acknowledgment

This work was supported by the National Genome Research Network (NGFN) of the Federal Ministry of Education and Research in Germany under contract number 0313323.

References

1. T. Akutsu, S. Miyano, and S. Kuhura. Identification of genetic networks from a small number of gene expression patterns under the boolean network model. In *Proceedings of the Pacific Symposium on Biocomputing*, pages 17–28, 1999.
2. T. Akutsu, S. Miyano, and S. Kuhura. Algorithms for identifying boolean networks and related biological networks based on matrix multiplication and fingerprint function. In *Proceedings of the fourth annual international conference on Computational molecular biology*, pages 8 – 14, Tokyo, Japan, 2000. ACM Press New York, NY, USA.
3. S. Ando, E. Sakamoto, and H. Iba. Evolutionary modeling and inference of gene network. *Information Sciences*, 145(3-4):237–259, 2002.

4. K. Deb, S. Agrawal, A. Pratab, and T. Meyarivan. A fast elitist non-dominated sorting genetic algorithm for multi-objective optimization: Nsga-ii. In *Proceedings of the Parallel Problem Solving from Nature VI Conference*, number 1917 in LNCS, pages 849–858, 2000.
5. N. Hansen and A. Ostermeier. Adapting arbitrary normal mutation distributions in evolution strategies: the covariance matrix adaptation. In *Proceedings of the 1996 IEEE Int. Conf. on Evolutionary Computation*, pages 312–317, Piscataway, NJ, 1996. IEEE Service Center.
6. J. Herz. Statistical issues in reverse engineering of genetic networks. In *Proceedings of the Pacific Symposium on Biocomputing*, 1998.
7. S. Imoto, T. Higuchi, T. Goto, K. Tashiro, S. Kuhara, and S. Miyano. Combining microarrays and biological knowledge for estimating gene networks via bayesian networks. In *Proceedings of the IEEE Computer Society Bioinformatics Conference (CSB 03)*, pages 104 –113. IEEE, 2003.
8. S. A. Kauffman. *The Origins of Order*. Oxford University Press, New York, 1993.
9. E. Keedwell, A. Narayanan, and D. Savic. Modelling gene regulatory data using artificial neural networks. In *Proceedings of the International Joint Conference on Neural Networks (IJCNN 02)*, volume 1, pages 183–188, 2002.
10. S. Kikuchi, D. Tominaga, M. Arita, K. Takahashi, and M. Tomita. Dynamic modeling of genetic netowrks using genetic algorithm and s-sytem. *Bioinformatics*, 19(5):643–650, 2003.
11. S. Liang, S. Fuhrman, and R. Somogyi. REVEAL, a general reverse engineering algorithm for inference of genetic network architectures. In *Proceedings of the Pacific Symposium on Biocomputing*, volume 3, pages 18–29, 1998.
12. Y. Maki, D. Tominaga, M. Okamoto, S. Watanabe, and Y. Eguchi. Development of a system for the inference of large scale genetic networks. In *Proceedings of the Pacific Symposium on Biocomputing*, volume 6, pages 446–458, 2001.
13. I. Ono, R. Yoshiaki Seike, N. Ono, and M. Matsui. An evolutionary algorithm taking account of mutual interactions among substances for inference of genetic networks. In *Proceedings of the IEEE Congress on Evolutionary Computation (CEC 2004)*, pages 2060–2067, 2004.
14. C. Pridgeon and D. Corne. Genetic network reverse-engineering and network size; can we identify large grns? In *Proceedings of the Computational Intelligence in Bioinformatics and Computational Biology (CIBCB 2004)*, pages 32–36, 2004.
15. M. A. Savageau. 20 years of S-systems. In E. Voit, editor, *Canonical Nonlinear Modeling. S-systems Approach to Understand Complexity*, pages 1–44, New York, 1991. Van Nostrand Reinhold.
16. C. Spieth, F. Streichert, N. Speer, and A. Zell. Iteratively inferring gene regulatory networks with virtual knockout experiments. In R. et al., editor, *Proceedings of the 2nd European Workshop on Evolutionary Bioinformatics (EvoWorkshops 2004)*, volume 3005 of *LNCS*, pages 102–111, 2004.
17. C. Spieth, F. Streichert, N. Speer, and A. Zell. Optimizing topology and parameters of gene regulatory network models from time-series experiments. In *Proceedings of the Genetic and Evolutionary Computation Conference (GECCO 2004)*, volume 3102 (Part I) of *LNCS*, pages 461–470, 2004.
18. C. Spieth, F. Streichert, N. Speer, and A. Zell. Utilizing an island model for ea to preserve solution diversity for inferring gene regulatory networks. In *Proceedings of the IEEE Congress on Evolutionary Computation (CEC 2004)*, pages 146–151, 2004.

19. F. Streichert, H. Ulmer, and A. Zell. Evaluating a hybrid encoding and three crossover operators on the constrained portfolio selection problem. In *Proceedings of the 2004 Congress on Evolutionary Computation*, pages 932–939, 2004.
20. D. Thieffry and R. Thomas. Qualitative analysis of gene networks. In *Proceedings of the Pacific Symposium on Biocomputing*, pages 77–87, 1998.
21. D. Tominaga, N. Kog, and M. Okamoto. Efficient numeral optimization technique based on genetic algorithm for inverse problem. In *Proceedings of the Genetic and Evolutionary Computation Conference (GECCO 00)*, pages 251–258, 2000.
22. D. Weaver, C. Workman, and G. Stormo. Modeling regulatory networks with weight matrices. In *Proceedings of the Pacific Symposium on Biocomputing*, volume 4, pages 112–123, 1999.
23. M. K. S. Yeung, J. Tegner, and J. J. Collins. Reverse engineering gene networks using singular value decomposition and robust regression. In *Proceedings of the National Academy of Science USA*, volume 99, pages 6163–6168, 2002.

High-Fidelity Multidisciplinary Design Optimization of Wing Shape for Regional Jet Aircraft

Kazuhisa Chiba[1], Shigeru Obayashi, Kazuhiro Nakahashi, and Hiroyuki Morino[2]

[1] Tohoku University, Katahira2-1-1, Aoba-ku, Sendai 980-8577, Japan
chiba@edge.ifs.tohoku.ac.jp
http://www.ifs.tohoku.ac.jp/edge/indexe.html
[2] Mitsubishi Heavy Industries, Ltd., Oye-cho 10, Minato-ku, Nagoya 455-8515, Japan

Abstract. A large-scale, real-world application of Evolutionary Multi-Criterion Optimization (EMO) is reported in this paper. The Multidisciplinary Design Optimization among aerodynamics, structures and aeroelasticity for the wing of a transonic regional jet aircraft has been performed using high-fidelity models. An Euler/Navier-Stokes (N-S) Computational Fluid Dynamics (CFD) solver is employed for the aerodynamic evaluation. The NASTRAN, a commercial software, is coupled with a CFD solver for the structural and aeroelastic evaluations. Adaptive Range Multi-Objective Genetic Algorithm is employed as an optimizer. The objective functions are minimizations of block fuel and maximum takeoff weight in addition to difference in the drag between transonic and subsonic flight conditions. As a result, nine non-dominated solutions have been generated. They are used for tradeoff analysis among three objectives. One solution is found to have one percent improvement in the block fuel compared to the original geometry designed in the conventional manner. All the solutions evaluated during the evolution are analyzed by Self-Organizing Map to extract key features of the design space.

1 Introduction

Recent researches on Multidisciplinary Design Optimization (MDO) have been conducted for aircraft design[1, 2]. Pure aerodynamic optimization shows wings with a low thickness-to-chord ratio and a high aspect ratio. These wings suffer undesirable aeroelastic phenomena from the low bending and torsional stiffness. Aerostructural interacted optimization is needed to overcome these phenomena and to perform a realistic aircraft design[3]. This multi-criterion optimization will provide a good application field for EMO.

The project to develop a more environmentally suitable, highly efficient transonic regional jet aircraft has been founded by Ministry of Economy, Trade and

Industry (METI) since 2003. Mitsubishi Heavy Industries, Ltd. (MHI) is the prime contractor for the project. The aim of this project is to build a demonstrator with advanced technologies, such as low drag wing design, light weight composite structures which are necessary for reduction of environmental burden. The initial aircraft geometry has been obtained from a conventional design method.

The objective of this study is to optimize the three-dimensional wing shape for the proposed regional jet aircraft using evolutionary multi-objective optimization with high-fidelity simulations as a collaboration between Institute of Fluid Science (IFS), Tohoku University and MHI. From the optimization results, tradeoff analysis has been performed among the three objectives. Moreover, by using a data mining technique, the aerostructural design knowledge for transonic regional jet aircraft has been obtained.

In the present study, high-fidelity simulation tools such as Reynolds-averaged Navier-Stokes solver for aerodynamics, NASTRAN, a versatile and high-fidelity commercial software, for structures and aeroelasticity are coupled together for MDO. Although the Euler/N-S solver may be still too expensive for the real-world design environment, it will predict complex and nonlinear flow phenomena such as shock wave and separation more accurately. Such nonlinearity will provide a severe test case for EMO. With the aid of rapid progress in computer hardware, the demonstration in this paper will become a standard design practice soon.

2 Multidisciplinary Design Optimization

2.1 Objective Functions

In this study, because the target range for the regional jet is given, the minimization of the block fuel derived from aerodynamics and structures is selected as an objective function instead of the range maximization commonly used for aircraft design. The block fuel is defined as the minimum fuel mass for the fixed range. In addition, two more objective functions are considered as the minimization of the maximum takeoff weight and the minimization of the difference in the drag coefficient between transonic and subsonic flight conditions.

2.2 Geometry Definition

The design variables define the aerodynamic geometry. Structural optimization and aeroelastic transformation are performed using NASTRAN under the given aerodynamic geometry after aerodynamics, structures and flutter are evaluated, the objective functions are calculated.

The design variables are related to airfoil, twist and wing dihedral. The airfoil is defined at three spanwise cross sections using the modified PARSEC[4] with nine design variables ($x_{up}, z_{up}, z_{xx_{up}}, x_{lo}, z_{lo}, z_{xx_{lo}}, \alpha_{TE}, \beta_{TE}$ and $r_{LE_{lo}}/r_{LE_{up}}$) per cross section shown in Fig. 1. The twists are defined at six spanwise locations, and then wing dihedrals are defined at kink and tip locations. An entire wing

3 Optimization Results

The population size is set to eight, and then roughly 70 Euler and 90 RANS computations are performed in one generation. It takes roughly one hour and nine hours of CPU time of NEC SX-5 and SX-7 one PE for Euler and RANS computations, respectively. The population is re-initialized at every five generations for the range adaptation. The total evolutionary computation of 16 generations is carried out so far. The evolution may not converge yet. However, the result is satisfactory, because several non-dominated solutions achieve significant improvements over the initial design. Furthermore, enough number of solutions have been searched so that the sensitivity of the design space around the initial design can be analyzed. This will provide useful information for designers.

All solutions evaluated are shown in Fig. 2, and Fig. 3 shows all solutions projected on two dimensional plane between two objectives, the block fuel and the drag divergence. As this figure shows that the non-dominated front is generated, there is tradeoff between the block fuel and the drag divergence.

Although the wing box weight, on the whole, tends to increase compared with that of the initial geometry, the block fuel can be also reduced. It means that aerodynamic performance can redeem the penalty due to the structural weight. An individual, indicated as 'optimized', on the non-dominated front shown in Fig. 3 is picked up, and then optimized geometry is compared with the initial geometry. Figure 4 shows the comparison of polar curves. Although the corresponding the drag minimization is not considered here, this figure shows that the drag coefficients tend to reduce on the whole. Comparing the polar curves at the constant lift coefficient for the cruising condition, the drag coefficient of the optimized geometry has been found to be reduced 5.5 counts. Due to the drag improvement, the block fuel of the optimized geometry can be decreased over one percent even with its structural weight penalty.

Next, the mechanism of the drag reduction is investigated. Figure 5 shows the comparison of the spanwise distributions of lift and drag coefficients of initial and optimized geometries. This figure shows that the drag decreases at the 35.0 % spanwise location. Figure 6 shows the comparison of the pressure distributions at the 35.0 % spanwise location. This figure shows that the variation of the leading-edge bluntness works to depress the shock wave on the upper wing surface, namely, to reduce the wave drag. In fact, the pressure drag coefficient is reduced by 5.6 counts. Figure 7 shows the comparison of shock wave visualized by the shock function F_{shock}[22] which is given as follows,

$$F_{\text{shock}} = \frac{\boldsymbol{V} \cdot \nabla P}{a \cdot |\nabla P|} \tag{1}$$

where $\boldsymbol{V}$ is velocity vector, P is pressure and a denotes the local speed of sound. The shock wave of the optimized geometry is weaker than the initial geometry in the vicinity of the 35.0 % spanwise location shown in Fig. 7. This fact signify the wave drag reduction. Moreover, the vorticity of wing wake of the

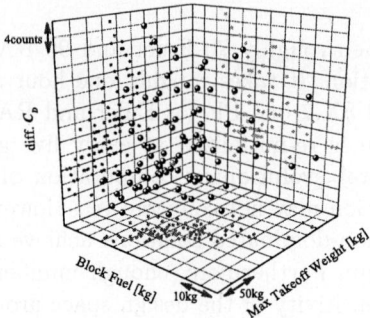 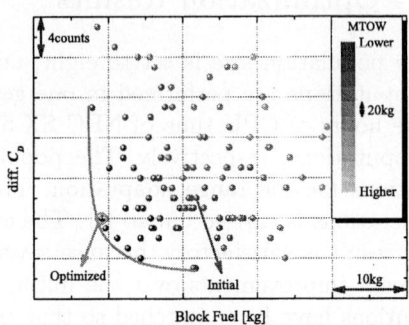

Fig. 2. All solutions plotted in three dimensional space of all objective function

Fig. 3. All solutions on two dimensional plane between block fuel and drag divergence

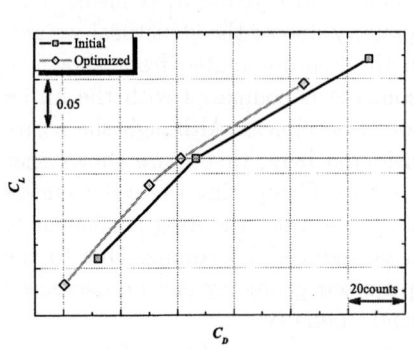 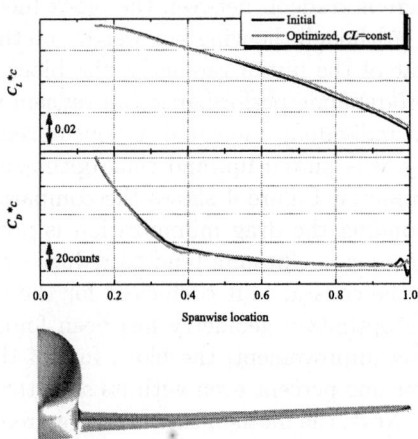

Fig. 4. Comparison of the polar curves between initial and optimized geometries

Fig. 5. Comparison of the lift and drag coefficients spanwise distributions between initial and optimized geometries

optimized geometry in the vicinity of the 35.0 % spanwise location is weaker than of initial geometry shown by helicity contours in Fig. 8. Therefore, these figures show that the shape near the 35.0 % spanwise location, namely, the shape in the vicinity of the kink location has been found effective to reduce the drag.

4 Data Mining

If the optimization problem has only two objectives, tradeoffs can be visualized easily. However, if there are more than two objectives, the technique to

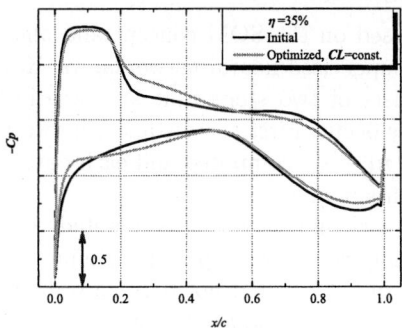

Fig. 6. Comparison of the pressure distributions and airfoil shapes between initial and optimized geometries at 35 % spanwise location

Fig. 7. Comparison of shock wave visualizations colored by entropy at the transonic cruise flight condition between initial (left) and optimized (right) geometries

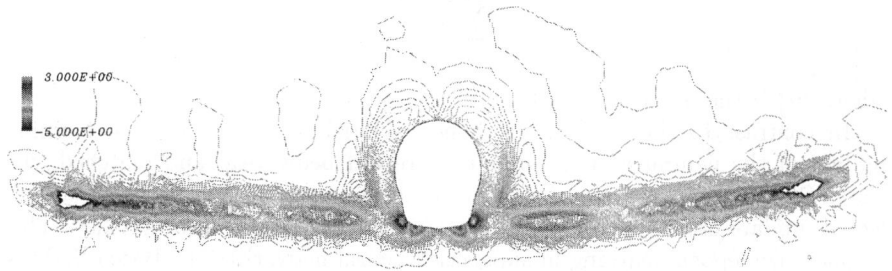

Fig. 8. Comparison of helicity contours of wing wake at the transonic cruise flight condition between initial (left) and optimized (right) geometries

visualize the computed non-dominated solutions is desired[23]. Therefore, in the present study, Self-Organizing Maps (SOMs) suggested by Kohonen[24] have been employed. SOM is not only the technique for the visualization but also the application tool for the intelligent compression of the information. In other words, SOM can be applied for the data mining technique to acquire the knowledge about design space. In this study, Viscovery® SOMine[25] produced by Eudaptics GmbH in Austria is employed.

4.1 Self-Organizing Map

Although SOMine is based on the SOM concept and algorithm, it employs an advanced variant of unsupervised neural networks, *i.e.* Kohonen's Batch-SOM.

The algorithm consists of two steps that are iteratively repeated until no more significant changes occur. First the distances between all data items $\{x_i\}$ and the model vectors $\{m_j\}$ are computed and each data item x_i is assigned to the unit c_i that represents it best.

In the second step, each model vector is adapted to better fit the data it represents. To ensure that each unit j represents similar data items as its neighbors, the model vector m_j is adapted not only according to the assigned data items but also in regard to those assigned to the units in the neighborhood. The neighborhood relationship between two units j and k is usually defined by a Gaussian-like function

$$h_{jk} = \exp\left(-\frac{d_{jk}^2}{r_t^2}\right) \tag{2}$$

where d_{jk} denotes the distance between the units j and k on the map, and r_t denotes the neighborhood radius which is set to decrease with each iteration t.

Assuming a Euclidean vector space, the two steps of the Batch-SOM algorithm can be formulated as

$$c_i = \arg\min \|x_i - m_j\| \tag{3a}$$

$$m_j^* = \frac{\sum_i h_{jc_i} x_i}{\sum_i h_{jc_i}} \tag{3b}$$

where m_j^* is the updated model vector.

In contrast to the standard Kohonen algorithm, which makes a learning update of the neuron weights after each record being read and matched, the Batch-SOM takes a 'batch' of data, typically all records, and performs a 'collected' update of the neuron weights after all records have been matched. This is much like 'epoch' learning in supervised neural networks. The Batch-SOM is a more robust approach, since it mediates over a large number of learning steps. Most important, no learning rate is required. The SOMine implementation combines four enhancements to the plain Batch-SOM algorithm[26]. In SOMine, the uniqueness of the map is ensured by the adoption of the Batch-SOM and the linear initialization for input data.

Much like some other SOMs[27], SOMine creates a map in a two-dimensional hexagonal grid. Starting from numerical, multivariate data, the nodes on the grid gradually adapt to the intrinsic shape of the data distribution. Since the order on the grid reflects the neighborhood within the data, features of the data distribution can be read off from the emerging map on the grid.

In SOMine, the trained SOM is systematically converted into visual information. The tool provides an extensive built-in capability for both pre-processing and post-processing as well as for the automatic colorcoding of the map and its components. SOMine is particularly useful in the determination of dependencies between variables as well as in the analysis of high-dimensional cluster distributions.

4.2 Cluster Analysis

Once SOM projects input space on a low-dimensional regular grid, the map can be utilized to visualize and explore properties of the data. When the number of SOM units is large, tofacilitate quantitative analysis of the map and the data, similar units need to be grouped, i.e., clustered. The two-stage procedure — first using SOM to produce the prototypes which are then clustered in the second stage — was reported to perform well when compared to direct clustering of the data[27].

Hierarchical agglomerative algorithm is used for clustering here. The algorithm stats with a clustering where each node by itself forms a cluster. In each step of the algorithm two clusters are merged: those with minimal distance according to a special distance measure, the SOM-Ward distance[25]. This measure takes into account whether two clusters are adjacent in the map. This means that the process of merging clusters is restricted to topologically neighbored clusters. The number of clusters will be different according to the hierarchical sequence of clustering. A relatively small number will be chosen for visualization, while a large number will be used for generation of codebook vectors for respective design variables.

4.3 Tradeoff Analysis and Data Mining of the Design Space

All of the solutions have been projected onto the two-dimensional map of SOM. Figure 9 shows the resulting SOM with 10 clusters considering the three objectives. Furthermore, Fig. 10 shows the SOMs colored by the three objectives and three characteristic parameters, respectively. These color figures show the SOM shown in Fig. 9 can be grouped as follows: Upper left corner corresponds to the designs with high block fuel and maximum takeoff weight. Lower left corner corresponds to the designs with low block fuel and high drag divergence. Figure 10(a) and Fig. 10(c) show that there is a tradeoff between these two objective functions. Lower center area corresponds to the designs with low block fuel. Right hand side corresponds to the designs with low drag divergence. As Fig. 10(a) is similar coloring to Fig. 10(b), there is not a severe tradeoff between the block fuel and the maximum takeoff weight. Lower right corner corresponds to the designs with low of all objectives. Extreme non-dominated solutions are indicated in Fig. 10(a) to (c). Because they are in the different clusters, the simultaneous optimization of the three objectives is impossible. However, the lower right cluster has relatively low values for all three objectives. This region of the design space may provide a sweet-spot for the present design problem.

Figure 10(d) shows the SOM colored by the drag coefficient at the cruising flight condition, the lower value exists on the lower right corner. As this area clusters the designs with the low of all objectives, this fact inspires that all objectives can be optimized simultaneously while the drag at cruising flight

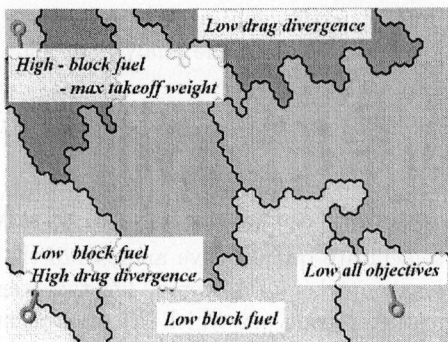

Fig. 9. SOM of all solutions in the three deimensional objective function space

condition is reduced. Furthermore, because the value of the maximum takeoff weight shown in Fig. 10(b) existing at a the similar location as in Fig. 10(d), the drag coefficient is found effective to reduce the maximum takeoff weight.

Figure 10(e) shows the SOM colored by the lift-to-drag ratio at the cruising flight condition, the lower value exists on the upper left corner. Because the higher value of the block fuel shown in Fig. 10(a) exists at a similar location, the lower lift-to-drag ratio is effective to increase the block fuel strictly. Furthermore, the higher value of lift-to-drag ratio exists on the lower area shown in Fig. 10(e). Because the lower value of the block fuel shown in Fig. 10(a) exist at a similar location, the higher lift-to-drag ratio is effective to decrease the block fuel. But, the higher value of the transonic lift-to-drag ratio is not necessarily effective to reduce the block fuel in Fig. 10(e) because the range of not only cruise but also of takeoff-to-landing is considered in this study.

Figure 10(f) shows the SOM colored by the angle between inboard and outboard on upper wing surface expressing the gull-wing at kink location. When this angle is greater/less than 180 deg, it means gull/inverted gull-wing. The characteristic inverted gull-wing shape is shown in Fig. 11. The location where the higher value exists shown in Fig. 10(f) corresponds to the position where the higher value of the drag coefficient at cruising flight condition shown in Fig. 10(d). However, when the angle is less than 180 deg, there is little correlation between Figs. 10(d) and (f). As it is known that the inverted gull-wing obtains structural weight increase, no-gull wing should be designed for structure and manufacture.

Finally, Fig. 12 shows the SOM colored by the characteristic design variables. Figure 12(a) shows the SOM colored by the design variable of PARSEC x_{up} at the 55.5 % spanwise location. When this value is increased in Fig. 12(a), the transonic drag is increased simultaneously shown in Fig.10(d). This means that the value of PARSEC x_{up} at the 55.5 % spanwise location has influence upon the transonic drag increase. Figures 12(b) and (c) show the SOMs colored by the design variables of PARSEC $r_{LE_{lo}}/r_{LE_{up}}$ at the 35.0 % spanwise location and PARSEC $z_{xx_{lo}}$ at the 55.5 % spanwise location, respectively. The decrease value

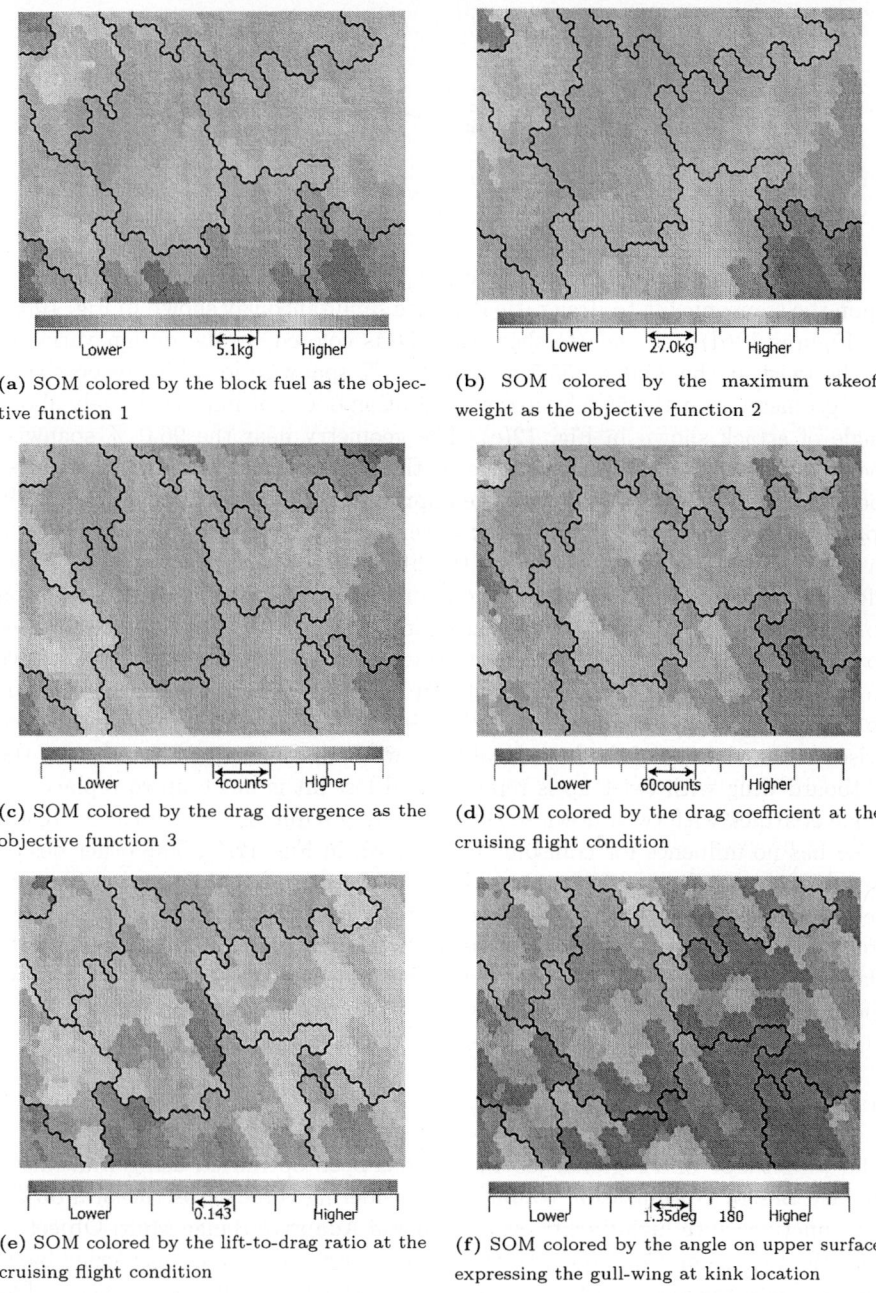

(a) SOM colored by the block fuel as the objective function 1

(b) SOM colored by the maximum takeoff weight as the objective function 2

(c) SOM colored by the drag divergence as the objective function 3

(d) SOM colored by the drag coefficient at the cruising flight condition

(e) SOM colored by the lift-to-drag ratio at the cruising flight condition

(f) SOM colored by the angle on upper surface expressing the gull-wing at kink location

Fig. 10. SOM colored by the objective functions and the characteristic values. The symbol × denotes the respective extreme non-dominated solutions in (a), (b) and (c)

Fig. 11. Visualization of a characteristic inverted gull-wing

of $r_{LE_{lo}}/r_{LE_{up}}$ and the increase value of $z_{xx_{lo}}$ in Figs. 12(b), (c) have influence upon the transonic lift-to-drag ratio decrease simultaneously shown in Fig. 10(e).

Figures 12(d), (e) and (f) show the SOMs colored by the design variables of the twist at the 35.0 %, 55.5 % and 96.0 % spanwise location, respectively. The geometry near the 55.5 % spanwise location does not improve largely about angle of attack shown in Fig. 12(e). The geometry near the 96.0 % spanwise location improves the twist up. In reverse, the geometry near the 35.0 % spanwise location improves the twist down. The improvement in the vicinity of the 35.0 % spanwise location is to restrain shock wave, namely, to reduce wave drag shown in Fig. 7. When the drag decreases, the lift may decrease simultaneously. The lift is increased to compensate for the reduction in the vicinity of kink location so that the angle of attack of the outboard wing is increased and roughly set to zero. It is noted that the angle of attack near the kink location is effective to the transonic drag especially as shown in Fig. 12(d). This fact corresponds to the phenomena visualized in Fig. 7. Specifically, as the shock wave which arises in the vicinity of the kink location is weaken, the angle of attack near the outboard wing with twist up is replaced and lost lift is made up to replace the angle of attack with twist down so that wave drag reduces. Twist up at outboard wing has no influence for transonic drag shown in Fig. 12(f). The other design variables are not found effective to reduce the objective functions and to increase aerodynamic performance as drag and lift-to-drag ratio at transonic cruise flight condition. Data mining technique using SOM has been found to be able to classify the design variables considering the influence upon the objectives and the aerodynamic performance.

5 Conclusion

The wing shape of a regional jet aircraft has been optimized using Multidisciplinary Design Optimization techniques considering three aerostructural objective functions with high-fidelity evaluation and Adaptive Range Multi-Objective Genetic Algorithm. Consequently, the objective function value considering block fuel has been reduced over one percent compared with the initial geometry. The geometry change at the kink location has been found effective for the drag reduction. The tradeoff information among three objective functions has been revealed, and a tradeoff has been found between the block fuel and the drag divergence. Moreover, data mining for the design space has been performed using Self-Organizing Map. Therefore, the particular design variables have been

High-Fidelity Multidisciplinary Design Optimization of Wing Shape

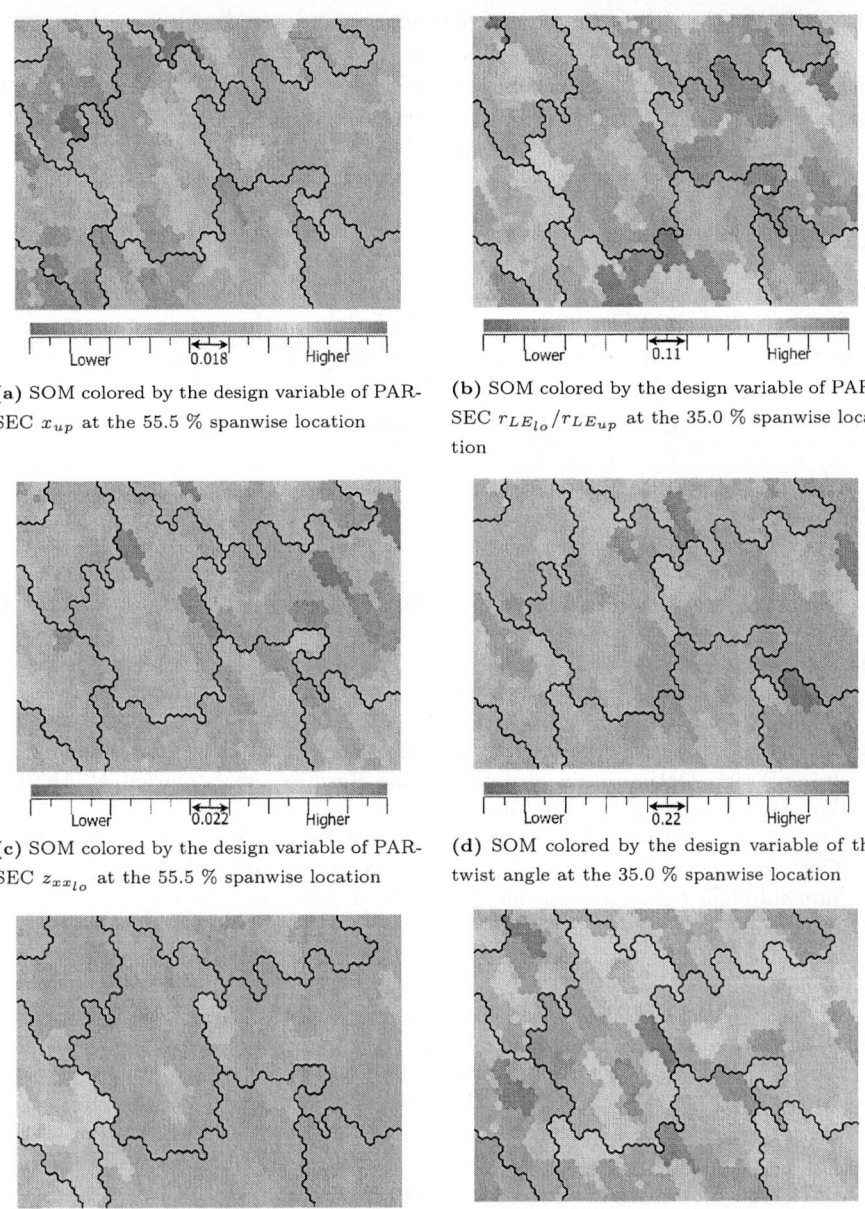

(a) SOM colored by the design variable of PARSEC x_{up} at the 55.5 % spanwise location

(b) SOM colored by the design variable of PARSEC $r_{LE_{lo}}/r_{LE_{up}}$ at the 35.0 % spanwise location

(c) SOM colored by the design variable of PARSEC $z_{xx_{lo}}$ at the 55.5 % spanwise location

(d) SOM colored by the design variable of the twist angle at the 35.0 % spanwise location

(e) SOM colored by the design variable of the twist angle at the 55.5 % spanwise location

(f) SOM colored by the design variable of the twist angle at the 96.0 % spanwise location

Fig. 12. SOM colored by the characteristic design variables

found effective to reduce the objective functions and aerodynamic performance. Detailed observation of SOM reveals there is a sweet-spot in the design space where the three objectives become relatively low. The data mining technique provides knowledge about the design space, which is considered an important facet of solving optimization problems.

Acknowledgements

We would like to thank Mr. Takano, Y., graduate student of Tohoku University for generating the CATIA STL and the mesh data of the initial geometry. The Navier-Stokes computations were performed using NEC SX-5 in the Institute of Fluid Science, Tohoku University, and NEC SX-7 in Super-Computing System Information Synergy Center, Tohoku University.

References

1. Kroo, I., Altus, S., Braun, R., Gage, P., and Sobieski, I., "Multidisciplinary Optimization Methods for Aircraft Preliminary Design," AIAA Paper 94-4325-CP, 1994.
2. Sobieszczanski-Sobieski, J. and Haftka, R. T., "Multidisciplinary Aerospace Design Optimization: Survey of Recent Developments," *Structural Optimization*, Vol. 14, No. 1, 1997, pp. 1–23.
3. Martins, J. R. R. A., Alonso, J. J., and Reuther, J. J., "High-Fidelity Aerostructural Design Optimization of a Supersonic Business Jet," *Journal of Aircraft*, Vol. 41, No. 3, 2004, pp. 523–530.
4. Oyama, A., Obayashi, S., Nakahashi, K., and Hirose, N., "Aerodynamic Wing Optimization via Evolutionary Algorithms Based on Structured Coding," *Computational Fluid Dynamics Journal*, Vol. 8, No. 4, 2000, pp. 570–577.
5. Murayama, M., Nakahashi, K., and Matsushima, K., "Unstructured Dynamic Mesh for Large Movement and Deformation," AIAA Paper 2002-0122, 2002.
6. Yamazaki, W., Matsushima, K., and Nakahashi, K., "Aerodynamic Optimization of NEXST-1 SST Model at Near-Sonic Regime," AIAA Paper 2004-0034, 2004.
7. Sasaki, D., Obayashi, S., and Nakahashi, K., "Navier-Stokes Optimization of Supersunic Wings with Four Objectives Using Evolutionary Algorithm," *Journal of Aircraft*, Vol. 39, No. 4, 2002, pp. 621–629.
8. Fonseca, C. M. and Fleming, P. J., "Genetic Algorithms for Multiobjective Optimization: Formulation, Discussion and Generalization," *Proceedings of the Fifth International Conference on Genetic Algorithms*, 1993, pp. 416–423.
9. Obayashi, S., Takahashi, S., and Takeguchi, Y., "Niching and Elitist Models for MOGAs, Parallel Problem Solving from Nature," *PPSN V, Lecture Notes in Computer Science*, Springer, Berlin, Heidelberg, New York, 1998, pp. 260–269.
10. Baker, J. E., "Reducing Bias and Inefficiency in the Selection Algorithm," *Proceedings of the Second International Conference on Genetic Algorithms*, 1987, pp. 14–21.
11. Eshelman, L. J. and Schaffer, J. D., "Real-Coded Genetic Algorithms and Interval Schemata," *Foundations of Genetic Algorithms 2*, Morgan Kaufmann, San Mateo, CA, 1993, pp. 187–202.

12. Deb, K., *Multi-Objective Optimization Using Evolutionary Algorithms*, John Wiley & Sons, Ltd., Chichester, 2001.
13. "MSC. website," URL: http://www.mscsoftware.com/ [cited 14 September 2004].
14. Ito, Y. and Nakahashi, K., "Direct Surface Triangulation Using Stereolithography Data," *AIAA Journal*, Vol. 40, No. 3, 2002, pp. 490–496.
15. Sharov, D. and Nakahashi, K., "A Boundary Recovery Algorithm for Delaunay Tetrahedral Meshing," *Proceedings of the 5th International Conference on Numerical Grid Generation in Computational Field Simulations*, 1996, pp. 229–238.
16. Ito, Y. and Nakahashi, K., "Improvements in the Reliability and Quality of Unstructured Hybrid Mesh Generation," *International Journal for Numerical Methods in Fluids*, Vol. 45, Issue 1, 2004, pp. 79–108.
17. Obayashi, S. and Guruswamy, G. P., "Convergence Acceleration of an Aeroelastic Navier-Stokes Solver," *AIAA Journal*, Vol. 33, No. 6, 1994, pp. 1134–1141.
18. Venkatakrishnan, V., "On the Accuracy of Limiters and Convergence to Steady State Solutions," AIAA Paper 93-0880, 1993.
19. Sharov, D. and Nakahashi, K., "Reordering of Hybrid Unstructured Grids for Lower-Upper Symmetric Gauss-Seidel Computations," *AIAA Journal*, Vol. 36, No. 3, 1998, pp. 484–486.
20. Dacles-Mariani, J., Zilliac, G. G., Chow, J. S., and Bradshaw, P., "Numerical/Experimental Study of a Wingtip Vortex in the Near Field," *AIAA Journal*, Vol. 33, No. 9, 1995, pp. 1561–1568.
21. Chiba, K., Obayashi, S., and Nakahashi, K., "CFD Visualization of Second Primary Vortex Structure on a 65-Degree Delta Wing," AIAA Paper 2004-1231, 2004.
22. Yamazaki, W., "Aerodynamic Optimization of Near-Sonic Plane Based on NEXST-1 SST Model," ICAS 2004-4.3.4, 2004.
23. Obayashi, S. and Sasaki, D., "Visualization and Data Mining of Pareto Solutions Using Self-Organizing Map," *EMO 2003, Lecture Notes in Computer Science*, Springer-Verlag Heidelberg, Faro, Portugal, 2003, pp. 796–809.
24. Kohonen, T., *Self-Organizing Maps*, Springer, Berlin, Heidelberg, 1995.
25. "Eudaptics website," URL: http://www.eudaptics.com [cited 16 June 2004].
26. Deboeck, G. and Kohonen, T., *Visual Explorations in Finance with Self-Organizing Maps*, London, Springer Finance, 1998.
27. Vesanto, J. and Alhoniemi, E., "Clustering of the Self-Organizing Map," *IEEE Transactions on Neural Networks*, Vol. 11, No. 3, 2000, pp. 586–600.

Photonic Device Design Using Multiobjective Evolutionary Algorithms

Steven Manos[1,2], Leon Poladian[1,3], Peter Bentley[4], and Maryanne Large[1]

[1] Optical Fibre Technology Centre, Australian Photonics CRC, 206 National Innovation Centre, Australian Technology Park, Eveleigh NSW 1430, Australia
[2] School of Physics, University of Sydney, NSW 2006, Australia
[3] School of Mathematics and Statistics, University of Sydney, NSW 2006, Australia
[4] Department of Computer Science, University College London, Gower Street, London WC1E 6BT, United Kingdom
s.manos@oftc.usyd.edu.au

Abstract. The optimization and design of two different types of photonic devices - a Fibre Bragg Grating and a Microstructured Polymer Optical Fibre is presented in light of multiple conflicting objectives in both problems. The fibre grating optimization uses a fixed length real valued representation, requiring the simultaneous optimization of four objectives along with variable bounds and a single objective constraint. This led to the human selection of a Pareto-optimal design which was manufactured. The microstructured fibre design process employs a new binary encoded variable length representation. An external embryogeny, or growth process is used to guarantee the creative generation of these complex designs which are automatically valid with respect to manufacturing constraints. Some initial results are presented for the case of two objectives which relate to the bandwidth and signal loss of a design.

1 Introduction

As more demands are made on telecommunications and other applications of optical fibres such as sensing, demand for the complexity and functionality of these devices increases. Fibre Bragg gratings (FBG) for optical filters and switching are an inherent part of such systems, offering highly tailorable optical filtering. Microstructured Polymer Optical Fibres (MPOF) are a more recent advent in the field of photonics, promising ease of manufacture along with functionality customization. Both these areas of research are served well since they can be manufactured along with the capabilities to characterize both FBGs and MPOFs. The design of both these devices is a complicated task, and forms an excellent set of technologically relevant problems for powerful design algorithms such as Evolutionary Algorithms (EA).

The design of FBGs using EAs has previously been explored using single objective techniques [1]. Typically the design goals have been simple (for example, bandpass filters), and EAs have proven very successful. In these cases the objective function was defined as the minimisation of the difference between the target spectrum and the design spectrum, using weights to increase the relative importance of regions of interest. The FBG used here is a 1-dimensional design problem, where the FBG features vary along the optical fibre. In this paper we consider the generalized case of multiple objectives by considering particular spectral features of interest, without reducing the problem to a single objective.

In the area of microstructured fibre design, there have been few examples of optimization technique usage. One possible reason is that the computational evaluation of traits for these types of structures can be expensive and difficult. Particular types of structures, such as hexagonal arrays of holes are used in silica due to the capillary stacking techniques used in manufacture, and this has led to most applications of microstructured fibre optimization considering these hexagonal arrays. MPOF technology on the other hand does not limit us to particular arrangements of holes, requiring the application of more general *design* techniques, as opposed to optimization using a fixed representation, which will open up the technology to more application areas. The MPOF design problem is 2-dimensional in nature, where the refractive index (placement of holes) varies across the optical fibre. In this paper we present a representation which was developed to effectively design these structures in an EA setting, along with some initial results using two conflicting objectives.

Real world engineering problems generally involve multiple objectives, and to simultaneously meet these most implementors will combine these multiple objective into a single one. In these cases, an a-priori decision is made about the relative importance of the objectives, emphasizing a particular type of solution. These techniques often require some problem-specific information, such as the total range each objective covers. In complex design problems such as those presented here, this information is rarely known in advance, making the selection of single objective weighting parameters difficult.

Instead of producing a single perfect design, multiobjective techniques consider all objectives simultaneously, resulting in a range of designs where the objectives are expressed to varying degrees (the non-dominated or *Pareto* set). Evolutionary techniques naturally lend themselves to multiobjective optimization, since they are inherently population based. Further if all the objectives can be simultaneously optimized, the whole non-dominated set converges to a single point, effectively achieving the same end results as single objective optimization.

FBGs are introduced in Section 2, along with the problem of interest and EA results in the two objective and full four objective cases. Section 3 introduces microstructured optical fibres, along with the representation used and MPOF design results.

2   Fibre Bragg Gratings

Permanent gratings in optical fibres were first demonstrated experimentally in the late 1970's. Since then, the theoretical and experimental aspects of FBG's have flourished, resulting in a multitude of applications. FBG's have been used as stand alone devices, for example, sensing applications for strain, temperature and voltage measurement. They have also been incorporated into fibre communications systems where they are used to combine, divide and filter digital light signals. The manufacture of FBG's has reached a level now where designs can be quickly and easily fabricated, making FBG design ideal for sophisticated design algorithms. An overview of FBG history and developments is available in [2].

The theoretical aspects of FBG's are well developed, and many *inverse* techniques to obtain a grating structure from a spectrum have been published. A recent example is presented in [3]. A common disadvantage of these methods is that the required transmission spectrum and other properties such as the group delay spectrum have to be specified over a large wavelength range. Further to that, the inverse algorithms do not allow the inclusion of other constraints, where the ability to specify important spectral features that should be present would be extremely useful. Finally, designs found using inverse techniques are often difficult to manufacture.

Experimentally, FBG's are produced by exposing a short length of optical fibre to an intense optical interference pattern, which results in a lengthwise modulation of the refractive index of the silica (glass) of the fibre. This high spatial frequency modulation is typically of the order of a μm, and forms an overall envelope profile which is cm's in length (Figure 1). The idea of *FBG design* is to find a profile which leads to the FBG exhibiting particular spectral properties of interest, where different wavelengths can be selectively reflected (or transmitted) to varying degrees.

Generally, the strongest interaction of the FBG with light occurs at the Bragg wavelength λ_B, defined by

$$\lambda_B = 2n_{\text{eff}} \Lambda \qquad (1)$$

where n_{eff} is the model effective index of the propagating wave and Λ is the grating period. For the purposes of this study we set $\lambda_B = 1.55 \mu m$, but the factor Λ has itself previously been used as a design variable [1]. Since n_{eff} is a tunable property of the optical fibre, and Λ can be controlled during the grating manufacture process with relative ease, they are not included as design variables.

One of the simplest FBG's consists of a uniform grating, which leads to a sinc-function like transmission spectrum. To smooth out these side ripples in the spectrum, a tapering, or *apodization* can be introduced at each end of the grating. This process of apodization refers to gradually changing the strength of the grating, rather than an abrupt change as in a uniform FBG. In addition to that, phase changes can be introduced in the grating to induce transmission peaks in the spectrum. Various functions are used to describe this type of FBG, where we have a rolloff on the ends and a constant central region. The raised cosine $\cos^2(z)$ is one of these commonly used functions.

2.1 FBG Design Parameters

The design parameters used in this problem are associated with details of the raised cosine function which describes the apodised profile. The most general form of the raised cosine grating structure is defined as

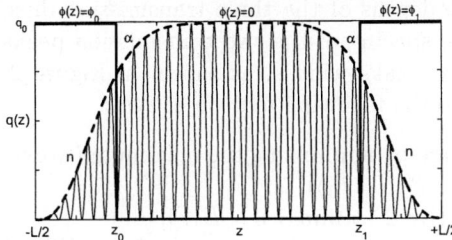
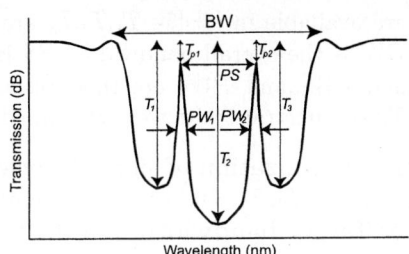

Fig. 1. Overview of the FBG design parameters and there contribution to various features of the profile

Fig. 2. Outline of the various spectral traits extracted using a peak-finding algorithm and used to evaluate the objectives of a design

Table 1. Outline of the parameter bounds

Design parameters	Parameter bounds
Grating strength q_0	$0.0 cm^{-1} < q_0 \leq 10.0 cm^{-1}$
Total length L	1.0 cm $\leq L \leq 15.0$ cm
Phase ϕ_0	$-\pi \leq \phi_0 < \pi$
Phase change positions $-z_0, z_0$	$|z_0| < L/2$
Curvatures α, n	$1.0 < \alpha, n < 10.0$

$$q(z) = q_0 \cos^\alpha \left(\frac{\pi}{2} \left| \frac{2z}{L} \right|^n \right) e^{i\phi(z)} \tag{2}$$

where $\phi(z)$ describes the phase over the length of the grating.

The function itself contains four design parameters: q_0 is the peak strength of the grating, L is the total length of the grating, n controls the curvature of the end drops and α controls the curvature of the ends. Typically $\alpha = 2$, but can be generalized to other values. The phase $\phi(z)$ can consist of a single or multiple steps. In the case where there are constant phase changes on each end of the grating, we have that $\phi(z) = \phi_0$ for $z < z_0$, $\phi(z) = \phi_1$ for $z > z_1$, otherwise $\phi(z) = 0$, where z_0 and z_1 are the locations of the ϕ_0, ϕ_1 phase changes.

In the most general case where all the above parameters are free design variables, we have a search space of 9 dimensions. In this study we make the design symmetrical about $z = 0$ by imposing $\phi_0 = \phi_1$ and $z_0 = -z_1$. The contribution of these parameters to the FBG design is shown graphically in Figure 1. The spectrum of the FBG was evaluated using the transfer matrix method.

2.2 FBG Design Objectives

The spectral characteristics of interest relate to the extraction of OTDM (Optical Time Domain Multiplexing) signals from a data stream. Further details on this are available in [4] [5]. T_1, T_2, T_3 are the depths of the three transmission dips, BW is the overall bandwidth, PS is the spacing of the two transmission peaks and PW_1 and PW_2 are their respective peaks widths, as shown in Figure 2. These spectral traits were combined into the following four objectives:

1. **Minimization of signal interference** by minimizing the maximum (worst) of T_1, T_2, T_3. Objective 1 → $\min(\max(T_1, T_2, T_3))$.
2. **Target bandwidth of 1nm.** Objective 2 → $\min |1.0 - BW|$.
3. **Target peak separation of 0.08nm** for optimal extraction of the OTDM 10GHz clock signal. Objective 3 → $\min |0.08 - PS|$.
4. **Increasing signal clarity** by minimizing the worst (largest) full width half maximum peak width. Objective 4 → $\min(\max(PW_1, PW_2))$.

We refer to these objectives for the remainder of the paper as T, BW, PS and PW respectively. An inequality constraint was also included to allow the consideration only of designs which had minimal signal loss (better than -0.5dB) from the two central transmission peaks, such that $\min(T_{p1}, T_{p2}) + 0.5 \geq 0$.

2.3 Application of the Multiobjective Evolutionary Algorithm

The Non-Dominated Sorting Genetic Algorithm (NSGAII) [6] was used as the selection mechanism, as it firstly provides an efficient method of elitist sorting and results in a well-spread final non-dominated set. Secondly, no external parameters need to be defined or tested. The 6 real valued design parameters used were bounded (Table 1), taking into consideration practical manufacturing limitations, such as the minimum length (L_{min}=1cm) and maximum grating strength (q_{0max}=10cm^{-1}). Simulated binary crossover (SBX) was used along with the polynomial mutation operator [7], and the design parameter bounds were enforced within these operators: any violations were simply repaired by setting them to the minimum or maximum bounds. A population size of 100, crossover probability of $p_c = 0.9$, mutation probability of $p_m = 0.2$ and random initial populations were used throughout.

The constraint defined in the previous section was dealt with by altering the definition of *domination*, which is used when sorting individuals according to their non-dominated level and deciding winners during tournament selection. Design **a** dominates **b** if any of the following are true

- **a** and **b** are infeasible, but **a** has a smaller constraint violation ($\mathbf{a}_c > \mathbf{b}_c$)
- **a** is feasible ($\mathbf{a}_c \geq 0$) and **b** is not ($\mathbf{b}_c < 0$)
- **a** and **b** are feasible and **a** dominates **b** in the usual Pareto-optimal sense

2.4 FBG Optimization Results

Bi-objective design is one of the most widely studied problems in multiobjective optimization. Given our objective space of 4-dimensions, along with unknown relationships between the objectives, a simpler bi-objective methodology was firstly used. A recent publication [8] examined the issue of high numbers of objectives

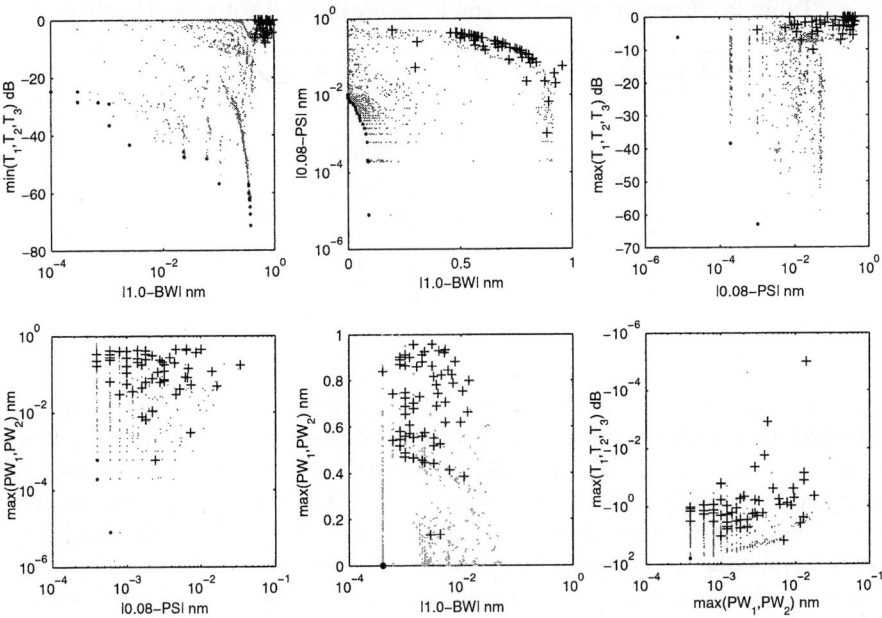

Fig. 3. All pairings of the 4 objectives outlined in Table 2. Crosses indicate the initial population, light grey points show all designs produced which satisfy the constraint, and the black points indicate the final non-dominated set after 100 generations

through the examination of pair-wise objective relationships. Objectives can be classified into harmonious and conflicting pairs. If a pairing is harmonious, which other objective suffers as a result, or do some objectives naturally follow and also improve? From these paired results we can also get some idea of the intrinsic dimensionality of the four objective non-dominated set. All 6 possible pairings were optimized over 100 generations, which was found sufficient to obtain a general idea of the non-dominated set. The SBX operator spread index $\eta_c = 5$ and polynomial mutation operator spread parameter $\eta_m = 10$ were used to facilitate a coarse parameter search. The results are shown in Figure 3, and the relationships summarized in Table 2. The most conflicting relationship was for the (BW, T) pairing, to a lesser degree followed by the (BW, PS), (PS, T) and (PS, PW) pairings. The final two pairings (BW, PW) and (T, PW) were harmonious, but in the process of optimizing for these objectives simultaneously, the other two objectives - respectively T, PS and PS, BW suffered. This indicates that there will be a non-dominated relationship when considering all four objectives simultaneously, resulting in a non-dominated set manifold where most of the variance exists in 2-dimensions as a result of the strongest conflict between BW and T.

The full 4 objective problem was then run for 1000 generations using a population size of 100, $\eta_c = 5$ and $\eta_m = 10$. The evolution of the design parameter space from the initial random population to the final non-dominated set is

Table 2. Summary of conflict and harmony in the 6 pairwise objectives

Objective 1	Objective 2	Type	Worst objectives
BW	T	Conflicting	-
BW	PS	Conflicting	-
PS	T	Conflicting	-
PS	PW	Conflicting	-
BW	PW	Harmonious	T, PS
PW	T	Harmonious	BW, PS

shown in Figure 4. To aid in the visualisation of the final non-dominated set, the ISOMAP [9] algorithm was used to reduce the dimensionality of the data. It was found that most of the variance (92%) was due to a 2-dimensional non-linear manifold embedded in the 4-dimensional space. Principle Component Analysis (PCA) was also attempted, but did not effectively realize the relationships between points due to their non-linear nature. The arrangement of non-dominated points on this 2-dimensional manifold is shown in Figure 5. Two clusters of

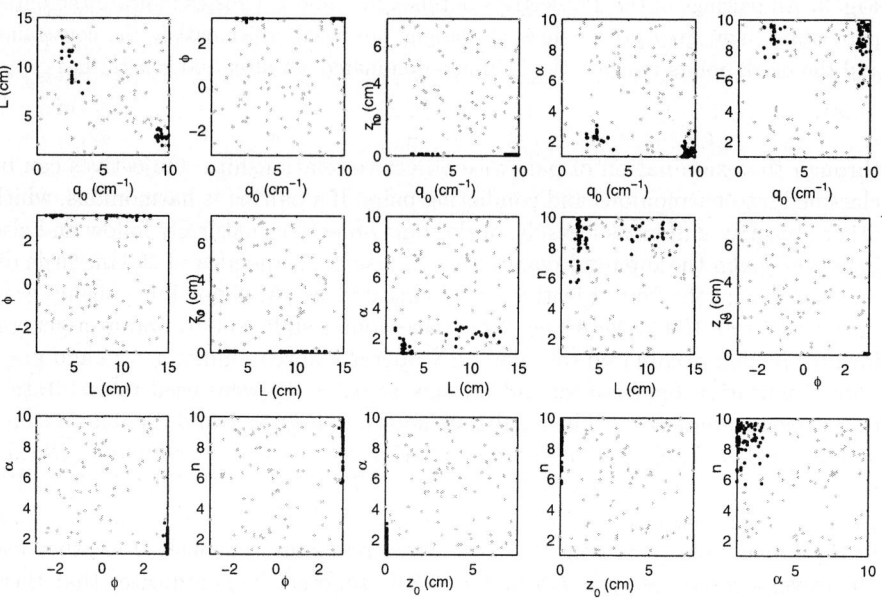

Fig. 4. Scatterplots in parameter space for the 4 objective FBG optimization problem. Light grey dots indicate the initial population, and black dots indicate the final non-dominated set after 1000 generations

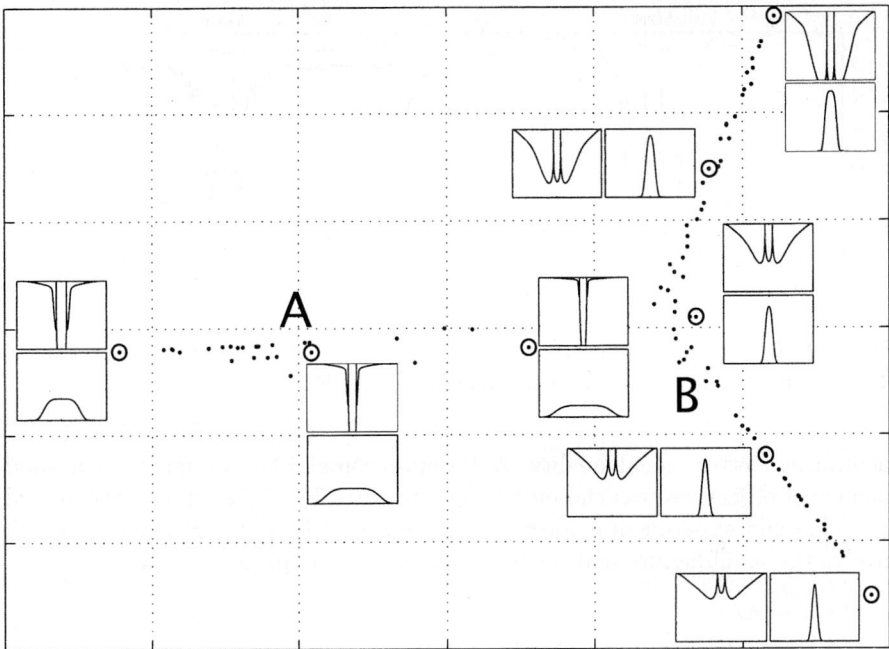

Fig. 5. Non-dominated designs of the four objective FBG problem represented as a flattened-out two dimensional surface. Two main clusters of designs, A and B are evident. Some of the designs (circled) have their spectra and FBG profile shown. Axes have been removed for clarity, and the plot bounds are: FBG spectrum - $\lambda \rightarrow [1549.5, 1550.5]$nm, $T \rightarrow [-70, 0]$dB, and FBG profile - $L \rightarrow [-5, 5]$cm, $q_0 \rightarrow [0, 11]$cm^{-1}

points are evident. Cluster A consists of designs which all excel in the PW objective, and mainly vary in terms of the three remaining objectives. Cluster B on the other hand contains designs with good BW which mainly vary in terms of the harmonious objectives PW and T but have good BW. As the transmission objective gets better from bottom to top, so does the peak width. This cluster is *locally non-dominated* due to the conflicting relationship between the BW, T and BW, PS pairings. In terms of the parameters, the designs in clusters A and B all converged to values of approximately $\phi = +\pi$, $\alpha = 1.0$, and $z_0 = 0.1$cm. The main variation overall occurs in the curvature n. In some runs they also converged to $\phi = -\pi$. The constraint handling method proved particularly efficient, since within a few generations only FBG designs within the constraint bounds were being produced.

Even though all of the designs in Figure 5 have good objectives to varying degrees, the two clusters represent two distinct types of FBGs found. Cluster A consists of longer but weaker gratings, whereas cluster B consists of shorter but stronger gratings (reflected in the groupings seen in Figure 4). As decision makers, we become more interested in cluster B since the second trough is wider,

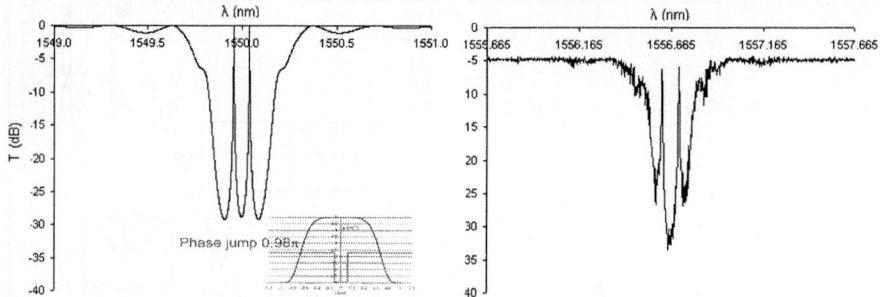

Fig. 6. Left: Theoretical spectrum from the Pareto optimal FBG shown in inset. **Right:** experimental spectrum of the manufactured FBG

facilitating better signal quality. A Pareto-optimal FBG design with a good balance of objectives was chosen for manufacture. The FBG profile, theoretical spectrum and experimental spectrum are shown in Figure 6. Further work will involve the manufacture and testing of other Pareto-optimal designs.

3 Microstructured Optical Fibres

Most optical fibres in use today are made of glass (silica) with solid cores. Deposited chemicals inside the core alter the refractive index profile of the glass and therefore its light guiding properties. Some previous work focused on the single objective optimization of such a fibre [10]. In recent years, silica holey fibres, or Photonic Crystal Fibres (PCF), which use air holes that run the length of the fibre, have presented themselves as an exciting new development in photonic technology. Even more recently has been the development of holey fibres in polymer - Microstructured Polymer Optical Fibres (MPOF), allowing the cheap and easy fabrication of arbitrary hole patterns in fibres. An overview of the technology is given in [11].

The refractive index profile in microstructured fibres is 2-dimensional - increasing the design complexity immensely. Many examples of holey fibres have so far focussed on arrangements of holes which relate to the manufacturing process, such as the stacking of capillaries in silica fibres which produce hexagonal array type structures. This has also been the basis of the numerical optimization of these structures - since the design can easily be parametized into a relatively small search space [12]. But what about the *generic design* of such structures - where no pre-disposition is to be made about the arrangement of holes? This is important given the large space of designs that can be manufactured using MPOF technology, for various applications from sensing to data transmission. We are no longer limited as much in terms of the creativity of designs, thus generic design sits very well with the MPOF technology.

Some of the recent fibre development work has focussed on the application of MPOF to high bandwidth data transmission applications. This section of the paper presents some initial work in using genetic algorithms to design such generic structures in a multiobjective setting.

3.1 Representation

Given the complexity of microstructured fibres - especially when considering in parallel constraints such as holes not overlapping and also conforming to manufacturing constraints (minimum wall thickness between holes), a flexible representation needs to be found. A general scheme was devised which can describe arbitrary hole patterns, automatically conforming to manufacturing constraints imposed through the use of an embroyogeny, where the fibres are grown. The only hard constraints built into the representation is that some symmetry $n_{symm} \geq 2$ is imposed, and that structures consist entirely of holes.

The process of the binary genotype $\rightarrow$ phenotype (holey fibre structure) conversion is outlined below.

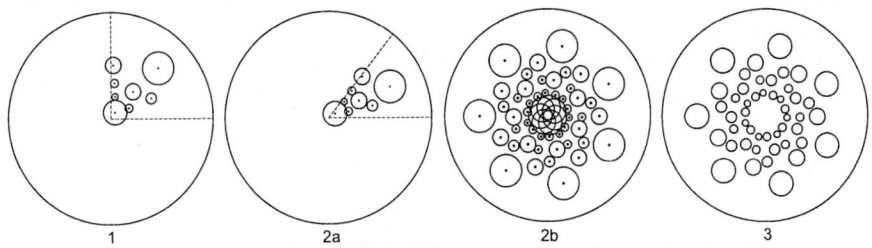

Step 1. The binary genotype is decoded into the symmetry n_{symm}, and N_h triplets of x_i, y_i, r_i values. The triplets describe the placement of holes in the first sector of the x, y plane, forming the raw structure with $n_{\text{symm}} = 4$. The binary versions of x_i, y_i are decoded to the real valued position of the hole. r_i is decoded to an integer value: each possible value refers to a user defined list of available hole sizes, which reflect the available drills used to produce the initial MPOF preform.

Step 2. The N_h hole positions x_i, y_i are symmeterized into the new symmetry n_{symm} (a symmetry of 7 is used in the above example). 2a. The holes are converted to polar coordinates r_i, θ_i, where θ_i values are scaled by $n_{\text{symm}}/4$. **2b.** ($n_{\text{symm}} - 1$) copies of the holes are made to complete the fibre.

Step 3. The holes are grown until manufacturing constraints prohibit further growth. We start with all holes having zero radius. The holes are grown in a step wise manner through the list of available hole sizes, and stop growing when they are within w_h of a neighboring hole or exceed their own maximum radius r_i. Growth states are updated in parallel at the end of every hole growth cycle, and the cycle is terminated when the growth state of every hole is false.

We can see that some holes are not influenced at all by surrounding holes, reaching the maximum allowed radius encoded in their genotype. Other holes

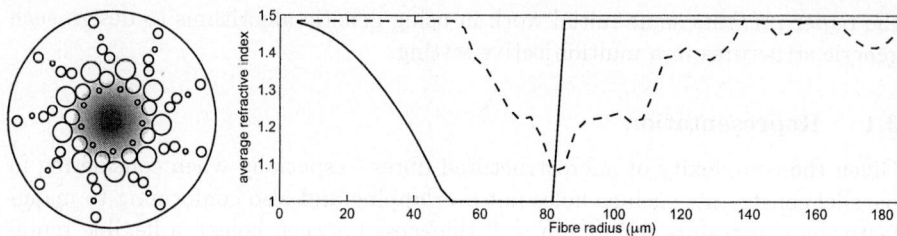

Fig. 7. Left: Randomly generated fibre design with the core fundamental mode field $|\psi_{01}|$ overlaid (graded region on the centre of the fibre), with $\gamma = 2.02 \times 10^{-6}$. **Right:** The solid line indicates the target average fibre profile, and the dashed line shows the average profile of the design to the left. The average profile objective is 0.14

never grow at all, but lay dormant in the genotype to appear in later generations during the evolutionary process. Most holes grow until inhibited by the surrounding phenotype. The advantage of using a growth scheme during the genotype → phenotype conversion is clear - we always automatically obtain manufacturable designs without the need for a penalty function or other constraint handling techniques. Each hole is of a size that can be inserted into the polymer, and the minimum wall thickness between holes guarantees structural stability.

3.2 Recombination and Mutation Operators

Single point crossover is used, where the crossover of two parents results in a single child being generated. Firstly a random point is chosen on parent 1. This in turn sets a constraint as to the selection of the crossover point in parent 2 since we have to preserve the length of the n_{symm} and hole x_i, y_i, r_i gene segments. Using this type of crossover, children often share the characteristics of the parents, and may become simpler or more complex structures depending on the two crossover locations.

Binary mutation with a probability of $p_m = 0.0035$ is used to mutate the binary strings. Two other types of mutation are also used - $p_a = 0.003$ refers to the frequency of hole addition to an individual, and $p_d = 0.003$ refers to the probability of hole deletion. No testing of different p_m, p_d, p_a values has currently been done to ascertain the best values to use for this type of representation.

3.3 MPOF Design Objectives

Two of the most important performance measures for transmission fibres relate to bandwidth and transmission loss [13]. Both of these parameters are computationally expensive to evaluate, but some approximations can be made.

Bandwidth. Polymer based fibres are typically highly multimodal, and a good approximation to bandwidth can be defined in terms of the fastest and slowest propagating modes in the fibre. Reducing this difference in velocity between these two extreme modes increases the bandwidth since adjacent light pulses in the fibre don't interfere as they spread while traveling down the fibre. Using a parabolic refractive index across the fibre approximately equalizes these mode velocities, effectively increasing the bandwidth. To apply this argument to holey fibres, small holes in the fibre microstructure are not completely resolved by the light, and we can approximate the whole structure by an averaged refractive index profile. So we can increase the bandwidth in a design by best suiting the average of the hole arrangement to a parabolic profile $n(r) = 1.48 - \alpha r^g$, where 1.48 is the refractive index of PMMA polymer at the wavelength $\lambda = 0.833 \mu m$ and $g = 2$. A small area of minimal refractive index ($n = 1.0$) is added to the outside of the core to help increase light confinement. The objective is to then minimize the mean difference between the target profile and the design average profile. The average profile of a randomly generated design is shown in Figure 7.

Transmission loss in MPOFs can be attributed, amongst other things, to material absorption and scattering due to surface roughness [13]. Material absorption in the polymer is well known and can be quantified. Scattering from surface roughness on a microscopic scale arises during the manufacture of the fibre preform, when holes are drilled into the polymer using a computer numerically controlled (CNC) mill. As light travels down the fibre, the interaction of the light with air-polymer interface of the holes results in the light scattering and coupling to lossy cladding modes, reducing signal intensity. A complete numerical evaluation of this overall loss is computationally expensive, but a simple approximation is to consider the overlap of the (normalized) fundamental mode ψ_{01} with the air-polymer interfaces. Reducing this overlap

$$\gamma = \sum_{i=1}^{N_h} \oint_{h_i} |\psi_{01}|^2 \, r_i d\theta \qquad (3)$$

effectively reduces the interaction of light with the air-polymer interface. Here N_h is the number of holes and r_i is the radius of hole h_i. The mode is evaluated at the wavelength $\lambda = 0.833 \mu m$. A numerical algorithm based on the Fourier decomposition method was used to solve Maxwell's scalar wave equation for the generated fibre structures. This particular implementation [14] was used since it has the ability to quickly evaluate the mode profiles of arbitrary MPOF structures, thus allowing complete automation for effective use in an evolutionary algorithm. An example of a fundamental mode is shown as a shaded region in the centre of the fibre in Figure 7.

3.4 MPOF Design

A random population of size 100 was used along with the NSGAII algorithm to order designs for subsequent selection into the next generation. Tournament selection ($k = 2$) was used to compare individuals for breeding based on their non-dominated levels and crowding distance measure. The algorithm was run over 5000 generations. The objective value calculations are computationally expensive, and for a run this long some parallelization was implemented. On shared

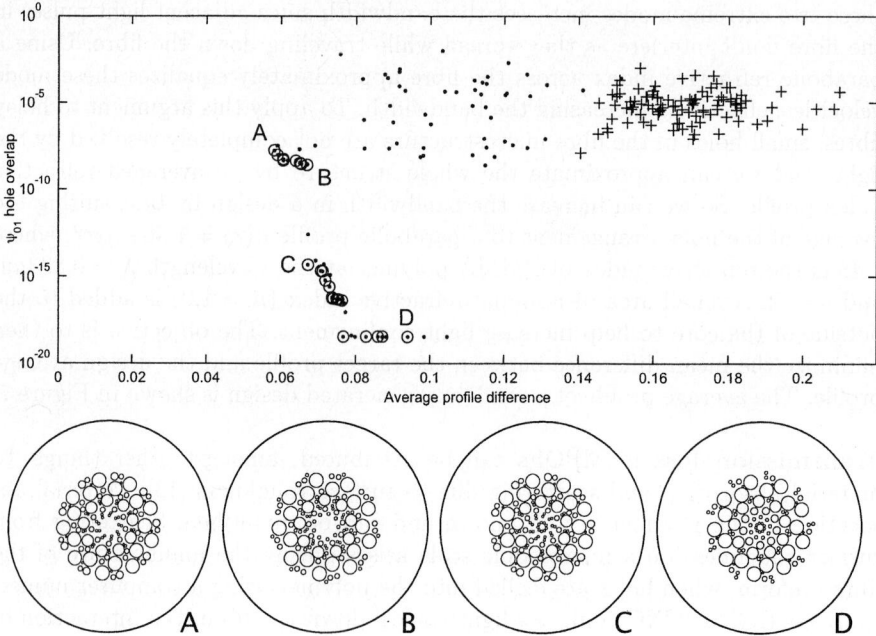

Fig. 8. Top: Population evolution with respect to the 2 design objectives. Crosses indicate the initial population, black dots the final population after 5000 generations and circles the final non-dominated set. **Bottom:** Four designs from the final non-dominated set, A,B,C and D are shown

memory architectures, OpenMP allows for trivial speedup given an objective evaluation loop. For example, in our C++ code it was achieved as follows:

```
#include <omp.h>
...
int i, popnsize = popn.size();
#pragma omp parallel default(shared) private(i)
{
   #pragma omp for schedule(dynamic)
   for(i=0; i<popnsize; i++)
      popn[i].evaluate_objectives();
}
```

Figure 8 shows the evolution from the initial random population to the final non-dominated set of designs. A good range of fibres are discovered. For example, A has a very good average index profile close to a parabolic index variation (0.0582) but the worst field hole overlap in the non-dominated set (1.7×10^{-8}). Design D is the other extremal solution, with an average index profile objective value of 0.094 but a very minimal overlap value of 3.32×10^{-19}. This is caused by the confining ring of small holes in the central region, which simultaneously conflict with the average index objective by not gradually changing the aver-

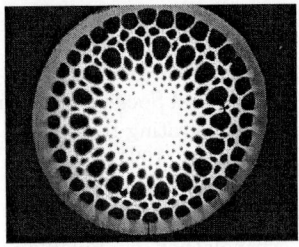

Fig. 9. Example of a fabricated MPOF with an average parabolic profile. The outer diameter of the fibre is approximately 220μm

age index. Designs B and C are intermediates on the non-dominated set, with objective values of $[0.0669, 2.61 \times 10^{-9}]$ and $[0.0672, 4.5 \times 10^{-15}]$ respectively.

An example of a manufactured MPOF with a good parabolic average index profile is shown in Figure 9. This particular design was not evolved using the EA discussed, but demonstrates the feasibility of manufacturing these designs. This MPOF design work is still in its initial stages. Future research will include the manufacture of these evolved MPOFs, followed by the experimental characterization of the bandwidth and loss associated with the objectives used.

4 Conclusion

In this paper we have demonstrated the successful application of evolutionary multiobjective algorithms to two quite different photonic design problems. Through the use of real parameter optimization we were able to identify Pareto-optimal FBGs, and choose a suitable design with a good balance of properties for manufacture. A more generic design approach to MPOFs demonstrated the effectiveness of a powerful and expressive representation along with EMO techniques to design complicated microstructured fibres. The creativeness of the evolutionary process went hand in hand with the breadth of manufacturable MPOF designs. Both these areas of design are still in early stages, where future work will involve fabrication and experimental characterization of evolved designs.

Acknowledgements. The authors would like to acknowledge the computational resources of Australian Centre for Advanced Computing and Communications (AC^3) and Sydney Vislab, Brian Ashton for the fibre Bragg grating manufacture, the Bandwidth Foundary, the Microstructured Polymer Optical Fibre research group at the Optical Fibre Technology Centre and helpful discussions with Professor Kalyanmoy Deb and Santosh Tirawi of the Kanpur Genetic Algorithms Laboratory.

References

1. Cormier, G., Boudreau, R.: Read-coded genetic algorithm for bragg grating parameter synthesis. Journal of the Optical Society of America, B **18** (2001) 1771–1776
2. Hill, K.O., Meltz, G.: Fibre bragg grating technology fundamentals and overview. Journal of Lightwave Technology **15** (1997)
3. Poladian, L.: A simple gratings synthesis algorithm. Optics Letters **25** (2000) 787–789
4. Attygalle, M., Ashton, B., Nirmalathas, A., Poladian, L., Padden, W.: Novel technique for all-optical clock extraction using fibre bragg gratings. In: OptoElectronics and Communications Conference, Shanghai, China. (2003) 13–16
5. Manos, S., Poladian, L., Ashton, B.: Novel fibre bragg grating design using multiobjective evolutionary algorithms. In: CLEO/IQEC, San Francisco, California, USA. (2004)
6. Deb, K., Pratap, A., Agarwal, S., Meyarivan, T.: A fast and elitist multiobjective genetic algorithm: Nsga-ii. IEEE Transactions on Evolutionary Computation **6** (2002) 182–197
7. Deb, K., Beyer, H.: Self-adaptive genetic algorithms with simulated binary crossover. Evolutionary Computation Journal **2** (2001) 197–221
8. Purshouse, R.C., Fleming, P.J.: Conflict, harmony and independence: Relationships in evolutionary multi-criterion optimisation. In: Proceedings of the Second International conference on Evolutionary Multi-Criterion Optimization (EMO). (2003) 16–30
9. Tenenbaum, J.B., de Silva, V., Langford, J.C.: A global geometric framework for nonlinear dimensionality reduction. Science **290** (2000) 2319–2323
10. Manos, S., Poladian, L.: Optical fibre design using evolutionary strategies. Engineering Computations **21** (2004) 564–576
11. van Eijkelenborg, M., Argyros, A., Barton, G., Bassett, I., Fellew, M., Henry, G., Issa, N., Large, M., Manos, S., Padden, W., Poladian, L., Zagari, J.: Recent progress in microstructured polymer optical fibre fabrication and characterization. Optical Fiber Technology **9** (2003) 199–209
12. Manos, S., Mitchell, A., Lech, M., Poladian, L.: Automatic synthesis of microstructured holey fibre designs using numerical optimisation. In: Australian Conference on Fibre Optic Technology (ACOFT), Sydney Convection Centre, Darling Harbour, Sydney, Australia. (2002)
13. Barton, G., van Eijkelenborg, M., Henry, G., Issa, N., Klein, K.F., Large, M., Manos, S., Padden, W., Pok, W., Poladian, L.: Characteristics of multimode microstructured pof performance. In: Plastic Optical Fibre (POF) Conference, Seattle, USA. (2003) 81–84
14. Poladian, L., Issa, N.A., Monro, T.: Fourier decomposition algorithm for leaky modes of fibres with arbitrary geometry. Optics Express **10** (2002) 449–454

Multiple Criteria Lot-Sizing in a Foundry Using Evolutionary Algorithms

Jerzy Duda and Andrzej Osyczka

AGH University of Science and Technology,
Faculty of Management, Dept. of Applied Computer Science,
ul. Gramatyka 10, 30-067 Kraków, Poland
{JDuda,AOsyczka}@zarz.agh.edu.pl

Abstract. The paper describes the application of multiobjective evolutionary algorithms in multicriteria optimization of operational production plans in a foundry, which produces iron castings and uses hand molding machines. A mathematical model that maximizes utilization of the bottleneck machines and minimizes backlogged production is presented. The model includes all the constraints resulting from the limited capacities of furnaces and machine lines, limited resources, customers requirements and the requirements of the manufacturing process itself. Test problems based on real production data were used for evaluation of the different evolutionary algorithm variants. Finally, the plans were calculated for a nine week rolling planning horizon and compared to real historical data.

1 Introduction

One of the authors has been working for a Polish foundry to develop the software which would help to improve shop production planning process. A weekly task for the planners at the operational level is to say how many castings for which orders will be produced on molding machines during all working shifts. The planning process is done manually with a little support of spreadsheets and basic MRPII/ERP (Material/Enterprise Resource Planning) related tools. It is a common practice not only in this particular foundry. The survey conducted by Van Voorhis and Peters [10] has shown that also in the USA only a few foundries used specialized software to assist the planning and scheduling process while the majority of them did it manually.

The production in small and medium iron foundries is often done in short series so the planners must take into consideration many orders, each for a different product. In the considered foundry there were about 100–200 active orders a week for 10 to 500 castings of various weight and iron grade.

The castings manufacturing process itself can be divided into following steps: designing a pattern, preparing molding sand and cores, making molds, melting and pouring hot iron and finally finishing operations. The patterns are prepared in a separated pattern shop and once they are made they can be used many times. If a casting requires cores, they are made in a core shop. Cores usually can be prepared earlier, even a few days before they are put in a mold. Next molding sand is compacted around a pattern in a flask thus creating a mold. Then hot iron, melted in electric furnaces is poured into the flasks, which are left to solidify. After several hours the castings are taken from the molds and they undergo cleaning and finishing operations in a dressing shop.

Operations for the pattern shop are planned independently of the main production process. Operations for the core shop and the dressing shop can be planned easily on the basis of a molding plan unless the plan includes an enormous number of castings which require many cores or a lot of time to be finished. This situation is very rare in the foundry so it will not be considered in the optimization model.

Thus a priority for the planners is to prepare an appropriate molding plan connected with a pouring schedule for the furnaces. These plans must be coordinated as melted iron cannot wait too long to be poured, and also the room for the molds waiting for pouring is limited. While building the plan many technological and organizational constraints have to be taken into consideration. The most significant are:

- capacities of furnaces and molding machines,
- the number, desired delivery date and cast iron grade of ordered castings,
- the number of different castings which can be produced during one shift (setup times are included in forming times),
- the number of flasks of various size available during a working shift,
- the minimum batch size a customer can accept.

The data for the optimization model can be collected nearly automatically from the existing production control system.

2 Mathematical Model

A mathematical model was formulated on the basis of the classical discrete capacitated lot-sizing problem with single level and multi item production on parallel machines. A similar approach can be found in Santos-Mezo et al. paper [8], which discusses a lot-sizing problem in an automated foundry. The model proposed in this paper, however differs in two main points. A commonly used minimization of an artificially built sum cost function has been replaced by two objective functions indicated directly by the planners. Another modification is that the equality constraint balancing inventory and demand has been changed into an inequality constraint.

The following symbols are used:
Decision variables:

x_{ijtz} – number of castings planned for order i to be manufactured on machine j during day t and shift z,
v_{htz} – number of heats of grade h during day t and shift z.

Data:

τ – week for which the plan is created,
k – number of working days in a week,
m_j – number of working shifts for machines type j,
n_j – number of active orders for machines type j,
C_P – daily melting capacity of the furnaces [kg],
W – weight of single heat [kg],

C_{Fj} – capacity of molding machines type j during a working shift [minutes],
w_{ij} – total iron weight needed to produce single i casting [kg],
a_{ij} – time of making a mold for casting i on machine j [minutes],
d_{ij} – ordered number of castings of type i to be produced on machine j,
γ – number of iron grades,
g_{ij} – iron grade for casting i, $g_{ij} \in \{1,...,\gamma\}$,
ω – number of flask types,
S_o – flask number of type o available during a working shift,
q_{ij} – flask type in which a mold for casting i is prepared, $q_{ij} \in \{1,...,\omega\}$,
κ_j – number of different castings which can be produced on machine type j during one working shift,
δ_{ij} – due week for castings of type i to be produced on machine j,
π^t_{ij} – penalty value for tardiness,
π^e_{ij} – penalty value for earliness.

First objective function:

$$\text{maximize} \sum_{j=1}^{l}\sum_{i=1}^{n_j}\sum_{t=1}^{k}\sum_{z=1}^{m_j}(\frac{x_{ijtz}w_{ij}}{kC_P} + \frac{x_{ijtz}a_{ij}}{km_j C_{Fj}}) \quad (1)$$

Second objective function:

$$\text{Minimize} \sum_{j=1}^{l}\sum_{i=1}^{n_j}((d_{ij}-\sum_{t=1}^{k}\sum_{z=1}^{m_j}x_{ijtz})(\tau-\delta_{ij})\pi^t_{ij}(\tau>\delta_{ij})(d_{ij}>\sum_{t=1}^{k}\sum_{z=1}^{m_j}x_{ijtz}))$$
$$+ \sum_{j=1}^{l}\sum_{i=1}^{n_j}((d_{ij}-\sum_{t=1}^{k}\sum_{z=1}^{m_j}x_{ijtz})(\tau-\delta_{ij})\pi^e_{ij}(\tau<\delta_{ij})) \quad (2)$$

Constraints:

$$\sum_{h=1}^{\gamma}v_{hzt}W \leq C_P, \quad t=1,...,k, \quad z=1,...,m_j \quad (3)$$

$$\sum_{i=1}^{n_j}x_{ijtz}a_{ij} \leq C_{Fj}, \quad j=1,...,l, \quad t=1,...,k, \quad z=1,...,m_j \quad (4)$$

$$\sum_{t=1}^{k}\sum_{z=1}^{m_j}x_{ijtz} \leq d_{ij}, \quad j=1,...,l, \quad i=1,...,n_j \quad (5)$$

$$\sum_{j=1}^{l}\sum_{i=1}^{n_j}(x_{ijtz}w_{ij}(g_{ij}=h)) \leq v_{htz}W, \quad h=1,...,\gamma, t=1,...,k, z=1,...,m_j \quad (6)$$

$$\sum_{i=1}^{n_j}(x_{ijtz}>0) \leq \kappa_j, \quad j=1,...,l, \quad t=1,...,k, \quad z=1,...,m_j \quad (7)$$

$$\sum_{j=1}^{l}\sum_{i=1}^{n_j}(x_{ijtz}(q_{ij}=o)) \leq S_o, \quad o=1,...,\omega, \quad t=1,...,k, \quad z=1,...,m_j \quad (8)$$

The first objective function (1) maximizes the utilization of both furnaces and molding machines as these are bottlenecks in the production system. Although the function (1) is to be maximized, for the sake of convenience it is transformed to minimization by multiplying it by −1. The second objective function (2) minimizes the penalty for not making as many castings as ordered by the customer on time or for making them earlier than in the week agreed with the customer. Constraints (3) and (4) are the capacity constraints for the furnaces and the molding machines, respectively. Constraint (5) ensures that no more production can be planned than ordered by the customer. In the classic lot-sizing model constraint (5) is an equality. However, in the considered foundry the sum of items ordered usually exceeds production capability so backlogging is a common practice and customers receive their castings in several batches. The penalty function (2) ensures that backlogging would is kept at a possibly low level. Inequality constraint (6) limits the weight of the planned castings of a particular cast iron grade to the weight of the metal which has to be melted. According to constraint (7) no more than κ_j different castings may be produced during one working shift. The last constraint (8) limits the flask availability.

The model is formulated as a nonlinear problem, however it is possible to change it into an integer programming formulation by entering additional binary variables.

3 Evolutionary Algorithm

A weekly plan for molding machines is coded in a single chromosome using integer gene values. A single gene represents quantity of castings planned for production or equals zero if the production for particular order during a given shift is not planned. This can be presented as the matrix shown in figure 1.

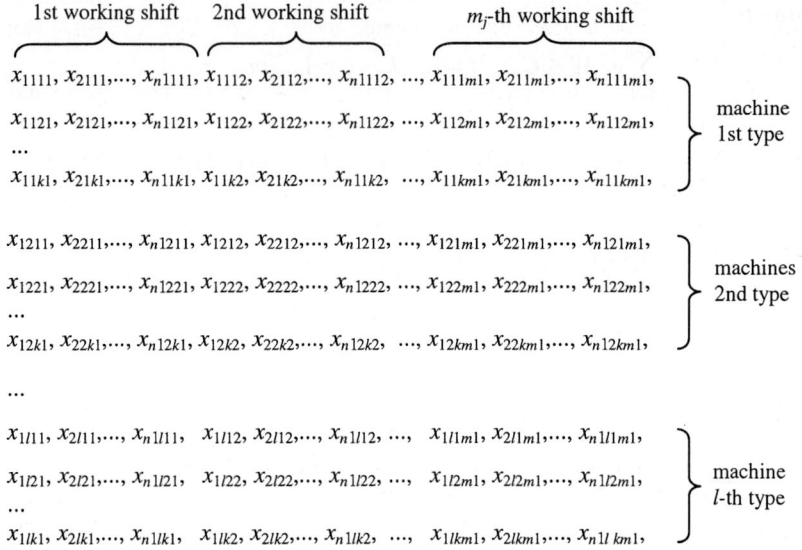

Fig. 1. Molding Plan Coded in a Chromosome

This chromosome structure leads to a situation where a correct plan should have a lot of zeroes, regarding constraint (7). To avoid keeping incorrect individuals in a population a simple repair algorithm is introduced. Whenever constraint (7) is violated for one of the machines and working shifts, the smallest lots planned so far are eliminated successively from the plan until the number of different lots which are allowed for production during one working shift is reached.

Note that there are no v_{htz} variables related to the pouring schedule in a chromosome. Instead of this a second repair algorithm is used. Its role is to keep molding plans always acceptable from a pouring schedule point of view. This means that there is enough hot iron for filling all the molds prepared. The idea of the algorithm is similar to the first repairing algorithm. If the maximum number of heats of a particular iron grade is exceeded than the lot with the minimum weight of castings is removed from the plan.

A new crossover operator which operates on working shifts is introduced. Two shifts in a plan are chosen randomly. Then the lot sizes in these shifts are swapped. The crossover operator is used with the probability of 80%. A mutation pool is being created using tournament selection together with an elitist mechanism. The simple uniform mutation is chosen experimentally as a mutation operator and is used for altering genes with the probability of 0.1%.

Many multicriteria evolutionary algorithms have been proposed in literature. The survey of them can be found in Coello Coello [1] or in Osyczka [7]. Among the algorithms proposed later, two are regarded as the most effective: NSGAII created by Deb et al. [2] and SPEA2 proposed by Zitzller et al. [11]. Both algorithms were used for generating the final approximation of Pareto front. Additionally, a slightly modified SPEA2 algorithm version has been tested. In the environmental selection process the best individuals regarding all the objective functions are copied into a mutation pool obligatorily. This modification will be denoted as SPEA2e (SPEA2 with extended elitism) in this paper.

The statistics proposed by Fonseca and Fleming [3] in the version implemented by Knowles and Corne [5] is used to test which multicriteria evolutionary algorithm performs better in terms of the presented model and test problems. Although the statistics fails in some cases [12], it enables us to compare two Pareto front approximation sets when a reference set is not known.

4 Test Problems

Test sets have been chosen from the production control computer system used in the foundry described in this study. The first test problem (*fixed1*) consists of 84 orders while the second (*fixed2*) has exactly 100 orders.

There are four molding lines in the considered factory, each consisting of two molding machines, one for making a cope and one for making a drag (top and bottom parts of a flask). However, there are only three types of molding machines. The type of machine which is used for making a mold for a particular casting is stated in its operation sheet. Tables 1, 2 and 3 shows detailed specification of the orders which are to be produced on machine types A, B, and C, respectively.

Table 1. Detailed Specification of *fixed1* Problem Orders for Machines Type A (Big Flasks)

order no.	flasks left to make	weight [kg]	forming time [min]	iron grade	due week	penalty coef.	order no.	flasks left to make	weight [kg]	forming time [min]	iron grade	due week	penalty coef.
1	282	61.2	30.5	4	-3	0.460	9	16	37.3	25.4	4	3	1.203
2	37	82.0	32.1	4	0	2.707	10	22	34.7	30.2	5	3	1.045
3	26	61.6	29.0	4	0	2.289	11	14	51.0	27.3	4	3	1.085
4	3	54.0	31.8	4	0	0.511	12	249	62.8	29.3	4	3	0.485
5	125	43.0	27.3	4	0	0.847	13	30	43.0	26.1	4	4	0.538
6	226	65.0	32.6	4	1	1.482	14	6	54.6	31.8	4	5	0.518
7	102	48.0	25.6	4	2	0.583	15	44	80.0	35.0	4	5	1.135
8	16	30.4	25.4	4	3	0.870	16	548	79.0	37.4	4	5	0.989

Table 2. Detailed Specification of *fixed1* Problem Orders for Machines Type B (Small Flasks)

order no.	flasks left to make	weight [kg]	forming time [min]	iron grade	due week	penalty coef.	order no.	flasks left to make	weight [kg]	forming time [min]	iron grade	due week	penalty coef.
1	32	24.0	14.2	5	-3	0.159	23	16	12.9	13.2	5	1	0.096
2	35	24.0	14.2	5	-3	0.159	24	37	12.8	13.2	5	1	0.126
3	231	18.0	11.8	2	-3	0.069	25	26	15.9	13.2	5	1	0.117
4	424	9.3	11.6	4	-3	0.183	26	24	21.4	15.1	5	1	0.207
5	8	3.4	5.5	4	-2	0.125	27	229	13.5	11.5	4	2	0.423
6	31	15.6	13.9	2	-2	0.128	28	8	6.8	14.3	5	3	0.083
7	404	15.1	14.2	4	-2	0.321	29	16	6.0	14.3	5	3	0.078
8	538	15.1	14.2	4	-1	0.321	30	31	1.8	3.6	5	3	0.093
9	432	16.2	15.3	5	0	0.458	31	6	10.4	13.9	5	3	0.161
10	44	14.3	12.7	4	0	0.108	32	11	9.1	13.9	5	3	0.133
11	28	18.1	14.3	4	0	0.153	33	16	10.8	13.9	5	3	0.158
12	83	25.0	13.9	4	0	0.831	34	5	10.9	14.0	5	3	0.030
13	91	25.0	13.9	4	0	0.831	35	5	13.1	14.0	5	3	0.049
14	212	10.4	12.1	2	0	0.217	36	19	15.2	13.1	2	3	0.183
15	159	12.2	12.1	2	0	0.238	37	10	13.9	12.3	2	3	0.055
16	4	9.0	12.6	5	0	0.209	38	112	9.6	13.3	5	3	0.072
17	47	13.8	12.0	5	0	0.310	39	458	12.2	12.7	5	3	0.212
18	16	12.4	13.9	5	0	0.169	40	32	12.6	13.0	5	3	0.209
19	16	11.6	13.9	5	0	0.147	41	184	23.8	12.1	4	5	0.348
20	16	11.0	13.9	5	0	0.161	42	52	8.3	13.0	5	5	0.370
21	16	12.0	13.9	5	0	0.175	43	59	4.2	11.5	5	5	0.138
22	133	12.0	12.3	5	1	0.094	44	545	28.9	13.2	4	5	0.431

The number of flasks to make is calculated as the number of castings ordered by the customers divided by the number of castings which fit in a single flask. Thus the weight and forming time refer to the whole flask, not to a single casting.

Table 3. Detailed Specification of *fixed1* Problem Orders for Machines Type C (Medium Flasks)

order no.	flasks left to make	weight [kg]	forming time [min]	iron grade	due week	penalty coef.	order no.	flasks left to make	weight [kg]	forming time [min]	iron grade	due week	penalty coef.
1	3	15.6	17.2	4	-3	0.209	13	36	6.9	2.9	5	2	0.203
2	257	18.2	15.1	4	-3	0.501	14	36	3.4	1.4	5	2	0.099
3	26	10.7	16.4	4	-2	0.239	15	83	5.8	2.4	5	2	0.171
4	25	10.8	18.1	4	-2	0.299	16	122	9.0	6.8	5	3	0.122
5	58	52.2	19.0	4	-2	0.652	17	96	23.6	16.7	5	3	0.148
6	196	29.6	17.7	4	-1	0.454	18	249	13.6	10.2	5	3	0.181
7	4	70.0	19.2	5	0	1.291	19	22	21.7	18.6	5	3	0.279
8	26	18.6	17.3	4	0	0.322	20	62	26.8	18.4	4	5	0.186
9	37	62.0	16.5	5	0	1.389	21	108	30.2	14.1	4	5	0.499
10	43	29.5	19.0	5	0	0.758	22	27	36.6	14.7	4	5	0.248
11	265	23.0	18.0	4	0	0.684	23	401	30.4	17.4	4	5	0.489
12	67	18.3	15.9	4	1	0.535	24	53	39.2	18.6	4	5	0.391

Due week is a week which has been agreed with the customer as a term of delivery. A negative number indicates that the remaining castings are overdue. A penalty coefficient for not making castings on time is calculated on the basis of the castings price and the customer's importance rating. A penalty value for earliness is set arbitrarily to 60% of the penalty coefficient value for tardiness, although this ratio can be set differently by the decision maker.

There are 3 working shifts for the lines of machine type A and C while there are only 2 working shifts for the lines of machine type B. A common practice in the considered foundry is that only two different castings can be produced during one working shift, so κ_1 and κ_3 are set to 2 and $\kappa_2=4$. The total daily capacity of the furnaces is 21 000 kg while a single pouring weighs 1400 kg, i.e. there are 15 heats a day. The number of flasks available for all molding machines during one working shift is limited to 50 big flasks, 100 medium and 120 small ones.

The goal for the two fixed planning horizon problems is to create a set of plans for a week which consists of 5 working days. The task for rolling horizon problem, presented later in this paper, is to build a series of weekly molding plans for 9 weeks, taking into account production quantities planned for previous weeks and the new orders appearing every week.

The second test set for fixed planning horizon (*fixed2*) and the set for the 9 week rolling horizon (*rolling1*) can be found at http://www.zarz.agh.edu.pl/jduda/foundry.

5 Results for a Single Week

Evolutionary algorithms for both test problems *fixed1* and *fixed2* were run for 30 000 generations. The size of the population and the size of the external sets were set to 50 individuals. The number of evaluated generations and the size of the population were chosen experimentally, as the best compromise between the solution quality and the computational time. It took about 10–15 minutes to generate a single set of plans, which is reasonable from a practical point of view. Calculations were repeated 20 times for every combination of the problem and algorithm type.

Figure 2 shows nondominated sets generated using NSGAII, SPEA2 and SPEA2e algorithms for the first test problem (*fixed1*). A nondominated set for a given MOEA was obtained by putting all the solutions from 20 runs together and choosing only nondominated ones. The solutions in a single set, however, did not differ from the solutions in the remaining 19 sets for more than 3% regarding each objective function.

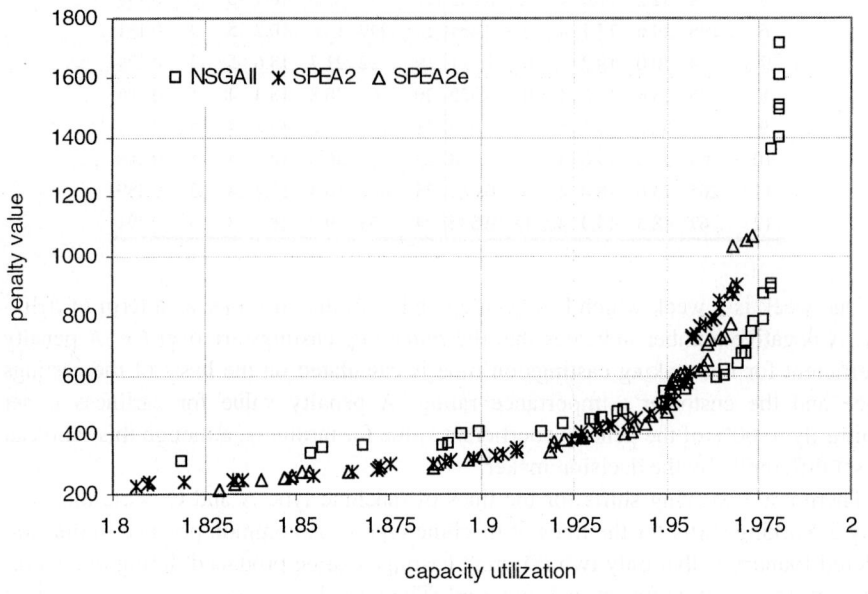

Fig. 2. Pareto Sets Achieved for Problem *fixed1*

It is worth noticing that the solutions generated by the NSGAII algorithm preferred the first objective function to the second one unlike SPEA2 algorithms. In 20 runs the algorithm achieved the highest overall machine and furnace utilization, however accompanied by the biggest penalty value.

Table 4 shows the area percentage of the solution space for which the tested algorithms are unbeaten by the others and the area percentage for which the tested algorithms beat all the others.

The calculations were done using Knowles and Corne algorithm at 0.95 confidence level with 500 lines generated. The area for which the tested algorithms remain unbeaten is the biggest for both versions of SPEA2 algorithms with a little superiority of the latter. However, SPEA2e algorithm beats all its rivals only in 0.5% of the solution space.

Table 4. Knowles Corne Statistic for *fixed1* Problem

	NSGAII	SPEA2	SPEA2e
unbeaten	42.6	81.2	99.8
beats all	0	0.2	0.5

Figure 3 illustrates the Pareto sets generated for *fixed2* problem. The NSGAII algorithm again gave the highest utilization level, thus confirmed its tendency to prefer the first objective function. This time the SPEA2 algorithm seems to prefer the second objective function to the first one. Only the SPEA2e treats both optimization criteria equally in this case. Nevertheless, these observations cannot be generalized to other problems without making additional tests and the use of additional metrics like for example generalization distance and error ratio proposed by Van Veldhuizen and Lamont [9].

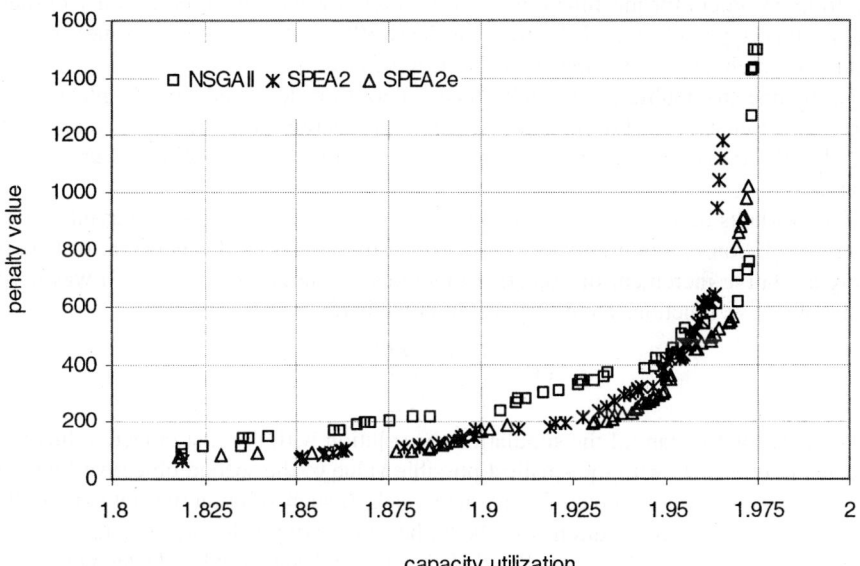

Fig. 3. Pareto sets achieved for problem *fixed2*

Fonseca and Fleming statistics, presented in Table 5, indicates that SPEA2e algorithm in the case of *fixed2* problem is unbeaten by the other algorithms, but there is no solution space where any of the algorithms beats the others.

The general conclusion is that all the tested multicriteria evolutionary algorithms perform well enough to be applicable in real-world production optimization. The choice of the algorithm will be easier if the planner regards one objective function a little more important that the other.

Table 5. Knowles Corne Statistic for *fixed1* Problem

	NSGAII	SPEA2	SPEA2e
unbeaten	54.7	81.2	100
beats all	0	0	0

To build the series of weekly plans for the rolling horizon problem (*rolling1*) the proposed modification of SPEA2 algorithm was chosen as it performed slightly better than the other algorithms.

6 Results for a Rolling Planning Horizon

In order to verify the effects resulting from the application of the proposed method the historical production data were compared against the plans created by the multicriteria evolutionary algorithm. The plans were build week after week within 9 weeks. Starting from 53 orders for machine type A, 129 orders for machine type B and 73 orders for machine types C (total 255 orders, at least 30% of them were overdue) in each following week new orders appeared (in the number of 3 to 21).

Each time the multicriteria evolutionary algorithm delivered weekly plans, only one was automatically chosen as the accepted production plan. After choosing a single plan the remaining orders were consequently altered by the planned production quantities.

Two variants of such automatic choice were considered. In the first variant a compromise solution in a min-max sense as defined by Osyczka [6] was taken. Instead of using a relative increment of objective functions, scalarized increment (9) was used, as the objective functions have very different value ranges.

$$f'_i(x) = \frac{f_i(x) - f_i^{min}}{f_i^{max} - f_i^{min}} \qquad (9)$$

In the second variant of the simulation the solution with the first objective function equaled at least 1.95 with the smallest possible value of the second objective function was taken as the plan accepted for the next week. If such solution did not exist in the obtained Pareto set, the solution with the highest utilization value was taken.

Table 6 shows the utilization level of the furnaces and molding machines calculated on the basis of the historical data and compared to the utilization attained in the simulation variants.

In the first variant the average utilization level of the furnaces equaled 90%, compared to 84%, which was observed in reality. In the second variant this utilization was even higher and equaled on average 93%, but the penalty value was bigger than the one calculated on the basis of historical data. Also the average utilization level of molding machines was 1% worse than in the first variant. In the plans obtained by evolutionary algorithm this utilization equaled 85% while in reality it was only 70%.

Table 6. Comparison Between the Results Obtained in the Simulation and the Historical Data

	historical	MOEA variant 1	MOEA variant 2
average furnaces utilization (weekly)	84%	90%	93%
lowest furnaces utilization	77%	80%	82%
average molding machines utilization	70%	84%	83%
lowest molding machines utilization	64%	73%	78%
average penalty value for tardiness	1880	1610	1790
average penalty value for earliness	230	390	410

It can be seen that in case of the simulation, the overdue production generally decreased. The biggest penalty value for the production which was made earlier than required by the customers as compared with the historical data might be seen as a disadvantage. However, this was not caused by increasing the penalty value for the overdue castings, which simply means that the time from order acceptance to its realization can be shortened.

In Table 7 the mini-max optimal solution obtained for the first week is shown. This solution may be viewed as the best compromise solution considering both criteria as equally important. The value of the first objective function (summarized utilization) is 1.91 (0.96 for furnaces and 0.95 for molding shop) while the value of the second objective function (summarized penalty) is 440 (373 for tardiness and 67 for earliness).

Table 7. The Exemplary Solution Obtained for the First Week (Order | Quantity)

day/shift	machine type A		machine type B		machine type C	
1/1	3|19	36|20	17|98	52|134	13|82	63|10
1/2	6|32	23|15	4|56	24|137	17|37	29|40
1/3	4|20	12|29			3|36	20|42
2/1	11|36	18|3	55|157	88|64	10|98	13|54
2/2	3|14	22|25	15|118	69|104	13|52	38|25
2/3	11|27	29|9			25|39	28|36
3/1	7|15	8|27	17|133	45|96	7|101	9|101
3/2	14|17	17|20	1|140	46|65	8|96	12|61
3/3	3|18	36|20			31|68	
4/1	8|27	16|19	34|93	44|132	5|35	40|40
4/2	2|16	21|14	1|94	16|106	16|31	35|41
4/3	8|27	19|17			15|68	21|5
5/1	5|13	18|24	4|98	53|144	17|78	
5/2	16|22	19|25	15|85	110|151	18|60	36|20
5/3	13|22	16|28			13|72	43|18

7 Conclusions

The results presented here look very promising for the future application not only in the foundry considered, but also in other similar manufactures. The model shown in this paper will be successively complemented with new technological and organizational constraints, especially resulting from the sequence of heats. Unfortunately, reliable data concerning the costs of iron grade changes were hard to collect because they were not present in the current computer system.

The multicriteria evolutionary algorithms together with the proposed repair algorithms prove to be a very effective optimization tool not only for standard test problems but also for real scale production optimization tasks. It is worth underlining that the simulation performed for a nine week rolling horizon can involve in a single run as many as 3125 variables and 345 constraints.

The introduction of additional objective functions also seems to be a very interesting alternative to the traditional approach with one objective function which optimizes usually artificially constructed sum of the production and relevant costs. This paper covered only two important aspects of operational production planning: how to maintain high utilization level of bottleneck machines and how to keep backlogged production as low as possible. The two objectives analyzed in this paper are very similar to the first two criteria proposed by Gravel et al. [4] for scheduling continuous casting of aluminum. This similarity, however not intentional, confirms that the proposed approach can be applied to a wider range of planning and scheduling problems in cast making companies of various kinds.

The main aim of the presented approach was to give the decision maker not a single plan which has to be implemented, but a set of plans from which she or he may choose the one which suits the best the current economical circumstances of the enterprise. The multicriteria evolutionary algorithms enable to obtain a wide spread of the solutions in a single run. This lets the decision maker to perform a quick what-if analysis before making the right planning decision.

Acknowledgments

This study was supported by the State Committee for Scientific Research (KBN) under Grant No. 0224 H02 2004 27 and PB 0808 T07 2003 25.

References

1. Coello Coello, C.A, Van Veldhuizen, D.A., Lamont, G.B., Evolutionary Algorithms for Solving Multi-Objective Problems, Kluwer Academic Publishers, New York (2002)
2. Deb, K., Agrawal, S., Pratab, A., Meyarivan T.: A Fast Elitist Non-Dominated Sorting Genetic Algorithm for Multi-Objective Optimization: NSGA-II. KanGAL report 200001, Indian Institute of Technology, Kanpur (2000)
3. Fonseca, C. M., Fleming, P.J., An Overview of Evolutionary Algorithms in Multiobjective Optimization. Evolutionary Computation, Vol. 3, 1 (1995) 1–16

4. Gravel M., Price, W.L., Gagné, C.: Scheduling continuous casting of aluminium using a multiple-objective ant colony optimization metaheuristic, European Journal of Operational Research, Vol. 143, 1 (2002) 218–229
5. Knowles, J.D., Corne, D.W.: Approximating the Nondominated Front using the Pareto Archived Evolution Strategy. Evolutionary Computation, Vol. 8, 2 (2000) 149–172
6. Osyczka, A.: Multicriterion Optimization in Engineering with Fortran Programs, John Wiley and Sons, New York (1984)
7. Osyczka A.: Evolutionary Algorithms for Single and Multicriteria Design Optimization, Physica Verlag, Heidelberg, New York (2002)
8. dos Santos-Meza, E., dos Santos, M.O., Arenales, M.N.: A Lot-Sizing Problem in An Automated Foundry. European Journal of Operational Research, Vol. 139, 3 (2002) 490–500
9. Van Veldhuizen, D.A., Lamont, G.B., Multiobjective Evolutionary Algorithm Research: A History and Analysis, Technical Report TR-98-03, Department of Electrical and Computer Engineering, Graduate School of Engineering, Air Force Institute of Technology, Wright-Patterson AFB, Ohio (1998)
10. Voorhis, T.V., Peters F., Johnson D.: Developing Software for Generating Pouring Schedules for Steel Foundries. Computers and Industrial Engineering, Vol. 39, 3 (2001) 219–234
11. Zitzler, E., Laumanns, M., Thiele, L.: SPEA2: Improving the Strength Pareto Evolutionary Algorithm. Technical Report 103, Computer Engineering and Networks Laboratory (TIK), Swiss Federal Institute of Technology (ETH), Zurich (2001)
12. Zitzler, E., Thiele, L., Laumanns, M., Fonseca, C.M., Grunert da Fonseca, V.: Performance Assessment of Multiobjective Optimizers: An Analysis and Review. IEEE Transactions on Evolutionary Computation, Vol. 7, 2 (2003) 117–132

Multiobjective Shape Optimization Using Estimation Distribution Algorithms and Correlated Information

Sergio Ivvan Valdez Peña, Salvador Botello Rionda,
and Arturo Hernández Aguirre

Center for Research in Mathematics (CIMAT),
Department of Computer Science,
A.P. 402, Guanajuato, Gto. 36000, México
{ivvan, botello, artha}@cimat.mx

Abstract. We propose a new approach for multiobjective shape optimization based on the estimation of probability distributions. The algorithm improves search space exploration by capturing landscape information into the probability distribution of the population. Correlation among design variables is also used for the computation of probability distributions. The algorithm uses finite element method to evaluate objective functions and constraints. We provide several design problems and we show Pareto front examples. The design goals are: minimum weight and minimum nodal displacement, without holes or unconnected elements in the structure.

1 Introduction

Shape optimization has been widely tackled by evolutionary algorithms. Genetic algorithms, (GAs), have been applied to shape optimization problems with some success, providing feasible solutions with acceptable fitness value [1, 3]. Nonetheless, GA based approaches present difficulties at finding solutions without holes or unconnected segments. This behavior can be explained by population diversity issues, which favor premature convergence and reduced search space exploration [2, 4].

In this paper, we present a multiobjective algorithm for shape optimization (MASO), which is based on estimation distribution concepts. The approach uses binary representation, and makes calls to an external finite element system to evaluate the fitness of candidate structures (individuals). MASO is related to univariate marginal distribution algorithms (UMDA) [6], and to Population Based Incremental Learning (PBIL) [7]. Therefore, every g generations, MASO estimates a (biased) probability distribution by sampling the current Pareto set. The new random population is generated with the updated distribution.

We have improved the algorithm's performance by using specific knowledge derived from the problem domain. This information, combined with the current Pareto set, provides better distribution estimations. Through experiments, we have observed enhanced exploration around promising areas, and less number of small holes and unconnected elements in the structure. Other approaches infer this relationship through the use of Bayesian probabilities [8, 9].

2 Problem Definition

The problem is to find the set of structures which fulfill design constraints (stress), and optimizes: total structure weight and, node displacement in one or more nodes (see Figure 1). Also, a minimum number of "objects or pieces" in the structure is desired. Another desired characteristic for the resulting structure is a minimum number of "small holes".

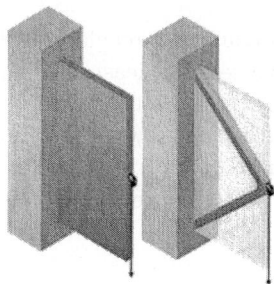

Fig. 1. Problem definition, initial search space and the minimum weight structure

In our problem, the design constraints are given by a maximum permissible Von Misses stress [10] (a standard criterion for mechanical design which represents the material resistance). The algorithm works on a delimited region (the whole piece) as it shown in Figure 1. The structure on the left side is the whole search region, that is, all structure's elements are present. The delivered design, shown on the right hand side, has minimum weight and minimum displacement (therefore, a member of the Pareto front). The design is achieved by "removing elements" from the structure, which is previously represented in discrete form for this purpose [11] (an "element" is one cell of the grid). Thus, binary representation is ad hoc for this problem. A "0" value represents a hole in the structure, while a "1" represents a given thickness value t. The discrete space and a representation example are shown in Figure 2.

3 Objective Functions and Constraints

As noted before, the design problem has two objective functions: the first objective is the minimization of the structure's weight, including the total number of "objects" needed to build the structure, and the number of "small holes". The

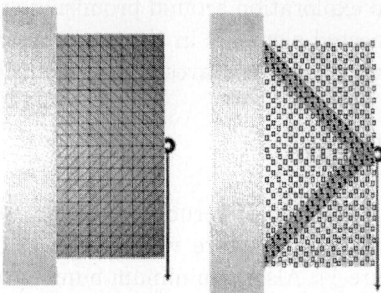

Fig. 2. Left: discrete search space; Right: representation of a structure configuration

second objective function accounts for displacement minimization at some nodes (user defined). An object is a set of at least two elements with one common side; likewise, a "small hole" is a non-present (0 value) element whose surrounded neighbors are present. Figure 3(a) shows a structure with 4 objects, and Figure 3(b) is a configuration example with 3 small holes.

The first objective function is expressed as follows:

Minimize:

$$F(x, O_n, O_b) = [1 + c_1 * (O_n - 1) + c_2 O_b] W(x) \tag{1}$$

Where:

$$W(x) = \sum_{i=0}^{n} w_i x_i \tag{2}$$

In Equation 1, c_1 is the object penalization constant. This constant modifies the function if the number of objects is greater than 1, otherwise it is 0. On the other hand, the constant c_2 penalizes holes whose size is exactly one element.

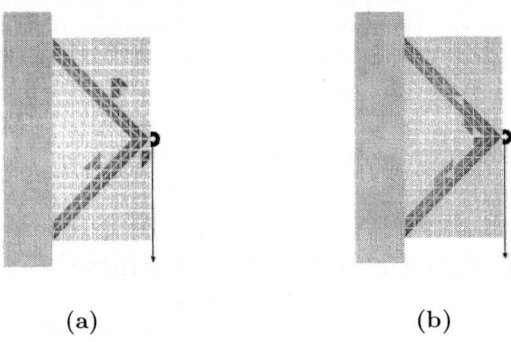

(a)　　　　　　　　　　(b)

Fig. 3. (a) 4-Object configuration, (b) 3-small hole configuration

In all experiments reported in Section 6, we used $c_1 = c_2 = 10$. Variable n is the total number of elements (cells in the grid) in the structure; O_n is defined by the number of objects; O_b is the number of small holes. Equation 2 models the structure weight, where w_i represents each element weight, and x_i is the bit value at the $i - th$ position (present, not-present).

The second objective function, Equation 3, minimizes the displacement at some specific nodes.

Minimize:

$$G(\delta, O_n, O_b) = [1 + c_1(O_n - 1) + c_2 O_b] \sum_{j=0}^{m} |\delta_j| \qquad (3)$$

Where $|\delta_j|$ is the absolute value of the displacement at the $j - th$ node. Finally, m is the number of nodes involved in the second objective function.

The design constraints, as we said before, represent the maximum Von Misses stress of each element (a standard mechanical criterion to evaluate the material resistance). Clearly, all elements must have a Von Misses stress value equal or lower than the maximum permissible for the material. Thus, in Equation 4, the sum of $\rho(\sigma)$ (first factor) represents the number of elements violating the Von Misses stress constraint. The second factor, represents the summation of the Von Misses stress of each element present in the structure.

$$H(x, \rho(\sigma), \gamma(\sigma)) = \left(\sum_{i=0}^{n} \rho_i(\sigma) \right) \left(\sum_{i=0}^{n} x_i \gamma_i(\sigma) \right) \qquad (4)$$

Where:

$$\rho_i(\sigma) = \begin{cases} 0 & if \ (\sigma_M - \sigma_i) \geq 0 \\ 1 & if \ (\sigma_M - \sigma_i) < 0 \end{cases} \qquad (5)$$

And:

$$\gamma_i(\sigma) = \begin{cases} 0 & if \ (\sigma_M - \sigma_i) \geq 0 \\ (\sigma_i - \sigma_M) & if \ (\sigma_M - \sigma_i) < 0 \end{cases} \qquad (6)$$

Where σ_M is the maximum permissible Von Misses stress, and σ_i is the Von Misses stress of each element.

4 Implementation

The probability vectors can be initialized in three different ways: a) with random numbers in the [0,1] interval, b) with all vector probability values equal to 0.5, and c) according to the algorithm in Section 4.1. The last approach was used for the experiments of this paper. The population is generated as described in Section 4.2, which is a common procedure used by EDA's with binary representation [7]. The objective and constraint functions are evaluated with triangular finite element standard routines [11]; the selection mechanism is Pareto dominance

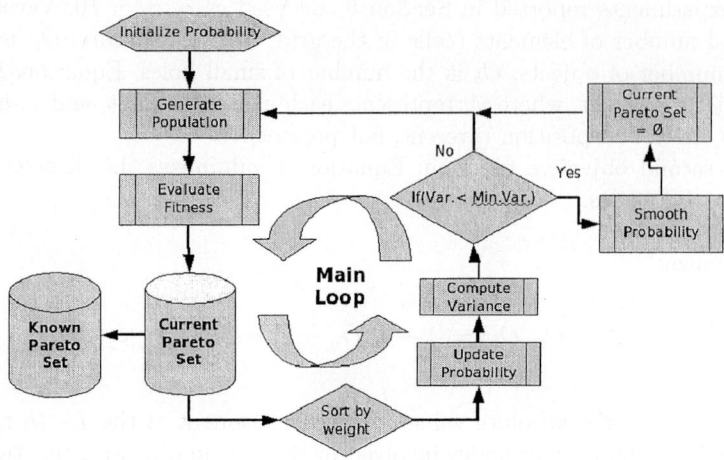

Fig. 4. Main loop of MASO depicting principal routines

taking the constraint as one more objective function. The main components of MASO are shown in Figure 4. In order to update the probability vectors, the current Pareto set is sorted by weight (see Section 4.3).

When the probability vectors have lost their exploration capacity (the variance is smaller than a threshold value, see Section 4.5), a membrane filter is applied to them in order to smooth the probability. The pseudo code of MASO algorithm is shown in Figure 5.

4.1 Probability Initialization

In Figure 6 we show an initialization procedure for probability vectors. Basically, we compute the stress values for "all-1" individuals, then we set to 0 the bits which represents the elements with low Von Misses stress value. A preset threshold is used in this decision. P and O are structure configurations; O in particular is the configuration with all bit values equal to 1, $VonMissesStress_i$ is the stress in the $i-th$ element, $InfProbLim$ and $SupProbLim$ are probability limits, and the function $Solve\_VonMisses(Configuration)$, solves the FE problem for a particular configuration.

4.2 Generation of the Population

The population is generated by a set of k distribution probability vectors called V_l. Every vector V_l generates q structure configurations called I_j. Thus, every bit from I_j is generated by a Bernoulli experiment with a success probability p_i.

```
Enter the next parameters:
μ: Smoothing Parameter. The weight of the gradient penalization in the membrane
filter (see [5] and Equation 9)
MinVar: Minimum Variance criterion to apply the membrane filter.
LSP,LIP: Superior and Inferior Probability Limit.
NV: Number of Probability Vectors.
NIM: Number of Individuals Generated by Every Probability Vector.
λ: parameter for the probability updating (learning rate,see subsection 4.3 ).
NG: Number of generations.

Probability_Initialization ();
k < − 0
i < − 0
while (i < NG ) {
  Generate_Population ();
  Evaluate_Objective_and_Constraints_Functions ();
  Select_Non-Dominated_Individuals ();
  Sort_Individualsbyweight();
  Update_Probability_Distributions ();
  Compute_Variance ();
  If (ComputedVariance < MinVar) {
    Smooth_Probability_Distributions ();
    Update_External_File ();
    k++; }
  i++; } end
```

Fig. 5. Pseudo-code of Multiobjective Algorithm for Shape Optimization

4.3 Updating Probability Distributions

The non-dominated individuals are used to compute and update the probability distributions. First, we find the non-dominated structures by treating the stress constraint as an additional objective. Thus, dominance is computed with three functions (two objectives plus the constraint). In this non-dominated temporal set there are feasible and infeasible individuals; the infeasible ones are sorted by total amount of constraint violation. All feasible individuals plus a small percentage of the unfeasible ones (those with a small constraint value) are used to update the probability distributions. For our experiments we used 10% of the infeasible structures. Before updating the distributions, all structures are sorted by one of the objectives; in our case we use the weight value to sort the structures.

Thus, being C_f the number of feasible non-dominated structures, and C_u the number of non-feasible non-dominated structures, ζ represents the percentage of non-feasible structures that are taken to update the probability vectors.

$$C_t = C_f + \zeta C_u \tag{7}$$

```
for i=1..NoElements O[i]=1 endfor
Solve_VonMisses(O);
Interval = (maxVonMisses-minVonMisses)/(NoElements)
threshold = maxVonMisses - Interval
Step 1:
For i=1..NoElements
   if (VonMissesStress_i < threshold)   P[i] = 0
   Else   P[i] = 1
endfor
if ( IsNotFeasible(P) ) {# True when is infeasible
   threshold = threshold - Interval
   go to: Step 1 }
Interval = (threshold - minVonMisses )/( NoVectors-1 );
For i= 0..NoVectors {
   For k=0..NoElements {
      if ( VonMissesStress_i < threshold)
         ProbabilityVector_{k,i} = InfProbLim;
      else
         ProbabilityVector_{k,i} = SupProbLim; }
   threshold = threshold - Interval; }
```

Fig. 6. Pseudo-code for probability initialization

For k probability vectors, the new probability vector will be:

$For\ j = 1..Number\ of\ Bits$

$$V_{l,j}^{t+1} = \lambda V_{l,j}^t + (1-\lambda) \sum_{i=(l-1)*Round(C_t/k)}^{(l)*Round(C_t/k)} I_{i,j}^b / Round(C_t/k) \qquad (8)$$

Where $V_l^{(t+1)}$ is the new probability vector for generation $t+1$, V_l^t is the probability vector at generation t, λ is a memory factor(learning rate, see [7]) that preserves the knowledge of the last distribution, I_i^b are the binary arrays (non-dominated individuals) that will be used to update the vector V_l. The $Round()$ function returns the nearest integer from the real result of C_t/k. As we can see, $\sum_{i=(l-1)*Round(C_t/k)}^{(l)*Round(C_t/k)} I_i^b / Round(C_t/k)$ is nothing else than the mean of a set of binary arrays. Note that the selection of non-dominated individuals is performed over the current population and the current Pareto set.

4.4   Smoothing the Probability Distributions

Due to the implicit reinforcement of the non-dominated individuals, Equation 8 takes the probability vectors to values close to "0", or to "1" . When this happens, the probability vectors have lost their exploration capacity. For MASO, the probability distribution vectors are smoothed, so predominant peaks are removed from the probability surface. Restarting the population with new probability distributions enhances exploration and local minima avoidance.

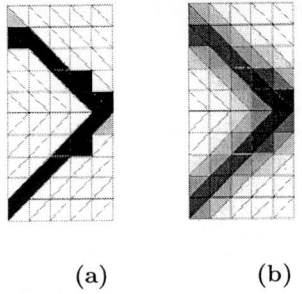

Fig. 7. Relationship between a structure and its probability distribution: (a) before applying the membrane filter, (b) after applying the membrane filter

Probability distributions are smoothed by a membrane filter [5]. Equation 9 finds a smooth function f which preserves the shape of the original distribution g, but will avoid the abrupt changes. The abrupt changes are penalized by the gradient ∇f.

$$U(f) = \int_\omega ([f(x,y) - g(x,y)]^2 + \mu ||\nabla f(x,y)||^2) d\omega \qquad (9)$$

Where $f(x, y)$ is the function which minimize the functional $U(f)$, $g(x, y)$ is the original probability distribution function, μ is a parameter which regulates how smooth f will be. In our experiments we used a μ value in the interval $0.35 - 0.5$. In this case, we approximate the gradient with the difference between the central element and each neighbor (we have three neighbors for a triangular mesh).

Figure 7 (a) shows a probability distribution (one vector V_l^t) before applying the membrane filter. Probabilities values are shown in gray scale color, where values close to 1 are shown in black, and probabilities close to 0 in white. In that figure we can observe that the probability distribution can generate a good approximated solution but, if we do not modify the probabilities they are not able of keeping search space exploration. In Figure 7 (b) we can observe how the membrane filter has spread the probability distribution over the neighbors.

We identify the poor search capacity of a probability vector by a variance measure. This measure is computed as explained in Section 4.5.

Once the smoothing process is finished, we must update the set of non-dominated individuals stored in the external file, (named *Known Pareto Set* in Figure 4). The new population is generated anew using the smoothed probability distributions; eventually the population will converge, the distributions smoothed again, and another *Current Pareto Set* will be generated and used to update the external file.

4.5 Variance Computation

We measure the variance of those locations whose probability vector are in the interval $[0.0002, 0.9998]$. The membrane filter is applied to every probability

```
ComputedVariance=0
For l=1.. NoVectors {
    Variance=0
    For=1..NoElementos {
        if (ProbabilityVector_{l,i} ≤ InfProbLim && ProbabilityVector_{l,i} ≥ SupProbLim)
            Variance=Variance+1; }
    if (Variance > ComputedVariance)
    ComputedVariance =Variance }
```

Fig. 8. Pseudo-code of variance computation algorithm

distribution if the computed variance is smaller than a threshold. Pseudo code of variance computation algorithm is shown in Figure 8.

5 Metrics

A MOEA convergence metric proposed by Deb and Jain is computed to measure the convergence and behavior of the algorithm [12].

5.1 Convergence Metric

A reference set P^* is determined from the union of 30 Pareto sets (therefore, Known Pareto set of each run). This metric takes a normalized value within $[0,1]$; near 0 means better [12]. The convergence metric is computed as follows:

1. Identify the no dominated set $F^{(t)}$ of $P^{(t)}$ (a population)
2. Form each point i in $F^{(t)}$, calculate the smallest normalized Euclidean distance to P^* as in Equation 10, f_k^{max} and f_k^{max} are the maximum and the minimum function values of k-th objective function in P^*

$$d_i \ min_{j=1}^{|P^*|} \sqrt{sum_{k=1}^{M} \left(\frac{f_k(i) - f_k(j)}{f_k^{max} - f_k^{min}} \right)} \qquad (10)$$

3. Calculate the convergence metric by averaging the normalized distance for all points in $F^{(t)}$:

$$C(P^{(t)}) = \frac{sum_{i=1}^{|F^{(t)}|} d_i}{|F^{(t)}|} \qquad (11)$$

Deb and Jain proposed to normalize the convergence metric by the maximum value (usually $C(P^{(0)})$) : $\hat{C}(P^{(t)}) = C(P^{(t)})/C(P^{(0)})$). To compute the convergence metric we calculate the normalized distance for the vectors in F^*, and if $|P^*| > |F^*|$ (we have more points in P^* than F^*) we set $|P^*| - |F^*|$ distances equal to 1.

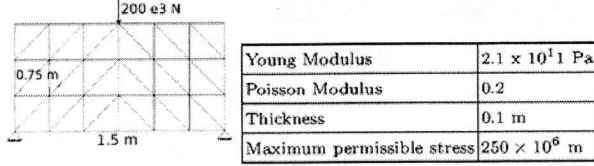

Fig. 9. Simply supported beam problem with a punctual load

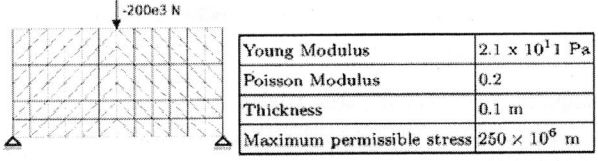

Fig. 10. Simply supported beam problem, discretized in 144 elements

6 Experiments

Two study cases are used to test our approach. In the first case we know the true Pareto set, whilst in the other it is unknown.

6.1 Experiment 1 (Description)

A simple structure with a punctual load is supported at the lower corners. The dimensions and load magnitude are shown in Figure 9. We chose this problem because the true Pareto set is easy to compute, therefore, comparisons are possible. Results are contrasted in Section 7.1.

6.2 Experiment 2 (Description)

Experiment 2 has the same conditions as experiment 1, as shown in Figure 10. But in this case the search space is discrete and consists of 144 elements. Since the true Pareto set is unknown, comparisons are made against a reference Pareto set obtained from 30 runs of our algorithm.

7 Results

7.1 Experiment 1. True Front Comparison

We compare results of 30 runs of our algorithm against the true Pareto front. There are 13 vectors in the true Pareto set. The average number of individuals found in the known Pareto set after 30 runs is 20.667, with an standard deviation of 1.3218 The average number of individuals from known Pareto set that belong to the true Pareto set is 12.6667 (out of 13), with a standard deviation of 0.479463. The average number of dominated individuals by the true Pareto

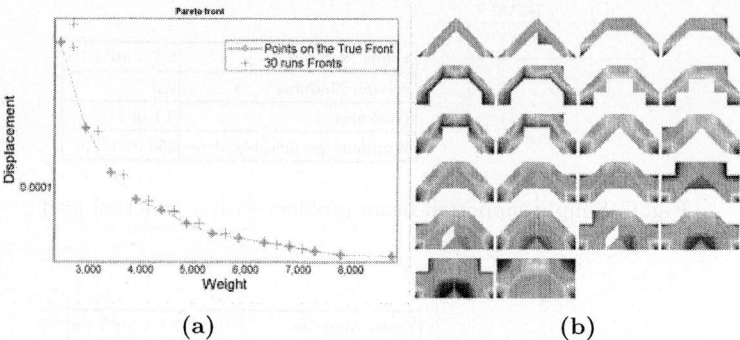

Fig. 11. (a) Diamonds represent the true Pareto front; crosses are vectors found in 30 runs (30 fronts are plotted but they overlap with each other). (b) Structures found by a typical run of MASO

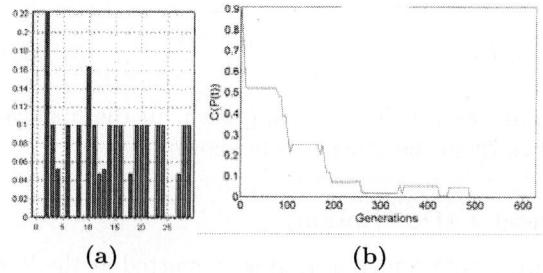

Fig. 12. (a) $C(P^{(t)})$ in the last generation, (b) Convergence graph of a typical run of Experiment 1

set is 0.066667, with a standard deviation of 0.253708. In Figure 11, we can see the true Pareto front and the non-dominated structures found in 30 runs (the dominance is check after either independent run, so in the graph we can see some dominated individuals but they are not from the same run). Note that all 30 runs found all the structures in the true Pareto set.

7.2 Experiment 1. Convergence Metric

We compute the convergence metric for the problem described in Section 6.1. Figure 12(a) shows the $C(P^{(t)})$ value for the last generation in 30 independent runs, the mean of 30 runs is 0.06783, with a standard deviation of 0.05586 (remember that smaller is better). On the right hand side, a convergence plot of a typical run is shown.

7.3 Experiment 2. Reference Front

Figure 13 shows the reference front and all the non-dominated individuals found in 30 runs. The crosses represent all the non-dominated individuals found in

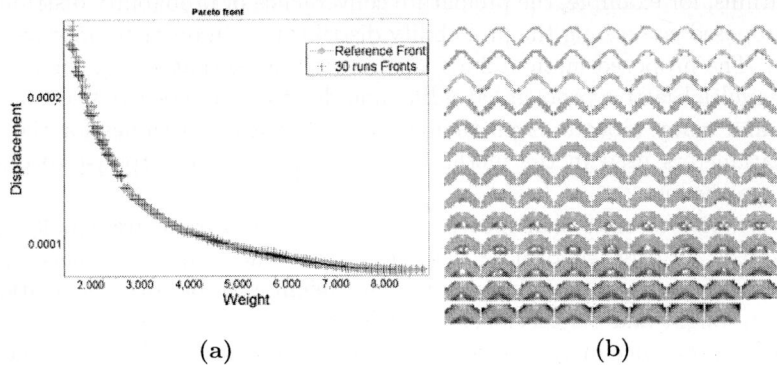

Fig. 13. (a) Diamonds are the reference front and the crosses represent the fronts from 30 runs, (b) Structures found by a typical run of MASO

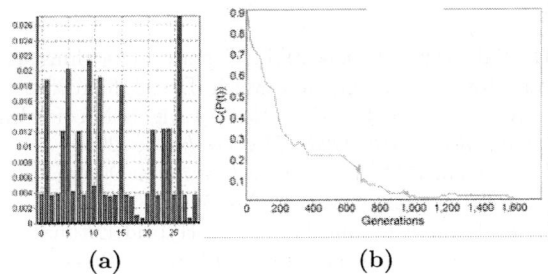

Fig. 14. (a) Graph of the convergence metric in 30 runs, (b) Convergence behavior

30 runs. Although some of them are dominated by the reference Pareto front (shown with diamonds), they were not dominated in their independent run.

7.4 Experiment 2. Convergence Metric

Figure 14(a) shows the $C(P^{(t)})$ value for the last generation. As we can observe, the value is very close to 0 implying good convergence behavior. Figure 14(b) shows the convergence behavior to the front for a typical run. A decreasing distance is clear in the graph; at some points there are some peaks which mean that the algorithm found some individuals that dominate more than one non-dominated individual in the previous generation. The average convergence is 0.008312, with a standard deviation of 0.007347, for 30 independent runs.

8 Conclusions

We proposed a new multiobjective optimization algorithm inspired by PBIL [7] and UMDA [6] algorithms. Several issues must be considered when using EDA algorithms, for example, the premature convergence of probability distributions. We proposed smoothing the probability distributions in order to improve exploration. The proposed method finds a set of structures that are optimal solution of the multiobjective problem, avoiding non-desired characteristics such as small holes and many objects. Even though we have suggested values for the different parameters used in the method, the algorithm is very robust to different parameter values.

It is worth to note that individuals on the Pareto front are equally spaced and spread all over. Also, note that the points that seem more spaced on the displacement axis, are equally spaced on the weight axis (because the relationship between weight and displacement is not linear).

The convergence metric makes evident the robustness of MASO. Even more, most vectors, 97.5% from the true Pareto set, were found by each run of a total of 30.

References

1. Chapman C.D., Saitou K., Jakiela M.J.. Genetic algorithms as an approach to configuration and topology design. Jou. Mech. Des. 116: 1005-11, (1994)
2. Deb K and Goell T. Multiobjetive Evolutionary Algorithms for Engineering Shape Optimization *KanGal report 200003 . Kanpur India*, (2000)
3. Kane, C. and Schoenauer M. Topological Optimum Design using Genetic Algorithms *Control and Cybernetics, Vol. 25 No. 5*, (1996)
4. Li, H., Zhang, Q., Tsang,E.P., and Ford, J.A. Hybrid Estimation of Distribution Algorithm for Multiobjective Knapsack Problem ,*In Proceedings of the 4th European Conference on Evolutionary Computation in Combinatorial Optimization, Coimbra, Portugal*, (2004)
5. Marroqun, J.L., Velasco,F.A., Rivera M. and Nakamura M. Gauss-Markov Measure Field Models for Low-Level Vision, *IEEE Trans. On PAMI, 23, 4: 337-348*, (2001)
6. Mühlenbein ,H. and PaaB, G. From reconbination of Genes to the estimation of distributions I. Binary parameters. *Parallel problem Solving form Nature (PPSN IV), 178-187*, (1996)
7. Shumeet Baluja. Population Based Incremental Learning: A Method for Integrating Genetic Search Based Function Optimization and Competitive Learning *School of Computer Science Carnegie Mellon University, Pittsburgh, Pennsylvania 1523. CMU-CS-94-163* , (1996)
8. Pelikan M. Goldberg D.E and Cantu Paz. Linkage problem, distribution estimation and bayesian networks *IlliGal Report No. 98013 Urbana Il University*, (1998)
9. Pelikan M. Goldberg D.E and Cantu Paz. BOA: The Bayesian Optimization Algorithm *Proceedings of the Genetic and Evolutionary Computation Conference (GECCO-99)*, (1999)
10. L.E Malvern, *Introduction to the Mechanics of a Continuous Medium*, Pretence Hall Inc. Englewood Cliffs New jersey, (1969).
11. O.C. Zienkiewicz y R.L. Taylor: *El Método de los Elementos Finitos*; Cuarta Edición, Volumen 2, Mc. Graw Hill-CIMNE. (1995).
12. Kalyanmoy Deb and Sachin Jain. Running Perfomance Metrics for Evolutionary Multi-Objective Optimization *KanGal report 2002004 . Kanpur India*, (2000)

Evolutionary Multi-objective Environmental/Economic Dispatch: Stochastic Versus Deterministic Approaches

Robert T.F. Ah King[1], Harry C.S. Rughooputh[1], and Kalyanmoy Deb[2]

[1] Department of Electrical and Electronic Engineering, Faculty of Engineering,
University of Mauritius, Reduit, Mauritius
{r.ahking, r.rughooputh}@uom.ac.mu

[2] Kanpur Genetic Algorithms Laboratory, Department of Mechanical Engineering,
Indian Institute of Technology, Kanpur, PIN 208 016, India
deb@iitk.ac.in

Abstract. Due to the environmental concerns that arise from the emissions produced by fossil-fueled electric power plants, the classical economic dispatch, which operates electric power systems so as to minimize only the total fuel cost, can no longer be considered alone. Thus, by environmental dispatch, emissions can be reduced by dispatch of power generation to minimize emissions. The environmental/economic dispatch problem has been most commonly solved using a deterministic approach. However, power generated, system loads, fuel cost and emission coefficients are subjected to inaccuracies and uncertainties in real-world situations. In this paper, the problem is tackled using both deterministic and stochastic approaches of different complexities. The Nondominated Sorting Genetic Algorithm – II (NSGA-II), an elitist multi-objective evolutionary algorithm capable of finding multiple Pareto-optimal solutions with good diversity in one single run is used for solving the environmental/economic dispatch problem. Simulation results are presented for the standard IEEE 30-bus system.

1 Introduction

The classical economic dispatch problem is to operate electric power systems so as to minimize the total fuel cost. This single objective can no longer be considered alone due to the environmental concerns that arise from the emissions produced by fossil-fueled electric power plants. In fact, the Clean Air Act Amendments have been applied to reduce SO_2 and NO_x emissions from such power plants. Accordingly, emissions can be reduced by dispatch of power generation to minimize emissions instead of or as a supplement to the usual cost objective of economic dispatch. Environmental/economic dispatch is a multi-objective problem with conflicting objectives because pollution is conflicting with minimum cost of generation. Various techniques have been proposed to solve this multi-objective problem whereby most researchers have concentrated on the deterministic problem.

Economic dispatch calculates the cost of generation based on data relating fuel cost and power output. This cost function is approximated by a quadratic equation with cost coefficients. In conventional economic dispatch the coefficients are assumed to be deterministic, but in real-world situations, these data are subjected to inaccuracies and uncertainties. These deviations are attributed to (i) inaccuracies in the process of measuring and forecasting of input data and (ii) changes of unit performance during the period between measuring and operation [1]. Thus, the operating point in practice will differ from the planned operating point and will thus affect the actual fuel cost. Similarly, emission coefficients may also be subjected to some deviations resulting in definite differences in practical systems.

There has been much research using the deterministic approach to solve the environmental/economic dispatch problem. Gent and Lamont [2] introduced the minimum-emission dispatch concept where they developed a program for on-line steam unit dispatch that results in the minimizing of NO_x emission. These authors introduced the mathematical representation of NO_x emission of steam generating units and used a Newton-Raphson convergence technique to obtain base points and participation factors. Zahavi and Eisenberg [3] proposed a dispatch procedure for power that meets the demand for energy while accounting for both cost and emission considerations. A tradeoff curve which present the decision maker with all possible courses of action (dispatch policies) for a given demand was introduced. Nanda et al. [4] presented an improved Box complex method for economic dispatch and minimum emission dispatch problems. Dhillon et al. [5] formulated the multiobjective thermal power dispatch using noncommensurable objectives such as operating costs and minimal emission. The epsilon-constraint method was used to generate non-inferior solutions to the multiobjective problem considering the operating cost as the objective and replacing emission objective as a constraint. More recently, multi-objective evolutionary algorithms have been applied to the problem at hand. Abido has pioneered this research by applying NSGA [6], NPGA [7] and SPEA [8] to the standard IEEE 30-bus system. In fact, it has been shown that NSGA-II can obtain minimum cost and minimum emission solutions comparable to Tabu search [9].

Not long after the introduction of the environmental consideration in the economic dispatch problem, researchers started considering stochastic approaches bearing in mind the uncertainties that are inherent in real-world situations. Viviani and Heydt [10] incorporated the effects of uncertain system parameters into optimal power dispatch. Their method employed the multivariate Gram-Charlier series as means of modeling the probability density function (p.d.f.) which characterizes the uncertain parameters. Parti et al. [1] extended the Lagrange multiplier solution method to solve the economic thermal power dispatch problem using an objective function consisting of the sum of the expected production costs and expected cost of deviations (a penalty term proportional to the expectation of the square of the unsatisfied load because of possible variance of the generator active power). Bunn and Paschentis [11] developed a stochastic model for the economic dispatch of the electric power. These authors used a form of stochastic linear programming method for online scheduling of power generation at 5 minute intervals taking into account the mismatch between dispatched generation and actual load demanded. Experimental results on real data demonstrated the efficiency of the approach compared to conventional deterministic linear programming model. Dhillon et al. [12] have used the weighted minimax technique to obtain trade-off relation between the conflicting objectives and fuzzy set theory is subsequently used to help the operator choose an optimal operating point. In another attempt, Dhillon et al. [13] solved the multiobjective stochastic economic dispatch problem whereby the weighted sum technique and Newton-Raphson algorithm are used to generate the non-inferior solutions considering expected operating cost and expected risk associated with the possible deviation of the random variables from their expected values. In their study, the random variables are assumed to be normally distributed and statistically dependent on each other, hence the deterministic objective functions have both variance and covariance terms. Recently, Bath et al. [14] presented an interactive fuzzy satisfying method for multi-objective generation scheduling with explicit recognition of statistical uncertainties in system production cost data. However, the multi-objective problem is converted into a scalar optimization problem and solved using weighted sum method. Hooke-Jeeves pattern search, evolutionary optimization and weight simulation methods are used to find the optimal weight combinations and fuzzy sets are used to obtain the 'best' solution from the non-inferior solutions set.

In this paper, both the deterministic and stochastic approaches are addressed. More precisely, the stochastic problem is considered in a unique way due to the nature of the problem when the load flow calculations determine the power generated by the slack bus. Thus, a reliability measure is used to test the power system under different stochastic considerations. The paper is organized as follows. The environmental/economic dispatch problem is defined in Section 2. Section 3 outlines the system parameters considered in this study. The simulation results of the deterministic approach are given in Section 4 while those of the stochastic approach are presented in Section 5. Based on these results, the main findings and some conclusions are outlined in Section 6.

2 Environmental/Economic Dispatch

The environmental/economic dispatch involves the simultaneous optimization of fuel cost and emission objectives which are conflicting ones. The deterministic problem is formulated as described below.

2.1 Objective Functions

Fuel Cost Objective. The classical economic dispatch problem of finding the optimal combination of power generation, which minimizes the total fuel cost while satisfying the total required demand can be mathematically stated as follows [15]:

$$C = \sum_{i=1}^{n} \left(a_i + b_i P_{Gi} + c_i P_{Gi}^2 \right) \text{ \$/hr} \qquad (1)$$

where

C: total fuel cost (\$/hr),
a_i, b_i, c_i: fuel cost coefficients of generator i,
P_{Gi}: power generated (p.u.) by generator i, and
n: number of generators.

NO$_x$ Emission Objective. The minimum emission dispatch optimizes the above classical economic dispatch including NO$_x$ emission objective, which can be modeled using second order polynomial functions [15]:

$$E_{NO_x} = \sum_{i=1}^{n} \left(a_{iN} + b_{iN} P_{Gi} + c_{iN} P_{Gi}^2 + d_{iN} \sin(e_{iN} P_{Gi}) \right) \text{ ton/hr} \qquad (2)$$

2.2 Constraints

The optimization problem is bounded by the following constraints:

Power Balance Constraint. The total power generated must supply the total load demand and the transmission losses.

$$\sum_{i=1}^{n} P_{Gi} - P_D - P_L = 0 \tag{3}$$

where
P_D: total load demand (p.u.), and
P_L: transmission losses (p.u.).

The transmission losses is given by

$$P_L = \sum_{i=1}^{N} \sum_{j=1}^{N} \left[\begin{array}{l} (r_{ij}/V_iV_j)\cos(\delta_i - \delta_j)(P_iP_j + Q_iQ_j) + \\ (r_{ij}/V_iV_j)\sin(\delta_i - \delta_j)(Q_iP_j - P_iQ_j) \end{array} \right] \tag{4}$$

where
N : number of buses
r_{ij} : series resistance connecting buses i and j
V_i : voltage magnitude at bus i
δ_i : voltage angle at bus i
P_i : real power injection at bus i
Q_i : reactive power injection at bus i

Maximum and Minimum Limits of Power Generation. The power generated P_{Gi} by each generator is constrained between its minimum and maximum limits, i.e.,

$$P_{Gimin} \leq P_{Gi} \leq P_{Gimax} \tag{5}$$

where
P_{Gimin}: minimum power generated, and
P_{Gimax}: maximum power generated.

2.3 Multiobjective Formulation

The multiobjective deterministic environmental/economic dispatch optimization problem is therefore formulated as:

$$\text{Minimize } [C, E_{NO_x}] \tag{6}$$

$$\text{subject to: } \sum_{i=1}^{n} P_{Gi} - P_D - P_L = 0 \quad \text{(power balance), and}$$

$$P_{Gimin} \leq P_{Gi} \leq P_{Gimax} \quad \text{(generation limits)}$$

3 System Parameters

Simulations were performed on the standard IEEE 30-bus 6-generator test system (Fig. 1) using the Elitist Nondominated Sorting Genetic Algorithm (NSGA-II) for both deterministic and stochastic approaches. Details of the algorithm of NSGA-II can be found in [16].

The power system is interconnected by 41 transmission lines and the total system demand for the 21 load buses is 2.834 p.u. Fuel cost and NO_x emission coefficients for this system are given in Tables 1 and 2 respectively.

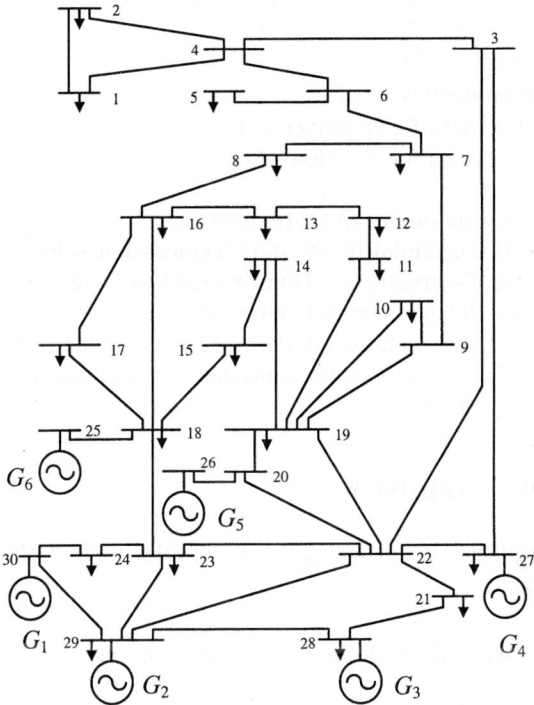

Fig. 1. Single-line diagram of IEEE 30-bus test system [8]

Table 1. Fuel Cost coefficients

Unit i	a_i	b_i	c_i	P_{Gimin}	P_{Gimax}
1	10	200	100	0.05	0.50
2	10	150	120	0.05	0.60
3	20	180	40	0.05	1.00
4	10	100	60	0.05	1.20
5	20	180	40	0.05	1.00
6	10	150	100	0.05	0.60

Table 2. NO_x Emission coefficients

Unit i	a_{iN}	b_{iN}	c_{iN}	d_{iN}	e_{iN}
1	4.091e-2	-5.554e-2	6.490e-2	2.0e-4	2.857
2	2.543e-2	-6.047e-2	5.638e-2	5.0e-4	3.333
3	4.258e-2	-5.094e-2	4.586e-2	1.0e-6	8.000
4	5.326e-2	-3.550e-2	3.380e-2	2.0e-3	2.000
5	4.258e-2	-5.094e-2	4.586e-2	1.0e-6	8.000
6	6.131e-2	-5.555e-2	5.151e-2	1.0e-5	6.667

In all simulations, the following parameters were used:
- population size = 50
- crossover probability = 0.9
- mutation probability = 0.2
- distribution index for crossover = 10
- distribution index for mutation = 20

The simulations were run for five different cases:
    Case D1: Deterministic - System is considered as lossless
    Case D2: Deterministic - Transmission losses are considered
    Case S1: Stochastic power generated
    Case S2: Stochastic power generated and system loads
    Case S3: Stochastic power generated, system loads, fuel cost and emission coefficients

4 Deterministic Approach

Using the deterministic parameters as given in Tables 1 and 2, the simulation results obtained are presented.

4.1 Case D1: Deterministic Without Transmission Losses

Fig. 2 shows a good diversity in the nondominated solutions obtained by NSGA-II after 200 generations.

Table 3 and 4 show the best fuel cost and best NO_x emission obtained by NSGA-II as compared to Linear Programming (LP) [15], Multi-Objective Stochastic Search Technique (MOSST) [17], Nondominated Sorting Genetic Algorithm (NSGA) [6], Niched Pareto Genetic Algorithm (NPGA) [7] and Strength Pareto Evolutionary Algorithm (SPEA) [8]. It can be deduced that NSGA-II finds comparable minimum fuel cost and comparable minimum NO_x emission to the last three evolutionary algorithms. To confirm that NSGA-II is able to obtain the Pareto front for the problem, the epsilon-constraint method [18] has been used as shown on the plot of Fig. 2. Genetic algorithm was used to solve the resulting single-objective problem.

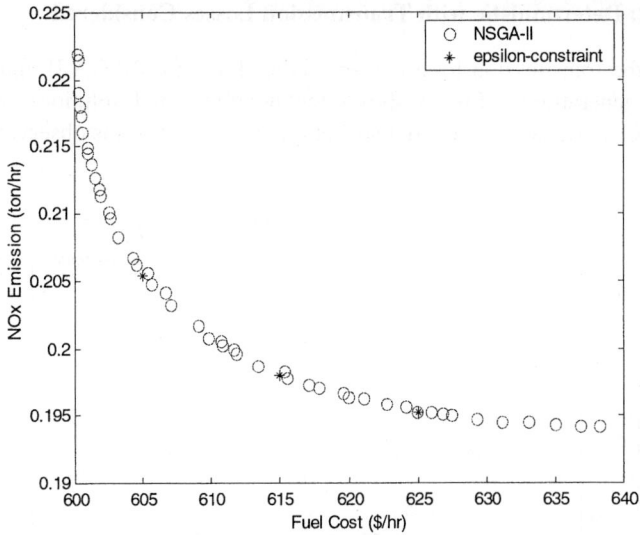

Fig. 2. Nondominated solutions for Case D1

Table 3. Best fuel cost

	LP [15]	MOSST [17]	NSGA [6]	NPGA [7]	SPEA [8]	NSGA-II
P_{G1}	0.1500	0.1125	0.1567	0.1080	0.1062	0.1059
P_{G2}	0.3000	0.3020	0.2870	0.3284	0.2897	0.3177
P_{G3}	0.5500	0.5311	0.4671	0.5386	0.5289	0.5216
P_{G4}	1.0500	1.0208	1.0467	1.0067	1.0025	1.0146
P_{G5}	0.4600	0.5311	0.5037	0.4949	0.5402	0.5159
P_{G6}	0.3500	0.3625	0.3729	0.3574	0.3664	0.3583
Best cost	**606.314**	**605.889**	**600.572**	**600.259**	**600.15**	**600.155**
Corresp. emission	0.22330	0.22220	0.22282	0.22116	0.2215	0.22188

Table 4. Best NO_x emission

	LP [15]	MOSST [17]	NSGA [6]	NPGA [7]	SPEA [8]	NSGA-II
P_{G1}	0.4000	0.4095	0.4394	0.4002	0.4116	0.4074
P_{G2}	0.4500	0.4626	0.4511	0.4474	0.4532	0.4577
P_{G3}	0.5500	0.5426	0.5105	0.5166	0.5329	0.5389
P_{G4}	0.4000	0.3884	0.3871	0.3688	0.3832	0.3837
P_{G5}	0.5500	0.5427	0.5553	0.5751	0.5383	0.5352
P_{G6}	0.5000	0.5142	0.4905	0.5259	0.5148	0.5110
Best emission	**0.19424**	**0.19418**	**0.19436**	**0.19433**	**0.1942**	**0.19420**
Corresp. cost	639.600	644.112	639.231	639.182	638.51	638.269

4.2 Case D2: Deterministic with Transmission Losses Considered

In this case, the transmission losses are considered and the NSGA-II algorithm was run for 200 generations. Fig. 3 shows the nondominated solutions obtained by NSGA-II for Case D2 where a good distribution of the solutions is observed.

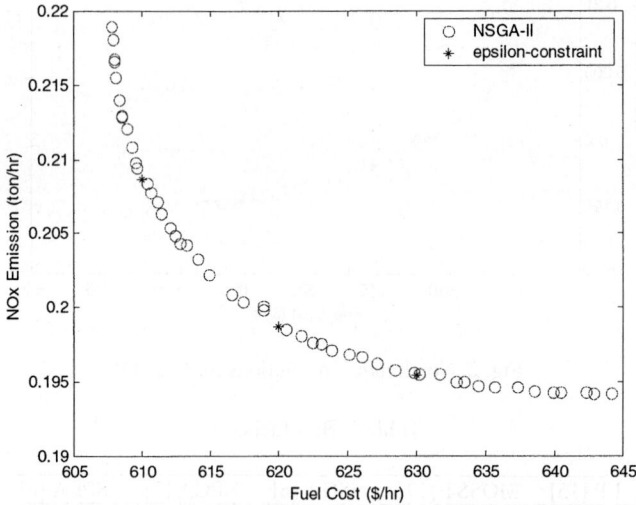

Fig. 3. Nondominated solutions for Case D2

The best fuel cost and best NO_x emission obtained by NSGA-II as compared to NSGA, NPGA and SPEA are given in Table 5 and 6. It is observed that NSGA-II again finds better minimum fuel cost and emission level than the other evolutionary algorithms.

Table 5. Best fuel cost

	NSGA [6]	NPGA [7]	SPEA [8]	NSGA-II
P_{G1}	0.1168	0.1245	0.1086	0.1182
P_{G2}	0.3165	0.2792	0.3056	0.3148
P_{G3}	0.5441	0.6284	0.5818	0.5910
P_{G4}	0.9447	1.0264	0.9846	0.9710
P_{G5}	0.5498	0.4693	0.5288	0.5172
P_{G6}	0.3964	0.3993	0.3584	0.3548
Best cost	**608.245**	**608.147**	**607.807**	**607.801**
Corresp. emission	0.21664	0.22364	0.22015	0.21891

Table 6. Best NO$_X$ emission

	NSGA [6]	NPGA [7]	SPEA [8]	NSGA-II
P_{G1}	0.4113	0.3923	0.4043	0.4141
P_{G2}	0.4591	0.4700	0.4525	0.4602
P_{G3}	0.5117	0.5565	0.5525	0.5429
P_{G4}	0.3724	0.3695	0.4079	0.4011
P_{G5}	0.5810	0.5599	0.5468	0.5422
P_{G6}	0.5304	0.5163	0.5005	0.5045
Best emission	**0.19432**	**0.19424**	**0.19422**	**0.19419**
Corresp. cost	647.251	645.984	642.603	644.133

Again, it can be deduced that the algorithm is capable of obtaining the Pareto front for the given problem as verified by the minimum of each objective and points obtained by the epsilon-constraint method in Fig. 3.

It has been shown that NSGA-II can obtain the Pareto front of the problem and it is therefore ideal for solving the multiobjective environmental/economic dispatch optimization problem which has conflicting objectives from the fact that the multiobjective approach yields multiple Pareto-optimal solutions in a single simulation run whereas multiple runs are required for the single objective approach with weighted objectives.

5 Stochastic Approach

Previous stochastic approaches involved the inclusion of deviational (recourse) costs to account for mismatch between scheduled output and actual demand in the formulation of the objective function [11], and conversion of stochastic models into their deterministic equivalents by taking their expected values and formulating the problem as the minimization of cost and emission plus additional objective for the expected deviation between generator outputs and load demand (unsatisfied load demand) [12, 13, 14]. The approach adopted in this paper is based on the reliability concept and simulations are performed to test the reliability of the stochastic system under different problem formulations. Decision variables P_{Gi} ($i = 1,...,6$) are assumed to be normally distributed with Mean P_{Gi} and Standard Deviation (SD) $\sigma_i = 0.1 P_{Gi}$. For each solution P_{Gi} ($i = 2,...,6$), 100 random instantiates having Mean P_{Gi} and SD σ_i are created within $2\sigma_i$. A good measure of system performance in the case of stochastic systems is its reliability [19]. We define reliability R as:

$$R = \frac{n}{m} \qquad (7)$$

and an additional constraint is included in the optimization problem:

$$R \geq R^{cr} \tag{8}$$

where R^{cr} is the required reliability which is 95.6% for which $Pr\{\mu_1 - 2\sigma_1 < P_1 < \mu_1 + 2\sigma_1\}$. Thus, Reliabilty R is calculated according to the number of cases for which P_1 is found to be within $2\sigma_1$.

In the stochastic approach, the objective functions are now reformulated as follows:

$$\text{Min.} \quad \overline{Cost} + 2\sigma_{Cost} \tag{9}$$

$$\text{Min.} \quad \overline{NO_x} + 2\sigma_{NOx}$$

subject to the following constraints:

$$\sum_{i=1}^{n} P_{Gi} - P_D - P_L = 0$$

$$P_{Gimin} \leq P_{Gi} \leq P_{Gimax}$$

$$R \geq R^{cr}$$

where $\overline{Cost}$, $\overline{NO_x}$, σ_{Cost} and σ_{NOx} are the Expected Cost and Expected NO$_x$ emission and SD of Expected Fuel Cost and SD of Expected NO$_x$ emission respectively.

Note that P_{GI} is calculated from the loadflow program and this satisfies implicitly the power balance constraint (equation 3).

The procedure used in this stochastic method is described as follows:

For each feasible solution $\overline{P}_j (j=2,...,n)$ obtained by NSGA-II,

Create m instantiates $P_j^{(i)}(j=2,...,n)$ by perturbing each $\overline{P}_j$ as N($\overline{P}_j$, σ_j) where m = 100 and σ_j = $0.1\overline{P}_j$.

Count the number of instantiates n for which
$P_1^{(i)} \in [P_1^{min}, P_1^{max}]$

where $P_1^{min} = \overline{P}_1 - 2\sigma_j \overline{P}_1 = 0.8\overline{P}_1$

and $P_1^{max} = \overline{P}_1 + 2\sigma_j \overline{P}_1 = 1.2\overline{P}_1$

Calculate $R = \dfrac{n}{m}$

Calculate Expected Cost $\overline{Cost}$ and Expected NO$_x$ $\overline{NO}_x$ and SD of Cost σ_{Cost} and SD of NO$_x$ σ_{NOx}

5.1 Case S1: Stochastic Power Generated with Fixed System Load

The multi-objective optimization problem is formulated as above with fixed total system load $P_D = 2.834$ p.u. Thus, power generated P_{Gi} are random variables. Fig. 4 shows the nondominated solutions for the stochastic case with fixed system load obtained as compared to the deterministic case.

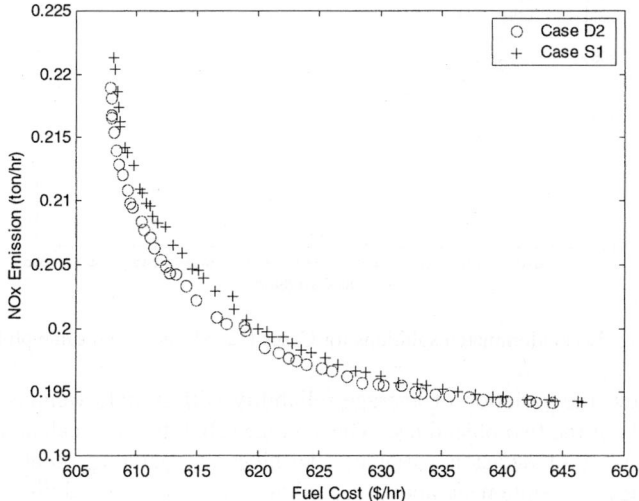

Fig. 4. Nondominated solutions obtained for stochastic power generated with fixed demand (Case S1) as compared to deterministic case (Case D2)

It can be inferred that for solutions excluding the minimum fuel cost and minimum NO_x emission (i.e. the two extreme points on the curve, optimum values of the two objectives would be generally worst than the deterministic case. In other words, for pseudo weights excluding (1, 0) and (0, 1), deterministic solutions always dominate stochastic ones.

5.2 Case S2: Stochastic Power Generated and System Loads

The multi-objective optimization problem is formulated as above but the individual loads on the system are treated as stochastic variables. Thus, power generated and system loads are random variables. Each of 21 loads is normally distributed with mean P_{Li} and $\sigma_i = 0.1 P_{Li}$. Power factor for each load is maintained as at the base load, i.e. ratio P_{Li} to Q_{Li} is constant.

Fig. 5 shows the nondominated solutions obtained for the three cases: deterministic (Case D2), stochastic power generated with fixed system load (Case S1), stochastic power generated and system loads (Case S2). It can be observed that the deterministic case shows that the minimum fuel cost obtained is no longer optimal when the decision variables are taken as stochastic. An interesting observation reveals that the minimum NO_x emission is not affected by stochastic considerations.

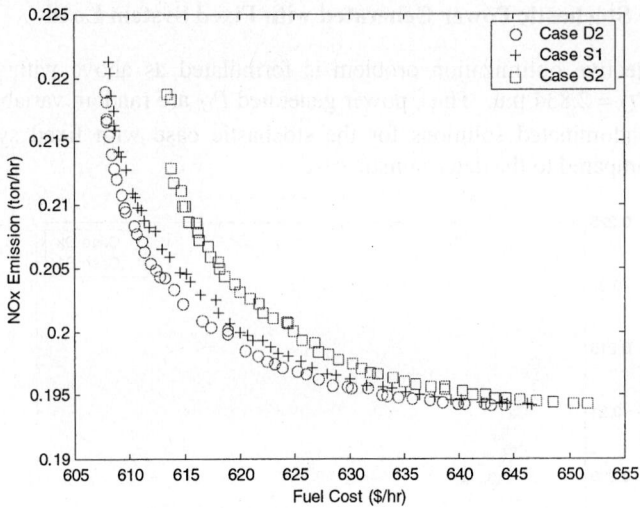

Fig. 5. Nondominated solutions for Cases D2, S1 and S2 on same plot

Fig. 6 shows the variation of average reliability with pseudo weights (1, 0), (0.5, 0.5) and (0, 1) of the two objectives. The average reliability was calculated over 100 runs for each set of decision variables (defining an operating point). This figure clearly verifies the statement above regarding the non-dependability of the NO_x emission on the nature of the decision variables for minimum emission level.

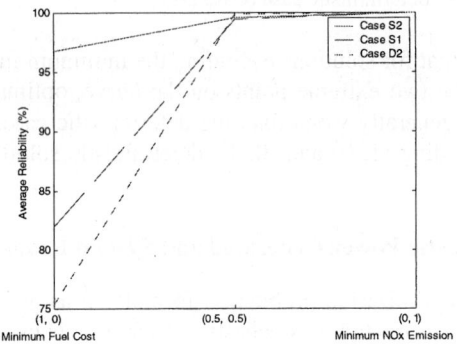

Fig. 6. Average reliability for different pseudo weights for Cases D2, S1 and S2

5.3 Case S3: Stochastic Power Generated, System Loads, Fuel Cost and Emission Coefficients

This case is similar to Case S2 but in addition the fuel cost and NO_x emission coefficients are considered as stochastic variables with mean as given in Tables 1 and 2 respectively and standard deviation as 0.1 of their respective means. Fig. 7 shows the nondominated solutions obtained for the stochastic case considering power generated, system loads, fuel cost and emission coefficients considered as random variables compared to the deterministic case as in Case D2.

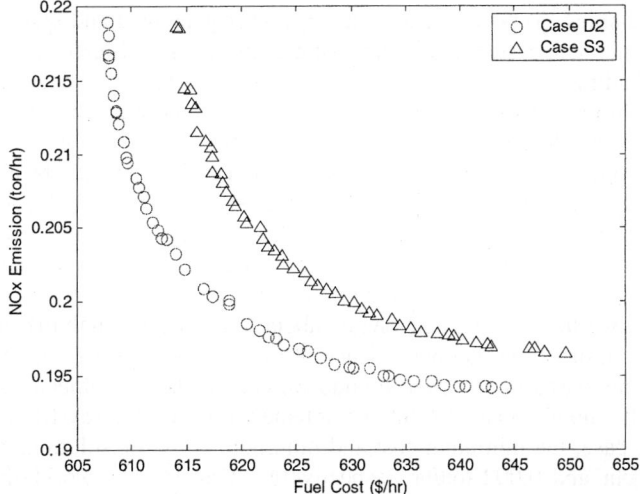

Fig. 7. Nondominated solutions for stochastic power generated, system loads, fuel cost and NO_x emission coefficients (Case S3) as compared to deterministic case (Case D2)

It can be observed that in this case the nondominated solutions obtained are shifted away from those of the deterministic case, that is, the solutions obtained when power generated, system loads, cost and emission coefficients are stochastic variables, are all dominated by the deterministic solutions. Therefore, higher cost and emission are expected when the power system is operated in real-world situation. The expected increase in minimum cost and minimum emission are about 6 $/hr and 0.002 ton/hr. Thus, a 1% increase has been obtained in both objectives when the standard deviation was taken as 10% of the mean value of the variables. It is to be noted that these figures are not negligible when the power system is operated over a year, corresponding figures would be $52,560 and 17.52 tons respectively.

6 Conclusions

In this paper, the multi-objective environmental/economic dispatch problem has been solved using the elitist Nondominated Sorting Genetic Algorithm. The algorithm has been run on the standard IEEE 30-bus system. Both the deterministic and stochastic approaches have been addressed.

In the deterministic problem, two cases have been studied: (i) the lossless system and (ii) when transmission losses are taken into consideration and given by the load flow solution. In the first case, the minimum cost and minimum emission solutions found by NSGA-II are better than those found by the conventional Linear Programming method. Moreover, these solutions are comparable, if not better than MOSST, NSGA, NPGA and SPEA reported from earlier studies. Considering the transmission losses, similar results were obtained thus confirming the superiority of NSGA-II as a fast evolutionary multi-objective algorithm.

For the stochastic problem, three cases with different complexities have been analyzed, all taking transmission losses into consideration with the following stochastic variables: (i) power generated, (ii) power generated and system loads, and (iii) power generated, system loads and cost and emission coefficients. The following interesting findings can be stated after comparison with the deterministic non-dominated solutions obtained. In the first case, the stochastic solutions obtained are dominated by the deterministic ones except for the two extreme solutions. This means that in practice, real-world operation cost and emission would be always higher except if the power system is operated at either its minimum fuel cost (economic dispatch) or minimum emission (environmental dispatch). In the second case, the minimum emission solution is not affected by stochastic considerations but all other solutions have higher cost for the same emission level. The minimum cost solution being higher than the deterministic one by about 6 $/hr. The reliability measure used in this study confirms the non-dependence of the emission level from comparison of operating points based on different pseudo-weights of the two objectives. The third case shows that nondominated solutions obtained are shifted away from those of the deterministic case, the minimum cost and minimum emission solutions being higher by about 6 $/hr and 0.002 ton/hr, respectively. Thus, in real-world situations, the power system would be operated at an operating point, which would have higher fuel cost and higher emission level than the calculated and planned operating point. In other words, the real-world (stochastic) operating point would always be dominated by the deterministic one.

References

1. Parti, S. C., Kothari, D. P., Gupta, P. V.: Economic Thermal Power Dispatch. Institution of Engineers (India) Journal-EL, Vol. 64 (1983) 126-132
2. Gent, M. R., Lamont, J. W.: Minimum-Emission Dispatch. IEEE Transactions on Power Apparatus and Systems, Vol. PAS-90, No. 6 (1971) 2650-2660
3. Zahavi, J., Eisenberg, L.: Economic-Environmental Power Dispatch. IEEE Transactions on Systems, Man, and Cybernetics, Vol. SMC-5, No. 5 (1975) 485-489
4. Nanda, J., Kothari, D. P., Lingamurthy, K. S.: Economic-Emission Load Dispatch through Goal Programming Techniques. IEEE Transactions on Energy Conversion, Vol. 3, No.1 (1988) 26-32
5. Dhillon, J. S., Parti, S. C., Kothari, D. P.: Multiobjective Optimal Thermal Power Dispatch. Electrical Power and Energy Systems, Vol. 16, No. 6 (1994) 383-389
6. Abido, M. A.: A Novel Multiobjective Evolutionary Algorithm for Environmental/ Economic Power Dispatch. Electric Power Systems Research, Vol. 65 (2003) 71-81
7. Abido, M. A.: A Niched Pareto Genetic Algorithm for Multiobjective Environmental/ Economic Dispatch. Electrical Power and Energy Systems, Vol. 25, No. 2 (2003) 97-105
8. Abido, M. A.: Environmental/Economic Power Dispatch using Multiobjective Evolutionary Algorithms. IEEE Transactions on Power Systems, Vol. 18, No. 4 (2003) 1529-1537
9. Ah King, R. T. F., Rughooputh, H. C. S.: Elitist Multiobjective Evolutionary Algorithm for Environmental/Economic Dispatch. IEEE Congress on Evolutionary Computation, Canberra, Australia, Vol. 2 (2003) 1108-1114

10. Viviani, G. L., Heydt, G. T.: Stochastic Optimal Energy Dispatch. IEEE Transactions on Power Apparatus and Systems, Vol. PAS-100, No. 7 (1981) 3221-3228
11. Bunn, D. W., Paschentis, S. N.: Development of a Stochastic Model for the Economic Dispatch of Electric Power. European Journal of Operational Research 27 (1986) 179-191
12. Dhillon, J. S., Parti, S. C., Kothari, D. P.: Stochastic Economic Emission Load Dispatch. Electric Power Systems Research, 26 (1993) 179-186
13. Dhillon, J. S., Parti, S. C., Kothari, D. P.: Multiobjective Decision Making in Stochastic Economic Dispatch. Electric Machines and Power Systems, 23 (1995) 289-301
14. Bath, S. K., Dhillon, J. S., Kothari, D. P.: Fuzzy Satisfying Stochastic Multi-Objective Generation Scheduling by Weightage Pattern Search Methods. Electric Power Systems Research 69 (2004) 311-320
15. Yokoyama, R., Bae, S. H., Morita, T., Sasaki, H.: Multiobjective Optimal Generation Dispatch based on Probability Security Criteria. IEEE Transactions on Power Systems, Vol. 3, No. 1 (1988) 317-324
16. Deb, K., Pratap, A., Agrawal, S., Meyarivan, T.: A Fast and Elitist Multiobjective Genetic Algorithm: NSGA-II. IEEE Transactions on Evolutionary Computation, Vol. 6, No. 2 (2002) 182-197
17. Das, D. B., Patvardhan, C.: New Multi-Objective Stochastic Search Technique for Economic Load Dispatch. IEE Proceedings. C, Generation, Transmission, and Distribution, Vol. 145, No. 6 (1998) 747–752
18. Haimes, Y. Y., Lasdon, L. S., Wismer, D. A.: On a Bicriterion Formulation of the Problems of Integrated System Identification and System Optimization. IEEE Transactions on Systems, Man, and Cybernetics 1 (3) (1971) 296-297
19. Deb, K., Chakroborty, P.: Time Scheduling of Transit Systems With Transfer Considerations Using Genetic Algorithms. Evolutionary Computation 6 (1) (1998) 1-24

A Multi-objective Approach to Integrated Risk Management

Frank Schlottmann[1,2], Andreas Mitschele[1,2], and Detlef Seese[2]

[1] GILLARDON AG financial software, Research Department,
Alte Wilhelmstr. 15, D-75015 Bretten, Germany
Frank.Schlottmann@gillardon.de
[2] Institute AIFB, University Karlsruhe (TH), D-76128 Karlsruhe
{mitschele, seese}@aifb.uni-karlsruhe.de

Abstract. The integrated management of financial risks represents one of the main challenges in contemporary banking business. Deviating from a rather silo-based approach to risk management banks put increasing efforts into aggregating risks across different risk types and also across different business units to obtain an overall risk picture and to manage risk and return on a consolidated level. Up to now no state-of-the-art approach to fulfill this task has emerged yet. Risk managers struggle with a number of important issues including unstable and weakly founded correlation assumptions, inconsistent risk metrics and differing time horizons for the different risk types. In this contribution we present a novel approach that overcomes parts of these unresolved issues. By defining a multi-objective optimization problem we avoid the main drawback of other approaches which try to aggregate different risk metrics that do not fit together. A MOEA is a natural choice in our multi-objective context since some common real-world objective functions in risk management are non-linear and non-convex. To illustrate the use of a MOEA, we apply the NSGA-II to a sample real-world instance of our multi-objective problem. The presented approach is flexible with respect to modifications and extensions concerning real-world risk measurement methodologies, correlation assumptions, different time horizons and additional risk types.

1 Introduction

In the recent study *Trends in risk integration and aggregation* [1] that has been conducted with 31 financial institutions worldwide the Working Group on Risk Assessment and Capital of the Basel Committee on Banking Supervision reports about two major trends in financial risk management. Firstly, the study has identified a strong emphasis on the management of risk on an integrated firm-wide basis. The second emerging trend comprises rising efforts to aggregate risks through mathematical models. At the end of the day banks are highly motivated to approximate their required capital base[1] that serves as a buffer against unexpected losses even more accurate.

---

[1] In internal banking models this is called economic capital.

While banks undertake high endeavors to gain an integrated sight of their entire business this aim in reality usually still rather resembles a mere vision. In real-world applications different types of risk are still assessed and controlled in a more silo-based manner[2], i.e. market risk is measured separately from credit risk etc. Assuming perfect correlation the resulting risk numbers are often just added up to get an aggregate risk measure. It is clear that this simple method only means a first step to true integrated risk management.

A multi-objective approach is obviously more appropriate under these circumstances. Thus, we propose a MOEA application which supports the silo-based approach currently adopted by many banks. Moreover, our approach allows the use of the Value-at-Risk which is also a commonly used risk measure in many financial institutions (and which we will explain in more detail below).

The remainder of this contribution is organized as follows: In the next section we give a short introduction to the key concepts which constitute integrated risk management. After that, we provide an overview of recent research in the area of integrated risk management and point out important obstacles in real-world applications. In the succeeding section we present our multi-objective approach which fits into current risk management practices and avoids some of the problems mentioned before. The application of a MOEA in our setting is then illustrated for a sample bank by applying the NSGA-II to recent market data. Finally, we give a conclusion and an outlook on possible future developments.

2 Integrated Risk Management

The Basel Committee on Banking Supervision [1] proposes the following definition: "An integrated risk management system seeks to have in place management policies and procedures that are designed to help ensure an awareness of, and accountability for, the risks taken throughout the financial firm, and also develop the tools needed to address these risks."

The core of such an integrated risk management[3] system is represented by an appropriate risk aggregation methodology. In the Basel Committee report [1] this is explained as follows: "Broadly, *risk aggregation* refers to efforts by firms to develop quantitative risk measures that incorporate multiple types or sources of risk. The most common approach is to estimate the amount of *economic capital* that a firm believes is necessary to absorb potential losses associated with each of the included risks."

Risk aggregation makes sense in a variety of different aggregation levels. To obtain a total bank risk measure the risk across different business units and risk types are summarized. Further possibilities include a measure for the total risk in one risk category or an aggregate measure by product or by business unit.

---

[2] Cf. Pézier [2] and Kuritzkes et al. [3].
[3] Similiar terms are *consolidated (financial) risk management* or *enterprise-wide risk management* (cf. Cumming & Hirtle [4]).

Such numbers facilitate internal comparisons across businesses and also between different companies and potential merging partners.

Cumming & Hirtle [4] proclaim two main goals that motivate integrated risk management. On the one hand with the safety-and-soundness concern the regulatory authority intends to maintain the stability of the international financial system by avoiding single bank crashes. To ensure that financial institutions hold sufficient amounts of capital to protect their risky positions, the Basel Committee on Banking Supervision has released the first Basel Accord in 1988 (with market risk amendment in 1995) and the new Accord (Basel II) that will probably become effective in 2007. On the other hand the bank's and its shareholders' perspective rather focuses on an efficient allocation of the scarce resource capital by having a more medium-term perspective. Through the integrated view at the entire institution the profitability of certain business lines can be analysed in a better way. Hence it is possible to distribute the available capital according to economically reasonable cost-benefit considerations to the different banking units.

2.1 Current Practice in Risk Management

Financial risks are inherent in financial markets and their management represents one of the main tasks in the business of financial institutions. Firstly, in this entire process the crucial risk types a certain institution faces have to be identified and defined firm-wide. In the second step the possible extents of these risks have to be quantified. As this involves high uncertainties, Alexander & Pézier [5] use the term *risk assessment* rather than *risk measurement*. Along with monitoring and reporting of the results comes the risk controlling function through trading and management action (cf. Alexander [6]). As identified by the Basel Committee on Banking Supervision [7] the main risk sources faced by banks are market, credit and operational risk. In the following section we introduce these key risk factors that we also incorporate into our model which is described later.

In general, **market risk** arises through adverse movements in the market prices of financial instruments. There exist a number of subcategories depending on the considered market factor, for instance *interest rate, equity* or *foreign currency risk*.

A prevalent method to measure market risk is Value-at-Risk (VaR). The formal definition by Frey & McNeil [8] which is derived from Artzner et al. [9] is as follows: Given a loss L with probability distribution $\mathbf{P}$, the Value-at-Risk of a portfolio at the given confidence level $\alpha \in]0,1[$ is represented by the smallest number l such that the probability that the loss L exceeds l is no larger than $(1-\alpha)$. Formally,

$$VaR_\alpha = \inf\{l \in \mathbf{R}, \mathbf{P}(L > l) \le 1 - \alpha\}. \tag{1}$$

Given a Value-at-Risk of USD 1m for a sample bank portfolio with respect to a risk horizon of 1 day and a confidence level of 99% the following conclusion may be drawn for instance: Within the next 100 days there should occur a maximum of one day with the loss on the current portfolio positions exceeding USD 1m.

The VaR calculation approaches are divided into in parametric (variance-covariance method) and non-parametric methods (historical simulation and Monte Carlo simulation). The main difference is that in the former statistical information is extracted from historical data and then employed into parameters for analytical formulae. The latter approaches perform a full valuation of the portfolio due to a number of risk factor scenarios.[4]

For instance, the historical simulation method which we will use later in our example is conducted as follows. Based on a chosen period of time (e.g. one year with 250 trading days) daily changes in the market risk factors are calculated and then applied to revalue the portfolio in its prevailing composition. By comparing the results with the current portfolio value we get a distribution of likely portfolio value changes within 1 day. Hence we are able to observe the desired quantile (e.g. 99%) directly to obtain the Value-at-Risk figure.

Even though VaR does not represent a coherent risk measure[5] it has still become a state-of-the-art methodology in the risk management of financial institutions. Furthermore in a very recent risk aggregation study performed by Rosenberg & Schuermann [10] the authors have found that the explanatory power of alternative risk measures[6] does not deviate strongly from the conclusions that can be drawn from a Value-at-Risk-based analysis. Due to these results and as we intend to present a real-world application we have decided to build our analyses upon the widespread risk measure VaR[7].

Compared to other risk types market risk measurement and management is rather well developed as extensive research has been carried out in this area (cf. Cumming & Hirtle [4]). Also long historical data sets are available for most of the instruments that are traded on financial markets. Last but not least typical returns of market instruments exhibit the convenient characteristic to resemble the standard normal distribution[8].

In the area of **credit risk** strongly intensifying efforts have been made both in research and practice. Credit risk concerns possible losses through unfavourable changes in a counterparty's credit quality. Within this category falls of course *default risk* in case a contractual partner is not capable to repay his debt anymore. But also possible depreciations in the bond value of an obligor through changes in his individual credit spread may lead to losses stemming from *spread risk* (cf. Crouhy et al. [12]).

---

[4] There are historical factor scenarios in the historical simulation method and simulated scenarios built on stochastic processes in the Monte Carlo approach.
[5] Cf. Artzner et al. [9] for details.
[6] Such as expected shortfall (also known as Tail-VaR), cf. Artzner et al. [9].
[7] For a more detailed illustration to the Value-at-Risk concept cf. [11].
[8] It has to be noted though that empirical data sets usually contain *fat tails*, i.e. the outer quantiles of the observed market distributions possess a higher probability density than assumed by the standard normal distribution. In our latter chosen approach (historical simulation method) fat tails are implicitly modelled through the empirical return distributions.

Not only the Basel Accords have increased banks' focus on credit risk but also competitive forces to establish adequate credit risk pricing systems. Nowadays there exists a number of credit risk models that have emerged as common practices. Also data availability[9] and risk management possibilities have improved substantially in recent years[10].

A risk type that has lately come into focus is **operational risk**, particularly through the new Capital Accord. The Basel Committee defines operational risk as 'the risk of loss resulting from inadequate or failed internal processes, people and systems or from external events. This definition includes legal risk, but excludes strategic and reputational risk' [7].

In the literature, there is a number of sophisticated approaches to assess operational risk, e.g. using extreme value theory (cf. Embrechts et al. [14]). In real-world applications however, many sophisticated approaches typically suffer from the absence of sufficiently given input data. Thus, we adopt the straightforward Basel Standardised Approach in our example which is not affected by this data problem. This approach will be introduced within section 4.

Further risks such as *business*, *reputational* and *strategic risks* are intentionally excluded by the Basel definition from operational risks. Based on the present state-of-the-art these risks are very hard if at all quantifiable. We therefore restrict our focus to market, credit and operational risk.

2.2 Research in the Area of Integrated Risk Management

In a very illustrative way Matten [15] describes the challenges that banks face when controlling their business on an integrated basis while entrapped between supervisory authorities and owners. The author recommends an economic profit concept that subtracts capital cost from profits to efficiently allocate capital within a bank.

Alexander & Pézier [5] propose a straightforward factor model to accomplish an aggregate risk assessment methodology. After the identification of the main bank-wide risk factors[11] aggregate risk measures can be calculated by applying certain correlation assumptions across the risk types. The authors demonstrate that the optimization of risk and return may be improved through the risk integration procedure.

Dimakos & Aas [16] present an approach to model the aggregate economic capital of a financial group taking into account pairwise interrisk correlations. Using a one year time horizon and a 99.97% confidence interval they find a reduction in the overall capital demand by around 20% compared to results obtained through the perfect correlation assumption.

Kuritzkes et al. [3] take the view of the supervisory authority to determine the possible extent of diversification benefits on the minimum capital require-

---

[9] Though still lacking behind market risk data due to comparably rare events.
[10] A detailed description of prevailing approaches is beyond the scope of this contribution (cf. e.g. Bluhm et al. [13]).
[11] These risk factors include for example interest rate, credit and equity risk.

ments within financial conglomerates. They suggest a building block approach that aggregates risk at three successive levels in the organisation and find that diversification effects are greatest within single risk factors while being smallest within different business lines.

Rosenberg & Schuermann [10] set forth a model that aggregates market, credit and operational risks through the use of copula functions which are a general concept for modelling dependencies between random variables. In their empirical analysis which has been performed with a wide number of publicly available data on financial institutions they find that simply adding up different risks overestimates total risk by more than 40%. The also popular assumption of a joint normality between risk factors underestimates risk by a similar amount. Besides their copula based method they also test a hybrid approximation that surprisingly achieves good results while still being easy to implement. The authors note that operational risk is not only relatively difficult to measure but it also deserves care when being aggregated.

Saita [17] takes a different perspective as he warns of overconfidence in the resulting risk aggregation numbers. Severe consequences may occur when wrong numbers serve as a basis for bonus payments for example. Apart from such model risk issues he addresses business risk and the varying definitions of capital as reference magnitude. Another contribution is a critical comparison of different aggregation approaches that have been proposed recently.

2.3 Obstacles for Integrated Risk Management

A main obstacle to integrated risk management are data problems. There may just be a lack of data as in the area of operational risk management for example. Due to insufficient data statistical methods fail to deliver clear statements in such cases. Sometimes there is enough data but in bad quality, i.e. with missing values or even wrong numbers.

Correlations also emerge in the context of inadequate data supplies. To take into account diversification effects in a reliable way stable correlations are required. If there is only short historical data available or if the parameters prove to be highly volatile empirically found correlations often fail. Additionally in crash situations correlations accross a variety of markets tend to move to one. These extreme cases also have to be taken into consideration when performing a sensible risk assessment. Inadequate correlation assumptions will very likely also lead to wrong incentives within the bank.

It has to be noted in this context that the supervisory authorities are still reluctant to accept internal assessments of correlation and thus diversification benefits. They rather make use of the supposedly conservative assumption of perfect correlation[12]. However the Basel Committee also tries to motivate banks to improve the reliability of their internal correlation estimates [1]. In the future

---

[12] Alexander & Pézier [5] show that risks can become worse than perfectly correlated if cascading effects are triggered. For example liquidity shortages across markets could worsen a market situation, leveraging correlation effects between two risk categories.

it may well be expected that the use of these proprietary estimates becomes more flexible when computing regulatory capital. Much more experience with correlations is still required though.

A further big challenge in respect of risk interdependencies lies in the correct distribution of diversification benefits between different entitites (e.g. business units or products). To motivate economically reasonable decisions the distribution should be deliberate.

It has already been mentioned above that risk measurement methodologies widely differ across different risk types. To account for the specific risk properties risk horizons usually range from 1 day (market risk) to 1 year (credit and operational risk). Also heterogeneous distributional assumptions hold, e.g. normal distributions for market and skewed distributions for credit risk. This prevents simply summing up the different obtained measures to get an overall risk figure.

Another problem are conceptional requirements. The best approach for integrated risk management seems to be a top-down process. In practical applications mostly bottom-up approaches are used, however. The reasons for this lie within the organisational structure of the institutions and these are hardly changeable.

It has become obvious that the aggregation of risks across these dimensions still turns out to be highly complicated. Unknown correlations between risks and business units, differing risk metrics and time horizons and data problems worsened by heterogenuous IT systems represent the main factors that make an implementation almost impossible. However, the following approach builds right upon these weaknesses of today's widespread structures of financial institutions and allows a bottom-up risk management as it is commonly performed.

3 A Multi-objective Approach to Integrated Risk Management

In the remainder, we consider a universe of $n \in \mathbf{N}$ investment opportunities (assets or asset classes). Any portfolio consisting of a subset of these assets is specified by an n-dimensional vector

$$X = (x_1, x_2, ..., x_n) \quad (2)$$

which satisfies the conditions

$$\sum_{i=1}^{n} x_i = 1 \wedge \forall i \in \{1, \ldots, n\} : x_i \in [0, 1]. \quad (3)$$

Each decision variable x_i represents the percentage of the bank's current wealth which is to be invested into investment opportunity $i \in \{1, \ldots, n\}$.

The following target functions reflect the usual objectives in a bank's silo-based approach of integrated risk (and return) management. Our first objective is the expected rate of return from a portfolio, given by

$$ret(X) := \sum_{i=1}^{n} x_i r_i \quad (4)$$

where r_i is the expected rate of return of investment opportunity i. This objective is to be maximized.

The second objective function is the Value-at-Risk of the portfolio due to changes of market prices (market risk), denoted by

$$mr(X) := VaR_{marketrisk}(X) \tag{5}$$

where the $VaR_{marketrisk}(X)$ is determined by one of the common calculation methods historical simulation, variance-covariance approach or Monte Carlo simulation. Usually, this objective is short-term oriented, e.g. measured on a time horizon of one or ten trading days, and to be minimized.

Our third objective function is the Value-at-Risk of the portfolio due to credit risk, i.e. defaults of obligors or other losses resulting from changing credit qualities of obligors. It is denoted by

$$cr(X) := VaR_{creditrisk}(X). \tag{6}$$

As mentioned in the first section, the $VaR_{creditrisk}$ is commonly calculated using one of the models CreditMetrics, CreditRisk+, CreditPortfolioView or similar approaches, cf. e.g. Bluhm et al. [13] for an overview of these models. A common time horizon for the calculations is one year, and this risk measure should be minimized.

The fourth objective which is relevant to our context is the required capital for operational risk compensation which we assume to be calculated according to the Basel Committee on Banking Supervision's Standardised Approach (cf. [7], p. 137ff). This yields a target function

$$or(X) := \sum_{i=1}^{n} x_i \beta_i \tag{7}$$

where β_i is specific for the business line in the bank which is affected by the investment $x_i > 0$ into opportunity i.

Summarizing the above definitions and restrictions as well as converting maximization of the ret function into minimization of $-ret$, we obtain the following problem setting:

$$f_1(X) := -ret(X) \tag{8}$$
$$f_2(X) := mr(X) \tag{9}$$
$$f_3(X) := cr(X) \tag{10}$$
$$f_4(X) := or(X) \tag{11}$$
$$X := (x_1, \ldots, x_n) \tag{12}$$
$$\forall i \in \{1, \ldots, n\} : x_i \in [0, 1] \tag{13}$$
$$\sum_{i=1}^{n} x_i = 1 \tag{14}$$

A portfolio X_2 is (weakly) dominated by a portfolio X_1 if the following condition is met:

$$\forall j \in \{1, \ldots, 4\} : f_j(X_1) \leq f_j(X_2) \wedge \exists k \in \{1, \ldots, 4\} : f_k(X_1) < f_k(X_2) \quad (15)$$

This is compatible to both the usual definition of dominated portfolios in the finance context and the common definition of dominated points in multi-objective optimization.

We assume that the bank is a rational investor, i.e. the bank is not going to invest in a dominated portfolio (cf. e.g. Markowitz [18]). Moreover, we assume that the bank's management prefers to choose from a whole set of individually optimal solutions, particularly by evaluating the trade-off between the desired expected rate of return and the different risks which have to be taken for the respective portfolio. Hence, we search for a set of non-dominated portfolios well-distributed in the four-dimensional objective function space $f_1(X)$ to $f_4(X)$ over the feasible search space which is specified by conditions (12) to (14).

The justification for the use of a heuristic algorithm builds upon the mathematical properties of the objective functions: According to Artzner et al. [9] and Gaivoronski & Pflug [20] the Value-at-Risk risk measure is a nonlinear and nonconvex function and has usually many local optima, hence f_2 and f_3 share this property which is problematic for conventional optimization approaches.[13] From the view of computational complexity, the problem of finding even a single feasible non-dominated point is **NP**-*hard* if the decision variables are restricted to integer values.[14]

Thus, we opt for a heuristic approach to compute approximation solutions. A MOEA is appropriate here since we search for a well-distributed approximation set in a restricted four-dimensional objective function space. In the literature, several different algorithms which actually implement a specific MOEA scheme are discussed, see e.g. Deb [21], Coello et al. [22] and many theoretical and empirical comparisons between the alternative approaches to evolutionary multi-objective optimization. In general, most of these MOEAs should be useful in our problem setting. It has to be pointed out here that it is not the goal of our work to propose a specific MOEA in our context as the best of all these algorithms.

However, for an illustrative example underlining the successful application of a MOEA to our real-world problem of integrated risk management, we have to choose an algorithm. Since the NSGA-II by Deb et al. [23] is an algorithm which has been successfully applied to many problem contexts in general, and more specifically, to other constrained portfolio optimization problems using less than four objective functions (cf. Schlottmann & Seese [24] for a general and [25] for

---

[13] If we assumed a Value-at-Risk measure for operational risk then this would also apply to f_4.

[14] This can be proven by reducing the standard *KNAPSACK* setting to a discrete version of our problem (cf. e.g. Seese & Schlottmann [19] for a formal analysis in the two-objective function case which can be generalised to more than two objectives). Since we assume real-valued decision variables, this does not apply directly here.

4 An Illustrative Example

a more specific overview of such studies), we have chosen this algorithm for our illustrative example in the following section.

We consider $n = 20$ investment opportunities for our sample bank with the characteristics shown in table 1. The historical market data range covers closing prices for the ten traded instruments from 15-MAY-2003 to 30-SEP-2004. The stocks are all traded on the Frankfurt stock exchange. For the 10 loans we assume the bank is hedged against market risk changes, i.e. interest rate risk is not relevant to these instruments. The loans are paying annual net interests. All calculations are based on a decision to be made by the bank's integrated risk manager on 30-SEP-2004.

Table 1. Investment opportunities for sample bank

Quantity	Category	Issuer/Obligor	Coupon	Maturity	Rating
1	Bond	German government (BUND)	6.250%	26-APR-2006	AAA
1	Bond	German government (BUND)	4.500%	04-JUL-2009	AAA
1	Bond	German government (BUND)	5.625%	20-SEP-2016	AAA
1	Bond	Deutsche Telekom (corporate)	8.125%	29-MAY-2012	BBB+
1	Bond	Volkswagen (corporate)	4.125%	22-MAY-2009	A-
1	Equity	BASF AG	-	-	-
1	Equity	Deutsche Bank AG	-	-	-
1	Equity	DaimlerChrysler AG	-	-	-
1	Equity	SAP AG	-	-	-
1	Equity	Siemens AG	-	-	-
2	Loan	Private Obligor	8.000%	30-SEP-2005	BB
2	Loan	Private Obligor	8.000%	30-SEP-2005	BB-
2	Loan	Private Obligor	8.000%	30-SEP-2005	B+
2	Loan	Private Obligor	8.000%	30-SEP-2005	B
2	Loan	Private Obligor	8.000%	30-SEP-2005	B-

For any given portfolio X, the expected rate of return $ret(X)$ is estimated from the historical time series for the ten traded instruments and from the expected annual net interest to be paid by the respective loan obligor.

Moreover, we assume the bank uses historical simulation for the calculation of the function value $mr(X)$ using a confidence level of 99% and a time horizon of 1 trading day. Furthermore, we assume the bank applies the CreditMetrics model by Gupton et al. [26] in the two-state variant described by Gordy [27] to determine $cr(X)$ for a confidence level of 99.9% and a one-year time horizon. At this point it has to be emphasized again that it is an important advantage of our multi-objective approach concerning real-world applications that different risk measures, distinct confidence levels and varying time horizons can be used within the search for non-dominated portfolios without adversely affecting the results.

The function value $or(X)$ is calculated according to the Basel Standardised Approach as specified within the previous section.

We apply the standard NSGA-II implementation provided by Kalyanmoy Deb to this problem instance. Using the genetic variation operators provided in this implementation (simulated binary crossover and a corresponding mutation operator for real-coded genes), we set the crossover probability to 0.8 and the mutation rate to $\frac{1}{n}$.

For the restriction of the decision variables specified in formula (3), we set the bounds for each real-coded gene in the NSGA-II to $[0, 1]$, respectively. In addition, we have to ensure that each portfolio X satisfies $\sum_{i=1}^{n} x_i = 1$. Since we have observed a worse empirical convergence of the algorithm when using an objective function penalty for infeasible individuals (in accordance to other studies in different application contexts), we opt for a simple repair algorithm: Immediately after performing crossover and mutation, every allele value x_j of an offspring individual is re-normed according to

$$\widetilde{x_j} = \frac{x_j}{\sum_{i=1}^{n} x_i} \tag{16}$$

and only the re-normed individuals $\widetilde{X} = (\widetilde{x_1}, \ldots, \widetilde{x_n})$ are considered in the succeeding steps of the NSGA-II.

The following figures 1 and 2 display the objective function values of the final individuals after 50000 population steps in the NSGA-II (100 individuals per population).

In figure 1, the three components concerning the market risk, the credit risk and the expected return of the respective approximated portfolio X are shown. For instance, the bank's risk manager can use this information straightforward to verify the current position of the bank against the drawn portfolios: Assume the bank has a current portfolio status Y. If the risk manager computes $f_i(Y)$ for $i = 1, .., 4$ he can immediately check the bank's current position in the three-dimensional plot. Of course, he also has to check the value $f_4(Y)$ against the objective function values in an additional figure (which we omit here) plotting operational risk and e.g. market risk against the expected return. If the bank's current position Y is dominated by a portfolio X he can directly observe the outcome of possible improvements and derive corresponding managing decisions to move the bank's risk-return profile into an improved position.

It is a striking advantage of the multi-objective view that the risk manager can see the consequences of different decision alternatives concerning different risk sources and the corresponding expected rate of return simultaneously. For instance, in figure 1 he can choose a portfolio that has e.g. high short-term risk (due to the risk horizon of 1 trading day in the market risk calculation) while having low medium-term risk reflected by the 1-year credit risk objective function value and yielding a high expected rate of return. Such portfolios are located in the upper left corner of the figure. If he does not desire a high short-term risk

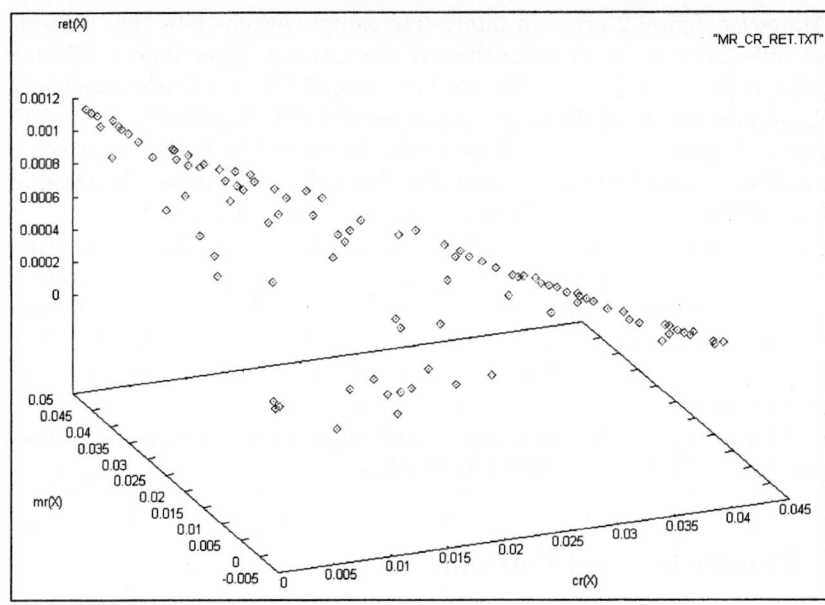

Fig. 1. Projection of mr(X), cr(X) and ret(X)

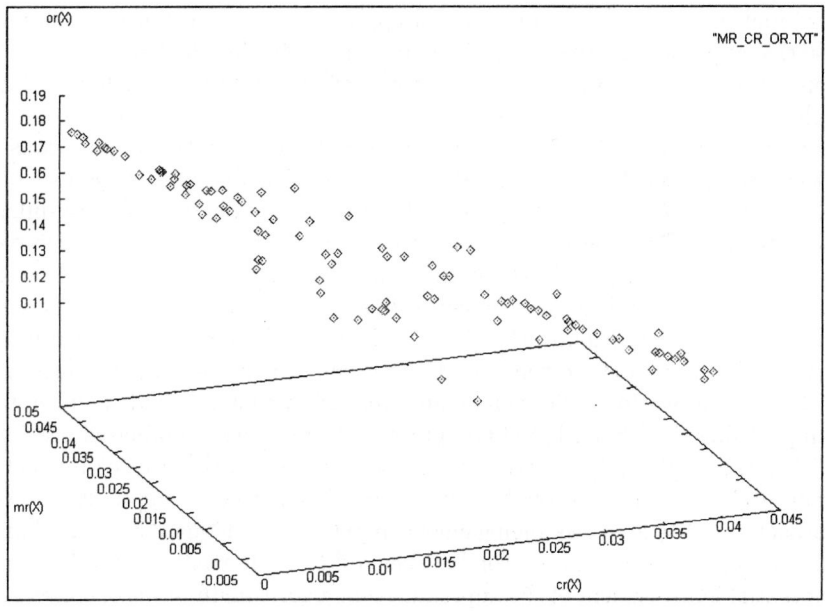

Fig. 2. Projection of mr(X), cr(X) and or(X)

(e.g. if the bank's short term risk limits are low), he can choose a portfolio in the lower right area having higher credit risk and so on.

Moreover, figure 2 gives an interesting sample insight into the trade-off between different sources of risk for the given investment opportunities. The mainly negative dependency between the market risk and the credit risk numbers is due to the immunization of the loans against market risk. The degree of operational risk to be taken by the bank is lower for the portfolios having relatively high credit risk and relatively low market risk. Note that in current real-world applications, this trade-off is usually not analyzed in such detail.

The preceding considerations represent a novel approach compared to the current state-of-the-art within the financial industry and the integrated risk management literature. Moreover, it has to be pointed out that the 20 asset example might seem small at first glance, however, we already mentioned that the risk manager can use more global risk factors representing whole asset classes instead of using single assets and the MOEA can of course process larger problems. Thus, the presented approach can be applied even to large portfolios in a top-down approach over different asset classes.

5 Conclusion and Outlook

The integrated management of different sources of risk is one of the largest challenges to the financial industry. Since each risk category has its own specific properties and is particularly measured on a distinct time horizon and an individual confidence level in real-world applications, we have proposed a multi-objective approach to integrated risk management in the previous sections. This approach does not require the aggregation of incompatible risk figures into a single number. Moreover, it does not necessarily require correlations between different risk types which are difficult to estimate in real-world applications due to the lack of data. Instead, the risk manager is provided with a number of solutions which he can use for an analysis of the trade-off between the different risk types and the expected rate of return. The manager can use a silo-based approach for the integrated risk-return management, which is currently standard in many financial institutions. Moreover, due to the use of a MOEA in the search of non-dominated portfolios, the Value-at-Risk which is commonly used in real-world applications, can be kept as a risk measure in the respective category. To illustrate a real-world application of our approach, we have provided an empirical example using the NSGA-II to find approximations of non-dominated portfolios.

For a thorough analysis of the MOEA performance in this area of application, a more detailed empirical study is necessary. Furthermore, our multi-objective approach to integrated risk management might be an adequate real-world application for an empirical comparison between different alternative MOEA schemes. A potential improvement of the approximation algorithm in terms of convergence speed could hybridize MOEAs and problem-specific knowledge, cf. the ideas used within the two-objective function approach presented recently in Schlottmann & Seese [28].

The problem of aggregating the multi-dimensional output to less dimensions still remains if the risk manager desires a single risk figure (although it seems not recommendable, cf. also Cumming & Hirtle [4]). However, a progress in the aggregation of different risk categories from the finance point of view can probably also be integrated into a refined MOEA approach due to its flexibility. In this case, the multi-objective approach would benefit from the development of new financial tools while still being attractive for analyzing the trade-off between different sources of risk and the expected rate of return as pointed out above. In addition, more objective functions of the bank which do not necessarily need to possess convenient mathematical properties might be incorporated quite easily into our MOEA-based approach in the future.

Acknowledgements

The authors would like to thank GILLARDON AG financial software for partial support of their work. Nevertheless, the views expressed in this chapter reflect the personal opinion of the authors and are neither official statements of GILLARDON AG financial software nor of its partners or its clients.

We are also grateful to Kalyanmoy Deb for providing the NSGA-II source code on the WWW, and to the anonymous referees for their helpful comments on an earlier version of this work.

References

1. Basel Committee on Banking Supervision: Trends in risk integration and aggregation, Basel Committee on Banking Supervision (2003)
2. Pézier, J.: Application-Based Financial Risk Aggregation Methods. Discussion Papers in Finance 11, ISMA Centre, University of Reading, UK (2003)
3. Kuritzkes, A., Schuermann, T., Weiner, S.: Risk Measurement, Risk Management and Capital Adequacy in Financial Conglomerates. Working Paper, Wharton Financial Institutions Center (2002)
4. Cumming, C.M., Hirtle, B.J.: The Challenges of Risk Management in Diversified Financial Companies. FRBNY Economic Policy Review (2001)
5. Alexander, C., Pézier, J.: Assessment and Aggregation of Banking Risks. ISMA Centre, University of Reading, UK. (2003) Presented at the 9th Annual Round Table of the International Financial Risk Institute.
6. Alexander, C.: The Present, Future and Imperfect of Financial Risk Management. Discussion Papers in Finance, ISMA Centre, University of Reading, UK (2003)
7. Basel Committee on Banking Supervision: International Convergence of Capital Measurement and Capital Standards - A Revised Framework, Basel Committee on Banking Supervision (2004)
8. Frey, R., McNeil, A.: VaR and expected shortfall in portfolios of dependent credit risks: conceptual and practical insights. Journal of Banking and Finance (2002) 1317–1334
9. Artzner, P., Delbaen, F., Eber, J.M., Heath, D.: Coherent Measures of Risk. Mathematical Finance (1999) 203–228
10. Rosenberg, J.V., Schuermann, T.: A General Approach to Integrated Risk Management with Skewed, Fat-Tailed Risks. Staff Report 185, Federal Reserve Bank of New York (2004)

11. Jorion, P.: Value at risk. 2 edn. McGraw-Hill, USA (2001)
12. Crouhy, M., Galai, D., Mark, R.: Risk Management. McGraw-Hill (2001)
13. Bluhm, C., Overbeck, L., Wagner, C.: An introduction to credit risk modeling. Chapman-Hall, Boca Raton (2003)
14. Embrechts, P., Kluppelberg, C., Mikosch, T.: Modelling extremal events for insurance and finance. Springer, Berlin (2003)
15. Matten, C.: Managing bank capital. John Wiley & Sons, Chichester (2000)
16. Aas, K., Dimakos, X.K.: Integrated Risk Modelling. NR Report, Norwegian Computing Center (2003)
17. Saita, F.: Risk Capital Aggregation: the Risk Managers Perspective. Working Paper, Newfin Research Center and IEMIF (2004)
18. Markowitz, H.: Portfolio selection. Journal of Finance **7** (1952) 77–91
19. Seese, D., Schlottmann, F.: The building blocks of complexity: a unified criterion and selected applications in risk management. (2003) Complexity 2003: Complex behaviour in economics, Aix-en-Provence, http://zai.ini.unizh.ch/www_complexity2003/doc/Paper_Seese.pdf.
20. Gaivoronski, A., Pflug, G.: Properties and computation of value-at-risk efficient portfolios based on historical data. (2002) Working paper, Trondheim University.
21. Deb, K.: Multi-objective optimisation using evolutionary algorithms. John Wiley & Sons, Chichester (2001)
22. Coello, C., Van Veldhuizen, D., Lamont, G.: Evolutionary Algorithms for solving multi-objective problems. Kluwer, New York (2002)
23. Deb, K., Agrawal, S., Pratap, A., Meyarivan, T.: A fast elitist non-dominated sorting genetic algorithm for multi-objective optimisation: NSGA-II. In Schoenauer, M., Deb, K., Rudolph, G., Yao, X., Lutton, E., Merelo, J., Schwefel, H., eds.: Parallel problem solving from nature, LNCS 1917. Springer, Berlin (2000) 849–858
24. Schlottmann, F., Seese, D.: Modern heuristics for finance problems: a survey of selected methods and applications. In Rachev, S., Marinelli, C., eds.: Handbook on Numerical Methods in Finance. Springer, Berlin (2004) 331–360
25. Schlottmann, F., Seese, D.: Financial applications of multi-objective evolutionary algorithms: Recent developments and future research. In Coello-Coello, C., Lamont, G., eds.: Handbook on Applications of Multi-Objective Evolutionary Algorithms. World Scientific, Singapore (2005) To appear.
26. Gupton, G., Finger, C., Bhatia, M.: CreditMetrics. Technical report, JP Morgan & Co., New York (1997)
27. Gordy, M.: A comparative anatomy of credit risk models. Journal of Banking and Finance **24** (2000) 119–149
28. Schlottmann, F., Seese, D.: A hybrid heuristic approach to discrete portfolio optimization. Computational Statistics and Data Analysis **47** (2004) 373–399
29. Hallerbach, W.G.: Capital Allocation, Portfolio Enhancement and Performance Measurement: A Unified Approach. Working Paper, Department of Finance, Erasmus University Rotterdam (2003)
30. Schlottmann, F., Seese, D.: A hybrid genetic-quantitative method for risk-return optimisation of credit portfolios. In Chiarella, C., Platen, E., eds.: Quantitative Methods in Finance 2001 Conference abstracts. University of Technology, Sydney (2001) 55 Full paper: http://www.business.uts.edu.au/finance/resources/qmf2001/Schlottmann_F.pdf.

An Approach Based on the Strength Pareto Evolutionary Algorithm 2 for Power Distribution System Planning

Francisco Rivas-Dávalos[1] and Malcolm R. Irving[2]

[1] Instituto Tecnológico de Morelia, Av. Tecnológico 1500, Morelia, Mich., México
frivasd2003@yahoo.co.uk
[2] Brunel University, Uxbridge, Middlesex, UB8 3PH, United Kingdom
malcolm.irving@brunel.ac.uk

Abstract. The vast majority of the developed planning methods for power distribution systems consider only one objective function to optimize. This function represents the economical costs of the systems. However, there are other planning aspects that should be considered but they can not be expressed in terms of costs; therefore, they need to be formulated as separate objective functions. This paper presents a new multi-objective planning method for power distribution systems. The method is based on the Strength Pareto Evolutionary Algorithm 2. The edge-set encoding technique and the constrain-domination concept were applied to handle the problem constraints. The method was tested on a real large-scale system with two objective functions: economical cost and energy non-supplied. From these results, it can be said that the proposed method is suitable to resolve the multi-objective problem of large-scale power distribution system expansion planning.

1 Introduction

A power distribution system is a network that consists of substations (electrical power source nodes), lines (electrical conductors connecting nodes and carrying power) and customers (power demand nodes). System planners must ensure that there is adequate substation capacity, line capacity and acceptable level of reliability to satisfy the power demand forecasts within the planning horizon. Planning these systems involves various tasks [1]; the main of these are: 1) To find the site of substations and lines, 2) To determine substations and lines sizes (substations and lines capacities) and 3) To determine the electrical power flow in substations and lines. These tasks have to be done simultaneously optimizing various objectives such as economical costs and reliability of the systems, and considering three main technical constraints: voltage drop limit, substation and line capacity limit and radial configuration (spanning tree configuration).

The vast majority of the developed planning methods consider only one objective function to optimize [2]. The objective function of these methods represents the economical costs of the system such as, investment, energy losses and interruption costs. However, there are other planning aspects that should be considered in the planning methods but they can not be expressed in terms of costs. For instances, environmental and social impact can be very important in some cases and they can not be expressed as economical costs. Reliability of the system is another planning aspect that have been expressed in terms of costs and considered in some planning methods but, it is required information about the economical impact of power interruptions on customers and suppliers. This information might be difficult to obtain in some cases. Therefore, some planning aspects to be considered need to be formulated as separate objective functions.

There are few multi-objective methods that have been proposed to resolve the problem of power distribution systems expansion planning with more than one objective function separately formulated. In [3], a planning method is proposed to optimize three objective functions: economical cost, energy non-supplied (a reliability index) and total length of overhead lines. This method generates a set of Pareto-optimal solutions using the ε-constrained technique. This technique transforms two objectives into constraints, by specifying bounds to them (ε), and the remaining objective, which can be chosen arbitrarily, is the objective function to optimize. In other words, the multi-objective problem is transformed into a single-objective optimization problem, which is resolved by classical single-objective algorithms. The bounds ε are the parameters that have to be varied in order to find multiple solutions.

Another planning method that uses the ε-constrained technique is reported in [4]. This method resolves the single-objective problems using a simulated annealing algorithm. The disadvantage of this technique is that the solution of the resulting single-objective problem largely depends on the chosen bounds ε. Some values of ε might cause that the single-objective problem has no feasible solution. Thus, no solution would be found. In addition, several optimization runs are required to obtain a set of Pareto-optimal solutions.

In [5], it is reported a planning method that uses the weighting technique to obtain non-dominated solutions. This technique consists in assigning weights to the different objective functions and combining them into a single-objective function. The Pareto-optimal solutions are identified by changing the weights parametrically with several optimization runs. One difficulty with this technique is that it is difficult to find a uniformly distributed set of Pareto-optimal solutions. In addition, many weight values can lead to the same solution and, in case of non-convex objective space, certain solutions can not be found.

In [6], a multi-objective optimization method based on genetic algorithms is presented. This method is able to find a set of approximate Pareto-optimal solutions in one single simulation run due to its population approach. The method is formulated to find the site and size of substations and lines optimizing two objective functions: economical cost and energy non-supplied. The drawback of this method is that the genetic algorithm has to be run several times in order to obtain solutions closer to the optimal ones. Moreover, the method uses genetic operators that generate many illegal solutions and its encoding technique has low heritability, making the algorithm inefficient and ineffective.

In this paper, we propose a new multi-objective planning method for optimal power distribution system expansion planning. The method is based on the Strength Pareto Evolutionary Algorithm 2 (SPEA2) [7]. The edge-set encoding technique [9] and the constrain-domination concept [11] were used to handle the problem constraints. The method was tested on a real large-scale system and some studies were carried out to analyze the effect of constraints and non-convex regions of the search space on the performance of the proposed method.

2   Problem Formulation

In this paper, the planning problem is formulated as the problem of selecting the number, site and size of substations and lines such that the investment cost, the cost of energy losses and the energy non-supplied index are minimum; maintaining the radiality of the network and at the same time not violating the capacity and voltage drop constraints in any part of the network. Fig. 1 shows a power distribution system for planning.

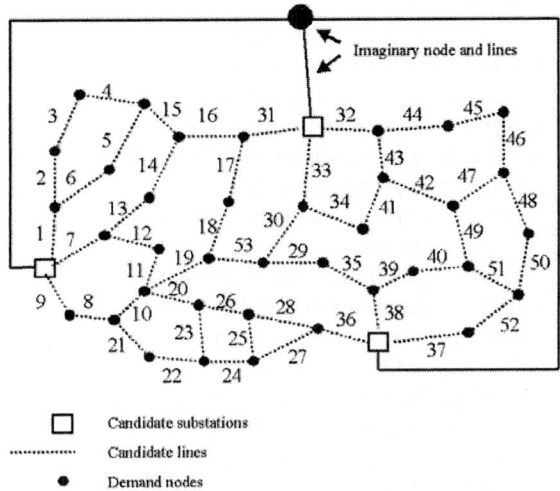

Fig. 1. A power distribution system for planning. The imaginary node and lines are used to manipulate problems with more than one substation and to represent the systems as spanning trees

The candidate substations and lines are the possible components to be selected. The candidate components and power demand nodes are known beforehand.

The mathematical formulation of the planning problem is expressed as follows:

$$\text{Minimize } F_{cost} = \sum_{t \in Nt} \sum_{s \in Ns} \{(FC_t)_s (X_t)_s + (Coeff)(PW)(3I_t^2 R_t)_s\} + \quad (1)$$

$$\sum_{l \in Nl} \sum_{c \in Nc} \{(FC_l)_c (X_l)_c + (Coeff)(PW)(3I_l^2 R_l)_c\}$$

$$\text{Minimize } ENS = \sum_{l=1}^{N_L} PF_l \, \lambda_l \, r_l$$

Subject to the conditions:

$V_{min} \leq V_j \leq V_{max}$ (Voltage drop constraint)
$I_l \leq Imax_l$ (Line capacity constraint)
$T_t \leq Tmax_t$ (Substation capacity constraint)
$\Sigma l = n-1$ (Radiality constraint)

Where:

F_{cost}	= Total economical cost (in Millions)
ENS	= Energy non-supplied index (in Megawatt-hour)
$(FC_t)_s$	= Investment cost of substation t to be built with size s
$(X_t)_s$	= 1 if substation t with size s is built. Otherwise, it is equal to 0.
I_t	= Current through substation t
R_t	= Resistance of the transformer in substation t
$(FC_l)_c$	= Investment cost of line l to be built with size c
$(X_l)_c$	= 1 if line l with size c is built. Otherwise, it is equal to 0.
I_l	= Current through line l
R_l	= Resistance of line l
Coeff	= Cost factor = (8760)(Cost of energy)(Loss factor)
PW	= Present worth factor = $[(1+d)^p-1]/[d(1+d)^p]$;
P	= planning years; d = discount rate
Nt	= Number of proposed substations
Ns	= Number of proposed sizes for substations
Nl	= Number of proposed lines
Nc	= Number of proposed sizes for lines
V_j	= Voltage in node j
PF_l	= Power flow on line l (in Megawatts)
λ_l	= Failure rate of line l (in failures/km*year)
r_l	= Failure duration of line l (in hours)
N_L	= Number of lines in the system
$V_{min,max}$	= Voltage drop limit (Permissible levels of voltage)
$Imax_l$	= Current capacity limit of line l
$Tmax_t$	= Power capacity limit of substation t
Σl	= Number of selected lines
n	= Number of nodes

3 A Multi-objective Planning Method for Power Distribution Systems

A multi-objective planning method is proposed to resolve the problem of power distribution system expansion planning. The method is based on Strength Pareto Evolutionary Algorithm 2 (SPEA2) [7].

3.1 SPEA2

SPEA2 uses a regular population and an archive (external set). The overall algorithm is as follows [7]:

Step 1 (Initialization): Generate an initial population P_o and create the empty archive $A_o = 0$. Set $t = 0$.
Step 2 (Fitness assignment): Calculate fitness values of individuals in P_t and A_t.
Step 3 (Environmental selection): Copy all non-dominated individuals in P_t and A_t to A_{t+1}. If size of A_{t+1} exceeds the archive size N_A then reduce A_{t+1} by means of a truncation operator; otherwise if size of A_{t+1} is less than N_A then fill A_{t+1} with dominated individuals in P_t and A_t.

Step 4 (Termination): If $t \geq G$ (where G is the maximum number of generations) or another stopping criterion is satisfied then set $\bar{A}$ (non-dominated set) to the set of the non-dominated individuals in A_{t+1}. Stop.

Step 5 (Mating selection): Perform binary tournament selection with replacement on A_{t+1} in order to fill the mating pool.

Step 6 (Variation): Apply recombination and mutation operators to the mating pool and set P_{t+1} to the resulting population. Increment generation counter ($t = t + 1$) and go to Step 2.

Fitness Assignment

The fitness assignment is a two-stage procedure. First, each individual i in the archive A_t and the population P_t is assigned a strength value $S(i)$, representing the number of solutions it dominates (the symbol $\succ$ corresponds to the Pareto dominance relation):

$$S(i) = |\{j \mid j \in P_t + A_t \wedge i \succ j\}| \qquad (2)$$

Second, the raw fitness of an individual i is determined by the strengths of its dominators in both archive and population:

$$R(i) = \sum_{j \in P_t + A_t, j \succ i} S(j) \qquad (3)$$

Additional density information is incorporated to discriminate between individuals having identical raw fitness values. The density information technique proposed in [7] is an adaptation of the k-th nearest neighbor method [8]. Thus, the fitness of an individual i is defined by:

$$F(i) = R(i) + D(i) \qquad (4)$$

Where $D(i)$ is the density information.

Environmental Selection

In the environmental selection, the first step is to copy all non-dominated individuals from the archive and population to the archive of the next generation A_{t+1}. In this step, there can be three scenarios: 1) The non-dominated set fits exactly into the archive ($|A_{t+1}| = N_A$), 2) The non-dominated set is smaller than the archive size ($|A_{t+1}| < N_A$) and 3) The non-dominated set exceeds the archive size ($|A_{t+1}| > N_A$).

In the first case, the environmental selection is completed. In the second case, the best $N_A - |A_{t+1}|$ dominated individuals in the previous archive and population are copied to the new archive. Finally, in the third case, an archive truncation procedure is invoked which iteratively removes individuals from A_{t+1} until $|A_{t+1}| = N_A$. The truncation procedure is as follows:

At each iteration, an individual i is chosen for removal if:

- $\sigma_i^k = \sigma_j^k$ for every value of k in the range $0 < k < |A_{t+1}|$ and $j \in A_{t+1}$ or
- $\sigma_i^q = \sigma_i^q$ and $\sigma_i^k < \sigma_i^k$ for every value of q in the range $0 < q < k$ and any value of k in the range $0 < k < |A_{t+1}|$.

Where σ_k^i is the distance(in the objective space) from an individual i to its k-th nearest neighbor[7].

3.2 The Proposed Multi-objective Planning Method

The proposed multi-objective planning method for power distribution system expansion planning is based on SPEA2. In the following sections, the main components of the method are described.

Encoding and Genetic Operators

It is proposed that the distribution system topologies be represented directly as sets of their lines (edge-set encoding technique), and special recombination and mutation operators be used. This encoding technique and genetic operators were proposed in [9] for the degree-constrained minimum spanning tree problem and they were adapted for power distribution system planning in [10]. For example, Fig. 1 shows a power distribution network with 37 demand nodes, 3 candidate substations and 53-numbered candidate lines. A potential solution for this network is encoded as the set of numbers that represent the lines that form the solution (Fig. 2). The imaginary node and lines are used to manipulate problems with more than one substation and to represent the networks as spanning trees. Therefore, each solution can be encoded with an array containing the lines of the solution.

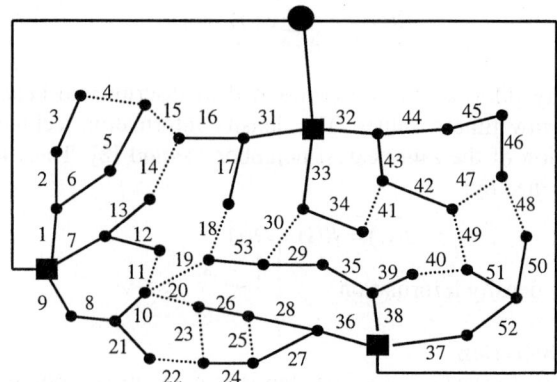

Encoded solution =
{9,8,21,10,1,2,6,5,3,7,12,13,31,32,17,16,33,34,44,45,46,43,42,37,52,50,51,
38,39,36,27,28,24,26,35,29,53}

Fig. 2. An example of a direct encoding of a solution

The recombination operator to create offspring consists on two steps: In the first step, a set of lines contained in both parents is selected to initialise the offspring. In the second step, lines are randomly and successively selected from the rest of the lines contained either in parent 1 or parent 2 (but not in both) to be included in the offspring (only lines that do not introduce cycles are included). Fig. 3 shows an example of a recombination operation for two parent solutions of the network in Fig 1. Figs. 3a) and 3b) are the parent solutions 1 and 2, respectively. Fig. 3c) is the offspring initialised with lines contained in both parents. In this phase, the offspring has components disconnected. In the second step, the disconnected components are connected with lines contained either in parent 1 or parent 2; as it is shown in figure 3d).

The mutation operator is described as follows (see Fig. 4): In a first step, a candidate line currently not in the offspring is randomly chosen and inserted in the offspring, e.g. in Fig. 4a) the line 30 (darker line) is inserted. A cycle will be formed with this action so, in a second step, a random choice among the lines in the cycle (triple lines) is then made (excluding the new line inserted and the imaginary lines), and the chosen one is removed from the offspring. In Fig. 4b) the line 41 is removed.

The recombination operation is controlled by a recombination probability parameter (Prc): the probability parameter is set to a real number in the range [0.0, 1.0] then, a real number is obtained by a random number generator. If the obtained number is smaller than Prc, the recombination is executed; otherwise one of the parents is randomly chosen and copied to the offspring population. Mutation operation is applied to every individual in the offspring population: exactly one line in an individual is changed. The size of the mating pool is equal to the population size and, in each iteration the old population is replaced with the offspring population.

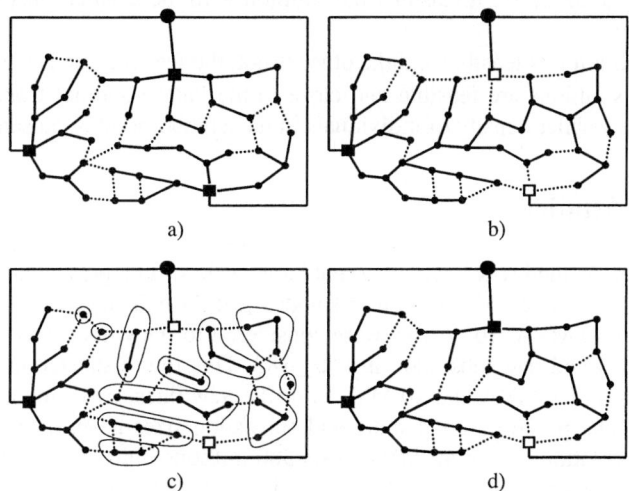

Fig. 3. An example of a recombination operation for two parent solutions of the network in Fig. 1

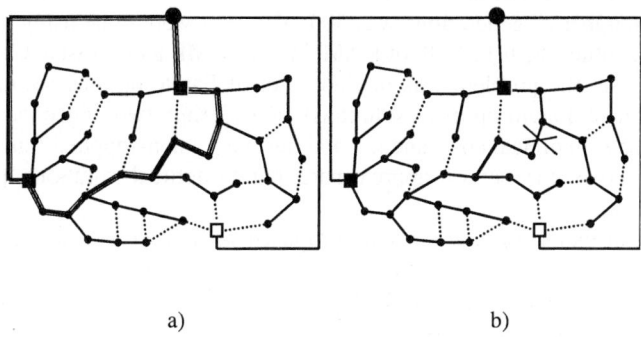

Fig. 4. An example of a mutation operation for the offspring in Fig. 3

Fitness Function

The fitness function is defined by the strategy formulated in SPEA2 algorithm. In this paper, the application of a different domination definition from the conventional one is suggested in order to handle the constraints of power system planning problems. This new concept of domination is called constrain-domination [11].

Selection Mechanism

The binary tournament selection is the selection mechanism used in the phase where solutions are selected for recombination (step 5 of SPEA2). This mechanism selects two solutions randomly and picks out the solution with better fitness value. Using the concept of constrain-domination in the fitness function formulation, one of the following scenarios is created each time this selection mechanism is applied:

- If both solutions are feasible, the solution closer to the Pareto-optimal front is chosen
- If both solutions are infeasible, the solution with the smaller constraint violation is chosen
- If one solution is feasible and the other is not, the feasible one is chosen
- If both solutions are feasible and close to the Pareto-optimal front, the solution with the smaller density of individuals in its neighborhood is chosen.

4 Case Studies

The proposed method was tested on a real large-scale system presented in [6] (Fig. 5).

The system has 45 existing demand nodes and 44 existing lines with one power substation of 40 MVA. 163 routes were considered for new lines to connect 137 new demand nodes and one substation in node 182. This future substation was proposed with two sizes of 8 MVA and 40 MVA. For the new lines, two conductor sizes were considered. The proposed conductors and substation sizes have different investment cost. The substation size of 40 MVA costs 300 millions (unit of money), whereas the other substation size costs 136 millions. Similarly, the conductor with the bigger size costs more but, it has less failure and failure duration rate than the other conductor.

The parameter values of the algorithm used to resolve the problem were: population size of 200; external archive size of 50; recombination probability of 0.8 and the maximum number of generations was 500. The tests were done using a PC compatible 1 GHz Pentium with 128 Mb of RAM, WindowsME and a Visual C++ compiler.

In this case, the problem was to find a set of Pareto-optimal solutions (or a set of approximate Pareto-optimal solutions) considering two objective functions to optimize: the economical cost function and the energy non-supplied function.

Fig. 6 shows the set of approximate Pareto-optimal solutions found by the proposed method.

Because of two substation sizes are proposed with different costs for the new substation, the Pareto front is divided into two fronts. The left front contains solutions with the new substation size of 8 MVA; and the other front contains solutions with the new substation size of 40 MVA.

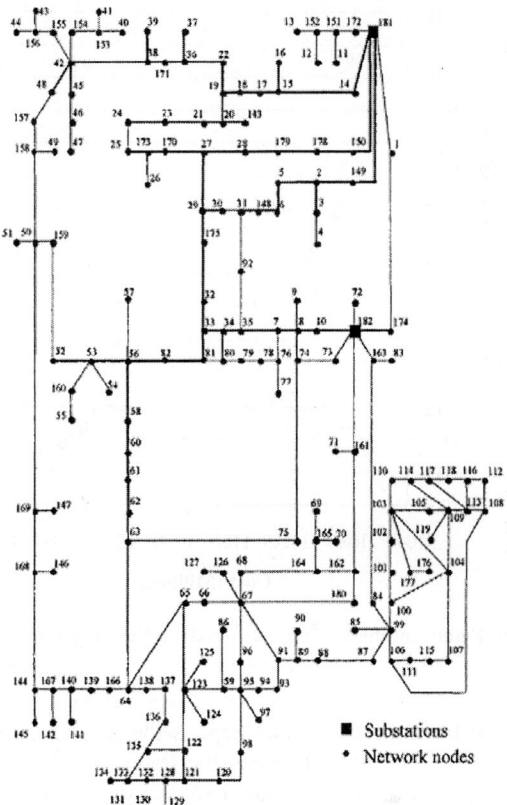

Fig. 5. Power distribution network for the case studies. The darker lines represent the existing lines and the proposed routes are represented by thin lines

Solutions with the substation size of 40 MVA provide more reliability (in terms of energy non supplied) than solutions with the substation size of 8 MVA because a bigger substation size can supplied more power; therefore, the total power demand is more equally shared between the existing and new substation and it experiences less interruption rate. However, in this case, the substation size of 40 MVA is more expensive. Similarly, in each front, there are solutions with better reliability than others but they have higher costs. This is because the energy non-supplied is a function of the configuration and the types of conductors in the system. In this case, the cheaper conductor has the higher failure rate.

The solutions shown in Fig. 6 produce conflicting scenarios between the both objective functions. If both objectives are equally important, none of these solutions is the best with respect to both objectives. However, this set of solutions can help the system planner to evaluate the solutions considering other criteria. The planner can assess the advantages and disadvantages of each of these solutions based on other criteria which are still important; and compare them to make a choice.

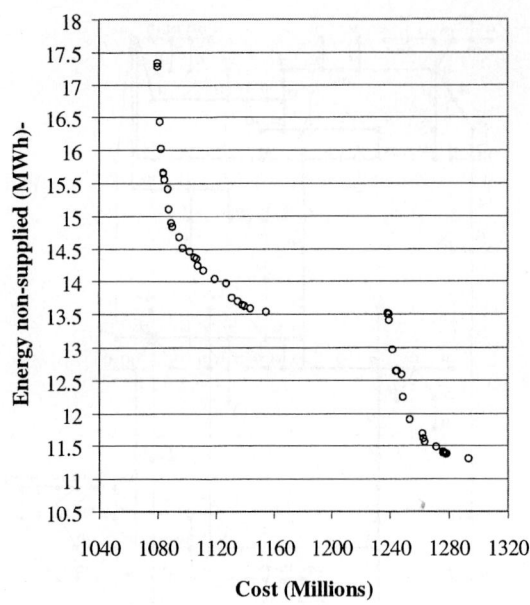

Fig. 6. Approximate Pareto-optimal solutions found by the proposed planning method to the problem of Fig. 5

Fig. 6 is different from the figure that depicts the solutions for the same problem reported in reference [6]. In this reference, the Pareto front is not divided into two fronts and it is not clear if the two proposed substation size have different fixed cost or not. Also, it is not mentioned which substation size was selected for each solution or what is the effect of the substation size on the set of non-dominated solutions.

In addition to the above study, more studies were carried out to analyze the effect of constraints and non-convex regions of the search space on the performance of the proposed method. These studies are reported as cases B, C, D and E. Case A corresponds to the original problem of reference [6](Fig. 6).

To analyze the effect of the constraints, the method was applied on the same problem of reference [6] with different levels of constraints. Fig. 7 shows the solutions found by the proposed method to the problem with the permissible level of voltage drop changed from the original 3.0 percent to 1.0 percent (case B). In Fig. 8, it is shown the solutions for the problem with the capacity limit of lines reduced by 50 percent (case C). Finally, Fig. 9 shows solutions to the problem with the proposed substation size of 40 MVA changed for a substation size of 9 MVA (the cost does not change) (case D).

Figs. 7 and 8 show that the proposed method was able to converge to one of the Pareto fronts previously found for the original problem. This Pareto front corresponds to the solutions with the new substation size of 40 MVA, which satisfy the new constraints. The other Pareto front of the original problem now lies on the infeasible region.

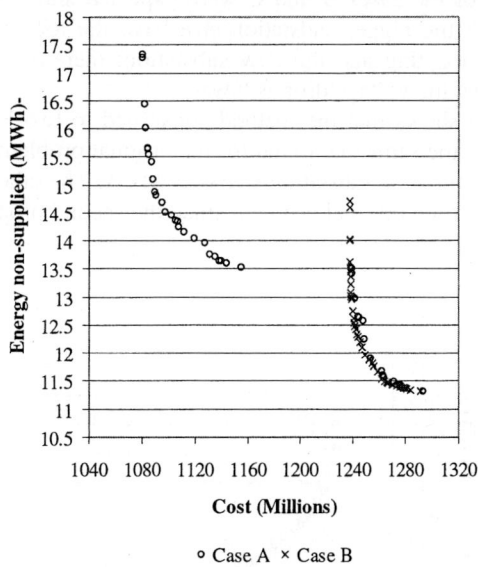

Fig. 7. Approximate Pareto-optimal solutions to the original problem (case A) and to the problem with the permissible level of voltage drop changed from 3.0 percent to 1.0 percent (case B)

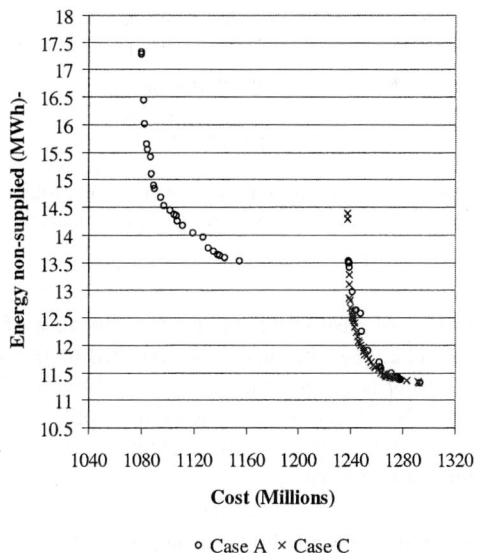

Fig. 8. Approximate Pareto-optimal solutions to the original problem (case A) and to the problem with the capacity limit of lines reduced by 50 percent (case C)

These solutions of the cases B and C were expected since, as it was mentioned early, solutions with the bigger substation size have the total power demand more shared between the existing and the new substation; therefore, the lines carry less amount of current and the voltage drop is lower.

In case D, Fig. 9 shows that the method converged to two Pareto fronts. One of these Pareto fronts is the same one found for the original problem, which corresponds to the solutions with the new substation size of 8 MVA. The other Pareto front is different from the one of the original problem since the second proposed substation size has been changed for a smaller one.

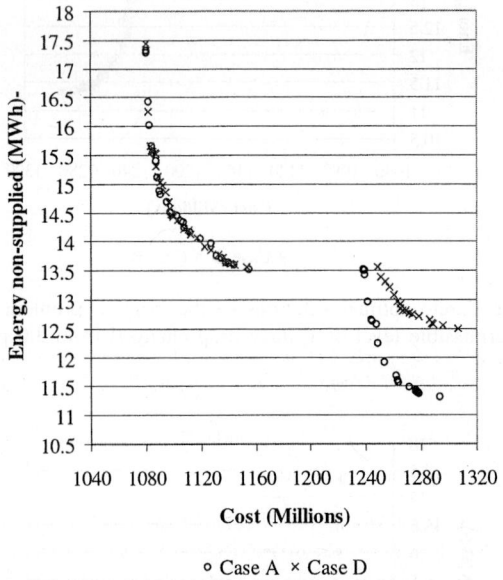

○ Case A × Case D

Fig. 9. Approximate Pareto-optimal solutions to the original problem (case A) and to the problem with the new proposed substation size of 40 MVA changed for a substation size of 9 MVA (case D)

Similarly, the solutions of case D were expected. Because of one of the proposed substation size was not changed, the method converged to the corresponding Pareto front. The other Pareto front is different from the one of the original problem because the substation size of 9 MVA has less capacity to supply energy; therefore, more power demand is satisfied by the existing substation and, as a consequence, the energy non-supplied index increases.

To analyze the effect of non-convex regions of the search space on the performance of the proposed method, the method was applied on the same original problem [6] with the cost of the new substation size of 40 MVA reduced from 300 millions to 200 Millions. Fig. 10 shows the solutions found by the method to this case (case E).

In this case E, there is a non-convex region in the Pareto front. The presence of several alternatives to build lines and substations with different economical and electrical characteristics can produce this type of scenarios.

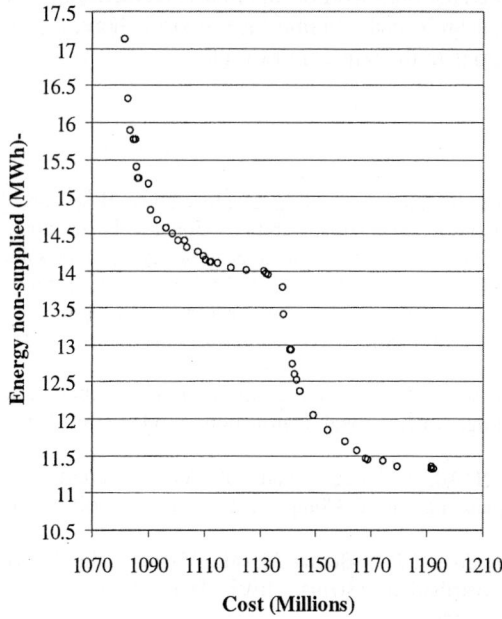

Fig. 10. Approximate Pareto-optimal solutions to the problem with the cost of the new proposed substation size of 40 MVA reduced from 300 millions to 200 millions (case E)

5 Conclusions

Traditionally, the planning problem has been formulated to minimize the economical costs of the system being treated. However, a distribution system involves other aspects such as reliability, environmental and social impact. If solutions to a planning problem are described only in terms of economical costs, it might be difficult to qualify the solutions. If instead the solutions are described in terms of other aspects, there would be more information available to help the planner to compare and select options.

Evolutionary algorithms (EAs) are ideal candidates to be applied on problems considering more than one objective since EAs work with a population of solutions; however, this property of EAs has been little exploited.

In this paper, a new multi-objective method for large-scale power distribution system expansion planning was introduced. The method is based on SPEA2 algorithm. The method has been tested on several multi-objective optimization problems. Some of these are presented in this paper. From the results we concluded that:

- The proposed method was able to find a set of approximate Pareto-optimal solutions, despite the complexity of these problems. One of the difficulties in these problems is that the Pareto-optimal front is not continuous.
- The constraints can cause complications for some planning methods to converge to the Pareto-optimal front and to maintain a diverse set of Pareto-optimal solutions. In these cases, it can be said that the proposed method was success in tackling these difficulties.

- Many multi-objectives optimization methods face difficulties in solving problems with non-convex search space. In one case reported here, the proposed method was able to find solutions in the non-convex region.

References

1. H. L. Willis, Power distribution planning reference book. New York: Marcel Dekker, 1997.
2. M. Vaziri, K. Tomsovic, A. Bose, and T. Gonen, "Distribution expansion problem: formulation and practicality for a multistage globally optimal solution," in Proc. Power Engineering Society Winter Meeting, vol. 3, 2001, pp. 1461 - 1466.
3. N. Kagan and R. N. Adams, "A Benders' decomposition approach to the multi-objective distribution planning problem," Electrical Power and Energy Systems, vol. 15, pp. 259 - 271, 1993.
4. M. T. Ponce de Leao and M. A. Matos, "Multicriteria distribution network planning using simulated annealing," International Transactions in Operational Research, pp. 377 - 391, 1999.
5. N. Kagan and R. N. Adams, "A computational decision support system for electrical distribution systems planning," Proc. Computer Systems and Software Engineering, pp. 133 - 138, May 1992.
6. I. J. Ramírez-Rosado and J. L. Bernal-Agustín, "Genetic algorithms applied to the design of large power distribution systems," IEEE Transactions on Power Systems, vol. 13, pp. 696 - 703, May 1998.
7. E. Zitzler, M. Laumanns, and L. Thiele, "SPEA2: Improving the Strength Pareto Evolutionary Algorithm," Swiss Federal Institute of Technology (ETH), Zurich, Switzerland. Technical report TIK-Report 103, May 2001.
8. B. W. Silverman, Density estimation for statistics and data analysis. London: Chapman and Hall, 1986.
9. G. R. Raidl, "An efficient evolutionary algorithm for the degree-constrained minimum spanning tree problem," Proc. 2000 Congress on Evolutionary Computation, vol. 1, pp. 104 - 111, July 2000.
10. F. Rivas-Davalos and M. R. Irving, "An efficient genetic algorithm for optimal large-scale power distribution network planning," in Proc. Power Tech Conference, IEEE Bologna 2003, vol. 3, pp. 797 - 801.
11. K. Deb, Multi-objective optimization using evolutionary algorithms, 1st ed. Chichester: John Wiley & Sons, 2001.
12. D. Pareto Vilfredo Federigo, Cours d'Économie politique, professé à l'Université de Lausanne: Lausanne, 1896.

Proposition of Selection Operation in a Genetic Algorithm for a Job Shop Rescheduling Problem

Hitoshi Iima

Kyoto Institute of Technology,
Matsugasaki, Sakyo-ku, Kyoto 606–8585, Japan
iima@si.dj.kit.ac.jp

Abstract. This paper deals with a two-objective rescheduling problem in a job shop for alteration of due date. One objective of this problem is to minimize the total tardiness, and the other is to minimize the difference of schedule. A genetic algorithm is proposed, and a new selection operation is particularly introduced to obtain the Pareto optimal solutions in the problem. At every generation in the proposed method, two solutions are picked up as the parents. While one of them is picked up from the population, the other is picked up from the archive solution set. Then, two solutions are selected from these parents and four children generated by means of the crossover and the mutation operation. The candidates selected are not only solutions close to the Pareto-optimal front but also solutions with a smaller value of the total tardiness, because the initial solutions are around the solution in which the total tardiness is zero. For this purpose, the solution space is ranked on the basis of the archive solutions. It is confirmed from the computational result that the proposed method outperforms other methods.

1 Introduction

Although many researchers have so far proposed various methods for solving scheduling problems in manufacturing systems, most studies deal with the determination of a schedule for conventional problems in which the condition is given in advance. However, in real manufacturing systems, alteration of problem condition such as change of due date and addition of job often obliges to revise a schedule worked out previously. Vieira et al. [1] presented definition appropriate for most applications of such rescheduling manufacturing systems, and reviewed various methods to solve the rescheduling problems. Moreover, genetic algorithms (GAs) [2] were also applied to some rescheduling problems [3,4,5]. The GA is an appropriate method to solve rescheduling problems, because diversity of population in the GA is useful in tracking change of problem condition [6].

The aim of most of these methods is to optimize only an original objective function, say the total tardiness. Consequently, the schedule obtained by such a method may be very different from that before the alteration. The difference of schedule incurs time and costs in re-preparing the processing for the case where the problem condition is altered after preparation of processing. As studies considering the schedule difference, Watatani and Fujii [7] defined a problem in which the objective function is a weighted sum of the makespan and the schedule difference, and applied the simulated annealing method to this problem. Abumaizar and Svestka [8] considered a problem for a breakdown of machine, and obtained a schedule with a small difference by rescheduling only the operations affected by the breakdown of machine.

This paper deals with a job shop rescheduling problem of minimizing both the total tardiness and the schedule difference in the case where the due dates of some jobs are altered, and a GA is proposed for obtaining the Pareto optimal solutions. Although a GA with a selection operation was applied to this problem [9] as a preliminary study, it is ineffective in instances with many jobs. In addition, the weight between the objective functions must be given in advance. In order to overcome these disadvantages, a new selection operation is proposed in this paper. In this operation, the solution space is ranked to select not only solutions close to the Pareto-optimal front but also solutions with a smaller value of the total tardiness.

The rest of the paper is organized as follows. Section 2 describes the rescheduling problem as well as the conventional scheduling problem before the alteration. Next, Section 3 presents the GA proposed for the rescheduling problem, and describes the new selection operation. The computational results can be found in Section 4, and the effectiveness of the proposed GA is investigated. Finally, Section 5 concludes the paper.

2 Problem Statement

2.1 Conventional Problem

At the beginning, a conventional problem P^* before alteration is described. In P^*, a set of I kinds of jobs J_i $(i = 1, 2, \cdots, I)$ is processed by using K machines M_k $(k = 1, 2, \cdots, K)$. A machine can process at most one job at a time. A job J_i should be completed by the due date D_i^*. Moreover, J_i consists of K operations O_{ij} $(j = 1, 2, \cdots, K)$. An operation O_{ij} is executed on $M_{R(i,j)}$ $(R(i,j) \in \{1, 2, \cdots, K\})$, which is given in advance, and its processing time is given as PT_{ij}. No preemption of operation is allowed. There exists a precedence constraint between operations belonging to a job, and the operations must be executed in the order of j. This constraint is often called the technological constraint. The total number of operations is denoted as Q $(Q = IK)$.

The problem P^* is to determine the completion time c_{ij}^* of O_{ij} in such a way that the total tardiness F_1^* should be minimized. The objective function F_1^* is formulated as

$$F_1^* = \sum_{i=1}^{I} \max(c_{iK}^* - D_i^*, 0). \qquad (1)$$

By applying a GA to P^*, the population at the final generation and the best solution (schedule) S^* can be obtained.

2.2 Rescheduling Problem

Consider the situation that the production has been prepared on the basis of the best schedule S^* obtained for the conventional problem P^*. After the preparation, some due dates are altered. The due date of J_i after the alteration is denoted as D_i. Since S^* may not be optimal for D_i no longer, it should be revised. In this situation, the second objective function is defined as the magnitude of difference between S^* and a schedule S revised.

The rescheduling problem P coped with in this paper is to determine the completion time c_{ij} of O_{ij} in such a way that both the total tardiness F_1 and the schedule difference F_2 should be minimized. The objective functions in P are formulated as

$$\min \begin{pmatrix} F_1 \\ F_2 \end{pmatrix} = \begin{pmatrix} \sum_{i=1}^{I} \max(c_{iK} - D_i,\ 0) \\ \sum_{i=1}^{I} \sum_{j=1}^{K} |a_{ij} \cap a_{ij}^*| \end{pmatrix} \qquad (2)$$

where a_{ij} and a_{ij}^* are the set of operations executed before O_{ij} on $M_{R(i,j)}$ in S and that after O_{ij} on $M_{R(i,j)}$ in S^*, respectively. Moreover, the symbol $|\ |$ means the number of elements. The schedule difference F_2 is given on the basis of Watatani's definition [7]. If there exists an operation belonging to both a_{ij} and a_{ij}^*, it means that the processing order of the operation and O_{ij} is reversed.

3 Application of Genetic Algorithm

GAs should be appropriately designed for respective optimization problems. In this section, a GA is designed for solving the rescheduling problem P. The population size and the final generation in this GA are denoted as N and G, respectively.

3.1 Individual Description

So far, many GAs have been proposed for solving job shop scheduling problems [10, 11, 12]. Recently Gen et al. [13], Ono et al. [14] and Shi et al. [15] proposed similar individual descriptions in which job numbers are sequenced, and obtained good solutions with less computational consumption.

The individual description proposed by Shi et al. is used in the proposed GA. This individual description is designed by utilizing the precedence constraint between operations effectively. The genotype is expressed by sequencing job numbers $\{i\}$ ($i = 1, 2, \cdots, I$) K times. The length of the sequence is equal to the total number Q of operations. The genotype represents basically a production order, and the schedule of operation corresponding to a gene is determined in turn according to the decoding procedure. The operation is uniquely determined from the precedence constraint. The completion time c_{ij} of each operation O_{ij} is calculated from the genotype in the following way.

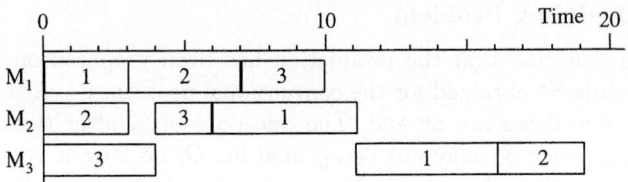

Fig. 1. Example of decoding genotype {213311223}

Table 1. Example of instance data ($I=3, K=3$)

	O_{11}	O_{12}	O_{13}	O_{21}	O_{22}	O_{23}	O_{31}	O_{32}	O_{33}
$R(i,j)$	1	2	3	2	1	3	3	2	1
PT_{ij}	3	5	5	3	4	3	4	2	3

Step 1 Set $m_k \leftarrow 0$ ($\forall k = 1, 2, \cdots, K$). The variable m_k represents the completion time of the operation to be executed last on the machine M_k.

Step 2 Set $j(i) \leftarrow 1$ ($\forall i = 1, 2, \cdots, I$). The variable $j(i)$ represents the operation number to be executed next in the job J_i.

Step 3 Set $\ell \leftarrow 1$. The variable ℓ represents the locus.

Step 4 The completion time of operation $O_{i_\ell j(i_\ell)}$ corresponding to the gene i_ℓ at the ℓ-th locus is given by

$$c_{i_\ell j(i_\ell)} = \max(m_{R(i_\ell, j(i_\ell))}, c_{i_\ell \; j(i_\ell)-1}) + PT_{i_\ell j(i_\ell)}. \tag{3}$$

Note that $c_{i0}=0$.

Step 5 Set $m_{R(i_\ell, j(i_\ell))} \leftarrow c_{i_\ell j(i_\ell)}$.

Step 6 If $j(i_\ell) < K$, set $j(i_\ell) \leftarrow j(i_\ell) + 1$.

Step 7 If $\ell = Q$, terminate this procedure. If not, set $\ell \leftarrow \ell + 1$ and return to Step 4.

The decoding procedure is explained by using an illustrative example. Figure 1 depicts the solution corresponding to genotype {213311223} in the instance shown by Table 1. At the beginning, the schedule of O_{21} corresponding to the first gene {2} is determined. Since $R(2,1)=2$ and $PT_{21}=3$, O_{21} is started at zero on M_2 and is completed at three. Next, the schedule of O_{11} and O_{31}, which correspond to the subsequence {13}, is determined in a similar way. Next, the fourth gene {3} is picked up. Since it is the second time this kind of gene is picked up ($j(3)=2$), the second operation O_{32} of J_3 corresponds to the gene. Since $m_2=3$ and $c_{31}=4$, O_{32} is started at four. Finally, the schedule of the remaining operations is determined in a similar way.

3.2 Initial Population

The initial population in GAs is generated randomly for optimization problems in general. Therefore the initial population is far from the Pareto optimal solutions,

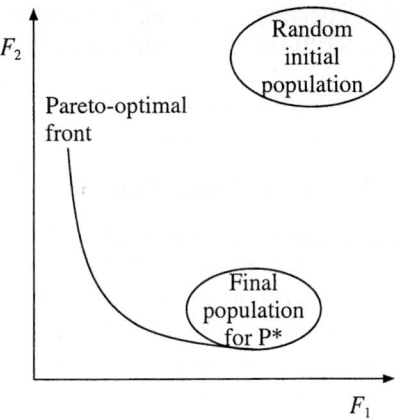

Fig. 2. Initial population of GA for a two-objective problem

and the population approaches them gradually by the exploration of GA, as shown in Fig. 2. On the other hand, the final population and the best schedule S^* obtained by applying a GA to the conventional problem P* can be utilized as the initial population for P. Since the schedule difference F_2 is zero for $S = S^*$, one of the Pareto optimal solutions is already obtained for P. Moreover, the other solutions in the final population are also closer to S^* than the random initial population, as shown in Fig. 2. Therefore, the final population is used as the initial population in the proposed GA. However, the diversity for the final population is lost, because a GA has been performed for P*. Thus, diverse initial solutions are generated by changing some solutions in the final population. These solutions changed are the non-dominated solutions in the final population in order to obtain a good solution set. The detail procedure to generate the initial population is as follows.

Step 1 Find the n' non-dominated solutions $y_1, y_2, \cdots, y_{n'}$ for P from the final population obtained for P*, and add these solutions to the initial population for P. Set $n \leftarrow 1$.

Step 2 Generate a solution by shifting a gene selected randomly in y_n just before another gene selected randomly, and add the solution to the initial population.

Step 3 If the N initial solutions are generated, terminate this procedure.

Step 4 Set $n \leftarrow n + 1$. If $n > n'$, set $n \leftarrow 1$.

Step 5 Return to Step 2.

Rescheduling may be invoked during the actual processing of jobs. In this case, the schedule of operations completed at the time is unchanged. The schedule of operations which are being executed at the time is also unchanged. The genes associated with their operations are removed from the initial population, and the operations of GA are applied to the remaining genes.

$$\begin{aligned}&\text{pa}^1: 112321332 &\rightarrow&\quad \text{ch}^1: 131221323\\&\text{pa}^2: 321213123 &\rightarrow&\quad \text{ch}^2: 213213132\end{aligned}$$

Fig. 3. Example of SPX

3.3 Crossover and Mutation Operation

As the crossover operation, the set partition crossover (SPX) proposed by Shi et al. [15] is used. SPX is effective from the viewpoint of the computation time. The procedure of SPX is explained as follows. In the following explanation, pa_ℓ^1 and pa_ℓ^2 are defined as the gene at the ℓ-th locus in two parents pa^1 and pa^2, respectively, and ch_ℓ^1 and ch_ℓ^2 are defined as the gene at the ℓ-th locus in two children ch^1 and ch^2, respectively.

Step 1 Separate the set of all genes into two subsets SU_a and SU_b, which are nonempty and exclusive.
Step 2 Set $\ell \leftarrow 1$, $\ell_1 \leftarrow 1$ and $\ell_2 \leftarrow 1$.
Step 3 If $pa_\ell^1 \in SU_a$, set $ch_{\ell_1}^1 \leftarrow pa_\ell^1$ and $\ell_1 \leftarrow \ell_1 + 1$. Go to Step 5.
Step 4 Set $ch_{\ell_2}^2 \leftarrow pa_\ell^1$ and $\ell_2 \leftarrow \ell_2 + 1$.
Step 5 If $pa_\ell^2 \in SU_b$, set $ch_{\ell_1}^1 \leftarrow pa_\ell^2$ and $\ell_1 \leftarrow \ell_1 + 1$. Go to Step 7.
Step 6 Set $ch_{\ell_2}^2 \leftarrow pa_\ell^2$ and $\ell_2 \leftarrow \ell_2 + 1$.
Step 7 If $\ell = Q$, terminate this procedure. If not, set $\ell \leftarrow \ell + 1$ and return to Step 3.

Figure 3 depicts an example of SPX, where $SU_a = \{1, 2\}$ and $SU_b = \{3\}$. As shown in this figure, SPX preserves the order of genes in each subset.

In the mutation operation, two genes are selected randomly, and each of them is shifted just before another gene selected randomly.

3.4 Flow of GA

Most of selection operations for multiobjective optimization problems are based on the multi-objective GA (MOGA) [16] proposed by Fonseca and Fleming. In MOGA the rank of solution is given on the basis of the number of solutions which dominate itself, and solutions close to the Pareto-optimal front tend to be selected. In a GA with such a selection operation, solutions out of edge of incumbent population are not actively explored, and the exploration progresses in the direction of arrow (a) in Fig. 4. Even if this GA is applied to P, it is hard to obtain Pareto optimal solutions with a smaller F_1 and a larger F_2 because the initial solutions are around S^*. In order to obtain all the Pareto optimal solutions, the exploration must progress in the direction of not only arrow (a) but also arrow (b).

In this paper, a new selection operation, the selection by area ranking (SAR), is used to obtain diverse Pareto optimal solutions. In SAR, solutions are selected by using the solution space ranked on the basis of the archive solution set X, which is the non-dominated solution set in the solutions explored by the incumbent generation. The solution space is ranked as follows.

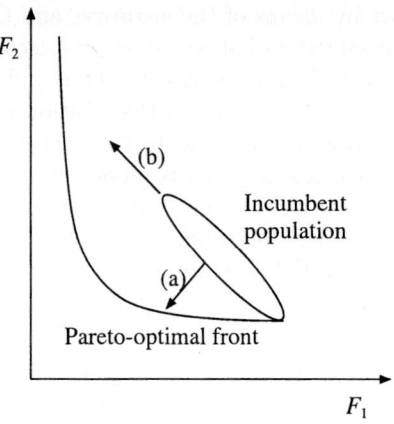

Fig. 4. Direction of exploration

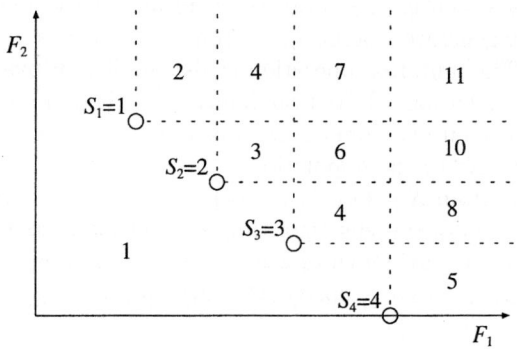

○ : Archive solution

Fig. 5. Example of area ranking

Step 1 Find the area in which all the solutions are not dominated by any solution of X, and set the rank of the area to one.

Step 2 Number all the solutions of X in the order of F_1, and set the score S_m of m-th solution to m ($S_m = m$).

Step 3 For the area in which the solution x is dominated by at least one archive solution of X, and set the rank of the area to $\sum_{m' \in X_{sub}} S_{m'} + 1$, where X_{sub} is the set of the archive solution numbers dominating x.

A solution with a small rank is close to the Pareto-optimal front, or has a small F_1. An example of area ranking is shown in Fig. 5 for $|X| = 4$.

At every generation in the proposed GA, two solutions are picked up as the parents. While one of them is picked up randomly from the population, the other is picked up randomly from X in order to explore an area around X intensively. The former solution is removed from the population. Next, four children are generated from the parents by means of the crossover and the mutation operation. Next, the two solutions with the smallest ranks are selected from the two parents and the four children, and then are added to the population. If the candidates for the second solution selected are plural, the solution with the minimum value of F_1 is selected from these candidates. In this selection, the population size increases, because only a single solution is picked up as a parent from the population. In order to prevent this, another solution selected randomly is removed from the population before the selection.

The flow of the proposed GA is as follows.

Step 1   Generate the initial population, and set $g \leftarrow 1$. The variable g represents the generation.
Step 2   Determine the archive solution set X from the initial population.
Step 3   As the parents, pick up a solution x_1 randomly from the population and a solution x_2 randomly from X. Remove x_1 from the population.
Step 4   Generate two solutions x_3 and x_4 by means of the crossover operation. The crossover operation is applied at every generation.
Step 5   Generate a solution x_5 from x_1 by means of the mutation operation. Similarly, generate a solution x_6 from x_2 by means of the mutation operation. The mutation operation is also applied at every generation.
Step 6   Remove a solution selected randomly from the population.
Step 7   Select two solutions from the solutions $\{x_1, x_2, \cdots, x_6\}$ by means of SAR, and add them to the population.
Step 8   Update X from $X \cup \{x_3, x_4, x_5, x_6\}$.
Step 9   If $g = G$, terminate this algorithm and output X as the answer. If not, set $g \leftarrow g + 1$ and return to Step 3.

The archive solution set X has no size limitation, because the size remains relatively small for this problem.

4   Computational Result

The proposed GA is applied to forty instances in order to evaluate its effectiveness.

4.1   Instance Data

The forty instances for the conventional problem P* are given by revising the benchmark instances la01–la40 [17]. Of the instance data, the machine number $R(i,j)$ and the processing time PT_{ij} of operation O_{ij} are the same as those of benchmark instance. The due date D_i^* of each job J_i is given by the total processing time $\sum_{j=1}^{K} PT_{ij}$ times a constant d. The constant d for each instance is given in such a way that the total tardiness F_1^* obtained by applying a GA to this instance is less than one hundred. The constant d, the number I of jobs, the number K of machines and the total number Q of operations are shown in Table 2.

Table 2. Instance data

Instance	d	I	K	Q	Instance	d	I	K	Q
1	2.10				21	1.85			
2	2.00				22	2.00			
3	2.10	10	5	50	23	1.80	15	10	150
4	2.20				24	1.85			
5	2.10				25	1.90			
6	3.00				26	2.30			
7	2.90				27	2.30			
8	2.80	15	5	75	28	2.30	20	10	200
9	3.00				29	2.40			
10	2.90				30	2.40			
11	3.40				31	3.20			
12	3.40				32	3.10			
13	3.30	20	5	100	33	3.30	30	10	300
14	3.80				34	3.17			
15	3.70				35	3.35			
16	1.50				36	1.65			
17	1.60				37	1.65			
18	1.55	10	10	100	38	1.63	15	15	225
19	1.55				39	1.60			
20	1.50				40	1.59			

A GA has been applied to these instances, and the final population and the best schedule S^* have been obtained. In order to generate the instances for the rescheduling problem P, the due date is altered for a few jobs on the basis of S^*. These jobs are selected randomly from the jobs such that the tardiness is zero in S^*. The number of the jobs selected is nearly ten percent of I. The due date D_i of each of the jobs $\{J_i\}$ is altered to $0.9c_i^*$, where c_i^* is the completion time for S^*. Hence, the jobs become tardy for $S = S^*$.

4.2 Comparison Among Procedures to Pick Up the Parents

In the proposed GA (called PGA), one of parents is picked up from the population, and the other is picked up from the archive solution set X. In order to evaluate the effectiveness of this procedure, PGA is compared with the following methods.

· GA-P : Both parents are picked up from the population.
· GA-X : Both parents are picked up from X.

In these GAs, there are two parameters: the population size N and the final generation G. The values of these parameters are decided through a preliminary calculation. The value of N is set to 500, and the value of G is set as shown in Table 3. Each GA is performed one hundred times with various random seeds for an instance.

The GAs are evaluated by the coverage metric [18] which means relation of domination between the solution sets obtained by two methods. The coverage

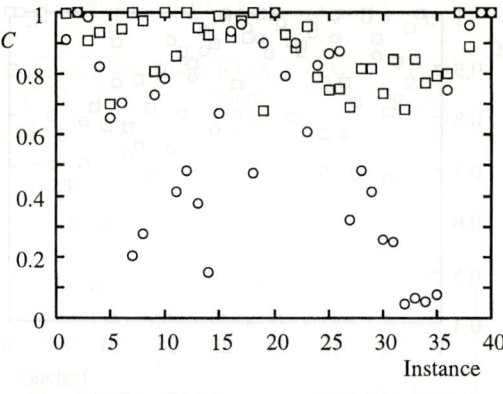

Fig. 8. Comparison between PGA and GA-M

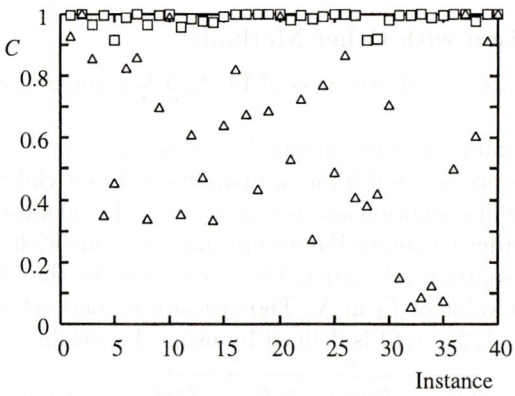

Fig. 9. Comparison between PGA and SPEA2

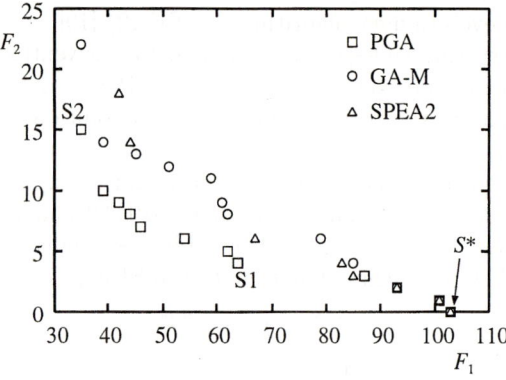

Fig. 10. Solution set obtained for Instance 8

a lower fitness value is better in SPEA2. By using ff^S, solutions with a smaller F_1 tend to be selected.

The population size N in GA-M and SPEA2 is as large as that in PGA. The final generation G is given in such a way that the computation time is the same, and is shown in Table 3. SPEA2 is a generational GA, and the distance between solutions in the population is calculated at every generation. Therefore, G in SPEA2 is smaller than those in the other GAs.

Figure 8 depicts the average coverage metric between PGA and GA-M. It is found from this figure that PGA is better than GA-M. In particular, PGA is much better than GA-M for Instances 11–15 and 26–35 with $I \geq 20$. Figure 9 depicts the average coverage metric between PGA and SPEA2. It is found from

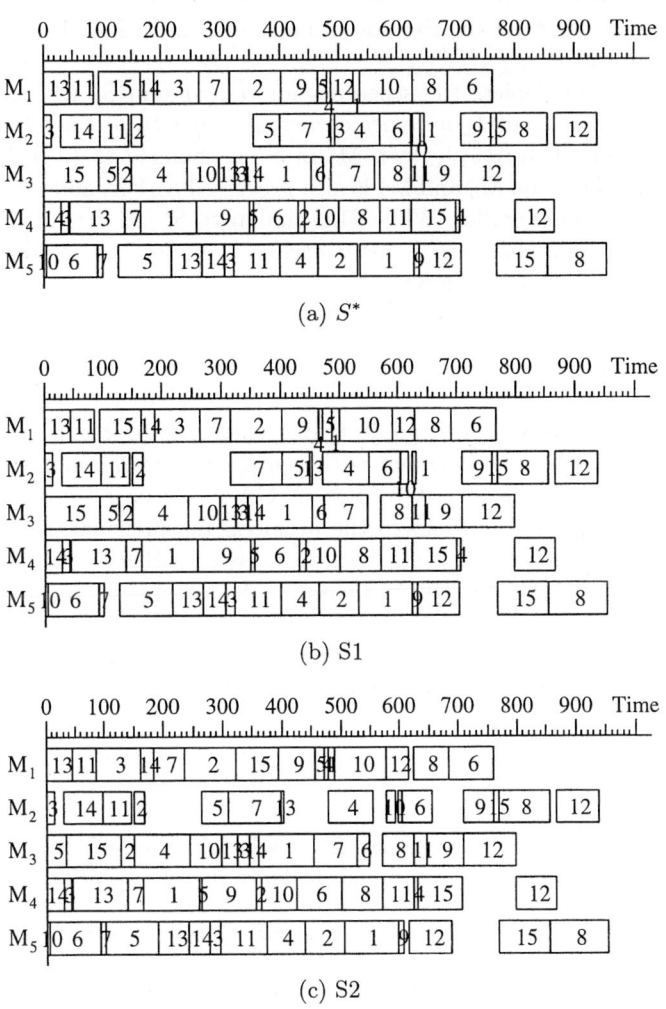

Fig. 11. Schedules obtained for Instance 8

this figure that PGA is better than SPEA2 for almost all instances. Since the value of G is relatively small in SPEA2, the search process may not converge still for the value given. PGA is more applicable than SPEA2, because a solution set should be obtained in short time for rescheduling problems.

Next, typical solution sets obtained by PGA, GA-M and SPEA2 are shown in Fig. 10 for Instance 8. These are results in a single trial, and each computation time is about 0.7 second by using a 2.4 GHz, Pentium IV PC, running under LINUX. As shown in this figure, the solutions obtained by PGA are closer to the Pareto-optimal front, and are distributed widely. A decision maker can confirm the total tardiness and the schedule difference of each schedule by the figure, and determine a desirable schedule in accordance with the situation at that time. The solutions obtained by GA-M and SPEA2 in $F_1 > 90$ are the same as those by PGA. Although the solutions obtained by GA-M in $F_1 < 90$ are distributed widely, they are far from the Pareto-optimal front. In SPEA2 few solutions are obtained in $F_1 < 80$.

Finally, Fig. 11 depicts the schedules for S^*, S1 and S2 in Fig. 10. In Instance 8 the due date of J_1 becomes earlier. In order to process J_1 earlier, the processing order $\{J_5 J_4 J_{12} J_1 J_{10}\}$ on M_1 in S^* is changed to $\{J_4 J_5 J_1 J_{10} J_{12}\}$ in S1. Consequently, J_1 on M_5 and M_2 is also processed earlier, and the completion time c_1 of J_1 becomes earlier. While S1 is similar to S^*, S2 is relatively different from S^*. In order to complete J_1 much earlier, the idle time at about time 100 on M_1 and M_5 in S1 is lost in S2.

5 Conclusion

This paper has dealt with a two-objective rescheduling problem after alteration of due date in a job shop. The aim of this problem is to minimize the schedule difference as well as the total tardiness. A GA has been proposed for obtaining the Pareto optimal solutions in the problem. In particular, a new selection operation by the area ranking has been proposed. In this operation the candidates selected are not only solutions close to the Pareto-optimal front but also solutions with a smaller value of the total tardiness. It is concluded from the computational result that solutions obtained by the proposed GA are closer to the Pareto-optimal front than those by other GAs.

References

1. Vieira, G.E., Herrmann, J.W., Lin, E.: Rescheduling manufacturing systems: A framework of strategies, policies and methods. Journal of Scheduling, 6 (2003) 39–62
2. Goldberg, D.E.: Genetic Algorithms in Search, Optimization and Machine Learning. Addison–Wesley (1989)
3. Lin, S., Goodman, E.D., Punch, W.F.III: A genetic algorithm approach to dynamic job shop scheduling problems. Proceedings of 7th International Conference on Genetic Algorithms (1997) 481–488

4. Bierwirth, C., Mattfeld, D.C.: Production scheduling and rescheduling with genetic algorithms. IEEE Transactions on Evolutionary Computation **7** (1999) 1–17
5. Branke, J., Mattfeld, D.C.: Anticipation in dynamic optimization: The scheduling case. Proceedings of Parallel Problem Solving from Nature VI (2000) 253–262
6. Branke, J.: Evolutionary Optimization in Dynamic Environments. Kluwer Academic Publishers, Norwell (2002)
7. Watatani, Y., Fujii, S.: A study on rescheduling policy in production system. JAPAN/USA Symposium on Flexible Automation **2** (1992) 1147–1150
8. Abumaizar, R.J., Svestka, J.A.: Rescheduling job shops under random disruptions. International Journal of Production Research **35** (1997) 2065–2082
9. Iima, H., Nakase R., Sannomiya, N.: Genetic algorithm approach to a multiobjective rescheduling problem in a job shop, Proceedings of 1st Multidisciplinary International Conference on Scheduling: Theory and Applications (2003) 422–437
10. Davis, L.: Job shop scheduling with genetic algorithms. Proceedings of First International Conference on Genetic Algorithms (1985) 136–140
11. Nakano, R.: Conventional genetic algorithm for job shop problems. Proceedings of Fourth International Conference on Genetic Algorithms (1991) 474–479
12. Fang, H.L., Ross, P., Corne, D.: A promising genetic algorithm approach to jobshop scheduling, rescheduling, and open-shop scheduling problems. Proceedings of Fifth International Conference on Genetic Algorithms (1993) 375–382
13. Gen, M., Cheng, R.: Genetic Algorithms & Engineering Design. John Wiley & Sons, Inc. (1997)
14. Ono, I., Yamamura, M., Kobayashi, S.: A genetic algorithm for job-shop scheduling problems using job-based order crossover. Proceedings of 1996 IEEE International Conference on Evolutionary Computation (1996) 547–552
15. Shi, G., Iima, H., Sannomiya, N.: A new encoding scheme for solving job shop problems by genetic algorithm. Proceedings of 35th IEEE Conference on Decision and Control (1996) 4395–4400
16. Fonseca C.C., Fleming P.J.: Genetic algorithms for multiobjective optimization: formulation, discussion and generalization. Proceedings of 5th International Conference on Genetic Algorithms (1993) 416–423
17. Lawrence S.: Resource constrained project scheduling: an experimental investigation of heuristic scheduling techniques (Supplement). Graduate School of Industrial Administration, Carnegie-Mellon University (1984)
18. Zitzler, E., Thiele, L.: Multiobjective evolutionary algorithms: A comparative case study and the strength Pareto approach. IEEE Transactions on Evolutionary Computation **3** (1999) 257–271
19. Zitzler, E., Laumanns, M., Thiele, L.: SPEA2: Improving the performance of the strength Pareto evolutionary algorithm. Technical Report 103, Computer Engineering and Communication Networks Lab, Swiss Federal Institute of Technology (2001)

A Two-Level Evolutionary Approach to Multi-criterion Optimization of Water Supply Systems

Matteo Nicolini

University of Udine, Faculty of Engineering,
Dipartimento di Georisorse e Territorio, via Cotonificio 114, 33100 Udine (Italy)
nicolini@dgt.uniud.it

Abstract. Purpose of the paper is to introduce a methodology for a parameter-free multi-criterion optimization of water distribution networks. It is based on a two-level approach, with a population of inner multi-objective genetic algorithms (MOGAs) and an outer simple GA (without crossover). The inner MOGAs represent the network optimizers, while the outer GA – the *meta* GA – is a supervisor process adapting mutation and crossover probabilities of the inner MOGAs. The hypervolume metric has been adopted as fitness for the individuals at the meta-level. The methodology has been applied to a small system often studied in the literature, for which an exhaustive search of the entire decision space has allowed the determination of all Pareto-optimal solutions of interest: the choice of this simple system was done in order to compare the hypervolume metric to two performance measures (a convergence and a sparsity index) introduced on purpose. Simulations carried out show how the proposed procedure proves robust, giving better results than a MOGA alone, thus allowing a considerable ease in the network optimization process.

1 Introduction

The problem of choosing the optimal combination of pipe diameters, in order to minimize the overall cost of a looped water distribution system (given a finite set of commercial available sizes), is proven to be NP-hard [1]. In the last decades, many authors have proposed several approaches based on different optimization techniques, mainly linear programming [2], [3], [4], [5], [6], [7] and non-linear programming [8], [9].

More recently, several researchers have applied genetic algorithms (GAs) to single-objective optimization of water supply systems, introducing some improvements with respect to the simple GA, [10], [11], [12], [13]. [14] applied GAs to optimal location of control valves, while [15] and [16] to leak detection and calibration problems; [17] used GAs for optimal scheduling of pipe replacement.

[18] have shown that networks designed taking into account only cost minimization (and in the case of just one loading condition) tend to branched config-

urations, as also pointed out by [19]. In a recent editorial, [20] stressed the need of adopting a multi-objective approach for the design of water supply systems.

These last years have seen an increasing number of applications of multi-objective optimization algorithms: generally, only two-objective problems have been considered, the first criterion being the total cost of the system and the second representing a measure of the network performance: [21] adopted for the first time a multi-objective algorithm for water network rehabilitation, minimizing cost and maximizing benefits; [22] considered the minimization of cost and of the maximum pressure deficit at nodes; [23] took into account the maximization of entropy or demand supply ratio, while [24] and [25] the maximization of the reliability of the system.

A multi-objective evolutionary algorithm (MOEA) has two main goals [26]: firstly, to find a set of solutions as close as possible to the Pareto optimal front; secondly, to find a set of solutions as diverse as possible. However, the performance of the algorithm is quite affected by crossover and mutation type and probability: as a result, many runs with different starting populations and parameter sets are usually performed in order to find a good population of non-dominated solutions.

In this paper, a different approach is proposed, consisting of a population of MOGAs at the inner level, and an outer single-objective GA (meta GA) controlling the MOGAs crossover and mutation probabilities. The fitness of each individual of the meta GA is given by the hypervolume (that is, the amount of the objective space dominated by the obtained non-dominated front, [27], [28]) obtained by the inner MOGA it represents.

This methodology reconsiders some ideas of [29] and [30], and is *non-self-adaptive* [31], thus basically different from the *self-adaptive* mechanism based on the inclusion of operators and control parameters within the individual representation, [32], [33].

In order to asses the validity of the hypervolume metric, it has been compared to two performance measures, namely a convergence and a sparsity index [34], which quantify the exploitation and exploration issues of the inner MOGAs.

The paper is organized as follows: in Section 2, the mathematical formulation of the problem is presented, together with the test problem adopted for the numerical analyses; Section 3 describes the performance metrics, while Section 4 the two-level approach; Section 5 presents the results obtained and Section 6 some concluding remarks.

2 Two-Objective Water Supply System Optimization

2.1 Mathematical Formulation

The problem is formulated as the minimization of the total cost of the network and the maximization of the minimum pressure level at nodes: for pressure level, we mean the deviation from the required pressure (see Figure 1 for an explanation), and hence both negative and positive values are allowed; however, in this work, the attention is focused only on negative values, indicating situations of pressure deficit (the maximum bound on the pressure level is then zero). The problem is constrained by continuity of mass at every node and energy conservation along every path in the system, giving:

$$\min f_1(d_1,\ldots,d_{N_p}) = \sum_{i \in D} \sum_{j=1}^{N_p} c(d_i) L_{ij} \tag{1}$$

$$\max f_2(d_1,\ldots,d_{N_p}) = \min_{k=1,\ldots,N_n} \left[\min(H_k - H_k^{req}, 0) \right] \tag{2}$$

s. to:
$$\sum_{i \in n_k} Q_i - \sum_{j \in m_k} Q_j = Q_{e,k} \tag{3}$$

$$\sum_{i \in p_j} h_{f,i} = \Delta E_j \tag{4}$$

$$H_k \geq H_k^{req} \tag{5}$$

$$d_{\min} \leq d_i \leq d_{\max} \tag{6}$$

in which

- $d_1,\ldots,d_{N_p}$ are the N_p (number of pipes in the network) decision variables;
- $c(d_i)$ and L_{ij} are, respectively, the cost per unit length and the total length of pipe j whose diameter is d_i;
- D is the set of the N_D available commercial diameters (whose minimum and maximum sizes are $d_{\min}$ and $d_{\max}$, respectively);
- N_n is the number of supply nodes in the system;
- H_k is the actual piezometric head at node k;
- H_k^{req} is the required pressure at node k;
- $Q_{i,k}$ and $Q_{j,k}$ respectively represent the n_k and m_k flows entering or leaving node k;
- $Q_{e,k}$ is the erogated flow at node k;
- p_j is the number of links belonging to path j;
- $h_{f,i}$ represents the energy loss in link i of path j;
- ΔE_j is the total energy loss along path j: for a closed loop, $\Delta E_j = 0$.

Equations (3) and (4) are guaranteed by the hydraulic simulator (in this work EPANET 2 [35]), to which the optimizer has been coupled. The following expression for the energy loss, h_f, has been adopted:

$$h_f = 10.668 \frac{Q^{1.852} L}{C_{HW}^{1.852} d^{4.871}} \tag{7}$$

in which Q is the discharge in the pipe (m³/s), L the length (m), d the diameter, and C_{HW} is the Hazen–Williams (adimensional) pipe roughness coefficient.

2.2 Test Problem Adopted for the Analyses

The two-loop network illustrated in Figure 2 has been considered, [2]: all links are 1000 m long, with a Hazen-Williams coefficient $C_{HW} = 130$. Nodal characteristics are also shown on the figure, while the available commercial diameters

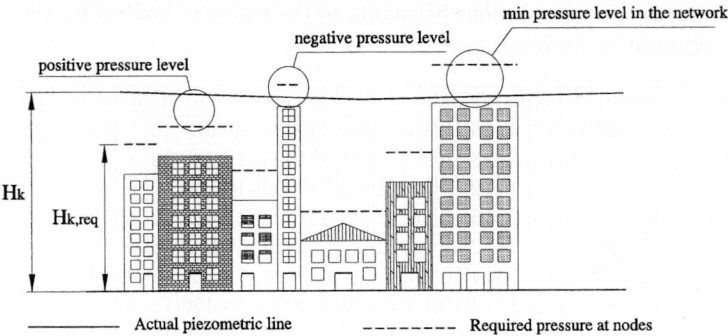

Fig. 1. Actual piezometric line and required pressures in a water supply system

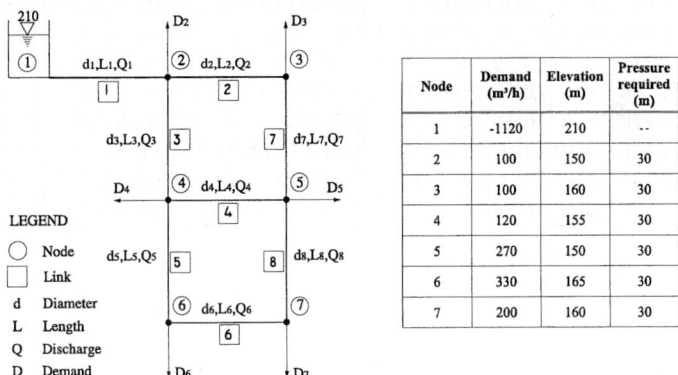

Fig. 2. Two-loop network adopted for the numerical analyses

(inches) are: 1, 2, 3, 4, 6, 8, 10, 12, 14, 16, 18, 10, 22, 24, and their respective costs per unit length ($/m): 2, 5, 8, 11, 16, 23, 32, 50, 60, 90, 130, 170, 300, 550.

The decision space consists of 14^8 configurations, and has been totally explored by exhaustive search (requiring nearly 50 hours of CPU time on a Pentium III 1GHz). In particular, the subsequent analyses have been focused only on the region of interest (ROI) of the objective space characterized by configurations having minimum pressure level at nodes not below -30 meters (Figure 3): this resulted in 647691 solutions inside the ROI, of which only 38 are Pareto-optimal, and are reported on Table 1.

A problem which arises when considering such multi-objective problem is the huge dimension of the Pareto front: there are actually non-dominated individuals characterized by not realistic pressure levels (extremely negative numbers). In order to avoid such (useless) configurations, the search has been biased towards the solutions inside the ROI through a bending of the Pareto front as indicated

Table 1. Pareto-optimal solutions belonging to the region of interest for the two-loop network optimization problem

n	f_1 (m)	f_2 (m)	d_1	d_2	d_3	d_4	d_5	d_6	d_7	d_8	n	f_1 (m)	f_2 (m)	d_1	d_2	d_3	d_4	d_5	d_6	d_7	d_8
1	419000	0.000	18	10	16	4	16	10	10	1	20	336000	-9.851	16	8	16	8	14	10	6	1
2	415000	-0.387	18	10	16	6	16	10	8	1	21	331000	-10.449	16	8	16	8	14	10	4	1
3	414000	-1.155	18	14	14	1	12	1	14	12	22	330000	-10.964	16	6	16	10	14	10	1	3
4	413000	-1.453	18	10	16	2	16	10	10	1	23	327000	-11.019	16	6	16	10	14	10	1	2
5	408000	-1.718	18	14	14	3	12	3	14	10	24	324000	-11.043	16	6	16	10	14	10	1	1
6	407000	-1.844	18	12	16	1	14	10	10	4	25	310000	-11.159	16	10	14	1	14	10	10	1
7	404000	-1.853	18	12	16	1	14	10	10	3	26	309000	-14.456	16	8	14	6	14	10	8	2
8	401000	-1.863	18	12	16	1	14	10	10	2	27	306000	-14.738	16	8	14	6	14	10	8	1
9	398000	-1.879	18	12	16	1	14	10	10	1	28	301000	-17.340	16	8	14	8	14	10	4	1
10	380000	-1.909	18	10	16	1	14	10	10	1	29	300000	-17.704	16	10	14	1	12	10	10	1
11	379000	-4.615	18	8	16	8	14	10	6	2	30	294000	-17.847	16	6	14	8	14	10	4	1
12	376000	-4.619	18	8	16	8	14	10	6	1	31	291000	-19.740	16	6	14	8	14	10	3	1
13	371000	-5.216	18	8	16	8	14	10	4	1	32	288000	-21.656	16	6	14	8	14	10	2	1
14	370000	-5.731	18	6	16	10	14	10	1	3	33	280000	-22.143	14	10	14	1	14	10	10	1
15	367000	-5.786	18	6	16	10	14	10	1	2	34	278000	-25.095	16	6	14	8	12	10	2	1
16	364000	-5.810	18	6	16	10	14	10	1	1	35	275000	-25.273	16	6	14	8	12	10	1	1
17	350000	-5.926	18	10	14	1	14	10	10	1	36	271000	-28.324	14	8	14	8	14	10	4	1
18	340000	-7.141	16	10	16	1	14	10	10	1	37	270000	-28.688	14	10	14	1	12	10	10	1
19	339000	-9.848	16	8	16	8	14	10	6	2	38	264000	-28.831	14	6	14	8	14	10	4	1

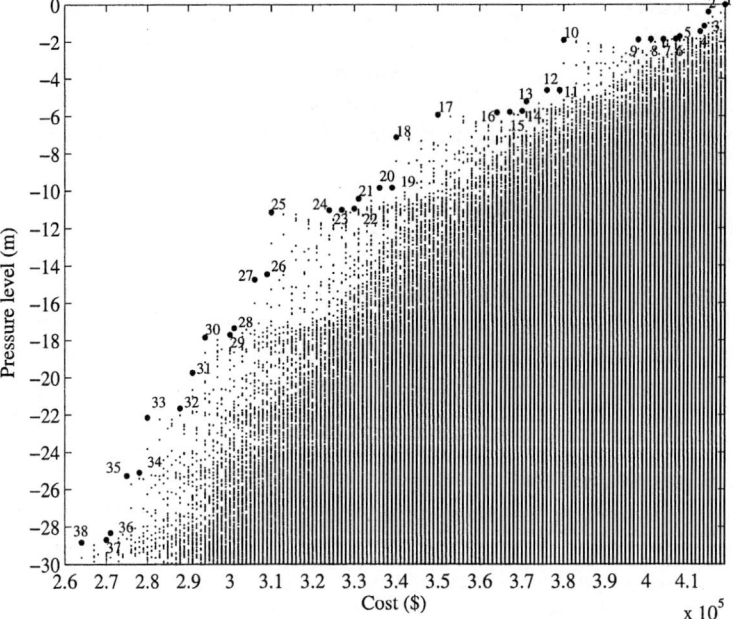

Fig. 3. Region of interest of the objective space with Pareto-optimal solutions evidentiated

in Figure 4, thus transforming all Pareto-optimal solutions outside the ROI into dominated individuals: mathematically, this has been achieved through a slight change in the first objective function, namely:

$$f_1(d_1,\ldots,d_{N_p}) = \begin{cases} \sum_{i \in D} \sum_{j=1}^{N_p} c(d_i) L_{ij} & \text{if } f_2(d_1,\ldots,d_{N_p}) \geq -30.0; \\ \sum_{i \in D} \sum_{j=1}^{N_p} c(d_i) L_{ij} + p\left[-f_2(d_1,\ldots,d_{N_p}) - 30.0\right] & \text{otherwise.} \end{cases} \quad (8)$$

in which $p > 0$ is a penalty factor.

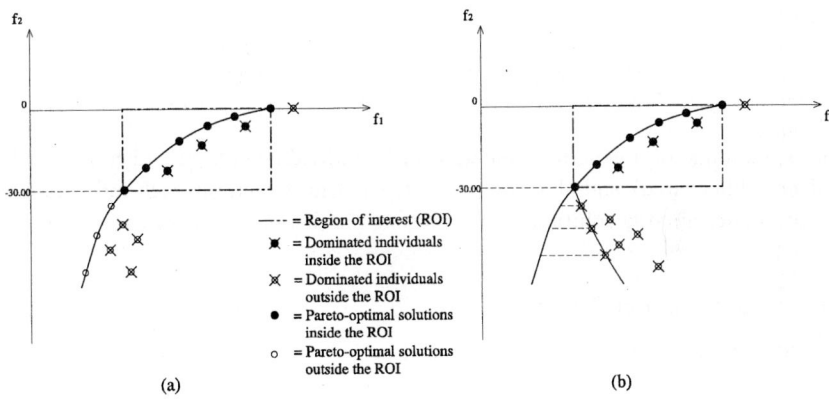

Fig. 4. Bending of the Pareto front in order to bias the search only on configurations of interest: (a), Pareto front with non-dominated solutions evidentiated; (b), preferred region of the frontier, with Pareto-optimal solutions outside the ROI being now dominated

3 Performance Metrics

Two kinds of performance measures have been considered:

1. the hypervolume, $\mathcal{HV}$, which quantifies with only one scalar the amount of criterion space dominated by the current non-dominated front, [26], [27], [28]. Since we knew the Pareto-optimal solutions (POS), it was decided to adimentionalize the metric with respect to its maximum value.
2. Two indices representing, respectively, the convergence towards the non-dominated solutions and the distribution of the individuals in the generic population along the front [34].

In the following, the convergence and sparsity indices are described in more detail.

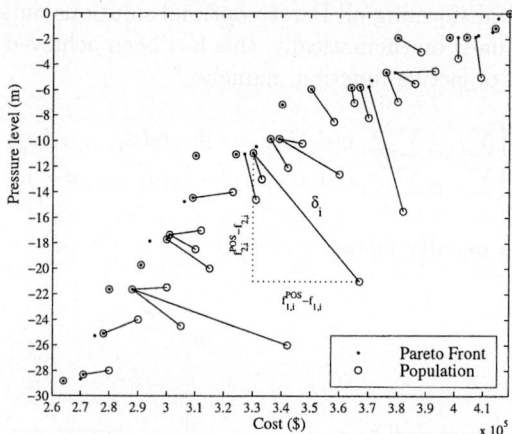

Fig. 5. Example of a generic population of 50 individuals (49 of which are inside the ROI here illustrated) and the Pareto front: solid lines connect each individual to its closest non-dominated solution, and represent the distance δ_i in equation (10)

3.1 Convergence Index

The convergence index, $\mathcal{CI}$, is expressed as:

$$\mathcal{CI} = \frac{N_f^{POS}}{N^{POS}}(1 - \overline{\delta}) \qquad (9)$$

where N_f^{POS} represent the number of POS inside the ROI found by the algorithm, $N^{POS} = 38$ (the size of P^*, P^* being the set of Pareto-optimal solutions reported on Table 1), and $\overline{\delta}$ is the average of the adimentionalized Euclidean distance values of all individuals inside the ROI from their nearest solution in P^*, given by:

$$\overline{\delta} = \frac{1}{N^{ROI}} \sum_i \delta_i = \frac{1}{N^{ROI}} \sum_i \sqrt{\left(\frac{f_{1,i}^{POS} - f_{1,i}}{f_{1,\max} - f_{1,\min}}\right)^2 + \left(\frac{f_{2,i}^{POS} - f_{2,i}}{f_{2,\max} - f_{2,\min}}\right)^2} \qquad (10)$$

where N^{ROI} is the number of individuals inside the ROI, $f_{1,i}$ and $f_{2,i}$ are respectively the cost and pressure level of the i-th individual, $f_{1,i}^{POS}$ and $f_{2,i}^{POS}$ the same quantities referred to the Pareto-optimal solution closest to the i-th individual, $f_{1,\max} = 419000$, $f_{1,\min} = 264000$, $f_{2,\max} = 0$, $f_{2,\min} = -30$ (these last values delimiting the ROI). The symbols are also represented in Figure 5.

3.2 Sparsity Index

The sparsity index, $\mathcal{SI}$, is expressed as:

$$\mathcal{SI} = \frac{N_r^{POS}}{N^{POS}}\left(1 - \frac{z_{\max}}{N^{POS}}\right)(1 - \sigma_{adim}) \qquad (11)$$

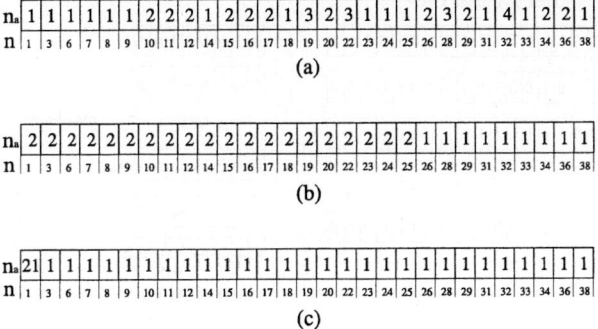

Fig. 6. Schematic representation of the distribution of the 49 individuals of the population around the $N_r^{POS} = 29$ reached. In (a), the actual distribution is reported (cfr. Figure 5), while in (b) and in (c), two examples of distributions characterized, respectively, by the minimum standard deviation (best distribution) and the maximum standard deviation (worst distribution). n_a is the number of individuals that have approached the n-th POS, whose progressive number is shown below (according to the numeration given on Table 1 and represented in Figure 3)

where N_r^{POS} represents the number of POS in P^* which have been approached (*reached*) by at least one individual in the population (of course, every POS found is also reached, but the contrary is not necessarily true), $z_{\max}$ is the maximum number of consecutive (adjacent) POS in P^* not reached, and σ_{adim} takes into account the actual distribution of individuals around the POS reached.

To understand the meaning of $z_{\max}$ and σ_{adim}, consider the example of before, in which 49 individuals are inside the ROI, and 29 POS have been reached: the actual distribution may be deduced from Figure 5 (counting the individuals around each POS), and is schematized in Figure 6 (a). Actually, the individuals inside the ROI may be distributed in many different ways around the POS reached: in particular, there will be (best) distributions characterized by the minimum standard deviation, as in Figure 6 (b), and (worst) distributions with the maximum standard deviation, as in Figure 6 (c). σ_{adim} is the adimentionalized standard deviation between these two extreme situations.

4 The Two-Level Approach

The methodology consists of a population of inner multi-objective GAs and an outer simple GA, adapting crossover and mutation probabilities of the inner level. Figure 7 shows a schematic representation of the inner and outer populations. In the figure, N_{in} and N_{out} are the inner and outer population sizes, respectively, while p_m and p_c the mutation and crossover probabilities.

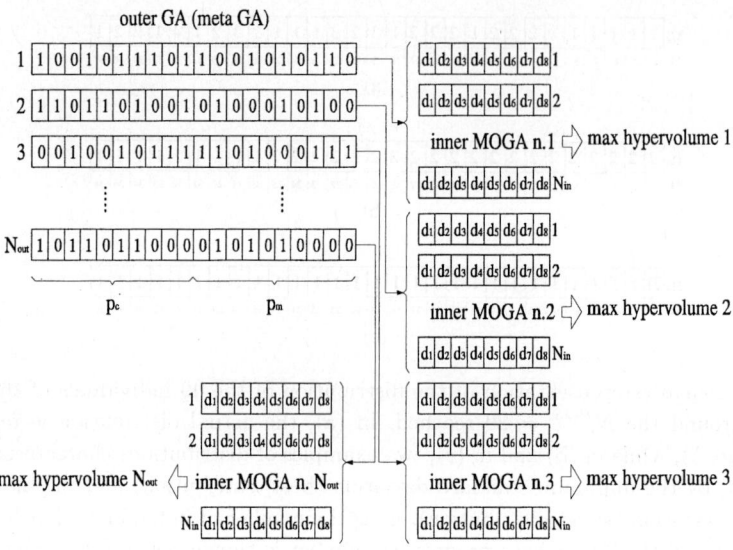

Fig. 7. Representation of inner and outer GAs for the two-loop network

4.1 The Inner GA

Two elitist multi-objective evolutionary algorithms have been implemented at the inner level with the aid of GAlib library [36], which has been modified in order to handle multi-criterion optimization problems:

1. The NSGA-II [37], a parameter-free NSGA based on the crowding distance of individuals; one problem of NSGA-II resides in the way elitism is performed, since it determines a deletion of solutions of non-elitist fronts and, as a result, the search process may suffer from stagnation or premature convergence.
2. The Controlled NSGA-II (CNSGA-II), introduced by [38], which, if compared to NSGA-II, performs elitism in a controlled manner, that is, instead of preserving all individuals of rank one, each front is allowed to have an exponentially decreasing number of solutions, thus forcing part of all non-dominated fronts to coexist in the new generation.

Every individual has a direct coding of commercial diameter values for each of the 8 decision variables, representing the 8 diameters to be assigned to the network. A population size of $N_{in} = 50$ individuals has been adopted, together with uniform crossover and adjacent mutation, allowing only changes to the nearest larger or smaller diameters.

4.2 The Outer GA

The meta GA is a simple GA with elitism, no crossover operator, a population size of $N_{out} = 5$, and a binary (gray) coding, mapping both mutation and probabilities from 0 to 1 with 3-digit precision (thus requiring 10 bits each), for a total of 20 bits chromosome representation. For every individual at the outer level, an inner multi-objective evolutionary algorithm is performed, and its fitness is represented by the (maximum) hypervolume $\mathcal{HV}$ reached during the evolution process.

5 Results

Numerical experiments have been divided in two phases: in the first, an assessment of the hypervolume metric has been performed, through its comparison with the convergence and sparsity indices previously introduced; in the second, the two-level approach has been tested.

Tables 2-5 report the results obtained with the NSGA-II and CNSGA-II, indicated as mean and standard deviation over ten runs with different initial populations. In particular, Tables 2 and 4 refer to the last generations of the evolution process, while Tables 3 and 5 to the best results achieved during the runs. Some considerations follow:

Table 2. Values of performance measures obtained with NSGA-II at last generation, for different crossover probabilities, p_c: μ is the mean and σ the standard deviation

	$\mathcal{HV}$		$\mathcal{CI}$		$\mathcal{SI}$		N_f^{POS}		N_r^{POS}		z_{max}	
p_c	μ	σ	μ	σ	μ	σ	μ	σ	μ	σ	μ	σ
0.00	0.9315	0.0836	0.3498	0.2812	0.6344	0.1918	13.40	10.65	26.70	5.95	1.90	0.70
0.25	0.9547	0.0694	0.4583	0.3028	0.6783	0.1683	17.50	11.47	28.20	5.04	1.80	0.60
0.50	0.9618	0.0673	0.5775	0.3232	0.7377	0.1615	22.00	12.22	29.80	4.64	1.30	0.46
0.75	0.9773	0.0472	0.6192	0.2950	0.7747	0.1399	23.60	11.16	31.30	3.98	1.40	0.49
1.00	0.9429	0.0718	0.4292	0.3200	0.5323	0.1528	16.40	12.13	29.80	5.56	1.50	0.50
Mean	0.9536	0.0679	0.4868	0.3044	0.6715	0.1629	18.58	11.52	29.16	5.03	1.58	0.55

Table 3. Best values of performance measures obtained with NSGA-II, for different crossover probabilities, p_c: μ is the mean and σ the standard deviation

	$\mathcal{HV}$		$\mathcal{CI}$		$\mathcal{SI}$		N_f^{POS}		N_r^{POS}		z_{max}	
p_c	μ	σ	μ	σ	μ	σ	μ	σ	μ	σ	μ	σ
0.00	0.9324	0.0824	0.3602	0.2900	0.6539	0.2008	13.80	10.99	27.40	6.48	1.90	0.70
0.25	0.9554	0.0681	0.4716	0.3239	0.6949	0.1824	18.00	12.29	28.80	5.47	1.70	0.64
0.50	0.9619	0.0671	0.6063	0.3345	0.7698	0.1694	23.10	12.65	31.10	4.93	1.30	0.46
0.75	0.9774	0.0470	0.6272	0.3021	0.7737	0.1440	23.90	11.42	31.40	4.08	1.40	0.49
1.00	0.9430	0.0717	0.4292	0.3200	0.5272	0.1474	16.40	12.13	29.80	5.56	1.50	0.50
Mean	0.9540	0.0673	0.4989	0.3141	0.6839	0.1688	19.04	11.89	29.70	5.31	1.56	0.56

Table 4. Values of performance measures obtained with CNSGA-II at last generation, for different crossover probabilities, p_c: μ is the mean and σ the standard deviation

	$\mathcal{HV}$		$\mathcal{CI}$		$\mathcal{SI}$		N_f^{POS}		N_r^{POS}		z_{max}	
p_c	μ	σ	μ	σ	μ	σ	μ	σ	μ	σ	μ	σ
0.00	0.9722	0.0534	0.5205	0.2552	0.6669	0.1179	19.90	9.66	28.60	3.72	1.80	0.40
0.25	0.9656	0.0622	0.4264	0.2369	0.6371	0.1325	16.30	8.98	27.20	4.56	1.80	0.60
0.50	0.9935	0.0082	0.6268	0.1635	0.7232	0.0317	23.90	6.16	30.30	1.01	1.40	0.49
0.75	0.9877	0.0096	0.5052	0.2122	0.7057	0.0702	19.30	8.00	29.70	2.05	1.60	0.49
1.00	0.9812	0.0110	0.4878	0.2102	0.6892	0.0814	18.70	7.94	31.00	2.05	1.90	0.54
Mean	0.9800	0.0289	0.5133	0.2156	0.6844	0.0868	19.62	8.15	29.36	2.68	1.70	0.50

Table 5. Best values of performance measures obtained with CNSGA-II, for different crossover probabilities, p_c: μ is the mean and σ the standard deviation

	$\mathcal{HV}$		$\mathcal{CI}$		$\mathcal{SI}$		N_f^{POS}		N_r^{POS}		z_{max}	
p_c	μ	σ	μ	σ	μ	σ	μ	σ	μ	σ	μ	σ
0.00	0.9734	0.0509	0.5235	0.2618	0.6892	0.1267	20.00	9.91	29.20	4.29	1.50	0.50
0.25	0.9667	0.0597	0.4395	0.2460	0.6656	0.1351	16.80	9.32	28.60	4.78	1.70	0.64
0.50	0.9940	0.0084	0.6614	0.1884	0.7517	0.0380	25.20	7.10	31.40	1.11	1.30	0.46
0.75	0.9879	0.0097	0.5338	0.2253	0.7235	0.0607	20.40	8.51	31.00	1.84	1.60	0.49
1.00	0.9812	0.0110	0.4904	0.2111	0.6852	0.0746	18.80	7.97	31.30	1.90	1.70	0.46
Mean	0.9807	0.0279	0.5297	0.2265	0.7030	0.0870	20.24	8.56	30.30	2.78	1.56	0.51

1. There is a good agreement between the hypervolume metric and the convergence and sparsity indices, also evidentiated by the number of Pareto-optimal solutions found or reached.
2. There may be situations in which hypervolume is high, although convergence and sparsity indices are not; this is due to the discrete character of the problem and to the actual distribution of Pareto-optimal solutions: looking at the front in Figure 3, it may be noted that there are some clusters of solutions which only marginally contribute to the hypervolume (a closer look at Table 1 reveals also that neighbour solutions on the objective space are characterized by changes in only one or two diameters).
3. CNSGA-II actually outperforms NSGA-II with respect to all the metrics, having also lower values of standard deviations, especially for the sparsity index; looking at Figure 8, which shows an example of the best evolutions of the two MOGAs, it may be observed that CNSGA-II exhibits both higher rapidity in reaching final values of performance measures, both lower oscillating evolutions, as confirmed by the comparison between Tables 4 and 5; NSGA-II, on the contrary, shows higher oscillating patterns (cfr. Tables 2 and 3).
4. There is a dependance of the two MOGAs with respect to crossover probability. Although mutation probability has been kept constant, some runs

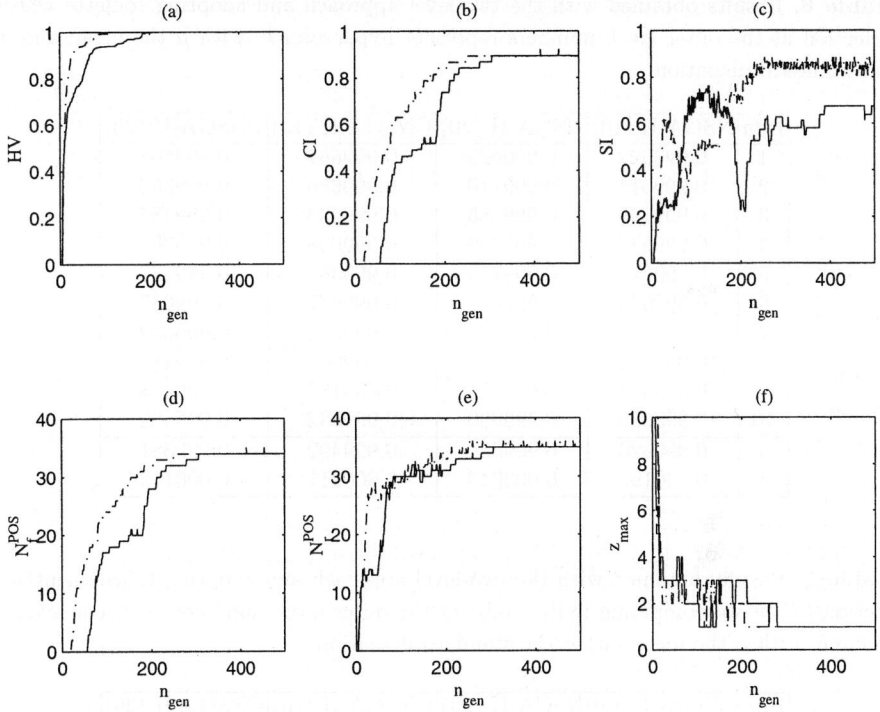

Fig. 8. Performance indices obtained with the best evolutions of NSGA-II (solid line) and Controlled NSGA-II (dash-dotted line): (a), hypervolume; (b), convergence index; (c), sparsity index; (d), Pareto solutions found; (e), Pareto solutions reached; (f) maximum number of consecutive POS not reached

performed confirmed the same behaviour, thus introducing the problem of the optimal choice of such values: this motivated the two-level approach.

Tables 6 and 7 report the results obtained with the two-level methodology, respectively adopting roulette wheel and tournament selection for the outer GA. Numbers represent the hypervolume obtained after 10 and 20 outer generations (indicated in parentheses in the Tables). CPU times were about 30 seconds for each outer generation. It may be observed that:

1. fitness obtained in one run is much higher than that achieved with a MOGA alone, even when several runs with different parameters are performed.
2. Very often results after 20 generations determine only a slight improvement with respect to those after 10.
3. NSGA-II and CNSGA-II present nearly the same performances, as well the tournament and roulette wheel selections.

Table 6. Results obtained with the two-level approach and adopting roulette wheel selection at the outer level: numbers represent hypervolume, with μ the mean and σ the standard deviation

Run	NSGA-II (10)	NSGA-II (20)	CNSGA-II (10)	CNSGA-II (20)
1	0.999622	0.999622	0.999667	0.999667
2	0.999514	0.999710	0.999656	0.999656
3	0.999190	0.999585	0.999434	0.999483
4	0.999633	0.999650	0.999658	0.999761
5	0.999673	0.999719	0.999382	0.999382
6	0.999634	0.999661	0.999687	0.999687
7	0.999673	0.999673	0.999652	0.999652
8	0.999122	0.999698	0.999064	0.999481
9	0.999504	0.999591	0.999183	0.999508
10	0.999630	0.999630	0.999532	0.999532
μ	0.999520	0.999654	0.999492	0.999581
σ	0.000190	0.000045	0.000211	0.000113

Table 7. Results obtained with the two-level approach and adopting tournament selection (with two competing individuals) at the outer level: numbers represent hypervolume, with μ the mean and σ the standard deviation

Run	NSGA-II (10)	NSGA-II (20)	CNSGA-II (10)	CNSGA-II (20)
1	0.999646	0.999646	0.999760	0.999760
2	0.999570	0.999570	0.999506	0.999573
3	0.999603	0.999622	0.999553	0.999583
4	0.999141	0.999547	0.999569	0.999677
5	0.999463	0.999615	0.999393	0.999564
6	0.999641	0.999664	0.999687	0.999820
7	0.999560	0.999618	0.999652	0.999652
8	0.999660	0.999660	0.999526	0.999554
9	0.999305	0.999522	0.999238	0.999267
10	0.999800	0.999800	0.999756	0.999756
μ	0.999359	0.999446	0.999564	0.999621
σ	0.000480	0.000484	0.000154	0.000148

6 Concluding Remarks

The paper has presented a two-level methodology for multi-criterion optimization of water distribution systems. Results show how such an approach, although requiring more computational effort than using a multi-objective genetic algorithm alone, is able to achieve very good performances in only one simulation run, thus proving its robustness and easing the network optimization process. In this work, a non-self adaptive procedure has been proposed: future research will be focused on the inclusion of mutation and crossover probabilities in the string representation, as in a fully self-adaptive mechanism.

References

1. Yates, D.F., Templeman, A.B., Boffey, T.B.: The Computational Complexity of the Problem of Determining Least Capital Cost Designs for Water Supply Networks. Engineering Optimization **7** (1984) 142-155
2. Alperovits, E., Shamir, U.: Design of Optimal Water Distribution Systems. Water Resources Research **13** (1977) 885-900
3. Quindry, G., Brill, E.D., Liebman, J.C.: Optimization of Looped Water Distribution Systems. Journal of Environmental Engineering **107** (1981) 665-679
4. Goulter, I.C., Lussier, B.M., Morgan, D.R.: Implications of Head Loss Path Choice in the Optimization of Water Distribution Networks. Water Resources Research **22** (1986) 819-822
5. Fujiwara, O., JenChaimahakoon, B., Edirisinghe, N.: A Modified Linear Programming Gradient Method for Optimal Design of Looped Water Distribution Networks. Water Resources Research **23** (1987) 977-982
6. Kessler, A., Shamir, U.: Analysis of the Linear Programming Gradient Method for Optimal Design of Water Supply Networks. Water Resources Research **25** (1989) 1469-1480
7. Bhave, P.R., Sonak, V.V.: A Critical Study of the Linear Programming Gradient Method for Optimal Design of Water Supply Networks. Water Resources Research **28** (1992) 1577-1584
8. El-Bahrawy, A., Smith, A.A.: Application of MINOS to Water Collection and Distribution Networks. Civil Engineering Systems **2** (1985) 38-49
9. Duan, N., Mays, L.W., Lansey, K.E.: Optimal Reliability-Based Design of Pumping and Distribution Systems. Journal of Hydraulic Engineering **116** (1990) 249-268
10. Simpson, A.R., Dandy, G.C., Murphy, L.J.: Genetic Algorithms Compared to other Techniques for Pipe Optimization. Journal of Water Resources Planning and Management **120** (1994) 423-443
11. Dandy, G.C., Simpson, A.R., Murphy, L.J.: An Improved Genetic Algorithm for Pipe Network optimization. Water Resources Research **32** (1996) 449-458
12. Savic, D.A., Walters, G.A.: Genetic Algorithms for Least-Cost Design of Water Distribution Networks. Journal of Water Resources Planning and Management **123** (1997) 67-77
13. Montesinos, P., Garcia-Guzman, A., Ayuso, J.L.: Water Distribution Network Optimization Using a Modified Genetic Algorithm. Water Resources Research **35** (1999) 3467-3473
14. Reis, L.F.R., Porto, R.M., Chaudry, F.H.: Optimal Location of Control Valves in Pipe Networks by Genetic Algorithm. Journal of Water Resources Planning and Management **123** (1997) 317-326
15. Vitkovsky, J.P., Simpson, A.R., Lambert, M.F.: Leak Detection and Calibration Using Transients and Genetic Algorithms. Computer Aided Civil and Infrastructure Engineering **15** (2000) 374-382
16. Meyer, R.W., Barkdoll, B.D.: Sampling Design for Network Model Calibration Using Genetic Algorithms. Journal of Water Resources Planning and Management **126** (2000) 245-250

17. Dandy, G.C., Engelhardt, M.: Optimal Scheduling of Water Pipe Replacement Using Genetic Algorithms. Journal of Water Resources Planning and Management **127** (2001) 214-223
18. Walters, G.A., Lohbeck, T.: Optimal Layout of Tree Networks Using Genetic Algorithms. Engineering Optimization **22** (1993) 27-48
19. Abebe, A.J., Solomatine, D.P.: Application of Global Optimization to the Design of Pipe Networks. In: Babovic and Larsen (Eds.), Hydroinformatics 1998, World Scientific, 989-996
20. Walski, T.M.: The Wrong Paradigm – Why Water Distribution Optimization Doesn't Work. Journal of Water Resources Planning and Management **127** (2001) 203-205
21. Halhal, D., Walters, G.A., Ouazar, D., Savic, D.A.: Water Network Rehabilitation with Structured Messy Genetic Algorithm. Journal of Water Resources Planning and Management **123** (1997) 137-146
22. Cheung, P.B., Reis, L.F.R., Formiga, K.T.M., Chaudry, F.H., Ticona, W.G.C.: Multiobjective Evolutionary Algorithms Applied to the Rehabilitation of a Water Distribution System: a Comparative Study. In: C.M. Fonseca et al. (Eds.), EMO 2003, LNCS 2632, 662-676 (2003) Springer
23. Formiga, K.T.M., Chaudry, F.H., Cheung, P.B., Reis, L.F.R.: Optimal Design of Water Distribution System by Multiobjective Evolutionary Methods. In: C.M. Fonseca et al. (Eds.), EMO 2003, LNCS 2632, 677-691 (2003) Springer
24. Tolson, B.A., Maier, H.R., Simpson, A.R., Lence, B.J.: Genetic Algorithms for Reliability-Based Optimization of Water Distribution Systems. Journal of Water Resources Planning and Management **130** (2004) 63-72
25. Prasad, T.D., Park, N.-S.: Multiobjective Genetic Algorithms for Design of Water Distribution Networks. Journal of Water Resources Planning and Management **130** (2004) 73-82
26. Deb, K.: Multi-Objective Optimization Using Evolutionary Algorithms. John Wiley & Sons (2001)
27. Knowles, J.D., Corne, D.W.: On Metrics for Comparing Nondominated Sets. Proceedings of the 2002 IEEE Congress on Evolutionary Computation (CEC 2002) 711-716
28. Fleischer, M.: The Measure of Pareto Optima In: C.M. Fonseca et al. (Eds.), EMO 2003, LNCS 2632, 519-533 (2003) Springer
29. Grefenstette, J.J.: Optimization of Control Parameters for Genetic Algorithms. IEEE Transactions Systems Man Cybernetics **16** (1986) 122-128
30. Bramlette, M.F.: Initialization, Mutation and Selection Methods in Genetic Algorithms for Function Optimization. In: R.K. Belew and L.B. Booker (Eds.), Proceedings of the Fourth International Conference on Genetic Algorithms, 100-107 (1991) Morgan Kaufmann
31. Spears, W.M.: Adapting Crossover in Evolutionary Algorithms. In: J.R. McDonnel et al. (Eds.), Proceedings of the Fourth Annual Conference on Evolutionary Programming, 367-384 (1995) The MIT Press
32. Back, T.: Self-Adaptation in Genetic Algorithms. In: F.J. Varela and P. Bourgine (Eds.), Proceedings of the First European Conference on Artificial Life, 263-271 (1992) The MIT Press
33. Srinivas, M., Patnaik, L.M.: Adaptive Probabilities of Crossover and Mutation in Genetic Algorithms. IEEE Transactions System Man Cybernetics **24** (1994) 656-666

34. Nicolini, M.: Evaluating Performance of Multi-Objective Genetic Algorithms for Water Distribution System Optimization. In: Liong et al. (Eds.), Hydroinformatics 2004, World Scientific, 850-857
35. Rossman, L.A.: EPANET 2 USERS MANUAL. U.S. Environmental Protection Agency, Cincinnati, Ohio (2000)
36. Wall, M.: GAlib: A C++ Library of Genetic Algorithm Components (v. 2.4.5). Mechanical Engineering Department, Massachussets Institute of Technology (2000)
37. Deb, K., Agrawal, S., Pratap, A., Meyarivan, T.: A Fast Elitist Non-dominated Sorting Genetic Algorithm for Multi-Objective Optimization: NSGA-II. In: Proceedings of the Parallel Problem Solving from Nature VI 849-858 (2000)
38. Deb, K., Goel, T.: Controlled Elitist Non-dominated Sorting Genetic Algorithms for Better Convergence. In: E. Zitzler et al. (Eds.), EMO 2001, LNCS 1993, 67-81 (2001) Springer

Evolutionary Multi-objective Optimization for Simultaneous Generation of Signal-Type and Symbol-Type Representations

Yaochu Jin, Bernhard Sendhoff, and Edgar Körner

Honda Research Institute Europe
63073 Offenbach/Main, Germany
yaochu.jin@honda-ri.de

Abstract. It has been a controversial issue in the research of cognitive science and artificial intelligence whether signal-type representations (typically connectionist networks) or symbol-type representations (e.g., semantic networks, production systems) should be used. Meanwhile, it has also been recognized that both types of information representations might exist in the human brain. In addition, symbol-type representations are often very helpful in gaining insights into unknown systems. For these reasons, comprehensible symbolic rules need to be extracted from trained neural networks. In this paper, an evolutionary multi-objective algorithm is employed to generate multiple models that facilitate the generation of signal-type and symbol-type representations simultaneously. It is argued that one main difference between signal-type and symbol-type representations lies in the fact that the signal-type representations are models of a higher complexity (fine representation), whereas symbol-type representations are models of a lower complexity (coarse representation). Thus, by generating models with a spectrum of model complexity, we are able to obtain a population of models of both signal-type and symbol-type quality, although certain post-processing is needed to get a fully symbol-type representation. An illustrative example is given on generating neural networks for the breast cancer diagnosis benchmark problem.

1 Introduction

Artificial neural networks are one of the most well known signal-type representations in cognitive and vision research. Neural networks are linear or nonlinear systems that encode information with connections and units in a distributed manner, which are more or less of biological plausibility. In contrast, symbol-type representations use meaningful symbols, and information is encoded by defining the relationship among various symbols. Several symbolic representation models have been developed, such as semantic networks, production systems (symbolic rules) and finite-state automata.

Symbolic representations and symbolic processing are believed to have several desirable features that are closely related to mental representations and cognition [12], namely, productivity, systematicity, compositionality and inferential coherence. Besides, increasing evidence has been found in cognitive neuroscience that the human brain does have mechanisms that are responsible for symbolic processing [13, 21, 4]. It is widely believed that symbolic systems are more transparent to human users than connectionist networks, which plays an important role when neural networks are employed in critical engineering applications.

For the above reasons, it is often necessary to extract symbolic or fuzzy rules from trained neural networks [3, 20, 11]. A common drawback of most existing rule extraction method is that the rules are extracted after a neural network has been trained, which incurs additional computational costs.

This paper attempts to generate signal-type and symbol-type models simultaneously using the multi-objective optimization approach. Multiple neural networks of various model complexities, instead of either a single signal-type or symbol-type model, will be generated using a multi-objective evolutionary algorithm combined with a local search, where accuracy and complexity serve as two conflicting objectives. It has been shown that evolutionary multi-objective algorithms are well suited and very powerful in obtaining a set of Pareto-optimal solutions in one single run of optimization [8, 7].

Training neural networks using evolutionary multi-objective optimization is not new in itself [30, 1, 2, 19]. However, the existing work focuses on improving the accuracy of a single network or an ensemble of networks. Generating an ensemble of fuzzy classifiers using evolutionary algorithms has also been studied in [16]. Objectives in training neural networks include accuracy on training data, accuracy on test data, number of hidden neurons, and number of connections. Note that a trade-off between the accuracy on the training data and the accuracy on test data does not necessarily mean a trade-off between accuracy and complexity.

Section 2 discusses very briefly the existing methods for controlling model complexity in the context of model selections in machine learning. Methods for converting signal-type neural networks to symbol-type rules in the area of neural networks will also be introduced. Section 3 shows that any formal neural network regularization methods can be treated as multi-objective optimization problems. The details of the evolutionary multi-objective algorithm, together with the local search method will be provided in Section 4. An illustrative example is given in Section 5, where a population of Pareto-optimal neural networks are generated for the breast diagnosis problem. It will be shown that among the models generated by the multi-objective evolutionary algorithm, those with a higher complexity are of more signal quality and those of a lower complexity are of more symbol quality.

2 Complexity Control and Rule Extraction

2.1 Model Selection and Complexity Control

The task of model selection is to choose the best model for a set of given data, assuming that a number of models is available. Several criteria have been proposed based on the Kullback-Leibler Information Criterion [6]. The most popular criteria are Akaike's Information Criterion (AIC) and Bayesian Information Criterion (BIC). For example, model selection according to the AIC is to minimize the following criterion:

$$AIC = -2\ log(\mathcal{L}(\theta|y,\ g)) + 2\ K, \tag{1}$$

where, $\mathcal{L}(\theta|y,\ g)$ is the maximized likelihood for data y given a model g with model parameter θ, K is the number of effective parameters of g.

The first term of Equation (1) reflects how good the model approximates the data, while the second term is the complexity of the model. Usually, the higher the model complexity is, the more accurate the approximation will be. Obviously, a trade-off between accuracy and model complexity has to be taken into account in model selection.

On the other hand, model selection criteria have often been used to control the complexity of models to a desired degree in model generation. This approach is usually known as regularization in the neural network community [5]. The main purpose of neural network regularization is to improve the generalization capability of a neural network by control its complexity. By generalization, it is meant that a trained neural network should perform well not only on training data, but also on unseen data.

2.2 Complexity Reduction in Rule Extraction

Generally, neural networks are signal-type representations that are difficult to understand for human users. Due to this reason, many efforts have been made to extract symbolic or fuzzy rules from trained neural network [3, 18, 11]. Two assumptions are often made during rule extraction from trained neural networks. First, units in the neural network are either maximally active or inactive. To meet this requirement, regularization techniques such as structural regularization [17], weight sharing [20] or network pruning [28] are usually implemented before rule extraction, which in effect reduces the complexity of neural networks. In fact, this assumption is also essential for the interpretability of the extracted rules. The complexity reduction procedure prior to rule extraction is also termed skeletonization [10]. The second assumption is that a label, or in other words, a meaning needs to be associated with each unit.

Rule extraction from trained neural networks can be seen as a model selection process that trades off between accuracy and interpretability, where preference is put on the interpretability of the model. Thus, the trade-off between accuracy and complexity in model selection also reflects a trade-off between accuracy and interpretability, in other words, a trade-off between signal-type and symbol-type of representations.

Existing methods for extracting symbolic rules from trained neural network can largely be divided into three steps: neural network training, network skeletonization, and rule extraction [11].

3 Multi-objective Optimization Approach to Complexity Reduction

3.1 Neural Network Regularization

Neural network regularization can be realized by including an additional term that reflects the model complexity in the cost function of the training algorithm:

$$J = E + \lambda \Omega, \qquad (2)$$

where E is an error function, Ω is the regularization term representing the complexity of the network model, and λ is a hyperparameter that controls the strength of the regularization. The most common error function in training or evolving neural networks is the mean squared error (MSE):

$$E = \frac{1}{N} \sum_{i=1}^{N} (y^d(i) - y(i))^2, \qquad (3)$$

where N is the number of training samples, $y^d(i)$ is the desired output of the i-th sample, and $y(i)$ is the network output for the i-th sample. For the sake of clarity, we assume that the neural network has only one output. Refer to [5] for other error functions, such as the Minkowski error or cross-entropy.

Several measures have also been suggested for denoting the model complexity Ω. A most popular regularization term is the squared sum of all weights of the network:

$$\Omega = \frac{1}{2} \sum_{k} w_k^2, \qquad (4)$$

where k is an index summing up all weights. This regularization method has been termed *weight decay*.

One weakness of the weight decay method is that it is not able to drive small irrelevant weights to zero, when gradient-based learning algorithms are employed, which may result in many small weights [25]. An alternative is to replace the squared sum of the weights with the sum of absolute value of the weights:

$$\Omega = \sum_{i} |w_i|. \qquad (5)$$

It has been shown that this regularization term it is able to drive irrelevant weights to zero [22].

Both regularization terms in equations (4) and (5) have also been studied from the Bayesian learning point of view, which are known as the Gaussian regularizer and the Laplace regularizer, respectively.

A more direct measure for model complexity of neural networks is the number of weights contained in the neural network:

$$\Omega = \sum_{i} \sum_{j} c_{ij}, \qquad (6)$$

where c_{ij} equals 1 if there is connection from neuron j to neuron i, and 0 if not. It should be noticed that the above complexity measure is not generally applicable to gradient-based learning methods.

A comparison of the three regularization terms using multi-objective evolutionary algorithms has been implemented [19]. Different to the conclusions reported [22] where gradient-based learning method has been used, it has been shown that regularization using the sum of squared weights is able to change (reduce) the structure of neural networks as efficiently as using the sum of absolute weights.

3.2 Multi-objective Optimization Approach to Regularization

It is quite straightforward to notice that neural network regularization in equation (2) can be reformulated as a bi-objective optimization problem:

$$\min \{f_1, f_2\} \tag{7}$$
$$f_1 = E, \tag{8}$$
$$f_2 = \Omega, \tag{9}$$

where E is defined in equation (3), and Ω is one of the regularization terms defined in equation (4), (5), or (6).

It is noticed that regularization is traditionally formulated as a single objective optimization problem as in Equation (2) rather than a multi-objective optimization problem as in equation (7). In our opinion, this tradition can be mainly attributed to the fact that traditional gradient-based learning algorithms are not able to solve multi-objective optimization problems.

3.3 Simultaneous Generation of Signal-Type and Symbol-Type Models

If evolutionary algorithms are used to solve the multi-objective optimization problem in Equation (7), multiple solutions with a spectrum of model complexity can be obtained in a single optimization run. Thus, if we choose a model of a higher complexity, the model will be of signal quality. On the contrary, if we choose a model of a lower complexity, the model will be of more symbol quality, assuming that a proper physical meaning can be associated with each neuron. In this sense, signal-type and symbol-type models can be generated simultaneously in one step using the multi-objective optimization approach to model selection.

4 Evolutionary Multi-objective Model Generation

4.1 Parameter and Structure Representation of the Network

A connection matrix and a weight matrix are employed to describe the structure and the weights of the neural networks. The connection matrix specifies the

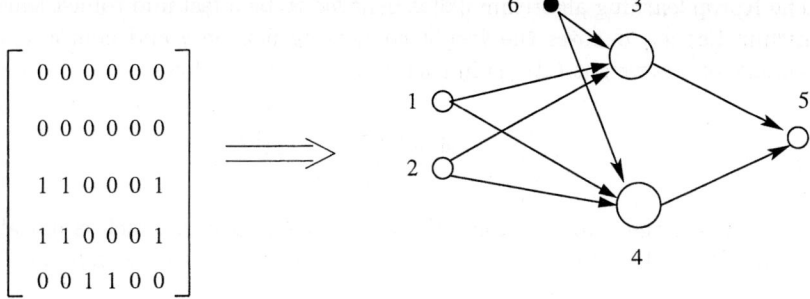

Fig. 1. A connection matrix and the corresponding network structure

structure of the network, whereas the weight matrix determines the strength of each connection. Assume that a neural network consists of M neurons in total, including the input and output neurons, then the size of the connection matrix is $M \times (M+1)$, where an element in the last column indicates whether a neuron is connected to a bias value. In the matrix, if element $c_{ij}, i = 1, ..., M, j = 1, ..., M$ equals 1, it means that there is a connection between the i-th and j-th neuron and the signal flows from neuron j to neuron i. If $j = M+1$, it indicates that there is a bias in the i-th neuron. Fig. 1 illustrates a connection matrix and the corresponding network structure. It can be seen from the figure that the network has two input neurons, two hidden neurons, and one output neuron. Besides, both hidden neurons have a bias.

The strength (weight) of the connections is defined in the weight matrix. Accordingly, if the c_{ij} in the connection matrix equals zero, the corresponding element in the weight matrix must be zero too.

4.2 Global and Local Search

A genetic algorithm has been used for optimizing the structure and weights of the neural networks. Binary coding has been used representing the neural network structure and real-valued coding for encoding the weights. Five genetic operations have been introduced in the global search, four of which mutate the connection matrix (neural network structure) and one of which mutates the weights. The four mutation operators are insertion of a hidden neuron, deletion of a hidden neuron, insertion of a connection and deletion of a connection [14]. A Gaussian-type mutation is applied to mutate the weight matrix. No crossover has been employed in this algorithm.

After mutation, an improved version of the Rprop algorithm [15] has been employed to train the weights. This can be seen as a kind of life-time learning (the first objective only) within a generation. After learning, the fitness of each individual with regard to the approximation error (f_1) is updated. In addition, the weights modified during the life-time learning are encoded back to the chromosome, which is known as the Lamarkian type of inheritance.

The Rprop learning algorithm [26] is believed to be a fast and robust learning algorithm. Let w_{ij} denotes the weight connecting neuron j and neuron i, then the change of the weight (Δw_{ij}) in each iteration is as follows:

$$\Delta w_{ij}^{(t)} = -\text{sign}\left(\frac{\partial E^{(t)}}{\partial w_{ij}}\right) \cdot \Delta_{ij}^{(t)}, \tag{10}$$

where $sign(\cdot)$ is the sign function, $\Delta_{ij}^{(t)} \geq 0$ is the step-size, which is initialized to Δ_0 for all weights. The step-size for each weight is adjusted as follows:

$$\Delta_{ij}^{(t)} = \begin{cases} \xi^+ \cdot \Delta_{ij}^{(t-1)}, & \text{if } \frac{\partial E^{(t-1)}}{\partial w_{ij}} \cdot \frac{\partial E^{(t)}}{\partial w_{ij}} > 0 \\ \xi^- \cdot \Delta_{ij}^{(t-1)}, & \text{if } \frac{\partial E^{(t-1)}}{\partial w_{ij}} \cdot \frac{\partial E^{(t)}}{\partial w_{ij}} < 0 \\ \Delta_{ij}^{(t-1)}, & \text{otherwise} \end{cases} \tag{11}$$

where $0 < \xi^- < 1 < \xi^+$. To prevent the step-sizes from becoming too large or too small, they are bounded by $\Delta_{\min} \leq \Delta_{ij} \leq \Delta_{\max}$.

One exception must be considered. After the weights are updated, it is necessary to check if the partial derivative changes sign, which indicates that the previous step might be too large and thus a minimum has been missed. In this case, the previous weight change should be retracted:

$$\Delta w^{(t)} = -\Delta_{ij}^{(t-1)}, \text{ if } \frac{\partial E^{(t-1)}}{\partial w_{ij}} \cdot \frac{\partial E^{(t)}}{\partial w_{ij}} < 0. \tag{12}$$

Recall that if the weight change is retracted in the t-th iteration, the $\partial E^{(t)}/\partial w_{ij}$ should be set to 0.

In reference [15], it is argued that the condition for weight retraction in equation (12) is not always reasonable. The weight change should be retracted only if the partial derivative changes sign and if the approximation error increases. Thus, the weight retraction condition in equation (12) is modified as follows:

$$\Delta w^{(t)} = -\Delta_{ij}^{(t-1)}, \text{ if } \frac{\partial E^{(t-1)}}{\partial w_{ij}} \cdot \frac{\partial E^{(t)}}{\partial w_{ij}} < 0 \text{ and } E^{(t)} > E^{(t-1)}. \tag{13}$$

It has been shown on several benchmark problems in [15] that the modified Rprop (termed as Rprop$^+$ in [15]) exhibits consistently better performance than the Rprop algorithm.

4.3 Elitist Non-dominated Sorting

In this algorithm, the elitist non-dominated sorting method proposed in the NSGA-II algorithm [9] has been adopted. Assume the population size is N. At first, the offspring and the parent populations are combined. Then, a non-domination rank and a local crowding distance are assigned to each individual in the combined population. In selection, all non-dominated individuals (say there are N_1 non-dominated solutions) in the combined population are passed to the offspring population, and are removed from the combined population. Now the combined population has $2N - N_1$ individuals. If $N_1 < N$, the non-dominated solutions in the current combined population (say there are N_2 non-dominated solutions) will be passed to the offspring population. This procedure is repeated until the offspring population is filled. It could happen that the number of non-dominated solutions in the current combined population (N_i) is larger than the left slots ($N - N_1 - N_2 - ... - N_{i-1}$) in the current offspring population. In this case, the $N - N_1 - N_2 - ... - N_{i-1}$ individuals with the largest crowding distance from the N_i non-dominated individuals will be passed to the offspring generation.

5 An Illustrative Example

To illustrate the feasibility of the idea of generating signal-type and symbol-type models simultaneously using evolutionary optimization approach, neural networks have been generated to solve the breast cancer benchmark problem in the UCI repository of machine learning database collected by Dr. W.H. Wolberg at the University of Wisconsin-Madison Hospitals [23]. Studies have been carried out to extract symbolic rules from trained neural network using the three-step procedure for rule extraction on this benchmark problem [29, 27]. The benchmark problem contains 699 examples, each of which has 9 inputs and 2 outputs. The inputs are: clump thickness (x_1), uniformity of cell size (x_2), uniformity of cell shape (x_3), marginal adhesion (x_4), single epithelial cell size (x_5), bare nuclei (x_6), bland chromatin (x_7), normal nucleoli (x_8), and mitosis (x_9). All inputs are normalized, to be more exact, $x_1, ..., x_9 \in \{0.1, 0.2, ..., 0.8, 0.9, 1.0\}$. The two outputs are complementary binary value, i.e., if the first output is 1, which means "benign", then the second output is 0. Otherwise, the first output is 0, which means "malignant", and the second output is 1. Therefore, only the first output is considered in this work. The data samples are divided into two groups: one training data set containing 599 samples and one test data set containing 100 samples. The test data are unavailable to the algorithm during the evolution.

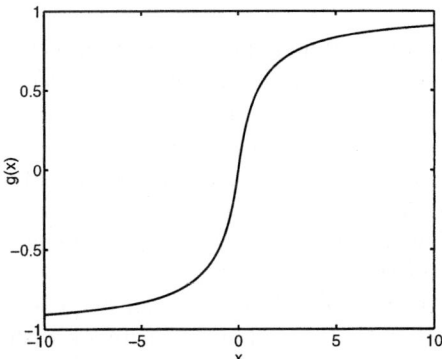

Fig. 2. The activation function of the hidden nodes

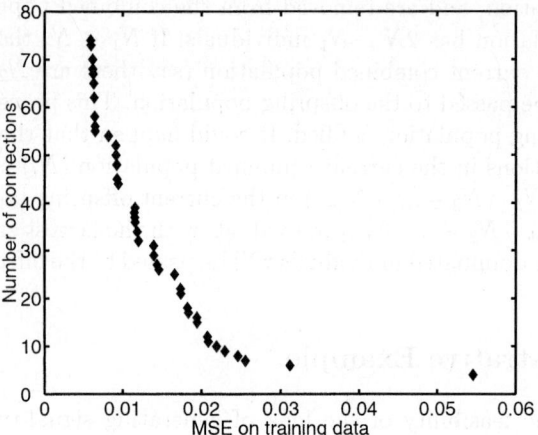

Fig. 3. The Pareto front containing 41 non-dominated solutions representing neural networks of a different model complexity

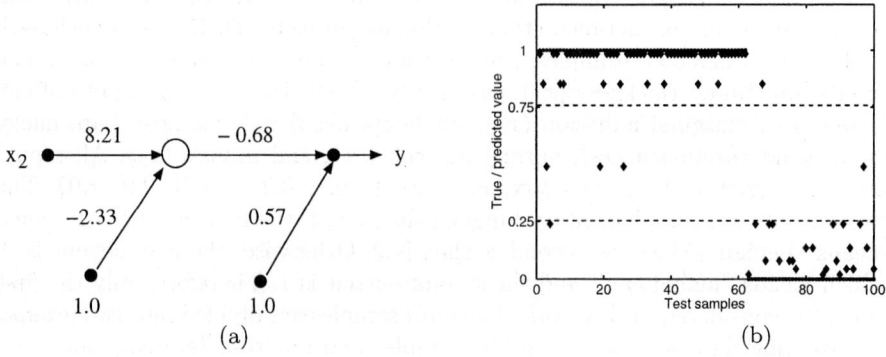

Fig. 4. The simplest neural network. (a) The structure; and (b) The prediction results on test data

The population size is 100 and the optimization is run for 200 generations. One of the five mutation operations is randomly selected and performed on each individual. The standard deviation of the Gaussian mutations applied on the weight matrix is set to 0.05. The weights of the network are initialized randomly in the interval of $[-0.2, 0.2]$. In the Rprop$^+$ algorithm, the step-sizes are initialized to 0.0125 and bounded between $[0, 50]$ during the adaptation, and $\xi^- = 0.2$, $\xi^+ = 1.2$, which are the default values recommended in [15] and 50 iterations are implemented in each local search.

Although a non-layered neural network can be generated using the coding scheme described in Section 3, a feedforward network with one hidden layer will be generated. The maximum number of hidden nodes is set to 10. The hidden

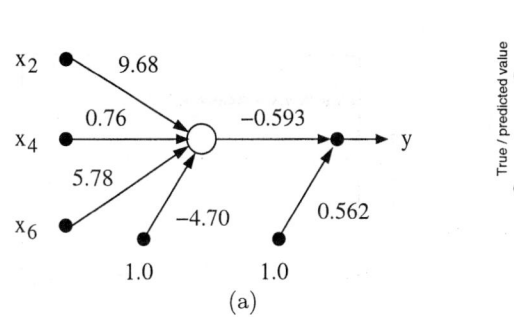
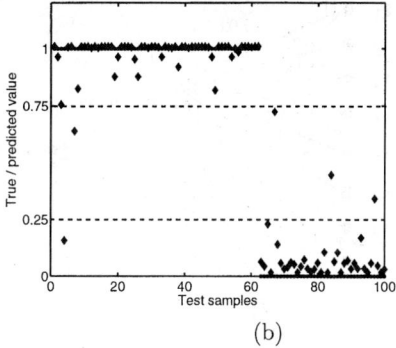

Fig. 5. The next simplest neural network. (a) The structure; and (b) The prediction results on test data

neurons are nonlinear and the output layers are linear. The activation function used for the hidden neurons is as follows,

$$g(z) = \frac{x}{1+|x|}, \qquad (14)$$

which is illustrated in Fig. 2.

In this study, the complexity measure defined in Equation (6) has been used as the objective describing the complexity of the neural networks. The non-dominated solutions obtained at the 200-th generation are plotted in Fig. 3. Note that many solutions in the final population are the same and finally 41 non-dominated solutions have been generated.

Among the 41 neural networks, the simplest one has only 4 connections: 1 input node, one hidden node and 2 biases, see Fig 4(a). The mean squared error (MSE) of the network on training and test data are 0.0546 and 0.0324, respectively.

Assuming that a case can be decided to be "malignant" if $y < 0.25$, and "benign" if $y > 0.75$, We can then derive that if $x_2 > 0.4$, which means that $x_2 \geq 0.5$, then "malignant" and "benign" if $x_2 < 0.22$, i.e., then $x_2 \leq 0.2$.

From such a simple network, the following two symbol-type rules can be extracted (denoted as MOO_NN1):

$$\text{R1: If } x_2 \text{ (clump thickness) } \geq 0.5, \text{ then malignant;} \qquad (15)$$
$$\text{R2: If } x_2 \text{ (clump thickness) } \leq 0.2, \text{ then benign.} \qquad (16)$$

Based on these two simple rules, only 2 out of 100 test samples will be misclassified, and 4 of them cannot be decided with a predicted value of 0.49, which is very ambiguous. The prediction results on the test data are presented in Fig 4(b).

Now let us look at the second simplest network, which has 6 connections in total. The connection and weights of the network are given in Fig. 5(a), and the prediction results are provided in Fig. 5(b). The MSE of the network on training and test data are 0.0312 and 0.0203, respectively.

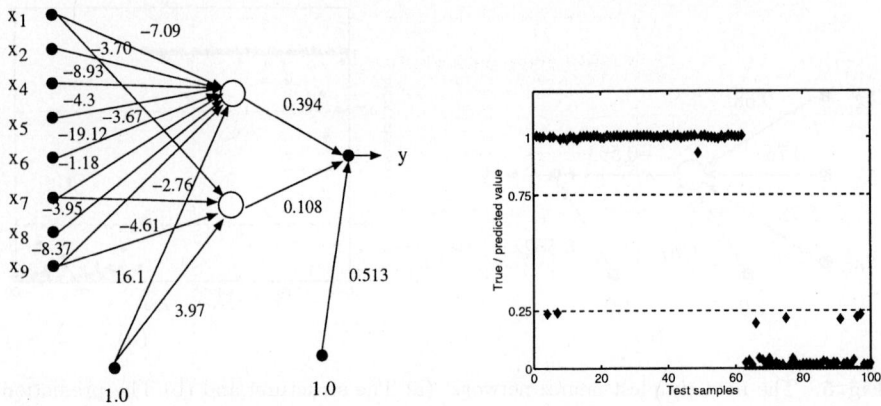

Fig. 6. The neural network with 16 connections. (a) The structure; and (b) The prediction results on test data

In this network, x_2, x_4 and x_6 are present. If the same assumptions are used in deciding whether a case is benign or malignant, then we could extract the following rules: (denoted as MOO_NN2)

R1: If x_2 (clump thickness) ≥ 0.6 or
x_6 (bare nuclei) ≥ 0.9 or
x_2 (clump thickness) $\geq 0.5 \wedge x_6$ (bare nuclei) ≥ 0.2 or
x_2 (clump thickness) $\geq 0.4 \wedge x_6$ (bare nuclei) ≥ 0.4 or
x_2 (clump thickness) $\geq 0.3 \wedge x_6$ (bare nuclei) ≥ 0.5 or
x_2 (clump thickness) $\geq 0.2 \wedge x_6$ (bare nuclei) ≥ 0.7, then malignant;
(17)

R2: If x_2 (clump thickness) $\leq 0.1 \wedge x_6$ (bare nuclei) ≤ 0.4 or
x_2 (clump thickness) $\leq 0.2 \wedge x_6$ (bare nuclei) ≤ 0.2, then benign; (18)

Compared to the simplest network, with the introduction of two additional features x_6 and x_4 (although the influence of x_4 is too small to be reflected in the rules), the number of cases that are misclassified has been reduced to 1, whereas the number of cases on which no decision can be made remains to be 4, although the ambiguity of the decision for the four cases did decrease.

The above two neural networks are very simple in structure. We have shown that for such networks of a low model complexity, rules of symbolic quality can be extracted. In the following, we will take a look at two neural networks obtained in the multi-objective optimization, which are of better accuracy but are of more signal-type quality.

The first network of relatively higher model complexity has 16 connections, whose structure and weights are described in Fig. 6(a). The prediction results

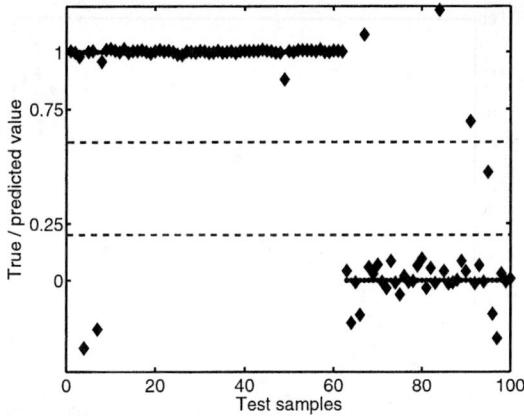

Fig. 7. The prediction results of the most complex neural network obtained in the simulation

Table 1. Comparison of Performance and Complexity

	No. Rules	Av. No. of Premises	Correct Classification Rate(%)
MOO_NN1	2	1	94
MOO_NN2	8	1.75	95
C4.5[25]	6	1.5	94.7
Ref.[23], 1e	1	4	97.2
Ref.[23], 2b	6	1.7	98.3
Ref.[25]	5	1.8	96.2

are plotted in Fig. 6(b). In this network, only x_3 is absent and there are 2 hidden nodes. The MSE on training and test data sets are 0.019 and 0.014, respectively. From Fig. 6(b), we can see that the classification accuracy is better: only two cases are mis-classified. However, extracting symbolic rules from the network becomes much more difficult. Besides, although the architecture of the two simple networks still exist in the current network, it not longer shows a dominating influence. Thus, the "skeleton" defined by the simple networks has been lost.

The most complex network obtained in the run has 74 connections. All input features are included in the network and the number of hidden nodes is 9. The MSE on the training data set is 0.0060, however, the MSE on the test data set increases to 0.066 with 5 samples misclassified and 1 undetermined. It seems that the network has over-fitted the training data and the understanding of the network is difficult. The prediction results of this neural network are provided in Fig. 7.

Table 1 lists the number of rules, the average number of premises in each rule and the correct classification rate on the test data of the rules obtained in this paper (MOO_NN1 and MOO_NN2), and those reported in [29] and [27]. It should be pointed out that the comparison is quite rough because the test data used in the three cases are different. Besides, samples with missing attribute values have

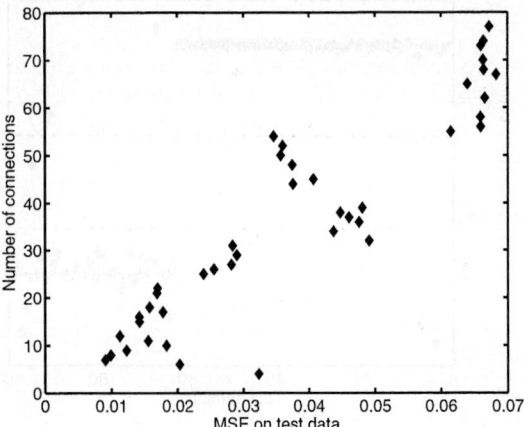

Fig. 8. The relationship between the accuracy on test data and complexity of the Pareto-optimal solutions

been discarded in [29] and [27]. In our simulation, the missing values are simply replaced with the mean of the non-missing values of this attribute. Finally, in our work, samples with a lower confidence (those undecidable) are not counted as "correct classification". The purpose of the comparison is to show that the performance of the symbol-type rules generated in our work is comparable to that of rules extracted using popular machine learning (e.g., C4.5 [24]) and rule extraction methods.

Finally, we take a look at the relationship between the accuracy on test data and the complexity of the neural networks, which is shown in Fig. 8. It seems that for this problem, most neural networks having a higher complexity perform poorly on the test data. The reason behind this phenomenon is still unclear.

6 Conclusions

Both signal-type and symbol-type representations are important for cognitive modeling, as well as for critical applications. This paper suggests a method for generating multiple models of both signal-type and symbol-type simultaneously using an evolutionary multi-objective optimization algorithm instead of extracting symbolic rules after a neural network has been trained, like in most existing rule extraction approaches. This idea has been embodied in generating neural network models for the cancer diagnosis benchmark problem.

A problem in the current method is that the "skeleton" of the most complex neural network is not fully similar to the most simple, symbol-type network. To achieve this property, some additional constraints need to be introduced in the multi-objective optimization, which should be further investigated.

References

1. H.A. Abbass. An evolutionary artificial neural networks approach for breast cancer diagnosis. *Artificial Intelligence in Medicine*, 25(3):265–281, 2002.
2. H.A. Abbass. Speeding up back-propagation using multiobjective evolutionary algorithms. *Neural Computation*, 15(11):2705–2726, 2003.
3. R. Andrews, J. Diederich, and A. Tickle. A survey and critique of techniques for extracting rules from trained artificial neural networks. *Knowledge Based Systems*, 8(6):373–389, 1995.
4. D. Badre and A. Wagner. Semantic retrieval, mnemonic control, and prefrontal cortex. *Behavior and Cognitive Neuroscience Reviews*, 1(3):206–218, 2002.
5. C. M. Bishop. *Neural Networks for Pattern Recognition*. Oxford University Press, Oxford, UK, 1995.
6. K.P. Burnham and D.R. Anderson. *Model Selection and Multimodel Inference*. Springer, New York, second edition, 2002.
7. C. Coello Coello, D. Veldhuizen, and G. Lamont. *Evolutionary algorithms for solving multi-objective problems*. Kluwer Academic, New York, 2002.
8. K. Deb. *Multi-objective Optimization Using Evolutionary Algorithms*. Wiley, Chichester, 2001.
9. K. Deb, S. Agrawal, A. Pratap, and T. Meyarivan. A fast elitist non-dominated sorting genetic algorithm for multi-objective optimization: NSGA-II. In *Parallel Problem Solving from Nature*, volume VI, pages 849–858, 2000.
10. W. Duch, R. Adamczak, and K. Grabczewski. Extraction of logical rules from backpropagation networks. *Neural Processing Letters*, 7:1–9, 1998.
11. W. Duch, R. Setiono, and J. Zurada. Computational intelligence methods for rule-based data understanding. *Proceedings of the IEEE*, 92(5):771–805, 2004.
12. J.A. Fodor and Z.W. Pylyshyn. Connectionism and cognitive architecture: A critical analysis. *Cognition*, 28(3):3–71, 1988.
13. J. Gabrieli, R. Poldrack, and J. Desmond. The role of left prefrontal cortex in langrange and memory. *Proceedings of the national Academy of Sciences*, 95:906–913, 1998.
14. M. Hüsken, J. E. Gayko, and B. Sendhoff. Optimization for problem classes – Neural networks that learn to learn. In Xin Yao and David B. Fogel, editors, *IEEE Symposium on Combinations of Evolutionary Computation and Neural Networks (ECNN 2000)*, pages 98–109. IEEE Press, 2000.
15. C. Igel and M. Hüsken. Improving the Rprop learning algorithm. In *Proceedings of the 2nd ICSC International Symposium on Neural Computation*, pages 115–121, 2000.
16. H. Ishibuchi, T. Yamamoto. Evolutionary multiobjective optimization for generating an ensemble of fuzzy rule-based classifiers. In *Proceedings of the Genetic and Evolutionary Computation Conference*, pages 1077–188, 2003.
17. M. Ishikawa. Rule extraction by successive regularization. *Neural Networks*, 13:1171–1183, 2000.
18. Y. Jin. *Advanced Fuzzy Systems Design and Applications*. Springer, Heidelberg, 2003.
19. Y. Jin, T. Okabe, and B. Sendhoff. Neural network regularization and ensembling using multi-objective evolutionary algorithms. In *Congress on Evolutionary Computation*, pages 1–8. IEEE, 2004.
20. Y. Jin and B. Sendhoff. Extracting interpretable fuzzy rules from RBF networks. *Neural Processing Letters*, 17(2):149–164, 2003.

21. A. Martin and L. Chao. Semantic memory and the brain: Structure and process. *Current Opinions in Neurobiology*, 11:194–201, 2001.
22. D.A. Miller and J.M. Zurada. A dynamical system perspective of structural learning with forgetting. *IEEE Transactions on Neural Networks*, 9(3):508–515, 1998.
23. L. Prechelt. PROBEN1 - a set of neural network benchmark problems and benchmarking rules. Technical Report 21/94, Fakultát für Informatik, Universität Karlsruhe, 1994.
24. J.R. Quinlan. *C4.5 Programs for Machine Learning*. Morgan Kaufmann, 1992.
25. R.D. Reed and R.J. Marks II. *Neural Smithing*. The MIT Press, 1999.
26. M. Riedmiller and H. Braun. A direct adaptive method for faster backpropgation learning: The RPROP algorithm. In *IEEE international Conference on Neural Networks*, volume 1, pages 586–591, New York, 1993. IEEE.
27. R. Setiono. Generating concise and accurate classification rules for breast cancer disgnosis. *Artificial Intelligence in Medicine*, 18:205–219, 2000.
28. R. Setiono and H. Liu. Symbolic representation of neural networks. *IEEE Computer*, 29(3):71–77, 1996.
29. I. Taha and J. Ghosh. Symbolic interpretation of artificial neural networks. *IEEE Transactions on Knowledge and Data Engineering*, 11(3):448–463, 1999.
30. R. de A. Teixeira, A.P. Braga, R. H.C. Takahashi, and R. R. Saldanha. Improving generalization of MLPs with multi-objective optimization. *Neurocomputing*, 35:189–194, 2000.

A Multi-objective Memetic Algorithm for Intelligent Feature Extraction

Paulo V.W. Radtke[1,2,*], Tony Wong[1], and Robert Sabourin[1,2]

[1] École de Technologie Supérieure,
Laboratoire d'Imagerie, de Vision et d'Intelligence Artificielle,
Department de Génie de la Production Automatisé,
1100, rue Notre Dame Ouest, Montréal, QC, Canada H3C 1K3
[2] Pontifícia Universidade Católica do Paraná,
Rua Imaculada Conceição 1155, Curitiba, PR, Brazil 80215-901
radtke@livia.etsmtl.ca

Abstract. This paper presents a methodology to generate representations for isolated handwritten symbols, modeled as a multi-objective optimization problem. We detail the methodology, coding domain knowledge into a genetic based representation. With the help of a model on the domain of handwritten digits, we verify the problematic issues and propose a hybrid optimization algorithm, adapted to needs of this problem. A set of tests validates the optimization algorithm and parameter settings in the model's context. The results are encouraging, as the optimized solutions outperform the human expert approach on a known problem.

1 Introduction

Image-based pattern recognition (PR) systems require that pixel information be first transformed into an abstract representation suitable for recognition, a process called feature extraction [1]. A methodology that extracts features for PR must select the most appropriate transformations and determine the spatial location of their application on the image. Related to the feature extraction process is the feature subset selection (FSS) operation [2]. FSS further refines the extraction process by selecting the most relevant features, within the extracted feature set, in order to reduce classifier's computation effort in the classification stage and improve recognition rate. A comparison of FSS methods in [3] indicates that genetic algorithm (GA) based approach performs better than traditional methods when the problem size is large (more than 50 features). In the context of isolated handwritten digits, Oliveira et al. applied a GA based FSS [4] to optimize classifier accuracy and feature set cardinality using a weighted vector. They postulated that a multi-objective genetic algorithm (MOGA) could further enhance the obtained results. Their postulate was later confirmed in

---

[*] Author for correspondence.

[5], where MOGA outperformed GA on the same problem. The superiority of MOGA in FSS is also confirmed by Emmanouilidis et al. using sonar and ionosphere data [6].

It is now understood that the advantage of MOGA lies in the inherent diversity of the optimized solution set, avoiding the population convergence to a single local optimum. However, the application of MOGA in FSS also faces a number of difficulties. In essence, classifier training is based on a finite set of labeled observations – the training set. The classifier may perform differently when presented to unknown observations, i.e., data not in the training set. This behavior is verified in [5] where the feature set optimized by a MOGA produces a recognition rate that is different for the optimization set compared to a set of unknown observations. This behavior can be explained by the fact that the input domain used in the MOGA optimization process does not match the one used in the classification stage of the recognition process. Thus, the corresponding objective spaces are also non matching because the same classifier is used in the optimization and the classification stages. Another difficulty arises when two or more feature sets sharing similar elements exist in the MOGA population. In the context of FSS, similar feature sets should yield comparable performances for a given classifier. If these feature sets also possess the same cardinality then the genetic selection operator is likely to emphasize the one with the highest recognition rate. Since the FSS problem has non matching objective spaces, the selected feature set may not perform adequately in the classification of unknown observations. Furthermore, the genetic selection operation is complicated by the dominance principle used in most Pareto-based MOGAs. The primary aim of the FSS operation is to reduce feature set cardinality while maintaining the highest possible recognition rate. This implies a mixed-integer objective space and standard dominance relationship can not be implemented directly. Due to the existence of the L_1 norm, special steps must be taken in order to ensure diversity on the Pareto-front. Finally, due to the non matching input domains and objective spaces, a non dominated feature set in the optimization stage is not necessarily non dominated in the classification stage.

Considering these aspects, we propose in Sect. 2 a methodology for feature extraction of isolated handwritten symbols formulated as an evolutionary multiobjective optimization problem (MOOP), supported by earlier experiments in [7]. Section 3 analyze the MOOP and verify the issues discussed in the context of isolated handwritten digits. Sections 4 to 6 describe an optimization algorithm adapted to the FSS problem and present a series of tests to verify its efficiency. Section 7 presents the conclusions.

2 The Intelligent Feature Extractor Methodology

Traditionally, human experts are responsible for the choice of the feature set. It is most often determined by using domain knowledge on a trial and error basis. We propose to use the domain knowledge in a methodology formulated as an

MOOP to genetically evolve a set of candidate solutions – the Intelligent Feature Extractor (IFE) methodology. The goal of this work is to help the human expert in defining representations (feature sets) in the context of isolated handwritten symbols.

2.1 IFE Concepts

The IFE methodology models handwritten symbols as features extracted from specific *foci* of attention on images using *zoning*. It is a strategy known to provide better results in recognition than features extracted from the whole image [8]. In the proposed IFE three operators are needed to generate representations: a *zoning operator* to define *foci* of attention over images, a *feature extraction* operator to apply transformations in zones, and a *feature subset selection* operator that removes irrelevant features. The domain knowledge introduced by the human expert lies in the choice of transformations for the feature extraction operator.

The operators are combined to generate a representation, as illustrated by Fig. 1. The *zoning operator* defines the zoning strategy $Z = \{z^1, \ldots, z^n\}$, where $z^i, 1 \leq i \leq n$ is a zone in the image I and n the number of zones. The pixels inside the zones in Z are transformed by the *feature extraction* operator in the representation $F = \{f^1, \ldots, f^n\}$, where f^i is the feature vector extracted from z^i. F has the irrelevant features eliminated by the *feature subset selection operator*, producing the representation $G = \{g^1, \ldots, g^n\}$, g^i being the feature subset of f^i.

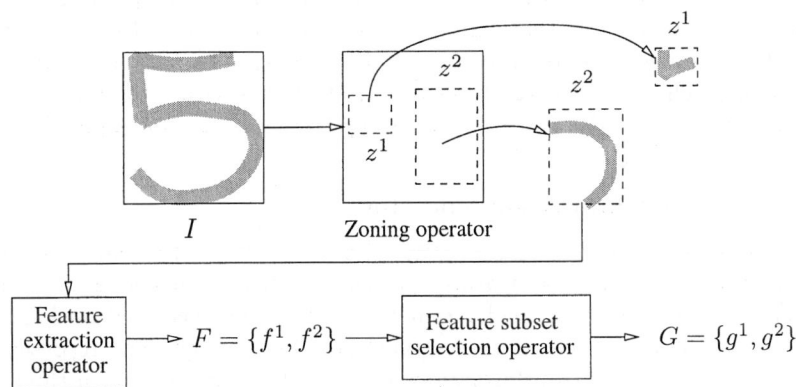

Fig. 1. IFE hierarchical structure

Candidate solutions are represented using a hierarchical genetic coding, with three different parts. Each part of the genetic coding is related to an IFE operator, as shown in Fig. 2. The parts are hierarchical in the sense that the coding in one part will determine the data manipulated by another. After optimization the result is a set of representations. The human expert can either select the representation with the highest accuracy, or use the result set to optimize an ensemble of classifiers (EoC) [9] for improved accuracy.

| Zoning | Feature extraction operators | Feature subset selection |

Fig. 2. IFE candidate solution coding

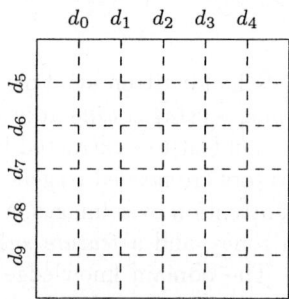

Fig. 3. Dividers zoning operator

2.2 Dividers Zoning Operator

To compare the IFE against the traditional human expert approach we consider a baseline representation known to achieve high accuracy with isolated handwritten digits [10]. The zoning on this representation can be defined as a set of three dividers, where the intersection of image borders and dividers defines zones as 4-sided polygons. Here we expand this concept into a set of 5 horizontal and 5 vertical dividers that can be either *active* or *inactive*. Figure 3 details the operator template, represented by a 10 bits binary string, each bit associated to a divider. This operator produces zoning strategies with 1 to 36 zones, and the baseline zoning in [10] can be obtained by setting d_2, d_6 and d_8 active.

2.3 Feature Extraction Operator

In [10], Oliveira et al. used a mixture of concavities, contour and surface transformations, extracting 22 features per zone – 13 for concavities, 8 for contour and 1 for surface. We consider that the performance achieved on handwritten digits supports the use of these transformations to optimize representations. With three different transformations the operator is encoded as a three bits binary string, where each bit indicates the state of the associated transformation. When all transformations are inactive, the zone becomes a missing part [11], a zone with no features extracted.

2.4 Feature Subset Selection Operator

The feature subset selection operator selects the most relevant features in the feature vector $F = \{f^1, \ldots, f^n\}$, creating a final representation $G = \{g^1, \ldots, g^n\}$. This task is performed with a binary string associated to each feature set f^i, where each bit indicates if the associated feature in f^i is active or not. Thus a 22 bits binary string is required to encode the feature extraction operator described in the previous section.

3 IFE Model

To verify the issues discussed in Sect. 1 we created a model, based on the dividers zoning operator with all 22 features extracted from each zone. In this model, only the dividers zoning operator is active, while the feature extraction and feature subset selection operators have fixed values – to extract all features from all zones. We calculated the entire objective space, evaluating the representations discriminative power on a *wrapper* approach [2] using actual classifier performance. We used the projection distance (PD) classifier [12], with the databases of digits in Table 1 to calculate representations error rate. The disjoint databases are extracted from the NIST SD19 database, a widely used database of isolated handwritten symbols, using the digits data sets *hsf_0123* and *hsf_7*. In this objective space, we minimize both the feature set cardinality and the error rate.

To train the PD classifier we use the *learn* database as the learning examples, and the *validation* database to configure classifier parameters during learning. The error rate during the IFE optimization is calculated with the trained classifier on the *optimization* database. In order to verify the generalization power of solutions optimized by the IFE, we use the *selection* and *test* databases to compare the error rates on unknown observations.

Table 1. Handwritten digits databases

Database	Size	Origin	Sample range
learn	50000	hsf_0123	1 to 50000
validation	15000	hsf_0123	150001 to 165000
optimization	15000	hsf_0123	165001 to 180000
selection	15000	hsf_0123	180001 to 195000
test	60089	hsf_7	1 to 60089

Figure 4 partially details the error rates of solutions using the model's precalculated objective space. Solutions A and C are dominated by solution B on the *optimization* database objective space – Fig. 4.a. In this context, solution B belong to the Pareto-optimal set. Solutions A and B have the same cardinality, but the later has lower error rate, and solution B outperforms solution C in both feature set cardinality and performance. Changing the context to the *selection* database to evaluate the error rate in Fig. 4.b, we have that solution B is dominated by solution A, and that solution C becomes non-dominated. Both solutions A and C belong to the Pareto-optimal set in the *selection* database context.

Because of the non matching objective function spaces, it is clear that the IFE needs a post-processing stage to analyze and select the optimized solutions

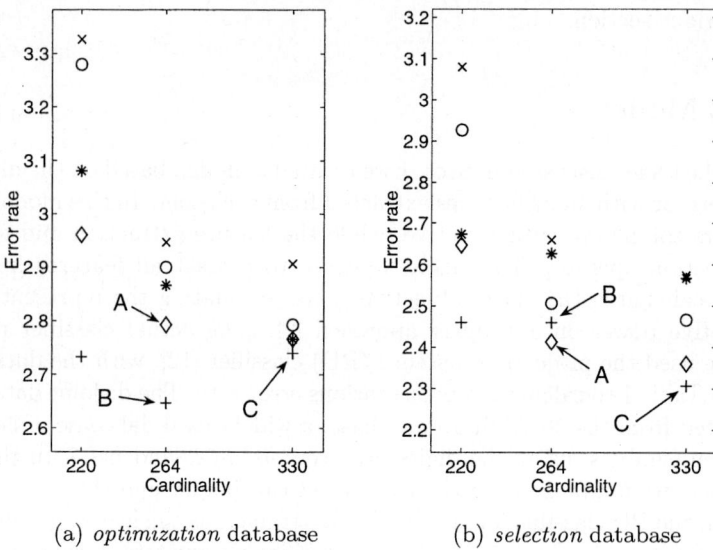

Fig. 4. Objective space – PD classifier

regarding the generalization power on unknown observations. Therefore we define the IFE optimization in two stages. The first stage optimizes solutions on the *optimization* database objective space. The second stage analyzes solutions archived by the optimization algorithm in the *selection* database objective space, where the IFE user selects one or more solutions, based on the error rate. This yields two requirements that must be satisfied by an MOGA algorithm for proper optimization of the IFE methodology:

1. Optimize the best solution for each cardinality value, which we call the *decision frontier* of the objective space.
2. Archive different levels of performance regarding solutions cardinality.

4 Multi-objective Memetic Algorithm

The Multi-Objective Memetic Algorithm (MOMA) combines a traditional MOGA with a local search (LS) algorithm, featuring modified selection and archiving strategies suitable for the IFE methodology. The combination of MOGA with LS is discussed in [13], and [14] demonstrates that hybrid methods outperform methods solely based on genetic optimization in some problems.

4.1 Concepts

To store the decision frontier defined in Sect. 3, we divide the objective functions in two categories, *objective function one* (o_1) in the integer domain, that defines the *slots* of our archive, and *objective function two* (o_2), which is optimized for each o_1 value. To archive different levels of performance, the slot S^l is a set of max_{S^l} solutions, associated to a possible value of o_1. For our IFE problem, o_1 is the feature set cardinality and o_2 is the error rate.

The archive is defined as $S = \{S^1, \ldots, S^j\}$, where j is the maximum number of slots. For solution X^i, $o_1(X^i)$ and $o_2(X^i)$ are the solution's values of o_1 and o_2. $B(S^l) = \{X^i \in S^l | o_2(X^i) = min(o_2(x)), \forall x \in S^l\}$ indicates the solution X^i in S^l with the best o_2 value, while $W(S^l) = \{X^i \in S^l | o_2(X) = max(o_2(x)), \forall x \in S^l\}$ indicates the opposite.

The decision frontier set P_S optimized by the MOMA algorithm is defined as $P_S = \bigcup_{l=1}^{j} \{B(S^l)\}$. We indicate that solution X^i is admissible into slot S^l as $X^i \bowtie S^l \equiv o_1(X^i) = o_1(x \in S^l)$, then $A(S^l, C) = \{c | c \in C \wedge c \bowtie S^l\}$ denotes the subset of solutions in C that are admissible in S^l.

To optimize the decision frontier, solutions are ranked for genetic selection by a *frontier ranking* approach. In the population P, the solution set belonging to the first rank is defined by $R^1 = \bigcup_{l=1}^{j} \{B(A(S^l, P))\}$. The solution set belonging to the second rank R^2 is obtained as the first rank of $P \backslash R^1$, and so on.

The decision frontier concept and the archive S are key elements for proper optimization of the IFE methodology. Combined they provide means to select solutions after optimization based on their generalization power on unknown observations. Selecting solutions by the decision frontier allows the optimization of solutions usually discarded by traditional Pareto based approaches. This need is justified to avoid optimization bias as indicated in Fig. 4, where solution C is dominated in the IFE model optimization objective space using the *optimization* database, but has better generalization power on unknown observations. The same principle justifies the need to store different levels of performance in the slot, solution A in Fig. 4 has better generalization power on unknown observations, but would be discarded from the archive on traditional approaches.

4.2 Algorithm Discussion

The MOMA algorithm is depicted in Fig. 5. It evolves a population P of size m, and archives good solutions found in the slots S, which are updated at the end of each generation. The population P is initialized in two steps. The first creates candidate solutions with a Bernoulli distribution, while the second generates individuals to initialize the slots. For each slot, we choose one random solution that is admissible in the slot and insert it in the population.

During genetic optimization, individuals in the current generation P_t are subjected to frontier ranking. Next a mating pool M is created by tournament selection, followed by crossover and mutation to create the offspring population P_{t+1}. In case of a draw in the tournament selection, one of the solutions is chosen randomly. To avoid genetic overtake, redundant individuals are mutated until the population has no redundant individuals as in [9].

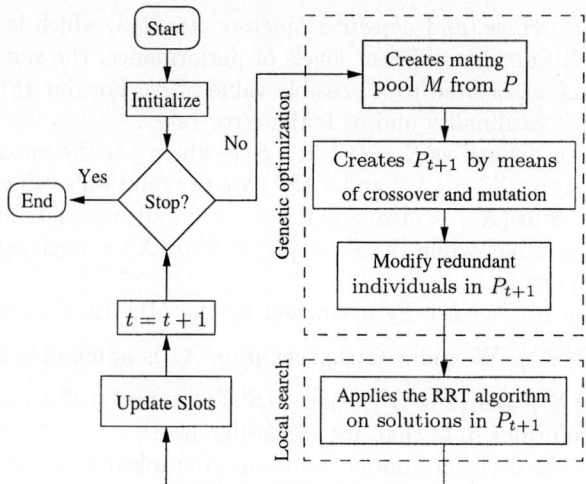

Fig. 5. MOMA algorithm

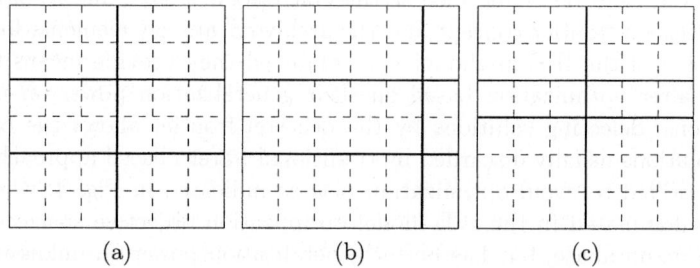

Fig. 6. Solution (a) and two neighbors (b and c)

After genetic optimization solutions are further improved by a LS algorithm. We choose the *Record-to-Record Travel* (RRT) algorithm [15], an annealing based heuristic. The RRT algorithm improves solutions by searching in its neighborhood for n potential solutions during NI iterations, allowing a decrease in the current performance of $a\%$ to avoid local optimal solutions.

Neighbors to solution X^i must have the same feature set cardinality and similar structure, which is achieved in the IFE model by modifying the zoning operator encoding. The model defined in Sect. 3 has all features extracted from all zones, and the feature extraction and feature subset selection operators are fixed. With the zoning operator dividers distributed in two groups, $g_1 = \{d_0, d_1, d_2, d_3, d_4\}$ and $g_2 = \{d_5, d_6, d_7, d_8, d_9\}$, to generate a neighbor we select a group to activate one divider and deactivate another. The solution in Fig. 6.a has solutions in Figs. 6.b and 6.c as two possible neighbors.

After the LS, the archive S is updated, storing good solutions from P_{t+1} in the slots as in Algorithm 1. Recall that max_{S^l} is the maximum number of solutions a slot can hold. At this point, we verify the stopping criterion, deciding

```
forall x^i ∈ P do
    Determines the slot S^l solution x^i relates to;
    if E(x^i) < E(W(S^l)) and x^i ∉ S^l then
        S^l = S^l ∪ {x^i};
        if |S^l| > max_{S^l} then
            S^l = S^l\{W(S^l)};
        end
    end
end
```

Algorithm 1. Update slot algorithm

if the algorithm should continue to the next iteration or stop the optimization process.

5 Experimental Protocol

To test the MOMA algorithm, we conducted three tests on the IFE model's objective space, optimizing only the IFE zoning operator, while keeping the remaining operators fixed to extract all 22 features from each zone. All tests used the *learn* and *validation* databases to train the PD classifier, and the *optimization* database to evaluate the error rate during the optimization process. The first test verifies that the genetic optimization has convergence properties in this type of problem. We achieve this by disabling the RRT algorithm with $NI = 0$. The second test evaluates the MOMA algorithm with a neighborhood subset best improvement strategy, while the third test uses a greedy first improvement strategy, where $n = 1$ and $a = 0\%$. We call these tests as *Test A*, *Test B* and *Test C*, respectively.

The usual genetic operators for the IFE are the *hierarchical single point crossover*, where the single point crossover is performed independently over each IFE operator in the chromosome with probability p_c, and the *hierarchical bitwise mutation*, which also performs the bitwise mutation independently over each IFE operator in the chromosome with probability $p_m = 1/L$, where L is the length in bits of the encoded operator being mutated. For the IFE model, genetic operations are restricted to the zoning operator, in order to extract all features from each zone.

We defined a set of values for each algorithmic parameter, using a *fractional design* approach [16] to obtain the 18 configuration sets in Table 2. During *Test A* and *Test C*, we replace columns in this table with specific values to achieve the desired effects.

Each configuration set is subjected to 30 runs of 500 generations in each test, and comparisons are made at the generation of convergence. One run is said to have converged when the optimized decision frontier set P_S can no longer be improved. Preliminary experiments indicated that 500 generations far exceed the number of generations required to converge *Test A*, our worst case scenario. Thus the following metrics are used to compare runs:

Table 2. Parameter values

#	m	p_c	p_m	a	n	NI
1	32	70%	1/L	5%	2	7
2	32	70%	1/L	5%	3	5
3	32	70%	1/L	5%	4	3
4	32	80%	1/L	4%	2	7
5	32	80%	1/L	4%	3	5
6	32	80%	1/L	4%	4	3
7	32	90%	1/L	2%	2	7
8	32	90%	1/L	2%	3	5
9	32	90%	1/L	2%	4	3

#	m	p_c	p_m	a	n	NI
10	64	70%	1/L	5%	2	7
11	64	70%	1/L	5%	3	5
12	64	70%	1/L	5%	4	3
13	64	80%	1/L	4%	2	7
14	64	80%	1/L	4%	3	5
15	64	80%	1/L	4%	4	3
16	64	90%	1/L	2%	2	7
17	64	90%	1/L	2%	3	5
18	64	90%	1/L	2%	4	3

1. *Unique individual evaluations* – how many unique individuals have been evaluated until the algorithm convergence, which relates to the computational effort.
2. *Coverage by the global optimal set* – percentage of individuals in P_S that are covered by solutions in the global optimal set [17], adapted to the decision frontier context. When P_S converges to the optimal set, the coverage is equal to zero.

Both metrics are fair as they hold the same meaning for all three tests. A final test evaluates representations optimized by the MOMA algorithm in the IFE model. We select a result set S, evaluate the error rate of these solutions with the *selection* database and calculate the decision frontier P_S. From this decision frontier we select a set of solutions for testing to compare with the baseline representation.

6 Results

The results for the MOMA tests are presented in Figs. 7 to 9. The horizontal axis on the plots relate to configuration sets in Table 2. Experiments 1 to 9 represent a smaller population – 32 individuals, while experiments 10 to 18 represent a larger population – 64 individuals. The box plots summarize the values attained in the 30 runs of each configuration set.

The results for *Test A* in Fig. 7.a indicate the convergence property of genetic operations alone, which is capable to optimize an approximation to the global optimal. The best coverage values where achieved by the larger population, which also explored better the objective space. The exploratory aspect is measured as the number of unique individual evaluations in Fig. 7.b.

To improve convergence and the objective space exploration, we use in *Test B* the complete MOMA algorithm. The RRT algorithm improved convergence and all runs but a few outliers in the smaller population converged to the optimal set. Objective space exploration in *Test B* is improved, as the number of unique individual evaluations in Fig. 8 is higher than in *Test A* – Fig. 7.b. This

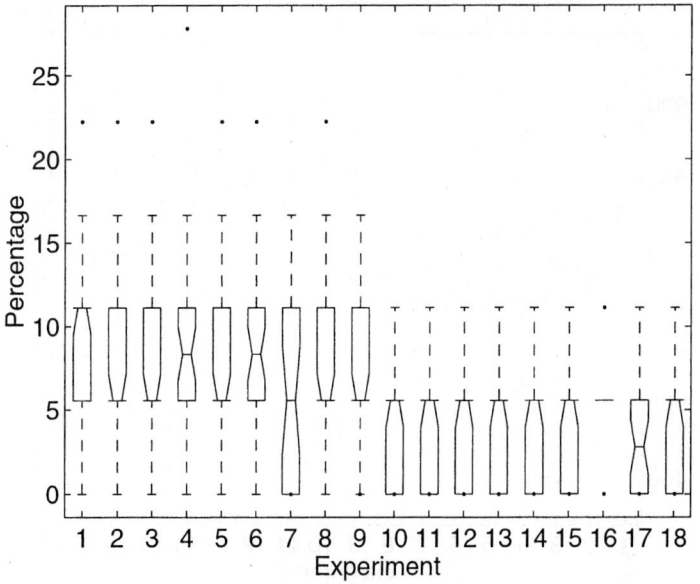

(a) Coverage – *Test A*

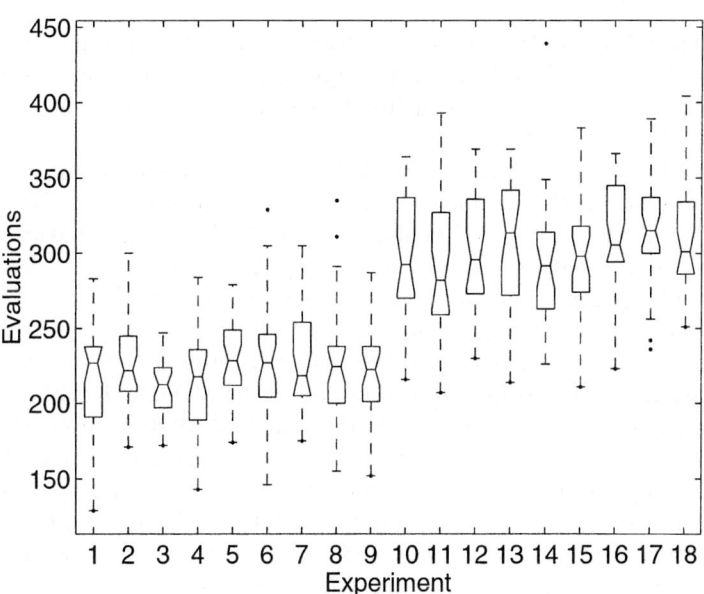

(b) Unique individual evaluations – *Test A*

Fig. 7. Results – *Test A*

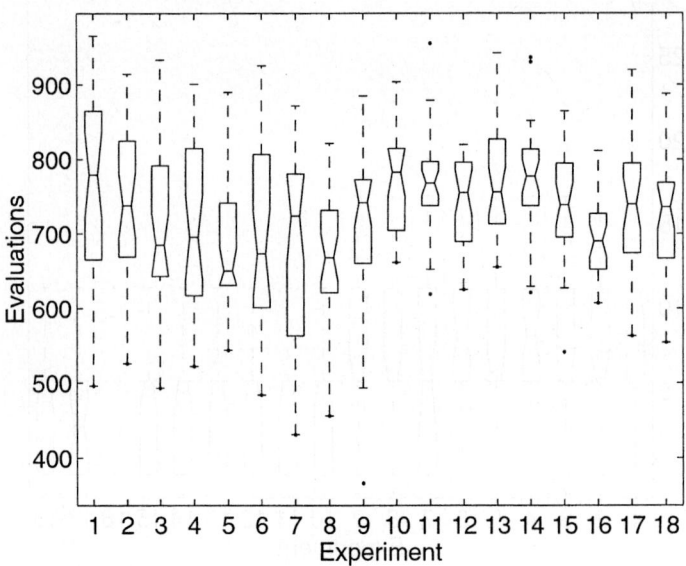

Fig. 8. Unique individual evaluations – *Test B*

improvement reflects in the convergence toward the global optimal set, which is better than in *Test A*. The LS helps to improve convergence when searching for better solutions, which may also helps the genetic algorithm to better explore the objective space.

In the IFE we are concerned with the error rate evaluation cost. Thus, it is desirable to restrain the number of unique individual evaluations by reducing the strength of the LS. *Test C* modifies the RRT algorithm behavior, using a greedy first improvement strategy. The convergence is similar to *Test B* – all runs but a few outliers converged to the optimal set. However, the number of unique individual evaluations in Fig. 9 is lower than in Fig. 8, which suggests that this improvement strategy is more suitable for the IFE problem optimization.

These results demonstrate the effectiveness of the MOMA algorithm with the IFE methodology, reaching solutions that traditional MOGA approaches could not. For better convergence with lower number of unique individual evaluations, the LS with the greedy first improvement strategy is most appropriate. As for the configuration parameters, configuration set 15 in Table 2 modified for the greedy improvement strategy ($n = 1$ and $a = 0\%$) is a good trade-off between convergence and number of unique individual evaluations.

Our final test evaluates a set of solutions optimized by the MOMA algorithm in the IFE model. We selected a random run from *Test C* and evaluated the error rate of solutions in S with the *selection* database. Then we arbitrarily selected solutions a to g from the decision frontier P_S calculated with the error rates

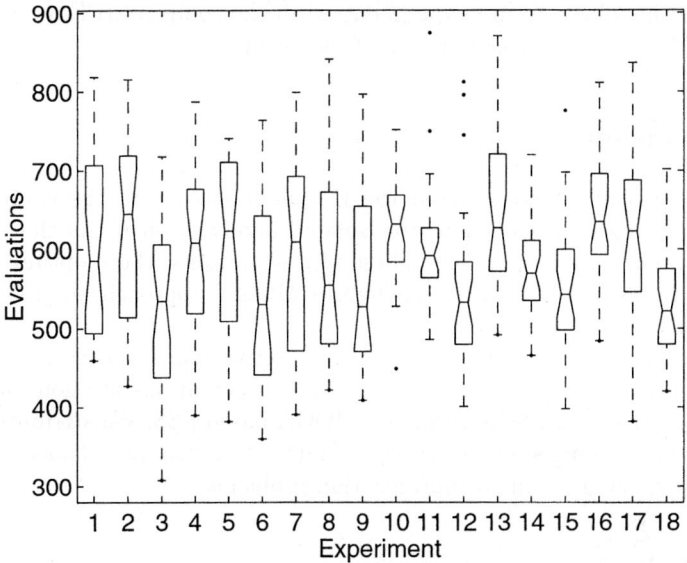

Fig. 9. Unique individual evaluations – *Test C*

Table 3. Representations comparison

Representation	features	zoning operator	e_{opt}	e_{sel}	e_{test}
Baseline	132	00100 01010	3.527%	3.010%	2.959%
a	110	00000 01111	3.587%	3.053%	3.272%
b	132	00000 11111	3.260%	2.987%	2.930%
c	176	00010 01101	3.153%	2.980%	2.981%
d	198	01010 01101	3.273%	2.993%	2.521%
e	220	00100 01111	2.733%	2.460%	2.438%
f	264	01100 01110	2.647%	2.460%	2.568%
g	330	00110 01111	2.740%	2.307%	2.180%

on the *selection* database, as discussed in Sect. 3. Finally we tested the selected solutions with the *test* database to compare with the baseline representation. The results are presented in Table 3, where the baseline representation was also trained and evaluated with the PD classifier using the same database set. The table details the feature set cardinality, the binary string associated to the zoning operator and the error rate in three databases, *optimization*, *selection* and *test* – e_{opt}, e_{sel} and e_{test}, respectively.

The results in Table 3 demonstrate that the IFE methodology is able to optimize and select solutions that outperform the traditional human expert approach in the domain of unknown observations – the *test* database. Representation *g*'s error rate is 26.33% lower than the baseline on the *test* database, which justifies the IFE methodology for actual applications. As the model is a subset of the complete methodology, future experiments with the complete IFE methodology are expected to achieve at least this performance level.

7 Conclusions

This paper presented and assessed the IFE methodology in the context of a model, generating representations for isolated handwritten digits that outperform the human expert approach. These representations where optimized with the proposed MOMA algorithm, an hybrid MOGA approach adapted to the objective space of the IFE problem.

The IFE model demonstrated that the objective space during optimization and the objective space with a set of unknown observations are non matching, which is also verified in the literature on MOGA based FSS. We attribute this to the supervised learning stage, based on a finite set of examples, hence the same behavior is expected in similar optimization problems.

Acknowledgments

The first author would like to acknowledge CAPES and the Brazilian government for supporting this research through scholarship grant BEX 2234-03/3.

References

1. Heutte, L., Paquet, T., Moreau, J.V., Lecourtier, Y., Olivier, C.: A structural/statistical feature based vector for handwritten character recognition. Pattern Recognition Letters **19(7)** (1998) 629–641.
2. John, G. H., Kohavi, R., Pfleger, K.: Irrelevant Features and the Subset Selection Problem. Proceedings of the International Conference on Machine Learning (1994) 121–129.
3. Kudo, M., Sklansky, J.: Comparison of algorithms that select features for pattern classifiers. Pattern Recognition **33(1)** (2000) 25–41.
4. Oliveira, L. S., Benahmed, N., Sabourin, R., Bortolozzi, F., Suen, C. Y.: Feature Subset Selection Using Genetic Algorithms for Handwritten Digit Recognition. Proceedings of the XIV Brazilian Symposium on Computer Graphics and Image Processing (2001) 362–369.
5. Oliveira, L. S., Sabourin, R., Bortolozzi, F., Suen, C. Y.: Feature Subset Selection Using Multi-Objective Genetic Algorithms for Handwritten Digit Recognition. Proceedings of the 16th International Conference on Pattern Recognition **I** (2002) 568–571.
6. Emmanouilidis, C., Hunter A., MacIntyre J.: A Multiobjective Evolutionary Setting for Feature Selection and a Commonality-Based Crossover Operator. Proceedings of the 2000 Congress on Evolutionary Computation (2000) 309–316.
7. Radtke, Paulo V. W., Oliveira, L. S., Sabourin, R., Wong, T.: Intelligent Zoning Design Using Multi-Objective Evolutionary Algorithms. Proceedings of the 7th International Conference on Document Analysis and Recognition (2003) 824–828.

8. Sabourin, R., Genest, G., Prêteux, F. J.: Off-Line Signature Verification by Local Granulometric Size Distributions. IEEE Transactions on Pattern Analysis and Machine Intelligence **19(9)** (1997) 976–988.
9. Ruta, D., Gabrys, B.: Classifier Selection for Majority Voting. Accepted to Information fusion (2004).
10. Oliveira, L. S., Sabourin, R., Bortolozzi, F., Suen, C. Y.: Automatic Recognition of Handwritten Numerical Strings: A Recognition and Verification Strategy. IEEE Transactions on Pattern Analysis and Machine Intelligence **24(11)** (2002) 1438–1454.
11. Li, Z.-C., Suen, C. Y.: The partition-combination method for recognition of handwritten characters. Pattern Recognition Letters **21(9)** (2000) 701–720.
12. Kimura, F., Inoue, S., Wakabayashi, T., Tsuruoka, S., Miyake, Y.: Handwritten Numeral Recognition using Autoassociative Neural Networks. Proceedings of the International Conference on Pattern Recognition **1** (1998) 152–155.
13. Knowles, J.D., Corne, D. W.: Memetic Algorithms for Multiobjective Optimization: Issues, Methods and Prospects. In Recent Advances in Memetic Algorithms. Krasnogor, N., Smith, J.E., and Hart, W.E. (eds). pp. 313-352, Springer, 2004.
14. Jaszkiewicz, A.: Do Multiple-Objective Metaheuristics Deliver on Their Promise? A Computational Experiment on the Set-Covering Problem. IEE Transactions on Evolutionary Computation **7(2)** (2003) 133-143.
15. Pepper, J., Golden, B. L., Wasil, E. A.: Solving the Traveling Salesman Problem With Annealing-Based Heuristics: A Computational Study. IEEE Transactions on Systems, Mand and Cybernetics – Part A: Systems and Humans **32(1)** (2002) 72–77.
16. Gunst, R. F., Mason, R. L.: How to Construct Fractional Factorial Experiments – ASQC basic references on quality control: v. 14. American Society for Quality Control – Statistics Division, 611 East Wisconsin Avenue, Milwaukee, Wisconsin 53202, USA (2001).
17. Zitzler, E., Thiele, L.: Multiobjective Optimization Using Evolutionary Algorithms – A Comparative Case Study. Parallel Problem Solving from Nature – PPSVN V, Springer Verlag (1998).

Solving the Aircraft Engine Maintenance Scheduling Problem Using a Multi-objective Evolutionary Algorithm

Mark P. Kleeman* and Gary B. Lamont

Air Force Institute of Technology,
Dept of Electrical and Computer Engineering,
Graduate School of Engineering & Management,
Wright-Patterson AFB (Dayton) OH, 45433, USA
{Mark.Kleeman, Gary.Lamont}@afit.edu
http://www.afit.edu

Abstract. This paper investigates the use of a multi-objective genetic algorithm, MOEA, to solve the scheduling problem for aircraft engine maintenance. The problem is a combination of a modified job shop problem and a flow shop problem. The goal is to minimize the time needed to return engines to mission capable status and to minimize the associated cost by limiting the number of times an engine has to be taken from the active inventory for maintenance. Our preliminary results show that the chosen MOEA called GENMOP effectively converges toward better scheduling solutions and our innovative chromosome design effectively handles the maintenance prioritization of engines.

Keywords: Multi-objective Evolutionary Algorithms, Scheduling Problem, Aircraft Engine Scheduling, Variable-length chromosome.

1 Introduction

Scheduling problems are a very common research topic. This is because, for efficiency reasons, our world relies heavily on schedules and deadlines. Aircraft engine maintenance is no exception. The United States Air Force has many planes that it must keep up and running. But with the downsizing that has occurred in recent years, the number of planes that are operational has become more critical. This means that every effort needs to be made to ensure that not only are the engines repaired in an efficient manner, but that their component's scheduled maintenance cycles are in sync so that the engine has fewer trips to the logistics maintenance center.

---

* The views expressed in this article are those of the authors and do not reflect the official policy of the United States Air Force, Department of Defense, or the United States Government.

Section 2 looks briefly at the types of scheduling problems that are researched and describes how each one behaves. Section 3 describes specifically the aircraft scheduling problem. It states why this problem is important, shows the maintenance flow, and gives the particulars as to how a system was implemented to solve the problem. Section 4 describes the algorithm used to implement the problem. Section 5 discusses our design of experiments and tells what metrics we used and why. Section 6 show the results of our experiments and looks at them analytically in an effort to discern any trends. Finally, the paper concludes with Section 7 where we summarize our findings and describe the future work we are doing on this problem.

2 Scheduling Problems

Scheduling problems can be formulated in many different ways. They are commonly abstracted into three different classes of problems: open shop, flow shop, and job shop. The definitions for these problems are described in [1].

2.1 Flow Shop

The flow shop scheduling problem consists of m machines and n jobs. All m machines are situated in a defined series. All n jobs have to be processed on each machine. All the jobs must follow the same routing along the series of machines. Once a job is completed on one machine, it is placed into the queue of the next machine in the series. Normally, jobs are removed from the queue on a *first in first out* (FIFO) basis, but this can be modified to fit the needs of the problem, such as higher priority jobs could be bumped to the front of the queue.

2.2 Job Shop

For the job shop problem, unlike the flow shop problem, each job has its own route to follow. With job shop problems, they can be modelled by having jobs visit any machine at most one time, or they can be created to allow for multiple visits to machines [1].

3 Aircraft Scheduling Problem

The U.S. Air Force has many aircraft in its arsenal. Its fleet is aging and it appears that plans are for the aircraft to be around for a while longer. But many of these engines are beyond their design life. In fact, 97% of all F100 engines and 84% of all F110 engines will be past their design life by the year 2010 [2]. One essential thing that the Air Force relies upon is a dependable fleet. If the Air Force does not have reliable aircraft, then they must procure redundant systems in order to ensure the success of any mission with a high degree of certainty. To achieve this type of dependability, the old "fix it when it breaks" mentality or the on condition maintenance (OCM) practice had to be revamped in order to overcome much of the uncertainty that is inherent in this type of repair mentality.

So now, aircraft engines are brought into the shop for two reasons: unscheduled maintenance and routine maintenance. Unscheduled maintenance occurs when an engine part breaks and requires repair or replacement. Routine maintenance occurs on predefined intervals, when the engine component is still operational, but history shows that the mean time to repair (MTTR) for that component is almost up. For dependability purposes, it is better to refurbish a component before a failure occurs instead of only repairing broken components. This type of mindset is typical of the reliability-centered maintenance (RCM) philosophy. A properly implemented RCM philosophy "systematically analyzes aircraft components/systems to identify and implement the best preventive maintenance policies" [3]. Scheduled maintenance is only part of the equation when it comes to RCM. Another goal of RCM is to reduce the amount of scheduled maintenance. While this is somewhat limited to the MTTR of specific components, it can also be influenced by the components used in the assembly of an engine. For example, it would be better to have two components that have a MTTR around 500 than it would be to have MTTR values of 500 and 1000 for the same two components.

3.1 Maintenance Flow

With RCM as our focal point, our problem is set-up in the following manner. When an engine comes into a logistics workcenter for repair, it is first logged into the system. Aircraft engines are commonly divided into smaller subcomponents which can be worked on individually and then recombined. For this problem, we divided the engine into five logical subcomponents: fan, core, turbine, augmenter, and nozzle. We assumed that the maintenance shop had one specific work area for each of the components. This is an example of the job shop problem, but with a twist. After all maintenance is completed on an engine, each engine component's mean time to repair (MTTR) is compared with other components on the engine. If there is a large disparity among the MTTRs then a component swap may be initiated with another engine in an effort to ensure the MTTRs of the components of a particular engine are similar. This is done so that the engine can have more time on wing (TOW) and less time in the shop.

Once the swaps are done, the engine is reassembled and tested as a whole to ensure functionality. This represents a small flow shop problem in that each engine has to have maintenance done first, followed by swapping and then testing. So the problem at hand is actually a hybrid of two scheduling problems.

Figure 1 shows an example of the flow for two engines. As you can see the problem has a certain flow to it, and the component repair is more of a job shop problem that is embedded into the flow.

3.2 Program Specifics

Static Scheduling. The program was written as a static scheduling problem, where the program receives an initial set of inputs and it outputs results based

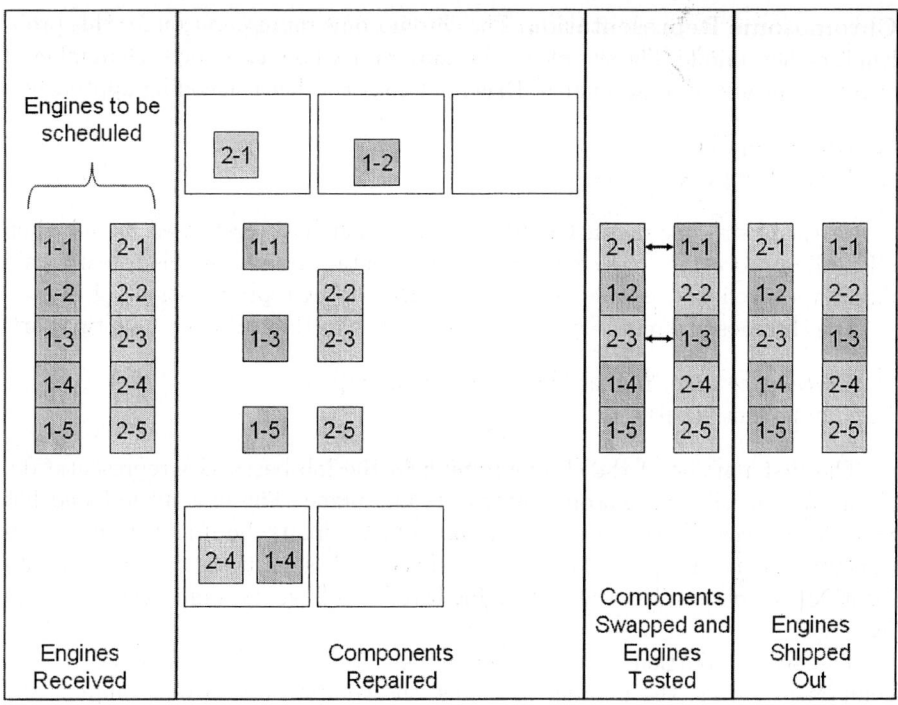

Fig. 1. Example of maintenance flow for two engines

on them. A dynamic scheduling problem allows the user to add additional items to the schedule while the program is running. For our problem, it is reasonable to assume that the logistics center is always alerted ahead of time as to what engines are coming and when. Since the shop is alerted a priori of the items to be scheduled, we felt that static scheduling was more appropriate for our problem domain.

Input File. Our input file is made up of all the information that may be useful for our scheduling problem. Some of the information is not used currently, but it was included for completeness and for future research. The first line of the file lists the n number of engines that need to be scheduled for repair. This is followed by n lines, where each line details the particulars of a specific engine. First the engine part number is listed. Then the estimated arrival time of the engine is listed. This is followed by the due date and weight. The due date is self explanatory, while the weight is a priority system for the program. A weight of one signifies a top priority, while a weight of three is a low priority job. The next ten numbers are divided into five sets of two, where each set represents one of the five components of the engine. The first number in a set is the time that is estimated for the component to be fixed. A zero in this spot means that the component is good and does not need repair work. The second number of each set is the MTTR.

Chromosome Representation. The chromosome representation for this problem is rather unique. The size of the chromosome varies based upon the number of engines that are to be scheduled. Bagchi [4] lists two basic encoding approaches:

1. Direct approach
2. Indirect approach

He then lists nine distinct representations that have been used for job-shop scheduling problems. The chromosome representation used in this research is a direct approach that is based on the "Job-Based GA representation" [4, 5].

The chromosome representation used can be effectively divided into two part:

1. Precedence order for the engines requiring repair
2. Component swaps

The first portion of the chromosome uses the job-based GA representation. Each allele in the chromosome represents an engine. The first allele listed has precedence over all other engines for component repairs. So if it has three components that need repair, they will be scheduled first in those respective repair locations. It follows that the last engine listed will have its components repaired last.

The second portion of the chromosome determines the precedence of the component swaps, the number of swaps, the components to be swapped, and the engines that the components are to come from. The most difficult part of this problem determining how many swaps to allow. By having a lot of swaps, you increase the problem complexity and if you increased the number of swaps beyond a reasonable limit, the problem efficiency is greatly reduced. For example, say I have three engines being repaired. Suppose that it would take four swaps in order to get the best answer. If you allow for 20 swaps, the search space is increased drastically by the additional 16 swaps that are unnecessary. Conversely, if you only allow for 4 swaps and you need 10, then there is no way that you can ever achieve the best answer because you do not allow for enough swaps. We concluded that as the number of engines is increased, the number of swaps would increase as well. So our chromosome was developed in order to accommodate the varying number of swaps based on the number of engines that are to be repaired. But we did not want to force the program to make swaps when none were needed. For example, suppose three engines go in for complete overhauls one week. When the repairs are done, none of the engines need to swap because all of the MTTR values are similar. By doing unneeded swaps, the makespan goes up needlessly. Therefore, the number of swaps can be zero or go as high as the total number of engines being repaired. Section 6 contains an analysis regarding the maximum amount of swaps and the average number of swaps recorded for individuals along the known Pareto front.

Figure 2 show a representation of a chromosome for four engines. Note that the first four alleles are for the engine scheduling precedence, while the last 12 alleles are for the swaps. There are four possible swaps. Each swap is represented by three alleles, the two engines that will be swapping components and the actual

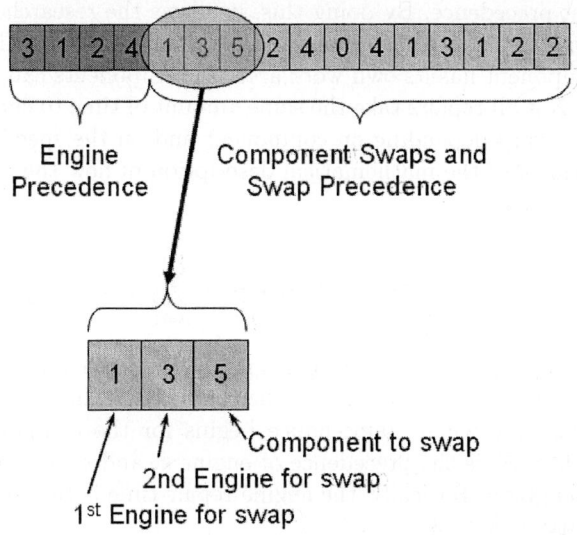

Fig. 2. Example of chromosome representation for four engines

component to be swapped. Note that for this example, there will be only three swaps because one of the swaps has a zero for the component, signifying no swap is to occur.

The formula for determining the chromosome length is found in the following equation:

$$L_{Chromo} = E + 3E = 4E \tag{1}$$

where L_{Chromo} is the length of the chromosome, and E symbolizes the number of engines to be scheduled. As you can see the chromosome length is four times the number of engines.

Fitness Functions Used. Another important aspect of the program are the fitness functions. As stated earlier, the goal is to develop a program that limits the time an engine is in the shop for repairs and to limit the number of times the engine has to go into the shop for scheduled maintenance. These are competing objectives because the first objective is to get the engines out as quickly as possible and the other objective is to match component MTTR values, which requires more time in the shop. It was determined that the best fitness functions to use for this problem were the makespan and the aggregate swap count.

The makespan, C_{max} is equivalent to the completion time of the last job to leave the system. For our problem this is not easily defined. First, every engine component that requires maintenance, is scheduled based on two things:

1. Its precedence defined in the chromosome
2. The arrival time of the engine

If an engine has not arrived in the shop yet and it has the highest precedence, all other components must wait until the work is done on the component with the highest precedence. By doing this, it allows the researcher to be more flexible in having higher priority items getting attention before lower priority items. Each component has its own work area, so components can be worked on simultaneously. Not all repairs take the same amount of time to complete, so finishing times may vary depending on component and on the specific component problem. Equation 2 is the mathematical description of how the repair time for each engine is calculate.

$$\forall e \in E, R_e = \substack{max \\ c \in C} \; t_{CompStart} \sum_{i=1}^{P_e} t_R(i) \qquad (2)$$

where e is an engine, E is the set of engines, R_e is the repair time for the engine, c is one of the five components of the engine, C is the set of components, $t_{CompStart}$ is the time when maintenance begins for the component with the highest precedence, P_e is the precedence of engine e, and $t_R(i)$ it the estimated repair time for engine i. Basically, the engine repair time is the completion time of the final component.

Next, component swaps are calculated based on two things:

1. The last (maximum) completion time for repair for either of the two engines involved in the swap
2. The precedence of the swap

Once these two items are determined, the swap occurs and the new completion time for both engine components is adjusted. If multiple swaps occur for a single engine, the precedence, as defined in the chromosome, determines which swaps occur first. Also note that different components have different swap times. This is because, in reality, not all component swaps take the same amount of time to perform. Equation 3 shows how the swap times are calculated

$$S_e = \substack{max \\ \forall e \in S} \; R_e + t_s(c) \qquad (3)$$

where S_e is the engine swap completion time, e is an engine, S is the set of two engines that are involved in the swap, R_e is the repair time for engine e, and $t_s(c)$ is the swap time for component c, the component to be swapped. Since not all engines will participate in swaps, we can conclude that $S_e \geq R_e$.

Finally, each engine is tested to ensure it is working properly before it is shipped back to the owner. Testing occurs after all repairs and swaps are done on an engine. Equation 4 describes how this is done.

$$T_e = S_e + t_{test} \qquad (4)$$

where T_e is the engine testing completion time, S_e is the swap completion time, and t_{test} is the time it takes to accomplish engine testing.

Finally, the makespan is calculated according to Equation 5.

$$C_{max} = \substack{max \\ \forall e \in Engines} \; T_e \qquad (5)$$

where C_{max} is the makespan for a chromosome representation, e is an engine, and T_e is the engine testing completion time. So the test completion time of the last engine to be tested is the makespan. This value is the fitness value for the first fitness function. So the lower the makespan value, the better the fitness for the chromosome.

The second fitness function is the aggregate swap count. The aggregate swap count is calculated after the makespan is calculated. The program compares the MTTR for all the components of an engine. If the MTTR is outside the provided tolerance levels for MTTR, then the aggregate swap count is increased. This is done for each engine and the summation of the results is the aggregate swap count.

$$ASC = \sum_{i=1}^{e} \sum_{k=1}^{c} tol_{ik_{MTTR}} \tag{6}$$

where the ASC is the aggregate swap count, e is the total number of engines, c is the total number of components (for this problem, $C = 5$), and $tol_{ik_{MTTR}}$ represents the total number of components of engine i that are outside the tolerance limits of the component k on engine i. These values are calculated in such a way so that they are counted only once. The total number of components whose MTTR are out of tolerance with others on the same engine are then summed together with all the other engines to get the aggregate swap count.

4 GENMOP Software Design

The problem was integrated into the General Multi-objective Parallel Genetic Algorithm (GENMOP). GENMOP was originally designed as a real-valued, parallel MOEA. This section briefly describes MOEAs and then discusses in more detail GENMOP operations.

4.1 Multi-objective Evolutionary Algorithms (MOEAs)

Evolutionary algorithms (EA) include genetic algorithms, evolution strategies (ES), and evolutionary programming (EP). EAs consist of a class of algorithms that use the concepts of genetics which enable them to explore the search space. In an evolutionary algorithm, there is a collection of individuals, each one is known as a chromosome. A group of chromosomes are created and compared to one another. This group is known as a population. Chromosomes consist of alleles that can be encoded into a variety of datatypes: binary, integer, real-valued, etc. These allele values are altered by EA operators such as mutation and recombination. Mutation works by inserting new genetic material into the population by modifying allele values. Recombination is accomplished by exchanging allele values between two or more individuals of the population. There are many varieties of genetic operators each with a different set of parameters that may be modified given a particular EA type [6]. After the chromosomes are modified, a selection process occurs where a new population is determined for the next generation. There many are ways to select one chromosome over another [7], but the main objective of the selection process is to steer the EA toward the solution. Regardless of the selection process used, all EAs have a fitness function that they assign to a chromosome. The selection process compares the fitness functions of the chromosomes in order to determine the new population.

A multi-objective EA (MOEA) differs from an EA in that there is more than one fitness function for each chromosome. This often creates a situation where there is a vector of answers that can be considered optimal. To determine which answer is best, the researcher must either weight the fitness values or must pick one point out of the group via an inspection process. The book by Coello Coello [8] explains many of the important aspects of MOEAs.

4.2 GENMOP

GENMOP is an implicit building block MOEA that attempts to find good solutions with a balance of exploration and exploitation. It is a Pareto-based algorithm that utilizes real values for crossover and mutation operators. The MOEA is an extension of the single objective GENOCOP algorithm [9, 10]. GENMOP was initially used to optimize flow rate, well spacing, concentration of injected electron donors, and injection schedule for bioremediation research [11, 12]. In this research, it was used to maximize percholate destruction in contaminated groundwater and minimize the cost of treatment. The algorithm was later used to optimize quantum cascade laser parameters in an attempt to find viable solutions for a laser that operates in the terahertz frequency range [13, 14].

The algorithm has four crossover methods and three mutation methods that are used. Each operator is chosen based upon an adaptive probability distribution, where the operators that produce the most fit individuals have their selection probability increased.

The algorithm flow is similar to most MOEAs. First, the input is read from a file. For our problem, a list of engines is read in, with each engine listing its arrival time, due time, priority (weight), and MTTR for each of its components. Next, the initial population is created and each chromosome is evaluated. The population is then ranked based on the Pareto-ranking of the individuals. Then a mating pool is created and only the most fit individuals are chosen to be in the mating pool. Crossover and mutation are performed on the members of the mating pool. The children created are then evaluated and saved. These children are then combined with the rest of the population $(\mu + \lambda)$. The population is then put into Pareto-rank order. The program then checks to see if the program has run through its allotted amount of generations. If it has, the program exits. If it has not, the program creates another mating pool and goes through the process again.

5 Design of Experiments and Testing

This section describes the computational system that executes GENMOP and then discusses our experimental method.

5.1 System Configuration

The systems used for testing were Linux Beowulf clusters. Each node runs dual Pentium III, 1 GHz chips. They are all running in a homogenous environment with Linux 7.3 as the operating system. While our current experimentation is done serially, future work will look into effective parallel processing for this problem.

5.2 Experimental Approach

The goals of this research were simple:

1. Develop an effective way to model the real-world aircraft engine scheduling problem.
2. Implement the model using GENMOP.
3. Determine if the number of swaps encoded in the chromosome is adequate.
4. Determine if the data shows trends that may require further study.

The first goal was difficult because we had to define some way of quantifying how costs are affected by engine swaps. Several different fitness functions were studied, but most of them did not capture what we were looking for. The aggregate swap count, it turns out, captures everything needed for this problem. Since the goal is decreasing costs by limiting the number of times an engine is in the shop, counting the out of tolerance variance in the MTTRs of each engine appears to be the most logical choice. This and other design decisions have been discussed earlier in the document.

The model implementation was another difficult task. Since the algorithm is a real-valued MOEA, and our problem is not a real-valued problem, some changes had to be made to the algorithm in order to accommodate our problem. GENMOP is discussed at greater length in Section 4.

As mentioned in section 3.2, our chromosome has encoded into it a limited number of swaps. These swaps are critical in limiting the amount of time an engine must be serviced over its lifetime. We had no initial idea of what a good number of swaps would be for each problem size, so we developed a baseline to test and see if we need to increase or decrease the number of swaps allowed per chromosome representation.

The last goal is to determine if the data shows some trends that can be analyzed further. For example, is it better to make large populations with a smaller number of generations or is it better to have smaller populations but more generations? With larger population sizes, we expect to get more exploration of the search space, while with more generations, we expect more exploitation of the good fitness values.

For these initial data results we looked at two instances: the five engine and ten engine scheduling problem. While the five engine problem may be a little small to be of practical use in the real world, the ten engine one is more in line with reality. Table 1 lists the testing parameters that we used.

The main metrics used are the average number of component swaps used for nondominated members and the Overall Nondominated Vector Generation

Table 1. Testing Parameters

Number of engines scheduled	Generation Size	Population size	Number of engines scheduled	Generation Size	Population size
5	10	10	10	10	10
5	10	100	10	10	100
5	10	1000	10	10	1000
5	25	25	10	25	25
5	100	10	10	100	10
5	100	100	10	100	100
5	100	500	10	500	100
5	100	1000	10	100	1000
5	1000	1000	10	1000	1000

(ONVG) metric [8, 15]. The average number of component swaps is used to determine if our maximum number of swaps is a limiting factor or if we are increasing our search space with too many swaps. This metric is different from the aggregate swap count fitness function. The metric pertains to the number of swaps utilized by the chromosome representation and the fitness function takes into account how many times an engine would have to be serviced because of varying MTTRs on the components. Our target is to have over half of the swaps utilized by the nondominated members, while at the same time have less than the maximum used.

The ONVG metric is one that we use with caution. We are using this metric to see which runs create the most number of nondominated points. But at the same time, we are also keeping an eye on the values of the these nondominated members. We must do this in order to avoid the situation where one instance generates a lot of nondominated points and another generates a fraction of the first instance, but has many points that dominate the first. This can give us contrary results. This is due to the fact that the ONVG is more of a diversity measure and less of a convergence measure.

Each iteration of the experiment was run 30 times in order for the central limit theorem to apply. This gives our data an approximately normal distribution [16].

6 Analysis

Tables 2 and 3 list the results from our five engine and ten engine experiments.

First, let's look at the average number of swaps. This value was derived from the swaps done by the members on the Pareto front. The analysis of the average number of swaps shows that the number of swaps for the five engine problem is about right. Of the five swaps allocated, 7 out of 10 instances utilized more than 50% of them. But with respect to the ten engine problem, only one of nine surpassed the threshold. This indicates that our baseline number of swaps is adequate for the 5 engine problem, but for 10 engines it may be limiting the efficiency of our system. This means our linear swap scaling factor of 1:1 was

Table 2. Testing Results for 5 Engines

Number of engines	Generation size	Population Size	Avg/Std Dev of Component Swaps	Avg/Std Dev of ONVG	Avg/Std Dev Makespan
5	10	10	2.04 / 0.93	2.63 / 1.73	971.5 / 39.4
5	10	100	3.43 / 0.93	4.3 / 1.53	941.5 / 78.2
5	10	1000	3.61 / 1.10	4.37 / 1.69	921.0 / 88.3
5	25	25	2.28 / 1.04	3.93 / 1.78	932.8 / 71.5
5	100	10	1.57 / 0.85	4.1 / 2.43	944.6 / 77.22
5	100	100	2.80 / 1.16	4.67 / 2.38	904.3 / 79.3
5	100	500	3.12 / 0.80	4.0 / 1.82	885.9 / 66.6
5	100	1000	3.54 / 0.92	4.4 / 2.46	889.5 / 67.5
5	500	1000	3.36 / 1.03	7.63 / 5.80	918.0 / 76.1
5	1000	1000	2.94 / 0.75	9.3 / 6.43	886.5 / 68.74

Table 3. Testing Results for 10 Engines

Number of engines	Generation size	Population Size	Avg/Std Dev of Component Swaps	Avg/Std Dev of ONVG	Avg/Std Dev Makespan
10	10	10	3.51 / 2.57	4.7 / 3.13	2165.5 / 83.1
10	10	100	5.6 / 2.72	4.33 / 2.20	2148.1 / 50.2
10	10	1000	7.32 / 1.74	3.8 / 1.16	2144.9 / 56.4
10	25	25	3.55 / 2.00	3.77 / 1.63	2142.2 / 43.5
10	100	10	3.72 / 1.79	6.13 / 4.51	2139.0 / 29.1
10	100	100	4.01 / 1.97	4.57 / 2.37	2119.1 / 32.7
10	100	500	4.03 / 2.30	5.29 / 4.11	2128.8 / 36.0
10	100	1000	4.45 / 1.98	4.33 / 1.97	2133.8 / 53.5
10	1000	1000	4.16 / 2.08	15.43 / 21.73	2105.5 / 4.53

too much. A more appropriate method would be to scale by a factor of one-half. Better still, would be to implement a variable scaling factor that increases the number of swaps in a chromosome when a certain threshold is reached, or decreases the number when a declining threshold is reached. This would make our chromosome lengths vary while the program is running.

Looking at the average ONVG, there really is no trend that one can decipher between the various instances, with the exception of the 10 engine, 1000 generation instance. That instance has a very high mean with an extremely high standard deviation. This is because of several runs that approached 70 members on the Pareto front. Most of these members had the same fitness values, but had different schedules.

As for the makespan, the trend is as one would expect, the average makespan of the members tends decrease as the number of generations and/or number of population members is increased. It is also interesting to note that the standard deviation tends to creep higher as the number of generations is increased. This can be expected to happen in a multi-objective problem. As population members

11. Knarr, M.R.: Optimizing an In Situ Bioremediation Technology to Manage Perchlorate-Contaminated Groundwater. Master's thesis, Air Force Institute of Technology, Wright-Patterson AFB, OH (2003)
12. Knarr, M.R., Goltz, M.N., Lamont, G.B., Huang, J.: *In Situ* Bioremediation of Perchlorate-Contaminated Groundwater using a Multi-Objective Parallel Evolutionary Algorithm. In: Congress on Evolutionary Computation (CEC'2003). Volume 1., Piscataway, New Jersey, IEEE Service Center (2003) 1604–1611
13. Keller, T.A.: Optimization of a Quantum Cascade Laser Operating in the Terahertz Frequency Range Using a Multiobjective Evolutionary Algorithm. Master's thesis, Air Force Institute of Technology, Wright-Patterson AFB, OH (2004)
14. Keller, T.A., Lamont, G.B.: Optimization of a Quantum Cascade Laser Operating in the Terahertz Frequency Range Using a Multiobjective Evolutionary Algorithm. In: 17th International Conference on Multiple Criteria Decision Making (MCDM 2004). Volume 1. (2004)
15. Schott, J.R.: Fault Tolerant Design Using Single and Multicriteria Genetic Algorithm Optimization. Master's thesis, Department of Aeronautics and Astronautics, Massachusetts Institute of Technology, Cambridge, Massachusetts (1995)
16. Milton, J.S., Arnold, J.C.: Introduction to Probability and Statistics: Principles and Applications for Engineering and Computer Science. Third edition edn. McGraw Hill (2002)

Finding Pareto-Optimal Set by Merging Attractors for a Bi-objective Traveling Salesmen Problem

Weiqi Li

School of Management, University of Michigan-Flint, 303 East Kearsley Street,
Flint, Michigan 48502, U.S.A.
weli@umflint.edu

Abstract. This paper presents a new search procedure to tackle multi-objective traveling salesman problem (TSP). This procedure constructs the solution attractor for each of the objectives respectively. Each attractor contains the best solutions found for the corresponding objective. Then, these attractors are merged to find the Pareto-optimal solutions. The goal of this procedure is not only to generate a set of Pareto-optimal solutions, but also to provide the information about these solutions that will allow a decision-maker to choose a good compromise solution.

1 Introduction

A multi-objective optimization seeks to optimize a vector of non-commensurable and often competing objectives. In other words, we whish to find a set of values for the decision variables that optimizes a set of objective functions. The general multi-objective combinatorial optimization problem can be formulated as:

$$optimize \quad f(x) = \begin{cases} f_1(x) = z_1 \\ f_2(x) = z_2 \\ \hbar \\ f_k(x) = z_k \end{cases} = z \in Z \tag{1}$$

$$subject\ to \quad x = (x_1, x_2, \hbar, x_n) \in X$$

where x is the decision vector, or *solution*, and $X \in \Re^n$ is the n-dimensional *decision space*, consisting of a finite set of feasible solutions. The objective function $f(x)$ maps x into $Z \in \Re^k$, the k-dimensional *objective space*, where k is the number of objectives. Whereas a single-objective problem is typically studied in decision space, multi-objective optimization is mostly studied in objective space. The image of a solution in the objective space is a point, $z = [z_1, z_2, ..., z_k]$. A point, z, is attainable if there exists a solution $x \in X$ such that $z = f(x)$. The set of all attainable points is denoted as Z. The ideal objective vector z^* is defined as $z^* = [optf_1(x), optf_2(x), ..., optf_k(x)]$, which is obtained by optimizing each of the objective functions individually. Normally, the ideal objective vector is not attainable because of the conflict among the objectives. Therefore, there will not exist a single optimal solution to the multi-objective combinatorial problem. Instead, we must look for "trade-off" solutions when dealing with a multi-objective optimization problem.

Objective vectors are compared according to the concept of Pareto-optimality and dominance relation. A partial ordering can be applied to solutions to the problem by the dominance criterion. A solution $x^a \in X$ is said to dominate a solution $x^b \in X$ if x^a is superior or equal in all objectives and at least superior in one objective. Mathematically, the concept of *Pareto optimality* is as follows [21]: assume, without loss of generality, a minimization problem, and consider two decision vectors, $x^a, x^b \in X$, then x^a is said to dominate x^b (often written as $x^a \hbar x^b$) if and only if

$$\forall i \in \{1,2,...,k\}: f_i(x^a) \le f_i(x^b) \ \land \\ \exists j \in \{1,2,\hbar,k\}: f_j(x^a) < f_j(x^b) \quad (2)$$

The solution x^a is said to be indifferent to a solution x^b, if neither solution is dominating the other one. When no a priori preference is defined among the objectives, dominance is the only way to determine if one solution performs better than the other does. The concept of Pareto optimality almost gives us a set of solutions called the *Pareto-optimal set*. The solutions in the Pareto-optimal set are also called *nondominated*, characterized by the fact that starting from a solution within the set, one objective can only be improved at the expense of at least one other objective being deteriorated. The curve formed by joining the Pareto-optimal solutions is known as a *Pareto-optimal front*. The goal of solving multi-objective problem is to find the Pareto-optimal set for the decision-maker to choose the most preferred solution. A solution selected by the decision-maker always represents a compromise between the different objectives.

The bounds on the Pareto-optimal set in the objective space can be defined by the ideal point and the nadir point [16]. The ideal objective vector, z^*, denotes an array of the lower bound of all objective functions. For each of the k conflicting objectives, there exists one different optimal solution. An objective vector constructed with these individual optimal objective values constitutes the ideal objective vector z^*. In general, the ideal objective vector corresponds to a non-existent solution. This is because the optimal solution for each objective function need not be the same solution. The nadir objective vector, z^{nad}, represents the upper bounds of each objective in the entire Pareto-optimal set.

The problem of finding the true Pareto-optimal set is NP-hard [5]. Thus, the goal of the multi-objective combinatorial optimization is to approximate the Pareto-optimal set. Over the years, the work of a considerable number of researchers has produced an important number of techniques to deal with multi-objective optimization problems [4], [7], [16], [22].

The TSP is the most well-known of all NP-hard combinatorial optimization problems. Multi-objective TSP is even harder than its corresponding single-objective version. Some researches have specifically treated the multi-objective TSP. Fischer and Richter [8] used a branch and bound approach to solve a TSP with two (sum) criteria. Gupta and Warburton [9] used the 2- and 3-opt heuristics for the max-ordering TSP. Sigal [20] proposed a decomposition approach for solving the TSP with respect to the two criteria of the route length and bottlenecking, where both objectives are obtained from the same cost matrix. Tung [23] used a branch and bound method with a multiple labeling scheme to keep track of possible Pareto-optimal tours. Melamed and Sigal [14] suggested an ε-constrained-based algorithm for bi-objective TSP. Ehrgott [6] proposed an approximation algorithm with worst case performance bound. Hansen [10] applied the tabu search algorithm to multi-objective TSP. Borges and Hansen [2] used the weighted sums program to study the global convexity for multi-objective TSP. Jaszkiewicz [11] proposed the genetic local search which combines ideas from evolutionary algorithms, local search with modifications of the aggregation of the objective functions. Paquete and Stützle [18] proposed the two-phase local search procedure to tackle bi-objective TSP. During the first phase, a good solution to one single objective is found by using an effective single objective algorithm. This solution provides the starting point for the second phase, in which a local search algorithm is applied to a sequence of different aggregations of the objectives, where each aggregation converts the bi-objective problem into a single objective one. Yan et al. [24] used an evolutionary algorithm to solve multi-objective TSP. Angel, Bampis and Gourvès [1] proposed the dynasearch algorithm which uses local search with an exponential sized neighborhood that can be searched in polynomial time using dynamic programming and a rounding technique. Paquete, Chiarandini and Stützle [17] suggested a Pareto local search method which extends local search algorithm for the single objective TSP to bi-objective case. This method uses an archive to hold non-dominated solutions found in the search process.

This study proposes a new search procedure to tackle multi-objective TSP. Fig. 1 sketches the schematic of the search procedure. This procedure incorporates the relationship between the problem presentation and data structure into the algorithm design. For a TSP with k objectives, this procedure first constructs the solution attractor for each of objectives individually. The solution attractor contains a set of the best solutions found for the corresponding objective. It is also reasonable to believe that the attractor consists of a large proportion of low-cost edges for the objective. These edges are in the extreme for that objective function. Then the procedure combines these k attractors in order to mix the edges contained in these attractors. Finally, the Pareto-optimal solutions can be found from these mixed edges.

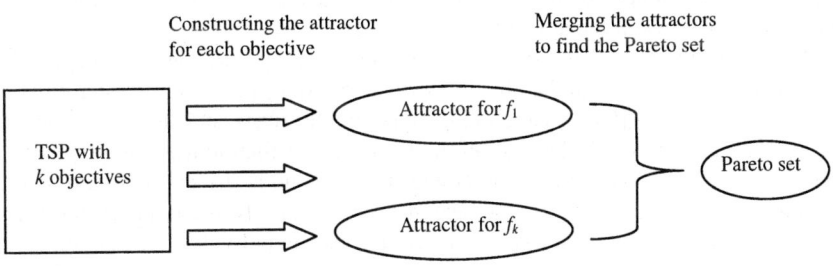

Fig. 1. Schematic of the search procedure

The remaining of the paper is organized as follows. Next section introduces a method for constructing a solution attractor for a single-objective TSP. Section 3 describes the proposed procedure for finding the Pareto-optimal solutions for multi-objective TSP and presents the computational results for a bi-objective TSP instance.

The goal of this procedure is not only to generate a set of Pareto-optimal solutions effectively, but also to provide information about these solutions that will allow a decision-maker to choose a good compromise solution. The final section concludes this paper.

2 Constructing Solution Attractor for TSP

Local search heuristics is a widely used general approach to find reasonable solutions to hard combinatorial optimization problems. Local search algorithms are simple to implement and quick to execute, but they have the main disadvantage that they are locally convergent.

When we apply a local search algorithm to the TSP, the common opinion about local optima is that the set of local optima forms a "big valley" structure in the solution space [3], [13], [15], [19]. In fact, it is a *solution attractor*, i.e., a set of fixed points, that drives the local search trajectories into the small region of the solution space [12]. The attractor is formed by the set of all local optimal tours. Since the global solution is a special case of local optimal solutions, the global tour is expected to be included in the attractor.

Li [12] suggests a procedure for constructing a solution attractor for single-objective TSP. For a TSP instance, the solution space contains the tours that the salesman may traverse. Li's procedure uses an $n \times n$ matrix E, called *hit-frequency matrix*, to record the number of hits on each edge by a set of local optimal tours. The hit-frequency matrix explores the information on edges and thus stores rich information about the solution attractor for the TSP instance. If we use a local search algorithm and generate all possible search trajectories for the TSP instance, when all search trajectories reach their local optima we could obtain the real attractor for the problem. Then, when all involved edges are recorded in the hit-frequency matrix, we should immediately recognize the attractor and easily identify the global optimal tour. Unfortunately, this "all possible search trajectories" scenario is unrealistic, due to the enormous amount experimental data required. A more realistic goal would be gathering a moderate sample of local optima to construct the solution attractor and infer statistical properties of the attractor.

We denote all edges that are contained in the global optimal tour as G-edges. When a local search process searches for an optimal solution, the search trajectory is constantly adjusting itself by disregarding unfavorable edges and trying to collect G-edges. If it successfully collects all of the G-edges, the final tour is the global optimal tour. If it only collects some of the G-edges, it ends up at a local optimum. If a tour contains none of the G-edges, the search process can always improve the tour by exchanging edges. Local heuristic algorithms cause individual search trajectories explore only a tiny fraction of the enormous solution space when n is large. Thus, it is difficult for a particular search trajectory to select G-edges globally, and a search trajectory often ends at a local optimum which contains most G-edges and some unfavorable edges. The more G-edges a local optimal tour contains, the closer to the global optimum it is. Local optimal tours are actually linked together by sharing the G-edges. When a solution attractor is constructed by a large number of local optimal tours, the attractor should consist of all G-edges and some unfavorable edges, called *noise*.

```
procedure TSP-Attractor(Q)
begin
  repeat
     s_i = Initial_tour();
     s_j = Local_Search(s_i);
     Update(E(s_j));
  until StoppingCriterion = M;
        A = Find_Core(E);
  Exhausted_Search(A)
end
```

Fig. 2. The procedure for contracting solution attractor of local search in TSP

Fig. 2 presents the procedure for constructing solution attractor for TSP. The procedure is very straightforward: randomly generating a sample of local optimal solutions to construct the attractor, and then removing the noise contained in the attractor to identify the core of the attractor. In the procedure, Q is a TSP instance. s_i is an initial solution generated by Initial_Tour(). s_j is a local optimum outputted by a local search process Local_Search(). E is the hit-frequency matrix to record M local optima. Each time when a search trajectory reach a local optimal solution s_j, the function Update() records the edges contained in s_j into E. Since a solution attractor contains G-edges and noise, the function Find_Core() is used to try to remove the noise. The remaining edges form the core of the attractor and are recorded into the matrix A. Finally, the matrix A is searched by an exhausted enumeration process Exhausted_Search() to generate all solutions in the attractor core. The hit-frequency matrix E plays an important role for collecting all information about the solution attractor. It acts as an input/output data table where the entry e_{ij} records the number of times that the corresponding edge is hit by the set of local optimal tours.

A solution representation is a mapping from the solution space of a possible solution to a solution space of encoded configuration within a particular data structure. For a TSP with n cities, there are many ways to represent a tour in the computer. One way is to make an ordered list of the cities to visit, with a return to the home city being implied. Another way is with an $n \times n$ matrix $E = [e_{ij}]$, such that $e_{ij} = 1$ if and only if city j follows city i in the tour. A tour therefore must always have exactly one "1" in every row and every column. The matrix E is an effective data structure that allows the local search process to maintain a functional link among the local optimal tours. The basic idea of the hit-frequency matrix E used in the procedure is to build a probabilistic representation of the solution attractor based on the M local optimal tours and then generate new candidate solutions based on the knowledge contained in the attractor.

The hit-frequency value in e_{ij} represents the probability that the corresponding edge in the TSP will be hit by a local optimal tour. If M is large enough, this value also can be viewed as the probability that the corresponding edge is a G-edge. For example, if an edge is hit by 73 percent of M local optimal tours, although each local optimum selects the edge based on its neighborhood structure, the edge is globally superior since the edge is reached by these individual optima from different search trajectories.

The hit-frequency matrix gives us important insights into the nature of the search space and provides an opportunity for us to concentrate the search in the region that contains the most promising solutions. When we restrict our attention to a smaller solution space represented by the attractor, the number of possibilities is no longer prohibitive.

The hit-frequency matrix E has capacity of learning. The basis of learning in the matrix is the generation of long-lived memory and statistics-based pattern. The matrix not only plays a fundamental role in the organization of memory, but also provides a powerful way of discovering the pattern in the searched edges. We can exploit this information to generate more local optimal tours, and even the global optimal tour.

Fig. 3 uses a simple 20-city TSP example to explain the attractor-construction procedure. This example generates $M = 100$ random initial tours. Since these initial tours are randomly produced, the edges should have an equal probability to be selected. The darkened elements in the matrix shown in Fig. 3(a) represents the union of the edges

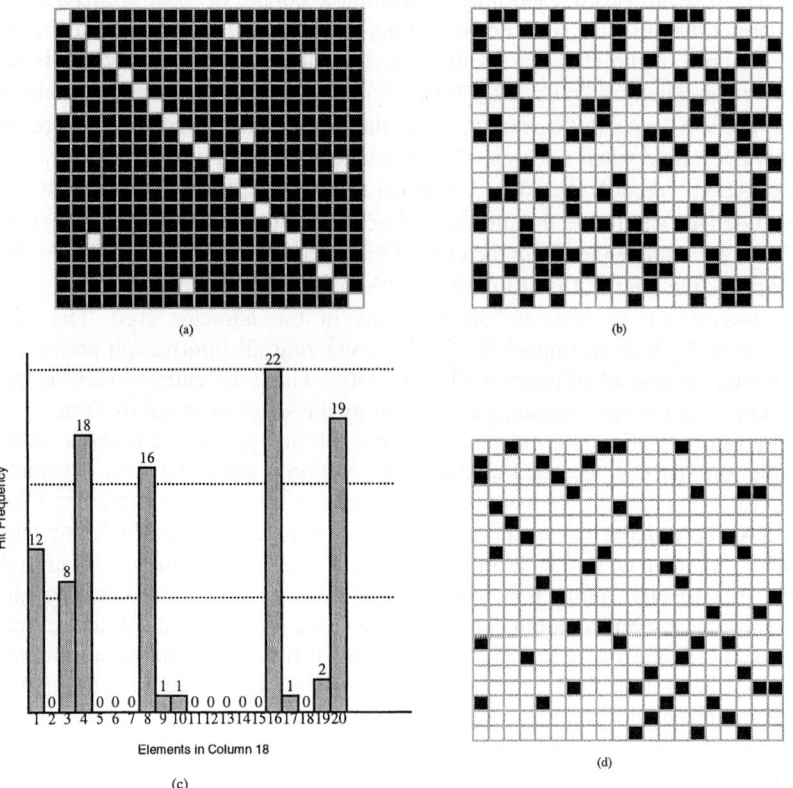

Fig. 3. A 20-city TSP example for illustrating attractor construction. (a) presents the union of the edges in the initial tours; (b) marks the union of the edges hit by the local optimal tours; (c) illustrates the clustering process for the column 18 of E; and (d) displays the attractor core

found in these initial tours. After applying the 2-opt local search algorithm to the initial tours, we obtain 100 local optimal tours. Fig. 3(b) marks the union of the edges hit by these 100 local optimal tours. Each of these marked elements also contains a value which is the number of hit by the local optimal tours.

It is interesting to see how the search space is reduced. During the local search process, the only thing the process is doing on a particular search trajectory is to replace bad edges with good ones. As result, the search process causes the elements in the matrix E to have unequal hit frequency. Good edges are selected by many search trajectories; bad edges are displaced and therefore contain low or zero hit frequency. After search trajectories reach their local optimal points, they leave their "final footprints" in E. The darken area in Fig. 3(a) is reduced to the one shown in Fig. 3(b), which exhibits the structure of solution attractor for the TSP. Comparing to the full solution space, the size of the attractor is very small.

The attractor constructed by local optima contains G-edges and noise. The function Find_Core() groups the edges into clusters in an attempt to remove the noise. A *cluster* is defined here as a set of edges that contain the hit frequency within a certain range. The cluster with the highest range of hit frequency constitutes the core of the attractor, while the cluster with the lowest range can be regarded as noise. In each column (or row) of the hit-frequency matrix E, the value of the maximum hit $MaxV$ is identified. Knowing that the range of possible value for an edge in that column can vary from zero to $MaxV$, we could divide this range into r equal portions. Our example chooses $r = 3$ to cluster the edges in each column. Fig. 3(c) illustrates the clustering process for the column 18 in E. Fig. 3(d) displays the cluster in which the edges are within the highest range of hit frequency. The darkened elements form the attractor core. In this way the attractor is further reduced into a core, an even smaller region. The most-hit edges in the core form the most promising region for search. Now it is possible to use an exhausted-enumeration algorithm to find all solutions in the core. In our example, the function Exhausted_Search() found 32 solutions in the core.

3 Finding Pareto-Optimal Set for a Bi-objective TSP

3.1 The TSP and Search Procedure

In a multi-objective setting, the TSP becomes even more difficult and complex. The general multi-objective TSP can be formulated as follows:

$$\min f(x) = \begin{cases} f_1(x) = c_1(n,1) + \sum_{i=1}^{n-1} c_1(i, i+1) = z_1 \\ f_2(x) = c_2(n,1) + \sum_{i=1}^{n-1} c_2(i, i+1) = z_2 \\ \hbar \\ f_k(x) = c_k(n,1) + \sum_{i=1}^{n-1} c_k(i, i+1) = z_k \end{cases} \quad (3)$$

subject to $x(1,2,\hbar,n) \in X$

where n is the number of cities, $c_q(i, j)$ is the cost between city i and j according to the q-th objective, $q = 1,...k$, and the decision variable x holds a cyclic permutation of the n cities. In practical applications the cost factors may correspond to distance, travel time, expenses, tourist attractiveness, energy consumed, degree of risk, or other relevant considerations for the tour. The goal is to find the "minimal" Hamiltonian circuit of the graph in terms of Pareto optimality. In a TSP, if the cost weights of the edges satisfy the triangle inequality, the problem is called the *metric TSP*. A special case is when the cities are points on the plan, and the cost weights are the Euclidean distances between the points. When the cost weights satisfy $c(i, j)=c(j, i)$, it is called the *symmetric TSP*, which has many practical applications.

The design of a test problem is always important in designing any new search algorithm. The context of problem difficulty naturally depends on the nature of problems that the underlying algorithm is trying to solve. In the context of solving multi-objective optimization problems, we are interested in designing the features that makes a problem difficult for the proposed multi-objective optimization algorithm. In this study, the test problem instance is designed based on several considerations. First, the size of problem should be large, since the TSP instances as small as 200 cities must now be considered to be well within the state of the global optimization art. The instance must be considerably larger than this for this study to be sure that the proposed approach is really called for. Second, the instance should be multi-modal, that is, with many local Pareto-optimal regions. Third, there is no any pre-known information related to the result of the experiment, since in a real-world problem one does not usually have any knowledge of the Pareto-optimal front. Fourth, the problem instance should be general, understandable and easy to formulate so that the experiments are repeatable and verifiable, but difficult to solve.

This study generates a general symmetric TSP instance, which consists of $n = 1000$ cities with two cost matrixes c_1 and c_2. The cost matrices are generated at random, where each cost element $c(i, j) = c(j, i)$ is assigned a random number in the interval [1, 1000]. This study uses two objectives primarily because of the ease in which two-dimensional Pareto-optimal front can be visually demonstrated.

Fig. 4 presents a general search procedure used in this study. Q is a TSP instance with a set of cost matrices $\{c_1, c_2,..., c_k\}$ with respect to objective $f_1, ..., f_k$. The function TSP_Attractor() finds the core of the attractor for each of objective functions respectively. This study uses the 2-opt algorithm [13], which is one of the earliest local search algorithms for the TSP.

The information about the attractor core for objective q ($q = 1,..., k$) is stored in matrix E_q, in which we mark the elements to represent the corresponding edges that are hit by the core. In this study, only 15 best solutions in each attractor core are selected and stored in E_q. And also the objective vector values $Z_q(z_1,...,z_k)$ for the best solution in the E_q is calculated. The vector $Z_q(z_1,...,z_k)$ can help us to determine the ideal objective vector z^* and nadir objective vector z^{nad}. After k attractor cores with respect to objective $f_1, ..., f_k$ are constructed, the function Merge() stores the union of all marked elements in the matrixes $E_1, ..., E_k$ into the matrix E. Then the function Find_Pareto-Set() finds all solutions in E through an exhausted search method, and outputs all *non-dominated solutions* into list L. Finally, the function Analize_Paroto_Set() analyzes the Pareto set and generate the information about each of the solutions.

```
procedure TSP_Pareto_Set(Q)
begin
    q = 1;
    while (q ≤ k) do
    E_q = TSP_Attractor(c_q);
    q = q + 1;
    end while
    E = Merge(E_1,…, E_k);
    L = Find_Pareto_Set(E);
    Analize_Pareto_Set(L);
end
```

Fig. 4. The procedure for searching Pareto-optimal set in multi-objective TSP

This procedure intuitively reflects the idea of finding solutions around the extreme ends of the Pareto-optimal front and then mixing their characteristics (edges) to find other trade-off solutions in the Pareto-optimal region. By considering each of objective functions separately, this procedure generates high-quality solutions in the solution attractor corresponding to that objective. The merge of these attractors will form a well-distributed set of the Pareto-optimal solutions, each of which takes part of high-quality edges from each of the attractors. Fig. 5 illustrates examples. Suppose that there are two local optimal tours (1,2,3,4,5,6) and (5,1,2,4,3,6) in the attractor 1 that corresponds to objective f_1 (see Fig. 5(a)), and there is one local optimum (5,3,1,4,2,6) in the attractor 2 corresponding to objective f_2 (Fig. 5(b)). If we merge these two attractors into one matrix, as illustrated in Fig. 5(c), we can identify other two solutions (1,2,4,5,3,6) and (1,4,2,3,6,5). The first solution takes four edges from attractor 1, one edge from attractor 2 and one common edge shared by both attractors. The second solution takes three edges from attractor 1, one edge from attractor 2 and two common edges. Even in the case in which two objectives are mutually exclusive (i.e., the two cost matrixes are mutually disjunctive), we will see that solutions in different attractors do not share edges, but we still can find the solutions that mix edges from attractor 1 and 2. For example, suppose that there are two solutions (1,3,5,2,6,4) and (1,3,6,4,2,5) in attractor 1 (Fig. 5(d)) and one solution (1,2,3,4,5,6) in attractor 2 (Fig. 5(e)). When we merge these two attractors into a matrix, as shown in Fig. 5(f), it is easy to see that no common edge is shared by both attractors. In addition to the three original solutions, we can find other four new solutions (1,2,3,5,6,4), (1,2,3,6,4,5), (1,3,4,2,5,6) and (1,3,4,5,2,6), each of them combines three edges from attractor 1 and three edges from attractor 2.

3.2 The Experiment Results

To guarantee the diversity of the sample of local optima, this study generated $M = 2000$ initial points, and, as a result, generated 2000 local optima for each objective. This experiment also relied heavily on randomization. All initial tours were randomly constructed. In the 2-opt local search, the algorithm randomly generated a solution in the neighborhood of the current solution. A move that gave the first improvement was chosen. The great advantage of first-improvement pivoting rule is to produce random

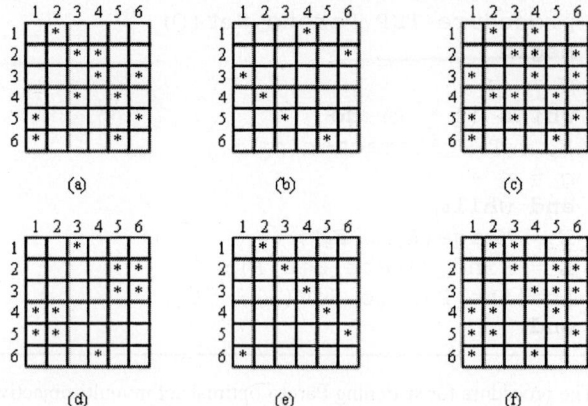

Fig. 5. Examples of merging two attractors

ized local optima. The local search process on each search trajectory terminated when no improvement had been achieved during 1000 iterations

This study used two 1000×1000 matrixes, E_1 and E_2, to store 15 best solutions taken from each of the attractor cores, respectively. Of course, the number of solutions selected from the attractor core affects the size of the Pareto-optimal set, the coverage of the set, and the computational resources needed to generate the set. Fig. 6 illustrates the solution points from each objective in the objective space. This figure also indicates the lower bound z^* and upper bound z^{nad} on the Pareto-optimal set to display the topology of the set.

Then these solutions were combined into the matrix E, in which 40.4% of the marked edges belong to attractor 1, 41.1% to attractor 2, and 18.5% are shared by both attractors. The fraction of edges that are common to both attractors can be defined as the overlap between the two attractors. This study then used an exhausted search algorithm to find all solutions in E. After discarding all dominated solutions, we obtained 31 Pareto-optimal solutions, as illustrated in Fig. 7.

For the purpose of comparison, this study also applied the aggregation approach to the same TSP instance. This study varied systematically the weights in the fashion of [0, 1], [0.04, 0.96], [0.08, 0.92], ..., [1, 0], and then use the 2-opt algorithm to search on the aggregated objective function. The search process in each run terminated when no improvement had been made during 1500 iterations. These multiple runs generated 26 points, among which 23 points were nondominated. These nondominated points are also displayed in Fig. 7. It is clear that the solutions generated by the proposed procedure maintain good properties of convergence and diversity. These solutions probably provide a more accurate description of the true Pareto-optimal set.

Mathematically, the multi-objective optimization problem is considered to be solved when the Pareto-optimal set is found, and all the Pareto-optimal points are equally acceptable solutions to the problem. However, it is not enough in many practical cases. The ultimate goal is to select the single *best compromise solution*. Selecting one solution out of the Pareto-optimal set calls for a decision-maker, who has better insight into the problem and can express preference relations between different

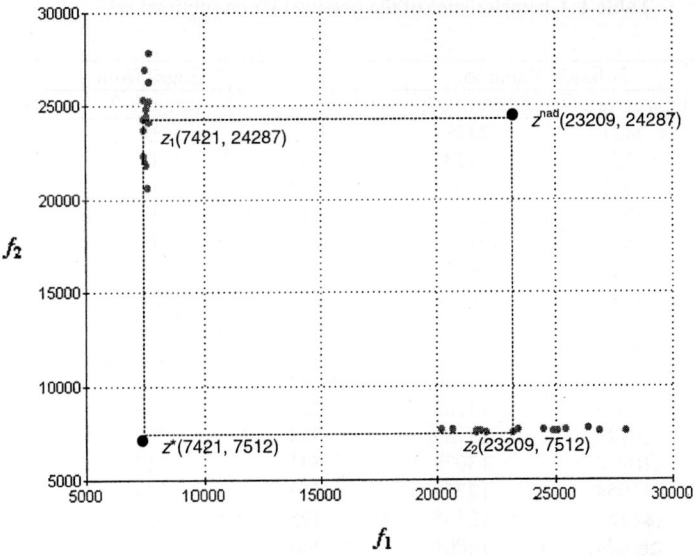

Fig. 6. Solutions in each of the attractor core

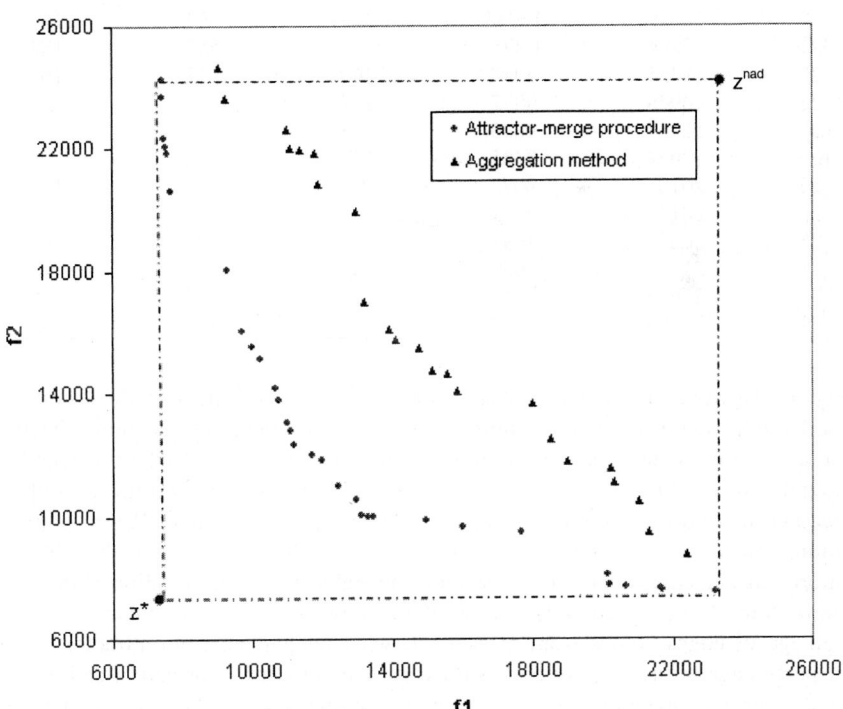

Fig. 7. The Pareto-optimal sets for the test TSP

Table 1. Characteristics of the obtained Pareto-optimal set

Tour #	Solution Value for		Edges from		
	Objective 1	Objective 2	Attractor 1	Attractor 2	Shared
1	7421	24287	972	0	28
2	7435	23731	967	0	33
3	7465	22343	934	0	66
4	7529	22060	919	0	81
5	7551	21838	906	0	94
6	7645	20629	893	0	107
7	9285	18021	744	97	159
8	9698	16001	698	153	149
9	9963	15530	661	172	167
10	10205	15120	634	214	152
11	10648	14205	587	248	165
12	10720	13789	574	262	164
13	10978	13075	531	312	157
14	11054	12775	512	334	154
15	11155	12335	495	353	152
16	11674	12001	446	399	155
17	11973	11847	433	401	166
18	12456	11007	409	425	166
19	12986	10548	374	452	174
20	13099	10048	329	507	164
21	13294	10001	286	552	162
22	13455	9987	260	577	163
23	14976	9877	227	615	158
24	15987	9645	205	644	151
25	17654	9465	147	702	151
26	20101	8073	14	871	115
27	20170	7719	0	901	99
28	20651	7697	0	906	94
29	21644	7639	0	917	83
30	21675	7553	0	933	67
31	23209	7512	0	964	36

objectives. However, finding a reliable important solution is difficult in the absence of any knowledge about the Pareto-optimal solutions. An important question related to this issue is how to present the Pareto-optimal solutions to the decision-maker in a meaningful way. This requires a multi-objective optimization algorithm not only to be capable of finding multiple and diverse Pareto-optimal (or near Pareto-optimal) solutions, but also to be able to provide necessary information about the obtained solutions. In our case, what would be more desirable is the information about each obtained tour. More specifically, we want to know that in a particular tour, what percentage of edges comes from attractor 1, what percentage from attractor 2, and what percentage are shared by both attractors. This kind of information can aid the decision-maker in arriving at a final decision. The characteristics of the solutions are essential decision elements when people look for the best compromise solution, and they are implicitly included in the common-sense notion of optimality.

The function `Analize_Pareto_Set()` calculated the distribution of edges for each obtained tour, as illustrated in Table 1. For instance, among 1000 edges in the tour 1, 28 edges are shared by both attractors, and other 972 edges are taken from attractor 1. For the tour 12, 164 edges are shared by both attractors, and among other edges, 574 edges are taken from attractor 1 and 262 edges are taken from attractor 2. It can be interpreted as: If we choose the tour 12, it will correspond to 57.4% preference of the objective f_1, 26.2% preference of the objective f_2, and 16.4% preference of both objectives at same time. However, if we choose the tour 1, 97.2% of this solution will satisfy the objective f_1, and 2.8% will satisfy both objectives simultaneously. No doubt, such information is useful to the decision makers for comparing multiple optimal solutions and choosing the best compromise solution, as and when required.

4 Conclusion

The multi-objective nature of most real-world problems makes multi-objective optimization a very important research topic. The increasing complexity of typical search space demands new strategies in solving multi-objective optimization problems. This study provides one possibility. This study shows that the use of simple local search, together with an effective data structure, can identify high quality Pareto-optimal solutions. Although the procedure was applied to a bi-objective TSP in this study, it can be expected that, with little modification, this procedure can be used to deal with the TSP with three or more objectives. Even if generalization cannot be claimed, this work provides a new search strategy that casts new light into multiple-objective optimization and might serve as a basis to build new algorithm for solving other multiple objective problems.

References

1. Angel, E., Bampis, E., Gourvès, L.: A Dynasearch Neighborhood for the Bicriteria Traveling Salesman Problem. In: Gandibleux, X., Sevaux, M., Sörensen, K., T'kindt, V. (eds.): Metaheuristics for Multiobjective Optimization, Lecture Notes in Economics and Mathematical Systems 535. Springer-Verlag Berlin (2004) 153-176.
2. Borges, P. C., Hansen, M. P.: A Study of Global Convexity for a Multiple Objective Traveling Salesman Problem. In Ribeiro, C. C., Hansen, P. (Eds.): Essays and Surveys in Metaheuristics. Kluwer Academic Publishers Norwell, MA (2002) 129-150.
3. Boese, K. D., Kahng, A. B., Muddu, S.: A New Adaptive Multistart Technique for Combinatorial Global Optimizations. Operations Research Letters 16:2 (1994) 101-113.
4. Collette, Y., Siarry, P.: Multiobjective Optimization – Principles and Case Studies. Springer-Verlag Berlin (2003).
5. Ehrgott, M.: Multicriteria Optimization. Springer-Verlag Berlin (2000).
6. Ehrgott, M.: Approximation Algorithms for Combinatorial Multi-Criteria Problems. International Transactions in Operations Research 7 (2000) 5-31.
7. Ehrgott, M., Gandibleux, X.: A Survey and Annotated Bibliography of Multiobjective Combinatorial Optimization. OR Spectrum 22:4 (2000) 425-460.

8. Fisher, R., Richter, K.: Solving a Multiobjective Traveling Salesman Problem by Dynamic Programming. Mathematische Operationsforschung und Statistik, Series Optimization 13:2 (1982) 247-252.
9. Gupta, A., Warburton, A.: Approximation Methods for Multiple Criteria Traveling Salesman Problems. In: Sawaragi, Y. (ed.): Towards Interactive and Intelligent Decision Support Systems: Proceedings of the 7th International Conference on Multiple Criteria Decision Making. Springer-Verlag Berlin (1986) 211-217.
10. Hansen, M. P.: Use of Substitute Scalarizing Functions to Guide a Local Search Based Heuristics: The Case of MOTSP. Journal of Heuristics 6 (2000) 419-431.
11. Jaszkiewicz, A.: Genetic Local Search for Multiple Objective Combinatorial Optimization. European Journal of Operational Research 137:1 (2002) 50-71.
12. Li, W.: Attractor of Local Search in the Traveling Salesman Problem. Journal of Heuristics, Forthcoming.
13. Lin, S.: Computer Solutions to the Traveling Salesman Problem. Bell Systems Technical Journal 44 (1965) 2245-2269.
14. Melamed, I. I., Sigal, I. K.: The Linear Convolution of Criteria in the Bicriteria Traveling Salesman Problem. Computational Mathematics and Mathematical Physics 37:8 (1997) 902-905.
15. Mezard, M., Parisi, G.: A Replica Analysis of the Traveling Salesman Problem. Journal of Physique 47 (1986) 1285-1296.
16. Miettinen, K. M.: Nonlinear Multiobjective Optimization. Kluwer Academic Publishers Dordrecht (1999).
17. Paquete, L. Chiarandini, M., Stützle, T.: Pareto Local Optimum Sets in the Biobjective Traveling Salesman Problem: An Experimental Study. In: Gandibleux, X., Sevaux, M., Sörensen, K., T'kindt, V. (eds.): Metaheuristics for Multiobjective Optimization, Lecture Notes in Economics and Mathematical Systems 535. Springer Berlin (2004) 177-199.
18. Paquete, L., Stützle, T.: A Two Phase Local Search for the Biobjective Traveling Salesman Problem. In: Fonseca, C. M., Fleming, P. J., Zitzler, E., Deb, K., Thiele, L. (eds.): Evolutionary Multi-Criterion Optimization, Proceedings of Second International Conference, EMO2003. Springer, Berlin (2003) 479-493.
19. Raidl, G. R., Kogydek, G., Julstrom, B. A.: On Weight-Biased Mutation for Graph Problems. In: Merelo-Guervós, J. J., Adamidis, P., Beyer, H. G., Fernáandez-Villacañas J. L., Schwefel, H. P. (eds.): Parallel Problem Solving from Nature: PPSN VII, LNCS2439. Springer Berlin (2002) 204-213.
20. Sigal, I. K.: Algorithm for Solving the Two-Criterion Large-scale Traveling Salesman Problem. Computational Mathematics and Mathematical Physics 34:1 (1994) 33-43.
21. Steuer, R. E.: Multiple Criteria Optimization – Theory, Computation and Application. John Wiley & Sons New York (1986).
22. Tan, K. C., Lee, T. H., Khor, E. F.: Evolutionary Algorithms for Multi-objective Optimization: Performance Assessments and Comparisons. Artificial Intelligence Review 17 (2002) 253-290.
23. Tung, C. T.: A Multicriteria Pareto-optimal Algorithm for the Traveling Salesman Problem. Asia-Pacific Journal of Operational Research 11 (1994) 103-115.
24. Yan, Z., Zhang, L., Kang, L., Lin, G.: A New MOEA for Multi-objective TSP and Its Convergence Property Analysis. In: Fonseca, C. M., Fleming, P. J., Zitzler, E., Deb, K., Thiele, L. (eds.): Evolutionary Multi-Criterion Optimization, Proceedings of Second International Conference, EMO2003. Springer Berlin (2003) 342-354.

Multiobjective EA Approach for Improved Quality of Solutions for Spanning Tree Problem

Rajeev Kumar, P.K. Singh, and P.P. Chakrabarti

Department of Computer Science and Engineering,
Indian Institute of Technology Kharagpur,
Kharagpur, WB 721 302, India
{rkumar, pksingh, ppchak}@cse.iitkgp.ernet.in

Abstract. The problem of computing spanning trees along with specific constraints is mostly NP-hard. Many approximation and stochastic algorithms which yield a single solution, have been proposed. In this paper, we formulate the generic multi-objective spanning tree (MOST) problem and consider edge-cost and diameter as the two objectives. Since the problem is hard, and the Pareto-front is unknown, the main issue in such problem-instances is how to assess the convergence. We use a multiobjective evolutionary algorithm (MOEA) that produces diverse solutions without needing a priori knowledge of the solution space, and generate solutions from multiple tribes in order to assess movement of the solution front. Since no experimental results are available for MOST, we consider three well known diameter-constrained minimum spanning tree (dc-MST) algorithms including randomized greedy heuristics (RGH) which represents the current state of the art on the dc-MST, and modify them to yield a (near-) optimal solution-fronts. We quantify the obtained solution fronts for comparison. We observe that MOEA provides superior solutions in the entire-range of the Pareto-front, which none of the existing algorithms could individually do.

1 Introduction

Computing a minimum spanning tree (MST) from a connected graph is a well-studied problem and many fast algorithms and analytical analyses are available [1, 2, 3, 4, 5, 6, 7, 8]. However, many real-life network optimization problems require the spanning tree to satisfy additional constraints along with minimum edge-cost. For example, communication network design problem for multicast routing of multimedia communication requires constructing a minimal cost spanning/Steiner tree with given constraints on diameter. VLSI circuit design problems aim at finding minimum cost spanning/Steiner trees given delay bound constraints on source-sink connections. Analogously, there exists the problem of degree/diameter-constrained minimum cost networks in many other engineering applications too (see [3] and the references therein).

Many such MST problem instances having a bound on the degree, a bound on the diameter, capacitated trees or bounds for two parameters to be satisfied simultaneously are listed in [3]. Finding spanning trees of sufficient generality and of minimal cost subject to satisfaction of additional constraints is often NP-hard [3, 4]. Many such design problems have been attempted and approximate solutions obtained using heuristics.

For example, the research groups of Deo et al. [5, 6, 7, 8] and Ravi et al. [3, 4] have presented approximation algorithms by optimizing one criterion subject to a budget on the other. In recent years, evolutionary algorithms (EAs) have emerged as powerful tools to approximate solutions of such NP-hard problems. For example, Raidl & Julstrom [9, 10] and Knowles & Corne [11, 12] attempted to solve diameter and degree constrained minimum spanning tree problems, respectively using EAs. All such approximation and evolutionary algorithms yield a *single* optimized solution subject to satisfaction of the constraint(s). Moreover, researchers have demonstrated superiority of one algorithm over other algorithms for a *particular* value of a constraint and did not assess the performance over entire range of the values.

We argue that such constrained MST problems are essentially multiobjective in nature. A multiobjective optimizer yields a set of all representative equivalent and diverse solutions; the set of all optimal solutions is the Pareto-front. Secondly, extending this constraint-optimization approach to multi-criteria problems (involving two or more than two objectives/constraints) the techniques require improving upon more than one constraints. Thirdly and more importantly, such approaches may not yield all the representative optimal solutions. For example, most conventional approaches to solve network design problems start with a minimum spanning tree (MST), and thus effectively minimize the cost. With some variations induced by ϵ-constraint method, most other solutions obtained are located near the minimal-cost region of the Pareto-front, and thus do not form the complete (approximated) Pareto-front.

In this work, we try to overcome the disadvantages of conventional techniques and single objective EAs. We use multiobjective EA to obtain a (near-optimal) Pareto-front. For a wide-ranging review, a critical analysis of evolutionary approaches to multiobjective optimization and many implementations of multiobjective EAs, see [13, 14] for algorithms and implementations, and [15] for various applications.

We use Pareto Converging Genetic Algorithm (PCGA) [16] which has been demonstrated to work effectively across complex problems and achieves diversity without needing *a priori* knowledge of the solution space. PCGA excludes any explicit mechanism to preserve diversity and allows a natural selection process to maintain diversity. Thus multiple, equally good solutions to the problem, are provided. Another major challenge to solving unknown problems is how to ensure convergence. Some multiobjective problems have a tendency to get stuck at local Pareto-front [16], therefore, we generate solutions using multiple tribes and merge them to ensure convergence. PCGA assesses convergence to the Pareto-front which, by definition, is unknown in most real search problems of multi-dimensionality, by use of rank-histograms [17]. We consider, without loss of generality, edge-cost and tree-diameter as the two objectives to be minimized, though the framework presented here is generic enough to include any number of objectives to be optimized. Initial results of this work were presented in other conferences [18, 19]. In this paper, we extend the work for larger problem instances, present a systematic approach to assess the convergence, and compare qualitatively and quantitatively the obtained solution-fronts from three well-known techniques, namely, One-Time-Tree Construction (OTTC) [7], Iterative Refinement (IR) [7], and Randomized Greedy Heuristics (RGH) [9] algorithms.

The rest of the paper is organized as follows. In Section 2, we include a brief review of the issues to be addressed for achieving quality solutions in the context of a MOEA. We describe, in Section 3, the representation scheme for the spanning tree and its implementation using PCGA. Then, we present results in Section 4 along with a comparison with other approaches. Finally, we draw conclusions in Section 5.

2 Multiobjective Evolutionary Algorithms : Issues and Challenges

EAs have emerged as powerful black-box optimization tools to approximate solutions for NP-hard combinatorial optimization problems. In the multiobjective scenario, EAs often find effectively a set of mutually competitive solutions without applying much problem-specific information. However, achieving proper diversity in the solution-set while approaching convergence is a challenge in multiobjective optimization, especially for unknown problems.

There exist many algorithms/implementations which have been demonstrated to achieve diverse and equivalent solutions [13, 14]. For diversity, some of the algorithms make explicit use of parameterized sharing, mating restriction and/or some other diversity preserving operator. Apart from its heuristic nature, the selection of the domain in which to perform sharing (variable (genotype) or objective (phenotype)) is also debatable. Any explicit diversity preserving mechanism method needs prior knowledge of many parameters and the efficacy of such a mechanism depends on successful fine-tuning of these parameters. Purshouse & Fleming [20] extensively studied the effect of sharing, along with elitism and ranking, and concluded that while sharing can be beneficial, it can also prove surprisingly ineffective if the parameters are not carefully tuned. Also, it is the experience of almost all researchers that proper tuning of sharing parameters is necessary for effective performance.

In particular to MOST problem where we use a special encoding [10], incorporation of such knowledge is not an easy task. There exist some other MOEAs, e.g., NSGA-II [21] and SPEA2 [22], which have now dispensed away with parameters for explicit niching. However, almost all the multiobjective evolutionary algorithms and implementations have ignored the issue of convergence and use some pre-determined metrics (e..g, number of generational runs) as the stopping criterion. Other common metric used is the distance metric which finds distance of the obtained solution front from the true Pareto front; this is trivially done for known problems. Such a metric is based on a reference and, a true reference is not known for unknown problems. A commonly practiced approach to determine the reference for unknown problems is to extract the reference from the best-solutions obtained so far, and the reference is incrementally updated with every generation in iterative refinement based algorithms.

Kumar & Rockett [17] proposed use of rank-histograms for monitoring convergence of Pareto-front while maintaining diversity without any *explicit* diversity preserving operator. Their algorithm is demonstrated to work for problems of *unknown* nature. Secondly, assessing convergence does not need *a priori* knowledge for monitoring movement of Pareto-front using rank-histograms. Some other studies have been done on combining convergence with diversity. Laumanns et al. [23] proposed an ϵ-dominance for getting an ϵ-approximate Pareto-front for problems whose optimal Pareto-set is *known*.

Many metrics have been proposed for quantitative evaluation of the quality of solutions [13, 14]. Essentially, these metric are divided into two classes:

- Diversity : Coverage and sampling of the obtained solutions across the front, and
- Convergence : Distance of the obtained solution-front from the *reference* front.

Some of the commonly used metrics are R-measure [24], S-measure (hyper-volume) [25], Generational distance (GD) [26], Spread measure [14, 27], and Convergence measure [28]. Some of these metrics (e.g., generational distance, volume of space covered, error ratio measures of closeness of the Pareto-front to the true Pareto front) are only applicable where the solution is known. In case of unknown nature, the metrics are sensitive to the choice of the reference. Other metrics (e.g. ratio of non-dominated individuals, uniform distribution) quantify the Pareto-front and can only be used to assess diversity. Knowles & Corne gave a detailed critical review of these measures in his paper [29], and recommended use of some of the metrics as stable measures. They have also shown the sensitivity of some of the metrics with respect to the arbitrary choice of the reference point/front.

The MOST problem is an NP-hard problem, the actual Pareto-front is not known. In Section 4, we will show that different algorithms give different shapes of the solution front; and interpretation of convergence and diversity from the metrics extracted from different shapes to be done meaningfully is not a straight forward task.

3 Design and Implementation

Evolutionary algorithm operators, namely, mutation and crossover imitate the process of natural evolution, and are instrumental in exploring the search space. The efficiency of the evolutionary search depends how a problem (in this case, a spanning tree) is represented in a chromosome and the reproduction operators are defined. There are many encoding schemes to represent spanning trees – see [10] for a detailed review and comparison. For example, one classic representation scheme is Prüfer encoding which is used by Zhou & Gen [30]. Raidl & Julstrom [10] and Knowles & Corne [12] have pointed out that Prüfer numbers have poor locality and heritability and are thus unsuitable for evolutionary search. Deo et al. suggested use of other variants of Prüfer mappings [8]. Recently, Raidl & Julstrom [10] proposed spanning trees to be represented directly as sets of the edges and have shown locality, heritability and computational efficiency of the edge sets for evolutionary search. (While writing this paper, we have come across a newer encoding scheme [31] for tree-based combinatorial optimization problems, which is shown to give superior performance on larger instances of dc-MST; we are currently using this encoding scheme for MOST and the results will be published elsewhere.) In all the results reported in this paper, we use edge-set scheme for representing spanning trees to exploring the search space.

Initial Population: We generate initial population based on random generation of spanning trees. We do not choose the cheapest edge from the currently eligible list of edges (as per Prim's algorithm) rather we select a random edge from the eligible list; this is done to un-bias the randomly generated population from the links found in MST. The other variants of generating initial trees could be based on One-Time-Tree Construction (OTTC) [7] and Randomized Greedy Heuristics (RGH) [9] algorithms.

Fitness Evaluation: We use Pareto-rank based EA implementation. The Pareto-rank of each individual is equal to one more than the number of individuals dominating it in the multiobjective vector space. All the non-dominated individuals are assigned rank one. The values of the two objectives to be minimized (cost and diameter) are used to calculate rank of the individual. Based on the two objectives rank of the individual is calculated. In this work, we calculate fitness of an individual by an inverse quadratic function of the Pareto-rank.

Other Genetic Operator: We select crossover operator to provide strong habitability such that the generated trees consist of the parental edges as far as possible. For generating valid trees, we include non-parental edges into the offspring tree. The crossover operator used in this work is a variant of the operator used by Raidl & Julstrom [9]. Raidl & Julstrom used the diameter information to know the center of the tree. Since we do not generate trees for a specific value of a constrained diameter, we do not have the diameter information to be embedded in the crossover. We start with an edge which is common in both parents as the start edge.

The mutation operators used in this work are again the variants of the operators used by Raidl & Julstrom and designed for edge-set encoding [9]. They designed all the four mutation operators based on the diameter information. In our case, we do not know diameter value, therefore, we adapted their mutation operators to work for diameter-independent values.

Ensuring Convergence: We compute *Intra*-island rank-ratio histogram for each epoch of the evolutionary evolution and monitor the movement of the Pareto-front. Since, this is a hard problem, it is likely that the improvement may get trapped in local minima. To ensure a global (near-) optimal Pareto-front, we use a multi-tribal/island approach and monitor the Pareto-front using *Inter*-island rank histogram. Our multi-island/tribal approach is essentially a test on convergence rather parallelizing the computational efforts as done by others, e.g., Cantu-Paz [32]. For details of computation of Intra-island rank-ratio and Inter-island rank histograms, see [16].

Algorithm: The PCGA algorithm [16] used in this work is a steady-state algorithm and can be seen as an example of (μ + 2) – Evolutionary Strategy (ES) in terms of its selection mechanism [13, 14]. In this algorithm, individuals are compared against the total population set according to a tied Pareto-ranking scheme and the population is selectively moved towards convergence by discarding the lowest ranked individuals in each evolution. In doing so, we require no parameters such as size of the sub-population in tournament selection or sharing/niching parameters. Initially, the whole population of size N is ranked and fitness is assigned by interpolating from the best individual (rank = 1) to the lowest (rank $\leq N$) according to some simple monotonic function. A pair of mates is randomly chosen biased in the sizes of the roulette wheel segments and crossed-over and/or mutated to produce offspring. The offspring are inserted into the population set according to their ranks against the whole population and the lowest ranked two individuals are eliminated to restore the population size to N. The process is iterated until a convergence criterion based on Intra-island rank-ratio and Inter-island rank histogram is achieved [16, 17]. A brief Pseudocode of the PCGA is included in Algorithm 1.

Algorithm 1 : Pareto Converging GA

1: *Input*: N - size of initial population and GA parameters
2: *Output*: a set of (near-) optimal solutions
3: *Algorithm*:
4: Generate an initial population of size N
5: Compute individual's objective vector
6: Pareto-rank the population and generate rank-ratio histogram
7: **while** *Intra*-island rank-ratio histogram does not satisfy stopping criterion **do**
8:     Select two parents using a selection scheme
9:     Perform crossover and mutation to generate two offsprings
10:     Compute objective vectors of offsprings
11:     Pareto-rank the population including offsprings
12:     Remove the two least fit individuals to keep the size N
13:     Generate rank-ratio histogram
14: **end while**
15: One while-loop for *Inter*-island rank-histogram satisfying stopping criterion
16: Output set of solutions

If two individuals have the same objective vector, we lower the rank of one of the individual by one; this way, we are able to remove the duplicates from the set of nondominated solutions without loss of generality. For a meaningful comparison of two real numbers during ranking, we restrict the floating-point precision of the objective values to a few units of precision; this is problem dependent and can be tuned by trial-and-error during few initial runs of the algorithm. Otherwise, this algorithm does not explicitly use any other diversity preserving mechanism. However, lowering the rank of the individual having the identical objective vector (with restricted units of precision) is analogous in some way to a sort of sharing/niching mechanism (in objective space) which effectively controls the selection pressure and thus *partly* contributes to diversity (For other factors that contribute to diversity, see [16]).

4 Results

We tested generation of dual objective spanning tree using our MOEA framework and selected benchmark data taken from Beasley's OR library[1]. The OR-Library is a collection of test data sets for a variety of Operations Research (OR) problems. We considered the Euclidean Steiner problem data which was used by previous researchers, e.g., Raidl-SAC. We considered datasets of up to 250 nodes for this work, and few representative results are included in rest of this Section.

---
[1] http://mscmga.ms.ic.ac.uk/info.html

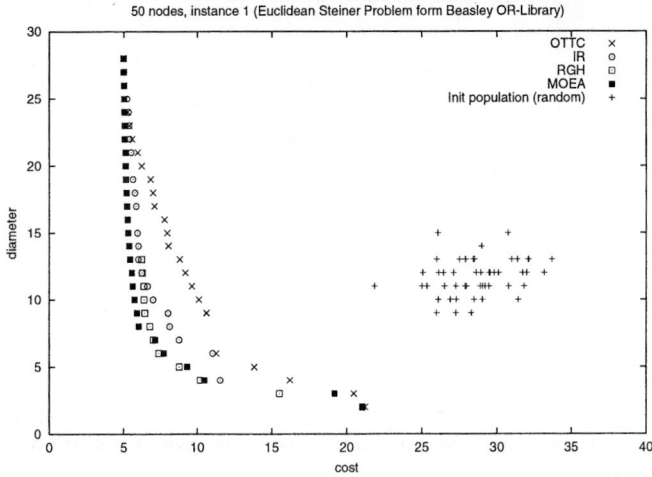

Fig. 1. Pareto front generated from evolutionary algorithm for a 50 node data. Initial population and other fronts generated from OTTC, IR and RGH algorithms are also shown in the plot

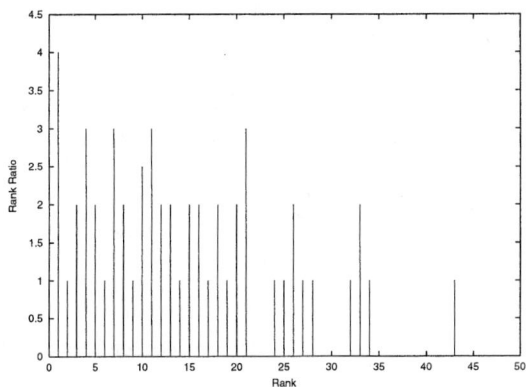

Fig. 2. Rank-ratio histogram computed from the initial population for 50 node data

For comparison, we also include results obtained from three well-known diameter constrained algorithms, namely, One-Time Tree Construction (OTTC) [7], Iterative Refinement (IR) [7] and Randomized Greedy Heuristics (RGH) [9] algorithms. All the three algorithms have been demonstrated for Beasley's OR data and few results included in their respective papers. All three algorithms are *single* objective single constraint algorithms and generate a single tree subject to the diameter constraint. Our MOST algorithm simultaneously optimizes both the objectives and generates a (near-optimal) Pareto-front which comprises a set of solutions. Therefore, we iteratively run all the three - OTTC, IR and RGH - algorithms by varying the value of the diameter constraint and generate sets of solutions to form the respective Pareto-fronts, for comparison with the Pareto-front

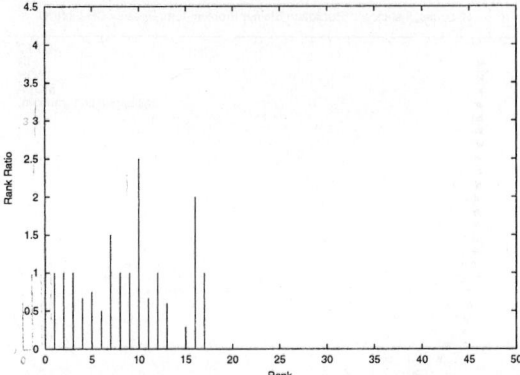

Fig. 3. Rank-ratio histogram computed from the population after first iteration/epoch for 50 node data. Movement of the solution-front can easily be seen with reduction of the histogram tail in comparison with Fig. 2

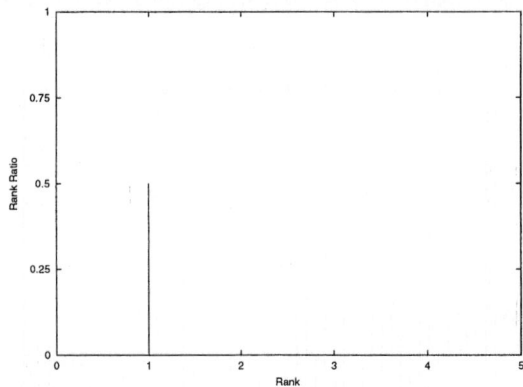

Fig. 4. Rank-ratio histogram at convergence for 50 node data

obtained from the proposed multiobjective evolutionary algorithm. For randomized algorithms, evolutionary and RGH, we have repeated experiments ten times to observe the variability due to randomization, and include here a single set of representative results obtained from the runs.

First, we include results obtained for 50 node data from all the four - OTTC, IR, RGH and our proposed MOEA - algorithms in Fig. 1. Initial population for the proposed MOEA is also shown in Fig. 1, and the corresponding intra rank-ratio histogram is shown in Fig. 2. The rank-ratio histogram after one iteration (epoch) is included in Fig. 3. The reduction in the size of the tail of the histogram indicates movement of the Pareto-front towards convergence which is substantial in this case. At convergence, the final rank-ratio histogram is depicted in Fig. 4. At this stage, all the entries are non-dominated and status of none of the individual was changed from non-dominated to dominated one in the past iteration/epoch. However, this necessarily does not mean that the Pareto-front as shown

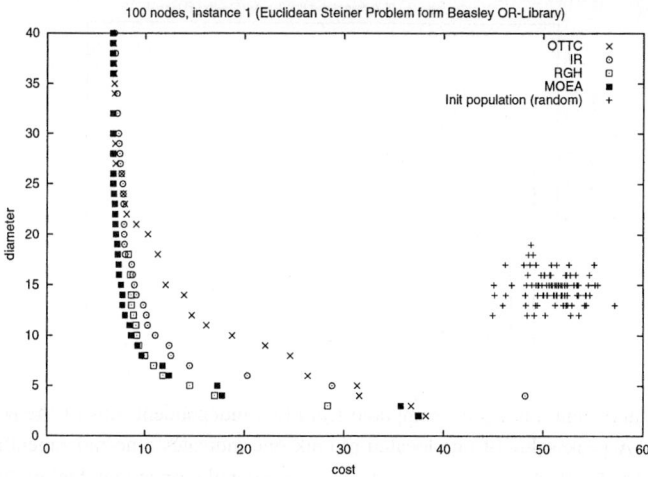

Fig. 5. Pareto front generated from evolutionary algorithm for a 100 node data. Initial population is also shown. Other fronts from OTTC, IR and RGH algorithms are also shown in the plot

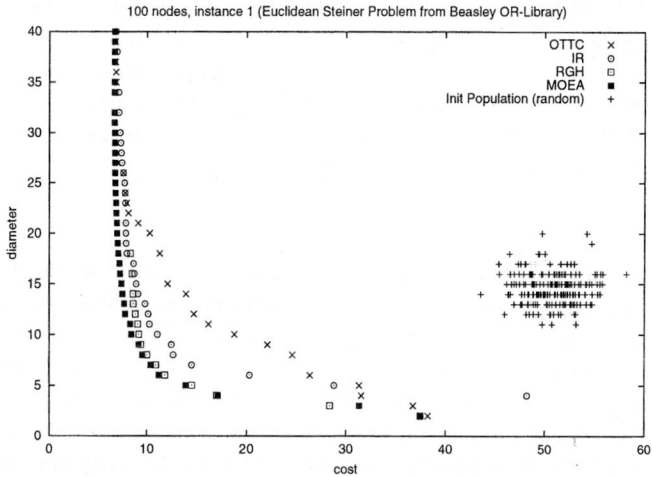

Fig. 6. Improved Pareto front generated from two tribes of evolutionary algorithm for the 100 node data; improvement in the lower and higher ranges of diameters are clearly visible. Initial population of one single tribe is shown. Other fronts from OTTC, IR and RGH algorithms are also shown in the plot

in Fig. 1 is necessarily optimal. Second, we also do not have *a priori* knowledge of the solution space, therefore, we are not in any position to know about the distance between the actual Pareto-front and the obtained solution-front. Therefore, we run MOEA for another run with another set of randomly initialized population, and get a solution set which was marginally superior to the previous one. On merging these two sets, we could

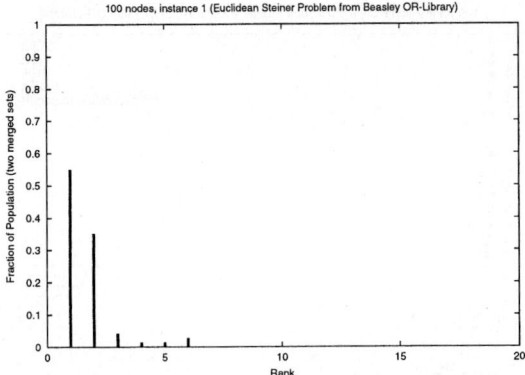

Fig. 7. Inter-Island Rank-histogram computed from two independent runs of the population for 100 node data. A peak value of one located at rank one indicates a no-movement state. In this case, a shift in the Pareto-front is indicated by non-zero population at rank higher than one

Table 1. Diversity and Convergence metrics for 100 node dataset

Algorithm	R-measure	S-measure	Spread	Convergence
OTTC	0.961	8788	0.489	0.072
IR	0.971	8730	0.818	0.050
RGH	0.977	8819	0.864	0.021
MOEA - 1 tribe	0.979	8950	0.769	0.011
MOEA - 2 tribes	0.980	8960	0.671	0.003

get little improvement in the Pareto-front. We may also run MOEA for the third run, and we may again get improvement. The movement of the Pareto-front with each additional run can be monitored on Inter-island rank-histogram (Fig. 7).

Next, we experimented for 100 node data. Results obtained from all the four algorithms are included in Fig. 5. The solutions obtained by MOEA are improved by running the algorithm again, and merging the obtained solutions to form a single Pareto-front. Results obtained from two randomly initialized runs of evolutionary algorithm for 100 node data to form an improved Pareto-front are included in Fig. 6. It can be seen from Fig. 6 that the solutions are improved in the lower and higher ranges of diameters. We plot the Inter-tribal rank-histogram for these two runs of the algorithms and include in Fig. 7. The movement of the solution front by merging the second set of solutions can easily be seen by the long-tail of the rank-histogram in Fig. 7. Otherwise, in a converged state, the rank-histogram should ideally have a single peak of normalized value one at rank one. This indicates that the solution quality is marginally improved by merging two tribes. This is possible because some of the solution points obtained from two tribes of MOEA were distinct and diverse in lower and higher ranges of diameter. This is the clear advantage of using the multi-tribal approach; the results could still be improved with a few more tribes. Such a multi-island/tribal approach is a test on convergence too.

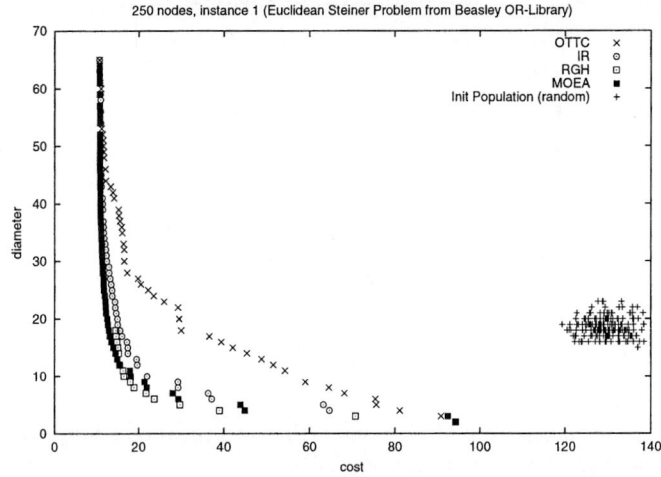

Fig. 8. Pareto front generated from evolutionary algorithm for a 250 node data. Initial population is also shown. Other fronts from OTTC, IR and RGH algorithms are also shown in the plot. Results could be improved by adding solutions from few more tribes

We also collected results from 250 node data; results are included in Fig. 8. Solutions obtained from MOEA are marginally sub-optimal compared to RGH algorithm in very low-range of diameter; this is obvious because MOEA is generic for any diameter values while RGH is tuned to the specific values. The quality of solutions can be further improved by merging solutions obtained from few more tribes, and having a test on convergence. Moreover, if genetic operators of MOEA are tuned to a specific value of diameter like RGH, it will give superior solutions.

Finally, we quantitatively evaluate the solution fronts obtained from each of the algorithms. We compute R-measure [24], S-measure [25], Spread [14] and Convergence measures [28], and include representative results for 100 node dataset in Table 1.

It can be observed from Figures 1, 5, 6 and 8, that this is indeed difficult to find the solutions in the higher range of diameter. In fact, RGH algorithm could not find any solution in higher range of diameter; we generated multiple sets of solutions with multiple runs of RGH algorithm with different initial values but none of the run could generate any solution in this range of diameter. It can also be observed from the figures that the solutions obtained form OTTC algorithm are good in lower and higher range of diameter, however, the results obtained from RGH are good only in the lower range of the diameter. Contrary to this, MOEA is able to locate solutions in the higher range of the diameter with almost comparable quality of the solutions obtained by OTTC. The solutions obtained by OTTC in the middle range are much sub-optimal and are inferior to the solutions obtained by MOEA. In the upper-middle range of diameters, RGH could not locate solutions at all, and the solutions located in this range by OTTC are much inferior to the solutions obtained by MOEA. Thus, quality of the solutions obtained by MOEA is much superior in this range, and comparable in higher range to those of OTTC. To reflect this by rank-histogram, we include, for example, in Fig. 9 the rank-histogram

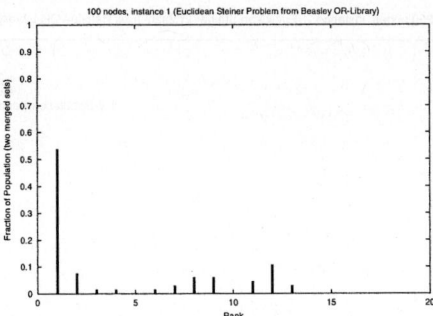

Fig. 9. Rank-histogram computed from OTTC and MOEA solution sets for 100 node data

of two solution-sets taken from MOEA and OTTC respectively; a long-tail reflects the inferior quality of solutions obtained from OTTC.

These are interesting observations, and are partly contrary to those reported by Raidl & Julstrom [9]. Raidl & Julstrom have shown that their technique works the best over all the other such techniques including OTTC. On looking from the plots in Figures 1, 5, 6 and 8, it can be observed that IR results are very close to RGH results in lower-diameter region, and little inferior in higher-diameter range to those of RGH. Moreover, IR could find competitive solutions in the entire range of the diameter which no other algorithm (barring MOEA) could do, is a significant achievement. However, since previous researchers could not visualize the solutions in the entire range of the diameter, their observations were biased. We reiterate that their conclusions were based on the experiments which they did for a particular value of the diameter and they could not evaluate the results over the entire range of diameter. In this work, since we could simultaneously obtain solutions for the entire range of the diameter, we could have a meaningful comparison of the existing algorithms too.

The above arguments are well supported by the metrics shown in Table 1. We are currently investigating the empirical behavior shown by these four algorithms, and how this knowledge can be used to further improve the solution-set by fine-tuning the evolutionary operators for the MOST problem.

5 Discussion and Conclusions

In this work, we demonstrated generating spanning trees subject to their satisfying the twin objectives of minimum cost and diameter. The obtained solution is a set of (near-optimal) spanning trees that are non-inferior with respect to each other. A network designer having a range of network cost and diameter in mind, can examine several optimal trees simultaneously and choose one based on these requirements and other engineering considerations.

To the best of our knowledge, this is the first work which attempts obtaining the complete Pareto front. Zhou & Gen [30] also obtained a set of solutions, they did not experiment on any benchmark data and, therefore, could not compare the quality of the solutions. It is shown by Knowles & Corne [12] that the front obtained by Zhou

& Gen [30], was sub-optimal. We attribute the sub-optimality due to their use of an EA implementation which was unable to assess convergence. Knowles & Corne [12] used a weighted sum approach and could get comparable solutions but their approach is sensitive to the selection of weight values.

The work presented in this paper presents a generic framework which can be used to optimize any number of objectives simultaneously for spanning tree problems. The simultaneous optimization of objectives approach has merits over the constrained-based approaches, e.g., OTTC, IR and RGH algorithms. It is shown that the constrained-based approaches are unable to produce quality solutions over the entire range of the Pareto-front. For example, the best known algorithm of diameter-constrained spanning tree is RGH which is shown to be good for smaller values of diameters *only*, and is unable to produce solutions in the higher range. Similarly, the other well-known OTTC algorithm produces sub-optimal solutions in the middle range of the diameter. MOEA could obtain superior solutions in the entire range of the objective-values. The solutions obtained by MOEA may further be improved marginally by proper tuning of evolutionary operators for the specific values of the objectives by introducing problem specific knowledge while designing evolutionary operators; such type of improvement, is however, difficult with an approximation algorithm.

Acknowledgements

Authors would like to thank anonymous reviewers for useful suggestions. A part of the research is supported from the Ministry of Human Resource Development (MHRD), Government of India project grant.

References

1. Garey, M.R., Johnson, D.S.: Computers and Interactability: A Guide to the Theory of NP-Completeness. San Francisco, LA: Freeman (1979)
2. Hochbaum, D.: Approximation Algorithms for NP-Hard Problems. Boston, MA: PWS (1997)
3. Marathe, M.V., Ravi, R., Sundaram, R., Ravi, S.S., Rosenkrantz, D.J., Hunt, H.B.: Bicriteria Network Design Problems. J. Algorithms **28** (1998) 142 – 171
4. Ravi, R., Marathe, M.V., Ravi, S.S., Rosenkrantz, D.J., Hunt, H.B.: Approximation Algorithms for Degree-Constrained Minimum-Cost Network Design Problems. Algorithmica **31** (2001) 58 – 78
5. Boldon, N., Deo, N., Kumar, N.: Minimum-Weight Degree-Constrained Spanning Tree Problem: Heuristics and Implementation on an SIMD Parallel Machine. Parallel Computing **22** (1996) 369 – 382
6. Deo, N., Kumar, N.: Constrained Spanning Tree Problems: Approximate Methods and Parallel Computations. DIMACS Series in Discrete Mathematics and Theoretical Computer Science, American Math Society **40** (1998) 191 – 217
7. Deo, N., Abdalla, A.: Computing a Diameter-Constrained Minimum Spanning Tree in Parallel. In: Proc. 4th Italian Conference on Algorithms and Complexity (CIAC 2000), LNCS 1767. (2000) 17 – 31
8. Deo, N., Micikevicius, P.: Comparison of Prüfer-like Codes for Labeled Trees. In: Proc. 32nd South-Eastern Int. Conf. Combinatorics, Graph Theory and Computing. (2001)

9. Raidl, G.R., Julstrom, B.A.: Greedy Heuristics and an Evolutionary Algorithm for the Bounded-Diameter Minimum Spanning Tree Problem. In: Proc. 18th ACM Symposium on Applied Computing (SAC 2003). (2003) 747 – 752
10. Julstrom, B.A., Raidl, G.R.: Edge Sets: An Effective Evolutionary Coding of Spanning Trees. IEEE Trans. Evolutionary Computation **7** (2003) 225 – 239
11. Knowles, J.D., Corne, D.W.: A New Evolutionary Approach to the Degree-Constrained Minimum Spanning Tree Problem. IEEE Trans. Evolutionary Computation **4** (2000) 125 – 133
12. Knowles, J.D., Corne, D.W.: A Comparison of Encodings and Algorithms for Multiobjective Minimum Spanning Tree Problems. In: Proc. 2001 Congress on Evolutionary Computation (CEC-01). Volume 1. (2001) 544 – 551
13. Coello, C.A.C., Veldhuizen, D.A.V., Lamont, G.B.: Evolutionary Algorithms for Solving Multiojective Problems. Boston, MA: Kluwer (2002)
14. Deb, K.: Multiobjective Optimization Using Evolutionary Algorithms. Chichester, UK: Wiley (2001)
15. Coello, C.A.C., Lamont, G.B.: Applications of Multiojective Evolutionary Algorithms. World Scientific (2004)
16. Kumar, R., Rockett, P.I.: Improved Sampling of the Pareto-front in Multiobjective Genetic Optimization by Steady-State Evolution: A Pareto Converging Genetic Algorithm. Evolutionary Computation **10** (2002) 283 – 314
17. Kumar, R., Rockett, P.I.: Assessing the Convergence of Rank-based Multiobjective Genetic Algorithms. In: Proc. 2nd IEE/IEEE Int. Conf. Genetic Algorithms in Engineering Systems: Innovations and Applications (Galesia 97). Volume I446., IEE, London, UK (1997) 19 – 23
18. Kumar, R., Singh, P.K., Chakrabarti, P.P.: Multiobjective Genetic Search for Spanning Tree Problem. In: Proc. 11th Int. Conf. Neural Information Processing (IcoNIP), LNCS 3316, Kolkata, India, Springer-Verlag (2004) 218–223
19. Kumar, R., Singh, P.K., Chakrabarti, P.P.: Improved Quality of Solutions for Multiobjective Spanning Tree Problem Using EA. In: Proc. 11th Int. Conf. High Performance Computing (HiPC), LNCS 3296, Bangalore, India, Springer-Verlag (2004) 494 –503
20. Purshouse, R.C., Fleming, P.J.: Elitism, Sharing and Ranking Choices in Evolutionary Multi-criterion Optimization. Research Report No. 815, Dept. Automatic Control & Systems Engineering, University of Sheffield (2002)
21. Deb, K.: A Fast Non-Dominated Sorting Genetic Algorithm for Multiobjective Optimization: NSGA-II. In: Proc. Parallel Problem Solving from Nature, PPSN-VI. (2000) 849 – 858
22. Zitzler, E., Laumanns, M., , Thiele, L.: SPEA2: Improving the Strength Pareto Evolutionary Algorithm. In: Proc. Evolutionary Methods for Design, Optimization and Control with Applications to Industrial Problems (EUROGEN). (2001)
23. Laumanns, M., Thiele, L., Deb, K., Zitzler, E.: Combining Convergence and Diversity in Evolutionary Multiobjective Optimization. Evolutionary Computation **10** (2002) 263 – 282
24. Hansen, M.P., Jaszkiewicz, A.: Evaluating the Quality of Approximations to the Non-dominated Set. Tech. Rep. IMM-REP-1998-7, Tech. Univ. Denmark (1998)
25. Zitzler, E.: Evoluationary Algorithms for Multiobjective Optimization: Methods and Applications. Ph.D. Thesis, Swiss Federal Institute of Technology, Zurich, Switzerland (1999)
26. van Veldhuizen, D.A.: Multiobjective Evolutionary Algorithms: Classifications, Analysis, and New Innovations. Ph.D. Thesis, Technical Report No. AFIT/DS/ENG/99-01, Air Force Institute of Technology, Dayton, OH (1999)
27. Schott, J.R.: Fault Tolerant Design Using Single and Multicriteria Genetic Algorithms. Masters Thesis, Department of Aeronautics and Astronautics, MIT, Massachusetts (1995)
28. Deb, K., Jain, S.: Running Performance Metrics for Evolutionary Multiobjective Optimization. In: Proc. 4th Asia-Pacific Conf. Simulated Evolution and Learning (SEAL 02), Singapore (2002) 13–20

29. Knowles, J., Corne, D.: On Metrics for Comparing Nondominated Sets. In: Proc. Congress Evolutionary Computation (CEC '02). Volume I., Piscataway, NJ, IEEE (2002) 711–716
30. Zohu, G., Gen, M.: Genetic Algorithm Approach on Multi-Criteria Minimum Spanning Tree Problem. European J. Operations Research **114** (1999) 141–152
31. Soak, S.M., Corne, D., Ahn, B.H.: A powerful new encoding for tree-based combinatorial optimisation problems. In: Parallel Problem Solving from Nature - PPSN VIII. Volume 3242 of LNCS., Birmingham, UK, Springer-Verlag (2004) 430–439
32. Cantu-Paz, E.: Efficient and Accurate Parallel Genetic Algorithms. Boston, MA: Kluwer (2000)

Developments on a Multi-objective Metaheuristic (MOMH) Algorithm for Finding Interesting Sets of Classification Rules

Beatriz de la Iglesia, Alan Reynolds, and Vic J Rayward-Smith

University of East Anglia, Norwich, Norfolk, NR4 7TJ, UK
{bli, ar, vjrs}@cmp.uea.ac.uk

Abstract. In this paper, we experiment with a combination of innovative approaches to rule induction to encourage the production of interesting sets of classification rules. These include multi-objective metaheuristics to induce the rules; measures of rule dissimilarity to encourage the production of dissimilar rules; and rule clustering algorithms to evaluate the results obtained.

Our previous implementation of NSGA-II for rule induction produces a set of cc-optimal rules (coverage-confidence optimal rules). Among the set of rules produced there may be rules that are very similar. We explore the concept of rule similarity and experiment with a number of modifications of the crowding distance to increasing the diversity of the partial classification rules produced by the multi-objective algorithm.

1 Introduction

Data mining is concerned with the extraction of patterns from large databases. One particular task of data mining which is attracting increased research attention is the extraction of classification rules. Partial classification, also known as nugget discovery, involves the production of accurate yet simple rules (nuggets) that describe subsets of interest within a database.

Recently, we have developed a multi-objective metaheuristic algorithm for the extraction of partial classification rules [7]. The problem of nugget discovery was formulated as a multi-objective optimisation problem by using some of the frequently used measures of interest, namely confidence and coverage of a nugget, as objectives to be optimised. NSGA-II was then used to perform the search for Pareto-optimal rules according to the defined objectives.

The approach was evaluated by comparison to another algorithm, ARAC [18] which is guaranteed to find all cc-optimal rules subject to certain constraints. The constraints may affect the number of attribute tests that are allowed in the antecedent of the rule, or the maximum cardinality allowed for any attribute that participates in a test. For small datasets, where constraints do not have to be applied, ARAC can deliver all the cc-optimal (i.e. Pareto-optimal) rules efficiently, so it provides a perfect point of comparison.

Results showed the strength of the new multi-objective approach for finding a good approximation to the Pareto front in a number of datasets. For the larger datasets, the multi-objective approach showed real advantage as it could find good sets of solutions in a fraction of the time, with predictable termination times, and without having to apply any restrictions to the number of attributes or their cardinality.

One question raised was whether the set of rules delivered may contain very similar rules or rules that appear to be different but match similar records. There may also be rules that are interesting because they cover different subsets of records, but which are dominated in terms of coverage and confidence and are, therefore, never found.

In this paper we investigate the quality of the rule sets obtained. In particular, we investigate various options for refining the quality of the rule sets obtained in order to deliver an *interesting* set of rules or nuggets. Defining interest in rule induction has been an area of research for some time. Most methods of measuring individual rule interest use a combination of confidence and support for the rule [14]. Considerations about rule novelty or surprise for individual rules are sometimes included [9, 10]. We study the novelty of the rules in relation to other rules within the set; that is, we would like to deliver a set of rules of high quality in terms of confidence and coverage, but also where rules are as diverse as possible with respect to other rules in the same set. This should increase the interest of the rule set, as opposed to the interest of the individual rules. We examine the interest of the rule sets obtained with our previous approach and attempt various modifications to improve our rule sets.

Section 2 covers the basic concepts and terminology used in the paper. Section 3 describes briefly the original multi-objective nugget discovery algorithm. Section 4 describes measures of rule dissimilarity and their applicability to the algorithm and introduces the concept of clustering rules for better interpretation of results. We describe some initial experimentation in section 5 and give our conclusions and ideas for further work in section 6.

2   Concepts and Terminology

2.1   Nugget Discovery

The task of partial classification [1] is also known as nugget discovery; it seeks to find patterns that represent "strong" descriptions of a specified class, even when that class has few representative cases in the data. For example, in insurance data, groups of people that constitute an unacceptably high risk are in a minority. However, if an insurer can identify such groups, with their defining characteristics, they may gain a competitive advantage.

Let Q be a finite set of attributes where each $q \in Q$ has an associated domain, $\text{Dom}(q)$. Then a record specifies values for each attribute in Q. A tabular database, D, is defined to be a finite set of such records. A classification tabular dataset is one in which a class attribute is present.

Rules that represent a partial classification are of the general form

$$\text{antecedent} \Rightarrow \text{consequent}$$

where the antecedent and consequent are predicates that are used to define subsets of records from the database D and the rule underlines an association between these subsets. In nugget discovery, the antecedent comprises a conjunction of Attribute Tests, ATs, and the consequent comprises a single AT representing the class description. The strength of the rule may be expressed by various measures, as described in section 2.2

Attributes may be described as ordinal or nominal (categorical). An ordinal attribute is defined as an attribute that has some explicit or implicit ordering, so numeric attributes are usually ordinal. Nominal attributes are those that are not ordinal, i.e. they have no implied ordering.

For a database with n attributes, the ATs for nominal attributes can be expressed in any of the following forms:

Simple Value: $AT_j = v$, where v is a value from the domain of AT_j, Dom_j, for some $1 \leq j \leq n$. A record x satisfies this test if $x[AT_j] = v$.

Subset of Values: $AT_j \in \{v_1, \ldots, v_k\}$, where $\{v_1, \ldots, v_k\}$ is a subset of values in the domain of AT_j, for some $1 \leq j \leq n$. A record x satisfies this test if $x[AT_j] \in \{v_1, \ldots, v_k\}$.

Inequality Test: $AT_j \neq v$, for some $1 \leq j \leq n$. A record x satisfies this test if $x[AT_j] \neq v$.

For a numeric attribute, ATs can take the following form:

Simple Value: $AT_j = v$, $1 \leq j \leq n$, as for categorical attributes.

Binary Partition: $AT_j \leq v$ or $AT_j \geq v$, for some $1 \leq j \leq n$, $v \in Dom_j$. A record x satisfies these tests if $x[AT_j] \leq v$ or $x[AT_j] \geq v$ respectively.

Range of Values: $v_1 \leq AT_j \leq v_2$ or $AT_j \in [v_1, v_2]$, for some $1 \leq j \leq n$ and $v_1, v_2 \in Dom_j$. A record x satisfies this test if $v_1 \leq x[AT_j] \leq v_2$.

Decision tree induction [4, 15] and rule induction algorithms [5, 6] are often used to extract partial classification rules. However, decision trees often have thousands of rules, with each rule covering only a few cases; hence their use as descriptive patterns is limited. Also, both decision tree and rule induction algorithms often fail to produce patterns for minority classes. Association rule algorithms have been adapted to find patterns in classification data [2, 3], but they are predominantly developed for categorical data and often apply restrictions to the syntax of the rules to keep the search feasible. They deliver *all* rules underlying a database, which can result in output of overwhelming size.

2.2 Strength of a Rule

Given a record, t, *antecedent*(t) (represented by a conjunction of ATs in nugget discovery) is true if t satisfies the predicate, *antecedent*. Similarly *consequent*(t) is true if t satisfies the predicate, *consequent*. Then the subsets defined by the antecedent or consequent are the sets of records for which the relevant predicate is true.

For a rule r, we define three sets of records.

$A(r) = \{t \in D | antecedent(t)\}$, (i.e. the set of records defined by the antecedent)
$B(r) = \{t \in D | consequent(t)\}$ (i.e. the set of records defined by the consequent)
$C(r) = \{t \in D | antecedent(t) \wedge consequent(t)\}$.

The support, $sup(M)$, for any conjunction, M, of ATs, is the number of records which satisfy M. Given a rule, r, we designate the antecedent of the rule r^a and the consequent r^c. Then, the support for the antecedent, $sup(r^a) = |A(r)| = a$ and the support for the consequent, $sup(r^c) = |B(r)| = b$ (i.e. the cardinality of the target class).

The *support* for r, $sup(r)$, is defined as $sup(r^a \wedge r^c) = |C(r)| = c$.

The *confidence* (also known as *accuracy*) of r, $conf(r)$, is defined as

$$conf(r) = \frac{sup(r)}{sup(r^a)} = \frac{c}{a}.$$

The support for a rule may be expressed as a proportion of the support for the consequent, this measure is referred to as *coverage*. In nugget discovery, it is often convenient and more intuitive to use this measure in place of rule support as we are interested in rules that represent a strong description of a predefined class.

The coverage of r, $cov(r)$, is defined as

$$cov(r) = \frac{sup(r)}{sup(r^c)} = \frac{c}{b}.$$

A strong rule may be defined as one that meets certain confidence and coverage thresholds. Those thresholds are normally set by the user (or the data owner) and are based on domain or expert knowledge about the data. Strong rules may be considered interesting if they are found to be novel and useful.

2.3 CC-Optimality

The complete set of strong rules that underlie a database may be very large and many rules may be very similar. In order to address this problem and present a concise set of rules, the cc-optimal (coverage-confidence optimal) subset of rules was proposed in [3]. The cc-optimal set is a set of rules where each rule is optimal with respect to coverage and confidence.

A partial ordering, $\leq_{cc}$, is defined on rules where $r_1 <_{cc} r_2$ if and only if:

Condition A- $cov(r_1) \leq cov(r_2) \wedge conf(r_1) < conf(r_2)$,
Condition B- $cov(r_1) < cov(r_2) \wedge conf(r_1) \leq conf(r_2)$.

Also, $r_1 =_{cc} r_2$ in the partial ordering if $cov(r_1) = cov(r_2) \wedge conf(r_1) = conf(r_2)$.

It is easy to see how the concept of cc-optimality fits in with a multi-objective approach as the cc-optimal set of rules are those that lie in the Pareto optimal front when the objectives to be optimised are confidence and coverage. In other words, if $r_1 <_{cc} r_2$, rule r_2 is said to dominate r_1. r_2 is said to be Pareto optimal (or cc-optimal) if and only if there is no other rule, r_i that dominates r_2.

Hence, we can use a number of algorithms, including multi-objective metaheuristics, to search the space of all possible rules and extract those that are cc-optimal, or close to cc-optimality.

The problem with cc-optimality as a criterion to choose rules is that, if two rules have the same confidence and coverage, only one of them may be kept in the cc-optimal set. However, the two rules could be very different either in attribute space or they could describe a different subset of records. In such cases it could be argued that the cc-optimal rule set may not be suitably representative of the most interesting rules underlying the database. In this paper, we investigate this claim, since that has been a doubt cast over our previous research. Intuitively, if this is the case, one would expect to find at least some rules of similar coverage and accuracy which cover different sets of records in the final Pareto front.

3 NSGA-II for Nugget Discovery

The algorithm NSGA-II [8] was applied to the problem of finding cc-optimal rules [7]. NSGA-II uses non-dominated sorting as a mechanism for introducing elitism in the search. It also uses a crowding operator to ensure diversity of solutions within the Pareto front.

A population of solutions is created and sorted into fronts according to non-domination with respect to the multiple objective functions. Solutions within the same front are then sorted according to crowding distance. Solutions that are non-dominated (i.e. those that belong to the first front) are given priority for reproduction. If two solutions are non-dominated, the solution that is least crowded has a higher priority for reproduction. The cycle of selection and reproduction using crossover and mutation creates a new pool which is merged with the initial pool, and the process is repeated again over a number of generations.

3.1 Implementation Details

The solution to be represented is a conjunctive rule or nugget following the syntax described in section 2.1. A binary string is used for this as follows. The first part of the string is used to represent the numeric fields or attributes. Each numeric attribute is represented by a set of Gray-coded lower and upper limits, where each limit is allocated a user-defined number of bits, p ($p = 10$ is the default). There is a scaling procedure that transforms any number in the range of possible values using p bits $[0, 2^p - 1]$ to a number in the range of values that the attribute can take.

The second part of the string represents categorical attributes, with each attribute having v number of bits, where v is the number of distinct values (or the number of labels) that the categorical attribute can take. If a bit assigned to a categorical attribute is set to 0 then the corresponding label is included as an inequality in one of the conjuncts.

In this work, we assume that the consequent of the rule is fixed and of the form of an attribute test, AT, on the class label, hence it does not need to be represented as part of the rule.

Random initialisation proved ineffective in a number of experiments. A more effective approach for this kind of problem is to use mutated forms of the default rule as initial solutions. The default is the rule in which all limits are maximally spaced and all labels are included. In other words, it predicts the class without any pre-conditions. This is the approach used in all experiments reported here, with a mutation probability of 1% which was set after parameter experimentation.

To evaluate a solution, the bit string is first decoded into a rule, and the data in the database, which has been previously loaded in memory, is scanned. For each record the values of the fields are compared against the rule, and the class is also compared. The counts of c (support for the rule) and a (support for the antecedent) are updated accordingly. The counts of b (support for the consequent) and d (cardinality of the database) are known from the data loading stage. Once all data has been examined, the measures of strength used as the objectives, in this case the coverage and confidence, are calculated for each nugget.

Parameter experimentation established the use of one-point crossover, with a crossover rate of 80%. The size of the population was set at 100 solutions.

The output of this algorithm can either be the best solutions found through the search (of which we keep a copy) or the final parents.

The rules obtained by this approach are a subset of the rules underlying the database and should be a good representation of the cc-optimal set. There is, of course, no way of guaranteeing optimality with any heuristic technique. In practical experimentation, however, when the implementation was tested on a number of standard databases against an algorithm (ARAC [18]) capable of finding all rules in the cc-optimal set, it performed very well and was shown to find a good approximation to the Pareto front of this set in each database tested (for details see [7]). The spread of solutions in the Pareto optimal front with respect to the objectives to be optimised appeared to be good, but it was difficult to know how close some of those rules may have been to one another in real terms, and which other rules (perhaps interesting rules) may have been side-stepped in the search for cc-optimal rules. In order to assess these factors, the set of rules produced by the algorithm needs to be analysed in terms of similarity of rules within the set.

4 Rule Dissimilarity

Rule dissimilarity can be measured in a number of ways. First, one could look at the specific syntactic difference between two rules, i.e. the difference in attribute space. This may be considered as testing the appearance of two rules.

Rules that appear to be different in attribute space may represent interesting concepts for the user. For example, they may represent different (alternative) characterisations of the same subset of records, perhaps by using attributes that are correlated.

On the other hand, we can simply examine the subset of records that is characterised by a rule. Rules that characterise different (non-overlapping) subsets of records may be considered dissimilar.

In the case of nugget discovery, we may be interested in encouraging diversity of solutions in terms of both their appearance and the population they characterise. However, the first concern must be to characterise as much of the target class as possible, so we will start by looking at dissimilarity in the sets of records that 'match' different rules. We leave rule appearance as an issue for further research.

4.1 Dissimilarity Measure

When trying to define the set of records that match a particular rule, it is possible to use $A(r)$, i.e. the set of records that match the antecedent of the rule r. Another possibility is to use $C(r)$, i.e. the set of records that match both the antecedent and consequent of the rule r. In all experimentation conducted we use the set $C(r)$ to calculate rule dissimilarity. In terms of their use for calculating distances, they are interchangeable by replacing C by A in the equation below.

If r_1 and r_2 are two arbitrary rules, we can define a dissimilarity measure as

$$d(r_1, r_2) = |C(r_1) \cup C(r_2)| - |C(r_1) \cap C(r_2)|$$
$$= |C(r_1) - C(r_2)| + |C(r_2) - C(r_1)|$$

This initial measure provides a count of records matching one and only one of the two rules. Dividing this measure by the number of records in the database, $|D|$ gives the simple matching coefficient [12].

Alternatively, we can use the Jaccard coefficient [11] on the sets of support for the rules and define

$$n(r_1, r_2) = d(r_1, r_2) / |C(r_1) \cup C(r_2)|.$$

Intuitively, two rules that are mutually supported by a thousand records and differ over only six are more similar than two rules that are mutually supported by no records and differ over five. Hence the Jaccard coefficient may be a better measure of rule dissimilarity for our purposes.

4.2 Clustering of Rules

In order to understand the results of applying distance metrics to the rules obtained by the NSGA-II algorithm, we proceed to apply a clustering algorithm to cluster similar rules together. This should help in the presentation of results.

Our recent research on suitable approaches to clustering rules presented a number of possible clustering algorithms [16, 17] for rule clustering. Here, we use two of the algorithms: Partitioning Around Medoids (PAM) and AGlomerative NESting (AGNES). Both algorithms work on a pre-prepared dissimilarity matrix which contains the distance between each pair of rules calculated using the Jaccard coefficient. The resulting clusters are based on these distances between rules, hence rules that appear in the same cluster should apply to the same or overlapping subsets of records.

5 Experimentation

For our experimentation, we are using the *Adult* dataset from the UCI repository [13]. Initial experiments used a set of rules produced by the NSGA-II algorithm for nugget discovery. The algorithm was applied to produce rules to describe the class "Income > 50 k". The AGNES clustering algorithm was applied to cluster the best rules found through the search. The clustering used the Jaccard coefficients on the sets of support for the rules. The hierarchy of rules produced was then cut at a point that lead to 8 clusters. The PAM clustering algorithm was also used to cluster the best rules obtained. The results are presented in figures 1 and 2 respectively. Both graphs show that rules that are close in terms of the values of their objective functions (coverage and confidence) are also close in terms of the set of records that support them. The clustering tends to be neatly distributed on the Pareto front, with little overlap of clusters within the front. Hence in the majority of cases, selecting sets of rules with similar coverage and confidence tends to deliver similar rules that describe a similar subsets of records. As we examine rules of different coverage/confidence we are likely to be finding rules that describe different subsets of records. A similar exercise was performed to cluster the rules in the final parent population, and this exhibited exactly the same characteristics. Other sets of rules produced from different databases provided similar clustering behavior.

5.1 Encouraging Diversity in Terms of Support Sets

It may be possible to encourage diversity of the rule set by using some measure of dissimilarity of rules as a third criterion to be optimised. However, since dissimilarity can only be measured in the context of other rules in the set, this will result in a less efficient evaluation procedure. Also, the application of dissimilarity to the non-dominated sorting may result in a new partial ordering of rules which does not reflect the requirements of nugget discovery. We consider diversity of rules in terms of support sets as a "second priority" objective, to be achieved once we can guarantee a pool of strong rules.

For our purpose, we decided to experiment with the crowding measure of NSGA-II [8]. The crowding measure in normal operation ensures that the population within the Pareto optimal front is as diverse as possible, so it acts as a secondary criterion for ordering rules. We first analysed the effect of not using a crowding measure for this application of NSGA-II for rule induction, so within the algorithm all rules where considered to be equally crowded at all times. The

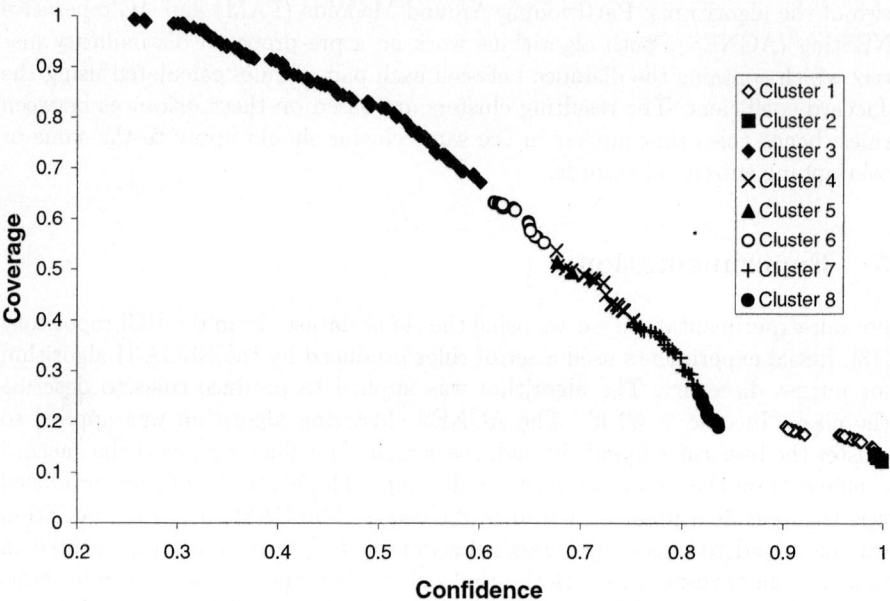

Fig. 1. Clustering of rules for Adult dataset using AGNES - 8 clusters

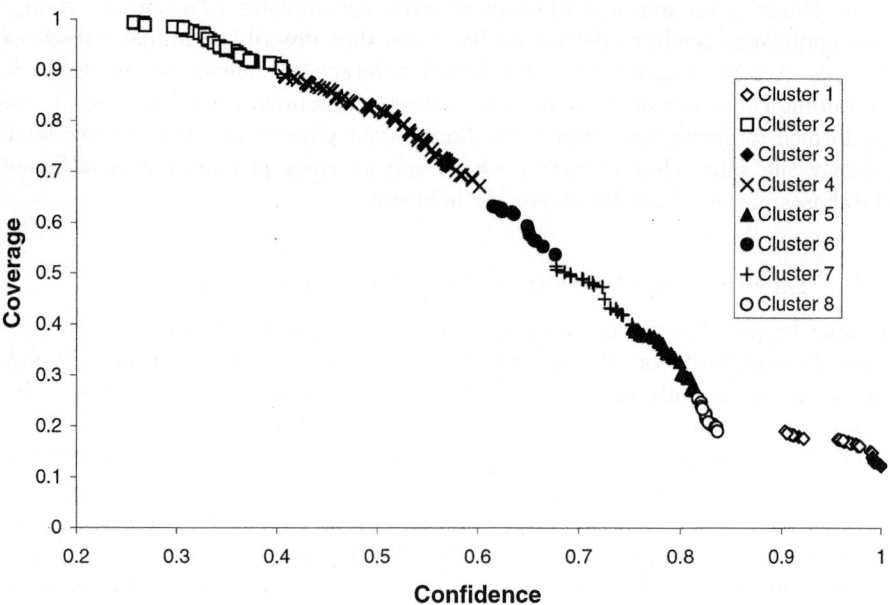

Fig. 2. Clustering of rules for Adult dataset using PAM - 8 clusters

Pareto front obtained by running our algorithm on the Adult data with equal crowding is shown in figure 3. In this section, we show the solutions in the fi-

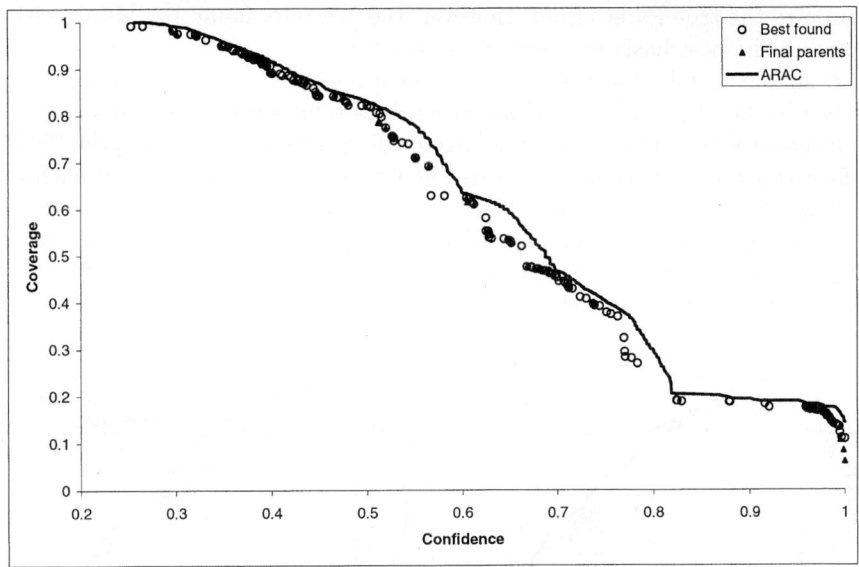

Fig. 3. Rules obtained with NSGA-II for the Adult database with equal crowding

nal population as well as the best solutions found during the search. To aid the analysis of results, the approximation to the Pareto front obtained by the ARAC algorithm is also plotted as a line. ARAC had to impose some restrictions on the search due to the size of this database, hence it can only give us an approxi-

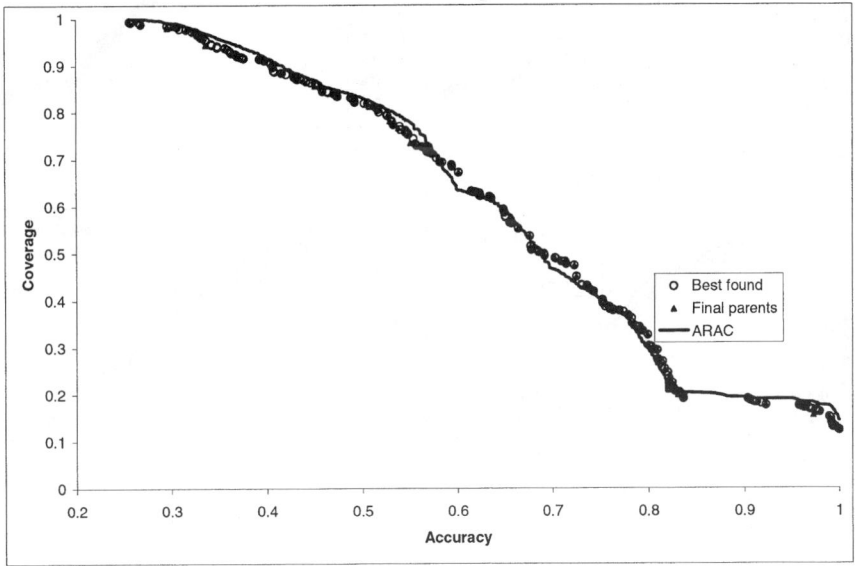

Fig. 4. Rules obtained with NSGA-II for the Adult database with standard crowding

mation to the true Pareto front. However, the 976 rules found by this algorithm represent the best basis for comparison of results.

A similar graph showing the results of using the crowding measure as described in the original NSGA-II algorithm is also presented in figure 4.

It can be seen that using the standard crowding measure proposed by the NSGA-II algorithm produces a much better spread of solutions in the Pareto optimal front, both in terms of the best solutions found as well as in terms of the final parents.

A number of approaches were tried to adapt the crowding measure to encourage diversity within the rule set in terms of support sets. We only discuss the most successful approach here.

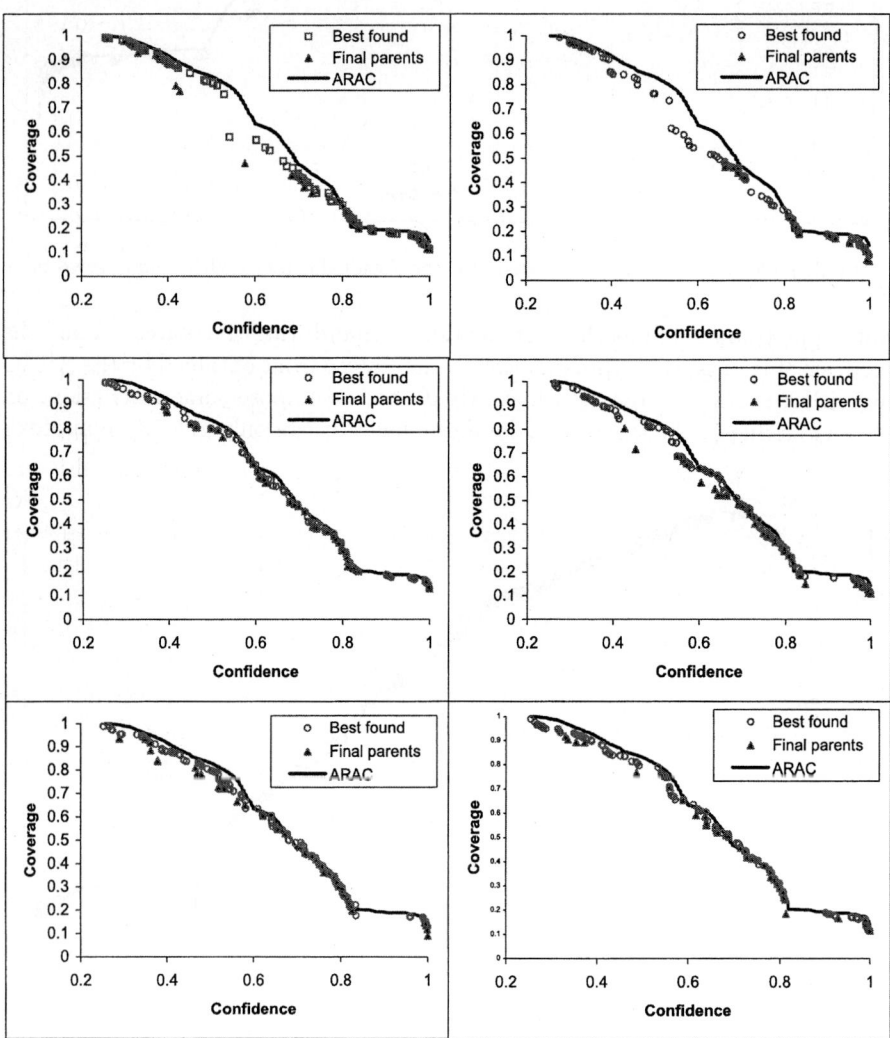

Fig. 5. Rules obtained with NSGA-II for the Adult database with new crowding mechanisms. From top to bottom graphs represent thresholds of 0.5, 0.2 and 0.1 respectively

The Jaccard dissimilarity measure was calculated for each pair of rules using the support set, C, for calculations. The crowding measure was then modified to be a count of the number of rules within a certain threshold distance, T, from the rule being examined, according to the Jaccard dissimilarity measure. Three values were tried for this distance threshold: 0.1, 0.2, 0.5. In the standard crowding measure, the crowding distance is calculated by using solutions within the same front. In our proposed approach it is possible to use the whole population to calculate the new crowding distance, as well as the solutions within the same front only. We experimented with both options.

The results are shown in figure 5. The left hand column of graphs shows the results using crowding based on the distance of the solutions in the front only, whereas the right column represents the results using the whole population. The first row of graphs represents the threshold value $T = 0.5$; for the second row $T = 0.2$; for the final row $T = 0.1$.

When crowding is calculated using threshold distances of 0.2 and 0.1 the spread of best found solutions seems to cover most of the Pareto front. However, the final parent solutions show less coverage of the front. A threshold of 0.5 produces poor coverage of the Pareto front. Some of these observations are expected: since we are no longer encouraging diversity as per the objective functions, some of that diversity will be lost in the population.

For each of the sets of final parents created with different crowding mechanisms (since there are always the same number of parents in each set), we calculate the sum of distances between rules. This is reported in table 1. The sum of distances increases in all cases with the new crowding mechanism with respect to standard crowding. The sum of distances also increases for equal crowding but the coverage of the front is poorer, with some areas not represented in the final parent population.

To further assess the increase in diversity of solutions in terms of support sets, we use PAM to cluster some of the rules produced and observe the degree of overlap. For this purpose, we choose the rules produced using the new crowding measure with the whole population and a threshold of 0.2. We use the support set, C, for dissimilarity calculations and clustering. We feed the clustering algorithm the set of best rules found. As before, we aim to produce 8 clusters for comparative purposes. The results of this process are shown in figure 6. There is some overlap now in clusters 5 and 2, so in the high confidence / low coverage

Table 1. Sum of distances between rules using different crowding mechanisms

Approach	Sum	Approach	Sum
Standard	4,570.51	Equal	5,057.4
Crowding based on distances by front		Crowding based on distances by population	
0.1	5,081	0.1	5,510.52
0.2	5,007.25	0.2	5,301.09
0.5	5,776.27	0.5	5,784.05

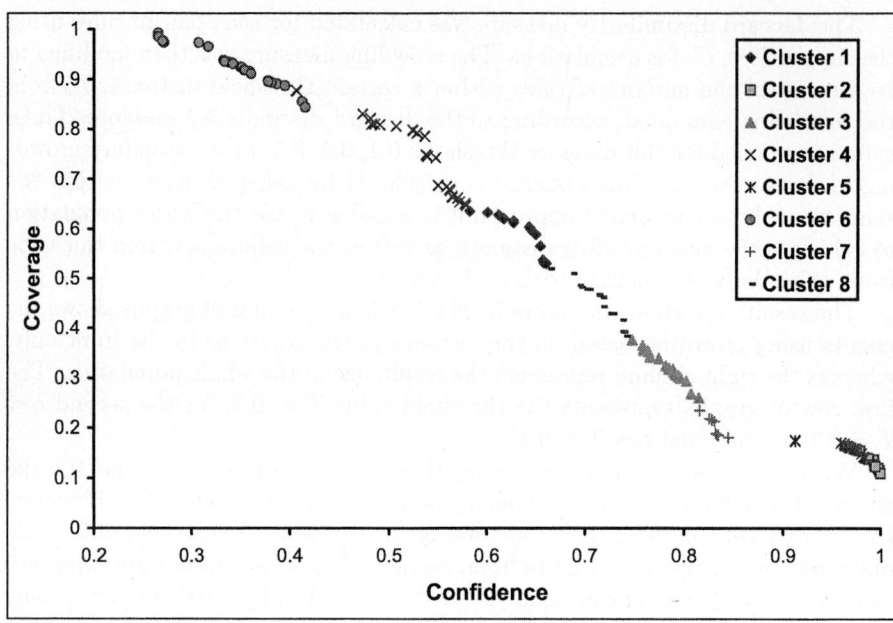

Fig. 6. Clustering of rules for Adult Database produced with dissimilarity crowding (PAM) - 8 clusters

area of the Pareto front we have managed to increase the diversity of solutions according to support sets. It seems reasonable that this is the area in which we have created diversity with our approach, as high coverage rules would apply to a high percentage of records within the population, and therefore finding alternative high coverage rules that apply to different sets of records is unlikely. Within the low coverage rules, there is obviously more scope for creating diversity as we have managed to encourage this with our new crowding measures. Hence we may now be able to present a set of rules which includes more diverse (in terms of support sets) rules of high accuracy.

6 Conclusions and Further Research

In this paper we have combined a number of innovative approaches to rule induction, exploiting the power of multi-objective metaheuristics to obtain interesting rule sets. In particular, we have experimented with the crowding mechanism in NSGA-II to improve the quality of rule sets obtained. We have also assessed the quality of rule sets obtained by using innovative approaches to cluster rules according to dissimilarity measures. This combined approach, when fully tested, may become a very powerful tool for rule induction.

We have shown that the rule sets obtained by NSGA-II in the standard implementation do not contain many cases of rules that are close in the objective space but far apart in terms of their support sets.

We have created a modification of NSGA-II by introducing the concept of rule dissimilarity in the crowding measure. This has allowed us to increase the diversity of rules in some areas of the Pareto front in terms of support sets.

The work presented here is only in its initial stages, and there is scope for extending it and improving in a number of ways. More experimentation is required to draw conclusive results. As further work, other measures of rule dissimilarity may be used to encourage diversity of the rule set. This may include considering the appearance of rules. Modifications of the NSGA-II algorithm to include our criteria may not be limited to the crowding measure, but may be more drastic using different selection criteria altogether. Experimentation with other multi-objective metaheuristics may also be beneficial.

References

1. S. Ali, K. Manganaris and R. Srikant. Partial classification using association rules. In D. Heckerman, H. Mannila, D. Pregibon, and R. Uthurusamy, editors, *Proceedings of the Third Int. Conf. on Knowledge Discovery and Data Mining*, pages 115–118. AAAI Press, 1997.
2. R. Bayardo and R. Agrawal. Constraint based rule mining in large, dense databases. *Data Mining and Knowledge Discovery Journal*, 4:217–240, 2000.
3. R. Bayardo and Agrawal R. Mining the most interesting rules. In *Proceedings of the 5th International Conference on Knowledge Discovery and Data Mining, (KDD 99)*, pages 145–152. AAAI Press, 1999.
4. L. Breiman, J. H. Friedman, R. A. Olshen, and C.J. Stone. *Classification and regression trees*. Wadsworth, Pacific Grove, CA, 1984.
5. P. Clark and T. Niblett. The CN2 induction algorithm. *Machine Learning*, 3:261–284, 1989.
6. W. Cohen. Fast effective rule induction. In *Proceedings of Twelfth International Conference on Machine Learning (ICML-95)*, pages 115–123. Morgan Kaufman, 1995.
7. Beatriz de la Iglesia, Graeme Richards, Mark S. Philpott, and Vic J. Rayward Smith. The application and effectiveness of a multi-objective metaheuristic algorithm for partial classification. *European Journal of Operational Research*, 2004, to appear.
8. Kalyanmoy Deb, Samir Agrawal, Amrit Pratab, and T. Meyarivan. A Fast Elitist Non-Dominated Sorting Genetic Algorithm for Multi-Objective Optimization: NSGA-II. In Marc Schoenauer, Kalyanmoy Deb, Günter Rudolph, Xin Yao, Evelyne Lutton, J. J. Merelo, and Hans-Paul Schwefel, editors, *Proceedings of the Parallel Problem Solving from Nature VI Conference*, pages 849–858, Paris, France, 2000. Springer. Lecture Notes in Computer Science No. 1917.
9. A. A. Freitas. On objective measures of rule surprisingness. In *Principles of Data Mining and Knowledge Discovery (Proc. 2nd European Symp., PKDD'98. Nantes, France),Lecturer Notes on Artificial Intelligence, 1510, 1-9*. 1998.
10. A. A. Freitas. On rule interestingness measures. *Knowledge-Based Systems Journal*, 12(5-6):209–315, 1999.
11. P. Jaccard. Étude comparative de la distribution florale dans une portion des Alpes et des Jura. *Bulletin de la Société Vaudoise de la Sciences Naturelles*, 37:547–579, 1901.

12. Leonard Kaufman and Peter J. Rousseuw. *Finding Groups in Data: An introduction to Cluster Analisys*. Wiley Series in probability and mathematical statistics. John Wiley and Sons Inc., 1990.
13. C. J Merz and P. M. Murphy. UCI repository of machine learning databases. Univ. California, Irvine, 1998.
14. G. Piatetsky-Shapiro. *Discovery, Analysis, and Presentation of Strong Rules*, chapter 13, pages 229–248. AAAI/MIT Press, 1991.
15. J. R. Quinlan. *C4.5: Programs for Machine Learning*. Morgan Kaufmann, San Mateo, CA, 1993.
16. A. P. Reynolds, G. Richards, B. de la Iglesia, and V. J. Rayward-Smith. Nugget clustering: A comparison of partitioning and hierarchical clustering algorithms. *TBA*, In preparation 2004.
17. A. P. Reynolds, G. Richards, and V. J. Rayward-Smith. The Application of K-medoids and PAM to the Clustering of Rules. In *Proceedings of the Fifth International Conference on Intelligent Data Engineering and Automated Learning (IDEAL'04). Lecture Notes in Computer Science No. 3177*, pages 173–178. Springer-Verlag, 2004.
18. G. Richards and V.J. Rayward-Smith. The discovery of association rules from tabular databases comprising nominal and ordinal attributes. *Intelligent Data Analysis*, 9(3), 2004.

Preliminary Investigation of the 'Learnable Evolution Model' for Faster/Better Multiobjective Water Systems Design

Laetitia Jourdan, David Corne, Dragan Savic, and Godfrey Walters

Schoolof Engineering, Computer Science and Mathematics, Harrison Building,
University of Exeter,Exeter EX4 4QF, United Kingdom
jourdan@lifl.fr, {D.W.Corne, D.A.Savic, G.A.Walters}@ex.ac.uk

Abstract. The design of large scale water distribution systems is a very difficult optimisation problem which invariably requires the use of time-expensive simulations within the fitness function. The need to accelerate optimisation for such problems has not so far been seriously tackled. However, this is a very important issue, since as MOEAs become more and more recognised as the 'industry standard' technique for water system design, the demands placed on such systems (larger and larger water networks) will quickly meet with problems of scaleup. Meanwhile, LEM (Learnable Evolution Model') has appeared in the Machine Learning literature, and provides a general approach to integrating machine learning into evolutionary search. Published results using LEM show very great promise in terms of finding near-optimal solutions with significantly reduced numbers of evaluations. Here we introduce LEMMO (Learnable Evolution Model for Multi-Objective optimization), which is a multi-objective adaptation of LEM, and we apply it to certain problems commonly used as benchmarks in the water systems community. Compared with NSGA-II, we find that LEMMO both significantly improves performance, and significantly reduces the number of evaluations needed to reach a given target. We conclude that the general approach used in LEMMO is a promising direction for meeting the scale-up challenges in multiobjective water system design.

1 Introduction

Fast optimization (in terms of using as few fitness evaluations as possible) is more and more essential when faced with real-world problems in which each fitness evaluation involves running an time-expensive simulation. However, there is of course a compromise between speeding up the optimization method, and ensuring good quality in the obtained solutions (see figure 1). One area in which this issue is paramount is in the design of large scale water distribution networks. This area contains various difficult and complex optimisation problems which invariably requires the use of time-expensive simulations within the fitness function. These problems are typically multi-objective (often trading off financial cost against requirements for pressure, speed of flow and other aspects of the

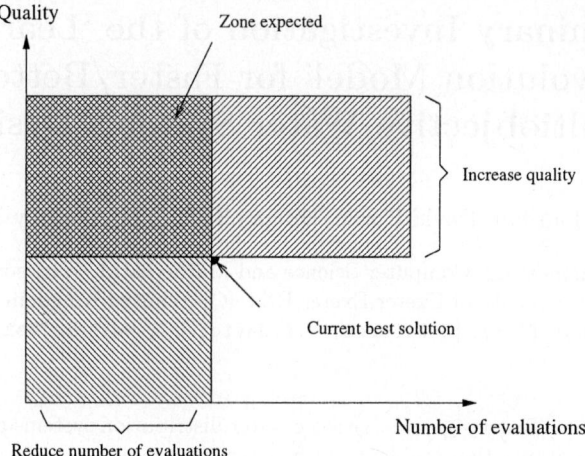

Fig. 1. Partition of the potential solutions

required design), and there is an increasing body of published research which addresses such problems, e.g.: [2, 26, 9, 23, 6, 4, 7, 1] However, the need to accelerate optimisation for such problems has not so far been seriously tackled. This is presumably because the problems involved are those of *design*, and sufficient time often exists to allow long optimisation runs before a final design is to be scrutinised and approved. However, as the quality of MOEA approaches causes them to be considered for ever larger problems, the time taken to find good solutions to such larger problems (e.g. a water design network with thousands of pumps, which is not uncommon) may turn out to be unacceptable with existing methods.

In this article, we address this scale-up problem by investigating a method based on the Learnable Evolution Model (LEM) [16, 17], which has appeared in recent years in the machine learning literature, but has so far been very little explored on real problems. The idea, which involves combining machine learning with evolutionary search, is a very generalised notion of which certain current trends in evolutionary computation (such as Estimation of Distribution Algorithms [15]) can be seen as specific instances. In particular, results to date on benchmark function optimisation problems show that LEM can save very significantly on the number of fitness evaluations needed to reach a certain target fitness. Our adaptation of this method to multiobjective optimisation, and in particular its application to water network design problems, is investigated here.

The remainder of the article is set out as follows. In section 2 we describe LEM and LEMMO (our version of this method). In section 3 we describe water systems design optimisation problems in general, and present the benchmark problems we use and briefly describe the associated fitness function simulator. In section 4, we describe the implementation of our genetic algorithm, the different variants of LEMMO and the metrics we use. Section 5 presents results from our LEMMO method, compared against NSGA-II (a well-known high-quality MOEA). Some concluding discussion is provided in section 6.

2 LEMMO

There are several publications that investigate the hybridisation of machine learning and evolutionary algorithms, e.g. Sebag [29, 27, 28, 10, 11, 12], while some works concentrate on statistical learning integrated within an EA [21, 15]. These and similar studies so far have concentrated on binary encodings and have rarely strayed from single-objective optimisation. However, a recent highly general framework for integrating machine learning and evolutionary search was proposed by Michalski, which we describe below.

2.1 LEM

The Learnable Evolution Model (LEM) [16, 17] integrates a symbolic learning component within evolutionary search; it seeks out rules (or other predictive models) explaining the differences between the better and worse performers in the population, and generates new individuals based on the templates specified in these rules. The LEM methodology has proven able to significantly improve efficiency on a variety of test problems. It is easy to describe the general shape of LEM. Given a machine learning method, and an EA, LEM consists of *Evolution* phases interrupted by *Learning* phases. During an *Evolution* phase, a normal EA is running. For the purposes of LEM, we can regard the result of one of these phases to be a collection of individuals with their associated fitnesses (i.e. all or a subset of those visited by the EA). During a *Learning* phase, this result of the most recent *Evolution* phase is used as training data by a machine learning method, which then learns a model which can discriminate between good and bad solutions. The details are highly general and there is as yet no theory concerning the best way to do this; however, in LEM work so far, a rule-induction method is used to learn a set of rules which are able to predict whether a genome's fitness is above a given threshold (a *positive* rule), and a similar set of rules is learned which capture *bad* solutions, whose fitness is therefore predicted to lie below a given threshold. These rules are then made use of in the generation of an initial population for the next *evolution* phase, and so it repeats. A particular instantiation of a LEM method involves deciding on what machine learning method to use, precisely how to use the rules to help generate the initial population of the next *Evolution* phase, and deciding when to switch between phases.

2.2 Adaptation to Multi-objective Genetic Algorithm

Our framework for incorporating LEM into MO search is described next. We first note that our framework can be used with any MO algorithm. The only requirement is that the MO algorithm maintains an up-to-date Pareto set after each generation.

We have designed LEMMO (Learnable Evolution Model for Multiobjective Optimization), which is designed to seek rules which discriminate between good and bad solutions from the multiobjective viewpoint. In LEM, this discrimination is based purely on thresholds associated with a single-objective fitness; the

```
R1: A3 <= 0 AND A7 <= 6 AND A18 <= 4
-> class bad
R2: A19 <= 4 AND A21 <= 1
-> class bad
R3: A7 > 6 AND A10<=5
-> class good
R4: A18 > 4
-> class good
```

Fig. 2. Example of produced rules by C4. and C4.5rules for continuous attributes

class of good solutions is simply called *good*, and the class of poor solutions is simply termed *bad*. In LEMMO, there are various possible ways to do this, and our Experiments section investigates some of these. In one method, for example, we only use individuals on the current approximation to the Pareto front as the *good* set, and individuals of rank three or below (in the non-dominated sorting sense) as the *bad* set; section 4 provides details of the different variants studied here.

The rules learned are then used to help create (and/or filter) new individuals, by seeking solutions that match the positive rules and don't match the negative rules. There are several ways to design and apply such a mechanism, and part of our ongoing research is to find the ideal ways of doing this for a range of water system design problems.

LEMMO is designed to use any machine learning method, but particularly attuned to using a rule induction algorithm, in order to produce rules to create new individuals. LEM publications so far have used the AQ learning algorithm. Although existing implementations, such as AQ11 [18] and AQ15 [19] handle noise with pre and post-processing techniques, the basic AQ algorithm heavily depends on specific training examples during search (the algorithm actually employs a beam search). Rather than using a sophisticated version of AQ, we decided in these initial experiments to use a rule induction algorithm with which we are more familiar, and which does not suffer from the lack of robustness to specific training examples as badly as basic AQ; we therefore used the well-known C4.5 decision tree induction algorithm [22], combined with 'C4.5rules' (which extracts rules from the resulting tree), which is provided with the C4.5 code distribution. An example of a rule generated in our application is given figure 2. The features in a rule (A3, A7 etc ...) refer to specific pipe connections in a water network, and the choices for such a feature are potential diameters for the pipe in question (see next section).

After learning a set of rules in the Learning stage of LEMMO, we generate several new matching individuals for each rule that is of the class good (called a *positive rule*). For example, an individual which matches the schema ******9**3*********** will match rule R3. Typically, only a few parts of the genotype are 'fixed' by a rule; the rest of the genotype is (in this study) filled

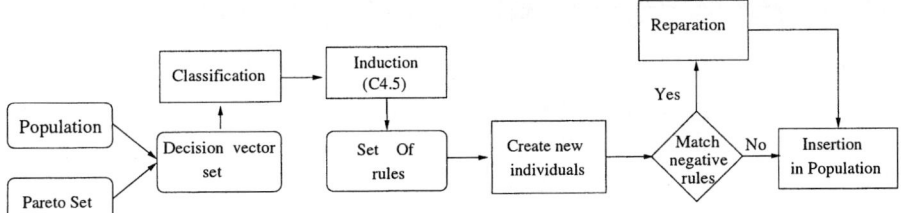

Fig. 3. Creation of new individuals using learned rules

in via one-point crossover of an arbitrary pair of selected parents, ensuring that the parts fixed by the rule are protected. Following this. we verify that the new generated individuals do not match any negative rules, repairing any individual which does. Figure 3 shows the entire process.

3 Application to Water System Design

Water distribution network modeling is used for a variety of purposes. Computer modeling of water distribution networks continues apace in the water industry as computers become increasingly powerful, and more complex systems need to be modeled. Open source software such as EPANET [24] is often used to model water systems, and indeed EPANET is what we used in the current research. In recent years, computational methods such as non-linear programming, dynamic programming and search techniques have been used to optimise the design, operation and rehabilitation of these networks, however MOEA methods, due to their general success in this arena, are becoming increasingly favoured.

Optimisation problems in this area are usually either the design of networks for new supply areas (design problems), modifying existing designs to meet new demands or other factors (rehabilitation problems), and modifying network parameters to ensure that they are accurate with respect to the real world (calibration problems). The test problems used in this article are of the first and second types respectively, and both are regularly used in the water systems research community.

We note that EPANET can be criticised for its suitability and accuracy in certain contexts. However, at this stage we feel that such problems outweighed by the fact that EPANET is freely available, and commonly used as a platform for this field of research.

3.1 NY Problem

The first test problem we use is the the New York Tunnels pipe network (NYT). The objective of the NYT problem was to determine the most economically effective design for addition to existing system of tunnels that constituted the primary water distribution system of the city of New York. Tunnel (pipe)diameters are considered as design variables. There are 15 available discrete diameters 36, 48,

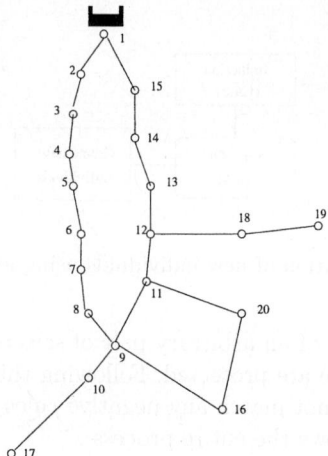

Fig. 4. The New York City problem

60, 72, 84, 96, 108, 120, 132, 144, 156, 168, 180, 192, 204 inches and one extra possible decision which is 'do nothing' option. The minimum head requirement at all nodes is fixed at 255 ft except for node 16, 17 and 1 that are 260, 272.8 and 300 ft respectively. All twenty one tunnels are considered for duplication. Supplying demand at an adequate pressure to consumers is the main constraint in the design of water distribution systems. Each pipe has a cost. The cost function is a non-linear function: $C = 1.1 \times D_{ij}^{1.24} \times L_{ij}$ in which cost C is in dollars, diameter D_{ij} is in inches, and lenght L_{ij} is in feet. A full enumeration of all possibilities would require: $16^{21} = 1.9342 \times 10^{25}$. From an optimisation perspective, the objective of the NYT problem is to modify the rehabilitated pipe diameters to meet the demands at the nodes. The current optimal solution for this is 38.64 million dollars and no pressure deficit although this can vary slightly depending on the modelling software and parameters used.

3.2 Hanoi Problem

Our second, larger test problem concerns a water distribution network in Hanoi, Vietnam, is considered in this study. The network [8] consists of one reservoir (node 1), 31 demand nodes and 34 pipes (see figure 5). The minimum pressure head required at each node is 30 m. The cost of commercially available pipe sizes (12, 16, 20, 24, 30, 40; in inches) was calculated using the equation provide in [8]): $C = 1.1 \times D_{ij}^{1.5} \times L_{ij}$.

4 LEMMO for Water Distribution Systems

Our comparative MOEA is NSGA-II [3]. That is, the comparative algorithm is NSGA-II, while the variants of LEMMO are all NSGA-II hybridised with

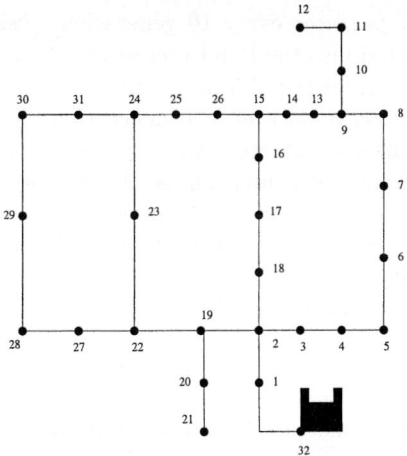

Fig. 5. Hanoi problem

phases in which rules are learned (by the process described above) leading to the generation of an initial population for the next phase of evolution (i.e. NSGA-II).

4.1 Implementation and Parameters

In all cases, the encoding is straightforward, with one individual containing one gene for each pipe, representing a choice of diameter for that pipe. In all cases we use as operator: one-point crossover and single-gene 'random new allele' mutation.

The population size is always 100, the crossover rate is 0.9, the mutation rate is 0.9, and in every trial of every experiment the maximum number of evaluations was set to 250,000.

4.2 LEMMO Variants

For all the explored schemes, the *Learning* is executed on decision vectors (here the size of the pipe) and the value of the objectives is used for the classification task. During the classification phase, each decision vector is put into a class (*bad or good*) according to different strategies that are explained below. This classification is re-done before each call of the induction algorithm in order to avoid conflict between rules.

The different LEMMO variants explored are:

- LEMMO-1: run *Learning* when there has been no change to the approximation to the Pareto front for two successive generations.
- LEMMO-fix1: run *Learning* every 10 generations, using the initial population of the previous *Evolution* phase as the *bad* set and the final population of that phase as the *good* set.

- LEMMO-fix2: run *Learning* every 10 generations, using the 20 individuals most recently inserted into the Pareto set as the *good* set, and the remaining individuals of the current population as the *bad* set.
- LEMMO-fix3: run *Learning* every 10 generations, using randomly chosen individuals from the current approximation to the Pareto set as the *good* set, and the remaining individuals of the current population as the *bad* set.
- LEMMO-fix4: run *learning* every 10 generations, using the best 30% of individuals found so far on one of the objectives (randomly chosen in each learning phase) as the *good* set, and the worst 30% of individuals found so far on that objective as the *bad* set.

4.3 Quality Measures

As suggested in [14], the S metric [30] and the R metrics [13] are generally better than other metrics in terms of their properties with respect to sensitivity to scaling, compatibility with common-sense 'outperformance' relations [13], cycle-inducing properties (i.e. some metrics may rate set A as better than set B, set B as better than set C and set C as better than set A, which is clearly undesirable), and computational overhead. We mainly use the S metric in this study (although it has a rather high computational overhead, this is not such a problem when $k = 2$ and the non-dominated sets are not overly large), but also supplement our results with the use of the R_{1R} metric. Regarding the S metric (details in Zitzler in [30]), its implementation requires a reference point Z_{ref}. In this work we calculate Z_{ref} by using the worst possible solutions in terms of both objectives. For the cost objective, we use the cost arising from using the largest (most expensive) diameter for each pie, and for head deficit objective we take the value resulting from using the minimum diameter for each pipe. When we use the R_{1R} metric (details in [13] and summarised in [14]), this requires a reference non-dominated set, for which we choose an arbitrary result from an NSGA run.

5 Results

In our first set of experiments the aim was to explore a number of parameter variants of our basic LEMMO method, in order to establish a feel for the balance required between learning and evolutionary search on the smaller of the two test problems. Note that even though the NYT and Hanoi water system test problems are relatively small in terms of design variables (compared to other real water system design problems), the computational cost of fitness evaluation remains very great, so we needed to limit the number of experiments. We therefore performed our initial experiments on the smaller of the two problems (NYT), and used a second set of experiments to begin to explore scale-up properties by testing the best variant from the first set of experiments on the larger problem (Hanoi).

5.1 Results on the NYT Problem

We did 30 trials each on the NYT problem with each of the five LEMMO variants and NSGA-II. The obtained approximate Pareto fronts are highly populated. Table 1 gives the S metric statistics, showing the percentage improvement (positive values) of the given LEMMO scheme over the corresponding value achieved by NSGA-II. For example, the median S metric value achieved by LEMMO-fix4 was 9.12% better than the median achieved by NSGA-II. We can see that the LEMMO methods, particularly LEMMO-fix4, led to better final approximate Pareto set quality than NSGA-II. The standard deviation column shows the percentage improvement of the given LEMMO scheme over the standard deviation for NSGA-II. It is notable that in each case the standard deviation is considerably reduced, suggesting that the LEMMO methods not only achieve better quality results, but do so more reliably than NSGA-II. Overall, LEMMO-fix4 seems to be the clear winner. For both LEMMO-fix3 and LEMMO-fix4, the improvement in the median and mean was found to be significant at a 95% confidence level based on a randomisation test [5]. For the $R1_R$ metric (see table 3), we observe that all the schemes are better than NSGA-II.

Table 2 presents statistics on the number of evaluations required to obtained the final Pareto front over the 30 runs for each scheme. That is, the number of evaluations at which further improvements to the Pareto front ceased; this is essentially a convergence measure. Meanwhile, table 2 shows the comparison of this measure between each of the LEMMO schemes and NSGA-II. For example, the median convergence time for LEMMO-fix1 was 16.28% faster than the median convergence time for NSGA-II. Generally, the LEMMO schemes are not always faster than NSGA-II, however this must of course be traded off against the fact that (as indicated above) the occasional slower convergence does lead to better results than NSGA-II. However, LEMMO-fix4 again turns out to be the best in this sense, achieving reliably faster convergence, and to better results. It is worth examining figure 6, which illustrates the evolution of the S metric value over time for LEMMO-fix1 and for NSGA-II. The LEMMO scheme is always ahead of NSGA-II in quality for a given time.

Table 4 presents the variation of the number of evaluations required to obtained different values of the normalized S metric between the scheme and NSGA-II over 30 runs. The first value 0.955 is chosen for this problem as it show the beginning of the convergence of the algorithm for this problem. The second value 0.968 is chosen as it is rather at the end of the convergence.

All the schemes are faster than NSGA-II to obtained a normalized Smetric greater than 0.968. We notice that LEMMO-fix4 is the less expensive scheme in comparison of NGSAII in median (-16.66%) and in mean (-20.58%) to obtain this value.

If we have a look to the solutions generated by LEMMO-fix4, we can notice that the best mono-objective solution obtained in [20, 25] is in all the obtained Pareto sets. The size of the pipes is the same than in [25].

Table 1. Quality assessment S metric (S(Scheme) - S(NSGAII))/S(Scheme)

Scheme	S Median	S Mean	S Min	S Max	S Std Dev.
LEMMO-1	7.83%	4.9%	7.3%	0.26%	-33.32%
LEMMO-fix1	8.28%	5.44%	3.4%	0.26%	-57.6%
LEMMO-fix2	8.19%	5.5%	2.04%	0.18%	-25.9%
LEMMO-fix3	8.14%	6.75%	8.84%	**0.42%**	-58.58%
LEMMO-fix4	**9.12%**	**6.8%**	**7.77%**	0.27%	-22.19%

Table 2. Statistic on number of evaluations for finding the final Pareto front over 30 runs for the different tested schemes (T(Scheme)-T(NSGAII)/T(Scheme))

Scheme	Median	Mean	Min	Max	Std Dev.
LEMMO-1	+14.07%	+6.02%	+29.8%	-1.84%	+4.11%
LEMMO-fix1	**-16.28%**	**-11.81%**	-10.04%	-0.93%	+1.26%
LEMMO-fix2	-8.97%	-4.54%	+20.69%	-0.00%	+5.22%
LEMMO-fix3	-14.69%	-7.38	-14.41%	-0.36%	+8.53%
LEMMO-fix4	-16.14 %	-8.45 %	+25.40%	-0.44%	-21.74%

Table 3. Quality assessment $R1_R$ metric with a front of NSGA-II for NY problem

Scheme	$R1_R$ Median	$R1_R$ Mean	$R1_R$ min	$R1_R$ max	$R1_R$ Std Dev.
NSGA-II	0.39172	0.44780	0.37525	0.62271	0.09460
LEMMO-1	0.38323	0.43119	0.35229	0.64671	0.09236
LEMMO-fix1	0.38024	0.41153	0.35229	0.64271	0.09044
LEMMO-fix2	0.38025	0.39476	0.35229	0.51597	0.05348
LEMMO-fix3	0.38124	0.40108	0.35229	0.65469	0.07496
LEMMO-fix4	0.38224	0.39680	0.36527	0.51497	0.04350

In tables 5 and 6, we show for the different possibilities of the parameters for each scheme the quality of the metrics. We observe that the choice of 10 was a good one.

5.2 Results on the Hanoi Problem

Having found that that the LEMMO schemes, but particularly LEMMO-fix4, seem capable of improving speed (and quality) of results, we did further experiments with LEMMO-fix4 on a larger problem, the Hanoi network [8,7]. The larger size of this problem (and other time constraints) mean that at the moment we can only report on results of 10 trial runs on this problem for each of NSGA-II and LEMMO-fix4. Table 7 shows the results in terms of the $R1_R$ metric (in this case, lower values are better). It is clear that LEMMO-fix4 achieved far better quality results than NSGA-II on this larger problem, and we find the comparisons between mean and median are significant at a 90% confidence level based on a randomisation test [5].

We also compared LEMMO-fix4 and NSGA-II on this problem in terms of their speed in attaining certain chosen values of the S metric (we chose the S

Table 4. Statistic on number of evaluation for finding a normalized Smetric greater than 0.955 and 0.968 over 30 runs for the different tested schemes (T(Scheme)-T(NSGA-II)/T(Scheme)) on NY problem

Scheme	$S_{norm} > 0.955$ Median	Mean	$S_{norm} > 0.968$ Median	Mean
LEMMO-1	0%	+16.16%	-16%	-6.8%
LEMMO-fix1	+10%	+10%	-10.10%	-5.01%
LEMMO-fix2	+16.66%	+16.66%	-7.8%	+2.6%
LEMMO-fix3	+6.25%	+14.28%	-13.51%	-10.81%
LEMMO-fix4	+10 %	0%	-16.66 %	-20.58%

Table 5. Impact of the parameters: $R1_R$ metric with a front of NSGA-II on NY problem

Scheme (parameter)	$R1_R$ Median	$R1_R$ Mean	$R1_R$ Std Dev.
LEMMO-fix1 (5)	0.38124	0.38503	0.02553
LEMMO-fix1 (10)	0.38024	0.41152	0.09044
LEMMO-fix1 (20)	0.38024	0.42664	0.12099
LEMMO-fix4 (5)	0.40118	0.47329	0.07234
LEMMO-fix4 (10)	0.38224	0.39681	0.0435
LEMMO-fix4 (20)	0.37924	0.39654	0.07272

Table 6. Statistic on number of evaluation for finding a normalized Smetric greater than 0.955 and 0.968 over 30 runs for the different tested schemes (T(Scheme)-T(NSGA-II)/T(Scheme)) on NY problem

Scheme (parameter)	$S_{norm} > 0.955$ Median	Mean	$S_{norm} > 0.968$ Median	Mean
LEMMO-fix1 (5)	+21.05%	+21.21%	+2.31	+4.65%
LEMMO-fix1 (10)	+10%	+10%	-10.10%	-5.01%
LEMMO-fix1 (20)	+16.66%	+16.66%	+13.51%	+2.38%
LEMMO-fix4 (5)	+6.25%	+16.66%	-14.28%	-7.89%
LEMMO-fix4 (10)	+10 %	0%	-16.66 %	-20.58%
LEMMO-fix4 (20)	+6.89%	+6.66%	-2.43%	+2.43%

Table 7. Quality assessment $R1_R$ metric with a front of NSGA-II for Hanoi problem on 10 experiments

Scheme	$R1_R$ Median	$R1_R$ Mean	$R1_R$ min	$R1_R$ max	$R1_R$ Std Dev.
NSGA II	0.05439	0.06871	0.00199	0.02745	0.07335
LEMMO-fix4	0.01247	0.01854	0.00598	0.03992	0.01261

metric for this for pragmatic reasons concerning our current implementation). We found that to obtain 0.711, LEMMO-fix4 required on average 27.5% fewer evaluations than NSGA II. Meanwhile, LEMMO-fix4 obtains the final S metric 0.7488 in 6 out of 10 experiments, but this is never achieved by NSGA-II within the 250,000 evaluations limit.

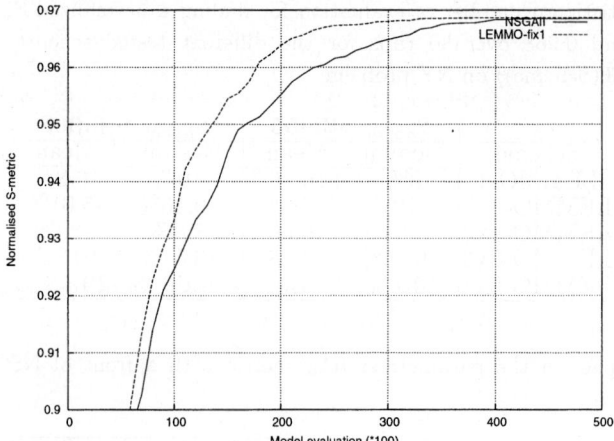

Fig. 6. Example of evolution of the normalized S-metric during the evolution against model evaluation on NY Problem

6 Conclusion

We have reported on initial research into ways to find good solutions to multiobjective water systems design problems in fewer evaluations, with the ultimate goal of enabling multiobjective EAs to be applied regularly to large scale such problems. In particular, inspired by impressive speedup results reported on other (but single-objective) problems we have started to explore whether the LEM method [16, 17] can be successfully used in this domain. Our adaptation of this technique, called LEMMO, and in particular the variant in which the learning method focuses on single-objective values (but alternates between objectives), has indeed been found to significantly improve both the speed and quality of solutions to two benchmark water systems design problems. The fact that LEMMO-fix4 was the better variant suggests that it is very hard to capture, in simple rules, characteristics of solutions which relate to their distance from the Pareto front (which was required of the other LEMMO variants); this indeed seems intuitively fair. We therefore recommend alternating single-objective based learning in such schemes applied to multiobjective problems, although of course the idea of *learning* correlates of distance from the Pareto front remains an intriguing research area.

We were particularly interested in the LEM framework as a way of reducing the required number of evaluations for these problems, however the achievement of better quality results was a welcome added bonus. Indeed it partly makes up for the fact that we would prefer to see more dramatic improvements in speed to achieve a given target. However, we feel that significantly further reduced relative numbers of evaluations will be achieved on larger problems (note that LEMMO-fix4 speedup on the NYT problem was around 16%, and on the larger Hanoi problem the speedup was around 28%), where the 'focussing' effect of the *Learning* phase becomes relatively more imperative. Meanwhile, considering that there is an immense space of possibilities for the way in which the *Learning* phase in LEMMO can be configured (e.g. different learning methods, different ways of generating new solutions from the learned models, and so on), it is very encouraging that these first few attempts have already led to promising results.

Acknowledgment

We are grateful to Joshua Knowles for supplying some of the metrics code, to the EPSRC for the support of Dr Jourdan and to the anonymous referees for their helpful comments.

References

1. P. B. Cheung, L.F.R. Reis, K.T.M. Formiga, F.H. Chaudhry, and W.G.C. Ticona. Multiobjective evolutionary algorithms applied to the rehabilitation of a water distribution system: a comparative study. In C.M. Fonseca, P.J. Fleming, E. Zitzler, K. Deb, and L. Thiele, editors, *Evolutionary Multi-Criterion Optimization (EMO-2003)*, pages 662–676, Springer-Verlag, LNCS 2632, 2003.
2. S.E. Cierniawski, J.W. Eheart, and S. Ranjithan. Using genetic algorithms to solve a multiobjective groundwater monitoring problem. *Water Resources Research*, 31(2):399–409, 1995.
3. K. Deb, S. Agrawal, A. Pratab, and T. Meyarivan. A fast elitist non-dominated sorting genetic algorithm for multi-objective optimization: Nsga-II. In *Proceedings of the Parallel Problem Solving from Nature VI Conference*, pages 849–858, 2000.
4. J.L. Dorn. and S.R. Ranjithan. Evolutionary multiobjective optimization in watershed quality management. In C.M. Fonseca, P.J. Fleming, E. Zitzler, K. Deb, and L. Thiele, editors, *Evolutionary Multi-Criterion Optimization (EMO-2003)*, pages 692–706, Springer-Verlag, LNCS 2632, 2003.
5. E. Edgington. *Randomization tests*. Marcel Dekker Inc., 1980.
6. M. Erickson, A.S. Mayer, and J. Horn. The niched pareto genetic algorithm 2 applied to the design of groundwater remediation systems. In E. Zitzler, K. Deb, L. Thiele, C. Coello Coello, and D. Corne, editors, *Evolutionary Multi-Criterion Optimization (EMO-2001)*, pages 681–695, Springer-Verlag, LNCS 1993, 2001.
7. K.T.M. Formiga, F.H. Chaudhry, .B. Cheung, and L.F.R. Reis. Optimal design of water distribution system by multiobjective evolutionary methods. In C.M. Fonseca, P.J. Fleming, E. Zitzler, K. Deb, and L. Thiele, editors, *Evolutionary Multi-Criterion Optimization (EMO-2003)*, pages 677–691, Springer-Verlag, LNCS 2632, 2003.
8. O. Fujiwara and D.B. Tung. A two-phase decomposition method for optimal design of looped water distribution networks. *Water Resources Research*, 26(4):539–549, 1990.
9. D. Halhal, G.A. Walters, D. Ouazar, and D. Savic. Multiobjective improvement of water distribution systems using a structured messy genetic algorithm approach. *ACSE Journal of Water Resources Planning and Management*, 123(3):137–146, 1997.

10. H. Handa, T. Horiuchi, O. Katai, and M. Baba. A novel hybrid framework of coevolutionary ga and machine learning. *International Journal of Computational Intelligence and Applications*, 2002.
11. H. Handa, T. Horiuchi, O. Katai, T. Kaneko, T. Konishi, and M. Baba. Coevolutionary ga with schema extraction by machine learning techniques and its application to knapsack problems. In *Proceedings of the 2001 Congress on Evolutionary Computation CEC2001*, pages 1213–1219, COEX, World Trade Center, 159 Samseong-dong, Gangnam-gu, Seoul, Korea, 27-30 May 2001. IEEE Press.
12. H. Handa, T. Horiuchi, O. Katai, T. Kaneko, T. Konishi, and M. Baba. Fusion of coevolutionary ga and machine learning techniques through effective schema extraction. In *Proceedings of the Genetic and Evolutionary Computation Conference (GECCO-2001)*, page 764, San Francisco, California, USA, 7-11 July 2001. Morgan Kaufmann.
13. M.P. Hansen and A. Jaszkiewicz. Evaluating the quality of approximations to the non-dominated set. Technical Report TR-IMM-REP-1998-7, Technical University of Denmark, 1998.
14. J. D. Knowles and D. W. Corne. On metrics for comparing non-dominated sets. In IEEE Service Center, editor, *Congress on Evolutionary Computation (CEC'2002)*, volume 1, pages 711–716, Piscataway, New Jersey, May 2002.
15. P. Larranaga and J. Lozano. *Estimation of distribution algorithms: a new tool for evolutionary computation*. Kluwer Academic Publishers, 2001.
16. R.S. Michalski. Learnable evolution model: Evolutionary processes guided by machine learning. *Machine Learning*, 38(1–2):9–40, 2000.
17. R.S. Michalski, G. Cervon, and K.A. Kaufman. Speeding up evolution through learning: Lem. In *Intelligent Information Systems 2000*, pages 243–256, 2000.
18. R.S. Michalski and J.B. Larson. Selection of most representative training examples and incremental generation of vl1 hypothesis: The underlying methodology and the descriptions of programs esel and aq11. Technical Report Report No. 867, Urbana, Illinois: Department of Computer Science, University of Illinois, 1978.
19. R.S. Michalski, I. Mozetic, J. Hong, and N. N. Lavrac. he multipurpose incremental learning system aq15 and its testing application to three medical domains. In *Proc. of the Fifth National Conference on Artificial Intelligence*, pages 1041–1045. PA: Morgan Kaufmann, 1986.
20. L.J. Murphy, A.R. Simpson, and G.C Dandy. Pipe network optimisation using an improved genetic algorithm. Technical report, Dept of civil and environment engineering, University of Adelaide, Australia., 1993.
21. M. Pelikan, D.E. Goldberg, and E. Cantu-Paz. Boa: The bayesian optimization algorithm. In W. Banzhaf, J. Daida, A.E. Eiben, M.H. Garzon, V. Honavar, M. Jakiela, and R.E. Smith, editors, *Proceedings of the genetic and evolutionary computation - GECCO-99*, pages 525–532, Morgan Kaufmann, 1999.
22. J. Ross Quinlan. *C4.5: programs for machine learning*. Morgan Kaufmann Publishers Inc., 1993.
23. P.M. Reed, B.S. Minsker, and D.E. Goldberg. A multiobjective approach to cost-effective long term groundwater monitoring using an elitist nondominated sorting genetic algorithm with historical data. *Journal of Hydroinformatics*, 3(2):71–89, 2001.
24. L.A. Rossman. Epanet, users manual. Technical report, U.S. Envi. Protection Agency, Cincinnati, Ohio, 1993.
25. D. Savic and G. Walters. Genetic algorithms for least-cost design of water distribution networks. *Journal of Water Resources Planning and Management*, 123(2):67–77, 1997.

26. D. Savic, G.A. Walters, and M. Schwab. Multiobjective genetic algorithms for pump scheduling in water supply. In D. Corne and J. Shapiro, editors, *AISB Workshop on evolutionary computation - selected papers*, pages 227–236, Springer-Verlag, LNCS 1305, 1997.
27. M. Sebag, C. Ravise, and M. Schoenauer. Controlling evolution by means of machine learning. In *Evolutionary Programming*, pages 57–66, 1996.
28. M. Sebag and M. Schoenauer. Controlling crossover through inductive learning. In Yuval Davidor, Hans-Paul Schwefel, and Reinhard Männer, editors, *Parallel Problem Solving from Nature – PPSN III*, pages 209–218, Berlin, 1994. Springer.
29. M. Sebag, M. Schoenauer, and C. Ravise. Toward civilized evolution: Developing inhibitions. In Thomas Bäck, editor, *Proc. of the Seventh Int. Conf. on Genetic Algorithms*, pages 291–298, San Francisco, CA, 1997. Morgan Kaufmann.
30. E. Zitzler. Evolutionary algorithms for multiobjective optimization: Methods and applications. Master's thesis, Swiss federal Institute of technology (ETH), Zurich, Switzerland, november 1999.

Particle Evolutionary Swarm for Design Reliability Optimization

Angel E. Muñoz Zavala, Enrique R. Villa Diharce,
and Arturo Hernández Aguirre

Center for Research in Mathematics (CIMAT),
Department of Computer Science,
A.P. 402, Guanajuato, Gto. CP 36000, México
{aemz, villadi, artha}@cimat.mx

Abstract. This papers proposes an enhanced Particle Swarm Optimization algorithm with multi-objective optimization concepts to handle constraints, and operators to keep diversity and exploration. Our approach, PESDRO, is found robust at solving redundancy and reliability allocation problems with two objective functions: reliability and cost. The approach uses redundancy of components, diversity of suppliers, and incorporates a new concept called *Distribution Optimization*. The goal is the optimal design for reliability of coherent systems. The new technique is compared against algorithms representative of the state-of-the-art in the area by using a well-known benchmark. The experiments indicate that the proposed approach matches and often outperforms such methods.

1 Introduction

The reliability of any device is very important for manufacturers and users. Larger reliability of the final product is desired but, the consequent rise in production cost has negative effects on the user's budget. Therefore, the design reliability optimization problem is phrased as reliability improvement at a minimum cost. The common sense perception of reliability is the absence of failures. Therefore, reliability is sometimes referred to as "quality in time dimension", because is determined by the failures that may or may not occur to the product during its life span. The design reliability problem is hard and challenging, mainly due to the interaction of many subsystems whose conflicting local goals must contribute to the overall performance. New product design involves the specification of performance requirements, the evaluation and selection of components that perform some defined function, and the determination of the system level architecture.

The problems of interest to us are the redundancy and reliability allocation problems. There are two conflicting goals in them: reliability, and cost. The allocation problem is characterized by a large combinatorial search space, ruled by multiple design constraints.

Several optimization approaches have previously been used to solve the reliability allocation problem [1]. We introduce a new approach based upon the

Particle Swarm Optimization (PSO) paradigm which was originally proposed by Kennedy and Eberhart [2]. Our approach, based on multi-objective optimization concepts (treats constraints as objective functions), includes: a selection criteria based on feasibility rules; a local-best PSO with ring topology organization; and two perturbation operators aimed to keep diversity.

The remainder of this paper is organized as follows. In Section 2, we introduce the problem of interest. Section 3 introduces our proposed approach. In Section 4, we describe a benchmark of 3 test functions. Section 5 provided a comparison of results with respect to techniques representative of the state-of-the-art in the area. Finally, our conclusion and future work are provided in Section 6.

2 System Reliability Optimization

As noted before, the redundancy and allocation problem deals with the optimization of reliability and cost. Frequently, this problem is described as optimal reliability design, subject to cost and weight constraints. Thus, cost and weight constraints are also optimized during the process. The optimization criterion could be posed in two different forms:

- Maximization of system reliability, subject to cost and weight constraints.
- Minimization of system cost, subject to reliability and weight constraints.

The reliability of a product at time t, $R(t)$, is the probability that it works like foresaw, during the time interval $(0, t]$; under several operational conditions and environment. The distribution of the time to failure of a product, determines its $R(t)$ reliability at time t.

$$R(t) = 1 - F(t) = P(T > t) \qquad t > 0 \qquad (1)$$

where T is the time to failure and $F(t)$ is the cumulative distribution function. $F(t)$ thus denotes the probability that the unit fails within the time interval $(0, t]$.

The reliability of a product with several components is calculated using its structure function. Figure 1 shows the three kind of structures used to model the system's components of this paper.

Each system has its own structure function. The general structure functions for series and parallel structures are shown in Equations 2 and 3, where n is the number of components. The Equation 4 presents the structure function of a k-out-of-n system with n identical components, where k is the minimum number of components required for a system to work.

- Series

$$R_S = \prod_{i=1}^{n} R_i \qquad (2)$$

- Parallel

$$R_S = 1 - \prod_{i=1}^{n}(1 - R_i) \qquad (3)$$

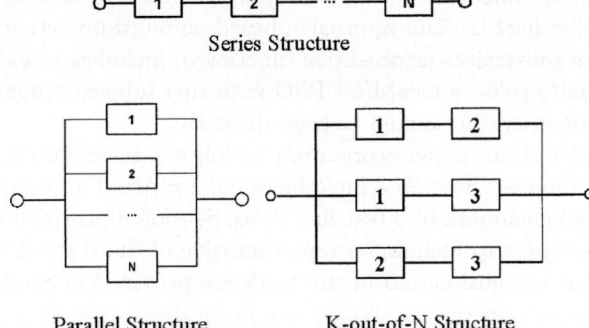

Fig. 1. Structures to model system components

- K-out-of-N

$$R_S = \sum_{i=k}^{n} \binom{n}{i} (R)^i (1-R)^{n-i} \quad (4)$$

The diversity of structures, resource constraints, and options for reliability improvement has led to the construction and analysis of several optimization models. We are interested in two general cases: redundancy allocation problem and reliability allocation problem. For the first case, there is a set of discrete components to choose from, whose characteristics are known (reliability or distribution function, cost, weight, etc.). The objectives of this combinatorial problem are the selection of components to use, and the corresponding redundancy levels. The redundancy allocation problem has been shown NP-hard by Chern [3]. For the second case, the component reliability or a vector of distribution parameters is treated as the design variables, and component's cost is a predefined increasing function of the component reliability.

As noted, our design optimization problem may appear in two forms:

P1

$$\text{Find } \mathbf{R} \text{ which maximize } R_s \quad (5)$$

subject to:

$$\sum_{i=1}^{n} C_i \leq c \quad (6)$$

$$\sum_{i=1}^{n} W_i \leq w \quad (7)$$

P2

$$\text{Find } \mathbf{R} \text{ which minimize } C_s \quad (8)$$

subject to:

$$R_s \geq r \quad (9)$$

$$\sum_{i=1}^{n} W_i \leq w \quad (10)$$

where $\boldsymbol{R}$ is the reliability vector of each component, $\boldsymbol{R} = (R_1, R_2, \ldots, R_n)$, C_i and W_i are the cost and the weight of the ith component, n is the number of components, and c, w, r are constants: $c \geq 0$, $w \geq 0$, and $0 \leq r \leq 1$.

There are several assumptions we must consider to solve the redundancy - reliability allocation problem:

- The state of the components and the system have only two options: good or bad.
- Failures of components are independent events.
- Failed components do not damage the system.
- Failed components are not repaired.
- The component's reliability or distribution function is known.

All redundancy and reliability allocation problems meeting these assumptions are denominated *coherent systems*. A system of components is said to be coherent if all its components are relevant and the state of the system is nondecreasing in each argument. One might get the impression that all systems of interest must be coherent, but this is not the case.

A component i is irrelevant if for all combinations of the state components, the state of the ith component does not affect the state of the system.

Here we will assume that a system will not run worse than before when we replace a component in failed state with one that is operating correctly. This is the same as requiring that the state of the system is nondecreasing in each argument.

3   The PESDRO Algorithm

A solution algorithm for this problem is based on PSO search. The PSO algorithm is a population-based search algorithm based on the simulation of the social behavior of birds within a flock. In PSO, individuals, referred to as particles, are "flown" through a hyperdimensional search space. PSO is a kind of symbiotic cooperative algorithm, because the changes to the position of particles within the search space are based on the social-psychological tendency of individuals to emulate the success of other individuals.

The feature that drives PSO is social interaction. Individuals (particles) within the swarm learn from each other, and based on this shared knowledge tend to become more similar to their "better" neighbors. A social structure in PSO is determined through the formation of neighborhoods. These neighborhoods are determined through labels attached to every particle in the flock (so it is not a topological concept). Thus, the social interaction is modeled by spreading the influence of a "global best" all over the flock as well as neighborhoods are influenced by the best neighbor and their own past experience.

Figure 2 shows some neighborhood structures that have been proposed and studied [4]. Our approach, PESDRO, adopts the **ring** topology. In the ring organization, each particle communicates with its n immediate neighbors. For instance, when $n = 2$, a particle communicates with its immediately adjacent neighbors as illustrated in Figure 2(b). The neighborhood is determined through

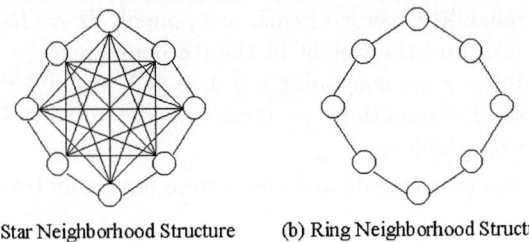

(a) Star Neighborhood Structure (b) Ring Neighborhood Structure

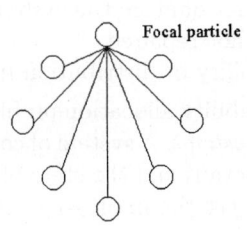

(c) Wheel Neighborhood Structure

Fig. 2. Neighborhood structures for PSO [4]

an index label assigned to all individuals. This version of the PSO algorithm is referred to as *lbest* (LocalBest). It should be clear that the ring neighborhood structure properly represents the LocalBest organization. It has the advantage that a larger area of the search space is traversed, favoring search space exploration (although convergence has been reported slower) [5,6].

PSO-LocalBest has been reported to excel over other topologies when the maximum velocity is restricted. PESDRO's results with and without restricted velocity reached similar conclusions noted by Franken and Engelbretch [7], thus, PESDRO algorithm incorporates this feature. Figure 3 shows the standard PSO algorithm adopted for our approach, where X_L and X_U are the limits of the search space, and n is the population size. The pseudo-code of **LocalBest** function is shown in Figure 4.

Constraint handling is embedded into the selection operator, and described by "feasibility and dominance" rules. These rules are: 1) given two feasible particles, pick the non-dominated one; 2) if both particles are infeasible, pick the particle with the lowest sum of constraint violation, and 3), given a pair of feasible and infeasible particles, the feasible particle wins.

The *speed* vector drives the optimization process and reflects the socially exchanged information. Figure 5 shows the pseudo-code of **Speed** function, where $c1 = 0.1$, $c2 = 1$, and w is the inertia weight. The inertia weight controls the influence of previous velocities on the new velocity.

3.1 Perturbation Operators

PESDRO algorithm makes use of two perturbation operators to keep diversity and exploration. PESDRO has three stages; in first stage the standard PSO

```
P_0 = Rand(X_L, X_U)
F_0 = Fitness ( P_0 )
PBest_0 = P_0
FBest_0 = F_0
Do While i¡MaxGenerations
  LBest_i = LocalBest ( PBest_i, FBest_i )
  S_i = Speed ( S_i, P_i, PBest_i, LBest_i )
  P_{i+1} = P_i + S_i
  F_{i+1} = Fitness ( P_{i+1} )
  For k = 0 To n
    < PBest_{i+1}[k], FBest_{i+1}[k] > = Best ( FBest_i[k], F_{i+1}[k] )
  End For
End Do
```

Fig. 3. Pseudo-code of PSO algorithm with local best

```
For k = 0 To n
  LBest_i[k] = Best(FBest_i[k-1], FBest_i[k+1])
End For
```

Fig. 4. Pseudo-code of **LocalBest**($PBest_i, FBest_i$)

```
For k = 0 To n
  For j = 0 To d
    r1 = c1 * U(0, 1)
    r2 = c2 * U(0, 1)
    w = U(0.5, 1)
    S_i[k, j] = w * S_i[k, j] +
                r1 * (PBest_i[k, j] - P_i[k, j]) +
                r2 * (LBest_i[k, j] - Pbest_i[k, j])
  End For
End For
```

Fig. 5. Pseudo-code of **Speed**(S_i, P_i, $PBest_i$, $LBest_i$)

algorithm [6] is performed, then the perturbations are applied in the next two stages.

The main algorithm of PESDRO is shown in Figure 6.

The goal of the second stage is to add a perturbation in a way similar to the so called "reproduction operator" found in *differential evolution* algorithm. This perturbation, called C-Perturbation, is applied all over the flock to yield a set of temporal particles $Temp$. Each member of the $Temp$ set is compared with the corresponding (father) member of $PBest_{i+1}$, so the perturbed version replaces the father if it has a better fitness value. Figure 7 shows the pseudo-code of the **C-Perturbation** operator.

In the third stage every vector is perturbed again so a particle could be deviated from its current direction as responding to external, maybe more promis-

```
P₀ = Rand(X_L, X_U)
F₀ = Fitness ( P₀ )
PBest₀ = P₀
Do While i¡MaxGenerations
  LBest_i = LocalBest ( PBest_i, FBest_i )
  S_i = Speed ( S_i, P_i, PBest_i, LBest_i )
  P_{i+1} = P_i + S_i
  F_{i+1} = Fitness ( P_{i+1} )
  For k = 0 To n
    < PBest_{i+1}[k], FBest_{i+1}[k] > = Best ( FBest_i[k], F_{i+1}[k] )
  End For
  Temp = C − Perturbation (P_{i+1} )
  FTemp = Fitness ( Temp )
  For k = 0 To n
    < PBest_{i+1}[k], FBest_{i+1}[k] > = Best ( PBest_{i+1}[k] , FTemp[k] )
  End For
  Temp = M − Perturbation (P_{i+1} )
  FTemp = Fitness ( Temp )
  For k = 0 To n
    < PBest_{i+1}[k], FBest_{i+1}[k] > = Best ( PBest_{i+1}[k] , FTemp[k] )
  End For
  P_i = P_{i+1}
End Do
```

Fig. 6. Main algorithm of *PESDRO*

```
For k = 0 To n
  For j = 0 To d
    r = U(0, 1)
    p1 = Random(n)
    p2 = Random(n)
    p3 = Random(n)
    Temp[k, j] = P_{i+1}[p1, j] + r * (P_{i+1}[p2, j] - P_{i+1}[p3, j])
  End For
End For
```

Fig. 7. Pseudo-code of **C − Perturbation**(P_{i+1})

sory, stimuli. This perturbation is performed with some probability on each dimension of the particle vector, and can be explained as the addition of random values to each particle component. The perturbation, called M-Perturbation is applied to every particle in the current population to yield a set of temporal particles *Temp*. Again, as for C-Perturbation, each member of *Temp* is compared with its corresponding (father) member of the current population, and the better one wins. Figure 8 shows the pseudo-code of the **M-Perturbation** operator. The perturbation is performed with probability $p = 1/d$, where d is the dimension of the decision variable vector.

```
For k = 0 To n
  For j = 0 To d
    r = U(0, 1)
    If r ≤ 1/d Then
      Temp[k, j] = Rand(LI, LS)
    Else
      Temp[k, j] = P_{i+1}[k, j]
  End For
End For
```

Fig. 8. Pseudo-code of $M - Perturbation(P_{i+1})$

These perturbations, in differential evolution style, have the advantage of keeping the self-organization potential of the flock as no separate probability distribution needs to be computed [8]. Zhang and Xie also try to keep the self-organization potential of the flock by applying mutations (but only) to the particle best (in their DEPSO system) [9]. In PESDRO, the self-organization is not broken as the link between father and perturbed version is not lost. Thus, the perturbation can be applied to the entire flock. Note that these perturbations are suitable for real-valued function optimization.

3.2 Important Aspects of PESDRO

Particle Evolutionary Swarm for Design Reliability Optimization (PESDRO) is an implementation of PSO to solve the redundancy and reliability allocation problem. We propose a new method that offers a variety of options to optimize each component of a system. Several approaches representatives of the state-of-the-art to solve the redundancy and reliability allocation problem, seek the optimum by applying only one approach, either redundancy or reliability, to all system's components. In this paper, we propose a combined approach for the optimization of each component, either by redundancy or by reliability allocation, but yielding a system with optimum reliability and costs.

Redundancy and diversity techniques are used in redundancy allocation problem.

- Redundancy: It is a technique that replaces one component by a subsystem formed by N equal components with a parallel structure. Subsystem's reliability is increased by each component allocated in parallel. Is important to mention that there is only one supplier for the component, therefore, a subsystem is conformed by n components with same reliability, cost and weight.
- Diversity: One component has several suppliers but it is not limited to choose only one. Instead, it is free to build a subsystem of N components with parallel structure, whose components come from different suppliers.

Reliability allocation problem could be solved by using the technique proposed in this paper, called distribution optimization.

- Distribution Optimization: The goal is to increase the component's reliability until a expected or required value is met. Component's cost is a predefined increasing function of the component reliability.

$$g(R_i) = a\left[\Delta R_i(t)\right]^b \tag{11}$$
$$\Delta R_i(t) = RF_i(t) - RI_i(t) \tag{12}$$

When the reliability function is independent of the time, a constant reliability value is used as the initial reliability $RI_i(t)$; and the final reliability $RF_i(t)$ is used to improve the reliability component until its desired value.

But, if the reliability function of the component is dependent of the time, then when a final reliability value $RF_i(t)$ is reached, we must find the optimal distribution parameters such that the distribution yields that value. We try to find the set of distribution parameters that minimize the distance with respect to the original distribution parameters, and yields a final reliability value $RF_i(t)$.

For instance, in Figure 9, an initial reliability value is at Weibull($\alpha = 70, \beta = 2$) that yields a reliability equal to $RI(50) = 0.60$, but the new parameters α, β must be determined for the yielded reliability curve $RF(50) = 0.77$. There are two main cases that may occur. In the first case, the distribution function has only one parameter, then the relation is one-one between the reliability function and its parameter; also there is one and only one parameter value that yields the desire reliability at time $t = 50$.

In the second case, the number of distribution parameters is at least two, $p > 2$, then there are an infinity distribution parameters set that yields the

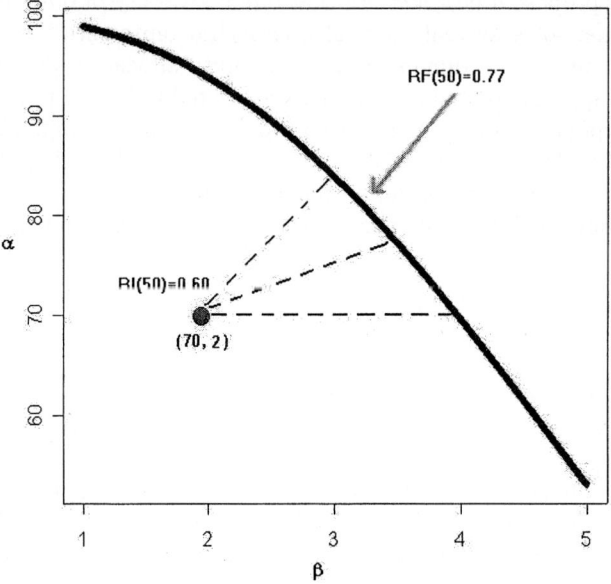

Fig. 9. Finding optimal distribution parameters

same reliability value. For this case, we can find the new distribution parameters set that has the minimum distance with respect to the original distribution parameters, and even standardize the distance of each parameter. Also, we can change only one parameter and set the other $p-1$ distribution parameters, then the first case appears.

Starting off from the mentioned techniques, a design optimization tool could be created to simultaneously solve both redundancy and reliability allocation problems. In PESDRO, we assign a sub-population to each component.

4 Experiments

PESDRO was evaluated using three well-known problems, the first two problems are of type **P1**, and the third one is type **P2**. The results are contrasted against a Genetic Algorithm, an Ant Colony System, and Tabu Search. The problems are explained next.

1. **Test case 1**: The first test problem was originally proposed by Fyffe, Hines and Lee in 1968 [10] and modified by Nakagawa and Miyazaki in 1981 [11]. Fyffe, Hines and Lee specified a system with 14 subsystems. For each subsystem, there are three or four component options. Component cost, weight and reliability are provided in Table 1. The objective is to maximize system reliability given constraints limits of 130 units of system cost, 170 units of system weight. The maximum number of components within a subsystem has been defined to be six ($n_{max,i} = 6$). Results are show in Table 4 and the analysis is presented in Section 5.1.
2. **Test case 2**: Coit and Liu in 2000 [12] proposed a system made of 14 k-out-of-n subsystems with $k_i \in \{1, 2, 3\}$. The problem is a modified version

Table 1. Component's data for Test case 1

	S_1			S_2			S_3			S_4		
i	R_{i1}	C_{i1}	W_{i1}	R_{i2}	C_{i2}	W_{i2}	R_{i3}	C_{i3}	W_{i3}	R_{i4}	C_{i4}	W_{i4}
1	0.90	$1	3	0.93	$1	4	0.91	$2	2	0.95	$2	5
2	0.95	$2	8	0.94	$1	10	0.93	$1	9			
3	0.85	$2	7	0.90	$3	5	0.87	$1	6	0.92	$4	4
4	0.83	$3	5	0.87	$4	6	0.85	$5	4			
5	0.94	$2	4	0.93	$2	3	0.95	$3	5			
6	0.99	$3	5	0.98	$3	4	0.97	$2	5	0.96	$2	4
7	0.91	$4	7	0.92	$4	8	0.94	$5	9			
8	0.81	$3	4	0.90	$5	7	0.91	$6	6			
9	0.97	$2	8	0.99	$3	9	0.96	$4	7	0.91	$3	8
10	0.83	$4	6	0.85	$4	5	0.90	$5	6			
11	0.94	$3	5	0.95	$4	6	0.96	$5	6			
12	0.79	$2	4	0.82	$3	5	0.85	$4	6	0.90	$5	7
13	0.98	$2	5	0.99	$3	5	0.97	$2	6			
14	0.90	$4	6	0.92	$4	7	0.95	$5	6	0.99	$6	9

Table 2. Component's data for Test case 2

i	k_i	λ_{i1}	C_{i1}	W_{i1}	λ_{i2}	C_{i2}	W_{i2}	λ_{i3}	C_{i3}	W_{i3}	λ_{i4}	C_{i4}	W_{i4}
1	1	0.001054	$1	3	0.000726	$1	4	0.000943	$2	2	0.000513	$2	5
2	2	0.000513	$2	8	0.000619	$1	10	0.000726	$1	9			
3	1	0.001625	$2	7	0.001054	$3	5	0.001393	$1	6	0.000834	$4	4
4	2	0.001863	$3	5	0.001393	$4	6	0.001625	$5	4			
5	1	0.000619	$2	4	0.000726	$2	3	0.000513	$3	5			
6	2	0.000101	$3	5	0.000202	$3	4	0.000305	$2	5	0.000408	$2	4
7	1	0.000943	$4	7	0.000834	$4	8	0.000619	$5	9			
8	2	0.002107	$3	4	0.001054	$5	7	0.000943	$6	6			
9	3	0.000305	$2	8	0.000101	$3	9	0.000408	$4	7	0.000943	$3	8
10	3	0.001863	$4	6	0.001625	$4	5	0.001054	$5	6			
11	3	0.000619	$3	5	0.000513	$4	6	0.000408	$5	6			
12	1	0.002357	$2	4	0.001985	$3	5	0.001625	$4	6	0.001054	$5	7
13	2	0.000202	$2	5	0.000101	$3	5	0.000305	$2	6			
14	3	0.001054	$4	6	0.000834	$4	7	0.000513	$5	6	0.000101	$6	9

Table 3. Component's data for Test case 3

i	1 ($k_1 = 4$)			2 ($k_2 = 2$)		
	R_{1j}	C_{1j}	W_{1j}	R_{2j}	C_{2j}	W_{2j}
S1	0.981	$95	52	0.931	$137	83
S2	0.933	$86	94	0.917	$132	96
S3	0.730	$80	32	0.885	$127	94
S4	0.720	$75	92	0.857	$122	93
S5	0.708	$61	41	0.836	$100	95
S6	0.699	$45	33	0.811	$59	63
S7	0.655	$40	98	0.612	$54	65
S8	0.622	$36	96	0.432	$41	49
S9	0.604	$31	83	0.389	$36	33
S10	0.352	$26	66	0.339	$30	51

of the original problem proposed by Fyffe, Hines and Lee [10] and modified by Nakagawa and Miyazaki [11]. For each subsystem, there are three or four component options. Component cost, weight and exponential distribution parameter (λ_{ij}) are given in Table 2. The objective is to maximize system reliability at a time of 100 hours given constraints limits of 130 units of system cost, 170 units of system weight. The maximum number of components within a subsystem has been defined to be six ($n_{max,i} = 6$). Results are show in Table 5 and the analysis is presented in Section 5.2.

3. **Test case 3**: Coit and Smith in 1996 [13] proposed a system made of 2 k-out-of-n subsystems with $k_1 = 4$ and $k_2 = 2$. For each subsystem, there are ten component options. Component cost, weight and reliability are given in Table 3. The objective is to minimize system cost given constraints limits of 650 units of system weight and 0.975 units of system reliability. The maximum number of components within a subsystem has been defined to be eight ($n_{max,i} = 8$). Results are show in Table 6 and the analysis is presented in section 5.3.

5 Comparison of Results

Because of the stochastic nature of PESDRO, 30 trials with 100 particles were performed. The best solution of each run is stored and later used to compute the statistics. PESDRO was compared to other algorithms used to solve the experiments explained in Section 4 (information regarding experiment conditions of other authors are not available as to perform an exact comparison).

5.1 Test Case 1

For this test, 30 runs with 3333 generations each were performed. Table 4 presents the corresponding results from the Ant Colony System of Liang and Smith (1999) [14], where 10 runs of the algorithm were performed. Also shown are the results of the TSRAP algorithm of Kulture-Konak, Smith and Coit (2003) [15] (10 trials of the algorithm were performed).

Table 4. Test case 1: A comparison of AS [14], TSRAP [15] and PESDRO

Reliability	AS 10 runs	TSRAP 10 runs	PESDRO 30 runs
Maximum	0.963510	0.970760	0.966950
Minimum	0.959090	-	0.952045
Average	0.962160	-	0.960001
Std. Dev.	0.001790	0.000490	0.003677

5.2 Test Case 2

For this test problem, 30 runs with 3333 generations each were performed. Table 5 presents the corresponding results from TSRAP of Kulture-Konak, Smith and Coit (2003) [15] (10 trials of the algorithm were performed).

Table 5. Test case 2: A comparison of TSRAP [15] and PESDRO

Reliability	TSRAP 10 runs	PESDRO 30 runs
Maximum	0.413450	0.403942
Minimum	-	0.341003
Average	-	0.370370
Std. Dev.	0.002625	0.015551

5.3 Test Case 3

For this last problem, 30 runs with 1666 generations each were performed. Table 6 presents the corresponding results from TSRAP algorithm of Kulture-Konak, Smith and Coit (2003) [15], where 20 trials of the algorithm were performed. Also shown are the results of the Genetic Algorithm of Coit and Smith (1996) [13](20 trials of the algorithm were performed).

Table 6. Test 3: A comparison of the GA [13], TSRAP [15] and PESDRO

Cost	GA 20 runs	TSRAP 20 runs	PESDRO 30 runs
Minimum	$727.00	$727.00	$727.00
Maximum	-	-	$728.00
Average	$727.25	$727.80	$727.27
Std. Dev.	-	-	0.449776

The results of PESDRO in the first two test problems are close to the TSRAP's results. For the last test problem, the results of PESDRO are competitive with the results of the TSRAP and the GA. PESDRO algorithm has a good performance because it works in two steps. First, PESDRO quickly finds feasible solutions by means of the perturbation operators (coarse search). At the same time, those operators help the algorithm to avoid stagnation at local optima. When a feasible solution is near the optimal, the search is mainly driven by the standard PSO operators, thus performing a fine search.

6 Conclusions and Future Work

In this paper a PSO approach was described and applied to the optimal design of redundancy and reliability allocation problems. PESDRO was demonstrated in three test problems with competitive results. A new technique is propose to solve reliability allocation problem. There are many applications for *Distribution Optimization* in system reliability optimization [16]. PESDRO provided results competitive with TSRAP in all problems, it is better than Ant System in Test problem 1, and in the average is better than TSRAP in problem 3. The proposed system can deal with 4 different probability distributions, and any combination of parallel, series, and K-out-of-N subsystems. This ability makes the system applicable to a broad kind of problems; this feature is not found in the other reviewed approaches: TSRAP, AS, or GA. Thus, the trade-off between generalization and specialization must be considered before any analysis of the results.

References

1. Kuo,W., Prasad,R.: An Annotated Overview of System Reliability Optimization. *IEEE Transactions on Reliability*, Vol. **49**(2) (June 2000) 176-187.
2. Kennedy,J., Eberhart,R.: Particle Swarm Optimization. *Proceedings of the IEEE International Conference On Neural Networks*, Vol. **4** (1995) 1942-1948.
3. Chern,M.: On the Computational Complexity of Reliability Redundancy Allocation in a Series System. *Operations Research Letters*, Vol. **11** (1992) 309-315.
4. Kennedy,J.: Small Worlds and Mega-Minds: Effects of Neighborhood Topology on Particle Swarm Performance. *IEEE Congress on Evolutionary Computation*, Vol. **3** (1999) 1931-1938.
5. Eberhart,R., Dobbins,R., Simpson,P.: Computational Intelligence PC Tools. Academic Press, (1996).

6. Kennedy,J., Eberhart,R.: The Particle Swarm: Social Adaptation in Information-Processing Systems. *New Ideas in Optimization*, McGraw-Hill (1999) 379-387.
7. Franken, N. and Andries P. Engelbrecht. Comparing PSO structures to learn the game of checkers from zero knowledge. In *Proceedings of the Congress on Evolutionary Computation 2003 (CEC'2003)*, Canberra, Australia, (2003) Vol. 1, pages 234-241
8. Storn, R. Sytem Design by Constraint Adaptation and Differential Evolution. *IEEE Trans. on Evolutionary Computation*, 1999, Vol. 3, No. 1, pp. 22 - 34
9. Zhang, W J., Xie XF. DEPSO: Hybrid Particle Swarm with Differential Evolution Operator. In *Proceedings of IEEE International Conference on Systems, Man and Cybernetics*, Washington D.C., USA, (2003) 3816-3821
10. Fyffe,D., Hines,W., Lee,N.: System reliability allocation and a computational algorithm. *IEEE Transactions on Reliability*, Vol. **17** (1968) 74-79.
11. Nakagawa,Y., Miyazaki,S.: Surrogate Constraints Algorithm for Reliability Optimization Problems with Two Constraints. *IEEE Transactions on Reliability*, Vol. **30** (1981) 175-180.
12. Coit,D., Liu,J.: System Reliability Optimization with k-out-of-n Subsystems. *International Journal of Reliability, Quality and Safety Engineering*, Vol. **7**(2) (2000) 129-143.
13. Coit,D., Smith,A.: Reliability Optimization of Series - Parallel Systems Using a Genetic Algorithm. *IEEE Transactions on Reliability*, Vol. **45**(2) (1996) 254-260.
14. Liang,Y., Smith,A.: An Ant System Approach to Redundancy Allocation. *Proceeding of the 1999 Congress on Evolutionary Computation, IEEE, Piscataway,N.y.*, (1999) 1478-1482.
15. Kulturel-Konak,S., Smith,A., Coit,D.: Efficiently Solving the Redundancy Allocation Problem Using Tabu Search. *IIE Transactions*, Vol. **35** (2003) 515-526.
16. Angel E. Muñoz Zavala. Optimal Design for Reliability. Master Thesis in Computer Science and Industrial Mathematics, Center for Research in Mathematics, 2004

Multiobjective Water Pinch Analysis of the Cuernavaca City Water Distribution Network

Carlos E. Mariano-Romero[1], Víctor Alcocer-Yamanaka[1], and Eduardo F. Morales[2]

[1] Mexican Instite of Water Technology, Paseo Cuauhnáhuac 8532,
Jiutepec, Mor, 62550, Mexico
cmariano@tlaloc.imta.mx
http://www.imta.mx/

[2] ITESM Campus Cuernavaca, Paseo de la Reforma 182-A,
Temixco, Mor, 62589, Mexico

Abstract. Water systems often allow efficient water uses via water reuse and/or recirculation. Defining the network layout connecting water-using processes is a complex problem which involves several criteria to optimize, frequently accomplished using Water Pinch technology, optimizing freshwater flowrates entering the system. In this paper, a multiobjective optimization model considering two criteria is presented: (i) the minimization of freshwater consumption, and (ii) the minimization of the cost of the infrastructure required to build the network that make possible the reduction of freshwater consumption. The optimization model considers water reuse between operations and wastewater treatment as the main optimization mechanism. The operation of the Cuernavaca city water distribution system was analyzed under two different operation strategies considering: leak reduction, operation of wastewater treatment plants as they currently operate, operation of wastewater treatment plants at design capacity, and construction of new infrastructure to treat 100 % of discharged wastewater. Results were obtained with $MDQL$ a multiobjective optimization algorithm based on a distributed reinforcement learning framework, and they were validated with mathematical programming.

1 Introduction

Water pinch technology (WPT) evolved out of the broader concept of process integration of materials and energy and the minimization of emissions and wastes in chemical processes. WPT can be seen as a type of mass-exchange integration involving water-using operations, which enables practicing engineers to answer important questions when retrofitting existing facilities and designing new water-using networks. There are three basic tasks in WP: a) Identification of the minimum freshwater consumption and wastewater generation in water-using operations (analysis), b) Water-using network design that achieves the identified flow rate targets for freshwater and wastewater through water reuse, regeneration, and recycle (synthesis), and c) Modify an existing water-using network to maximize water reuse and minimize wastewater generation through effective process changes (retrofit).

Nowadays most WPT problems are formulated as non linear highly restricted programming problems [2]; [13]; [14]. Important efforts have been made in order to make mathematical models more robust and applicable to real situations [1]; [7]; [9].

In general, WPT minimizes freshwater flow rate entering the system, using the mass balance and the contaminants concentrations at the inlet and outlet in all water-using operations as restrictions. Because of the diverse types of water-using operations, treatment effectiveness and cost, and types of contaminants, the criteria for efficient use of water is inherently non linear, multiple and conflicting [1]; [9]; [13]. Some of the criteria that easily arise are: equipment cost minimization, maximization of reliability (amount of contaminant captured in treatment plants), and minimization of wastewater production.

This paper presents a methodology that exploits specific features of the water and wastewater minimization problem. The formulation extends the domain of WPT analysis with elements of capital cost of the required pipe work. Consequently, the optimization is made based on cost efficient networks and networks featuring freshwater consumption. The methodology involves two criteria: the minimization of freshwater consumption and the infrastructure costs. Two techniques are used to solve the multiobjective optimization problem stated for the design of water-using systems: 1) weighted aggregation considering variation in the weight coefficients in order to construct the Pareto set [19], and 2) *MDQL*, which is a heuristic approach based on the exploitation of the knowledge generated during the search process.

The proposed multiobjective optimization model was applied for the case of the water distribution network in the city of Cuernavaca. An operation analysis considering two different strategies was performed: 1) reduction of leaks in the network and operation of wastewater treatment plants as they currently operate, and 2) reduction of leaks in the network, operation of wastewater treatment plants at their design capacity, and construction of new treatment infrastructure to reach 100 % wastewater treatment.

Section two presents the mathematical formulation for the bi-objective optimization problem and its description. In section three the function aggregation method and the MDQL heuristic approach are described. Section four describes the application case. Section five contains a discussion on the obtained results, general conclusions and future research directions.

2 Mathematical Formulation

The mathematical model describing a water demanding process considers two main components: a) freshwater sources available to satisfy demands, and b) water-using operations described by loads of contaminants and concentration levels. A case with two sources and two operations is sketched in Figure 1. The design task is to find the network configuration that minimizes the overall de-

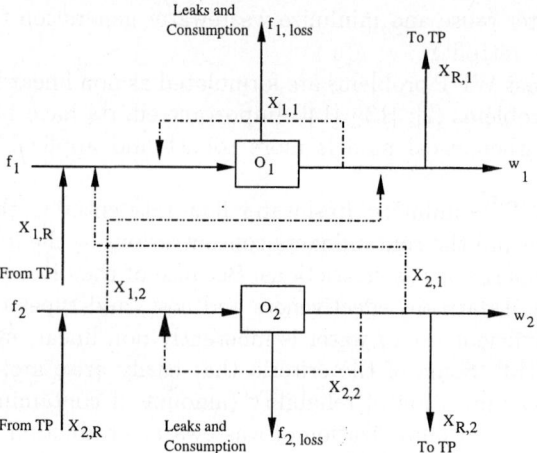

Fig. 1. Block diagram of a water-using system with one two sources and two operations

mand for freshwater, $\sum f_i$, (and consequently reduce wastewater volume $\sum W_i$) compatible with minimum investment cost. In order to complete the design task the optimization problem is stated in terms of low freshwater consumption, a suitable network topology for water reuse, $X_{i,j}$, and low investment cost. Also, water treatment through treatment plants (TP) is considered, $X_{R,i}$, as a unitary operation, so other operation can reuse water from TP, $X_{i,R}$.

Unitary operations demanded water, O_i, are defined through their contaminant loads, required flow rates and allowable minimal and maximal contaminant concentrations at influxes and discharges.

Objective functions for freshwater consumption minimization and for infrastructure cost minimization are represented by Equations 1 and 2.

$$Min Z_1 = F_1 = \sum_j cst_j + TPC, \qquad (1)$$

$$Min Z_2 = F_2 = \sum_i f_i \qquad (2)$$

Where: F_1 is the total cost of the distribution network considering the connection of freshwater sources to unitary operations receiving water directly, and the connection for reusing water between unitary operations. The total distribution network cost is composed by the sum of the partial costs, cst_j, of the pipe segments used for connecting freshwater sources to unitary operations and unitary operations to unitary operations, and TPC (the treatment plant construction cost that applies only for new treatment infrastructure).

F_2, is the total freshwater demanded by the system, obtained by the partial demands of freshwater from each of the unitary operations in the system. Partial demands from unitary operations, say operation O_i, are represented as, f_i. That is f_i is the partial freshwater demand of operation O_i.

2.1 Infrastructure Cost

Evaluation of the first objective function, F_1, depends only on the pipe segment costs in the network. These costs are represented as cst_j, (see Equation 3) and depend on three variables: a) pipe length, L_j; b) cost per unit length, PC_j; which depends on the pipe diameter required to transport the demanded flow of water, D_j; and c) a cost factor, CF_j, related to pipe materials required to resist corrosive effects of contaminants. It is important to note that one of the objectives of this work is to demonstrate the benefits obtained by the solution of the multiobjective approach, compared with those obtained with the single objective approach. It is for this reason that some considerations regarding the hydraulic behavior of the network are not included.

$$cst_j = L_j \times PC_j \times CF_j \qquad (3)$$

As mentioned in the previous paragraph, the variable PC_j depends on the pipe diameter, $D_j = f(Q_j)$, which is obtained calculating the minimum diameter required to transport the water flow through the pipe. The minimum diameter, D_{min_j}, is obtained through Equation 4; deduced from the definition of flow ($Q = velocity/area$) considering maximum velocities of water in pipes of 2.5 m/s. D_{min_j} is approximated to the closest mayor commercial diameter. Table 1-(b) shows diameters and cost per unit length for commercial pipes considered in this work. The data in Table 1 is only demonstrative and can be substituted with real data from local markets.

$$D_{min} = 0.714\sqrt{Q} \qquad (4)$$

where: D_{min} is the minimal pipe diameter in mm required to transport flow rate Q ; $Q \in \{f_i, X_{i,j}, W_i\} \forall i,j$ and is given in m^3/s.

In a similar manner, the factor CF_j is related to the capacity of pipe segments to resist corrosive effects due to the presence of contaminants in water flows. Values for the CF_j factor are included in Table 1-(a), calculated considering local prices in Mexico for non corrosive pipes.

Finally the treatment plant construction cost considered in this work is 10 $/l$, that is the construction cost in monetary units per liter of treatment capacity for the plant or plants.

2.2 Freshwater Demand

To guarantee steady state conditions in the system, it is necessary to restrict the objective functions by the mass balance between unitary operations, and by the maximum and minimum allowed contaminant concentrations on the influxes and discharges of operations [14].

The flowrate required in each unitary operation is related to the mass load of contaminants ($\Delta m_{i,k,tot}$) discharged by operations. This is described in Equation 5.

$$f_i = max_c \frac{\Delta m_{i,k,tot}}{c_{i,k,out}^{max} - c_{i,k,in}^{max}} \qquad (5)$$

Table 1. Cost factors for pipes resistant to abrasive effects of contaminants *(a)* and cost per unit length for commercial diameter pipes *(b)*

(a)		(b)	
Contaminant concentration (mg/l)	CF	Diameter (mm)	PC $/m
$0 \leq c \leq 50$	1.25	99	4.8
		150	5.0
		200	8.9
$50 < c \leq 100$	2.0	250	12.9
		300	17.7
		350	23.6
$100 < c \leq 150$	2.0	400	25.6
		450	34.1
$150 < c \leq 200$	3.0	500	40.9
		610	42.6
		762	45.9
$200 < c \leq 500$	5.0	838	54.6
		1,016	69.9
		1,118	83
$500 < c$	10.0	1,219	94
		1,372	110

where f_i is the freshwater flow rate for operation O_i; $\Delta m_{i,k,tot}$ is the total mass transfer for each contaminant, k, to the water used at operation O_i (this term is also known as contaminant mass charge [3] and is expressed in kg/h); $c_{i,k,out}^{max}$ and $c_{i,k,in}^{max}$ are the maximum allowed concentration of contaminant k on the discharge and influx of operation O_i, in mg/l respectively.

The optimization model depends on the mass balance between all inlets and all outlets of water to the operation O_i. According to Figure 2, the expression for the mass balance has the form shown in Equation 6.

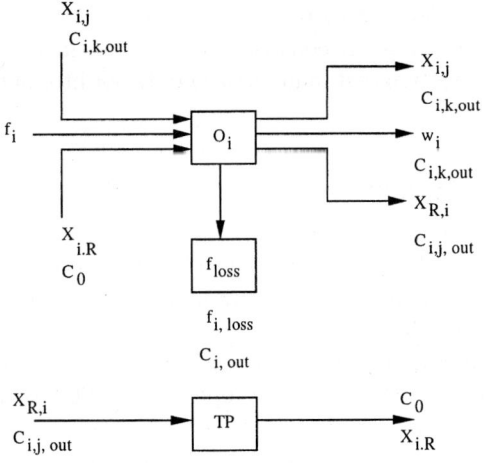

Fig. 2. General structure

$$f_i + \sum_{j \neq i} X_{i,j} + X_{i,R} - f_{i,loss} - W_i - \sum_{j \neq i} X_{j,i} - X_{R,j} = 0 \quad (6)$$

where, $X_{i,j}$ is the reusable water flow rate from other operations, say O_j, in operation O_i; $X_{i,R}$ is the treated water from the wastewater treatment plants that can be used in operation O_i; $f_{i,loss}$ is the portion considered as water loss in the operation or water consumption by the operation; W_i is the wastewater flow rate from operation O_i; $X_{j,i}$ is the reusable water flow rate from operation O_i in operations O_j; and $X_{R,i}$ is the portion of the discharged water from operation O_i that receive treatment. All flowrates are represented in m^3/h.

Several (k) different contaminants can be considered in the optimization model. This consideration requires the definition of constraints to restrict the concentration of contaminants at the inlets and outlets of operations, in order to guarantee that water influxes will not affect the operation performance, and to avoid the violation of environmental or operation standards. The satisfaction of this constraints will determine the quantities of fresh and reused water to supply to operations. Contaminant concentration constraint at the influx of the ith operation, $c_{i,k,in}$ is defined by Equation 7.

$$c_{i,k,in} = \frac{\sum_{j \neq i} X_{i,j} c_{j,k,out} + c_{k,0} X_{i,R} - f_{i,loss} c_{i,k,in}^{max}}{\sum_{j \neq i} X_{i,j} + f_i + X_{i,R} - f_{i,loss}} \leq c_{i,k,in}^{max} \quad (7)$$

where, $c_{i,k,in}$ is the concentration of contaminant k at the influx of operation O_i; $c_{j,k,out}$ is the concentration of contaminant, k, at the discharge of operation O_j, $c_{k,0}$ is the concentration of contaminant k in the treated water, $c_{i,k,in}$ is the maximum allowable concentration of contaminant k at the influx of operation O_i. Concentrations are expressed in mg/l.

The same way, contaminant concentration constraint at the outlet of jth operation, $c_{j,k,out}$ is defined by Equation 8.

$$c_{j,k,out} = c_{i,k,in} + \frac{\Delta m_{i,k,tot}}{\sum_{j \neq i} X_{i,j} + f_i + X_{i,R} - f_{i,loss}} \leq c_{i,k,out}^{max} \quad (8)$$

Finally, non negativity constraints are established according to the following equations.

$$X_{i,j} \geq 0;$$
$$f_i \geq 0;$$
$$L_j \times PC_j \times CF_j \geq 0.$$

3   Solution Method

In this sense we propose the use of two techniques especially designed to solve optimization problems with more than one criterion. The first, uses an aggregated function constructed with the use of weight coefficients representing the relative importance of the two objective functions. The resulting optimization problem is solved by the reduced gradient method in order to avoid penalty parameters [4] for five combinations of weights to construct the Pareto set. The second approach is an heuristic based on the solution of Markov decision processes known as *MDQL* [16]. MDQL is capable of exploiting the knowledge acquired during the solution process, and has been tested on several benchmark problems showing good performance [16], [18] and more recently [19].

3.1 Aggregated Function

This approach is probably the most known and simplest way to solve this type of problems. Some of the first references on it are [11] and [25]. The main idea behind this approach is the construction of a weighted function resulting from the combination of the m objective functions with the use of weight coefficients. The weighted function is then used on a single objective optimization problem. In general terms it is proposed that the weight coefficients, p_i, to be real values such that $p_i \geq 0 \forall i = 1, \ldots, m$. It is also recommended to use normalized weight coefficients, so $\sum_{i=1}^{m} p_i = 1$. More precisely, the multiobjective optimization problem is transformed to the problem stated in Equation 9, which will be called in the successive the "weighted problem".

$$min \sum_{i=1}^{m} p_i \cdot F_i \qquad (9)$$

where, $p_i \geq 0 \forall i = 1, \ldots, k$ and $\sum_{i=1}^{k} p_i = 1$.

This approach guarantees the optimality of the Pareto set if the weighted coefficients are positive or the solution is unique [12] [20]. Pareto set construction is made with the variation of the weight coefficients values, solving the weighted problem as many times as the number of variations of the weight coefficients can be configured.

The resulting problem after the application of the weighted aggregation approach to the two objective functions presented in section 2, takes the form presented in Equation 10.

$$F = p_1 \sum_{i} f_i + p_2 (\sum_{j} cst_j + TPC) \qquad (10)$$

Solution of the weighted problem in Equation 10 is made through the reduced gradient method with the use of the GAMS/MINOS program [10]. Weight coefficients combinations used (p_1, p_2) are: $(0.1, 0.9)$, $(0.25, 0.75)$, $(0.5, 0.5)$, $(0.25, 0.75)$ and $(0.9, 0.1)$.

3.2 Multiple Objective Distributed Q-Learning(MDQL)

In order to efficiently solve optimization problems with more than one objective function it is desirable to use population based approaches, that is, approaches

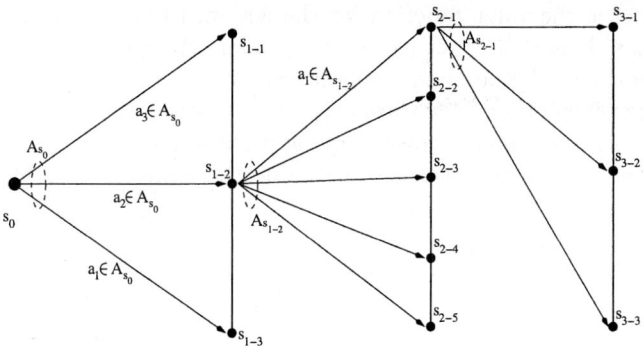

Fig. 3. Variable space division for MDQL

with the capability to generate more than one solution concurrently. Moreover, it is necessary to apply the dominance optimality criterion to evaluate the generated solutions. This is the main hypothesis of much of the recently developed approaches designed to efficiently solve multiobjective optimization problems based on evolutionary computation. Taking advantage of some of the characteristics of evolutionary approaches, it has been previously established that optimization problems can be solved considering search processes as a Markovian decision process [8]. Furthermore successful application of reinforcement learning to multiobjective optimization problems was first presented in [15]; extended and improved in [16] and [18].

MDQL considers a group of agents searching a terminal state, s_t, in an environment formed by a set of states, $\mathcal{S}$. The set of states, or environment, is constructed dividing variable ranges in the parameter space in fixed number of parts, considering that all decision variables can be discretized in a finite number of divisions. Minimum and maximum limits for divisions are considered states, as illustrated in Figure 3. An environment with these characteristics allows to the agents to propose values for each one of the decision variables in the problem.

For each state, $s \in \mathcal{S}$, a set of actions, $\mathcal{A}_s$, is settled, see Figure 3. All actions in states, $a \in \mathcal{A}_s$ have an associated value function, $Q(s,a)$, indicating the goodness of the action to complete a task.

The search mechanism for an agent in MDQL operates when an agent located in a state selects an action based on its value function, $Q(s,a)$. Most of the time the agent selects the best evaluated action (the action with the higher estimated value for $Q(s,a)$), and occasionally a random action with a probability $\epsilon \approx 0$. Action value functions are updated depending on how useful an action can be to an agent to reach a terminal state. This behavior is adjusted with the help of a reward value, $r \in \Re$, and the value function for the best evaluated action in the future state reached by the agent after the execution of the selected action, $Q(s',a')$. This update rule is expressed in Equation 11.

$$Q(s,a) \leftarrow Q(s,a) + \alpha \left[r + \gamma \max_{a' \in \mathcal{A}'_s} Q(s',a') - Q(s,a) \right] \quad (11)$$

where $Q(s,a)$ is the value function for the action, $(0 \leq \alpha \leq 1)$ is the learning step, $(0 \leq \gamma \leq 1)$ is a discount parameter r is an arbitrary reward value, $r \in \Re$, s' and a' are the next state and the best evaluated action for s' respectively.

As an agent explores the state space, $Q(s,a)$ estimations improve gradually, and, eventually, each $\max_{a' \in A'_s} Q(s',a')$ approaches: $E\left\{\sum_{n=1}^{\infty} \gamma^{n-1} r_{t+n}\right\}$ [22]. Here r_t is the reward received at time t due the action chosen at time $t-1$. Watkins and Dayan [24] have shown that this Q-learning algorithm converges to an optimal decision policy for a finite Markov decision process.

In *MDQL* there is a group of agents, instead of a single agent, interacting with the environment described above, and since the task for the agents is the construction of the Pareto set, the original *Q-Learning* [24] algorithm must be adapted. Main adaptations considered in *MDQL* are listed below.

- Decision variables in the environment have a predefined order, as illustrated in 3, the agents move in the decision variables space obeying this order, so the definition of the values for the decision variables is made in the same order by all the agents.
- A 'map' is constructed with a copy of the action values for each of the available environments (one for each of the solutions in the Pareto set, and solutions violating constraints).
- When all the agents finish a solution (define values for all the decision variables), all solutions are evaluated using the Pareto dominance criterion. Environments for non dominated solutions and solutions that violate any constraint remain in memory to be used in future episodes (see previous item).
- Agents are assigned randomly to the environments in memory.
- Action values update is made in two stages. The first is when agents make a transition using the 'map' of the environment (copy of action value functions) in which all agents working in the sane environment show their experience updating value functions [17]. Action values update is made considering the following criteria: (i) non dominated solutions receive a positive reward and (ii) solutions violating any constraint receive a punish, calculated as a function of the magnitude of the violated constraints [16] and [18]. Finally, the original action (those used to construct 'map's) value functions are updated considering the same criteria (second stage), 'maps' are destroyed and new solutions incorporated.

Reported Pareto fronts obtained with *MDQL* are the best after ten executions of the algorithms with the same execution parameters.

4 Cuernavaca City Water Distribution System, México

Water uses are classified in five categories [6]. Table 2 includes the values for the freshwater demand by each of the operations. It is relevant to note that part of the demanded water is consumed by the operation itself, other part can not be register and is considered as a loss caused by leaks occurring along distribution systems which is about 43.41% [21]. The rest is declared as wastewater and

Table 2. Freshwater demand and inflow and outflow limit concentration for all operations. Current situation for Cuernavaca city

Operation O	Water demand f_i l/s	BOD_5			TSS		
		$c_{i,A,in}^{max}$ mg/l	$c_{i,A,out}^{max}$ mg/l	$\Delta m_{i,tot}$ kg/h	$c_{i,B,in}^{max}$ mg/l	$c_{i,B,out}^{max}$ mg/l	$\Delta m_{i,tot}$ kg/h
Urban & Public	3,003	0.00	220.00	1,767.74	0.00	220.00	1,403.07
Services	16.19	0.00	220.00	9.53	0.00	200.00	7.56
Agriculture	593.0	50.00	350.00	449.57	50.00	300.00	449.57
Multiple	2.24	0.00	220.00	1.32	0.00	220.00	1.05
Industrial	47.58	0.00	874.00	85.57	0.00	371.00	36.32
Self Service	1.36	0.00	220.00	0.60	0.00	240.00	0.73

supposedly is discharged with the effluents to the receiving water bodies, in this case the Apatlaco river.

Two contaminants indexes are considered, five day biochemical oxygen demand (BOD_5) and total suspended solids (TSS). Wastewater treatment plants treat 339.15 l/s to BOD_5 and TTS mean concentration of 50 mg/l according to the data reported in the literature [3].

Values for both water quality indexes, $c_{i,k,out}^{max}$, were established using the information obtained from some studies performed to evaluate the degree of contamination in the Apatlaco river [5]. For both contaminants, the concentration in the freshwater supplied to the system is considered to be zero, see Table 2.

There are 15 wastewater treatment plants in Cuernavaca city, ten for the treatment of municipal wastewater and five for the treatment of industrial wastewater. The total treated wastewater flowrate is 364.15 l/s (339.65 municipal and 24.50 industrial [5]. In general the 10 municipal wastewater treatment plants operate at 66.35% of their design capacity and the five industrial wastewater treatment plants at 71.01% of their design capacity.

In order to verify how the system performance can be improved two different strategies were evaluated. The first is the operation of the treatment plants at their current operation capacity, that is 339.65 l/s, and the reduction of leaks in the distribution from 43% to 25%. The second strategy considers the operation of the existing treatment plants at their design capacity, 511.86 l/s, a leak reduction program to decrease the non accounted water from 43% to 25%, and the construction of new treatment facilities to reach 100% of treatment covering. MDQL operation parameters used for all test cases were: $\alpha = 0.1$, $\gamma = 0.01$, $r = 1$ for non dominated solutions and $r = -1$ for solutions violating constraints. Otherwise, for both problems decision variables are: six fresh water flow rates, f_i; reusable water flow rate in each operation including treatment plants, $X_{i,j}$, and waste water flow rates, W_i. Each variable discreticized to have increments of 0.1 l/s. Computational cost is $O(k^2)$, being k the number of agents in the problem and is related with the number of objective functions evaluations. More detail can be consulted in [18].

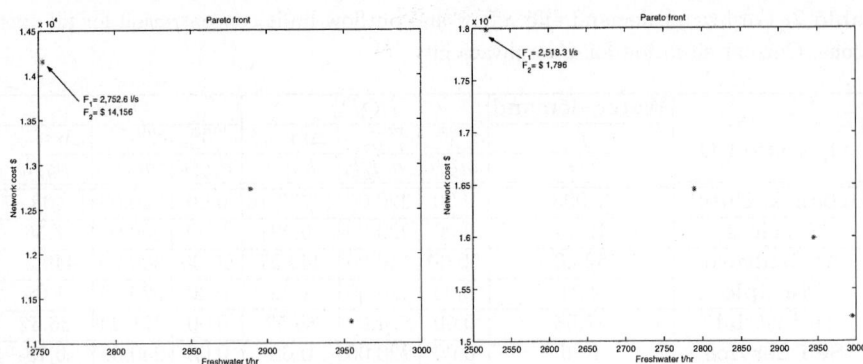

Fig. 4. Cuernavaca city distribution system results for the first strategy (left) and second strategy (right) both Pareto fronts obtained with MDQL

4.1 Results for the First Strategy

In this strategy leaks reduction from 43% to 25% and operation of treatment plants keeped in their current operation levels are considered (339.65 l/s).

Results are presented in Figure 4 (left). Since the agriculture is the most demanding operation in Cuernavaca city, the main change in the operation of the Cuernavaca city water distribution network is that water supplied for the agriculture can be supplied from three different sources: the wastewater treatment plants, freshwater and wastewater from the urban and public sector, saving water that cab be used to: a) increment of the irrigated area, and/or b) reduction of freshwater sources exploitation with a benefit to the environment.

The upper left most solution in Figure 4 (left) has a total demanded freshwater flowrate 2,752.6 l/s, compared with the current demand which represents a decrement of approximately 24.87 %, that is, 911.57 l/s of the amount of water taken from the sources. Freshwater savings represent approximately 28.74 millions of m^3 per year that could increase water availability in Cuernavaca valley aquifer from eight millions of cubic meters to 36.74 millions of cubic meters.

4.2 Results for the Second Strategy

The second operation analysis strategy considers the operation of existing wastewater treatment plants to their design capacity, leak reduction to 25 %, and construction of new treatment facilities.

The Pareto front obtained for this test case is presented in Figure 4 (right). The upper left solution with the lowest freshwater demand of 2,581.3 l/s and cost of \$1,796 requires the construction of two new wastewater treatment plants. The first with a capacity to treat 81.38 l/s configured to receive 4.6 % of the discharged wastewater from the urban and public sector, and 53 % of the discharged water by multiple sector. The second proposed wastewater treatment plant capacity is 1,130.44 l/s and could receive the 95.4 % resting discharged wastewater by the urban and public sector.

Similar to the results found with the first strategy, demanded water by agriculture is satisfied with the total of the municipal treated water, and with 81.38 l/s coming from the new plant proposed in the design. Industrial water is treated independently from the existing industrial treatment plants.

Water savings arise since a considerable flow of freshwater is not longer supplied to the agriculture sector. Freshwater savings represent 34.15 millions of cubic meters per year, savings that could represent an increment in the freshwater availability of the aquifer from eight to 44.13 millions of cubic meters per year.

As can be also appreciated in Figure 4 (right) that the upper left solution is the lowest freshwater flowrate demand solution, compared with Pareto solution found with the two strategies analyzed. It is also true that Pareto solutions cost are the highest but, at least intuitively, solutions into this Pareto set are more efficient solutions since all discharged water by the Cuernavaca city water distribution system is treated and contamination levels are the lowest. Qualitative efficiency is measured in terms of the remaining contaminant concentration in discharged wastewater to the reception bodies, this parameter is not included in the optimization model, but according to the environmental standards (included in the model) solutions for both strategies are feasible and do not violate them.

4.3 Function Aggregation Comparison

Figure 5 is a comparison of Pareto fronts obtained with the two strategies evaluated the Cuernavaca city water distribution network. This figure also includes five solutions for the two strategies obtained with the function aggregation and reduced gradient approach. Pareto fronts obtained with $MDQL$ and mathematical programming are close to the two analyzed strategies, so it can be said that the obtained results are valid and can be used to make decisions over the real-world problem stated in this paper. It also can be appreciated from Figure 5

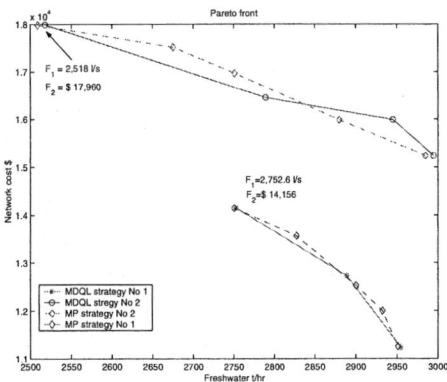

Fig. 5. Comparison of the Pareto fronts obtained for the two strategies evaluated to perform the Cuernavaca city distribution system operation

that solutions on the extremes of the Pareto fronts are similar, presumably because that solutions correspond to the end extremes of the Pareto fronts for the mathematical model.

This conjecture can be enlarged with the comparison of the solution on the upper left corner in Figure 5 (second strategy) $(2,518\ l/s)$ with the solution for the same strategy, but with the sole criterion of freshwater minimization (ideal vector) presented in [3], for which the total demanded freshwater flowrate was $2,586.96\ l/s$. The difference between the single objective and the bi objective solutions can be partially attributed to the weight factors used in the bi objective optimization. This comparison permits the validation of $MDQL$ for the solution of the bi-objective optimization model for the design of water using systems presented in this paper, considering the convergence properties of the function aggregation approach [20] and previous results on similar problems [19].

5 Conclusions

A water pinch optimization model that considers more than one criteria was presented. The model considers the reuse of wastewater from operations, wastewater treatment, consumption flowrates and leaks in the system, and the combination of this mechanisms for the optimization of two objective functions. The reduction of freshwater demands is possible with the guarantee that the quality of the water served to the different users do not violate ecological and sanitary norms. The bi-objective optimization model operates considering mass balances between operations, freshwater sources, wastewater treatment plants, and wastewater disposal effluents.

Model solution permits the verification of its behavior, consistency and completeness [19]. Mathematical programming for the solution of a weighted aggregated function of criteria was used as a mean of comparison, selected on the basis of previous results and convergence properties reported in the literature [20]. The objectives of this work were completely satisfied, it can be said that proposed model is complete and represents the behavior of real water distribution systems as the Cuernavaca city distribution system.

The quality, number, and distributions of solutions along the Pareto fronts obtained with $MDQL$ seems better compared with the those obtained with mathematical programming. The main difference is that $MDQL$ solutions were obtained on a single run without the definition of weight coefficients. So, based on this analysis it can be said that $MDQL$ is more competitive, especially because the quality of solutions (approximation of the Pareto front). It is also possible to note that combination of weight coefficients is, sometimes a tedious work decision makers are not totally convinced to do, especially for the preferences definition.

Solution for water pinch problems represent important technical challenges that are only partially solved in the industry. Results presented here can help as a sample of how real applications may be solved with the participation of multidisciplinary teams involving researches from different communities, as in this case.

Finally it is important to say that more work can be made with the optimization model. In this order of ideas constraints implementation to optimize the processes is one of the future activities, this implementation could help in the selection of more efficient processes, for example if wastewater treatment technology is selected in terms of the type of contaminants, the mass remotion could be made more effective and the system more efficient if the proper process is selected and optimized in terms of cost and efficiency. Another important aspect to implement is the cost function, which need to be extended in order to quantify operation costs, reuse costs, and other economic factors affecting the operation of a system with the characteristics.

References

1. Alva-Argaez, A *An Automated design tool for water and wastewater minimization*, PhD Thesis, Department of Process Integration University of Manchester IST, UK 1999.
2. Alcocer, V. and Arreguín, F., Minimización de agua de primer uso en procesos industriales a través de técnicas de optimización. *Memorias del XX Congreso Latinoamericano de Hidráulica*, La Habana, Cuba, september 2002, (in spanish).
3. Arreguín, C., F., Alcocer, Y. V. Modelación sistémica del uso eficiente del agua. *Revista Ingeniería Hidráulica en México*. Vol. XIX, No. 3, july, 2004.
4. Beightler, Ch., Phillips, D., and Wilde, D., *Foundations of Optimization*, Prentice Hall International Series in Industrial and Systems Engineering, 1979.
5. National Water Commission, *Clasification estudy in the Apatlaco river*, Sanitation and Water Quality Department from the Morelos state office, 1996 (In Spanish).
6. National Water Commission, *Hydro-geological update of the aquifer of Cuernavaca, state of Morelos* 1999, CNA 2000, (In Spanish).
7. Coetzer, D., Stanley, C. and Kumana, J. Systemic Reduction of Industrial Water Use and Wastewater Generation, In *Proc. of AIChE National Meeting*, Houston, March 1997.
8. Dorigo, M., and Stützle, T. (2004) *Ant Colony Optimization*. MIT press, July 2004.
9. Galan, B., and Grossmann, I.E., Optimal Design of Distributed Wastewater Treatment Networks, *Ind. Eng. Chem. Res.*, 37:4036, 1998.
10. GAMS/MINOS http://www.gams.com General Algebraic Modeling System Development Corporation, (2001).
11. Gass S. and Saaty T., *The conputational Algorithm for the Parametric Objective Function*, Naval Research Logistics Quarterly, 2 (1955).
12. Das, I., and Dennis, J. A Closer Look at Drawbacks of Minimizing Weighted Sums of Objectives for Pareto Set Generation in Multicriteria Optimization Problems, *Structural Optimization* Vol 14 no. 1, pp 63-69, 1997
13. Kuo, W.C. and Smith, R. Efluent treatment system design, *Chem Eng. Sci.*, 52:4273, 1997.
14. Mann, G.J. and Liu, Y.A. *Industrial Water reuse and wastewater minimization* McGrawHill Eds., 1999.
15. Mariano, C. and Morales, E. (1999) MOAQ: An Ant-Q algorithm for multiple objective optimization problems. In W. Banzhaf, J. Daida, A.E. Eiben, M.H. Garzon, V. Honavar, M. Jakiela and R.E. Smith (Eds.) *Proceedings of the Genetic and Evolutionary Computation Conference (GECCO-1999)*, Vol 1 pp. 894-901. San Francisco, Morgan Kaufmann.

16. Mariano, C. and Morales, E., (2000) A new approach for the solution of multiple objective optimization problems based on reinforcement learning, *Lecture Notes in Artificial Intelligence, Proceedings of the Mexican International Conference on Artificial Intelligence*, pp. 212-223, Acapulco, Mex, April, 2000.
17. Mariano, C. and Morales, E., (2001) A new updating strategy for reinforcement learning based on Q-learning, In P. Flach and L. de Raeldt (eds.), *Springer Verlag, Lecture Notes in Artificial Intelligence,vol 2167: 12th European conference on Machine Learning*, pages 324-335, 2001.
18. Mariano, C., (2001)*Reinforcement learning in multiobjective optimization* PhD Thesis in computer Science, Instituto Tecnolgico y de Estudios Superiores de Monterrey, Campus Cuernavaca, March, 2002, Cuernavaca, Mor., México.
19. Mariano, C., Alcocer, V., and Morales, E., (2005) Diseño de sistemas hidráulicos bajo criterios de optimización de puntos de pliegue y múltiples criterios, *Revista de Ingeniería hidráulica en México*, to appear in Vol XX, No, 3, september, 2005.
20. Miettinen K., *Nonlinear Multiobjective Optimization*, Kluwer Academic Publishers, (1999)
21. Ochoa, L., Bourguett, V. *Integral reduction of leaks*, Mexican Institute for Water Technology, National Water Commission, September 1998, (In Spanish)
22. Putterman, M. *Markov Decision Processes Discrete Stochastic Dynamic Programming*, Wiley Series in Probability and Mathematical Statistic, (1994)
23. Sutton R., and Barto G. *Reinforcement Learning An Introduction*, MIT Press, (1998).
24. Watkins, C., Dayan, P., (1992) Q-Learning, *Machine Learning*, 3:279-292, 1992.
25. Zadeh L., (1963) Optimality and Non-Scalar-Valued Performance Criteria, *IEEE Transactions on Automatic Control*, 8, (1963), 59-60.

Multi-objective Vehicle Routing Problems Using Two-Fold EMO Algorithms to Enhance Solution Similarity on Non-dominated Solutions

Tadahiko Murata and Ryota Itai

Department of Informatics, Kansai University
2-1-1 Ryozenji, Takatsuki 569-1102, Osaka, Japan
murata@res.kutc.kansai-u.ac.jp
http://www.res.kutc.kansai-u.ac.jp/~murata/

Abstract. In this paper, we focus on the importance of examining characteristics of non-dominated solutions especially when a user should select only one solution from non-dominated solutions at a time, and select another solution due to the change of problem conditions. Although he can select any solution from non-dominated solutions, the similarity of selected solutions should be considered in practical cases. We show simulation results on vehicle routing problems that have two demands of customers: Normal Demand Problem (NDP) and High Demand Problem (HDP). In our definition the HDP is an extended problem of NDP. We examined two ways of applying an EMO algorithm. One is to apply it to each problem independently. The other is to apply it to the HDP with initial solutions generated from non-dominated solutions for the NDP. We show that the similarity of the obtained sets of non-dominated solutions is enhanced by the latter approach.

1 Introduction

Although we have many approaches in EMO (Evolutionary Multi-criterion Optimization) community [1, 2] recently, there are few research works that investigate the similarity of obtained non-dominated solutions. Deb considered topologies of several non-dominated solutions in Chapter 9 of his book [3]. He examined the topologies or structures of three-bar and ten-bar truss. He showed that neighboring non-dominated solutions on the obtained front are under the same topology, and NSGA-II can find the gap between the different topologies. While he considered the similarity of solutions in a set of non-dominated solutions from a topological point of view, there is no research work relating to EMO that considers the similarity of solutions in different sets of non-dominated solutions from that point of view.

We employ the Vehicle Routing Problem (VRP) to consider the similarity in different sets of solutions. The VRP is a complex combinatorial optimization problem, which can be seen as a merge of two well-known problems: the Traveling Salesman Problem (TSP) and the Bin Packing Problem (BRP). This problem can be described as follows: Given a fleet of vehicles, a common depot, and several customers scattered geographically. Find the sets of routes for the fleet of vehicles. Many research works [4, 5, 6, 7, 8] on the VRP try to minimize the total route cost that is calculated using the distance or the duration between customers. Several hybrid algorithms have been proposed to improve the search ability of genetic algorithms [4, 5]. The research works in [6, 7, 8] are related to multi-objective optimization. Tan *et al.* [6] and Saadah *et al.* [7] employed the travel distance and the number of vehicles to be minimized. Chitty and Hernandez [8] tried to minimize the total mean transit time and the total variance in transit time.

In this paper, we employ an EMO algorithm, NSGA-II [9], to our vehicle routing problems with minimizing the number of vehicles and the maximum routing time among the vehicles. It should be noted that we don't employ the total routing time of all the vehicles, but use the maximum routing time among the vehicles. We employed it in order to minimize the active duration of the central depot. We consider two problems with different demands. One problem has a normal demand of customers. The other has a high demand. We refer the former problem and the latter problem as NDP and HDP, respectively. We define the demand in the HDP as an extended demand of the NDP in this paper. For example, we assume that the demand in the HDP is a demand occurring in a high season such as Christmas season. In that season, the depot may have an extra demand as well as the demand in the normal season. In order to avoid a large change of each route from the depot, a solution (i.e., a set of route) in the HDP should be similar to a solution in the NDP. This situation requires us to consider the similarity of solutions on different non-dominated solutions in multi-objective VRPs.

In order to find a set of non-dominated solutions in the HDP that is similar to a set of non-dominated solutions in the NDP, we apply a two-fold EMO algorithm to the problem. In a two-fold EMO algorithm, first we find a set of non-dominated solutions for the NDP by an EMO algorithm. Then we generate a set of initial solutions for the HDP from the non-dominated solutions for the NDP. We apply an EMO algorithm to the HDP with initial solutions that are similar to those of the NDP problem.

We organize this paper as follows: Section 2 gives the problem model for multi-objective VRPs. The outline of our two-fold EMO algorithm is described in Section 3. We define a measure of the similarity between solutions in Section 4. A small example of our multi-objective VRP is also shown in Section 4. Section 5 presents the extensive simulations and compares results of the two-fold EMO algorithm and those obtained individually for the HDP and NDP. Conclusions are drawn in Section 6.

2 Multi-objective Vehicle Routing Problems

The domain of VRPs has large variety of problems such as capacitated VRP, multiple depot VRP, periodic VRP, split delivery VRP, stochastic VRP, VPR with backhauls, VRP with pick-up and delivering, VRP with satellite facilities, and VRP with time windows. These problems have the basic architecture of the VRP except their own constraints. Their constraints are arisen in practical cases. Please see for the detail of the VRP problem in [10].

The objective of the basic problem is to minimize a total cost is described as follows:

$$\text{Minimize } \sum_{m=1}^{M} c_m , \tag{1}$$

where M is the number of vehicles each of them starts from the depot and is routed by a sequence of customers, then return to the depot. The cost of each vehicle is denoted by c_m and described as follows:

$$c_m = c_{0,1} + \sum_{i=1}^{n_m-1} c_{i,i+1} + c_{n_m,0}, \tag{2}$$

where $c_{i,j}$ means the cost between Customers i and j. Let us denote 0 as the index for the depot in this paper. Equation (2) indicates the sum of the cost between the depot and the first customer assigned to the m-th vehicle (i.e., $c_{0,1}$), the total cost from the 1st customer to the n_m-th customer (i.e., $\sum_{i=1}^{n_m-1} c_{i,i+1}$), and the cost between the final customer n_m and the depot. Each vehicle is assigned to visit n_m customers, thus we have $N = \sum_{m=1}^{M} n_m$ customers in total. The aim of the VRP is to find a set of sequences of customers that minimizes the total cost. Each customer should be visited exactly once by one vehicle.

While the total cost of all the vehicles is ordinarily employed in the VRP, we employ the maximum cost to be minimized in this paper. When the cost $c_{i,j}$ is related to the driving duration between Customers i and j in Equation (2), the total cost c_m for the m-th vehicle means the driving duration from the starting time from the depot to the returning time to the depot. In order to minimize the activity duration of the depot, the maximum duration of the vehicles should be minimized since the depot should wait until all the vehicles return to the depot. We also consider the minimization of the number of vehicles in our multi-objective VRP. The objectives in this paper can be described as follows:

$$\text{Minimize } \max_m c_m, \tag{3}$$

$$\text{Minimize } M. \tag{4}$$

When we have a solution with $M = 1$, our problem becomes the traveling salesman problem, the TSP. In that case, however, the other objective to minimize the maximum driving duration in Equation (3) can be reduced to at most the optimal value of the TSP with only one vehicle. That is, the maximum driving duration can not be reduced as in the case of using multiple vehicles. On the other hand, the maximum driving duration becomes minimum when the number of vehicles equals to the number of customers (i.e., $M = N$). In that case, each vehicle needs to visit only one customer. The driving duration for each vehicle in (2) can be described as follows:

$$c_m = c_{0,[1]_m} + c_{[1]_m,0}, \tag{5}$$

where $[k]_m$ denotes the index of the customer who is the k-th customer visited by the m-th vehicle. The maximum driving duration in (5) over M vehicles becomes the optimal value of that objective in the case of $M = N$. We have the trade off between these two objectives: the minimization of the maximum driving duration and the minimization of the number of vehicles.

We consider two problems with different demands: the NDP and the HDP. In the NDP, a normal demand of customers should be satisfied. On the other hand, extra demands should also be satisfied in the HDP. In this paper, we increase the number of customers in the HDP. That is, $N_{NDP} < N_{HDP}$, where N_{NDP} and N_{HDP} are the number of customers in the NDP and the HDP, respectively. We can obtain a set of non-dominated solutions for each problem. We refer a set of non-dominated solutions for the NDP as Ψ_{NDP}, and that for the HDP as Ψ_{HDP}. These two sets of non-dominated solutions can be obtained by applying one of multi-objective algorithms such as EMO algorithms. But if we apply the algorithm to each of the NDP and the HDP independently, we can not expect to obtain a set of solutions with similar routes in the HDP to that obtained for the NDP. Before introducing a measure of similarity of a solution set in Section 4, we describe a two-fold EMO algorithm for our multi-objective VRP in the next section.

3 Two-Fold EMO Algorithm for Multi-objective VRPs

In this section, we show the employed coding scheme to find a set of non-dominated solutions using the genetic operations to apply an EMO algorithm to our multi-objective VRPs. Then we show how we apply a two-fold EMO algorithm to obtain a similar set of solutions in the NDP and the HDP.

3.1 Coding Scheme to Describe Solutions

We code a solution of the VRPs by a permutation of N customers, and we split it into M parts as shown in Fig. 1. There are eight customers in Fig. 1, and they are served by one of three vehicles. The first vehicle denoted v_1 in the figure visits three customers in the order of Customers 1, 2, and 3. It is noted that the depot is not appeared in the coding. Each route is divided by a closed triangle. Therefore the driving duration for v_1 is calculated by $c_{0,1} + c_{1,2} + c_{2,3} + c_{3,0}$. Fig. 2 shows an example of three routes depicted on the map of eight customers and the depot.

3.2 Genetic Operations

We employ the cycle crossover [11] as a crossover operator in this paper. Fig. 3 shows an example of generating an offspring from selected two parents by the crossover. From this figure we can find that the number and the location of splits are changed by this crossover. In this figure, the generation of the offspring is started from Customer 3 in Parent 1. Another offspring is also generated by starting from Parent 2.

As for the mutation, we employ two kinds of operators in order to modify locations of splits and the order of customers in a selected route. Fig. 4 shows examples of these mutations. It should be noted that the order mutation itself does not affect the two objectives (i.e., the maximum driving duration and the number of vehicles). But it can be useful to increase the variety of solutions when it is used with the cycle crossover and the split mutation.

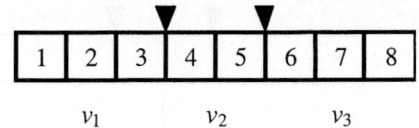

Fig. 1. An example of eight customers visited by three vehicles. Each triangle shows the split between the routes for vehicles

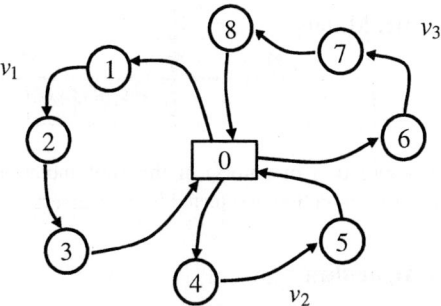

Fig. 2. An example of eight customers visited by three vehicles. Each triangle shows the split between the routes for vehicles

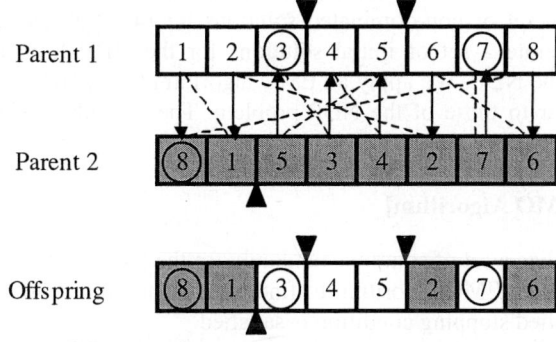

Fig. 3. An example of the cycle crossover. Customer 3 in Parent 1 is randomly chosen and inherited to an offspring. Then Customer 3 in the other parent is found (the dotted line). The customer in Parent 1 locating in the same position of Customer 3 in Parent 2 is inherited to the offspring (i.e., Customer 4). This operation is repeated until returning to Customer 3 in Parent 1. One of remaining customers is chosen from Parent 2 (Customer 8 in this figure). Repeat these operations until all customers are inherited to the offspring. It is noted that a split denoted by a closed triangle is also inherited to the offspring when a customer that is the final one in a route is inherited (Customers 3 and 5 from Parent 1, and Customer 1 from Parent 1 in this figure).

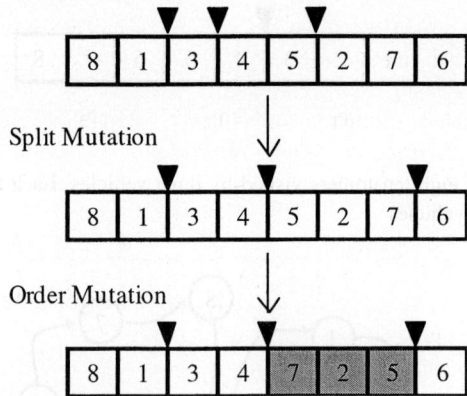

Fig. 4. Examples of two mutation operators. In the split mutation, locations of splits are changed randomly. In the order mutation, a selected route is inversed its order of customers

3.3 Two-Fold EMO Algorithm

In our multi-objective VRP, we have two problems, the NDP and the HDP. Since the HDP has extra demands of customers with the demands of the NDP, we have two approaches to search a set of non-dominated solutions for each of the NDP and the HDP. One approach is to apply an EMO algorithm individually to each of them. The other is to apply a two-fold EMO algorithm to them. In the two-fold EMO algorithm, first we find a set of non-dominated solutions for the NDP by an EMO algorithm. Then we generate a set of initial solutions for the HDP from the non-dominated solutions for the NDP. We apply an EMO algorithm to the HDP with initial solutions that are similar to those of the NDP problem. The procedure of the two-fold EMO algorithm is described as follows:

[Two-Fold EMO Algorithm]

Step 1: Initialize a set of solutions randomly for the NDP.
Step 2: Apply an EMO algorithm to find a set of non-dominated solutions until the specified stopping condition is satisfied.
Step 3: Obtain a set of non-dominated solutions for the NDP.
Step 4: Initialize a set of solutions for the HDP using a set of non-dominated solutions of the NDP.
Step 5: Apply an EMO algorithm to find a set of non-dominated solutions until the specified stopping condition is satisfied.
Step 6: Obtain a set of non-dominated solutions for the HDP.

In Step 4, we initialize a set of solutions as follows:

Step 4.1: Obtain a set of non-dominated solutions of the NDP.
Step 4.2: Specify a solution of the set.
Step 4.3: Insert new customers randomly into the solution.
Step 4.4: Repeat Steps 4.2 and 4.3 until all solutions in the set of non-dominated solutions of the NDP are modified.

It should be noted that the number of vehicles of each solution is not changed by this initialization. The number of vehicles of each solution is changed by the crossover operation. Using this initialization method, we show that the similarity between non-dominated solutions for the NDP and those for the HDP can be increased. Before showing the simulation results we explain a measure of the similarity of solutions in the next section

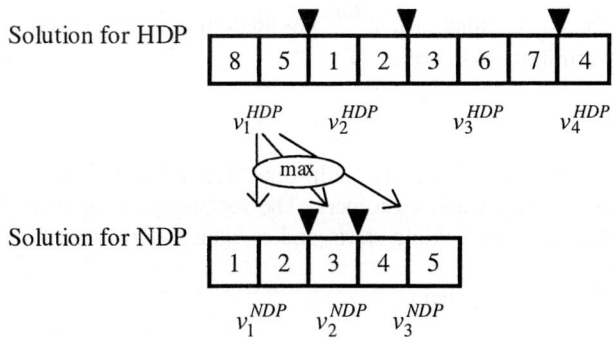

Fig. 5. A set of example solutions for the NDP and the HDP

4 Measure of Similarity Between Sets of Non-dominated Solutions

4.1 Similarity Measure

We define a similarity measure to compare non-dominated solutions obtained for the NDP and the HDP. Figure 5 shows a set of example solutions to be compared. Suppose that we have five customers in the NDP and eight in the HDP. The five customers in the NDP are denoted by 1, 2, ..., 5. The other three customers inserted in the HDP are denoted by 6, 7 and 8. When we obtain a solution with three vehicles for the NDP and one with four for the HDP, the similarity of a solution in the HDP to one in the NDP can be calculated as follows:

$$s(v_i^{HDP}) = \max_j rsr(v_i^{HDP}, v_j^{NDP}), \quad i = 1,...,M_{v_i^{HDP}}, \tag{6}$$

where $s(v_i^{HDP})$ is the similarity of the i-th vehicle v_i^{HDP} of a solution for the HDP to vehicles v_j^{NDP} ($j = 1,...,M_{v_j^{NDP}}$) of a solution for the NDP, and $rsr(v_i^{HDP}, v_j^{NDP})$ is the ratio of the same route in v_i^{HDP} and v_j^{NDP}. The number of vehicles in these solutions is denoted by $M_{v_i^{HDP}}$ and $M_{v_j^{NDP}}$, respectively. In the example of Fig. 5, $M_{v_i^{HDP}} = 4$ and $M_{v_j^{NDP}} = 3$. We calculate $rsr(v_i^{HDP}, v_j^{NDP})$ as follows:

$$rsr(v_i^{HDP}, v_j^{NDP}) = |R_{v_i^{HDP}} \cap R_{v_j^{NDP}}| / |R_{v_j^{NDP}}|, \tag{7}$$

where R_{v_i} shows a set of routes for Vehicle v_i, and $|R_{v_i}|$ indicates that the number of routes of Vehicle v_i. For example, $R_{v_1^{HDP}}$ consists of routes {0 to 8, 8 to 5, 5 to 0}. The number of routes of v_1^{HDP} is three. When we compare the routes of v_1^{HDP} to the routes of vehicles in the solution obtained for the NDP, we can obtain the similarity of v_1^{HDP} using (6) as follows:

$$s(v_1^{HDP}) = \max_j rsr(v_1^{HDP}, v_j^{NDP}),$$
$$= \max\{rsr(v_1^{HDP}, v_1^{NDP}), rsr(v_1^{HDP}, v_2^{NDP}), rsr(v_1^{HDP}, v_3^{NDP})\},$$
$$= \max\{0/3, 0/2, 1/3\} = 1/3. \tag{8}$$

As shown in the above equations, v_1^{HDP} has no common routes with v_1^{NDP} and v_2^{NDP}, and 1/3 as its similarity to the route of v_3^{NDP}.

4.2 Example

We show a small example of our multi-objective vehicle routing problem. Fig. 6 shows a map of a depot and customers. The rectangle in the map shows a central depot from which every vehicle starts and to which it returns. Open circles show

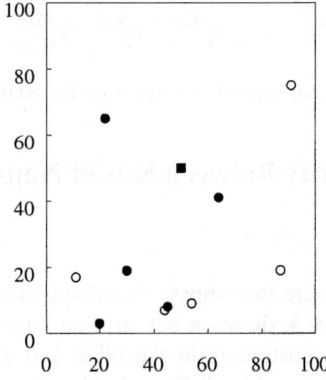

Fig. 6. The locations of customers in a vehicle routing problem with five customers in the NDP, and ten in the HDP

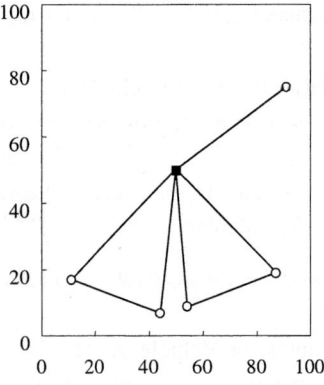

Fig. 7. A solution with three vehicles for the NDP

customers in the NDP, and closed circles show those added in the HDP. Fig. 7 shows an example solution with three vehicles for the NDP with five customers. Figs. 8 and 9 show solutions with a high maximum similarity and a low maximum similarity. From these figures, we can see that the solution for the HDP in Fig. 8 has the same routes (thick routes) in Fig. 7 more than the one in Fig. 9. If we can obtain a solution with a high similarity, we can reduce the effort of drivers of vehicles to be informed of routes among customers.

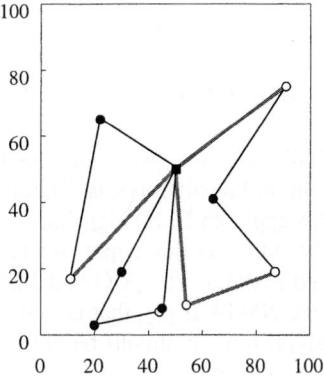

Fig. 8. A solution with three vehicles with a high similarity

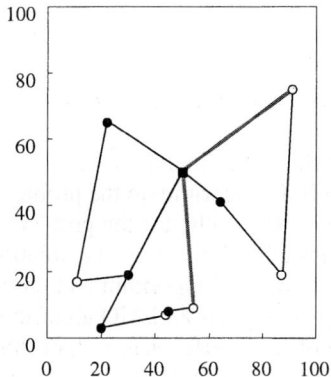

Fig. 9. A solution with three vehicles with a low similarity

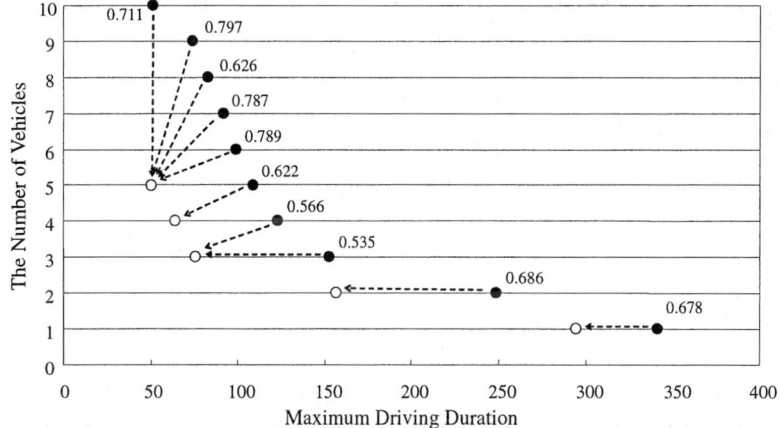

Fig. 10. Sets of Pareto-optimal solutions for the NDP and the HDP in Fig. 6

When we apply an exact search algorithm to find optimal Pareto set for the NDP and the HDP, we obtained the Pareto front depicted in Fig. 10 (In this paper, we employed an exhaust method to examine all possible solutions in the problem. We could employ the exhaust method since the problem is small). The open circles in Fig. 10 show the Pareto solutions for the NDP with the five customers in Fig. 6 with respect to the maximum driving duration and the number of vehicles. The closed circles show the Pareto solutions for the HDP. Each dotted arrow shows the corresponding solution in the NDP, that has the maximum similarity to each solution in the HDP. For example, the solution with four vehicles in the HDP has the routes with the maximum similarity with the solution with three vehicles in the HDP. This figure shows that not all solutions in the NDP are similar to solutions obtained in the HDP. We can increase the number of vehicles according to the increase of demands of customers in the HDP with a small effort of drivers to be informed of new routes between customers.

5 Simulation Results by Two-Fold EMO Algorithm

We show the simulation result on the NDP and HDP problems. We apply our two-fold EMO algorithm to the problem with 100 different initial solution sets. That is, we obtain 100 results for the problem in Fig. 6. We also apply an EMO algorithm to the HDP individually from the solution obtained for the NDP. In this paper, we refer a two-fold EMO algorithm and an individually applied EMO as 2F-EMO and I-EMO, respectively. As an EMO algorithm, we employed the NSGA-II [9], that is known as one of high performance algorithm among EMO algorithms. It should be noted that any EMO algorithm can be used in a 2F-EMO and an I-EMO.

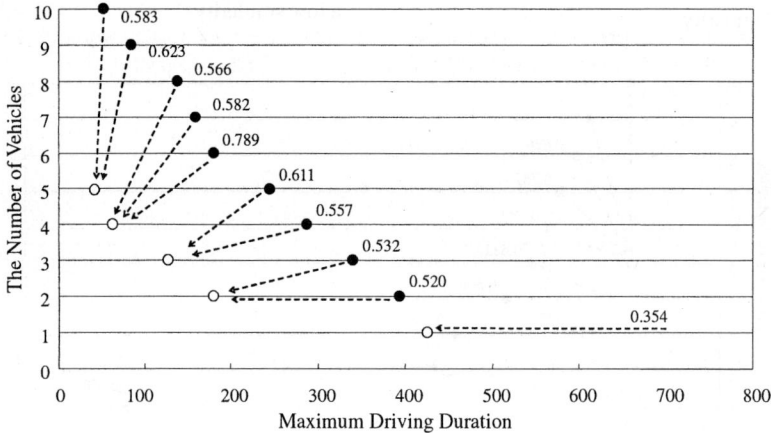

Fig. 11. Similarities of solutions obtained by a two-fold EMO algorithm (2F-EMO). Solutions of the HDP are obtained by an EMO algorithm with initial solutions generated from the solutions obtained in the NDP

Figs. 11 and 12 show examples of non-dominated solutions obtained by a 2F-EMO and by an I-EMO. In these figures, we compare each set of non-dominated solutions obtained by a 2F-EMO or an I-EMO for a HDP to a set of non-dominated solutions for a NDP. A value attached to each solution of non-dominated solutions for the HDP means a maximum similarity of that solution calculated by (8). By comparing these figures, we can see that the obtained sets of non-dominated solutions are not so different each other. However, when we measure the maximum similarity of each solution for the HDP to a set of solutions for the NDP, we can see the difference between those sets of solutions. That is, while the average value of the maximum similarity for each solution obtained by a 2F-EMO in Fig. 11 is 0.5717, that obtained by an I-EMO in Fig. 12 is 0.3858. These values show that the set of non-dominated solutions obtained by a 2F-EMO has higher similarity to the set obtained for the NDP.

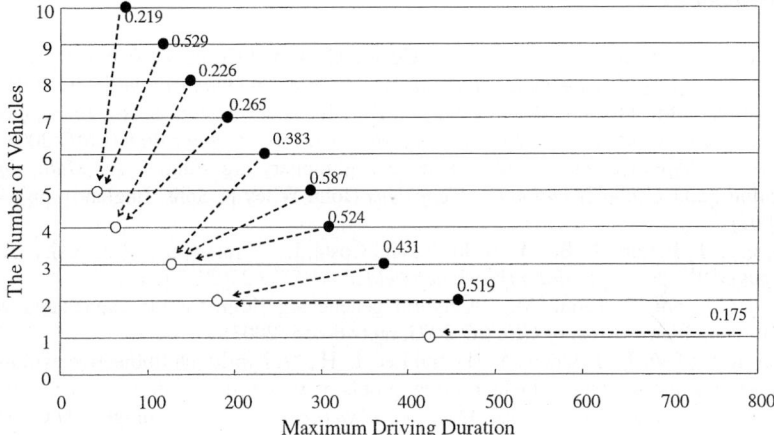

Fig. 12. Similarities of solutions obtained by an individually applied EMO algorithm (I-EMO). Solutions of the HDP are obtained by an EMO algorithm applied individually

Table 1. The average maximum similarity over 100 trials

# of vehicles	10	9	8	7	6	5	4	3	2	1	Ave.
2F-EMO	0.41	0.39	0.43	0.41	0.38	0.33	0.31	0.20	0.13	0.05	0.304
I-EMO	0.37	0.44	0.38	0.35	0.36	0.28	0.25	0.19	0.13	0.04	0.277

Table 1 shows that the average simulation results over 100 trials on the NDP and the HDP shown in Fig. 6. It summarizes the average maximum similarity for the solutions of each number of vehicles. We can see that the maximum similarity tends to become small for the solutions with the small number of vehicles. In a solution with a small number of vehicles, it has a small similarity since each vehicle should visit many customers.

6 Conclusion

In this paper, we consider the similarity of sets of non-dominated solutions that are obtained for a vehicle routing problem and its variant. When the number of customers are increased in a vehicle routing problem, it is better to reduce the effort of drivers to be informed of routes among customers when the new vehicle routing problem is solved. In this paper, we have just considered the influence of using initial solutions generated by a solution obtained for the former problem. Simulation results show that it is better to use initial solutions for an EMO algorithm to solve the increased problem. We are tackling to develop an algorithm to increase the similarity using genetic operations such as crossover and mutation.

References

1. Zitzler, E., Deb, K., Thiele, L., Coello Coello, C. A. and Corne, D. (eds.), *Proc. of First International Conference on Evolutionary Multi-Criterion Optimization* (EMO 2001).
2. Fonseca, C. M., Fleming, P. J., Zitzler, E., Deb, K., and Thiele, L. (eds.), *Proc. of Second International Conference on Evolutionary Multi-Criterion Optimization* (EMO 2003).
3. Deb, K., "Applications of multi-objective evolutionary algorithms," In: *Multi-objective Optimization Using Evolutionary Algorithms* (John Wiley & Sons, England), pp.447-479 (2001).
4. Tavares, J., Pereira, F. B., Machado, P., and Costa, E., "Crossover and diversity: A study about GVR," *Proc. of AdoRo (Workshop held at GECCO 2003)*, 7 pages.
5. Berger, J., and Barkaoui, M., "A hybrid genetic algorithm for the capacitated vehicle routing problem," *Proc. of GECCO 2003*, pp.646-656 (2003).
6. Tan, K. C., Lee, T. H., Chew, Y. H., and Lee, L. H., "A hybrid multiobjective evolutionary algorithms for solving vehicle routing problem with time windows," *Proc. of IEEE International Conf. on Systems, Man, and Cybernetics 2003* (Washington D.C., U.S.A., Oct. 5-8, 2003), pp.361-366 (2003).
7. Saadah, S., Ross, P., and Paechter, B., "Improving vehicle routing using a customer waiting time colony," *Proc. of 4th European Conf. on Evolutionary Computation in Combinatorial Optimization*, pp.188-198 (2004).
8. Chitty, D. M., and Hernandez, M. L., "A hybrid ant colony optimisation technique for dynamic vehicle routing," *Proc. of GECCO 2004*, pp.48-59 (2004).
9. Deb, K., Pratap, A., Agarwal, S., and Meyarivan, T., "A fast and elitist multiobjective genetic algorithm: NSGA-II," *IEEE Trans. on Evolutionary Computation*, **6** (2), pp.182-197, 2002.
10. Lenstra, J. K., and Rinnooy Kan, A.H.G., "Complexity of vehicle routing and scheduling problems", *Networks*, 11, pp. 221-227 (1981).
11. Oliver, I. M., Smith, D. J., and Holland, J. R. C., "A study of permutation crossover operators on the traveling salesman problem," *Proc. of 2nd International Conference on Genetic Algorithm and Their Application*, pp.224-230 (1987).

Multi-objective Optimisation of Turbomachinery Blades Using Tabu Search

Timoleon Kipouros[1], Daniel Jaeggi[1], Bill Dawes[2], Geoff Parks[1], and Mark Savill[3]

[1] Engineering Design Centre, Department of Engineering,
University of Cambridge, Cambridge CB2 1PZ, United Kingdom
[2] Computational Fluid Dynamics Laboratory, Department of Engineering,
University of Cambridge, Cambridge CB2 1PZ, United Kingdom
[3] Computational Aerodynamic Design Group, Department of Aerospace Sciences,
Cranfield University, Cranfield MK43 0AL, United Kingdom

Abstract. This paper describes the application of a new multi-objective integrated turbomachinery blade design optimisation system. The system combines an existing geometry parameterisation scheme, a well-established CFD package and a novel multi-objective variant of the Tabu Search optimisation algorithm. Two case studies, in which the flow characteristics most important to the overall performance of turbomachinery blades are optimised, are investigated. Results are presented and compared with a previous (single-objective) investigation of the problem.

1 Introduction

The optimisation of airfoil designs is a challenging, computationally expensive, highly constrained, non-linear problem. As with most real-world problems, there are multiple (usually conflicting) performance metrics that an engineer might seek to improve in optimising, for example, the design of turbomachinery blades, wings or other aerodynamic surfaces. This suggests a multi-objective approach, a notion that is reinforced by the recognition that any consideration of robustness – the retention of performance over a range of operating conditions, in the face of geometry changes (e.g. through creep) etc. – must also inevitably entail multiple objectives.

Despite this obvious motivation, multi-objective aerodynamic optimisation seems to have been somewhat overlooked. However, two recent studies in particular have embraced multi-objective optimisation and show the possible benefits compared to single-objective optimisation with a composite objective function.

Gaiddon et al. [9] perform multi-objective optimisation on a supersonic missile inlet. They compare a number of optimisation algorithms using both composite and multiple objective functions, and conclude that "performing real multi-objective optimization and finding a Pareto front is the only effective way to find a set of designs satisfying several performance criteria in an industrial context".

Nemec et al. [17] perform multi-objective optimisation on both a single and a multi-element 2-D aerofoil. Their integrated approach combines a Newton-

Krylov adjoint CFD code, a b-splines-based parameterisation scheme and both a gradient-based optimiser and a Genetic Algorithm (GA). They obtain good results on some simple test problems.

The multi-objective integrated design system used in the present work has been developed and described by Kipouros et al. [16] building on the single-objective integrated design optimisation system (BOS3D) developed by Harvey [11] and described by Dawes et al. [6]. The system combines an existing, efficient and flexible geometry parameterisation scheme, a well-established CFD package and a novel multi-objective variant of the Tabu Search (TS) optimisation algorithm for continuous problems [13]. The system can readily be run on parallel computers, which can substantially reduce wall-clock run times – a significant benefit when tackling computationally demanding design problems.

In previous work [16] the performance of this system has been investigated considering a compressor blade design test case. The effectiveness of the multi-objective optimisation procedure was verified and the expected trade-offs between the chosen objectives confirmed. In the work presented in this paper we use our system to tackle more realistic turbomachinery design test cases, taking advantage of the greater computational power offered by exploiting its parallel processing capabilities.

2 Description of the Integrated System

Fig. 1 presents a flow diagram showing the stages of the process executed by our integrated multi-objective turbomachinery blade design optimisation system. The first stage is the parameterisation of the initial blade design, input through an initial CAD geometry together with boundary conditions for the flow solution. The geometry is parameterised using a Partial Differential Equation approach [3], giving a compact but flexible representation of the design, in a design vector comprising 26 variables. This design vector is the input to the main loop of the design system, which consists of the flow simulation and optimisation processes. On receipt of a new design vector, a computational mesh is automatically generated from the geometry specification, and then a detailed CFD analysis (blade to blade) is performed. The mesh is a 3D structured grid consisting of $21 \times 87 \times 23$ nodes in each direction. The flow simulation is performed by a CFD code solving the 3D Navier-Stokes equations, and this routine returns all the necessary metrics that describe the flow around the blade [5]. Based on this evaluation, the optimisation routine generates a new design vector that is meshed and evaluated, and this process continues until a stopping criterion is met.

At the end of the optimisation process, the best design vectors identified and their associated flow solutions are converted into a single file, in the final stage of representation. This stage is accomplished by using Non-Uniform Rational B-Splines (NURBS) [18]. The optimal geometries can then be examined in detail through, for instance, contour plots.

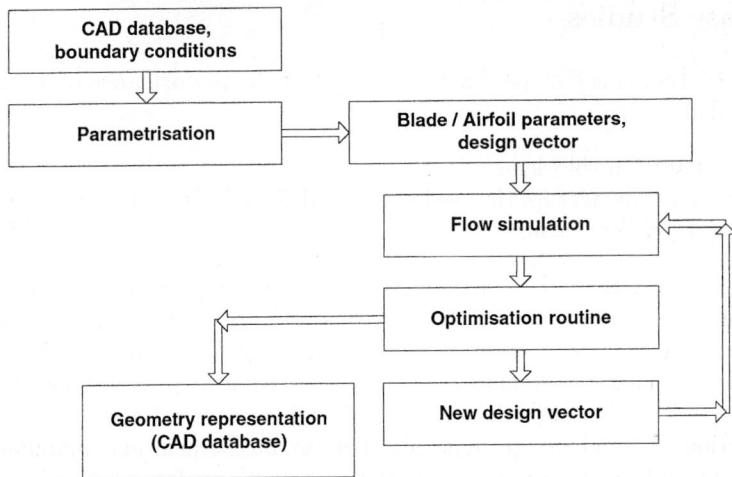

Fig. 1. The structure of the integrated design optimisation system

At the heart of our system is a multi-objective variant [13] of the well-established Tabu Search optimisation algorithm [10]. There has been substantial recent interest in developing multi-objective optimisation algorithms – the vast majority of which concerns multi-objective GAs [7]. GAs have been applied to aerodynamic design problems, but, although able to locate optimal designs, the method proved to be sensitive to constraint handling schemes and required significantly more computational time than a gradient-based method [1]. This is not surprising. The highly constrained nature of most aerodynamic design optimisation problems suggests that algorithms, like GAs, that routinely make large changes to solutions may experience difficulties in trying to negotiate feasible space on such problems, and that optimisation algorithms which progress by making small changes are likely to be more effective. Harvey attributes the local search paradigm at the heart of TS as being one of the reasons for its effectiveness in his work [11]. The execution of a local search at each iteration of the algorithm also offers the potential benefit of being able to estimate the robustness of the current design to variations in geometry without the need for additional flow solutions [2]. For these reasons we have opted to use a multi-objective TS variant in our work.

Our multi-objective TS variant takes as its starting point the single-objective TS implementation of Connor and Tilley [4]. This uses a Hooke and Jeeves local search algorithm (designed for continuous optimisation problems) [12] coupled with short, medium and long term memories to implement search diversification and intensification, as prescribed by Glover and Laguna [10]. This algorithm and analysis of its performance on benchmark constrained optimisation problems from the literature are described in detail in a companion paper [14].

3   Case Studies

The most important flow parameters that affect the performance of a turbomachinery blade are:

- flow separation (blockage),
- losses (any flow feature that reduces the efficiency of the turbomachine), and
- deviation in flow turning.

Here we seek to find, starting from a gas turbine compressor guide vane specification, a blade geometry that efficiently gives a good pressure rise at a particular flow coefficient. Thus, a global performance measure of a given blade geometry is needed for the optimisation process. Efficiency is only one of a number of possible design objectives when undertaking detailed aerodynamic shape optimisation. A good design must also respect mechanical and manufacturing constraints while achieving the required aerodynamic performance (with respect to flow turning, separation and good off-design performance etc.).

The design optimisation of compressor blade geometries has previously been studied by Harvey [11] from a single-objective perspective. In our multi-objective test cases we retain Harvey's objective function (equation 1) as an essential 1D (throughflow) measure of blade performance. This is a normalised function including penalty function terms for specific flow characteristic and geometry constraints. This objective function considers the span-averaged blockage for a given mass flow rate:

$$f_1 = \frac{B}{B_0} + 250\left(1 - \frac{\dot{m}}{\dot{m}_0}\right)^2 + 0.4 max^2\left(0, 1 - \frac{R_{LE}}{R_{LE,0}}\right) \\ + 500 max^2\left(0, 1 - \frac{\Delta\theta}{\Delta\theta_0}\right) + 0.5 max^2\left(0, 1 - \frac{C}{0.015}\right) \quad (1)$$

In equation 1, B represents the blockage, the extent to which viscous forces restrict the effective flow area in a blade passage, which is probably the most critical quantity in high-speed compressor design. Then, $\dot{m}$ is the mass flow rate; R_{LE} is the minimum radius of the leading edge of the blade; $\Delta\theta$ is the mass-averaged flow turning; and C measures the tip clearance of the blade. The zero subscripts identify the equivalent quantities for the datum blade geometry, the initial design in the optimisation – a real compressor blade design shown in Fig. 2. Harvey established suitable values for the weightings for each of the penalty terms through extensive testing [11].

In highly loaded compressors, the flow tends to separate from the blade under conditions of low mass flow. Flow separation acts as a blockage in the flow path, which limits pressure recovery.

The mass flow associated with the design should be equality constrained for two reasons. First, if it was not, the inlet dynamic head from the rotor would vary, which is not modelled by the boundary conditions. Second, if the axial

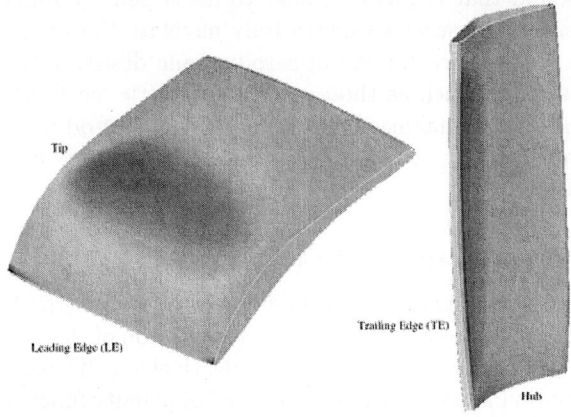

Fig. 2. The datum blade geometry showing its axial velocity distribution

velocity drops, then the inlet static pressure must be higher (since inlet pressure and flow angles are prescribed), so that the static pressure rise across the stator will be lower (outlet pressure is fixed), and the blade row will not be an effective diffuser. Equally, it is important that, if the mass flow is fixed, the flow turning in the stage should not be reduced during optimisation, otherwise the static pressure recovery will not be sufficient. Therefore, control of the flow turning is achieved by treating it as a penalty term.

In addition, there are two terms in equation 1 describing the geometrical constraints on the blade. The first limits the sharpness of the blade's leading edge, while the second allows a weighted penalty factor to trade off aerodynamic performance against mechanical proximity. The objective function value is penalised when the blade design has less than 1.5 cm clearance. Both these penalty terms reflect a concern for robust aerodynamic performance from the design, since these geometric characteristics are closely related to the off-design performance of the blade:

1. A sharp leading edge produces a high velocity profile at the front of the blade and, in addition, such a design may well have a flat section at its front. Such a geometry results in a velocity peak at off-design conditions, which means that the velocity distribution produced has a high probability of early transition, laminar or turbulent separation, and consequently poor off-design performance.
2. The tip clearance is responsible for the deviation created on the end walls. In particular, the secondary flows are known to create over-turning very close to the end walls and a region of under-turning some distance away from them. The effects of this deviation on the exit velocity profile and on the inlet incidence angles to the next blade row can be very large and can substantially increase the incidence losses on the next blade row.

Harvey [11] found that it was necessary to use a penalty function approach with these constraints in order to successfully navigate the highly constrained, nonlinear search space characteristic of aerodynamic design optimisation problems. Other constraints, such as those on the geometric feasibility of blade designs and on their operational feasibility (a design which produces unsteady flow patterns is not acceptable), are handled as hard constraints – designs violating them are not accepted.

3.1 First Multi-objective Test Case

In our first test case, we introduce a second objective $Brms$ (the RMS variation in blockage), which provides an additional representation of the spatial variation, and hence homogeneity, of the blockage – effectively a 2D measure of blade performance (equation 2), subject to the same set of penalty function constraints:

$$f_2 = \frac{Brms}{Brms_0} + 250\left(1 - \frac{\dot{m}}{\dot{m}_0}\right)^2 + 0.4max^2\left(0, 1 - \frac{R_{LE}}{R_{LE,0}}\right) \quad (2)$$
$$+ 500max^2\left(0, 1 - \frac{\Delta\theta}{\Delta\theta_0}\right) + 0.5max^2\left(0, 1 - \frac{C}{0.015}\right)$$

The same penalty terms are included in the formulation of each objective in order to ensure that the final Pareto front identified contains only designs that are (near-)feasible.

3.2 Second Multi-objective Test Case

In our second test case, we introduce an objective function which quantifies losses associated with the design. This is defined in terms of the rate of entropy generation [8], subject again to the same set of penalty function constraints:

$$f_3 = \frac{EntropyRate}{EntropyRate_0} + 250\left(1 - \frac{\dot{m}}{\dot{m}_0}\right)^2 + 0.4max^2\left(0, 1 - \frac{R_{LE}}{R_{LE,0}}\right) \quad (3)$$
$$+ 500max^2\left(0, 1 - \frac{\Delta\theta}{\Delta\theta_0}\right) + 0.5max^2\left(0, 1 - \frac{C}{0.015}\right)$$

In a turbomachine the isentropic efficiency is defined as the ratio of the actual work to the isentropic work for a work-producing device (such as a turbine) and the ratio of the isentropic work to the actual work for a work-absorbing device (such as a compressor). The only factors that change this efficiency are departures from isentropic flow. These may arise due either to heat transfer or to thermodynamic irreversibility. For most turbomachines the flow is close to adiabatic (no heat transfer) and so only entropy creation by irreversibilities contributes significantly to the loss of efficiency, which means that entropy generation rate is the only rational measure of overall loss in an adiabatic machine. Any irreversible flow process creates entropy and thus reduces the isentropic efficiency. The sources of loss in a turbomachine can be categorised as profile loss, secondary (or end wall) loss, and tip leakage loss. In many machines the three are comparable in magnitude, each accounting for about 1/3 of the total loss.

4 Results

4.1 First Test Case

The optimisation was initiated from the datum geometry shown in Fig. 2. Fig. 3 shows the progress in objective-space of the search performed by our multi-objective TS algorithm over the 1350 iterations of the run, using the control parameters specified in Table 1. See [14] for a detailed explanation of these parameters. The values of these control parameters were chosen based on experience, but it should be noted that studies in [14] show that the algorithm's performance is relatively insensitive to the control parameter settings.

The optimisation was run on an 8-node parallel PC cluster of 2.8GHz Pentium 4 machines in order to reduce wall-clock run times by exploiting our system's parallel capabilities. The CFD flow solution required for evaluation of a single candidate design takes 3 minutes on a single node of the cluster, and up to 52 CFD evaluations (on average 32 evaluations) are required at each iteration of the optimisation algorithm. Thus, on our cluster a 1350 iteration run takes 270 hours (just over 11 days).

Fig. 4 shows the set of Pareto-optimal solutions found during the search. The geometry shown in Fig. 5 represents a compromise design from the middle of the Pareto front, which is clearly quite different to the datum design. The large changes made to the blade geometry during the search demonstrate the flexibility of the geometry management system used. A high twist along the span is the main characteristic of this blade. Reassuringly, this optimised geometry

Table 1. Tabu Search Parameter Settings

Parameter	Value	Description
intensify	25	Intensify search after *intensify* iterations without adding to the Medium Term Memory (MTM)
diversify	75	Diversify search after *diversify* iterations without adding to the MTM
reduce	95	Reduce step size and restart after *reduce* iterations without adding to the MTM
n_stm	15	Short Term Memory size – the last *n_stm* visited points are tabu
n_regions	4	In the Long Term Memory each variable is divided into *n_regions* to determine which regions of the search space have been under-explored
SS	1%	Initial step size as percentage of variable range
SSRF	0.5	Factor by which step sizes are reduced on restarting

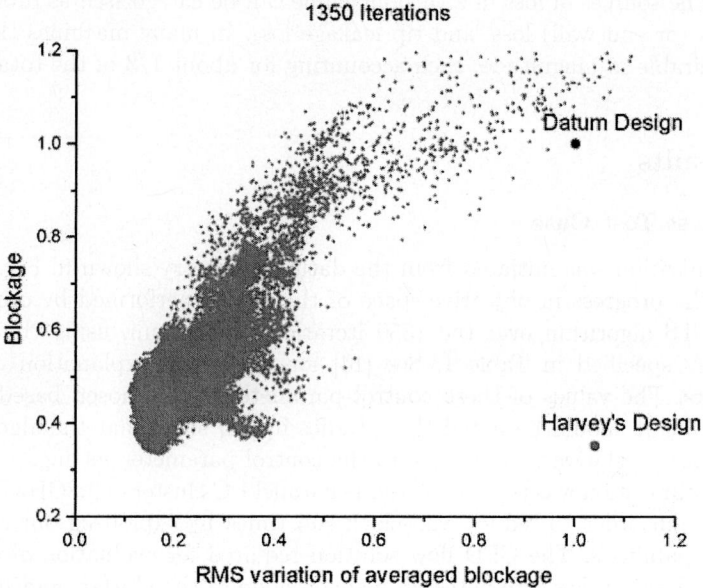

Fig. 3. The optimisation search pattern for the first blading test case

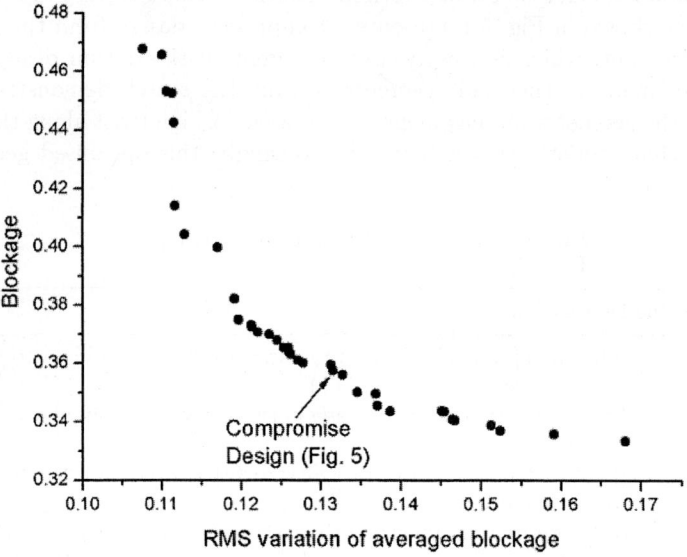

Fig. 4. The Pareto front found for the first test case

has a similar leading edge (LE) profile to the single-objective optimised design found by Harvey, shown in Fig. 6 [11].

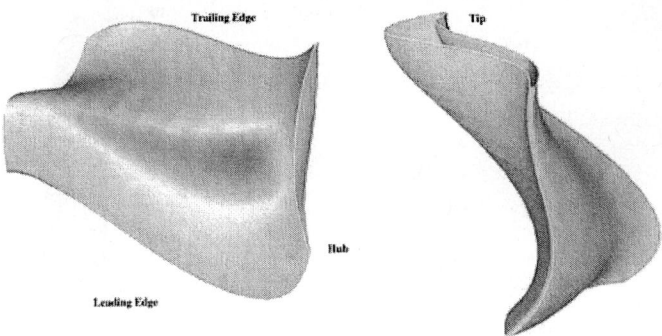

Fig. 5. Test case 1: The optimised geometry for a compromise point on the Pareto front (Fig. 4)

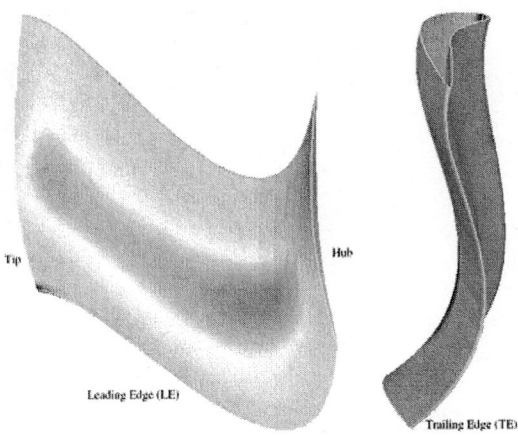

Fig. 6. The optimised geometry for the single-objective test case [11]

It can be seen that the performances of the optimised geometries lying at the low blockage end of the Pareto front (Fig. 4) have matched or slightly exceeded that of the optimised geometry found by Harvey (Fig. 6). This has been achieved in an optimisation run that is actually shorter in terms of the number of CFD flow solutions required than that reported by Harvey [11] (using the same initial design), even though we are tackling a multi-objective optimisation problem. In effect the additional information provided to the designer by revealing the trade-off between the main objective (blockage) and the secondary objective (the RMS variation in blockage) costs nothing because the quality of the best designs found with respect to the primary objective have not been compromised at all by the switch to multi-objective optimisation.

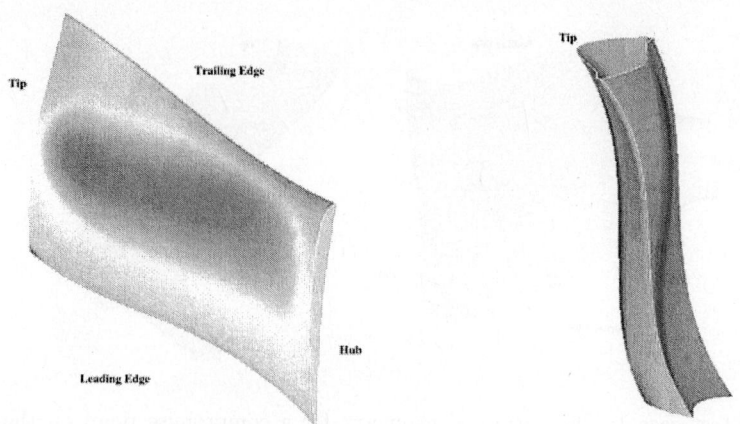

Fig. 7. Test case 2: The optimised geometry for lowest blockage (Fig. 11)

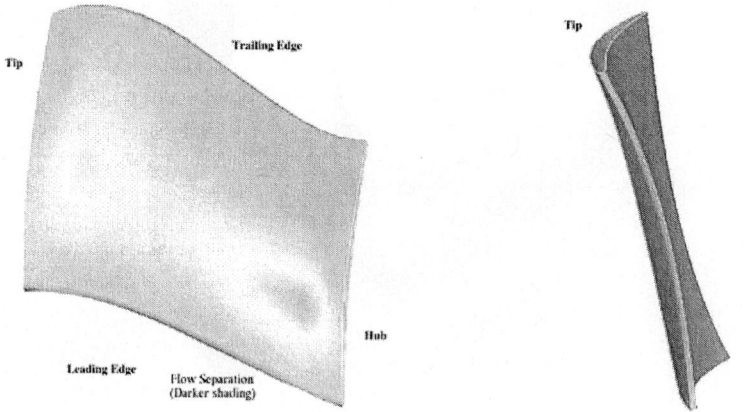

Fig. 8. Test case 2: The optimised geometry for lowest entropy generation (Fig. 11)

4.2 Second Test Case

Optimised geometries found for our second test case are presented in Figs. 7, 8, 9 and 10, which show respectively the lowest blockage design, lowest entropy generation design and two compromise geometries from the trade-off surface (Fig. 11). These blades are again quite different to the datum geometry used to initiate the optimisation, and in addition there are significant differences between them. The LE of these geometries is similar, but differs from the LE of the single-objective optimised blade (Fig. 6). Furthermore, there are considerable differences in the trailing edge (TE) between the multi-objective and single-objective designs.

It is worth remarking that compromise geometry A (Fig. 9) displays geometrical characteristics from both the lowest blockage (Fig. 7) and the lowest

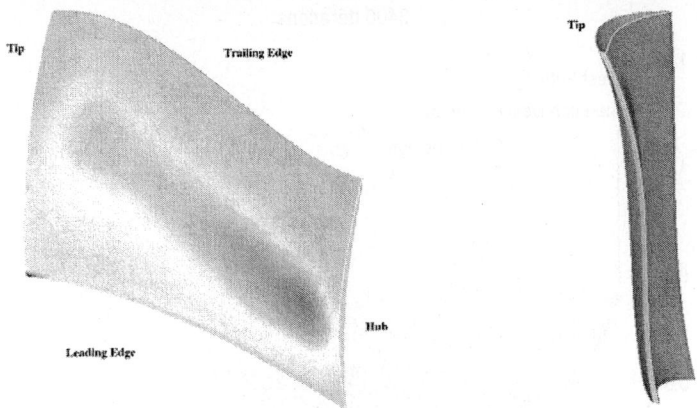

Fig. 9. Test case 2: The optimised geometry for compromise point A on the Pareto front (Fig. 11)

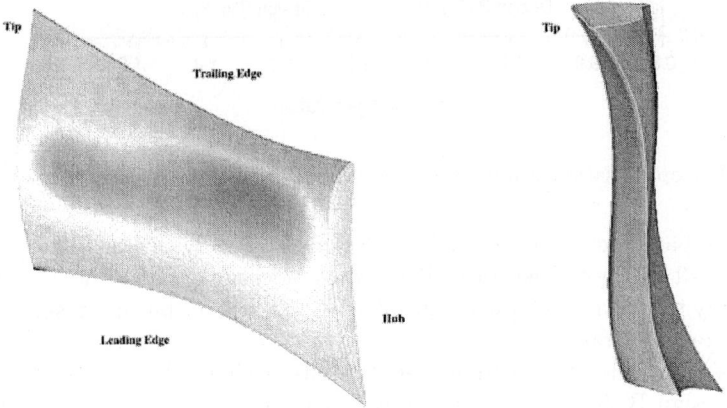

Fig. 10. Test case 2: The optimised geometry for compromise point B on the Pareto front (Fig. 11)

entropy generation (Fig. 8) designs. However, the big difference between these blades is in the tip profiles. For the lowest blockage design (Fig. 7) and compromise design B (Fig. 10), for which the blockage is almost as low, there is a rapid change in the tip camber, which results in a thick profile, whereas for the geometries in Figs. 8 and 9 there is a smooth change in the camber of the tip profile. This geometrical characteristic is shared with the blade in Fig. 6.

As regards aerodynamic performance, all the blades have good, i.e. smoothly varying, axial velocity distributions. However, there is noticeable flow separation along the TE of the blade optimised for lowest entropy generation.

The Pareto front shows that significant performance improvements are achievable. For instance, compromise design B (Fig. 10) reduces the blockage signif-

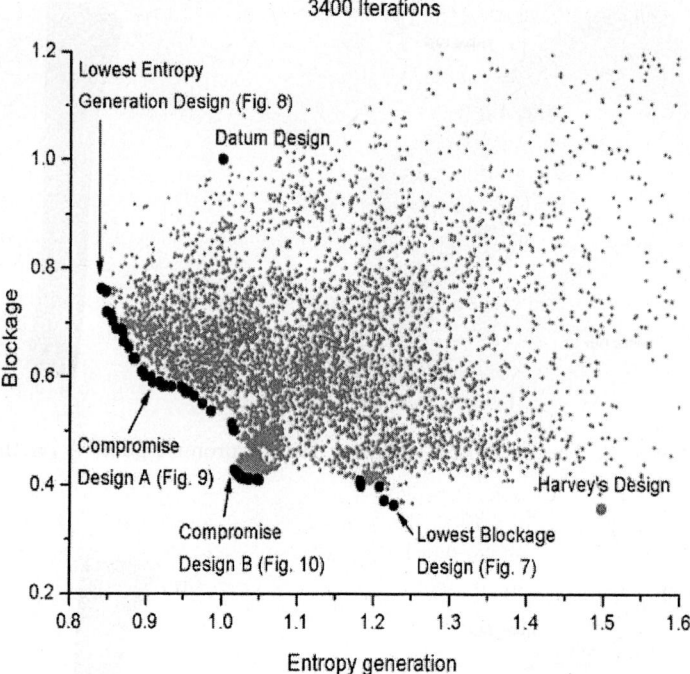

Fig. 11. The optimisation search pattern and the Pareto front for the second test case

icantly compared to the initial datum design (Fig. 2) with very little increase in the rate of entropy generation. The lowest blockage design (Fig. 7) performs comparably to Harvey's blockage-optimised blade (Fig. 6) but has a significantly lower entropy generation rate.

Interestingly the Pareto front also exhibits a sharp "elbow" around compromise design B. Relative to this design, it is possible to reduce the blockage objective but only at the cost of quite large increases in entropy generation. It is also possible to reduce the rate of entropy generation but only at the cost of significantly increased blockage. In this case the designer might find it quite straightforward to select a good compromise design.

5 Conclusions and Future Work

The foregoing test cases demonstrate that our multi-objective integrated turbomachinery design optimisation system can successfully tackle realistic real-world problems, negotiating the highly constrained, nonlinear search space, and presenting the designer with a range of designs showing the trade-offs between the objectives under consideration.

In both test cases the performance of the designs found match or exceed the performance of the optimised blade identified in an earlier single-objective study. The computational effort required to solve these multi-objective problems was no more than that required to solve the single-objective problem and, in addition to equally good or better designs, the designer is also presented with helpful information about the trade-offs between the objectives of interest. This demonstrates very clearly the value of multi-objective optimisation.

The factors influencing efficiency of turbomachinery blades and the trade-offs between them are extremely complex and therefore this area needs further investigation. The next steps will be to define additional loss objectives for the design problem examined above. Thus, new objective functions will evaluate individually the profile losses and the secondary losses in order to improve understanding of the trade-offs between them in design. These investigations will require the tackling of three- and four-objective problems and will therefore also allow us to test the effectiveness of our multi-objective TS variant on these higher dimension problems.

To improve further the effectiveness of our multi-objective TS variant a more sophisticated system for selecting the design variables to be modified in search, based on the work of Kellar [15], will be developed. This will seek to identify the variables that have the greatest impact on the performance of the current design and prioritise them at each local search iteration. It is hoped that this will improve the wall-clock performance of the system substantially.

Acknowledgements. The first author gratefully acknowledges the support of the Embiricos Foundation and the Cambridge European Trust. The second author acknowledges the support of the UK Engineering and Physical Sciences Research Council (EPSRC) under grant number GR/R64100/01.

References

1. Aly, S., Ogot, M., Pelz, R.: Stochastic Approach to Optimal Aerodynamic Shape Design. Journal of Aircraft. **33** (1996) 956–961
2. Asselin-Miller, C. S.: Robust 2D-Aerofoil Design: Proof of Concept. 4th Year Project Report. University of Cambridge, Department of Engineering (2003)
3. Bloor, M. I. G., Wilson, M. J.: Efficient Parameterisation of Generic Aircraft Geometry. Journal of Aircraft. **32** (1995) 1269–1275
4. Connor, A. M., Tilley, D. G.: A Tabu Search Method for the Optimisation of Fluid Power Circuits. IMechE Journal of Systems and Control. **212** (1998) 373–381
5. Dawes, W. N.: Development of a 3D Navier-Stokes solver for application to all types of turbomachinery. ASME Conference Paper 88-GT-70. ASME Gas Turbine Conference, Amsterdam. (1988)
6. Dawes, W. N., Kellar, W. P., Harvey, S. A., Dhanasekaran, P. C., Savill, A. M., Cant, R. S.: Managing the Geometry is Limiting the Ability of CFD to Manage the Flow. 33rd AIAA Fluid Dynamics Conference, Orlando, Florida. AIAA-2003-3732. (2003)
7. Deb K., Multi-Objective Optimization using Evolutionary Algorithms. John Wiley & Sons Ltd., Chichester UK (2001)
8. Denton, J. D.: Loss Mechanisms in Turbomachines. Journal of Turbomachinery. **115** (1993) 621–656

9. Gaiddon, A., Knight, D. D., Poloni, C.: Multicriteria Design Optimisation of a Supersonic Inlet Based upon Global Missile Performance. Journal of Propulsion and Power. **20** (2004) 542–558
10. Glover, F., Laguna, M.: Tabu Search. Kluwer Academic Publishers, Boston MA (1997)
11. Harvey, S. A.: The Design Optimisation of Turbomachinery Blade Rows. Ph.D. Dissertation. University of Cambridge (2002)
12. Hooke, R., Jeeves, T.: Direct Search Solution of Numerical and Statistical Problems. Journal of the ACM. **8** (1961) 212–229
13. Jaeggi, D. M., Asselin-Miller, C. S., Parks, G. T., Kipouros, T., Bell, T., Clarkson, P. J.: Multi-objective Parallel Tabu Search. In: Yao, X., Burke, E., Lozano, J-A., Smith, J., Merelo-Guervos, J., Bullinaria, J., Rowe, J., Tino, P., Kaban, A., Schwefel, H-P. (eds.): Parallel Problem Solving from Nature – PPSN VIII. Lecture Notes in Computer Science, Vol. 3242. Springer-Verlag, Berlin (2004) 732–741
14. Jaeggi, D. M., Parks, G. T., Kipouros, T., Clarkson, P. J.: A Multi-objective Tabu Search Algorithm for Constrained Optimisation. In: 3rd Int. Conf. Evolutionary Multi-Criterion Optimization. Lecture Notes in Computer Science, Vol. 3410. Springer-Verlag, Berlin (2005) 490–504
15. Kellar, W. N.: Geometry Modeling in Computational Fluid Dynamics and Design Optimisation. Ph.D. Dissertation. University of Cambridge (2002)
16. Kipouros, T., Parks, G. T., Savill, A. M., Jaeggi, D. M.: Multi-objective Aerodynamic Design Optimisation. In: Giannakoglou, K. C., Haase, W. (eds.): ERCOFTAC Design Optimization: Methods and Applications Conference Proceedings. On CDRom (2004) Paper ERCODO2004_239
17. Nemec, M., Zingg, D. W., Pulliam, T. H.: Multipoint and Multi-objective Aerodynamic Shape Optimization. AIAA Journal. **42** (2004) 1057–1065
18. Rogers, D. F.: An Introduction to NURBS With Historical Perspective. Morgan Kaufmann Publishers, San Francisco CA (2000)

Author Index

Aguilera-Contreras, Miguel Angel   235
Aguirre, Hernán E.   355
Ah King, Robert T.F.   677
Alba, Enrique   443
Alcocer-Yamanaka, Víctor   870
Alvarez-Benitez, Julio E.   459

Barone, Luigi   280
Basseur, Matthieu   120
Bentley, Peter   636
Berry, Adam   77
Beume, Nicola   62
Bortolozzi, Flávio   592
Bosman, Peter A.N.   428
Botello Rionda, Salvador   664
Brizuela, Carlos A.   206

Carrasquero, Néstor   221
Chakrabarti, P.P.   811
Chiba, Kazuhisa   621
Clarkson, John   490
Coello Coello, Carlos A.   505
Corne, David   841

Dawes, Bill   897
Day, Richard O.   296
de la Iglesia, Beatriz   826
de Moura Oliveira, P.B.   165
Deb, Kalyanmoy   47, 150, 311, 677
Du, Haifeng   474
Duda, Jerzy   651

Ehrgott, Matthias   33
Emmerich, Michael   62
Emperador, José M.   576
Everson, Richard M.   459

Fieldsend, Jonathan E.   459
Filipič, Bogdan   520
Fleming, Peter J.   14
Fonseca, Carlos M.   250

Galván, Blas   221, 576
Gamenik, Jürgen   191
Gandibleux, Xavier   33
Gong, Maoguo   474

Greiner, David   576
Grunert da Fonseca, Viviane   250
Guo, Shao Dong   135
Gupta, Himanshu   150
Gutiérrez, Everardo   206

Handl, Julia   547
Haubelt, Christian   191
Hernández Aguirre, Arturo   664, 856
Hingston, Phil   280
Huband, Simon   280
Hughes, Evan J.   176

Igel, Christian   534
Iima, Hitoshi   721
Irving, Malcolm R.   707
Ishibuchi, Hisao   265, 341, 370
Itai, Ryota   885

Jaeggi, Daniel   490, 897
Jeong, Min Joong   561
Jiao, Licheng   474
Jie, Luo Yang   135
Jin, Yaochu   752
Jourdan, Laetitia   841

Kaige, Shiori   341, 370
Kang, Lishan   108
Kipouros, Timoleon   490, 897
Kleeman, Mark P.   782
Knowles, Joshua   176, 547
Kobayashi, Takashi   561
Kollat, Joshua B.   386
Köppen, Mario   399
Körner, Edgar   752
Kumar, Rajeev   811

Lamont, Gary B.   296, 782
Large, Maryanne   636
Leyva-Lopez, Juan Carlos   235
Li, Weiqi   797
Limbourg, Philipp   413
Liu, Yong   108
Lu, Bin   474
Luna, Francisco   443
Lygoe, Robert J.   14

Manos, Steven 636
Mariano-Romero, Carlos E. 870
Min, Yang Shu 135
Mitschele, Andreas 692
Morales, Eduardo F. 870
Morino, Hiroyuki 621
Morita, Marisa 592
Muñoz Zavala, Angel E. 856
Murata, Tadahiko 885

Nakahashi, Kazuhiro 621
Narukawa, Kaname 265, 341, 370
Naujoks, Boris 62
Nebro, Antonio J. 443
Nickolay, Bertram 399
Nicolini, Matteo 736
Nojima, Yosuke 341

Obayashi, Shigeru 621
Oliveira, Luiz S. 592
Osyczka, Andrzej 651

Paquete, Luís 250
Parks, Geoff 490, 897
Poladian, Leon 636
Purshouse, Robin C. 14

Radtke, Paulo V.W. 767
Rayward-Smith, Vic J. 826
Reed, Patrick M. 386
Reyes Sierra, Margarita 505
Reynolds, Alan 826
Rivas-Dávalos, Francisco 707
Robič, Tea 520
Rughooputh, Harry C.S. 677

Sabourin, Robert 592, 767
Salazar, Daniel 221
Savic, Dragan 841

Savill, Mark 897
Schlottmann, Frank 692
Seese, Detlef 692
Sendhoff, Bernhard 752
Seynhaeve, Franck 120
Shang, Ronghua 474
Shukla, Pradyumn Kumar 311
Singh, P.K. 811
Solteiro Pires, E.J. 165
Speer, Nora 607
Spieth, Christian 607
Streichert, Felix 92, 607

Talbi, El-Ghazali 120
Tanaka, Kiyoshi 355
Teich, Jürgen 191
Tenreiro Machado, J.A. 165
Thierens, Dirk 428
Tiwari, Santosh 47, 311

Ulmer, Holger 92

Valdez Peña, Sergio Ivvan 664
Vamplew, Peter 77
Vicente-Garcia, Raul 399
Villa Diharce, Enrique R. 856

Walters, Godfrey 841
While, Lyndon 280, 326
Winter, Gabriel 576
Wong, Tony 767

Yao, Shuzhen 108
Yoshimura, Shinobu 561

Zeleny, Milan 1
Zell, Andreas 92, 607
Zeng, Sanyou 108

Lecture Notes in Computer Science

For information about Vols. 1–3298

please contact your bookseller or Springer

Vol. 3418: U. Brandes, T. Erlebach (Eds.), Network Analysis. XII, 471 pages. 2005.

Vol. 3412: X. Franch, D. Port (Eds.), COTS-Based Software Systems. XVI, 312 pages. 2005.

Vol. 3410: C.A. Coello Coello, A. Hernández Aguirre, E. Zitzler (Eds.), Evolutionary Multi-Criterion Optimization. XVI, 912 pages. 2005.

Vol. 3406: A. Gelbukh (Ed.), Computational Linguistics and Intelligent Text Processing. XVII, 829 pages. 2005.

Vol. 3404: V. Diekert, B. Durand (Eds.), STACS 2005. XVI, 706 pages. 2005.

Vol. 3403: B. Ganter, R. Godin (Eds.), Formal Concept Analysis. XI, 419 pages. 2005. (Subseries LNAI).

Vol. 3398: D.-K. Baik (Ed.), Systems Modeling and Simulation: Theory and Applications. XIV, 733 pages. 2005. (Subseries LNAI).

Vol. 3397: T.G. Kim (Ed.), Artificial Intelligence and Simulation. XV, 711 pages. 2005. (Subseries LNAI).

Vol. 3393: H.-J. Kreowski, U. Montanari, F. Orejas, G. Rozenberg, G. Taentzer (Eds.), Formal Methods in Software and Systems Modeling. XXVII, 413 pages. 2005.

Vol. 3391: C. Kim (Ed.), Information Networking. XVII, 936 pages. 2005.

Vol. 3388: J. Lagergren (Ed.), Comparative Genomics. VIII, 133 pages. 2005. (Subseries LNBI).

Vol. 3387: J. Cardoso, A. Sheth (Eds.), Semantic Web Services and Web Process Composition. VIII, 147 pages. 2005.

Vol. 3386: S. Vaudenay (Ed.), Public Key Cryptography - PKC 2005. IX, 436 pages. 2005.

Vol. 3385: R. Cousot (Ed.), Verification, Model Checking, and Abstract Interpretation. XII, 483 pages. 2005.

Vol. 3382: J. Odell, P. Giorgini, J.P. Müller (Eds.), Agent-Oriented Software Engineering V. X, 239 pages. 2005.

Vol. 3381: P. Vojtáš, M. Bieliková, B. Charron-Bost, O. Sýkora (Eds.), SOFSEM 2005: Theory and Practice of Computer Science. XV, 448 pages. 2005.

Vol. 3379: M. Hemmje, C. Niederee, T. Risse (Eds.), From Integrated Publication and Information Systems to Information and Knowledge Environments. XXIV, 321 pages. 2005.

Vol. 3378: J. Kilian (Ed.), Theory of Cryptography. XII, 621 pages. 2005.

Vol. 3376: A. Menezes (Ed.), Topics in Cryptology – CT-RSA 2005. X, 385 pages. 2004.

Vol. 3375: M.A. Marsan, G. Bianchi, M. Listanti, M. Meo (Eds.), Quality of Service in Multiservice IP Networks. XIII, 656 pages. 2005.

Vol. 3374: D. Weyns, H.V.D. Parunak, F. Michel (Eds.), Environments for Multi-Agent Systems. X, 279 pages. 2005. (Subseries LNAI).

Vol. 3372: C. Bussler, V. Tannen, I. Fundulaki (Eds.), Semantic Web and Databases. X, 227 pages. 2005.

Vol. 3368: L. Paletta, J.K. Tsotsos, E. Rome, G.W. Humphreys (Eds.), Attention and Performance in Computational Vision. VIII, 231 pages. 2005.

Vol. 3366: I. Rahwan, P. Moraitis, C. Reed (Eds.), Argumentation in Multi-Agent Systems. XII, 263 pages. 2005. (Subseries LNAI).

Vol. 3363: T. Eiter, L. Libkin (Eds.), Database Theory - ICDT 2005. XI, 413 pages. 2004.

Vol. 3362: G. Barthe, L. Burdy, M. Huisman, J.-L. Lanet, T. Muntean (Eds.), Construction and Analysis of Safe, Secure, and Interoperable Smart Devices. IX, 257 pages. 2005.

Vol. 3361: S. Bengio, H. Bourlard (Eds.), Machine Learning for Multimodal Interaction. XII, 362 pages. 2005.

Vol. 3360: S. Spaccapietra, E. Bertino, S. Jajodia, R. King, D. McLeod, M.E. Orlowska, L. Strous (Eds.), Journal on Data Semantics II. XI, 223 pages. 2004.

Vol. 3359: G. Grieser, Y. Tanaka (Eds.), Intuitive Human Interfaces for Organizing and Accessing Intellectual Assets. XIV, 257 pages. 2005. (Subseries LNAI).

Vol. 3358: J. Cao, L.T. Yang, M. Guo, F. Lau (Eds.), Parallel and Distributed Processing and Applications. XXIV, 1058 pages. 2004.

Vol. 3357: H. Handschuh, M.A. Hasan (Eds.), Selected Areas in Cryptography. XI, 354 pages. 2004.

Vol. 3356: G. Das, V.P. Gulati (Eds.), Intelligent Information Technology. XII, 428 pages. 2004.

Vol. 3355: R. Murray-Smith, R. Shorten (Eds.), Switching and Learning in Feedback Systems. X, 343 pages. 2005.

Vol. 3353: J. Hromkovič, M. Nagl, B. Westfechtel (Eds.), Graph-Theoretic Concepts in Computer Science. XI, 404 pages. 2004.

Vol. 3352: C. Blundo, S. Cimato (Eds.), Security in Communication Networks. XI, 381 pages. 2005.

Vol. 3350: M. Hermenegildo, D. Cabeza (Eds.), Practical Aspects of Declarative Languages. VIII, 269 pages. 2005.

Vol. 3349: B.M. Chapman (Ed.), Shared Memory Parallel Programming with Open MP. X, 149 pages. 2005.

Vol. 3348: A. Canteaut, K. Viswanathan (Eds.), Progress in Cryptology - INDOCRYPT 2004. XIV, 431 pages. 2004.

Vol. 3347: R.K. Ghosh, H. Mohanty (Eds.), Distributed Computing and Internet Technology. XX, 472 pages. 2004.

Vol. 3346: R.H. Bordini, M. Dastani, J. Dix, A.E.F. Seghrouchni (Eds.), Programming Multi-Agent Systems. XIV, 249 pages. 2005. (Subseries LNAI).

Vol. 3345: Y. Cai (Ed.), Ambient Intelligence for Scientific Discovery. XII, 311 pages. 2005. (Subseries LNAI).

Vol. 3344: J. Malenfant, B.M. Østvold (Eds.), Object-Oriented Technology. ECOOP 2004 Workshop Reader. VIII, 215 pages. 2005.

Vol. 3342: E. Şahin, W.M. Spears (Eds.), Swarm Robotics. IX, 175 pages. 2005.

Vol. 3341: R. Fleischer, G. Trippen (Eds.), Algorithms and Computation. XVII, 935 pages. 2004.

Vol. 3340: C.S. Calude, E. Calude, M.J. Dinneen (Eds.), Developments in Language Theory. XI, 431 pages. 2004.

Vol. 3339: G.I. Webb, X. Yu (Eds.), AI 2004: Advances in Artificial Intelligence. XXII, 1272 pages. 2004. (Subseries LNAI).

Vol. 3338: S.Z. Li, J. Lai, T. Tan, G. Feng, Y. Wang (Eds.), Advances in Biometric Person Authentication. XVIII, 699 pages. 2004.

Vol. 3337: J.M. Barreiro, F. Martin-Sanchez, V. Maojo, F. Sanz (Eds.), Biological and Medical Data Analysis. XI, 508 pages. 2004.

Vol. 3336: D. Karagiannis, U. Reimer (Eds.), Practical Aspects of Knowledge Management. X, 523 pages. 2004. (Subseries LNAI).

Vol. 3335: M. Malek, M. Reitenspieß, J. Kaiser (Eds.), Service Availability. X, 213 pages. 2005.

Vol. 3334: Z. Chen, H. Chen, Q. Miao, Y. Fu, E. Fox, E.-p. Lim (Eds.), Digital Libraries: International Collaboration and Cross-Fertilization. XX, 690 pages. 2004.

Vol. 3333: K. Aizawa, Y. Nakamura, S. Satoh (Eds.), Advances in Multimedia Information Processing - PCM 2004, Part III. XXXV, 785 pages. 2004.

Vol. 3332: K. Aizawa, Y. Nakamura, S. Satoh (Eds.), Advances in Multimedia Information Processing - PCM 2004, Part II. XXXVI, 1051 pages. 2004.

Vol. 3331: K. Aizawa, Y. Nakamura, S. Satoh (Eds.), Advances in Multimedia Information Processing - PCM 2004, Part I. XXXVI, 667 pages. 2004.

Vol. 3330: J. Akiyama, E.T. Baskoro, M. Kano (Eds.), Combinatorial Geometry and Graph Theory. VIII, 227 pages. 2005.

Vol. 3329: P.J. Lee (Ed.), Advances in Cryptology - ASIACRYPT 2004. XVI, 546 pages. 2004.

Vol. 3328: K. Lodaya, M. Mahajan (Eds.), FSTTCS 2004: Foundations of Software Technology and Theoretical Computer Science. XVI, 532 pages. 2004.

Vol. 3327: Y. Shi, W. Xu, Z. Chen (Eds.), Data Mining and Knowledge Management. XIII, 263 pages. 2005. (Subseries LNAI).

Vol. 3326: A. Sen, N. Das, S.K. Das, B.P. Sinha (Eds.), Distributed Computing - IWDC 2004. XIX, 546 pages. 2004.

Vol. 3325: C.H. Lim, M. Yung (Eds.), Information Security Applications. XI, 472 pages. 2005.

Vol. 3323: G. Antoniou, H. Boley (Eds.), Rules and Rule Markup Languages for the Semantic Web. X, 215 pages. 2004.

Vol. 3322: R. Klette, J. Žunić (Eds.), Combinatorial Image Analysis. XII, 760 pages. 2004.

Vol. 3321: M.J. Maher (Ed.), Advances in Computer Science - ASIAN 2004. XII, 510 pages. 2004.

Vol. 3320: K.-M. Liew, H. Shen, S. See, W. Cai (Eds.), Parallel and Distributed Computing: Applications and Technologies. XXIV, 891 pages. 2004.

Vol. 3319: D. Amyot, A.W. Williams (Eds.), Telecommunications and beyond: Modeling and Analysis of Reactive, Distributed, and Real-Time Systems. XII, 301 pages. 2005.

Vol. 3318: E. Eskin, C. Workman (Eds.), Regulatory Genomics. VIII, 115 pages. 2005. (Subseries LNBI).

Vol. 3317: M. Domaratzki, A. Okhotin, K. Salomaa, S. Yu (Eds.), Implementation and Application of Automata. XII, 336 pages. 2005.

Vol. 3316: N.R. Pal, N.K. Kasabov, R.K. Mudi, S. Pal, S.K. Parui (Eds.), Neural Information Processing. XXX, 1368 pages. 2004.

Vol. 3315: C. Lemaître, C.A. Reyes, J.A. González (Eds.), Advances in Artificial Intelligence – IBERAMIA 2004. XX, 987 pages. 2004. (Subseries LNAI).

Vol. 3314: J. Zhang, J.-H. He, Y. Fu (Eds.), Computational and Information Science. XXIV, 1259 pages. 2004.

Vol. 3313: C. Castelluccia, H. Hartenstein, C. Paar, D. Westhoff (Eds.), Security in Ad-hoc and Sensor Networks. VIII, 231 pages. 2005.

Vol. 3312: A.J. Hu, A.K. Martin (Eds.), Formal Methods in Computer-Aided Design. XI, 445 pages. 2004.

Vol. 3311: V. Roca, F. Rousseau (Eds.), Interactive Multimedia and Next Generation Networks. XIII, 287 pages. 2004.

Vol. 3310: U.K. Wiil (Ed.), Computer Music Modeling and Retrieval. XI, 371 pages. 2005.

Vol. 3309: C.-H. Chi, K.-Y. Lam (Eds.), Content Computing. XII, 510 pages. 2004.

Vol. 3308: J. Davies, W. Schulte, M. Barnett (Eds.), Formal Methods and Software Engineering. XIII, 500 pages. 2004.

Vol. 3307: C. Bussler, S.-k. Hong, W. Jun, R. Kaschek, D. Kinshuk, S. Krishnaswamy, S.W. Loke, D. Oberle, D. Richards, A. Sharma, Y. Sure, B. Thalheim (Eds.), Web Information Systems – WISE 2004 Workshops. XV, 277 pages. 2004.

Vol. 3306: X. Zhou, S. Su, M.P. Papazoglou, M.E. Orlowska, K.G. Jeffery (Eds.), Web Information Systems – WISE 2004. XVII, 745 pages. 2004.

Vol. 3305: P.M.A. Sloot, B. Chopard, A.G. Hoekstra (Eds.), Cellular Automata. XV, 883 pages. 2004.

Vol. 3303: J.A. López, E. Benfenati, W. Dubitzky (Eds.), Knowledge Exploration in Life Science Informatics. X, 249 pages. 2004. (Subseries LNAI).

Vol. 3302: W.-N. Chin (Ed.), Programming Languages and Systems. XIII, 453 pages. 2004.

Vol. 3300: L. Bertossi, A. Hunter, T. Schaub (Eds.), Inconsistency Tolerance. VII, 295 pages. 2005.

Vol. 3299: F. Wang (Ed.), Automated Technology for Verification and Analysis. XII, 506 pages. 2004.